Withdrawn
University of Waterloo

Sedimentation and Tectonics in Rift Basins

JOIN US ON THE INTERNET VIA WWW, GOPHER, FTP OR EMAIL:

WWW: http://www.thomson.com
GOPHER: gopher.thomson.com
FTP: ftp.thomson.com
EMAIL: findit@kiosk.thomson.com

A service of I(T)P®

Sedimentation and Tectonics in Rift Basins Red Sea:- Gulf of Aden

Edited by

Bruce H. Purser
Emeritus Professor of Sedimentology
Université de Paris Sud
France

and

Dan W. J. Bosence
Professor of Carbonate Sedimentology
Royal Holloway
University of London
UK

Much of the research presented in this book was supported by the SCIENCE Programme of the European Community.

CHAPMAN & HALL
London · Weinheim · New York · Tokyo · Melbourne · Madras

Published by Chapman & Hall, an imprint of Thomson Science, 2–6 Boundary Row, London SE1 8HN, UK

Thomson Science, 2–6 Boundary Row, London SE1 8HN, UK

Thomson Science, Pappelallee 3, 69469 Weinheim, Germany

Thomson Science, 115 Fifth Avenue, New York, NY 10003, USA

Thomson Science, Suite 750, 400 Market Street, Philadelphia, PA 19106, USA

First edition 1998

© 1998 Chapman & Hall

Thomson Science is a division of International Thomson Publishing

Typeset in 10/12pt Times by Scientific Publishing Services (P) Ltd, Chennai, India

Printed in Great Britain by the University Press, Cambridge

ISBN 0 412 73490 7

Apart from any fair dealing for the purposes of research or private study, or criticism or review, as permitted under the UK Copyright Designs and Patents Act, 1988, this publication may not be reproduced, stored, or transmitted, in any form or by any means, without the prior permission in writing of the publishers, or in the case of reprographic reproduction only in accordance with the terms of the licences issued by the Copyright Licensing Agency in the UK, or in accordance with the terms of licences issued by the appropriate Reproduction Rights Organization outside the UK. Enquiries concerning reproduction outside the terms stated here should be sent to the publishers at the London address printed on this page.

The publisher makes no representation, express or implied, with regard to the accuracy of the information contained in this book and cannot accept any legal responsibility or liability for any errors or omissions that may be made.

A catalogue record for this book is available from the British Library

Library of Congress Catalog Card Number: 97-69705

∞ Printed on acid-free text paper, manufactured in accordance with ANSI/NISO Z39.48-1992 (Permanence of Paper).

Contents

Contributors   X
Preface   XV

SECTION A. INTRODUCTION

A1. Organization and scientific contributions in sedimentation and tectonics of rift basins: Red Sea–Gulf of Aden
B.H. Purser and D.W.J. Bosence   3

Significance of the Red Sea and Gulf of Aden as a natural laboratory for rift basin studies, 3
The organization of this volume, 4
Some major contributions of this volume, 4

A2. Stratigraphic and sedimentological models of rift basins
D.W.J. Bosence   9

Introduction, 9
Rift basin models, 10
Stratigraphic response to different geotectonic rift-basin models, 11
Application of rift models to the Gulf of Suez, Red Sea and Gulf of Aden, 14
Rift sub-basins and facies models, 18
Sequence stratigraphic concepts in rift basins, 21
Conclusions, 25

SECTION B. GEOPHYSICAL, MAGMATIC AND STRUCTURAL FRAMEWORK

B1. Geophysical studies on early tectonic controls on Red Sea rifting, opening and sedimentation
R. Rihm and C.H. Henke   27

Introduction, 29
Crustal domains of the Red Sea inferred from seismic data, 30
Gravity field of the Red Sea, 37
Magnetic trends in the southern and central Red Sea and onshore Yemen, 41
Geometry of Red Sea margins, 41
Plate tectonic reconstruction of the Red Sea, 45
Discussion and conclusions, 48

B2. Pre-, syn- and post-rift volcanism on the south-western margin of the Arabian plate
G. Chazot, M.A. Menzies and J. Baker   50

Introduction, 50
Chronology of magmatism, 50
Arabian plate magmatism, 53
Conclusions, 55

B3. Tectonic and sedimentary evolution of the eastern Gulf of Aden continental margins: new structural and stratigraphic data from Somalia and Yemen
P.L. Fantozzi and M. Sgavetti   56

Introduction, 56
General geologic setting, 57
Stratigraphic framework of the study areas, 59
Tectonic setting of the areas, 61
Discussion, 73
Conclusions, 74

B4. Structure, sedimentation, and basin dynamics
during rifting of the Gulf of Suez
and north-western Red Sea
*W. Bosworth, P. Crevello, R.D. Winn Jr.
and J. Steinmetz*   77

Introduction, 78
Regional setting, 78
Fault block evolution and stratigraphic
   development, 82
Conclusions, 90

B5. Rift development in the Gulf of Suez
and the north-western Red Sea:
structural aspects and related sedimentary
processes
*C. Montenat, P. Ott d'Estevou, J.-J. Jarrige
and J.-P. Richert*   97

General setting, 97
Structural framework, 102
Tectonic evolution, 104
Structural stages and sedimentary processes, 107
Initial doming versus uplift of the rift
   shoulders, 114
Short comparisons with other tectono-sedimentary
   rift sequences, 115
Conclusion, 116

SECTION C. EARLY RIFT SEDIMENTATION AND TECTONICS

C1. Pre-rift doming, peneplanation or subsidence
in the southern Red Sea? Evidence from the
Medj-zir Formation (Tawilah Group)
of western Yemen
*A.-K. Al-Subbary, G.J. Nichols, D.W.J. Bosence
and M. Al-Kadasi*   119

Introduction, 119
Definition and distribution of uppermost Tawilah
   Group sediments, western Yemen, 121
Discussion, 130
Conclusions, 133

C2. Sedimentary evolution of early rift troughs
on the central Red Sea margin,
Jeddah, Saudi Arabia
M.A. Abou Ouf and A.M. Gheith   135

Introduction, 135
Geological setting of the Usfan and Shumaysi
   Formations, 136
Methods, 139
The Usfan Formation, 139
The Shumaysi Formation, 140
Discussion, 141
Conclusions, 144

C3. The sedimentary record of the initial stages
of Oligo-Miocene rifting in the Gulf of Suez
and the northern Red Sea
*C. Montenat, F. Orszag-Sperber, J.-C. Plaziat
and B.H. Purser*   146

Introduction, 146
Geodynamic framework, 146
Early rift sediments, 147
Climate, 154
Synsedimentary deformation related
   to early rifting, 155
Evolution of relief during early rifting, 156
Asymmetry of rift initiation in space and time, 157
Discussion and conclusions, 160

SECTION D. SYN-RIFT SEDIMENTATION AND TECTONICS

D1. Rift-related sedimentation and stratigraphy,
southern Yemen (Gulf of Aden)
*F. Watchorn, G.J. Nichols
and D.W.J. Bosence*   165

Introduction, 165
Previous work, 168
Regional stratigraphy, 168
Tectonic architecture, 168
Pre-rift stratigraphy, 171
Rift sediments, 172
Discussion, 185
Conclusions, 189
Appendix, 189

D2. Miocene isolated platform and shallow-shelf carbonates in the Red Sea coastal plain, north-east Sudan
J.H. Schroeder, R. Toleikis, H. Wunderlich and H. Kuhnett 190

Introduction, 190
Previous work, 191
Location of outcrops and geological framework, 192
Field and laboratory work, 193
Facies analysis, 193
Biostratigraphy, 201
Discussion, 204
Conclusions, 210

D3. Stratigraphy of the Egyptian syn-rift deposits: correlations between axial and peripheral sequences of the north-western Red Sea and Gulf of Suez and their relations with tectonics and eustacy
J.-C. Plaziat, C. Montenat, P. Barrier, M.-C. Janin, F. Orszag-Sperber and E. Philobbos 211

Introduction, 211
Overview of stratigraphic evolution, 211
The early-rift strata, 212
The open-marine syn-rift strata (Group B), 215
The outcropping evaporites series (Group C_2), 219
The post-evaporitic unit (Group D), 220
Conclusion, 222

D4. Extensional tectonics and sedimentation, eastern Gulf of Suez, Egypt
K.R. McClay, G.J. Nichols, S.M. Khalil, M. Darwish and W. Bosworth 223

Introduction, 223
Tectonic setting and geological history of the Gulf of Suez, 224
Stratigraphy of the eastern margin of the Gulf of Suez, 231
Tectono-stratigraphic evolution of the eastern margin of the Gulf of Suez, 237

D5. Carbonate and siliciclastic sedimentation in an active tectonic setting: Miocene of the north-western Red Sea rift, Egypt
B.H. Purser, P. Barrier, C. Montenat, F. Orszag-Sperber, P. Ott d'Estevou, J.-C. Plaziat and E. Philobbos 239

Introduction, 240
The Middle Miocene syn-rift platforms of the north-western Red Sea, 241
Discussion, 263
Conclusions, 270

D6. The tectono-sedimentary evolution of a rift margin carbonate platform: Abu Shaar, Gulf of Suez, Egypt
N.E. Cross, B.H. Purser and D.W.J. Bosence 271

Introduction, 271
Structural setting, 274
Regional syn-rift stratigraphy, 276
Depositional sequences, facies and environments, 277
Discussion, 291
Conclusions, 295

D7. Miocene coral reefs and reef corals of the south-western Gulf of Suez and north-western Red Sea: distribution, diversity and regional environmental control
C. Perrin, J.-C. Plaziat and B.R. Rosen 296

Introduction, 296
Age and regional setting of reef development, 297
Occurrence, geometry and development of coral reefs, 298
Discussion, 312
Conclusions, 318

D8. Miocene periplatform slope sedimentation in the north-western Red Sea rift, Egypt
B.H. Purser and J.-C. Plaziat 320

Introduction, 320
The mid-Miocene slopes and their sediments, 323
Specific slope features and sedimentary dynamics, 333
Discussion and conclusions, 343

D9. The tectonic significance of seismic sedimentary deformations within the syn- and post-rift deposits of the north-western (Egyptian) Red Sea coast and Gulf of Suez
J.-C. Plaziat and B.H. Purser 347

Introduction, 347
Deformation processes and morpho-structural influence on seismite types, 349
Chronological distribution of seismicity in the syn-rift series, 351

Discussion: relationships between seismites and tectonic history, 365
Conclusion, 366

SECTION E. SYN-RIFT CARBONATE DIAGENESIS

E1. Syn-rift diagenesis of the Middle Miocene carbonate platforms on the north-western Red Sea coast, Egypt
B.H. Purser   369

Introduction, 369
The principal diagenetic attributes of the mid-Miocene carbonates, 371
Discussion, 386
Conclusion, 387

E2. The dolomitization and post-dolomite diagenesis of Miocene platform carbonates: Abu Shaar, Gulf of Suez, Egypt
N. Clegg, G. Harwood and A. Kendall   390

Geological setting, 390
Methods, 391
Diagenesis, 392
Conclusions, 402

SECTION F. EVAPORITES AND SALT TECTONICS

F1. A review of the evaporites of the Red Sea–Gulf of Suez rift
F. Orszag-Sperber, G. Harwood, A. Kendall and B.H. Purser   409

Introduction, 409
The temporal distribution of the evaporites (Miocene–Recent), 410
Age of the Miocene evaporite formations, 411
Miocene evaporitic facies and sedimentation, 414
Relationships between Miocene evaporite sedimentation and rift structure, 418
Evaporite tectonics, 422
Post-Miocene evaporites, 423
Discussion and conclusions, 425

F2. Post-Miocene sedimentation and rift dynamics in the southern Gulf of Suez and northern Red Sea
F. Orszag-Sperber, B.H. Purser, M. Rioual and J.-C. Plaziat   427

Introduction, 427
'Pliocene' sedimentation and tectonics, 429
Discussion and conclusions, 444

F3. Salt domes and their control on basin margin sedimentation: a case study from the Tihama Plain, Yemen
D.W.J. Bosence, M.H. Al-Aawah, I. Davison, B.R. Rosen, C. Vita-Finzi and E. Whitaker   448

Introduction, 448
Geological setting of south-eastern Red Sea salt domes, 449
Sedimentology of Salif cover rocks (Kamaran Limestone and Al Milh Sandstone), 457
Discussion concerning the role of salt diapirism in controlling margin sedimentary evolution, 460
Conclusions and implications, 462
Appendix, 463

SECTION G. POST-RIFT AXIAL SEDIMENTS AND GEOCHEMISTRY

G1. Axial sedimentation of the Red Sea transitional region (22°–25° N): pelagic, gravity flow and sapropel deposition during the late Quaternary
M. Taviani   467

Introduction, 467
Geological and hydrological setting of the study area, 468
Axial zone sediments, 469
Discussion, 476

G2. Sedimentation, organic chemistry and diagenesis of cores from the axial zone of the southern Red Sea: relationships to rift dynamics and climate
P. Hofmann, L. Schwark, T. Brachert, D. Badaut, M. Rivière and B.H. Purser   479

Introduction, 479
The mineralogical and textural properties of the axial zone of the Red Sea, 481
Organic geochemical properties and origins of sapropels, 491
Diagenetic properties and origin of lithified crusts, 497
Comparisons with other areas of the Red Sea and Gulf of Aden, 501

Discussion and conclusions, 502
Appendix: analytic methods, 503

G3. Metalliferous sedimentation in the Atlantis II Deep: a geochemical insight
G. Blanc, P. Anschutz and M.-C. Pierret — 505

Introduction, 506
Mineralogy of the lithological units, 507
Geochemical history of the Atlantis II Deep, 517

SECTION H. POST-RIFT COASTAL SEDIMENTS, REEFS AND GEOMORPHOLOGY

H1. Present-day sedimentation on the carbonate platform of the Dahlak Islands, Eritrea
F. Carbone, R. Matteucci and A. Angelucci — 523

Introduction, 524
Environmental parameters, 524
General geological setting, 524
Local structural pattern, 526
Sedimentary patterns, 529
Discussion and conclusions, 533

H2. Quaternary marine and continental sedimentation in the northern Red Sea and Gulf of Suez (Egyptian coast): influences of rift tectonics, climatic changes and sea-level fluctuations
J.-C. Plaziat, F. Baltzer, A. Choukri, O. Conchon, P. Freytet, F. Orszag-Sperber, A. Raguideau and J.-L. Reyss — 537

Introduction, 537
Geographic and climatic context, 538
Chronology of Quaternary deposits and problems of dating, 539
Regional synthesis and discussion, 563
Conclusion, 572

H3. Post-Miocene reef faunas of the Red Sea: glacio-eustatic controls
M. Taviani — 574

Introduction, 574
Late Quaternary faunal turnovers, 579
Conclusions, 582

H4. Modern Red Sea coral reefs: a review of their morphologies and zonation
W.-C. Dullo and L. Montaggioni — 583

Introduction, 583
Siliciclastic control, 584
Modern reefs of the Aqaba–Eilat Gulf areas, 585
Tectonic control, 587
Salt diapir control on reefs, 592
Sea-level control and reef terraces, 593
Conclusions, 594

H5. Tectonic geomorphology and rates of crustal processes along the Red Sea margin, north-west Yemen
I. Davison, M.R. Tatnell, L.A. Owen, G. Jenkins and J. Baker — 595

Introduction, 595
Geologic history, 596
Stratigraphy, 598
Geomorphology of north-west Yemen, 599
Rates of crustal processes, 605
Discussion, 610
Conclusions, 611

References — 613
Index — 653

Contributors

M.A. Abou Ouf
Dept of Marine Geology
Faculty of Marine Science
King Abdulaziz University
Jeddah
Saudi Arabia

M.H. Al-Aawah
Dept of Geology
University of Sana'a
Sana'a
Republic of Yemen

M. Al-Kadasi
Dept of Geology
University of Sana'a
Sana'a
Republic of Yemen

A.K. Al-Subbary
Dept of Geology
Royal Holloway, University of London
Egham
Surrey TW20 0EX
UK

A. Angelucci
Dipartimento di Scienze della Terra
Universita 'La Sapienza'
P A Moro 5
00185 Rome
Italy

P. Anschutz
Département de Géologie et Océanographie (DGO)
URA CNRS 197
Université Bordeaux 1
Avenue des Facultés
33405 Talence Cedex
France

D. Badaut
Laboratoire de Géologie
Muséum National d'Histoire Naturelle
Paris
France

J. Baker
Dept of Geology
Royal Holloway, University of London
Egham
Surrey TW20 0EX
UK

F. Baltzer
Département des Sciences de la Terre
URA 723
Bâtiment 504
Université de Paris Sud
91405 Orsay Cedex
France

P. Barrier
Institut Géologique Albert de Lapperent (IGAL)
Institut Polytechnique Saint-Louis
13 Boulevard de l'Hautil
F 95092 Cergy-Pontoise Cedex
France

G. Blanc
Departement de Géologie et Océanographie (DGO)
URA CNRS 197
Université Bordeaux 1
Avenue des Facultés
33405 Talence Cedex
France

D.W.J. Bosence
Dept of Geology
Royal Holloway, University of London
Egham
Surrey TW20 0EX
UK

W. Bosworth
c/o Marathon Oil Company
Cairo Office
PO Box 1228
Houston
TX 77251-1228
USA

T. Brachert
Institut für Geowissenschaften
Universität Mainz
Germany

F. Carbone
c/o Dipartimento di Scienze della Terra
Universita 'La Sapienza'
P A Moro 5
00185 Rome
Italy

G. Chazot
Sciences de la Terre
Ecole Normale Supérieure
69364 Lyon Cedex 07
France

A. Choukri
Sciences de la Terre
Ecole Normale Supérieure
69364 Lyon Cedex 07
France

N. Clegg
School of Environmental Sciences
University of East Anglia
Norwich NR4 7TJ
UK

O. Conchon
Département des Sciences de la Terre
URA 723
Bâtiment 504
Université de Paris Sud
91405 Orsay Cedex
France

P. Crevello
Dept of Petroleum Geosciences
University of Brunei
Bandar 2028
Darassalam
Brunei

N.E. Cross
Badley Ashton Associates
Winceby
Horncastle
Lincolnshire LN9 6PB
UK

M. Darwish
Geology Department
Faculty of Science
Cairo University
Cairo
Egypt

I. Davison
Dept of Geology
Royal Holloway, University of London
Egham
Surrey TW20 0EX
UK

W.-C. Dullo
GEOMAR Research Centre for Marine
Geosciences
Wischhofstrasse 1–3
Geb 4
D-24148 Kiel
Germany

P.L. Fantozzi
Dept of Earth Sciences
University of Siena
Via delle Cerchia 3
53100 Siena
Italy

P. Freytet
Département des Sciences de la Terre
URA 723
Bâtiment 504
Université de Paris Sud
91405 Orsay Cedex
France

A.M. Gheith
Dept of Marine Geology
Faculty of Marine Science
King Abdulaziz University
Jeddah
Saudi Arabia

C.H. Henke
Terrasys
Eiffestrasse 78
20537 Hamburg
Germany

P. Hofmann
Geologisches Institut
Universität Köln
Germany

M.-C. Janin
Université Pierre et Marie Curie
Laboratoire de Micropaléontologie
CNRS
4 place Jussieu
75252 Paris Cedex 05
France

J.J. Jarrige
Elf-Aquitaine
CSTJF
Avenue Larribau
64108 Pau Cedex
France

G. Jenkins
Dept of Earth Sciences
University of Wales
PO Box 914
Cardiff CF1 3YE
UK

A. Kendall
School of Environmental Sciences
University of East Anglia
Norwich NR4 7TJ
UK

S.M. Khalil
Dept of Geology
Royal Holloway, University of London
Egham
Surrey TW20 0EX
UK

H. Kuhnett
Universität Bremen
Fachbereich 5
Geowissenschaften
Postfach 330 440
28334 Bremen
Germany

R. Matteucci
Dipartimento di Scienze della Terra
Universita 'La Sapienza'
P A Moro 5
00185 Rome
Italy

K.R. McClay
Dept of Geology
Royal Holloway, University of London
Egham
Surrey TW20 0EX
UK

M.A. Menzies
Dept of Geology
Royal Holloway, University of London
Egham
Surrey TW20 0EX
UK

L. Montaggioni
URA 1208 CNRS
Université de Provence
Centre de Sédimentologie et Paléontologie
3 Place V Hugo
F-13331 Marseille Cedex 3
France

C. Montenat
Institut Géologique Albert de Lapperent
(IGAL)
Institut Polytechnique Saint-Louis
13 Boulevard de l'Hautil
F 95092 Cergy-Pontoise Cedex
France

G.J. Nichols
Dept of Geology
Royal Holloway, University of London
Egham
Surrey TW20 0EX
UK

F. Orszag-Sperber
Département des Sciences de la Terre
URA 723
Bâtiment 504
Université de Paris Sud
91405 Orsay Cedex
France

P. Ott d'Estevou
Elf-Aquitaine
CSTJF
Avenue Larribau
64108 Pau Cedex
France

L.A. Owen
Dept of Earth Sciences
University of California
Riverside
California 92521
USA

C. Perrin
Laboratoire de Paléontologie
Museum National d'Histoire Naturelle
8 rue Buffon
75005 Paris
France

E. Philobbos
Dept of Geology
University of Assiut
Faculty of Sciences
71516 Assiut
Egypt

M.-C. Pierret
Centre de Géochimie de la Surface (CGS)
UPR 6251
Université Louis Pasteur
1 Rue Blessig
67084 Strasbourg
France

J.-C. Plaziat
Département de Sciences de la Terre
Bâtiment 504
Université de Paris Sud
91405 Orsay Cedex
France

B.H. Purser
Laboratoire de Petrologie et Paléontologie
Bâtiment 504
Université de Paris Sud
91405 Orsay Cedex
France

A. Raguideau
Département des Sciences de la Terre
URA 723
Bâtiment 504
Université de Paris Sud
91405 Orsay Cedex
France

J.-L. Reyss
Centre des Faibles Radioactivités
Laboratoire mixte CNRS/CEA
Domaine CNRS
F-91198 Gif-sur-Yvette Cedex
France

R. Rihm
Geomar Foschungszentrum für Marine
Geowissenschaften
Wischhofstrasse 1–3
24148 Kiel
Germany

M. Rioual
Départment des Sciences de la Terre
URA 723
Bâtiment 504
Université de Paris Sud
91405 Orsay Cedex
France

M. Riviére
Laboratoire de Petrologie et Paléontologie
Bâtiment 504
Université de Paris Sud
91405 Orsay Cedex
France

B.R. Rosen
Dept of Palaeontology
Natural History Museum
Cromwell Road
London SW7 5BD
UK

J.H. Schroeder
Technische Universität Berlin
Institut für Angewandte Geowissenschaften II
Sekr. EB 10
Ernst-Reuter Platz 1
01587 Berlin
Germany

L. Schwark
Geologisches Institut
Universität Köln
Germany

M. Sgavetti
Institute of Geology
University of Parma
Viale delle Scienze 78
43100 Parma
Italy

J. Steinmetz
Montana Bureau of Mines & Geology
Butte
Montana 59701–8997
USA

M.R. Tatnell
Dept of Geology
Royal Holloway, University of London
Egham
Surrey TW20 0EX
UK

M. Taviani
Instituto di Geologia Marina
CNR
Via Gobetti 101
40129 Bologna
Italy

R. Toleikis
Technische Universität Berlin
Institut für Angewandte Geowissenschaften II
Sekr. EB 10
Ernst-Reuter Platz 1
01587 Berlin
Germany

C. Vita-Finzi
Dept of Geological Sciences
University College London
Gower Street
London WC1E 6BT
UK

F. Watchorn
Dept of Geology
Royal Holloway, University of London
Egham
Surrey TW20 0EX
UK

E. Whitaker
Dept of Geology
Royal Holloway, University of London
Egham
Surrey TW20 0EX
UK

R.D. Winn Jr.
Dept of Geology
University of Papua New Guinea
PO Box 414
University
NCD
Papua New Guinea

H. Wunderlich
Technische Universität Berlin
Institut für Angewandte Geowissenschaften II
Sekr. EB 10
Ernst-Reuter Platz 1
01587 Berlin
Germany

Preface

The Red Sea–Gulf of Aden is an active maritime rift system which is unique in the world. Interest in rifts is considerable because they are vital in terms of understanding the fundamental mechanisms of plate movements and their related seismicity, and because the sedimentary basins which result can be important in terms of hydrocarbon and metalliferous potential. Furthermore, the Red Sea–Gulf of Aden rift system covers about 20° of latitude and therefore is also a favourable site for the study of past climatic changes through a particularly rich sedimentary record.

In recent years, the European Community programme SCIENCE (Stimulation of the Cooperation and Interchange among European Scientists) has financed several research groups working in the important field of rifting. Thus, from 1988 to 1994, four rift projects were implemented which strengthened the collaboration between the research teams, developed competences, and reinforced the mobility of researchers, thus contributing to the understanding of the Red Sea–Gulf of Aden rift system.

This volume, comprising eight sections, presents results of extensive research carried out by 62 scientists based mainly in different European and Middle East countries. It represents a valuable reference for all potential users (students, geoscientists and climatologists), since it covers most aspects of rifting mechanisms, starting with the geophysical, magmatic and structural framework of rifts, through early rift sedimentation and tectonics, syn-rift sedimentation, diagenesis of rift sediments and, finally, post-rift sedimentation and geochemistry. The originality of this volume is that, for the first time, rift evolution is treated from many angles. The compilation includes items of varying scale ranging from 'broad-brush' plate tectonics to microscopic sedimentary components, all relating to a common factor – rift dynamics and evolution. This compilation is also based both on classical onshore field studies and on offshore oceanographic methods, giving a balanced picture of rift dynamics and its sedimentary expression. It also discusses both the superficial and deep-crustal expressions of a given set of geological events.

Finally, the Red Sea–Gulf of Aden volume represents a genuine step forward in the knowledge of the sedimentary processes within rift systems. Although discussed mainly in terms of a specific rift-type basin, the results may be readily extrapolated to other parts of the world and thus be utilized as a comparative model for the better understanding of rift-type basins in general. In particular, this volume provides a unique source of new data readily accessible to all researchers working on the plate tectonics of new ocean basins and the economically important sedimentary fill of rifts.

Gilles Ollier
European Commission
Directorate General XII
Science, Research and Development

Section A
Introduction

Chapter A1
Organization and scientific contributions in sedimentation and tectonics of rift basins: Red Sea–Gulf of Aden

B. H. Purser and D. W. J. Bosence

This volume comprises thirty-two chapters arranged in eight sections examining the various controls on the sedimentological evolution of the Red Sea–Gulf of Aden rift system. The book synthesises and provides new case studies of land-based and marine research carried out by 62 scientists based largely in Europe and the Middle East. This introduction outlines the significance of the Red Sea–Gulf of Aden to studies of rift basins, the organization, and the major contributions of this volume.

SIGNIFICANCE OF THE RED SEA AND GULF OF ADEN AS A NATURAL LABORATORY FOR RIFT BASIN STUDIES

The scale and geography of the Red Sea–Gulf of Aden rift system

The rift system extends from the Gulf of Aden to the northern end of the Gulf of Suez – a distance of about 3500 km. Because the Red Sea is essentially oriented north-north-west–south-south-east, it covers 20° degrees of latitude and thus a considerable range of climates ranging from tropical oceanic in the Gulf of Aden, to temperate intracontinental settings in the northern Red Sea and Gulf of Suez. These major geographic variations are also associated with varying degrees of aridity relating partly to increased monsoon effects towards the Indian Ocean. Since the late Oligocene this rift system has included a wide range of continental and marine environments, with present depths locally attaining 2200 m. The hydrographic factors are also influenced by the presence of the Bab el Mandab sill (present depths about 140 m) whose presence has influenced marine faunas, notably during lowering of global sea level.

This wide range of climatic and hydrographic variations, together with a complex tectonic history, are reflected in the great variety of sediments whose depositional environments include continental, coastal, open shelf, deep slope and oceanic settings. The Red Sea–Gulf of Aden rift system therefore is justly considered as the world's best natural laboratory within which one may study the tectonic and climatic controls on a wide range of sedimentary environments.

The sedimentary expression of rift evolution

The Red Sea–Gulf of Aden rift system, since its initiation in the late Oligocene, has evolved from a series of shallow continental depressions to a relatively deep, oceanic trough. Its history therefore is recorded in continental, transitional and marine environments, relating to different stages of structural evolution. This rift basin, therefore, is not limited to lacustrine, intracontinental or oceanic settings but reflects extensional tectonism in a range of major sedimentary regimes.

Evolutionary stages of the rift system

Various stages of rift evolution are recorded in different parts of the rift; the Gulf of Aden segment has evolved from rift to drift and is now in the post-rift phase, the Red Sea is in an early post-rift stage with incipient ocean basin formation, while the southern parts of the Gulf of Suez are still in the syn-rift phase. Thus, virtually all phases of rift geodynamics may be examined within this system.

Sedimentation and Tectonics of Rift Basins: Red Sea–Gulf of Aden. Edited by B.H. Purser and D.W.J. Bosence. Published in 1998 by Chapman & Hall, London. ISBN 0412 73490 7.

Variable geotectonic settings of the rift system

The system comprises both active, plume-related rifting in the southern Red Sea as well as passive rifting in the Gulf of Aden, northern Red Sea and Gulf of Suez. These basic differences are reflected in the nature and timing of major unconformities in the rift basins and occurrence of thick, pre- and syn-rift volcanics in the southern Red Sea.

Economic implications

Rift studies presented in this volume are based on outcrop, remote sensing and subsurface data and as such, some provide integrated three-dimensional pictures of rift basin evolution essential for exploration and development of rift basin siliciclastic and carbonate reservoirs. This is particularly significant for providing analogues for exploration in the deeply buried, rifted North Sea and Atlantic margins.

The exceptional quality of outcrops

Because of arid conditions and considerable peripheral uplift, outcrops of pre-, syn-, and post-rift strata are often exceptionally well exposed; this being particularly true of the north-west Red Sea and Gulf of Suez, and the Gulf of Aden. Large outcrops cut by numerous deep wadi systems often enable three-dimensional study of reservoir-scale sediment geometries and the evolution of fault and fracture systems. A more humid climate would certainly have masked many of the attributes of rift sedimentation and tectonics.

The above factors combine to make the Red Sea–Gulf of Aden one of the world's best natural laboratories in which to study the geological evolution of extensional basins.

THE ORGANIZATION OF THIS VOLUME

The 32 individual contributions comprising this volume cover most aspects of sedimentation and rifting. Following a review of rift basin stratigraphy section (A), section B reviews the magmatism and structural evolution of the Red Sea–Gulf of Aden rift that form an important framework for the subsequent sedimentological chapters. Eight further sections (C to H) treat specific phases or aspects of rift evolution from pre-rift to post-rift environments. The geographic and stratigraphic localities of these studies within the rift basin (Figures A1.1 and A1.2) should help selective reading. References cited in individual contributions are all grouped in a single list at the end of the volume.

Although most aspects of rifting are treated, this volume should not be regarded as a synthesis of the sedimentary evolution of the rift basin. Rather, it is an attempt to describe, analyse and compare the many processes related to rifting and thus may help in the understanding of other comparable sedimentary basins. Readers concerned with palaeogeographic evolution of the region are advised to consult the recent World Bank-sponsored project (Crossley et al. 1992).

This volume is concerned mainly with the relationships between sedimentation and rifting. In spite of the fact that sedimentation in the Red Sea–Gulf of Aden basins is an extensional basin, many authors also stress the importance of climatic and other factors unrelated to rifting. Thus the volume tries to give a balanced picture of all the major processes concerned with sedimentation within the rift system.

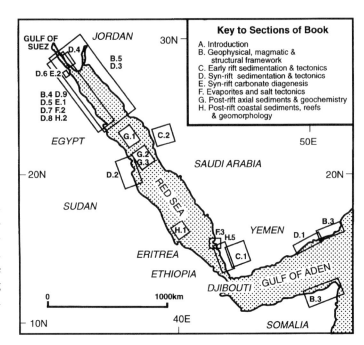

Figure A1.1 Map of the Red Sea and Gulf of Aden showing the location of the various contributions (chapters) to this volume. Numbers correspond to the chapters.

SOME MAJOR CONTRIBUTIONS OF THIS VOLUME

Geophysical, structural, and magmatic framework

Geophysical data on the nature and timing of basin formation
The abrupt ocean–continental crust transition along the western margin of the Red Sea suggests that the early basin had a transform margin that was subsequently (22 Ma) dominated by extensional tectonics. Axial oceanic crust may have started at 10–12 Ma in the Red Sea but with rapid dike injection in the axial troughs starting at about 5 Ma.

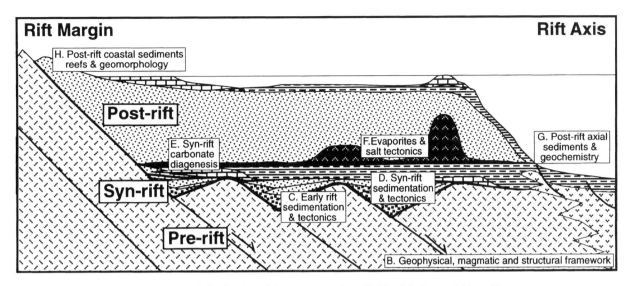

Figure A1.2 Cartoon showing the geological distribution of the various sections (B–H) of the book within a rift system.

Plume-related magmatism dominates the eastern margin of the Red Sea
Magmatism in Yemen and Saudi Arabia is linked to upwelling mantle plumes beneath Afar and ocean-floor spreading in the Red Sea and Gulf of Aden. $^{40}AR/^{39}Ar$ dating reveals three phases of volcanism related to rift basin evolution: the pre-rift, large igneous province in Yemen and alkaline volcanism in Saudi Arabia, syn-rift dikes, plutonic and hyperbyssal rocks, and the very extensive Saudi Arabian and Yemeni post-rift alkaline province.

New outcrop, Landsat, well and seismic data reveal insights into structural evolution
Structural and stratigraphic reconstructions of opposing margins of the Gulf of Aden indicate west-north-west–east-south-east trending half-graben separated by transfer zones developed in the early- to mid-Oligocene. Transfer zones coincide with later transform faults and half-graben (< 20 Ma) with sea-floor spreading centres. In the younger, syn-rift, Gulf of Suez, detailed analyses reveal a multiphase structural evolution with inherited north-north-east trending transfer faults in the earliest Miocene.

1. Pre- and early-rift environments

Cretaceous and Eocene sediments are well developed in Yemen as the Tawilah Sandstone which has much in common with the continental parts of the Nubian Sandstone of Egypt, the latter being the principal oil reservoir in the Gulf of Suez. In Egypt the syn-rift unconformity is often a flat, slightly angular discordance generally lacking conglomerates or other erosional features.

Inheritance of Precambrian structural trends
Rejuvenation of pre-rift fault systems within the underlying crystalline basement has strongly influenced the geometry of early- and syn-rift structural trends and thus the distribution and the nature of sediments within the rift.

The absence of pre-rift doming
This is indicated by the nature of the syn-rift unconformity, by the preservation of pre-rift sediments within the axial parts of the Gulf of Suez and northern Red Sea, and by the frequently fine-grained textures of early syn-rift sediments in the Gulf of Suez and in the Gulf of Aden.

Early syn-rift morpho-structural framework
Initial strike-slip faulting was often oblique and discontinuous with respect to the subsequent north-north-west–south-south-east axis in the Gulf of Suez. It was associated with low continental relief whose depressions (half-graben) were filled with alluvial fan, playa and flood-plain sediments. Comparable facies also occur in Yemen, Somalia, northern Sudan, in Saudi Arabia and in Egypt. Lacustrine deposits generally are poorly developed.

Magmatic activity
The main phase of magmatic activity (31–26 Ma) predates extension in the southern Red Sea. Basalts in the north-west Red Sea and Gulf of Suez are younger (25 Ma) and are interbedded within the continental, early-rift sediments.

2. Syn-rift environments

First marine transgression
Latest Oligocene (Chattian) faunas are recorded from the Gulf of Aden, Red Sea and the Gulf of Suez while Aquitanian microfaunas, identified from the north-west Red Sea (Gebel Honkorab), post-date local block faulting. The regional distribution of dated marine sediments is insufficient to confirm the frequently proposed northwards propagation of the rift system, e.g. opening from the south (Indian Ocean). The extensive post-31 Ma magmatism and uplift in the southern Red Sea also argues against a connection through to the Gulf of Aden. Numerous reefs of mid-Miocene (Langhian–Serravallian) age which have Mediterranean coral faunas confirm independence, at least during the mid-Miocene, of the Red Sea from the Gulf of Aden.

Major phase of rifting
Probably diachronous on a regional scale, the major subsidence and creation of tilt-blocks occurred during the early Miocene both in the Red Sea and Gulf of Suez. In the latter region it is expressed by the formation of north-north-west–south-south-east oriented tilt-blocks and half-graben containing up to 4000 m of Lower to Middle Miocene open marine sediment. In the Gulf of Aden where rift sedimentation starts in the early Oligocene, the marginal relief was lower and sediment supply higher, and the classic half-graben fills are not developed.

The tilt-block/half-graben model
Early Miocene tilt-blocks, notably in north Sudan, north-west Red Sea (Egypt) and southern Gulf of Suez, constitute nuclei for the accumulation of shallow marine sediments whose geometries, depositional sequence boundaries and thickness variations are controlled to a large extent by the structural attributes and tectonic rejuvenation of each block. However, this tilt-block model does not apply to syn-rift sedimentation in the Gulf of Aden (southern Yemen and Somalia) which is characterized by low relief and little wedging of strata into half-graben.

Carbonate platform initiation and development
Carbonate platforms are often regarded as post-rift features. However, in the north-western Red Sea (Sudan and Egypt), shallow marine carbonates with multiple reef bodies, develop in the syn-rift phase near the crests of tilt-blocks. Accretion, controlled by accommodation space generated by synsedimentary tilting, varies according to the tilt direction, axial plunge and position of the block with respect to the rift periphery.

Siliciclastic sedimentation and cross-faulting
Topographic corridors created by cross-faults and/or transfer zones tend to funnel siliciclastic sands and gravels towards the axis of the rift where they accumulate as localized shallow marine fans and turbidites.

Evaporite sedimentation and rift evolution
Unlike many rift basins, the evaporites of the Red Sea rift system are not limited to the initial marine transgression but occur throughout the syn- and post-rift phases. The principal mid- to late Miocene evaporites partly predate the well-known Messinian evaporites of the Mediterranean basin. Increasing salinity was caused by progressive isolation from the parental Mediterranean by a topographic barrier situated north of the Gulf of Suez.

Salt-tectonics
Rejuvenation of pre-evaporite extensional fault systems has stimulated diapirism and other evaporite remobilization structures in the southern Gulf of Suez and the Red Sea. The bathymetric relief created by halokinetic movements has strongly influenced late and syn-rift sedimentation; carbonate platforms have formed on the diapirs whose flanks are onlapped by siliciclastics.

Diagenetic modifications of syn-rift carbonates
Most carbonates have been dolomitized or replaced by secondary sulphate. Isotopic signatures suggest that dolomitization is related to heated sea waters whose movements are favoured by the steep bathymetric relief permitting the penetration of sea water into the sedimentary cover, and by high thermal gradients stimulating convectional circulation.

Diachronism of the syn- to post-rift transition
The age of the post-rift unconformity varies from the Gulf of Aden where it is late Miocene to the Red Sea where it occurs at the end of the Pliocene. Therefore, the Gulf of Aden and the Red Sea are in the post-rift phase. The southern parts of the Gulf of Suez have undergone subsidence and some block rotation since the Miocene and are still in the syn-rift stage, while the central and northern parts of the Gulf of Suez appear to be more stable. In sum, the rift system has evolved to a greater degree within the Gulf of Aden than in the northern segments.

Tectonic control of syn-rift stratigraphy
In Yemen, Sudan and in Egypt, lithostratigraphic boundaries, marked by angular unconformities, generally coincide with major changes in sedimentary environment recording major changes in the architecture of the rift. These boundaries may be correlated for several hundreds of kilometres, although they may not be synchronous. On a smaller scale, depositional sequence boundaries may result either from eustatic sea-level fluctuation or, more frequently, to tilting of individual blocks and cannot be correlated from one tilt-block to another.

Tectonic instability recorded by synsedimentary deformation
Most early, syn- and post-rift sediments exhibit stratiform deformation whose character and frequency depend on the diagenetic state of the sediment and its proximity to inclined surfaces. These deformations may be formed *in situ* or may result from intrastratal shearing and sliding. Hydroplastic and brittle (breccia) deformations probably record seismic events. These 'seismites' are particularly frequent near tectonically controlled stratigraphic boundaries.

3. Post-rift environments

Ocean-floor spreading and drifting
In the Gulf of Aden ocean-floor spreading and dike injection occurred at about 20 Ma. In the southern Red Sea scarps bordering the axial trough cut both Miocene evaporites ('reflector S') and much of the Plio-Quaternary cover. Local cells of basaltic extrusion started at 10–12 Ma and major drifting appears to have begun about 5 Ma, e.g. early Pliocene. There is no axial magmatism in the northern Red Sea although peri-axial scarps are well developed. Because drifting is limited to the north-west by the Aqaba transform zone, there is no deep axial zone in the Gulf of Suez.

Deep axial sedimentation and diagenesis
Post-rift sediments within the axial zone have multiple origins: iron and manganese oxides and anhydrite are related to axial brines while argillaceous clays (smectite, palygorskite) relate both to detrital (continental) sources and ocean-floor diagenesis. Relatively abundant (up to 40%) detrital quartz and feldspar have been delivered by fluviatile and aeolian processes while pelagic carbonates are composed of globigerinid foraminifera and pteropods. Both the pelagic carbonates and the siliciclastic sands occur in the form of turbidites derived from the steep escarpments flanking the axial trough.

Peripheral reef terraces and structural stability
Pre-Quaternary (Pliocene) strata outcropping along the north-west coast of the Red Sea dip towards the rift axis and are truncated by a series of subhorizontal reef terraces dated by U/Th isotopes. The most extensive, 125 000 Y (Eemian) terrace has a relatively constant altitude (about 8 m) along 400 km of Red Sea coast indicating tectonic stability. However, within the south-western parts of the Gulf of Suez it varies in height, and is locally tilted to 20 m, indicating that this segment of the rift is tectonically active and still in the syn-rift phase.

Peripheral uplift
Fission track and geomorphological studies both in north Yemen and in Egypt show that peripheral uplift relating to syn-rift isostatic compensation, is still active. Uplift of the rift shoulders occurred mainly during the Miocene.

4. Non-tectonic factors affecting sedimentation within the rift basin

Arid climates have favoured the coexistence of carbonates and siliciclastics
Although syn-rift carbonates are better developed on the crests of tilt-blocks and siliciclastics are deposited either in the axes of half-graben or within cross-fault depressions, both facies are often interbedded, both on highs and in lows; carbonates and clastics are not necessarily antagonistic. This intimate association seems to be related to weathering in dry climates which produces little argillaceous sediment. Coarse clastics are delivered during intermittent periods of flash-flooding and the rapid return to clear waters favours renewed carbonate sedimentation.

Winds today, in the Red Sea and probably in the past, are predominantly from the north-west and therefore are parallel to the rift axis. Within mid-Miocene and Pliocene syn-rift sediments this polarity is expressed by the leeward accretion of some shelf and slope deposits.

Glacial eustacy
Global eustatic effects expressed by sequence boundaries are difficult to demonstrate within the syn-rift sediments because of repeated structural movements. However U/Th and strontium isotope dates illustrate that glacio-eustacy, and locally salt tectonics, controlled sea-level changes in the Quaternary. On open marine shelves and in deep axial environments, the 19 000 BP glacial low is recorded by isotopically positive lithified crusts often associated with sapropels. These crusts formed in hypersaline waters whose reducing environments favoured the preservation of organic matter. These stagnant bottom waters were related to the barrier-effects of the Bab el Mandab sill which almost isolated the Red Sea during the last glacial maximum. This isolation also resulted in the extinction of many Red Sea faunas whose present character is related to post-glacial Indian Ocean influx.

ACKNOWLEDGEMENTS AND COLLABORATION

The research presented in this volume has been supported by industry and government grants mentioned in individual contributions. Research by the French team, mainly in Egypt, was carried out initially under the project 'GENEBASS' (Genèse des Bassins Sédimentaires), funded by French oil companies ELF and

TOTAL-CFP and the Centre National des Recherches Scientifiques (CNRS). Practical help has been offered by the Compagnie Générale de Géophysique (CGG). British funding has come mainly from the Royal Society, National Environmental Research Council (NERC), British Council and the following oil companies: British Petroleum, Canadian Occidental, Marathon, Arco British, Conoco, Mobil North Sea, Petrobras and Sun Oil Britain.

Funding for research by German colleagues was also provided by the German Research Foundation while Italian research was funded partly by the Italian National Research Council.

In 1990 a consortium of European researchers formed the group 'RED SED' which was funded for 4 years by the European Economic Community 'Science Program'. Many results presented in this volume are derived from this EC-funded research. The RED SED project has also organized international meetings in Cairo (1992) and in Sana'a (1995) which concerned sedimentation and rifting in the Red Sea and Gulf of Aden.

Research in the Red Sea and Gulf of Aden by European workers has been carried out in collaboration with researchers of the bordering states: Egypt (University of Assiut), Saudi Arabia (King Abdulaziz University (Jeddah), Sudan (University of Port Sudan), and Yemen (University of Sana'a) and considerable support has been given by the Egyptian General Petroleum Corporation (EGPC). This help and collaboration is gratefully acknowledged.

Finally, publication of this volume has been helped considerably by European Community funding (Program 'Science') which is also acknowledged.

Chapter A2
Stratigraphic and sedimentological models of rift basins

D. W. J. Bosence

ABSTRACT

This chapter reviews models for the stratigraphic and sedimentological evolution of rift basins at the basin scale and at the half-graben or facies model scale. Different models have been presented for the geotectonic evolution of rifts (pure shear, simple shear, heterogeneous stretching and volcanic- or plume-related rifts) and these are reviewed. The different stratigraphic signatures of each of these geotectonic models are discussed in terms of the nature and the occurrence of pre-rift strata, syn-rift unconformity, syn-rift strata, post-rift unconformity and post-rift strata. This simple stratigraphic analysis reveals that each geotectonic model should have a unique stratigraphic signature that has not been previously documented. The stratigraphic signatures of the Gulf of Suez, Red Sea and Gulf of Aden are reviewed and compared with the geotectonic models previously proposed for these basins. It is concluded that the pure shear model is applicable to all basins and that the southern Red Sea is modified by plume-related volcanism.

Rift basins are reviewed at the half-graben or facies model scale where recent advances in structural and sedimentary geology have revealed a complex and evolving three-dimensional template for erosion, sediment production, sediment transport and deposition. New facies models for the tectono-sedimentological evolution of rift basins are proposed in the light of the three-dimensional structural models. Recent work applying sequence stratigraphic concepts to rift basins is reviewed which illustrates that the recognition and nature of depositional sequences at a variety of scales gives important information on the tectono-sedimentary evolution of rift basins. These include the timing and style of fault movement, the complexity of relative sea-level changes on rotating fault blocks and transfer zones, and the temporal and spatial isolation of clastic and carbonate facies associations in depositional sequences.

INTRODUCTION

The stratigraphy of rift basins is controlled by a variety of tectonic, climatic, magmatic and sedimentary processes which all leave distinctive signatures in the sedimentary record. A number of different geotectonic models of extensional basin formation have been proposed and applied to different rift basins. The stratigraphic signatures of these different rift process models have not been investigated in such detail and this is addressed in this chapter. The basin-scale stratigraphic history of the Gulf of Suez, Red Sea and Gulf of Aden is also reviewed in the light of new data presented in this volume and compared with the various geotectonic models that have been proposed for these basins.

At the sub-basin scale the evolving three-dimensional history of half-graben and transfer or accommodation zones is crucial to the understanding of the development of facies and depositional sequences. Previously, largely two-dimensional facies models have been presented for half-graben basin evolution and these are reviewed and revised as three-dimensional models of facies evolution through time to illustrate the different stages of rift basin evolution.

Finally, the application of sequence stratigraphic concepts to extensional basins is discussed. Important characteristics of the tectono-sedimentary evolution of

---

Sedimentation and Tectonics of Rift Basins: Red Sea–Gulf of Aden. Edited by B.H. Purser and D.W.J. Bosence. Published in 1998 by Chapman & Hall, London. ISBN 0412 73490 7.

half-graben that affect depositional sequences are the rotation of fault blocks which leads to synchronous relative sea-level rise and fall in hangingwall and footwall locations respectively, the temporal and spatial isolation of carbonate and clastic facies associations in depositional sequences, and the rates of fault-slip in relation to rates of sediment supply and regional or global sea-level change.

RIFT BASIN MODELS

During the 1970s and 1980s a number of geotectonic models of rift basins were proposed and applied to basins such as the North Sea, Gulf of Suez and the early stages of evolution of the Atlantic Ocean margins. These models have been broadly classified by Sengor and Burke (1978) into passive rift models (pure shear, simple shear and heterogeneous stretching models, below), in which the extension is generated by plate movements external to the rift basin and the input of asthenospheric material is a passive response to lithospheric thinning. Alternatively, active rifting occurs where extension relates to the involvement of hot asthenosphere through decompression melting and volcanic plumes.

In the pure shear model (McKenzie 1978) the crust is uniformly and instantaneously extended by homogeneous, pure shear by faulting and rotation of strata in the brittle upper crust and ductile deformation in the lower crust and the formation of a symmetric rift basin (Figure A2.1(a)). Such symmetric rift basins are defined by two rift border faults dipping towards each other as was proposed by early works on the Rhine graben (reviewed in Einsele, 1992) and the Red Sea (Lowell and Genik, 1972). A two-stage history of basin evolution is proposed based on an isostatic response to the initial stretching and thinning of the brittle crust leading to subsidence followed by the longer period of thermal cooling of the lithosphere and subsidence.

An asymmetric graben structure was subsequently proposed by Wernicke and Burchfiel (1982) and Wernicke (1985) and illustrated in deep seismic reflection profiles (e.g. Beach, 1986; Blundell *et al.*, 1991). These studies indicate that extension may take place by simple shear along an intracrustal detachment surface (Figure A2.1(b)). The low-angle detachment results in a proximal or footwall margin with thinning and associated subsidence of rotated fault blocks in the brittle upper crust, but with little or no thinning of the mantle lithosphere. The distal or hangingwall margin of the basin in this model is affected mainly by thinning of the lithospheric mantle which will undergo thermal expansion (McKenzie, 1978), resulting in uplift and erosion which is then followed by thermal subsidence. Therefore the proximal mechanically rifted area is laterally sepa-

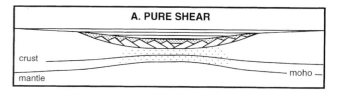

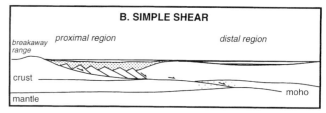

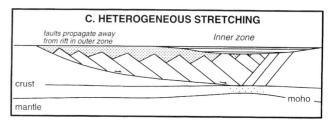

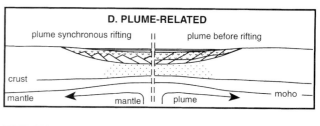

Figure A2.1 Models of the formation rift basins. (a) Pure shear model (after McKenzie, 1978). (b) Simple shear model (after Wernicke and Burchfiel, 1982). (c) Heterogeneous stretching model (after Coward, 1986). (d) Plume-related model (after White and McKenzie, 1989).

rated from the distal thermally subsiding basin (Figure A2.1(b)).

The heterogeneous stretching model of Coward (1986) involves an upper crustal zone of rotated fault blocks which propagate laterally away from the rift on a low-angle lithospheric detachment (Figure A2.1(c)). This is accompanied by extension and thinning of the underlying lower ductile crust and lithospheric mantle. In this case a period of uplift and erosion of fault blocks will occur over the thinned ductile zone from thermal expansion prior to thermal cooling and subsidence.

Many rift basins may be affected by small amounts of magmatic activity throughout their history and in response to lithospheric stretching (Storey *et al.*, 1992). However some rift margins may be locally or temporally associated with mantle plumes and the eruption of large igneous provinces such as the Deccan Plateau of India, the Parana of Brazil and the Yemen–Ethiopia–Eritrea volcanic province in the southern Red Sea (White and

McKenzie, 1989). Such associations have given rise to the active, volcanic- or plume-related rift models (Figure A2.1(d) (White, 1987; White and McKenzie, 1989), or 'open-system extensional model' of Leeder (1995). These are considered to give rise to very different rift basin histories than those generated by lithospheric stretching alone and have been modelled by Braun and Beaumont (1989a). Magmatism occurs when a rifted and thinned continental lithosphere passes over a region of hot asthenosphere or plume. Decompression and partial melting from relatively small rises in temperature lead to massive outpourings of basaltic lava (White and McKenzie, 1989) which is broadly synchronous with rifting. Plume-related magmatism is thought to have an important effect on subsidence, and uplift of extensional basins through the addition of lower density igneous material at the base of the crust (underplating), dynamic support by the upwelling plume and the reduction in density as the asthenosphere cools following volcanism. Depending on the amount of stretching of the lithosphere and the thermal anomaly, it has been proposed that a plume-related rift may 'remain near to or above sea-level as a margin rifts' (White and McKenzie, 1989). If the plume is present prior to rifting then it may cause uplift of up to 1 km or more over a broad 1500 to 2000 km dome. Because of the nonlinear effects of plumes they may be limited both spatially and temporally as indicated by the Yemen margins of the Red Sea and Gulf of Aden (Chazot *et al.*, this volume). Following rifting and magmatism the margin will subside thermally as the asthenosphere cools. However, because of magmatic underplating or residual thermal effects they may not subside as much as rifts generated dominantly by mechanical stretching.

STRATIGRAPHIC RESPONSE TO DIFFERENT GEOTECTONIC RIFT-BASIN MODELS

Classification of rift basin stratigraphy

Rift basin fills have been classified in accordance with the different stratigraphic stages in their tectono-sedimentary evolution. The different usages of these important stratigraphic stages and their bounding unconformities are discussed below.

Pre-rift strata (Figure A2.2)
Falvey (1974) initially introduced the terms infrarift for strata deposited prior to rifting but subsequent authors have largely followed Grow *et al.* (1983) and Badley *et al.* (1984) by using pre-rift for these strata. The upper surface of pre-rift strata is the syn-rift unconformity or a superimposed post-rift unconformity (see below).

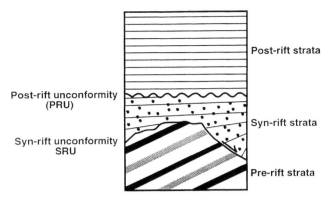

Figure A2.2 Terminology of rift-basin stratigraphy used in this chapter.

Syn-rift unconformity (Figure A2.2, SRU)
The unconformity between the pre-rift and the syn-rift is either a widespread or local erosion surfaces that are readily defined by an unconformity surface on rotated fault blocks with onlap of syn-rift strata (e.g. Schlische and Olsen, 1990). Surprisingly this unconformity does not appear to have been previously named and is named here as the syn-rift unconformity (SRU) as it is generated prior to and during rifting and underlies syn-rift strata.

Syn-rift strata (Figure A2.2)
Syn-rift strata are sediments deposited during active mechanical extension and subsidence (Badley *et al.*, 1984; 1988; Grow *et al.*, 1983) and are equivalent to the 'rift' phase of Falvey (1974) and the 'synrift megasequence' of Hubbard (1988). Syn-rift strata are recognizable by repeated thickening (fanning) of strata into hangingwall basins near faults, by facies variations (e.g. footwall derived fans) adjacent to faults, soft sediment deformation structures and by fewer drag effects (i.e. compactional) adjacent to faults (Prosser, 1993). Care should be exercised in using the term syn-rift because one part of a basin may show the above features but another part may be tectonically inactive during the same period of time. Thus the term syn-rift is used either in a local sense only where there are stratigraphic features indicating syndepositional fault movement, or it may be used in a more general time-stratigraphic sense to indicate a period of time of basin evolution generally characterized by mechanical extension. Because of the difficulty of identifying true syn-rift from post-rift in some seismic sections Frostick and Steel (1993, p. 122) suggest that for convenience of description authors should 'refer to the entire infill between adjacent rotated blocks as syn-rift'; however, this seems a retrograde step and such strata might be better referred to simply as rift strata or rift basin fill.

Post-rift unconformity (Figure A2.2, PRU)

This surface was initially labelled the 'break-up unconformity' by Falvey (1974) as it was considered to be generated by the onset of sea-floor spreading. However, this usage cannot be applied to continental basins or to failed maritime rifts such as the North Sea, which have not experienced sea-floor spreading but do have strata deposited after extensional fault movement has ceased. Grow et al. (1983) and Badley et al. (1984, 1988) therefore use the term post-rift unconformity for the erosional surface which marks the base of strata deposited during thermal or post-rift subsidence phase of basin evolution.

Post-rift strata (Figure A2.2)

Post-rift strata are those overlying the post-rift unconformity and are deposited during the stage of post-rift thermal subsidence (Badley et al., 1984; Grow et al., 1983). Hubbard (1988) uses the terms 'passive margin' or the 'passive wedge' for these strata. Although this stratigraphic stage is labelled as post-rift it should be remembered that the basin may still be classified as a rift basin until it evolves either, seamlessly over time into a mature passive margin of an oceanic basin, or as a failed rift. The post-rift phase is commonly marked by thick, onlapping and offlapping, stratigraphic sections which may be initially sawtooth-shaped if there are remnants of the half-graben topography unfilled from the syn-rift phase. Subsidence is controlled by cooling and increase in density of the lithosphere and asthenosphere and will occur over a larger area than subsidence generated by mechanical extension. Maximum subsidence will occur over sites of the most thinned mantle lithosphere, and differential subsidence may be accommodated by normal fault movement which is planar and vertical (Bosworth et al., this volume). Subsidence is increased by sediment and water loading (Bott, 1992).

While these terms are of use in subdividing the package of rift basin fill it should be appreciated that in detail there may be many episodes of fault movement and quiescence during the syn-rift phase (e.g. Hubbard, 1988; Bosence et al., 1996) and, during the evolution of a rift, faulting may be both episodic and diachronous with different syn-rift to post-rift phases (e.g. Tankard and Balkwill, 1989, Figure 5). The terms defining stratigraphic fills do not imply facies associations or palaeoenvironments and are applicable to rifts which are essentially continental in their origin (e.g. Himalya-Baikjal rift and Rhine rift) or to those that evolve from continental to marine environments (e.g. Red Sea–Gulf of Aden, Reoconcavo, Brazil).

Subsidence and rift basin geometry

The first-order stratigraphic response to continental crustal extension is subsidence and filling of accommodation space. Subsidence occurs through isostatic readjustments to extended and thinned continental crust, through sediment loading of the basin but in part is compensated for by thermal uplift. Bott (1992) gives an example of the different components of subsidence as rifted continental margins evolve into passive margins which have undergone 30 km of crustal thinning. This yields 5.2 km of submarine subsidence in the absence of other factors. Thermal cooling and subsidence will follow an initial phase of decompression melting induced by lithospheric stretching (McKenzie, 1978). Bott (1992) calculates that a relatively rapid 3 km of thermal uplift will follow from the initial 30 km of thinning (above) reducing the effective subsidence to 2.2 km but that this would be followed by about 3 km of slow thermal subsidence so that the net effect is 5.2 km of subsidence. Sediment loading will amplify this response and if average sediment density is taken as 2630 kg/m^3 and accommodation space in the syn-rift and post-rift stages is filled to sea level then sediment thickness will be 3.2 times the original water depth resulting in up to 16.6 km of sediment. Conversely, plume-related rifted margins will be expected to either remain at sea level or be uplifted by up to 1 km (White and McKenzie, 1989) and have very little sedimentary infill.

Viewed at this scale, syn-rift to post-rift basin-fills are saw-shaped in cross-section and thin to a sediment starved basin depocentre and to the rift basin margin. The stratigraphic fill is characterized by aggrading and prograding geometries with surfaces dipping into the enlarging accommodation space in the basin centre. The lower surface of the basin is sawtooth-shaped in section resulting from burial of the half-graben topography (Figure A2.1). Post-rift subsidence is more widespread and post-rift strata onlap outer marginal areas of the rift basin as has been modelled by Cochran (1983a) and may be spatially separated from the mechanically rifted region.

Basin-scale stratigraphic responses to models of rifting

The stratigraphic divisions defined and discussed above provide a very useful terminology for the first-order description and analysis of rift basins. These subdivisions of basin fill can be used to characterize the evolving stratigraphy that develops in response to the different tectonic and magmatic controls within the various models proposed for rifting. The stratigraphic responses to the various geotectonic rift models have not been previously investigated in any consistent way, and this is presented below and in Figures A2.3. and A2.4. For each model schematic stratigraphic panels illustrate the different stratigraphic relations and degree of development of rift-related strata and unconformities in axial and lateral portions of the rift basin.

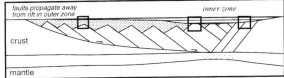

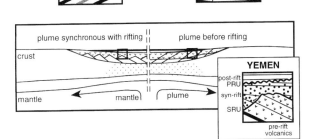

Figure A2.3 Models of rifting and predicted nature of stratigraphic fills in rift basins resulting from (a) pure shear, and (b) simple shear.

In the pure shear model (Figure A2.3(a)) there is no uplift predicted prior to rifting and the rift basin will be locally floored by a syn-rift unconformity developed over the footwall highs of rotated fault blocks. Hangingwall areas may show little stratigraphic break from the pre-rift to the syn-rift. Syn-rift strata may be well developed in basins with rapid sediment supply in response to the accommodation space generated by lithospheric stretching and thinning. The post-rift unconformity may be a prominent surface which will extend laterally over the rift shoulders, and post-rift strata will occur in accommodation space generated by the thermal subsidence phase.

The simple shear model is predominantly characterized by differing distal and proximal regions (Figure A2.3(b)). The syn-rift unconformity is locally developed in proximal areas in response to stretching, fault-block rotation, and footwall uplift as discussed above. However, a regional syn-rift unconformity will be present in the distal region as this area will be uplifted in response to mantle heating and upwelling (Wernicke, 1985). Syn-rift strata will be well developed in the proximal area as accommodation space is generated by the subsiding and rotating fault blocks over the lithospheric detachment. No syn-rift strata are present in distal sites as this area is undergoing uplift and erosion. The post-rift unconfor-

Figure A2.4 Models of rifting and predicted nature of stratigraphic fills in rift basins resulting from (a) heterogeneous stretching, and (b) mantle plume.

mity is present in both distal and proximal regions and in the former is likely to represent a long-lived break in deposition superimposed on the extensive syn-rift unconformity. Post-rift strata are thin in proximal areas because of limited upper mantle thinning and consequent thermal subsidence. Subsidence in this area will also be controlled by the isostatic adjustments caused by erosion in the breakaway range and any addition of magma to the base of the crust (Wernicke and Tilke, 1989). Distal post-rift strata are expected to be widespread in response to thermal subsidence of this area. Basin margin footwall uplift (e.g. rift flank uplift) will relate to a combination of unloading and flexural isostacy which are both synchronous with axial extension and subsidence (Braun and Beaumont, 1989b).

The heterogeneous stretching model (Figure A2.4(a)) also has a very distinctive combination of unconformities and basin fills, and part of the evidence for the erection of the model was based on the occurrence of a regionally extensive syn-rift unconformity around the North Sea Central Graben (Coward, 1986). This may develop locally from a combination of footwall uplift, fault-block rotation but more regionally from isostatic uplift in response to thinning of the lower crust and/or lithospheric mantle. It therefore differs from the simple shear model where mechanical and thermal subsidence are spatially separated. Syn-rift strata may or may not be present in the inner zone of regional uplift but will be well preserved in the outer zone (Figure A2.4(a)) particularly in response to laterally propagating half-graben. A widespread post-rift unconformity overlain by post-rift strata are present in this model which may be superimposed on or cut the syn-rift unconformity in the formerly uplifted inner zone. Thinly developed post-rift strata are expected in areas overlying laterally propagating syn-rift sub-basins as they are not underlain by thinned lower crust and upper mantle.

Plume-related (Figure A2.4(b)) rift basins are distinctive because of their volcanic stratigraphy and their unique subsidence/uplift history. Stratigraphic relations are largely controlled by the timing of when the plume affects the stretching margin (White and McKenzie, 1989). Two hypothetical examples are given in Figure A2.4(b) for when the plume is synchronous with rifting and when the plume is prior to rifting. If synchronous to rifting then a local syn-rift unconformity will be developed on footwall blocks as in the pure shear model (above) which are covered by thick syn-rift volcanics. The post-rift unconformity is followed by relatively thin post-rift strata because of the predicted limited thermal subsidence of these margins (White and McKenzie, 1989). If, however, the plume starts prior to rifting then a 1000–2000 km broad 1–2 km high dome is predicted to form with associated extensional tectonism. This will generate a significant regional syn-rift unconformity followed by thick volcanics and limited post-rift thermal subsidence. The stratigraphy in the south-eastern Red Sea (Yemen) differs from these two predicted patterns as the volcanics are part of the pre-rift (Figure A2.4(b) inset and discussion below). The crustal doming associated with plumes should generate a pronounced radial surface drainage pattern. If the doming is pre-rift then the possibility exists that radial palaeocurrents may be preserved in pre-rift strata. However, pre-rift doming cannot be inferred from present-day radial drainage patterns over some ancient plume-related rifts as has been suggested by Cox (1989) as these patterns may relate to lithospheric doming at any stage of the rifting or post-rifting history.

APPLICATION OF RIFT MODELS TO THE GULF OF SUEZ, RED SEA AND GULF OF ADEN

Of the three areas, the Gulf of Suez (Figure A2.5) has been studied in by far the greatest detail both by academic researchers and by oil company geologists. The Red Sea (Figure A2.5) has not been studied in such detail and here is considered in two areas: the southern portion that was affected by mantle plume volcanism and by initial phases of sea-floor spreading; the northern portion which was not affected by such magmatism and is discussed here together with the Gulf of Suez. The Gulf of Aden has an earlier geological evolution from the Red Sea and today represents a well established ocean basin but the details of its stratigraphy are still poorly known and so sections are not illustrated.

Gulf of Suez (Figure A2.5) and northern Red Sea

Pre-rift strata and syn-rift unconformity
Pre-rift Paleocene and Eocene strata form east–west facies belts and facies deepen and thicken southwards into the Gulf away from the east-north-east–west-south-west Syrian Arc (Kuss, 1992). These strata then thin

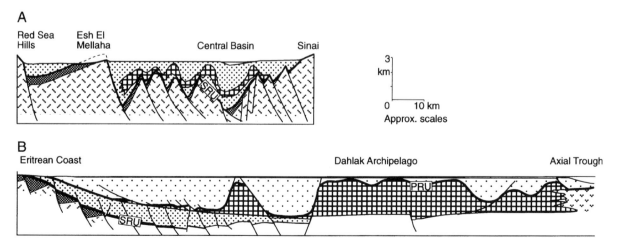

Figure A2.5 Schematic cross-sections through (a) the southern Gulf of Suez (after Bosworth, 1995), and (b) southern Red Sea (after Mitchell *et al.*, 1992) illustrating basin morphology and major unconformities and stratigraphic intervals (not accurately scaled).

southwards on to the Kharga Arch (east-north-east–west-south-west) in the northern Red Sea and therefore appear not be related to any extensional or transtensional tectonics of the essentially Neogene rifting in this area (Patton *et al.*, 1994). The syn-rift unconformity shows only slight angularity where it overlies the Eocene (Figure A2.5; Montenat *et al.*, this volume) and there is no major hiatus on the eastern margin of the Gulf (McClay *et al.*, this volume). These relationships indicate there is no evidence for widespread pre-rift doming in this region (Patton *et al.*, 1994).

Syn-rift strata
Onset of syn-rift sedimentation was earlier on the eastern margin of the rift (Chattian) than on the western margin (Aquitanian) (Dullo *et al.*, 1983; Purser and Hotzl, 1988; Montenat *et al.*, this volume). The earliest sediments on the rift margins (e.g. Abu Zenima Formation) are generally fine grained, following thin, local, basal conglomerates, and indicate no major topographic relief at this time (Montenant *et al.*, this volume). However, subsurface data from the rift axis indicates thick conglomerates and considerable relief (Bosworth, 1995). The earliest marine syn-rift is of Chattian age in the Midyan region of the north-east Red Sea (Dullo *et al.*, 1983) and the early syn-rift Nukhul Formation (Aquitanian) thins northwards through the proto Gulf of Suez rift basin (Patton *et al.*, 1994). Syn-rift strata occur as classic wedge-shaped half-graben fills within the basin (Figure A2.5) in response to a number of locally variable pulses of north-east–south-west extension (e.g. 'mid-clysmic event'; Garfunkel and Bartov, 1977). Hangingwall basins in the late Oligocene to early mid Miocene are characterized by clastics and marls and footwall crests may be carbonate platforms. In mid to late Miocene the basin becomes partially isolated and thick evaporites form (Figure A2.5). Data in this volume (McClay *et al.*; Purser *et al.*) indicate that coarse clastic material derived from uplifted marginal rift shoulders appears to occur synchronously with mechanical extension and subsidence. Fission track studies (Omar *et al.*, 1989) suggest rift shoulder uplift began at 22 Ma±1 which is at, or soon after, the onset of extension. With time the central and southern parts of the Gulf narrowed with post-Miocene subsidence and sedimentation concentrating in the axial parts of the basin (Figure A2.5; Bosworth, 1995).

The early (late Oligocene–early Miocene) stages of rifting in the northern Red Sea and Gulf of Suez are now considered to be similar (Bartov *et al.*, 1980). Left-lateral shear on the Aqaba–Dead Sea fault system, which accommodates the continued extension of the Red Sea, is thought to have started in the middle Miocene (Bartov *et al.*, 1980).

Post-rift unconformity and strata
The major phase of basin extension and subsidence is considered by many (e.g. Bosworth, 1995; Patton *et al.*, 1994) to have ended by the Langhian, but rift shoulder escarpments remain, minor fault movements and unconformities continue through to the present day. Some workers (e.g. Orszag-Sperber *et al.*, this volume) refer to the Pliocene of the north-western Red Sea as post-rift; it is marked by an unconformable lower surface; away from salt structures it is relatively undeformed; it onlaps the basin margins and shows downlap toward the rift axis. Because of these varying signatures it can best be concluded that the region is probably entering the post-rift phase but that this will be dominated by structures generated by salt tectonics (Figure A2.5).

Discussion
Earlier workers have suggested that active rift models are appropriate for the Gulf of Suez because of the high heat flow and the high rift shoulder uplift (Steckler, 1985). The absence of extensive volcanism (restricted to minor volcanics in early phase of syn-rift) and a significant and widespread pre-rift unconformity argue

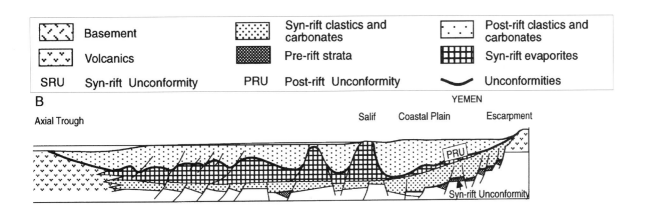

against pre-rift updoming. This suggests that passive rifting models may be more appropriate (McClay et al., this volume). The rift shoulders are symmetric to the basin and were formed synchronous with rifting and therefore they are most likely to have formed through flexural isostacy (Braun and Beaumont, 1989b). There is little asymmetry to the basin, except at the half-graben scale, and no evidence of a large offset or axial area marked by pre-rift doming as is predicted by the simple shear and heterogeneous-stretching models. The basin has yet to move into a post-rift phase.

Therefore the stratigraphic signature of the Gulf of Suez sub-basin, in particular the nature of the pre-rift unconformity, and the symmetry of the basin and the rift flanks suggest that McKenzie's (1978) pure shear model may be the most appropriate.

Southern Red Sea

Pre-rift strata and syn-rift unconformity

The pre-rift strata of the southern Red Sea comprise continental clastics with a regional palaeoslope to the north-west and oblique to the present-day margins of the Red Sea (Al-Subbary et al., this volume). Slow rates of deposition are indicated by extensive laterite palaeosols. Similar pre-rift strata are found in Saudi Arabia (Abou-Ouf and Geith, this volume). These siliciclastic units are conformably overlain by 1.0 to 1.5 km thick, plume-related, basaltic-silicic volcanics of Yemen, Ethiopia and Eritrea (Chazot et al., this volume). These were erupted over 2 m.y. from 30.9 to 26.5 Ma (late Rupelian to Chattian) and were followed by extensional faulting (Davison et al., 1994).

The syn-rift unconformity in the southern Red Sea is poorly known. On the uplifted eastern marginal escarpment erosion is down to pre-rift volcanics and sedimentary rocks. On the coastal plain and shelf syn-rift strata onlap pre-rift volcanics on the Yemen margin (Davison et al., 1994; Mitchell et al., 1992) and on pre-rift Mesozoic strata on the Ethiopian–Eritrean margin (Mitchell et al., 1992) (Figure A2.5). This unconformity is poorly imaged on the sub-salt seismic profiles and published well data are very limited.

Syn-rift strata

Syn-rift strata outcrop in northern Sudan where the earliest marine deposits are late Oligocene in age (Schroeder et al., this volume). Tilted strata of mid Miocene (N13) age have also been dated. These stratigraphic data complement that from the Thio-1 well offshore Eritrea that encountered marine shales with late Oligocene (NP25) foraminifera (Hughes et al., 1991). Syn-rift strata initially thicken away from the coast but then thin toward the rift axis (Figure A2.5) and subsurface data indicate restricted marine muds with marginal continental to coastal clastics of lower to mid Miocene age (Crossley et al., 1992). These are followed by mid to upper Miocene evaporites. Maximum recorded thickness of the syn-rift in Thio-1, where it is not thickened by salt diapirism, is about 3 km. There are few onshore outcrops of syn-rift strata. Seismic images of the sub-salt are generally of poor quality and there are differing views on the upper boundary of the syn-rift. In the World Bank project, evaporites are considered as part of the syn-rift (Figure A2.5) throughout the Red Sea (Hughes and Beydoun, 1992; Mitchell et al., 1992) while (Davison et al., 1994) interprets extensional faults from offshore Yemen as terminating below the salt and the salt is considered post-rift. On the eastern margin volcanism continues into the syn-rift as evidenced by Aquitanian (21–24 Ma) aged dikes in Saudi Arabia and 21.4 to 22.3 Ma plutonic rocks in Yemen (Chazot et al., this volume).

Post-rift unconformity and strata

The post-rift unconformity is taken by Hughes and Beydoun (1992) and Mitchell et al. (1992) as the top mid-late Miocene salt, and synchronous with the onset of rapid sea-floor spreading at 5 Ma. The first indications of sea-floor spreading in the Atlantis-II deep are at 12 to 10 Ma (Coutelle et al., 1991; Rihm and Henke, this volume). This post-rift unconformity has erosional truncation of the Miocene salt, followed by onlap and downlap of Pliocene to Recent clastics (Warden and Desset Fms) and then carbonates (Shagara or Dhunishub Fms) (Figure A2.5; Mitchell et al., 1992). Alternatively, Davison et al. (1994) place the post-rift at the base of the salt where most extensional faults terminate and rotated strata are truncated (Mitchell et al., 1992, Figure 7; Davison et al., 1994, Figure 11).

Thicknesses of post-rift strata are very variable because of salt loading (Figure A2.5) but reach 3 km of post-rift clastics in the Abbas-1 well on the Yemen coastal plain (Heaton et al., 1995; Bosence, this volume). Post-rift volcanics are concentrated on the eastern margin of the rift stretching from Yemen to Syria to form one of the largest alkaline volcanic provinces in the world (Chazot et al., this volume).

Discussion

This southern portion of the Red Sea clearly involves active rifting from plume-related magmatism (Menzies et al., 1992). In addition there is considerable asymmetry to the Red Sea in that the western margin is narrow and steep, the volcanism is almost exclusive to the eastern side and the most significant uplift is on the eastern side. This has led Wernicke (1985) and Voggenreiter and Hötzl (1989) to propose a simple shear model with an east dipping low angle detachment under the Arabian plate. However this intracrustal detachment has not been identified and there is no stratigraphic evidence for pre-rift uplift on this margin as is predicted in the simple shear model. Similarly, the timing of volcanism, uplift and rifting do not fit White and McKenzie's (1989) predicted models of plume-related rifting in which the plume is either prior to, or synchronous with, rifting (Figure A2.4). On the south-eastern margin of the Red Sea there is no evidence of the predicted 1 to 2 km doming prior to volcanism as the pre-rift stratigraphy ends with continental siliciclastics conformably overlain by 31–26 Ma (Baker *et al.*, 1996) plume-related volcanics (Al Subbary *et al.*, this volume). This is followed by extensional faulting (27 to 10 Ma; Davison *et al.*, 1994) with rotation of volcanics and earlier strata which are onlapped by syn-rift clastics in the subsurface of the Yemeni coastal plain (inset Figure A2.4). The 3 km of post-volcanic subsidence and basin fill are not as predicted by the White and McKenzie (1989) model (see above). Pre-volcanic marine strata now occur at 2.6 km altitude on the rift shoulder indicating significant syn- to post-volcanic uplift from lithospheric heating, magmatic addition to the crust, surface extrusion of lava, crustal flexure and erosion (Davison *et al.*, this volume).

Recent geophysical investigations by Makris and Rihm (1991) and Rihm and Henke (this volume) on the steep and narrow western margin of the Red Sea suggest that this margin originated as a series of strike-slip pull-apart basins in the early history of this oblique rift basin. Subsequent extension by pure shear, rather than simple shear, is then envisaged on the eastern margin and this accounts for the present-day asymmetry of the basin (Rihm and Henke, this volume). To summarize, the tectonostratigraphic data indicate that the most appropriate model for the southern Red Sea is that prior to rifting the south-eastern margin was affected by plume-related volcanism but was not predated by updoming, and that subsequent rifting occurred through pure shear with enhanced rift shoulder uplift.

Gulf of Aden

Pre-rift strata and syn-rift unconformity
Pre-rift Palaeogene strata form widespread outcrops in both south-eastern Yemen and northern Somalia (Bott *et al.*, 1992; Fantozzi and Sgavetti, this volume). These pass upwards from Paleocene to early Eocene shelf carbonates (Umm Ar Radumah, Yemen, and Auradu Fm. Somalia) to restricted marine and evaporitic units (Jeza and Rus Fms, Yemen and Taleh Fm., Somalia), and finally the mid Eocene Habshiya Fm. (Yemen) and Karkar Fm. (Somalia) with more marine facies to the Indian Ocean to the east (Beydoun, 1964; As Saruri, 1995). There is a general absence of late Eocene strata in the Gulf of Aden basin, with the exception of the Daban sub-basin in Somalia (Abbate *et al.*, 1993b), eastern Socotra and Dhofar (Samuel *et al.*, in press) and early Oligocene–Miocene syn-rift sediments rest unconformably on eroded pre-rift strata. Away from basement highs where the syn-rift Shihr Group rests on Mesozoic and Precambrian basement (Beydoun, 1964; Bosence *et al.*, 1996) the syn-rift unconformity erodes down to a remarkably consistent level within the underlying Habshiya Formation (Watchorn *et al.*, this volume).

Syn-rift strata
Wells offshore from Yemen and Somalia reveal that the oldest marine strata are middle Oligocene (Hughes *et al.*, 1991) and that these overlie marginal marine evaporitic sands and muds similar to those of the onshore, syn-rift Shihr Group. The predominant onshore lithologies of the Shihr Group are fine-grained marginal marine and continental sediments. These accumulate is response to basin-wide subsidence throughout the late Oligocene and do not show dramatic thickening and tilting into half-graben structures (Watchorn *et al.*, this volume). Adjacent to basement highs steeper, fault controlled basin margins occur with a complex interfingering of high energy marine and continental facies (Bosence *et al.*, 1996). Offshore, in contrast, there are many tilted fault blocks with <2 km thick syn-rift strata recording deepening upward sequences through the Oligo-Miocene (Bott *et al.*, 1992). The rift shoulders supply sediment to the marginal areas throughout rifting and clasts record a gradual unroofing history of the underlying pre-rift strata (Bosence *et al.*, 1996; Watchorn *et al.*, this volume).

Post-rift unconformity and strata
Offshore Yemen latest Miocene to recent post-rift sediments rest unconformably on tilted Shihr Group (Bott *et al.*, 1992; Brannan *et al.*, in press). These sediments vary from deep-water pelagic oozes through to shallower siliciclastic and carbonate lithologies and may be up to 2 km thick. These strata contrast with the Red Sea post-rift in that they are not affected by salt diapirs because of a continuous connection between the Gulf of Aden and the Indian Ocean. Onshore, Watchorn *et al.* (this volume), working to the east of Mukalla, date the base of the essentially horizontal post-rift strata at 16 to 22 Ma. These sequences record the continuing uplift of the rift margin through to the present day. The dates broadly correspond to the earliest dates for sea-floor spreading in the outer portions of the Gulf of Aden at around 20 Ma (Sahota *et al.*, 1995; Brannan *et al.*, in press).

Discussion
The Gulf of Aden is an oblique rift that shows a large degree of symmetry in the geological history and stratigraphic record in the two margins (Fantozzi and Sgavetti, this volume). The west-north-west–east-south-

east basin has developed in response to north-east–south-west extension of the Afro-Arabian plate. The basin has passed from rift to drift with ocean floor spreading dating back to the early Miocene. The area was not affected by widespread, long-lived pre-rift updoming. Typically, only Rupelian strata are missing. Rift shoulders have been symmetric positive features throughout the basin's history. While the western end of the basin has been affected by plume-related volcanism around Afar and south-west Yemen the remainder of the basin shows only minor, and mainly post-rift volcanism. The area of post-rift subsidence is broadly coincident with syn-rift thermo-mechanical subsidence but post-rift onlap has not extended as far as the preserved limits of the syn-rift basin. Of note is the observation that in Yemen the basin margins show regional sag rather than fault-related subsidence which is commoner basinwards of the current shoreline. Together, these features all suggest that McKenzie's pure shear model is the most appropriate for the Gulf of Aden.

RIFT SUB-BASINS AND FACIES MODELS

The half-graben model

Most workers consider that the essential local-scale structural unit of the different types of rift basins are asymmetric half-graben (Figure A2.6; Bosworth, 1985; Gibbs, 1984; Rosendahl, 1987, Wernicke, 1985). Rift basins are also characterized by zones of varying polarity of half-graben tilting (Bosworth, 1985). For example, the Gulf of Suez is characterized by large-scale domino-style faults which can be divided into northern and southern provinces with easterly dipping faults and westerly dipping strata and a central province of westerly dipping faults and easterly dipping strata (Bosworth, 1994a; Bosworth et al., this volume). These provinces of opposing dip, or polarity are separated by structural accommodation zones (Figure A2.6) with a complex internal structure (Bosworth, 1985; McClay et al., this volume). Many outcrop and seismic surveys have shown that on a smaller scale major rift faults are limited to a few tens of kilometres in length and that laterally the strain may be taken up by other offset faults with an intervening structural ramp or relay (Bosworth et al., this volume; McClay et al., this volume). Where faults break through this offset with an oblique trend to link extensional faults then these are referred to as transfer faults (Bally, 1981; Gibbs, 1984; Morley, 1995). Those oblique faults cutting across both half-graben and extensional faults are referred to as cross-faults or cross-strike faults (Morley, 1995). Therefore a complex three-dimensional structural control on sedimentation exists which itself evolves through time as fault tips propagate, half-graben subside, and footwall areas rise. It is therefore useful to consider the effects that the evolving three-dimensional models have on the sedimentological evolution of rifts.

Tectono-sedimentary facies models

This complex four-dimensional picture of evolving rift structure (Figures A2.6–A2.8) contrasts with earlier essentially two-dimensional tectono-sedimentary models of Leeder and Gawthorpe (1987), Frostick and Read (1990), and Surlyk (1990). Although these earlier models display information in three dimensions they do not take into account and incorporate three-dimensional processes associated with varying fault displacement and accommodation zones (cf. Gawthorpe et al., 1994; Lambiase and Bosworth, 1995). Any stratigraphic or sedimentological study should be undertaken in association with a full structural analysis. The converse is also true as stratigraphic relations (e.g. thickness changes and onlap surfaces) and sedimentological responses (e.g. footwall-derived fans and unconformities) give important information on the timing of fault initiation and propagation which is essential to understanding the evolving three-dimensional rift structure (see Lambiase and Bosworth, 1995; McClay et al., this volume). This has led Leeder (1995) to question the complexity of some structural analyses when synchroneity of fault movement is unproven.

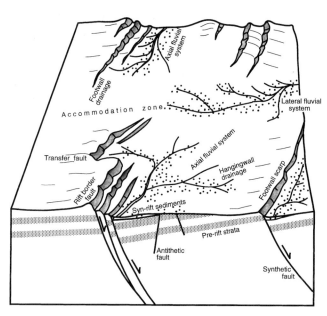

Figure A2.6 Tectono-sedimentary model of three-dimensional structure and sedimentary environments in early stages of syn-rift evolution of rift basins.

Syn-rift sedimentology

With the onset of rifting a complex, fault-related topography develops which not only provides an intricate template for sediment accumulation but also controls many erosional and sedimentological processes (Figure A2.6). The extended crust will undergo regional subsidence, footwall crests will uplift (Jackson, 1987), undergo erosion, and become local source areas for clastic sediments. However, areas away from faults, or in fault-related hangingwall sub-basins may continue to receive sediment. Therefore the syn-rift basal unconformity may be either localized to footwall highs in rift basins resulting purely from lithospheric extension of the upper crust, or, the unconformity will be regional in extent in plume-related basins or those involving thinning of the lower crust and/or lithospheric mantle.

As rifting continues into the syn-rift stage the complex surface topography which develops in response to extensional fault propagation, fault surface geometry, magnitude of fault slip, hard and soft linkage transfer zones, accommodation zones and polarity of the half-graben may all control clastic source areas, development of drainage patterns, and patterns of sediment accumulation. This early continental stage of rift evolution has been recently reviewed in detail by Lambiase and Bosworth (1995). Drainage basins will therefore have a complex and changing relationship with structure in addition to their inherited antecedent drainage pattern. Drainage may initiate off footwall crests, hangingwall dip slopes and pass axially into hangingwall sub-basins to eventually cut through the structural grain at transfer faults or accommodation zones. If rates of erosion, transport and deposition are high in relation to rates of fault movement then drainage patterns may disregard structure and cut across minor and basin bounding faults (Nichols and Daley, 1989, Figure 8; McClay *et al.*, this volume). However, rapidly extending rift basins will have a geomorphology closely following the structural template and this main phase of rift basin evolution tends to result in underfilled accommodation space (Figures A2.7, A2.9) indicating that rates of fault-related subsidence exceed rates of clastic sediment supply. This has led Prosser (1993) to suggest that true syn-rift strata (as defined above) may not be as thick as has been commonly identified and much of this strata may be of post-rift origin. Later in rifting (Figures A2.7, A2.8), and into the post-rift stage, topography will be reduced by erosion of footwall areas and sedimentation in hangingwall areas and accommodation zones.

An important sedimentological event in the subsidence history of a rift basin is marine flooding. This can occur at any stage and is controlled by subsidence, eustacy, pre-rift topography and location of the rift with respect to adjacent marine basins. In the Red Sea to Gulf of Aden basins marine flooding occurs early within the syn-rift phase. Complete marine flooding effectively isolates footwall areas from clastic sedimentation and in tropical climates they may become sites of carbonate sediment production (Figures A2.8, A2.9). The sedimentology and depositional sequences developed within

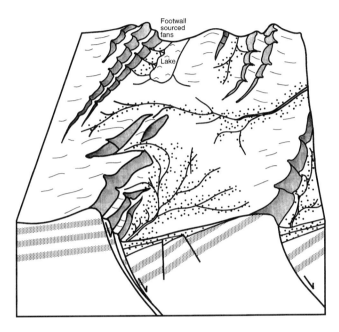

Figure A2.7 Tectono-sedimentary model of three-dimensional structure and sedimentary environments in middle stages of syn-rift evolution.

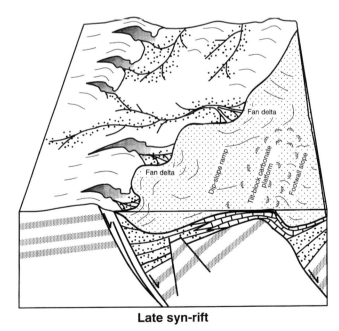

Figure A2.8 Tectono-sedimentary model of three-dimensional structure and sedimentary environments in late stages of syn-rift evolution of rift basins and following marine flooding.

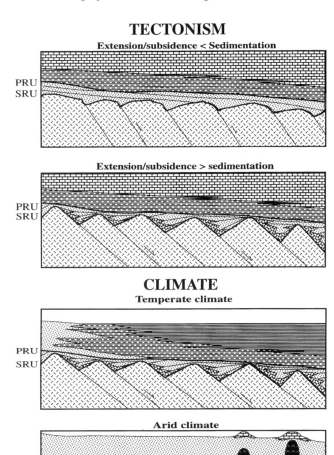

Figure A2.9 Schematic cross-sections of rift basin margins to illustrate major effects of tectonic, climatic and magmatic controls on rift stratigraphies.

these tilt-block carbonate platforms will have a first order control determined by whether the faults are basinward dipping, which results in a steep seaward-facing platform margin, or whether the faults are landward dipping and a gently sloping shelf is formed (Purser et al., this volume). In the former case a rimmed platform forms on seaward-facing footwall highs with a bypass or escarpment margin (Figure A2.8; Read, 1985). The hangingwall area will have a ramp-like profile and may interfinger downslope with laterally or axially derived clastic sediments (Figure A2.8; Cross et al., this volume). If the fault block is rotated seawards, on a land dipping fault, then initially a ramp will occur which may steepen up through time with differential shallow water production to develop into a progradational rimmed carbonate platform. The footwall margin facing the shoreline is unlikely to develop into a rimmed platform because of the restricted coastal waters and the likeli-

hood of burial from shoreline derived clastics (Purser et al., this volume). Because carbonates are likely to fill accommodation space to a sea-level datum progressive rotation of the fault block will result in shallowing-upward stratigraphic units which are repeatedly thickened into the hangingwall sub-basin (Cross et al., this volume). Footwall sites are characterized by erosion surfaces, formed in response to footwall uplift, and by shallow marine, shelf margin facies.

Submarine clastic sediment will be transported along topographic lows such as fan deltas in the Gulf of Suez (McClay et al., this volume) and in the extensional Aegean Gulf seasonal run-off maintains active alluvial fans, braid deltas, fan deltas and submarine fans (Leeder, 1995). Submarine channels, either parallel or normal to the rift axis, are fed by these fluvial systems and generate submarine fans at the base of slope (Figure A2.8). The combination of steep slopes and seismic activity results in mass movement and slump scars, slumps and debris flows are common (Papatheodorou and Ferentinois, 1993). Deep marine areas become sites for shales, pelagic and redeposited facies (Hofmann et al., this volume).

During marine rift-basin formation when the basin is still narrow, such as the Red Sea–Gulf of Suez (Orszag-Sperber, this volume) and the southern Atlantic Cretaceous rift (Mohriak et al., 1989), the basin may become isolated by uplift or falling global sea levels and in arid climate settings thick evaporites may develop (Figure A2.9). These evaporitic fills can be significant in the thickness and rapidity of basin fill (e.g. around 2 km in 6 m.y. in the southern Red Sea (Heaton et al., 1996), but also later on because of the structural and sedimentological effects of salt diapirism. While the structural effects have been well documented (Davison et al., 1996) the sedimentological effects, such as the formation of salt-cored offshore islands, siting of carbonate platforms over salt domes, and the influence of salt diapirs on clastic sediment pathways have not been investigated in such detail (but see Bosence et al.; Carbone et al., Orszag-Sperber et al., this volume b). The effects of diapirism can be overwhelming and mask the syn-rift to post-rift transition as in the case of the north-western Red Sea (Orszag-Sperber et al., this volume b).

Rift-shoulder uplift, generated by footwall uplift adjacent to major border faults and/or the isostatic response of adding asthenospheric melt to the base of the crust during lithospheric stretching (White and McKenzie, 1989), or flexural isostacy of the lithosphere (Braun and Beaumont, 1989 a,b) may increase local sediment supply into the basin margins (Davison et al., and McClay et al., this volume). Shoulder uplift may be long-lived and in the Gulf of Aden there is evidence of rift shoulder-derived clastics from its inception in the late Oligocene through to the present day. However on a

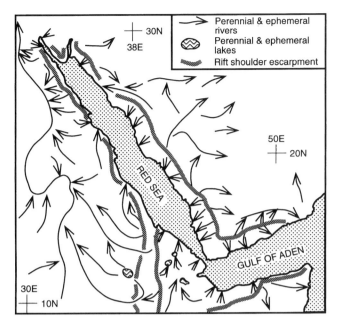

Figure A2.10 Present-day drainage pattern around the Red Sea and Gulf of Suez illustrating deflection of main drainage systems away from the rift shoulders. (After Frostick and Reid, 1989).

regional scale drainage may be directed away from the rift to external basins as occurs in the Red Sea today (Figure A2.10). This may partially explain why the central parts of the Red Sea are underfilled despite massive evaporite precipitation since the major phase of extension during the early Miocene.

Post-rift sedimentology

Unconformities are likely to develop around rift shoulders. Stratigraphic relations around such unconformities give useful data on the phase of thermal subsidence of the rift shoulders as they become progressively buried by the post-rift basin fill as documented in the North Sea (Badley *et al.*, 1984). Strata may initially accumulate as downlapping units away from any uplifted rift shoulders and subsequently will mainly accumulate in parallel stratified units which will onlap and bury any remaining syn-rift topography as in the North Sea (Figure A2.9; Badley *et al.*, 1984). Large-scale compactional drapes may be present against buried topographic relief.

Post-rift marine facies in axial areas will be pelagic and hemipelagic muds with redeposited facies adjacent to remnant half-graben topography. Economically important black shales or sapropels may occur in topographically isolated sub-basins or when the entire basin becomes restricted from open ocean conditions (Hofmann *et al.* and Taviani, this volume a). High sediment supply will result in progradational clastic shelf margins as exemplified by the north-east shelf of the USA (Hubbard, 1988). In the northern North Sea, Steel (1993) has documented a number of widespread post-rift 'megasequences' whose accommodation space was thought to have been generated by periodic increases in rates of thermal subsidence. These 200–1000 m thick 'megasequences' are progradational in style with maximum progradation during intervals of minimum subsidence with low rates of addition of accommodation space.

Unusually low rates of sediment supply result in starved shelves which may have unburied pre-rift strata outcropping on the sea floor and in an extreme case upper mantle is exposed (Boillot *et al.*, 1989).

In tropical areas where aggradation of tilt-block carbonate platforms is expected to keep pace with subsidence carbonate shelf areas will persist through the post-rift into passive margin stages. The carbonate platforms of the Bahamas are the world's most impressive example of this as they originated in the Jurassic on rotated half-graben and portions of these early platforms have persisted through to the present day as shallow-water carbonates which are thousands of kilometres long, hundreds of kilometres wide and up to eight kilometres thick (Sheridan, 1981).

Basins which had a significant earlier syn- or post-rift evaporitic stage will have axial and shelf areas dominated structurally and stratigraphically by salt diapirism. This may have a complex and intimate relation with extensional structures of the rift (Heaton *et al.*, 1995) and give rise to a complex of diapirs, salts walls, canopies and rafts (Figure A2.9). If this occurs during the post-rift stage as is the case in the southern Red Sea then the post-rift unconformity can be masked by stratigraphic responses to salt movement rather than to the slower regional subsidence (Figure A2.8; Bosence *et al*, Orszag-Sperber *et al.*, this volume b).

In conclusion there are a large number of independent and interdependent controls on rift-basin stratigraphy which broadly fall under the three main headings of tectonic, climatic and magmatic processes (Figure A2.11). These controls may either be restricted to the different phases of rift-basin evolution, or be longer lived and affect all stages of rift-basin evolution.

SEQUENCE STRATIGRAPHIC CONCEPTS IN RIFT BASINS

At the basinal-scale Hubbard (1988) was one of the first to carry out a detailed regional study of a rift-basin margin which critically assessed the earlier claims of Vail *et al.* (1977) that Mesozoic to Tertiary passive margin stratigraphy was controlled by global second- and third-order sea-level cycles. Hubbard (op. cit.) clearly demonstrated that cycles of the rifted margins of the North Atlantic, South Atlantic and Arctic Oceans were not synchronous, and therefore not global, and that they were controlled by rates of basin subsidence, sediment input and long-term tectono-eustatic cyclicity.

Recently, Dolson *et al.* (1996) and Ramzy *et al.* (1996) have used good quality subsurface data from the Gulf of Suez to modify and apply sequence stratigraphic methods to a syn-rift stratigraphy. They show that in a rift setting subsidence and uplift can occur synchronously within a small area so that regional or 'global' sea-level changes are overwhelmed by tectonic effects. Thus, in contrast to passive margin stratigraphies, the syn-rift stratigraphy of rift basins does not always show basinwide erosion surfaces and the wedge-shaped stratigraphic geometries of half-graben is so prevalent that the presence of highstand systems tracts, transgressive systems tracts and lowstand wedges is difficult to resolve. These authors use three-dimensional seismics, abundant well data and high resolution biostratigraphy in order to identify local stratigraphic stacking patterns, hiatuses and condensed sections, maximum flooding surfaces to erect a local rift basin model (Figure A2.12).

Prosser (1993) reviewed the structural controls on rift basin stratigraphy and suggests a scheme whereby systems tracts are related to stages in the structural evolution of a rift basin (e.g. 'rift climax systems tract'), rather than to sea-level changes. It remains to be seen whether this additional terminology for systems tracts is accepted as an improvement on what is already available.

The smaller-scale field based analysis of depositional sequences and the erection of genetic stratigraphies in syn-rift strata is proving an important tool in basin analysis (Dart *et al.*, 1994; Gawthorpe *et al.*, 1994; Bosence *et al.*, 1996; Howell and Flint, 1996) as sequence and/or parasequence boundaries may give precise timings for tectonic movement, and stratigraphic geometries give information on the varying sediment supply routes and generation of accommodation space. This has resulted from detailed fieldwork and establishment of high resolution sequence stratigraphies (e.g. Cross *et al.*, this volume) together with tectono-sedimentary computer modelling (Hardy and Waltham, 1992).

When considering the application of sequence stratigraphic concepts to rift-basin fills it is essential to appreciate that accommodation and depositional sequences can be variously altered by fault-block rotation, by variations in clastic and carbonate sediment supply, and by varying rates of regional (basinwide geotectonic or eustatic) and fault-related sea-level change in half-graben fills.

Depositional sequences and fault block rotation

Of particular importance here is the straightforward geometric constraint that fault-related movement in a half-graben is rotational so that hangingwall subsidence, and relative sea-level rise, is synchronous with footwall uplift, and a relative sea-level fall (e.g. Bosence *et al.*,

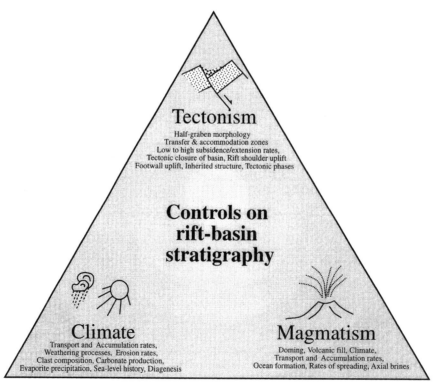

Figure A2.11 Summary of main tectonic, climatic and magmatic controls on rift basin stratigraphy.

1994; Cross et al., this volume). At the same time there may be little or no sea-level change in response to this fault movement in the adjacent accommodation zone (Gawthorpe et al., 1994). This contrasts significantly with shelf-wide more-or-less synchronous, relative sea-level changes which characterize passive margin basins (Hubbard, 1988).

If footwall uplift is synchronous with a hangingwall subsidence then footwall sequence boundaries will correspond with hangingwall flooding surfaces, and strata deposited during transgression and sea-level highstands. Because this pattern differs from conventional sequence stratigraphic concepts (Vail et al., 1977; Van Wagoner et al., 1988) differing terminologies have been applied by different authors to describe the sequence stratigraphy of half-graben basins.

Rosales et al. (1994) considered that tectonic tilting was the main control on sedimentation on Cretaceous carbonate platforms developed on fault blocks in northwest Spain. Their correlations suggest contemporaneity between coarser grained and slumped deposits in the hangingwall, as lowstand deposits, with emergent karstic surfaces and cave infills on the footwall. Transgressive and highstand deposits in the hangingwall are deeper-water carbonate or siliciclastic facies while footwall areas have aggrading shallow reefal facies. However, if the strata accumulated during tectonic tilting it would be expected that the emergent footwall karst would correlate with the transgressive and deeper deposits of the hangingwall. Their correlations indicate that this is not the case, which suggests that the deposition of strata in each case was post-tilting and was a passive fill of the wedge-shaped accommodation space during a regional relative sea-level rise. The pronounced onlap at the bases of the depositional sequences in the hangingwall dip slopes support this alternative interpretation.

Gawthorpe et al. (1994) consider this problem with respect to hangingwall-sourced versus footwall-sourced clastic depositional systems from the Quaternary of Central Greece. Here they recognize the contemporaneity of footwall uplift giving rise to footwall unconformities, and hangingwall subsidence giving rise to highstand systems tracts. Transgressive systems tracts are not developed because of the generally high rates of sea-level rise relative to sediment supply rates in this study area (Gawthorpe et al., 1994). This uses the terminology of systems tracts in a very local and time-transgressive sense so that within one half-graben basin a lowstand systems tract will pass laterally into a highstand systems tract. This usage departs somewhat from the original definition of systems tracts as 'a linkage of contemporaneous depositional systems' (Brown and Fisher, 1977; Van Wagoner et al., 1988). In half-graben settings contemporaneous depositional systems have different stratigraphic stacking patterns while similar depositional systems are time-transgressive or diachronous (Figure A2.12).

An alternative approach, which is favoured by this author (Bosence et al., 1994, 1996; Cross et al., this volume), and is also used by Howell and Flint (1996) is to abandon the concept of systems tracts and to simply identify and describe depositional systems in terms of stacking patterns and their bounding surfaces and facies which are then interpreted in terms of local, relative sea-level changes. This method also removes the problem of attempting to link particular stratigraphic geometries (e.g. retrogradation) with particular portions of a relative sea-level curve (e.g. rising sea level), when sea-level rise can result in progradation, aggradation or retrogradation depending on rates of sediment supply.

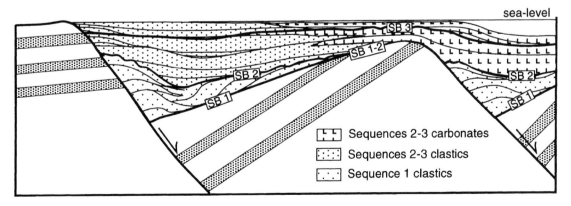

Figure A2.12 Sequence stratigraphic model for half-graben sub-basins (after Gawthorpe et al., 1994; Bosence et al., 1994; Dolson et al., 1996; Cross et al., this volume). Sequence 1 has synchronous onlap, aggradation and downlap in the hangingwall with erosion and development of sequence boundary (1–2) in the footwall in response to fault-block rotation. Sequence 2 initially shows onlap from hangingwall to footwall sites followed by aggradation with clastics in hangingwall basin and limestones (L) preferentially sited on the basin-ward footwall crest. Sequence Boundary 3 develops in response to fault-block rotation with synchronous footwall uplift and erosion (SB3) and hangingwall subsidence and increased clastic deposition.

Clastic versus carbonate sediment supply

It is now well established that clastic and carbonate depositional systems respond quite differently to changes in sea level and in their resultant depositional sequences (Schlager, 1992; Handford and Loucks, 1993; Bosence et al., 1994). Carbonate systems contrast with clastic systems in that their rates of supply (production) are highest during transgression and highstand while clastic systems have the highest supply rates during falling and low sea-level stands. Therefore in the mixed clastic and carbonate systems of half-graben basins there will be a complex interplay between depositional sequences and clastic and carbonate lithologies.

Firstly, it should be remembered that when fault blocks are submerged below sea level then carbonate platforms are favoured in footwall highs which may be isolated from clastic supply routes concentrated in hangingwall sub-basins and transfer zones (Figure A2.8). The two systems would be expected to behave differently in response to extensional fault movement and block rotation. Clastic systems in the hangingwall sub-basins would first respond with a flooding surface followed by aggrading footwall sourced fan deltas (Dart et al., 1994) and prograding submarine fans (Ferentinos et al., 1988). In tropical climates the synchronous response on the footwall carbonates will be rapid shallowing or emergence, following relative sea-level fall, and the development of a rapid shallowing-upward sequence or erosional sequence boundary. Carbonates on the hangingwall dip-slope are likely to behave in a similar way to hangingwall sourced clastics in that they will initially drown, but then prograde out from shallow areas resulting in a shallowing-upward depositional sequence.

In tropical settings and following marine flooding during any long periods of tectonic quiescence carbonates would be expected to dominate over clastics. Clastic supply will gradually decrease through time as the hinterland topography is reduced and carbonates will prograde out from footwall crests and also become dominant in axial areas if the basin is fully marine.

Carbonate and clastic depositional sequences may also be separated in time in fault-related settings. In a recent study from the northern margin of the Gulf of Aden (Bosence et al., 1996) it is shown that different parts of depositional sequences are characterized by different lithological associations. Depositional sequences were found to have erosion surfaces overlain by alluvial-fan conglomerates and sandstones. These were interpreted to correspond to relative lowstand periods when fault-related uplift in a footwall setting resulted in erosion followed by increased clastic supply. Subsequent transgression from regional subsidence resulted in a marine flooding surface followed by progradation of shallow-marine carbonates in transgressive and highstand periods. Subsequent sea-level fall produced the next sequence boundary. These depositional sequences therefore have a distinct tectonic signature with the association of sequence boundaries followed by thick basal conglomerates and then marine flooding and progradational carbonates.

Although few studies have been undertaken on this topic it is clear that clastic and carbonate depositional systems will respond in different ways to a similar tectonic movement and because of the different controls on their production and distribution they can be separated spatially or temporally in depositional sequences.

Rates of regional and fault-related sea-level change

The rates and magnitudes of fault movement are important controls on accommodation within half-graben in relation to rates of sediment supply and rates of regional sea-level change. Data from actively extending basins today indicate fault-related subsidence to occur as instantaneous seismic events resulting in throws of up to about five metres (Leeder, 1995). Slip rates averaged over thousands of years from Gulf of Patras, Greece, are 2–5 m/ka (Chronis et al., 1991), from Utah, USA, are 2 m/ka (Machette et al., 1991) and from the Pisia fault, Greece, are 0.5 m/ka (Machette et al., 1991; Roberts et al., 1993). Uplift in the footwall immediately adjacent to extensional faults is around 10 to 15% of the total displacement (Jackson and McKenzie, 1983).

These average slip rates are up to two orders of magnitude higher than rates of both thermal subsidence of rifted margins, and are at least an order of magnitude higher than rates for first- and second-order eustatic sea-level changes. Therefore, fault-related subsidence will normally override thermal subsidence and eustatic changes and produce a distinct tectonically generated sea-level signal against a background relative sea-level rise or fall generated by the latter processes. However, in glacial periods, rates of sea-level change may be higher (maxima of 10–14 m/ka) within the Quaternary in which case the background subsidence or uplift may be fault-related and the higher frequency cycles related to glacio-eustacy (Gawthorpe et al., 1994).

Rates of erosion and clastic supply are exceedingly variable but evidence from many rift basins indicates that during periods of maximum extension basins are under-filled (for review see Prosser, 1993). This suggests that accumulation rates of rift basin clastic sediments are less than the range of average slip rates (0.5 to 5 m/ka) described above. At the half-graben scale Dart et al. (1994) and Gawthorpe (1994) demonstrate that at sites of maximum slip rates near the centre of fault segments then sequences are stacked in aggradational packages dominated by highstand systems tracts, but that in other sites with lower subsidence rates such as transfer zones and hangingwall dip-slopes coeval sequences are more progradational and regressive (Figure A2.12).

The average slip rates quoted above are similar to rates of shallow-water carbonate production on carbonate platform margins (1–4 m/ka; Bosence and Waltham, 1990). This suggests that carbonate platform margins on footwall crests that are cut by extensional faults might either drown or keep pace with fault movement. Both these events occurred on Gebel Abu Shaar, in Egypt, where footwall areas are either truncated by the block-bounding Esh Mellaha Fault or at other periods there is evidence that the carbonate margin overgrew and sealed the block-bounding fault (Cross *et al.*, this volume; Burchette, 1988). With respect to the effect of fault movement on carbonate platforms on footwall highs these are more affected by footwall uplift and emergence than by rates of downward block subsidence as discussed above and footwall areas are typically sites of shallow, shelf-margin facies associations.

CONCLUSIONS

1. During the 1970s and 1980s a number of geotectonic models were proposed for the origin and evolution of rift basins. The predicted stratigraphic signatures of these models have not received so much attention and are reviewed here. Using a simple classification based on the nature and the occurrence of pre-rift strata, syn-rift unconformity, syn-rift strata, post-rift unconformity and post-rift strata it is shown that each geotectonic model has a distinct stratigraphic signature. This stratigraphic classification may be applied to the analysis of basins which are poorly exposed, or with few wells, or are poorly imaged on seismic data due to depth or poor penetration through evaporite lithologies.
2. A review of the evolution of the Gulf of Suez, Red Sea and Gulf of Aden sub-basins in the light of the proposed stratigraphic classification and new data presented in this volume indicates that McKenzie's (1978) pure shear model is the most appropriate geotectonic model for these rift basins. While early phases of the oblique Red Sea and Gulf of Aden rifts may be characterized by pull-apart basins and that the southern Red Sea experiences pre-rift plume-related volcanism.
3. Previous two-dimensional facies models for rift basins are revised in terms of current three-dimensional (four-dimensional including time) structural models. These indicate a complex evolving structural template which controls the erosion, transport routes and deposition of clastic sediments in continental through to marine settings, and also the isolation of footwall areas as sites of carbonate platform growth.
4. The restriction of early rift basins from ocean waters in an arid climate setting leads to extensive evaporite formation. This has a pronounced effect on subsequent basin history as salt diapirs rise, deform the subsequent stratigraphy and because salt diapirs may generate sites for new, offshore carbonate platforms which in turn control subsequent clastic sediment pathways.
5. The application of sequence stratigraphic concepts to the analysis of extensional basins is a significant new development and likely to improve our understanding of the evolution of rift basins. Important characteristics of the tectono-sedimentary evolution of half-graben that affect depositional sequences are the rotation of fault blocks which leads to a synchronous hangingwall relative sea-level rise and footwall sea-level fall, the temporal and spatial isolation of carbonate and clastic facies associations within depositional sequences, and the relative rates of fault-slip, fault-block rotation, sediment supply and regional and global sea-level changes.

ACKNOWLEDGEMENTS

This review is the outcome of research undertaken during receipt of grants from the Royal Society, European Union Science Programme 'REDSED', British Petroleum Exploration and the British Council in Yemen and Saudi Arabia. This financial support for fieldwork in Egypt, Yemen and Saudi Arabia has done much to clarify my views on rift sedimentation. I also acknowledge with thanks discussions and comments on earlier drafts of this paper from Ian Davison, Gary Nichols, Ken McClay, Martin Menzies and Fred Watchorn from Royal Holloway University of London and from Bruce Purser (Université de Paris Sud, Orsay) who also introduced me to the remarkable outcrops of tilt-block carbonate platforms of the Gulf of Suez.

SECTION B
Geophysical, magmatic and structural framework

This section contains five invited contributions from authors who have carried out extensive recent research on the geophysics, magmatism and structural geology of the Gulf of Aden, Red Sea and Gulf of Suez. These chapters consider the geophysics of basin evolution, the relationships between magmatism and basin evolution on the Arabian margin, and the structural setting and structural evolution of the Gulf of Aden, northern Red Sea and Gulf of Suez.

As well as providing up-to-date reviews on these subjects these contributions act as a framework for many of the sedimentological studies that follow in this volume. Many of the features of the sedimentology of the basin relate to these underlying controls.

B1 Geophysical studies on early tectonic controls on Red Sea rifting, opening and segmentation
R. Rihm and C. H. Henke

B2 Pre-, syn- and post-rift volcanism on the south-western margin of the Arabian plate
G. Chazot, M. A. Menzies and J. Baker

B3 Tectonic and sedimentary evolution of the eastern Gulf of Aden continental margins: new structural and stratigraphic data from Somalia and Yemen
P. L. Fantozzi and M. Sgavetti

B4 Structure, sedimentation and basin dynamics during rifting of the Gulf of Suez and north-western Red Sea
W. Bosworth, P. Crevello, R. D. Winn and J. Steinmetz

B5 Rift development in the Gulf of Suez and north-western Red Sea: structural aspects and related sedimentary processes
C. Montenat, P. d'Ott Estevou, J.-J. Jarrige and J.-P. Richert

Chapter B1
Geophysical studies on early tectonic controls on Red Sea rifting, opening and segmentation

R. Rihm and C. H. Henke

ABSTRACT

Seismic, gravity and magnetic data are combined with results of structural geology to elucidate the earliest stages of rifting in the Red Sea. The abrupt continent–ocean transition at the Egyptian and Sudanese Red Sea, compared to other margins, suggests that it was created as a transform margin. Following a major plate kinematic re-orientation around 22 Ma, the transform plate boundary became extensional and the Red Sea basin started to open. Formation of oceanic crust by sea-floor spreading is not restricted to the bathymetrically defined axial graben and deeps, but may have started at spreading rates below 5 mm/year, as early as 10–12 Ma. The geometry of the initial rift line is the result of the regional stress field interfering with older lithospheric lineaments that acted as stress guides. The rift follows the path of least resistance by jumping from one available zone of structural weakness to another. This minimizes the energy needed for break-up of the continents. Harmonization of local variations of spreading ridge orientation with the overall regional extension thus determines the original rift geometry. Such a control of old continental lineaments over the segmentation pattern of a developing spreading ridge has probably not only occurred in the Red Sea, but, in a much broader sense, can be applied to the segmentation of mid-ocean ridges in general.

INTRODUCTION

The Red Sea is one of the best natural laboratories on earth for studying the early generation of oceanic crust and lithosphere. The geometry of the Red Sea coastlines and escarpments (Plate 1) was already noted by Wegener (1915), who quoted it as support of his theory of continental drift. A large number of models for the evolution of this basin have since been proposed (e.g. Makris et al., 1991a; Bosence, this volume). The models are in general agreement that the last stage of evolution in the Red Sea is of a small ocean basin with a well organized sea-floor spreading centre in its southern axial graben. There are, however, substantial controversies relating to the initial stages of the development of the main trough and the way the northern and central parts of the basin are spreading. Several general models for mechanisms and early stages of oceanization (i.e. initial generation of oceanic type of crust) following continental break up have been applied to the Red Sea (Figure B1.1) such as opening through sea-floor spreading (McKenzie et al., 1970; Girdler and Styles, 1974; Le Pichon and Francheteau, 1978; Le Pichon and Gaulier, 1988) or gradual focusing of individual magmatic intrusions into organized sea-floor spreading (La Brecque and Zitellini, 1985). These models imply oceanic crust in the entire Red Sea basin except for a possible small portion of continental crust related to an initial continental rift (which remains in those reconstructions after 'closing' the Red Sea by 'removing' all oceanic crust). Restriction of oceanic crust to the axial graben was postulated by the various concepts of diffuse extension (e.g. Cochran, 1983b), of punctiform initiation of sea-floor spreading (Bonatti, 1985), and the simple lithospheric shear model (Wernicke, 1985; Voggenreiter et al., 1988). The alternative pull-apart model of shear controlled early rifting (Makris and Rihm, 1991) considered a more complex distribution of oceanic and continental domains, which is discussed below. The observation of the present-day distribution of crustal

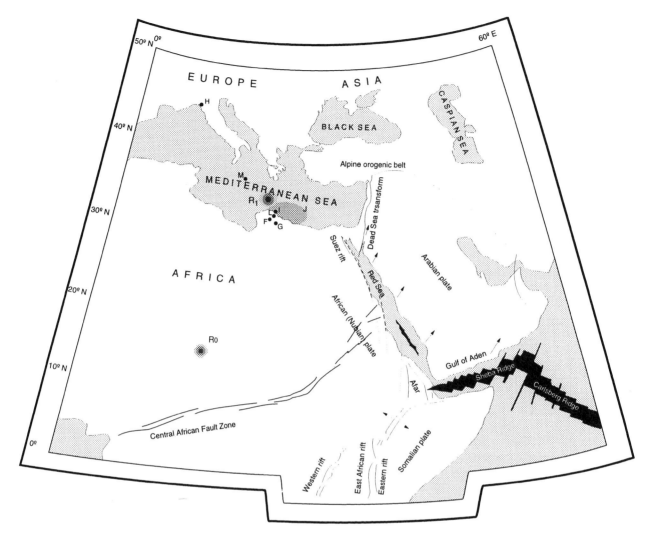

Figure B1.1 Plate kinematic setting of the Red Sea and adjacent regions after Rihm, 1996, in parts adapted from Izzeldin, 1982; Bonatti, 1987; Makris and Rihm, 1991. Poles of rotation for the recent opening of the Red Sea plot in the Mediterranean region: L = Laughton, 1996; M = McKenzie *et al.*, 1970; F = Freund, 1970; G = Girdler and Darracot, 1972; H = Hall, 1979; I = Izzeldin, 1982; J = Joffe and Garfunkel, 1987; R = Rihm, 1996. The latter analyses (stippled areas J and R_1) indicate considerable variation of direction and rate of motion for different times and areas of the Red Sea. The arrows on the Arabian plate are equivalent to a rotation of 4° around pole R_1. R_0 is a reconstruction for the pole of rotation describing the initial strike-slip motion in the Red area, the orientation of which is illustrated by the dashed/dotted lines on the western Red Sea flank.

domains within the Red Sea and the reconstruction of the early rift geometry suggest a causal relationship between pre-existing continental lineaments, crustal evolution during basin opening and, consequently, the development of a segmented pattern for the evolving mid-ocean ridge. The latter has implications for the architecture and geometry of the global mid-ocean ridge system (Rihm, 1996).

CRUSTAL DOMAINS OF THE RED SEA INFERRED FROM SEISMIC DATA

North-western Red Sea

The distribution of crustal domains in the Red Sea was evaluated from a number of deep seismic profiles (Figure B1.2). One of the most striking crustal features in the Red Sea is evident from a comparison of two

Figure B1.2 Location map of seismic profiles in the Red Sea region. The following sources (labelled in the map after the onshore survey area or the research vessel used in the Red Sea, respectively) were used for the definition of crustal types in and around the Red Sea: Afar (Berckhemer *et al.*, 1975; Makris and Ginzburg, 1987); Egypt (Makris *et al.*, 1988a; Rihm *et al.*, 1991), southern Saudi Arabia (USGS) (Healy *et al.*, 1982; Mooney *et al.*, 1985), STEFAN E, Saudi Arabian part (Makris *et al.*, 1983, 1988a; Rihm *et al.*, 1991), STEFAN E, Egyptian part (Rihm, 1984, 1989; Makris *et al.*, 1988a; Rihm *et al.*, 1991), MINOS (Gaulier *et al.*, 1988, these ESP profiles are shown in the insert map in the lower left corner), CONRAD (Rihm *et al.*, 1991), SONNE 53 (Rihm, 1989; Egloff *et al.*, 1991).

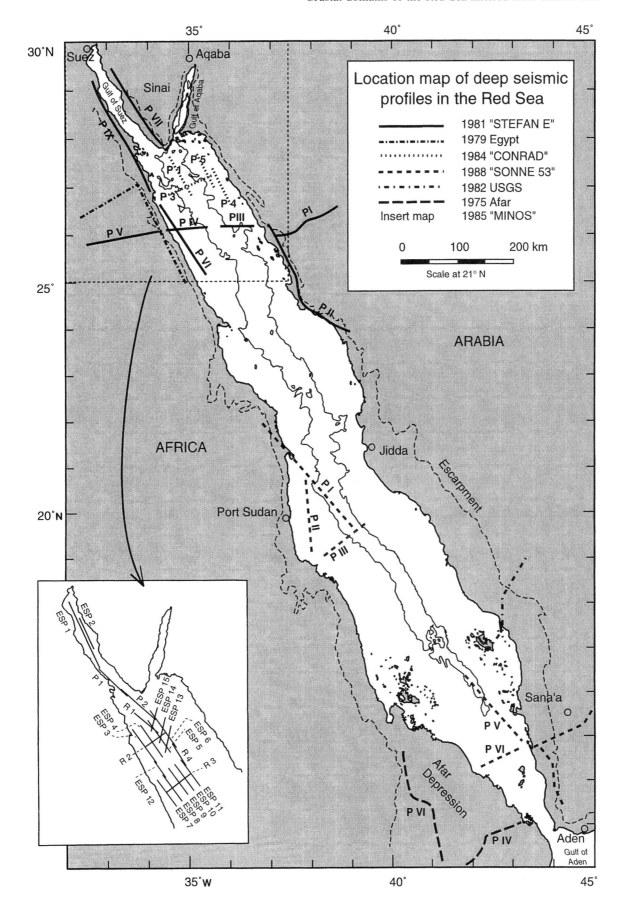

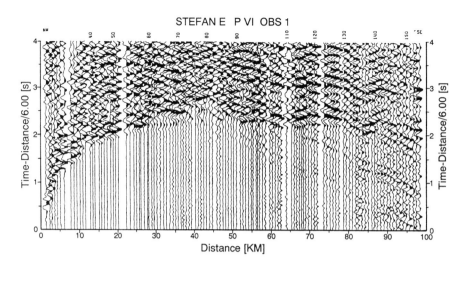

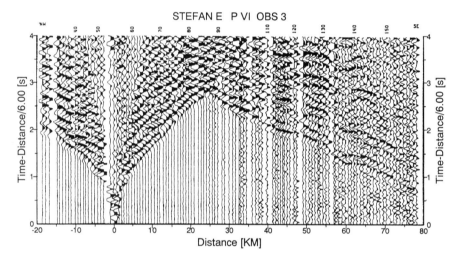

Figure B1.3 Comparison of seismogram sections of OBS positions 1 and 3, 24 km apart on STEFAN E profile VI, 1981 (Rihm, 1989; Rihm *et al.*, 1991). For location see Figure B1.2. The sections are reduced with a velocity of 6 km/s and normalized to maximum amplitudes. At OBS 1, first arrivals from 30 to 50 km offset have an apparent velocity of around 6 km/s and are interpreted as Pg-waves from the upper continental crust. Pn-arrivals from the upper mantle appear as first arrivals at an offset of about 45 km. At OBS 3, only sedimentary velocities around 3.5–4 km/s are observed up to 25 km offset, where this travel-time branch is directly overtaken by Pn- (mantle) arrivals of 7.5 km/s.

adjacent OBS (ocean bottom seismograph) sections situated along STEFAN E profile VI on the western flank of the northern Red Sea (Figure B1.3): the velocity–depth structure turns over a distance of 24 km from stretched continental crust observed at OBS Pos. 1 to something very close to oceanic at OBS Pos. 3. Reversed observation and seismic modelling at OBS positions 6 to 10 of the same profile (Figure B1.4) confirms this crustal structure to extend over the remaining profile length (some 70 km), indicating substantial portions of oceanic-type of crust as close as 30 km off the Egyptian coast (Rihm, 1984, 1989). ESP (expanded spread) profiles recorded parallel to the coastline (north-north-west–south-south-east) across the entrance from the Gulf of Suez into the northern Red Sea (MINOS cruise, see Figure B1.2) give a similarly abrupt crustal change from stretched continental to oceanic structure (Gaulier *et al.*, 1988). Land recordings of sea shots across the Egyptian margin show a similar result (Rihm *et al.*, 1991) and reveal one of the steepest continent–ocean transitions ever recorded on the globe (Figure B1.5).

Central western Red Sea

Further south, in the central Red Sea, the Saudi-Sudanese Red Sea Commission collected multi-channel seismic, gravity and magnetic data (Izzeldin, 1982, 1987). These surveys have established the correlation of oceanic crust from the axis towards both flanks of the central Red Sea 20 to 40 km beyond the oldest recognizable magnetic anomalies (Figure B1.6).

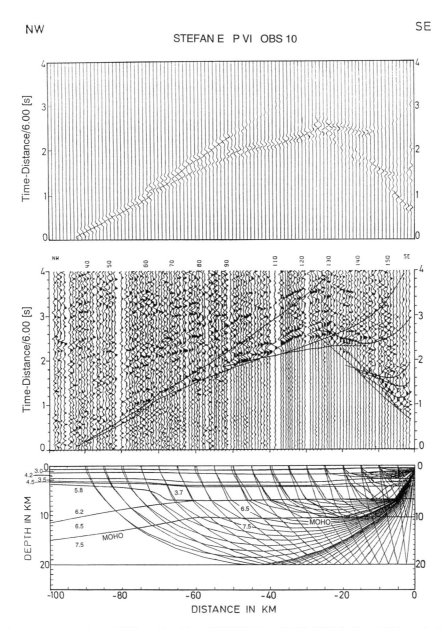

Figure B1.4 Observed seismogram section of OBS position 10 on STEFAN E profile VI, 1981 (in the middle), synthetic seismogram section (above) and ray tracing model (below), including locations of OBS used for computation (Rihm, 1989; Rihm et al., 1991). The travel-time pattern of OBS 10 resembles that of OBS 3 (Figure B1.3). Seismic modelling was performed for all OBS positions after the method of Cerveny and Psencik (1981) and reveals that across an approximate 15 km wide transition zone located near OBS Position 3, the sediment thickness increases rapidly, the continental upper crustal layer terminates and the upper mantle rises to a depth of only 10.5 km bsl.

Deep seismic data were collected offshore Sudan with the aim of identifying the nature of the crust across this margin (Rihm, 1989; Egloff et al., 1991; for locations see Figure B1.2). The structure of the margin was resolved in some detail along three profiles (SONNE 53 – profiles I, II and III), defining the domains of stretched continental crust along the margin (Figure B1.7) and the distribution of crustal types across the margin (Figure B1.8). In summary, the results off Sudan (Egloff et al., 1991) resemble those obtained off Egypt, i.e. a sharp continent–ocean transition with a slightly wider zone of stretched continental crust (60 km off Sudan vs. 30 km off Egypt) and significantly less sedimentary coverage on top of the oldest part of the oceanic crust (< 2000 m in some places compared to 5000 m off Egypt).

Eastern Red Sea

The eastern flank of the Red Sea reveals a different structural style, characterized by stretching of continental crust extending over much larger distances landward of the continent–ocean transition, as seismic data

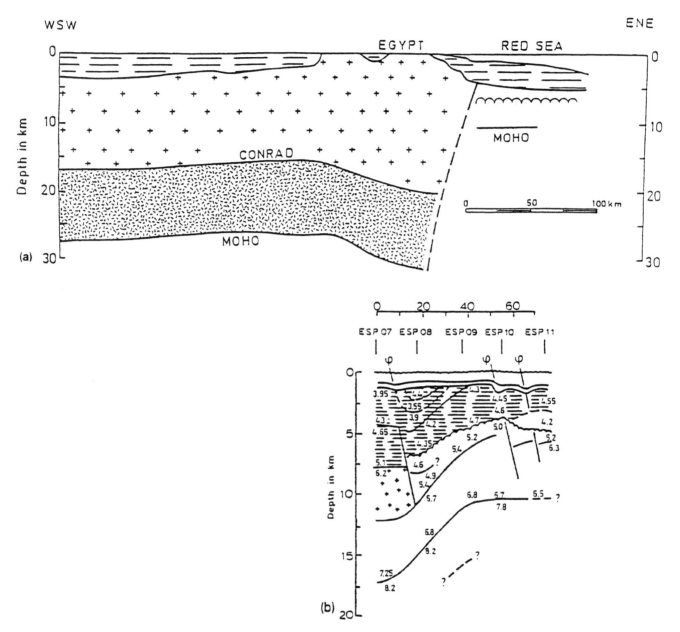

Figure B1.5 (a) The crustal section along 'STEFAN E' profiles IV and V (Rihm, 1989; Rihm *et al.*, 1991) illustrates the steepness of the Egyptian continental margin. (b) The sharp continent–ocean transitions on the western Red Sea flank are confirmed by ESP profile R3 (MINOS cruise, Gaulier *et al.*, 1988). The crossing points with profiles 7–11 are annotated. For locations see Figure B1.2.

indicate (Makris *et al.*, 1983, 1988a; Cochran and Martinez, 1988; Rihm *et al.*, 1991; Richter *et al.*, 1991; Egloff *et al.*, 1991). Although the relatively steep topographic escarpment along the edge of the Arabian plate (Plate 1) is associated with a significant reduction of crustal thickness (Healy *et al.*, 1982; Mooney *et al.*, 1985), the eastern flank of the northern Red Sea off Saudi Arabia and of the Southern Red Sea off Yemen (Figure B1.9) are both stretched continental crust from the coast to the axial region of the Red Sea.

Figure B1.6 Top of oceanic crust interpreted from 7 multichannel seismic (MCS) lines running perpendicular to strike across the entire width of the Red Sea between axial latitudes 19°N and 21°N (Izzeldin, 1987). Locations of the profiles are shown in the lower part of the figure. The upper part of the figure shows the correlation of the top of oceanic crust obtained from the seismic profiles 17–29, plotted with different signatures on top of each other and centred at the Red Sea axis at 0 km. The arrows mark the limits of magnetic anomaly 3, the oldest identifiable anomaly (*c.* 5 Ma, Izzeldin, 1987). All seven profiles have a relatively uniform variation of geometry with distance from the axis, and the depth scatter is around ± 1 km. The correlation shows that oceanic crust extends up to 40 km further off axis than indicated by magnetics and bathymetry.

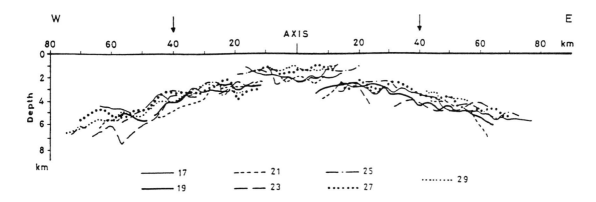

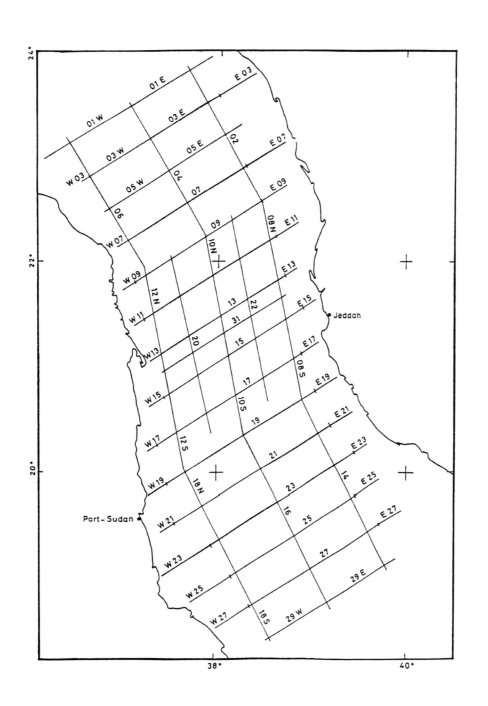

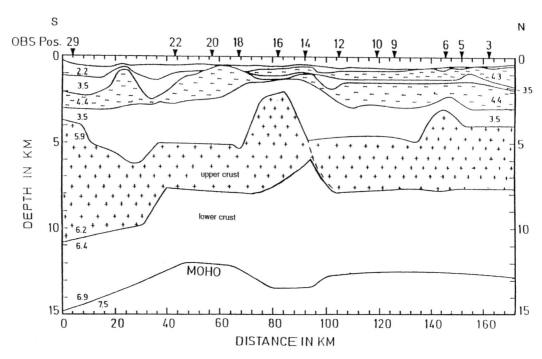

Figure B1.7 Crustal section of 'SONNE' profile II off Sudan parallel to the coast (Egloff et al., 1991; for location see Figure B1.2). The sedimentary sequence features marked halokinetic effects. The crystalline portion of the continental crust is thinned from > 30 km (Egloff et al., 1991, not shown on figure) to 12–7 km and strongly tectonized (signatures for evaporites and continental basement same as in Figure B1.8).

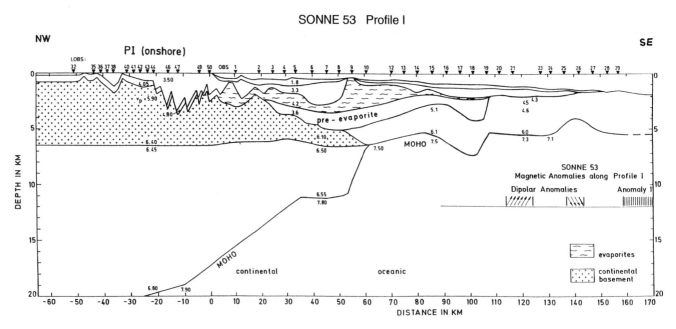

Figure B1.8 Onshore–offshore crustal section 'SONNE' profile I extending from the Sudanese coastal area into the Suakin Deep of the Red Sea axis (Egloff et al., 1991; for location see Figure B1.2). Locations of seismic land stations (LOBS) and ocean bottom seismographs (OBS) constraining the transect are annotated. The coast is 1 km seaward of LOBS 50. The boundary between strongly tectonized continental crust and oceanized crust coincides with the area of maximum sedimentary cover. The thinnest continental and the oldest oceanic crust lie under up to 2 km of pre-evaporitic sediment in the area of highest salt accumulation. Such old oceanized type of crust extends approximately 50 km along the profile, before younger oceanic crust covered with significantly less sediment (≤1 km) occurs at 1 to 2 km shallower depths. Further along profile I towards the Red Sea axis, oceanic crust of decreasing age and decreasing sedimentary cover is found, which increasingly correlates better with anomalies of the magnetic field.

Synthesis of the above-mentioned and other deep seismic data published elsewhere (refraction and wide-angle reflection: Berckhemer et al., 1975; Ginzburg et al., 1981; El-lsa et al., 1987; Makris and Ginzburg, 1987; Gaulier et al., 1988; Marzouk, 1988; Rihm, 1989; Prodehl and Mechie, 1991) and MCS seismic profiles (Izzeldin, 1982; Beydoun, 1989, 1991; Bunter and Abdel Magid, 1989), illustrates a complex pattern of crustal domains with both across-axis as well as along-axis variations (Figure B1.10). This pattern suggests a large number of individual pieces of oceanized crust, which are separated by first- and second-order discontinuities and by remnant pieces of continental crust. A series of elongate patches of oceanic type of crust, covered by one or several kilometres of sediment is found along the western margin of the northern and central Red Sea (Makris and Rihm, 1991). Young oceanic crust that has been produced by sea-floor spreading since c. 5 Ma floors the axial trough of the southern and central Red Sea and Red Sea Deeps. The same type of crust, however, extends considerably (20–40 km) further off the axis than indicated by both bathymetric deeps or high-amplitude, high-frequency linear magnetic anomalies (Girdler and Styles, 1974; Izzeldin, 1982, 1987; Makris and Rihm, 1991). Stretched continental crust is found elsewhere in the Red Sea basin, i.e. along the entire eastern flank, across the southern Red Sea except for the axial graben, and over (small) portions of the western flank of the northern and central Red Sea. The asymmetric geometry of these different crustal domains is shown in Figure B1.10. This has been mapped on the basis of the above-cited data sets and from interpretation of morphology and structural geology (Mohr, 1975; Bäcker et al., 1975; Garson and Krs, 1976; Vail, 1983, 1985, 1988; Guiraud et al., 1985; Berhe, 1986; Dixon et al., 1987, 1989; Garfunkel et al., 1987; Pallister et al., 1988; Sultan et al., 1988; Schandelmeier and Pudlo, 1990; Henke, 1995), and petrology data (Coleman et al., 1983; Coleman, 1984; Betton and Civetta, 1984; Altherr et al., 1988; Coleman and McGuire, 1988; Hart et al., 1989; Camp and Robool, 1989; Camp et al., 1987, 1991, 1992; Coleman, 1993) and will be discussed together with the results of other geophysical methods in the last section of this paper.

GRAVITY FIELD OF THE RED SEA

Obervations and interpretation of the gravity field and computation of two-dimensional gravity models (Makris et al., 1975; Gettings, 1977; Izzeldin, 1982; Akamaluk, 1989; Henke, 1989, 1995; Makris et al., 1991c) confirm the seismic results discussed above. The map of Bouguer gravity of the Red Sea (Plate 2 redrawn after Makris et al., 1991c) displays in the southern part the structure of stretched conjugate continental margins with substantial sedimentary cover (strong negative to low positive values) separated by a narrow oceanic graben (strong positive linear anomaly). Along the western limit of the Afar Depression (east of 40°E) a positive relative anomaly meets the Red Sea coast (at 41°E) and roughly follows the north–south oriented Marda Line (for location see Figure B1.10), delineating the zone of recent rifting in Afar. This zone is partly, as Plate 1 illustrates, below sea level. Similar, but less pronounced are north–south oriented positive relative anomalies correlated with the Baraka and Onib Hamisana Sutures on the western flank (annotated as lithospheric lineaments in Figure B1.10). In general, the image becomes more complicated in the central and northern Red Sea, reflecting the northward increase of discrepancy between the orientations of rift direction and plate motion as the pole of rotation is approached, and the resulting complexity of crustal domains. (The issue of obliquity of spreading is further discussed in the last two sections).

En-echelon structures are increasingly significant from south to north, with the axial zone turning from a linear north-north-west–south-south-east oriented spreading centre (15°N to 20°N) into small oblique segments of Aqaba orientation (north–south to north-north-east–south-south-west, e.g. 24°N–26°N at 36.5°E; 25.5°N–27°N at 35.5°E; and 26.5°N–28°N at 34.5°E). Simultaneously, the central positive anomaly widens, correlating with the bathymetry of a wider, shallow trough with individual isolated deeps.

Three 2-dimensional density profiles have been computed across the Red Sea at different latitudes (Figure B1.11). The results are in good agreement with the seismic models of asymmetric distribution of crustal domains in the northern Red Sea (Figures B1.11(a) and B1.5(a)). They also agree with symmetric stretching of the continental crust in the southern Red Sea associated with axial continental break up and seafloor-spread, young oceanic crust, as far south as latitude 16.5°N (Figures B1.11(b) and B1.9) and continuous continental crust across the entire Red Sea trough at 14.5°N (Figures B1.11(c) and B1.9). The southernmost profile illustrates that the maximum attenuation splits into two branches, correlating with two relative minima of Moho depth (i) along the Red Sea axis and (ii) along the western flank of the Afar Depression.

MAGNETIC TRENDS IN THE SOUTHERN AND CENTRAL RED SEA AND ONSHORE YEMEN

Measurements of the magnetic field provide evidence for sea-floor spreading in the axial southern graben of the Red Sea (Allan, 1970; Laughton et al., 1970; Hall, 1970,

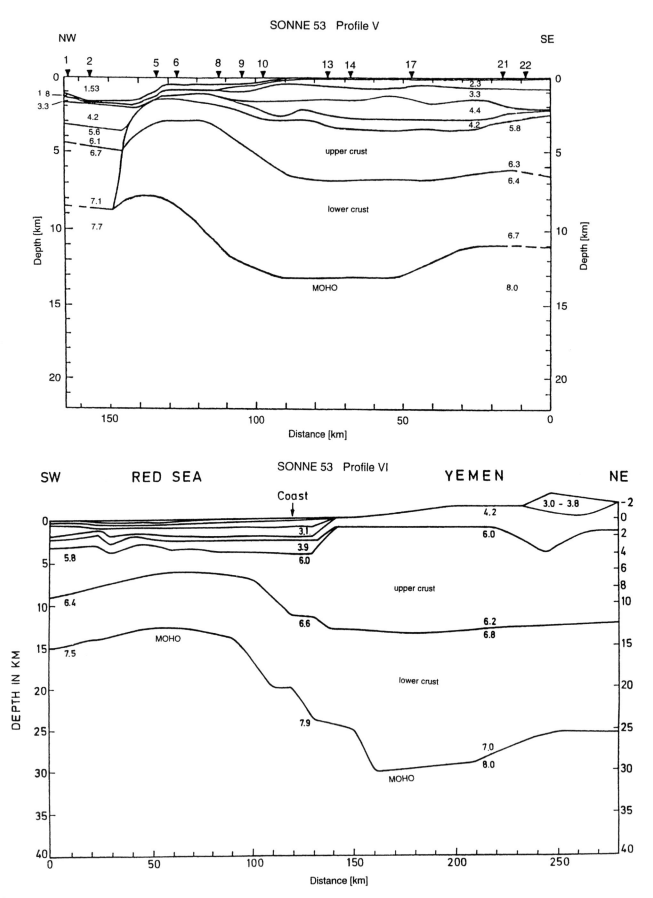

Figure B1.9 Crustal sections of 'SONNE' offshore-profile V (top), extending SE–NW from the coast of Yemen at Hodeida across the shallow eastern flank (water depths 20–50 m) into the axial graben of the Red Sea and onshore/offshore-profile VI (bottom) crossing the entire Yemen continental margin from Yemen highland into the Red Sea axis, where no oceanic crust has yet been produced (Egloff et al., 1991; for location see Figure B1.2). The profiles confirm that in the southern Red Sea only the axial graben is underlain by oceanic crust, whereas the flanks of the southern Red Sea and the Tihama coastal plain are underlain by stretched continental crust with β-values (McKenzie, 1978) exceeding 3. Thinning of the crust from the Yemen highland towards the Red Sea occurs in several major steps across the escarpment.

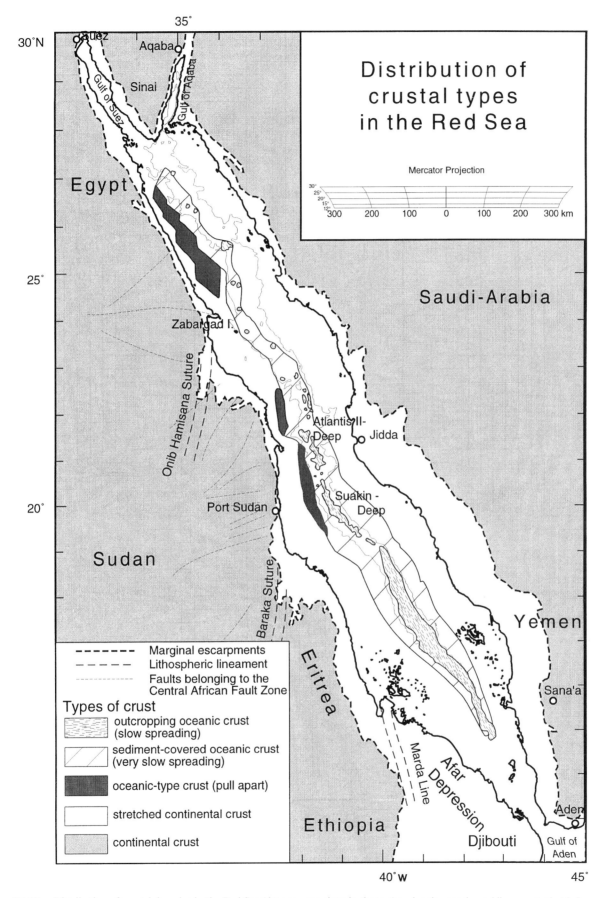

Figure B1.10 Distribution of crustal domains in the Red Sea (data sources given in the text) and major continental lineaments that influenced the development of the initial Red Sea rift. For discussion see the text.

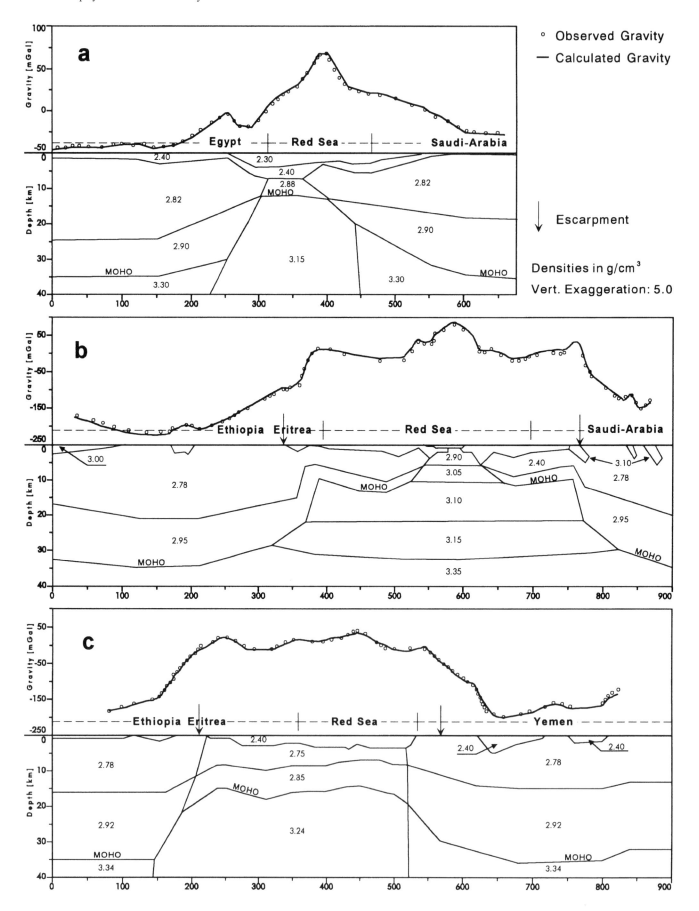

1979; Girdler and Styles, 1974; Roeser, 1975; Izzeldin, 1982; MOMR, 1990). A magnetic total intensity map of the southern and central Red Sea area (Plate 3; data sources in figure caption) shows strong dipolar anomalies resulting from the recently emplaced oceanic crust.

Another spectacular feature of the map is a series of north-west–south-east trending positive anomalies in the main trough of the southern Red Sea and onshore Yemen. The onshore structures correlate with the well-documented direction of Cretaceous to Miocene fault systems and rifts in central and eastern Yemen (e.g. Huchon *et al.*, 1991; Watchorn, 1995; Davison *et al.*, this volume), the most prominent of which is the Marib Graben. In western Yemen this trend of rifting has not been reported, but might be buried beneath the Yemen flood basalts. These rifts seem to be linked with the unsuccessful propagation of the Carlsberg Ridge across Arabia, which was deviated through the Gulf of Aden (Manighetti *et al.*, 1995). Here the large-scale plate kinematic direction is maintained by the en echelon pattern of north-west–south-east oriented spreading segments of the Sheba Ridge (Sahota *et al.*, 1995). The identically oriented magnetic anomalies running in the offshore areas of the eastern Red Sea both along and parallel to portions of the coastline (c. 16°–17°N), have not been previously imaged. Their occurrence and location suggest that the related faults have been reactivated during rifting and have been utilized for the accommodation of stretching on the eastern flank, possibly accompanied by injection of dikes. These north-west–south-east structures have therefore been preserved on the eastern margin of the southern Red Sea during opening where they provide additional evidence for the hypothesis that the orientation of initial rifting is controlled by the directions of pre-existing continental lineaments (Rihm, 1996).

On the eastern flank of the central Red Sea (19°–21°N), the north-west–south-east trend is preserved, suggesting continuation of this trend from the Carlsberg Ridge across the entire south-eastern part of the Arabian Peninsula into the central Red Sea. On the conjugate (western) flank, however, the magnetic anomalies are more discontinuous and south of 22°N correlate better with the locations of the proposed north–south oriented pull-apart basins. These are thought to have developed on the western flank of the northern and central Red Sea during the early stages of opening (Makris and Rihm, 1991).

GEOMETRY OF RED SEA MARGINS

Figure B1.12 illustrates that the western Red Sea margin is characterized by narrower continent–ocean transitions (< 50 km) than other passive continental margins such as the European North Atlantic margin (> 300 km). It is also significantly narrower than the Arabian Red Sea margin (> 100 km). A compilation of strongly simplified continent–ocean transitions from margins where deep seismic information is available suggests classification of passive margins into three categories of steepness, defined by $\delta\beta$, the gradient of McKenzie's (1978) stretching factor β across the margin (Rihm, 1996; Figure B1.13). The categories indicate whether rifting and opening of these margins has been controlled by one or other of the following factors:

1. Stretching ($\delta\beta$: 0.01–0.02); examples are the Norwegian margins into the Norwegian–Greenland Sea (Olafsson *et al.*, 1991; Goldschmidt-Rokita *et al.*, 1994), off Ireland (Makris *et al.*, 1988b) and off Biscay (Montadert *et al.*, 1979) into the North Atlantic, or off southern Argentina into the South Atlantic (Lohmann *et al.*, 1995).

2. Oblique motion ($\delta\beta$ around 0.05); e.g. the East Greenland margin (Weigel *et al.*, 1995), the Antarctic margin off Wilkes Land (Bohannon and Eittreim, 1991), the Siberian margin into the Arctic Ocean (Sorokin, personal communication, 1994), the north-west African margin off Mauritania (Fritsch *et al.*, 1978), the Arabian margins off southern Saudi Arabia (Healy *et al.*, 1982; Mooney *et al.*, 1985), and off Yemen (Egloff *et al.*, 1991).

3. Strike-slip motion ($\delta\beta$ well above 0.1, up to 0.2): the African margins into the northern Red Sea off Egypt (Rihm, 1989; Rihm *et al.*, 1991) and off Sudan (Egloff *et al.*, 1991), and the West African margin off Ghana (Edwards *et al.*, 1996).

The narrow continent–ocean transition off Ghana (only 50 km from full thickness continental crust to

Figure B1.11 Three 2-D density models across the Red Sea and adjacent coastal areas crossing the axis at latitudes 25.5°N, 16.5°N, and 14.5°N, respectively. Vertical and horizontal scales of all three profiles are identical. The coastlines are indicated by vertical lines, the escarpment is marked by arrows. All density lines are constrained by deep seismic data (Rihm *et al.*, 1991; Egloff *et al.*, 1991). Locations of density transects (see insert map of Plate 2) correlate with locations of seismic profiles 'STEFAN E' PI, III, IV and V (transect a), and 'SONNE' profile VI (transect c); transect (b) crosses 'SONNE' profile V near 16.5°N; for location of seismic profiles see Figure B1.2. Density models were in parts adapted from Akamaluk, 1989; Henke, 1989, 1995 and Makris *et al.*, 1991b. (a) Northern Red Sea. Note the contrast of the attenuated continental margin of the eastern Red Sea flank and the abrupt continent–ocean transition of the Egyptian margin. (b) Southern Red Sea across active spreading centre. Symmetric structure of the basin with stretched continental crust on both flanks and recent emplacement of oceanic crust in the axial trough. (c) Southern Red Sea, c. 220 km south of profile b. Stretched continental crust across the entire width of the Red Sea basin (including the Afar Depression; note the large distance between coastline and escarpment on the western flank) showing two minima of crustal thickness correlating with (i) the prolongation of the axial graben, and (ii) the western edge of the Afar Depression.

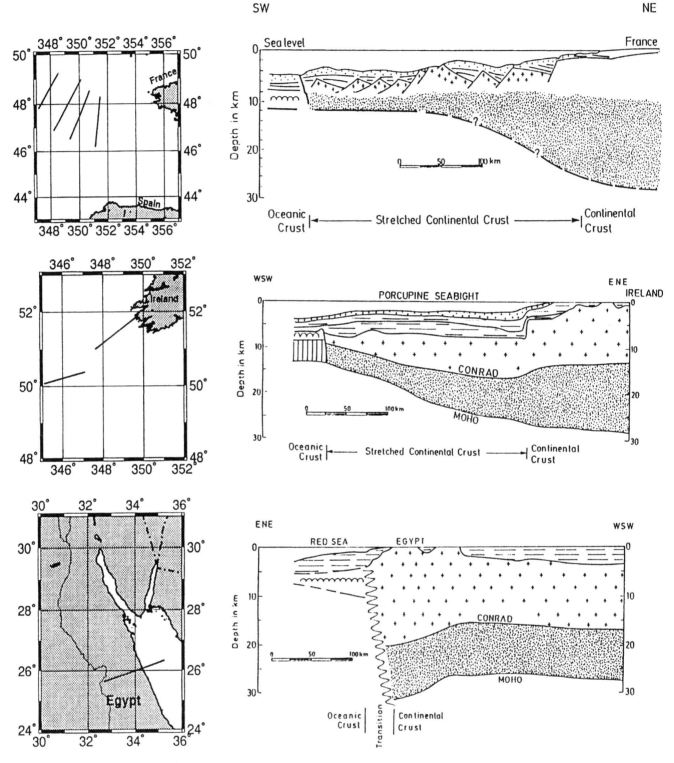

Figure B1.12 Crustal sections across three passive continental margins at identical scale (Rihm *et al.*, 1991). The top section (Montadert *et al.*, 1979) is a schematic profile across the Biscay margin based upon the four MCS profiles shown in the location map (from W to E: 202, 205, 207, 209). The middle section (Makris *et al.*, 1988b) is a deep seismic (ocean bottom seismograph) profile from the North Atlantic Ocean off Ireland. In both examples the continental shelf extends over several hundred kilometres, whereas the western continental margin of the northern Red Sea (lower section, based on offshore and onshore deep seismic data (Rihm *et al.*, 1991), note reversed orientation) appears to be 'cut off'.

Geometry of Red Sea margins 43

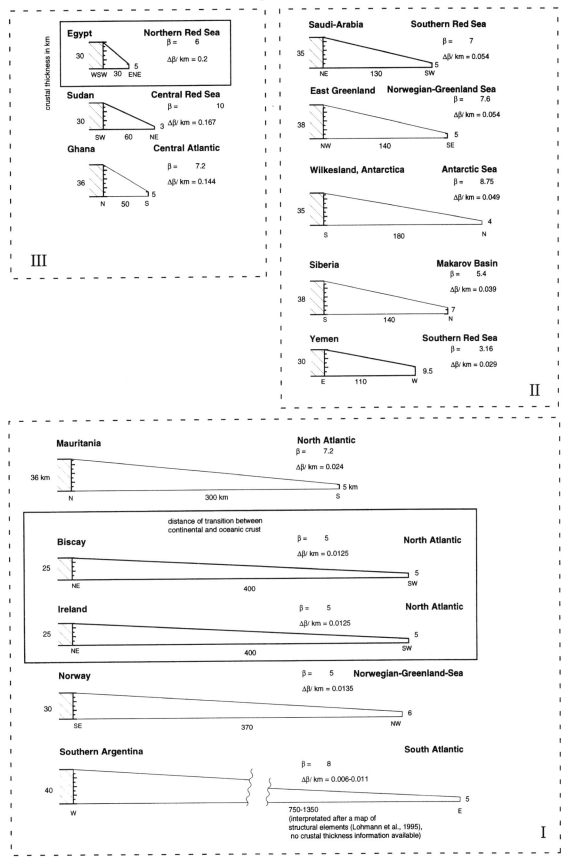

Figure B1.13(a)

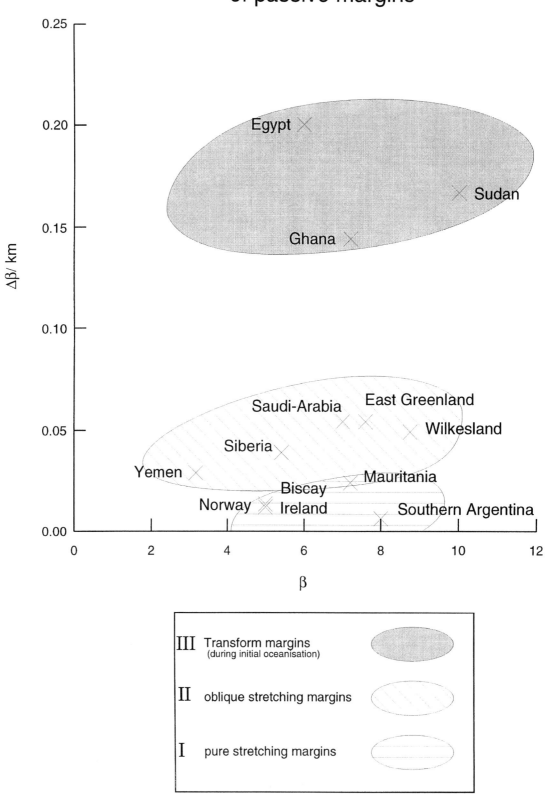

5 km thick oceanic crust) is clearly related to its tectonic position at a transform margin, which has been sheared during its entire development.

PLATE TECTONIC RECONSTRUCTION OF THE RED SEA

The geometry of the western margin of the Red Sea suggests that in its initial stage the plate boundary between Africa and Arabia was conservative along most of the northern and central Red Sea, i.e. the rift started as a transform fault in these areas. Rifting in a plate with uniform strength would lead to symmetric thinning and, finally, linear break up perpendicular to the direction of external stress. In reality, however, the geometry of an evolving rift is linked to the tectonic history of the region and controlled by the interplay of both the extensional stress field induced by plate motion and pre-existing lithospheric structures such as suture zones and major fault zones, acting as stress guides (Dixon *et al.*, 1987; Makris and Rihm, 1991; Rihm, 1996). The spatial discrepancy between such pre-existing directions of stress guides and the orientation theoretically required by the external plate kinematic stress field has to be accommodated by a secondary mechanism. In the case of the Red Sea, the line of initial rifting, as activated in late Oligocene, was located along a zone of structural weakness created during the Pan-African (late Precambrian) period of island-arc accretion (e.g. El Shazly, 1977; Schandelmeier *et al.*, 1987; Bosworth, 1994b), which defined the first-order orientation of the rift. As secondary mechanism, another pre-existing fault system, the – roughly east–west oriented – Central African Fault Zone (CAFZ) was utilized for adjusting the first-order orientation to the regional plate kinematic pattern. This adjustment happened in such a way that the rift line jumped at appropriate locations along the CAFZ from one branch of the first-order lineaments (e.g. Onib Hamisana suture zone) to another (e.g. Baraka suture zone). The result is a rift orientation following over several hundred kilometres first-order lineaments of reduced lithospheric strength which may in parts be oriented slightly oblique with respect to the theoretical orientation of a rift solely depending on the plate tectonic environment. These long segments of oblique orientation are offset by short, roughly perpendicular segments and also preferably occur where lineaments of suitable orientation already exist (Rihm, 1996).

Detailed reconstruction of the history of the African–Arabian plate boundary (Makris and Rihm, 1991) has distinguished four consecutive stages of dominant processes characterizing the evolution of the Red Sea.

1. A continental rift developed during the initial separation of Arabia from Africa, which occurred as primarily a conservative (left-lateral strike-slip) motion. The pole of rotation (R_0) is given in Figure B1.1. This sense of motion is well established for faults adjacent to the northern and central Red Sea coasts (Stern, 1985; Dixon *et al.*, 1987; Sultan *et al.*, 1988; Schandelmeier and Pudlo, 1990). Similar N20°E trending left-lateral strike-slip faults occur onshore in Yemen (Huchon *et al.*, 1991), and in the (pre-) Pan-African suture zones of the central and southern Red Sea (Stern *et al.*, 1986; Berhe, 1986; Kröner *et al.*, 1987). Such lithospheric discontinuities, which are preferentially reactivated during rifting, defined the geometry of the initial continental rift. They also determined where the African and Arabian plates were decoupled: (a) across axis by wrench faulting inducing nucleation of pull-apart basins, and (b) along axis at the major east–west offsets of the rift line (Figure B1.14).

2. Pronounced magmatism, which progressed from the southern Red Sea region (peak activity 29–26 Ma) to the central (25 Ma) and northern Red Sea (20 Ma) (e.g. Izzeldin, 1982, 1987; Menzies *et al.*, 1992; Coleman, 1993) preceded and accompanied a re-orientation of plate motion at the beginning of the Suez stage (pole R_1 of Figure B1.2). Rifting was intensified (Moretti and Colletta, 1987; Steckler *et al.*, 1988) and the separation of the master faults of the pull-apart basins increased, resulting in coalescence and oceanization of the pull-apart domains (segments 1 and 3 of Figure B1.14), whereas stretching of the continental crust continued in the areas that had previously been extensional (segments 2, 4 and 5).

3. Stagnation of extension in the Gulf of Suez at around 15 Ma (Steckler *et al.*, 1988; Bosworth, 1995) coincided with the onset of strike-slip motion in the Gulf of Aqaba, which has been dated not older than 14 Ma

Figure B1.13 (a) Compilation of strongly simplified continent–ocean transitions (morphology across the margin is approximated as straight line; continental and oceanic crustal thickness is defined equal at the continent–ocean boundary) across margins, where deep seismic or sufficient structural information is available. Crustal thickness data from Rihm *et al.*, 1991 (Egypt); Egloff *et al.*, 1991 (Sudan and Yemen); Edwards *et al.*, 1996 (Ghana); Mooney *et al.*, 1985 and Prodehl and Mechie, 1991 (Saudi Arabia); Weigel *et al.*, 1995 (East Greenland); Bohannon and Eittreim, 1991 (Antarctica); Sorokin, 1994, personal communication (Siberia); Fritsch *et al.*, 1978 (Mauritania); Montadert *et al.*, 1979 (Biscay); Makris *et al.*, 1988b (Ireland); Olafsson *et al.*, 1991 and Goldschmidt-Rokita *et al.*, 1994 (Norway), and Lohmann *et al.*, 1995 (Argentina). The three margins of Figure B1.12 have a solid frame. **(b)** Classification of continental margins. Based on the ratio of $\delta\beta$ vs β (β is the stretching factor, McKenzie, 1978), three distinct classes of margins can be identified, where rifting and early opening has been controlled by I – stretching, II – oblique motion, and III – strike-slip motion. Further details in the text.

46 Geophysical studies on early tectonic controls on Red Sea

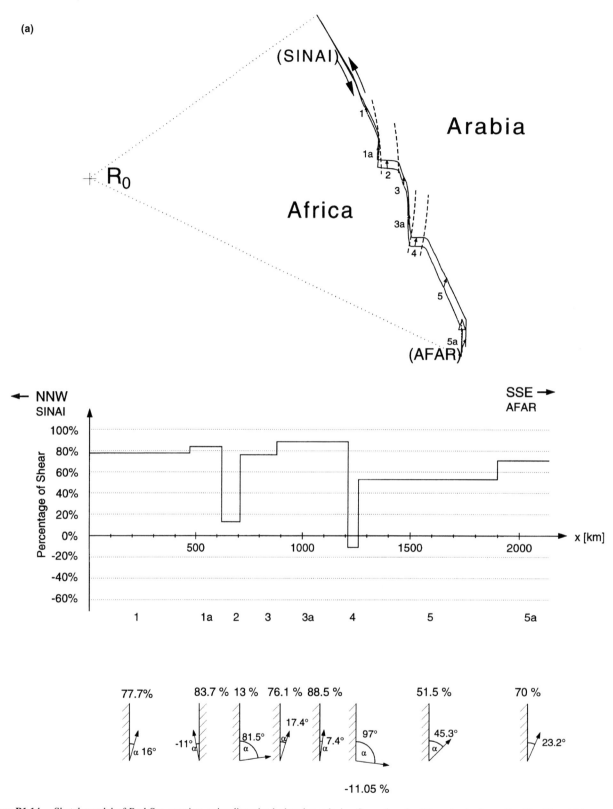

Figure B1.14 Sketch model of Red Sea opening, using linearized plate boundaries. Several major lithospheric discontinuities (Pan-African suture zones: Onib Hamisana, Baraka, and the Central African Fault Zone, see Figure B1.12) deviated the actual rift geometry from the 'default' direction perpendicular to plate motion. (a) Initial separation along a conservative plate boundary (strike-slip dominated (e.g. Shimron, 1990; Makris and Rihm, 1991), pole of rotation R_0, Figure B1.12), created an abrupt plate boundary along Segments 1 and 3. Local transtension resulted in nucleation of en echelon arranged pull-apart basins, possibly separated by some transpressional features. Along Segments 2 and 4 the sense of motion was extensional with percentages of shear below 20% (see middle panel), whereas along Segment 5 it was oblique at about 45° (see vectors of plate motion on lower panel), which seems to be not sufficient for a complete detachment of the lithospheric plates. Across the segment boundaries, differential motion separated crustal domains which could develop independently to the north and the south. (b) At the beginning of the Suez stage, rifting was intensified and accommodated by stretching of the continental crust in the Gulf of Suez (50% of total separation from 19 to 15 Ma, up to 5 mm/year, Moretti and Coletta, 1987; Steckler et al., 1988) and in the previously extensional areas of the Red Sea rift (Segments 2, 4). Along Segment 5 the shear

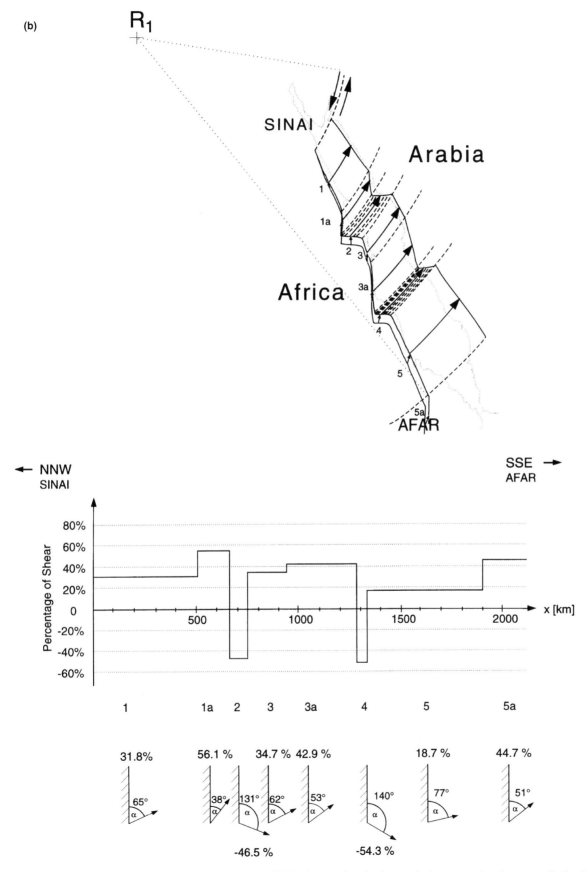

component fell below 20%, implying further stretching of the crust which had not yet been broken up. In the transtensional segments (1, 3), where the continental plates were already detached, continuous motion of the Arabian plate was possible without stretching the African side. Subsequently, extension was accommodated by strike-slip motion in the Gulf of Aqaba with further oceanization in the pull-apart domains and stretching of the interjacent segments and the eastern flank. The eastern Red Sea margin represents the stretched continental margin of the drifting Arabian plate. Its development and shape has been determined by the connection to the African plate through the areas, which had not been subjected to pull-apart evolution (and thus were not separated at an early stage) and by the tectonic setting of Arabia as the moving plate and Africa as the stable plate.

(Bayer et al., 1988). This rearranged configuration did not change the direction of plate motion in the Red Sea, and hence, the existing pull-apart basins were further oceanized by rapidly uprising magma, as the Arabian plate moved north-westward. Stretching of continental crust continued in the adjacent domains and on the eastern Red Sea margin.

4. Finally, sea-floor spreading has formed an axial graben in the southern Red Sea and a series of isolated deeps in the central and northern Red Sea. Reconstructions of the history of the Atlantis-II Deep, based on swath bathymetry (Pautot, 1983) and deep-tow magnetic data (Coutelle et al., 1991), for this most advanced of the Red Sea Deeps, have demonstrated the continuity of continental trends as structural control for opening.

Accretion of oceanic crust at the present axis of the Red Sea has started significantly earlier than indicated by magnetics and morphology (Figure B1.6), possibly as early as 12–10 Ma and at slower rates of motion than the recent ones.

DISCUSSION AND CONCLUSIONS

The cause for the high steepness of large portions of the African Red Sea margin has been explained by left-lateral strike-slip motion during the early stages of rifting in the Red Sea around pole of rotation R_0 of Figure B1.1 (Rihm, 1996). This motion, which has a direction more similar to the opening between Arabia and Somalia (Jestin et al., 1994) than to present-day extension of the Red Sea, resulted in the generation and rapid oceanization of linearly arranged pull-apart basins (Makris and Rihm, 1991; Rihm, 1996). Spatial distribution of these basins is inferred by the geometry of the strike-slip zone, which itself depends on the distribution of major old structural lineaments such as sutures or shear zones. The early Red Sea rift, mainly in the Egyptian and Sudanese coastal areas, was affected by the interplay of the external stress field resulting from plate kinematics with pre-existing continental lineaments. The most effective of the latter are identified as the old ($\geq$ Pan-African) suture zones of Onib Hamisana and Baraka. These intersect with the Red Sea coast at around 23°N and 20°N, respectively (Vail, 1983; Stern et al., 1986; Pallister et al., 1988). Individual branches of the Central African Fault Zone (CAFZ), which crosses the entire African continent in an east to east-north-east direction, meet the Red Sea coast in Sudan and southern Egypt (Garson and Krs, 1976; Guiraud et al., 1985; Schandelmeier and Pudlo, 1990; Bosworth, 1992, 1994b). Arabia therefore was separated from Africa where pull-apart basins developed and were rapidly oceanized as Arabia moved away. Both continents were still connected in the intervening segments by stretched continental crust. Because the Arabian plate moved away from the stable African plate, the eastern Red Sea flank was formed by pure shear through stretching, thinning and diffuse extension. As a consequence, the eastern and western flanks are asymmetrical. Different relative velocities of the conjugate plates with respect to the underlying asthenosphere have been described for other conjugate margins such as Norway–Greenland (Vogt et al., 1982) or Australia–Antarctica (Weissel et al., 1977; Bohannon and Eittreim, 1991) and can be attributed to slow migration of the lithosphere over a fixed zone of asthenospheric upwelling. In the case of the Red Sea, the north-easterly direction of migration has promoted intrusive processes (such as the oceanization of the pull-apart basins in the northern and central Red Sea by basaltic intrusions) on the western flank and stretching of the eastern flank (Dixon et al., 1989; Makris and Rihm, 1991). The intrusion of dikes along the coastal areas of Saudi Arabia and Yemen is consistent with this hypothesis, because it occurred mainly from 24 to 20 Ma (Coleman et al., 1977, 1979; Capaldi et al., 1987a; Pallister, 1987; Du Bray et al., 1991; Manetti et al., 1991; Mohr, 1991; Davison et al., 1994) and hence clearly pre-dates the main phases of opening of the Red Sea.

Crustal generation by sea-floor spreading has formed the axial Red Sea trough and this has been dated around 5 Ma at rates of 7–8.5 mm/year by correlation of linear magnetic anomalies (Allan, 1970; Girdler and Styles, 1974; Roeser, 1975; Hall et al., 1977; Izzeldin, 1982, 1987). Ocean bottom seismic results (Egloff et al., 1991) and reflection seismic mapping (Izzeldin, 1982, 1987; Figure B1.6), however, show the oceanic crust to extend off the axial graben up to 40 km towards the flanks beyond the oldest recognizable magnetic anomalies. This result suggests that axial formation of oceanic crust may have started already around 10 Ma, if spreading rates have remained constant, or as early as 11–12 Ma at spreading rates around 5–7 mm/year (Izzeldin, 1987). The latter reconstruction would also explain the lack of clear linear magnetic anomalies, which are not likely to be produced at such slow spreading (Girdler and Styles, 1974; Roeser, 1975). It also is in good agreement with recent reconstructions of sea-floor spreading in the Gulf of Aden based on magnetic, gravity and seismic data (Sahota et al., 1995), which demonstrate that oceanic crust has been produced continuously at the eastern Sheba Ridge since around 20 Ma and around 15 Ma in the western Gulf of Aden.

In a broader plate kinematic context, the development of the Gulf of Aden may be the result of deviation of the north-west propagating Carlsberg Ridge (northern end of Central Indian Ridge) influenced by the configuration of the existing continental lineaments available as stress guides. In consequence, the Red Sea may have been initiated as a transform fault of the Gulf of Aden during its westward propagation. This model is supported by the similar results obtained for opening of the Gulf of Aden (Jestin *et al.*, 1994) and the early motion in the Red Sea (Rihm, 1996 and this paper). The beginning of substantial extension in the Red Sea would then coincide with a re-adjustment of plate motion to the large-scale kinematic pattern, which is consistent with the extension at the Carlsberg Ridge. This re-adjustment may have induced the East African rift system, which accommodates the differential motion resulting from recent directions of opening in the Gulf of Aden and the Red Sea.

ACKNOWLEDGEMENTS

We acknowledge the tremendous effort, J. Makris and all colleagues at the Geophysics Institute, Hamburg University in collecting and working on data from the Red Sea, that were published in Makris *et al.* (1991b) and provided a major base for this paper. Participation in the Sana'a Conference was enabled by a DFG grant (477/1265/94). We are grateful for careful editing of an earlier version of the manuscript and valuable comments by B. Purser and D. Bosence and an anonymous reviewer.

Chapter B2
Pre-, syn- and post-rift volcanism on the south-western margin of the Arabian plate

G. Chazot, M. A. Menzies and J. Baker

ABSTRACT

Cenozoic magmatism in Yemen and Saudi Arabia is inextricably linked to upwelling mantle plumes beneath Afar and to sea-floor spreading in the Gulf of Aden (<20 Ma) and the Red Sea (<5 Ma). The precise timing of these magmatic events is crucial to our understanding of the mantle dynamics associated with continental break up and the initiation of sea-floor spreading. To this end it is apparent that $^{40}Ar/^{39}Ar$ dating is the only appropriate technique because of the erroneous conclusions forthcoming from whole-rock K-Ar dating, largely because of secondary processes that may have affected the volcanic rocks. Available $^{40}Ar/^{39}Ar$ data define three main episodes of Cenozoic volcanism on the south-western Arabian plate. Pre-rift volcanism was dominated by the eruption of a large igneous province (i.e. LIP) in Yemen and smaller volume alkaline volcanism in Saudi Arabia. Syn-rift volcanism was characterized by intrusion of a major tholeiitic dike swarm in Saudi Arabia parallel to the Red Sea and intrusion of sub-volcanic plutonic and intrusive rocks in western Yemen. Post-rift volcanism was notable by eruption of one of the largest alkaline provinces on earth.

INTRODUCTION

Surface uplift and extension of the Afro-Arabian plate and formation of the Gulf of Aden and the Red Sea were accompanied by widespread magmatism extending over $>1 \times 10^6$ km^2 of the western Arabian plate, from Yemen in the south to Syria and Jordan in the north. Many studies have focused on the processes related to the opening of the Red Sea and Gulf of Aden (i.e. volcanism, rifting and erosion). Dixon et al. (1987, 1989) pointed out that reactivation of pre-existent, basement lineaments can greatly influence the location and orientation of the rifting zones. The Shabwah-Balhaf graben parallels Najd lineaments in south-west Arabia and apatite fission track work in Yemen (Menzies et al., 1996b) supports Tertiary reactivation of lineaments that parallel the margins of this graben. The main problems to be resolved regarding magmatism in south-western Arabia relate to the relative timing of magmatism, crustal extension, surface uplift and erosion on volcanic (i.e. southern Red Sea) and non-volcanic (i.e. central and northern Red Sea and Gulf of Aden) margins (Davison et al., 1994; Menzies et al., 1992, 1994a,b). Also of some importance is the relevance of this timing to theoretical models of the evolution of active or passive models of continental rifting (Bosence et al., this volume). Herein we will outline a general tectonic framework related to development of magmatism on the Arabian plate but we will not attempt to review the different hypotheses put forward to explain the tectonic and magmatic events which have affected this area. We will review the timing and nature of this magmatism and relate the main magmatic phases to the tectonic events which have affected the whole region.

CHRONOLOGY OF MAGMATISM

K-Ar and Ar-Ar techniques

The absolute timing of magmatism is extremely important in the study of continental rifting because it can provide an absolute datum against which one can assess the relative timing of extension, surface uplift and erosion. In the last twenty years much of the age dating of volcanic rocks has involved whole rock K-Ar techniques (e.g. Dixon et al., 1989; McGuire and Bohannon, 1989; Menzies et al., 1990; Davison et al., 1994). More than one hundred K-Ar 'dates' have been published on the volcanic rocks from the Yemen large igneous province (i.e. LIP) (also called flood basalts, traps and the Yemen Volcanics) (Civetta et al., 1978; Capaldi et al., 1983, 1987a,b; Menzies et al., 1990; Huchon et al., 1991; Manetti et al., 1991; Al Kadasi, 1994). Many of the volcanic rocks have experienced secondary processes in the form of contamination with upper and lower crustal rocks (Baker et al., 1994) and surface alteration with groundwater. All of these factors influence the K-Ar system so that the K-Ar 'dates' bear little or no relationship to the actual primary crystallization age of the rock in question. This can be adequately demonstrated by several combined K-Ar and $^{40}Ar/^{39}Ar$ studies on volcanic rocks from Yemen and Saudi Arabia. For example, a detailed K-Ar study of the basal volcanic rocks of the Yemen LIP indicated that volcanism commenced from the Eocene to the late Miocene (Al Kadasi, 1994). A parallel $^{40}Ar/^{39}Ar$ study of exactly the same rocks revealed a period of tightly constrained volcanism in the late Oligocene. $^{40}Ar/^{39}Ar$ data showed that volcanism commenced around 29–31 Ma (Baker et al., 1994, 1996) and that volcanism continued to 26 Ma (uppermost flows) over 2500 m of section. This result is vastly different to the whole rock K-Ar database which indicated that volcanism began between 10 and 66 Ma ago (Al Kadasi, 1995). Even if one uses the $^{40}Ar/^{39}Ar$ data to screen the K-Ar data, such that those samples affected by secondary processes are eliminated, the K-Ar data have little meaning. Furthermore, more than 75% of the K-Ar data are shown to be in error by 2–35 m.y. Similarly, a recent review of c. 130 K-Ar analyses of volcanic rocks from Saudi Arabia (Camp and Roobol, 1991) gave 'dates' of 15–42 Ma with a few older 'dates' (Pallister, 1987; Gettings and Stoeser, 1981; Schmidt et al., 1982). However, when compared with $^{40}Ar/^{39}Ar$ data it is apparent that the main magmatic event occurred in a much shorter time, between 21 and 24 Ma (Féraud et al., 1991; Sebai et al., 1991) again demonstrating that K-Ar data can lead to erroneous conclusions. Consequently in many of these LIPs, K-Ar data cannot contribute to our understanding of the timing of magmatism in relation to extension, surface uplift/erosion. Consequently, we will primarily discuss $^{40}Ar/^{39}Ar$ data and mention will be made of K-Ar data when no other data are available, but with the cautionary statement that these data can be unreliable due to secondary processes (e.g. contamination and alteration).

A recent integration of apatite fission track, $^{40}Ar/^{39}Ar$ and field data indicates that the Red Sea margin exposed in western Yemen evolved in response to surface uplift and volcanism followed by extension and erosion (Menzies et al., 1996a,b). From these data it is apparent that eruption of several thousand metres of the Yemen LIP from 31 to 26 Ma predated break up and erosion on the Yemen margin by c. 5 Ma (i.e. pre-rift volcanism) (Plate 4). The period of erosion and break up on the Yemen margin was associated with emplacement of intrusive and plutonic rocks from 26 to 16 Ma (Baker et al., 1996; Féraud et al., 1991; Zumbo et al., 1995) (i.e. syn-rift volcanism) (Figure B2.1). More recently (< 10 Ma) the Yemen margin has been affected by a period of post-erosional volcanism that postdated break up and erosion of the margin and the transition from continent to ocean (i.e. post-rift) (Plate 4). It is on this basis that we interpret the volcanism along the eastern margin of the Red Sea and in south-western Arabia in terms of pre-, syn- and post-rift activity. Recent considerations of erosion on the uplifted Red Sea margins (Menzies et al., 1996b) indicate that extension and erosion were essentially synchronous. The major period of extension/erosion to have affected the Red Sea margins occurred at < 26 Ma.

Magmatic phases

Pre-rift magmatism (> 26 Ma) includes those extrusive (i.e. sub-aerial volcanic flows) and intrusive (i.e. dikes, sills and plutonic) rocks that formed prior to widespread extension and erosion on the margins of the Gulf of Aden or the Red Sea. Such magmatism may have been synchronous with surface uplift (Menzies et al., 1996a,b) which crustal cooling data (i.e. apatite fission track analyses) indicate most likely happened in the Oligo-Miocene. The only sedimentological evidence for base-level changes at this time is within the Tawilah Formation. The gradual change from the marine to fluvial sedimentation, palaeosol development and sub-aerial volcanism is consistent with a base level change on the order of tens of metres. This is interpreted as the earliest expression of surface uplift (Menzies et al., 1996a,b). Pre-rift magmatism was initiated by deep mantle processes (e.g. plumes) and presumably the pre-rift lithosphere was thicker and colder than it is today (i.e. low heat flow : shield geotherm).

Syn-rift magmatism (26–20 Ma) includes extrusive and intrusive rocks that formed during extension and enhanced erosion on the margins of the Gulf of Aden or the Red Sea. Syn-rift magmatism was driven by shallow mantle processes in the asthenosphere and presumably the syn-rift lithosphere had a thickness and temperature not very different from those existing today (i.e. high heat flow : oceanic geotherm).

Post-rift magmatism (≪ 20 Ma) includes extrusive and intrusive rocks that formed after the main period of extension and erosion on the margins of the Gulf of Aden or the Red Sea. In many cases, post-rift magmatism occurred once the rift shoulder had been uplifted, extended and eroded. Presumably post-rift magmatism was driven by shallow mantle processes and the post-rift lithosphere was thin and hot (oceanic/ridge geotherm) similar to that observed on the margins of the Gulf of Aden and the Red Sea today.

Pre-rift magmatism

Saudi Arabia

The earliest expression of Cenozoic volcanism in Saudi Arabia (Plate 5) occurs at Harrat (i.e. plateau) Hadan, the expression of a large north–south trending volcano comprising olivine alkali basaltic lava flows. These lavas have been dated by $^{40}Ar/^{39}Ar$ methods at 27–28 Ma (Sebai et al., 1991). In southern Saudi Arabia, the Harrat As Sirat (Plate 5) is composed of alkaline basalts that have K-Ar 'dates' in the range 30–20 Ma (Du Bray et al., 1991). Other poorly constrained K-Ar 'dates' and field relationships indicate similar 'ages' for the Harrat Harairah, Harrat Ishara, the lower part of Harrat Uwayrid and isolated basalts north of Harrat Rahat. Given the uncertainties in K-Ar data, outlined earlier, no conclusions can be drawn from these K-Ar data other than to say that such volcanism is probably ≫20 m.y. ago (Camp and Roobol, 1991) and as such may be pre- or syn-rift.

Yemen

The largest volcanic province in Arabia was erupted in Yemen (Plates 4, 5) before the main rifting episode that formed the Gulf of Aden and the Red Sea (Menzies et al., 1992; Davison et al., 1994). In the central and northern part of the Yemen LIP volcanostratigraphic and $^{40}Ar/^{39}Ar$ chronostratigraphic studies (Baker et al., 1994, 1996) have established that basaltic magmatism began at 30.9 Ma with a cumulative thickness of 1000–1500 m of basaltic rocks. Throughout western Yemen an important switch from basaltic to bimodal magmatism occurred at 29 Ma and this period of combined basaltic and silicic magmatism lasted from 29.1 to 26.5 Ma with a cumulative thickness of c. 1000 m. Formation of the Yemen LIP appears to have ended between 26.9 and 26.5 m.y. ago, evident as an erosional unconformity at the top of the basalt–rhyolite/ignimbrite (i.e. bimodal) units (Baker et al., 1996). In the southern part of the Yemen LIP, Zumbo et al. (1995) reported $^{40}Ar/^{39}Ar$ ages of 28.9 and 26.5 Ma for two basalts at the bottom and the top of a volcanic section respectively. These $^{40}Ar/^{39}Ar$ data reiterate the suggestion that volcanism ceased around 26.5 Ma (Baker et al., 1996). However, the lowermost flow in this section is by no means basal since it coincides in age with the period of bimodal volcanism in northern Yemen (i.e. 29–26 Ma). Overall bimodal volcanism in Yemen occurred before the main phase of extension and break up (Figure B2.2; (Davison et al., 1994) when the erosional response to surface uplift appears to have been largely suppressed judging from the lack of major erosional unconformities and interflow sediments.

Syn-rift magmatism

Saudi Arabia

Following the pre-rift eruption of scattered alkaline volcanic rocks, a major period of tholeiitic dike intrusion occurred along the Arabian coast. The Tihama Asir and Al Lith magmatic complexes (Plate 5) were intruded along north-west basement faults (e.g. Najd). These magmatic complexes comprise important volcanic sequences, intrusive rocks (i.e. feeder dike swarms) and associated plutonic bodies. A system of gabbroic dikes (10–100 m thick) is evident along the coast from Saudi Arabia to Sinai and a recent $^{40}Ar/^{39}Ar$ study of this magmatism has revealed that it occurred between 21 and 24 Ma (Féraud et al., 1991; Sebai et al., 1991). This is a very short period of magmatic activity when one considers that magmatism of this type extends along the Red Sea rift margin for >1000 km. Since this dike swarm parallels the orientation of the Red Sea it is considered to have been coeval with formation of the proto-Red Sea (Plate 5).

Yemen

Pre-rift eruption of the Yemen LIP was followed by a period of syn-rift volcanic activity associated with lower eruption rates and emplacement of plutonic rocks (e.g. gabbro–granite–syenite) (Chazot and Bertrand, 1993; Blakey et al., 1994) (Plates 4, 5). $^{40}Ar/^{39}Ar$ dates indicate that the plutonic rocks of western Yemen were intruded from 21.4 to 22.3 Ma (Zumbo et al., 1995). This time period is substantiated by Rb-Sr dates from granites (Blakey et al., 1994). $^{40}Ar/^{39}Ar$ plateau ages on intrusive rocks (Zumbo et al., 1995), cutting the LIP, range from 25.4 to 16.1 Ma suggesting that LIP formation straddled the boundary between pre-rift and syn-rift and was active for a longer period than the preserved volcanostratigraphy of the pre-rift LIP. Although LIP volcanism appears to have finished around 26 Ma, a significant part of the LIP may have been removed during Oligo-Miocene erosion (Menzies et al., 1996a,b) between 26 and 19 Ma. This time interval is marked by an unconformity at the top of the LIP where volcanic rocks with ages of 26 Ma are overlain unconformably by a volcanic unit with an age of 19 Ma (Baker et al., 1996). This emphasizes the fact that syn-rift volcanic activity was associated with a period of enhanced erosion and widespread extension in contrast to the earlier pre-rift period of volcanism (26–31 Ma) when erosion and extension was minimal (Menzies *et al.*, 1992; Davison *et al.*, 1994). It is important to note that the time range for this large intrusive event in Yemen corresponds to the main magmatic event recorded along the coast of Saudi Arabia between 21 and 24 Ma (Féraud *et al.*, 1991; Sebai *et al.*, 1991). It should be mentioned that extension and erosion, apparent in the Yemen LIP (26–20 Ma), coincided with rifting in the Gulf of Aden prior to sea-floor spreading (*c.* 20 Ma). Clearly a significant change in the volcanic and tectonic history of western Yemen occurred around *c.* 26 Ma probably driven by the opening of the Gulf of Aden and north-eastwards movement of Arabia away from Africa.

Verification of the syn-rift status of magmatism in western Yemen is forthcoming from a recent integration of $^{40}Ar/^{39}Ar$ data (Baker *et al.*, 1996) with apatite fission track analyses (Menzies *et al.*, 1996a,b). Combined chronological data indicate that: (a) volcanism (31–26 Ma) predated a major period of continental extension (break up) (< 26 Ma); (b) erosion (i.e. erosion/denudation) largely took place in < 26 Ma, indicating that surface uplift, which must precede erosion, occurred > 26 m.y. ago; (c) Tertiary reactivation of lineaments of Gondwana and Jurassic age was important, and (d) the volcanic margin in western Yemen evolved in response to surface uplift and volcanism followed some 5 m.y. later by extension and erosion.

Post-rift magmatism

Saudi Arabia

One of the largest alkaline volcanic provinces in the world (Figure B2.2) erupted after the main period of rifting and erosion in the Gulf of Aden and the Red Sea. Fissural alkaline volcanism extended over more than 1500 km from the Al Birk volcanic field in the southern part of Saudi Arabia (Plate 5) to Syria and Jordan in the north. For most of the harrats (Figure B2.2), the eruptive fissures were oriented north–south, but in some cases (e.g. Harrat Uwayrid) the direction of the fissures was parallel to the previously described Oligo–Miocene dikes. $^{40}Ar/^{39}Ar$ data obtained on these formations (Sebai, 1989) indicate ages of < 5 Ma. Most of the K-Ar dates are also < 5 Ma (Camp and Roobol, 1992).

Yemen

The alkaline province in Saudi Arabia continues south into Yemen (Plates 4, 5) where post-rifting and post-erosional volcanism is more diversified, but less voluminous and widespread, than in Saudi Arabia. One of the main volcanic features is the Aden Volcanic Line that extends along the south coast from Perim Island in the west to Aden in the east (Plate 4). This volcanic line comprises six important volcanic centres with basaltic and highly differentiated rocks (Gass and Mallick, 1968; Cox *et al.*, 1969; Cox *et al.*, 1970, 1977; Mallick *et al.*, 1990; Chazot, 1993). The location of these volcanic centres on a line along the south coast suggests a strong structural control on volcanism. In addition, several volcanic centres exist near Sana'a, Sada'a, Dhamar and Jabal An Nar. Unfortunately, no precise $^{40}Ar/^{39}Ar$ data exist for these rocks so the 'ages' are based on K-Ar data and cannot be confirmed. In the area of Sana'a, K-Ar data indicated a period of volcanism around 10 m.y. (Capaldi *et al.*, 1983; Manetti *et al.*, 1991) and in the case of the Aden Volcanic Line, K-Ar dates (Mallick *et al.*, 1990) indicate that volcanism lasted from 11 to 5 Ma, with an age progression from west (Perim Island) to east (Aden). However, this needs to be confirmed as spatial time progressions have been found elsewhere in the pre-rift volcanic rocks, using K-Ar data (Al'Kadasi, 1994), but $^{40}Ar/^{39}Ar$ data have shown them to be invalid (Baker *et al.*, 1996).

Fissural basaltic eruptions created several volcanic provinces (Plate 4) in more recent times. Five post-erosional alkaline volcanic fields were erupted; Sana'a, Marib, Dhamar, Bir Ali, Ataq and Shuqra (Cox *et al.*, 1977). Their state of preservation indicates that they are very young with an historical eruption in Sana'a around AD 200–500. Again no $^{40}Ar/^{39}Ar$ data exist for this episode, ages cannot be confirmed and only a few K-Ar ages are available on these volcanic rocks (Huchon *et al.*, 1991). Plio-Quaternary volcanism in Yemen around the Masila High of the Hadramaut (Watchorn 1995, this volume) is far less important in volume and extent, than in Saudi Arabia, and the orientation of the eruptive fissures is also different.

ARABIAN PLATE MAGMATISM

It is beyond the scope of this paper to review in detail the origin and evolution of pre-, syn- and post-rift magmatism on the Arabian plate since the late Oligocene. Two main sources can contribute to basaltic volcanism at the surface of the earth, the mantle and the crust, and in south-western Arabia mantle sources include possible mantle plumes beneath Afar, the asthenosphere (convecting upper mantle) that underlies the present-day Gulf of Aden and Red Sea ridges and the lithospheric mantle which underlies the crust. Crustal sources can contribute during crustal underplating at the crust–mantle boundary (Moho) or during magmatic evolution in shallow fractionating magma chambers in the upper crust (Camp and Roobol, 1992). Since many of the rocks erupted at the surface in the Arabian Plate are basaltic an initial mantle component is believed to have been involved but a later overprint, from interaction with crustal rocks, has contaminated many of these rocks. Detailed studies of volcanism in Djibouti (an area of thin crust, high heat flow with a shallow low velocity zone) define the unique chemical characteristics of the Afar mantle plume (Vidal et al., 1991; Deniel et al., 1994). Basalts erupted along the spreading ridge axes of the Red Sea (Eissen et al., 1989; Volker et al., 1993) and the Gulf of Aden (Schilling et al., 1992) define the isotopic composition of the asthenosphere (i.e. convecting upper mantle). In addition, studies of lithospheric mantle xenoliths found in volcanic rocks at Bir-Ali, Ataq, Marib (Yemen) (Plate 4) (Chazot et al., 1996a, b) and in Saudi Arabia can help us define the chemistry of the shallow sub-crustal mantle. Similarly, the nature of the crust, that may have contributed to magmatism, is forthcoming from the study of crustal rocks exposed at the surface or brought to the surface as xenoliths in recent volcanic rocks.

The earliest expression of continental rifting probably began in the Gulf of Aden at c. 35 Ma (Watchorn 1995, this volume) and some c. 10 m.y. later in the Red Sea and the Gulf of Suez (Bosworth 1995, this volume). The latter would indicate near synchronous development of the Gulf of Aden and the Red Sea along its entire length in under 5 m.y. This is supported by the age of the extensive dike swarm in the Red Sea (Plate 5). This opening or extension relates to movement of the Arabian plate away from the African plate, westward propagation of the Sheba ridge through the Gulf of Aden and upwelling of the Afar plume beneath Afar. Initiation of sea-floor spreading in the Gulf of Aden and the Red Sea is another key point which is difficult to resolve. In the Gulf of Aden, Courtillot (1980, 1982) and Cochran (1981, 1982) have used oceanic magnetic anomalies to demonstrate that formation of oceanic crust started at c. 10–12 Ma. This is in contrast to recent palaeomagnetic work in the Gulf of Aden (Sahota et al., 1995) that revealed oceanic crust with an age of c. 20 Ma. In the Red Sea, it is generally accepted that sea-floor spreading occurred between 5 and 6 Ma (Izzeldin, 1987) although rift development preceded this by 10 m.y. The recent work in the Gulf of Aden (Sahota et al., 1995) negates the suggestion that oceanic crust may have formed at the same time in the Red Sea and the Gulf of Aden (Le Pichon and Francheteau, 1978; Le Pichon and Gaulier, 1988). In both the Gulf of Aden and the Red Sea, sea-floor spreading post-dated the earliest expressions of rifting by some 15 m.y. (Courtillot et al., 1987b).

Pre-rift volcanism

Pre-rift volcanism was erupted through thick and cold lithosphere. To what extent the lithosphere had retained its initial architecture (i.e. pre-Gondwana) is not known, but it is obvious that the formation of major extensional basins in the Jurassic (Balhaf graben) must have been associated with a certain amount of lithospheric heating and thinning. This process would have converted pre-existent Gondwana lithosphere. Volcanic activity on the Gulf of Aden and Red Sea margins predated Miocene extension and erosion and the vast volumes of material erupted supports the generally accepted view that LIP volcanism in Yemen (and in Ethiopia) was related to impingement of mantle plume(s) beneath the Afro-Arabian lithosphere (White and McKenzie, 1989; Richards et al., 1989; Hofmann et al., 1995). In Yemen, Oligocene magmatism (Plate 4) was initiated by the Afar mantle plume and transfer of magmas through the lithosphere led to the involvement of either the lithospheric mantle (Chazot, 1993; Chazot and Bertrand, 1993) or the continental crust in magma production (Chiesa et al., 1989; Baker et al., 1994). Pre-rift alkaline volcanism in Saudi Arabia may have had a similar origin but is yet to be studied in detail for both its geochemistry and age.

Syn-rift volcanism

Syn-rift volcanism was erupted through lithosphere undergoing extension as a result of the separation of Arabia from Africa. The dike swarms in Saudi Arabia were intruded into continental lithosphere that had been heated and thinned after a period of plume activity in the preceding >8 m.y. The location and orientation of syn-rift magmatism and the restriction in age indicates that syn-rift volcanism in Saudi Arabia was related to a major episode of continental rifting along the present Red Sea. One could argue from the orientation of pre-rift and syn-rift magmatism that the plume that was responsible for much of the pre-rift magmatism helped orient subsequent rift propagation in the Red Sea (Plate 5). Syn-rift magmas are chemically distinct from pre-rift magmas indicating a change in source due to a changing thermo-tectonic regime. The syn-rift magmas of Saudi Arabia are believed to have originated in the asthenosphere with no involvement of the Afar plume (Camp and Roobol, 1992). Syn-rift volcanism in Yemen takes the form of emplacement of gabbro–granite–syenite plutons (Blakey et al., 1994), intrusion of dikes (Zumbo et al., 1995) and extrusion of volcanic rocks. The exact source of these rocks is not known but is believed to be complicated and to have involved heterogeneous mantle and crustal sources.

Post-rift volcanism

Post-rift magmatism (Plate 5) was erupted through the Arabian lithosphere after separation from the African plate. The architecture of the lithosphere was little different from that today and post-rift volcanism (besides postdating widespread extension) also postdated surface uplift and erosion on the Red Sea margin. Although post-rift magmatism in Saudi Arabia is located well beyond any possible influence of the Afar plume under Ethiopia, the volcanic rocks have the signature of the Afar plume (Altherr et al., 1990). Camp and Roobol (1992) suggested that convective flow emanating from the Afar plume had produced an elongated and extended lobe of hot mantle beneath Arabia thus permitting the involvement of Afar sources in magmatism at this location. Post-rift magmatism in Yemen is closer to the Afar plume, and volcanic centres along the Aden line and elsewhere in Yemen have a contribution from the plume and the asthenosphere (Chazot, 1993). The only exception is the Bir Ali volcanic field, east of Aden, which is probably out of the 'sphere of influence' of the Afar plume (Chazot, 1993).

CONCLUSIONS

Three main periods of volcanic activity were associated with break-up of the Afro-Arabian plate and the formation of volcanic and non-volcanic rifted margins. Overall volcanic rocks within any volcanic margin are crucial 'time indicators' in that they can be precisely dated using $^{40}Ar/^{39}Ar$ techniques. Application of these absolute dates to the evolution of margins of the Red Sea and Gulf of Aden can provide valuable information about:

1. Volcanism. The lowermost and uppermost volcanic rocks within a particular volcanic stratigraphy can be dated thus defining the period of volcanic activity (e.g. Yemen LIP, Baker et al., 1996). Individual volcanic flows within a sedimentary sequence can provide a valuable constraint on the age of that particular stratigraphy. Sills and dikes postdate the sediments and as such do not directly constrain the age of the stratigraphy.

2. Extension. Volcanic flows that cap faults within any rock sequence can provide a 'minimum' age on the timing of a period of rifting. Volcanic flows in the hangingwall of domino fault blocks can provide a 'maximum' age for a period of rifting (e.g. Red Sea margin, Davison et al., 1994). Similarly dike swarms can constrain the timing of a period of extension (e.g. Red Sea coast of Saudi Arabia, Sebai 1989; Feraud et al., 1991; Sebai et al., 1991; Zumbo et al., 1995). However caution must be exercised as dikes can be repeatedly used over a period of time, such that the age of a dike rock could constrain the latest intrusive episode but not necessarily the earliest. In some cases extension can also be dated using apatite fission track analyses because it appears that much of the erosion (i.e. crustal cooling) on the Red Sea, Gulf of Aden and Gulf of Suez margins was initiated by extension (Menzies et al., 1996a).

3. Erosion. Volcanic flows that overlie erosional unconformities can provide vital information on the timing of a period of erosion (e.g. north-east Sana'a, Baker et al., 1996).

ACKNOWLEDGEMENTS

The authors would like to thank the Association of Commonwealth Universities (Baker), the European Union (Chazot), and NERC (Menzies and Hurford) for funding geological research in Yemen. In addition field studies and overseas laboratory visits have been made possible by the Industrial Association of the Department of Geology, Royal Holloway, British Petroleum and the Royal Society. Invaluable logistical support was provided by the University of Sana'a, in particular Drs Mohamed Al Kadasi and Abdulakarim Al Subbary. Vincent Courtillot, Bruce Purser and Dan Bosence are thanked for their comments on an earlier version of this manuscript.

B3
Tectonic and sedimentary evolution of the eastern Gulf of Aden continental margins: new structural and stratigraphic data from Somalia and Yemen

P. L. Fantozzi and M. Sgavetti

ABSTRACT

The purpose of this work is to reconstruct the tectonic and stratigraphic evolution of the eastern sectors of these young margins. New structural and stratigraphic data were collected for the two continental margins facing the Gulf of Aden, in northern Somalia and southern Yemen, by integrating field work, aerial photo analysis and satellite multispectral data processing. A detailed stratigraphic correlation of pre-rift, Mesozoic to Eocene strata was constructed through north-eastern Somalia (Migiurtinia region), and selected sections were correlated between the two sides of the Gulf. The analysis of the deformation of these strata leads to the recognition of west-north-west–east-south-east trending half-graben and tectonic depressions, separated by structural highs interpreted as transfer zones. The transfer zones appear to be located on the landward extension of transform faults in the oceanic setting of the Gulf, and the tectonic depressions are approximately parallel to the oceanic ridge. Oligo-Miocene syn-tectonic basins developed within the depressions, and their patterns reflect their tectonic evolution of the eastern sector of the Gulf of Aden. The age of the deposits sealing the faults cutting Eocene strata confirm a early-middle Oligocene age for the beginning of rifting. The restoration of the pre-drift setting leads to a matching of the continental structures, such as basins and transfer zones, on the two sides of the Gulf. The comparison between the tectonic and stratigraphic features of both continental margins and the oceanic setting of the Gulf of Aden, indicates that a progressive crustal extension brought about the formation of oceanic spreading centres which correspond to the pre-existing syn-rift basins.

INTRODUCTION

The Gulf of Aden is a young ocean in which the formation of oceanic crust can be dated back to the magnetic anomaly 5 (11 Ma), east of 54°E and not earlier than of 3–4 Ma, west of this longitude (Cochran, 1981). According to Sahouta (1995) the oldest oceanic crust of Gulf of Aden is 20 Ma. This age is in agreement with a continental rifting of Lower Oligocene (see below) and a continental drifting of 11 m.y. ago. We think that the Cochran (1981) chronology of oceanic drifting of Gulf of Aden is still valid because it is related to an organized system of magnetic anomalies rather than only a radiometric age of oceanic basalts. The Gulf of Aden structure is characterized by an axial oceanic Sheba Ridge segmented by several transform faults expressed by approximately north-east–south-west fracture zones (Laughton, 1966b; Laughton and Tramontini, 1968; Laughton et al., 1970; Girdler and Styles, 1978; Cochran, 1981; Girdler, 1991) (Figure B3.1).

The Arabian and Somali continental margins of the Gulf of Aden record the initial phase of crustal extension which has previously been reported as late Oligocene–early Miocene (e.g. Cochran, 1981; Bosellini, 1986, 1992; Abbate et al., 1988). Several papers have provided considerable data on the stratigraphic succession and the overall tectonic evolution of northern Somalia (Azzaroli, 1958; Azzaroli and Fois, 1964; Canuti and Marcucci, 1968; Ducci and Pirini, 1968; Barnes, 1976; Bruni and Fazzuoli, 1977, 1980; Merla et al., 1979; Altichieri et al., 1982; Bosellini, 1986, 1989; Abbate et al., 1988; Luger et al., 1990; Boeckelmann and Schreiber, 1990; Ali Kassim Mohamed, 1991, 1993) and of southern Yemen (Beydoun, 1966; Greenwood

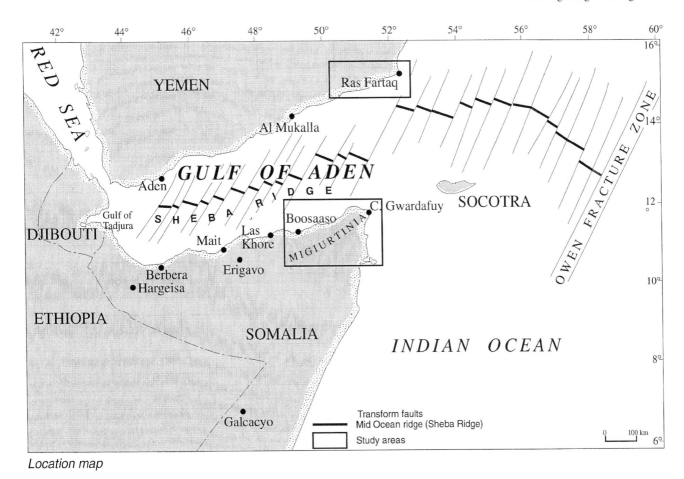

Figure B3.1 Schematic tectonic framework of the Gulf of Aden and location of the study area.

and Bleackley, 1967; Schüppel and Wienholz, 1990; Watchorn et al., this volume). The reconstruction of the pre-rift palaeogeography has been generally based on the overall continuity of certain geologic features and facies belts on the two margins (e.g. Beydoun, 1978; Bosellini, 1986, 1992; Abbate et al., 1988).

In this chapter, the authors present new stratigraphic and tectonic data that support and better constrain the time–space evolution of the rifting of the Gulf of Aden. Specifically, (1) we present a detailed stratigraphic scheme through north-eastern Somalia (Figure B3.3), from Las Khoreh to the Indian Ocean; (2) we analyse the geometry and tectonic significance of syn-rift basins along northeast Somalia and south-east Yemen; (3) we correlate the tectonic elements recognized on both the southern and northern margins of the Gulf of Aden with the transform faults and ridges that characterize this young ocean.

This enables us to restore the geometry and distribution of the Oligo-Miocene basins and their spatial relationships during the rifting phase. The data used include detailed field work integrated with analyses of aerial photos and satellite multispectral images. Field data are extensively documented in Fantozzi (1993,

1996), and in the 1:200 000 geological map of northeastern Somalia (Fantozzi et al., 1993). Methodological criteria for the construction of a stratigraphic correlation framework from aerial photos and for the lithological interpretation of multispectral remote sensing data are reported in Sgavetti (1992) and Sgavetti et al. (1995) and Ferrari et al. (1996), respectively. The transition from continental to oceanic rifting, of the Gulf of Aden, is discussed in detail in Fantozzi (1996).

GENERAL GEOLOGIC SETTING

On the two margins of the Gulf of Aden, pre-Palaeozoic metasediments and Palaeozoic granites outcrop extensively in the Hargeisha and Mait–Las Koreh areas in Somalia, and in the Aden–Al Mukalla (Figure B3.1) area in Yemen, and are overlain via a major unconformity by Jurassic to Quaternary sediments.

The Mesozoic to Eocene sediments of northern Somalia and southern Yemen were deposited in an overall eastward deepening basin, as documented by westward shallowing facies tracts. To the west, the basin was characterized by morphological highs exposing basements rocks, on which the Jurassic strata are often absent and the entire Mesozoic section is locally strongly reduced in thickness (e.g. Beydoun, 1970). Not considering the intense faulting affecting the entire area, the overall regional dip of the strata is toward the Indian Ocean, where Mesozoic strata are rarely exposed in outcrop.

The Jurassic to Eocene strata have been grouped into various formations, clearly correlatable on both sides of the Gulf of Aden, and representing the pre-rift succession (e.g. Beydoun, 1966, 1970; Abbate et al., 1974; Bruni and Fazzuoli, 1980; Bosellini, 1989, 1992; Cherchi et al., 1993).

Post-Eocene strata are laterally discontinuous and were deposited in downfaulted blocks, above an unconformity related to the beginning of tectonic evolution the Gulf of Aden. According to Cochran (1981), this evolution was characterized by shoulder uplift in both northern Somalia and southern Arabia simultaneously with faulting. The timing of the rifting of the continental margins is generally reported as lower Oligocene (Macfadyen, 1933; Azzaroli, 1958; Bosellini, 1989), whereas another phase of faulting is only inferred to occur at the Oligo-Miocene boundary (Abbate et al., 1988). From the literature, in north-east Somalia stratigraphic dates are restricted to the Indian Ocean coast and Boosaaso area (Figure B3.1), where Oligocene conglomerates were recognized, with provenance from uplifted western areas (Azzaroli, 1958). Westward, a stratigraphic transition is reported from middle Eocene to upper Oligocene in the Daban basin, in Somalia (Daban Series, Abbate et al., 1988), and in the Habban-Al Mukalla (Figure B3.1) area in south Yemen (Schüppel and Wienholz, 1990). Finally, Miocene marine carbonate deposits (Dubar Series, Macfadyen, 1933) unconformably overlie older Tertiary and Mesozoic strata and, locally, the basement.

The present study areas are from north-east Somalia, from west of Las Khore to the Indian Ocean, and a more limited area located in south-east Yemen, from Ras Sharmah to Ras Fartaq (Figures B3.1, B3.2, B3.7). The two areas are bordered by Oligo-Miocene continental margins on the north and south Gulf of Aden

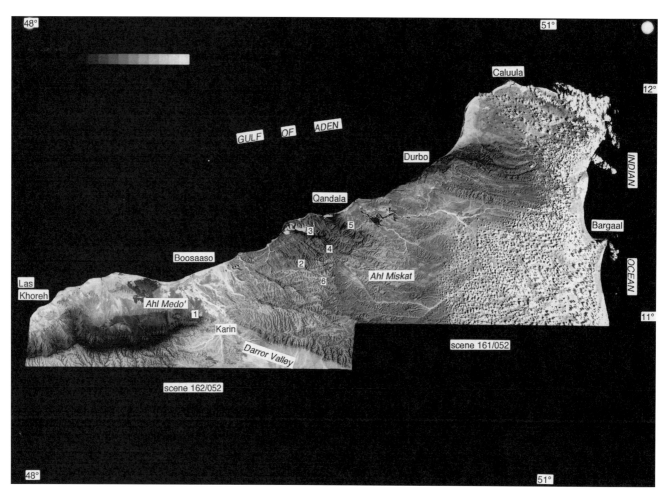

Figure B3.2 Mosaic of Thematic Mapper scenes of the northern Somalia study area. Colour composite of 321 (RGB) bands.

respectively. The evolution of the Eastern Gulf of Aden seems strongly controlled by the tectonic structures of both study areas, and in particular the development of small, onshore syn-tectonic basins in fault-controlled depressions. The area studied in Somalia comprises the entire north-east sector of the Somali plate, bounded to the east by the continental margin formed during the rapid spreading of the Indian Ocean in Late Cretaceous and Paleocene (Bosellini, 1986). The evolution of this margin is recorded within the Paleocene and Eocene stratigraphic succession, that indicates an overall transgression from the east, which today is expressed by the general eastward downwarping of the area.

STRATIGRAPHIC FRAMEWORK OF THE STUDY AREAS

The basement of the two areas consists mainly of low-grade metasediments and late Pan-African granitic intrusions. These are exposed in the Boosaaso and Ras Antara areas in north-east Somalia, and in the Ras Sharwayn area in south-east Yemen.

The Jurassic to Eocene strata in general consist of continental and shallow marine siliciclastic and carbonate deposits to the west, passing eastward into clastic and pelagic marly sediments of the Indian Ocean domain. These strata outcrop persistently throughout both study areas (Figures B3.2, B3.3, Plates 6, 7), and

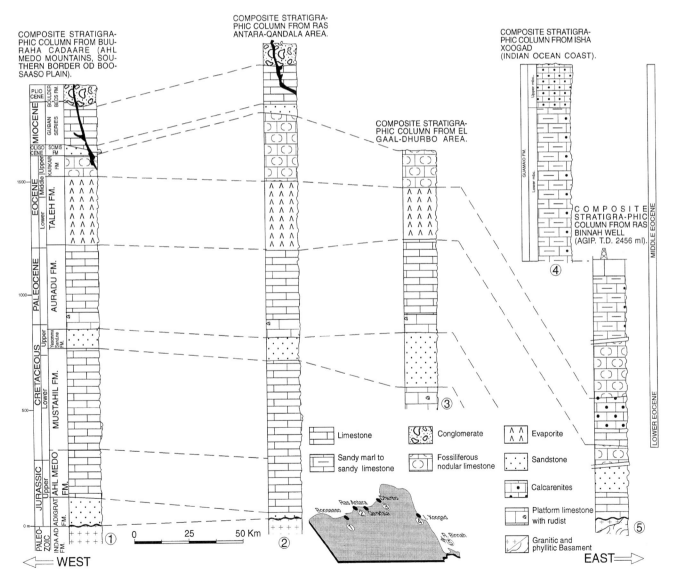

Figure B3.3 Lithostratigraphic framework north-east Somalia, from Las Khoreh to the Indian Ocean. After Fantozzi (1993).

are part of formations described by Bosellini (1989, 1992; Cherchi *et al.*, 1993).

In north-east Somalia, the recognized stratigraphic units include, from base to top: (1) continental conglomerates and sandstones (Jurassic Adigrat Sandstone) discontinuously overlying the basement, as the basal clastic unit of Somalia; (2) a succession that we place in the Ahl Medo Formation; the lower part comprising shallow marine or lagoonal mixed siliciclastic and carbonate deposits. These strata represent the Ahl Medo Formation of Bosellini (1989), which have been deposited in a fault-bound basin and are time equivalents of the Adigrat Sandstone; the middle part of Ahl Medo Formation consists of a carbonate platform succession that represents the major Jurassic transgression (Uarandab Transgression, Bosellini, 1989; Cherchi *et al.*, 1993) related to the East Africa–Madagascar rifting. This corresponds to the Hamanlei Formation, Bihen Limestone and Gawan Limestone of previous authors (Abbate *et al.*, 1974, 1988; Bruni and Fazzuoli, 1977, Merla *et al.*, 1979; Bosellini, 1989). The upper part of this formation is represented by a regressive succession and contains surfaces of subaerial exposure. (3) Late Barremian–Cenomanian shallow water, cyclic carbonates are included into the Mustahil Formation (Figure B3.3) (Cherchi *et al.*, 1993). (4) Continental and nearshore siliciclastic strata are part of the upper Cretaceous Yesomma Formation. (5) Transgressive Maastrichtian-Palaeocene massive or thick-bedded shallow-water platform limestones, that represent the lower part of the Auradu Formation (Macfadyen, 1933); these strata comprise at least four shallowing-up cycles, and are transitional upwards into 6. (6) Evaporitic facies of the Taleh Formation of Lower-?Middle Eocene, which includes of the Allakajid beds (Macfadyen, 1933). (7) Cyclic, shelf carbonates representing the Eocene Karkar Formation. (8) Marls, psammitic limestones and bioclastic limestones of the Gumaio Formation outcrop along the Indian Ocean coastline (Marchesini and Rocca, personal communication). These are laterally equivalent to parts of the Taleh and to Karkar Formations.

Lithological observations, measured sections, aerial photographs and data analysis of remote sensing, and biostratigraphical data and from Azzaroli (1958), Bosellini (1989), Cherchi *et al.* (1993) and Fantozzi (1993), have been integrated to obtain the lithostratigraphic correlation shown in Figure B3.3. In aerial photographs, strata and stratal surfaces are primarily expressed by the lithological contrast between adjacent strata on the land surface. In multispectral satellite images, strata with different lithological properties are expressed by characteristic responses, referred to as spectral image facies (Plates 6, 7; Ferrari, 1993; Chiari *et al.*, 1994; Ferrari *et al.*, 1996). The lithologic units recognized in the field provide the ground truth for the packages of strata prominently expressed in aerial photographs and in the multispectral images, and were traced from the Ahl Medo area to the Indian Ocean (Figure B3.2).

Aerial photo-interpretation was also applied to analysis of the stratal patterns in the Ahl Medo–Qandala area, where strata consist of continental, shelf and restricted platform environments with carbonate and evaporitic deposits. Through this analysis, a number of prominent stratal surfaces and characteristic stratal patterns were recognized and traced (Figure B3.4), resulting in a correlation framework based on surfaces approximating stratal surfaces and therefore having a relative chronostratigraphic significance. Within this framework the multispectral image facies can be directly integrated, and lithologically significant units thus can be correlated along prominent surfaces. As a result, the formations recognized in this sector, which are often characterized by lateral facies changes, have been subdivided into minor units bounded by surfaces, traced both within and between formations, and within which lateral facies transitions can be evaluated. One of the most striking examples is the prominent surface (2 in Figure B3.4) that separates thick-bedded limestones and dolostones of the lower Taleh Formation from the overlying thin-bedded aphanitic and very fine dolostones in the upper Taleh Formation. Another example is the thin unit (1 in Figure B3.4) recognized within the Auradu Formation and easily traceable through the entire area.

The final stratigraphic framework and map (Fantozzi *et al.*, 1993) obtained by combining both large-scale and minor stratigraphic units formed the base for delineating the pre-rift stratigraphy of the area and for a first-order geometric characterization of the tectonic structures.

Yemen pre-rift

In the pre-rift strata of south-east Yemen we recognize:

1. Mixed siliciclastic and carbonate strata, equivalent to the Amran Group (Beydoun, 1970). These strata should overlie the conglomeratic Kohlan Formation, but this is not recognized in the study area.
2. Mainly carbonate succession, corresponding to the Barremian–Aptian Qishn Formation (Beydoun, 1970) and the Ras Fartaq Formation (Wetzel and Morton, unpublished data, 1948; Beydoun, 1966, 1970; dated Albian–Turonian in Bosellini, 1989). This passes westward into the siliciclastic Harshiyat Formation (Beydoun, 1970; Bosellini, 1989).
3. Continental strata containing several hardground sections and levels with vegetal fragments that represent the widely occurring Mukalla Formation.

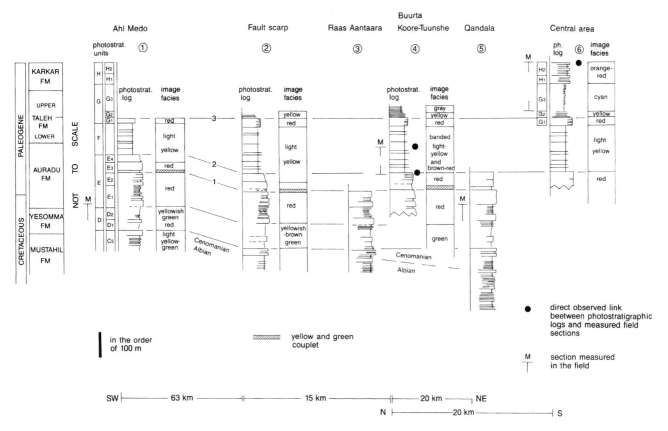

Figure B3.4 Photostratigraphic correlation framework of pre-rift strata through Ahl Medo'–Ahl Miskat area. The columnar sections (location in Plate 6(a)) were derived from the erosional profile of the strata in aerial photos and were correlated to lithological units and biostratigraphic data in the field. Strata correlation was based on prominent, laterally persistent stratal surfaces clearly traceable in aerial photos. 1, 2: prominent correlation surfaces. Slightly modified after Sgavetti *et al.* (1995)

4. Marly carbonate and micritic carbonates which are part of the upper Cretaceous Ras Sharwayn Formation.
5. Thick, shallow-water marine limestones representing the Paleocene Uma er Raduma Formation (Beydoun, 1970). Within these units selected sections were measured and correlated with the units outcropping in Northern Somalia (Figure B3.5). This correlation is discussed later in the chapter.

Somalia and Yemen syn- and post-rift

The Oligo-Miocene syn-rift and post-rift strata exposed in the study areas infill small fault-controlled basins aligned along the Gulf of Aden coasts and located in the Boosaaso, Qandala, El Gal and Aluula areas in Northern Somalia, and in the Sayut, Ras Uqab and Qishn areas in Southern Yemen (Figures B3.6 to B3.8). In north-east Somalia, the syn- and post-rift strata we have measured include: (1) continental and lagoonal deposits corresponding to the Scushuban Formation (Azzaroli, 1958) and Daban Series (Abbate *et al.*, 1988); (2) mainly bioclastic carbonate, marine strata representing the Miocene Dubar Series (Macfadyen, 1933); (3) continental conglomerate deposits representing the Upper Conglomerates of Azzaroli (1958) and Boulder Beds of Macfadyen (1933). In the Boosaaso and Qandala basins, we observed Oligocene deposits clearly sealing the faults affecting the Eocene strata. These deposits, contain *Austrotrillina asmariensis*, Adams, suggesting a Rupelian–Chattian age (Ali Kassim, 1991, 1993; Fantozzi, 1992) thus documenting an Oligocene age for the initial stages of the rifting. In south-eastern Yemen, the base of the syn-rift infill of the basins is represented by Oligo-Miocene strata of the Shir Group (Beydoun, 1968, 1970). The sedimentary geometry of these units within different syn-rift basins is discussed in a following section.

TECTONIC SETTING OF THE AREAS

The general structure of north-east Somalia is characterized by tectonic depressions elongated in a west-north-west–east-south-east direction, which separate complex

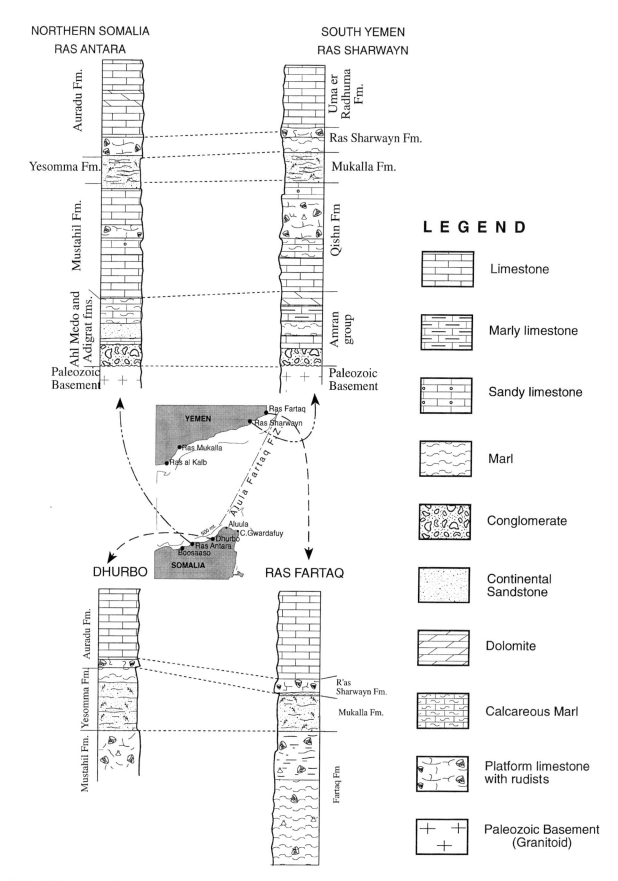

Figure B3.5 Lithostratigraphic correlation of pre-rift strata through the Gulf of Aden.

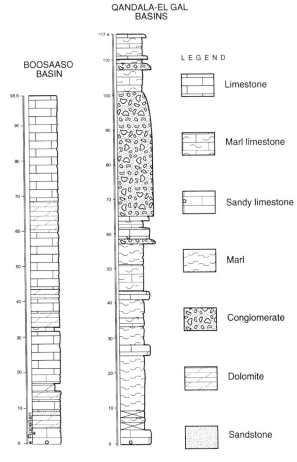

Figure B3.6 General lithostratigraphic sections of the Oligo-Miocene strata (Guban Series) in the Boosaaso and Qandala–El Gal basins.

structural highs (Figures B3.7, B3.8). They form a system of half-graben bounded by west-north-west–east-south-east, north-west–south-east normal faults. These faults affect the pre-rift strata with displacements in the order of kilometres. The faults bordering contiguous half-graben systems sometimes dip in opposite directions, so that a 'transfer zone' or 'accommodation zone' develops in the transition area between contiguous half-graben systems (Figures B3.9, B3.10; Bosworth, 1985; Bosworth et al., 1986; Rosendahl et al., 1986; Morley, 1990); syn- and post-rift Oligo-Miocene basins developed within the structural depressions.

In north-east Somalia the following basins and structural highs can be easily identified from west to east (Figure B3.9):

- the Ahl Medo Mountains structural high
- the Boosaaso Basin
- the Ras Antara–Ahl Miskat Mountains structural high
- the Qandala–El Gal basin
- the Ahl Bari Mountains structural high
- the Aluula Basin

The most significant examples of transfer zones are in the Boosaaso and Quandala–El Gal areas and these will be described in a following section. The Ahl Bari Plateau is characterized by the presence of numerous normal faults and associated roll-over folds (Figures B3.8, B3.9). The structural style of this area, completely different from the rest of north-east Somalia, can be explained by facies changes within the sedimentary sequences from the north-west (Dhurbo area) to the Indian Ocean coast. In the Dhurbo area, the Cretaceous–Eocene sediments consist of platform limestone, followed by more than 200 m of sandstone and by Paleocene–Eocene platform limestones. To the east and north-east, this sequence passes laterally to a monotonous succession of marls and marly limestones. These 'plastic' horizons may have favoured the detachment of the Paleocene–Eocene platform limestone, producing listric normal faults and large-scale roll-over folds.

In south-eastern Yemen, the following basins and structural highs are identified, from east to west, in the area between Ras Fartaq and Ras Sharmah facing northeastern Somalia (Figure B3.7):

1. the Ras Fartaq structural high
2. the Oligo-Miocene Qishn Basin
3. the Ras Sharwayn structural high
4. the Oligo-Miocene basin immediately to the west of Ras Sharwayn
5. the Sayut structural high
6. the Oligo-Miocene Sayut Basin
7. the transition area between the Sayut Basin and the large Oligo-Miocene basin outcropping between Al Mukalla and Ras Sharmah

Tectono-stratigraphic setting of the syn-rift basins

The overall depositional strike of the syn-rift basins is west-north-west–east-south-east and the stratal dip is toward the Gulf of Aden for almost all basins. In general, these strata are only slightly deformed, except in the vicinity of major faults where the continental deposits may sometimes be involved in roll-over folds. The integrated analysis of stratal patterns and tectonic structure of the basins and highs assists in the accurate definition of the transfer zones, of which the most interesting are located in the Boosaaso and Qandala–El Gal areas.

The sedimentary fill of the Boosaaso Basin is shown in Figure B3.11. In the lower part of the succession in the south, lenses of coarse-grained sandstones and conglomerates containing basement fragments are pres-

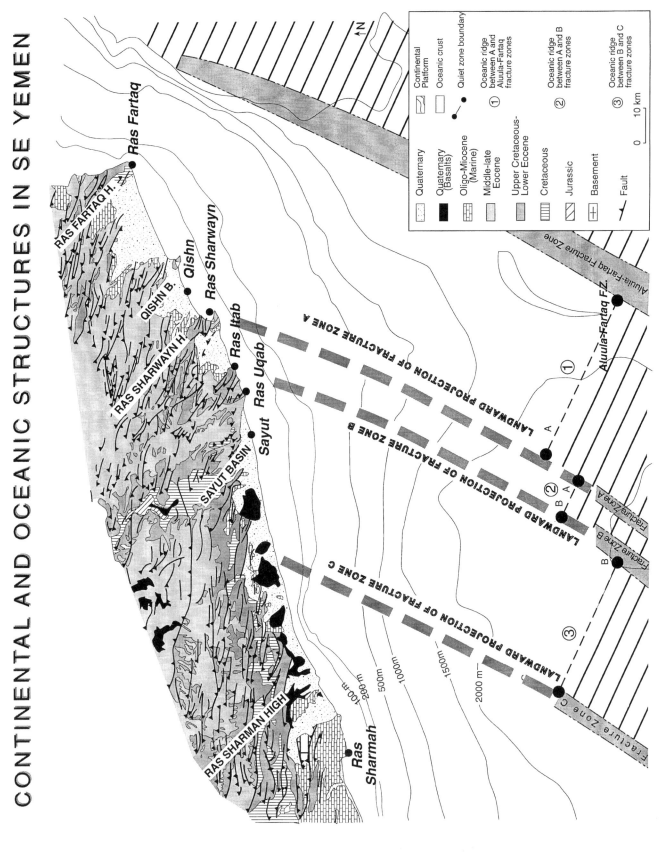

Figure B3.7 Tectonic sketch map of south-east Yemen showing the location of continental structural setting as well as oceanic structures and their landward extrapolation.

Tectonic setting of the areas 65

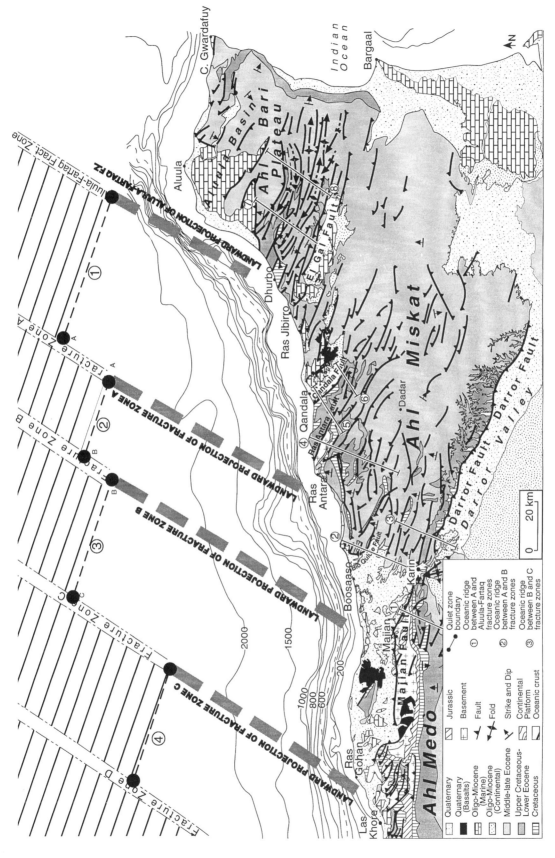

Figure B3.8 Tectonic sketch map of north-east Somalia showing the location of continental structural setting as well as oceanic structures and their landward extrapolation.

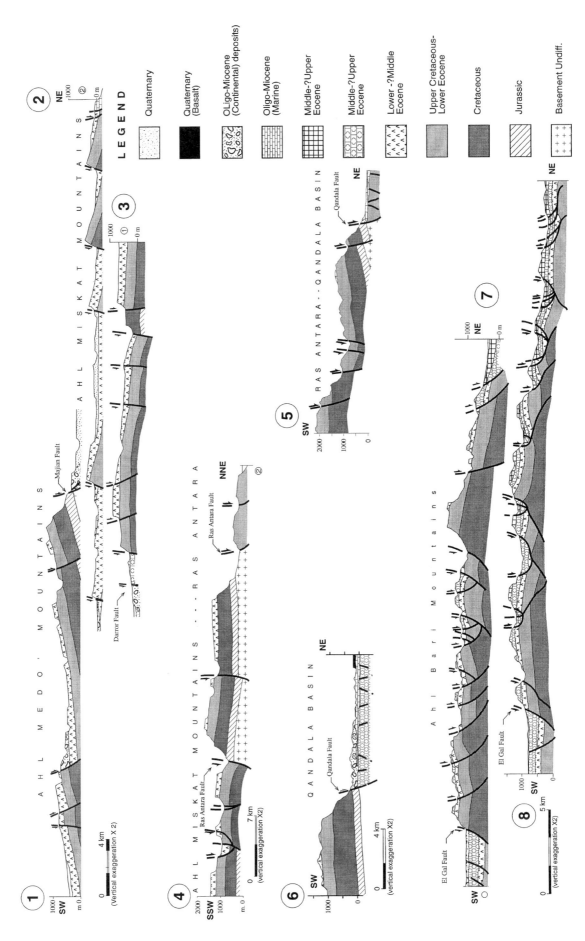

Figure B3.9 Geologic profiles across the structural basins and highs discussed in the text. See location in Figure B3.8.

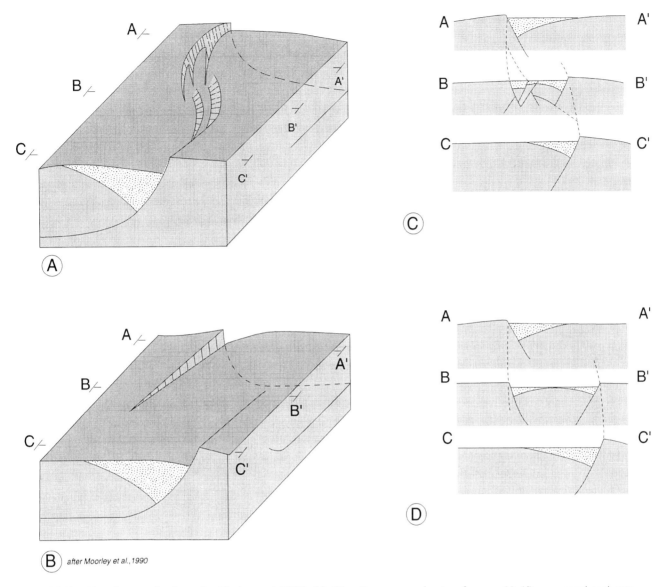

Figure B3.10 Transfer zone sketches, after Morley *et al.* (1990). (a), (b): colinear converging transfer zone; (c), (d): superposed conjugate converging zones.

ent locally, and have been dated Rupelian–Chattian (Oligocene Shimis sandstones of Azzaroli, 1958; Ali Kassim, 1991, 1993; Fantozzi, 1992). This age represents the beginning of rifting in this area. The overall facies pattern indicates, from south to north, a restricted shelf and lagoonal complex with fluvial influences, passing seaward to a north-westerly prograding marine succession, separated by a barrier formed by prevalent organogenic limestones (Figure B3.11). Alluvial fan conglomerates are interbedded in both the Oligo-Miocene lagoonal and marine strata.

In the Boosaaso area the following features indicate that a transfer zone developed between the structural domains of the Ahl Medo and those of the Ahl Miskat (Figure B3.8):

1. The normal faults which delineate the Ahl Medo and Ahl Miskat half-graben converge in this area.
2. The normal faults bordering the Ahl Medo and Ahl Miskat half-graben have the same strike but opposite dip: north-north-east in Ahl Miskat and south-south-west in Ahl Medo.
3. In the south-eastern sector of the Boosaaso Basin the basal contact between the Oligo-Miocene and the pre-rift strata is exposed, showing the pre-rift strata occur in a north-north-east-dipping half-graben. It is possible that this half-graben system, dipping toward the Gulf of Aden, continues under the Boosaaso Basin. This structural setting is not consistent with the large northward displacement of the Ahl-Medo border fault (Majian Fault). Structural constraints suggest the presence of a transfer zone between the Ahl Medo and Ahl

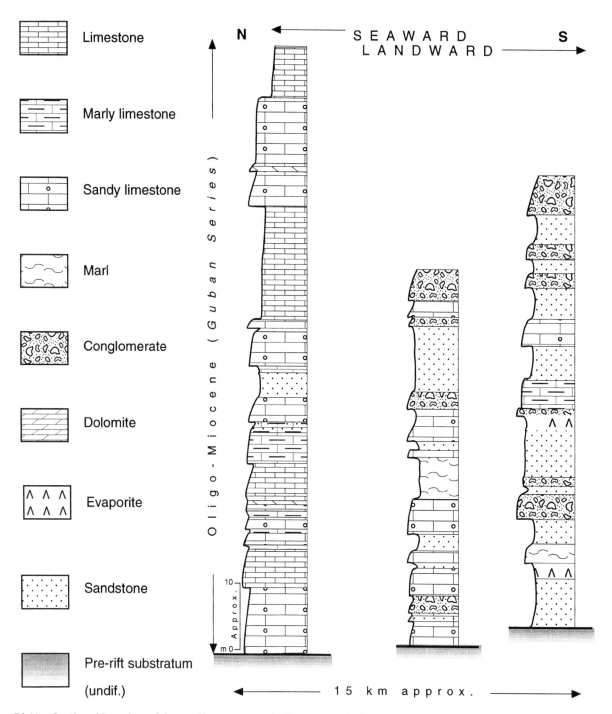

Figure B3.11 Stratigraphic sections of the syn-rift strata across the Boosaaso Basin, in an overall depositional dip direction, from north to south.

Miskat. Figure B3.12 shows the reconstruction of the geometry of the complex structure formed by the association of the Boosaaso Basin and the Ahl Medo-Karin transfer zones.

The well-exposed Oligo-Miocene infill of the Qandala basin indicates the northward depositional dip of the basin (Figure B3.13). The conglomerates and sandstones, outcropping near the Qandala fault, are interbedded with increasingly more abundant marine deposits toward the Gulf of Aden coast, having a depositional trend similar to that of the Boosaaso Basin. The Qandala Basin is bound to the west by the Ras Antara and Qandala faults.

The Ras Jibirro structural high separates the Qandala and El Gal basins. The structural setting of this zone, extending from Ras Jibirro for about 40 km southward, is characterized by small horst and graben in which the pre-rift sediments outcrop. In this area the vertical

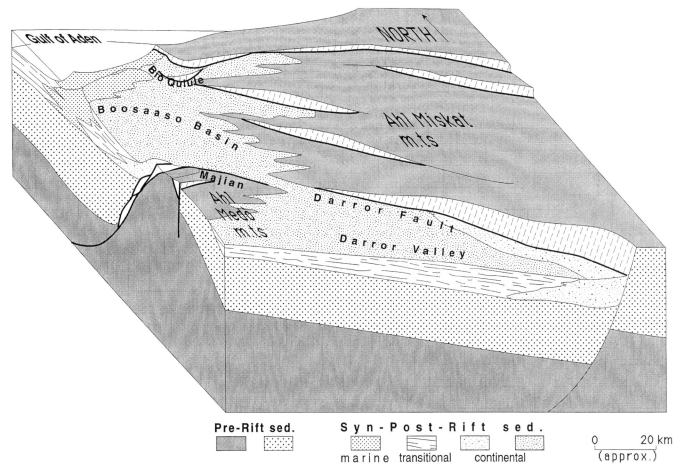

Figure B3.12 Schematic block-diagram of the Boosaaso Basin and transfer zone between Ahl Medo and Ahl Miskat areas (courtesy L. Carmignani).

displacement of the Qandala and El Gal faults diminishes markedly.

The widely exposed Oligo-Miocene strata of the El Gal Basin indicate a south-eastward depositional dip (Figure B3.11), in contrast with the Qandala Basin. Southward, the transition between continental and marine deposits indicates maximum marine ingression during the Miocene. As with the Qandala Basin, the El Gal Basin is characterized by an asymmetric geometry bounded at its north-east margin by the El Gal Fault.

From the above discussion, we summarize that the Boosaaso Basin may be considered to be a transfer zone analogous to those described from the East African Rift (Rosendahl *et al.*, 1986) and to those described in the classification scheme of Morley *et al.* (1990) (Figure B3.10). The Ras Antara area, where the granitic basement crops out, is also related to the development of a transfer zone. The structure of the Qandala–El Gal Basin comprises two tectonic depressions the outer limits of which are represented by two west-north-west–east-south-east oriented faults, separated by the Ras Jibirro high (Figure B3.13). The structural setting of the Qandala–El Gal Basin is in agreement with the classification of the transfer zone of Morley *et al.* (1990) and is sketched in Figure B3.10(b).

Correlation between north-eastern Somalia and south-eastern Yemen

The structure of north-eastern Somalia shows a consistent relationship with the structure of the continental platform between Ras Jibirro and Ras Gohan (Somalia). Along this transect, the continental platform is characterized by north-east–south-west oriented slopes and west-north-west–east-south-east oriented basins. The slopes are always located on the lateral projection of the transform faults to the continent and coincides with the transfer zones observed on the continental margins (Figure B3.8). The basins are always located at the projection of the oceanic ridges into the continent and coincide with the Oligo-Miocene syn-rift basins developed in the continental margin. This indicates that the basins, which developed during the initial stages of continental rift, were those in which the spreading sites originated during the later stages of oceanic rifting. The

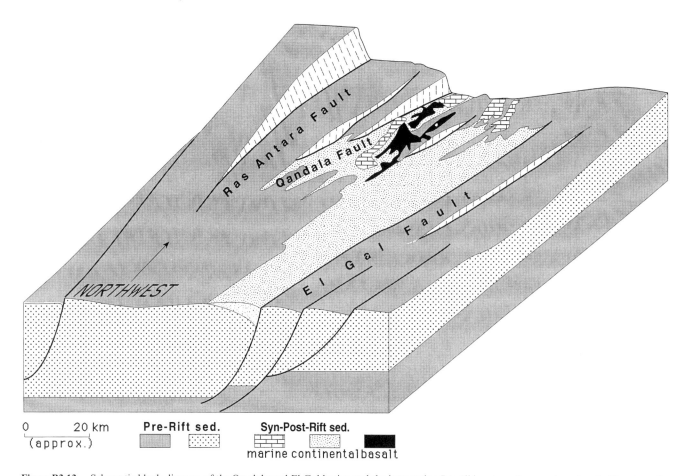

Figure B3.13 Schematic block-diagram of the Qandala and El Gal basins and the intervening Ras Jibirro transfer zone (courtesy L. Carmignani).

transition areas between one structurally depressed area and another are sites of accommodation zones which, during the phases of oceanic rifting, guided the development of the subsequent fracture zones. Such a model can be verified by restoring the pre-drift geometry of both margins of the Gulf of Aden, in order to obtain a complete picture of the pre- and syn-drift geometry.

For an ocean basin where well-developed transform faults extend to steps in the continental shelf, the opening pole of the ocean may be obtained by matching structural lineaments, coastlines and isobaths of the continental margins.

For this reconstruction, we have compared selected sections on the two sides of the Gulf, correlating the most significant pre-rift stratigraphic markers. The sections are located in Ras Antara (northern Somalia) and Ras Fartaq (southern Yemen), and in Dhurbo (northern Somalia) and Ras Sharwayn (southern Yemen) areas, and their correlation is shown in Figure B3.5.

In each section pairs, striking similarities are immediately perceptible, as has been observed already in other areas by previous authors (e.g. Beydoun, 1970; Bosellini, 1986, 1989; Abbate et al., 1988). Among the most easily correlatable stratigraphic units are the Upper Cretaceous Yesomma Sandstone and Mukalla Formation, representing a widespread continental interval, and the laterally persistent Paleocene calcareous Auradu and Uma er Raduma Formations.

New stratigraphic evidence from the study areas permit a more accurate correlation of the sections. Both the Ras Antara and Ras Sharwayn sections expose the basement. In the Ras Sharwayn section, Wetzel and Morton (Beydoun, 1970) reported the presence of granites apparently intrusive in the metasedimentary complex, and with an absolute age of 467 ± 25 Ma. In the Ras Antara section, the intrusive younger granite (S.O.E.C., 1954) has similar petrochemical characteristics and a comparable age of about 500 Ma. In the Dhurbo section, a new outcrop of Cretaceous strata has been recognized and assigned to the Mustahil Formation. Within these strata, a characteristic horizon consisting of often well-preserved rudists was observed. An identical reefal horizon with abundant rudists was observed in a corresponding stratigraphic position in the Ras Fartaq section.

The Jurassic strata outcropping in both Ras Antara and Ras Sharwayn areas, and the Cretaceous strata outcropping in both Dhurbo and Ras Fartaq areas are faulted, and Oligo-Miocene strata onlap the fault planes. Similar relationships between Oligo-Miocene strata and more or less intensely faulted Mesozoic and Eocene strata are observed in the various half-graben and structural depressions, within which stratal arrangement and facies patterns record the syn- and post-rift evolution.

These correlations, combined with transform fault directions, isobaths, coastlines and tectonic features of both continental margins, enables the pre-drifting restoration of the Gulf of Aden shown in Figures B3.14 and B3.15. This reconstruction fits the kinematic constraints of the Afro-Arabian rift system as described by Laughton et al. (1970), Le Pichon and Francheteau (1978), Cochran (1981). The pre-drift restoration of the continental margins matches the two coastlines in a north-east–south-west direction with a range of 30–50 km in between. According to Cochran (1981, 1982), the entire width of the magnetic quiet zone is 160 km and consequently the average extension factor ($\beta = 300-500\%$) can be calculated. This value is larger than a $b = 3.6$ calculated for a finite model of extension (Cochran, 1982). The discrepancy between the extension data proposed in this chapter and the data of Cochran (1982) is possibly related to some uncertainty regarding the amount of dike injection predicted by Cochran's model for the magnetic quiet zone. An increase in the magma injected into a magnetic quiet zone leads to a reduction in the amount of continental crust involved in the stretching process. Consequently, this would lead to a reduction of the width of a 160 km magnetic quiet zone and to a reduction in the b factor proposed in this chapter.

The syn-rift tectonic structure
Based on the reconstruction in Figure B3.15, the original extent of the major continental syn-rift structures of the margins of the Gulf of Aden can be estimated. From east to west we observe:

1. The normal fault which delimits the western sector of the Ras Fartaq structural high can be linked to the normal fault which delimits the Dhurbo structural high.

2. The entire Ras Jibirro transfer zone is connected with the structures in the central area of the Qishn Basin (Ras Darjah area).

3. The north-north-west projection of the major Qandala Fault corresponds to the fault bordering the Qishn Basin to the west.

4. The intermediate area between the Qandala and Ras Antara Faults is the extension of the relief located

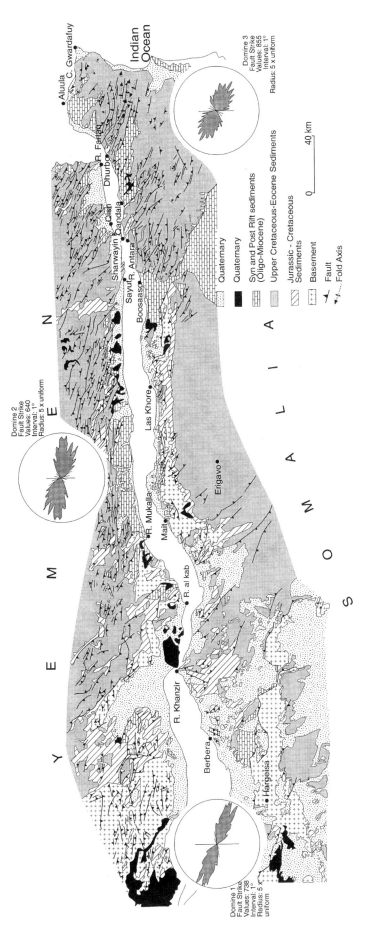

Figure B3.14 Reconstruction of the syn-rift setting of the continental margins of the Gulf of Aden, based on geological marine data (Cochran, 1981, 1982) and structural and stratigraphic constraints.

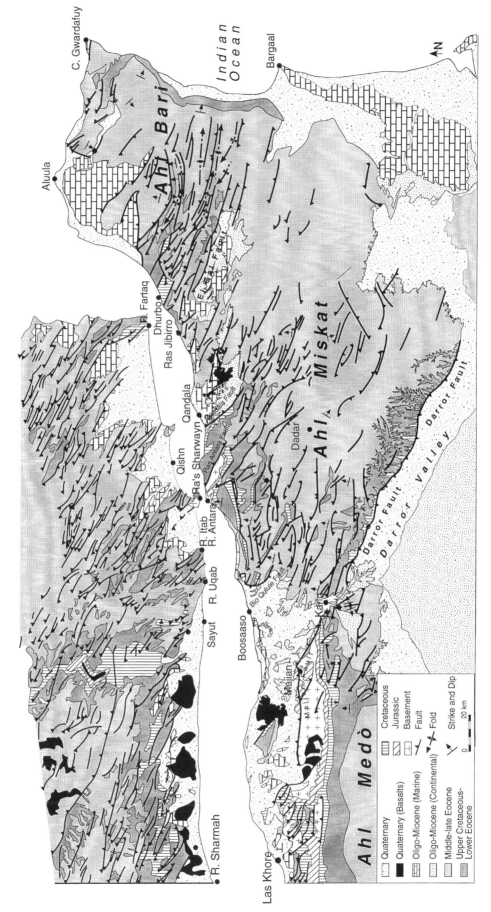

Figure B3.15 Detail of the Gulf of Aden syn-rift reconstruction shown in Figure B3.12, relative to the Las Khoreh–Cape Guardafui area.

between the western edge of the Qishn Basin and the Ras Sharwayn structural high.

5. The Ras Sharwayn and Ras Antara faults are collinear but are displaced in opposite directions, forming a transfer zone in which the transform fault 'A' subsequently develops.

6. The small Oligo-Miocene basin to the west of Ras Sharwayn extends into the Oligo-Miocene strata located between the area north of Bio Qulule and the beginning of the Ras Antara Gulf.

7. The Oligo-Miocene Boosaaso Basin is adjacent to the basin between Sayut and Ras Sharmah. The normal faults bordering the Sayut Basin to the north and northeast fit the normal fault system located between Bio Qulule and the Darror Valley in Somalia.

8. The fault east of Las Khore (Ras Adado) is an important structure which causes the basement to outcrop. It is a transfer zone in which the transform fault 'C' will develop, possibly originating between the north-north-east-dipping normal fault system (including the Majian and Las Khoreh Faults in Somalia) and the south-south-west-dipping fault system (including the faults west of Ras Sharmah in Yemen).

DISCUSSION

The data presented show that the geometry of the oceanic structures in the Gulf of Aden are strongly controlled by the continental rifting of the Afro-Arabian plate. This relationship has been proposed by many authors (Laughton et al., 1970; Cochran, 1981, 1982; Withjack and Jamison, 1986; Abbate et al., 1988; Bosellini, 1989; Fantozzi, 1992, 1993). Several authors have also considered the effects of continental rift structures on oceanic structures in different regions, including the Bay of Biscay (De Charpal et al., 1978; Montadert et al., 1979), the West African coast (Rosendahl et al., 1986), the east coast of South America, the Gulf of Suez and the Red Sea (Cochran and Martinez, 1988). In these regions transfer zones segmenting the rift were observed.

The relationship between transfer zones of continental margins and oceanic structures is a recurrent theme both in the proposed models for the evolution of passive margins and in many examples of evolved or incipient oceanic margins (Bally, 1981; Wernicke and Burchfiel, 1982; Gibbs, 1984; Bosworth, 1985; Lister et al., 1986; Milani and Davidson, 1988; Lister et al., 1986). In the rifting area, the axis of accommodation zones are not parallel with respect to the basins' axis; this degree of non-coaxiality might result in strain softening and focusing of late transverse structures. During continued crustal stretching, transfer areas can result as sites of major strain concentration, whereby syn-rift basins and later oceanic expansion centres originate prevalently in basin areas. The present-day northern Red Sea may represent a situation similar to the pre-drift setting of the Gulf of Aden proposed in this chapter (Figures B3.15, B3.16). This represents an advanced rift setting

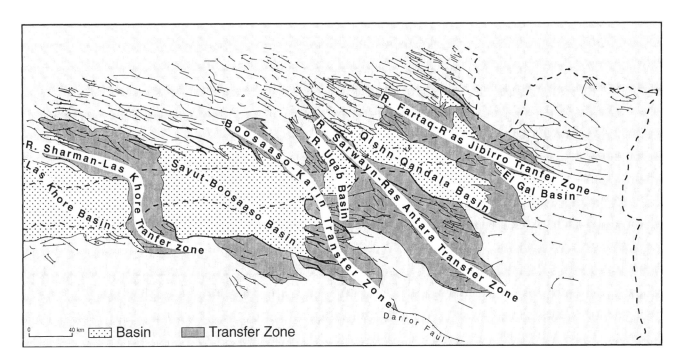

Figure B3.16 Reconstruction of the Early Oligocene syn-rift setting of the southern Yemen and northern Somalia continental margins, showing the basin areas and transfer zones.

analogous to that studied by Cochran and Martinez (1988) and Favre and Stampfli (1992) for the Red Sea and compares with Bosworth's 'transitional basin' (1993). In the northern Red Sea, according to Cochran and Martinez (1988), a rift is developing and oceanic expansion is incipient; this area is characterized by crustal extension, concentrated in a lowered axial area, and by basaltic intrusions. As crustal stretching continues, the intrusions evolve into limited expansion centres that subsequently coalesce to form oceanic ridges.

In summary, in the restoration shown in Figure B3.16, three important structurally depressed areas are reconstructed, in which the syn-rift sequences were deposited. These areas are the Qishn–Qandala Basin, the basin between Ra's Sharwayn and Ra's Uqab including its continuation into Somalia between Ras Antara and the Boosaaso Basin, and the Sayut–Boosaaso Basin. These three basins contain Chattian-Rupelian deposits (Oligocene). The basins are delimited by transfer areas which also extend into the two future continental margins. From the reconstruction (Figure B3.16), it is also evident that:

1. The Qishn–Qandala Basin is in a position which was subsequently occupied by the oceanic ridge located between the Alula Fartaq Fracture Zone and Fracture Zone 'A'.

2. The Oligo-Miocene basin between Ra's Sharwayn and Ra's Uqab, including its prolongation in north-eastern Somalia, occupies the area which will later be occupied by the oceanic ridge between transform faults 'A' and 'B'.

3. The Oligo-Miocene Boosaaso Basin and its extension into the Las Khore and Sayut Basins, occupies an area in which the segment of oceanic ridge between transform faults 'B' and 'C' later developed.

4. The Ra's Jibirro transfer zone is an area intensely deformed by small north-west–south-east oriented faults having no strike-slip component. The Alula–Fartaq Transform Fault has formed in this area of weakness.

5. Between the Ra's Sharwayn and Ras Antara faults, which are parallel, collinear and both displaced for several kilometres in opposite senses, a transfer zone has formed in which the transform fault 'A' develops.

6. The transform fault 'B' developed in the area of the Boosaaso–Karin Transfer Zone.

7. Between the Las Khore Fault and the fault system east of Ras Sharmah, a transfer zone formed in which transform fault 'C' later developed.

To summarize, from the data resulting from the restoration of the pre-rift tectonics of the eastern continental margins of the Gulf of Aden (Figures B3.15, B3.16) we propose that the Oligocene basins represent centres of persistent crustal extension controlling the positioning of the centres of oceanic spreading, in agreement with the findings of Cochran and Martinez (1988) for the northern Red Sea.

In all the examples of transfer zones reconstructed by restoring the position of the two continental margins, no transcurrent structures or structures cross-cutting the major west-north-west–east-south-east-oriented fault system were found on shore. As already observed in the East African rift and in other continental rifts characterized by transfer zones, transcurrent faults are very rarely found in the transfer zones. Even in the most advanced continental rifts, transcurrent faults separating half-graben systems with opposite dip have not been recorded (Bosworth, 1985; Faulds et al., 1990; Davison et al., 1994). It seems that the transfer of displacement took place along normal faults, generally obliquely oriented with respect to the rift direction and frequently displaying a curved trace (splay or relay fault system), analogous to those described in the eastern sector of the Boosaaso Basin (Figure B3.12).

CONCLUSIONS

According to Cochran (1981, 1982), the magnetic anomaly sequence extends back to 10 Ma east of long. 44°E and the sea-floor spreading began almost simultaneously along the entire length of the trough. Thus, on the basis of marine geological data, the whole Gulf of Aden shows a unitary evolution. Based on new field data from north-eastern Somalia and south-eastern Yemen together with previous oceanic data it is possible to reconstruct the pre-rift geometry of this portion of the Afro-Arabian continent and its evolution during Tertiary rifting and breakup (Figure B3.17).

1. During Late Eocene–Early Oligocene a north-east–south-west extension stress field affected the Afro-Arabian platform. In agreement with Cochran (1982), this direction of extension was related to a east-north-east displacement of the Indian and Arabian plates.

2. During the early Oligocene (Rupelian), the Afro-Arabian platform underwent an important phase of continental rifting which led to the formation of small west-north-west–east-south-east-oriented basins separated by transfer zones and structural highs.

3. In these syn-rift basins, progressive crustal extension brought about the formation of oceanic spreading centres which correspond to the pre-existing syn-rift basins.

4. The progressive separation of the Somalia and Yemen continental margins led to the formation of an oceanic basin.

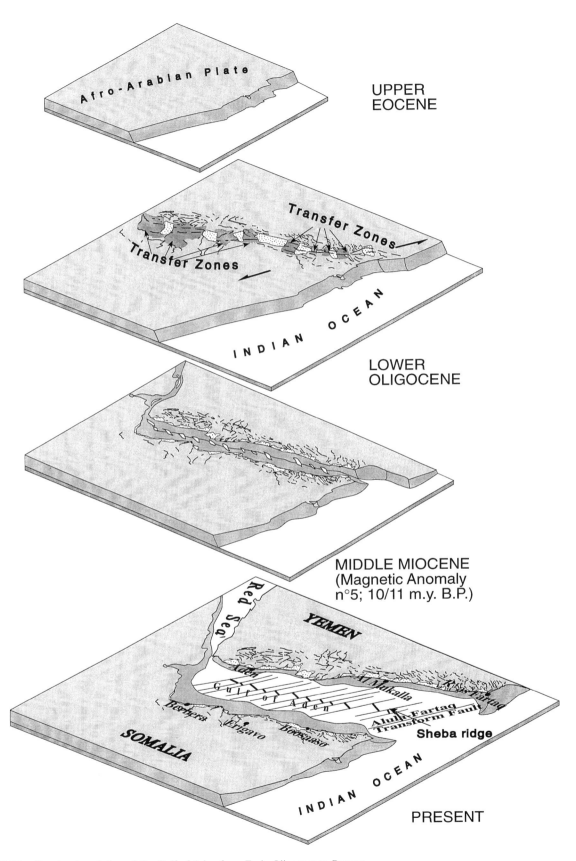

Figure B3.17 Structural evolution of the Gulf of Aden from Early Oligocene to Present.

ACKNOWLEDGEMENTS

The field data used in this study were collected during 1988–1990 missions organized by L. Carmignani, Abdirahman Hilowe Mohamed, P. Fantozzi, Ali Kassim Mohamed and with the co-operation of the Somali National University. Thematic Mapper images of Figures B3.2 and 3.4(a) were processed by M.C. Ferrari at Jet Propulsion Laboratory, Pasadena, California, and at Telespazio s.p.a., Rome, respectively. Aerial photos at 1:50 000 scale were provided by Somali Ministry of Mines. Financial support by Ministero Università e Ricerca 60% to University of Siena, and by Agenzia Spaziale Italiana to University of Parma.

Chapter B4
Structure, sedimentation, and basin dynamics during rifting of the Gulf of Suez and north-western Red Sea

W. Bosworth, P. Crevello, R. D. Winn Jr. and J. Steinmetz

ABSTRACT

The Gulf of Suez formed as the northern segment of the late Oligocene–early Miocene Red Sea rift. Rare occurrences of basaltic dikes that cut and flows that are interstratified with the oldest syn-rift strata, a basal red bed sequence (Abu Zenima–Nakheil formations), suggest that rifting had initiated prior to ~25 Ma. The oldest palaeontologically datable strata are of Aquitanian age (Nukhul Fm.), and are older than ~21 Ma. Nearby basal strata in north-west Saudi Arabia are Chattian age (~27 Ma). Apatitie fission-track data suggest that the earliest phase of Gulf of Suez and northern Red Sea rifting may have begun ~34 Ma, but there is no dated sedimentary record of this. The areal distribution of the basal syn-rift units was largely controlled by the geometry and timing of movement of the early rift-fault system, which was strongly influenced by pre-existing structures in the Pan-African basement complex. In the early Burdigalian, the Gulf of Suez and northern Red Sea entered a period of rapid subsidence and increase in marine water depths, resulting in the widespread deposition of *Globigerina* marls (Rudeis and Kareem formations) in axial areas. Fault density decreased during the Burdigalian to Langhian, and most extension was accommodated by movement on a few large, basin- or block-bounding faults. By the Serravalian, connection between the Gulf of Suez and Mediterranean basins became restricted and sedimentation rapidly changed to laterally continuous evaporites (Belayim and South Gharib Formations). In the late Tortonian, renewed influx of normal marine water occurred from the southern Red Sea, and deposition switched to mixed evaporite–marginal marine settings (Zeit Fm.). The Zeit sediments differentially loaded the underlying South Gharib massive salt, driving diapirism and intra-sediment slump faulting that were generally focused along Middle Miocene fault trends. Basinal sedimentation patterns during the Messinian, Pliocene and Quaternary were largely controlled by the geometry of the evolving salt ridges. Subsidence became increasingly focused along the axis of the rift during this time.

Analysis of fault geometries, fault kinematics and sedimentation patterns indicates that the early Nukhul phase of rifting occurred during north-east–south-westerly directed extension. The existence of pre-existing north-north-east-trending basement structures, however, caused local development of N10–20°E transfer faults, and local rhombic basin geometries that other workers have interpreted as pull-apart basins. As rifting progressed, the N10–20°E faults were abandoned and younger transfer faults became predominantly rift-normal. In the Middle Miocene, the relative movement between Africa and Sinai shifted to the Gulf of Aqaba transform boundary, and extension rates across the Gulf of Suez decreased dramatically. During the late Pleistocene, the regional stress field changed to a N15°E extension direction and a new system of faults began to evolve.

The early phase of pre-Nukhul and Nukhul extension in the Gulf of Suez was accompanied by relatively minor basin subsidence. Between 19 and 16 Ma, within the Rudeis Fm., subsidence rates increased several fold. This increase in subsidence, and accompanying sedimentation rate, followed quickly after the onset of significant rift shoulder uplift, which according to apatite fission-track analysis occurred at ~23–21 Ma. Many workers have therefore inferred that Gulf of Suez

Sedimentation and Tectonics of Rift Basins: Red Sea–Gulf of Aden. Edited by B.H. Purser and D.W.J. Bosence. Published in 1998 by Chapman & Hall, London. ISBN 0412 73490 7.

rifting began gradually, and culminated in the Rudeis. Our observations suggest, however, that the early Nukhul faulting and stratal rotation reflect significant regional extension during the Aquitanian. The extension was distributed across broad regions and accommodated by movement on many, closely spaced faults. Eventually, distinct master faults evolved and footwall uplift to the larger structures became a principal factor controlling erosion and regional sedimentation patterns. Initially, subsidence rates were relatively low, perhaps reflecting the inherent strength of the pre-rift lithosphere. The onset of rapid subsidence eventually occurred a few million years later, and most of the basin was flooded with deep marine waters. The Gulf of Suez is therefore best modelled with a non-instantaneous relationship between extension and subsidence.

INTRODUCTION

The early histories of continental rifts are often poorly known, partly because interpretation of the early history is greatly complicated by syn-rift and post-rift evolution of the basin. An equally confounding situation is that the syn-rift fill of most rifts is of continental origin, and its palaeontological dating is imprecise. The presence of volcanic rocks can greatly help in this respect, but many rifts formed without significant volcanism. The precise relationships between such parameters as horizontal extension, vertical subsidence, and rift-shoulder uplift are therefore often difficult to constrain.

The Gulf of Suez and Red Sea (Figure B4.1) is a rift that initiated in a continental plate interior situated at, or near, sea level (Garfunkel and Bartov, 1977; Sellwood and Netherwood, 1984; Steckler, 1985; Coleman and McGuire, 1988). As a result, marine waters reached most of the sub-basins within this rift very early in their development, depositing a stratigraphic sequence accurately datable with microfossils.

We have studied excellent exposures of the early syn-rift fill of the southern Gulf of Suez and northern Egyptian Red Sea, and have integrated our results with the abundant subsurface data that exist in the Gulf. In this chapter we present our interpretation of the syn-rift sediment response to crustal extension, and propose a new model for the kinematics of the early rift. We also suggest that standard models of instantaneous basin subsidence do not satisfy the observed geologic relationships.

REGIONAL SETTING

Age of rift initiation

Stratigraphy of the syn-rift fill of the offshore northwestern Red Sea and southern Gulf of Suez are very

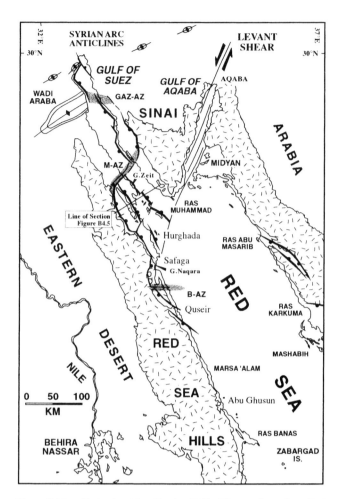

Figure B4.1 Tectonic setting for the Gulf of Suez and northern Red Sea rift basin. Dash pattern is crystalline basement. Stipple pattern is the area of a rift accommodation zone. GAZ-AZ = Galala-Abu Zenima Accommodation Zone; M-AZ = Morgan Accommodation Zone; B-AZ = Brothers Accommodation Zone.

similar, and the two basins are therefore generally interpreted to have originated as a single, sedimentologically connected rift system (Tewfik and Ayyad, 1984; Miller and Barakat, 1988). The oldest syn-rift strata exposed along the Egyptian Red Sea coast, the central and southern Suez coasts, and drilled in the Gulf of Suez subsurface are unfossiliferous, but are intruded by basaltic dikes or overlain by flows that have been K-Ar dated between 20 and 25 Ma (Steen, 1984; Roussel, 1986; Bosworth, 1995). The onset of rifting for this entire region is therefore constrained to have been no later than about 25 Ma.

Reworked Chattian age foraminifera in the basal Nukhul Fm. of a well drilled at Hurghada (Figure B4.1; El Shinnawi, 1975) prove that late Oligocene marine strata were deposited in the area of the southern Gulf. It is not known, however, whether these marine strata were deposited in a fault-bounded rift basin, or a proto-rift sag. Along the northern Saudi Arabian Red Sea margin at Midyan (Figure B4.1), benthic foraminifera of Chattian age (planktonic zone N2 of Blow, 1969) have been found in carbonate beds of the early syn-rift Musayr Fm. (Dullo et al., 1983; Bayer et al., 1988; Abou Out and Gheith, this volume). In a pre-rift configuration, the Midyan region restores to a position essentially adjacent to Hurghada. The stratigraphic limits to initiation of rifting in the northernmost Red Sea can therefore reasonably be set at a minimum age of 27 Ma (approximately the end of the N2 zone), and certainly somewhat earlier due to the presence of the undated underlying red beds at Midyan (Sharik Fm.) and elsewhere.

Apatite fission-track studies suggest that a phase of basement unroofing began in the rift shoulder of the southern Red Sea of Egypt and in a small area of the north-western Gulf of Suez at ~34 Ma (Steckler and Omar, 1994; Omar and Steckler, 1995). Omar and Steckler interpret this as part of a regionally synchronous, Red Sea rift system pulse of extension. No definitive stratigraphic or structural evidence presently exists to ascertain the relationship between this rift event and the main phase of extension that began several million years later.

Main rift events

The main period of horizontal extension in the Gulf of Suez lasted from the late Oligocene–Aquitanian into the Middle Miocene, to about 14 Ma (Garfunkel and Bartov, 1977; Angelier, 1985; Steckler, 1985). Two phases of deformation can often be differentiated: (1) early normal and strike-slip faulting that was strongly influenced by pre-existing basement structures such as west-north-west-trending Pan-African shear zones ('Duwi' trend; Montenat et al., 1986a; Jarrige et al., 1990; Sultan et al., 1992; Montenat et al., this volume) and north-north-east-striking steep faults ('Aqaba' trend; Jarrige et al., 1990; McClay et al., this volume; Montenat et al., this volume); (2) later normal faulting that was principally parallel to the rift-axis ('Clysmic' trend; Hume, 1921; Robson, 1971). These early phases of the rift history were marked by rapid rotation of strata and significant growth of the syn-rift section into active faults. Subsidence rates reached their maximum between 19 and 16 Ma (Steckler, 1985; Evans, 1988; Steckler et al., 1988). In the early Serravallian (~14 Ma), the connection between the Gulf of Suez and Mediterranean basins became severely restricted, causing the Gulf of Suez and northern Red Sea to shift from predominantly open and marginal marine conditions to an evaporitic, silled-basin setting (Hassan and El-Dashlouti, 1970; Robson, 1971; Orszag-Sperber et al., this volume). Tortonian deposition in the Red Sea and southern and central Gulf consisted largely of halite along basin axes, and gypsum along the margins. The subsequent loading and flowage of the salt produced extensive salt walls and rimming synclinal troughs. Most of the salt walls more closely reflect the geometry of underlying Middle Miocene age faults, rather than that of the main Early Miocene fault systems (Bosworth, 1995). Localization of salt flowage may either have been triggered by late movement on the Middle Miocene faults, or by instabilities induced where salt deposition covered fault scarps (palaeotopography) in the basin floor.

The age of the end to evaporite deposition is not precisely known due to the sparsity of datable microfossils in these units. Sedimentation and subsidence rates are therefore speculative. It appears, however, that renewed subsidence occurred in the late Miocene to Pliocene (Steckler, 1985; Evans, 1988; Steckler et al., 1988). Regionally, only very small amounts of post-early Serravallian rotation of strata can be observed, suggesting that the rate of horizontal extension was low (Bosworth, 1995). Based on subsidence modelling, Steckler et al. (1988, Table 1) interpreted the late Burdigalian (Rudeis), early Serravallian (Kareem) and late Serravallian (Belayim) opening rates at the vicinity of Hurghada (Sinai triple junction) to be approximately 0.48, 0.18 and 0.05 cm/yr, respectively. The early Serravallian drop in the rate of Gulf of Suez extension corresponds in time to the final shift of the Sinai–Africa plate boundary to the Gulf of Aqaba transform system, leaving Suez as an essentially aborted rift (Steckler and ten Brink, 1986; Steckler et al., 1988). The synchroneity of the restriction of the Suez–Mediterranean seaway with the onset of the main phase of Aqaba faulting suggests that these two events are geodynamically related.

Microearthquakes and some teleseismic events continue to occur along a few major normal faults in the southern Gulf, indicating that horizontal extensional stresses are still present in this region (Daggett et al., 1986; Jackson et al., 1988). Localized uplift of Pleistocene coral terraces similarly reflects continued deformation in the vicinity of the Aqaba transform–Red Sea rift plate boundary intersection (Andres and Radtke, 1988; Bosworth and Taviani, 1996). The present-day extension direction in the southern Gulf of Suez is north-north-east–south-south-west, approximately parallel to the slip-direction of the Aqaba transform (Bosworth and Taviani, 1996).

Rift geometry

The Gulf of Suez and Egyptian northern Red Sea rift consists of large-scale half-graben that alternate in polarity along the basin axis (Figure B4.1; Moustafa,

1976; Bosworth, 1985, 1995; Moretti and Colleta, 1987; Coffield and Schamel, 1989; Jarrige *et al.*, 1990). Accommodation zones (AZ) that link the half-graben are located at Wadi Araba in the northern Gulf (Galala-Abu Zenima AZ), the north end of Gebel el Zeit in the central Gulf (Morgan AZ), and near Quseir on the Red Sea coast (Brothers AZ) (Figure B4.1). The accommodation zones contain complex arrays of nested and conjugate normal faults linked by ramp-relays (distributed deformation or 'soft' transfer) and local cross-faults (localized deformation or 'hard' transfer). In exposures of the Morgan AZ on the western side of the Gulf of Suez, pre-existing north-north-east-striking basement fault were locally reactivated as transfer faults (Moustafa and Fouda, 1988). None of the accommodation zones, however, are defined by individual cross-faults that run from rift shoulder to shoulder.

Cross-structures smaller than accommodation zones segment the rift at a variety of scales, producing apparent or real offsets in rift-trend normal faults. Based on outcrop and subsurface data, transfer faults have been mapped with lengths of a few tens of metres to kilometres, and follow predominantly east–west, north-east–south-west, and north-north-east–south-south-west orientations (see also Montenat *et al.*, this volume). In the southern Gulf of Suez, the transfer fault–rift-trend fault pattern generally produces right-stepping, gently north-west-plunging fault-block geometries.

Rift shoulder uplift is very pronounced in the southern Gulf of Suez and northern Red Sea, with present-day elevations commonly between 600 and 1000 m in the Eastern Desert and Sinai (Figure B4.1), and individual basement peaks locally exceeding 2 km. Geologic and apatite fission-track data show that pre-rift (pre-late Oligocene) uplift was minimal, and probably less than the magnitude of eustatic seal-level fall during the Chattian (Garfunkel and Bartov, 1977; Kohn and Eyal, 1981; Steckler, 1985; Garfunkel, 1988; Omar *et al.*, 1989). Erosion of the pre-rift sedimentary section was correspondingly small. The general absence of early Oligocene strata can be accounted for solely by eustatic effects. Exhumation of the rift shoulder at ~34 Ma as suggested by fission-track data can be interpreted as evidence that a rift escarpment began to form at that time (Omar and Steckler, 1995). However, rapid uplift and development of large elevation differences between rift shoulder and rift axis did not commence until sometime after about 22 Ma, when faulting and basin subsidence were well underway (Omar *et al.*, 1989).

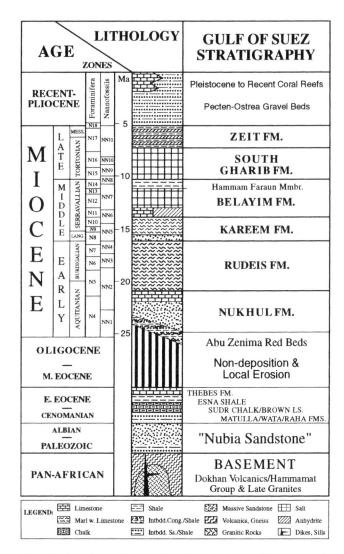

Figure B4.2 Generalized stratigraphy and microfossil zonations of the southern Gulf of Suez. Modified from Evans (1988), Richardson and Arthur (1988) and Bosworth (1995).

Stratigraphy

Pre-rift

The pre-rift section in the southern Gulf of Suez consists of a lower sandstone, overlain by interbedded shale, sandstone, and limestone (Figure B4.2). The lower sandstone is generally referred to as the 'Nubia' (e.g. Sellwood and Netherwood, 1984; Evans and Moxon, 1988). The base of the Nubia, in places, consists of several tens of metres of interbedded red siltstone and sandstone that is undated, but may correlate in part with lithologically similar Paleozoic to Triassic strata in Sinai (Perry, 1986; Van Houten *et al.*, 1984; Allam, 1988). The upper Nubia consists of several hundred metres of quartzose, dominantly cross-bedded, non-marine sandstone that is referred to the Naqus (Allam, 1988) and/or Malha formations (e.g. Kerdany and Cherif, 1990). The

Malha is overlain by a widespread shallow marine sequence of Cenomanian to Eocene age, consisting of the Raha, Wata, Matulla, Brown Lime, Sudr, Esna, and Thebes formations (Said, 1962).

Along the northern Egyptian Red Sea coast, the pre-rift stratigraphy is very similar to that of the southern Gulf of Suez. The basal Nubia Sandstone is undated and generally about 200 m thick (Said, 1990b). The Cenomanian to Santonian formations of the Gulf of Suez are absent, but are partly represented by the poorly fossiliferous Quseir Variegated Shales (Youssef, 1957). The late Senonian to Eocene stratigraphies are nearly identical between the two areas, but with different formational names (Said, 1990b; Kerdany and Cherif, 1990).

Syn-rift
The syn-tectonic, dominantly clastic, Early to Middle Miocene sediments of the southern Gulf of Suez are divided in the subsurface into the Nukhul, Rudeis and Kareem Formations (Figure B4.2; Ghorab et al., 1964; Nat. Strat. Sub-Comm., 1974). These terms have also been applied to offshore Red Sea wells (Tewfik and Ayyad, 1984; Miller and Barakat, 1988). Other workers have proposed a similar stratigraphy for exposures along the rift margin and at fault-blocks such as Gebel el Zeit (Ghorab and Marzouk, 1967; Purser et al., 1990; Darwish and El-Azabi, 1993).

The Nukhul consists of a heterolithic assemblage of sandstone, conglomerate, carbonate, shale and evaporite. Four members have been defined based on well penetrations at the Shoab Ali field (Saoudi and Khalil, 1986). The basal unit, the Shoab Ali Member, consists of sandstone with intercalated red shale and occasional coaly carbonaceous material. The unit is unfossiliferous, but based on lithology and stratigraphic position is probably correlative with the Aquitanian age Abu Zenima Fm. of the northern Gulf (Sellwood and Netherwood, 1984). Similar units along the Red Sea coast from Safaga to south of Quseir (Figure B4.1) are informally called 'red beds', or have been placed within the Ranga Fm. (Gindy, 1963; Issawi et al., 1971). These continental deposits are the basal part of the Group A of Montenat et al. (1986a; also Plaziat et al., this volume). At Gebel Duwi near Quseir, basal syn-rift conglomerate, sandstone, and lacustrine sediments are referred to the Nakheil Fm. (Akkad and Dardir, 1966). The Shoab Ali member is overlain by the laterally interfingering Ghara anhydrite member and carbonate and sandstone of the Gharamul and October Members containing benthonic foraminifera of Early Miocene age (Saoudi and Khalil, 1986). Throughout the Gulf of Suez rift basin, the datable part of the Nukhul Fm. predominantly belongs to Blow's (1969) early Aquitanian N4 foraminifera zone, although a few authors continue the unit into the very earliest Burdigalian (Scott and Govean, 1985; Richardson and Arthur, 1988).

The Rudeis Fm. (Figure B4.2) is predominantly *Globigerina* marl that is cyclically interbedded with fine-grained, deep-water limestone. Its age extends from the late Aquitanian to the Burdigalian, but its base is probably time-transgressive (Richardson and Arthur, 1988, and references therein). Short-lived, basin-scale relative sea-level changes resulted in the deposition of thin, laterally continuous, marine evaporites and more areally restricted thin, turbidite sandstones at approximately the Burdigalian–Langhian boundary. Marl deposition generally resumed in most sub-basins through the Langhian. The thin evaporites and sand markers, or, where absent, the occurrence of Langhian faunas in marls, are commonly picked as the base of the Kareem Fm. (Figure B4.2). The basal Kareem evaporites and sandstones were not generally deposited along the rift-margins or on the crests of fault blocks, which comprise most of the presently outcropping areas. The Rudeis–Kareem boundary cannot therefore be recognized in the field (see further discussion in Purser et al., this volume).

In the early Serravallian, at about 14 Ma, deposition in the southern Gulf of Suez and northern Red Sea shifted to predominantly anhydrite and salt of the Belayim Fm. (Figure B4.2; Hassan and El Dashlouti, 1970). The base of the Belayim is often an angular unconformity, with Belayim evaporite resting directly on Rudeis marl or older formations. The uppermost part of the Belayim Fm. is a regionally extensive, normal marine shale referred to the Hammam Faraun Member. It is commonly the only datable part of the post-Kareem Miocene stratigraphic section. The latest Serravallian age (N14) of the Hammam Faraun, and the general lack of erosion at its top, show that the overlying South Gharib salt deposition probably began in the Tortonian.

Post-rift
In the southern Gulf of Suez and offshore northern Red Sea, the South Gharib Fm. consists predominantly of massive halite, with minor stringers of anhydrite (Figure B4.2). Whether the South Gharib is truly a syn- or post-rift deposit is somewhat arbitrary, because without palaeontologic age control it is impossible to accurately estimate extension rates during deposition of the salts. A significant number of faults cut the underlying Belayim Fm. (discussed below; also Bosworth, 1995, Figure 10), and hence we consider the Belayim part of the syn-rift succession. It is difficult, however, to determine how many of these faults were active in the Tortonian, due to the flowing nature of the salt and the lack of correlatable marker beds.

The South Gharib Fm. is overlain by interbedded evaporite, shale, sandstone and carbonate of late

Miocene (Zeit Fm.) to probably Pliocene age (Figure B4.2). These in turn are capped by extensive gravels and lesser *Ostrea–Pecten* coquinas and marls which are only broadly Pliocene to Pleistocene in age (Hume *et al.*, 1920; M. Taviani, personal communication). Post-Miocene units are best exposed along the Red Sea coast, where they are referred to the Gabir, Shagra and Tubia formations (Akkad and Dardir, 1966; Khedr, 1984; Said, 1990b).

FAULT BLOCK EVOLUTION AND STRATIGRAPHIC DEVELOPMENT

Most basins and fault systems in the southern Gulf of Suez and northern Egyptian Red Sea margin experienced generally similar, or at least systematically coherent, structural and stratigraphic histories. No single locality, in outcrop or subsurface, however, completely illustrates the details of the rift's evolution. To accomplish this, we describe the geological relationships at key areas, moving chronologically through the Early to Middle Miocene development of the basin.

Rift initiation (Aquitanian): south Gebel el Zeit

Gebel el Zeit is a large, intra-rift fault block that exposes granitic basement, an excellent section (~650 m) of pre-rift strata, and a thin (~150 m), crestal section of Nukhul to lowermost South Gharib syn-rift rocks (Plate 8; Perry, 1986; Montenat *et al.*, 1986a and this volume; Evans, 1988; Allam, 1988; Bosworth, 1995 and references therein). The north-east slopes of the gebel are a slightly eroded fault-line scarp, separating the basement block from the offshore East Zeit basin. The south-west flank of the gebel dips into the Gemsa basin, which attains a maximum stratigraphic thickness of approximately 7 km.

Gebel el Zeit is separated by a saddle (Sarg el Zeit) into the main northern range and a smaller complex to the south referred to as South (Ganoub) or Little Gebel el Zeit (Plate 8). Earliest syn-rift deformation is best observed at South Zeit. The large-scale structure of South Zeit is produced by the main fault at the coast, and a second large, rift-trend normal fault that exposes basement and pre-Miocene strata in its footwall (Plate 8). The hanging-wall of this second fault contains a large, north-west-plunging syncline exposing South Gharib, Belayim and Kareem Formation evaporite and marl. The basement–Nubia Sandstone contact and bedding planes within the pre-Miocene on the footwall dip on the average 45° SW.

The pre-Miocene series at South Zeit is overlain unconformably by Nukhul conglomerate, sandstone and carbonate. The thickness of the Nukhul is extremely irregular, frequently attaining a maximum adjacent to some of the early, syn-depositional cross-faults (Figure B4.3). The unconformity at the base of the Nukhul dips 30–35° south-west.

Samples from the uppermost Nukhul carbonate beds at North Zeit (plate 8) contain the large benthic foraminifera genus *Miogypsinoides*, indicating N4 Zone of Blow (1969) and therefore an early Aquitanian age (~21–25 Ma) (Bosworth, 1995). The base of the Rudeis marl locally contains *Globigerina ciperoensis*, which is only as young as the early part of N5 (~20–21 Ma; latest Aquitanian). In other sections, the basal Rudeis is as young as N7–N8 (~16–17 Ma; late Burdigalian), with abundant reworked older faunas (A. Sadek, personal communication; also Montenat *et al.*, this volume). The top of the Nukhul in outcrop is therefore locally unconformable, with up to ~4 Ma of section missing, but elsewhere appears to be sedimentologically and faunally transitional into the overlying Rudeis *Globigerina* marls. In general, the Nukhul and Rudeis strata in wells to the south-west of Gebel el Zeit (Gemsa basin; Figure B4.4) are conformable in dipmeter records.

The pre-Miocene and Nukhul strata at South Zeit are cut by numerous, small, rift-parallel normal faults and cross-faults, most of which are healed by overlying Rudeis marls (Plate 8). Over much of South Zeit, and similarly at North Zeit (Bosworth, 1995, Figure 6), the upper Rudeis–Kareem marls and base of the Belayim Fm. are unfaulted. The intense internal deformation of the South Zeit block was therefore restricted to the early phase of the rift history (pre-deposition of exposed Rudeis marls). This argument is supported by the strong control that faults played in localizing Nukhul sediment accumulations and determining facies distribution (Figure B4.3), whereas most of the overlying Rudeis is composed of laterally continuous marl beds. Clearly, deformation at Gebel el Zeit continued after this early phase, as all the younger stratigraphic units are rotated and/or uplifted (Evans and Moxon, 1988; Bosworth, 1995). At South Zeit, the large normal fault that cuts down the axis of the gebel offsets strata as young as basal South Gharib Fm. (Plate 8). The block-bounding fault along the coast and the cross-faults that terminate the southern ends of both North and South Zeit also experienced up to kilometres of movement after the deposition of the Belayim Fm. (Bosworth, 1995, Figure 15). However, the spacing between normal faults in the pre-Miocene and Nukhul exposure is generally several tens to a few hundred metres, while active faults by upper Rudeis to Belayim time were generally separated by kilometres (Plate 8).

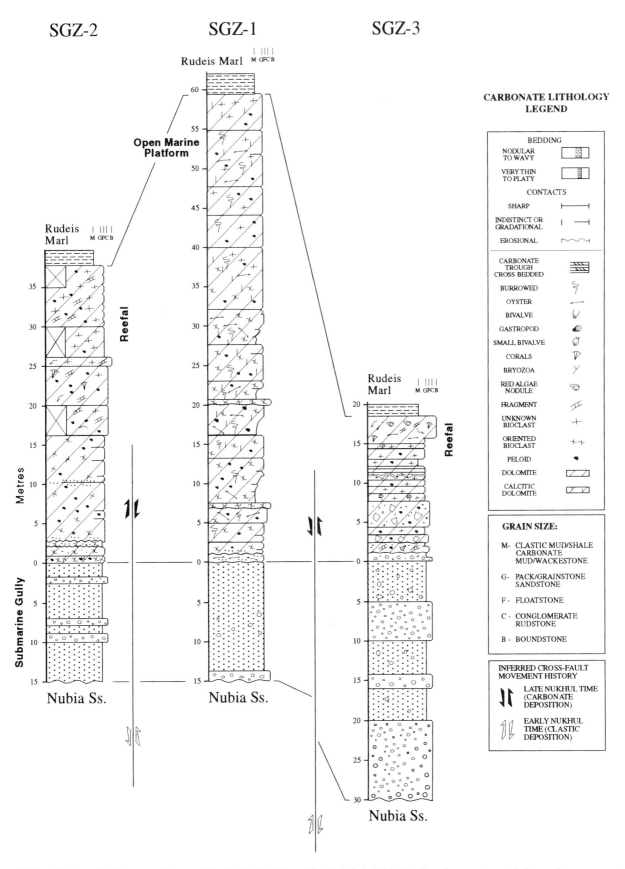

Figure B4.3 Facies and thickness variations within the Nukhul Fm. at South Gebel el Zeit. Stratigraphy was strongly influenced by movement on cross-faults, which commonly display changes in sense of throw through time (double arrows).

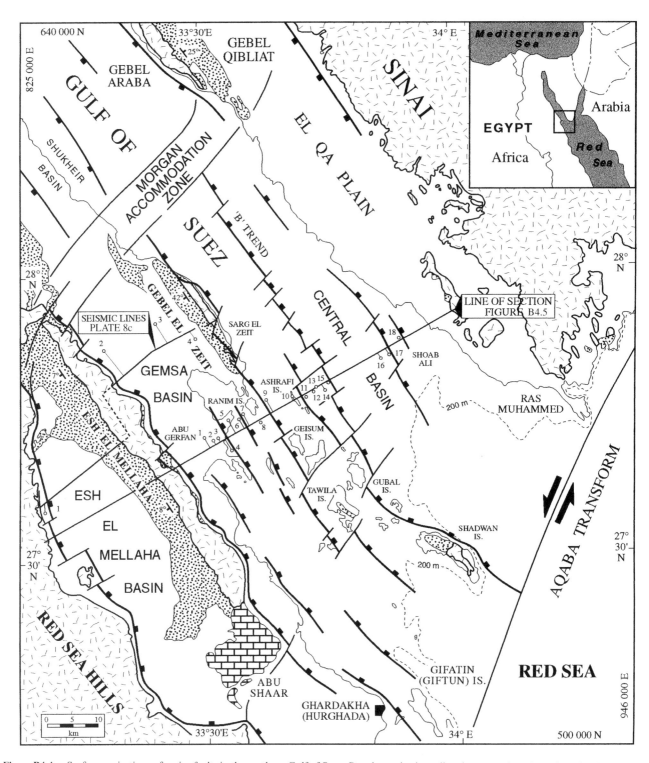

Figure B4.4 Surface projections of major faults in the southern Gulf of Suez. Based on seismic, well and outcrop data. Location of regional cross-section (Figure B4.5) and Esh el Mellaha–Gemsa basin seismic profile (Figure B4.6) are given. The main pre-rift and syn-rift sedimentary outcrops are shown by stipple pattern. Large carbonate complex at Abu Shaar el Qibli is shown by block pattern. Dash pattern is crystalline basement (after Bosworth, 1995).

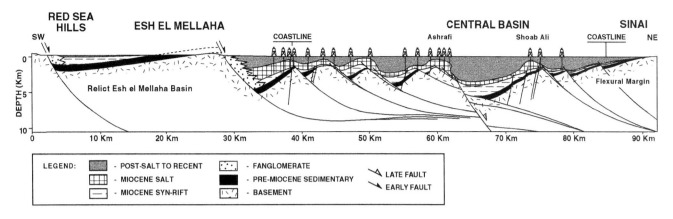

Figure B4.5 Regional cross-section from the Red Sea Hills to Sinai. Location is given in Figures B4.1 and B4.4 (after Bosworth, 1995).

Main phase rift subsidence (Burdigalian): southern Gulf of Suez subsurface

The main phase of syn-rift sedimentation and structuring, which occurred during the Burdigalian deposition of Rudeis marls, is not well-exposed in outcrop, and generally requires a more synoptic view to appreciate its overall style. This is best accomplished by considering subsurface data from an area of close well control, such as the area south of Gebel el Zeit (Figure B4.4). The cross-section in Figure B4.5 extends from rift shoulder-to-shoulder, running through Ras el Bahar, Zeit Bay, Ashrafi, and Shoab Ali fields.

The regional cross-section across the southern Gulf can be divided into three principal Burdigalian age structural provinces: 1. an abandoned, partially exposed basin adjacent to the Red Sea Hills (Esh el Mellaha basin; Figure B4.6), 2. the main, active fault trends of the central (axial) rift, and 3. the eastern flexural or ramp margin at Sinai. Uplifted rift-shoulder basins similar to the Esh el Mellaha basin are also found along the northern Egyptian Red Sea coast, such as at Gebel Duwi, and have been referred to as 'relict basins' (Bosworth, 1994a). These basins contain thick sections of the basal red beds and/or Aquitanian age strata (Nukhul Fm.), but no or very thin younger syn-rift units. The central rift includes all the hydrocarbon-producing fault blocks of the basin, which are at least partially sourced by the thick Rudeis *Globigerina* marls. The flexural margin begins at Shoab Ali (Figure B4.5), and is a region of progressively decreasing syn-rift stratal rotation as the exposed basement complex of Sinai is approached.

The structural geometry of the central rift and eastern flexural margin has been discussed in detail elsewhere (Perry and Schamel, 1990; Bosworth, 1995). The principal attribute of Burdigalian, main phase rifting and subsidence is the tremendous growth that is focused at a few, regionally continuous normal fault systems such as the Ashrafi–Shadwan and Ranim–Tawila trends (Figure B4.4; Bosworth, 1995). Distributed fault deformation was much less significant in the Burdigalian rift than it had been in the Aquitanian rift. Many of the large normal faults active during the Burdigalian have been reached by the drill bit. The fault planes are generally gently listric in profile, and dip as shallow as 23° northeast (Bosworth, 1995).

End main phase rifting (Serravalian): Gebel Naqara

Gebel Naqara forms the high basement range west and south-west of Safaga (Figures B4.1, B4.7). The outcrops along the lower slopes of Gebel Naqara provide an excellent example of the structural and stratigraphic attributes of Middle Miocene carbonate–terrigenous clastic facies distribution developed on rotated fault blocks adjacent to the uplifted border fault complex of the north-western Red Sea rift. Previous work in the area is limited essentially to a large mapping project conducted by GENEBASS (e.g. Montenat *et al.*, 1986a; Thiriet *et al.*, 1986).

The general structural geometry of the rift shoulder between Safaga and Quseir is controlled by large, down-to-the-east normal faults, striking between north–south and north-west–south-east. At Gebel Naqara, the block-bounding fault consists of two to three parallel segments, producing steps in the basin border geometry (Figure B4.7(b)). Each of the normal faults at Naqara is segmented along-strike by cross-faults, striking up to 20° north or south of east–west. Several of these structures are well-exposed. Where this is the case, they consist of a discrete fault, rather than a zone of broadly distributed deformation. We did not observe any evidence for significant strike-slip movement on the cross-faults. Some of the cross-faults and rift-parallel step-faults are healed by Middle Miocene carbonates (age dating discussed below), demonstrating that move-

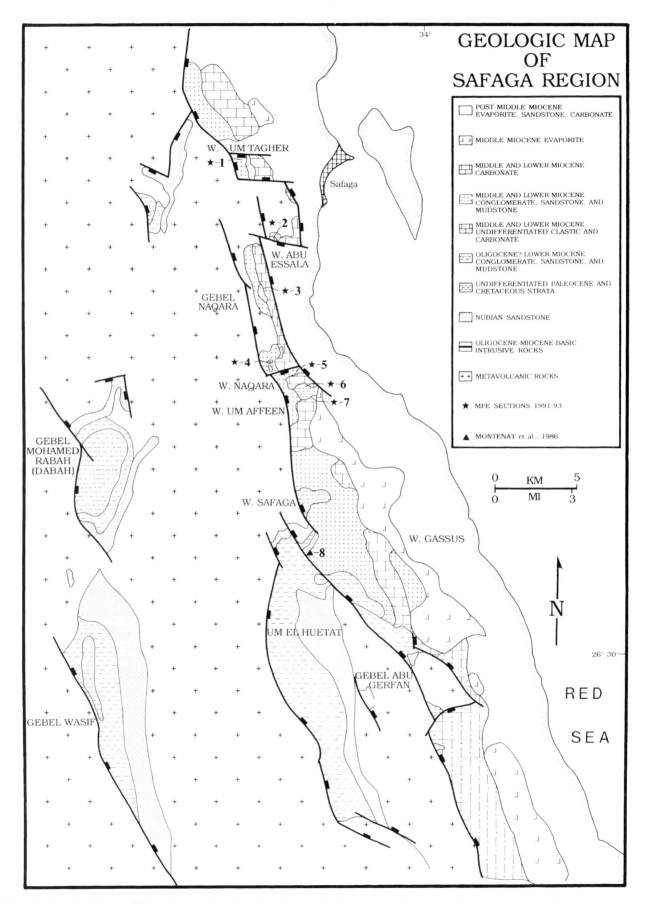

Figure B4.6 Gebel Naqara field study area. General geological map of Safaga region, with location of measured sections. After Montenat et al. (1986a).

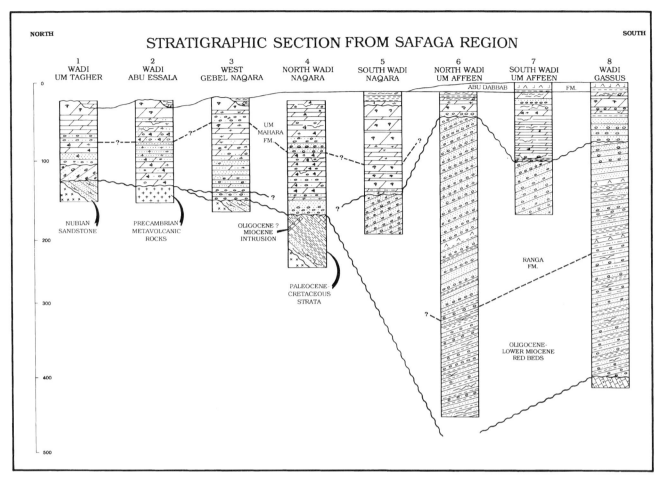

Figure B4.7 (a) Measured stratigraphic sections. Symbols as in Figure B4.3.

ment occurred during Early Miocene rifting (Figure B4.8). In general, we interpret the cross-faults to have formed synchronously with normal faulting, acting as relays between normal faults, analogous to the transfer faults observed at Gebel el Zeit. The extensive healing of faults by the Middle Miocene strata, which is also present at Gebel el Zeit and in the subsurface fault trends (Figure B4.5), indicates that the rift underwent a second decrease in structural complexity, similar to that observed at the end of the Aquitanian.

The sedimentary succession at Naqara consists of nearly 100 m of mixed terrigenous siliciclastics and reefal platform dolomites unconformably overlying Nubia Sandstone and metavolcanic basement (Figure B4.7(a)). Very poorly exposed, patchy remains of Late Cretaceous pre-rift formations are also present, as just north of Wadi Naqara. The basal siliciclastic depositional sequence is transitional and conformable with the overlying mixed siliciclastic–carbonate sequence. The reefal platforms form plateaux at ±300 m. The carbonate plateaux are a sporadic but locally spectacular feature of the Red Sea margin from Abu Ghusun to the Esh el Mellaha basin north of Hurghada (Figure B4.1). Similar plateaux are present along the north-eastern Red Sea margin in Saudi Arabia.

The biostratigraphic age for the reef complex of Gebel Naqara was determined by dating two samples from neptunian fissures in the seaward margin of the gebel adjacent to a transfer fault slope canyon (locality GN-6 in Figure B4.7(b)). The fissures are overlain by the reef complex. Age-diagnostic planktonic foraminifera species *Orbulina universa*, *O. suturalis*,/*Praeorbulina* sp. indicate that the fissure-fill sediments and at least the basal reef carbonates are Serravallian, and thus correlate to either the Kareem or Belayim Formations. We have also determined planktonic foraminifera-based Serravallian–Langhian ages for the Gubal Island subsurface carbonate platform (Steinmetz and Crevello, unpublished data). James *et al.* (1988, p. 557) reported that the carbonate platform at Abu Shaar (Figure B4.4) is 'probably latest early Miocene, based on calcareous nannoplankton,' which makes the platform correlative with upper Rudeis Fm. *Globigerina* marls. However, the calcareous nannoplankton to which James *et al.* were

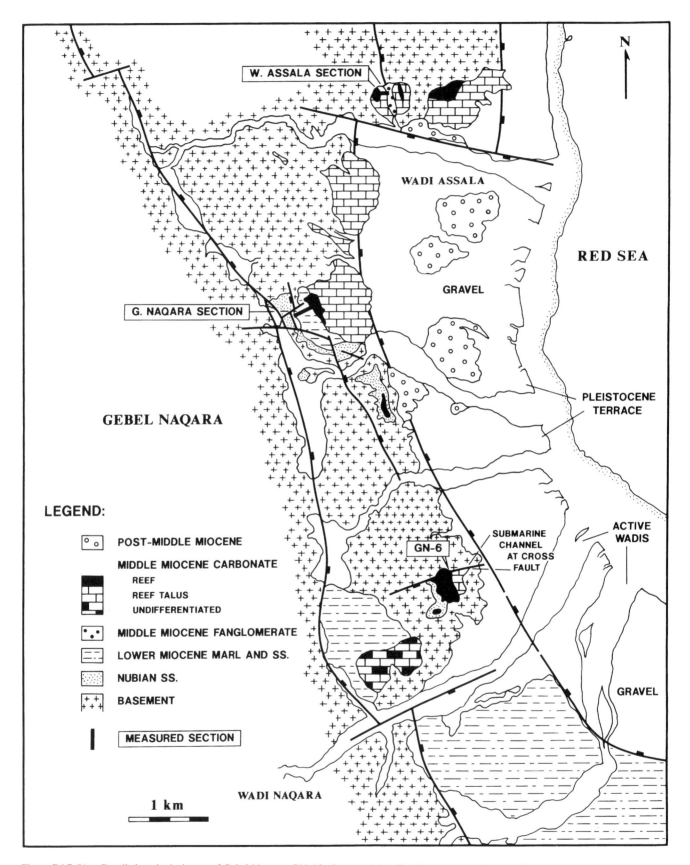

Figure B4.7 (b)   Detailed geological map of Gebel Naqara. GN-6 is the sample locality that contained Serravallian foraminifera.

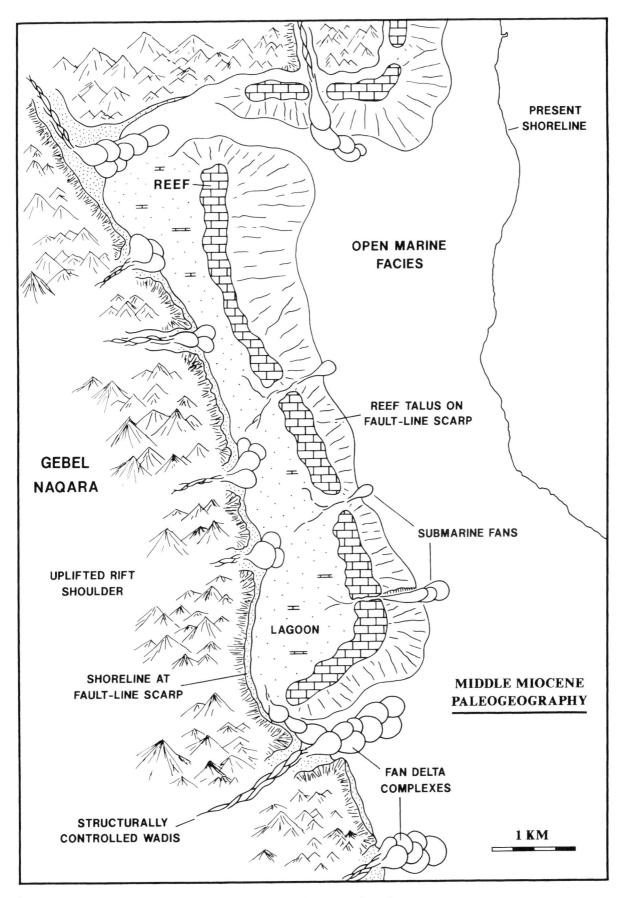

Figure B4.7 (c) Middle Miocene palaeogeographic reconstruction of Naqara reef complex.

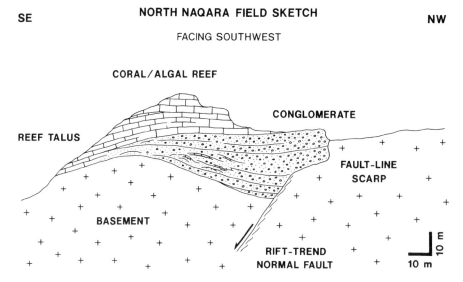

Figure B4.8 Field sketch of Middle Miocene reef complex healing a rift-parallel step fault at north Gebel Naqara. Location is labelled 'W. Assala section' in Figure B4.7(c).

referring were actually recovered from basinal carbonates in well bores in the Zeit Bay area (Cofer et al., 1984). Our data, which consist of direct determinations of fossils recovered from two of the platform carbonate complexes, suggest that carbonate platforms developed more or less synchronously throughout the Gulf–northwestern Red Sea during the Middle Miocene, following a prolonged period of detrital clastic sedimentation of the Rudeis Fm. As evidenced by Gebel el Zeit, an earlier period of platform development occurred in the Aquitanian during late Nukhul deposition (Evans and Moxon, 1988; Bosworth, 1995; Purser et al., this volume).

Serravallian sea levels are generally believed to have been somewhat higher than at present, but not by as much as 300 m (Haq et al., 1987). The present elevation of the Naqara carbonate platform (and the similar ±200 m platform at Abu Shaar) is therefore at least partially the result of later uplift along the margin of the Red Sea basin.

The lithofacies and stratigraphy of the depositional sequences above the Nubia are variable. The siliciclastics are interpreted as aqueous deposits of alluvial fan-braided streams and flood plain environments. Poorly sorted, pebbly and clayey sandstones are interpreted to be deposits of high-flow capacity alluvial channels and flood plains. The continental siliciclastic strata were deposited on the landward, western margins of rotated fault blocks. The siliciclastics pass upward into Miocene mixed carbonate–terrigeneous marine–fan delta strata. Figure B4.7(c) shows a Middle Miocene palaeogeographic reconstruction of the Naqara area, and Figure B4.9 is a schematic block diagram of our depositional model for the structurally controlled Naqara reef platform. The Middle Miocene reef geometry directly reflected the underlying pattern of intersecting rift-parallel and rift-normal faults.

CONCLUSIONS

Early to Middle Miocene basin dynamics

Well control for the central rift and north-eastern flexural margin is quite good in the southern Gulf of Suez (Figure B4.5). Age dating of the Esh el Mellaha basin fill is less certain. However, a well drilled in the northern part of the basin penetrated ~600 m of post- to syn-rift section prior to reaching the pre-rift strata (Plate 8c). Seismic data quality in the Esh el Mellaha basin is excellent, and the well can readily be tied to outcrop and other parts of the basin. The basal ~150 m of the syn-rift is Early Miocene marine strata (Nukhul–Rudeis Formations). The shallower section is undifferentiated Miocene, with the exception of the top 30 m which were unfossiliferous and possibly post-Miocene. This well confirmed that in the Early Miocene, the entire rift basin (Esh el Mellaha basin, central rift, and flexural margin), with the exception of the crests of some fault blocks, was reached by marine waters.

Following the Early Miocene, the area of active sediment accumulation narrowed through time, culminating with post-Zeit deposition focused in the Shoab Ali basin. Figure B4.10 shows sediment thicknesses from the regional southern Gulf of Suez cross-section broken out into four time intervals: 1. late Oligocene to early Middle Miocene (Nukhul, Rudeis, and Kareem Formations), 2. late Middle to early Late Miocene (Belayim and South Gharib Formations), 3. latest Miocene (Zeit

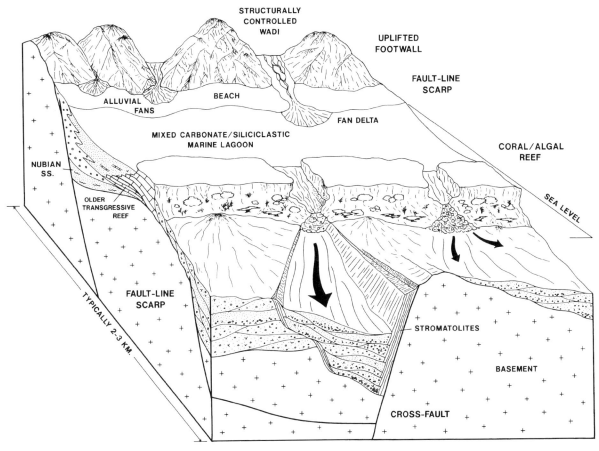

Figure B4.9 Block diagram of reconstructed Serravallian palaeoenvironments in the Naqara reef complex. Active cross-faults controlled the location of submarine channels, which eroded into basement. The cross-faults were then lined by growth of stromatolites (see Figure B4.7 (c)).

Fm.), and 4. post-Miocene (Pliocene to Recent formations). The narrowing of the rift through time is not restricted to the south, but also occurred in the central Gulf of Suez half-graben (Moretti and Colleta, 1987).

The precise timing of abandonment of the Esh el Mellaha basin is not known (see discussions in Perry and Schamel, 1990, and Bosworth, 1995). Middle to Late Miocene evaporites have not been found thus far in the basin. During the post-Miocene the basin was predominantly non-marine and did not experience any significant subsidence (Plate 8c). Our Serravallian age interpretation for at least the upper part of the Abu Shaar carbonate platform, and their basinal equivalents, locates these strata at roughly the boundary between Figures B4.10(d) and (e). No younger carbonate build-ups are known from outcrop along the Esh el Mellaha range. The Serravallian may have been the time of the last significant marine presence in the Esh el Mellaha basin. Post-Serravallian deformation of Esh el Mellaha was minimal, as Abu Shaar strata are rotated no more than about 2° to the south-west (Cross *et al.*, this volume).

Our structural and sedimentologic observations indicate that the early Nukhul period of rifting at Gebel el Zeit was one of active faulting, turbidite deposition, and significant differential uplift between the Gemsa basin and the eroding basement crest of the Zeit fault block. Many authors have argued that the early Gulf of Suez displayed only very limited structural relief (e.g. Garfunkel and Bartov, 1977), and low-relief, shallow marine interpretations have been suggested for the lower Nukhul Fm. at Gebel el Zeit (e.g. Evans, 1988). Outcrops of early syn-rift strata in Sinai near the central Gulf of Suez eastern border fault complex suggest that the central rift margin was not significantly eroded in the Aquitanian (Sellwood and Netherwood, 1984; McClay *et al.*, this volume). It is probable that not all segments of the rift experienced precisely the same structural, topographic and sedimentologic early rift histories. However, our Gebel el Zeit observations are from an intra-rift fault block, distant from the rift shoulder. It is possible that the early rift, although without great escarpments along its margins, displayed significant intra-rift topography. Our observations from the Nu-

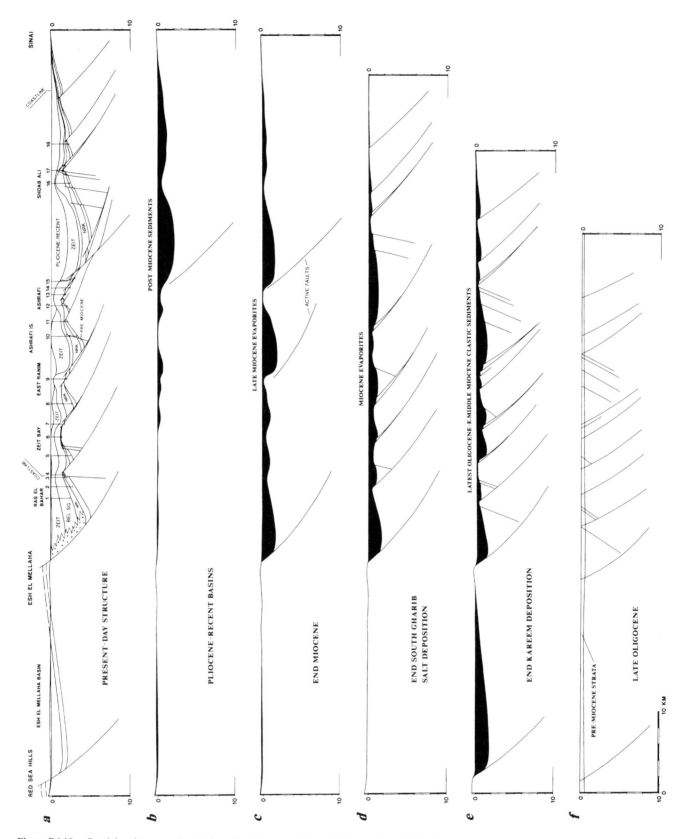

Figure B4.10 Spatial and temporal evolution of sub-basins in Figure B4.5, southern Gulf of Suez. Black shading represents sediment that accumulated during each indicated time interval. NRK = Nukhul–Rudeis–Kareem formations; BEL-SG = Belayim–South Gharib formations. Stipple pattern = alluvial fan and reef talus along the Esh el Mellaha fault; see zone of poor seismic data in Figure B4.6 (after Bosworth, 1995).

khul at Gebel el Zeit would suggest that the early rift floor was broken into a large number of fault-block defined depocentres.

We have calculated rotation rates for each formation at Gebel el Zeit (Figure B4.11). Although theoretically there is not a simple, linear relationship between rotation rate and horizontal extension, the fastest period of structuring does appear to have taken place prior to and during deposition of the Nukhul (~27–21 Ma), with about 15° total rotation, or 2.5° of rotation per million years. Assuming an age of ~24 Ma for the base of the Nukhul as exposed at Gebel el Zeit, this factors into 10° of pre-Nukhul rotation (3.3°/Ma) and 5° of syn-Nukhul rotation (1.7°/Ma) (Figure B4.11). During Rudeis time, this slowed to 1.4°/Ma, and, within the limitations of the outcrop data, remained subdued and relatively constant until sometime during the post-Miocene when a significant increase in the rate of rotation again occurred. For comparison, the cumulative rotation of strata at Gebel el Zeit and that encountered in nearby wells, normalized such that 100% equals the present pre-rift stratal dip (or total rotation) at each locality, are plotted against time in Figure B4.12. In many subsurface fault blocks, rapid rotation started in Nukhul time and continued at about the same rate through deposition of the Rudeis, Kareem, and Belayim Formations. By the end of the Middle Miocene (~10 Ma), block rotation ceased in blocks such as Sarg el Zeit and Tawila (Figure B4.12). During the Late Miocene to Recent, rotation was restricted to progressively fewer areas, such as Gebel el Zeit (see further discussion in Bosworth and Taviani, 1996).

The Gulf of Suez basin has been particularly amenable to subsidence modelling due to the relatively good age control provided by marine microfossils. Studies from Wadi Gharandal in Sinai (Evans, 1988), the central Gulf (Steckler, 1985; Evans, 1988), southern Gulf (Steckler et al., 1988) and Gemsa basin (Evans, 1988) all confirm that the major phase of basin subsidence commenced 20–19 Ma with the onset of Rudeis Fm.

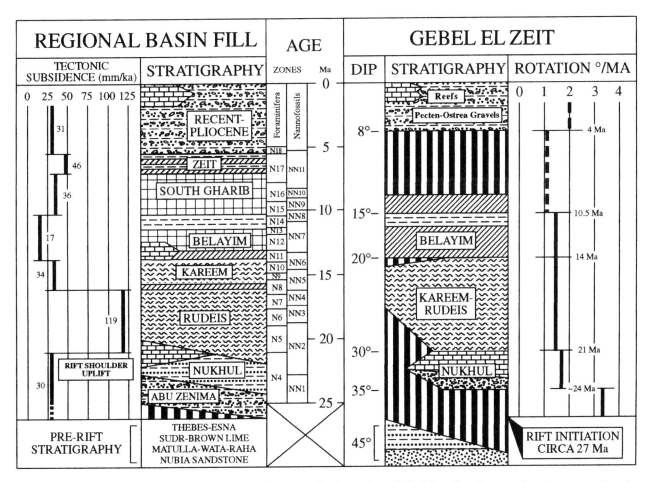

Figure B4.11 Rotation rates for Gebel el Zeit, and subsidence rates for the southern Gulf of Suez. Rotation rates for units younger than the Hammam Faraun Shale (dashed) are only approximate estimates due to lack of age control. Tectonic subsidence rates are calculated from data in Steckler et al. (Table 1, 1988) and are averaged across the entire Gulf of Suez basin. The ages of stratigraphic boundaries used by Steckler et al. are slightly different than those we have followed, which are taken from the extensive compilation by Richardson and Arthur (1988). We have adjusted the Steckler et al. data to match our time boundaries. Lithologic symbols are the same as those in Figure B4.2.

marl deposition, and continued until the Middle Miocene (~16 Ma). Using data from an analysis of a regional cross-section by Steckler et al. (1988), we have calculated average rates of tectonic subsidence for each of the units in Figure B4.12. It is apparent that the period of maximum subsidence lagged behind rapid Nukhul age faulting and rotation by several million years.

Further insight into Gulf of Suez rift dynamics can be gained by analysis of the timing and magnitude of rift shoulder uplift (Steckler, 1985). Analysis of apatite fission-track ages and track-length measurements has shown that the main unloading of the basement complex in the rift shoulder of the western Gulf did not begin until ~23–21 Ma (Omar et al., 1989). Omar et al. interpreted these results as indicating that uplift accompanied, or slightly preceded, the onset of rapid extension and concomitant tectonic subsidence marked by the deposition of early Rudeis marls. The apatite data also confirmed the absence of significant regional unroofing prior to the onset of rifting (Garfunkel and Bartov, 1977).

Additional fission-track data from the Red Sea hills of southern Egypt, a small area of the west margin of the central Gulf of Suez, and south-western Saudi Arabia indicate a regional period of exhumation starting at ~34 Ma (Steckler and Omar, 1994; Omar and Steckler, 1995). This suggests that the Oligocene South Turkana (Kenya)–Southern Sudan rift system (Morley et al., 1992; Hendrie et al., 1994; Bosworth, 1994b) may have reached to the area of the Miocene Red Sea basin. However, unlike in East Africa, no structural or stratigraphic record for this period of rifting has yet been identified in Egypt. It is possible that although rifting propagated essentially instantaneously throughout the Red Sea region at ~34 Ma as proposed by Omar and Steckler (1995), the amount of extension was too small to leave a readily recognizable rift basin signature. We therefore assume that the main phase of extension and block rotation commenced at ~27 Ma, when the first stratigraphic section was preserved (see discussion above).

Our field data from Gebel el Zeit suggest that the connection between extension, subsidence, and uplift is more complex than envisioned in the standard McKenzie (1978) model of lithospheric extension. About one-third of the total block rotation at Gebel el Zeit occurred between the start of the main phase of rifting (~27 Ma) and about 21 Ma. Well data from across the rift confirm that rate of rotation during deposition of the Nukhul and, by inference, rate of extension, were at least as great as that during Rudeis deposition, when the rate of subsidence was nearly an order of magnitude greater (Figure B4.11; Steckler et al., 1988). Nukhul facies observed in outcrop, together with fault relationships, demonstrate that the early southern Gulf of Suez was a tectonically active, marine setting, with significant topographic relief within the basin itself, although probably lacking large, rift margin escarpments. Marked rift shoulder uplift followed the onset of rifting by a few million years, and was accompanied by abandonment of peripheral basins such as Esh el Mellaha. This in turn was followed by a dramatic increase in the rate of sediment accumulation along the active, central basins (Figure B4.11).

The structural and topographic profiles of several active and failed rift basins have recently been successfully modelled by assigning a finite strength to the continental lithosphere during the time of rifting (Weissel and Karner, 1989; Kusznir and Egan, 1990; Ebinger et al., 1991). If the proto-Gulf of Suez rift lithosphere possessed significant strength, then horizontal extension (cause) and subsidence (vertical response) should never have been instantaneously in equilibrium. The question is whether or not this disequilibrium could produce a time delay of sufficient duration to be geologically resolvable. Our data suggest that the delay is, in fact, measurable. If our interpretation of the early extensional history of the Gulf of Suez is representative of rift basins in general, then future refinements in mechanical rift models must also incorporate a non-instantaneous relationship between extension and initial subsidence.

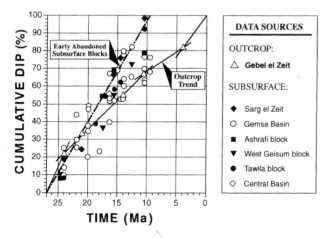

Figure B4.12 Plot of cumulative dip versus time, normalized to 100%, for Gebel el Zeit outcrop and well data (dipmeter) from adjacent basins of the southern Gulf of Suez. Solid line follows outcrop trend with rapid initial sructuring (pre-Nukhul, ~27–24 Ma), main phase of deformation with moderate rotation rates (Nukhul to Belayim, 24–10 Ma), Late Miocene post-rift (10–4 Ma), followed by late, renewed rotation (~4–0 Ma). Dashed line is history of some large subsurface fault blocks that were abandoned by the end-Middle Miocene, heralding the onset of post-rift subsidence. Other subsurface blocks display intermediate histories. Scatter in data at ~20 Ma reflects uncertainties in palaeontologic age-dating; scatter in data at younger times is an indication of real differences in rotation histories.

Relationship between reef platform development and rift fault system evolution

Three periods of extensive shallow-water carbonate deposition can be recognized throughout much of the northern Red Sea and southern Gulf of Suez: 1. Aquitanian reefs and platform facies of the Nukhul Fm., 2. Serravallian reefs and platform facies of the late Kareem–Belayim Formations (and their age equivalents along the rift shoulder), and 3. uplifted coral terraces and living coral reefs of the Pleistocene to Recent. Formation of these reef platform complexes was undoubtedly dependent upon the coincidence of several key factors, including climate, absolute sea level (connectivity to Mediterranean Sea–Indian Ocean) and rates of sea-level change. However, there is also an apparent relationship between the timing of platform development and the evolution of fault systems within the rift basin.

As discussed above, outcrop relationships at Gebel el Zeit (Plate 8) and along the Egyptian Red Sea margin in the Gebel Naqara–Gebel Mohamed Rabah region (Figure B4.6; also unpublished mapping, 1:20 000 scale, Red Sea Phosphate Company), Gebel Ambagi–Gebel Duwi region near Quseir (Figure B4.1; Greene, 1984), and Ras Honkorab near Abu Ghusun (Figure B4.1; Montenat et al., 1986b) indicate that the early rift (Chattian–Aquitanian) was structurally and stratigraphically complex, and marked by distributed faulting on a variety of scales. Sedimentation at this time was predominantly clastic facies (generally marine in the Gulf of Suez), with local evaporite deposition (e.g. Ghara member). In the upper Nukhul Fm., deposition shifted to carbonates. At approximately the same time, many small- and intermediate-scale faults were abandoned, and the next phase of deposition was marked by laterally continuous marl beds of the Rudeis Fm. (early Burdigalian). The overall nature of the Nukhul–Rudeis contact is variable, ranging from essentially continuous deposition to a small angular discordance of up to ∼5°. Jarrige et al. (1990) interpreted the early rift stress field to have been a strike-slip regime, with the maximum horizontal stress parallel to the rift-axis (see also Montenat et al., this volume). They suggested that this changed to an extensional regime, with the minimum horizontal stress generally perpendicular to the rift, in the Burdigalian.

The characteristics of the Kareem–Belayim transition in many respects resemble those of the upper Nukhul–Rudeis. Lithofacies within the Kareem are generally much more variable than those of the upper Rudeis. Although still predominantly *Globigerina* marls, the upper Kareem also contains abundant immature sandstone and conglomerate that was deposited in submarine channel and fan settings. These are generally located on the downthrown sides of faults, and are interpreted here to indicate a brief period of intensified structuring at about 16–14 Ma. This structural activity in the Kareem rift may have initiated at the 'mid-Clysmic event' (Garfunkel and Bartov, 1977). Despite the evidence for faulting and erosion of fault-block crests, the Kareem was a period of relatively slow subsidence, similar to the Nukhul (Steckler et al., 1988). This structural pulse was followed by the formation of the carbonate platforms at Gebel Naqara, Gubal Island and probably Abu Shaar, where the sedimentary overlapping of many faults can be observed in outcrop or inferred from subsurface data. The Serravallian is also a time of one of the most important stress field changes in the Gulf of Suez region, with the onset of the main phase of movement along the Gulf of Aqaba transform plate boundary (Steckler et al., 1988). Steckler et al. also inferred that the regional stress field shifted from north-north-east extension to east-north-east extension in the Middle Miocene.

An increase in subsidence rates is generally recognized in Gulf of Suez data for the Late Miocene (South Gharib and Zeit Formations) (Steckler, 1985; Evans, 1988; Steckler et al., 1988). The tectonic significance of this increase is not well understood, although Steckler et al. have suggested that it may reflect minor adjustments in the partitioning of Red Sea extension between the newly formed Aqaba transform boundary and the failing Gulf of Suez rift. The nature of tectonism in the Pliocene is even less well-constrained, because very few published palaeontologic dates exist for this part of the post-rift fill. It is possible that Pliocene and Late Miocene subsidence and tectonic environments were similar. In the Pleistocene, a dramatic change in stress field occurred, shifting back from east-north-east or north-east extension to north-north-east extension, with activation of a new orientation of normal faults (Bosworth and Taviani, 1996). Again, this period of structural reorganization was followed by the development of extensive fringing carbonate reefs and broad reef platforms.

Based on these structural and stratigraphic observations and interpretations, we suggest that the main Gulf of Suez–Red Sea shallow marine carbonate complexes formed during the end of active tectonic phases, when fault blocks were undergoing minimum local movements. The upper Nukhul (Aquitanian) and Belayim (Serravallian) carbonate events also occurred at times in the rift history when the size of fault blocks was increasing, due to the abandonment and healing of faults (Plate 8, Figure B4.8; Jarrige et al., 1990; Bosworth, 1995). This provided a more stable environment with respect to sea level, and enabled carbonate facies to develop into broad platforms.

ACKNOWLEDGEMENTS

Mahmoud Atta, Adham Gouda, Emad Kadry and Mahmoud Galal assisted with our field work at Gebel el Zeit. Discussions with Mohammed Darwish, Terry Engelder, Andrew Evans, Chris Hathon, Samir Khalil, Ken McClay, Jim Pendergrass, Bruce Purser, Mahmoud Raslan, Ali Sadek, Marco Taviani, Nazih Tewfik, Barry Wood and Amgad Younis were very helpful. W. M. Abdel Malak and Ali Sadek made identifications of foraminifera and calcareous nannofossils for us. Thoughtful reviews of the manuscript were provided by Dan Bosence, Ian Davison and Bruce Purser. We thank Marathon Oil Company for permission to publish this paper.

Chapter B5
Rift development in the Gulf of Suez and the north-western Red Sea: structural aspects and related sedimentary processes

C. Montenat, P. Ott d'Estevou, J.-J. Jarrige and J.-P. Richert

ABSTRACT

Field studies including extensive geological mapping carried out in the Gulf of Suez–north-western Red Sea provide original data concerning the geometry, kinematics and sedimentation of the rift, as follows:

- Structural trends inherited from the Precambrian–Palaeozoic basement determine four main groups of faults N140–N160 (Clysmic), N00–N20 (Aqaba), N100–N120 (Duwi) and N40–N60 (cross-faults).
- The association of these different trends during rifting gives a 'zigzag' structural pattern resulting in a complex geometry of fault blocks.
- The rift suffered a polyphased structural evolution, including an initial phase of wrenching, followed by different episodes of extensional tectonics. The arrangement of fault blocks is closely related to the initial episode of strike-slip faulting (number and size of blocks, direction of dip, etc.).
- Sedimentation is directly influenced by the structural evolution, enabling the definition of tectonic–sedimentary sequences which express both the structural and sedimentological originalities of rift evolution.

These features are compared with those of other rift systems in order to evaluate the possible broader scale importance of the proposed structural evolution.

INTRODUCTION

This contribution summarizes and modifies slightly the results from field surveys performed by the 'Genebass' research group in the Gulf of Suez and the north-western Red Sea (Figure B5.1b). The work focuses on the structural aspects of Miocene rift development near its margin, insofar as they influence sedimentary processes. The features and events which characterize the tectonic evolution of the rift are reviewed with respect to sedimentation and basin dynamics. Neogene stratigraphy does not utilize the classical lithostratigraphic nomenclature based mainly on subsurface well data, because sometimes it is difficult to correlate the subsurface formations with outcropping series (see discussion in Plaziat et al., this volume and Figure B5.4). Therefore a simple, more informal terminology, based on tectono-sedimentary units was introduced by the 'Genebass' research team. It includes four groups referred to as Groups A to D, easily recognizable in the field, separated by regional angular unconformities (Montenat et al., 1986 and Figure B5.3). The correlation between these groups and the classical formations are discussed elsewhere in this volume (Plaziat et al.), and summarized in Figure B5.4. In conclusion, comparison with other rift systems in the world are made in order to point out the more basin aspects of rift evolution.

GENERAL SETTING

The Gulf of Suez and Red Sea rift was initiated between the African and Arabian shields near the end of the Oligocene. It is a part of a complex and diversified Neogene geotectonic system which includes (Figure B5.1a):

1. the Taurus and Zagros orogenic belts to the north and the north-east;

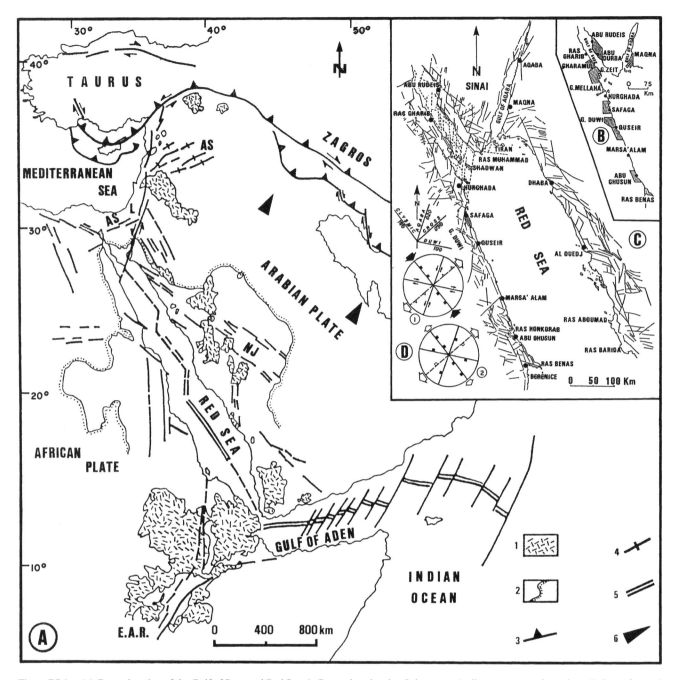

Figure B5.1 (a) General setting of the Gulf of Suez and Red Sea. 1. Cenozoic volcanics; 2. basement/sedimentary cover boundary; 3. thrust front; 4. fold axis; 5. oceanic spreading zone; 6. relative movement of the Arabian shield. AS – Syrian folded arc; E.A.R. – East Africa rifts; NJ – Nadj fault system. (b) Surveyed areas, including geological mapping (hatching), by the GENEBASS group (maps published in Montenat (ed), 1986. (c) Fault pattern of the Gulf of Suez and the north-western Red Sea. (d) Major fault trends and main stages of structuration: (1) initial stage of strike-slip faulting; (2) extensional faulting.

2. the Levant or Aqaba wrench fault and the Syrian arc folds, directly connected with the rift to the north;
3. the oceanic spreading zone of the Gulf of Aden to the south-east;
4. the Ethiopian rift to the south.

Red Sea rifting succeeded the formation of the folded Syrian arc in the late Cretaceous to Eocene (Bartov *et al.*, 1980; Chorowicz and Lyberis, 1987).

The lithostratigraphy of the rift includes three main units (Figure B5.2).

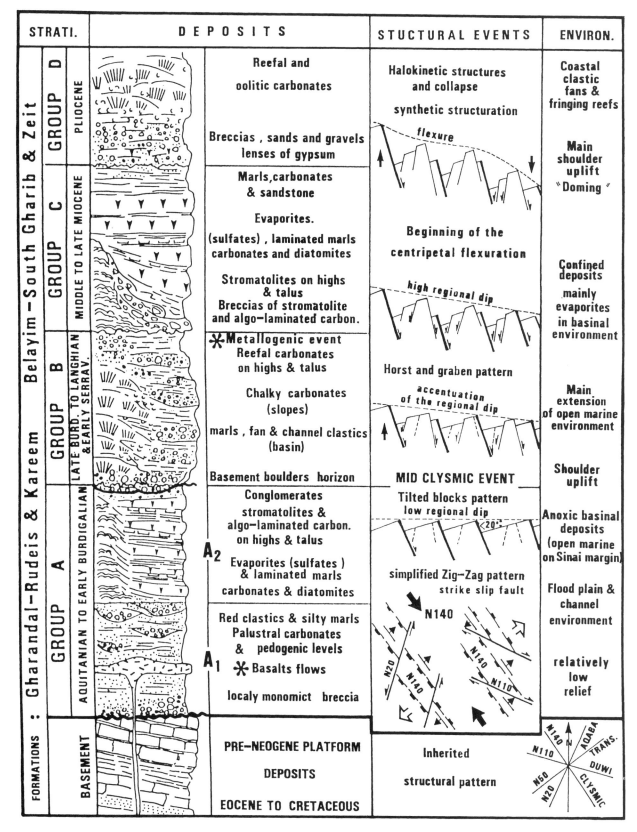

Figure B5.2 Pre-rift sedimentary cover and syn-rift deposits in the Safaga area (after Thiriet et al., 1986).

1. The basement series which outcrops both along the marginal relief (shoulders) of the rift and within a number of large fault blocks along the margins (i.e. Abu Durba, Gharamul, Mellaha, Ras Honkorab, Ras Benas, etc.). They include various crystalline, metamorphic, volcanic and sedimentary rocks related to several Precambrian orogenic cycles and lower to middle Palaeozoic magmatic events (dikes and batholiths; cf. Burollet et al., 1982). The Pan-African tectono-magmatic event (about 500 Ma) played an important part in the structuring of this polyphased basement. The Cenozoic rift evolved partly by reactivation of the various categories of deformation within the basement.

2. The pre-rift sedimentary cover rests unconformably on the basement (Figure B5.2). The sequence begins with a continental terrigenous series, including the Cretaceous Nubian sandstones and older units (lower Palaeozoic to Jurassic in the northern part of the Gulf). These sandstones are overlain by shaley and carbonate marine deposits of late Cretaceous to middle Eocene or locally late Eocene ages on the Sinai edge (Said, 1962; Garfunkel and Bartov, 1977). The Cretaceous transgression advanced southward and an open marine sedimentation thus existed earlier in the north, with Cenomanian in the Gharamul area of the Gulf of Suez, or with Campanian at Safaga and Quseir in the northern Red Sea. These peri-continental deposits display notable variations in thickness, facies and synsedimentary deformations related to tectonic movements which occurred during late Cretaceous and Eocene (see example at Gharamul, south of Ras Gharib in Ott d'Estevou et al., 1986a). Tectonic and palaeogeographic features are oriented basically east–west to north-east–south-west and related to Tethyan influence. They do not relate to the later rift configuration and cannot be used to support the existence of a Mesozoic 'proto-rift', as proposed by Sellwood and Netherwood (1984).

The thickness of the pre-rift cover varies from about 2000 m in the north of the Gulf (Sinai margin) to 400–800 m in the south (Safaga and Quseir). To the south of Quseir, the absence of pre-rift sedimentary cover, on both Egyptian and Arabian margins, results from erosional processes which probably occurred during Palaeogene times (see below).

3. The syn-rift and post-rift deposits (late Oligocene to Plio-Quaternary age) display similar characteristics in the Gulf and the Red Sea margins, at least as far south as Sudan (Montenat et al., 1990). They are characterized by spectacular variations in facies and thickness, and are

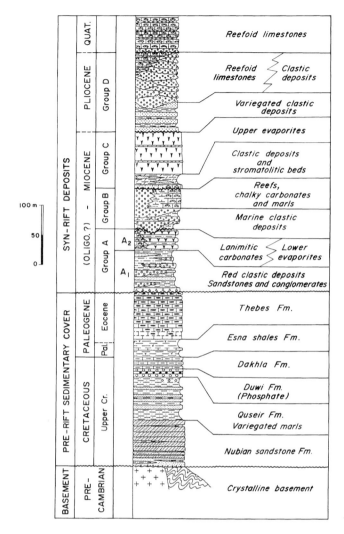

Figure B5.3 Schematic representation of the successive sedimentary and structural events recorded on the Gulf of Suez and north-western Red Sea margins (after Montenat et al., 1986).

affected by syndepositional tectonics. Strata are separated by regional unconformities which occur in all studied areas (Figure B5.1b) and may be correlated from site to site. The discordances delimit four major tectonic–sedimentary units, termed Groups A to D (Montenat et al., 1986a) (Figure B5.3). These outcropping units have offshore, axial equivalents, corresponding to thicker and more uniform deposits (Figure B5.4; see discussion concerning the correlation of these units in Plaziat et al., this volume). Correlations with the series observed on the Saudi Arabian margin have been proposed (Purser and Hotzl, 1988; see also Montenat et al., this volume for group A deposits).

Figure B5.4 Chrono-diagram of tectonic, sedimentary and magmatic events which characterize the Gulf of Suez and north-western Red Sea evolution (hatching: stratigraphic gap on the western margin). For details and discussion concerning stratigraphic correlations, see Plaziat et al., this volume.

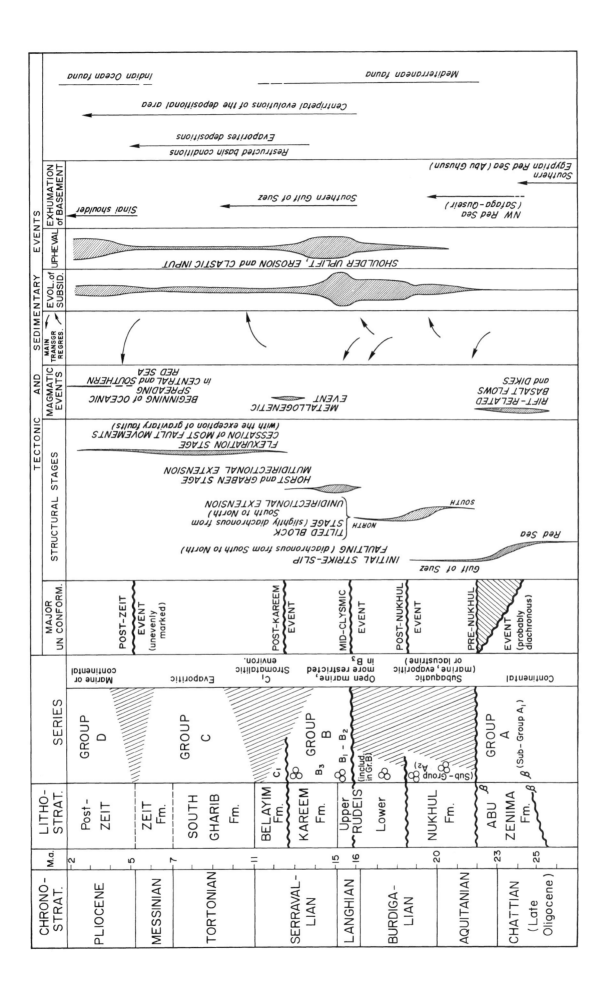

Groups A to C (late Oligocene to late Miocene) are syn-rift deposits, while the situation of Group D sediments (Pliocene) requires more comment. In the Red Sea, the Pliocene deposits are contemporaneous to the beginning of oceanic spreading, so that they can be referred to as post-rift deposits. On the contrary, during the same Pliocene interval, the Gulf of Suez remains in a more or less active rift situation, generating syn-rift deposits.

Group A deposits include two sub-units: (1) a lower Red series (A_1, late Oligocene to Aquitanian), including red terrigenous sediments and rare basaltic dikes and flows (Wadi Nukhul and Abu Zenima on the Sinai edge; south of Quseir; Marsa Shinab on the Sudanese margin); available K-Ar radiometric ages range between 26 and 22 Ma; (2) an upper series (A_2, late Aquitanian to early Burdigalian) including evaporites and related anoxic sediments (Abu Ghusun and Safaga areas) or open marine carbonates and clastics (Nukhul Formation in the Abu Rudeis area on the Sinai edge). Biostratigraphic and sedimentary data concerning the deposits related to the A group are discussed in Plaziat *et al.* and Montenat *et al.* elsewhere in this volume. Group A deposits are weakly unconformable to conformable with the pre-rift sedimentary cover. To the south of Quseir, they lie directly on the basement.

Group B are open marine deposits unconformably overlying Group A strata or the basement. They are late Burdigalian to Langhian (and possibly Serravallian) in age. Biostratigraphic data are reviewed in Plaziat *et al.* (this volume). They display various facies related to distinct but spatially related environments (reef platform and talus, basinal planktonic muds, clastic fans and channels), as a result of the structural morphologies inherited from the substratum.

Group C (Serravallian to late Miocene) unconformably overlies the older groups or the basement. This represents the major evaporitic event which occurred during rift development. The deposits are thin at outcrop (some hundred metres) and consist mainly of sulphates, locally associated with stromatolitic carbonates. The thickness increases significantly toward the central part of the basin where halite was deposited (Richardson and Arthur, 1988) (Figure B5.19).

Group D (Pliocene–Pleistocene) sediments overlie the underlying evaporitic deposits, unconformably or conformably, depending on the structural position. Important continental siliciclastic sedimentation (sands, conglomerates) reflects reactivation and uplift of the peripheral basement. Open marine carbonates including calcarenites, reefs and peri-reefal sediments, were deposited in areas protected from detrital sediments. In many places, Group D sediments were affected by halokinesis.

STRUCTURAL FRAMEWORK

The general north-west–south-east trend of the Gulf of Suez and Red Sea rift, in fact, comprises a complex structural pattern which directly influences depositional processes (transit of clastics, nature and geometry of sedimentary bodies).

The inherited structural pattern

Four major tectonic trends comprise the structural framework of the rift (Figure B5.1c, d) (Jarrige *et al.*, 1986).

1. North-west–south-east (N140°–160°) faults, the so-called 'Clysmic' trend of Hume (1921), are parallel to the axis of the basin. They commonly occur on the margins where they often border large fault blocks.
2. North-north-east–south-south-west (N10°–20°) 'Aqaba' trend, parallel to the Levant or Aqaba or wrench fault, is well developed in the Gulf of Suez and in the southern extension of the Gulf of Aqaba (Safaga area, Thiriet *et al.*, 1986). Along the Red Sea margins this trend is not well developed, except locally (for example, on the Sudanese margin; Montenat *et al.*, 1990).
3. The sub east–west (N90°–120°) or 'Duwi' trend (Jarrige *et al.*, 1986b) comprises large corridors of faults which occur locally along the rift margins (Quseir or Ras Banas area, on the western margin; Nadj fault system on the Arabian shield; Delfour, 1979) (Figure B5.1b).
4. The 'cross' trend (N40°–60°), perpendicular to the rift axis, occurs only in the Gulf of Suez when it plays a minor role in the rift structure (with the notable exception of the Gharamul area, near Ras Gharib; Ott d'Estevou *et al.*, 1986a) (Figure B5.8).

The four fault trends are clearly inherited from polyphased basement structures (Jarrige *et al.*, 1986a). The Cenozoic rifting rejuvenated various mechanical weaknesses of the basement indicated by intrusions of dikes or elongated batholiths, metamorphic foliation, brittle fractures, large Precambrian shear zones, etc. (various examples of such inherited structures are given in Montenat, 1986).

The zigzag fault pattern

The different fault trends are linked to compose the rift fault-block pattern. Clysmic, Aqaba and Duwi trends are predominant whereas the cross trend is recessive. In a given area, the Clysmic faults are generally present and associated with another major, Aqaba or Duwi faults, resulting in the formation of a 'zigzag' fault pattern (Garfunkel and Bartov, 1977; Jarrige *et al.*, 1986a) which controls the Cenozoic structure. Such an arrangement is observed at various scales, for example, as a part of the margin (decakilometric scale), as a fault block (kilometric scale) or, sometimes, at outcrop (metric scale) (Figures B5.5 to B5.9).

This characteristic pattern of the rift margin is illustrated by the following examples:

- the Abu Durba–Abu Rudeis area, on the Sinai margin, where Clysmic and Aqaba faults predominate (Ott d'Estevou *et al.*, 1986b); (Figure B5.5);
- the Abu Ghusun–Ras Honkorab area (South Egypt Red Sea margin) where Clysmic and Duwi faults predominate (Montenat, 1986b); (Figure B5.6);
- The Safaga area, at the Gulf of Suez–Red Sea transition, display a complicated mosaic of fault blocks and a large variety of shapes and block size (Thiriet *et al.*, 1986; (Figure B5.7).

Detailed studies demonstrate that large tilt-blocks, initially bounded by major Clysmic faults, are cut by other secondary faults. These later faults define secondary blocks within the first-order Clysmic block, as illustrated by the Gebel Zeit block at the southern end of the Gulf of Suez (Prat *et al.*, 1986; Figure B5.9).

Importance of the inherited structural trends

Considering this complex pattern and the movements recorded by the various fault trends during the Miocene, the structure cannot be the result of a simple extension inducing the creation of neoformed faults. Thus, the Gulf of Suez and Red Sea rift is not a purely neoformed Cenozoic structure but a recombination and reactivation of pre-existing discontinuities during the Tertiary. The size, geometry and complexity of the blocks depend directly on the nature and density of the discontinuities susceptible to rejuvenation. For example, in the Abu

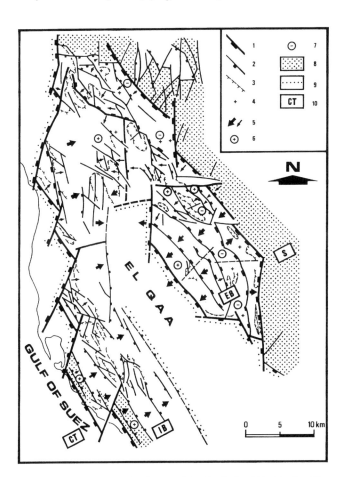

Figure B5.5 Tectonic sketch map of the Sinai edge (Abu Durba–Abu Rudeis area, Gulf of Suez eastern margin), showing a typical 'zigzag' fault pattern formed by prevailing Aqaba and Clysmic faults, which generate a complex mozaic of blocks. 1. and 2. major and minor faults with downthrown side, 3. bedding with dip, which delineates anticline and syncline structures, 4. horizontal dip, 5. tilt of major and minor blocks (large or small arrow), 6. horst, 7. graben, 8. basement, 9. limit of recent depression, 10. morphostructural units: CT, central trough; IB and EB, internal and external benches; S, shoulders (after Jarrige *et al.*, 1990, slightly simplified). For location see Figure B5.1b.

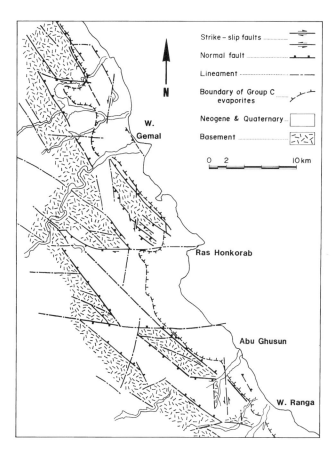

Figure B5.6 Structural sketch map of the Red Sea margin in the Abu Ghusun area (see location on Figure B5.1(b). The fault pattern is formed by predominant Clysmic and Duwi faults which draw a harpoon-shaped limit of the western shoulder and delimit rhombic blocks. Note the dispersal of Aqaba faults which display relatively small segments. The group C evaporites (lower boundary is represented) coat and seal the fault pattern. Slight reactivations of faults are indicated by morphological lineaments (for example Ras Honkorab) (after Montenat *et al.*, 1986b, modified).

104 Rift development in the Gulf of Suez and the north-western Red Sea

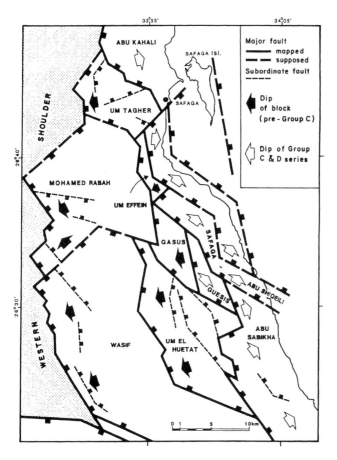

Figure B5.7 Fault-block pattern of the Red Sea margin in the Safaga area (see location on Figure B5.1(b)). Note the difference between the general westward dipping of the external blocks, inherited from the tilted block stage and the basinward dipping of evaporites (Group C) and post-evaporitic series (Group D), related to the flexure stage (after Thiriet *et al.*, 1986).

Ghusun–Ras Honkorab area, the foliated metamorphic basement favoured the creation of numerous small tilt-blocks (Ras Honkorab), while the adjacent intrusive diorite gave rise to large and massive fault blocks (Abu Ghusun; Montenat, 1986b).

The specific character of the fault-block pattern observed in the Gharamul area (Figure B5.8), where the cross-trending faults are markedly developed, is determined by the presence of the Dara granitic batholith (500 Ma) intruded along the cross-direction (Ott d'Estevou *et al.*, 1986a). On the Sudanese margin, north of Port Sudan, the predominance of faults oriented north–south coincides with the presence of large north–south trending granitic plutons in the basement (Montenat *et al.*, 1990).

The comparison with other rift systems (East African Gregory rift, South Atlantic African margin, Rhine graben, etc.) leads to a similar conclusion: the reactivation of numerous inherited structures and, correspondingly, the constitution of a zigzag fault pattern, may be a general rule in rift evolution. Oblique or transverse faults are often of a special importance (Reyre, 1984) as they divide the rift into segments which record quite different tectonic and sedimentary histories (see below).

TECTONIC EVOLUTION

Polyphase rifting

As previously stated, the different Cenozoic sedimentary units referred to as Groups A to D are separated from each other by regional unconformities (Figure B5.3).

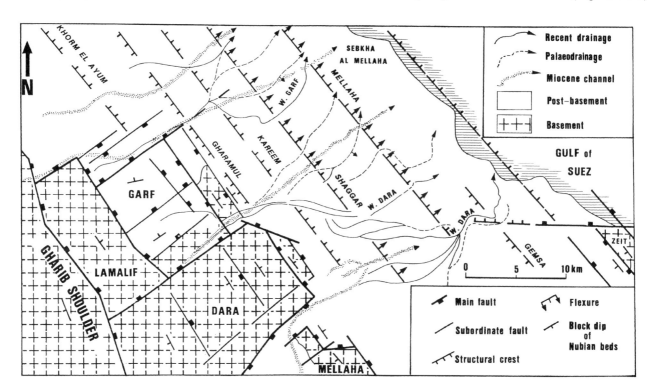

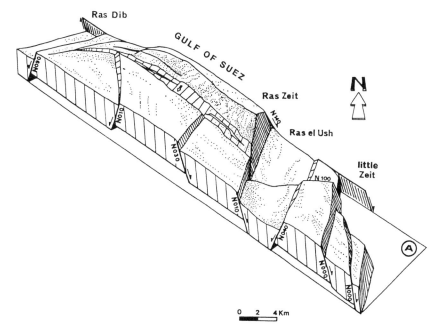

Figure B5.9 Schematic block-diagram of the Gebel Zeit stripped of its Miocene series (see location on Figure B5.1(b)). Firstly, the Zeit was a large tilt-block dipping south-westward, coupled with the Mellaha block to the South (see Figure B5.10). It was split subsequently into a mosaic of minor blocks bounded by Aqaba, Duwi and Clysmic faults, during the horst and graben stage. Note the formation of cross-trending graben (Ras el Ush) usable for the transit of clastics basinward (after Prat *et al.*, 1986).

Many examples of faults successively sealed during the sedimentation of the different units are known from field or subsurface observations. Such features clearly indicate successive stages of structural evolution. Moreover, superimposed deformations have been recorded within the Oligo-Miocene series. Two sets of movements have been observed on some fault planes, or one set of faults cutting another (Figure B5.10). Such occurrences evidenced in the different studied areas, are coherent, and involve two main tectonic phases in the rift evolution (Jarrige *et al.*, 1986b, 1990).

The first phase, contemporaneous with deposition of unit A_1 is typically exposed in the Wadi Sharm el Qibli (south of Quseir) by fracturing a 24.9 Ma basalt flow and surrounding red sediments (top Oligocene). This section shows: (1) sinistral movement of the north–south to N30 trending Aqaba fault; (2) dextral movement of the N105 to N130 trending Duwi fault; (3) normal movement of the north-west–south-east clysmic faults. Some strike-slip fault planes have recorded superimposed normal movements related to the second phase (Jarrige *et al.*, 1986a).

The second phase, from unit A_2 (lower evaporites) and Group B onwards, is characterized by normal faulting without any evidence of strike-slip movement. The extensional tectonics begin with the creation of tilt-blocks, followed by the formation of a horst and graben tectonic pattern during this latter stage of extensional tectonics. Clysmic faults have large downthrows (several hundreds to a thousand metres) and other fault trends also recorded normal movements, giving rise to a complex breaking up of larger fault blocks (Figure B5.9).

Successive stages of palaeostress

The compilation and processing of a great number of microstructural measurements, in various areas of the Gulf of Suez (Abu Rudeis–Abu Durba on the Sinai edge; Gharamul, Gebels Zeit and Mellaha on the western margin; Figure B5.10) and the north-western Red Sea (Safaga, Quseir and Abu Ghusun zones) have enabled us (Ott d'Estevou *et al.*, 1989b) to determine the mean principal palaeostress axes ($\sigma 1$, $\sigma 2$ and $\sigma 3$ are

Figure B5.8 Structural sketch map of the Gharamul-Dara area (Gulf of Suez; location on Figure B5.1(b)). The recessive cross-trending faults (about N45°) are clearly expressed in this area (Dara block) during the rifting and are still exposed. Note the relation between the network of Miocene channels used for the submarine transit of Group B clastics and the drainage of Quaternary and modern wadis. The flexures affect Group C evaporites, directly above buried Clysmic faults (after Ott d'Estevou *et al.*, 1986a). Due to the scale of the sketch map these flexures are represented in a simplified manner as straight lines.

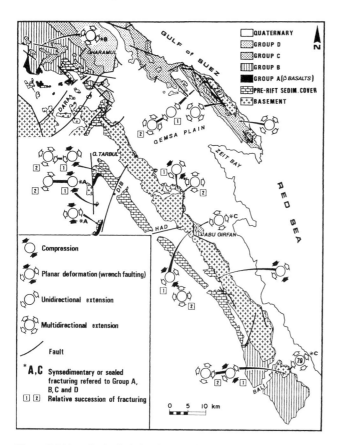

Figure B5.10 Geological sketch map of Gebel Zeit and Mellaha (location on Figure B5.1(b)). The sites of microstructural analysis give examples of successive stages of deformation: initial stage of wrench faulting and subsequent stages of extensional faulting (after Ott d'Estevou et al., 1989).

respectively referred to as maximum, intermediate and minimum compressional stresses).

The first stage of deformation which displays specific strike-slip movement, is related to a stress system characterized by north-west–south-east compression and north-east–south-west extension (respectively $\sigma 1$ and $\sigma 3$ in the horizontal plane). Fold structures (en echelon folds) with steeply dipping axes are induced by wrenching. These are located along or at the tips of Aqaba strike-slip faults (or more seldom associated with Duwi faults). Such features are well exposed in the Abu Rudeis–Abu Durba area (Ott d'Estevou et al., 1989b). These faults are sealed by Group B deposits (Ott d'Estevou et al., 1986b; Jarrige et al., 1990). Thus, the initial stage of rifting is characterized by a compressional strike-slip regime, lacking evidence of transtensional regime. The wrench tectonics which occurred during the lower half (A_1) of Group A are probably diachronous, being older (Oligocene) in the Red Sea and younger (Aquitanian–early Burdigalian) in the Gulf of Suez (see discussion in Montenat et al., 1988 and in Montenat et al., elsewhere in this volume). On the other hand, it appears that wrench tectonics were more accentuated to the north (Gulf of Suez) than in the south (Red Sea).

The second stage began during the upper half of Group A (A_2) and is clearly evidenced within Group B deposits. It is characterized by palaeostress tensors with vertical maximum stress. A true extensional regime ($\sigma 1$ vertical, $\sigma 3$ oriented north-east–south-west) controlled the formation of tilt-blocks, predominantly bounded by north-west–south-east Clysmic faults (A_2 and beginning of Group B; late Aquitanian and Burdigalian, for example, Abu Ghusun, Safaga; see discussion of structural data in Ott d'Estevou et al., 1989b).

The axes of minimum stress ($\sigma 3$), related to the direction of extension, vary from N 20 to N 85, with a maximum between N 50–N 60, perpendicular to the rift axis. Such variations are related to the influence of major faults on the regional stress field (deviation of palaeostress tensors); the hypothesis of a varying direction of extension during rift development (Gautier and Angelier, 1986; Chorowicz and Lyberis, 1987; Giannerini et al., 1988) is not supported by chronological data.

A multidirectional extension has progressively replaced the previous tectonic regime from the upper part of Group B (Langhian–Serravallian) and still prevails within Group D Plio-Pleistocene series. Nevertheless, from the Group C interval onwards, faulting is drastically reduced and replaced by a general flexure of the margin toward the basin axis (for example, the Safaga area, Thiriet et al., 1986, and Figure B5.19). The flexure is more marked in the north-western Red Sea than in the Gulf of Suez.

Comparisons with other rift developments

The polyphased rifting of the Gulf of Suez and Red Sea basin, with gradual change from compressional to extensional tectonics, may be compared with other rift evolutions. For example, the East African rift, located close to the Red Sea and initiated approximately at the same time, recorded early Oligocene–early Miocene strike-slip tectonics: north-east–south-west compression ($\sigma 1$ horizontal) associated with north-west–south-east extension ($\sigma 3$) (Chorowicz et al., 1979, 1987). Subsequently, a true extensional regime developed ($\sigma 1$ vertical and north-west–south-east direction of extension); later kinematic evolution of both rifts reveals relevant analogies (Jarrige et al., 1990).

The late Eocene–Oligocene North European rift recorded a similar evolution. At the beginning of the late Eocene, the Rhine graben was initiated in a compressional strike-slip regime (submeridian horizontal $\sigma 1$) which induced a slight east–west extension ($\sigma 3$ horizontal). This extension became progressively stronger, resulting in the graben opening at the beginning of the Oligocene (Villemin et al., 1984).

The Ales basin (France), a southern segment of the West European rift, provides a precise chain of events which characterize the rift evolution, detailed by Fredet (1987). The region suffered a strong compressional deformation, with predominant folding (sub north–south direction of compression) by the end of Lutetian to early Bartonian times (Pyrenean–Provencal phase). This evolved into a compressional strike-slip tectonic (horizontal north-north-east–south-south-west compression and induced west-north-west–east-south-east extension) during the middle part of the Bartonian. By the end of the late Eocene, a west-north-west–east-south-east to north-west–south-east trending extension predominated, giving birth to the Ales graben, which was filled with Oligocene deposits.

The above examples are related to a common tectonic scenario, although the regional context may differ as follows.

- the early compressional stage changes to compressional strike-slip deformation associated with a slight perpendicular extension ($\sigma 1$ and $\sigma 3$ horizontal);
- the direction of extension, previously minor, becomes dominant in the extensional regime;
- the final stage may involve multidirectional extension.

The successive tectonic stages occur as a continuum of deformation. This results mainly from a simple permutation of $\sigma 1$ and $\sigma 2$ stress axes, while $\sigma 3$ keeps a constant orientation, as discussed in Ott d'Estevou et al. (1989b) for the Red Sea data.

STRUCTURAL STAGES AND SEDIMENTARY PROCESSES

The successive structural styles closely influence sedimentation in this area. Both tectonics and sedimentation are associated to define a tectonic–sedimentary sequence, representative of rift evolution during the Cenozoic (Figure B5.11).

Initial stage of rifting

As shown above, the initial stage of rifting is characterized mainly by strike-slip movement along Aqaba (sinistral) and Duwi (dextral) faults. The conjunction of these two major systems, which determine a pseudo-conjugate system *sensu* Angelier (1979), results in the partition of large rhomb-shaped blocks (Figure B5.12, B5.13) which correspond to the maximal extent of the area subjected to rifting (Ott d'Estevou et al., 1989b; see also Richardson and Arthur, 1988). These large blocks were probably slightly tilted with little subsidence.

The basal A_1 Red series was deposited in an alluvial plain environment cut by minor channels which supply proximal fans (see Montenat et al., this volume). Predominance of fine-grained detritus suggests a low relief hinterland. Pedogenic horizons with calcareous nodules, weakly developed palustral episodes, frequent desiccation features and good preservation of feldspars in sands, all indicate a relatively arid climate (Plaziat et al., 1990). These terrigenous deposits covered an extensive low-relief region with shallow, poorly drained depressions.

Sediments deposited in such a flat depositional environment record a range of synsedimentary deformations such as convoluted structures, breccias, sedimentary dikes, clay diapirs, liquefaction, which are regarded as seismites, induced by strong and repeated earthquake shocks (Plaziat et al., this volume). Such seismic occurrences indicate a tectonic activity. Synsedimentary deformations directly related to tectonic movements are occasionally intense, especially when located along Aqaba strike-slip faults (e.g. Gebel Tarbul at the north-western end of the Mellaha block (Prat et al., 1986; Figure B5.14); Gebel Mohamad Rabat, east of Safaga (Thiriet et al., 1985)), giving rise, for example to the formation of depocentres located within synsedimentary synclines.

In summary, the strike-slip deformation which characterizes the initial stage of rifting has important consequences for the rift evolution. However, fault movement did not generate high relief fault blocks and therefore did not exert a major influence on sedimentation.

Magmatic activity including tholeiitic basalt flows, dikes and sills occurred during this early stage. In the Gulf of Suez and the northern Red Sea these volcanics occur as scattered, small-size magmatic bodies; many dikes are clysmically oriented and concentrated on the Sinai margin (Garfunkel and Bartov, 1977; Patton et al., 1994). The setting of the volcanics is probably related to tension gashes generated in the strike-slip regime (north-west–south-east trending gashes, parallel to the horizontal direction of compression; Ott d'Estevou et al., 1986b). In this case, the small volume of volcanics contrasts with large and repeated outflows occurring in parts of the southern Red Sea (for instance, Yemen volcanics groups; Davison et al., 1994).

Tilt-blocks and sediments

Within the upper part of Group A (A_2, late Aquitanian–early Burdigalian), antithetic normal movements on Clysmic-trending faults became dominant, while strike-slip movements stopped, resulting in the creation of a system of large antithetically tilted blocks. The antithetic tilt-block pattern is widely present at the beginning of

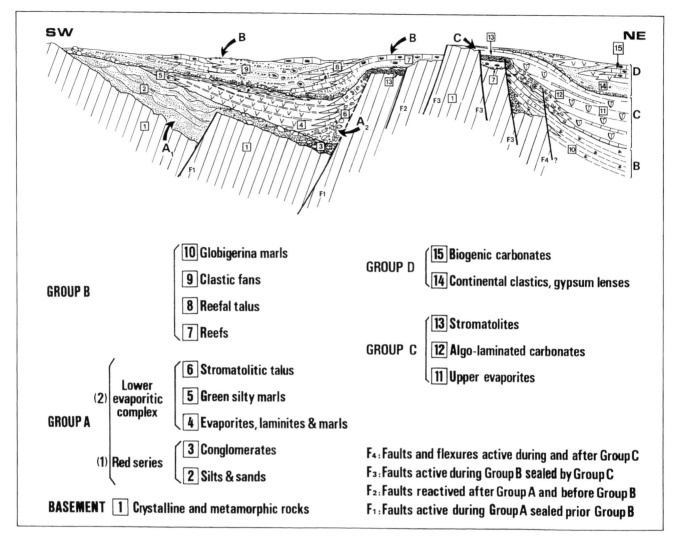

Figure B5.11 Example of relations between the different sedimentary units (Group A to D). Synthetic section from outcropping series of the Abu Ghusun area (location on Figure B5.1(b)). Note the unconformities and successive sealing of faults. The limits of the basement is a schematic illustration of the two kinds of extensional structures: antithetic tilt-blocks in the moderately subsiding external (SW) part, and synthetic faulting toward the axis of the trough, inducing a higher rate of subsidence (after Montenat *et al.*, 1986b).

the fault-block evolution. It occurred both in the Gulf of Suez (e.g. Gebels Zeit and Mellaha or Abu Rudeis) and Red Sea areas (Safaga; Ras Honkorab near Abu Ghusun for instance; various example are described in Montenat, 1986) (Figures B5.15, B5.19). The blocks are often limited by Clysmic-trending faults and locally complicated by the interference of other trends of faults, resulting in a zigzag pattern (see above) (Figure B5.15). Different groups of tilt-blocks with opposite dips are separated from one another by Aqaba or Duwi faults which previously acted as strike-slip faults, at the boundaries of the aforesaid large rhomb-shaped blocks (Jarrige *et al.*, 1986a) (Figures B5.12, B5.16). These are transfer faults, accommodation zones or twist zones of previous authors (Bosworth, 1985) or doglegs of Harding (1984). It is noted that such tectonic boundaries between groups of tilt-blocks are obviously inherited from the initial stage of wrenching (Ott d'Estevou *et al.*, 1987).

This stage of structuring corresponds with a notable change in sedimentation. The alluvial red deposits were succeeded by sediments related to confined permanent water conditions: anoxic lacustrine laminated carbonate or marine evaporites (lower evaporitic episode) associated with fetid carbonates and diatomites (Ras Honkorab; Safaga). Moreover, these subaquatic anoxic deposits accumulated in marked structural depressions. For example, in the Ras Honkorab area, North of Abu Ghusun, the evaporites accumulated in half-graben while the crests of tilt-blocks and fault-scarps are draped

Structural stages and sedimentary processes   109

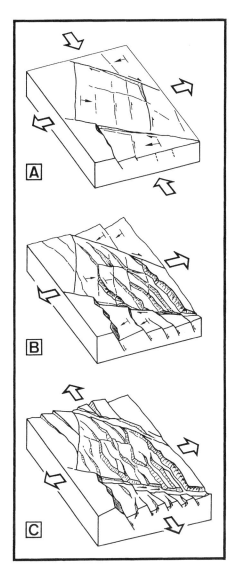

Figure B5.12 Block-diagrams representing the polyphase tectonic evolution of the rift, inspired from examples of the Gulf of Suez (after Ott d'Estevou *et al.*, 1989). (a) Wrench-faulting stage, resulting in the partition of large rhombic panels slightly subsident (combination of Duwi dextral and Aqaba senestral faults; Clysmic faults are little apparent) (after Ott d'Estevou *et al.*, 1989). (b) Tilt-block stage. The panels are cut by antithetic Clysmic normal faults. Note the tilting inversion of the blocks on both sides of Duwi faults. (c) Horst and graben stage. The pre-existing tilt-blocks split into a mosaic of smaller blocks. The different fault trends act with normal throw; synthetic Clysmic faults generate strongly subsident areas (compare with Figure B5.17).

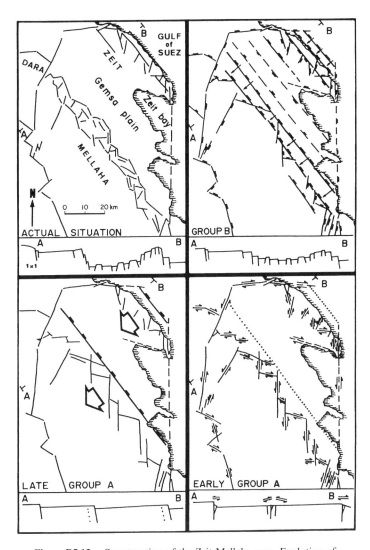

Figure B5.13 Structuration of the Zeit-Mellaha area. Evolution of the fault pattern and related cross-sections (the profiles A–B refer to the basement). The deep structure of the Gemsa plain is interpreted from subsurface data. Note the typical zigzag fault pattern of the present day Mellaha crest of block. Compare with Figure B5.12 theoretical model (after Prat *et al.*, 1986).

by dolomitized algal mats or stromatolitic domes, brecciated on palaeoslopes (Figure B5.15). The planktonic microfauna (*G. primordius* zone) from marls located in the upper part of the evaporites indicates that the tilt-block pattern developed as early as the Aquitanian (Montenat *et al.*, 1986b; see also Montenat *et al.*, elsewhere in this volume) and was still active during the early Burdigalian (Figure B5.4).

Age and amplitude of rotation of tilt-blocks

Generally, the rotation of the large tilt-blocks does not exceed 20°–25°, a similar indication being deduced from seismic profiles (Figure B5.19) (Thiriet *et al.*, 1986). Most of the rotation of large blocks was completed before deposition of Group B sediments. For example, the large Gebel Mellaha block (southern end of the Gulf) is tilted south-westward at about 10°. The marine erosional surface, which truncated the crest of the block just before the deposition of Group B sediments, has retained a subhorizontal position (<5°) up to the present (Purser *et al.*, this volume). Such a disposition is commonly observed in other parts of the Gulf of Suez and the northern Red Sea (Gharamul, Safaga, Quseir,

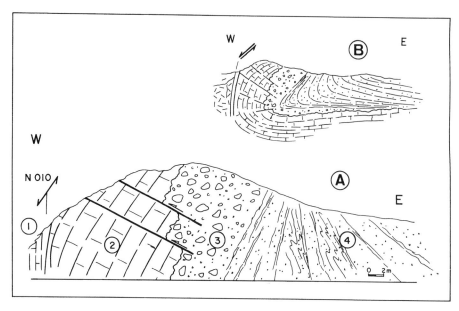

Figure B5.14 Example of synsedimentary deformation of the clastics (A_1 sub-group) related to contemporaneous wrench faulting along N010 sinistral fault at Gebel Tarbul (see location on Figure B5.10). (a) Outcropping structure (after photograph). 1. fault zone, 2. Eocene overturned limestones (Thebes Fm.), 3. basal breccia of the red clastic formation (Eocene silex and blocks of limestone) (overturned), 4. red clastics composed of reworked Nubian sandstones. The deposits show gradual synsedimentary discordance (from overturned to normal dip of beds) and indications of slumps to syndepositional deformation. (b) Synthetic view of the structure.

Abu Ghusun; various illustrations in Purser *et al.*, this volume). For that reason it is not possible to agree with the hypothesis of a gradual and continuous tilting (i.e. rotation) of blocks during the rift evolution, as proposed by Bosworth *et al.*, this volume. In some places, dips of basement blocks exceeding 20° to 25° occur locally for a short distance and are related to bending along fault planes (Gebel Zeit, for instance).

The principal rotation of the tilt-blocks therefore occurred for a short time (Late Aquitanian–Burdigalian; Figures B5.3, B5.4) and was of a moderate amplitude. The continuation of the rift structures results mainly from other tectonic processes (horst and graben pattern; see below). The completion of the tilt-block pattern, when the Group B sedimentary interval began, has several consequences.

1. The Group B series rest unconformably on the underlying rocks including basement, pre-rift sedimentary cover and Group A deposits. This unconformity and the associated erosional surface are widely repre-

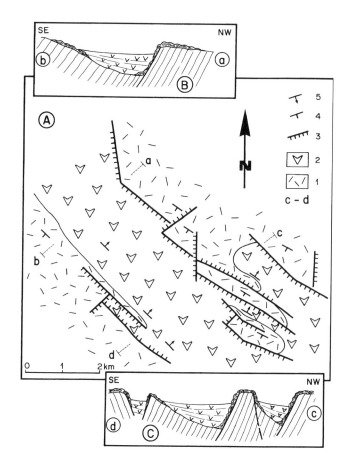

Figure B5.15 Syn-depositional tilt-block structures (evaporites, A_2 unit, Ras Honkorab area). (a) Schematic map of block-faulting which occurred during the deposition of evaporitic A_2 unit (drawn after geological map in Montenat *et al.*, 1986b). 1. basement, 2. evaporites, 3. fault (faulting contemporaneous with the deposition of A_2 unit), 4. foliation of metamorphic basement, 5. dip of evaporites beds, c–d location of sections. (b) and (c) Sections, see location on map.

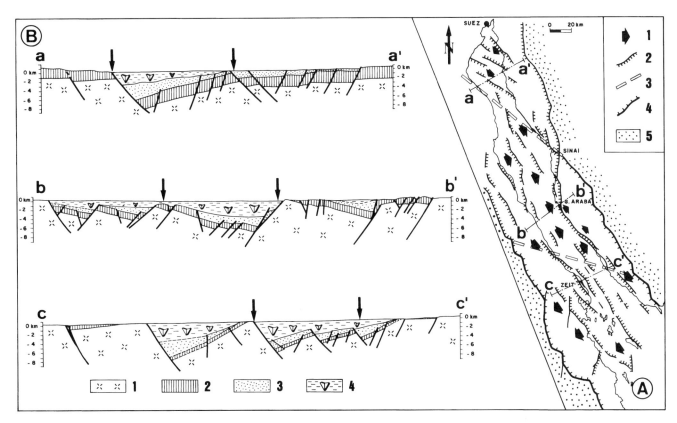

Figure B5.16 Successive group of tilt-blocks with opposing dips. The homogeneous tilt provinces are separated by tectonic boundaries, the so-called transfer faults, dog-legs accommodation, hinge or 'twist' zones, which are the effects of faults (Aqaba or Duwi trends) which acted as strike-slip faults during the early stage of the rift-structuration. (a) The tilt provinces in the Gulf of Suez (after Jarrige et al., 1990). 1. dip of block, 2. fault, 3. tectonic boundary of tilt provinces, 4. major fault bounding the rift shoulders, 5. rift shoulder. (b) Simplified cross-sections in the different tilt provinces of the Gulf of Suez (adapted from Patton et al., 1994). 1. basement, 2. pre-rift sedimentary cover, 3. syn-rift deposits (pre-evaporites of Group C), 4. evaporites (Group C) and recent deposits. Arrows: limits of the offshore part of the Gulf. Locations of sections are shown on figure (a).

sented and referred to as the 'mid-Clysmic event' (Garfunkel and Bartov, 1977), identified by subsurface data in the offshore trough as intra-Burdigalian in age (Figure B5.4).

2. Group B sedimentation recorded the various morphostructural and erosional features of the underlying fault blocks (abraded crest of block, steep talus on fault scarp or gently sloping flank of tilt-blocks; channels controlled by Aqaba or Duwi trending faults which separate different groups of fault blocks).

3. The rejuvenation of fault blocks (horsts and graben) and consequent reactivation of erosion during Group B sedimentation, provided a large amount of clastics (including material of the basement) for Group B submarine channel and fans. The clastics supplied from the rift margins may be funnelled toward the axis of the basin via structural corridors controlled by Duwi, Aqaba, or occasionally cross-trending faults, oblique to the tilt-block pattern (discussion in Purser et al., this volume; Figure B5.8).

Horst and graben structural stage and the maximum subsidence

As noted above, continuation of extensional tectonics and concomitant accentuation of the subsidence during Group B sedimentation are not due to persistent rotation of tilt-blocks along curved listric faults as inferred by Moretti and Colletta (1987) (see also Bosworth et al., this volume). At outcrop there is no evidence of such major listric faults affecting the basement. On the other hand, the listric faults that are present seem to be of superficial origin, of relatively minor size, and are considered to be related to gravitational processes (Ott d'Estevou et al., 1987; Jarrige et al., 1990; Figure B5.19).

The subsequent stage of the rift structure during Group B consists in the break up of pre-existing tilt-blocks which are dislocated by predominantly synthetic faults (flank of block), associated with a longitudinal cutting or cleavage of many block crests (Figures B5.9, B5.17). During this stage, all fault trends have large

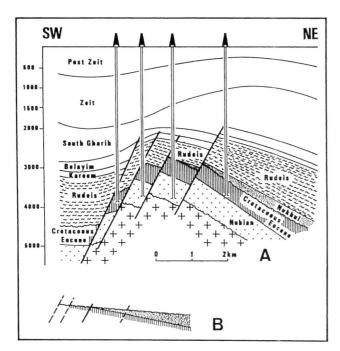

Figure B5.17 Example of cleavage of a crest of tilt-block which occurred during the deposition of the '*Globigerina* marls' (upper Rudeis Formation, i.e. Group B). Note the important downthrow recorded during the horst and graben stage. (a) July field, offshore Gulf of Suez, south-east of Ras Gharib, after Brown (1980). (b) Reconstitution of the tilted block before cleavage (Nukhul deposits).

vertical throw (thousand metres during Group B sedimentation). As a result, the former large tilt-blocks are split into a mozaic of smaller blocks, leading to a complicated horst and graben system (Figures B5.12, B5.13).

Reefs are built on structural highs (horst and crest of tilt-block) and developed large carbonate talus covering fault-scarps (Purser *et al.*, this volume) while pelagic muds (*Globigerina* marls) accumulated in highly subsident graben. At the exit of the previously described transverse tectonic corridors, clastic discharges resulted in detrital fans sandwiched between marls (for instance, Gharamul area, Ott d'Estevou *et al.*, 1986a; Abu Ghusun area, Montenat *et al.*, 1986b). The horst and graben pattern evolved during the Group B interval and shows various synsedimentary effects (Figure B5.18). The formation of the horst and graben pattern does not exclude a slight rotation of some blocks (<5°, for example Gebel Esh Mellaha) but this tilting is of minor importance during this kinematic stage of structuring (Figure B5.3).

The predominance of synthetic faults resulted in the increase of the regional basinward dip, which marks the beginning of the centripetal migration of subsidence during Group B and coincides with the maximum subsidence of the rift (Moretti and Colletta, 1987) (Figure B5.4).

Flexure and centripetal evolution

Most of the faults active during the formation of the horst and graben pattern were sealed during deposition of Group C evaporites (Figures B5.17, B5.19). In many places the evaporitic episode is preceded by the deposition of dolomitized algal mats and stromatolitic carbonates, which coat pre-existing morphologies including fault-scarps and these seal the previously active faults (spectacular examples are seen in Gebel Mellaha; Prat *et al.*, 1986; Purser and Plaziat, this volume; Figure B5.18).

Group B sediments and older rocks, including the basement, are unconformably onlapped by the evaporites. For the main part, these evaporites are not related to playa or sabkha deposits and have generally been deposited under subaquatic basinal conditions. Large tectonic features such as fault-scarps (tens of metres high) or reef talus, were buried under the fine-grained sulphate sediment (Orszag-Sperber *et al.*, this volume).

The block faulting previously discussed was succeeded by a general flexure of the margins towards the axial part of the basin. This flexuring is illustrated by:

1. The emergence of large parts of the rift margins which subsequently remain outside the area of marine syn-rift sedimentation.
2. The thickening of the evaporites basinwards, clearly recorded along clysmically-oriented flexures (Gharamul, Safaga, Abu Ghusun, etc.), and the location of salt deposits in the axial part of the trough (Patton *et al.*, 1994; Figure B5.19).
3. Synsedimentary sliding of the evaporites as evidenced on seismic lines by numerous and classical examples of listric faults and rollover structures (for example Safaga offshore margin, Thiriet *et al.*, 1986; Figure B5.19).

These features indicate migration of subsidence towards the basin centre, which coincides with a gradual reduction of the marine domain, which tends to be limited to the axial part of the rift.

During Plio-Pleistocene Group D, the outstanding event is the beginning of halokinetic movements which strongly deformed the Plio-Pleistocene sediments (Orszag-Sperber *et al.*, this volume). Halokinesis due to the mobilization of Group C evaporites seems to be especially located directly above buried major faults (Mart and Rabinovitz, 1987). The different stages of the rift structures are synthetized on Figure B5.20.

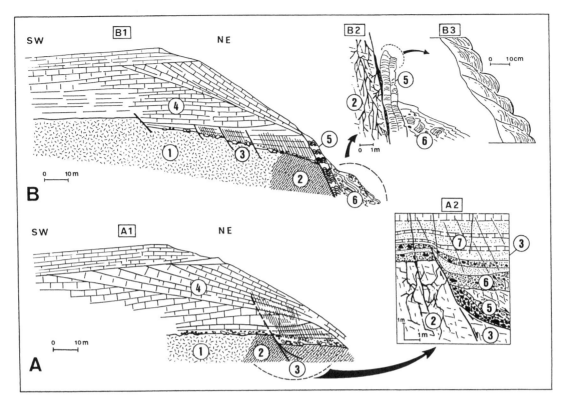

Figure B5.18 Fault-block and sediments. Two field sections of the crest of the Mellaha block (Bir Abu Shaar zone) illustrating synsedimentary faulting and sealing of fault. A_1 – General setting. 1. volcanic basement (crest of block), 2. strongly fractured basement, cut by sedimentary dikes (filled with Group B micritic sediment), 3. fault zone active at the beginning of Group B sedimentation (detail in A_2), 4. peri-reefoid calcarenites (Group B) with internal truncations and talus, induced by faulting (major fault bounding the crest of the block to the north-east). A_2 – detail of the active fault zone, 2 and 3 *idem* A_1, 5. brecciated basement with internal scars of sliding, 6. basal conglomerate, 7. sandy carbonates, gradually sealing the fault and affected by jointing. B_1 – General setting. 1, 2, 3 and 4 *idem* A_1. Note the successive synthetic faults which are satellites of the major fault bounding the block to the north-east and active during the same time (illustration of the cleavage of a crest of block). B_2 – 2. *idem* B_1, 5. stromatolitic coating plastered on faultscarp and sealing previous fault (beginning of Group C), 6. collapsed stromatolitic material and evaporites (Group C). A_3 – detail of the stromatolitic coating. Note the absence of tilting of Group B deposits, although they were affected by subsequent faulting (horst and graben stage).

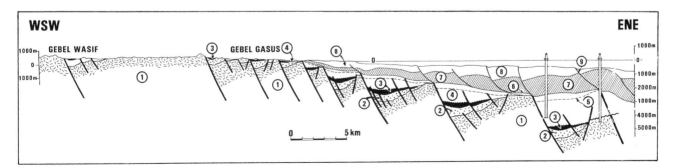

Figure B5.19 Simplified cross-section of the western Red Sea margin near Safaga including onshore and offshore data. The section shows the dislocation of large tilt-blocks into second order smaller blocks and the general flexure of the margin which result in a substantial thickening of the Group C evaporites basinwards. The evaporites suffered gravitary deformations resulting in the formation of listric faults and associated rollover structures.

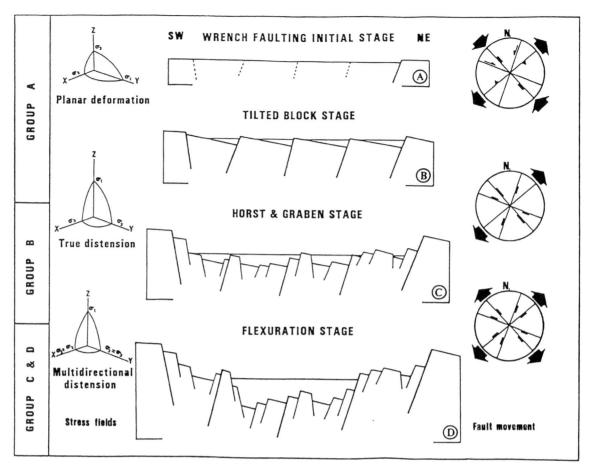

Figure B5.20 Schematic illustration of the different stages of structuration of the Gulf of Suez. Northern Red Sea rift. Note the centripetal evolution of the depositional area (after Montenat et al., 1988).

INITIAL DOMING VERSUS UPLIFT OF THE RIFT SHOULDERS

It has often been considered that an initial stage of thermal doming preceded the rifting and prefigured its subsequent geometry. According to that point of view, the rift resulted from collapse of the domed and thinned crust. However, more recent work indicates that there is no evidence of pre-rift uplift (doming) and related erosional phase. Heybroek (1965), Metwalli et al. (1978), previously recognized that the pre-rift sedimentary cover is preserved in the blocks lying in the central trough of the Gulf (Figure B5.16). These Cretaceous–Paleogene rocks were not eroded prior to the rift initiation and, in most cases, the contact with overlying Group A deposits is conformable. The stratigraphic gap between the pre-rift and the syn-rift stages is not important (see Montenat et al., elsewhere in this volume): the pre-rift sedimentary cover includes late Eocene deposits on the Sinai margin (Garfunkel and Bartov, 1977; Ott d'Estevou et al., 1989a) and a Rupelian age is assigned for the first syn-rift marine deposits in the Midyan Peninsula (Purser and Holtz, 1988).

The first sediments of the rift sequence (A_1 Red series) indicate a relatively low-lying environment (flood plain-type) lacking notable relief. The evidence presented indicates that the uplift of the rift shoulders occurred subsequently. It started at the beginning of Group B and is recorded by a sudden and massive influx of clastics, including in many places, a horizon with large basement boulders at the base of the unconformable Group B sediments. This horizon has been observed also on the Sudanese margin (Montenat et al., 1990). On the Sinai margin the basement is deeply buried under a thick pre-rift cover which was exhumed later (Sellwood and Netherwood, 1984; Figure B5.4). According to Kohn and Eyal (1981), at least 3000 m of uplift has occurred since 9 Ma on the Sinai shoulder. Generally the uplift was reactivated by the end of Group C onwards. The vertical difference between the axis of the rift of Suez and its shoulders is between 6000 and 7000 m and very steep slopes with youthful morphologies indicate recent uplift (for example Gebel Gharib, 1750 m, in the Gulf; Figure B5.16).

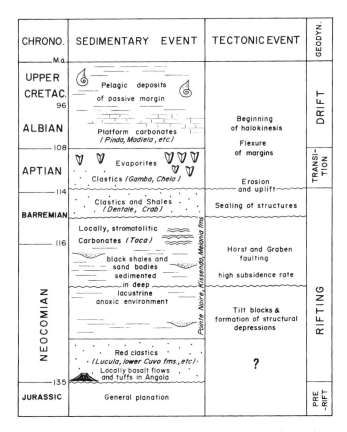

Figure B5.21 Compilation of the major sedimentary and tectonic events which characterize the evolution of the South Atlantic Cretaceous rift (African margin) for comparison with the Gulf of Suez tectonic–sedimentary sequence. Several important and classical sedimentary units are mentioned in italic.

SHORT COMPARISONS WITH OTHER TECTONO-SEDIMENTARY RIFT SEQUENCES

It is useful to compare the above-mentioned tectono-sedimentary sequence with that recorded by other rift systems. An instructive example is given by the West Africa Atlantic margin (from Gabon to Namibia) which was part of the South Atlantic rift during the early Cretaceous (Schlumberger, 1983).

This domain is well documented from subsurface data provided by intensive oil prospecting (see reviews in Reyre, 1984; Teisserenc and Villemin, 1990), although little is known concerning the detailed structural history, due to the scarcity of field data.

The tectono-sedimentary evolution of the South Atlantic rift is synthesized in Figure B5.21. The most significant elements of comparison are as follows.

1. The South Atlantic rift is founded on a fault pattern clearly inherited from Precambrian structures: fault blocks are mainly bounded by north-north-west–south-south-east and north-north-east–south-south-west trending faults which are combined with transverse north-east–south-west to sub east–west trending faults, resulting in the formation of a zigzag fault pattern (Reyre, 1984). The latter transverse faults divide the margin into segments each of which comprise different associations and geometries of fault blocks. Some transverse fault zones play an important role in the subsequent stages of the margin evolution (late Cretaceous and Tertiary) as a preferential exit for clastic transit ('faisceau N'Komi' on the Gabon margin for instance; Reyre, 1984).

2. The rift was initiated in a zone affected by peneplanation and underwent minor subsidence since late Palaeozoic time; thus it was not preceded by a general doming. Nevertheless, a giant thermal doming related to a mantle plume was centred on the southern end of the future rift (i.e. the Walvis ridge, located initially close to the Namibian and southern Brazil edges) and emitted important basalt flows into the peripheral areas (Baumgartner, 1974). A similar thermo-magmatic dome possibly existed close to the south-eastern end of the Red Sea rift (Chazot et al., this volume) where the volcanic emissions were especially important during the early rifting stage (Yemen volcanic group, for example; Davison et al., 1994).

3. The sedimentary sequence in the South Atlantic rift begins with early Cretaceous red terrigenous sediments, including palustral episodes. They reflect poorly drained alluvial plains, with little hinterland relief, under relatively arid conditions as suggested by the presence of arkosic sands. Such a depositional environment is similar to that prevailing at the beginning of Gulf of Suez and Red Sea rifting (Abu Zenima or Abu Ghusun fms, or sub-group A_1). Scarce basaltic outflows and tuffs (Brognon and Verrier, 1966) complete the similarity with the first sedimentary episode of the northern Red Sea. There are no available data concerning the tectonic regime which occurred during this time in the South Atlantic zone.

4. The subsequent deposition of subaquatic sediments (early Neocomian–Barremian) is related to the creation of structural depressions, formed by north-north-west–south-south-east to north–south trending tilt-blocks where anoxic dark lacustrine shales, silts and channelized sands were deposited. These black shales are an important potential source rock. Dolomitized lacustrine carbonates including coquina banks, algal mats and stromatolitic platform deposits (e.g. Toca formation, for example; Schlumberger, 1983) are a lateral equivalent to the black shales and are limited to palaeo-highs formed by crests of blocks. The depositional environment is compared by the authors to present-day deep anoxic lakes of the East Africa Rift (Reyre, 1984). Obvious similarities exist between these deposits and various anoxic sediments such as laminated carbonates, evapo-

rites and diatomitic marls or microbial and stromatolitic carbonates, observed at Safaga and Ras Honkorab (sub-group A_2).

5. According to Reyre (1984), the structural pattern is composed of two categories of fault blocks:
   a) antithetic tilt-blocks are located in the inland (peripheral) zone, where a moderate subsidence is recorded and is related to a low rate of crustal stretching; these blocks are mainly tilted eastward (see also Brice et al., 1982);
   b) synthetic tilt-blocks prevail in basinward locations, in a highly subsident zone where up to several thousand metres of anoxic lacustrine sediments were deposited. This subsident area corresponds to a very thinned and stretched crust (i.e. zone of 'surextension' of Reyre, 1984), which develops toward the rift axis.

Two kinds of extensional structures are also observed in the north-western Red Sea where a synthetic block faulting is superimposed to the antithetic tilt-block pattern and is responsible for the creation of a steep regional dip and the formation of a strongly subsiding trough (see above). Unlike the north-western Red Sea, it is not clearly established whether or not the two kinds of Atlantic structures are contemporaneous.

Teisserenc and Villemin (1990) support a polyphase rift structure in the South Gabon sub-basin. They describe an initial episode of moderate extensional tectonics, followed by more accentuated tectonic activity: 'Deposition of the Melania Formation (upper part of Neocomian–Barremian black shales and sandstones) was contemporaneous with graben and horst block faulting. Abrupt thickness changes are found. Sedimentation can be different from one tectonic block to the next'. In addition, the authors noted that the episode of horst and graben faulting corresponds to the maximum subsidence of the rift sequence (Figure B5.21). A polyphase rift structure is also supported by Brice et al. (1982).

By mid-Barremian times, tectonic activity gradually ceased and the last syn-rift deposits (e.g. upper Melania beds and terrigenous Crabe and Dentale formations) are unconformably draped on the underlying structures. At the same time, the westward flexure of the margin started. Rift structures are sealed and covered by clastic deposits (e.g. Gamba or Chela formations), associated with a regional erosional surface and are overlain by Aptian evaporites which thicken westwards.

The evaporitic episode is regarded either as a transitional stage or as the early phase of drifting (Teisserenc and Villemin, 1990; Brice et al., 1982). The same question is also under discussion for the Red Sea (see above the limit between syn-rift and post-rift deposits). As early as Albian times, salt tectonics began to influence sedimentation on the Atlantic margin. This early stage of halokinesis may be compared with the salt structures which affect the Plio-Pleistocene series (Group D) of the Red Sea. The comparison ends at this stage as, the north-western Red Sea today is in a juvenile stage of passive margin development.

CONCLUSION

Detailed field studies and additional subsurface data provide accurate information concerning: 1. the geometries and kinematics of the rift, on one hand; 2. the succession of sedimentary episodes contemporaneous with the structuring, on the other hand. From the integration of these data, it is possible to define a tectono-sedimentary sequence which characterizes the evolution of the Gulf of Suez and north-western Red Sea rift.

The brief comparison, with other rift systems, either within an intracontinental context or evolving towards an oceanic passive margin, is instructive. It appears that a number of similar features occur such as inherited structures, formation of the zigzag fault pattern, initial strike-slip structural stage, polyphase extensional structures, analogy of sedimentary sequences suggested by the comparison with the Cretaceous South Atlantic rift, etc. These features may be regarded as typical attributes of rift dynamics and useful guides for economic prospecting. However, such studies of rifts require more in-depth investigations. These comparative data may also be useful for the interpretation of ancient rifts influenced by subsequent orogenic cycles; the early Jurassic Tethyan rift, for instance, lends itself to instructive comparison and analogies with the north-western Red Sea rift (Dumont and Grand, 1987).

ACKNOWLEDGEMENTS

The authors wish to thank Dan Bosence for improving the manuscript. They also acknowledge financial support from the Centre National de Recherche Scientifique, ELF – Aquitaine and TOTAL – Compagnie Française de Pétroles (Group GENEBAS).

SECTION C
Early rift sedimentation and tectonics

These three chapters focus on the earliest rift-related sediments in the Red Sea and Gulf of Aden. Late pre-rift and early syn-rift continental clastic sediments provide information on the geomorphological and structural evolution of the basin during its initiation and early stages.

Critical to this aspect of basin evolution is the nature and timing of the syn-rift unconformity at the base of the rift basin fill which, in this case, is found to be diachronous both within the basin as a whole and locally within the Gulf of Suez.

C1 Pre-rift doming, peneplanation or subsidence in the southern Red Sea? Evidence from the Medj-zir Formation (Tawilah Group) of western Yemen
A.-K. Al-Subbary, G. Nichols, D. W. J. Bosence and M. Al-Kadasi

C2 Sedimentary evolution of the early rift troughs of the central Red Sea margin, Jeddah, Saudi Arabia
M. A. Abou Ouf and A. M. Geith

C3 The sedimentary record of the initial stages of Oligo-Miocene rifting in the Gulf of Suez and the northern Red Sea
C. Montenat, F. Orszag-Sperber, J.-C. Plaziat and B.H. Purser

C1
Pre-rift doming, peneplanation or subsidence in the southern Red Sea? Evidence from the Medj-zir Formation (Tawilah Group) of western Yemen

A.-K. Al-Subbary, G. J. Nichols, D. W. J. Bosence and M. Al-Kadasi

ABSTRACT

The behaviour of the lithosphere prior to rifting and rift-related volcanism in the southern Red Sea area is recorded in the stratigraphy underlying the Yemen Volcanic Group. The Medj-zir Formation is dated as Paleocene to Oligocene in age and is the upper part of the Cretaceous to Paleogene Tawilah Group of Yemen. The Medj-zir Formation is well exposed over a large area of western Yemen, forming a unit which varies in thickness from 30–60 m in the west to 60–75 m in the east. The Yemen Volcanic Group overlies the Medj-zir with a sharp, but conformable boundary. The lowest volcanics are dated as 30–32 Ma. The Medj-zir Formation consists of clastic and subordinate carbonate facies which indicate deposition in fluvial, shallow marine and lacustrine–lagoonal environments. Three members are recognized: the basal Zijan Member comprises fluvial facies and shallow marine sandstones and mudstones; the Kura Member is made up of fluvial channel and overbank facies, the latter including well-developed ferruginous paleosols; the Lahima Member, at the top, is fine-grained, including gastropod-rich limestones deposited in a lacustrine or brackish lagoonal environment. These facies and transitions between them represent deposition in a coastal plain to shallow marine shelf setting which was subjected to minor fluctuations in sea level. It is noteworthy that there are no angular unconformities or erosional hiatuses within the Medj-zir Formation anywhere in western Yemen. Depositional hiatuses of unknown duration are represented by well-developed paleosols indicating long periods of tectonic stability. The contact with the overlying volcanics is conformable in all examined sections and there are no volcanic horizons or clasts in the Medj-zir Formation.

The evidence from the sedimentology and stratigraphic relationships in the Medj-zir Formation shows that the Paleocene to Oligocene period prior to volcanism and rifting in the southern Red Sea the region was tectonically stable. During the Paleogene western Yemen was part of a very extensive alluvial to coastal plain extending out from a source area in central Africa to the Tethyan margins of North Africa, southern Arabia and the Horn of Africa. The topography of this alluvial plain was not affected by a prevolcanic phase of doming as has been predicted in current models for plume-related rifting.

INTRODUCTION

There are two main constraints to establishing the evolution of sedimentary environments in response to rifting in the southern Red Sea. Firstly, the outcrops in western Yemen (Figure C1.1) are dominantly volcanic, which until recently (Baker *et al.*, 1996a) have been poorly dated, and postvolcanic exhumation and uplift have affected the entire margin (Menzies *et al.*, 1992; Davison *et al.*, 1994) so that there are very few surface outcrops of syn-rift sediments. Secondly, the published subsurface data from the southern Red Sea is sparse and of poor quality. Work undertaken in this area emphasizes the poor imaging of subsalt seismic stratigraphies (Mitchell *et al.*, 1992, p. 192) and the paucity of wells penetrating the early syn-rift or pre-rift (Hughes and Beydoun, 1992; Crossley *et al.*, 1992). Thus Crossley *et al.* (1992) conclude (p. 171) that 'understanding of the

Sedimentation and Tectonics of Rift Basins: Red Sea–Gulf of Aden. Edited by B.H. Purser and D.W.J. Bosence. Published in 1998 by Chapman & Hall, London. ISBN 0412 73490 7.

120 Pre-rift doming, peneplanation or subsidence

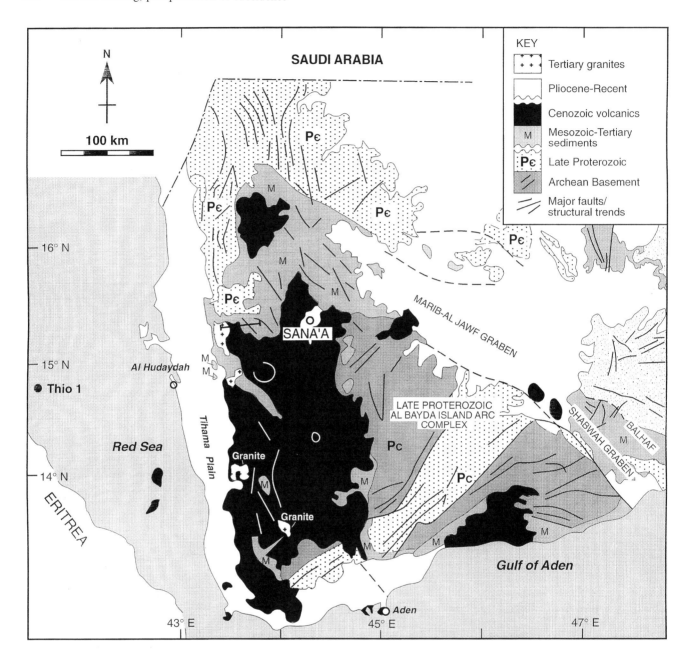

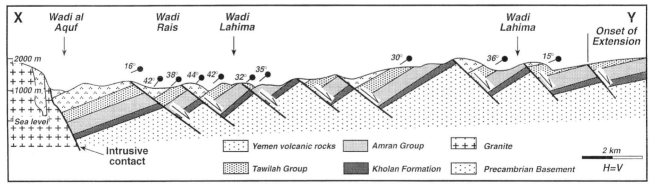

Figure C1.1 Geological map of western Yemen with an east–west cross-section or rift shoulder map and location of Thio 1 well in southern Red Sea (after Davison *et al.*, 1994).

relationships between sedimentation and tectonic history remains weak' and that 'many unknowns remain'. The one well that does penetrate the Oligocene (Thio 1, offshore Eritrea, Figure C1.1) indicates that this area had a marginal marine environment during the late Oligocene (Hughes and Beydoun, 1992) and this has been used to provide a date for late Oligocene rifting in the southern Red Sea (Hughes et al., 1991). However, a rifting unconformity was not penetrated in this well so that while it provides the earliest known marine flooding of the area it cannot be tied to the tectonic development of the basin.

Recent research on the extensive outcrops in western Yemen of the Cretaceous to Tertiary Tawilah Group sediments and the overlying Yemen Volcanic Group have shown that whereas previously the contact was thought by some to have been unconformable (cf. Civetta et al., 1978; El-Nakhal, 1988), recent works (Menzies et al., 1991, 1995; Davison et al., 1994) have shown that this junction is conformable and that the onset of volcanism over western Yemen was broadly synchronous between 30 and 32 Ma. The volcanic sections (c. 2.5 km thick) do not contain any major unconformities but are themselves faulted into domino-style tilt-blocks (Figure C1.1; Davison et al., 1994). The timing of the onset of this rifting must therefore be post-volcanic (< 26 Ma) and may be related to the crustal extension associated with the intrusion of granites along the Great Escarpment between 22.3–21.4 Ma (Zumbo et al., 1995) and c. 25 Ma (Blakey et al., 1994).

If extension started at, or soon after, 26 Ma on the rift shoulder then this correlates well with recent work (Toleikis and Schroeder, 1995; Schroeder et al., this volume) on the benthic foraminifera of limestone outcrops further north on the coastal plain of Sudan (Abu Imama Formation) that show fully marine strata of zones N4–P22 giving an age of about 24–26 Ma (cf. Haq et al., 1988). Similarly the earliest recorded marine strata from the central Red Sea in Thio 1 well is about 25 Ma ('penetration of zone NP25', Hughes et al., 1991, p. 357). Therefore, current evidence suggests rifting started at around 26 Ma and the earliest evidence for marine flooding within the southern Red Sea is at 26–24 Ma.

The best information from western Yemen on these early phases of rifting comes from well-exposed stratigraphic sections of the uppermost Tawilah Group which are conformably overlain by 30–32 Ma basalts and therefore must be within the pre-rift phase of this part of the Red Sea basin. Data from these sections are therefore essential to understanding the basin evolution of this area and are crucial to the question as to whether the basin was affected by a phase of pre-rift updoming (Lowell and Genik, 1972) or whether there was no pre-rift updoming (Bohannon et al., 1989; Coleman and McGuire, 1988; Davison et al., 1994).

This chapter describes in detail the diverse suite of sedimentary rocks which accumulated in western Yemen prior to the 32–26 Ma volcanism of the Yemen Volcanic Group. The data come from extensive outcrops in this area which have been studied as part of a University of Sana'a/Royal Holloway, University of London collaborative project. Eighteen sections have been studied (Al Kadasi, 1994; Al-Subbary, 1995) throughout the Sana'a area (Figures C1.1, C1.2) and this has led to a revision of the lithostratigraphy of these strata and the recognition and erection of three new members within the Medj-zir Formation of the Tawilah Group (Figure C1.3). These comprise:

(top conformable with the Yemen Volcanic Group dated at 30–32 Ma)
Lahima Member – lacustrine mudstones, cherts and freshwater limestones.
Kura' member – fluvial sandstones with well-developed ferricrete paleosols.
Zijan Member – alluvial plain and shallow-marine sandstones.
(base conformable on the Ghiras Formation of the Tawilah Group)

Below we describe and interpret the occurrence, nature and environment of deposition of these strata and use these data to demonstrate that there is a conformable sequence from the late Mesozoic to the Oligocene in western Yemen. The sedimentary environments indicate changes from shallow marine to fluviatile conditions with mature quartz arenites being derived from northern Ethiopia–southern Sudan to the southwest across the area of the present-day Red Sea. These formed an extensive alluvial plain that was subjected to long periods of soil formation in a tropical to subtropical environment which then changed to lacustrine–lagoonal environments prior to the onset of subaerial volcanism.

DEFINITION AND DISTRIBUTION OF UPPERMOST TAWILAH GROUP SEDIMENTS, WESTERN YEMEN

Medj-zir Formation

The Tawilah Group is a unit of sedimentary rocks up to 400 m thick which occurs in western Yemen (Plate 9(a)) and is subdivided into two formations: the Ghiras Formation and the Medj-zir Formation (Al Subbary et al., 1993). Both formations occur in all the sections of the Mesozoic outcrops of Yemen that we have investigated (Figure C1.2) with the exception of the Medj-zir Formation which does not occur to the north in the Jebel Al Jahili section (loc. 18 south of Sana'a; Figure C1.2). The Medj-zir accounts for between a quarter to a third of the thickness of the measured Tawilah sections (Al Subbary, 1995). Formerly the Tawilah had been assigned formation status with the Ghiras and Medj-zir as members (El-Nakhal, 1988). Al Subbary et al. (1993)

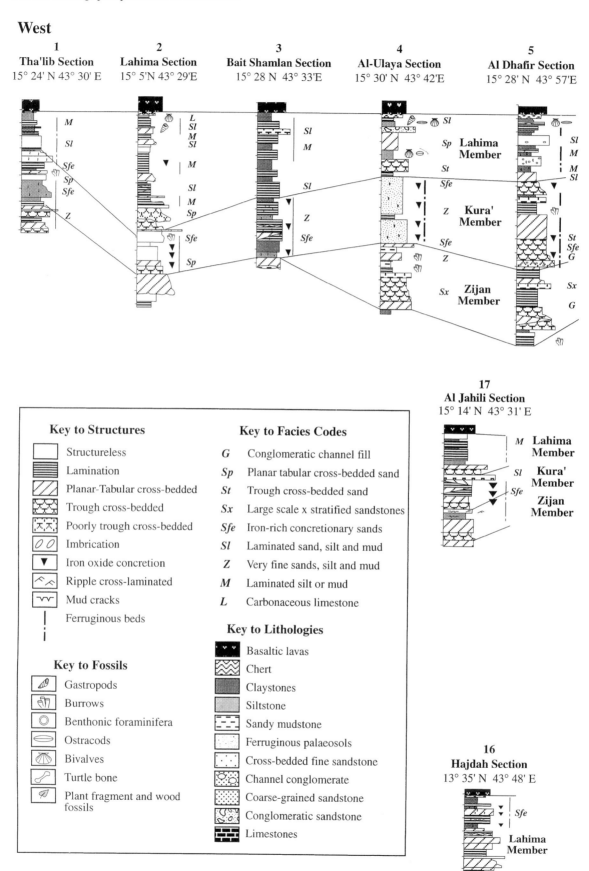

Figure C1.2 Logged sections through the Medj-zir Member of the Tawilah Group in western Yemen. (All logs have a granulometric scale: c – clay, s – silt, f – fine, m – medium, c – coarse, p – pebble and b – boulder.)

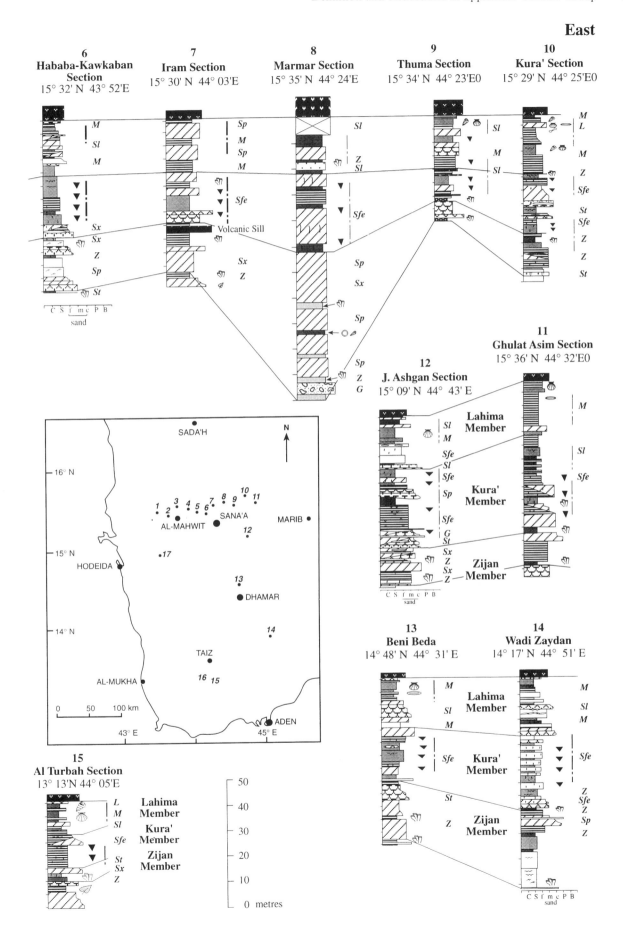

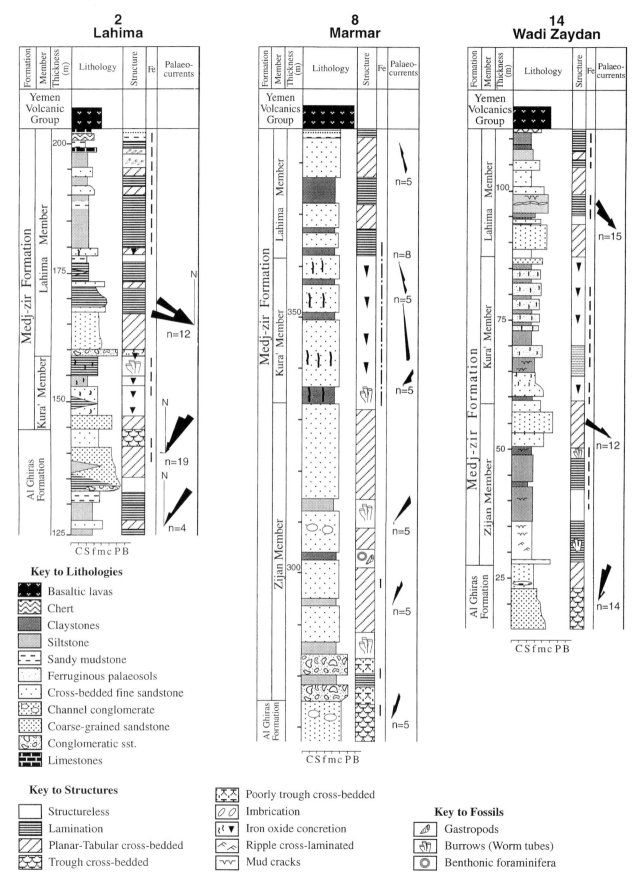

Figure C1.3 Three sedimentological logs from the west-east-southeast cross-section through the Medj-zir Formation illustrating members, lithologies, structures, palaeocurrents, facies and facies associations used in the construction of Figure C1.2.

raised the Tawilah to group status with the Ghiras and Medj-zir as formations on the basis that they are widespread mappable units in western Yemen.

The Medj-zir Formation varies in thickness from zero at the Jebel Al Jahili section (loc. 18) to 115 m in the Jebel Marmar type section (location 8, Figure C1.3). The unit comprises medium- to fine-grained sandstones, sometimes with large-scale cross-bedding. The sands and associated claystones are locally fossiliferous with trace fossils and microfauna (Al Subbary, 1993). Some horizons are conglomeratic and fine-up to sandstones. The upper beds frequently contain ferricrete horizons. The interpreted environments range from shallow marine to fluviatile, to flood plains with iron-rich soils and lacustrine areas. These are detailed below.

The age of the Medj-zir is considered to be early Tertiary; Paleocene to Eocene by Geukens (1960), Paleocene by Geukens (1966), Paleocene by El Nakhal (1988). The recent radiometric dating of the overlying, conformable basal flows of the Yemen Volcanic Group at 30–32 Ma (Baker *et al.*, 1996) indicates that the upper parts of the Medj-zir may be as young as early Oligocene.

The Medj-zir Formation is subdivided into three members: from base to top, the Zijan, Kura and Lahima (Al Subbary, 1995) – and these occur in all of the outcrops we have examined. The essential lithostratigraphic data of these members is summarized below and in Figures C1.2 and C1.3.

Zijan Member
The Zijan Member (Al Subbary, 1995) attains 60 m in thickness at the type locality in Zijan village (15° 29′ 20″ N and 44° 25′ 08″ E) but is more commonly 20–30 m except in the most western (locations 1, 2, and 3, Figure C1.2), southern (location 16, Figure C1.2) and northern outcrops (location 18, south of Sana'a, Figure C1.1) where the member is absent. The member thickens considerably to the south-east. The boundaries are conformable with the underlying Ghiras Formation and the overlying Kura' Member.

The sediments of the Zijan Member are brown, mainly massive, medium to fine-grained sandstones which are interbedded with greenish-grey to brown carbonaceous rich siltstones (Plate 9). Ferricretes are present and are composed of very dark, red to black iron oxides. These iron-rich beds are confined to the fine-grained sediments.

Kura' Member
This member (Al Subbary, 1995) is conformable with the underlying Zijan Member, or the Ghiras Formation where the Zijan is absent, and also conformable with the overlying Lahima Member. The unit is 33 m thick at the type locality at Jabal Kura, to the east of Wadi Al-Sir (15° 29′ 20″ N, 44° 25′ 08″ E) and is commonly 20–30 m thick throughout the study area. The member is present in all the studied sections of the Medj-zir and shows a gradual thickening to the south-east (Figures C1.2, C1.3).

The lower part of the member comprises very fine-grained, variable coloured (often yellowish) sandstones and siltstones with lenticular units of dark grey/pink claystones. The middle layers are characterized by dark-brown very fine sandstones, ferruginous claystone and yellowish brown to mottled pink mudstone horizons with iron concretions. Upper sections are dark-brown to deeply reddish-coloured very fine sandstones and ferruginous claystones with concretionary iron-rich zones up to 1 m thick (Plate 10).

Lahima Member
This member conformably overlies the Kura' Member and is conformably overlain by volcanic tuffs and lavas of the Yemen Volcanic Group (Figure C1.3). The type locality is Wadi Lahima (Al Subbary, 1995), west-south-west of Al-Mahwit (15° 25′ 37″N, 43° 29′ 19″E) where the unit is 45 m thick but it is commonly about 20 m throughout the rest of the outcrop. The south-eastern sections are thicker than the sections in the centre of the east–west transect (Figure C1.2), with the exception of the locally thickened section at the type locality. The member thins to the south to 10–15 m at locations 15 and 16 (Figure C1.2) and is absent at Hajdah (location 16) where the Yemen Volcanics rest conformably on the Kura' Member.

The Lahima Member (Figures C1.3, C1.4) consists of laminated mudstones, siltstones (locally with chert) and very fine sandstones with a variety of structures including current ripples, planar lamination with heavy minerals and some desiccation cracks. The siltstones and claystones contain gastropods, ostracods and bivalves. Iron concretions occur within some of the fine sediments. Locally, the uppermost part of this member are dark-grey limestones with lenticular horizons rich in gastropods. None of these lithologies are laterally continuous with the exception of the lacustrine mudstones in the east and south-east (Figure C1.2).

Jihana Member of Yemen Volcanic Group, western Yemen

The type section of the Jihana Member (Al Kadasi, 1994) is at Jihana (Wadi Ashgan, 44° 34′ 00″E, 15° 09′ 3″N; locality 12, Figure C1.2) where it comprises 20–45 m thick beds of volcanic ashes, lithic tuffs and lithic crystal tuffs. The lower contact of the Jihanah Member varies from sharp to conformable with the Lahimah Member. The upper contact is sharp and conformable with the basal flows of the *c.* 2500 m thick Yemen

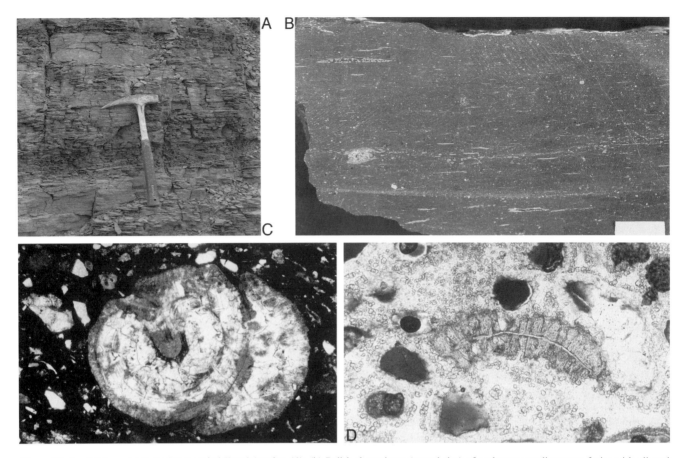

Figure C1.4 (a) Lacustrine mudstone Jabal Kura' (section 10). (b) Polished specimen (actual size) of carbonaceous limestone facies with aligned shells and bone fragments (left), Wadi Lahima (section 2). (c) Thin section of carbonaceous limestone facies with gastropod (centre) and quartz (field of view 2 mm). (d) Thin section of bioclastic grainstone from carbonaceous limestone facies with *Microcodium* and intraclasts of micrite (field of view 2 mm).

Volcanic Group. The main lithotypes found within the Jihana Member (Al Kadasi, 1994) are medium- to fine-grained lithic-crystal and crystal-lithic tuffs containing irregularly shaped mafic boulders, lithic fragments and pyroxene crystals. Chlorite patches (after lithic fragments or pyroxene crystals) and calcite also occur. These lithic crystal tuffs may be mixed with volcanic ashes and irregular-shaped siltstone beds.

The presence of sub-angular to sub-rounded quartz with angular clinopyroxene suggest volcanic eruptions contemporaneous with sedimentation. Similarly, the presence of a fine layering of clinopyroxene crystals interbedded with fine volcanic ashes suggests reworking of volcanic air-fall deposits.

Facies of the Medj-zir Formation

Sedimentological analysis of eighteen outcrop sections from a major east–west traverse together with isolated outcrops to the north and south (Figures C1.1, C1.2) results in the recognition of nine principal facies within the Medj-zir Formation. The facies are described below and the designations used here are adapted from the coding scheme of Miall (1977, 1978). They are subsequently grouped into facies associations and interpreted in terms of depositional environments.

Conglomeratic channel-fill facies (G)

This facies occurs rarely within the Zijan and Kura' Members (Figure C1.2) as thin units associated with the trough cross-bedded sandstone facies and very fine sandstone, siltstone and mudstone facies (see below). It consists of lenses of medium to coarse sandstone and pebbly sandstone, sandy conglomerate and clast-supported conglomerate with subordinate amounts of fine sand and slit (Plate 9(b)). The sandstones are well to very well rounded and poorly sorted. Intraformational clasts, including sandstone and mudstone clasts, pebbles, and large blocks up to 65 cm across are common to abundant. Pebbles are well rounded and the sizes observed range up to 5 cm in diameter. Individual lenses of sand and gravel range from 1 to 5 m thick. Internal structures are characterized by large-scale, low to moderate angle bed sets, parallel and planar stratification, scour and fill structures, and medium- to small-scale, 5 to 25 cm thick trough cross-stratification (Plate 9(b)). Poorly developed trough cross-beds record variable flow directions but a north-eastward vector predominates. Silicified wood is recorded at the top of this facies.

Trough cross-bedded sandstone (S_t)
The facies occurs only locally within the Zijan Member and rarely in the Lahima Member (Figure C1.2). The sandstone beds comprise white to yellowish light-grey, medium- to very coarse-grained, moderately well-sorted to moderately sorted quartz arenite, commonly with quartz granules, small pebbles and mud clasts. The sandstone beds are generally thin, < 1 m, trough cross-bedded in the sets ranging from 10 to 40 cm in thickness. Locally they are also horizontally laminated beds and normally graded. Bases are erosive and cross-cutting while bed tops are not well preserved. The quantity and size of the pebbles generally decrease upwards. Fossils are absent. Trough cross-bedding shows that the direction of flow was to the north-east with subordinate flow to the north-west.

Planar-tabular cross-bedded sandstone (S_p)
This facies is found locally in the Zijan and Kura' Members (Figures C1.2, C1.3) and is particularly well developed at location 8 (Figure C1.3) as 4–8 m thick units. The sandstones are pale, medium- to coarse-grained, moderately sorted quartz arenites. Planar-tabular cross-beds occur mostly in sets of 10 to 25 cm; large-scale sets of 30 to 100 cm also occur (Plate 9(c)). Horizontal bounding surfaces are common. The general trend of palaeocurrents is to the north-west.

Concretionary iron-rich sandstone facies (S_{fe})
Concretionary ironstones developed in sandstone beds dominate the Kura' Member in all areas and are common in the basal parts of the Lahima Member throughout the study area. These show features typical of paleosols developed in sandstones such as very dark red to black ironstone concretions and mottling, plant and root remains. XRD analysis of the concretions indicates that kaolinite is the main clay mineral and the ferruginous material is hematite (Al Subbary, 1995). The concretions generally concentrate towards the tops of beds of sandstone, siltstone and mudstone (Plate 10(c)). Individual iron-rich zones show a well-developed pedogenic horizonation and can be divided into three subzones according to colour and abundance of the iron concentration (Figure C1.5). The lower subzone is usually greenish mottled with whitish and yellowish patches, and in thin section sands are seen to be bioturbated and have iron-rich patches. This subzone grades upwards into a pinkish coloured subzone in the middle with coated lithoclasts with circumgranular cracks. The lithoclasts are multigenerational showing successive reworking of lithified layers. Small areas of haematite cement occur between the grains. The upper subzone comprises concretionary, iron-rich zones up to 1 m thick. Very dark, red to black ferruginous glaebules occur which may amalgamate into a continuous ferricrete pavement. In thin section, opaque iron oxide can be seen to cement intergranular pore space, coating grains and infilling fractures in grains and replacing the rock (Menzies *et al.*, 1991).

Large-scale cross-stratified sandstones (S_X)
The large-scale cross-bedded facies occurs only within the Zijan Member and is commonly associated with the very fine sandstone, siltstone and mudstone facies. The facies is most common in the central region of the outcrop (Figure C1.2). These sands are mainly light brown, medium-grained, large-scale planar cross-bedded sandstone in units which range from 3 to 15 m in thickness (Plate 10(b)). Commonly, the unit is composed of two or three sets, the boundaries of which are nearly parallel and horizontal. This facies has distinctive large-scale tabular cross-strata, typically dipping between 15° and 25°. At one locality (location 5, Figure C1.2) there is a compound set with superimposed (0.1 to 0.5 m thick) cross-beds with foresets dipping at less than 15°. Simple tabular sets consisting of foresets, up to 25 cm thick, that are inclined to the west are common. Mudstone intraclasts are abundant within the beds and are concentrated at the tops of beds. *Thalassinoides* burrows are found on some of the bed bases. Palaeocurrent directions measured at several exposures show a bimodal distribution with means to the north-west and to the north-east.

Siltstone with very fine sandstone and mudstone facies (Z)
This facies always occurs in association with sandstone facies S_x (above) where it forms units up to 20 m thick, and is particularly well developed in the eastern sections of the Zijan Member (Figures C1.2, C1.3). The facies occurs only rarely in the Kura' and Lahima Members. It is characterized by mudstones and siltstones interbedded with fine-grained ferruginous sandstone (Figure C1.5(a)). The beds are 20–60 cm thick with common horizontal lamination and minor ripple cross-lamination (mean palaeocurrent to 020°). Colours vary from yellowish-grey to light-grey to a dark-brown and pink colour for the iron-rich beds. The base of many beds have 3 cm diameter branching and anastomosing burrows. At location 8 (Figure C1.3) at the base of the Zijan Member a thin, dark-grey calcareous mudstone contains a marine microfauna dominated by foraminifera (El-Nakhal, 1988; Al-Subbary, 1995).

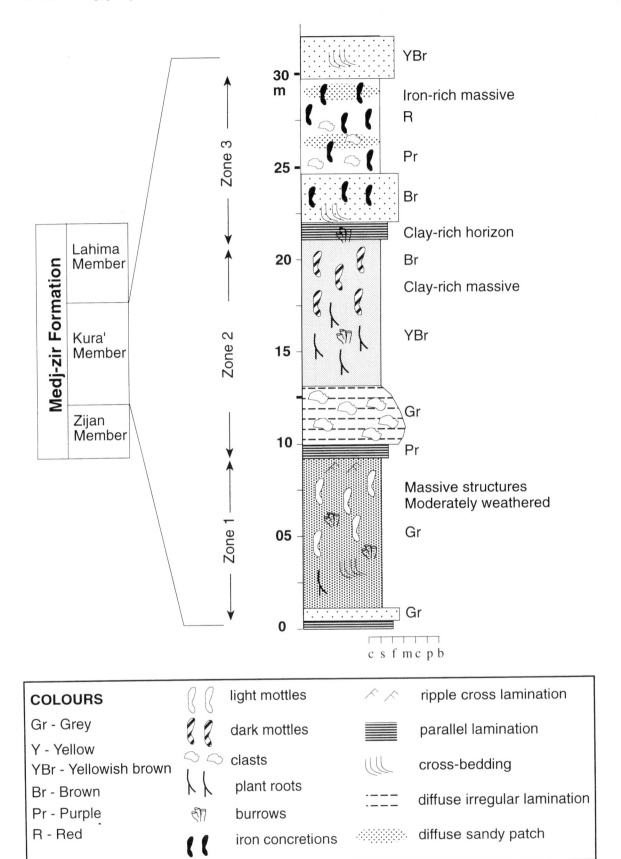

Figure C1.5 Log of ferricrete paleosols from the Kura, Wadi Kura (locality 10, Figure C1.2).

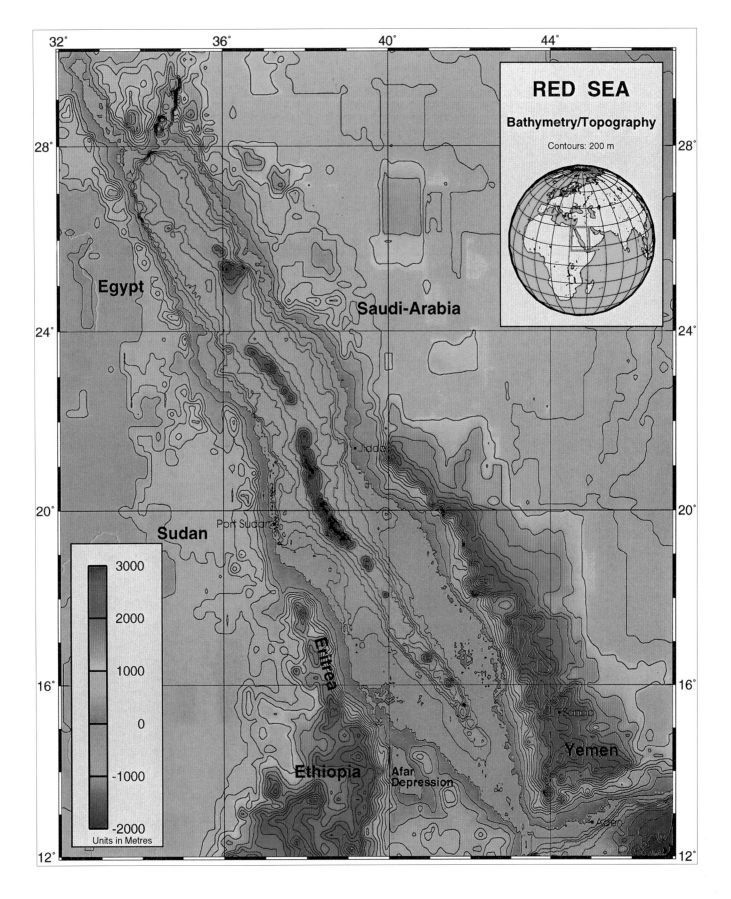

Plate 1 Bathymetry and topography of the Red Sea area, displayed with the use of GMT software (Wessel and Smith, 1991). Data compiled from several detailed local surveys (same sources as gravity data, see text) and ETOP-05-grid (National Geographic Data Center, Boulder, Colorado). The maximum elevation in the Ethiopian and Yemen highlands and the maximum water depths in the Red Sea axial areas exceed the maximum values of the scale.

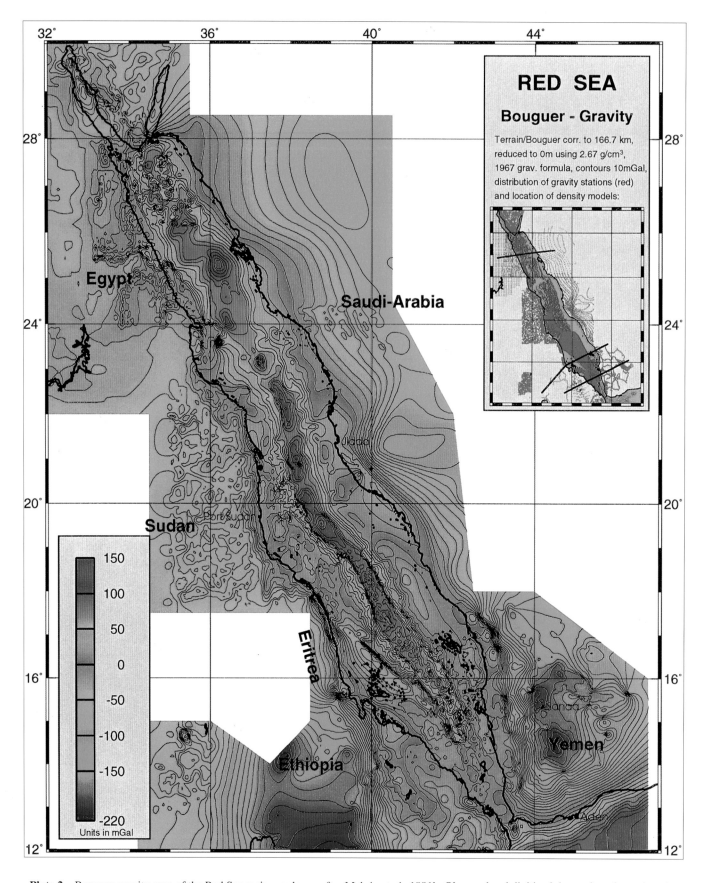

Plate 2 Bouguer gravity map of the Red Sea region, redrawn after Makris *et al.*, 1991b. Observed and digitized data points shown in red colour in the legend, uniformally reduced to 0 m with 2.67 g/cm³. Maximum values of >100 mGal occur in the axial region of the Red Sea, minima exceeding -200 mGal are found on the Ethiopian and Yemen plateaus. Note the relative negative anomalies on the flanks of the southern Red Sea and the relative positive anomalies on the western flank correlating with the western edge of the Afar Depression and with old north–south striking continental lineaments in Sudan and Egypt (Baraka and Onib Hamisana Suture Zones).

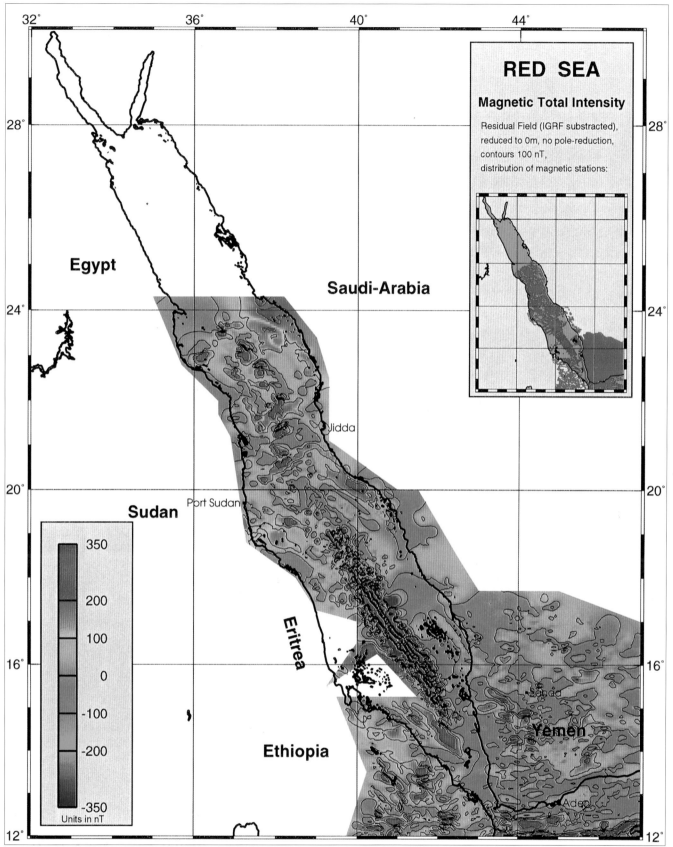

Plate 3 Total intensity residual magnetic map of the central and southern Red Sea and adjacent coastal regions. Observed and digitized data points shown in red colour in the legend, uniformally reduced to 0 m. Data sources are Allan, 1970; Laughton *et al.*, 1970; Hall, 1970, 1979; Hall *et al.*, 1977; Girdler and Styles, 1974; Roeser, 1975; Izzeldin, 1982; and MOMR, 1990. Strong dipolar linear magnetic anomalies correlate with the sea-floor spreading zone of the axial graben of the Red Sea. Note the longer wavelength – lower amplitude north-west–south-east striking anomalies onshore Yemen, on both flanks of the southern Red Sea and on the eastern flank of the central Red Sea. Further discussion in the text.

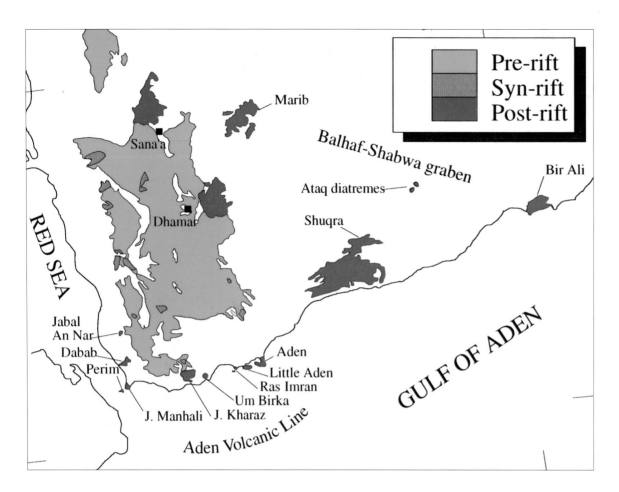

Plate 4 Pre- (>26 Ma); syn- (26–20 Ma) and post-rift (<20 Ma) magmatic activity in Yemen. Note that most volumetrically significant eruptive episode was pre-rift on the southern Red Sea margin in Yemen. Syn-rift magmatism may have occupied a greater aerial extent than is shown because there is evidence of removal of part of the volcanic stratigraphy during erosional processes between 26–19 Ma. Post-rift magmatism is represented by (a) main post-erosional volcanic fields at Sana'a, Marib, Dhamar, Ataq-Shuqra and Bir Ali; (b) the Aden Volcanic Line (Perim to Aden), and (c) Jabal An Nar. (See text for references and discussion.)

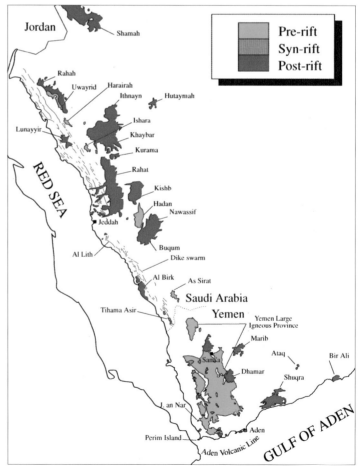

Plate 5 Pre-, syn- and post-rift volcanism in southwest Arabia. Pre-rift is dominated by the large igneous province (LIP) in Yemen and scattered alkaline volcanism in Saudi Arabia. Note that the spatial distribution of pre-rift volcanism mimics the orientation of the proto-Red Sea and that the major amount of magmatic activity is in the south perhaps due to proximity to the Afar plume/hot spot in Ethiopia. (See text for references and discussion.) Syn-rift volcanism in south-west Arabia is dominated by intrusion of a major dike swarm from Jordan to Yemen that is parallel to the proto-Red Sea. Emplacement of plutonic rocks in Saudi Arabia and Yemen was associated with this intrusive activity. See text for references and discussion. Post-rift volcanism in south-west Arabia is dominated by eruption of a large alkaline province stretching from Syria and Jordan to Yemen. Note that the most volumetrically significant volcanism is in Saudi Arabia. (See text for references and discussion.)

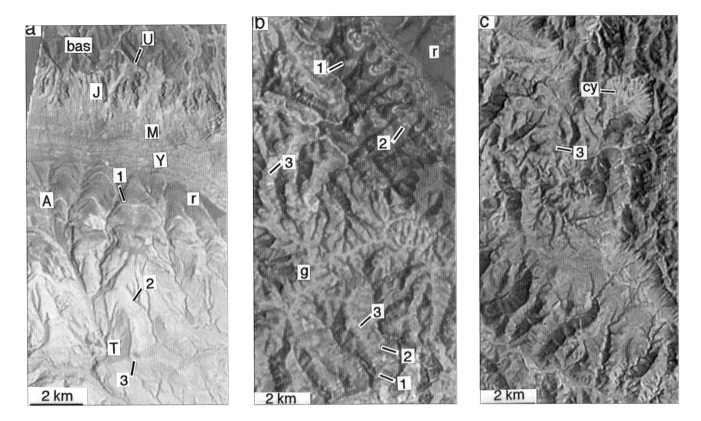

Plate 6 Thematic Mapper image of the Ahl Medo'–Ahl Miskat area in northern Somalia. Processing: 741 (RGB) colour composite. Note examples of multispectral signatures of specific lithological units (r, c). r: red response of limestones with lichen coating; c: cyan (blue) response of overall evaporitic deposits (dolostones and gypsum). 1–6: location of the photostratigraphic logs of Plate 7.

Plate 7 Details of Thematic Mapper 541 (RGB) band composite of (a) Ahl Medo' (northern Somalia) and (b) Sayut (southern Yemen) areas. Note the similarity in multispectral responses of the pre-rift Cretaceous (Cr) and Paleocene-Eocene (P) strata. The cyan facies on top of the Sayut section (c) is strikingly similar to the evaporitic facies exposed in the Ahl Miskat area (c in Plate 6 (a)).

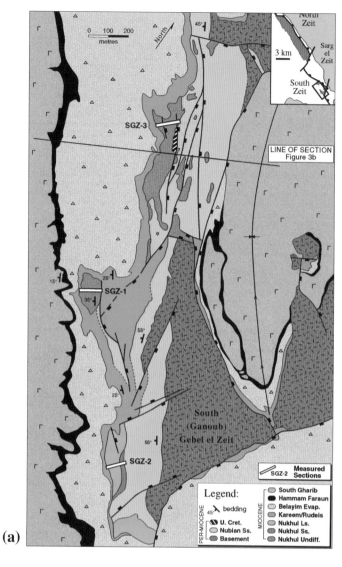

(a)

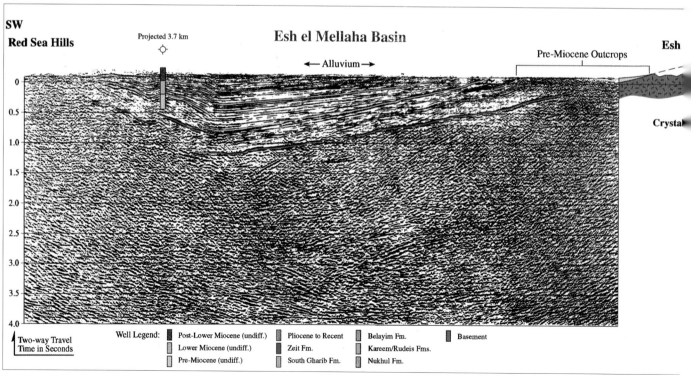

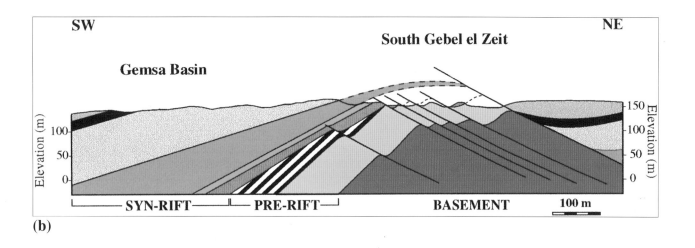

Plate 8 Detailed outcrop map (a) and cross-section (b) through South (Ganoub) Gebel el Zeit illustrating structural geometry at the crest of the highly rotated fault block. Rift trend faults are spaced tens to hundreds of metres; most cut through Nukhul Fm. (Aquitanian) and are truncated by the basal Rudeis unconformity or die out within the Rudeis marls (Burdigalian) (after Bosworth, 1995). (c) Seismic profile from the Red Sea Hills to Gabel Zeit, crossing the Esh el Mellaha and Gemsa basins SW Gulf of Suez. Migrated data. Location is given in Figure B4.4.

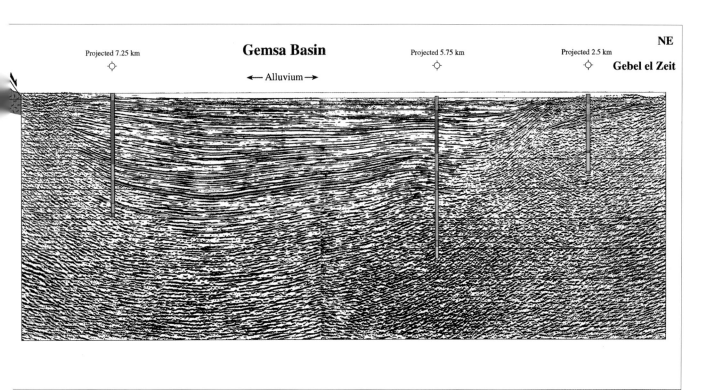

Plate 9 A. General view of Tawilah Group, Jabal Assim (section 10) indicating conformable boundaries between Ghiras (G) and Medzzir (MZ) and Yemen Volcanic Group. (YVG). B. Conglomeratic channel fill facies Jabal Al Ulaya (section 4) C. Planar-tabular crossbedded sandstone facies, Ghulat Asim (section 11).

Laminated sandstones (S_l)

This facies is confined to the Lahima Member of western Yemen where it is most abundant in the lower levels of this unit (Figure C1.2) forming sections up to 10 m in thickness. It comprises fining-upward cycles with each cycle starting with a lower erosion surface overlain by tabular to lenticular sandstone units from 2.5 to 7.0 m thick. The sandstones are moderately well-sorted quartz arenite weathering from a white and pale-yellow colour to a distinctive dark brown. In some localities (locations 9, 10 and 11, Figure C1.2) the sandstones are thin (0.3 to 1 m), have a sheet-like geometry, and have sharp but irregular lower and upper contacts. The sandstones commonly show horizontal lamination, ripple cross-lamination, and medium-scale sets of tabular cross-bedding. Fine-grained sandstones pass gradationally upwards into maroon to grey siltstone, forming laterally impersistent beds up to 50 cm thick. These are characterized by flaser and lenticular bedding and are interbedded with mudstone. These rocks are locally extremely fossiliferous and assemblages comprise small, well-sorted and preserved gastropod shells (Plate 10(d)) which are particularly well developed in sections 2, 10 and 16 (Figure C1.2). The gastropod-rich beds are frequently silicified into beds or nodules of chert.

The gastropods are of two genera. The smaller specimens are *Tarebia* (Thiarinae) which ranges from the Paleocene to Recent. The larger specimens (Plate 10(d)) are identified as *Coptostylus* (Thiarinae) which ranges from the Paleocene to late Oligocene (personal communication, P. Nuttall, 1992, and P. Jeffrey, 1992; Natural History Museum London).

Mudstone (M)

This facies is interbedded with the fluvio-lacustrine sandstones (above) and also confined to the Lahima Member (Figure C1.2). The rocks consist of very dark grey to black, laminated claystone, carbonate rich mudstone, muddy limestone (Figure C1.4(a)) and a single 0.3 m thick laminated, black shaly limestone horizon which is recognized at the top of the Lahima Member. Beds are thin and sometimes exhibit small desiccation cracks. Siltstones and fine-grained sandstones are rare. Black laminated mudstones locally contain abundant ostracods, bivalves and sparse lacustrine gastropods. At sections 2, 10 and 16 (Figure C1.2) the siltstone, shale and limestone interbedded with fine sandstone contain abundant fresh-water bivalves and well-preserved small gastropods. The siltstone, which is parallel laminated and lenticular bedded, is sometimes ferruginous and contains concentrations of fine plant fragments and ripple lamination.

Limestones (L)

Thinly bedded, lenticular dark coloured micritic (Figure C1.4(b)) and sparry limestones are present only at the top of profiles in the south and west, sections 2, 15 and 16 (Figure C1.2). They are well-laminated and organic rich, containing a variety of bioclasts; molluscs (Figure C1.6(c)), ostracods and *Microcodium* fragments (Figure C1.4(d)) set in a micritic matrix or calcite spar cement. Textures range from micrites to grainstones. The micrites are finely laminated and carbonaceous and locally contain pyrite. The beds occur immediately below, and are conformable with the first flow of the Yemen Volcanic Group at each of the sections (Figure C1.2).

Depositional environments of the Medj-zir Formation

Fluvial facies association

The trough cross-bedded sandstones (facies S_t) and planar-tabular cross-bedded sandstones (facies S_p) are associated in units ranging from 5 to 12 m. These units have a sharp erosional base and fine upwards and the size of cross-sets also diminishes upwards. The uncommon conglomerate lithofacies (G) occurs sporadically at the bases of these fining-up units. A sandy to gravelly braided stream setting is the probable environment of deposition for this association of lithofacies (cf. Cant and Walker, 1978; Allen, 1983; Miall, 1988; Rust and Gibling, 1990; Godin, 1991). The well-developed concretionary ironstones (facies S_{fe}) in the Kura' Member are consistent with pedogenic ferricretes (Fe-rich laterites) formed within fluvial channel and overbank facies (cf. Allen, 1965; Rust, 1978; Bown and Kraus, 1981; Wright, 1994). Complete channel-fill successions are overprinted by ferruginous paleosols in several sections (5 and 10) while a finer-grained sandstone facies acted as the host for pedogenic processes in other places (sections 6 and 9).

Shallow marine facies association

Marine fossils (foraminifera) are uncommon, occurring only at a few localities in the Zijan Member of the Medj-zir Formation (Al Subbary et al., 1993; Al Subbary, 1995). However, shallow marine depositional facies are more widespread. The large-scale cross-stratified sandstones (facies S_x) occur interbedded with siltstones (facies Z), both of which contain *Thalassinoides* trace fossils which are common in shallow marine environments (Inden and Moore, 1983). Moreover, the large-scale cross-beds occur in tabular sets which have non-erosive bases and are not confined within channels. They are interpreted as the deposits of shallow marine bars or sand waves which may have been driven by tidal currents (Stride, 1970; Anderton, 1976; Nio, 1976).

Lacustrine facies association

Horizontal and cross-laminated sandstones and siltstones (facies S_l) and the interbedded dark mudstones (facies M) occur primarily in the Lahima Member. Scours representing localized channels occur locally but in general these two facies are thin to medium bedded with ripples structures. Rare, thin-bedded limestones (facies L) occur towards the top of certain sections. Bioturbation is common, as is a shelly fauna dominated by gastropods. *Tarebia* and *Coptostylus* are well known in the European Tertiary where they are typically associated with mixed assemblages of brackish-water molluscs indicating tolerance of a wide range of salinities. Such deposits are invariably shallow water, with lagoonal and fluvial environments indicated. Freshwater environments are confirmed by the *Microcodium*.

DISCUSSION

This work represents the first analysis of facies and palaeoenvironment for the upper part of the Tawilah Group. Whereas these sediments had previously been regarded as a sequence of sands of fluviatile and shallow marine origin, there is clearly a wide range of siliciclastic, carbonate, ferruginous and siliceous lithologies within the Medj-zir Formation. Many of these facies are unusual, as are their associations, and these sections merit more detailed study. For the purposes of this contribution the lithostratigraphic, facies and palaeoenvironmental classification presented above forms the database for the following discussion of the palaeoenvironmental and tectono-sedimentary evolution of the pre-rift stage of the south-eastern Red Sea.

1. Revised, lithostratigraphy, chronology and regional correlation

The lithological diversity of the upper part of the Tawilah Group, the Medj-zir Formation, and its subdivision into three lithologically distinct members in western Yemen has not been previously recognized. The fact that these members can be recognized at most locations allows correlation to be made on the basis of lithofacies characteristics. Correlation throughout western Yemen as depicted in Figure C1.2 provides a useful tool for subsequent stratigraphic and palaeoenvironmental analysis. The Tawilah Group is also recognized as a major stratigraphic unit of continental and shallow marine sands which extends across eastern Yemen (Al Subbary *et al.*, 1993; Watchorn *et al.*, this volume) and northwards into southern Saudi Arabia as the Tawilah Formation (Blank *et al.*, 1986) and also as a time-equivalent to much of the Cretaceous Nubian sandstones of Ethiopia, Sudan and Egypt (Al Subbary, 1995).

The basal part of the Medj-zir Formation has been shown previously to be of Paleocene to Eocene age on the basis of its marine microfauna (Geukins, 1960, 1966; El Nakhal, 1988; Al Subbary *et al.*, 1993). The age ranges of the gastropods found within the Lahima Member of the Medj-zir Formation are consistent with these previous dates, but do not provide any further constraints. *Tarebia* has a range from Paleocene to Recent and, more significantly, *Coptostylus* ranges from the Paleocene to the late Oligocene (P. Nuttall, British Museum Natural History, personal communication, 1990). The only other chronological data in this section come from the lava flows lying on top of the Medj-zir with a date of 30–32 Ma or late Oligocene. The conformable nature of the boundary between the Medj-zir and the Yemen Volcanic Group suggests that the topmost Medj-zir Formation deposits may be as young as mid-Oligocene. However, if rates of deposition were extremely slow the Lahima Member could be as old as Paleocene.

2. Evolution of pre-rift environments

The timing of rifting in the southern Red Sea basin cannot be closely constrained because of the absence of outcrops of syn-rift strata and the lack of well penetration or published data on this interval in offshore areas. However, available data suggest the onset of rifting started at around 26–24 Ma some 5–6 m.y. after the Medj-zir to Yemen Volcanic boundary (see Introduction above). Therefore the sedimentary rocks of the upper part of the Tawilah Group and the majority of the Yemen Volcanics can be regarded as pre-rift, and the former gives valuable data on the evolving palaeogeography prior to volcanism and subsequent rifting of this margin. The contemporaneity of the basal volcanic flows and the fact that the preceding stratigraphic boundaries are broadly parallel to this surface are the best evidence available that the members were deposited at different time intervals and that they are not diachronous (except for the base of the Zijan Formation, see below). Therefore the evolution of this area can be discussed with respect to the three members of the Medj-zir Formation.

Zijan Member

The sandstones, siltstones and mudstones of the Zijan Member formed in a complex of fluvial and shallow marine environments. In the western and south-western sections (localities 1, 2, 3 and 17, Figure C1.6) the Zijan Member is absent. In the central part of the main east–west transect (localities 4, 5 and 6) the lower part of this unit is fluvial, passing east (section 8) into a mixture of fluvial and shallow marine sandy facies (Plate 10). In all these sections the upper part of the Zijan Member

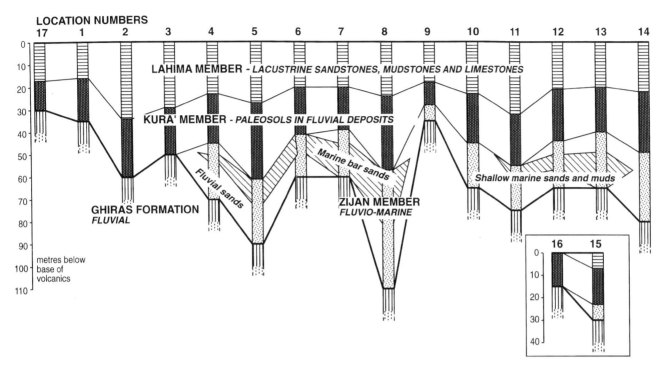

Figure C1.6 Simplified logs (from Figure C1.2) illustrating main depositional environments and thickness changes in the Medj-zir Formation, western Yemen. (For locations see Figure C1.2.)

comprises large-scale cross-bedded sandstones deposited in a shallow marine environment as sandy bars. In eastern and south-eastern exposures (10 to 14) the Zijan Member is wholly marine consisting of a mixture of bioturbated fine sandstones and sandy bar deposits (Figure C1.6).

These relationships are consistent with the west to east palaeoflow in fluvial deposits of the underlying Ghiras Formation (Al Subbary et al., 1993) and the associated fluvial sediments in the Medj-zir Formation (Figure C1.3). Facies distributions and palaeocurrents therefore indicate a transition from fluvial conditions in the west to a shallow marine environment in the east (Figure C1.7(a)). The absence of the Zijan Member in the west is attributed to the diachronous boundary between the wholly continental sandstones of the Ghiras Formation and the mixed fluvial and marine facies of the Medj-zir Formation. The base of the Medj-zir Formation is defined by the lowest appearance of bioturbated sandstones indicating marine or paralic conditions: in a transgressive succession this boundary would be diachronous. Time-equivalent beds of the Zijan Member are likely to be fluvial deposits at the top of the Ghiras Formation.

Kura' Member
The sandstones and mudstones of the Kura' Member were largely deposited in fluvial channel and overbank environments. Extensive ferruginous paleosols formed within these deposits to create a unit of superimposed ferricrete horizons. The thickness of the Kura' Member is relatively constant at around 20 m thick, but varying between 10 and 30 m (Figure C1.6). Thick, well-developed ferricrete paleosols develop under conditions of slow sedimentation and slow subsidence. The area must have had very low relief and been very stable for the long periods of time required (up to a million years; cf. estimates in Wright, 1994; Kraus and Bown, 1986) for the formation of mature laterites. Humid and organic-rich conditions are required for ferricrete formation because of the requirement of varying Eh/Ph levels for iron mobilization (Bown and Kraus, 1981). During the Eocene Yemen lay more or less on the equator and it is likely that humid equatorial conditions prevailed. The well-developed ferricretes in the Kura' Member indicate that this was a period of very slow to negligible aggradation and negligible subsidence or uplift.

The fluvial deposits and paleosols represent a return to fluvial conditions throughout the area. This may have been due to either progradation of the alluvial plain or a relative fall in base level of a few tens of metres: the absence of erosion at the base of the Kura' Member indicates that the magnitude of relative base level fall could not have been of any greater magnitude.

Lahima Member
The Lahima Member lies conformably on the paleosols of the Kura' Member and is conformably overlain by

132 Pre-rift doming, peneplanation or subsidence

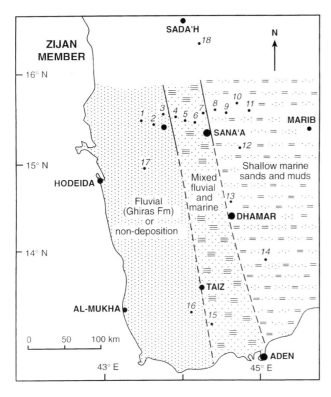

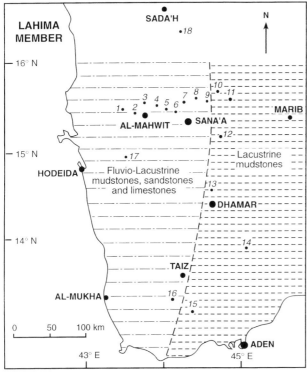

Figure C1.7 Palaeogeographic maps of western Yemen for (a) the Zijan Member, and (b) the Lahima Member of the Medj-zir Formation.

volcanics of the Jihanah Member. Thickness of the Lahima Member varies little across the area being around 20 m in most localities: it is only 7 m thick in

section 17 and just over 30 m thick at localities 2 and 11 (Figure C1.2). It is apparently absent in the most south-westerly section (16).

Lacustrine mudstones and sandstones occur in discontinuous layers and lenses. Lenses of limestone occur locally and only at the top of the Lahima Member. There are no consistent trends in the distribution of facies in the member. Lateral facies changes are commonly recorded in shallow lacustrine depositional environments. A broad, low relief area with localized lacustrine sub-environments is envisaged. The south-western part of the area may have been a region of nondeposition or of slow sedimentation in deposits amalgamated with the underlying Kura' Formation. Palaeogeographically, the area preserved the low relief indicated by the ferricretes of the Kura' Member but in addition there were a number of apparently isolated lakes and/or brackish lagoons in the area (Figure C1.7(b)). This indicates a rise in the water-table and base level compared to the conditions of formation of the ferricretes when there was essentially no deposition.

To summarize, the pre-rift environments evolve from sand-dominated fluvial and shallow marine deposits of the Zijan Member to fluvial sand with ferricrete paleosols of the Kura' Member to fluvio-lacustrine sediments of the Lahima Member. Facies and facies association boundaries trend north-north-east–south-south-west (Figure C1.7) through the central and southern sections but the area just to the north remains positive and no sediments of the Medj-zir have been recognized here. The absence of intermediate outcrops makes it impossible to determine if previous sediments were eroded or whether sediments never accumulated in this area.

Correlation with eastern Yemen

In time-equivalent strata in eastern Yemen there is accumulation of a much thicker and more marine-dominated succession with the Paleocene to early Eocene marine carbonate shelf Umm ar Radhumah Formation and marginal marine–evaporitic Jeza and Rus Formations. These are followed in the middle Eocene by the Habshiya Formation which shows east to west facies changes (As Saruri, 1995) from open marine carbonates in eastern Yemen to terrestrial facies (including ferricrete paleosols) in the Mayfa'ah Member in the Balhaf graben. There is a large (c. 200 km) gap in exposure between the Paleogene exposures of western Yemen and those of the Balhaf area (Figure C1.1) but there are two indications in the western Yemen sections described in this chapter (Figure C1.6) of a marine depocentre to the east of the country. Firstly, the shallow marine facies association in the Medj-zir Formation is only well developed in the eastern part of the present study area. Secondly, all palaeocurrent data in fluvially deposited strata indicate an easterly palaeoslope. A palaeogeography for the Paleogene is presented in Figure C1.8 indicating the fluvial to marginal marine facies associations of the Medj-zir in the east, and the marginal marine and carbonate shelf deposits of eastern Yemen were separated by a large island of Precambrian rocks including the A1 Bayda complex (Figure C1.1).

Tectono-sedimentary evolution

In western Yemen the evolving low relief and easterly sloping fluvial plains and lacustrine areas gave way between 32 and 30 Ma (Baker et al., 1996) to basaltic volcanism which continued until about 26 Ma. The constancy of fluvial palaeocurrents through the three members and the stable position of the eastern depocentre both indicate an absence of any change in depositional topography prior to volcanism. The only evidence for possible regional uplift within the Medj-zir is at the base where the Zijan Member thins out towards the west. Note this does not occur at the top of the Medj-zir and prior to the volcanics. There is no indication of widespread erosion within this western stratigraphy.

We therefore agree with Bohannon et al. (1989) that there is no evidence of 1–2 km doming and uplift which are predicted in the plume-related rifting model of White and McKenzie (1989) as there are no forced regressions, no widespread or local erosion surfaces and no fluvial incision prior to volcanism. The only evidence for changes in palaeotopography come from the fluviolacustrine sediments overlying laterites which suggests a base-level rise and not fall as would result from uplift.

This area of Yemen in the Paleogene was part of a very broad alluvial to coastal plain with a source area in central Africa. The plain stretched out to the Tethyan margins of north Africa, southern Arabia and the Horn of Africa and shows no palaeotopographic effects around the centre of the plume in the southern Red Sea area. Rifting and extension are considered to have started at 21–24 Ma (Davison et al., this volume; Menzies et al., 1995). These dates for the onset of extension tie in well with the recent biostratigraphic work of Hughes et al. (1991) and Toleikis (1995) and Schroeder et al. (this volume) which indicate that marine environments of the early Red Sea were present in central and western areas (Sudan) at around 25 Ma. Surprisingly this marine deposition was taking place synchronously and within 200–300 km of the volcanism in western Yemen. The same relationship is seen to the east in Yemen where small normal marine syn-rift carbonate platforms were also forming at this time (Bosence et al., 1996).

The fact that the shallow marine sandstones of the Zijan Member now occur at altitudes of around 2 km and that there is no pre-rift uplift indicates that this 2 km uplift must have occurred either during or after volcanism within the area and not prior to volcanism as has been previously suggested.

CONCLUSIONS

The Medj-zir Formation consists of siliciclastic, ferruginous and subordinate carbonate facies which indicate deposition in fluvial, shallow marine and lacustrine–lagoonal environments. The facies and facies transitions

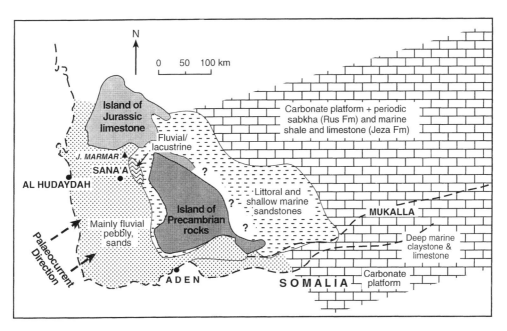

Figure C1.8 Palaeogeography of southern Arabia during the Paleogene (modified from Bott, 1992).

indicate deposition in a coastal plain to shallow marine shelf setting which was subjected to minor fluctuations in sea level.

There are no angular unconformities or erosional hiatuses within the Medj-zir Formation anywhere in western Yemen. Depositional hiatuses of unknown duration are represented by well-developed ferricrete paleosols indicating long periods of tectonic stability. The contact with the overlying volcanics is conformable in all sections examined.

The evidence from the sedimentology and stratigraphic relationships in the Medj-zir Formation shows that the Paleocene to Oligocene period prior to volcanism and rifting in the southern Red Sea the region was tectonically stable. There is no evidence for doming, peneplanation or significant subsidence in the pre-rift stratigraphy.

Chapter C2
Sedimentary evolution of early rift troughs of the central Red Sea margin, Jeddah, Saudi Arabia

M. A. Abou Ouf and A. M. Gheith

ABSTRACT

Field work, sedimentological investigations and biostratigraphical data from the Jeddah area provides new information on the Tertiary depositional history of the Red Sea in the central coastal area of western Saudi Arabia.

Sedimentary rocks outcropping in the coastal plain around Jeddah are subdivided into two major units which both unconformably overlie the Precambrian crystalline basement: the Usfan Formation probably of late Cretaceous to Eocene age and the Shumaysi Formation of late Oligocene to early Miocene.

The pre-rift Usfan Formation outcrops in the Haddat Ash Sham area north-east of Jeddah and the Shumaysi Formation overlies the granodioritic basement rocks in Wadi Shumaysi, east of Jeddah. The Usfan is represented by a succession of fluviatile sandstones and siltstones (Nubian type), which pass upward into littoral marine facies (phosphatic and carbonate lithologies), and supratidal facies (dolomitic marl and shales with gypsum) indicating shoreline sedimentation. These are followed by a thick red-bed sequence of fluvial sands and conglomerates of the lower Shumaysi Formation.

The presumed syn-rift Shumaysi Formation in Wadi Shumaysi is divided into three units. The lower is a section of fluviatile sandstones and conglomerates. The middle beds are fluvio-lacustrine silts and shales with two oolitic ironstone deposits. The upper unit comprises shales, siltstones, sandstone tuffs and freshwater–marine cherty limestones. Igneous activity contributes a lava at the top of the upper Shumaysi and tuffs to the basin-fill.

These two sequences occur in separate troughs in the Jeddah region and their stratigraphic range from the late Cretaceous–Eocene through to the early Miocene and provides information about the early rifting history of the central Red Sea coastal margin. They represent some of the earliest rift sediments of the Red Sea basin.

INTRODUCTION

Field and laboratory studies carried out on the sedimentary sequences outcropping in the coastal plain near Jeddah provide data on the early sedimentary environments of the central Red Sea and constrain the timing of the early stages of rifting within the basin. Beydoun (1988) records that there are no widespread late Eocene sediments bordering the Red Sea and Gulf of Aden. He considered that they were either not deposited or were removed by erosion because Arabia was emergent except for a narrowing seaway between the Mediterranean and the Indian Ocean along the area of north Syria, north Iraq and parts of south-west Iran to western Oman.

The Cenozoic sediments of the Jeddah region, although still not well dated, are likely to preserve the earliest stages of the structural evolution and sedimentation in the Red Sea rift system. The Paleogene sediments of the Jeddah region either rest unconformably on, or are faulted against the Precambrian basement. Unfortunately fossils are rare in these largely continental facies and many of the sections await precise biostratigraphic and/or isotopic dating. The outcrops described in this chapter come from two isolated troughs (Figure C2.1): the Suqah trough to the north-east of

Jeddah and the Shumaysi trough to the east of Jeddah (Moore and Al-Rehaili, 1989). Both troughs contain north-east to north-north-east dipping strata and are parallel to the present-day Red Sea rift escarpment and coast.

Spencer (1987) divides these Paleogene sections into two formations: the lower Shumaysi (which he further divides into three members, Khusluf, Mataah and Haddat ash Asham) and the upper Khulays Formation of Oligocene–Miocene age. He did not recognize the Usfan Formation as a distinct unit but considers that it is probably a lateral equivalent of the Shumaysi Formation (Spencer, 1987, p. 8).

Moore and Al-Rehaili (1989) in their mapping of the Makkah quadrangle recognize three formations: the Hadat ash Sham, the Usfan and the Shumaysi which they regard as partial time-equivalents spanning the Paleocene to early Eocene.

Earlier work in the Jeddah region by Brown *et al.* (1962), Al Shanti (1966), Yamani (1968), Moltzer and Binda (1981) and Basahel *et al.* (1982), does not add significantly to the discussion concerning the broad stratigraphic setting of these deposits.

Our field investigations suggest that the traditional (i.e. Karpoff 1957a,b) subdivision of these sections is more practicable than the later revisions and we recognize the earlier Usfan Formation of Karpoff which occurs only in the Suqah trough (Figure C2.1) and the later Shumaysi Formation which outcrops in Wadi Shumaysi as well as the Suqah trough (Figure C2.1). This is informally subdivided here into the lower, middle and upper units. The stratigraphic relations between these two formations are seen in Wadi Ash Sham in the Suqah trough where lower Shumaysi lithologies are seen to rest with no angular discordance on top of the Usfan Formation (Plate 11A).

The Usfan Formation was considered by Karpoff (1957a) to be of late Maastrichtian to Eocene age on the basis of molluscs and shark teeth within a limestone unit towards the top of the formation. A glauconitic bed immediately underlying the limestone was dated by Brown (1970) using K-Ar dating as 43–55 Ma but he considered there may have been some argon loss from the sample and that it might be older. Basahel *et al.* (1982) considered that the lower portions of the Usfan, and in particular a phosphatic bed contained Maastrichtian nautiloids and baluchicardid molluscs. A broad stratigraphic range from the Maastrichtian to the early Eocene has been suggested by the most recent work on the Usfan by Hughes and Filatoff (1995).

The Shumaysi Formation was believed by Karpoff (1957a) to be of Oligocene–Miocene age. Brown *et al.* (1962) gave an Eocene age. Micropalaeontological work by Moltzer and Binda (1981) assigned the middle Shumaysi Formation to the early Eocene (Cuisian).

Dating (K-Ar) of the lava flow capping the Shumaysi Formation to the east of Jeddah by Brown (1970) gave ages of 32 ± 2 and 25 ± 3 Ma indicating a late Oligocene–Miocene age. The Shumaysi Formation is not recognized in the recent work by Hughes and Filatoff (1995).

The principal aim of this chapter is to present stratigraphic logs of these two formations for the first time and to discuss their environments of deposition, which, on the basis of their Paleogene ages, span the early stages of rifting of the central Red Sea.

GEOLOGICAL SETTING OF THE USFAN AND SHUMAYSI FORMATIONS

The geology of the Jeddah area is strongly controlled by extensional faulting parallel to the Red Sea which forms a number of north-north-west–south-south-east fault blocks and half-graben basins or troughs filled with Tertiary sediments. Local faulting and fracturing has taken place in both the crystalline basement and in the Tertiary sedimentary rocks.

Rocks of the Usfan Formation occur in the north-western end of the Suqah Trough near the village of Usfan and in Wadi Ash Sham (Figure C2.1). The rocks are either faulted against the Precambrian or rest unconformably on the Precambrian (Karpoff, 1957b). The beds either dip to the east at around 40°–50° in the Usfan area or are dipping at 20°–30° (Plate 11(a)) to the north and north-east; or are vertical to overturned in Wadi Ash Sham (Plate 11(b)). The formation is unconformably overlain by the subhorizontal late Miocene to Pliocene basalts of the Hammah Formation (Moore and Al-Rehaili, 1989) (Plate 11(b)). The Usfan Formation is reported to be 215 m thick (Vail *et al.* in Moore and Al-Rehaili, 1989).

The Paleogene sedimentary rocks to the east of Jeddah comprise the Shumaysi Formation (Karpoff, 1957a). Outcrops tend to be in isolated low hills surrounded by rocky desert or wadi sands and gravels. Good outcrops occur in Wadi Fatima and its tributary Wadi Shumaysi (Figure C2.1, Plate 12). The total thickness of the Shumaysi is given by Al Shanti (1966) as ranging between 80 and 200 m. Moore and Al-Rehaili (1989) give thicknesses of 74 to 183 m. The Shumaysi Formation unconformably overlies the Precambrian crystalline rocks which are strongly weathered, and is conformably overlain by a basalt flow. In general, the strata dip north-easterly at between 15° and 30° which probably relates to an unmapped basinward dipping fault on the eastern margin of the Shumaysi Trough. Wadi Shumaysi is believed to be formed along a north-western regional fault parallel to the Red Sea (Al Shanti, 1966).

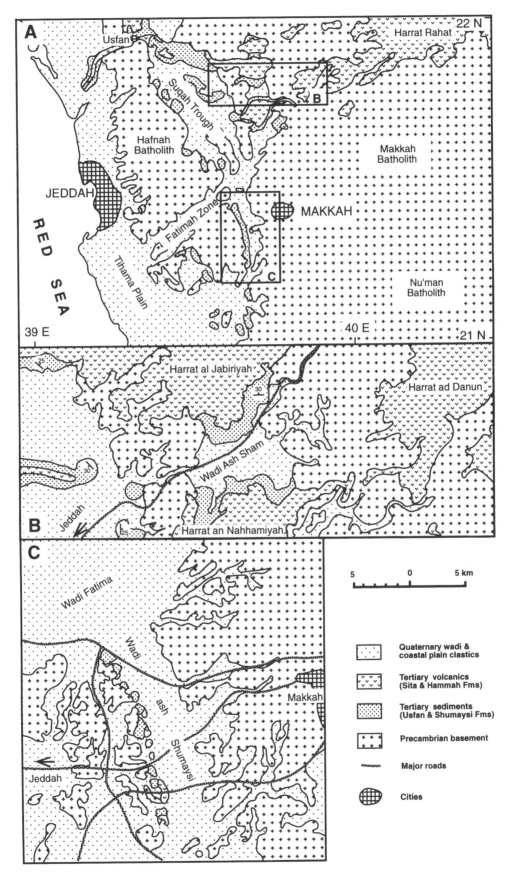

Figure C2.1 Maps of Jeddah area indicating: (a) location of study area, (b) geology of Usfan, Hadat Ash Sham area, and (c) Wadi Shumaysi (after Moore and Al-Rehaili, 1989).

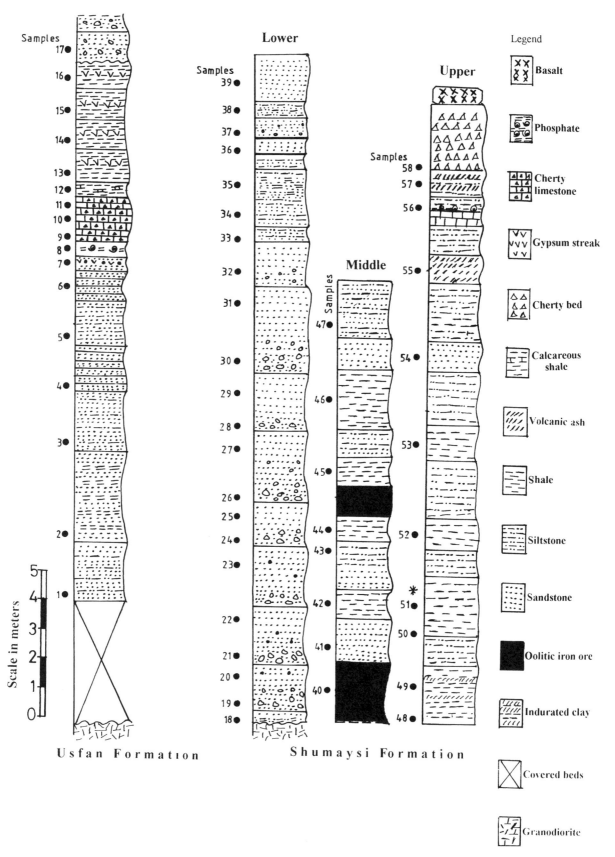

Figure C2.2 Logged sections and sample positions from Usfan Formation (Hadat Ash Sham, illustrated in Plate 11(a)) and lower, middle and upper Shumaysi Formation from Wadi Shumaysi (Plate 12(a)–(c)). First occurrence of marine fossils marked with asterisk.

METHODS

The study area lies on the western coast of Saudi Arabia and includes Wadi Shumaysi between Jeddah and Makka and the Haddat Ash Sham area north-east of Jeddah (Figure C2.1). Field work involved logging and description of sections and collection of samples (Figure C2.2).

Laboratory study included textural and mineralogical analysis of sedimentary rocks. Mechanical analysis of the sediments was carried out using methods of Folk (1968) and Folk and Ward (1957). Clay minerals were identified by X-ray diffraction and oriented slides were prepared by the sedimentation technique described by Carver (1971). The reflections used for identification of the different clay mineral groups and their occurrence in samples from the Shumaysi Formation are given in Table C2.1. Peak area ratios were used for a semiquantitative estimate of the amount of a mineral present.

Heavy minerals were separated, identified and counted and these data, together with percentage opaques are given in Figure C2.3.

THE USFAN FORMATION

Rocks from part of the Usfan Formation have been logged at Haddat Ash Sham (Plate 11 (a), Figure C2.2). The lower beds are formed of reddish-brown, thinly laminated, fine-grained sandstones of 'Nubian' type. Sandstone beds are locally rippled. Investigation of the heavy mineral suite in six sandstone samples indicates a dominance of opaques (average 78%), rutile (average 29%) and tourmaline (average 25%). Minor amounts of staurolite, hornblende, augite, epidote and zircon are also present. A 0.5 m thick, brownish to greenish coloured arenaceous phosphatic and glauconitic bed is found in the upper part of the section. This horizon was dated using K-Ar by Moore and Al-Rehaili (1989) as 55.2 ± 1 Ma and 42.8 ± 1 Ma (early to mid Eocene).

This is overlain by a widespread unit of limestone which has been examined at two localities: in the Haddat Ash Sham area it is a gently dipping (Plate 11a) bed about 1.5 m thick and the other is at Usfan where it stands as a vertical wall and is 4.5 m thick. The 0.1 to 1.5 m beds of limestone are dominantly bioclastic

Table C2.1 Clay mineral analyses from the Shumaysi Formation.
(a) Reflections used for identification of clay minerals

Clay mineral group (001)	Untreated 20°	d(Å)	Glycolated 20°	d(Å)	Heated (550°) 20°	d(Å)
Montmorillonite	6.4	13.8	5.3	16.7	8.8	10.04
Illite	8.2	10.8	8.2	10.8	8.6	10.20
Kaolinite	12.0	7.37	12.0	7.37	–	–

(b) Percentages of clay minerals from shales of the middle unit of the Shumaysi Formation. For location see Figure C2.2

S. No.	Kaolinite %	Illite %
41	100	–
42	100	–
43	100	–
44	100	–
45	100	–
46	33	67
47	12	88

(c) Percentages of clay minerals from shales of the upper unit of the Shumaysi Formation. For location of samples see Figure C2.2

S. No.	Kaolinite %	Montmorillonite
48	25	75
49	100	–
50	100	–
51	100	–
52	100	–
53	54	46

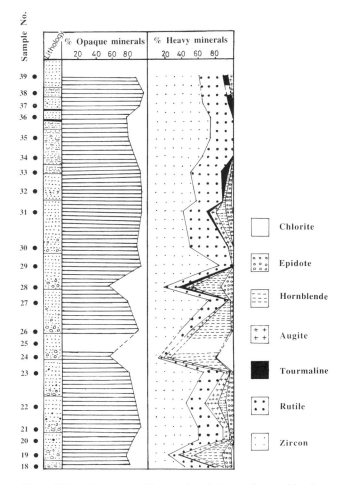

Figure C2.3 Occurrence of heavy minerals in very fine sand fraction of lower unit of the Shumaysi Formation.

with abundant casts, moulds and shell fragments of bivalves (*Cardita*) and gastropods. Thin-section study indicates bioclastic wackestones, packstones and some molluscan rudstones.

The limestone is overlain by a 0.5 m thick bed of hard dolomitic marl with a rippled upper surface. A 5 m thick succession of green-brown limonitic shale with thin veins of gypsum (Plate 11(b)) caps the Usfan succession at Wadi Ash Sham. An erosive contact is followed by a very thick (several hundred metres) layer of red beds which are unconformably overlain by basalts of the Hammah Formation (Plate 11(a)). These red beds are very similar to, and are considered to be part of, the lower Shumaysi Formation (see below) but await further study.

The lower part of the Usfan Formation with unfossiliferous, iron-oxide-rich sandstones and siltstones is considered to have been formed in a low to moderate energy fluviatile environment. The heavy mineral suite indicates derivation from a metamorphic source area such as the surrounding areas of Precambrian metamorphic basement. The dominance of tourmaline and rutile suggest mineralogical maturity. The overlying succession of a phosphatic and glauconitic bed, limestone, dolomite marl, and shales and gypsum indicate an initial transgression with starved sedimentation followed by moderately agitated fully marine conditions for deposition of the limestones. This is followed by a regression as the marine fauna disappears and restricted dolomitic marls and gypsiferous shales are deposited.

THE SHUMAYSI FORMATION

On the basis of our logging and subsequent analysis of the Shumaysi Formation in the Wadi Shumaysi area we have subdivided the formation into the lower, middle and upper informal lithostratigraphic units.

Lower Shumaysi Formation

The lower Shumaysi unconformably overlies the Precambrian basement in Wadi Shumaysi (Figure C2.1) and has a considerable lateral variation in thickness and lithology of individuals units. The basal part of the lower Shumaysi Formation consists predominantly of reddish to brownish beds of conglomeratic sandstones alternating with fine-grained sandstone units (Plate 12(a)). Gravel content of these sediments is up to a maximum of 53%, while mud content does not exceed 26%. Finer-grained pink to violet coloured silts and sands also occur in the lower Shumaysi Formation.

The lower Shumaysi Formation exhibits many sedimentary structures including planar bedding, graded bedding, cross-bedding (Plate 12(b), ripple cross-lamination, convolute lamination, lenticular and wavy bedded sands and muds, and trace fossils. Graded bedding and fining-upward units are the most common features present (Figure C2.2). Sequences start with thick beds of medium- to coarse-grained pebbly sandstones. The pebbles comprise rounded ironstone, siltstone and fine orthoquartzite sandstone clasts and quartz pebbles. The sands are poorly sorted. Overlying the pebble sands are finer-grained sandstone beds as the sequence fines upward.

Fine-grained sandstone and siltstone units with abundant ripple-drift cross-lamination occur and the sands are well sorted. Trace fossils are common on the bases of beds and bioturbation may destroy lamination in the fine-grained units. Deformed bedding such as convolute and slump structures are also present. The upper parts of this unit contain fine-grained sandstone and siltstone with lenticular, wavy and flaser-bedded intervals.

The fine sandstones from the lower Shumaysi are characterized by a heavy mineral suite of ultrastable minerals (Figure C2.3): opaques, zircon (euhedral, prismatic, bipyramidal), rutile (yellow to red, prismatic, bipyramidal), hornblende (dark-green, elongate, prismatic). Augite, epidote, tourmaline and chlorite are present in small amounts.

The coarse grain-size of these sediments and their deposition as conglomerate-lined channel fills and lenticular and fining-upward conglomerate to sandstone sequences, lacking marine fossils, all indicate deposition within a fluvial system. The absence of significant overbank deposits and the abundance of channel fills and lenticular bedding units indicate a braided fluvial system (Allen, 1965). The planar cross-bedded sandstones are attributed to high stage deposition within transverse channel bars while the small-scale cross-bedded finer sands may have formed within the falling stage. The plane-bedded fine sandstone and interlaminated lithologies are attributed to sedimentation in a late state, and falling or low waters over the channel bars (Coleman, 1969). The liquefaction may have been induced by fluidization of grains following emergence and reflooding (Reineck and Singh, 1975). The mineralogically mature heavy mineral suite indicates a long period of weathering and transport, from a deeply weathered acid to intermediate igneous source area, prior to deposition of these sands.

Middle Shumaysi Formation

This unit is distinguished by two oolitic ironstone beds separated by alternating shales and fine-grained sandstone (Figure C2.1(b)). The unit varies in thickness from 8 m to 15 m. Beds of siltstone, shale and occasional fine-grained sandstone occur with a number of biogenic sedimentary structures; trace fossils are very common, as are silicified wood, leaves and plant rootlets and vertebrate tracks. The only body fossils detected by the authors are some bivalve moulds, and teeth of fish and reptiles.

The two distinctive oolitic ironstones are separated by a fining upward sandstone unit. The ironstones are predominantly goethite with minor amounts of hydrohaematite and haematite with iron contents varying from 42 to 48% (Moore and Al-Rehaili, 1989). The clay mineral analysis of seven fine fractions separated from mudstone samples (Table C2.1) prove the dominance of kaolinite, while illite is recorded in only two samples near the top of the section.

The association of fine-grained sediments with these preserved biotas indicates low-energy, freshwater environments in a humid setting. Fluviatile overbank and lacustrine environments have been proposed by Moltzer and Binda (1981) possibly similar to those around the modern Lake Chad. The iron of the middle Shumaysi Formation was probably derived from erosion of the surrounding Precambrian basement by humid climate weathering and transport in solution as bicarbonate and subsequent precipitation in agitated waters as ooids. The predominance of kaolinite indicates humid periods which lead to intensive leaching and chemical weathering (Singer, 1984). The occurrence of illite, with a sharp, narrow peak, reflects a high degree of crystallinity and an authigenic origin.

Upper Shumaysi Formation

This is the most lithologically varied sequence of the Shumaysi Formation with shales, siltstones, sandstones, tuffs, limestones and cherts recorded (Plate 12(c)(d), Figure C2.2). Thickness varies between 51 m and 92 m and the section in Wadi Shumaysi is capped by a basaltic lava (Sita Formation) dated as 20.1 ± 7 and 25.3 ± 3 Ma (Moore and Al-Rehaili, 1989). The dominant facies include laminated shales, varicoloured and enriched with volcanic tuffs (Plate 12(c)). Fossiliferous calcareous shales rich in molluscs and bivalves occur with iron concretions and fissures filled with authigenic silica. Siltstone and fine-grained sandstone beds are also encountered. The sandstones are poorly sorted quartz arenites with abundant chert nodules. The uppermost beds are silicified limestones (Plate 12(d)) with an abundant marine to non-marine molluscan fauna and freshwater algae (charophyte oogonia).

Identification of clay minerals from six clay fractions shows a dominance of kaolinite in four samples while montmorillonite is important from samples from the base and the middle of the succession (Table C2.1).

The fauna was studied by Cox and by Sohl and Taylor (in Al Shanti, 1966) from different areas and it was concluded that ages were either Oligocene or early Eocene respectively. We have carried out micropalaeontological preparations from shales from the base of this unit and found two planktonic foraminiferal fossils (*Globigerinoides* spp.) which give a late Oligocene to early Miocene age. Moltzer and Binda (1981) record *Ammobaculites* from this unit, also indicating marine conditions.

The presence of a marine fauna at the base passing upwards to marine and non-marine molluscs and charophyte algae at the top indicates deposition of the upper Shumaysi Formation in shallow marine through to freshwater environments.

The predominance of kaolinite in the Shumaysi clays indicates a relatively warm and tropical climate acting on a terrestrial source and the tuffs and montmorillonite clays indicate episodes of volcanic activity within the upper part of the Shumaysi.

DISCUSSION

Stratigraphic and basin setting of Paleogene sections around Jeddah

There have been different opinions concerning the stratigraphic status of the Usfan and the Shumaysi Formations and the relations between these two units. While early workers (e.g. Karpoff, 1957a,b) proposed that the two units were distinct, and were given separate names, subsequent workers have either considered the two units as synonomous and refer the entire section to the Shumaysi Formation (e.g. Spencer, 1987), or that the two formations are distinct lithologically but are, in part, laterally equivalent to one another (Moore and Al-Rehaili, 1989). Most recently Hughes and Filatoff (1995) provide useful data on the Usfan Formation but make no mention of the Shumaysi Formation. Our logs indicate that although there are some lithological similarities between the two formations, such as the lower fluviatile dominated sections in both areas, there are important differences such as the marine phosphorite and limestone at the top of the Usfan, and the oolitic ironstones, and the freshwater limestones and cherts of the Shumaysi Formation. We are therefore in agreement with the original work of Karpoff (1957a,b) and more recently Moore and Al-Rehaili (1989) that the Paleogene strata of this area can be subdivided into two lithologically distinct formations.

Although these continental and marginal marine strata are difficult to date accurately there is evidence to support the view that the Shumaysi Formation is younger than the Usfan Formation. In Wadi Ash Sham glauconites from the Usfan have been dated as early as mid Eocene (55–43 Ma; Moore and Al-Rehaili, 1989). The molluscs from the limestone and the palynomorphs from the continental facies (Hughes and Filatoff, 1995) are both long-ranging (late Cretaceous to Eocene) biotas and provide little help in constraining the age. The Usfan in this area is erosively overlain by what is interpreted in this chapter to be the lower unit of the Shumaysi Formation and these are unconformably overlain by late Miocene to Pliocene basalts. In the Wadi Shumaysi sections the upper units of the Shumaysi Formation have biostratigraphic data suggesting a late Oligocene to early Miocene age, and these are capped by an apparently conformable lava flow (Sita Formation) of the same age range dated as 20.1 ± 7 and 25.3 ± 3 Ma (Moore and Al-Rehaili, 1989). The earlier molluscan and palynological biostratigraphy are inconclusive and suggest either an Eocene or an Oligocene age. In conclusion we believe that the Usfan Formation is both lithologically distinct and older (late Cretaceous to Eocene) than the Shumaysi Formation (late Oligocene to early Miocene) and forms part of the pre-rift stratigraphy.

In the southern Shumaysi trough the younger Shumaysi Formation rests on the Precambrian basement and Oligo-Miocene sediments and basalts are tilted, but in the north-west of the Suqah trough the older Usfan sediments rest on the basement, and are erosively overlain by the younger Shumaysi Formation. This indicates a long-lived basin margin syn-rift unconformity within this area. These rocks were then tilted to the north-east and eroded prior to the late Miocene eruption of the subhorizontal late Miocene basalts of the Hammah Formation.

Comparison with other Tertiary sequences along the eastern Red Sea margin

The early Tertiary sedimentary successions outcropping on the eastern margin of the Red Sea document the early stages of structural and sedimentary evolution of the rift system. Comparison of the sedimentary sequences studied here in the central part of Red Sea coastal plain (Jeddah region) with those previously studied in the northernmost Midyan region (Dullo *et al.*, 1983; Purser and Hötzl, 1988), the northern Ayunah to Yanbu (Jado *et al.*, 1989) and the southern Jizan area (Schmidt *et al.*, 1982; Blank *et al.*, 1986) enable this work to be considered in its regional context (Figure C2.4).

Northern Saudi Arabia

The Midyan area in northern Saudi Arabia has been most recently studied by Dullo *et al.* (1983) and by Clark (1985) (Figure C2.4). The oldest Tertiary unit, the Jabal Tayran Formation, rests unconformably on the eroded Precambrian basement. The basal conglomeratic section of Dullo *et al.* (1983) (Wadi al Hamdh Member) is interpreted as a fluvial deposit. This member is interpreted to be of early Oligocene age because it is overlain by marine limestones of late Oligocene age. The limestones (Wadi al Kils Member) are shallow marine bioclastic packstones and coral bafflestones with associated limestones with late Oligocene foraminifera (Dullo *et al.*, 1983). These are followed by a succession of marginal and shallow marine sandstones, siltstones and marls with intercalations of reefal limestones of early Miocene age. They correlate these marls and reefs with the *Globigerina* marls of the Gulf of Suez. The Al Bad Formation follows firstly with shales and gypsum and then gypsum, anhydrite dolomite and some limestones and shales occur. Foraminifera in the lower part of this unit give middle Miocene ages and those from higher give late Miocene (Tortonian) ages. The section is unconformably overlain by the Ifal Formation which comprises sandstones and conglomerates that have not been dated, but are assumed to be of late Miocene and Pliocene age.

Further south in north-west Saudi Arabia, Jado *et al.* (1989) have found similar evidence for late Oligocene flooding with dolomitized reef limestones of late Oligocene to Miocene age succeeded by Miocene siliciclastic sediments with limestone interbeds which become evaporitic up-section.

Wells offshore from this area which penetrate the basement (Al Kurmah-1, Barqan-1, and An Numan-1) have shales and sandstones labelled as 'Rudeis' in Mitchell *et al.* (1992), which is taken here to be of Burdigalian to Langhian in age, overlying basement. Barquan-1 well has the earlier 'Nukhul', which we take to be Aquitanian in age, sandstones and shales overlying basement. These are interpreted to be restricted marine muds by Crossley *et al.* (1992).

When compared with the Jeddah region there are clear differences as there are no equivalents of the late Cretaceous to early Eocene? deposits of the Usfan Formation in the northern area. The late Oligocene–early Miocene Shumaysi Formation on present evidence appears to be age-equivalent to the Jabal Tayran Formation and shows a broadly similar palaeoenvironmental evolution from basal fluvial passing up to shallow marine deposits. However, there the similarity ends and the limestones from the two areas differ as do the occurrences of volcanic tuffs and lava in the Jeddah area.

Hughes and Beydoun (1992) suggest that the Midyan sections may relate to a marine connection to a 'proto-Jordon valley' rather than being connected to the early Red Sea. Similarly Jones and Racey (1994) show no marine sediments of Oligocene age in this area in their review of Cenozoic palaeogeographies of the Arabian plate. The ages and palaeoenvironmental correlations presented here, together with new evidence of late Oligocene marine conditions in Sudan (Schroeder *et al.*,

Discussion 143

this volume), suggest that both the central and northern areas of the Red Sea were probably transgressed by marine waters in the late Oligocene.

Southern Saudi Arabia
The oldest syn-rift deposits of south-west Arabia are represented by the informal Jizan group of sedimentary and volcanic rocks (Schmidt and Hadley, 1984). This informal group is divided into five informal formations (Ayannah sandstone, Ad Darb, Baid, Liyyah and Damad) which are considered to be of late Oligocene–early Miocene age on the basis of palaeontological and radiometric dating. The Jizan group rests unconformably on Precambrian, Cambro-Ordovician (Wajid Formation), Jurassic (Amran Formation), and Cretaceous (Tawilah Formation) rocks. Intrusive igneous rocks of the Tihamat Asir complex have K-Ar ages of late Oligocene to early Miocene (24.3 ± 1 to 20.6 ± 0.6).

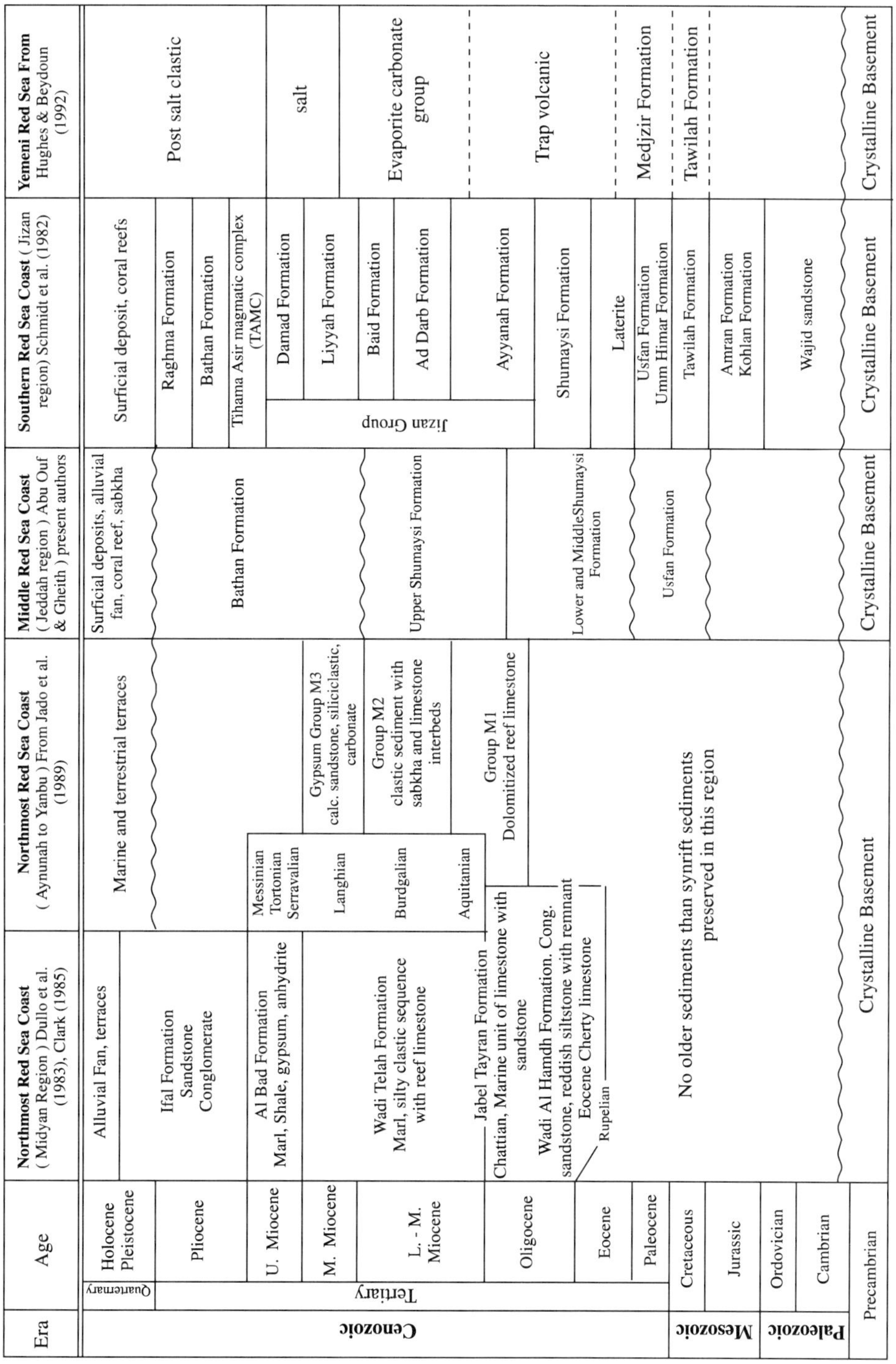

Figure C2.4 Regional stratigraphic correlation of Tertiary successions along the eastern margin of the Red Sea.

Mid to late Miocene shales and evaporites occur in and around the salt domes of the Tihama Plain e.g. Jizan (Blank et al., 1986).

The Tawilah Formation is 10–30 m thick and by comparison with occurrence and age in Yemen (see Al Subbary et al., this volume) is considered to be late Cretaceous to early Tertiary in age. The Tawilah comprises a sequence of cross-bedded quartz-arenite sandstones and conglomerates with rhyolitic tuffs in the upper portions. The upper surface is weathered into mottled sandstones with haematite concretions as is also the case at the top of the Tawilah Group in Yemen (Al Subbary et al., this volume). The Tawilah Formation shows some lithological and stratigraphic similarities to the Usfan Formation of the Jeddah region with the exception of the marine limestones. Most of the lower unit of the Shumaysi Formation appears to resemble the informal Ayyanah formation (sandstones with conglomerates) of the Jizan group (personal observation). The informal Baid formation consists of laminated varicoloured shales and volcanic tuffs with inorganic precipitated silica and appears lithologically similar to the upper Shumaysi Formation of the Jeddah region.

Tectono-sedimentary evolution

The occurrence of the Usfan and Shumaysi Formations unconformably overlying Precambrian basement within fault-bound troughs parallel with the present Red Sea suggests that their deposition was intimately related to evolution of the rift basin. While the stratigraphic relations have yet to be studied in detail, and integrated with a structural analysis, the current work provides some useful insight into the geological evolution of the central area of the Red Sea. Unfortunately this work cannot be tied into the recent biostratigraphic scheme of Hughes and Filatoff (1995) because their data are not tied to outcrop sections and new names have been introduced.

The Usfan Formation represents initially fluvial conditions followed by a short-lived marine transgression and regression. The age of this unit is not well constrained. Faunal data suggest a late Cretaceous age while K-Ar dating, which has been shown elsewhere to be unreliable (Chazot et al., this volume) suggests a much younger, Eocene age. These age bands suggest that the Usfan is a pre-rift sequence equivalent to the fluvial dominated Tawilah Group of Yemen or the Nubian, Esna, or Thebes units of the Gulf of Suez. Unlike the Tawilah Group of Yemen the sands of the Usfan appear to be locally derived from the Precambrian basement which together with its unconformable relation with the Precambrian, suggest considerable pre-Paleogene uplift and erosion. Whether or not this relates to early stages of rifting has yet to be established.

The Shumaysi Formation, which is unconformable on Precambrian basement, covers a larger area as it occurs in both the Suqah and the Shumaysi troughs. Here it is considered to be an early syn-rift deposit. Its late Oligocene to early Miocene age indicates an early fluvial and later marginal marine and freshwater deposits on this eastern margin of the early Red Sea rift. The sequence is contemporaneous with the similar fluvial to shallow marine Jabal Tayran Formation described from the Midyan area in north-western Saudi Arabia. Together, these two sections indicate early uplift and erosion (basal unconformity), locally derived siliciclastics in the earliest syn-rift phase followed by marine flooding and/or lacustrine conditions in low relief coastal plains. Tuffs and a lava flow within the Shumaysi indicate the onset of volcanism in this area which becomes much more important and widespread further to the south in the age-equivalent Jizan group of south-western Saudi Arabia.

The late Oligocene age of the marine sediments within the upper part of the Shumaysi Formation agree with recent results from Sudan (Schroeder et al., this volume) and Hughes and Beydoun (1992) offshore Ethiopia that marine flooding of the central part of the Red Sea basin occurred at this time.

The lava flows within and on top of the Shumaysi Formation are important in establishing a time framework of rifting. The tilted late Oligocene–early Miocene sediments and lava of the Shumaysi section are the youngest tilted strata seen in the sections studied. Unconformably overlying the tilted Shumaysi Formation are horizontal basalts of the Hammah Formation dated as late Miocene to Pliocene (see Chazot et al., this volume). This indicates that extensional faulting and tilting in the area had finished by late Miocene times and that the main phase of extension in the central Red Sea must have been during the mid Miocene.

CONCLUSIONS

Pre-rift and early syn-rift deposits occur in fault-bound troughs, parallel to the Red Sea margin in the vicinity of Jeddah, Saudi Arabia. The earlier Usfan Formation occurs only to the north-east of Jeddah, rests unconformably on the Precambrian basement and comprises fluvial sands and conglomerates capped by a transgressive–regressive unit of limestones, dolomites and gypsiferous shales. K-Ar dating and biostratigraphic data give a large age range from Maastrichtian to the Eocene. The younger Shumaysi Formation outcrops to the east of Jeddah where it rests unconformably on the Precambrian basement and also to the north-east of Jeddah where it rests erosively on the Usfan Formation. The Shumaysi comprises a lower fluvial unit of sandstones and conglomerates, a middle freshwater shale and siltstone unit with oolitic ironstones, and an upper sequence, formed in marine to freshwater conditions, of shales, silts and cherty limestones. Palaeontological and radiometric dating both indicate a late Oligocene to early Miocene age.

Regional comparisons indicate that age equivalents of the Usfan Formation do not occur in northern Saudi Arabia, but that pre-rift strata of this age do occur in the Gulf of Suez (Nubian, Esna and Thebes Formations). To the south, the Usfan is probably laterally equivalent to the Tawilah Formation of the Jizan area and the Tawilah Group of Yemen. Similar age equivalents of the Shumaysi Formation occur in the Jabal Tayran Formation of the Midyan area and sedimentary sections within the volcanic, siliciclastic rocks of the Jizan group of southern Saudi Arabia.

The late Oligocene–early Miocene marine transgression within the upper Shumaysi unit indicates marine conditions in the central Red Sea at this time and supports recent reports of late Oligocene marine conditions in Sudan and offshore Ethiopia.

There was a mid Miocene phase of extension and tilting subsequent to early rift basin sedimentation as indicated by late Miocene–Pliocene flat-lying basaltic lavas unconformably overlying the Shumayusi Formation.

ACKNOWLEDGEMENTS

The authors wish to express their deep thanks to Dan Bosence for fruitful and critical discussions in the field and for assistance with writing and compilation of this chapter. We thank the British Council (Jeddah) for a travel grant for M.A. Abou Ouf to visit the UK to work on this chapter.

Chapter C3
The sedimentary record of the initial stages of Oligo-Miocene rifting in the Gulf of Suez and the northern Red Sea

C. Montenat, F. Orszag-Sperber, J.-C. Plaziat and B. H. Purser

ABSTRACT

The late Oligocene–early Miocene strata in the Gulf of Suez and the northern Red Sea were deposited in response to the morphotectonic conditions prevailing during the initial stages of rifting.

The Precambrian basement and pre-rift Mesozoic and Eocene cover have been eroded but there is no evidence of doming prior to initial syn-rift sedimentation. The early-rift Abu Zenima Formation of the eastern and central Gulf of Suez together with the equivalent Group A clastics of the north-west Red Sea are composed of both fine and coarse continental, reddish clastics. The conglomeratic units are derived from the uplifted Eocene carbonates and subsequently from Precambrian basement, and were deposited as both proximal alluvial fans and in channels. These sediments occur predominantly in structural depressions oriented oblique with respect to the axis of the rift. These sub-basins were formed by initial post-Eocene strike-slip movements and then extensional block tilting. They are locally filled by marine evaporites or open marine deposits indicating a marine transgression.

The transition from continental to marine conditions generally occurs in response to block tilting and appears to be slightly diachronous. The earliest marine sediments are Aquitanian at Ras Honkorab (north-west Red Sea) while the earliest marine facies (Nukhul) in the Gulf of Suez ranges from Aquitanian to Burdigalian in age. Differences in radiometric ages of associated basalts tend to confirm this slight diachronism.

Differences between the evolution of the Arabian and African rift margins is reflected in the relative intensity of volcanism and by the thickness and ages of early rift sediments. Marine facies are markedly thicker both in Sinai and in the Midyan area of Saudi Arabia where initial open-marine sedimentation is Chattian, as opposed to Aquitanian to Burdigalian on the opposing African margin. This regional synthesis thus confirms both a longitudinal and transverse asymmetry of the north-west Red Sea–Gulf of Suez rift.

INTRODUCTION

This chapter focuses on the sedimentary aspects of rift initiation (late Oligocene–early Miocene) in the Gulf of Suez and the north-west Red Sea and the lithostratigraphic units concerned are reviewed with respect to their age and sedimentological characters.

Additional data concerning chronological and tectonic aspects are dealt with in chapters by Bosworth et al., Montenat et al., and Plaziat et al. (this volume).

GEODYNAMIC FRAMEWORK

Pre-rift history

The Precambrian basement belongs to several superimposed orogenic cycles, the latest being the 'Panafrican' tectonomagmatic phase (about 500 Ma), and by early and late Palaeozoic magmatic events. These resulted mainly in the intrusion of granitic plutons and dikes (Gebels Dara and Gharib in Gharamul area, Gulf of Suez; Ott d'Estevou et al., 1986a).

The pre-rift sedimentary cover, unconformably overlying the Precambrian basement, attains 2000 m in the northern Gulf of Suez and in general is a transgressive sequence. It is topped with Eocene carbonates of the Thebes Formation (Said, 1962). This Eocene outcrops in the Nile Valley as well as on the Red Sea coast and extended over the Arabo-African platform before its breakup. It is thus independent from the rift formation. Moreover, the facies and thickness are arranged in east–west belts and, in agreement with Garfunkel and Bartov (1977) there is no reason to relate these sediments to a proto-rift (Kostandi, 1959).

Contact between pre-rift and early syn-rift sediments

The pre-rift sediments covered a wide area of the Gulf of Suez, during the initiation of the rift. Industrial wells located in the central part of the Gulf, show a similar amount of erosion of pre-rift strata as their equivalent on the periphery (Richardson and Arthur, 1988). This indicates, at least in this area, that the rifting was not preceded by a general doming and a widespread erosional phase (Garfunkel, 1988; Bohannon et al., 1989).

The pre-rift cover also is preserved in the north-west part of the Red Sea, as far as 30 km south of Quseir (Figure C3.1) (Orszag-Sperber and Plaziat, 1990). Further south (Abu Ghusun), the cover, initially probably thinner (Orszag Sperber and Plaziat, 1990; Patton et al., 1994), was completely eroded prior to the deposition of the early rift sediments. At an equivalent latitude, along the Arabian margin, it appears to have been more deeply eroded (Purser and Hötzl, 1988).

The first sediments referable to the early syn-rift depositional phase (late Oligocene, see below) rest unconformably on the pre-rift cover. In most cases, the discordance is less than 10° or barely perceptible. Locally, close to the north–south strike-slip faults (Quseir area; Roussel et al., 1986), the first beds of the syn-rift sequence overlie, with an angular discordance of 5°, the tilted Eocene.

When overlying the pre-rift cover the first syn-rift clastics begin with a submonomictic breccia of angular cherts together with blocks of limestone, supplied from the underlying weathered and karstified cherty limestones of the Thebes Formation (Middle Eocene). The variable thickness of the breccia reflects the filling of an irregular surface. These clastics are overlain by well-sorted sands provided by the erosion of the Nubian sandstones. This succession of chert breccia and sandstone is typically illustrated by the Nakheil formation (Figure C3.2) (Wadi Nakheil, west Quseir), which rests conformably on the Thebes limestones of the Gebel Duwi (Jarrige et al., 1986).

The contact between the pre-rift and the early, syn-rifting sequences suggest a long period of planation and weathering before the deposition of the syn-rift clastic sediments.

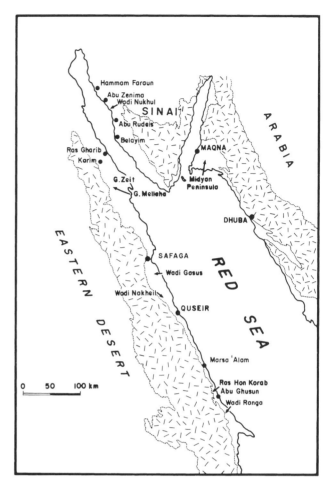

Figure C3.1 Location of sites mentioned in the text.

EARLY RIFT SEDIMENTS

Lithostratigraphic nomenclature

The rift basin is initiated by the deposition of late Oligocene–early Miocene clastic sediments. These sediments were deposited within tectonically controlled depressions which do not always coincide with the down-faulted N140 blocks formed later during rift development (i.e. west-north-west–east-south-east trending Hamadat corridor, south of Quseir; Roussel et al., 1986). For that reason, the term of 'proto-rift' was proposed, to refer to this early stage of sedimentation (Orszag-Sperber and Plaziat, 1990; Plaziat et al., 1990).

The early rift sediments discussed in this contribution are considered part of the Abu Zenima and Nukhul Formations in the Gulf of Suez, late Oligocene to early Miocene in age (see Plaziat et al., this volume).

In the north-western Red Sea, the lithostratigraphic nomenclature is complicated by numerous local formation names. Here, lithologic units equivalent to those of the Gulf of Suez are the Abu Ghusun and Nakheil Formations and probably the lower Um Abbas and

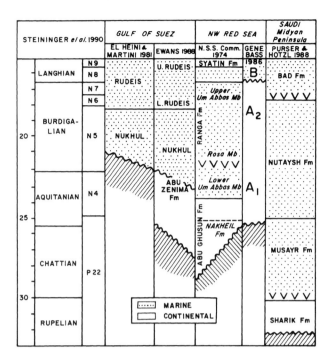

Figure C3.2 Stratigraphic nomenclature of early rift deposits in the northern Red Sea, Gulf of Suez and Midyan Peninsula (Saudi), after NSS Comm. (1974); Montenat et al. (1986); Evans (1988); Purser and Hötzl (1988).

Rosas Members, belonging to the Ranga Formation (Figure C3.2). On the Saudi margin (Midyan Peninsula), the Musayr and Nutaysh Formations occupy the same interval, between the late Oligocene (Chattian) and the early Miocene (Aquitanian–early Burdigalian) (Purser and Hötzl, 1988; also see discussion below and figures).

A more simple terminology, based on tectono-sedimentary units (Montenat et al., 1986) comprises four major units, Groups A, B, C and D, each separated by a regional unconformity. The Group A (late Oligocene–early Burdigalian), relates to the early stage of the rifting and corresponds approximately to the above-mentioned formations. This group is made of two sub-groups: the reddish continental clastics A_1 and the subaquatic deposits A_2 (Figure C3.2). Group A is overlain, locally with unconformity, by the marine Group B (Plate 13).

Early rift deposits in the Gulf of Suez

Eastern margin
Well-known sections are exposed in the Abu Rudeis area (Garfunkel and Bartov, 1977; Ott d'Estevou et al., 1986b; Evans, 1990; Patton et al., 1994), specially at Abu Zenima and Wadi Nukhul. In these sections, two major facies are observed (Figure C3.3):

- reddish or variegated continental clastics, at the base, locally associated with basalts, referred to as the Abu Zenima Formation (i.e. the A_1 sub-group).

- yellowish marine detrital deposits, in the upper part, referred to as the Nukhul Formation (the A_2 sub-group).

The base of the early-rift series is of particular interest at Wadi Nukhul (Ott d'Estevou et al., 1986a), where pre-rift strata consist of marly limestone (Middle Eocene) and grey-beige marls (10 m) containing a planktonic-rich assemblage of Late Eocene (*Globigerinapsis semiinvoluta* zone). These open marine marls are overlain, without any notable transition, by lacustrine limestones (0.80 m). A large basalt dike, oriented N140 (Clysmic) cut these sediments (Figure C3.3). The Abu Zenima Formation is unconformably overlain by marine bioclastic detrital sediments, sandstones, marly sandstones and silty marls of the Nukhul Formation. These are interpreted as littoral sandbars with large-scale oblique bedding and ripples at the base, passing upwards into marly sand-bodies related to gravity deposits (grain flows with massive reworking of Eocene microfauna) alternating with planktonic silty marls upwards. This sedimentary sequence clearly indicates the increase in water depth.

At Abu Zenima, these marine clastics begin with a conglomerate overlying an erosional surface at the top of a basalt flow which overlies the variegated clastics of the Abu Zenima Formation.

Locally (Gebel Ekma, north Abu Durba), these marine detrital Nukhul deposits exhibit intense syn-depositional deformations, such as various gravity features (slumps, olistoliths, mud-flows and debris-flows), rapid variations in thickness and synsedimentary faulting. Planktonic and large benthic foraminifera indicate an Aquitanian–early Burdigalian age of the Nukhul Formation (El Heini and Martini, 1981; Ott d'Estevou et al., 1986a; Steininger et al., 1990).

A marine unit, comprising calcarenites and reefoid carbonates, unconformably overlies the Nukhul Formation and may rest directly on the Eocene pre-rift cover. These unconformable deposits, widely outcropping on the Sinai margin, contain a late Burdigalian microfauna and are referred to as the Lower Rudeis Formation (Garfunkel and Bartov, 1977). The discordance between the Nukhul and overlying Rudeis Formation (i.e. 'Post Nukhul event' of Evans, 1988) is the upper limit of deposits discussed in this chapter.

Western margin
Group A deposits related to the early stages of rifting are limited to scattered outcrops on the western margin. They consist mainly of reddish continental clastics correlated with the Abu Zenima Formation (sub-group A_1) of the eastern margin. In the Gharamul area (Ott d'Estevou et al., 1986a) these deposits rest with a slight unconformity on Eocene limestones. They include a basal breccia of angular unconsolidated blocks of

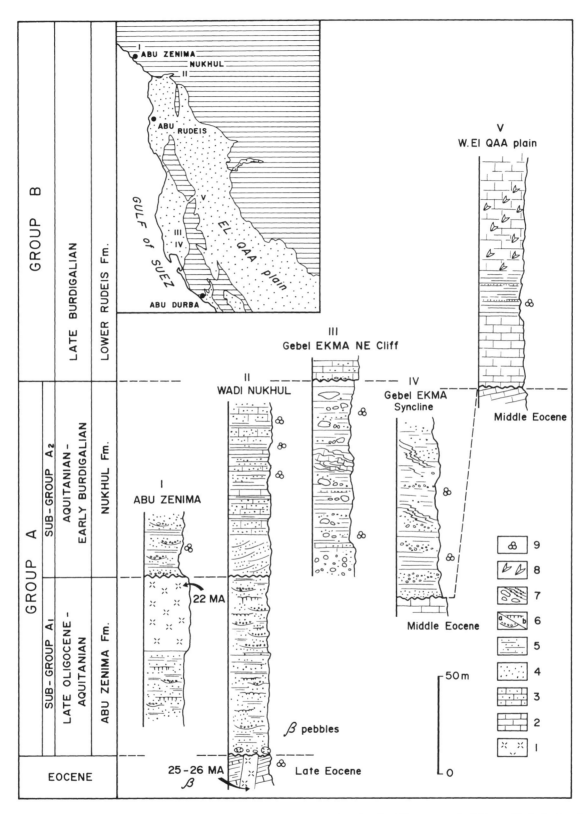

Figure C3.3 Correlation of early-rifting deposits outcropping on the Sinai margin (after Ott d'Estevou et al., 1986, modified). 1 – basalt; 2 – carbonate; 3 – sandy-carbonate; 4 – sandstone; 5 – sandy marls; 6 – a. channel, b. paleosol; 7 – olistolith and slumps; 8 – gypsum replacing carbonate; 9 – micropalaeontological data.

Eocene cherts and limestones, located in discontinuous depressions. Alternating beige to reddish sands or sandstones, and yellowish sandy marls compose the bulk of the strata. Laminated carbonate and gypsiferous marls, locally interbedded with the clastics suggest temporary playas within a flat alluvial plain. The same kinds of deposits outcrop at the north-west extremity of the Mellaha block (Gebel Tarbul; Prat et al., 1986).

Outcrops related to the marine Nukhul Formation are absent, except for biogenic carbonates at Gebel el Zeit dated as Aquitanian (Evans, 1988). This single example can be explained by the specific position of the Gebel el Zeit, as an uplifted block situated near the axial part of the Gulf.

Axial part of the Gulf
Continental sediments, corresponding to the Abu Zenima Formation, are poorly developed or absent in the industrial wells (Richardson and Arthur, 1988; Evans, 1990). Fluviatile sandstones and shales occurring only in the Southern Gulf are considered part of the Shoab Ali Member and regarded as the basal unit of the Nukhul Formation (Saoudi and Khalil, 1984); however these could be as well correlated with the continental clastics of the Abu Zenima Formation.

In most cases, in the sub-surface, the first sediments overlying the pre-rift strata are of marine origin, including reefoid and bioclastic limestones (Gharamul Member), marine sandstones and conglomerates (October Member) laterally replaced by anhydrite and marls (Ghara Member) in more restricted environments (Saoudi and Khalil, 1984). The total thickness of these deposits is highly variable, from 0 to 700 m (Richardson and Arthur, 1988), depending on the syndepositional structure of the substratum.

Early syn-rift deposits of the north-western Red Sea

Egyptian margin
Basically, two kinds of sediments are exposed:

- reddish continental clastics (sands, sandstones and conglomerates) are similar to the Abu Zenima Formation (A_1 sub-group). They are widely distributed and refer to the Abu Ghusun and Nakheil Formations and/or? to the Lower Um Abbas Member of the Ranga Formation (Plaziat et al., this volume).
- evaporites interbedded with marls, which correspond to the A_2 sub-group and are correlated with the Rosa Member, an evaporite unit of the Ranga Formation (Plaziat et al., this volume).

Abu Ghusun and Nakheil Formations (A_1 sub-group)
North of Quseir (Wadi Nakheil), the base of the reddish continental clastics (Abu Ghusun Formation, about 200 m) includes conglomerates composed of cherts and sandstone pebbles with a sandy matrix (Nakheil Formation) (Roussel et al., 1986; Roussel, 1986; Thiriet et al., 1986; Figure C3.5). However, 250 km further south, at Abu Ghusun-Ras Honkorab, Precambrian granite pebbles appear at the base of the sequence (Montenat et al., 1986).

In Wadi Gasus (south Safaga) as with the Abu Ghusun area, series of mudstones and sands occur in the proximal continental deposits (Figures C3.4(b), C3.5(c)). These are mainly debris-flows (Figure C3.4(b)) and mud-flows (Figure C3.5) lacking basal erosional surfaces, intercalated within finer, flood-plain mudstones. Mudstones are often thinly bedded, with frequent desiccation cracks (Plate 14(b)). In distal sections, sands and muds are deposited as fining-upwards sequences with marked basal erosional surfaces (Figure C3.5(e)). Depositional structures and textures suggest wadi-type channel sedimentation grading downstream into flood-plain sediments with rare mud-flows.

Local coarsening and thickening upward metre-thick sequences suggests playa-type deposits. These may be associated with sands showing horizontal, millimetric lamination of aeolian origin and limestones deposited in lakes (fish scales in thinly bedded limestone at Abu Ghusun) or palustrine environments (massive nonlaminated beds with root-traces at Wadi Gasus).

Grain-size distribution of the sediments (Figure C3.6) suggests that the particules are current-transported in channels, through the flood plain. Deposition by decantation appears only in lakes and in the abandoned channels. In the upper part of the sequence, sedimentation becomes more dominated by fluviatile deposits, with the appearance of crystalline basement pebbles indicating the uplift of the periphery. The presence of synsedimentary deformational structures suggests an increase in seismicity (Plaziat et al., 1990, this volume) (Figure C3.3(d)).

Subaquatic deposits (sub-group A_2)
Locally, the continental clastics are overlain by a succession of gypsum beds alternating with marls, diatomites and laminated carbonates corresponding to the sub-group A_2 (Ras Honkorab, north of Abu Ghusun and Wadi Gasus, south Safaga; Montenat et al., 1986; Thiriet et al., 1986) and tentatively correlated with the Lower Umm Abbas Member.

At Wadi Gasus (Safaga), the evaporite series (20 to 30 m) (see also section on evaporites in Orszag-Sperber et al., this volume) includes several beds of laminated crystalline gypsum (1 to 5 m each in thickness), alternating with yellowish silty marls, conglomerates and sandstones, with graded bedding, flutemarks and loadmarks, suggesting gravity-flow deposition. The western edge of this evaporite basin is indicated by the development of laminated dolomicrite or dolomicrosparite

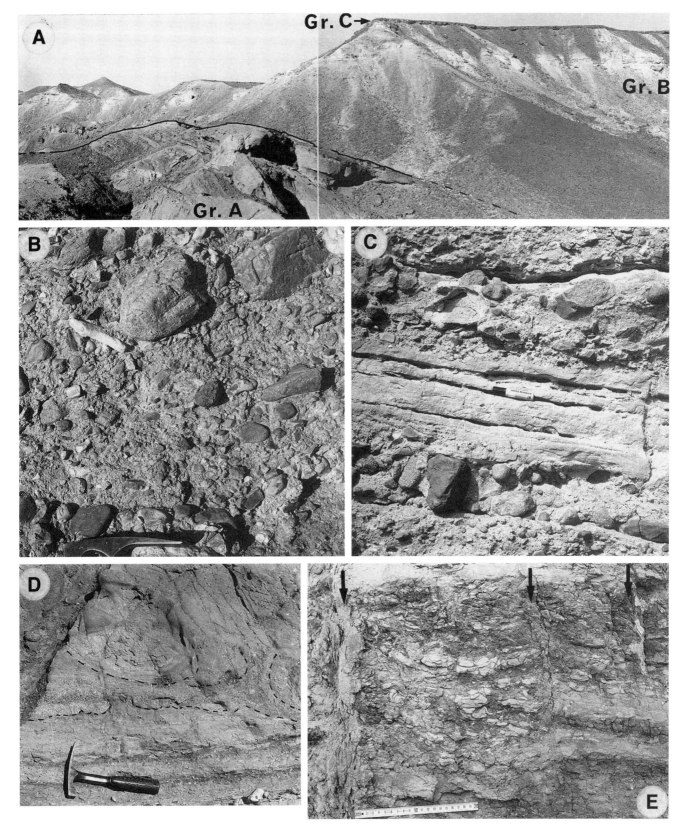

Figure C3.4 Outcrops and details of the Group A deposits. (a) Discordance between Group A continental sandstones and Group B marine limestone (left) and sandstones, followed by stromatolites. The plateau is topped by laminated dolomite marking the end of Group B (Wadi Sharm el Qibli, S Quseir). (b) Debris-flow with muddy matrix. These poorly sorted clasts include slightly rounded Eocene and Precambrian pebbles typical of proximal alluvial fan deposition (N Wadi Asal, S Quseir). (Hammer for scale). (c) Alternating bedded sands and poorly sorted gravels with sandy matrix, interpreted as wadi deposits (Wadi Sharm el Bahari, S Quseir). (d) Hydroplastic deformation with interbedded muddy sandstones and gravelly mudstones initially deposited as mud flows, parts of which were subsequently remobilized during earthquakes (Wadi Sharm el Bahari). (e) Sandy flood plain mudstones whose bedding is cut by deep desiccation cracks filled with sand (arrows) (Mohamed Raba Basin, SW Safaga).

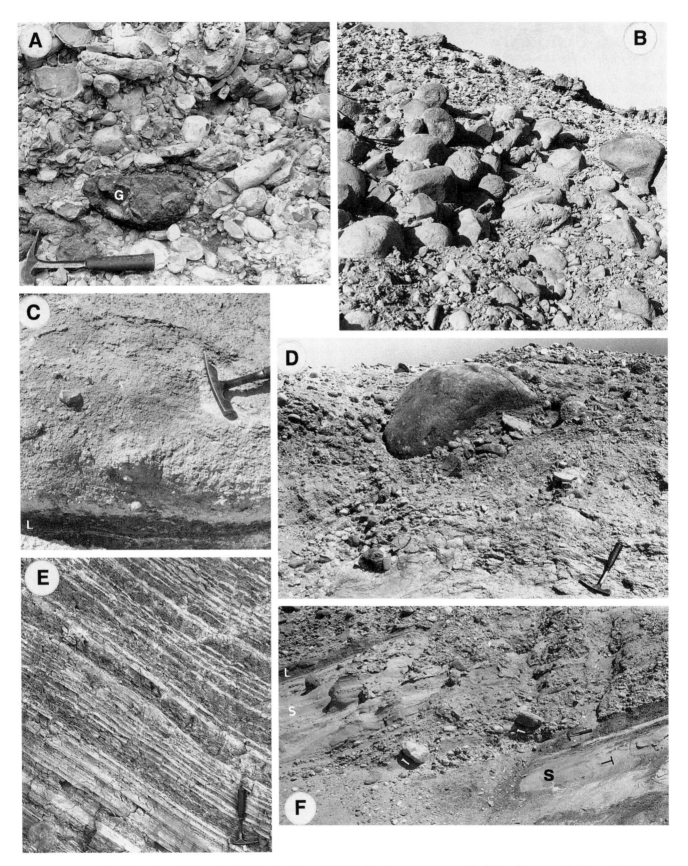

Figure C3.5 Facies of the Group A deposits (Abu Zenima Formation). (a) Conglomerate composed of chert from pre-rift (Eocene Thebes Formation) and sandstones (Cretaceous Nubian Formation), with sandy matrix. This facies is typical of the Nakheil Formation (Gebel Atshan, W Quseir). (b) Conglomerate with rounded granite blocks (30–50 cm), and muddy sand matrix (Wadi Asal, S Quseir). (c) Muddy sand-flow deposit with scattered Eocene chert cobbles, overlying flood-plain mudstones (L), in the lower part of section at Wadi Gasus (S Safaga). (d) Muddy, sandy conglomerate with large (80 cm) blocks. The muddy sands at the base are deformed by the overlying, 2 m thick debris flow (Hamadat depression, S Quseir). (e) Interbedded mud and sand with upwards fining granulometry and bed thickness; subaquatic distal deposits at Wadi Gasus. (f) Extreme variability of proximal Group A deposits. The lower finer deposits include a sandy bed inclined to the south, intercalated between mudstones (L). These are overlain by a series of sand-flow deposits which include blocks inclined in a downstream direction. Each flow exceeds 1–5 m in thickness and lacks an erosional base. A 20 cm thick muddy horizon (L) separates these sand-flows from a poorly sorted conglomeratic bed (Eocene chert cobbles and granite boulders). Hammer for scale. (Hamadat depression).

which locally display a vuggy fabric due to dissolution of gypsum crystals. These carbonates exhibit a cauliflower microstructure related to microbial growth. These peripheral carbonate deposits rest conformably on the continental clastics (A_1) or unconformably, on the pre-rift sedimentary cover (Thiriet, 1987).

At Ras Honkorab, the evaporites are thicker (100 m) than at Safaga. The deposits are located within a system of N120 trending depression (Figure C3.7(b)), between the Gebels Honkorab and Rafa el Tanish (Figure C3.7(a); Montenat et al., this volume). Evaporites lie directly on metamorphic basement, the reddish continental series (A_1) is exposed only in the southern area (Montenat et al., 1986). The facies pattern of the evaporites is controlled by syndepositional faulting. Basinal deposits composed of several episodes of laminated microcrystalline gypsum (< 10 m) interbedded with laminated green marls, diatomaceous beds and laminated dolomicrites, are best developed within half-graben, while the crest of blocks, fault-scarps and fault-scarp breccias are coated with vuggy, brecciated dolomite, showing traces of algal lamination. Stromatolitic growths are located on the crest of blocks. Smallest half-graben (decametric) are filled only with laminated carbonates.

At Ras Honkorab, as well as at Safaga, the evaporites are contemporanous with tilt-block faulting, as is clearly shown in the Ras Honkorab area (Montenat et al., 1986, this volume; Ott d'Estevou et al., 1987).

At Wadi Gasus (south Safaga), the gypsum beds, dipping at about 25°, are conformable on the underlying continental beds, these latter exhibiting frequent syn-sedimentary deformation structures reflecting the first phases of tilting. This tilting may have led to the transition from subaerial to subaquatic sedimentation. The resulting subaquatic gypsum has been locally displaced to form large (10 m) lenses associated with olistolithes (6 m) of consolidated marls containing planktonic microfauna (Lower Miocene, Thiriet, 1987).

At both Ras Honkorab and Wadi Gasus, these evaporites are covered unconformably by subsequent open-marine deposits (Group B) (Thiriet, 1987; Montenat et al., 1986). These evaporites represent the first marine ingression on the north-western margin of the rift, and thus may be correlated with the Aquitanian–early Burdigalian Nukhul Formation, which includes an evaporite Ghara Member (Saoudi and Khalil, 1984).

Sudanese margin

On the Sudanese coast, north of Port Sudan, the syn-rift sedimentary sequence begins with a succession similar to its Egyptian equivalent (Montenat et al., 1990; Figure C3.8).

- Red shales and silty or sandy clays interbedded with numerous basalt flows and some basaltic tuff layers (cf. A_1 sub-group or Abu Ghusun Formation).
- Alternation of evaporites (laminated white crystalline gypsum beds, 1 to 5 m in thickness) and green marls associated with laminated diatomites, which rest conformably on the red beds (cf. A_2 sub-group).

These deposits record tilt-block extensional tectonics before burial by marine calcarenites and clastics probably of Early to Middle Miocene age, correlateable with

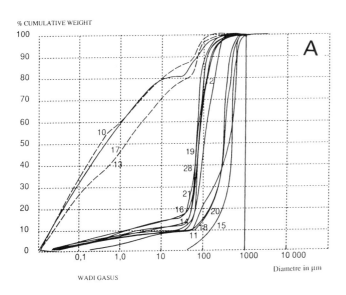
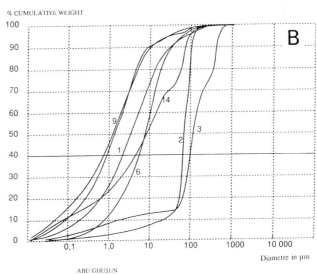

Figure C3.6 Grain-size distribution curves in the Abu Ghusun (Zenima) Formation (sub-group A1). (a) Wadi Gasus: curves 10, 13, 17 show an hyperbolic facies indicating a deposition of particles by settling in a calm subaquatic environment (lake, abandoned part of a channel). The other curves showing a parabolic facies indicate particles transported and deposited by a traction current. Note the relative abundance of fines particles (< 100 μm) (after Roussel, 1986). (b) Abu Ghusun: grain-size distribution indicative of deposition in a flood plain, of not well sorted particles, except for curves 2 and 3.

the marine group B deposits of Egypt (Figure C3.9; Montenat *et al.*, 1990).

Saudi Arabian margin

Outcrops in the Midyan region extend from Dhuba in the south, to Maqna (Gulf of Aqaba), including the Midyan Peninsula (Figure C3.1). The lithostratigraphic succession (Dullo *et al.*, 1983; Le Nindre *et al.*, 1986; Purser and Hötzl, 1988) is more complete and diverse than that of the Egyptian margin and can be summarized as follows (see details in Purser and Hötzl, 1988).

- Coarse clastic continental sediments of the Sharik Formation (?Rupelian).
- Yellowish to grey marine limestone, locally reefoid and intercalated sandstones of the Musayr Formation (Chattian microfauna, late Oligocene). The Musayr Formation conformably overlies the Sharik Formation, or rests directly on the faulted crystalline basement (Jabal Musayr). At Jabal Tayran an evaporitic episode (35 m) marks the transition between the Sharik and Musayr Formations (Figures C3.2, C3.11).
- Marine terrigenous deposits (sandstones, marls and siltstones) and intercalated reef limestone of the Nutaysh Formation (Aquitanian to Early Burdigalian, i.e. contemporaneous to the Nukhul Formation). These overlie the upper Oligocene Formation (Musayr Formation) with slight unconformity.

The pronounced unconformity separating the Nutaysh Formation and the overlying terrigenous marine deposits of the Bad Formation (Middle to Upper Miocene) can be related to the post-Nukhul event of Evans (1990).

CLIMATE

Numerous sedimentary and diagenetic features indicate a relatively arid climate during the early stage of rift sedimentation. The important role of debris-flow sediments and the general architecture of the deposits are indicative of alluvial-fan systems. Highly weathered materials are systematically mixed with feldspathic sandstones.

In susceptible muddy alluvial sediments there are deep desiccation cracks (Plate 14) extending down to 1 m in massive or layered mudstones. However, the most significant climatic expression is the weak development

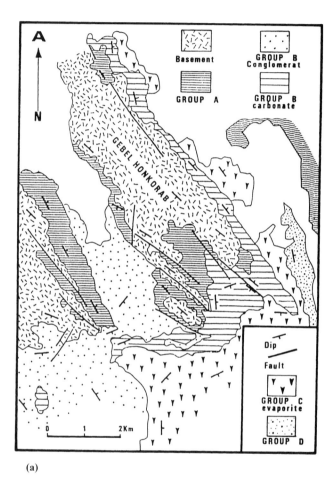

(a)

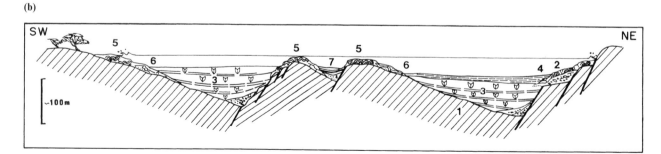

(b)

Figure C3.7 (a) Simplified geological map of the Ras Honkorab area, north Abu Ghusun. The small fault blocks, with graben filled by evaporites of the sub-group A2, are overlain unconformably by marine Group B deposits. (b) Syntectonic deposition of evaporites (sub-group A2) in the Ras Honkorab area, N Abu Ghusun. 1. faulted basement; 2. fault-scarp breccias coated with dolomitized stromatolitic talus; 3. evaporites interbedded with diatomites, marls and anoxic laminated carbonates; 4. green silty marls with *Globigerina* onlapping the talus; 5. stromatolitic build-up; 6. slumped and brecciated stromatolite material; 7. small half-graben filled with anoxic laminated carbonate.

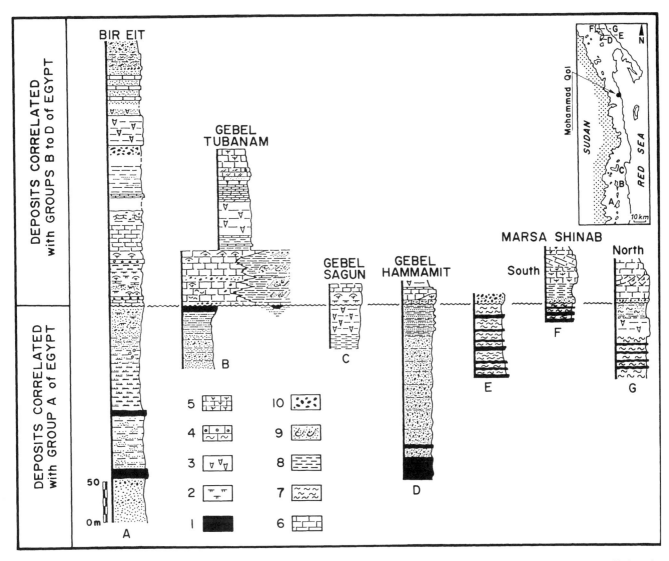

Figure C3.8 Correlation of syn-rift deposits outcropping on the Sudanese margin, north of Port Sudan, with special reference to early-rift deposits (equivalent to Group A) (after Monetant et al., 1990, modified). 1. basalt flow; 2. palustrine carbonate and paleosol; 3. evaporites; 4. oolitic and coquina limestone; 5. reefal carbonate, carbonate; 6. reddish marl and silt; 8. marls; 9. sandstone; 10. conglomerate.

of alluvial paleosols which are partially responsible for the preservation of mud cracks. In low lying areas reoxidized gleys (ferruginous mottling) and Fe-Mn nodules (north Wadi Abu Ghusun) and palustrine limestones (Mohamed Raba, Gebel Ashtan or Gebel Tarbul to the north-west Gebel Mellaha) are also developed. The same type of sedimentation occurs along the Sudanese Red Sea coast (Montenat et al., 1990). These attributes suggest a semi-arid, dry tropical climate, with a rainy season sufficiently long to account for active alluvial, flash-flood transportation terminating in overflows into closed basins. The occurrence of large granite boulders (2 m) in sandy deposits, on gentle slopes (Figure C3.5(b),(d), Plate 14(a)), at a distance of several kilometres from the basement sources, clearly implies flash-flood transportation. Silicified woods support this climatic interpretation, in agreement with the existence of ephemeral lakes (for example in the Gharamuls, Ott d'Estevou et al., 1986a).

SYNSEDIMENTARY DEFORMATION RELATED TO EARLY RIFTING

An overview of early rift tectonics

The structure of the Gulf of Suez and north-western Red Sea rift is controlled, at various scales, by four major trends of faults: N00–20 (Aqaba), N100–120 (Duwi), N140 (Clysmic) and N40–60 (Cross), inherited from the structure of the Precambrian basement (Garfunkel and Bartov, 1977; Jarrige et al., 1986). The combination of these different fault trends results in a zigzag fault pattern which characterizes the morphostructure of the

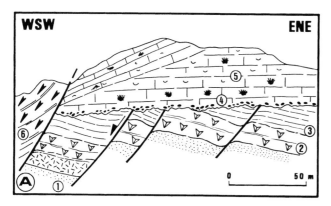

Figure C3.9 Example of syn-rift deposits outcropping at Marsa Shinab, N Port Sudan. 1. red series and basalt flow (to compare with the A1 sub-group on the Egyptian margin); 2. evaporites; 3. green marls (sub-group A2); 4. major unconformity and boulder horizon with basement material; 5. biogenic carbonates locally reefal, with syndepositional truncation (WSW part) (marine Group B); 6. evaporites (Group C). After Montenat et al., 1990)

rift and directly influences the distribution of syn-rift sediments (see Montenat et al., this volume).

Sealed structures and polyphased fault movements indicate at least two major stages of deformation related respectively to compressive strike-slip faulting and extensional tectonics which occurred during the early rifting phase of sedimentation (i.e. group A deposits; Montenat et al., 1986). The transition from one pattern to another may be slightly diachronous within that interval (Oligocene pro-parte-early Burdigalian) (Figure C3.11).

Syndepositional deformation structures related to strike-slip movements

Strike-slip movements along N00°–20° Aqaba faults (sinistral) and N100°–120° Duwi faults (dextral) are responsible for the preferential accumulation of sediments (red continental clastics of the Abu Zenima and related formations) as a succession of discontinuous depocentres located within these tectonic corridors. Such depocentres resemble highly structured synclines and record spectacular syndepositional unconformities (Mohamed Raba, near Safaga, Thiriet et al., 1985; Gebel Tarbul, north-west of the Gebel Esh Mellaha, Montenat et al., this volume). The strike-slip faults and these depocentres are oblique to the general orientation (Clysmic) of the rift, and include the N110°–120° Duwi-trending Hamadat corridor, south Quseir (Roussel et al., 1986; Orszag-Sperber and Plaziat, 1990; Patton et al., 1994).

On the Sinai margin, strike-slip deformation frequently affects the Nukhul Formation and relates mainly to N00°–20° Aqaba faults. Strike-slip movements produce folds of kilometric scale, often with high-angle dip axis, which evolved contemporaneously with sedimentation (Gebel Ekma, north of Abu Durba).

These structures are sealed by late Burdigalian deposits (Rudeis Formation) (Ott d'Estevou et al., 1986b).

First appearance of the tilt-block pattern

At Ras Honkorab, the initiation of the tilt-block pattern is clearly recorded during the deposition of the evaporite unit (A_2 sub-group) (Figure C3.7(a),(b)). This large-scale fault-block pattern was preceded by the formation of numerous faults (mainly N140°, south-west dipping faults of decimetric to metric scale) within the red clastics of the Abu Ghusun Formation (Montenat et al., 1986). These generate a minor antithetic fault pattern, possibly related to strike-slip motion, controlled by the prevailing Duwi fault trend.

EVOLUTION OF RELIEF DURING EARLY RIFTING

The composition of the clastic materials within the rift fill provides information concerning the rocks which existed on the rift shoulders.

Pre-rift unroofing and weathering

The pre-rift sedimentary cover was originally very variable in thickness. It is thinning from 2000 m in the north of the Gulf of Suez to several hundred metres to the south, in the northern Red Sea. As noted by Patton et al. (1994) this thinning also reflects an early unroofing of the basement on the rift margins in the southern area (south of Quseir) where the pre-rift sedimentary cover was totally eroded prior to deposition of the earliest syn-rift sediments, while the basement was exhumed later in time in the northern Gulf of Suez.

Obviously, it is easier to study the evolution of the erosion and marginal reliefs in the area where the pre-rift sedimentary cover was preserved at the beginning of the rift history.

As previously noted, in the areas devoid of pre-rift sedimentary cover, the crystalline basement was deeply weathered before being covered by the first syn-rift clastics (Abu Ghusun). When the pre-rift sedimentary cover is present, the weathering of Eocene cherty limestones gave rise to eluvial deposits such as the sporadic accumulation of autochtonous chert breccias at the base of the Abu Zenima or Nakheil Formations.

History of uplift and erosion

The occurrence of pebbles derived from the crystalline basement together with their size increase up section (Figure C3.10), suggests progressive increase in reliefs

(A_2 reddish deposits in Safaga and Quseir areas). At the same time, the alluvial cones lacking channels in the first hundred metres of sections, and developed a channelized fluviatile system with deep incision and coarse in-filling in the upper part of the sections.

Within the lower part of the above-mentioned reddish formations, the detrital material is derived only from the Eocene (cherts as blocks and sand-size detritus). This indicates initial low relief whose altitude did not exceed the thickness of the Eocene limestones (less than 200 m; Garfunkel, 1988; Patton et al., 1994). However, the red clastics are rapidly enriched in well-sorted, sandy material derived from the underlying Nubian sandstones. The exhumation of these sandstones indicates erosion of about 150 m (west of Safaga) to about 500 m (Gebel Duwi, west of Quseir) of the pre-rift cover during the deposition of the red clastics (A_1 subgroup). This erosion tends to be greater in certain areas located close to strike-slip faults (Mohamed Rabah, west Safaga; Thiriet et al., 1985; Gebel Tarbul, north Gebel Mellaha, Montenat et al., this volume).

The composition of the upper part of the red clastics, indicate exposure of the basement (Quseir, Safaga, Gharamuls) as reworked pebbles and cobbles of granite and metamorphic rocks mixed with debris originating from the pre-rift sedimentary cover. The total removal of the pre-rift sedimentary cover, at least locally, suggests an uplift of 300–800 m in the Quseir and Safaga areas.

In Sinai, studies based on the composition of the early rift sediments by Ott d'Estevou et al. (1986b), Garfunkel (1988), Evans (1990), as well as the fission track studies of basement apatite age analysis (Kohn and Eyal, 1981; Omar et al., 1989), show a northward decrease in age and amplitude of uplift with unroofing beginning at about 17 Ma in the southern Sinai, and at 15 Ma in the northern part. This corresponds approximately with deposits of marine Rudeis sediments.

Patton et al. (1994) indicate that some of the best data in support of a northward younging of the rift of Suez are derived from isopach and facies data of the Nukhul Formation (Saoudi and Khalil, 1986); the Shoab Ali Member (lower part of the Nukhul Formation) exists only in the southern part of the Gulf. These strata are mainly composed of reworked Nubian sandstones.

ASYMMETRY OF RIFT INITIATION IN SPACE AND TIME

Indications of north–south diachronism of rift initiation

Chronological data
The comparison of different stratigraphic sections from the Gulf of Suez and the north-western Red Sea reveals the following points (Figure C3.11).

1. At Abu Zenima, the red clastics (Abu Zenima Formation) are dated between 26–25 Ma and 22 Ma (Late Oligocene–Aquitanian pro-parte). The overlying Nukhul Formation begins after 22 Ma (within the Aquitanian) while it ends within the early Burdigalian, around 18 Ma (see above).

2. At Sharm el Bahari, near Quseir, the major part of the red clastics (Abu Zenima or Abu Ghusun Formations) underlies a basalt flow dated at 24.9 Ma (Roussel et al., 1986), and thus is older than the dates from Sinai.

3. At Ras Honkorab, north of Abu Ghusun, the lower evaporites (A_2 sub-group, part of the Nukhul Formation) are Aquitanian in their upper part (Montenat et al., 1986; Plaziat et al., this volume).

Thus, from these data, the first syn-rift deposits are diachronous and young slightly from south to north. However this diachronism appears relatively minor (less than 5 Ma) and should be supported by more precise data.

Structural data
The succession of tectonic patterns from strike-slip to extensional which characterize an early rifting stage also shows a diachronism. Data from south to north include the following.

- In the southernmost area (Abu Ghusun and Ras Honkorab) the strike-slip regime is weakly perceptible and occurred essentially prior to deposition of the red clastics sub-group A_1 (Montenat et al., 1986). On the other hand, initiation of the tilt-block stage related to the extensional tectonics is clearly recorded during the deposition of the lower evaporites (sub-group A_2).
- In the Quseir area, deformation related to strike-slip tectonics is clearly recorded by the red clastics and associated basalt (Jarrige et al., 1986), but this does not affect younger deposits.
- In the Safaga area, the red clastics (A_1 sub-group) have been affected by compressive strike-slip deformation (Thiriet et al., 1985, 1986), while the deposition of the lower evaporites (A_2 sub-group, probably Aquitanian in age) is contemporaneous with initiation of tilt-blocks (Thiriet et al., 1986; Orszag-Sperber and Plaziat, 1990).
- Northward, on the Sinai margin, the compressive strike-slip regime is clearly expressed up to the Nukhul Formation (Aquitanian pro-parte–early Burdigalian) and the related deformations are sealed by Late Burdigalian deposits (Lower Rudeis Formation; Ott d'Estevou et al., 1986b; Jarrige et al., 1990). On the contrary, southwards, on the Sudanese margin, the syn-rift sedimentary sequence does not record evidence of strike-slip tectonics (Montenat et al., 1990).

Thus, the strike-slip tectonics increase in intensity from south to north and are recorded in progressively younger sediments towards the north.

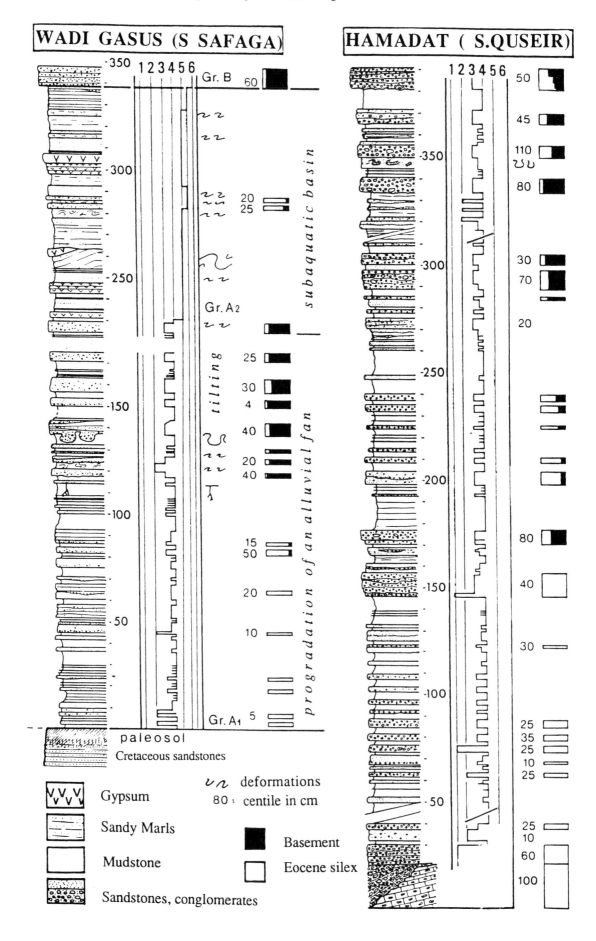

Asymmetry of rift initiation in space and time   159

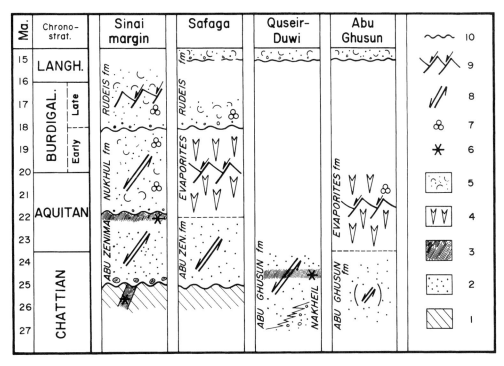

Figure C3.11 Sketch illustrating the diachronism of sedimentary and tectonic events from south (Red Sea) to north (Gulf of Suez). 1. basement; 2. reddish clastics; 3. basalt; 4. evaporites; 5. marine deposits (Nukhul Formation); 4. evaporites; 6. radiometric data; 7. biostratigraphic data; 8. strike-slip tectonic pattern; 9. tilt-block, extensional pattern; 10. unconformity.

Jarrige *et al.* (1990) have also noted that the increasing importance of strike-slip tectonics northwards coincides with the proximity of the Syrian foldbelt which is oriented in the same direction.

Transverse asymmetry of the rift initiation

Depositional asymmetry
As previously noted, the stratigraphic units related to early rifting differ on either side of the Gulf of Suez as well as the north-western Red Sea (Figure C3.11).

- On the western margin, early rift deposits (Group A) are weakly developed or absent. When present, they include continental lacustrine or restricted marine sediments (lower evaporites). Open marine sediments do not occur on this margin before the late Burdigalian–Langhian transgression (Group B; Plaziat *et al.*, this volume).

- On the eastern margin (Sinai edge and Midyan Peninsula), early rift deposits are thicker, rather widely distributed and contain open marine deposits (Nukhul Formation or its equivalent Nutasyash Formation) are predominant. The earliest marine syn-rift deposits (Musayr Formation of the Midyan area, Chattian in age) also occur on the eastern side of the rift (Figure C3.12).

This asymmetry is maintained during deposition of the Lower Rudeis Formation (late Burdigalian–early Langhian) which is well developed on the Sinai margin, but not represented on the western margin where it corresponds to an important stratigraphy gap (Plaziat *et al.*, this volume). However, more homogeneous marine sedimentation, on both rift margins occurs during upper Rudeis sedimentation (late Langhian).

Structural asymmetry
According to Purser and Hötzl (1988), Oligocene deposits of the eastern side (Midyan Peninsula) are closely related to pre-existing fault blocks. Moreover, block faulting was especially active during deposition of the Nutaysh Formation (equivalent to the Nukhul Formation). Thus, block-faulting appears to be earlier and more active on this eastern margin. This may be linked to the earlier Chattian marine transgression and

Figure C3.10 Sections through Group A sediments. The sections at Wadi Gasus (S Safaga) and at Gebel Hamadat (S Quseir), show distal and proximal continental deposits respectively. In both sections, basement (Precambrian) material, deeper erosional channel, and synsedimentary deformational structures, all appear progressively in the upper part of both sections. At Wadi Gasus, tilting of the deposits coincided with the flooding of the graben by marine waters which deposited gypsum and siliciclastic slopes deposits exhibiting slope structure. 1–6: interpretation of the environment of deposition. 1. debris-flow; 2. mud-flow; 3. fluviatile channels; 4. flood-plain; 5. permanent but restricted marine basin; 6. submarine delta. ▪ 25 = centile, i.e. the size above which only 1% of the pebbles is found. (After Orszag-Sperber and Plaziat, 1990).

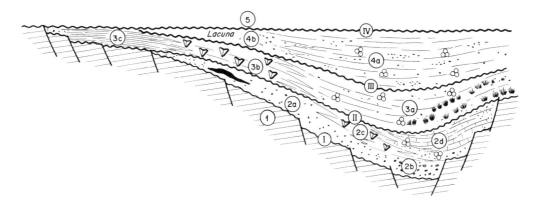

Figure C3.12 Schematic sketch of the Gulf of Suez–north Red Sea, showing asymmetry of the depositional area during late Oligocene–Early Miocene. 1. Substratum with pre-rift sedimentary cover widely preserved (Gulf of Suez) or eroded (Midyan Peninsula). The fault pattern is not related to a precise geometry; it just indicates an earlier and stronger block faulting (Oligocene) on the Arabian margin. 2a. Abu Zenima Formation (including basalts); 2b. Sharik Formation; 2c. evaporites (Jabal Tayran Formation); 2d. Musayr Formation. 3a. Open marine Nukhul and Nutaysh Formations (reefs on Saudian margin); 3b. evaporites (Aquitanian) Safaga; 3c. laminated lacustrine carbonate (W Safaga; N Gebel Esh Mellaha). 4a. Open marine deposits (Lower Rudeis Formation); 4b. marine clastics deposits of Safaga or gap. 5. Transgressive Langhian deposits (Upper Rudeis Formation or Group B). Major unconformities: I. basal unconformity (diachronous), II. pre-Nukhul unconformity, III. post-Nukhul event, IV. mid-Clysmic event. Note the merging of various unconformities on the western margin.

unroofing of the pre-rift sedimentary cover on the western margin.

Magmatic data: east–west asymmetry and north–south diachronism

Although magmatism is generally regarded as typical of the early phase of rifting, basalts are relatively rare along the north-western Red Sea margin. Several basalt flows occur on the Sudanese margin (Montenat *et al.*, 1990), a single flow is intercalated in Group A sediment at Sharm el Qibli (Egypt) and a basalt sill exists in the Gharamuls (south-western Gulf of Suez) (Ott d'Estevou *et al.*, 1986a).

In contrast, a widespread system of dikes and smaller bodies occur along the north-eastern margin of the Red Sea–Gulf of Suez. Large dikes were emplaced mainly parallel to the Red Sea–Gulf of Suez (Clysmic direction), from the Sinai Peninsula (Abu Durba, Sharm el Sheikh, Santa Catarina, Mutarish and Sheikh Atiya) to Yemen (Baldridge *et al.*, 1991, Figures 1, 2; Chazot, this volume).

The basaltic magmatism started earlier in the south of the Red Sea (30–29 to 20 Ma in Yemen, Davison *et al.*, 1994) than to the north (26–25 to 22 Ma) in the Gulf of Suez and the Red Sea (Steinitz *et al.*, 1981; Ott d'Estevou *et al.*, 1986a).

DISCUSSION AND CONCLUSIONS

1. The late Oligocene–early Miocene sediments of the Gulf of Suez, and the north-western Red Sea document the sedimentary and tectonic events which occurred during the early stage of rift sedimentation.

2. The preservation of the pre-rift (Eocene) sediments in the Gulf of Suez and on the western Red Sea (up to 30 km to the south of Quseir) indicate that doming was not significant during the early phases of rifting.

3. Rift development was preceded by a stage of planation and weathering. Thus the rifting began in an area of poorly drained and low relief hinterland.

4. The evolution from flood-plain deposits (sub-group A_1) to subaquatic environments (sub-group A_2) during early rifting may be correlated with the tectonic evolution of the area. The development of subaquatic deposits coincides with the initiation of the tilt-block faulting which generated structural depressions favouring marine deposition. Due to the geometry of the marginal dipping antithetic tilt-blocks, which favours the creation of restricted environments, sediments are generally fine-grained anoxic deposits. Such features often appear to be related to rift initiation (for example, Cretaceous South Atlantic rift; Reyre, 1984; Montenat *et al.*, this volume).

5. Composition of the clastic material within the rift fill and the chronological data indicate asymmetry and diachronism of rift initiation as follows:

- slight south–north and an important east–west decreasing age of the first syn-rift deposits;
- strike-slip faulting increases in intensity and is recorded in younger sediments from south to north;
- most of the basaltic magmatism is located on the eastern side of the rift;
- progressive unroofing and uplift from south to north.

ACKNOWLEDGEMENTS

The authors gratefully acknowledge Dan Bosence who improved the manuscript. They also thank the Centre National de la Recherche Scientifique, Elf Aquitaine and Total-Compagnie Française des Petroles (Group GENEBASS) for their financial support.

Section D
Syn-rift sedimentation and tectonics

This, the largest section in the book, comprises nine chapters focusing on the syn-rift sedimentology of the entire rift basin from the eastern Gulf of Aden to the Gulf of Suez.

The first five chapters are broader-scale treatments of the sedimentological and stratigraphic evolution of particular areas and the relationships to the structural evolution of the basin margins.

More detailed investigations from the Miocene of the north-western Red Sea and western Gulf of Suez are presented in the final four chapters and these cover more specific topics such as tilt-block carbonate platforms, coral reefs, slope deposits and synsedimentary deformation.

D1 Rift-related sedimentation and stratigraphy, southern Yemen (Gulf of Aden)
F. Watchorn, G. J. Nichols and D. W. J. Bosence

D2 Miocene isolated platform and shallow-shelf carbonates in the Red Sea coastal plain, north-east Sudan
J. H. Schroeder, R. Toleikis, H. Wunderlich and H. Kuhnert

D3 Stratigraphy of the Egyptian syn-rift deposits: correlations between axial and peripheral sequences of the north-western Red Sea and Gulf of Suez and their relations with tectonics and eustacy
J.-C. Plaziat, C. Montenat, P. Barrier, M.-C. Janin, F. Orszag-Sperber and E. Philobbos

D4 Extensional tectonics and sedimentation, eastern Gulf of Suez, Egypt
K. McClay, G. J. Nichols, S. M. Khalil, M. Darwish, and W. Bosworth

D5 Carbonate and siliciclastic sedimentation in an active tectonic setting: Miocene of the north-western Red Sea rift, Egypt
B. H. Purser, P. Barrier, C. Montenat, F. Orszag-Sperber, P. Ott d' Estevou, J.-C. Plaziat and E. Philobbos

D6 The tectono-sedimentary evolution of a rift margin carbonate platform: Abu Shaar, Gulf of Suez, Egypt
N. E. Cross, B. H. Purser and D. W. J. Bosence

D7 Miocene coral reefs and reef corals of the south-western Gulf of Suez and north-western Red Sea: distribution, diversity and regional environmental controls
C. Perrin, J.-C. Plaziat and B. R. Rosen

D8 Miocene periplatform slope sedimentation in the north-western Red Sea rift, Egypt
B. H. Purser and J.-C. Plaziat

D9 The tectonic significance of seismic sedimentary deformations within the syn- and post-rift deposits of the north-western (Egyptian) Red Sea coast and Gulf of Suez
J.-C. Plaziat and B. H. Purser

Chapter D1
Rift-related sedimentation and stratigraphy, southern Yemen (Gulf of Aden)

F. Watchorn, G.J. Nichols and D.W.J. Bosence

ABSTRACT

Syn- and post-rift successions exposed on land in a 200 km stretch along the northern margin of the Gulf of Aden in southern Yemen can be divided into three depositional sequences. Sequence 1 is a patchily developed and highly variable alluvial plain and saline playa facies assemblage of probable Oligocene age. Sequence 2 comprises fine-grained fluvial sands grading into marginal marine, coastal sabkha-type deposits with some local reefal beds. It has a cumulative thickness of between one and two kilometres and displays evidence of upward gradation to isolated depocentres. Sequence 3 facies reflect the current physiography of the margin with alluvial sediments interfingering with coastal clastic and carbonate facies.

These depositional sequences and their relationship to the rift structure indicate that rift flank uplift occurred before any significant regional extension. The main control on sedimentation was a series of basement highs which compartmentalized the rift margin into numerous sub-basins. Thick sedimentary packages were deposited between the high regions while thin and patchy deposits characterize the highs. Three phases of tectonism are indicated: at 35 Ma the onset of rifting occurred followed by an acceleration in subsidence at around 30 Ma; the rifting climax occurred between 20 and 18 Ma. Sequences 1 and 2 were deposited during extension but do not show growth relationships to faults or any synsedimentary deformation. Sequence 3 represents a post-rift stage which began at about 18 Ma and was characterized by regional uplift, tectonic quiescence and rift shoulder erosion. The facies and geometry of the depositional units (sequence 1 and 2) suggest that sediments accumulated passively in a low relief marginal marine/coastal system dominated by regional subsidence. This is not typical of classical facies models of rift fill and is interpreted to be due to sediment passively infilling earlier formed, fault-generated, accommodation space.

INTRODUCTION

The Gulf of Aden is a juvenile ocean which began rifting during the late Eocene (between 42 and 35 Ma) and completed oceanization by the early Miocene (18.5 Ma) (Bott et al., 1992; Sahota et al., 1995). In the west the margin is volcanic and highly affected by the activities of the Afar plume (Davison et al., 1994). Up to 2000 m of Oligo-Miocene volcaniclastic sediments–flows are present offshore and onshore Aden (Tard et al., 1991). Eastward the volcanics die out and in the region studied the passive margin is a steep, narrow, sediment-poor, non-volcanic continental margin (Bott et al., 1992; Tard et al., 1991). Along both the Yemeni and the conjugate Somali margin, rift-related deposits are limited to a number of discrete embayments separated by structural (basement) highs (Figures D1.1, D1.2).

The Hadhramaut region affords the largest continuous rift exposures along the Gulf of Aden, covering approximately 2700 km^2 in total (Figure D1.2). Physiographically, the region is presently a low relief coastal plain covered in gravel terraces shed from a deeply dissected, 1500–2000 m high rift shoulder. Pre-rift strata are exposed in the footwalls of the major faults (Figures D1.2, D1.4). The Oligocene to Recent syn-rift and post-rift succession, the Shihr Group (Beydoun, 1964; Beydoun and Sikander, 1992), occurs as patchy outcrops on the coastal plain.

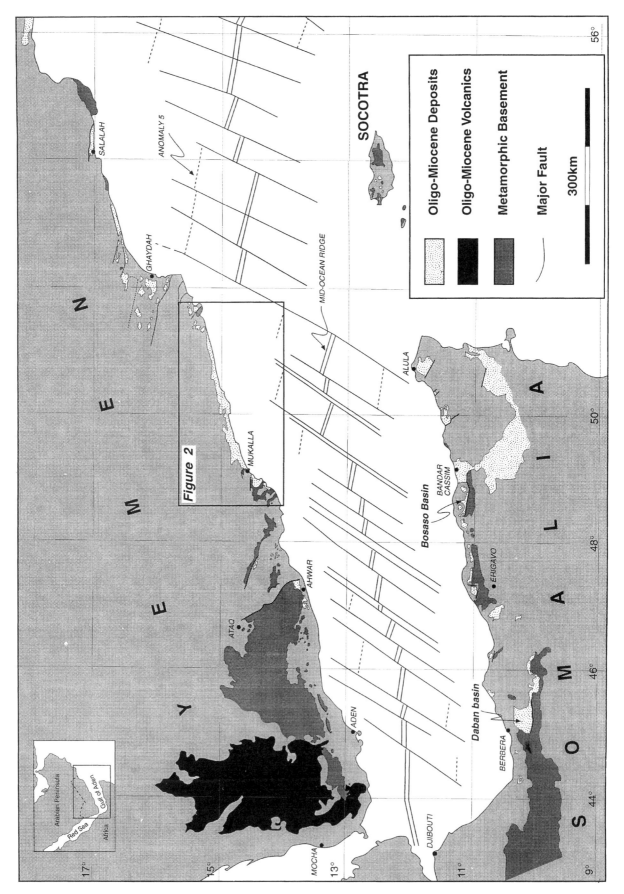

Figure D1.1 Outline map of Oligo-Miocene depocentres and basement exposures of the Gulf of Aden and location of study area.

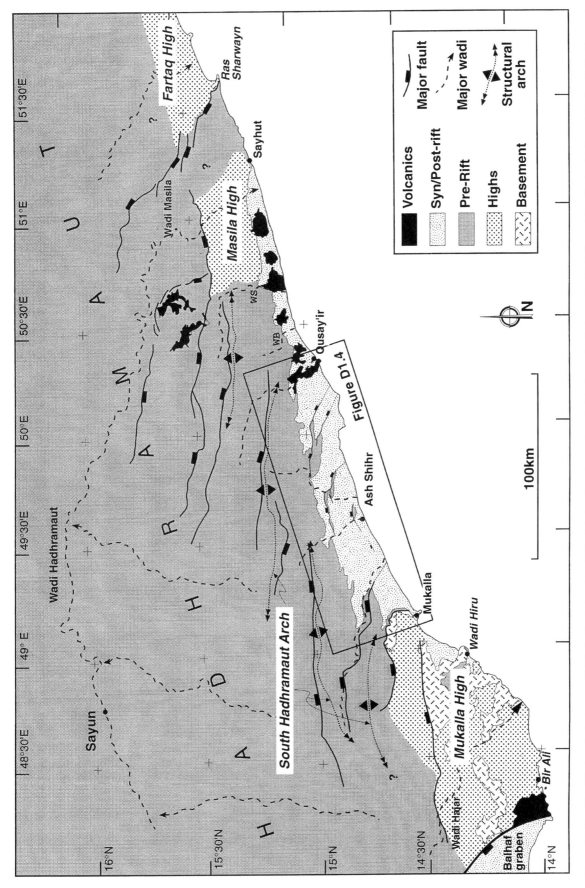

Figure D1.2 Simplified geological map of Mukalla-Sayhut region. Rift-related sediments are deposited in the region between the Mukalla and Ras Fartaq highs and the South Hadhramaut Arch. WB = Wadi Bidish, WS = Wadi Shakhawi. Note location of Wadi Hiru (see discussion).

The data and interpretations of the Shihr Group presented here are based on structural and stratigraphic mapping of the area supplemented by information from aerial photographs, Landsat TM and SPOT images. Facies analysis was performed with reference to Reading (1986) and Tucker and Wright (1990). Proprietary seismic sections were examined for the purpose of cross-checking surface-based interpretations and have been incorporated into the conclusions. Age constraints have been provided by Sr isotope dating (Appendix, data from Reeder, 1994).

PREVIOUS WORK

Most previous work in the area has been regional in scope (Beydoun, 1964, 1978; DGME, 1986; Haitham and Nani, 1990; Bott *et al.*, 1992). The stratigraphy erected by these authors is summarized in Figure D1.3. Schüppel and Weinholz (1990) and Bosence *et al.* (1996) have completed detailed studies on the Tertiary rift sediments. Basic mapping and stratigraphic work has also been carried out by Agip and the East German government (AGIP, internal reports, 1977–1984; DGME, 1986). Investigations on the Mesozoic basin configuration and hydrocarbon framework also allude to the development of the Gulf of Aden (Ellis *et al.*, 1996; Redfern and Jones, 1995). Studies from offshore Yemen have concentrated on defining the spreading history and ocean–continent configuration of the margin and not on the age, nature or spatial configuration of the sediments (Cochran, 1981; Sahota *et al.*, 1995). Poor seismic reflection quality precludes detailed mapping of the offshore sediments. Thus, the offshore database is limited to scattered exploration wells (15 in total) and the 3 DSDP wells (Bott *et al.*, 1992).

REGIONAL STRATIGRAPHY

The lowermost Oligocene, represented by the Ghaydah Formation offshore Yemen and the Bandar Harshau Formation offshore Somalia, is a mixed marginal marine–sabkha–estuarine sequence deposited in isolated sub-basins. Hughes *et al.* (1991) have shown that by the middle Oligocene a deep proto-Gulf of Aden existed at least as far east as Al Hami (Figure D1.4). A westward shoaling probably reflects both earlier palaeogeography (As Saruri, 1995; Al Subbary *et al.*, this volume) and differential uplift due to the Afar plume. Deep marine globigerinid oozes of Miocene age (Hami Formation) are present in the central Gulf and reflect the development of a narrow and deep early oceanic trough. Nearshore equivalent beds of the Miocene Hami Formation are fossiliferous silty and sandy marls overlain by chalky limestones deposited in outer neritic–bathyal depths. On seismic lines upper Miocene reflectors are generally undeformed. Pliocene deltaic deposits fed by the major wadi systems (such as Wadi Masila, which has produced a fan over 2000 m thick) constitute the early stages of passive margin blanketing (Sarar Formation).

Rift sediments of northern Somalia and Dhofar, the conjugate and adjacent margins, are better documented (Bosellini, 1989; Abbate *et al.*, 1993b; Osman, 1992; Platel and Roger, 1989; Roger *et al.*, 1989; Fantozzi *et al.*, this volume). In Somalia, small early Oligocene (Chattian–Repelian) syn-rift basins are filled with lacustrine and continental clastics and evaporites and minor interbedded volcanic horizons (Fantozzi *et al.*, this volume). The Daban Basin (Figure D1.1) in western Somalia records a continuous succession of alternating shallow marine–lagoonal, lacustrine and continental sediments throughout the Eocene and Oligocene (Abbate *et al.*, 1993b). The Boosaaso Basin in Migiurtania (Figure D1.1) records a dual phase of rifting with a minor early Oligocene rifting event accumulating terrestrial sands and a major early Miocene event accumulating variable lagoonal, lacustrine and continental clastics and carbonates (Oligocene Shimis Sandstone and lower Miocene Mait Group of Osman, 1992). These syn-rift deposits are unconformably overlain by Pliocene wadi conglomerates and volcanics. In Dhofar a very different (and much more complete) marine-dominated succession is encountered (Roger *et al.*, 1989). Here deposition continued throughout the upper Eocene (Priabonian) although lacustrine limestones mark a regional regression and uplift. The first evidence for rifting is in the middle Oligocene when a marked deepening of the basin is indicated by a thick succession of turbidites and olistoliths. Deep marine conditions continued until the early Miocene (Burdigalian) with shallow marine carbonates deposited with a marked unconformity on the underlying bathyal sediments. This unconformity represents the break-up unconformity and the Dhofar region has been progressively elevated since the early Miocene.

TECTONIC ARCHITECTURE

The studied Shihr Group exposures are confined between the Mukalla high to the west, the Masila–Fartaq high to the east and the South Hadhramaut Arch (*sensu lato*) to the north (Figure D1.2; Beydoun, 1964, 1969). Syn- and post-rift sediments are in fault contact with basement to the west (at Mukalla) and unconformable on Tertiary limestones to the east and along the South Hadhramaut Arch. Consistent eastward-directed wadi systems suggest a regional eastward basement slope across the region. The pre-rift stratigraphy indicates that the highs are at least Jurassic and

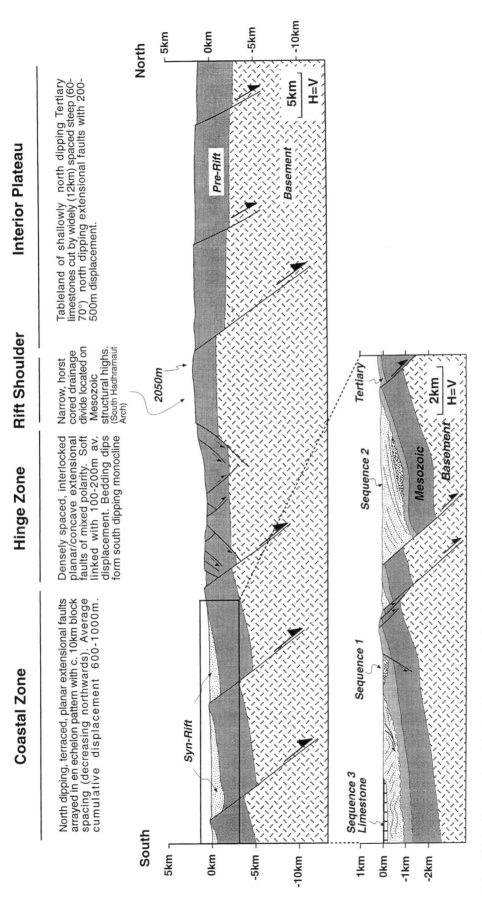

Figure D1.2 Cross-section of the Hadhramaut region based on surface mapping and proprietary seismic data. Insert highlights the asymmetrical nature of rift sedimentation. Note that the overall wedge shape of the rift sediments is not supported by stratal geometry which is everywhere subparallel.

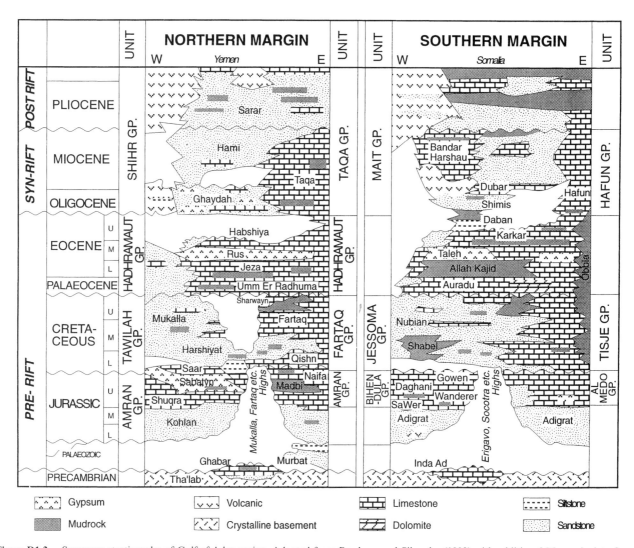

Figure D1.3 Summary stratigraphy of Gulf of Aden region. Adapted from Beydoun and Sikander (1992) with additional Mesozoic data from authors' observations, Redfern and Jones, 1995; Ellis *et al.*, 1996 and Beydoun, 1964; and additional Oligo-Miocene data from authors' own observations, Bott *et al.*, 1992; Osman, 1992; Abbate *et al.*, 1993b and Roger *et al.*, 1990.

possibly Palaeozoic in age and they were reactivated early in the evolution of the rift. Pre-rifting structural trends, principally the north-west–south-east Najd wrench trend and the north-west–south-east to east–west Jurassic rift trend have clearly influenced the initial location and geometry of the rifting. Restored cross-sections indicate that the total strain across the margin is low, with an average β-value of 1.15. Stretching is primarily effected by a complex extensional fault network and subordinately by monoclinal flexing–downwarping; no strike-slip transfer faults segmenting the margin have been identified.

The South Hadhramaut Arch forms the shoulder of the rift and is a composite drainage divide which is systematically breached eastwards along strike. The arch is coincident with the footwalls of the major basin-margin faults. These define a left-stepping, en echelon system of concavo-planar, north-dipping (50–65°) faults with variable throws of *c.* 250–700 m. Deformation in the interior plateau is localized along discrete, curvilinear fault zones formed by linkage of numerous individual fault segments. Displacements are in the range of 100–350 m. Fault trends are predominantly east–west, but eastwards they become markedly oblique to the rift margin, trending north-west–south-east.

Between the arch and the coastal plain is a 10 to 15 km wide, intensely faulted, hinge-zone (Figure D1.2(a)). Consistently south-dipping beds along the margin indicate this hinge-line to be a south vergent monocline with 1500–2000 m of relief. Interlocking planar faults of varying polarity and strike form a dense network of 'soft-domino' style faults. Fault throws are *c.* 100–300 m but have steep along-strike displacement gradients. North-throwing faults predominate along the southern margin and south-throwing faults along the northern margin of this hinge zone.

Pre-rift stratigraphy

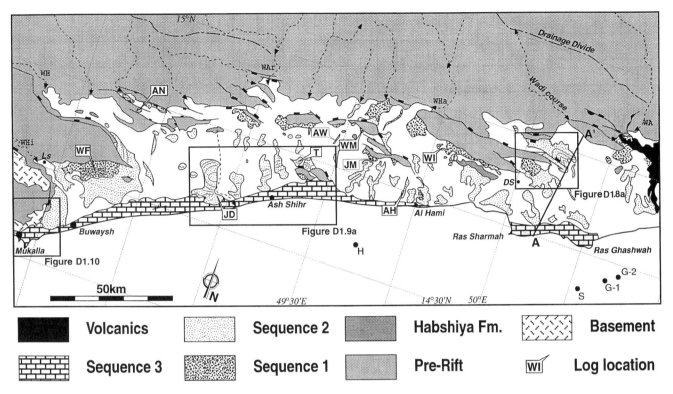

Figure D1.4 Simplified geological map of rift sediments. Only the coastal (i.e. raised beach) exposures of Sequence 3 are shaded: wadi gravels and gypsum accumulations are white. Boxed letters indicate log locations. WF = Wadi Filik; AN = An Naq'ah; JD = Jebel Dhabba; T = Tibelah; AW = Al Wasit; WM = Wadi Marun, JM = Jebel Marun; AH = Al Hami; WI = Wadi Idthik. Offshore wells located are H = Hami-1, S = Sharmah, G-1 = Ras Ghashwah-1, G-2 = Ras Ghashwah-2. Major wadi systems also located and abbreviated as follows: WHi = Wadi Himmum; WH = Wadi Huwayra; WAr = Wadi Arif; WHa = Wadi Harad; WA = Wadi Assid. DS = Town of Ad-Dis-Sharqiah, Ls = Lusb village. A-A' = location of cross-section of Plate 16(a).

The coastal plain is crossed by a few major east–west trending, northerly dipping fault zones. These fault systems are commonly terraced planar faults (in a zone up to 750 m wide), with cumulative displacements of approximately 750–1000 m. The fault zones are composed of 5–10 km long segments which link laterally in complex, 1–2 km wide accommodation zones. Because of the predominance of ductile shales and evaporites in the Shihr Group many faults at depth are expressed on the surface as broad, complex zones of folding and minor faulting rather than discrete fault strands. The basal evaporitic shales also act as detachment horizons on the large fault blocks producing very tight synclines with steep to subvertical limbs.

It is worth noting that although the surface exposures reveal a simple deformation style in the Shihr Group, seismic profiles show a considerable complexity of internal stratigraphic geometries and deformational styles which are only hinted at on the surface.

PRE-RIFT STRATIGRAPHY

Basement

Basement exposures are confined to the region west of Mukalla. A tripartite lithostratigraphic subdivision was erected by Beydoun (1964). Proterozoic metavolcanics (Tha'lab Group) are overlain by unmetamorphosed late Proterozoic clastics, carbonates and volcanics (Ghabar Group); both are intruded by 400 Ma granitoids. No Gondwanan sequences have been recorded anywhere in south Yemen and it is likely that the region remained an area of positive relief throughout the Palaeozoic (Beydoun, 1978).

Amran Group (Jurassic)

The Jurassic is seen only in the Mukalla high (Wadi Hajar region) and in the Masila High (Wadi Masila). This group consists of marine and marginal marine carbonates and clastics of considerable, but variable thickness. They were deposited in narrow, high relief north-west–south-east trending graben. These graben have been the focus of considerable hydrocarbon exploration recently and their complexity is just beginning to be understood (Ellis *et al.*, 1996; Redfern and Jones, 1995). Significantly, these graben systems are markedly oblique to the trend of the Gulf of Aden.

Tawilah Group (Cretaceous–Paleogene)

This dominantly sandy unit occurs throughout Yemen and is unconformable on the underlying Amran Group. It varies from continental facies in the west to marine in the east and represents a period of slow epicontinental subsidence (Al Subbary *et al.*, this volume; Beydoun, 1978). Exposures in the Hadhramaut are widespread and evidence for both continental and shallow-marine siliciclastic sedimentation is present. Four marine carbonate horizons, progressively thicker eastward, are present across the area studied (Beydoun, 1964, 1978).

Hadhramaut Group (Tertiary)

Tertiary sedimentary rocks cover much of south-eastern Yemen and have been interpreted to either disconformably (Beydoun, 1964) or unconformably (our observations) overlie the Tawilah Group. This thick (255–1770 m) pre-rift succession comprises the Umm Er Radhuma Formation (Paleocene–early Eocene), the Jeza Formation (early-mid Eocene), the Rus and Habshiya Formations (mid Eocene) and the Hamara and Rimah Formations (late Eocene–Oligocene) (DGME, 1986; Beydoun, 1964). The Umm Er Radhuma Formation is cliff-forming and comprises foraminiferal grainstones, packstones and wackestones. The overlying Jeza Formation consists of calcareous paper shales and well-bedded micritic limestones. The Rus Formation is a thick (*c.* 300 m), homogeneous gypsum unit and the Habshiya Formation consists of various platform carbonate facies dominated by nummulitic packstones and mudstones with subsidiary gypsum. Upwards a number of thin sandy horizons indicate increasing clastic input.

RIFT SEDIMENTS

We have subdivided the Shihr Group into three sequences of distinct facies assemblages separated by regional unconformities. Sequences 1 and 2 are loosely classed as syn-rift, which is defined here as all sediment accumulated between initiation of tectonism and final oceanic spreading. There is no evidence within either sequence for active fault-controlled sedimentary accumulation and these sequences may actually have formed by passive infill of accommodation space (see discussion below; Prosser, 1991). Sequence 3 represents the post-rifting (drift) evolution of the margin.

An informal stratigraphy has been adopted: formal stratigraphic definitions await more comprehensive field mapping and more detailed dating. A summary map of the distribution of the sequences is shown in Figure D1.4: sequence 1 deposits are preferentially exposed on the dip-slopes of the major north-dipping faults, and sequence 2 in the hangingwalls of the faults. Sequence 3 forms a blanketing veneer over the two underlying sequences. Logs of sequence 1 deposits are shown in Figure D1.6 and of sequence 2 facies in Figure D1.7 (a key to the logs is given in Figure D1.5).

Sequence 1

These sediments were deposited with a marked angular unconformity on the Habshiya, Jeza and Umm Er Radhuma Formations. This sequence forms an areally (and volumetrically) small portion of the exposed rift sediments (Figure D1.4). Facies relationships are complex with large thickness variations and a marked localization of certain facies. The age of the base of the sequence is poorly constrained: the youngest pre-rift exposures are ≈42 Ma (Lutetian, Habshiya Formation) while the oldest Shihr Group deposits are at least Chattian (35 Ma) (Beydoun, 1964; Hughes *et al.*, 1991). The upper age of the sequence is also poorly controlled. The contact between sequence 1 and sequence 2 is not exposed but is inferred to be an unconformity because the former always occurs with steeper dips than the latter.

Two main facies are recognized: a siliciclastic (alluvial) facies association generally lies directly on pre-rift and is overlain by a dominantly carbonate (marginal marine) facies association (Figure D1.6 – compare lower and upper deposits of Wadi Idthik and Jebel Marun). A distinct evaporitic association is found in sequence 1 at Wadi Filik.

Siliciclastic association
Description
The Wadi Idthik log (lower section) is an example of a relatively fine-grained association (Figure D1.6). Basal exposures (not shown here) are reddened, well-sorted and rounded, unstratified gravel and boulder conglomerates. Typical field exposures of these conglomerates (and intervening rhizolith horizons) are shown in Plate 15(a),(e). Clast composition is overwhelmingly (approximately 90%) Umm Er Radhuma limestone with a minor component of chert and other Tertiary limestones. The matrix is generally coarse-granular red sand. Bed thicknesses vary from 1 to 5 m and have both planar and scoured bases. Imbrication and coarse sand lenses are occasionally present in the thinner beds. All conglomerate horizons display a large amount of lateral variation. Marls and calcretes occur associated with these conglomerates along with less common tufa horizons. The maximum measured thickness encoun-

Rift sediments

Lithology

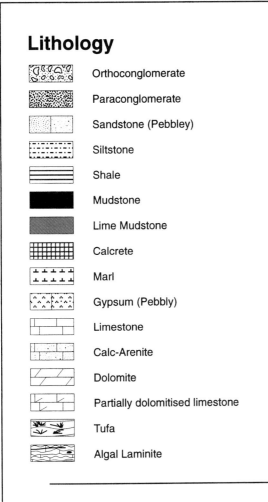

- Orthoconglomerate
- Paraconglomerate
- Sandstone (Pebbley)
- Siltstone
- Shale
- Mudstone
- Lime Mudstone
- Calcrete
- Marl
- Gypsum (Pebbly)
- Limestone
- Calc-Arenite
- Dolomite
- Partially dolomitised limestone
- Tufa
- Algal Laminite

Grain size/Texture

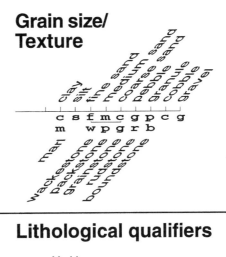

Lithological qualifiers

- Muddy
- Silty
- Nodular
- Pebbly (clastic)
- Pebbly (carbonate)
- Coated grain
- Fe Ferruginous
- Chk Chalky
- Gyp Gypsiferous
- Gl Glauconitic
- Fault

Biogenic structures/fauna

- Foraminifera (undiff.)
- Gastropod
- Bivalve
- Molluscan bioclasts
- Brachiopod
- Solitary coral
- Colonial coral
- Echinoderm
- Echinoderm spine
- Algae (undiff.)
- Algal oncolites
- Bioturbated (weak, moderate, strong)
- Burrowed (undiff.)
- Burrow (vertical)
- Rhizolith
- Chert bed

Sedimentary structures

- Trough cross-bedding
- Planar tabular cross-bedding
- Asymptotic tabular cross-bedding
- Low angle cross-bedding
- Overturned foreset
- Disturbed foreset
- Reactivation surface
- Graded foreset
- Channel
- Wavy planar bedding
- Planar bedding
- Asymmetric ripple cross lamination
- Symmetric ripple cross lamination
- Desert Rose
- Imbrication
- Mudcrack

Figure D1.5 Key to sedimentary logs of Figures D1.6 and D1.7.

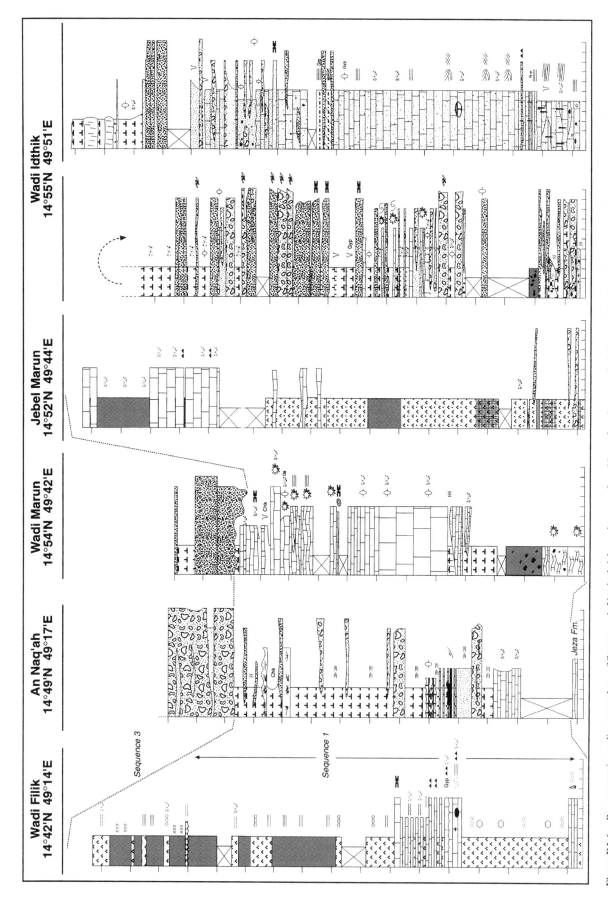

Figure D1.6 Representative sedimentary logs of Sequence 1. Vertical tick marks every 5 m. Unconformity on underlying early Tertiary is low-angle with muted topography and rarely exposed. Note that the two Wadi Idthik logs are not contiguous, but are from the same, conformable succession. Wadi Filik. d) Bleached white alternating tufas, marly paleosols and pebbly alluvial beds sharply overlain by sheet flood derived gravel conglomerates, Wadi Idthik (not shown in Fig. 6.). Exposure ≈20 m high, 60 m long. e) Unfossiliferous, poorly sorted, intraclastic, thick-bedded grainstone lower Wadi Marun f) Mixed gypsum conglomerate facies similar to basal beds of Wadi Marun. Hammer for scale. 4 km north of Tibelah.

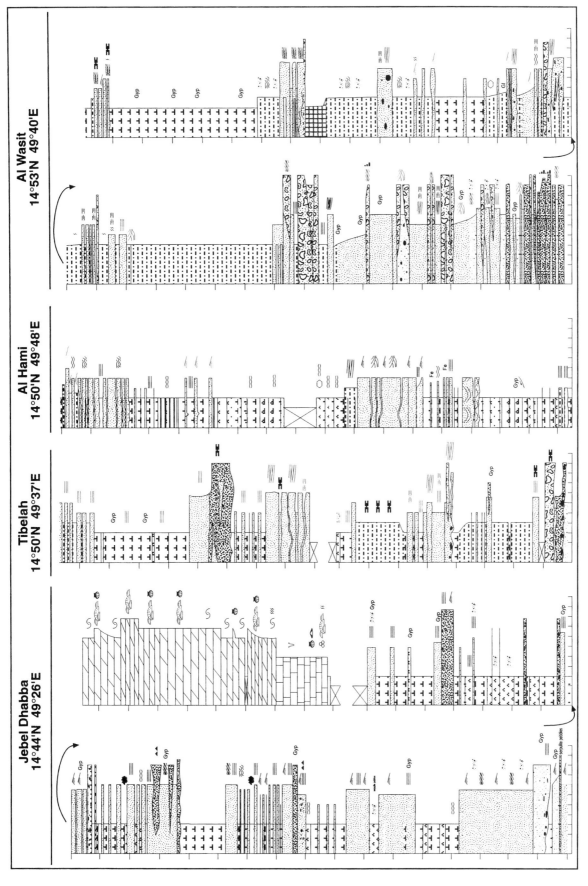

Figure D1.7 Representative sedimentary logs of Sequence 2. Vertical tick marks every 5 m. Underlying unconformity with Sequence 1 is not exposed: Sequence 2 units generally lie on eroded Habshiya Fm. Commonly overlain by Sequence 3 gravels.

tered was about 125 m at the entrance of Wadi Harad where a clear upward gradation from thick boulder conglomerates to thick reddened paleosols with minor gravel conglomerate horizons is seen. The calcretes are mottled, deeply reddened and vertically burrowed in beds up to 2 m thick. Red, subvertical rhizolith stringers are common and generally define distinct horizons. These fine beds increase in frequency and thickness upwards.

Interpretation
The coarse, structureless nature of these deposits suggests deposition as sheet-flood events in proximal parts of alluvial fans. Close spatial relationships with present-day wadi systems may indicate the longevity of these systems. The calcretes formed in areas between the small, isolated fans in a semi-arid–arid climate.

Carbonate association
Description
Limestone beds form the bulk of the exposures mapped in sequence 1 and are illustrated in logs measured at An Naq'ah, Jebel Marun, Wadi Marun and Wadi Idthik (upper) (Figure D1.6). They are predominantly rubbly and variably micritized biopackstones and skeletal grainstones with distinctive pink mottling. Granule to pebble clasts of chert, white wackestones and grainstones and rounded quartz are common. Algal oncolites occur and algal laminites bind the sediments. No macro- or microfauna were recovered. Bioturbation is rare and limited to coarser beds. Typical thicknesses are 40–60 m and are relatively constant across the region.

At Wadi Idthik (Figure D1.6, Plate 15(d)) alternating thickly laminated and low-angle cross-bedded, white grainstones with scours and lime-mud rip-up clasts occur. *Skolithos* burrows are common in the lower part of the section while *Thalassinoides* is common at the top. In the uppermost 25 m a number of thin conglomerate beds of Umm Er Radhuma clasts are seen. At Jebel Marun (Figure D1.6) red, mottled, 2–5 m thick beds of structureless coarse gypsum occur interbedded with irregular lenses of angular limestone granule and pebble conglomerates. These facies are also exposed at Hummum (Figure D1.8(a)) and at Wadi Arif (Plate 15(f)).

Westward around Wadi Arif and An Naq'ah, muddier deposits predominate and limestone beds are not common (Figure D1.6). Here laminated marls, tufas and numerous conglomeratic interbeds are present. Tufas are consistently present across the region as white, thin (generally 1–3 m) crudely laminated units (occasionally containing upright, encrusted reed horizons). Five kilometres west of Ad Dis-Sharqiah two spectacular circular tufa mounds (approximately 30 m diameter, 10–15 m thick) composed of thick and uneven, superimposed encrusted reed layers with a central cavity are present (at 'T' in Figure D1.4).

At Wadi Arif a coralgal framestone unit occurs, bedded within 5–10 m thick units of structureless, coarse, open framework limestone conglomerates (Plate 15(b)). The framestones comprise faviid corals (?*Porites*) and molluscan debris in a coarse grey bioclastic wackestone matrix.

Interpretation
This facies association is interpreted as saline playa deposits fringing a low relief marine shoreline. Much of the gypsum in the marginal playa is sourced by redeposition of gypsum from the Rus Formation. The occasional conglomerate lenses are interpreted as sheet flood deposits. Around An Naq'ah, these playa deposits are less gypsiferous, muddier and contain numerous tufa beds. This reflects both a different substratum (Jeza Formation marls rather than Rus Formation evaporites) and indicates more frequent flooding of the coastal plain. Eastward at Wadi Idthik, the calcarenites are interpreted as beach deposits. The tufa mounds at Ad Dis-Sharquiah are considered to be deposits of CO_2 rich fossil springs (note that there are a number of present-day geothermal springs nearby).

The bedded limestones are interpreted as peritidal marine beds from their abundance of algal debris but lack of any other fauna. The regional upward gradation to bedded limestones reflects transgression. The intimate interbedding of coral limestones and conglomerates at Wadi Arif suggests a fan delta development and colonization by framework building corals (e.g. Hayward, 1982; Bosence et al., 1996).

Evaporite association
Description
Figure D1.6 shows 85 m of the 300 m section at Wadi Filik. It is composed of laterally continuous beds of mixed gypsum, limestones and marl in repeated 10–15 m thick coarsening-up cycles (Plate 15(c)). The marls and lime mudstones contain mudcracks and horizontal burrows and display continuous, very thin (algal?) lamination. Thin conglomerates occur and several 1–2 m thick grainstone horizons contain rare gastropods and bioclastic debris. Gypsum occurs as thin, continuous laminae, nodular white gypsum arranged in thickening-up beds and as replacive, brown, unbedded gypsum.

Interpretation
This association has the characteristics of an ephemeral, evaporitic lake or lagoon. The basal marls and conglomerates are interpreted as a playa mudflat. The repeated marl–gypsum cycles are taken to represent repeated flooding and desiccation of the water body.

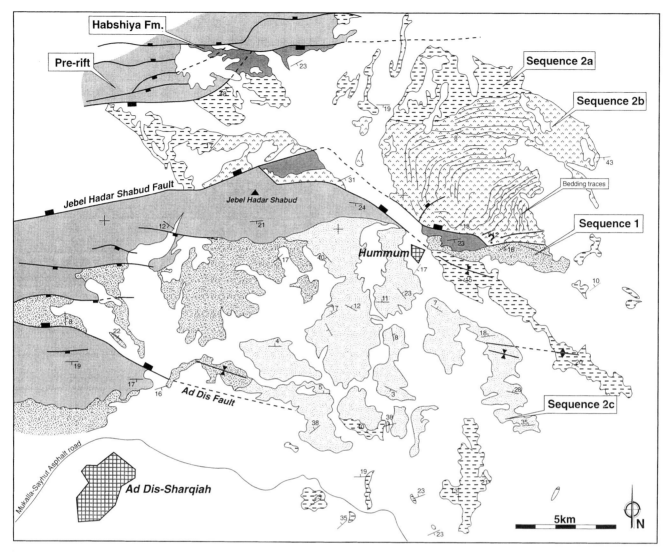

Figure D1.8 (a) Geological map of Ad-Dis-Hummum region, based on field mapping and Landsat TM interpretation, highlighting close correlation between observed facies and fault architecture. Sequence 2a = lower facies of Sequence 2, sequence 2b = gypsum facies of Sequence 2, and Sequence 2c = middle facies of Sequence 2.

Sequence 2

Sequence 2 comprises the bulk of the onshore exposures (Figure D1.4). It unconformably overlies sequence 1 of the Shihr Group, Umm Er Radhuma, Jeza and Habshiya Formations. The total thickness is difficult to gauge because of alluvial cover and post-depositional faulting, but is of the order of 1000–1500 m and is relatively constant along strike. The age of the base of the unit is poorly constrained but is estimated at around 30 Ma. The upper age limit is between 21.1 Ma and 17.4 Ma (see Jebel Dhabba below; Table D1.1).

There is a clear upwards trend from laterally persistent, alluvial dominated facies to spatially limited, coastal salina–sabkha facies. This is best illustrated by comparing the log at Al Wasit, which typifies the basal 300–500 m, with all the other logs of Figure D1.7 which illustrate the top of the sequence.

Basal facies assemblage

Two basal facies associations are recognized: a coarse sandy and conglomeratic facies association and a thicker silty association.

Sandstone–conglomerate facies association
This mixed sandstone–conglomerate facies is generally deposited unconformably on the Habshiya Formation, and occurs across the entire region in a unit about 50 m thick. It is typified in the lower part of the Al Wasit log (Figure D1.9). The sandstone–conglomerate ratio varies widely and nonsystematically along strike from 30% to 90% conglomerate.

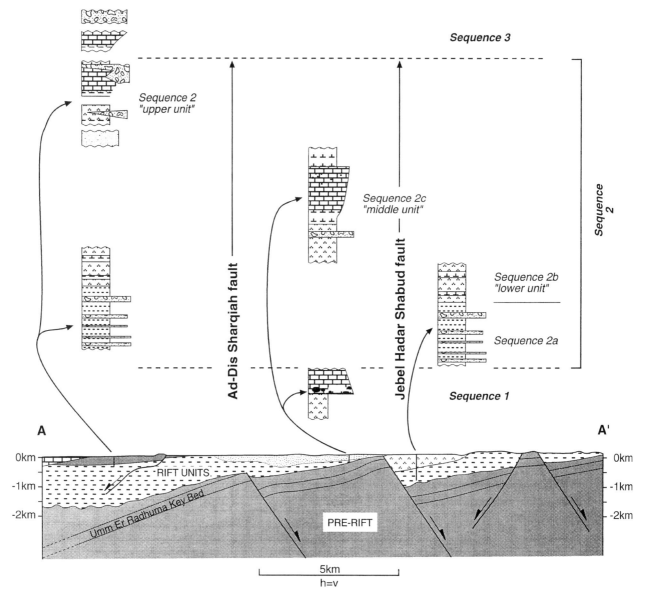

Figure D1.8–*Contd.* (b) Schematic cross-section and associated sketch stratigraphy from Ad Dis-Hummum-Ras Sharmah region. See A–A' of Figure D1.4 for location. Stratal geometry and thickness based on nearby seismic reflection line.

The sandstones are reddened, coarse and pebbly lithic arenites–wackes with numerous silt and gypsum-rich horizons and also distinct slit lenses. Trough cross-bedding is pervasive in the coarser sand beds and a general fining-up pattern in 3–5 m thick units is apparent. North-west of Jebel Marun these coarse sandstones are almost exclusively developed with only minor lenses and beds of polymict conglomerates.

The conglomerates are lensoid and weakly sub-planar bedded with occasional deep (2–3 m) scours. Sparse palaeocurrent data indicate a general south-easterly trend to these channels. Most beds, however, are 1–2 m thick with sub-planar, nonerosive bases. Lithologically they are polymict (limestone clasts of all Tertiary Formations), rounded, moderately sorted conglomerates with a coarse lithic matrix.

Interpretation

The facies is interpreted as the deposits of a braided river system. The conglomerates, with their crude stratification, open framework and repeated scouring are channel bar deposits while the sandy units are interpreted as waning flood channel deposits. The lateral persistence of the conglomerates suggest a broad braidplain existed. Some interfingering with overlying laminated silts–marls shows that the braid-plain fed into a subaqueous environment.

Table D1.1

Sample number	Organism	Initial $^{87/86}$ Sr ratio	Range (+/−0.000018)	Min.–max.age my		Age	Longitude	Latitude	Sequence
DBSY31	Ostreid	0.70891	0.708928/0.708892	6.5–10.0	8.2	Tortonian	14°42.39′N	49°28.42′E	3
DBSY32	Ostreid	0.708912	0.708930/0.708894	6.4–10.0	8.2	Tortonian	14°42.39′N	49°28.42′E	3
DBSY35	Ostreid	0.708771	0.708789/0.708753	14.4–16.3	15.3	Langhian	14°42.19′N	49°29.04′E	3
DBSY66	Ostreid	0.708569	0.708677/0.708641	16.7–18.2	17.4	Burdigalian	14°43.24′N	49°30.57′E	3
DBSY67	Matrix	0.708569	0.708677/0.708641	16.7–18.2	17.4	Burdigalian	14°43.24′N	49°30.57′E	3
DBSY155	Ostreid	0.70842	0.708438/0.708402	20.7–21.6	21.1	Aquitanian	14°38.45′N	49°11.18′E	2

Silty facies association
Red silty marls constitute the majority of the exposures of sequence 2 and are illustrated by the upper part of the Al Wasit log (see also Plate 16 (a), (c)). They are unfossiliferous, undisturbed and planar laminated in 15–25 m thick units. Mudcracks are occasionally present. In thicker units the marls are capped by distinctive 1–3 m thick mottled purple-red calcretes. Interbedded horizons of conglomerates and sandstones are common throughout. The sandstones are lithic arenites, 0.1–5 m thick, normally graded with ripple cross-laminated and low angle trough cross-bedding. Silt drapes are common. The conglomerates are 1–3 m thick units of planar erosively based, pebble-cobble, polymict limestone conglomerates with weakly developed imbrication. In Wadi-Huwayra a 15 m thick olistolith bed containing lithified, gypsum-veined blocks of similar lithologies occur.

Interpretation
This is interpreted as a sub-littoral, hydrologically open lacustrine facies. The numerous channelized sand and conglomerate beds are interpreted as storm deposits. A crude, repeated, fining-up pattern to the facies is characteristic and represents periodic deepening–recharge of the lake which may have been related to tectonics. These fining-up units are always capped by calcretes, indicating subaerial exposure. Desiccation cracks at other levels in the succession indicate other periods of exposure. Numerous disturbed and contorted laminae in the sand beds point to either dewatering or gravitational instabilities. Abundant gypsum and a lack of body fossils may indicate hypersaline and possibly stratified lake waters.

The petrography of the silts and marls indicates principally a Tertiary source; however, the presence of abundant feldspar and quartz and some glauconite clasts point to a subordinate Mesozoic source.

Upper facies assemblage
The upper facies in sequence 2 vary with geographic location and are described below in terms of the facies typical of particular localities.

Mukalla. A 120–150 m thick, unbedded coralline limestone forms Jebel Mukalla and rests unconformably on basement and Qishn Formation limestones (Plate 16(f), Figure D1.10). The basal 10 m contains angular boulders of brown dolomite and pink lithic wackestones of Habshiya and Harshiyat Formation provenance. This isolated unit is located on a basement horst and probably formed as an isolated carbonate platform. This unit was correlated with the Fuwwah Formation and dated as lower Miocene (DGME, 1986) and misinterpreted as Umm Er Radhuma Formation by Beydoun (1964).

Jebel Dhabba. Jebel Dhabba is a prominent flat-topped, coastal hill which exposes a 160 m thick mixed clastic–gypsum–dolomite facies assemblage (Figures D1.4, D1.7, D1.9, Plate 16(e)).

Deposits of mixed silts exposed at the base grade up to an alternating sequence of nodular gypsum, pebbly conglomerates, coarse lithic sandstones and brown marls (Figure D1.7, Plate 16(b)). These lithologies form a number of fining-up units. The bases are channelized with pebble-cobble conglomerate and planar-laminated, ripple cross-laminated and trough cross-bedded coarse lithic arenite infills. These clastics are overlain by alternations of planar laminated/rippled sands and nodular gypsum–marl couplets which increase upwards. Desert rose gypsum and gypsum pebble conglomerates (with occasional serpulid encrustations) are common. The sand beds are rippled and planar laminated, with rare convolute lamination.

The above is capped by a 45 m thick, coarse shelly dolospartite with a number of 1–2 m thick horizons of coral (*Porites*), molluscan and forminiferal rudstones. The upper 15 m of these dolostones contain a number of patch reefs and form a set of low-angle, downstepping, sigmoidal clinoforms (Figure D1.9).

Interpretation
The lower unit is interpreted as shallow marine shore-face sands with interbeds of nodular gypsum representing deposition within a coastal salina–sabkha environment during relative sea-level falls.

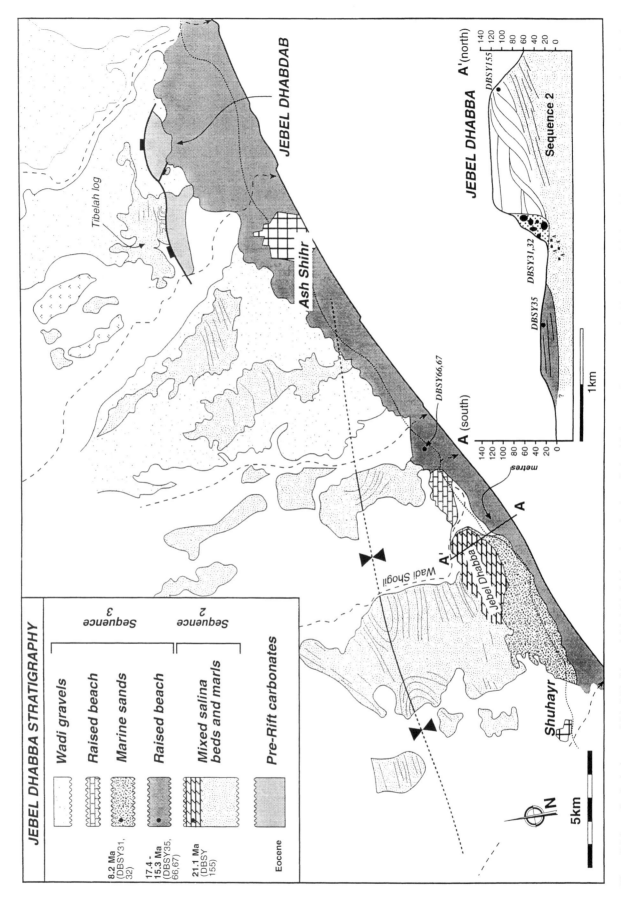

Figure D1.9a Geological map of Jebel Dhabba and Jebel Dhabdab regions showing relationship between Sequence 3 raised beaches and deformed Sequence 2 marls. Insets show sketch stratigraphy and cross-section of Jebel Dhabba. Sr dated samples are as given in Appendix.

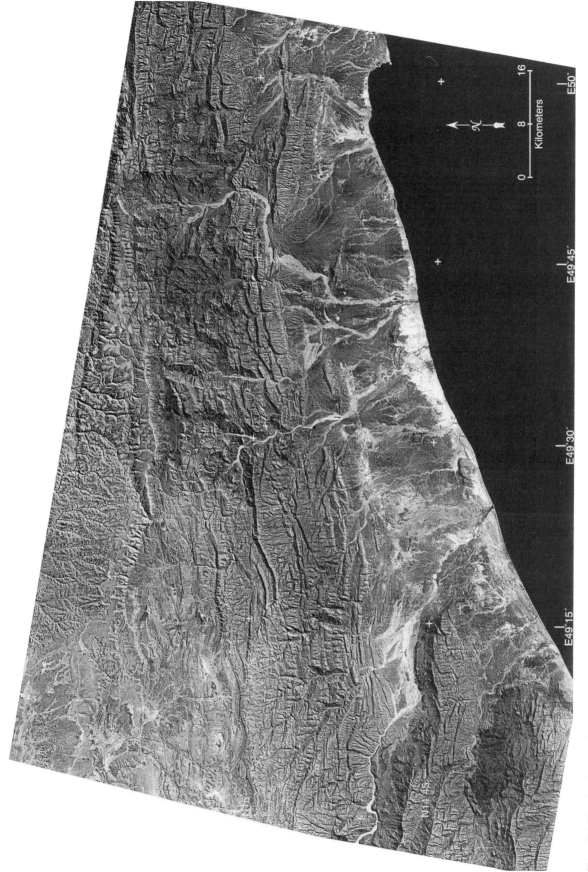

Figure D1.9b Landsat TM image of area shown in Figure D1.4 indicating oblique rift structures in northern pre-rift strata and rift sediments along coastal zone. The area of Figure D1.9a lies between 49° 20′ and 49° 49′E.

Abruptly overlying these marine sands are three fining-up units interpreted to be deposits of an arid coastal playa environment. Interbedded channelized conglomerates are interpreted as flash-flood deposits. The fining-up units are formed by periodic water inundation and progressive desiccation.

The isolated horizons of reefal debris indicate deposition as patch reefs. The clinoforms suggest southward (i.e. basinward) progradation of the platform. The lack of any similar facies nearby suggests this was an isolated reefal buildup but may reflect preferential preservation due to dolomitization. The reefal beds of Jebel Mukalla are of a similar Chattian–Aquitanian age (Beydoun, 1964; Figure D1.10, see below), as are some of the dolomitized reefs in the Wadi Hiru area (Bosence et al., 1996).

Tibelah-Jebel Dhabdab. Poorly exposed, varicoloured marls cover a 7–8 km wide region centred on Jebel Dhabdab (Figures D1.4, D1.9). They consist of repeated, 5–10 m thick deposits of brown, grey, red and purple marls with thin interbeds of recrystallized gypsum and occasional 1–5 m thick horizons of ripple cross-laminated and low angle, trough cross-bedded, lithic arenites. The marls are generally featureless with occasional planar lamination being the only sedimentary structures evident. Lenses and beds of polymict gravel conglomerate are present in some of the sand beds. Conglomerate units are 0.5–2 m thick, composed of rounded Tertiary limestone clasts, and are generally trough and low angle, cross-bedded.

This unit can be crudely divided into four coarsening-up cycles. The base of each cycle is interpreted as a deepening event with the basal beds of marls and mudstones suggesting deposition in a quiescent lagoonal environment. No sedimentary structures indicate subaerial exposure. These beds coarsen up to erosively based conglomerate and sand beds which are interpreted as proximal channels in a subaqueous delta system. The sands contain a large proportion of heavy mineral grains and strained quartz which are probably derived from Mesozoic siliciclastics.

Al Hami. Thinly interbedded, laminated gypsum, nodular gypsum and green-brown marls form the majority of the exposure. They are extensively replaced by secondary gypsum and quartz pseudomorphs after gypsum. Relict planar lamination is common as are rare desiccation cracks. Interbedded with these 20–30 m thick units are 2–5 m horizons of trough cross-bedded and ripple cross-laminated lithic arenites. They are full of gypsum clasts and display numerous silt and marl drapes. Large-scale (1 m) dewatering convolution of laminae is occasionally present. Palaeocurrents indicate a crude northward palaeoflow.

The gypsum–marl couplets are interpreted to be the products of repeated flooding and desiccation of a sabkha coastal plain (cf. Warren and Kendall, 1985). Interbedded sand beds are interpreted as fluvial channels. Extensive secondary alteration of the gypsum and marl beds is apparent in thin section with a typical 'chevron' style deformation of the large gypsum crystals.

Wadi Himmum
This unit of conglomerates occurs 12 km north of Mukalla and was deposited unconformably on sequence 1 and Tertiary limestones (Figure D1.4, Plate 16(d)). Clasts present are cobble-gravel in size and are composed of metavolcanics (Tha'lab Group), megacrystic granite, shelly quartz wackes (Harshiyat Formation) and white wacke–packstones (Umm Er Radhuma Formation). Foraminifera within one limestone clast gives a Sr age of 24 Ma (early Aquitanian - DBSY144, Appendix). Conglomerates predominate in channelized units with interbedded coarse sand beds containing lenses of cobbles. The thickness is estimated to be 50–70 m and forms a north–south swathe adjacent to the main border fault in the area.

The channelled form, good sorting, weak imbrication and abundant sand lenses of the Wadi Himmum conglomerates suggest deposition as fluvial channel bars. The provenance of the clasts provides vital clues in the interpretation of the geological history. The presence of Chattian limestone blocks, probably from equivalent deposits to Jebel Mukalla (see above) indicates that this unit post-dates sequence 2. However, the beds are strongly tilted (southwards), unlike sequence 3, and are therefore interpreted as a late Miocene sediment pulse, possibly generated by movement of the main border fault. Additionally, we can infer that basement was exposed to the west at this time and also that there must have been an extensive suite of coral limestones parallel to the main border fault which are not preserved but act as a source for the conglomerates.

We suggest that these conglomerates are structurally controlled. Wadi Himmum parallels both the main east–west trending border fault and the east–west strike of the pre-rift throughout its upper reaches (Figure D1.4). At Lusb village the fault abruptly changes strike to south-south-east–north-north-west, while the pre-rift continues to strike east–west. The Mukalla conglomerates were deposited in a pattern that radiates radially away from this divergence point. This is interpreted to be caused by a sudden decrease in stream gradient associated with this point.

Hummum-Sharmah
There are a number of localized facies in this area, a summary of which is given in Plate 16(a).

Lower beds
Gypsum beds abruptly and conformably lie on red silts and sands of facies association 1 and have a strong

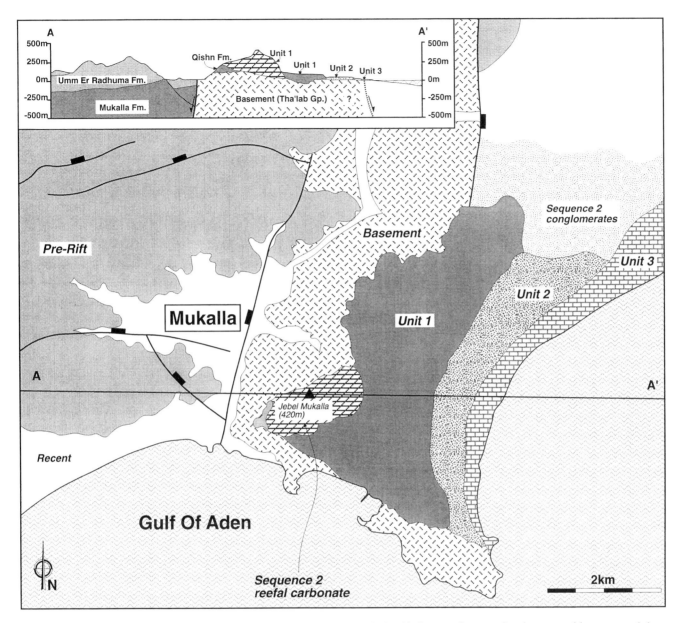

Figure D1.10 Geological map and cross-section of Jebel Mukalla showing relationship between Sequence 3 carbonates and basement and the terraced nature of carbonate deposits.

spectral response making this unit particularly visible on Landsat TM images. These beds are seen only in the immediate hanging wall of the Jebel Hadar Shabud fault and have lateral extent of 5 km and are at least 650 m thick. Towards the east the beds are mostly gypsiferous, mudcracked siltstones which grade upward to thick horizons of alternating nodular greenish gypsum, prominent ribs of black (?algal)/white banded gypsum and brown lime mudstone with occasional ripple cross-laminated, gypsum-rich siltstones. The gypsum beds show evidence of replacive–displacive growth of gypsum and the mudstones have extensive mudcracks infilled with crystalline gypsum.

The continuous nature of the gypsum and marl beds implies a quiescent, dominantly subaqueous, lacustrine or lagoonal depositional environment; mudcracks indicate periodic desiccation. The marls are interpreted as periods of low lake level while the rare rippled silt beds probably represent storm deposits ('underflows'), and indicate the low relief, fine-grained nature of the catchment area. The laminated gypsum and algal beds are interpreted to have formed by evapotranspiration following water recharge. This is interpreted as a coastal salina with the tilt-block of Jebel Hadar Shabud forming the marine barrier and recharge periodically from the east (cf. Lake McLeod, Western Australia in Al-Sharhan and Kendall, 1994).

Middle beds

The middle part of the succession at Hummum-Sharmah is up to 300 m thick and consists of a highly differentiated and laterally variable sequence of alternating gypsum, calcarenites, marl and conglomerate. Homogenous reddened, laminated gypsum beds full of scattered limestone pebbles are overlain by a thick sequence of brilliant white calcarenites. Upwards, some interbeds of laminated red marls appear and the white calcarenite grades up to marls with a number of thick limestone conglomerate lenses.

The fine, gypsiferous beds are interpreted as being the products of a broad, dominantly saline, mudflat with thick gypsum units formed in brine pools or salt pans and the sand horizons as widespread storm-derived, sheet-flood deposits. The calcarenites are interpreted to be littoral beds of small coastal salinas. The age of this unit is unknown: in places it is steeply tilted and unconformably overlies limestone conglomerates of sequence 1 and red marls of sequence 2 and is overlain by undeformed sequence 3 gypsum and limestone beds.

Upper beds

The upper beds are exposed further south in the region of Ras Sharmah and Ras Ghashwah (area not covered in Plate 16) and are composed of a mixed sequence of gypsum, limestone and conglomerates. These beds rest unconformably on salina beds of sequence 2 and are overlain unconformably by sequence 3 wadi gravels and limestones. Total thicknesses are between 20 and 75 m. The distribution of the facies is complex. The basal facies is a coarse, poorly sorted, planar laminated lithic arenite with numerous floating gravel clasts. Unconformably on this facies is a mixed nodular gypsum, gypsum sand and channelized conglomerates. The uppermost facies is a distinctive, repeated, coarsening-up facies with peloidal muds grading into bioclastic grainstones and capped by sub-planar stratified, muddy conglomerates. These beds all contain a rich fauna of echinoids, forams and coralline algae.

These upper deposits show a gradual increase in marine influence. The basal sand beds are interpreted as braided fluvial sheet bars, while the gypsum–conglomerate facies are interpreted as playa deposits at the edge of an alluvial fan. The carbonates show a clear progression from low energy offshore muddy carbonates, possibly lagoonal in origin, to high energy, foreshore grainstone shoals and ending in conglomeratic beach limestones.

Each of the above facies assemblages is confined to a single fault block (Plate 16(a)). Although no evidence exists for deposition in an active extensional regime, the fault blocks must have acted as morphological barriers which localized deposition (Plate 16(a)). The middle facies assemblage over-steps the Jebel Hadar Shabud fault trace at Hummum providing unequivocal evidence for the relative timing of sedimentation on these blocks and also suggesting that the Jebel Hadar Shabud fault scarp was more or less peneplaned prior to deposition. This sequential deposition on different fault blocks reflects different episodes of fault movement implying a progressive oceanward migration in faulting through time. This suggests a progressive narrowing of the rift with time (see Dart et al., 1995). Note that by correlating the prominent carbonate benches of Ras Sharmah and Ras Ghashwah with those of Jebel Dhabba and Buwaysh, all deformation and sedimentation must have occurred prior to mid Miocene times.

Sequence 3

The sediments comprising this sequence represent the post-rifting phase of the evolution of the rift margin. They generally reflect present-day sedimentary environments and distributions and form a thin horizontal veneer covering the underlying sequences. This suggests that the region had reached its present form by the beginning of deposition of the sequence. The restriction of marine beds of sequence 3 age to a narrow strip 100 m–1 km inland of the present coastline further implies that the coastal plain attained its present-day width early and that lateral shoreline migration had little effect in this area since the Mid Miocene. $^{87/86}$Sr dating places the base of this sequence between 21.1 Ma and 17.4 Ma and coastal deposition continues to the present day (see Appendix and below).

Two regions (Jebel Dhabba and Jebel Mukalla) are described in detail because of their importance in unravelling the middle-late Miocene evolution of the margin.

Jebel Dhabba

A summary stratigraphic column is shown in Figure D1.9 (see also Plate 16(e)). As well as the siliciclastic beds and sequence 2 deposits, two carbonate levels are present. These are coarsening-up raised beach deposits, grading from bioclastic lime mudstones at the base to burrowed, bioclastic packstones–rudstones at the top. There is an abundant fauna of gastropods, corals, molluscs and echinoids. An older unit forms a plateau at 25 m and a younger unit forms a 10 m high ledge. A third unit is a shallow marine sandstone which interfingers laterally with a chaotic cliff breccia on the side of Jebel Dhabba (cross-section, Figure D1.9). A distinct west–east reduction in grain size is apparent, from coarse clean sands along the western margin of the hill to marls on the eastern margin. The sandstones are well-sorted, well-rounded, polymict, clean coarse lithic arenites: 1–2 m cross-bed sets indicate a westerly palaeoflow. The conglomerates are unsorted, polymict boulder conglomerates composed of dolosparite clasts.

A series of $^{87}Sr/^{86}Sr$ dates have been obtained from pristine shell material collected from these deposits (Appendix). The location of the samples (and ages) are also shown in Figure D1.9. Only the upper two ledges have been dated. The older raised beach (Figure D1.9) is laterally extensive and can be correlated east to Ras Ghaydah and west to Buwaysh and Jebel Mukalla. Dates recorded from this raised beach are 15.3 Ma at Jebel Dhabba, 17.4 Ma at Dhabba east and 17.4 Ma at Riyan (Figure D1.9). The marine sands are dated at 8.2 Ma from ostreids embedded in the sediments (Plate 16(e)).

These data show that the shoreline has been located in the area at various levels since 21 Ma. All the sequence 3 beds 'wrap-around' Jebel Dhabba and imply that the Jebel has been a topographic high (?island) since middle Miocene times. This interpretation is further supported by the dolomitization, minor karstification and peneplanation of the upper surface of the Jebel. The three marine horizons are limestone and do not show a linear and progressive uplift pattern, but rather indicate alternating deposition, uplift, subsidence and erosion. This fluctuating history is interpreted here as a result of interaction between higher-order sea-level variations (glacio-eustatic?) superimposed on progressive uplift of the rift margin.

Jebel Mukalla
The stepped nature of the topography surrounding Jebel Mukalla reflects the successive deposition of a number of different limestone deposits (Figure D1.10, Plate 16(f)). There are three topographic levels with associated flat-lying carbonates which are all lithologically similar, shallow-water reefal rudstones and boundstones with a rich coralline and molluscan fauna. Each carbonate unit is flat lying, rests unconformably on all underlying sequences (and earlier units – by definition) and each successively younger unit is topographically lower, forming a downstepping stratigraphic geometry. Unit 1 is approximately 30 m thick, Unit 2 is at least 25 m thick and Unit 3 is approximately 15 m thick (Figure D1.10).

The three carbonate units seal the underlying border fault (Figure D1.10). Correlation of Unit 1 with similar beds at Buwaysh-Jebel Dhabba beds implies progressive uplift without faulting since at least 17 Ma. These beds also record the post-tectonic uplift of the margin, showing about 120 m of movement during at least three uplift pulses. Furthermore, it can be implied that about 300 m of uplift occurred within the early Miocene (i.e. after deposition of the sequence 2 reefal carbonate and prior to the deposition of sequence 3, unit 1). This progressive, punctuated uplift contrasts with that seen in Jebel Dhabba and suggests that regional uplift of Jebel Mukalla was rapid enough to always overprint any sea-level variation.

Other facies
Numerous alluvial terraces are present in the area. At least five levels are present with the highest at 220 m (35 m above the present wadi level). Although individual terraces are traceable along distances of 5 to 10 km it is not possible to map them across the region. Terrace sediments are composed of well rounded, moderately sorted, unbedded, monomict limestone conglomerates. All show a distinctive red coloration and a coarse, gypsum-rich matrix. Chert and iron concretions often form a crude desert armour.

Redeposited gypsum beds are areally limited and occur where gypsum-rich underlying beds are being eroded. They form a flat-bedded, thin, draping veneer, 1 to 5 m thick, and contain clasts of the locally exposed rocks. The region north of Jebel Dhabdab shows the best development of this facies.

A number of basalt flows are present east of Wadi Assid and as yet remain undated (Figure D1.2). Some have flowed down present-day wadis, damming them in places (Wadi Assid and Wadi Shakhawi) and are overlain only by windblown sand and modern wadi gravels. They are mostly fissure fed, although three minor volcanic cones have been identified.

Present-day wadis have braided, sinuous to anastomosing courses. Two levels of wadi are seen: the topographically lower antecedent drainage systems of the major wadis and the consequent streams superimposed on the coastal plain.

DISCUSSION

Regional correlation

Offshore Gulf of Aden
Correlation between Shihr Group Sequences 1 and 2 with wells offshore Hadhramaut is good. The lower part of the Ghaydah Formation (Bott et al., 1992) comprises sabkha-type, marginal marine to fluvial deposits similar to those interpreted in Sequences 1 and 2. However, the upward gradation, with apparent conformity, to deeper marine conditions of the upper Ghaydah and Hami formations is not recognized onshore. The age span of Oligocene–Miocene (?Rupelian–Burdigalian) dates for Sequence 1 and 2 deposits correlates well with the reported Oligocene Ghaydah and Oligocene to early Miocene Hami Formation dates (Haitham and Nani, 1990; Bott et al., 1992). The westward gradation to predominantly carbonate facies (Taqa Formation) is not encountered onshore, and is interpreted to represent eastward linkage with open marine conditions of the Indian Ocean (Haitham and Nani, 1990; Bott et al., 1992).

The strong unconformity between Sequences 2 and 3 is replicated offshore where there is onlap by the Sarar Formation (middle Miocene–Recent; Haitham and

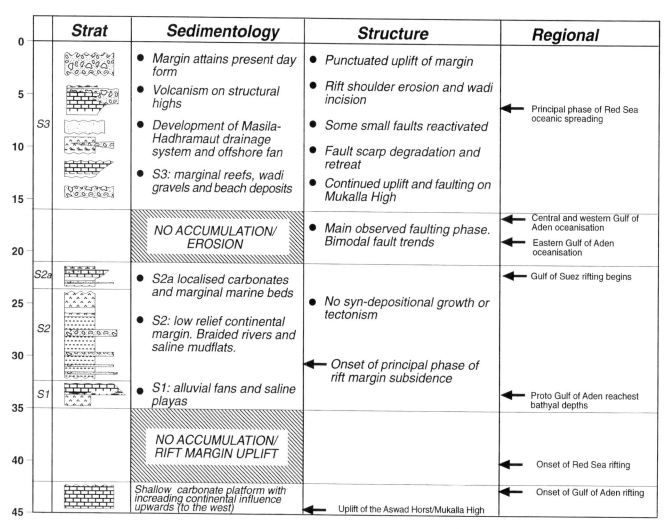

Figure D1.11 Tables summarizing the key depositional and tectonic events of Gulf of Aden evolution.

Nani, 1990). The Sarar is thickest (1126 m) in the region of Wadi Masila, reflecting its source from the Masila-Hadhramaut drainage system (Figure D1.2). Away from these major sediment input points the Sarar Formation is a thin unit of clastic submarine turbidites with occasional carbonate beds, in many ways very similar to the deposits of Sequence 3.

Somalia

Comparison with Somalia shows considerable facies similarities with the Boosaaso region, where the Oligocene-aged fluvial Shimis sands interfinger with marine limestone beds (Osman, 1992). The unconformably overlying Mait Group (confusingly also referred to as Daban series, Guban series and Dubar series) composed of reefal limestones and fanglomerates, bears some similarities to the uppermost deposits of Sequence 2, especially in the region of the Mukalla basement high. The limestones have similar early Miocene ages, although the Mait Group also encompasses beds up to middle Miocene age (Osman, 1992). Further west in Somalia, the Oligocene–Miocene Daban basin is a half-graben filled with basal lagoonal beds but progressively upward more continental ephemeral and perennial lacustrine deposits reflect the westward shoaling of the proto-Gulf of Aden (Abbate et al., 1993b; Fantozzi et al., this volume). Similarly, in western Yemen, freshwater lacustrine beds are intercalated within subaerial fissure-fed basalt lavas and ignimbrites of Oligocene age (Davison et al., 1994).

Controls on facies

Apart from the pervasive influence of extension and subsidence, discussed below, a number of other factors control the type and distribution of the exposed rift facies. The nature of the pre-rift substratum, the relief of depositional basins and surrounding hinterland, and climate are most important. Climate is probably the most important factor as the high temperatures and evaporation rates resulted in abundant and widespread evaporitic facies formed either in sabkhas or in shallow-water bodies, principally lagoons. However, unlike the Red Sea (Orszag-Sperber et al., this volume) the Gulf of Aden never became a deep evaporite basin.

The influence of pre-rift lithology can be seen both in the general predominance of carbonate and evaporite rift facies and in more specific examples of recycling of pre-rift deposits. The overall paucity of coarse-grained facies is a function of a number of factors, specifically: low topographic relief (that is, minimal rift shoulder uplift and constant bevelling of land surface) sediment by-pass (most of the sediment is directed landwards from the nearby rift shoulder and thence into the Gulf of Aden by the Hadhramaut-Masila drainage system: Figure D1.2) and the pre-rift lithologies (predominantly limestones providing gravel but little sandy detritus). The coarse material is all derived from lower Tertiary limestones exposed on the rift shoulder. There are no pebbles or cobbles of Cretaceous sands (except in the Mukalla region where deposits are very proximal) suggesting complete disaggregation of sandstone has occurred. Mesozoic clasts become more important in the higher Sequence 2 sands and silts.

Sedimentation and tectonics

The overall geometry of the rift fill (Sequences 1 and 2) shows a characteristic wedge form (Figure D1.2(a)), typical of many half-graben rift systems (e.g. Nottvedt et al., 1995; Prosser, 1991). Seawards this wedge-shape becomes less pronounced as sediment progressively blankets both hangingwall dip-slopes and fault-block crests. However, the detailed geometry of the sedimentary strata lack any progressive rotation or onlap and show no evidence for soft-sediment deformation, slumping or disturbance. Some doubt about the detailed stratal geometry exists because of the poor quality of the seismic data, the paucity of hangingwall exposures and the large amount of post-depositional erosion.

The character and distribution of the facies in these deposits along the margins of the Gulf of Aden rift do not conform to any of the classic models for extensional basin sedimentation (cf. Leeder and Gawthorpe, 1987). Deposition occurred in a low-relief setting with limited supply of coarse clastic detritus; fine-grained, evaporitic, coastal facies are dominant throughout the main syn-rift depositional phase (Sequence 2). Where coarse-grained facies occur they are associated with either antecedent drainage systems or uplift-controlled consequent streams and are not footwall scarp-related. The supply of fine-grained material from biochemical and evaporitic precipitation kept pace with subsidence to fill available accommodation space along the margin. Throughout the 10–15 Ma period of tectonism the shoreline remained in essentially the same position. In the deeper parts of the basin (offshore) sediment supply was from larger, regional drainage systems which by-passed the margin.

These sediment geometries and facies distributions indicate that the classical models of syn-rift deposition do not fit this rift margin (see Leeder and Gawthorpe, 1987; Gawthorpe and Hurst, 1993; Prosser, 1993; Nottvedt et al., 1995). The observed parallel, nondivergent stratal geometries and overwhelmingly fine-grained marginal marine deposits indicate a low-relief, continually subsiding basin.

The evolution of the sediment fill can be considered in terms of major phases of fault movement which separate major phases in sediment accumulation. Thus the observed stratal geometries are due to passive infill of accommodation space formed by a combination of continuous basin-wide subsidence and fault-related topography. The continuous deposition of marginal marine–coastal sediments in Sequence 2 shows that sediment accumulation kept pace with subsidence throughout the upper Oligocene. There must have been a steep continental shelf at this time as all the offshore wells record continuous deep marine sedimentation throughout the Oligocene and Miocene (see above; Bott et al., 1992). Differential compaction across tilt-blocks may have been a significant factor in creating additional accommodation space, but this remains as yet unqualified. Preliminary interpretation of offshore seismic data confirms the importance of compactional effects. This flexural subsidence style of sedimentation was partially alluded to by Bott et al. (1992, p. 221), with respect to the Mio-Pliocene section.

Tectonic implications

Pre-rifting uplift

There is a time lag of at least 7 Ma between the latest pre-rift and earliest syn-rift deposits (i.e. late Eocene (Priabonian) during which the region was apparently undergoing nondeposition and/or erosion. A fundamental question in the evolution of the rift margin is whether the margin underwent uplift during this time or whether the lack of preserved sediment may be due to some other factor, such as later erosion or eustatic sea-level fall. Indeed, Haq et al. (1987) indicate a 60–75 m eustatic sea-level fall in the Priabonian, which could account for the nondeposition.

Our mapping shows that the syn-rift unconformity has eroded to a remarkably consistent depth across and along the margin as the Habshiya Formation, the uppermost pre-rift unit, is a relatively constant thickness of 150–250 m across the region. The original thickness of the Habshiya is not known and total erosion is unquantifiable.

West of the study area, the Habshiya Formation indicates that this region had evolved into a number of segregated, north–south and east–west trending facies assemblages suggesting a regional retreat of the sea to the east and the formation of a proto-Gulf of Aden marine embayment with fringing terrestrial clastics and marginal marine sediments by late Eocene times (DGME, 1986; As Saruri, 1995). This has also been suggested for northern Somalia (Abbate et al., 1993b). In Dhofar, Roger et al. (1989) have shown that marine sedimentation continued throughout the Eocene. Our observations of the Habshiya, which has been dated only as middle Eocene in the region studied (Beydoun, 1964), shows variable shallow-marine carbonate facies with only a minor influx of clastic sediments upwards and no major facies divisions.

Sparse apatite fission-track work along the Yemeni and Somali margins indicates systematic variations in ages along the margins. Palaeozoic and Mesozoic ages are recorded from the palaeohighs (e.g. Mukalla high = 238 Ma) suggesting that they were not deeply buried prior to Tertiary rifting and that they have not undergone significant exhumation. Two other clusters of dates at 40 Ma and 20–23 Ma can be related to rifting and suggest a two stage uplift/exhumation process (Abbate et al., 1993a; Menzies et al., 1996a).

All these data suggest that uplift of the rift shoulders began prior to rifting with a pronounced west–east slope.

Basement highs
Basement highs are important features along both Gulf of Aden margins and appear to have exerted a considerable influence on both the architecture of rifting, the nature of the rift basins and the form (and content) of the feeder drainage systems (Figures D1.1, D1.2). The highs are regions of minimum strain along the margin (Watchorn, unpublished data). They are also the sites with minimum accumulated thickness and width of rift-related sediments. Bosence et al. (1996) have described the Wadi Hiru sub-basin (on the Mukalla High; Figure D1.2): they document a stacked sequence of interfingering and alternating alluvial fans, fan deltas and reefs which reflect repeated faulting and sea-level rises/falls over the past 25 Ma. These coarse, high-energy coastal facies are in striking contrast to the low-energy coastal facies described here from east of Mukalla. The presence of the highs appears to be a significant control on the nature of rift sedimentation. The Mukalla fault is the boundary between the two regions (Figure D1.10). It is an extensional fault with a highly oblique (070°) trend to the rift margin.

Constraints on timing of faulting
Our facies mapping constrains regional deformation and subsidence to have occurred in three main phases: prior to Sequence 1, between Sequences 1 and 2 and between Sequences 2 and 3 (Figure D1.9). It is also apparent, however, that local tectonic movements have occurred at intervening times (see Hummum region above).

Evidence for the early tectonism is based on the regional unconformity of Sequence 1 on tilted Hadhramaut Group beds. Coarse conglomeratic facies and rapid facies variations indicate tectonism in the form of differential uplift along the margin. This phase constitutes the main rifting unconformity and occurred before deposition of Sequence 1 and tilted the pre-rift deposits to form accommodation space. The timing of this event is poorly constrained, ranging between late Eocene (42 Ma) and early Oligocene (35 Ma).

Following the first phase of tectonism, a thick sedimentary succession formed: outcrop and seismic reflection profiles indicate up to 2500 m of subsidence. Time constraints on this depositional phase are poor with initiation at about 30 Ma (see above) and continued sedimentation to about 21 Ma (Figure D1.9). As with the earlier phase, no faults related solely to this phase were observed: all seem to have been reactivated in the final regional deformation. Thus, evidence for this phase is deduced from the unconformable contact with Sequence 1 and the large, fault-controlled thickness of accumulated sediment.

The third tectonic event produced the fault pattern exposed today and occurred after the deposition of Sequence 2 and prior to the deposition of Sequence 3, between 21 and 17.5 Ma. The most recent work on the sea-floor spreading history of this portion of the Gulf of Aden indicates that oceanization occurred about 16.5 Ma ago (Sahota et al., 1995). The close temporal correlation between the marked Sequence 2/3 unconformity and earliest oceanic crust may suggests that this is the so-called 'break up unconformity' (*sensu* Braun and Beaumont, 1989b). Offshore up to 1.1 s TWT of relief can be seen on the unconformity surface. The commonly observed time-lag between the break-up unconformity and oldest sea-floor spreading anomaly (cf. Tankard et al., 1989) is reversed: here the unconformity pre-dates the magnetic anomalies. Rather than use the genetic term 'break-up' (which in any case is under a considerable amount of dispute), here we term this surface the post-rift unconformity.

No regional tectonic events have occurred since the lower Miocene, as evidenced by the undeformed nature of Sequence 3 deposits, although some minor, localized faulting has occurred (Wadi Masila). It is worth pointing out that although many of the exposed faults appear to have fresh fault scarps, field inspection indicates that these scarps have retreated up to 300 m back from the fault plane by linear scarp retreat. The region is seismically inactive with no recorded historical or instrument seismicity.

CONCLUSIONS

1. Sedimentation associated with break up can be divided into three major sequences; the first two correspond to the onset and climax of rifting and subsidence respectively and the third sequence reflects initial passive margin sedimentation.
2. Uplift along the rift flanks and in the west occurred prior to the onset of rifting.
3. Rift-related basins along the northern margin of the Gulf of Aden are located away from basement and structural highs. The tectonic, subsidence and sedimentary histories of the highs and inter-high areas are markedly different.
4. The sedimentary facies and geometry of the depositional units do not conform to 'typical' rift models of sedimentation. Sedimentary facies indicate a predominantly marginal marine, arid, low-relief depositional environment throughout the rift phase.

ACKNOWLEDGEMENTS

This work forms part of a doctoral thesis funded by Canadian Occidental Petroleum who also provided logistical and scientific, support. D.W.J. Bosence and G.J. Nichols were supported by EU Science Project REDSED (No. SCI.CT 91.07190). Jerry Sykora, Eric Bolton, Andy Pearson and Ken McClay are thanked for advice, data and assistance.

APPENDIX

Dating the depositional units

Method

Original calcitic or aragonitic shells were collected for $^{87/86}$strontium isotope dating (Faure, 1977). These shells were checked petrographically and all heavily bored, recrystallized, or neomorphosed shells were discarded. The samples were cleaned of matrix and cement prior to analysis by thermal ionization mass spectrometer. Table D1.1 lists the samples, locations, taxonomic assignments, and $^{87/86}$strontium ratios for the material collected from this study.

To obtain ages from strontium isotope ratios it is necessary to compare analysed ratios with ratios obtained from chronostratigraphically constrained samples, e.g. from Beets (1992) and Hodell *et al.* (1991) from the Gulf of Aden at DSDP sites 588, 558A and 722 (see Bosence *et al.*, this volume). The latter results have been standardized to those of Beets (1992) to correct for their SRM standard value of 0.710235. No standardization was required for our data (with a precision $+/-$ 0.000018) because our results were within six decimal places of Beets' standard value. The spread of published ratios and dates have been enveloped to show minimum and maximum ratios and dates. The derived ratios are given together with their analytical errors to show how age ranges are obtained graphically (see Bosence *et al.*, this volume for a worked example). Ages are subsequently quoted as the median value (Table D1.1).

Chapter D2
Miocene isolated platform and shallow-shelf carbonates in the Red Sea coastal plain, north-east Sudan

J. H. Schroeder, R. Toleikis, H. Wunderlich and H. Kuhnert

ABSTRACT

Detailed sedimentologic, petrographic and micropalaeontologic studies of surface outcrops north of Port Sudan revealed:

1. Miocene carbonates of the Red Sea coastal plain of the Sudan occur in two different depositional settings: (a) isolated platforms characterized by massive margins and concave, closely crenulate talus slopes and (b) shallow shelf deposits characterized by horizontally and conformably deposited bioclastic sediments which, due to structural movement, now form tilt-blocks.
2. Both outcrop types are characterized by respective lateral patterns of facies and microfacies distribution; both were deposited in shallow water, at slightly elevated temperatures, in normal marine to hypersaline conditions.
3. Biostratigraphically these carbonates range from the Oligocene–Miocene boundary to at least late Serravallian, i.e. N4 to N13 planktonic foraminiferal zones. Deposition on the oldest isolated platform carbonates began in latest Oligocene; others followed. The shallow-shelf carbonates are younger than all platforms dated. These findings suggest considerable local differentiation during Miocene sedimentation.

INTRODUCTION

During the past decade geophysical and structural investigations have revealed that the Red Sea rift is a complex system with a most intriguing history. Makris et al. (1991b) recently summarized this development. With this progress sedimentological studies have not kept pace; realizing the deficiency, B. H. Purser initiated and subsequently coordinated the RED SED Programme.

Independently the Special Research Project 'Geoscientific problems in arid and semiarid regions' included young coastal basins in its programme. The present subproject began along the coast of northern Somalia and intended to provide a comparison of the Somali Gulf of Aden and the Sudanese Red Sea margin. Political events cut off promising work in Somalia and the Sudanese coastal plain became the focal point of this project. The objectives were:

- To obtain a more detailed and differentiated picture of Tertiary carbonate sedimentation from existing surface outcrops by characterizing facies, microfacies and their lateral and vertical distribution. In this way the depositional environments were to be characterized.
- To correlate Miocene carbonate rocks litho- and biostratigraphically.

This chapter is a synthesis of the dissertations of Wunderlich (in prep.) and Toleikis (in prep.) with the respective databases complemented by work of Kuhnert (1993 and unpublished report). In view of space allotted for this contribution, it focuses on the sedimentology of the carbonates, leaving details on underlying and overlying sediments as well as sedimentological and diagenetic details for more specific contributions.

Sedimentation and Tectonics of Rift Basins: Red Sea–Gulf of Aden. Edited by B.H. Purser and D.W.J. Bosence. Published in 1998 by Chapman & Hall, London. ISBN 0412 73490 7.

PREVIOUS WORK

Sudanese coastal plain

The Sudanese coastal plain has been repeatedly studied by petroleum exploration and the basic sedimento-stratigraphic framework resulting from the AGIP campaign was presented by Carella and Scarpa (1962) and Sestini (1965). This was followed by a University of Khartoum MSc thesis of Aboul-Basher (1975), contributing new surface observations. A major effort was the World Bank programme on the hydrocarbon potential of the Red Sea and Gulf of Aden region, which in the Sudan yielded important small-scale maps, a national review, and regional syntheses (Robertson Research International, 1987; Beydoun, 1989; Bunter and Abdel Magid, 1989; Crossley et al., 1992). The offshore exploration activity of the French company TOTAL included a short campaign of Montenat et al. (1990) to outcrops in the Sudanese coastal plain applying concepts developed in Egypt.

Preludes to the present project included a diagenetic study of Tertiary carbonates by Schroeder (1985) during his directorship of the Sudanese National Institute of Oceanography in Port Sudan (now the nucleus of the Red Sea University).

Red Sea–Gulf of Aden area

The best-known pre-rift and syn-rift sediments of the Red Sea are in the Egyptian coastal plain, where they have been studied intensively by a French/Egyptian team (Purser and Philobbos and their respective co-

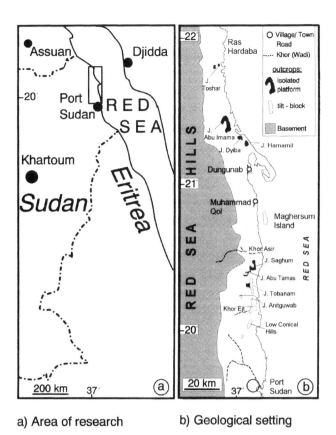

a) Area of research   b) Geological setting

Figure D2.1 Location of area and major outcrops studied.

workers; e.g. Purser et al., 1987; Purser and Hötzl, 1988); the carbonate platforms along the Gulf of Suez have become textbook examples.

Table D2.1 Miocene carbonate outcrops in the Red Sea coastal plain, north-east Sudan

Location	Position Longitude/Latitude	Dimensions NS/EW (km)	Type (Dip)	Subjacent clastics (m)	Carbonates (m)	Subsequent evaporites (m)
J. Toshar	36°43′E 21°55′N	2.5 × 0.7	isolated platform, closed	25	36	–
J. Abu Imama	37°16′E 21°22′N	5.5 × 3.5	isolated platform, open	51	60	–
J. Dyiba	37°16′E 21°02′N	3.5 × 2.8	isolated platform, closed	52	30	12
Maghersum Is.	37°16′E 20°48′N	10.0 × 4.0	tilt-block (12–17°W)	151	68	–
J. Saghum	37°07′E 20°25′N	5.0 × 3.8	isolated platform, closed	11	38	–
J. Anitguwab	37°05′E 20°15′N	1.4 × 0.6	tilt-block (20°W)	–	14	–
J. Tobanam E	37°08′E 20°18′N	9.0 × 1.0	tilt-block (14°E)	7	14	–
J. Tobanam W	37°07′E 20°20′N	3.0 × 3.0	isolated platform, closed	56	37	c. 25
Khor Eit	37°09′E 20°09′N	6.0 / 9.6*	tilt-block (12–18°E)	352	56	c. 200
L. Conical Hills	37°09′E 20°04′N	16.0** / 1.0	tilt-block (2–5°E)	6	22	–

J = Jebel = hill or mount(ain); Is. = island; L = low; *quasi-linear extent of khor; ** irregular chain of individual hills.

In Somalia marine rift sediments occur respectively near the towns of Las Koreh and Candala; especially at the latter location spectacular outcrops have been observed by the senior author in reconaissance missions for the project mentioned above.

LOCATION OF OUTCROPS AND GEOLOGICAL FRAMEWORK

The Red Sea coastal plain of the Sudan extends for about 600 km from the Sudanese–Egyptian to the Sudanese–Eritrean border (Figure D2.1(a)). Its width ranges from few to about 25 km. On its western side, the coastal plain is bounded by the Red Sea Hills, a mountain chain with peaks rising to more than 2000 m, composed of a large variety of Precambrian rocks. The rift boundary, in general, is sharp, but in detail it is highly serrated due to young erosion and deposition. Tertiary outcrops occur only in the northern portion of the coastal plain, from about 30 km north of Port Sudan to the Egyptian border. They consist either of isolated hills, several kilometres in horizontal dimension and tens to a few hundred metres in height, or of wadis (in the Sudan called khors) dissecting the coastal plain (Figure D2.1(b), Table D2.1).

Stratigraphically, according to the nomenclature presently used in the Sudan, these rocks constitute the Abu Imama Formation, while the subjacent clastics are part of the Maghersum Formation. For reference and comparison relevant stratigraphic scales are presented in Table D2.2.

Table D2.2 Compilation of terminology concerning Miocene rocks in the Sudan

Age				Carella and Scarpa (1962)	Sestini (1965)	Robertson Research (1986)	Bunter and Abdel Magid (1989)	
m.y.	Holocene				Abu Shagara Fm	unnamed carbonates	Abu Shagara Group	Shagara Fm
	Pleistocene			Abu Shagara Fm				
2	Pliocene	Late	Piacenian		upper siliciclastic group			Wardan Fm
4		Early	Zanclian					
6	Miocene	Late	Messinian	Dungunab Fm	Dungunab Fm	Zeit Fm		Dungunab Fm
8			Tortonian			South Gharib Fm		
10								
12		Middle	Serravallian	Abu Imama Fm	Abu Imama Fm	Belayim Fm	Belayim Fm	Hammam Faraun
14								Feiran
								Sidri
16			Langhian					
18		Early	Burdigalian		Khor Eit Fm	Kareem Fm	Maghersum Group	Baba
20								Kareem Fm
22			Aquitanian	Maghersum Fm	Maghersum Fm	Rudeis Fm		Rudeis Fm
24								
26	Oligocene	Late	Chattian	Hamamit Fm	Hamamit Fm	Nukhul Fm		Hamamit Fm
28								

FIELD AND LABORATORY WORK

Fieldwork

Fieldwork began with a reconaissance campaign in 1991 followed by expeditions in 1992 and 1993. During the first, at the classic locations of Khort Eit, Jebel (= J. = 'hill' or 'mountain'). Tobanam, J. Saghum, J. Dyiba, J. Abu Imama and Maghersum Island sections were measured and sampled. During the second mission additional profiles were studied and supplementary samples taken. Less conspicuous outcrops were also visited, and either short profiles or individual oriented samples were taken. This study is based mainly on 36 profiles.

Laboratory work

For the purpose of microfacies analysis, polished slabs and thin sections were prepared for petrographic investigation. Staining techniques (Dickson, 1966) were used to establish the distribution of carbonate minerals, while carbonate contents and calcite/dolomite ratios were analyzed using carbonate bomb and X-ray diffractometer.

For palaeontological studies samples were disintegrated with H_2O_2 and thermally; specimens were picked from the fraction $> 63\,\mu m$. In thin sections used for microfacies analysis fossil contents were also investigated. Oriented sections of individual foraminifera were prepared for species identification. Some specimens were investigated and illustrated by scanning electron microscope. Nanoplankton were searched for in smear slides and sporomorphs in the insoluble residue following HCl/HF-treatment; however, no useful fossil material of either was recovered.

FACIES ANALYSIS

Outcrop morphology and general framework

Positions, dimensions, sedimentary characteristics and thicknesses of sedimentary rocks from the major outcrops are presented in Table D2.1; in Figure D2.1(b) different types of outcrops are marked. Two types of outcrop morphologies occur: irregular rings or portions thereof and tilt-blocks; they correspond to the depositional settings of isolated platforms and shallow shelves, respectively.

Isolated platforms

Isolated platforms, less than one kilometre to several kilometres in lateral extent, are composed of interior lagoon, rim and outer slope, of which generally only the latter two are preserved (Figures D2.2, D2.3, D2.4, D2.6). The rim is either massive or horizontally bedded; it is crenulate in outline. The talus is characterized by beds varying in dip between 0° and 30°; dips are steepest near the rim and decrease downward, thus providing a concave profile. Individual layers comprising the slope are traced in cross-cutting outcrops from the rim to the shelf floor (Figure D2.6); they vary downslope in thickness, dip and composition.

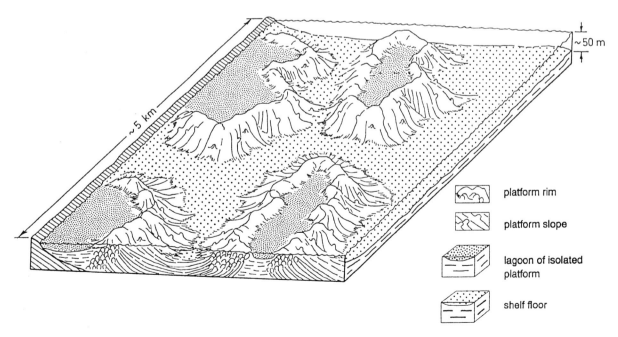

Figure D2.2 Palaeogeographical reconstruction of the two types of Miocene isolated carbonate platforms in the Sudanese coastal plain. Fringing-type opens toward the coast, and closed-rim type.

194 Miocene isolated platform and shallow-shelf carbonates

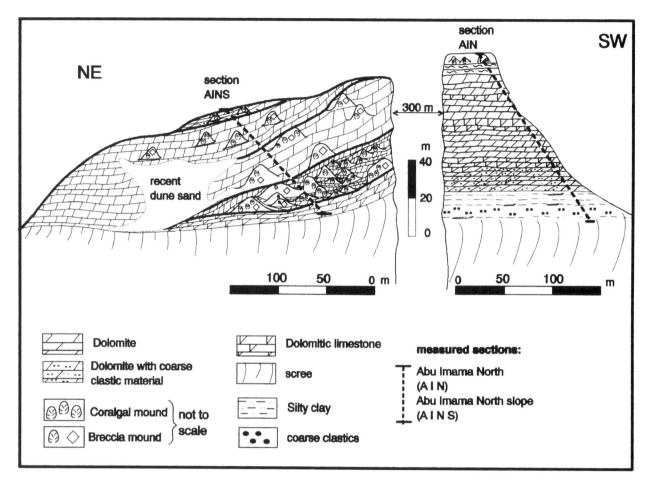

Figure D2.3 Section across the northern portion of J. Abu Imama platform as seen in outcrop. Note at the south-west rim the thick clastic base and horizontal bedded to massive carbonate rocks, and at the north-east slope rocks are variously dipping with mounds and toe of the slope reaching below the base of the carbonates in south-west.

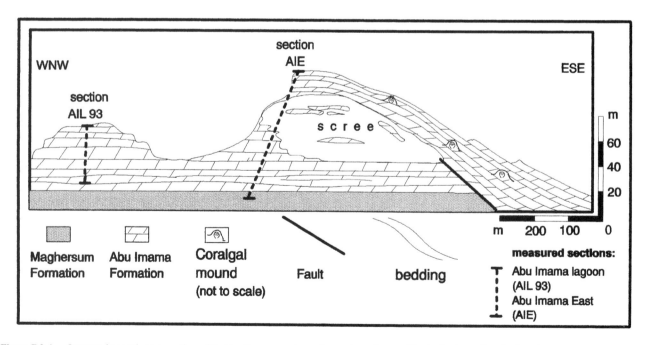

Figure D2.4 Outcrop in south-east portion of J. Abu Imama; note on the east-south-east side the fault cutting the hangingwall covered by slope sediments including small mounds.

A characteristic of the slopes are wart-like mounds of one to three metres in horizontal and vertical dimensions, which are irregularly distributed (Figures D2.3, D.2.6). They either are *in situ* coralgal mounds or breccia piles. These are smaller, yet similar in nature to those described by Purser *et al.* (1987) from the slopes of Abu Ghusun, Egypt, which now are interpreted to be olistoliths (Purser, personal communication, 1996).

As suggested by present outcrop configuration, the synsedimentary relief between the rim of the platform and toe of the slope ranges from 10 metres as observed at Abu Imama to 50 metres as observed at Jebel Tobanam. In one location (Abu Imama south-east) a relation between sedimentary platform development and synsedimentary faulting was observed (Figure D2.4): The platform or possibly an early ramp stage (*sensu* Read, 1985), was cut by a fault, and the fault-plane served as basis of the slope layers draped on to it.

The interior lagoon generally has been eroded (Figure D2.3); where preserved (J. Abu Imama south-east) it consists of horizontal layers of bioclastic carbonates.

In plan, the present outline of the rims suggests that there were two types of platform, one open toward the west, i.e. toward the coast and thus probably corresponding to a fringing reef, the other with a completely closed rim suggests a rather stubby atoll (Figure D2.2). This difference, readily apparent in aerial photograph and outcrop, may be due, partly, to selective erosion.

There are no outcrops of sediment deposited between these platforms, i.e. on a shelf or basin floor; slope configurations and dimensions suggest that the floor was no more than several tens of metres deep. The general paucity of clastic components in the slope sediments indicates low terrigenous input.

Examples of different morphologies are the Jebel Abu Imama Complex for a platform open toward the coast, and Jebel Dyiba as well as Jebel Saghum for a closed-rim platform.

Shallow shelves

Outcrops of the tilt-block type comprise layers originally deposited horizontally having consistent attitude and parallel bedding planes. Present dip angles vary between 3° and 18°, dip directions are either east or west; as a result outcrops are elongate in the north–south direction and tend to be relatively narrow (for data see Table D2.1). Some tilt-blocks are dissected by a system of step faults. A good example of a tilt-block is the Khor Eit complex (Figure D2.11), which thanks to erosion by the east–west trending wadi (= 'khor' in Sudanese usage) and the dip of about 10–12°E, exhibits the most complete stratigraphic section of the region; if linked with the low conical hills in the south and Jebel Tobanam (east) in the north it is traced for almost 40 km. The other example is Maghersum Island, less accessible, but with spectacular outcrops.

Subjacent sediments

Platform and carbonate-shelf deposits rest on siliciclastics. In the tilt-blocks, deposition of coastal plain to nearshore clastics of tens of metres thickness conformably underly carbonate deposition suggesting a gradual transgression. These clastics were deposited on alluvial plains and fans, in littoral and shallow marine environments (Wunderlich (in prep.) studied these in detail). At the base of the isolated carbonate platforms there are horizontally bedded clastics, with coarse conglomerates, sands and silts. Intermittently there are marine deposits including marine fossils such as bivalves and foraminifera. Within the clastics at J. Abu Imama *in situ* coral colonies, several decimetres in size, occur (Kuhnert, 1993), and at Jebel Toshar there exist *in situ* coralgal mounds several metres in diameter and about one metre in height; both suggest that the intensity of clastic input fluctuated considerably.

The clastics provided pedestals of limited lateral extension for the platform; their morphology, as indicated by the platform slopes, which topographically reach several tens of metres farther down than the clastics–carbonate transition below the platform rim (J. Abu Imama and J. Dyiba outcrops; Figure D2.3). Thus carbonates became established on the tops of a pre-existing relief, which may have been determined by faulting or by differential erosion. The latter suggestion is supported by comparison with the present coastal plain which exhibits relics of alluvial plains, eroded by a system of channels and left as flat-topped hills. Following drowning they would provide suitable pedestals for the platforms (Figure D2.4(c); Schroeder and Mansour, 1994). This suggestion is supported by the erosional unconformity visible at the small outcrop at J. Abu Tamas (Figure D2.1), immediately south of Jebel Saghum, where carbonates cover basement, the only contact of this type.

At Jebel Dyiba volcaniclastics and volcanic sills are intercalated within the clastic sediments; the volcanic activity may also have contributed to pre-carbonate platform relief.

Carbonate components and ecology

Although most of the rocks studied are dolomitized, in many units the original carbonate components were recognized. As listed in Figure D2.10, biogenic components include sessile and vagile benthos. The former is composed of corals, coralline red algae, vermetid gastropods, serpulid worms, encrusting foraminifera

and bryozoa as well as microbial crusts. The bioclastic components include pelecypods, gastropods, a variety of benthic foraminifera (nummulitids, alveolinids, miliolids) and planktonic ones (*Orbulina* sp., *Praeorbulina* sp., *Globigerinoides* sp. etc.), echinoderms, and green algae (e.g. *Halimeda* sp.); skeletal fragments of sessile benthos are incorporated into the bioclastic sediment. The nonskeletal components include peloids as well as normal, superficial, multiple and micritic ooids. In general, organisms represented in carbonate components suggest shallow, open marine photic environments.

More specific ecological information was obtained from distribution of foraminifera (Toleikis, in prep.; mainly based on Reiss and Hottinger (1984), Hart (1987) and Murray (1991). In general, two communities of benthic foraminifera were distinguished:

A. *Sorites orbicularis*, *Archaias* sp. and *Peneroplis* spp. represent the innermost shelf and lagoons with water depths less than 10 m (Murray, 1991).
B. *Ammonia beccari*, *Amphistegina* spp., *Acervulina* sp., *Borelis melo*, *Bolivina* spp., *Challengerella bradyi*, *Calcarina* sp., *Elphidium* spp., some miliolids as well as *Textularia* spp. including those encountered in group A (although markedly less abundant) indicate the shallow marine environments in the order of 5–50 m (Reiss and Hottinger, 1984).

Water temperatures are deduced by comparison with recent foraminifera: representatives of communities A and B (above) indicate temperatures between 20 and 30°C; in particular *Ammonia* sp., *Amphistegina* sp., *Archaias* sp., *Borelis* sp., *Calcarina* sp., *Cancris* sp., *Elphidium* sp., *Heterostegina* sp., and miliolids are relevant (Hart, 1987).

Salinities are interpreted also by comparison with Holocene foraminifera (Murray, 1981). The above-mentioned foraminifera suggest normal marine salinities with a marked tendency toward hypersaline conditions. Thus, Miocene correspond to Holocene conditions in the Sudanese Red Sea (38–41% with higher values in lagoonal settings; Schroeder, 1982).

Facies distribution

As revealed by the study of 28 measured columnar sections in both isolated platforms and shallow shelves studied, the respective lateral distributions of carbonate sediments follow characteristic patterns which were modified from general facies pattern proposed by Wilson (1975).

The isolated platform is characterized by the presence of a well-developed platform rim (Figures D2.3, D2.5–D2.7), which constitutes a high-energy environment, where carbonate sediment is produced, destroyed/brecciated and moved inward into the lagoon as well as outward down the slope. The schematic distribution of sedimentary environments, of carbonate components and resultant textures is presented in Figure D2.10. Conditions vary considerably between different environments, and correspondingly the lithologies vary laterally within tens and hundreds of metres. Detailed lithological profiles demonstrating vertical and horizontal variations are presented by Wunderlich (in prep.); examples of characteristic microfacies types are presented in Figure D2.11.

In contrast, the shallow shelf deposits comprising the tilt-blocks are more uniform and can be traced laterally for several kilometres without significant change, notably along the discontinuous ridge from the low conical hills across Khor Eit and along to Jebel Tobanam (Figures D2.8, D2.9). As readily seen in the respective sections and summarized in Figures D2.11 and D2.12, there are considerable facies variations including relatively high-energy shoals with oolitic grainstones, open and protected portions of the shelf, and conversely, depressions such as wide shallow basins; the common denominator is that all environments are shallow, i.e. within the photic zone (Figure D2.13).

The lack of directional trends and the absence of shelf edge or slope deposits suggest a relatively broad shelf, modified locally by ridges or patches which provided protection and reduced circulation or conditions suitable for biostromes. Although outcrops in general are linear along the escarpments (Figure D2.8), perpendicular khor sections (Khor Eit, Maghersum Island) suggest relatively uniform three-dimensional facies geometry.

Post-carbonate sedimentary history

Carbonate sediments in both settings are followed by evaporites; at this stage only field observations are available. Evaporites cover slope deposits of isolated platforms. At the northern side of J. Dyiba in the toe of

---

Figure D2.5 Isolated platforms in the coastal plain of north-east Sudan: outcrops. J. Abu Imama N: platform rim viewed from (now eroded) interior of the platform. Height of outcrop about 70 m; lower two-thirds of outcrop consist of subjacent clastics, upper third consists of massive carbonates (see also Figure D2.3). (b) J. Abu Imama platform: eastern rim viewed from exterior, i.e. from east; note variously dipping slope beds and almost horizontal platform beds at right; height above coastal plain about 50 m. (c) Holocene alluvial plain north of Khor Eit with residual hills of alluvial plain (height about 20 m); note flat tops, uniform elevations and talus slopes. In Miocene, such hills in the subjacent Maghersum clastics upon transgression may well have formed the base for isolated carbonate platforms.

Figure D2.6 Slopes of isolated platforms. (a) J. Abu Imama platform: northern slope viewed from outside (person in circle for scale). Note variously dipping layers, near-horizontal layers at tip of toe (1) and mound (M). (b) Depositional dip-slope surface of isolated platform at J. Tobanam SE. Note wart-like mounds (M).

Figure D2.7 Isolated platforms: characteristic rock-types in outcrop. (a) Top portion of platform rim composed of biogenic framework, brecciated portions and internal sediment (J. Abu Imama N, top cliff about 20 m high); for log see Figure D2.3. (b) Coral-algal framework at rim of the platform; corals partly dissolved (J. Abu Imama N, near top). (c) Biohermal growth at platform rim: corals (largely dissolved) covered by thick crenulate crust of corallinacean algae (J. Abu Imama N, near top). (d) Rhodolith floatstone from inner rim of platform (J. Abu Imama N, near base of carbonate section on lagoonal side).

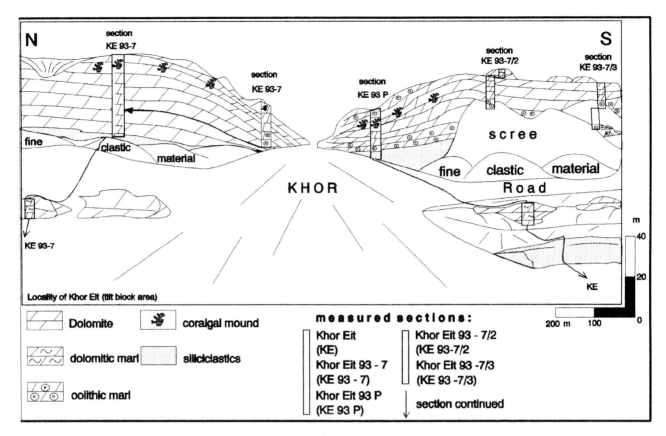

Figure D2.8 N–S extending outcrop of shallow shelf deposits in tilt-block extending from J. Tobanam to Khor Eit.

the platform slope, a sequence of dolomitic and calcitic beds contains conformable layers of gypsum. In contrast, at the southern-western side of Jebel Dyiba deeply incised valleys were filled by massive evaporites above clearly exposed erosional unconformities; similarly at Jebel Tobanam west, an erosion surface precedes dominantly evaporite deposition.

In Khor Eit a transition from carbonates to evaporites is observed; the evaporites, soon after deposition, became involved in large-scale sliding and slumping forming a chaotic layer of several metres thickness. This deformation is considered to represent the initiation of block-tilting.

In this area subsequently further evaporites were deposited. North-east of Khor Eit the top 50 m portion of economically viable gypsum deposit is mined in the open pit of Eit Mine; below the pit another 150 m portion has been explored.

BIOSTRATIGRAPHY

The palaeontological work aims to biostratigraphically date and correlate the lithological units observed and to provide a framework for the sedimentological and palaeogeographic evolution of the area. In spite of intensive dolomitization of the rocks, a variety of benthic and planktonic foraminifera were recovered, which provide stratigraphic markers (Figures D2.14–D2.16; Toleikis and Schroeder, 1995; Toleikis, in prep.). In particular, Miogypsinids identified according to Drooger (1966), represent the oldest microfossils in Miocene sediments of this area. Further time markers were obtained using planktonic biozones of the Mediterranean and Paratethyian; *Praeorbulina transitoria*, *P. sicana* and *P. glomerosa* were found in the platform carbonates. These species were also reported by El Heini and Martini (1981) from three boreholes at the western margin of the Gulf of Suez.

The data presented in Figures D2.14 and D2.15 are summarized as follows:

1. Marine carbonate sedimentation began near the Oligocene–Miocene boundary. The oldest carbonate deposits found occur in the lower parts of the isolated platforms; the youngest in tilt-blocks. At J. Abu Imama some of the oldest planktonic foraminifera were found in subjacent clastic deposits (Figure D2.15).
2. The respective earliest carbonates on various platforms differ in age. Implications of these observations are as follows.

Figure D2.9 (a) Tilt-block of originally horizontally bedded shelf carbonates. View south along west escarpment of J. Tobanam–Khor Eit block; height of escarpment is about 40 m. (b) Biostromal facies in shelf carbonates; isolated coral colonies *in situ* (Khor Eit, southern flank E of Khor).

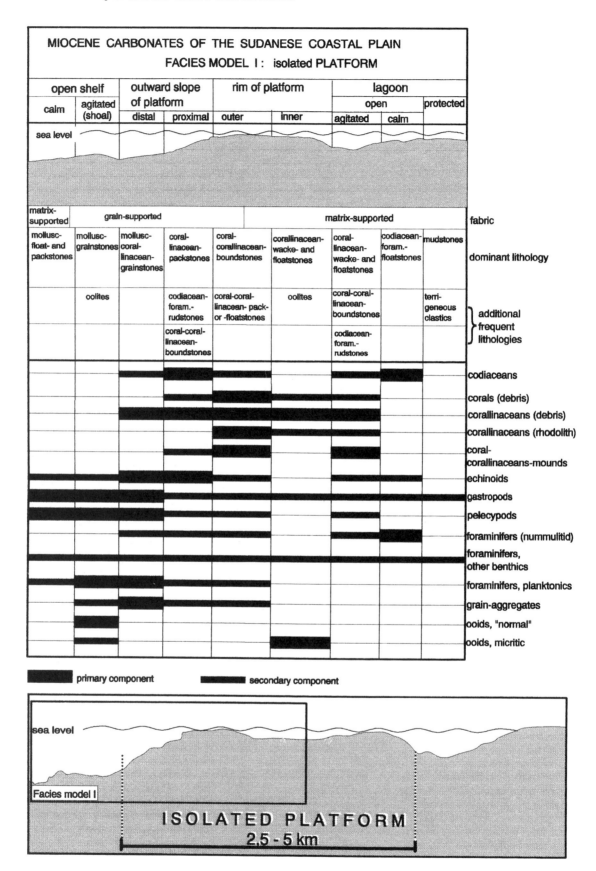

Figure D2.10 Facies model and distribution of microfacies types as well as major components in Miocene isolated carbonate platforms at the Sudanese coast.

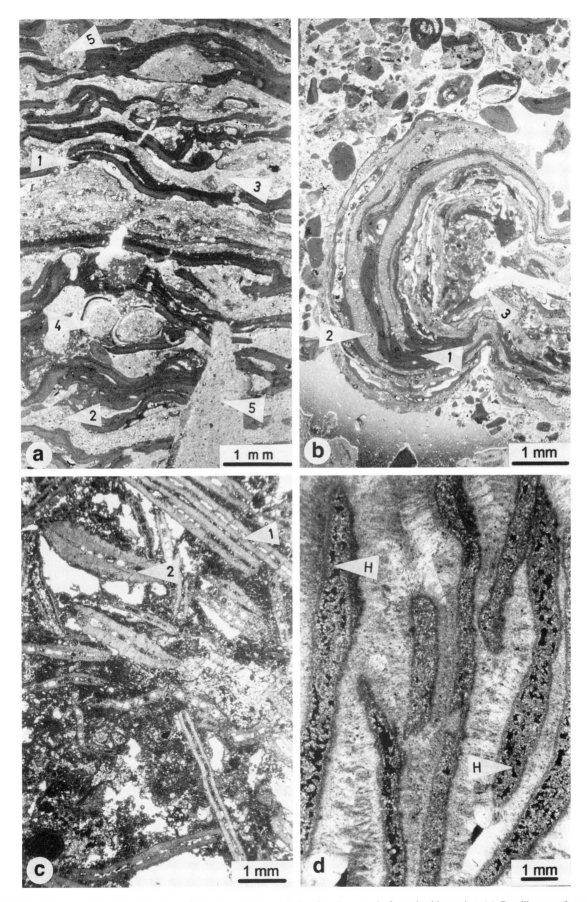

Figure D2.11 Characteristic microfacies types in Sudanese Miocene isolated carbonate platforms in thin section. (a) Corallinacean-foraminiferal bindstone: dark layers red algal crusts (1); light grey layers: encrusting foraminifer *Acervulina* sp. (2) in oblique section; sediment between crusts (3); vermetid gastropod (4); bioerosion by pelecypods (5) filled with sediment (J. Abu Imama E, sample SD 91 TAI 3M). (b) Rhodolith (see Figure D2.7(d)) composed of alternating crusts of coralline red algae (1) and *Acervulina* sp. (2) with cement in intercrustal pores (3); (J. Abu Imama, near base of carbonate section on lagoonal side; sample 93 20.3-1). (c) Foraminiferal rudstone with oblique and axial sections of *Planostegina* sp. Banner and Hodginson (1) and axial section of *Miogypsinoides bantamensis* (2; see Figure D2.16(g)) (J. Abu Imama section AICH 93, middle portion; Sample AICH 93–21). (d) Codiacean rudstone consisting exclusively of *Halimeda* sp. (H) with early marine cement (Jebel Dyiba platform on proximal slope; Sample D 92/202–3h).

The lithostratigraphic boundary between clastics and carbonates has no regional or local biostratigraphic significance as the carbonates are diachronous. Their deposition began in various portions of the Sudanese coastal plain at different times. Even on a given platform, carbonate sedimentation did not begin simultaneously in all parts. For example, at Jebel Abu Imama (Figure D2.7) marine carbonates developed first, probably in a ramp phase (Read, 1985), while the platform rim developed later.

The onset of carbonate deposition on the isolated platforms indicates favourable local conditions in the regional transgressive history of the shelf rather than a single transgressive episode. These conditions possibly include the existence of inherited relief and reduction in clastic input. The latter ecologically permitted or increased growth of organisms and sedimentologically prevented the carbonates from being produced or diluted by clastic components. Shallow-shelf deposition, represented by tilt-blocks, reflects similar sets of conditions prevailing over larger areas.

It also appears that carbonate deposition ended at various times before and during the onset of evaporite deposition. The evaporites were subject to later erosion and their record was largely destroyed.

DISCUSSION

Nature of the isolated platforms

Carella and Scarpa (1962) described the platforms (e.g. J. Abu Imama and J. Dyiba) as eroded and faulted anticlines. This notion prevailed up to 1988. On the Geological Map of the Sudan 1 : 1 000 000 compiled by Robertson Research for the Geological Research Authority of the Sudan, these carbonate complexes are again depicted as anticlines, some of them plunging. Montenat et al. (1990) first recognized their nature as documented here, i.e. that the dipping beds are original sedimentary slopes of isolated platforms.

Sedimentology and facies distribution

Sedimentological patterns in the Miocene of the Sudanese coastal plain resemble present ones (Schroeder and Mansour, 1994). This applies to the subjacent clastics of the Maghersum Formation (Wunderlich, in prep.), where one finds the complete inventory of coastal plain, beach and shallow sea. The resemblance covers the general conditions with respect to depth, temperature and salinity of the marine carbonates as well as to particular facies types. For example, the shelf biostromes in Khor Eit (Figure D2.9(b)) characterized by individual scattered coral colonies, closely resemble the coral zone of the present fringing reef (Schroeder and Nasr, 1981). The oolitic shoals are similar to those found on top of Holocene barrier reefs, for example on Shaab Baraya located north-east of Maghersum Island (Schroeder, unpublished observation).

For sedimentology and facies distribution, generally the situation in the Egyptian coastal plain as outlined by Purser et al. (1990) is closest for comparison. The major differences are the width of the coastal plain (narrow in Egypt and wide in Sudan) and the availability of basement relief (present in Egypt, virtually absent in Sudan). In Egypt, basement frequently forms platform cores, and the relief is more pronounced (about 100 m) and slopes are steeper (up to 50°); in the Sudan only one example has been observed (J. Tamas; Figure D2.1).

Biostratigraphy and palaeogeography

Either expressly or by implication many previous authors working on Miocene rocks in the Sudanese Red Sea region attributed ages to the lithofacies and their boundaries. Micropalaeontological studies indicate that this is not correct and suggest a relatively early onset of local marine deposition in the central Red Sea, i.e. at the Oligocene–Miocene boundary. These results differ from those obtained in Egypt; Purser et al. (1987, 1990) place the earliest syn-rift carbonates in Early Burdigalian, as do Montenat et al. (1990) with respect to Sudanese carbonates. According to Purser (personal communication, 1996) recent studies of the Jebel Zeit area yielded Aquitanian age. Sun and Esteban (1994) placed the lower limit of carbonate settings in the Gulf of Suez and Red Sea region at the lower boundary of the N8 zone. Our studies indicate that it coincided with the lower boundary of the N4 zone, i.e. with the Oligocene–Aquitan boundary.

Evans (1988) has described laterally discontinuous marine carbonate deposits of N4 age from the Gulf of Suez, which contains mainly planktonic foraminifera, while according to Crossley et al. (1992) by Early Miocene (20 m.y. BP) marine conditions were established throughout the length of the Red Sea rift system. In the map of the respective period these authors show widespread marine muds and coastal clastics, but no carbonates. Only in the Gulf of Suez are minor carbonate buildups mentioned; at the next time slice (15 m.y. BP) these build ups, although local in the Gulf of Suez, are attributed reservoir potential. Recently, Bosworth (1995) referring to unpublished work of W. M. Abdel Malak reported that the dolomitized limestones of the Nukhul Formation contain foraminifera of the N4 zone.

In their 'Simplified regional lithostratigraphical variations in Tertiary nomenclature in the Red Sea', Hughes and Beydoun (1992) illustrate carbonate lithologies at

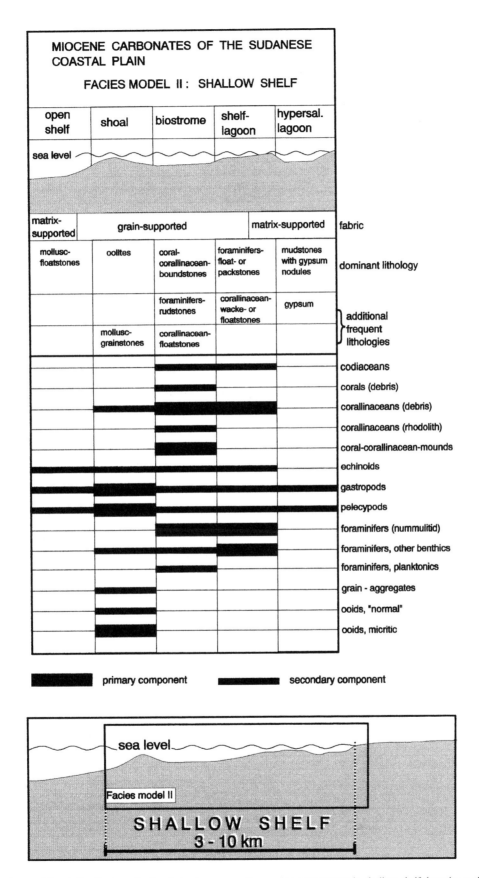

Figure D2.12 Facies model and distribution of microfacies types as well as major components in shallow shelf deposits at the Sudanese coast.

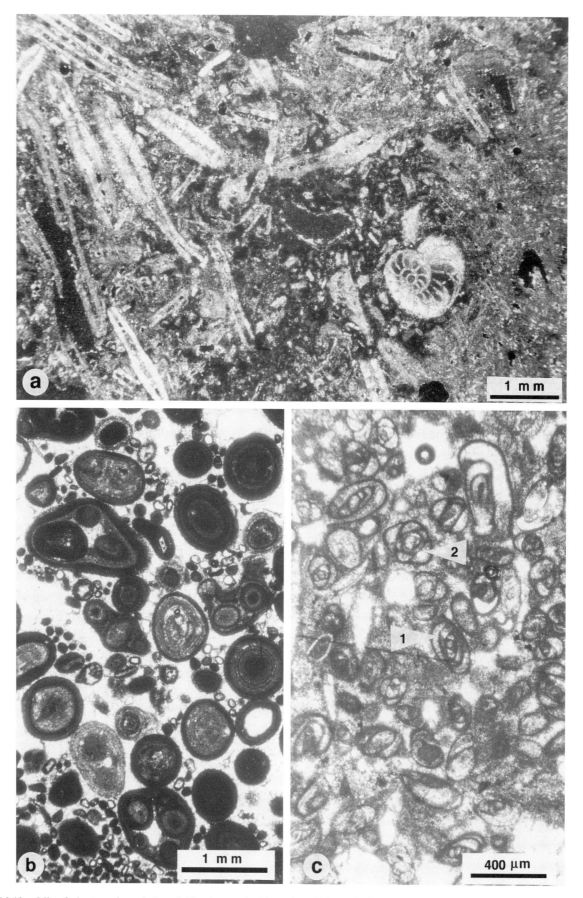

Figure D2.13 Microfacies types from shallow shelf carbonates in thin section. (a) Foraminiferal rudstone with axial (1) and equatorial (2) section of *Planostegina* sp. (Khor Eit, top of section; sample KE 93 VIII). (b) Dolomitic ooid grainstone with single and multiple ooids (Khor Eit, upper portion of sections VII /2, VII/3, P & JEN 93 II; sample JEN 93 II-2). (c) Grainstone containing miliolid foraminifera exhibiting axial and equatorial sections of *Biloculina* sp. (1) and *Quinqueloculina* sp. (2).

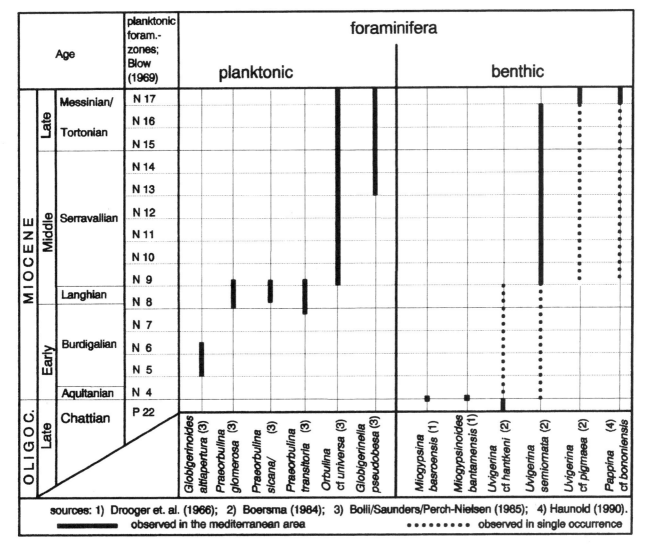

Figure D2.14 Taxonomy and stratigraphic range of benthic and planktonic foraminifera used for correlating Miocene carbonates.

the base of Miocene both from Egypt and Sudan (Hamamit Fm.); the latter are overlain by clastics of the Maghersum Formation. However, the early Miocene carbonates studied by the present authors clearly succeed Maghersum clastics.

The fact that the Sudanese isolated platforms and tilted shelf carbonates are of different ages suggests that they should be distinguished stratigraphically. We submit these results to the competent Sudanese national authorities for consideration when their stratigraphic terminology is reviewed.

On the basis of our study we envisage a relatively differentiated sedimentary history in the Sudanese coastal region. Such compartmentalization may well be limited to the marginal portion of rift system, while in the central parts may have a more consistent history.

Structural considerations

At Jebel Tobanam isolated platform and shallow shelf sediments, i.e. carbonate sediments of different ages and depositional settings, were observed in juxtaposition, separated by a major fault oriented subparallel to the coast. The fault determines the position of the western escarpment of the Khor Eit–J. Tobanam tilt-block. It has been illustrated by Sestini (1965) and interpreted as a normal fault. Makris and Henke (1992) in their pull-apart model have suggested left-lateral strike-slip faults extending roughly in a north–south direction in this region.

The respective distributions of isolated platforms and tilt-blocks, in particular the lack of remnants of (younger) shelf carbonates in the area where platforms are found, support the suggestion of lateral movement.

208 Miocene isolated platform and shallow-shelf carbonates

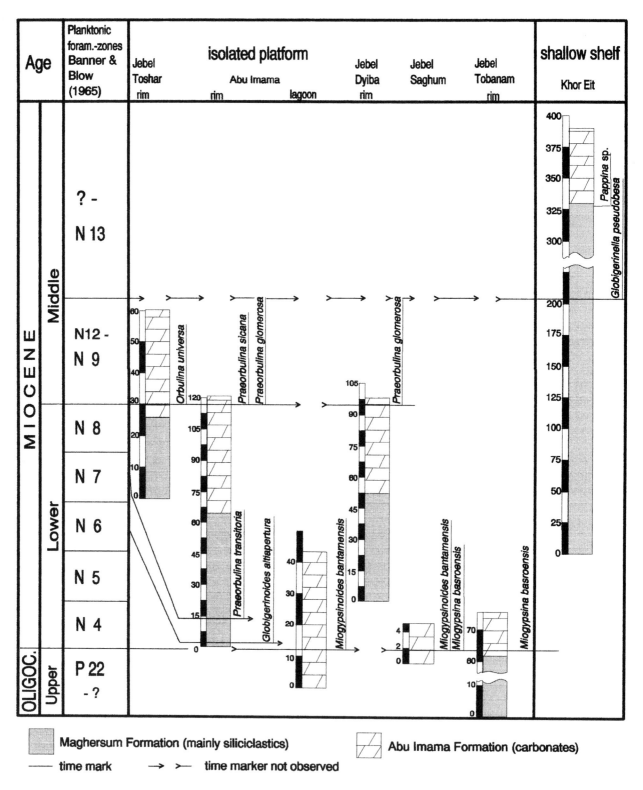

Figure D2.15 Stratigraphic correlation of various outcrops of Miocene carbonate deposits at the Sudanese coast.

Figure D2.16 Selected foraminifera of stratigraphic value in the Miocene carbonate rocks of the sudanese coastal plain (scanning electron micro photography); for stratigraphic range see Figure D2.15. (a) and (b) *Uvigerina semiornata*: (1) apertural view; (2) side view (J. Abu Imama platform; sample Al 4). (c) *Pappina cf. bononiensis*, side view (Khor Eit; tilt-block; sample KE 84). (d) and (e) *Globigerinoides altiapertura*; (4) spiral side; (5) enlarged part of the surface of the chamberwall (J. Abu Imama platform; sample Al 4). (f) *Praeorbulina glomerosa*, spiral side (J. Dyiba platform; sample JDA 3). (g) *Miogypsinoides bantamensis*, horizontal section of the test showing the arrangement of the embryonic apparatus, the spiral and the following chambers (J. Abu Imama Platform: sample AIL 5).

Discussion 209

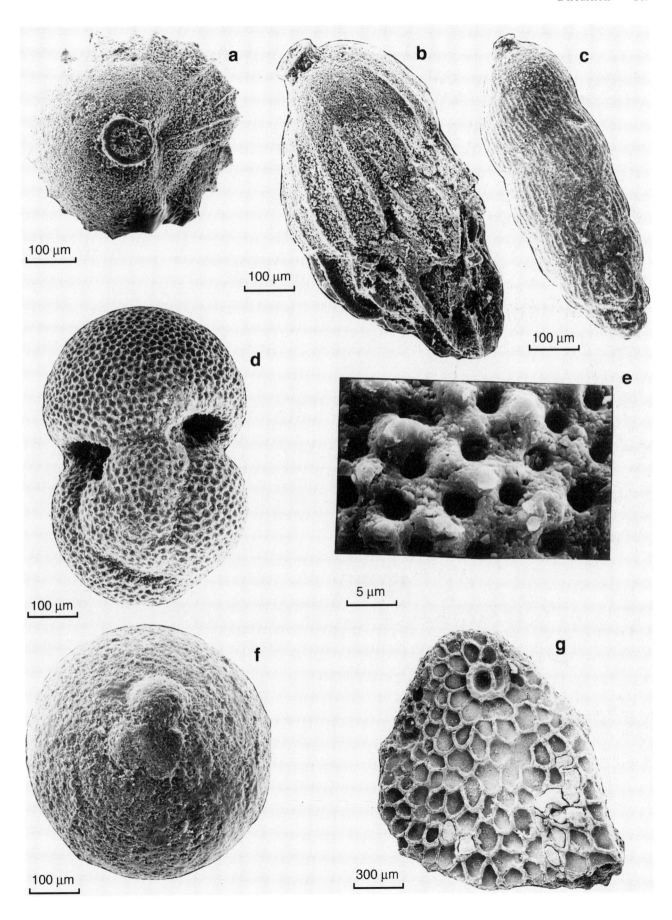

Conversely, the tectono-geophysical consideration of Makris *et al.* (1991b) nourishes speculation that isolated platforms and shallow shelves may have formed in different structurally defined compartments, i.e. in different basins. More detailed tectonic and geophysical work in the Sudanese region is required.

CONCLUSIONS

1. Miocene carbonates of the Red Sea coastal plain of the Sudan occur in two different depositional settings:

(a) Isolated platforms, several hundred metres to several kilometres in width and of a few to several tens of metres in relief, characterized by massive margins and concave, closely crenulate talus slopes. Their preservation is due to the hardness of the rim, which resisted various stages of erosion during the subsequent geological history.

(b) Shallow-shelf deposits characterized by horizontally and conformably deposited layers which are laterally continuous for several tens of kilometres, which due to structural movement now form tilt-blocks dipping either east or west.

2. Both basic types are characterized by a respective assemblages of facies and microfacies types; both were deposited in shallow water, at slightly elevated temperatures in normal marine to hypersaline conditions.

3. Biostratigraphically these carbonates range from Zone N4 to at least N13. The isolated platform carbonates are older, their deposition beginning at the Oligocene–Miocene boundary, however, their ages vary. In contrast, the shallow-shelf carbonates are distinctly younger. These findings suggest considerable local and regional differentiation during Miocene sedimentation.

ACKNOWLEDGEMENTS

This study is a result of a subproject of the Special Research Project 69 'Geoscientific problems of arid and semiarid regions' funded by the German Research Foundation; leadership of E. Kitzsch and administrative guidance of U. Thorweihe and H. W. Linke are gratefully acknowledged. Funds received from the EC Science Programme 'RED SED' are acknowledged. We thank the Geological Research Administration of Sudan (GRAS) collectively and many colleagues in Khartoum and Port Sudan specifically for professional partnership and administrative assistance.

B. H. Purser (Paris) and E. Philobbos (Assiut) guided (in Egypt) in the field and joined in many discussions; L. Hottinger (Basel) and W. Piller (Vienna) generously shared their palaeontological experience with R. Toleikis. C. v. Engelhardt helped with preparatory work, B. Dunker with drafting, H. Glowa with photography. Student assistants involved in this project included R. Schaaf, T. Schermutzki, M. Schumann, L. Seiffert and O. Stieghorst. G. Schirrmeister and C. Reinhold shared in discussions and critically read the manuscript. Thorough reviews by B. H. Purser and D. Bosence of an earlier version submitted helped considerably to improve the chapter. We thank all of them for their assistance.

Chapter D3

Stratigraphy of the Egyptian syn-rift deposits: correlations between axial and peripheral sequences of the north-western Red Sea and Gulf of Suez and their relations with tectonics and eustacy

J.-C. Plaziat, C. Montenat, P. Barrier, M.-C. Janin,
F. Orszag-Sperber and E. Philobbos

ABSTRACT

Dating of late Oligocene basalt in the continental early-rift deposits and micropalaeontological data (planktonic foraminifera, nanoplankton) recently obtained from Gebel el Zeit give limited but reliable correlations with the plankton-rich subsurface formations. The Burdigalian deposits equivalent to Nukhul and lower Rudeis formations appear to be rarely represented in the peripheral sequences where the sediments related to the major Langhian marine transgression generally lie unconformably either on Aquitanian evaporitic syn-rift unit or on pre-rift strata and basement morphologies. The upper pelagic marls, underlying the upper evaporitic unit at Gebel el Zeit, appear to be Serravallian in age. Thus a Middle Miocene age is proposed for most of the marine pre-evaporite strata of the western rift margin. Because of limited biostratigraphic correlations, subdivisions of the informal group stratigraphy are retained for general description of regional stratigraphy.

INTRODUCTION

Stratigraphic studies of the north-western Red Sea–Gulf of Suez region have not been carried out with such detail as structural research. This probably results from the importance of this region in developing rift models, as the outcrop and subsurface data offer a relatively complete reconstruction of tectonic events. However, these models require a stratigraphic chronology and exact correlations between the different regions and especially between the peripheral and axial sequences. Transgressive and regressive trends, hiatuses and coarse sediment influxes may have tectonic or eustatic origins.

Thus, any model of rift evolution should be chronostratigraphically constrained and the detailed palaeogeographies, including geometric organization of the deposits, require precise correlations. Unfortunately, this has not been achieved in the Red Sea–Gulf of Suez basin and this contribution is an attempt to simplify a complex terminology and to draw attention to the relative precision of the current stratigraphic usage.

Much of biostratigraphic data have been published at least ten years (El Heiny and Martini, 1981; Steininger et al., 1990; Montenat et al., 1986; Evans, 1988; Richardson and Arthur, 1988; Sellwood and Netherwood, 1984; etc.). Several important contributions have been added more recently (Ouda and Massoud, 1993; Philobbos et al., 1993). With this background complemented by new data in this chapter we discuss the previous correlations and propose local changes and a general assessment of eustatic versus tectonic influences on sediment distribution.

OVERVIEW OF STRATIGRAPHIC EVOLUTION

There is general agreement on the late Oligocene initiation of the northern Red Sea and Gulf of Suez rifting, and on its multi-staged evolution (references cited above). We have emphasized the importance of the initial late Oligocene to early Miocene stage of rifting (named proto-rift by Orszag-Sperber and Plaziat, 1990 and early-rift by Hughes and Filatoff, 1995), characterized by strike-slip faulting (Thiriet et al., 1985; Jarrige et al., 1986) and a marine transgression at first limited to the axial part of the Red Sea and north-eastern Gulf of Suez. Subsequent rift stages developed from early to middle Miocene times. Various forms of extensional deformation, a wide marine transgression and significant shoulder uplift (Montenat et al., 1986) are the major events which took place successively during this interval.

The Aqaba–Dead Sea transform fault began to dissociate the Gulf of Suez from the Red Sea by the end of Middle Miocene (16–12 Ma; Bartov et al., 1980), when rifting is said to cease in the northern Red Sea, followed by sea-floor spreading in its southern part. The Gulf of Suez then underwent a different evolution during Late Miocene–Quaternary times (i.e. a prolonged rifting with a low extension rate or a Pliocene and Quaternary renewed rifting: Orszag-Sperber et al. and Plaziat et al., this volume; Hughes et al., 1992). Thus, the north-western Red Sea stratigraphy may be quite different from that of the Gulf of Suez, and this is the main subject of our discussion. The regional stratigraphic framework established for the Gulf of Suez is largely based on subsurface studies as well as western Sinai outcrops (EGCP committee, 1964). This lithostratigraphic nomenclature is the standard reference even for the Red Sea (Hughes and Beydoun, 1992). While this is reasonable for the axial areas, with sequences deposited at depths similar to those of the Gulf of Suez during Miocene times, it is not suitable for the peripheral sequences. For the African margin a lithostratigraphic chart has been proposed by the National Stratigraphic Sub-committee of Geological Sciences of Egypt (NSSC, 1974), later completed by Samuel and Saleeb-Roufaeil, Philobbos et al. (1989 and 1994) and Purser and Philobbos (1993). Independently, the GENEBASS research group introduced an informal, more simple terminology, based on tectono-sedimentary units (Montenat et al., 1986; Purser et al., 1990): comprising Groups A to D which are separated by regional angular unconformities.

Within the deeper marine sequence of the Gulf of Suez, below the main evaporite deposits referred to the Late Miocene (Tortonian + ? Messinian; see Orszag-Sperber et al., this volume), three major discontinuities (hiatuses) have been demonstrated, the tectonic origin of which is fairly well established (Figure D3.1). The 'Post-Nukhul event' (Evans, 1988) in the early Miocene, coincides with the beginning of active subsidence in the Gulf of Suez and, moreover, is characterized by block-tilting and faulting. The 'Mid-Clysmic event (MCE)' (Garfunkel and Bartov, 1977), within the late Early Miocene, marked by coarse sediment influx, was interpreted as indicating uplift of the axial blocks as well as the rift shoulders. The coincidence of the MCE with a change in regime of extensional deformation has been emphasized by Montenat et al. (1988). In the offshore region the deposits located below and above the MCE have been referred to as the Lower and Upper Rudeis formation. The 'Post Kareem event' (Evans, 1988), on the contrary, is characterized by a pronounced unconformity followed by a marked change in sedimentation (widespread and continuous accumulation of evaporites). It terminates the period of important fault movement and high subsidence rates (Moretti and Colletta, 1987). The origin of the important restriction limiting connexion with the Mediterranean is related to minor tectonic rise of the Suez region, a part of the Syrian arc. However, we must also take into account any eustatic changes in sea level that may occur during these tectonic events (see chronostratigraphic diagram of the syn-rift event in Montenat et al., this volume).

The major transgressive sequences culminate at the beginning of Early Miocene, Middle Miocene and Pliocene times, respectively (Haq et al., 1987). These highstands are especially influential in the peripheral areas which register only major transgressions. These areas are also susceptible to the slightest regressions (high frequency fluctuations). It is thus difficult to utilize the basinal data for a detailed chronologic subdivision of the peripheral north-western Red Sea series.

This chapter compares the available dates obtained from outcrops with the EGCP and NSSC stratigraphic charts and with the synthetic correlation charts of Haq et al. (1987) and Steininger et al. (1990), which differ to some extent.

THE EARLY-RIFT STRATA

These strata comprise a large variety of rocks: coarse immature conglomerates, reddish alluvial mudstones, gypsum, sandy marls and sandstones related to open-marine environment (locally with evidence of gravity flows), basalt dikes and flows. This tectono-sedimentary unit, named 'Group A' in our previous publications, may be subdivided into A_1, a basal continental unit with basalts, and A_2 typified by gypsum associated with marine marls (the lower Miocene evaporite) and correlated with the lowermost marine deposits of the Sinai outcrops ('First marine series' in Ott d'Estevou et al., 1986a).

The A_1 sub-group

This continental, frequently reddish unit is the Abu Zenima Fm. of the Western Sinai and the Abu Ghusun Fm. of the north-western Red Sea. A very coarse, localized facies, consisting of residual cherts and limestone pebbles, is termed Nakheil Fm., after the outcrop of Wadi Nakheil near Quseir (Red Sea). These conglomerates grade upward to Abu Ghusun-type beige and red sandstone and should be better considered as a basal member of the Abu Ghusun Fm.

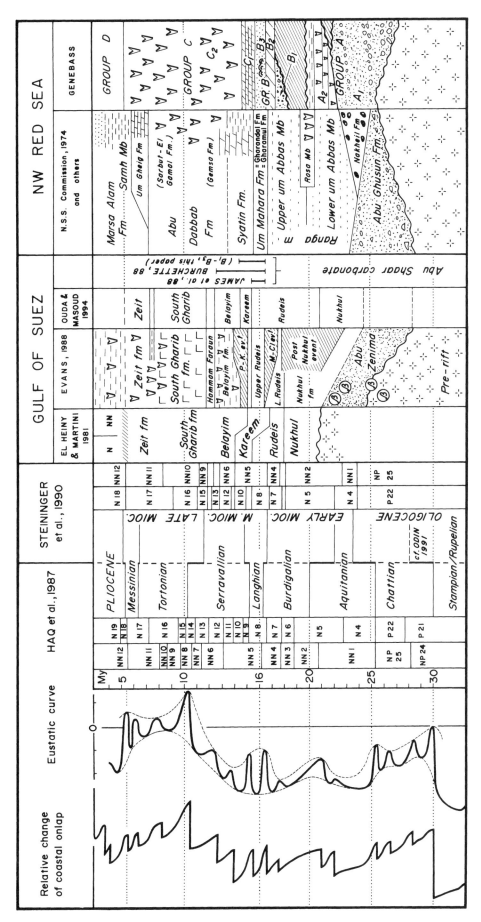

Figure D3.1 Table showing correlation between the basinal Gulf of Suez and the marginal, north-western Red Sea lithological units from Late Oligocene to Early Pliocene, based on diverse sources and related to the chrono- and biostratigraphic charts of Haq et al. (1987) and Steininger et al. (1990). The only common reference between these two official charts is the radiometric chronology (million years scale), while the biostratigraphic (N–NN) and the chronostratigraphic (stages) are different. We present the older Haq et al., stratigraphic scale because it is the most widely accepted base for eustatic curves. However, we utilize the biostratigraphic chart by Steininger et al. (1990) and show that the respective correlations by El Heiny and by Martini (1981) have some discrepancies concerning the controversial Middle Miocene interval. It is clear from the Rudeis, Kareem and Belayim Formations have variable age determinations. In agreement with Evans (1988), we suggest that the Abu Shaar platform developed during a relatively short time interval, benefitting from the highest Miocene sea levels. The influence of brief lowstands, visible on the eustatic curve for that period (or of higher frequency) are probably important but difficult to demonstrate. These must have been intermingled with hiatuses related to tectonic events in the basinal Gulf of Suez and should be represented in the marginal areas. In the north-western Red Sea, the complex lithostratigraphic frame has been oversimplified and correlated according to the informal GENEBASS system (Montenat et al. (1986). Hiatuses (hatched symbol) are located only in the Evans and Genebass systems: Post-Nukhul event, mid-Clysmic event and post-Kareem event.

These deposits unconformably overlie Cretaceous or Eocene rocks although the angularity is often weakly expressed in the Gulf of Suez. In the north-western Red Sea, to the South of Mersa Alam, the Abu Ghusun Fm. rests directly on Precambrian basement.

On the Sinai margin, in the type section of Abu Zenima, the red beds are capped by a basalt flow K/Ar dated at 22 Ma (21.95 ± 0.5 Ma and 22.15 ± 0.5 Ma; Ott d'Estevou et al., 1986b). The basalt underlies early Burdigalian marine deposits (see below). In Wadi Nukhul, the basal reddish strata begin with a conglomerate including basalt pebbles reworked from a large, nearby basalt dike whose radiometric K/Ar ages are 25 ± 0.5 Ma and 26 ± 0.6 Ma (Ott d'Estevou et al., 1986b). These ages pre-date the base of the Abu Zenima Formation which should be dated at between 25–26 and 22 Ma (i.e. later half of Chattian–early Aquitanian). No additional chronological data are available concerning the Gulf of Suez. The 24.7 (± 0.6) Ma basalt of Gebel Monsill in the Gharamul (near Ras Gharib) is not associated with sediments (Ott d'Estevou et al., 1986a).

On the Red Sea margin, the basalt flow of Sharm el Bahari and Sharm el Qibli near Quseir is interbedded in the upper part of the alluvial fan–playa sediments referred to as the Abu Ghusun Fm. Its radiometric age of 24.9 ± 0.6 Ma (Roussel et al., 1986) and 22.6 ± 0.5 Ma (El Haddad, 1984) suggests a late Oligocene–Aquitanian age for the major part of this unit.

The Abu Ghusun Fm. outcropping in the type area (Wadi Ranga–Ras Honkorab to the north of Abu Ghusun) is not dated. However, the subsequent evaporitic unit (A_2) contains Aquitanian planktonic microfauna near its top (see below). For that reason, the underlying red strata are obviously older, and possibly late Oligocene in age.

The Abu Zenima and Abu Ghusun formations are mainly affected by post-depositional tilting. They also occasionally display synsedimentary normal faults (prevailing Clysmic N140° trend, which is the Gulf of Suez axis orientation, *Clysma* being the Greek name for Suez) of small (decimetric to metric) amplitude in the Abu Ghusun area. In places, these deposits exhibit synsedimentary deformation related to strike-slip movements of the sub-meridian Aqaba trend of fault such as at Gebel Tarbul, north-western end of the Mellaha block and south of Safaga (Prat et al., 1986; Thiriet et al., 1985).

The A_2 sub-group

1. The lowermost marine strata of the Sinai margin. The continental, red Abu Zenima Fm. is unconformably covered by mixed marls and sandy marine sediments. This unconformity is especially well-exposed in Wadi Nukhul. In many places these marine deposits overstep the underlying red clastics to rest directly on the pre-rift Cretaceous or Eocene rocks (e.g. Gebel Ekma). They contain a planktonic association of late Aquitanian–early Burdigalian age (*Globigerinoides trilobus trilobus, G. trilobus immaturus, G. trilobus altiaperturus, G. subquadratius, Globigerina* gr. *praebulloides, G.* gr. *ciperoensis, Globorotalia obesa*) frequently associated with large benthic foraminiferas (*Miogypsina intermedia, M. complanata mauretanica, Sphaerogypsina* sp., *Heterostegina costata*) reworked from a shallow marine environment. In certain localities, these sediments show important syndepositional instability (olistoliths, slumps, etc., Gebel Ekma; see sections of the lowermost marine parts in Montenat et al., this volume).

This 'First marine series' of Ott d'Estevou et al. (1986b) correlates with the Gulf of Suez subsurface marine deposits referred to the Nukhul Fm. in the Gulf of Suez which also belongs to the late Aquitanian–early Burdigalian. These open marine deposits do not outcrop on the western margin of the Gulf and northern Red Sea. They are generally represented by a stratigraphic break or locally by a restricted marine facies: marls with rare planktonic microfaunas and gypsum beds deposited in small marginal graben. This is the Rosa member of the Ranga Fm. in Philobbos et al. (1993).

2. The lower evaporites. South of Safaga, the reddish alluvial-fan deposits (A_1) are conformably covered by a tilted 100 m evaporitic unit (A_2) with several gypsum beds, intercalated between silty marls, slumped sandstones and laminated diatomites (Wadi Gasus; Thiriet et al., 1986; Orszag-Sperber and Plaziat, 1990; Orszag-Sperber et al., this volume). In this region the evaporites are replaced westward (i.e. towards the basin edge) by laminated dolomicrites (algal mats) which overstep the underlying red clastics (A_1) to lie unconformably on late Cretaceous–Eocene rocks. The evaporites of Wadi Gasus are topped with slumped and reworked sands and marls which contain early Burdigalian planktonic foraminifera (Thiriet, 1987). The age given by this microfauna either post-dates the evaporites or corresponds to the upper limit of the A_2 sub-group.

These lower evaporites at Abu Ghusun (Ras Honkorab, Montenat et al., 1986c) include several gypsum beds about ten metres thick, interbedded with laminitic unfossiliferous carbonates and diatomaceous marls. Green marls located close to the top of the evaporitic unit yielded an Aquitanian planktonic microfauna: *Globigerinoides trilobus primordius, G. quadrilobatus, G. praebulloides, G.* gr. *ciperoensis*. The deposition of the evaporitic series is contemporaneous with the initiation of the tiltblock structural pattern (Orszag-Sperber et al., 1994; Montenat et al., this volume).

In this region a laminitic dolomicrite is associated with the evaporite unit, underlying locally the gypsum beds but more frequently blanketing the basement horst topographies. South of Wadi Gemal, it separates A_1 from A_2 deposits (Purser and Philobbos, 1994).

3. The lacustrine limestones. This significant but very localized facies occurs west of the northern Esh Mellaha (Prat et al., 1986). The laminated deposit with only charophyte remains suggests a permanent, stratified body of (saline?) water. It reflects a more humid environment than the playas of the A_1 upper deposits of Wadi Sharm el Qibli (Roussel et al., 1986) and it may represent a westernmost expression of graben subsidence during A_2 unit.

The intra-Group A or pre-Nukhul event

A_1 and A_2 deposits may be separated by a marked unconformity (Wadi Nukhul) or an erosional surface. The contact may also appear conformable (Wadi Gasus near Safaga). In other places the A_2 deposits overstep the underlying reddish unit and rest unconformably on pre-rift rocks. This geometric diversity may be related to contemporaneous block-tilting which resulted in either conformable (hangingwall, i.e. bottom of half-graben) or unconformable (footwall, flank and crest of tilted blocks) contacts. It is worthy of note that the block rotation correlates with local marine transgression and, in a more general way, with the accentuation of subsidence and a major change in facies.

Conclusion

1. Two successive suites of sediments were deposited during the initial stage of rifting: alluvial fan-to-playa deposits (the A_1 or Abu Zenima and Abu Ghusun Fms) and the A_2 subaquatic deposits: open marine ('First marine series' or Nukhul Fm.), restricted marine ('Lower evaporites' or Rosa Member of Ranga Fm.) or anoxic lacustrine.

2. This general sedimentary sequence is clearly related to rift evolution. Outcrops of the Nukhul Fm. are limited to the Sinai margin, suggesting an asymmetry of the rift basin. The same sedimentary sequence (red beds including numerous basalt flows and upper evaporites) also outcrops on the Sudanese margin, north of Port-Sudan (Montenat et al., 1990).

3. The chronological data suggest that the transition of the A_1 red beds to the A_2 lowermost marine unit may be diachronous, being younger in the north (Montenat et al., 1988). Consequently, the event involving block-tilting (and known in the basin deposits as the 'Post-Nukhul event') would also be diachronous. This important question still requires confirmation from more detailed chronological information before it can be solved (see discussion in Patton, 1994 and Montenat et al., this volume).

4. The African deposits referred to as sub-groups A_1 and A_2 show distinct syndepositional deformation. The A_1 series and associated basalts are contemporaneous with transpressive strike-slip faulting (initial stage of deformation; Jarrige et al., 1990; Montenat et al., this volume), whereas the A_2 marine evaporites are contemporaneous with extensional tilt-block faulting, evidenced in the Abu Ghusun and Safaga regions. On the other hand, it must be emphasized that the A_2 marine unit of the Sinai edge (Abu Rudeis area), of late Aquitanian–early Burdigalian age, was also affected by strike-slip deformation. This suggests that the wrench fault regime occurred a little later in the northernmost part of the rift, in agreement with the diachronism suggested above (see Montenat et al., this volume).

THE OPEN-MARINE SYN-RIFT STRATA (GROUP B)

This unit is characterized by: 1. open marine conditions, 2. abundant coarse marine clastics related to uplift of the rift shoulders and of intrabasinal tilt-blocks, 3. marked angular unconformity between the Group B deposits and their tilted substratum (Group A beds to basement rocks), 4. the onlap disconformity of Group C evaporites on Group B platform margin morphologies.

Depending on the local setting within the African Red Sea and Gulf of Suez margins, this group is either rich in carbonate (platforms and talus fringing the uplifted fault blocks) or mainly composed of coarse terrigenous siliciclastics (fan deltas = fanglomerates of Gilbert deltas) or even planktonic marls (hangingwall of fault blocks). It is frequently difficult to elucidate the precise stratigraphic relations between these contrasting facies. Therefore, Group B concerns the diverse marine deposits (open-marine as well as somewhat restricted) preceding the bulk of evaporites referred to as Group C.

Apart from the undivided pelagic marl facies (see below), this group comprises three units whose definition depends on their environment of deposition more than on their lithology. They are regional stratigraphic units (Laffitte et al., 1972) rather than formal lithostratigraphic formations. These informal depositional units are best illustrated in the South Quseir region and at Abu Shaar (Figures D3.2, D3.3).

B_1 open-marine unit

This detrital unit unconformably onlaps the erosional topographies of basement and A_1–A_2 resulting from block faulting and tilting. Faulting and associated erosion are responsible for 20 m deep gullying adjacent to 100 m highs as at Sharm el Qibli (south of Quseir). Such complex topography results in localized sedimentary units such as small juxtaposed fan deltas infilling submarine valleys and subsequently, overflowing as extensive terrigenous sheets. During this initial stage, there is little evidence of open-marine conditions in the sand and conglomerate deposits but a few red algal fragments and scarce coral patches. Where morphology is simpler, as in the large tilt-blocks such as Abu Shaar (Esh Mellaha), coarse terrigenous material is limited to the island shorelines and is replaced seaward by mixed terrigenous and bioclastic sands and by reefs (Figure D3.3; see Cross et al., this volume). The carbonate fraction has been generally overemphasized. It tends to mask the mixed nature of these units. For example, the Kharasa Member of the Abu Shaar plateau is not a typical Bahamian carbonate platform.

B_2 Open-marine unit

This unit comprises detrital carbonate facies mainly developed at the basinward periphery of mixed platforms. It may be difficult to differentiate from B_1 in the reefal areas. These characteristics of shelf-margin wedge are susceptible to high frequency fluctuations of sea level as progradation and associated coral growth are affected by relative sea level falls and rises. Tectonics may either reduce or exaggerate eustatic influences which complicates the interpretation of erosional surfaces as well as reef downstepping or backstepping (see Perrin et al., this volume).

This B_2 unit is the most obviously open-marine stage of the peripheral margin. This suggests that it is the expression of the early Mid-Miocene major highstand. This is not in contradiction with major lowering, because on the Haq et al. (1987) eustatic curves, the Mid-Miocene is characterized by an alternation of high- and low-level sea stands (Figure D3.1).

These shallow-water facies are unsuitable for precise biostratigraphic dating and their age has only been estimated by reference to the underlying planktonic marls (uppermost B_1).

B_3 more restricted marine unit

This unit is a mixed terrigenous-dominated sequence similar to but finer grained than B_1. It comprises sands (frequently with basement gravels) and muds. Oolitic sands with wave ripples, unfossiliferous carbonate muds and the frequent large, stromatolites indicate a restricted peritidal environment that alternates with extensive but thin reefal horizons (e.g. at Abu Shaar, Figure D3.3), or narrow coral fringes on the shallow-marine fans filling transfer-fault corridors (e.g. at Sharm El Bahari, Figure D3.2). These recurrent environmental changes do not grade progressively from open-marine to evaporitic conditions but end abruptly with the dolomitic laminites of the following C_1 unit. This B_3 unit has not been dated by biostratigraphic methods. Micropalaeontologic data only come from the lateral equivalent marls reviewed below.

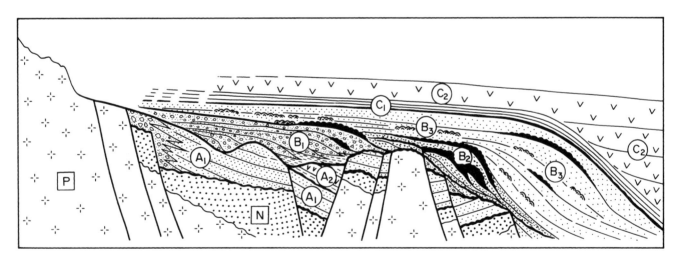

Figure D3.2 Generalized chronological and spatial relationships between the diverse structural settings of the north-western Red Sea coast and the proposed informal lithostratigraphic units. P = Precambrian basement, N = pre-rift cover: Nubian sandstones, Upper Cretaceous–Paleocene–Eocene strata. Syn-rift units: A_1 = continental alluvial fans-playa deposit with interbedded basalt flows, Abu Ghusun Fm.; A_2 = marine marls with gypsum beds, Rosa Mb of the Ranga Fm.; B_1 = coarse terrigenous fan-deltas with local intercalated and capping reefs; B_2 = shelf-edge reefs and associated mixed marine sediments; B_3 = finer terrigenous deposits with widespread stromatolitic episodes and marginal coral biostromes; C_1 = dolomitic microbial laminites encrusting the platform and peripheral slope; C_2 = onlapping bedded gypsum with fine terrigenous intercalations.

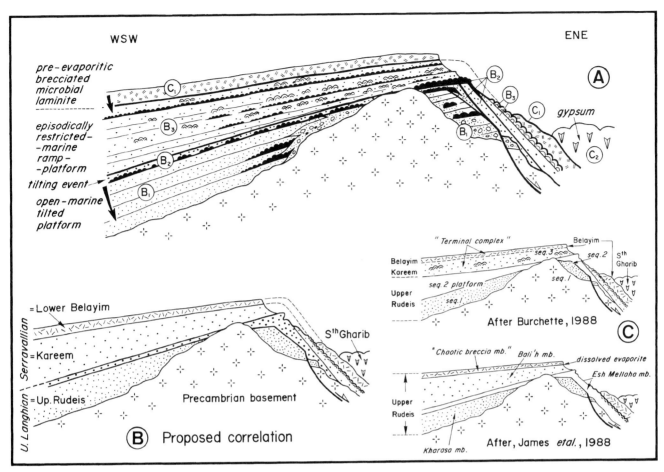

Figure D3.3 Major lithostratigraphic units of the Abu Shaar platform and proposed correlations with the axial Gulf of Suez and W Sinai terminology: A, according to the proposed subdivisions of groups A to C, the different units defined on the eastern margin of the tilt-block have been correlated with those on the western ramp (hangingwall); B, proposed correlation with the basinal lithostratigraphic units; C, the initial correlations by James *et al.* (1988) and Burchette (1988) related to the proposed geometric interpretation. For detailed sequence stratigraphic subdivisions see Cross *et al.* (this volume).

Biostratigraphic dates

The correlations of units (B_1) and (B_2) with planktonic assemblages is proposed as follows. In Wadi Gasus (south of Safaga), the unconformity between A_2 marine evaporites and the marine clastic sediments is weakly developed. The lowermost detrital deposits (northern bank of Wadi Gasus) yielded a late Burdigalian microfauna (Thiriet, 1987): *Globigerina* gr. *praebulloides*, *Globigerinoides trilobus immaturus*, *G. subquadratus*, *G.* gr. *bisphaericus*. The same level includes large blocks of semilithified marls probably reworked from the underlying, tilted evaporites (A_2).

This is the only reliable late Burdigalian age for the western Egyptian margin. This lower clastic unit is overlain by upper Group B_1–B_2 deposits including reef, bioclastic or chalky carbonates, planktonic-rich marls and minor clastic materials (referred to as Bc in Thiriet, 1987). In many places (eastern part of the area South of Safaga), the B_2 part of Bc oversteps the lower clastics and lies directly on the pre-rift (including the basement) and buries fault scarps on the eastern flank of Gebel Gasus. In Wadi Guesis, the marls associated with carbonate lenses, of the upper Bc series, contain *Globigerinoides sacculifer*, *G.* gr. *trilobus*, *G.* gr. *subquadratus*, *Praeorbulina transitoria*, *P.* gr. *glomerosa* and questionable *Orbulina* sp., which probably indicate a late Langhian age (Thiriet, 1987).

Other reliable micropalaeontologic data are rare but consistent. At the extreme northern limit of the studied area, at Gharamul (Wadi Garf) within a conglomeratic unit referable to B_1 as well as to B_2, *Globigerinoides trilobus* and *Orbulina* sp. indicate a late Langhian age (Ott d'Estevou *et al.*, 1986a). In the southernmost area of Abu Ghusun (Wadi Gemal), a marl layer directly below biogenic limestone, yielded *Globigerinoides quadrilobatus trilobus*, *G. sicanus*, *G. subquadratus*, *Praeorbulina glomerosa*, *Orbulina universa*, *O.* cf. *suturalis*, *Globorotalia obesa*, which characterize the late Langhian (Philobbos *et al.*, 1993).

To sum up, the available biostratigraphic data concerning the open marine deposits of the western Egyptian margin indicate a late Langhian age, with the exception of a single late Burdigalian determination.

If we accept a late Burdigalian age for B_1, it implies either an important hiatus before B_2 which is late Langhian (or early Serravallian), or a low sedimentation rate in contradiction with the observed style of deposition. On the other hand, a late Burdigalian age correlates with Lower Rudeis Formation (late Burdigalian–early Langhian) and late Langhian with Upper Rudeis Formation. Lower and Upper Rudeis are separated by the Mid-Clysmic tectonic event while the Lower Rudeis succeeds the Post-Nukhul tectonic event (Figure D3.1). For that reason, we consider that the Post-Nukhul and Mid-Clysmic events merge on the western periphery of the rift, constituting the major unconformity recorded between groups A and B. In this hypothesis the Lower Rudeis Formation has no (or few) representative deposits in these western peripheral outcrops, the clastic (Bd) unit of South Safaga being an exception, overstepped by late Langhian sediments. From this point of view, we agree with the correlation proposed by Burchette (1988) for the Gebel Esh Mellaha–Abu Shaar tilted platform units (B_1 and B_2 units) with the Upper Rudeis Formation (Figure D3.3).

On the contrary, marine deposits of the Lower Rudeis formation, dated as late Burdigalian, outcrop extensively on the eastern margin of the Gulf of Suez (Sinai edge), again indicating the asymmetry of rift sedimentation.

The upper, more or less restricted marine sequence B_3, lacking biostratigraphic markers, may be correlated with the basinal sequence only by means of its relative position above B_2 (late Langhian) and below evaporites and by the sporadic restriction of the marine environment. In the axial Gulf of Suez, below the main evaporites (South Gharib and Zeit Fms) two partly evaporitic formations (Kareem and Belayim Fms) have been distinguished in the subsurface and these are terminated by a more marine episode (= Hamman Faraun Member). The alternative correlations of B_3 are either with the Kareem Fm. or with both Kareem and Belayim formations. This problem is not easy to solve because of the absence of biostratigraphic data from the rift periphery.

The late Langhian age (N9 *Orbulina suturalis* planktonic zone) of marls intercalated in the lower part of the evaporitic sequence of Saudi Arabia (Makna, south of the Gulf Aqaba, in Le Nindre *et al.*, 1986) is the only indication of a Middle Miocene age for so-called South Gharib evaporites. However, considering their position, the correlation matches better with the Kareem Formation, and the foraminifera may even be suspected of reworking (see below).

In the Gebel el Zeit section, below the thick gypsum (C_2) unit, the plankton-rich marls have variable thickness (1–100 m) which indicates that they fill a complex erosional/structural morphology. Macrofauna, including the pteropod *Vaginella lapuguyensis*, suggests water depth exceeding 100 m. *Orbulina suturalis*, *Globigerinoides trilobus trilobus*, *G. sacculifer*, *G. obliquus obliquus* and *Dentoglobigerina altispira* of late Langhian age (*Orbulina suturalis* zone) have been identified by M. Toumarkine (personal communication, 1990) in a basal sample. A higher sample yielded *Praeorbulina glomerosa*, *Globigerinoides trilobus trilobus*, *G. trilobus sacculifer*, *G. bisphaericus*, *G. obliquus obliquus*, *G. ruber*, *Globoquadrina venezuelana*, *G. dehiscens*, *G. boroenoenensis*, *Dentoglobigerina altispira*, *Globorotalia obesa*, *G. mayeri* and *G. siakensis*, indicative of an early Langhian (N8 *Praeorbulina glomerosa* zone). This reversal of biostratigraphic ages suggests reworking of microfauna. Another micropalaeontological study by J. Cravatte (in Prat *et al.*, 1986) yielded *Orbulina suturalis*, *O. universa*, *Globorotalia* gr. *mayeri*, which is in good agreement with the late Langhian age proposed by Arafa (1982).

The calcareous nanoplankton recently studied by M.-C. Janin (Figure D3.4) also indicate an apparent stratigraphic reversal associated with major reworking: among the 79 morphs identified, 38 belong to Paleogene and Cretaceous beds, and only 30 are typical for the Neogene. The NN7 *Discoaster kugleri* biozone, indicative of the late Serravallian, is recorded even in the lower six metres of the marls, followed by levels with apparent Langhian assemblages characterized by *Sphenolithus heteromorphus* (NN5), sometimes mixed with late Serravallian markers (*D. kugleri, Catinaster* spp.). All assemblages include high amounts of Cretaceous–Paleogene forms, and it is probable that most Langhian material is also reworked. However, Langhian and Serravallian nanoflora have many forms in common, thus a Serravallian age would be expressed only in a few levels, related to episodes of less restricted marine environment compatible with the occurrence of the Serravallian marker group *Discoaster/Catinaster*, only common in open marine conditions.

The uppermost marls, situated immediately beneath the evaporites, show obvious indications of anoxic deposition: laminated marls, pyritic granules, presence of fish remains (*Bregmaceros*), diatom levels, etc., although some planktonic assemblages are still present.

As indicated above, the '*Globigerina* marls' of Gebel el Zeit, generally referred to as late Burdigalian–late Langhian age (e.g. Prat *et al.*, 1986) belong to the late Serravallian. Reworking of older (Langhian) pelagic deposits are recorded but the existence of late Burdigalian marls within the sequence is not confirmed and appears doubtful. We still have no precise data concerning the Miocene so-called Nukhul carbonate which is the substratum of the marls. As it has been deeply eroded and reworked (olistoliths) in the pelagic marls, it was possibly an ancient deposit (early Miocene ?) but this is questionable (cf. Moretti and Colletta, 1987).

The late Serravallian age of the upper part of the 'Globigerina marls' (in Gebel el Zeit and possibly in other sites) correlates which the age of the marine Hamman Faraun member in Sinai, at the top of the Belayim Fm., i.e. just below the main evaporitic formations (South Gharib and Zeit Fms), in a similar stratigraphic position as at Gebel el Zeit.

Conclusion

According to the available biostratigraphic data, we confirm the Middle Miocene age of most of the marine deposits, including coral reefs, of the north-western coast of the rift. However, the episodically restricted B_3 unit paradoxically would correspond to the plankton-rich marls of Gebel el Zeit. The rarity of reliable open-marine biostratigraphic assemblages, as evidenced by nanoplankton in the pelagic marls, is nevertheless in good agreement with the fluctuating environments of B_3 observed in the higher parts of peripheral fault-blocks.

THE OUTCROPPING EVAPORITES SERIES (GROUP C_2)

The geometric relations between the thick gypsum beds named Group C_2 and the underlying series is not usually well exposed. The weathered sulphate (dissolved and dehydrated at the surface) has frequently slid over the older outcropping rocks and is often located in low-lying areas lacking fresh outcrop.

Group C rests unconformably on the older rocks (including the basement), generally onlapping and overlapping the frontal slope of platforms but sometimes with a high angular discordance. Various topographic highs such as fault blocks and fault scarps, reef talus and channels, are buried by the upper evaporite beds which levelled the uneven substratum. Buried carbonate bodies, such as reef talus, have occasionally suffered sulphatization, i.e. carbonate replacement by massive gypsum (Orszag-Sperber et al., 1986) causing

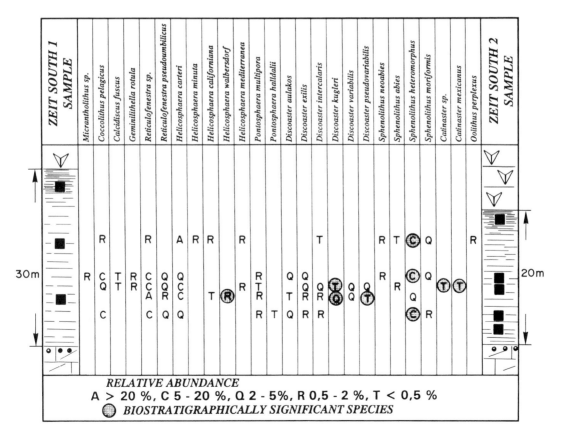

Figure D3.4 New biostratigraphic data from two sampled sections through the pelagic marls at Wadi Kabrit, Gebel el Zeit. Nanoplankton identification by M.-C. Janin.

possible stratigraphic confusion with true evaporites. Because of the progressive onlap on to structural highs (up to 50 m visible height) it is often impossible to observe the most ancient evaporite beds, deposited at greater depths and now buried below the Quaternary plain deposits. For example, immediately west of Quseir (northern side of Wadi Ambagi) horizontal gypsum layers clearly onlap the B_2–B_3 slope sediments, extending on to the platform culmination (Figure D3.5 and Perrin *et al.*, this volume, Figure D7.7). As the lowermost part of the sulphatized talus is not exposed, the real base of the evaporites (Abu Dabbab Formation) is unknown.

Furthermore, the upper surface of the evaporites is also difficult to study, because of erosion and weathering in the highest outcrops. The boundary with the post-evaporitic 'Pliocene' series, where it is visible, shows sliding and deformation which suggests a final seismic event, possibly associated with a general emergence and hiatus implying lack of Messinian deposits south of the Suez isthmus.

In most outcrops, Group C sulphates (C_2) are preceded by a 10 to 25 m thick laminitic microbial unit (unfossiliferous, vuggy dolomitized carbonates), comprising stratiform mechanical breccias (Plaziat *et al.*, 1990, this volume) referred to here as C_1. These deposits are still influenced by the morphology of the substratum. At Abu Shaar, they covered the footwall and the hangingwall of a tilt-block above the B_3 stromatolitic unit. Laminated carbonates plastered the external slope and flat top of the platform and after seismically induced sliding, piled up in depressions (up to 50 m thick in the Abu Ghusun area where they fill a pre-existing channel; cf. Montenat *et al.*, 1986).

The upper massive and banded gypsum logically corresponds to the South Gharib and Zeit Formations of the subsurface. However the preceding, partly evaporitic Kareem and Belayim Formations of basinal settings may possibly be correlated with the unexposed basal evaporites as well as with the B_3 and laminite (C_1) platform deposits. As the present authors suggest a correlation between the B_3 restricted marine unit with the Kareem Formation it is logical to correlate the highly restricted C_1 laminites with the lower part of (or the whole) Belayim Formation. This differs from Burchette's proposal (1988) of an equivalence of his 'stromatolitic sequence', including the uppermost 'dolomitic vuggy carbonates' (the C_1 laminites), with the Belayim Formation, and a 'karstification phase' of the B_2 reefs with the Kareem Formation (cf. Figure D3.3).

According to our hypothesis, the correlation between the C_1 microbial carbonates and the lower Belayim Formation does not imply that the main time interval during the deposition of the basinal Belayim Formation is represented by a hiatus on the peripheral margin. Rather, it is likely to correspond to the lower evaporitic interval or is not deposited at the foot of the B_1–B_3 highs.

THE POST-EVAPORITIC UNIT (GROUP D)

The comprehensive Group D has not been precisely dated and biostratigraphically separated from the underlying Miocene. It is generally referred to as Pliocene and Quaternary (Orszag-Sperber *et al.*, this volume, b). As noted above, the frequent breccias and sliding dislocation of the Group C evaporites suggest instability generated by emersion of alternating layers of mud and gypsum. The middle Messinian Mediterranean lowstand may explain these disturbances, being responsible for a complete desiccation of the Red Sea, unless the Miocene Bab el Mandeb isthmus was broken and a Plio-Quaternary communication with the Indian Ocean during the Messinian or earlier was established (Orszag-Sperber *et al.*, this volume, b).

Certain southern sections (north and south of Mersa Alam) conversely suggest continuous submarine flooding, the uppermost gypsiferous deposits grading into fine subaquatic terrigenous deposits. Intercalated carbonate muds, with stunted molluscs, are indicative of very restricted marine environments. These are followed, after a deformation horizon, by open-marine calcarenites, oolites and coarse terrigenous deposits occasionally including patch reefs.

This varied style of the Mio-Pliocene boundary may be explained by an alternative hypothesis. The contrast may be related to topographic differences induced by tectonics: the uplifted sites showing an emergent boundary (e.g. south of Safaga) while the subsiding areas would have registered a continuity. In this case, the transitional post-evaporitic deposits would be either latest Miocene or Pliocene in age and the uppermost evaporites of the Red Sea would belong to the Messinian. At present, these interpretations remain unconfirmed.

The authors prefer to retain the term 'Pliocene' for the whole post-evaporite marine deposits (restricted and open-marine, with oyster banks and coral reefs) that generally slope towards the axial Red Sea (5° to 25°, Figure D3.5), this dip being considerably higher than that of the overlying Quaternary terraces. The reefal limestones have a contrasting characteristic Indo-Pacific fauna (molluscs: *Tridacna*, *Pinctada*, *Lambis*; echinids: *Clypeaster reticulatus*, *Laganum*; *Acropora* dominated corals), demonstrating the new biogeographic connection of the Red Sea.

Except for these lithified deposits, associated with synsedimentary halokinetic domes, it is not possible to assume a Pliocene age of the uppermost layers which may be Pleistocene in age (cf. Taviani, this volume, b) despite the general angular unconformity observed

Figure D3.5 Structural and sedimentary relations between the syn-rift deposits and their respective substratum, northern bank of Wadi Ambagi (Quseir). (a) Marine Gr. B platform and reef on a block of pre-rift Eocene limestone; (b) onlapping and overstepping C2 gypsum on the B2 platform edge. The section shows the peripheral clinoforms which terminate in a dark bed rich in reefal faunas; (c) general view towards the north-west showing east-dipping 'Pliocene' beds in foreground, Miocene evaporites and basement horst with narrow Mid-Miocene shelf (in background) and pre-rift formations of Duwi block on horizon.

along the Red Sea, south of Quseir, between 'Pliocene' and unquestionable Quaternary terraces. The assumed Quaternary sequence, with its reefal terraces associated with alluvial fans, is discussed by Plaziat *et al.* (this volume). However the authors use inverted commas in order to suggest the remaining doubts on the limits of this 'Pliocene'.

CONCLUSION

The sedimentological objective, namely the discrimination between tectonic and eustatic influences on the sediment distribution, is limited by poor biostratigraphic data. The correlation chart (Figure D3.1) relating the basinal Gulf of Suez series and the peripheral Red Sea is an attempt to integrate the stratigraphy of both domains. The synchronous (as far as it is possible to ascertain) environmental shifts expressed by relative sea-level changes or ecological deterioration must logically complement this tentative correlation.

The Langhian transgression, the highest proposed Neogene highstand (Haq *et al.*, 1987), may well explain the open-marine carbonate episode (Group B) which covers the topographic highs of the rift periphery. However, the Mid-Clysmic Event possibly results from a combination of a tectonic event related to block-faulting and a short but important regression at the end of the Burdigalian, prior to the observed Langhian flooding. Similarly, the major early Serravallian transgression may account for the sustained high-stand (B_3–C_1) on the platforms of Abu Shaar and South of Quseir. However, the progressive environmental restriction from B_2 to C_1 may be interpreted to be the result of local tectonics, north of Suez. Combined with this tectonic upheaval the major eustatic regression placed at the end of the Serravallian (Haq *et al.*, 1987) would therefore play a threshold role in the final tectonic isolation (one-way influx of sea water from the Mediterranean) which brought about the thick axial halite deposits of the South Gharib and Zeit formations (Evans, 1988). On the other hand, the repeated environmental oscillations within B_3 cannot be explained by only one regressive event and it seems more likely to relate them to high frequency eustatic oscillations, of limited vertical range, such as those demonstrated in the Tortonian–Messinian of the Mediterranean (Pomar and Ward, 1994).

Predictably, the tentative correlations between the subsurface, basinal formations and the littoral deposits of the western periphery reveal the discontinuous character of the latter. Groups A to D are separated by hiatuses of variable amplitude due to a combination of tectonic and eustatic events. During early and middle Miocene times (Groups A and B) the hiatuses seems to be larger on the south-west than on the north-east margin. There appears to be an east to west merging of Post-Nukhul and Mid-Clysmic events and a correlative lacuna of the Lower Rudeis sediments on the western side, while such deposits develop on the Sinai edge, between the Morgan and Galala-Abu Zenima accommodation zones. A similar eastern development is noted concerning the marine Nukhul series, emphasizing the fundamental asymmetry of this part of the rift.

The alleged south to north diachronism of rifting within the northern Red Sea cannot be resolved because of the lack of precise dates of the first graben. However, encouraging preliminary results of this integrated study of biostratigraphic, tectonic and eustatic data, should not overlook the insufficient nature of biostratigraphic dates from outcrops on the Egyptian margin. We would suggest that future progress in the reconstruction of the Red Sea rift will come mainly from new stratigraphic investigations.

D4
Extensional tectonics and sedimentation, eastern Gulf of Suez, Egypt

K. R. McClay, G. J. Nichols, S. M. Khalil, M. Darwish and W. Bosworth

ABSTRACT

The eastern margin of the Gulf of Suez displays spectacular interrelationships between rift-border fault systems and sedimentation patterns. The structural and sedimentological evolution of a 100 km long segment of the well-exposed eastern margin of the Gulf of Suez rift has been studied by detailed field mapping, structural and sedimentological analysis together with analysis of Landsat TM and SPOT data. The eastern margin of the Gulf of Suez between Hammam Faraun in the north and Gebel Araba in the south is formed by a series of north-west trending, south-west dipping rift-border faults that are segmented and offset in an en-echelon pattern together with a second set of segmented, north-west trending, south-west dipping, major intra-rift faults that form the 'coastal fault system'. In cross-section the faults are typically south-west dipping, planar domino – to gently curving listric in shape. The pre-rift stratigraphy includes crystalline Pan-African basement rocks (largely metavolcanics, migmatites and granites) and a pre-rift sedimentary succession of clastics and carbonates that ranges in age from the Cambrian through to the Eocene. Syn-rift clastics and carbonates range from the Oligocene through to the Recent and typically show depositional patterns which relate to the geometries of the extensional fault systems. Shallow marine Miocene syn-rift clastics were deposited in fault-bounded sub-basins at the rift margin and fanglomerate bodies with well-developed progradational wedge geometries formed in the hangingwalls of major rift-border faults. Continued movement along the major fault systems has deformed the pre-rift and syn-rift sequences leading to the present-day structural configurations.

INTRODUCTION

The Gulf of Suez is the north-western arm of the Red Sea rift system and partly separates the Sinai Peninsula from the remainder of Egypt. It is a Cenozoic rift about 300 km long and up to 80 km wide (Figures D4.1 and D4.2). Uplift of both flanks of the rift exposes a relatively complete stratigraphy from the Precambrian to the Quaternary with rift sedimentation continuing to the present day. Exposure is excellent, with an arid climate and active erosion allowing a detailed examination of the structures, sedimentological and stratigraphical relations to be undertaken.

The Gulf of Suez has been the focus of hydrocarbon exploration for over 100 years (E.G.P.C., 1996), and a number of significant discoveries have been made including the Belayim (2.3 billion BOE), Morgan (>1 billion BOE), July and October fields (>700 M BOE each). The structure and stratigraphy of the area have been summarized in a number of publications: Robson (1971) reviewed the geology of the eastern side of the Gulf; the National Stratigraphic Sub Committee (1974) synthesized the Miocene rock stratigraphy of the Gulf of Suez; Gawthorpe et al. (1990) concentrated on the Miocene sedimentary history, and various aspects of the structure and stratigraphy are described in Barakat et al. (1986, 1988); Khalil and Meshref (1988); Said (1990b), Darwish (1992, 1994); El Barkooky (1992); Darwish and El Araby (1993); Patton et al. (1994) and Schutz (1994). It is worth mentioning that the published offshore data are quite limited, and a comprehensive tectono-stratigraphic review of the evolution of the whole of the Gulf of Suez has yet to be undertaken (but see Bosworth, this volume). In this chapter we review our recent studies on

Sedimentation and Tectonics of Rift Basins: Red Sea–Gulf of Aden. Edited by B.H. Purser and D.W.J. Bosence. Published in 1998 by Chapman & Hall, London. ISBN 0412 73490 7.

224   Extensional tectonics and sedimentation

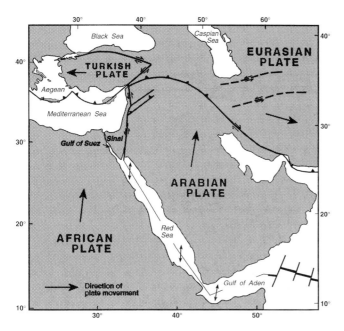

Figure D4.1   Tectonic setting of the Gulf of Suez (after Hempton, 1987).

the structure and stratigraphy of the central eastern part of the Gulf of Suez.

TECTONIC SETTING AND GEOLOGICAL HISTORY OF THE GULF OF SUEZ

Plate tectonic setting

The Gulf of Suez–Red Sea–Gulf of Aden rift system (Figure D4.1) has long been recognized as having been formed by the separation of Arabia from Africa (e.g. Coleman, 1974a, 1993; Hempton, 1987) and is perhaps the best modern example of continental rifting and incipient ocean formation. The rift system was formed by the anticlockwise rotation of Saudi Arabia away from Africa about a pole of rotation in the central or south-central Mediterranean Sea (cf. McKenzie et al., 1970; Freund, 1970; Le Pichon and Francheteau, 1978; Morgan, 1990; Meshref, 1990). Extension decreases westwards along the Gulf of Aden and northward along the Red Sea, consistent with, and constraining the pole of opening. At the northern end of the Red Sea the opening is divided between the Gulf of Suez and predominantly sinistral shear along the Gulf of Aqaba–Dead Sea transtensional system (Figure D4.1) (Freund, 1970; Ben Menahem et al., 1976; Abdel Khalek et al., 1993).

Geological setting

The Gulf of Suez trends north-west–south-east and is characterized by north-west–south-east striking major extensional faults both at the rift borders and within the

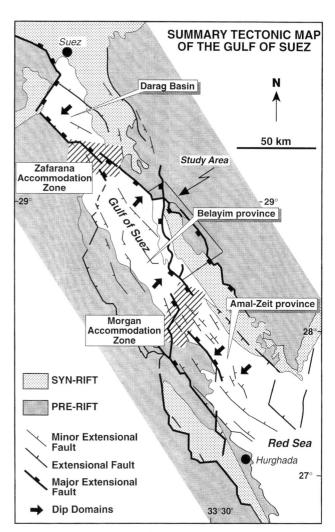

Figure D4.2   Tectonic map of the Gulf of Suez (compiled from various sources quoted in the text and from the authors' own observations).

rift basins (Figure D4.2). There are three distinct depocentres within the Gulf of Suez – the Darag basin at the northern end, the central basin or Belayim province in the middle, and the southern Amal-Zeit province (Figure D4.2). Each basin is asymmetric with a dominant dip direction. The northern Darag basin has a dominant dip towards the south-west whereas the strata in the central (Belayim) province dip dominantly to the north-east. In the southern province (Amal-Zeit) the dominant dip direction is to the south-west. Structurally complex accommodation zones separate these three distinct dip domains (Figure D4.2).

The eastern side of the Gulf of Suez, in the central dip domain, is shown in the Landsat Thematic Mapper image in Plate 17 with the regional geology and structure summarized in Figures D4.3 and D4.4. A summary stratigraphy of the eastern side of the Gulf of Suez is shown in Figure D4.5. The pre-rift units vary

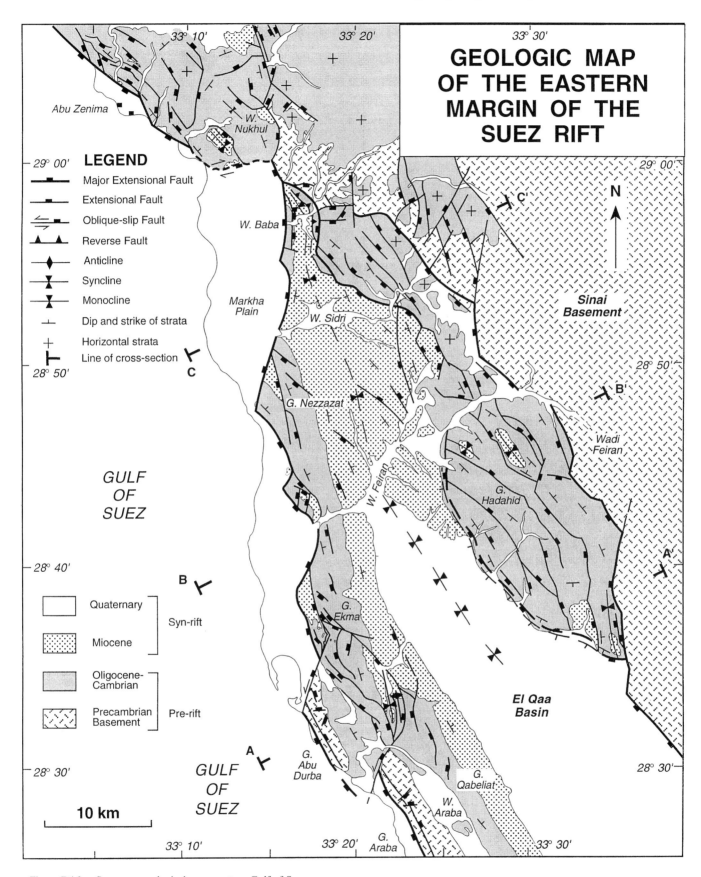

Figure D4.3 Summary geological map, eastern Gulf of Suez.

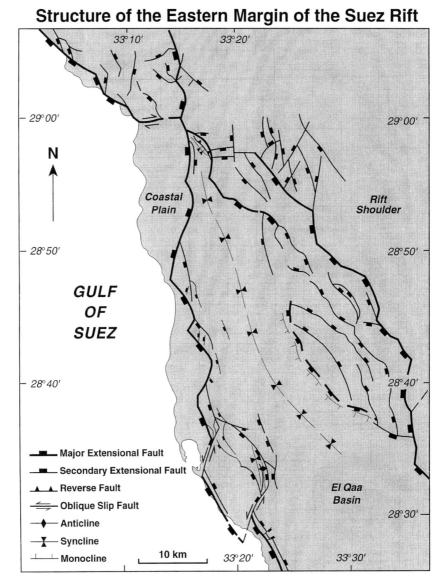

Figure D4.4 Principal structural features of the central eastern part of the Gulf of Suez.

from Precambrian crystalline basement, Lower Palaeozoic sediments (Lower Cambrian), Upper Palaeozoic (Carboniferous–Permian), Mesozoic sediments of the southern margin of Tethys (Triassic–Cretaceous) and Paleocene through to Eocene strata (Figure D4.5). The syn-rift strata extend from the Upper Oligocene through to the present day (Figures D4.5 and D4.6).

The pre-rift stratigraphy shows a relatively clear variation along the length of the central section of the eastern margin of the Gulf, with a southwards thinning of most of the marine units. In this eastern section there appears to be only minor variations in the pre-rift thicknesses between areas affected by the Cenozoic rifting and regions beyond the influence of Cenozoic extensional faulting to the east. Rifting appears to have commenced during the mid to late Oligocene with limited igneous activity and deposition of a continental red bed to shallow marine early syn-rift succession in the Late Oligocene and Earliest Miocene (the Tayiba Red Beds, Figure D4.5).

Timing of rifting

The development of the Gulf of Suez rift has been the subject of considerable debate in that there is conflicting evidence for the timing of uplift and extension in the Red Sea area, a problem associated with many rift systems (Sengor and Burke, 1978). It has been suggested that Late Cretaceous alkaline magmatism was associated with regional uplift and doming of the Red Sea prior to extension (Resseter and Nairn, 1980), and Cretaceous doming in the area now occupied by the Red Sea is

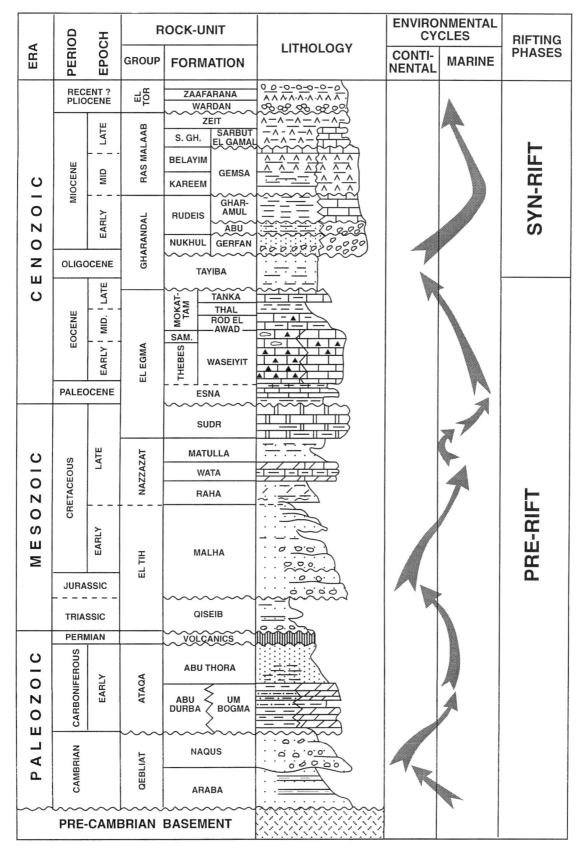

Figure D4.5 Summary stratigraphy, Gulf of Suez (after Darwish, 1993; personal communication).

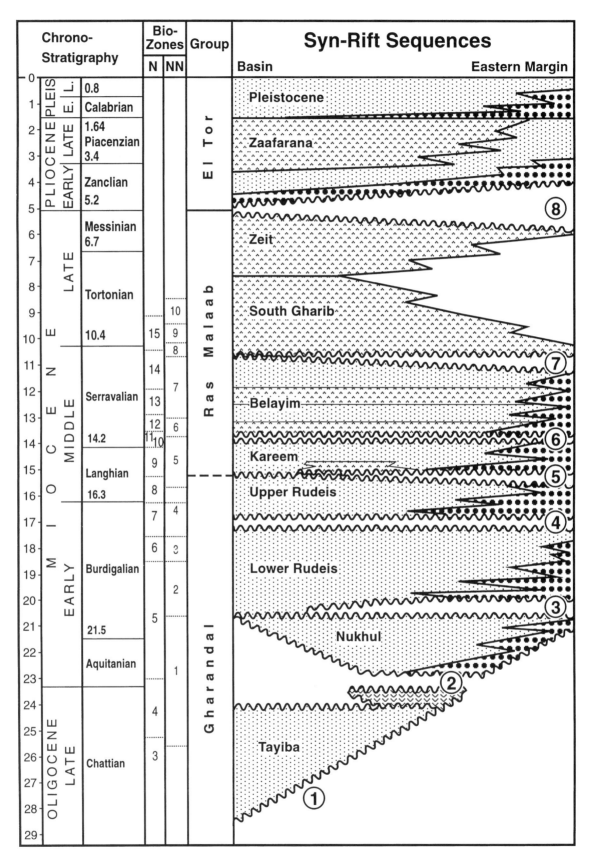

Figure D4.6 (a) General syn-rift stratigraphy for the central eastern part of the Gulf of Suez. (Compiled from the authors' own research and from sources quoted in the text.) Major unconformities are labelled 1–8. These are (1) Basal Tayiba unconformity; (2) Basal Nukhul unconformity; (3) Lower Rudeis unconformity; (4) Mid-Rudeis unconformity; (5) Mid-Clysmic unconformity; (6) Basal Belayim unconformity; (7) Basal Gharib unconformity; (8) Basal Pliocene unconformity.

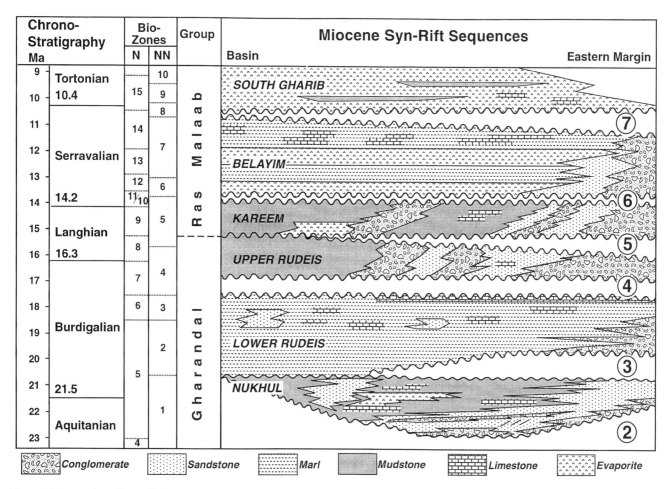

Figure D4.6 – *contd.* (b) Detailed Early–Mid-Miocene syn-rift stratigraphy (compiled from Patton *et al.*, 1994; Schutz, 1994; and the authors' own mapping and section measurements).

indicated by sediment transport directions (Ward and McDonald, 1979). Most workers, however, do not consider the initiation of rifting to be this early. Meneisy (1986) identifies a late Eocene–early Oligocene magmatic event (40 ± 10 Ma) in the Arabian plate which he associates with Red Sea doming and extension. Kohn and Eyal (1981) used fission-track data to determine the uplift history of Sinai and deduced that uplift preceded extension by a few million years. Steckler (1985), however, maintained that uplift and extension were synchronous, on the basis of a stratigraphic analysis of the Gulf of Suez. Omar and Steckler (1995) analysed fission-track data from the Red Sea and concluded that the Red Sea rift system initially opened along its entire length in the late Oligocene with a second major pulse of extension in the Early Miocene.

There is now general agreement that major faulting associated with the Gulf of Suez rifting began in the latest Oligocene to early Miocene (Robson, 1971; Steckler, 1985; Patton *et al.*, 1994; Schutz, 1994), at approximately the same time or slightly later than the initial Red Sea rifting (Coleman, 1974a; Gass, 1977).

Subsidence in the Gulf of Suez and northern Red Sea was affected by the development of the Gulf of Aqaba and Dead Sea transtensional systems (cf. Figure D4.1) (Quennell, 1958, 1984; Freund, 1970; Freund *et al.*, 1970; Moretti and Chénet, 1987; Moretti and Colletta, 1987). The sinistral offset of Miocene sedimentary and volcanic rocks along the Gulf of Aqaba shear has been taken to indicate about 45 km of post-Miocene movement (Freund *et al.*, 1970) with an additional 62 km of movement in the early Miocene, possibly related to initial faulting in the Gulf of Suez and Red Sea rift. The timing of this earlier movement has been questioned by Bartov *et al.* (1980) who suggested that the extension was all Middle Miocene or later. The cumulative evidence (Dewey and Sengor, 1979; Abdel Khalek *et al.*, 1993) now indicates that extension in the northern Red Sea was initially achieved by extension in the Gulf of Suez quickly followed by rifting in the northern Red Sea (late Oligocene–early Miocene) and the shear on the Dead Sea–Gulf of Aqaba system occurred later (Pliocene).

Subsidence history

Cross-sections across the Gulf of Suez (Moretti and Colletta, 1987; Patton et al., 1994; Bosworth, 1994a) generally show an asymmetric graben system formed by domino-style extensional faults with most of the subsidence focused in the central section of the rift. Subsidence patterns (Moretti and Colletta, 1987; Moretti and Chénet, 1987) across the rift reveal a distinctly three-stage tectonic evolution for subsidence in the Gulf of Suez. An initial major phase of tectonic subsidence occurred between 22 and 16 Ma (early Miocene) with especially rapid subsidence since 20 Ma. During this period, it appears that the Suez rift was linked with the northern Red Sea (Garfunkel and Bartov, 1977; Mart and Hall, 1984). In the second stage, between 16 and 13 Ma, the tectonic subsidence slowed during deposition of the Kareem and Belayim Formations (Garfunkel and Bartov, 1977). One interpretation of this stage is the initiation of sinistral strike-slip motion along the Aqaba–Levant fault with the subsequent opening of the Red Sea coupled to the Aqaba–Dead Sea rift system. In a third stage, from middle–late Miocene to the present, a second major period of subsidence occurred. Tectonic subsidence is still active in the central part of the rift and normal faults are found in late Quaternary deposits associated with uplift of Quaternary beach deposits along the footwalls of major fault blocks (Ball, 1939; Abu Khadrah and Darwish, 1986). The rate, however, is slower than during the early Miocene and is likely to be significantly lower than the separation rate between the Arabian and African plates within the Red Sea (1.6 cm/y, Hempton, 1987). Away from the central trough there is a 20 km wide transition zone where tectonic subsidence has almost ceased. Onshore, the rift flanks show a continuous history of uplift since extension commenced. An average value for the amount of extension is 34–36 km ($\beta = 1.56-1.62$) (e.g. Patton et al., 1994; Schutz, 1994) although the amount of extension and thinning increases to the south in the region of Gebel Zeit ($\beta = 1.9-2.0$; Bosworth, 1994a). The Gulf of Suez is underlain by thinned continental crust and no oceanic crust is developed.

The present-day tectonic subsidence has two main characteristics: 1. the uplift of the northern part of the Gulf and 2. the narrowing and deepening of the rift trough (Moretti and Colletta, 1987). It appears that significant tectonic subsidence does not occur on the western margin north of Wadi Araba (north of the Zafarana accommodation zone; Figure D4.2). The Wadi Araba area corresponds to a large Syrian arc fold structure of crustal scale that formed during the Late Cretaceous–Eocene inversion of the southern margin of the Tethyean ocean basin. This transition zone where tectonic subsidence is presently negative corresponds to the narrowest part of the Gulf. In the central and southern parts of the Gulf, subsidence is still active, although in a restricted area (Moretti and Chénet, 1987).

Rift flank uplift

One of the most impressive tectonic features of the Gulf of Suez is the relief of the rift flanks: elevations on average exceed 1000 m and on Sinai exceed 2600 m. Fission-track dating on samples from the Precambrian basement on the western (Omar et al., 1989) and eastern sides of the Gulf (Kohn and Eyal, 1981) generally indicate that there must have been about 5 km of uplift/exhumation relative to present-day sea level. Such large movements require significant heat input into the crust and mantle. Two endmember models for rifting that may have caused these large vertical movements are: 1. active rifting, in which extension is a response to upwelling asthenosphere beneath the rift zone and 2. passive rifting in which extension is in response to remote plate boundary forces. Heat flow data indicate a hotter crust than would be expected simply from passive lithosphere stretching (Morgan et al., 1985), and uplift of the rift margins of the Gulf of Suez is significantly in excess of that predicted by lithospheric thinning during extension (Steckler, 1985). Both of these features suggest active rifting. However, there is no convincing evidence that rifting was preceded by major doming, which is the main characteristic associated with active rifting. One mechanism for the rift flank uplift is by flexural isostasy which assumes the lithosphere has lateral strength and flexes like an elastic beam during loading. The thinning of the rift acts as a negative load upon the lithosphere generating isostatic uplift of the rift flanks. Flexural isostasy produces permanent topographic uplift and is consistent with the amplitude and wavelength of the topography seen around the Gulf of Suez.

Structural evolution of the Gulf of Suez

The Gulf of Suez is dominated by four major fault trends that have been recognized by many researchers (e.g. El Tarabili and Adawy, 1972; Moretti and Chénet, 1987; Colletta et al., 1988; Meshref, 1990; Patton et al., 1994; Schutz, 1994). The dominant fault system is the north-west–south-east 'Clysmic' trend which is the orientation of the major basin-bounding extensional faults in the Gulf (Figures D4.2, D4.3 and D4.4). The second major fault orientation is that of the Aqaba trend (N 10°–20°E) found predominantly in the southern Gulf where the influence of the Gulf of Aqaba–Dead Sea transtensional system becomes apparent. In particular the north-north-east trending faults commonly form the linkages between the north-west rift faults producing the characteristic rhomboidal fault pattern seen at the rift border fault systems (Figures D4.3 and D4.4). Two other relatively minor cross-fault orientations – the 'Duwi' (Montenat et al., 1986, 1988) and 'Cross Fault' (Robson, 1971; Patton et al., 1994) trends have been identified and these appear to be at least partly controlled by structural grains in the basement.

The three major dip domains in the Gulf of Suez (Figure D4.2) are separated by two accommodation zones marking the change in regional dip – the Zafarana–Abu Zenima accommodation zone to the north and the Morgan accommodation zone to the south (Figure D4.2). These accommodation zones appear to be wide (up to 20 km), somewhat poorly defined zones of complexly faulted blocks of variable dips and interlocking 'flip-flop' conjugate fault systems. Within each dip domain the depocentres have a dominant half-graben asymmetry such that one margin of the basin is characterized by only one or two major rift border faults which have very large throws (commonly in the order of 2.5 to 5 km) and the other margin has a distinctly flexural character with a number of extensional faults each having only relatively small throws (hundreds of metres maximum).

Within each dip domain the styles of extensional faulting may be more complex. In the central dip domain on the eastern margin of the Gulf of Suez in the vicinity of the Gebel Araba to Wadi Baba fault blocks the rift margin faulting is complex with both the northwest and north-north-east fault trends developed (Figures D4.3 and D4.4). This produces zig-zag fault traces and rhomboidal fault blocks. Some of the north-north-east trending faults (particularly the fault that separates the Araba block from the Abu Durba block – Figure D4.3) are transfer faults with both strike-slip and extensional displacements. These faults accommodate displacement transfer between systems of like-dipping extensional faults and most probably result from reactivation of an older basement structural grain. As a result 'trap door' fault systems developed with elevated footwall corners of individual fault blocks forming the prominent exposed fault blocks at the surface (e.g. Gebel Araba – Figure D4.3) and corresponding hangingwall depocentres developed in the opposite corners of the fault-block system. In detail this structural style controls the local basin development within the individual tilted fault blocks of the rift margin and has a major influence on the styles and expressions of clastic deposition in the rift system.

Across the southern dip domain in the Gulf of Suez, (Figure D4.2) the regional balanced cross-sections of Bosworth (1994a) reveal the spatial and temporal evolution of this part of the rift. The stratigraphic sections were approximately area balanced at each sequential step and satisfactory restorations were obtained with low-angle detachments or shear zones placed at depths of 10–15 km. Bosworth's restorations indicate that active sediment accumulation in the southern Gulf has narrowed through time (Bosworth, this volume). The narrowing of the rift also seems to have affected the central Gulf sub-basin (Moretti and Colletta, 1987). This changing subsidence pattern has been interpreted by Perry and Schamel (1990) as reflecting a shift in the position of the detachment breakaway.

STRATIGRAPHY OF THE EASTERN MARGIN OF THE GULF OF SUEZ

The stratigraphy of the eastern side of the Gulf of Suez is summarized in Figure D4.5 with the stratigraphic relationships for the syn-rift stratigraphy on the eastern margin of the Gulf of Suez shown in greater detail in Figures D4.6 (a) and (b). Figure D4.7 shows seven regional cross-sections through the eastern margin of the Gulf of Suez as located on the geological map of Figure D4.3. Plate 18 shows field photographs of the exposures found in the map area.

Pre-rift stratigraphy

The oldest rocks of the area are Precambrian metamorphic and igneous basement which form the mountainous region of the Sinai. Uplift in the footwall of the rift bounding faults expose Precambrian basement close to the coast on both sides of the Gulf of Suez (Plate 18(b)). Erosion of this basement has provided clastic detritus to the rift, as metamorphic and igneous clasts are found in proximal deposits from the middle Miocene to the present day. It is also very likely that structural trends in the basement rocks have provided lines of weakness which have been exploited by the later Cenozoic extensional faults.

The Lower Palaeozoic

The Araba and Naqus Formations (Hassan, 1967; Said, 1971; El Barkooky, 1992) of the Qebliat Group (Figure D4.4) lie with a sharp, distinct planar unconformity on the Precambrian basement. These two distinctive red and white sandstone units are up to 130 m and 380 m thick respectively and well exposed and developed at Gebel Araba–west Qebliat region (Plate 18(a)). In the north of the Wadi Fieran–Belayim area the Cambrian sequences are marine in character whereas to the south they are more continental.

Upper Palaeozoic

A long hiatus separates the Cambrian sandstones from the overlying Carboniferous strata but there is no angular unconformity between the two along the east

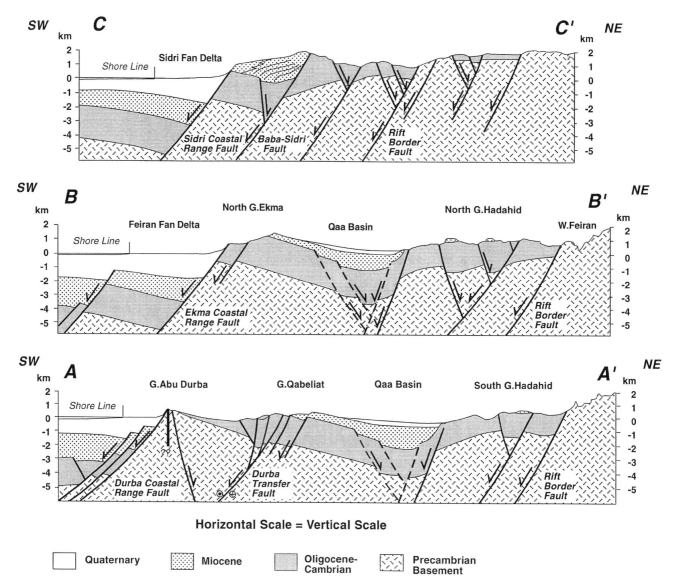

Figure D4.7 Structural cross-sections, eastern Gulf of Suez (locations on Figure D4.3)

side of the Gulf of Suez. The Umm Bogma Formation (Kostandi, 1959; Klitzsch, 1990) is a grey, sandy dolomitic limestones and dolostones 40 m thick of Carboniferous (Visean) age. To the south in the Abu Durba area are fossiliferous black shales and mudstones of the Abu Durba Formation (Hassan, 1967; El Barkooky, 1992). These lie beneath the brown sandstones of Abu Thora (Weissbrod, 1969; Kora, 1984), a thick succession (up to 200 m) of buff to brown weathering medium-grained sandstones of shallow-marine origin. The Carboniferous sandstones are capped and intruded by Permian basaltic flows and sills.

Triassic, Jurassic and Early Cretaceous
Clastic sedimentation continued into the Mesozoic as the rather thin, poorly fossiliferous strata of the Qiseib and Malha formations of the El Tih Group (Barakat et al., 1986). The Qiseib Formation (Abdallah et al., 1963; Barakat et al., 1986) is Permo-Triassic in age and consists of unfossiliferous red beds which are in places clearly continental (paleosol profiles) and in other regions shallow marine, with evidence of bidirectional (tidal) cross-stratification. The Qiseib Formation is about 300 m thick. The Malha Formation (Abdallah et al., 1963) is variable from 30 to more than 150 m thick and consists of white to pale yellow and pink fluvial sandstones with clay intercalations and kaolin pockets.

Depositional facies are more varied in the Upper Cretaceous strata (Figure D4.4), consisting of a succession of shales, sandstones and limestones, which are mostly shallow marine in origin. Cenomanian strata

(Raha Formation; Ghorab, 1961) are dominantly green marls and shales with sandy intercalations and thin limestones. The Raha Formation is 80–100 m thick and decreases in thickness southwards. The overlying Turonian (Wata Formation; Ghorab, 1961) contains a thicker limestone which serves as a useful marker across the region. It is also up to 100 m thick, decreasing southwards. A return to shales and sandy shales marks the Coniacian–Santonian Matulla Formation (Ghorab, 1961) which is glauconitic in places. The thickness of this formation ranges from 120 to 170 m, and it decreases in thickness south of Wadi Belayim. The Campanian to Maastrichtian beds of the Sudr Formation (Ghorab, 1961) are a fine-grained limestone, the Sudr Chalk. It is composed of snow-white, hard, poorly bedded chalk and chalky limestone which attains 100–120 m in thickness. The total thickness of Upper Cretaceous strata in about 500 m. The first three units are included in the Nezzazat Group (Steen and Helmy, 1982).

Paleocene–Eocene
The Esna Shales (Beadnell, 1905) are a dark greenish shale which is easily picked out above the white Maastrichtian chalks and below Eocene limestones. It therefore forms a useful marker horizon. The unit is about 35 m thick and is composed of light green to grey, soft shale with mirror carbonate thin beds, gypsum veins and brownish iron concretions in the middle part. The Paleocene–Eocene boundary is found to be within the upper 3 m (Said, 1990a; Zico et al., 1993). The Eocene outcrops are in the eastern side of the Gulf and are more than 500 m in thickness; they are entirely made up of carbonates with occasional shale interbeds.

There are two distinctive facies suites characterizing the Eocene sequences in the Gulf of Suez region. The Nile Valley facies is well developed south of the Feiran-Belayim area and is represented by the Thebes Formation (70 to 130 m of well-bedded, grey/white limestone with abundant chert in certain horizons; Said, 1960) and the Samalut Formation, a thick-bedded, highly fossiliferous limestone, 90 to 110 m thick (Bishay, 1966). In some instances, sporadic outcrops of the Mokattam Formation (Middle to Upper Eocene) occur (Said, 1990a). The Central Gulf of Suez facies are found between Wadi Feiran in the south and Wadi Sudr in the north. This facies suite includes the following formations: Waseiyit (Lower to Middle Eocene), Rod El Awad, Thal (Middle Eocene) and Tanka (Upper Eocene) Formations. The distribution, characteristics and boundary relationships are discussed in detail within different studies including Moon and Sadek (1923), Viotti and El Demerdash (1968), Barakat et al. (1988), Abul Nasr (1990), and Said (1990a). In many pre-Miocene blocks the deep erosion has resulted in syn-rift facies lying directly on middle Eocene or older rocks.

Syn-rift stratigraphy

The Oligo-Miocene stratigraphy illustrated in Figure D4.6 has been built up using both onshore exposures and offshore well data. A period of extensive erosion clearly preceded syn-rift sedimentation as the Oligo-Miocene strata lie unconformably on units which range in age from the Precambrian basement to the Eocene. There are rapid facies changes in the Miocene units which can be related to proximal deposition along the flanks of the rift and deeper water facies in the subsurface along the axis of the rift.

Late Eocene–Oligocene transition
Tayiba Formation
The Tayiba Formation (Hume et al., 1920) is a succession of red and white sandstones and siltstones which are considered to be late Eocene–Oligocene in age (Hermina et al., 1989). The formation is dominated by continental deposits of fluvial sandstones and poorly developed paleosols. In the northern part of the area the lower part of the red beds are found to contain shallow marine fossils (Abul Nasr, 1990). The Tayiba Formation is not widely exposed and the relationship between sedimentation and rifting during the late Oligocene is unclear.

Oligocene basic dikes and flows
The Tayiba Formation also includes Oligocene lavas, dikes and sills of basaltic composition. The dikes are almost vertical and have a dominant north-north-west trend (Plate 18(c)). Some of these dikes are dated by whole-rock K/Ar dating at 27 Ma whereas others appear to be younger, ranging between 24 and 21 Ma (Meneisy, 1990). The basaltic flows are found in isolated fault blocks in the Hammam Faraun and Abu Zenima region and are in part underlain by altered and weathered volcanic ash. These basic rocks are the only exposed rift-related igneous activity found in the Gulf of Suez.

Miocene
Research on the Miocene sequences in the Gulf of Suez region dates back to the end of the last century. Sadek (1959) divided the Miocene of the Gulf of Suez into three main groups: lower, middle and upper Miocene based upon characteristic facies changes within the Miocene strata.

In 1974 the National Stratigraphic Sub-Committee introduced comprehensive lithostratigraphic schemes for the Miocene of the Gulf of Suez; one for the open marine facies and the other for the non-marine and coastal facies. Most of the Miocene outcrops on the eastern side of the Gulf of Suez represent the open marine suites and include the Nukhul and Rudeis formations of the Lower Miocene Gharandal Group, as well as the Kareem, Belayim, South Gharib and Zeit formations (of the Middle to Upper Miocene Ras Malaab Group; Figure D4.6).

Proximal Miocene facies include coarse-grained alluvial deposits and fan deltas which are exposed in a number of places onshore. Distal facies are not so well exposed in the central sub-province (Wadi Nukhul and Wadi Gharandal areas). The Miocene in the centre of the Gulf is known from drilling to be evaporitic and deep-water clastics, mostly turbidites. The exposures of the Miocene in the eastern part of the Gulf of Suez largely belong to the Nukhul, Rudeis, Kareem, Belayim formations plus the lower part of South Gharib Formation.

Nukhul Formation
The Nukhul Formation (e.g. Plate 18(d)) is a succession of mainly shallow marine deposits which include calcareous conglomerates, sandstones and marls. In the type section in Wadi Nukhul there is a transitional contact with the underlying Tayiba Formation consisting of red beds passing up into pebble conglomerates and calcarenites. The lowest beds of the Nukhul Formation are considered by some workers to be continental on the basis of poorly developed paleosols interbedded with cross-bedded sandstones and conglomerates. Further up-section the calcareous sandstones are wave-rippled and siltstones contain marine fauna (oysters). These beds are overlain by pale green calcareous mudstones which are abruptly terminated by a unit of cross-bedded calcareous sandstones deposited in lower and upper shoreface environments. Other exposures are composed of massive or horizontally stratified gravelly, coarse calcareous sandstones with angular chert fragments, pebbles of white, Thebes Formation limestone and broken pieces of oyster shell; flute and groove marks indicate deposition as gravity flows of sandy and gravelly debris.

Adjacent to pre-existing topography (e.g. Wadi Baba, Figure D4.3 and Plate 18(d)) proximal Nukhul Formation deposits are coarse conglomerates of locally derived material, principally limestones and chert pebbles from the adjacent ridge of Thebes Formation. These conglomerates are clast-supported and crudely stratified but with a random orientation of clasts in the sandy and muddy matrix. The limited extent of exposures of the Nukhul Formation suggest that this unit was only locally developed adjacent to footwall highs along the rift flank (Figure D4.10(a)). These highs included a ridge of limestone which was emergent and created a narrow gulf in the adjacent hangingwall behind which some deposition occurred in the Wadi Baba area. Further to the south the absence of Nukhul Formation is assumed to reflect non-deposition in this part of the area. Erosion of Nukhul Formation deposits prior to deposition of the overlying Rudeis Formation is considered unlikely because lowers beds of the latter are fine-grained, low-energy deposits.

Rudeis Formation
The early Miocene to earliest middle Miocene Rudeis Formation is much more widespread than the Nukhul Formation. This formation is divided into two parts by a sharp change in facies from a fine-grained marl and sandstone succession in the lower part to a coarse glauconitic fossiliferous sandstone succession in the upper part. Such a change in facies coincides with the 'mid-Rudeis event' which, in some localities, is a marked unconformity.

In the northern part of the area the lower Rudeis deposits are marls with thin sandstones (Figure D4.8(b)). The area of deposition had evidently expanded since Nukhul Formation times and in the south, around Wadi Feiran, sandstones of Rudeis age directly overly middle Eocene limestones. These sandstones have a quartzose composition which suggests derivation from Cretaceous siliciclastic strata because the younger parts of the pre-rift stratigraphy are marls and limestones. The same sandy facies have been found in the subsurface in the El Qaa depression. Shallow marine reefal carbonate deposition occurred directly on tilted Middle Eocene nummulitic limestones (Samalut Formation) in the southwestern part of the area, the Araba and Abu Durba blocks.

In the upper part of the Rudeis Formation, the deposits which occurred after the mid-Rudeis event are the earliest, extensive coarse syn-rift sediments in the area (Figure D4.8(c)). The most extensive area of coarse conglomeratic sedimentation (proximal facies) is in Wadi Sidri, where conglomerate deposition persisted from Rudeis Formation times through to the Kareem Formation. They are known as the Abu Alaqa Conglomerates (Garfunkel and Bartov, 1977) and are mostly poorly sorted, matrix-supported or clast-supported with a crude stratification interpreted as the deposits of shelf-type fan deltas (Gawthorpe et al., 1990). Sandy beds also occur, and algal limestones are present, principally at the tops of the 'cycles' of sedimentation separated by unconformities. In the lower parts of this upper Rudeis succession the clasts are mainly of Eocene and upper Cretaceous limestone, Palaeozoic and Mesozoic sandstones, followed by Precambrian basement higher up section in conglomerates of the Kareem Formation.

In Wadi Baba the ridge of pre-rift limestone strata was evidently breached in upper Rudeis times as conglomerates rich in chert clasts occur above and to the west of this ridge. Further north, another area of gravel deposition marks another fan delta which initially developed adjacent to a major fault in Rudeis times and

Stratigraphy of the eastern margin of the Gulf of Suez 235

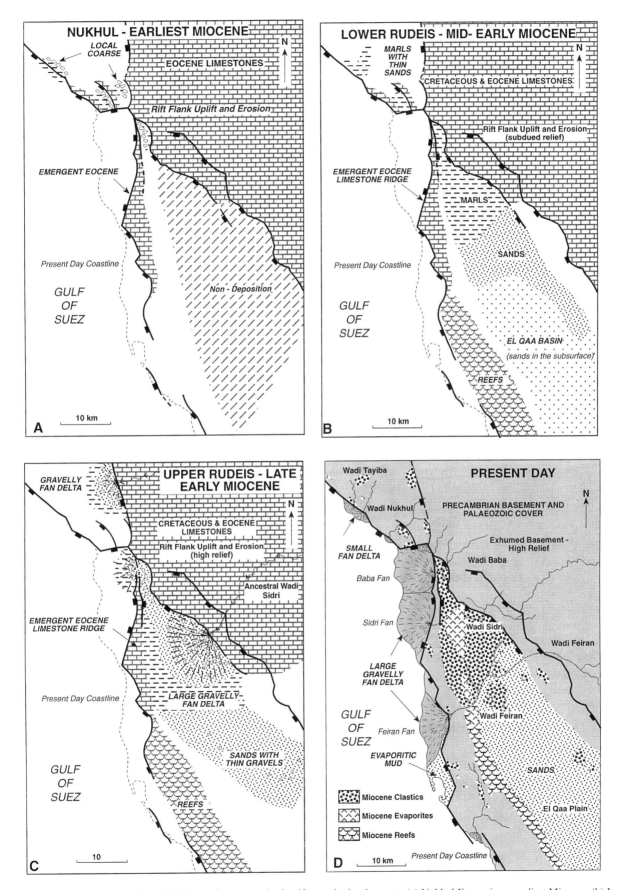

Figure D4.8 The distribution of syn-rift facies at four stages in the rift margin development: (a) Nukhul Formation – earliest Miocene; (b) Lower Rudeis Formation – mid-early Miocene; (c) Upper Rudeis Formation – latest early Miocene; (d) present-day showing the Miocene outcrops.

persisted into Kareem Formation times. The lower part of the succession is a limestone breccia of redeposited reefal debris and overlying beds are poorly sorted, polymict, pebble to boulder conglomerates with depositional dips of up to 20°. These are considered to be the topsets and foresets of Gilbert-type fan deltas. To the south-west, reef formation continued on the Araba and Abu Durba blocks.

Kareem Formation
The early-middle Miocene Kareem Formation (Ghorab et al., 1964) is locally unconformable on top of the Rudeis Formation. The mid-Clysmic event demarcates the boundary between the two formations. On the basin margin conglomerates of Abu Alaqa Group continued to develop. Shelly calcarenites and patch reef limestones make up most of the exposed sections, with a subordinate proportion of polymict conglomerates. In exposures near the coast in the northern part of the area the first Middle Miocene evaporites denote the abrupt change in facies and basin restriction during early Kareem Formation time. This is followed by the open-marine dark grey shales (Shagar Member), well-defined Middle Miocene marker unit in the subsurface offshore sequences. It is worth mentioning that in some areas the Kareem facies at the basin margins is developed earlier than in the open basin. Accordingly it is considered to be time transgressive from the latest Early Miocene to the Middle Miocene.

Belayim Formation
Late Middle Miocene deposition is dominated by evaporites, marls and reefal limestones and conglomeratic sandstone of the Belayim Formation (Ghorab et al., 1964). The reefal facies of the Nullipore rocks (also referred to as the Hammam Faravn Member) is extensively developed in most of the emergent and submerged blocks. These strata are generally rather poorly exposed on the flanks of the rift and, along with the Late Miocene strata, are better known from subsurface data where the facies are more distal to the rift margins. In the subsurface the whole evaporite-bearing formations are penetrated, these include Belayim, South Gharib and Zeit formations. The first two are generally anhydrite–bearing while the latter is dominated by salt and subordinate anhydrite with thin mudstone interbeds.

Pliocene
In the subsurface a section of clastics and thin evaporites succeed the Zeit Formation. In the surface outcrops the fifth gypsum unit of Moon and Sadek (1923) or its equivalent South Gharib Formation occurs in fault blocks which have conglomeratic and sandstone sequences of Pliocene age deposited against them. These sequences are well exposed in the area of Wadi Gharandal. It is fossiliferous with different molluscan and echinoid remains of Indo-Pacific affinities. In addition to the Pliocene fossil remains, the conglomeratic sequence of the Wardan Formation includes re-worked gypsum lithoclasts, chert, sandstone and limestone detritus. In the subsurface a considerable thickness of more than 1000 m is encountered in the wells. This clastic sequence is overlain by an evaporite series (Zafarana Formation) that is dominated by anhydrite at the base with salt in the middle and grades upwards into pisolitic limestones and pebbly sandstones at the top.

Quaternary deposits
Quaternary deposits cover the flat and topographically low areas as well as the coastal plains in the area. Quaternary alluvium and wind-blown sands cover the wadi floors. On the coastal plains the deposits include loose to moderately consolidated coarse clastics derived from the older pre-rift rocks that form the surrounding topographic highs. These coarse clastics mainly form fan deltas on the coast (Figure D4.8(d)). A single large gravelly fan delta has developed at the mouth of Wadi Feiran and flanks the lagoon at Belayim. The Baba Plain consists of two coalesced fan delta bodies, one issuing from the mouth of Wadi Sidri, the other supplied by detritus from Wadi Baba. The gently sloping shoreface indicates that these are low-angle, shelf-type fan deltas (Wescott and Ethridge, 1990) which have built out into the present-day Gulf of Suez. In addition, a series of raised beaches are developed at different altitudes ranging from a few metres on the coastal plains up to more than 60 m on the marginal coastal ranges (e.g. Hammam Faraun, Abu Durba and Gebel Araba). Most of these raised beaches include patch coral reefs and oyster banks (Abu Khadrah and Darwish, 1986).

TECTONO-STRATIGRAPHIC EVOLUTION OF THE EASTERN MARGIN OF THE GULF OF SUEZ

The spectacular exposures along the eastern margin of the Gulf of Suez graphically illustrate the complex interactions between tectonics, sedimentation and stratigraphy in a young rift that has not undergone post-rifting thermal relaxation and subsidence. Prior to the initiation of the Gulf of Suez rift the area had a long and relatively stable history, with only some extensional faults developing during the Jurassic–Early Cretaceous Tethyean extension and subsequent inversion of these extensional fault systems in the Late Cretaceous through to the Middle Eocene. Our mapping and analysis of the stratigraphic, sedimentological and structural relationships along the central eastern margin of the Gulf of Suez rift has identified four phases of rift development from the late Oligocene to the present. There is a clear and strong structural control on the sedimentation patterns with the development of individual fault-bounded basins with local depocentres and syn-rift stratigraphy. The fault patterns within the rift margin reflect stages in the tectonic development whereas uplift of the rift flank determined the nature and extent of the syn-rift deposits. There is a strong hinterland control on the clast composition within the syn-rift sequences.

Pre-rift history

The Precambrian units form a metamorphic and igneous basement which outcrops widely over the Sinai Peninsula. The Precambrian structural fabric is likely to have played a role in determining the location and orientation of Cenozoic tectonic features. Cambrian fluvial and shallow-marine clastic sedimentation occurred on a peneplained surface to form the oldest pre-rift sedimentary unit. A long hiatus, with some local erosion of the Cambrian succession, separates this from Carboniferous shallow-marine carbonates, shales and sandstones. Permian basaltic volcanics are locally preserved and formed the substrate for lateritic soil development in the Triassic and are seen at the base of the coarse, siliciclastic red beds of the Qiseib Formation.

The Lower Cretaceous rocks are also fluviatile, up to the Cenomanian stage when a marine transgression marked the onset of a long period of shallow-marine sedimentation which persisted up to the late Eocene. The lower formations of this suite are mainly sandstones and shales with carbonates more abundant in Late Cretaceous strata, notably the Sudr chalk. The Paleocene is marked by a distinctive shale which separates the Cretaceous limestones from the succession of shallow-marine carbonate facies of Eocene age.

This pre-rift succession is now completely exposed in tilted fault blocks along the margins of the Gulf of Suez. It is remarkable for the absence of any major angular unconformities between the base of the Cambrian strata and the Eocene limestones (Plate 18(a)), although there are clearly several major depositional hiatuses within the succession, notably at the bases of the Carboniferous, Triassic and Lower Cretaceous strata (Figure D4.5).

These pre-rift lithologies formed the hinterland bedrock which was eroded to supply the syn-rift clastic deposits: the limestones of the upper part of the pre-rift succession yielded little sandy detritus and it was only in areas where erosion on the rift flanks reached older siliciclastic rocks in the later stages of rift development that significant amounts of quartz sand were deposited in the basin.

Rift development

Four stages of rift development, each of which developed a characteristic tectono-stratigraphic sequence, can be identified on this eastern margin of the Gulf of Suez. Each stage is characterized by regionally significant unconformities and sedimentary sequences that reflect fault activity at the rift margin (Figure D4.6).

Stage 1 – Late Oligocene–Earliest Miocene: Tayiba Phase
Red-beds of Oligocene age and basalt dikes and flows dated at between 24 and 21 Ma mark the onset of rift development in the Gulf of Suez. The red beds are apparently limited in distribution and are generally fine-grained; little rift flank topography appears to have been developed at this stage and the igneous activity may have been in response to only a small amount of extension across the area.

Stage 2 – Earliest Miocene: Nukhul Phase
A limited amount of rift-flank topography is indicated by localized conglomerates of early Miocene age – the Nukhul Formation (Figures D4.6, D4.8(a)) made up of limestone and chert clasts eroded from Eocene strata. These Nukhul Formation strata were deposited mainly in shallow-marine environments in fault controlled sub-basins in the hangingwalls of domino-style tilt blocks.

Stage 3 – Mid-Miocene–Late Miocene: Mid Rudeis to Zeit Phase
The first major influx of coarse clastic detritus occurred after the mid-Rudeis event (unconformity 4 – Figure D4.6(b)). This is a widespread unconformity which is overlain by limestone conglomerates in many parts of the area. These units are spectacularly developed as the deposits of large, gravelly fan deltas which formed at the mouths of the ancestral drainage systems (Figures D4.7, D4.8(c)). In the southern part of the area, uplift of the rift flank was greater, exposing Cretaceous and older sandstones which supplied detritus to areas of shallow-marine, sandy deposition in the early Miocene. A progressive erosion, or unroofing/exhumation history, of the rift flank is evident from the clast types found in the upper parts of the fanglomerate successions. Continued and renewed phases of extension on faults along this eastern margin of the rift is indicated by syndepositional unconformities within the conglomerate sequences (Figure D4.6(b)). The Middle to Late Miocene period in the Gulf of Suez is marked by some continued coarse deposition on the rift margins but the dominant facies are marls and evaporites of the basin centre.

Stage 4 – Pliocene–Recent

Renewed extension and fault-block movement is evident from the coarse conglomeratic facies of Pliocene age that are deposited unconformably (unconformity 8 – Figure D4.6(a)) on the underlying strata along the rift margins.

The present-day pattern of sedimentation in this area is dominated by large gravelly fan deltas which have formed at the mouths of the major wadis (Baba, Sidri, Feiran – Figure D4.8(d)). Drainage patterns which developed during the early stages of rifting have apparently persisted and remain the main wadi systems feeding the margins of the Gulf and the development of the rift border faults with associated relay structures has exerted little control on the major wadis. One exception is the capture of the Feiran drainage by the system which formerly issued from Wadi Sidri: this drainage has been diverted north along a relay between two fault strands to form a fan which coalesces with the Wadi Baba fan on the Baba plain (Figure D4.8(a) and (b)).

In conclusion, our research shows that this eastern margin of the Suez rift developed as a series of fault-bounded sub-basins each with a distinctive tectono-stratigraphic history. Syn-rift sediments are characterized by coarse clastic facies at the rift margins with finer-grained clastics deposited in the more open marine conditions towards the centre of the rift. The composition of the syn-rift clastics was strongly controlled by the exhumation histories of the pre-rift sequences on the rift flanks.

ACKNOWLEDGEMENTS

Research summarized in this paper was supported by the Fault Dynamics Project (sponsored by Arco British Ltd., BP Exploration, Conoco (UK) Limited, Mobil North Sea Limited, Petrobras UK, and Sun Oil Britain). S. Khalil's fieldwork was supported by Marathon Egypt, Limited. Fault Dynamics Publication No. 69.

D5
Carbonate and siliciclastic sedimentation in an active tectonic setting: Miocene of the north-western Red Sea rift, Egypt

B. H. Purser, P. Barrier, C. Montenat, F. Orszag-Sperber, P. Ott d'Estevou, J. -C. Plaziat and E. Philobbos

ABSTRACT

This study, based on extensive French and European Community field projects, compares shallow-marine sedimentation on fault blocks from the south-western margin of the Gulf of Suez and the north-western Red Sea. These synsedimentary blocks have been examined on various scales. On a regional scale, the fault blocks are grouped according to their distance from the western periphery, this factor determining not only the importance of siliciclastic facies relative to carbonates but also the structural behaviour of each block. Blocks attached directly to the periphery have thin, fairly variable sedimentary covers. Others separated from the periphery by one or more half-graben have somewhat thicker sequences often with better developed carbonate platforms, while one example (Gebel el Zeit), situated close to the axis of the Gulf of Suez rift, has very limited coarse siliciclastics and thin shallow-marine carbonates due to relatively high rates of subsidence.

Structural parameters have also been the main factor determining local sediment geometries. Virtually all tilt-blocks are characterized by a variable asymmetry expressed by the direction of tilting (towards the axis or periphery of the rift), vertical relief, and orientation of each structural unit with respect to the general north-north-west–south-south-east axis of the rift. These morpho-structural attributes determine not only the relative importance of carbonate and siliciclastic facies but also the occurrence of specific facies, such as reefs.

The complexity of the sedimentary cover on any given block depends mainly on its tectonic stability. All blocks, irrespective of their position in the rift, exhibit the effects of synsedimentary tilting, generally in the form of multiple reef systems and localized depositional sequences. However, these are best developed at Gebel el Zeit where a series of angular unconformities, stratigraphic gaps and synsedimentary erosion of both pre- and syn-rift sediments are clearly visible. Structural instability on many blocks has also resulted in the collapse of platform margins while major north-east–south-west fault systems crossing certain platforms have also contributed to their demise. These fault systems, oblique with respect to the dominant 'Clysmic' north-north-west–south-south-east trend, also served as corridors for coarse siliciclastic discharges exported from the periphery towards the axis of the rift.

Although tectonic movements have dominated mid-Miocene sedimentation, they have not been the sole factor. In spite of marked lithostratigraphic variation among all blocks, an overall palaeoenvironmental change occurs from early-mid Miocene open-marine conditions to mid and late Miocene restriction, culminating in evaporite precipitation. This change may be related either to global sea-level movement, or to regional tectonic adjustments which culminated in the separation of the Red Sea and Mediterranean. Climatic conditions relating both to temperature and wind directions appear to have been significant factors influencing the thickness and geometry of carbonate platform sediments. Relatively thin carbonate buildups and modest platform dimensions may have resulted from the relatively high latitudes of the area during the Miocene, while a predominant north-westerly wind direction, parallel to the axis of the narrow basin, may have contributed to the asymmetric development of reefs and marine sands.

The mid Miocene sediments exposed along the north-western margin of the Red Sea rift clearly demonstrate the intimate relationships between tectonics and sediment geometry, both on regional and on local scales. Because these geodynamics concern the relatively early phases of rifting, they illustrate the tectonic, eustatic and climatic factors governing the distribution of carbonate and siliciclastic sediments with respect to the *first* phases of individual platform development.

INTRODUCTION

The north-western margin of the Red Sea and Gulf of Suez (Egypt) rift is limited by a dissected escarpment, the Red Sea Hills, whose altitudes locally attain 2200 m. The escarpment is composed of Precambrian metamorphics, granites and volcanics, and has slopes that descend rapidly to about 250 m to form an irregular coastal plain ranging in width from 1 to 15 km. The coastal region is traversed by numerous large wadis exposing excellent sections across a series of structural blocks and their related sedimentary cover. In some areas structural units may be studied in three dimensions. These consist of tilted fault blocks which are generally elongate parallel to the main axis of the rift, the 'Clysmic direction' of Hume (1921). This structural frame, progressively established and rejuvenated during early phases of rifting, has strongly influenced both continental and marine palaeoenvironments, and thus has played a predominant role in determining the composition and geometry of the syn-rift, Miocene sediments.

The nature and objectives of this contribution

The history of the Red Sea rift system, which began in the late Oligocene, includes many tectonic phases which are recorded by the nature and geometry of the syn-rift sediments. This contribution discusses only the mid Miocene, marine phase. Earlier studies of these syn-rift sediments include those of Akkad and Dardir (1966), Ghorab and Marzook (1967) and Issawi *et al.* (1971) together with doctoral studies, mainly from the universities of Assiut (El Haddad, 1984), Orsay (France) (Roussel *et al.*, 1986), and IGAL (France) (Prat *et al.*, 1986) and Thiriet *et al.* (1986). Detailed structural and sedimentological studies were carried out by the French group 'GENEBASS' whose results have been summarized in a series of thematic volumes: 'Documents et Travaux', Institut Géologique Albert de Lapparent (IGAL) 10, 1986; 'Notes et Memoires' de TOTAL, Compagnie Française des Pétroles, 21, 1987; *Bulletin de la Société Géologique de France*, 3, 1990. More recent studies including those carried out under the European Community project RED SED, in collaboration with the University of Assiut, have been summarized in the *Bulletin of the Geological Society of Egypt*, Special Publication No. 1, 1993. Publication of a thematic volume of *Tectonophysics*, 153, 1988, the World Bank project published in the *Journal of Petroleum Geology*, vol., 15, 1992, and the *Geological Society of London*, Special Publication, 80, 1995, have also been major contributions.

The above publications have been utilized extensively in the preparation of this contribution. However, more recent field-studies by the present authors on individual fault blocks provide a database for the present discussion. In particular, the north-western Red Sea region offers a unique opportunity to examine the early stages of mixed carbonate–siliciclastic platform development which are rarely visible in older rift systems. Through discussion of a series of examples the authors demonstrate the temporal and geometric relationships between the major carbonate, siliciclastic and evaporite facies in the region.

The present study concerns only outcrops from the north-west, peripheral parts of the 350 km wide rift system (Figure D5.1). However, the early phases of rifting involving the creation of fault blocks, probably was a regional phenomenon. The structural evolution, relative sea-level changes and climatic factors, all affecting the composition and geometry of these syn-rift sediments, produce a series of sedimentary 'models'.

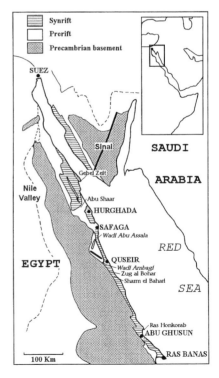

Figure D5.1 Map of the north-western Red Sea and Gulf of Suez showing the localization of the areas studied.

Factors potentially influencing sedimentation on Miocene fault blocks

Structural controls

These may be grouped into two categories based on scale. On a broad scale, the position of the structural unit within the rift will be a determining factor in the composition of its sedimentary cover. Blocks situated near the periphery may be buried by terrigenous materials, those closer to the axis may be sites of carbonate or deeper marine, marly sedimentation.

On a local scale, within a given fault block, the following parameters are considered (Figure D5.2).

- The tilt of half-graben blocks exerts a major influence on sedimentation and is expressed by an important asymmetry in sediment composition and geometry relating to the footwall and dip-slope sectors.
- The axial plunge (pitch) of a block similarly results in a lateral variation in bathymetry and accommodation space which will determine both sediment composition and thickness.
- The relief (amplitude) of a block may influence the composition of its sedimentary cover, low-relief blocks being more susceptible to burial by terrigenous materials.
- The orientation of a fault block with respect to the rift periphery may be fundamental. Virtually all positive elements are oriented parallel to the rift axis. However, fault-related depressions (transfer zones) are oblique with respect to the axis, those formed relatively early in rift evolution having been inherited from Precambrian fracture systems (Jarrige et al., 1986; Montenat et al., this volume).
- The stability of the tilt-block during sedimentation will considerably influence both peripheral (e.g. footwall) collapse and the number and geometry of shallow platform depositional sequences.

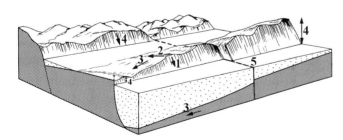

Figure D5.2 Diagram of tilted fault blocks showing their principal attributes influencing Miocene sedimentation, and terminology employed in this contribution: footwall scarp (1); dip-slope (2); axial plunge of block (3); relief (amplitude) relative to adjacent sea floor (4); oblique fault or transfer zone (5).

Relative sea level

The position of sea level relative to the crests of shallow platforms will determine the geometry of depositional sequences. This is particularly true for reefs and related carbonate sand bodies. Within an active tectonic setting relative sea level may be influenced by repeated movements of the block, as discussed by Cross et al. (this volume). Tectonically driven relative sea-level changes may be expressed by sequence boundaries whose character varies from one part of the structure to the other.

Although tectonic instability may be the major control of relative sea-level fluctuation, global eustatic movements occur, the Burdigalian–Serravallian being a period of global sea-level rise (Haq et al., 1987). Separation of the two controlling factors (tectonic and eustatic) is not usually possible.

Climatic factors

Perhaps the most important climatic factor affecting carbonate platform sedimentation is wind intensity and direction. On the oceanic Bahamian platforms, Hine et al. (1981), Grammer and Ginsburg (1992) and others have shown that peripheral slope sedimentation attains a maximum along leeward flanks which are supplied by detritus swept from shallow platform tops. In the north-western Red Sea remarkably constant north-north-west wind is parallel to the axis of the rift, as may expected to be the case in narrow sedimentary basins. Because wave movement strongly influences both reef growth and distribution of bioclastic sands, sedimentation on, or around, any north-north-west–south-south-east oriented block may be expected to be asymmetric, with reefs on the windward and sands on the south-south-east (leeward) slopes of the block.

THE MIDDLE MIOCENE SYN-RIFT PLATFORMS OF THE NORTH-WESTERN RED SEA

As already noted (Purser et al., 1990; Purser and Philobbos, 1993), the structural framework upon which the mid Miocene sediments were deposited comprises tilted blocks oriented parallel to the axis of the rift. While certain blocks (half-graben) are structurally simple (e.g. Esh Mellaha and Abu Shaar), others result from the structural disintegration of larger blocks, discussed by Montenat et al. (this volume). This disintegration results in series of small, closely spaced blocks (e.g. Ras Honkorab) and thus a more complicated sedimentation. The north-west–south-east structural pattern is further complicated by the presence of structural depressions which are oblique to the predominant north-west–south-east system. While certain of these important corridors may be regarded as 'transfer' or 'accommodation zones', most do not correspond to this classical notion. Their orientation, to a large extent, is conditioned by the rejuvenation of Precambrian fracture systems (Jarrige et al., 1986). Their ages and orientation are variable, some preceding the development of the dominant north-west–south-east direction, and others post-dating this clysmic trend, their evolution being discussed by Montenat et al. (this volume). Most act as important corridors for the transit of terrigenous detritus across the predominant north-west–south-east structural pattern towards the axis of the rift.

The eight examples presented in this contribution include both positive blocks (horsts) which are referred to as 'tilt-block platforms', as well as structural depressions (graben, half-graben and transfer zones). These latter are filled with shallow marine fans whose subtidal relief has favoured Miocene reef development. There exists an overall lithological evolution from open marine, often coral-rich facies near the base, to stromatolitic and oolitic carbonates in the upper third. Progressive restriction indicated by algal laminites culminates in late Miocene evaporites (Figure D5.3). This overall sequence never exceeds 150 m at outcrop and is clearly exposed on the Abu Shaar platform (Figure D5.1) situated at the entrance to the Gulf of Suez (Cross et al., this volume). Because all structural elements are progressively buried, the age of the marine sedimentary cover on any block will depend on its relief relative to the rising mid Miocene global sea level.

The following examples are considered essentially in terms of their structural geometry and occurrence with respect to the rift periphery and are grouped into three categories:

- tilt-block platforms attached to the rift periphery
- tilt-block platforms adjacent to, but detached from, the rift periphery
- tilt-block platforms independent of the rift periphery

Tilt-block platforms attached to the rift periphery

Wadi Ambagi, Quseir: a tilted shelf

A linear basement scarp situated 3 km to the south-west of the port of Quseir (Figure D5.4(a)) is fringed by a relatively continuous carbonate platform composed mainly of clinoform deposits whose present (visible) relief ranges from 50 to 70 m (Roussel, 1986). These beds are attached to a steep surface representing an eroded Miocene fault scarp which is terraced and overlain by outliers of Miocene carbonate and marine siliciclastic gravels (Figure D5.4(b), (c)). The terrace is gently inclined towards the north-east, and was a Miocene shoreline. To the west, basement rises gently to altitudes averaging 120 m.

The narrow shelf is composed mainly of coarse slope detritus inclined at angles averaging 20°. While biogenic

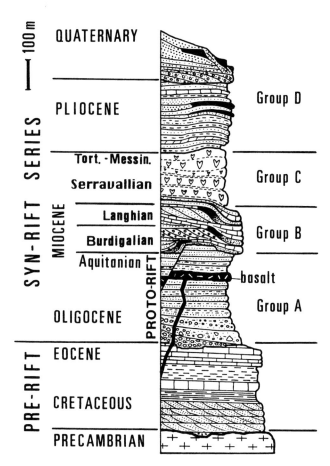

Figure D5.3 General stratigraphic log or pre- and syn-rift sediments of the north-western Red Sea coast (after Montenat et al., 1986).

elements predominate, an important terrigenous fraction occurs in the form of rounded basement pebbles. These are generally concentrated into thin (10 cm) beds suggesting grain-flow. Numerous spectacular blocks (Figure D5.5(b)) occur down-slope from the reef platform. The clinoform beds grade upwards into reefal carbonates including massive corals often in growth position. Onlapping relationships between the peripheral reef and the horizontal lagoonal or beach carbonates indicate that this reef had a slight relief above the adjacent subhorizontal platform surface (Figure D5.4(b)).

The shelf fringing the basement scarp may be traced continuously for a distance of 6 km (Figure D5.4(a)). At its northern extremity, near Wadi Ambagi, it is composed solely of slope material with no horizontal platform (Figure D5.4(b)). However, it widens progressively towards the south attaining a maximum width of nearly 1000 m in the vicinity of Wadi el Aswad (Figure D5.4(c)). This progressive widening is associated with the development of a multiple reef-terrace system, the single platform in the north passing to four platforms some 3–4 km to the south, with younger

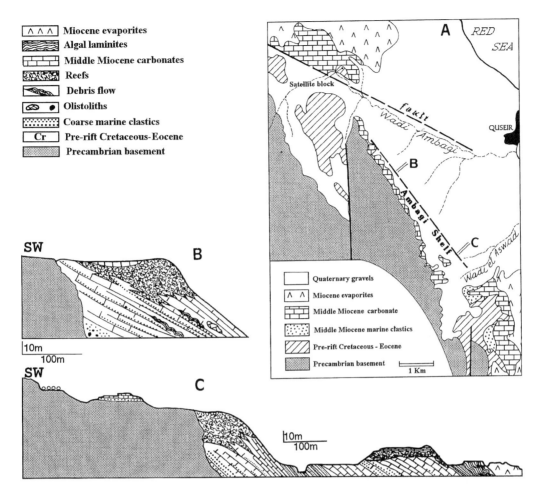

Figure D5.4 Ambagi Shelf, Quseir: A. Simplified geological map of the Wadi Ambagi area showing structural context of the Ambagi shelf. The basement tilt-block plunges towards the north (modified after Roussel *et al.* (1986). B. Section across northern end of shelf showing single reef-terrace. C. Section across southern part of Ambagi shelf showing multiple reef-terraces suggesting repeated uplift of this part of the shelf.

platforms at lower altitudes. The lateral variation in the number of reef terraces can be explained only in terms of local tectonic adjustments, with repeated lowering of relative sea level towards the south, i.e. structural uplift. This uplift was probably facilitated by faults flanking both sides of the basement ridge (Figure D5.4(a)), this latter tending to plunge towards the north.

Conclusion
The Ambagi Shelf illustrates the relatively simple relationships between basement relief and both carbonate and siliciclastic sedimentation. The Ambagi Shelf and adjacent areas also record the effects of synsedimentary structural movements, notably the successive rejuvenation of the plunge of a fault block resulting in an increase in the number of reefal sequences towards the uplifted end of the block.

Tilt-block platforms adjacent to, but detached from the rift periphery

Numerous fault blocks are separated from the rift margin by one or more half-graben, many formed during the initial stages of rift evolution. The distance from the peripheral terrigenous sources and the number of intervening dip-slope depressions, potential sediment traps, are important factors influencing the composition of sediments deposited on and around these blocks.

Zug al Bohar platform: a block with basinwards-tilt
This relatively small basement high (Figure D5.6) is located 17 km south of Quseir. It is situated only 4 km to the east of the major basement block of Gebel Zarib from which it is separated by two narrow graben. The 'inner graben' (Figure D5.7) contains down-faulted pre-rift Nubian Sandstone whose existence is confirmed by Miocene slope deposits which dip westwards into this depression. The 'outer graben' is filled completely with Miocene carbonates and coarse siliciclastics which

Figure D5.5 Field aspects of the Wadi Ambagi area: A. Prograded reef-shelf near southern end of Ambagi Shelf. B. Detail of slope deposits, Ambagi Shelf, showing large reef olistolith and mass-flows. C. Detail of grain-flow deposits showing numerous erosional discontinuities (hammer shows scale).

overlap the culmination of the Zug al Bohar platform linking it to the 'western high'. Miocene sedimentation in the Zug al Bohar area therefore concerns both fault blocks and intervening depressions.

The fault blocks are composed of both Precambrian granite and Cretaceous Nubian Sandstone faulted into contact prior to the early Miocene transgression. This composite basement unit, in common with virtually all blocks situated south of Quseir, is tilted north-east towards the rift axis. This tilting influenced Miocene sedimentation as the steeper footwall (south-west) flank acted as a barrier to detritus derived from the west, whereas the gently inclined, north-easterly dip-slope favoured lateral accretion of reefs and sands.

The thickest, earliest reef developed on the dip slope of the eastern block (Figure D5.7). It aggraded during relative sea-level rise and also prograded some 300 m towards the north-east forming a massive lens 25 m

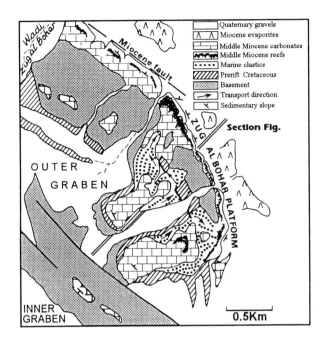

Figure D5.6 Map of the Zug al Bohar area showing multiple basement blocks, siliciclastic accretion (arrows) and carbonate reef-platform. Blocks are tilted towards the north-east.

thick comprising favid corals in growth position and is flanked to the east by a series of well-bedded talus deposits. This reef was interrupted by a clysmic fault (Figure D5.7, F3), whose exposed fault-plane was extensively bored by molluscs. The subsequent, but considerably thinner, coral biostromes formed shortly after this faulting and have also been affected by a second fault overlain by well-bedded slope deposits against which the late Miocene evaporites were deposited.

The reef complex formed on the north-eastern front of Zug al Bohar block attains maximum development near the northern end of the platform (Figure D5.6). This spectacular exposure coincides with a swing (towards the north-west) of the platform margin, the resulting promontory probably having been exposed to north-westerly wind-driven currents stimulating reef growth. The main reef body decreases in volume towards the south-east where it is replaced by a series of siliciclastic sands deposited on a lee-slope. The mixed carbonate and siliciclastic sediments deposited on the platform attain a maximum thickness of about 30 m.

Coarse, marine terrigenous deposits, about 20 m in thickness, fill the 'outer graben', its lower parts being piled against the footwall (south-westerly) scarp of the adjacent platform. This is indicated by divergent cross-bedding directions (Figure D5.6). However, the upper part of this shallow marine fan overflows the structural depression and the sands extend on to the Zug al Bohar platform.

The filling of the graben and contemporaneous aggradation of the platform during relative sea-level rise have led to the fusion of the two structural elements to form a single, essentially carbonate, platform. This has involved an increase in the width from 600 m to about 2 km. This amalgamated platform is capped by a laminated algal micrite which is folded and brecciated, probably by synsedimentary seismic instability (Plaziat and Purser, this volume). These highly restricted carbonates cap the summit of the 'western high' and today attain altitudes of about 120 m. Together with the underlying open-marine carbonates and siliciclastics, they record a considerable rise (at least 100 m) of relative sea level prior to the onset of Miocene evaporitic sedimentation.

Conclusion

The Zug al Bohar and adjacent units demonstrate the significance of basinwards (north-eastern) tilt of fault blocks prior to platform sedimentation. The gentle slope facilitates the lateral accretion of reefs and related

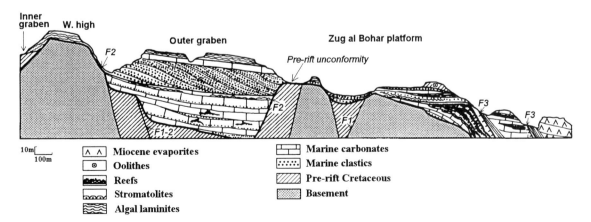

Figure D5.7 North-east–south-west section across the Zug al Bohar blocks showing infilling of the half-graben ('outer graben') and overflow of siliciclastics on to the adjacent Zug al Bohar tilt-block platform.

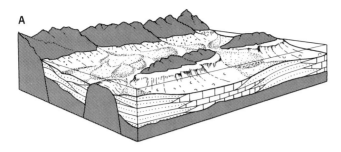

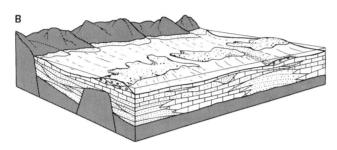

Figure D5.8 Cartoons showing evolution of siliciclastic and carbonate sedimentation on the Zug al Bohar blocks: (A) early stages, with siliciclastic sedimentation in graben and carbonates on seawards side of tilt-block; (B) late stage, with tilt-block drowned and replaced by emergent carbonate shoals.

platform sediments. Structural instability of the seawards (north-east) margin of the tilt-block platform, as on the Ambagi shelf, has led to the formation of multiple reef bodies. The area also shows the potential of structural lows to trap siliciclastic sediments. However, when the amplitude of these structural units is relatively modest – here about 50 m – their filling leads to the flooding of the adjacent platform with peripherally derived clastics (Figures D5.7, D5.8).

The Abu Ghusun platform: a tilt-block disrupted by cross-faulting

The Abu Ghusun block, situated in the southern part of the Egyptian Red Sea coast (Figure D5.1), is composed of Precambrian volcanic basement. Together with a second block situated immediately to the south-west, it has been capped by continental Group A sediments and strongly tilted towards the north-east, i.e. towards the axis of the rift (Figure D5.9). The outer (Abu Ghusun) block is located about 3 km from the periphery of the rift. The main axis of the block plunges towards the south-east exposing basement rocks on the up-plunge extremity. However, this plunge is cut by a major north-east–south-west oriented fault approximately coincident with the wadi axis. This important cross-fault cuts the platform into two segments which are separated by massive siliciclastics deposited in the fault-related depression. The area has been mapped in detail by Montenat *et al.* (1986) and by Philobbos *et al.* (1993).

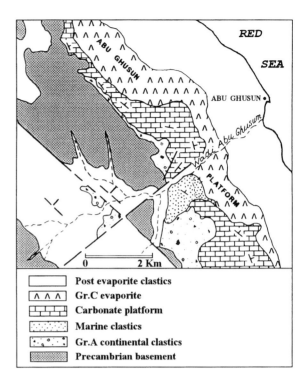

Figure D5.9 Simplified map of the Abu Ghusun area (modified after Montenat *et al.*, 1986) showing the south-east plunging basement block and its carbonate platform, cut by a north-east–south-west oblique fault-line depression (transfer fault). This oblique depression is filled laterally by slope deposits and axially by conglomerates.

The tilt-block situated to the north-west of Wadi Abu Ghusun supports a Miocene carbonate platform which has formed unconformably on the tilted, early syn-rift (Group A) sandstones (Figure D5.10). These reefal carbonates attain maximum thickness of about 40 m

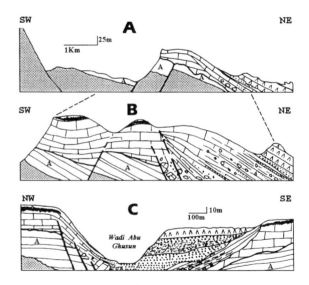

Figure D5.10 Simplified sections across the Abu Ghusun tilt-blocks. Multiple phases of Miocene faulting have severely dislocated the platform: A, General north-east–south-west section; B, detail of section A; C, longitudinal (north-west–south-east) section across the oblique fault-line depression and its Miocene filling.

clearly exposed in high cliffs flanking the wadi. Northwards, the subhorizontal platform thins rapidly and disappears about 2 km north-west of the wadi where its existence is recorded by a series of talus deposits (discussed by Purser and Plaziat, this volume). This thinning is probably related in part to the south-eastern plunge of the block, its northern extremity having been close to or above mid-Miocene sea level. As such, the underlying Precambrian volcanics provided a nearby source for detrital materials which fed the adjacent north-eastern dip slopes. These are rich both in coarse terrigenous sands and conglomerates and in numerous large (10 m) reef olistoliths. The abundance of coarse detritus of reefal origin records the synsedimentary destruction of the platform due to a high degree of tectonic-instability.

The existence of important tectonic movements postdating the initial tilting of the block is amply confirmed in the steep slopes bordering Wadi Abu Ghusun where synsedimentary sliding and faulting are clearly exposed (Figures D5.10, D5.11). Massive platform carbonates have been disrupted and megablocks 25 m in diameter have gravitated into fine-grained, open-marine marly sediments situated near the foot of the slope. The creation of a major re-entrant coincident with Wadi Abu Ghusun is also confirmed by the progressive change in strike of platform slope deposits (Figure D5.9).

The segment of the Abu Ghusun platform situated to the south of the wadi also demonstrates the presence of the adjacent mid-Miocene depression. A palaeocliff composed of at least 50 m of massive platform carbonate is flanked (to the north-west) by a series of large (5 m) olistoliths embedded in marls (Figure D5.11(c)).

Conclusions

The Abu Ghusun tilt-block platform illustrates the destructive effects of synsedimentary faulting on carbonate platforms (Figure D5.12). Much of the initial platform was destroyed during the mid-Miocene, notably at its up-plunge extremity where the greater part has disintegrated into large olistoliths whose preservation on the adjacent slopes is the principal evidence of a reef-platform's existence. Further south, at Wadi Abu Ghusun, the presence of a major cross-fault, developed and rejuvenated during marine sedimentation, has not only contributed to the disintegration of the platform but has created a corridor for great volumes of coarse terrigenous materials derived from the adjacent rift periphery (Figure D5.12).

The Abu Shaar–Esh Mellaha platforms:
blocks tilted towards the rift periphery
Near the entrance to the Gulf of Suez, Abu Shaar el Qibli is situated about 20 km from the peripheral escarpment from which it is separated by the Esh Mellaha Basin (Figure D5.13). This spectacular platform, exposing about 130 m of Middle Miocene carbonates and siliciclastics, has been the subject of a number of studies discussed by Cross et al. (this volume). For this reason only certain specific aspects relating mainly to synsedimentary tectonic movements will be presented.

In common with most Miocene platforms of the north-western Red Sea coast, Abu Shaar is a block of Precambrian basement bound along its north-eastern edge by a major, basinwards-dipping clysmic fault. In contrast to numerous blocks situated south of Quseir (see previous examples), Abu Shaar is tilted towards the periphery of the rift (Figure D5.14(a)). This direction is important in terms of delivery of peripheral clastics up the gentle (3°) dip slope. The main axis of the tilt-block plunges to the south-east (Figure D5.14(b)). This results in an elevated basement (the Esh Mellaha range) to the north-west, a major source of siliciclastic detritus, and a relatively thick accumulation of platform carbonates towards the structurally lower, south-eastern extremity of the block.

Initial basement relief prior to the onset of mid-Miocene burial was important, basement in the nearby Myos Hormos well being in the order of 2000 m (oral information, TOTAL Oil Co.). However, at outcrop, only the terminal parts of the transgressive marine series are exposed (Figure D5.14). These sediments, the basal parts of which have been correlated with the Langhian by Plaziat et al. (this volume), onlap an inclined Precambrian basement surface. Virtually the entire mid-Miocene series is composed of very shallow marine facies with numerous small reefs, algal stromatolites and beach deposits (Figure D5.17). This series therefore records a progressive rise in relative sea level whose amplitude was at least 130 m, and possibly considerably more. This is clearly in excess of the global eustatic rise of mid-Miocene sea level (about 100 m; Haq et al., 1987) and therefore must include an important element of synsedimentary subsidence, in part due to block rotation (Cross et al., this volume).

The Miocene sediments onlapping the inclined Precambrian basement (Figure D5.15(a)) range from mixed carbonate and arkosic gravels locally rich in corals, red algae, bryozoa and other open marine biota associated with local beach deposits towards the base, to very shallow platform facies, also with beaches, towards the top (Figure D5.17). As elsewhere on the Red Sea coast, the platform series terminates in a 20 m thick microbial laminite exhibiting folding and brecciation discussed by Plaziat and Purser in this volume. Miocene sediments deposited on the platform can be subdivided into depositional sequences which are clearly visible along the footwall escarpment (Plate 19).

248 Carbonate and siliciclastic sedimentation in an active tectonic setting

Figure D5.11 Field-views of the Wadi Abu Ghusun area: A. Panorama across the southern flank of the fault-line depression showing platform scarp (a), (top right); proximal platform slope deposits with olistoliths (b); distal platform slope marls (c); lense of dark coloured conglomerates filling the depression (d); late Miocene evaporites sealing the fault-line depression (e). B. Detail of northern flank of depression showing massive carbonate platform sediments (a) sliding on an inclined surface (b). The shearing has created a series of imbricated blocks (c). C. Detail of erosional base of conglomeratic filling; distal slope marls with olistoliths (arrow) (a); conglomerates (b). D. Block of coral in distal slope sediments; notice piling of sediment on upslope side (knife gives scale = arrow). E. Detail of conglomerates filling fault-line depression; bedded lower part (a) suggest a high flow-regime, the massive upper parts (b) are probably mass-flow deposits.

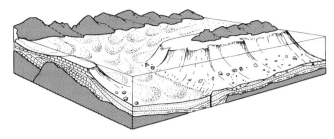

Figure D5.12 Cartoon of the Abu Ghusun platform and the oblique fault-line depression, the latter favouring the funnelling of coarse clastics towards the axis of the rift.

Synsedimentary tilting

In addition to the excessive thickness of the very shallow-marine transgressive series (see above) with respect to global sea-level rise, the mid-Miocene cover decreases in dip from 3–4° at the base to 1–2° at the top, confirming synsedimentary tilting of the block (Figure D5.15(c)). Although not all depositional sequence boundaries are due to structural tilt, the contact between the two main stratigraphic units of James *et al.* (op cit.) – the Kharasa and Bali members – records such movements.

Evolving platform morphologies recorded by variation in siliciclastic supply

The systematic study of siliciclastic sand bodies and their cross-bedding directions reveals a progressive change upwards through the platform sediments (Figure D5.16). These siliciclastics are well exposed within Wadi Bali which traverses the Abu Shaar platform (Figure D5.15). The south-eastern orientation of the cross-bedding in the basal arkosic sands indicates provenance from the north-west while their feldspathic mineralogy and moderate rounding suggest a fairly close source. The Abu Shaar block rises progressively to the north-west, culminating at Esh Mellaha, its immediate prolongation (Figure D5.13). This area was almost certainly above mid-Miocene sea level, constituting a local source for the noncarbonate fraction deposited on the platform. Relatively constant cross-bedding dips are oriented parallel to the axis of the emergent tilt-block.

Sand units somewhat higher in the sequence (Figure D5.16(b)) were probably deposited as shallow marine bars whose variable cross-bedding directions indicate a weaker slope relative to that of the preceding morphology. The uppermost units (Figure D5.16(c), (d)) differ markedly from the preceding sands and are composed of rounded quartz gravel and pebbles of basement rocks. These units exhibit large-scale foresets and are interpreted as shallow-marine fan-delta sediments. Their north and north-eastern-oriented dips imply progradation from the south or south-west, i.e. across the Esh Mellaha basin. The delivery of coarse siliciclastics up the dip slope of the block towards the axial culmination of the platform, at first sight appears illogical. However, this could readily be achieved if the early structural slope had aggraded to a near-horizontal surface.

The delivery of coarse siliciclastics onto the Abu Shaar platform records its evolution from an inclined ramp-like morphology associated with an active long-shore current from the north-west, to a horizontal, shallow marine shelf directly attached to the rift periphery. Successive tilting resulted in morphologies which oscillated between the two systems.

Synsedimentary erosion of the footwall margin

The footwall (north-eastern) margin of the Abu Shaar platform is characterized by slope deposits (Figure D5.17) whose dips average 25°. These relatively steep clinoforms are cut by steeper surfaces whose slopes locally attain 70°. These scarps (Figure D5.17), probably initiated by faulting, have been modified by erosion. Their extremely high angles are conditioned by early lithification as discussed by Aissaoui *et al.* (1986)

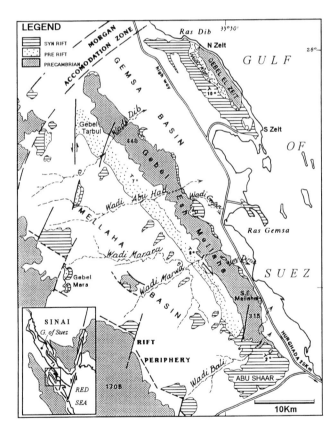

Figure D5.13 Generalized map of the south-west Gulf of Suez region showing the Abu Shaar, Esh Mellaha and Gebel el Zeit tilt-blocks and intervening basins. Mid-Miocene sediments on the platform crests are about 130 m thick, those in the adjacent half-graben basins exceed 2000 m (map based on Prat *et al.*, 1986). Note the irregular footwall (NE) edge of Esh Mellaha due to Miocene collapse.

250   Carbonate and siliciclastic sedimentation in an active tectonic setting

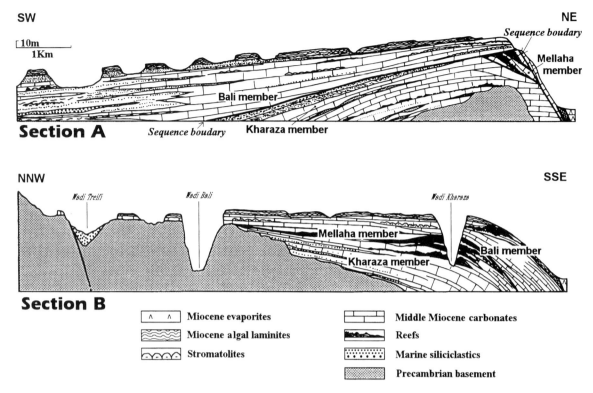

Figure D5.14 Sections across the Abu Shaar tilt-block platform showing main stratigraphic units and facies. Depositional sequence boundaries are most clearly developed along the footwall crest (north-east). Upward decrease in dip and onlapping relationships indicate repeated tilting to the south-west: A. north-east–south-west cross-section, B. north-north-west–south-south-east longitudinal section parallel to the axial plunge (modified after Cross et al., this volume).

and by Purser (this volume). This collapse, perhaps aided by sedimentary steepening, may have been stimulated by seismic instability relating to movements along the adjacent fault-system bounding the block.

Peripheral erosion along the footwall is one of the main characteristics of Abu Shaar and Esh Mellaha tilt-blocks. It has several important implications relative to the geometry of the platform cover.

- collapse of peripheral reef and carbonate sand facies leads to the anomalous situation where muddy inner platform facies fringe parts of the platform;
- stratigraphy and geometry of the slope deposits are complicated by the creation of numerous steep by-pass surfaces facilitating the basinward transit of sediments swept from the platform;
- important quantities of periplatform sediments, partly in the form of olistoliths, have probably accumulated at the base of the footwall scarp (Burchette, 1988).

Conclusion

Progressive burial of a gently tilted structural block by open-marine Miocene sediments resulted in its evolution from an offshore platform isolated from continental terrigenous contamination, to a subhorizontal, pericontinental shelf across which important quantities of coarse detritus, originating from the periphery of the rift, were delivered. This platform/shelf was affected repeatedly by synsedimentary structural tilting and by active, fault-related erosion along its footwall margin.

The Esh Mellaha tilt-block: a dismantled Miocene carbonate platform

The Abu Shaar tilt-block is the southern prolongation of the more extensive Esh Mellaha block from which it is offset by a north-north-east–south-south-west transfer fault (Figure D5.13). The two blocks, although probably segments of a common tilt-block, exhibit a marked geomorphological difference; the Esh Mellaha tilt-block

Figure D5.15 Field-views of Abu Shaar and Esh Mellaha tilt blocks: A. View across Wadi Bali (towards the south-east) showing mid-Miocene sediments (a) on the dip-slope, onlapping the inclined top of Precambrian basement (b). B. Slope deposits (a) flanking dark Precambrian basement (b), south-east end of Esh Mellaha. Notice that dips in the slope deposits follow the embayment in the footwall escarpment. C. Mixed open-marine carbonates and arkosic sands of Kharasa member, in Wadi bali; notice upwards decrease in dip. D. Sharp erosional base to arkosic sand unit (arrow). E. Detail of arkosic sand with sharp base (arrow) suggesting channel-fill. F. Quartz gravels of unit D (Figure D5.20) whose foresets prograde to the north-east, i.e. towards the footwall crest of the block.

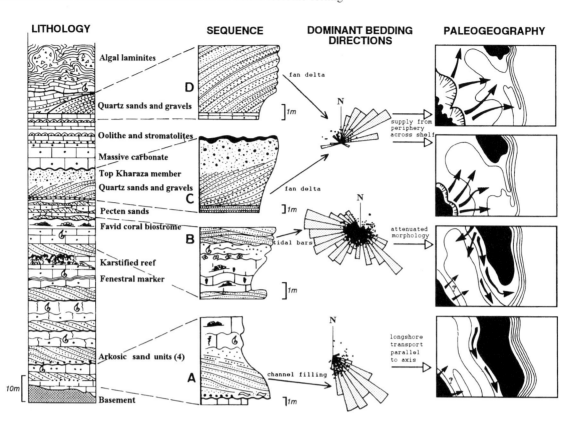

Figure D5.16 Vertical evolution of the main siliciclastic sand and gravel units on the dip-slope (W) of Abu Shaar tilt-block and inferred palaeogeography.

Figure D5.17 Panorama of footwall scarp, NE Abu Shaar platform showing the well-defined sequence boundaries; subhorizontal Kharasa member (a) is terminated by sharp, inclined surface (b). The reefal slope deposits (c) of the inclined Mellaha member are cut by another erosional limit (c1) which locally forms a bypass surface (arrow). The platform units (d, e, and f) pass into equivalent slope units (d1, e1 and f1).

exposes mainly Precambrian basement, virtually all its Miocene cover having been eroded.

The crest of the Esh Mellaha block is a spectacular, north-west–south-east oriented ridge extending without interruption for a distance of 65 km. As such, it is one of the longest structures in the Gulf of Suez and, possibly, within the entire rift system. It is tilted antithetically towards the south-west and the resulting transverse asymmetry is expressed by a steep, somewhat irregular, footwall escarpment and by a dip-slope comprising a planar pre-Miocene surface inclined at about 3° towards the south-west. This exhumed unconformity (Figure D5.18) is underlain by pre-rift sediments dipping at angles of about 8° to the south-west.

The internal structure of the Esh Mellaha tilt-block is complicated by the presence of numerous fault lines of which N10, N120 as well as N140° (Clysmic) directions predominate. These lineaments, few of which affect the dip-slope cover, appear to be pre-rift in age (Montenat *et al.*, this volume). However, along the north-eastern footwall escarpment both Clysmic and oblique faults affect the pre-rift sedimentary cover and are rejuvenated locally, downfaulting (to the east) the syn-rift Miocene series. The steeply inclined footwall escarpment, although having a general north-west–south-east trend, is highly irregular, being modified by a series of lunate embayments (Figure D5.13). That these latter are not the products of a capricious modern erosion is clearly demonstrated by the Miocene talus deposits whose dips locally follow this embayed morphology (Figure D5.15(b)) which is probably the consequence of Miocene footwall collapse. Most of the syn-rift sediments deposited on the Esh Mellaha tilt-block have been eroded during and subsequent to the Miocene.

Southern extremity of Esh Mellaha block
It is marked by a clearly exposed lense of fairly well-sorted conglomerate composed of transported cobbles of reworked Eocene chert whose source is located on the nearby south-western slopes of the block. The conglomerate fills a steep-sided, relatively narrow (1 km) palaeo-valley cut into Precambrian basement (Figure D5.19). The valley would appear to be oriented between north-north-east and east. The factors determining the localization of this Miocene incision, which cuts across the predominant north-west–south-east structural pattern, are probably tectonic. The presence of a major cross-trend between Esh Mellaha and Abu Shaar blocks, suggested by different structural tilting of these two segments and by the presence of north-north-west-oriented fault-planes, may have facilitated fault-line erosion leading to the discharge of coarse clastics across the north-west–south-east Clysmic trend.

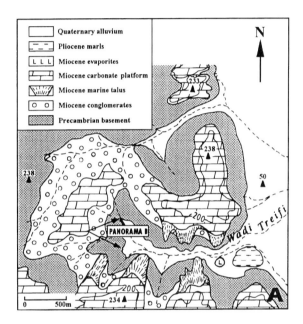

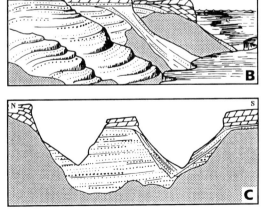

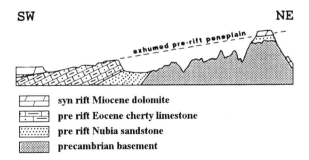

Figure D5.18 Simplified section across the Esh Mellaha block showing the multiphased tilting of the pre-rift sediments, the pre-rift unconformity and the syn-rift (Miocene) cover.

Figure D5.19 Miocene conglomeratic infilling of an oblique (north-east–south-west) depression at Wadi Treifi. This narrow incision cuts across the predominant north-west–south-east tilt-block system and probably follows a transfer zone separating the Esh Mellaha and Abu Shaar blocks.

The conglomeratic valley-fill measures about 80 m in thickness. Although including several paleosols near its base, subhorizontal bedding suggests subaqueous rather than subaerial slope deposition. The massively bedded conglomerates grade upwards into coarse arkosic gravels and sands with scattered coral (*Stylophora*) colonies in growth position and terminate in horizontally bedded carbonates which are correlated with the upper parts of the mid-Miocene series at Abu Shaar.

Wadi Um Dirra
Situated on the north-eastern escarpment about 18 km from the south-eastern extremity of the tilt-block, this segment of the footwall zone is a wide embayment within the escarpment which coincides with the presence of scattered outliers of Miocene Group B carbonates. Beige-coloured Miocene dolomites lie immediately in front of the dark-grey basement against which they appear to be faulted. This outlier of massively bedded, syn-rift carbonate is rich in molluscs and corals. However, it is neither a reef nor a slope deposit, its bedding and muddy textures being typical of platform-top environments.

A second outlier of pre- and syn-rift sediments is preserved on the crest of the footwall escarpment about 500 m north of Wadi Um Dirra (Plate 19(b)) where a flat surface limiting Precambrian basement is overlain by 25 m of Nubian Sandstone. These pre-rift sediments are overlain by Group B carbonates which form two distinct units, culminating at an altitude of about 300 m. The lower, 15 m thick unit is rich in quartz and molluscs and is followed by 20 m of beige-coloured dolomite whose south-east-dipping beds downlap on to the preceeding unit. These remnants indicate that the Esh Mellaha tilt-block supported an extensive carbonate platform which has been almost totally demolished both by Miocene collapse related to footwall instability, and probably by post-Miocene erosion.

Conclusion
The Esh Mellaha tilt-block, probably the longest (65 km) in the south-western Gulf of Suez rift, is a linear ridge of Precambrian basement almost devoid of sedimentary cover which has been dismantled, in part, by Miocene tectonic instability.

Ras Honkorab platforms: multiple, steeply inclined tilt-blocks
The rift periphery is situated some 15 km to the south-west of Gebel Honkorab tilt-block, from which it is separated by a series of narrow structural depressions. The positive blocks are tilted towards the axis of the rift and are characterized by marked plunge of their long axes towards the south-east (Figure D5.20). This plunge creates a broad structural low situated to the south-east, causing a major embayment in the Miocene shoreline. In addition to their axial plunge, the blocks are characterized by considerable vertical relief attaining present altitudes of 213 m at Gebel Honkorab.

The structural depressions preserve remnants of Group A continental siliciclastics as well as evaporites dated as Aquitanian (Montenat et al., 1986). These evaporites record the initial phase of marine flooding of the rift periphery (Orszag-Sperber et al., this volume). They are approximately contemporaneous with, or slightly younger than, a remarkable, 10 m thick brecciated algal laminite unit. These laminated dolomicrites encrust subvertical basement relief to altitudes approaching 100 m, locally capping the top of adjacent basement highs. Because these laminites must have been deposited under subaquatic conditions, their vertical distribution has important implications in terms of relative sea level. Stratigraphically, they clearly underlie adjacent Group B conglomerates and biogenic carbonates.

Lowering of relative sea level, probably due to structural uplift of the block, was followed by the development of a carbonate shelf fringing the basin-ward, north-eastern side of the Honkorab block. This shelf decreases in altitude northwards from Ras Honkorab towards Sharm el Luli, i.e. along the north-eastern flank of the block. The gentle inclination, oblique to the shelf periphery, is expressed by its bedding; gently inclined (10°) foresets prograde northwards along the shelf terminating in a multiple reef system opposite Sharm el Luli. A profile across this same shelf (Figure D5.21) indicates that progressive lowering coincides with the downstepping of four fringing reefs, described by Purser et al. (1996). This is thought to record the repeated uplift of the Honkorab block.

Inclination of the Honkorab shelf (Figure D5.22(e)) indicates the existence of a Miocene depression, still existing today as the Sharm el Luli embayment. A structural control to this embayment is suggested by the presence of an east-north-east-oriented fault shown by Montenat et al. (1986) to cut the basement immediately to the west-north-west of Sharm el Luli.

In the vicinity of Gebel Honkorab silty marls are overlain by a well-developed carbonate platform some 120 m in thickness, reflecting increasing accommodation space on the structurally lower parts of the plunging axis. Sediment geometry thus evolves along the axis of the block from a narrow, down-stepping shelf in the vicinity of Sharm el Luli, to a subhorizontal platform located on the axis of the block in the vicinity of Ras Honkorab (Figure D5.21). Tilting of the block is indicated by its slope deposits which dip at 15° along the north-eastern dip-slope and at about 25° along the south-western footwall (Figure D5.22(a)). These steep footwall slopes indicate the existence of a bathymetric depression coincident with the small graben situated

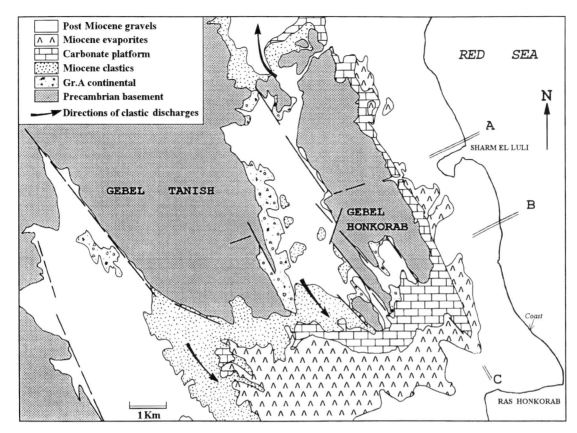

Figure D5.20 Simplified geological map (after Montenat *et al.*, 1986) of the Ras Honkorab area showing the multiple basement blocks which plunge towards the south-east. Arrows indicate the directions of terrigenous discharges. The location of the three cross-sections (Figure D5.21) are also indicated.

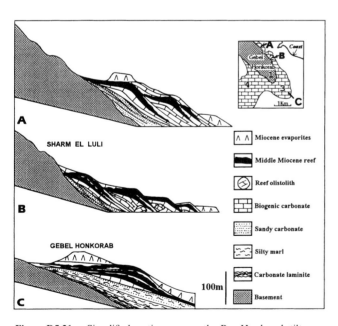

Figure D5.21 Simplified sections across the Ras Honkorab tilt-block, the base of each section being present sea level. Notice multiple, downstepping reefs and the more complete sequence (C) at the south-eastern plunging, end of the block where a small carbonate platform replaces the narrow reef-shelves (A and B).

immediately to the south-west of Gebel Honkorab. Unlike Zug al Bohar where peripheral graben were filled completely with terrigenous and other carbonate detritus, those at Ras Honkorab remained unfilled (Figure D5.22). This preservation seems to have been caused both by excessive heights of the intervening horsts and, probably, by the repeated uplift during mid-Miocene sedimentation.

Siliciclastic discharges

As at Zug al Bohar, coarse gravels and conglomerates are concentrated within the series of structural depressions situated between Gebel Honkorab and the rift periphery. However, unlike Zug al Bohar, these coarse clastics have not filled the depressions and thus have not overlapped the adjacent tilt-block platforms. Pure carbonate sedimentation was therefore continued along the basinwards margin of the block. The conglomerates have been blocked by the Honkorab high which caused a deviation of the discharge to structurally lower exits situated at the north and south extremities of the block.

These conglomerates have been deposited as shallow-marine fans. The small graben situated immediately to the south-west of Ras Honkorab is virtually devoid of

256 Carbonate and siliciclastic sedimentation in an active tectonic setting

Plate 10 A. Very fine sandstone, siltstone and mudstone facies, Hababa-Kawkaban (section 6). B. Giant cross-bedded sandstone facies (Jebel Al Dafir, NW Sana'a). C. Concretionary iron-rich sandstone facies, Jebel Kura (section 10). D. Fluvio-lacustrine sandstone facies with gastropods *Coptostylus* in a coarse-grained sandstone, Jebel Kura (section 10). 2cm scale.

Plate 11 Outcrops of the Usfan Formation and overlying units from Usfan and Hadat Ash Shaam areas in Saudi Arabia.
A. North-easterly dipping light-coloured shales, sandstones and limestone (middle of section) of the Usfan Formation, overlain by coarse, purple/brown sandstones and conglomerates of the lower Shumaysi Formation. Grey scree at top of hill is unconformably overlying basalts of the Hammar Formation (estimated 60 m of section). (Wadi Hadat Ash Shaam). B. Contact between lower Usfan Formation (green shales and gypsum veins) and Shumaysi Formation conglomeratic sandstone (2 m thick ledge). (Wadi Hadat Ash Shaam). C. Contact between subvertical Usfan Formation (younging to right) and basalts of the Mio-Pliocene Hammar Formation, near Usfan.

Plate 12 Outcrops of Shumaysi Formation from Wadi Shumaysi, Saudi Arabia.
A. North-easterly dipping lower and middle units of the Shumaysi Formation. 2 m thick oolitic ironstone bed arrowed. B. Lower unit of Shumaysi Formation with large-scale cross-bedded sandstones and conglomerates which unconformably overlie Precambrian granodiorites in this area. C. Section through upper Shumaysi Formation shales, siltstones and fine sandstones capped by limestones and cherts at top of hill. Makkah road section, Wadi Shumaysi. D. Coarse bioclastic limestone and siltstones partially replaced by chert at top of Shumaysi Formation. Hammer for scale. Makkah road section, Wadi Shumaysi.

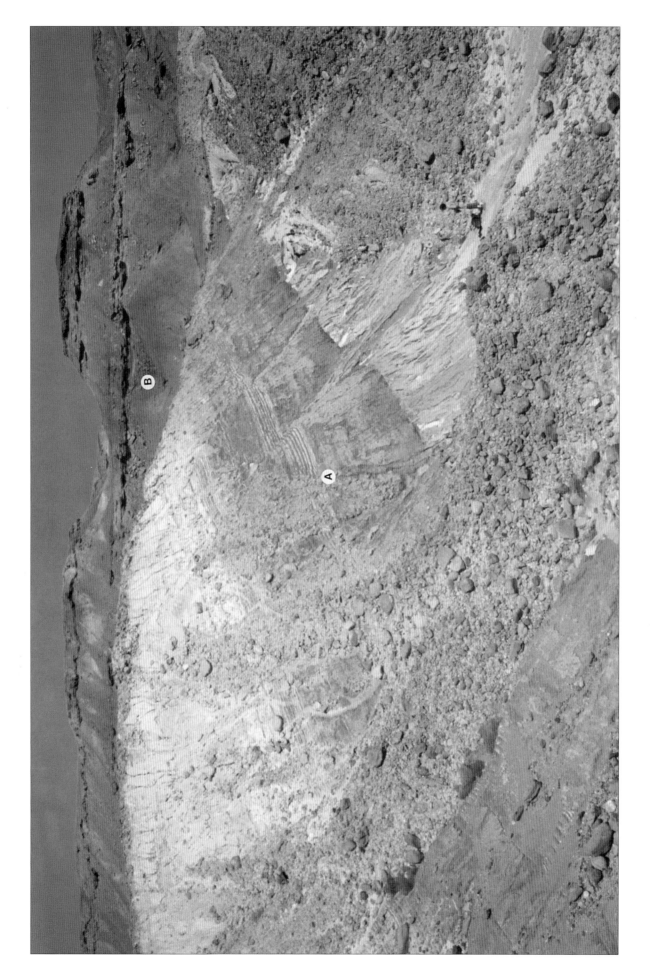

Plate 13 Group A continental deposits. Unconformity between folded Group A and subhorizontal, marine Group B (Mid-Miocene) beds. Group A consists of distal continental facies rich in red-brown floodplain mudstones and quartz sands. Towards the top an alternation of well-bedded sands and muds represents playa sedimentation. Wadi Sharm el Qibli, NW Red Sea Coast.

Plate 14 A. Typical continental facies of Group A. The lower part (1) is bedded floodplain silts. An erosional surface (dotted line) is overlain by unsorted mass-flow conglomerate with sandy matrix (2 and 3); boulders (granite) measure 1 m in diameter. Wadi Sharm el Bahari. B. Well-bedded floodplain silts and sands including surfaces with desiccation cracks (a) and mottled paleosols (b). Wadi Sharm el Bahari, NW Red Sea Coast.

Plate 15 Representative field photographs of constituent facies of Sequence 1, SE Yemen. a) Mixed rhizolith/marl palaeosol horizon overlain by weakly imbricated channelized boulder conglomerates, Wadi Idthik. Hammer at right-centre for scale. b) Coralgal framestone facies, Wadi Arif. Hammer for scale. c) Coarsening and shallowing up unit typical of basal exposures of evaporite assemblage. Unit = 12 m thick, hammer for scale. d) Bleached white alternating tufas, marly palaeosols and pebbly alluvial beds sharply overlain by sheet flood derived gravel conglomerates, Wadi Idthik. Exposure = 20m high, 60m long. e) Unfossiliferous, poorly sorted, intraclastic, thick-bedded grainstone, lower Wadi Marun. f) Mixed gypsum conglomerates facies similar to basal beds of Wadi Marun. Hammer for scale. 4km north of Tibelah.

Plate 16 Representative field photographs of constituent facies of Sequence 2 and 3, SE Yemen. a) Al Wasit region – typical wadi-edge exposures of reddish silty marl facies with numerous thin sand interbeds. Extensive sinuous, low-angle gypsum veining (fibrous, stretched fill) is typical and is interpreted to be emplaced during rift shoulder uplift following final oceanization. Outcrop height = 15 m. b) Interbedded nodular gypsum and recessive brown marl beds from basal 20–30 m of Jebel Dhabba log (Figure D1.7) Note hammer in foreground for scale. c) Interbedded gypsiferous silts and marls with prominent ribs of pebbly gypsum containing occasional mudcracks. Hummum region (Figure D1.8) Outcrop height = 12 m. d) Steeply dipping alluvial fan conglomerates containing mixed basement and Tertiary clasts, 2 km east of Lusb village. e) Aerial view of Jebel Dhabba looking north. Note dolomitized flat-top of Sequence 2 skirted by white limestone plateau in foreground (containing 8.2 Ma ostreids: see Appendix). Prominent bedding traces of alternating gypsum marl of Sequence 2 visible in background. f) South-west-directed aerial view of Jebel Mukalla showing prominent basement block, capped by Sequence 2 reefal carbonates on right middle ground. Stepped Sequence 3 limestone terraces visible in centre and left middle ground. The main terrace is considered (on basis of field mapping) equivalent to 15.3–17.4 Ma terraces mapped further east.

Plate 17 Landsat TM image, eastern Gulf of Suez.

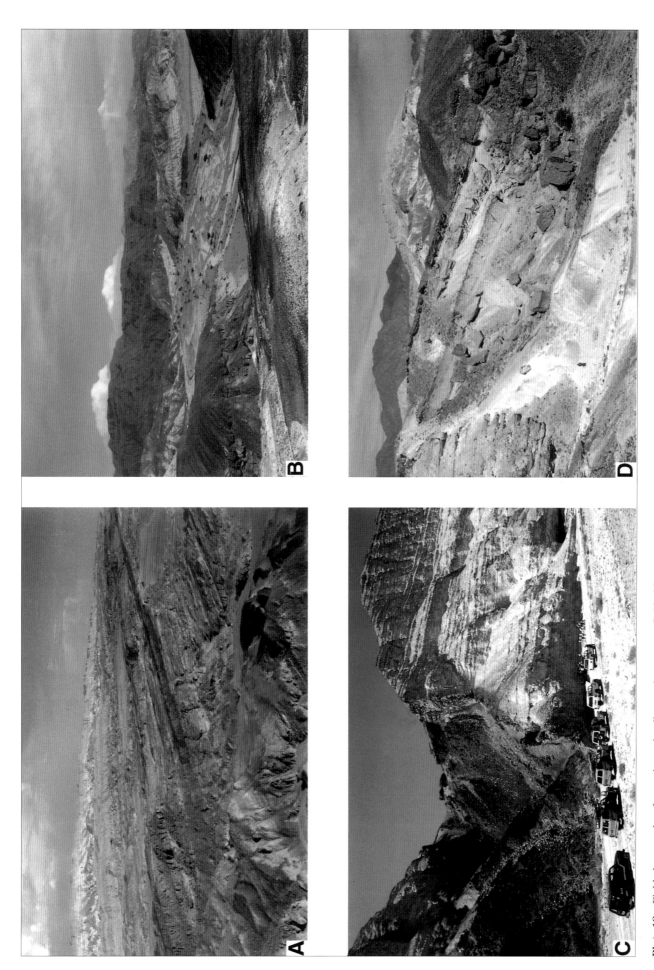

Plate 18 Field photographs of tectonics and sedimentation, eastern Gulf of Suez. (a) View of the Gebal Araba tilted fault block (looking to the south-east) showing the pre-rift stratigraphy from Cambrian Naqus Formation in the foreground to the ridge of Eocene Thebes limestone (Gebel Qabeliat) in the background. (b) Wadi Baba (looking east) showing the rift border fault and Precambrian basement in the background with syn-rift Nukhul and Rudeis clastics in the middle ground. c) Oligocene basic dyke intruding mid Miocene Rod el Awad Formation limestone in Wadi Nukhul. d) East dipping extensional fault between Rod el Awad Formation to the left and sun-rift early Miocene Nukhul Formation (brown) to the right. Looking north to Wadi Baba.

Plate 19 Panoramas of the Abu Shaar and Esh Mellaha blocks, SW Gulf of Suez. A. Footwall escarpment of the Abu Shaar tilt-block at the eastern entrance to Wadi Bali; Precambrian basement (a) is overlain by mid-Miocene, platform-edge carbonate (b) which pass into slope deposits (c); transfer zone between Abu Shaar and Esh Mellaha (d). B. Footwall crest and escarpment of the Esh Mellaha tilt-block showing Precambrian basement (a), pre-rift Nubian sandstone (b) and two mid-Miocene platform sequence (c and d) forming the small outlier near Wadi Um Dirra.

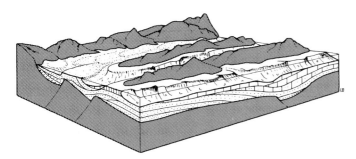

Figure D5.23 Cartoon of the Ras Honkorab blocks showing the relationships between the morpho-structural framework, carbonate, and coarse siliciclastic sediments.

siliciclastics and has been occupied by marine carbonates (Figure D5.22(b)). As one progresses westwards, i.e. towards the rift periphery, successive structural depressions are increasingly filled with coarse siliciclastics, and the peripheral carbonate platform does not enter these embayments. These structural depressions have functioned as major conduits for the transit of coarse terrigenous sediments.

Conclusion
The series of blocks comprising the Ras Honkorab area has favoured the separation of carbonate and siliciclastic regimes. Carbonates were developed along the relatively steep, north-eastern dip slopes of the Honkorab block where they were protected from terrigenous contamination (Figure D5.23). They are locally down-stepped in response to synsedimentary uplift. This stepped shelf passes laterally to a subhorizontal platform formed on the plunging axis of the block. The general morphology of this Miocene shelf-platform is dictated by the relatively high structural relief; carbonates have penetrated locally into depressions indicating that these structural lows have never been totally filled. Most conglomerates were deposited near the exits of the western structural depressions, or have been deviated by the positive blocks.

Blocks independent of the rift periphery

Gebel el Zeit: a series of tectonically unstable blocks buried under open-marine marls

The North and South Zeit blocks are located in the south-western part of the Gulf of Suez (Figure D5.1) and are situated about 45 km from either periphery of the Suez rift. They are bordered to the west by the Gemsa Basin and the north-east by the Suez Basin, both depocentres containing more than 4000 m of syn-rift Miocene sediments. Gebel el Zeit has been mapped by Prat et al. (1986) and by Bosworth (1995), and is discussed by Montenat et al. and Bosworth et al. (this volume). Only the essential sedimentary attributes are discussed in this comparative study.

The syn-rift sedimentary cover of the two Zeit structures differs markedly from that of the more peripheral platforms (Figures D5.24, D5.25). Basal, shallow-marine carbonates with scattered corals, large foraminifera, and molluscs comprising the Nukhul Formation have been considered as uppermost Aquitanian or lower Burdigalian (zone N5) by Evans (1988). These carbonates attain a maximum thickness of about 30 m, forming discontinuous lenses affected by synsedimentary faulting (Figure D5.24(a)). They have been tilted and buried by pelagic marls of the Rudeis and Kareem formations (Figure D5.26(a), (c)) which thicken from 50 m at outcrop to some 3000 m within the adjacent Gemsa Basin (Bosworth, 1995). Pelagic sedimentation reflects a rise in relative sea level which continued into the Serravallian before the appearance of South Gharib evaporites (Richardson and Arthur, 1988; Plaziat et al., this volume). Further tilting of the blocks has occurred before, and possibly during, evaporite sedimentation; the evaporites unconformably overlie all preceeding units, including Precambrian basement (Figure D5.24). Carbonate sedimentation was not re-established until the onset of the South Gharib evaporites when a series of lenticular masses of algally laminated, dolomicrites cap local palaeohighs. These laminites, formed in restricted marine environments, could be the equivalents of the microbial laminites which terminate the mid-Miocene succession on many of the peripheral platforms where they immediately preceed evaporite sedimentation.

Syn-rift sedimentation on the Gebel el Zeit structures is dominated by relatively deep marine facies; shallow-marine (Nukhul) carbonates, developed temporarily before middle Miocene subsidence, were rapidly drowned. Because of distance (40 km) from peripheral sources, and excessive depth, coarse siliciclastic sedimentation was very limited. However, a lensoid body of quartz sand and gravel (Figures D5.25 and D5.26(d)) is

Figure D5.22 Field-views of the Ras Honkorab area: A. Panorama along the axis of the Ras Honkorab block which rises to the north-west (horizon); Precambrian basement (a) is overlain by algal laminites and silty marls (b). Reefal carbonates (c) have steeper sedimentary dips along the south-western flank (c¹); Sharm el Luli (e). B. Carbonate clinoforms and reef within the half-graben immediately to the west of the Honkorab block. C. Prolongation of photo B showing slope deposits on plunging block (a) and carbonate and evaporite series on the axis of plunging Honkorab block (b). D. Down-stepping reefs (arrows) in the Sharm el Luli embayment. E. View towards the south-west at Sharm el Luli showing the reef-terraces (arrows) which are also inclined parallel to the axis of the basement block (dark), the multiple slope directions (photos D and E) being related to a small structural depression coincident with the wadi axis.

258 Carbonate and siliciclastic sedimentation in an active tectonic setting

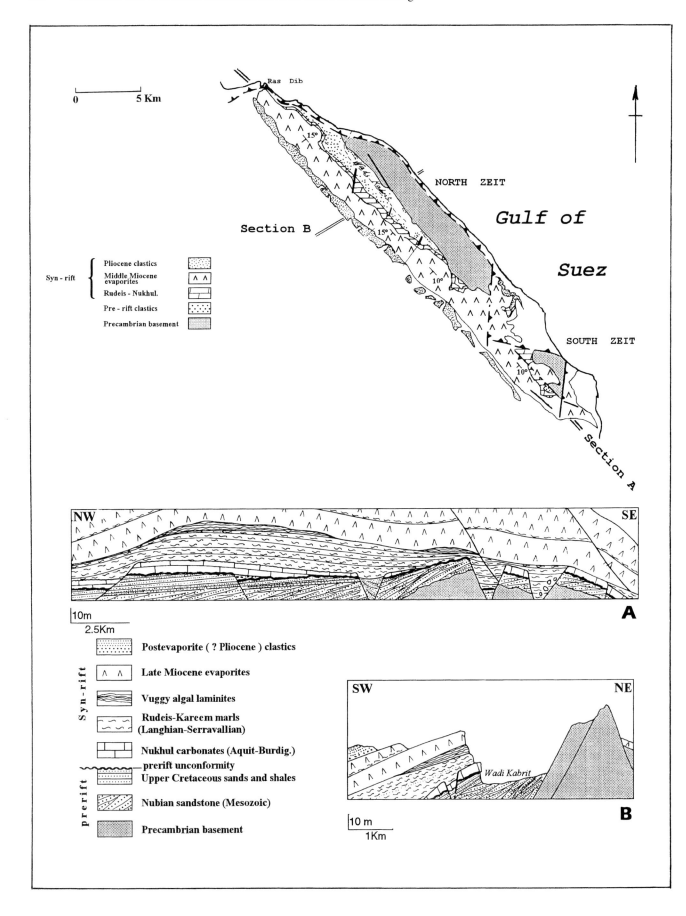

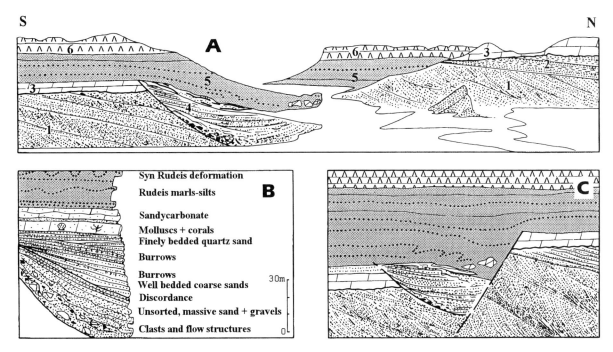

Figure D5.25 Geometry of local, early Miocene, sand body which fills oblique fault-line depression, North Zeit. A. General field-sketch showing pre-rift Nubian sandstones (1) and Cretaceous silts (2), the pre-rift unconformity being overlain by shallow marine Nukhul carbonates (3). These syn-rift carbonates are cut by a steep (35°) surface overlain by the sand body (4), the depression being sealed by Rudeis–Kareem marls (5). The marls are locally deformed and overlain by mid-late Miocene evaporites (6). B. Detailed log of sand body. C. Simplified diagram of sand body showing stratigraphic relationships.

concentrated along a fault-line depression situated near the south-eastern extremity of Wadi Kabrit where it is sealed by pelagic (Kareem) marls. These 25 m thick sands are composed mainly of reworked Nubian Sandstone which forms the pre-rift cover over much of Gebel el Zeit. These marine sands therefore have been derived from local sources following early Miocene tilting of the block.

Unlike the more peripheral platforms, structural movement of the Gebel el Zeit structures has continued not only during early- and mid-Miocene sedimentation, but also during upper Miocene evaporite and post-evaporite deposition (Orszag-Sperber et al., this volume), affecting even the peripheral Quaternary terraces (Plaziat et al., this volume). The present altitude of Gebel el Zeit (456 m), somewhat higher than the more peripheral blocks (Esh Mellaha, 398 m), is due mainly to these more recent movements. This marked structural instability has resulted in the development of multiple, tectonically controlled sequence boundaries and in the relatively steep structural dips (10–20°) affecting the syn-rift cover (Figure D5.24). Repeatedly affected by faulting, block rotation and subsidence, sedimentation at Gebel el Zeit records the dynamics relating to the more axial parts of the rift.

Conclusion

Gebel el Zeit, a major structural element located near the present axis of the Gulf of Suez rift, differs markedly from the peripheral blocks both in terms of block movement and in the lithological composition and stratigraphy of its syn-rift strata. Because of subsidence, the incipient carbonate platform was quickly drowned. Although the marked thinning of the pelagic Rudeis–Kareem marls clearly indicates the positive nature of the Miocene block, this bathymetric high must have been located at water depths in excess of those required for carbonate platform development.

The considerable rise of relative sea level during the mid-Miocene drowned the Gebel el Zeit structures relatively early in rift development (Figure D5.27). However, the more peripheral highs, including Esh Mellaha and Abu Shaar, had greater relief and thus remained near sea level until fairly late in the mid-Miocene (Langhian–Serravallian). Subsurface data (Bosworth, 1995, this volume) indicate that these highs

Figure D5.24 Map and sections of the Gebel el Zeit structures (modified after Prat et al., 1986, and Bosworth, 1995) showing the multiple stratigraphic discordances and the relatively steep dips (somewhat exaggerated in sketch).

Figure D5.26 Field-views of Gebel el Zeit area. A. Panorama towards the south at South Zeit showing Precambrian basement (a), Nubian sandstones (b), lenticular 'Nukhul' carbonates (c), pelagic Rudeis–Kareem marls (d) and south-west-dipping Miocene evaporites (e). B. Pre-rift Cretaceous silts (a) cut by pre-rift unconformity (arrows) overlain by Nukhul conglomerates (b) and 'Nukhul' carbonates (c), near Wadi Kabrit, N. Zeit. C. Ferruginized surface on top of syn-rift 'Nukhul' carbonates (a), olistoliths of 'Nukhul' dolomite (b) embedded in Rudeis–Kareem pelagic marls (c), topped by Miocene evaporites (d), at South Zeit. D. Cross-bedded sand body shown on Figure D5.25.

are flanked to the south-west (Gemsa Basin) by a thick development of Rudeis marl.

Carbonate and siliciclastic sedimentation within oblique structural depressions

The preceding examples have concerned Miocene sedimentation on and around tilt-blocks, generally associated with the development of carbonate platforms. However, these tilt-blocks are also the sites of siliciclastic sedimentation, notably on the gently inclined dip slopes or where the emergent footwall crest has created a local source for littoral siliciclastic deposition; on most tilt-blocks carbonate and siliciclastic sediments are intimately associated. In addition, relatively thick accumulations of fairly pure, generally coarse, siliciclastics occur within two distinct types of structural depression.

1. The axial parts of half-graben. These trap or funnel clastics, the adjacent footwall crests generally serving as topographic barriers which prevent transit towards the rift axis.
2. Accommodation zones, transfer zones, and cross-faults. The multiple half-graben (tilt-blocks) constitute segments which are interrupted by oblique fault-

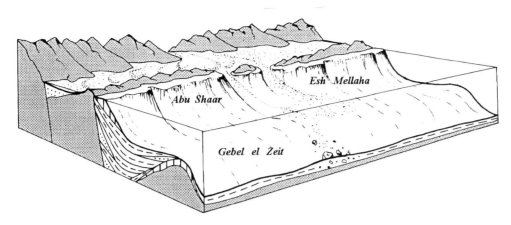

Figure D5.27 Cartoon showing the relationships between the Esh Mellaha and Zeit tilt-blocks during the early and mid Miocene.

systems whose importance in terms of siliciclastic sedimentation is well known (Gibbs, 1984; Frostick and Read, 1989; Evans, 1990; Nelson *et al.*, 1992; Lambiase and Bosworth, 1995; and others).

While the precise distinction between 'accommodation', 'transfer' and 'cross-fault' zones is not always clear, all these structural elements tend to create local topographic or bathymetric lows which affect the larger-scale half-graben structure of the rift. In continental settings, these lows favour the access of fluviatile systems and the development of fan deltas (Frostick and Read, 1989; Collier and Gawthorpe, 1995). Accommodation zones may even cut the rift periphery permitting penetration of extra-rift fluviatile systems (Lambiase and Bosworth, 1995). Certain transfer zones may coincide with the lateral offsetting of adjacent tilt-blocks and thus create passages between adjacent half-graben basins (Nelson *et al.*, 1992).

In the Gulf of Suez rift lenticular marine sand bodies have accumulated in certain north-west–south-east oriented half-graben associated with the intersection of cross-faults where they form important oil reservoirs (Morley, 1995). Others, exposed along the eastern periphery of the rift, have formed shallow-marine cones located near the exits of cross-faults (Sellwood and Netherwood, 1984; Richardson and Arthur, 1988).

Within the north-western Red Sea, numerous shallow-marine cones, generally associated with north–south, N40° or N120°-oriented fault-line depressions have been described by Purser *et al.* (1987), Philobbos *et al.* (1993) and by El Haddad *et al.* (1994). In addition to those traversing the tilt-blocks discussed above, an additional example occurs at Wadi Sharm el Bahari.

Wadi Sharm el Bahari: a shallow-marine cone supporting an incipient reef-platform
Situated 30 km south of Quseir, the Sharm el Bahari area, unlike most of the north-western Red Sea coast, lacks the more typical tilt-block morphology, being a relatively wide (11 km) coastal plain modified by a system of monoclinical cuestas dipping towards the adjacent shore. The absence of visible structural elements is the consequence of the south-eastern plunge of basement blocks, including Gebel Zarib (Figure D5.28), situated somewhat to the north. Together with the Hamadat half-graben, the axes of these plunging structural elements are oriented N120°, somewhat oblique to the main N140° axis of the rift.

The Hamadat half-graben, formed relatively early in the history of the rift (Roussel, 1986), was a structural corridor in which have been deposited the late Oligocene (Group A) continental series and intercalated basalt flows (Orszag-Sperber *et al.* and Montenat *et al.*, this volume).

Middle Miocene sediments include an important, shallow-marine, conglomerate cone at the base followed by a series of mixed shallow-marine clastics and carbonates. The cone is well exposed along the south-western side of Wadi Sharm el Bahari (Figure D5.30(c)) where it is composed mainly of round Eocene chert nodules and scattered molluscs. The cone has large-scale (30 m) foresets prograding mainly towards the south-east, i.e. parallel to the axis of the Hamadat graben. Progradation of the somewhat finer conglomerates and gravels comprising the upper parts of this 50 m thick cone gradually swings from south-east to north-east suggesting that the earlier south-eastern structural control was progressively attenuated (Figure D5.29). As the initial north-west–south-east depression was filled, detritus overflowed to the north-east, prograding towards the main north-north-west–south-south-east axis of the rift.

The bathymetric relief on the shallow marine cone, clearly expressed by its slopes and by the amplitude of the onlapping series (Figure D5.29) was in the order of 30 m, the toe of the cone passing laterally into marls and

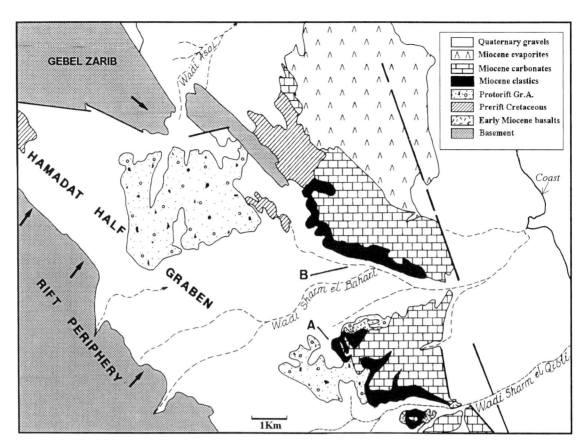

Figure D5.28 Simplified geological map of the Sharm el Bahari area (after Roussel *et al.*, 1986) showing the oblique (north-west–south-east), early Miocene, Hamadat half-graben which has influenced the geometry of both early Miocene (Group A) continental, and mid Miocene (Group B) shallow-marine conglomeratic cones; note the position of sections shown in Figure D5.29.

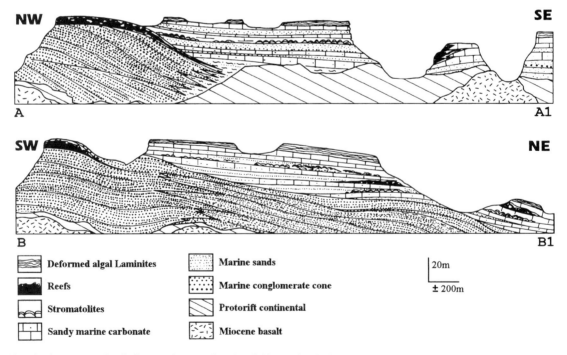

Figure D5.29 Sections across the shallow marine cone(s) at Wadi Sharm el Bahari. A. North-west–south-east section oriented parallel to the Hamadat half-graben showing the conglomeratic cones whose relief has favoured the installation of a small patch reef, progressively onlapped by shallow-marine carbonates. B. North-east–south-west section showing the north-eastern progradation of the upper parts of the cone and onlapping carbonates.

silt. The positive sedimentary relief appears to have favoured the installation of a 10 m thick coral-algal patch-reef, localized on the culmination of the cone. Following its senility, the cone was progressively buried by onlapping clays and sands whose carbonate fraction gradually increased upwards culminating in a series of stromatolitic ridges (Figures D5.29, D5.30(b)) whose axes are oriented at high angles to the foresets (dip) of the underlying cone.

Conclusion

The Sharm el Bahari system differs from the classical transfer or cross-trends discussed above. The oblique, early Miocene depression, probably related to underlying Precambrian basement fracture systems (Jarrige et al., 1986), was created before and rejuvenated after the development of the predominant north-west–south-east Clysmic tilt-block system. This oblique structural control was attenuated by filling of the depression being finally lost by overflow across the plunging axis of the adjacent basement block.

DISCUSSION

The middle Miocene sediments of the north-western margins of the Red Sea are characterized by their extreme variability, most aspects of which have been demonstrated by the preceding examples. These variations result from the complicated interplay of many factors of which syn-rift tectonics is the most important. However, structural control is complicated both by movements of global sea level and by climatic factors, including palaeolatitude, under which Miocene sediments have been generated. Thus, it is not surprising that each of the areas studied exhibits its own specific sedimentary style. These multiple variations nevertheless follow some common patterns which may be summarized below.

General attributes of mid-Miocene sedimentation

Facies variability

The marine sediments of the north-western Red Sea rift margin are an intimate association of siliciclastic and carbonate facies. This is particularly true of the mid-Miocene (essentially Langhian–Serravallian) sediments where pure carbonates rarely exceed 10 m in thickness and carbonate–silicate transitions are rapid and frequent. This variability is readily explained in terms of local and regional tectonics and also by climatic factors.

Size and frequency of structural units

The distribution of carbonate and siliciclastic facies is best understood in terms of structural framework. The relatively complicated sedimentary patterns result essentially from a fairly closely spaced system of pre- and synsedimentary faults of variable orientation, and the depositional morphology is relatively young. Structural blocks range from 0.5 km to 10 km in width and 5 km to 65 km in length and the carbonate platforms are correspondingly small. Their limited dimensions are also the result of repeated synsedimentary faulting and tilting resulting in inclined sedimentary substrates and relatively narrow, shallow-marine culminations. While this morphology tends to favour certain facies, notably reefs, it precludes the creation of extensive inner-platform seas or lagoons. However, because the north-western Red Sea tilt-blocks record the earliest stages of platform development, with time, the complicated patterns associated with platform initiation may be progressively simplified as adjacent highs amalgamate.

Thickness of syn-rift sediments

Syn-rift sediments within the eight examples presented never exceed 150 m and in many cases are less than 50 m; they barely qualify for the term 'carbonate platform'. The extreme thinness of the platform cover can be explained, in part, in terms of 'production surfaces' which are very limited (5–10 km^2). The shallow-marine substrates, favourable for carbonate production, are situated close to structurally controlled escarpments down which sediment is exported. Most areas examined are located close to the rift periphery and thus tend to be affected by uplift which has limited accommodation space. Only in areas more remote from the periphery do sediment thicknesses become important. However, these occur in subsident structural lows where shallow-marine facies may be replaced by deeper-marine marls. It is also possible that the absence of well-developed carbonate platforms was determined by climatic factors, discussed in a subsequent paragraph.

Synsedimentary tectonic modification and destruction of carbonate platforms

Virtually all examples discussed show the effects of repeated tectonic movements involving change in relative sea level. These movements are expressed both by multiple reef bodies, in general relating to the lowering of relative sea level, as at Wadi Ambagi, Zug al Bohar and at Ras Honkorab, and by the massive destruction of footwall zones. The formation of typical by-pass margins at Abu Shaar and Esh Mellaha are related to peripheral collapse leading to the development of subvertical escarpments. These gravitational movements are probably stimulated by structural instability of adjacent bounding-faults. Most platforms are also cut by important Miocene cross-faults whose movements cause the local disintegration of the platform.

Figure D5.30 Field-views of the Sharm el Bahari and Wadi el Aswad (Quseir) shallow marine cones. A. Sharm el Bahari showing prograding cone (a) onlapped by shallow-marine carbonates (b). B. Stromatolitic ridges in upper parts of onlapping Miocene carbonates (see Figure D5.29B). C. North-west–south-east section showing south-eastern prograding cone (a and b) onlapped by carbonates (c and d). D. Detail of Figure D5.29, section B showing upper part of cone (a) prograding to the north-east, onlapped by stromatolitic carbonates (b). E. Shallow-marine conglomeratic cone at Wadi el Aswad (Quseir) with three interstratified reefs (a, b, and c).

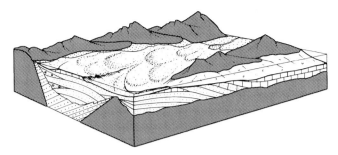

Figure D5.31 Cartoon showing the geometry of the Sharm el Bahari shallow-marine cone and related carbonates.

Properties common to most tilt-block platforms

Although each structure has its own sedimentary peculiarities, nevertheless there exists a common, overall sedimentary evolution possibly relating to global sea-level movement. Dating of the Miocene strata is limited. While the more axial areas such as Gebel el Zeit and nearby wells have early- and mid-Miocene (Langhian–Serravallian) microfaunas, the more peripheral, shallow-marine carbonates and siliciclastics appear to be somewhat younger. Rare biostratigraphic dates from Wadi Gemal (Philobbos et al., 1993), at Safaga (Thiriet et al., 1986), and at Abu Shaar (Cofer et al., 1984) all suggest Langhian to Serravallian ages of marly facies situated immediately below, or lateral to, the platform carbonates. Thus, the shallow-marine sediments deposited on the peripheral parts of the rift, notably on the positive blocks, are probably of mid-Miocene age. Within this somewhat uncertain stratigraphic framework, discussed by Plaziat et al. (this volume), all structural blocks were onlapped. Middle Miocene sediments generally range from shallow, open-marine carbonates and clastics with red algae, corals and benthic foraminifera to relatively pure carbonates with stromatolites, thin oolitic beds and other peritidal facies (Figure D5.32). Increasing restriction is reflected by the progressive appearance of algally laminated dolomicrites often with scattered gypsum or celestite pseudomorphs, and eventually by the development of massive late-Miocene evaporites.

This general sedimentary evolution is regional in extent, but the exact nature of these controls is uncertain. Progressive restriction occurred during a period of global sea-level rise. Therefore, it may be related to regional structural changes which culminated in the separation of the Red Sea and Mediterranean domains.

Within the Miocene series as a whole, major angular discontinuities limiting the four lithological Groups (A–D) record regional structural events relating to rift evolution, discussed by Montenat et al. (this volume). These boundaries may be followed over most of the north-western Red Sea and southern Gulf of Suez. However, in addition, there exist numerous discontinuities whose number and style (angular or disconformable) vary from structure to structure precluding any obvious regional correlation. They have been caused by local tectonic movements relating to specific structural units.

Structural controls of Miocene sedimentation

The importance of this predominant factor is best appreciated by considering, on various scales, the series of examples already discussed.

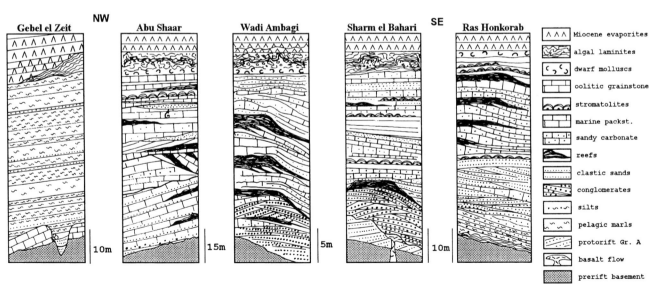

Figure D5.32 Simplified logs through the mid-Miocene series on various tilt-blocks showing similarities and especially the differences; notice the general upwards evolution from open-marine, partly siliciclastic facies at the base, to peritidal and restricted facies at the top; these sections may not be strictly synchronous.

Individual structural units

a. Tilt-block platforms

The morpho-structural relief, well defined during mid-Miocene sedimentation, has exerted a major control on facies composition and distribution, this being clearly illustrated in all areas examined. In common with most sedimentary basins, irrespective of age or tectonic style, the structurally low areas are characterized by terrigenous facies which tend to contrast with the carbonate platforms formed on and around the crests of tilt-blocks (Figure D5.33(a), (b)). However, this generalization is not without exception. Tilt-blocks such as Ras Honkorab and Wadi Ambagi, indeed, are the sites of a predominantly carbonate sedimentation. However, other blocks, including Zug al Bohar and also parts of Abu Shaar, are periodically enriched in siliciclastics, notably when the intervening dip-slope depression has been filled (Figure D5.33(c)).

Dip-slope basins, although generally the sites of coarse terrigenous deposition, occasionally can be sites of relatively pure carbonate sedimentation. If siliciclastics are not delivered to the depression the resulting accommodation space is partially filled with carbonates, as is the case at Ras Honkorab (Figure D5.33(d)). In other localities, fault-related depressions oriented obliquely to the main north-west–south-east pattern, may constitute corridors where shallow-marine cones develop. These form a positive bathymetric relief by purely sedimentary processes, this relief favouring the nucleation of small reefs, as is the case at Sharm el Bahari.

b. Peripheral versus axial sides of the block

On any given block there is generally a marked sedimentary asymmetry between opposing slopes. Because most structures tend to be oriented more or less parallel to the main axis of the rift, their flanks face the periphery or the axis, respectively. In general, the western flanks of any given block, oriented towards the periphery of the rift, are buried in siliciclastic detritus while the opposing eastern flanks, often protected from terrigenous burial by the intervening crest, are sites of carbonate deposition (Figure D5.34). The general tendency is true irrespective of whether the block is tilted towards or away from the periphery.

c. Block tilting and its influence on sedimentation

The orientation of tilt-blocks has a strong influence both on siliciclastic and on carbonate sedimentation. There are two obvious possibilities.

Firstly, if the steep footwall escarpment faces the somewhat deeper environments of the rift axis, as is the case for all blocks in the southern Gulf of Suez region (Gebel el Zeit, Abu Shaar, etc.), reefs and associated sediments will be concentrated along this footwall crest. The gently inclined dip slope facilitates the onlapping of siliciclastic detritus up the dip-slope ramp towards the footwall crest of the block, as clearly demonstrated at

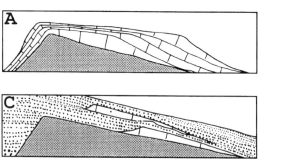
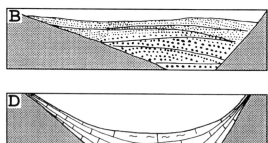

Figure D5.33 Sketch comparing sedimentation on positive and negative parts of tilt-blocks. A. Carbonates often dominate on footwall crests while coarse siliciclastics fill half-grabens (B). However, where upstream half-graben are filled, clastics may overflow on to the footwall crest (C). Inversely, when half-graben are not filled with clastics, they may be occupied by carbonates (D).

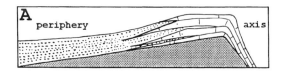

Figure D5.34 Comparison of sedimentation on peripheral (western) and axial (eastern) sides of tilt-blocks. A. The western sides, notably low-angle dip-slopes, are generally dominated by clastic sediments while the eastern sides (facing towards the axis of the rift) favour carbonate sedimentation, especially where clastics are blocked by footwall escarpments.

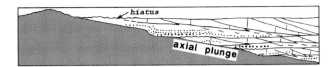

Figure D5.35 The plunge of the main axis of the block creates greater accommodation space favouring thicker sequences. The elevated up-plunge culmination may be a local source for siliciclastics.

Abu Shaar (Figure D5.34(a)). Thus, dip-slope surfaces, especially when oriented towards the peripheral sources, are dominated by siliciclastic sedimentation.

Secondly, if the footwall escarpment faces the periphery of the rift, as is the case for all blocks situated south of Quseir, the supply of terrigenous clastics from peripheral sources tends to be blocked by the relatively steep footwall escarpment (Figure 5.34(b)). The opposing dip slope, protected from terrigenous supply, is generally dominated by carbonates whose seawards accretion is facilitated by the relatively low angle of the structural ramp, as at Zug al Bohar.

d. Axial plunge (pitch)

The long axis (Figure D5.2) of virtually all blocks is not horizontal and its plunge, together with fault-block rotation, imparts a double asymmetry to the block. This axial plunge has important, if fairly obvious, implications mainly because of the accommodation space created (Figure D5.35). On most blocks, including Abu Shaar, Ras Honkorab, and Abu Ghusun, the most complete sections are localized near the lower extremity of the plunging axis; at Abu Shaar and Ras Honkorab the thickest reef development occurs on the down-plunge end of the platform. The up-plunge segment of the axis may coincide with emergence and thus non-deposition or erosion. At Abu Shaar, Ras Honkorab and Abu Ghusun, the structurally higher parts of the axis coincide with exposure of Precambrian basement, these footwall islands creating an important local source of siliciclastic detritus. Because this source is located near the culmination of the block (platform), relatively coarse detritus may readily accumulate on the structurally high areas following moderate longshore transport.

e. The vertical relief of the tilt-block

Depending on the palaeobathymetry of its culmination, the vertical relief of a block determines not only the type of sediment on its culmination but also its efficiency in terms of structural barrier. Where the culmination was above sea level (as at Wadi Ambagi) positive blocks favoured carbonate production leading to the formation of fringing shelves. More substantial carbonate platforms developed if the culmination was below sea level – as was the case for much of Abu Shaar. However, the supply of terrigenous detritus to the platform has depended partly on its relief; where the structurally based platform is bordered by relatively high footwall escarpment, or where the culmination was above sea level (as at Ras Honkorab), the supply of allochthonous terrigenous detritus is blocked (Figure D.36(a)). If, on the contrary the structural relief is modest, siliciclastics, having rapidly filled adjacent depressions, may overflow on to the culmination of the platform, as at Zug al Bohar (Figure D5.36(b)).

f. Fault-related depressions oblique to the main northwest–south-east (Clysmic) tilt-block system

Virtually all tilt-block segments are traversed or terminated by cross-faulting which forms morpho-structural corridors favouring the lateral transit of siliciclastics towards the axis of the rift (Figure D5.37). Local relief created by certain shallow-marine cones has favoured installation of patch reefs. While these oblique depressions have much in common with transfer zones, certain aspects differ.

Firstly, most fault-related depressions have been created by rejuvenation of pre-rift fracture systems affecting underlying Precambrian basement (Montenat et al., this volume). They have been active before, during and subsequent to the development of the north-west–south-east tilt-block system, i.e. their development is not necessarily part of the predominant north-north-west–south-south-east trend.

Secondly, these oblique trends may coincide with the offsetting of the axes of adjacent tilt-blocks, as between Esh Mellaha and Abu Shaar. Others appear not to offset adjacent blocks and their intersection with tilt-block platforms may be associated with the disintegration of these latter, as at Wadi Abu Ghusun.

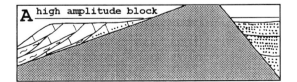
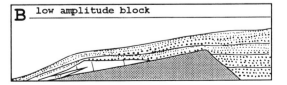

Figure D5.36 The elevation (amplitude) of a block influences sediment facies. A. A relatively high block, especially when emerging above sea level, favours a contrast in facies, while a low block (B) may be buried in siliciclastics or marls and carbonates may be limited.

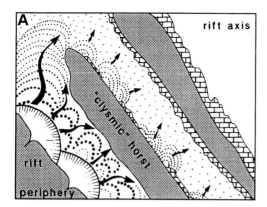
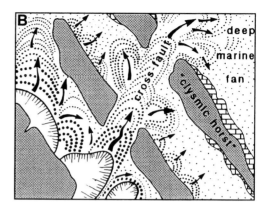

Figure D5.37 The presence of oblique, generally fault-related depressions favour the lateral transit of siliciclastics towards adjacent half-graben and, eventually, the rift axis. A. Clysmically oriented tilt-blocks tend to block siliciclastics favouring development of carbonates. However, where these blocks plunge, the topographic low may facilitate the overlow of siliciclastics. B. Cross-faults or transfer zones favouring transit of siliciclastics, may also disrupt the carbonate platforms.

Thirdly, oblique, fault-line depressions funnel clastics across the north-west–south-east tilt-block system (Figure D5.37). However, because these depressions seem to be shallow (100–200 m), they tended to fill. Sediment overflowed the crests of adjacent tilt-blocks especially where the main axis of the block plunged, forming a morpho-structural saddle.

Fourthly, because Middle Miocene sediments were deposited in shallow-marine settings, sedimentation within these shallow oblique depressions, although predominantly terrigenous, nevertheless included patch-reefs and other carbonates.

Relationship between sedimentation and structural stability
As already noted, many tilt-blocks are characterized by important Miocene erosion related mainly to tectonic instability. In addition, the repeated structural readjustments give specific sequence stratigraphic signatures to syn-rift strata. This results mainly from variation of relative sea level. Each structural unit has its own set of sedimentary packages, often limited by sequence boundaries which can not be correlated from structure to structure. On a given block, such as Abu Shaar, depositional sequences mainly record repeated tilting of the block, as discussed in detail by Cross *et al.* (this volume). Their tectonic controls are confirmed not only by the angular relationships between adjacent sequences, but also by the general increase in dip towards the base of the series.

Synsedimentary tectonics are also recorded by angular unconformities and by the sedimentary sealing of fault-planes, these features being well developed at Gebel el Zeit (Figure D5.24). Structural instability is also expressed by the multiplicity of reef-terraces clearly seen at Wadi Ambagi where the lateral evolution along the front of the Ambagi shelf involves an increase from one to four terraces over a distance of about 5 km. The large number of shallow-marine sequences within the mid-Miocene sediments of the north-western Red Sea therefore records synsedimentary tilting either across the main axis, or along its plunge.

Relationships between sedimentation and position of the block within the rift
Certain blocks are attached, almost directly, to the rift periphery (Wadi Ambagi) while others are separated from the periphery by one or more sub-basins (Zug al Bohar, Abu Ghusun, Abu Shaar). Finally, other structures are separated from the rift periphery by a series of half-graben (Ras Honkorab) while Gebel el Zeit is located close to the axis of the Gulf of Suez rift. Obviously, the tectonic and sedimentary history of the block, related to the overall evolution of the rift, depends partly on its position in the rift system (Figure D5.38). In general, blocks situated close to the periphery have been affected by moderate subsidence and thus the

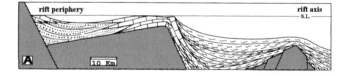
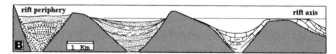

Figure D5.38 Sedimentation on a given tilt-block is influenced both by its situation with respect to the rift periphery, and by the presence of adjacent blocks. A. Distance from the rift periphery and bathymetry of the culmination may determine the texture and mineralogy (carbonate or siliciclastic) of sediments deposited on the culmination. B. A series of half-graben will tend to favour the trapping of coarse clastics within the more proximal graben and accumulation of carbonates within the distal depressions and culminations.

accommodation space on the culmination of the block has been limited; the shallow-marine cover tends to be thin (<50 m). Blocks such as Abu Shaar, situated in intermediate positions, have somewhat thicker (130 m) sequences, the adjacent half-graben containing 1000–2000 m of sediment. This situation contrasts markedly with that at Gebel el Zeit situated about 40 km from the rift periphery; it was drowned during the mid-Miocene (Langhian–Serravallian). As a result, shallow-marine sediments are thin and carbonate platform development has not kept pace with subsidence. Although a palaeohigh existed throughout much of the Neogene, Gebel el Zeit often has been bathymetrically deep, sediment on its crest including up to 100 m of pelagic marls.

The stability of the block, expressed by sedimentary discontinuities, also depends on its location. Blocks situated close to the periphery, although affected by moderate synsedimentary tilting, tend to retain their initial, subhorizontal or low (2–5°) structural dips and marked angular unconformities and associated erosion is not frequent. Blocks situated near the axis of the rift, such as Gebel el Zeit, on the contrary, have been affected by repeated faulting and rotation during Neogene sedimentation. This structural instability is clearly registered as a series of unconformities.

Relationship between sedimentation and adjacent blocks
Sedimentation on any given platform and its adjacent structural depressions is also influenced by the number and distance of adjacent structures (Figure D5.38(b)). Relatively isolated blocks such as Gebel el Zeit tend to be influenced mainly by local factors discussed in the preceding paragraphs. However, in a series of fairly closely spaced structures, notably those at Ras Honkorab, interdependent sedimentation is related to the barrier effect of the upstream block. Thus, structural depressions situated nearer the peripheral source areas are filled with coarse clastics which may overflow on to the positive blocks. On the contrary, adjacent depressions situated further to the east, i.e. somewhat more remote from the periphery of the rift, receive finer detritus and may be sites of marly or even carbonate sedimentation. The axial parts of the more eastern external graben at Ras Honkorab have been colonized by scattered branching corals and domed stromatolites. These external depressions tend to remain unfilled, the accommodation space permitting the development of slope deposits on both sides of the adjacent structural high (Figure D5.38(b)).

Nonstructural controls of Miocene sedimentation

During active rifting, tectonics, although predominant, was not the sole factor controlling sediment mineralogy and geometry; latitude and local climatic factors have also been important.

Possible latitudinal effects on Miocene sedimentation
The early- and mid-Miocene sedimentary sequences, notably those on the structural highs, are relatively thin and carbonate platform development is very modest. This may be due to a number of factors, including limited accommodation space. As already noted, reduced carbonate production may well have been controlled by the limited areas of production on the small tilted blocks. However, platforms of the north-western Red Sea are not markedly smaller than those elsewhere in the world, notably many of the smaller Pacific atolls and certain eastern Bahamian platforms. Furthermore, the time available for carbonate buildup has not been negligible; because certain platforms were formed during the early Miocene and persisted until the Serravallian, the period of sedimentation could have been in the order of 10 m.y.

Naturally, the thickness and geometry of any platform will depend on the accommodation space available. While many of the platforms examined in the north-western Red Sea area have been situated close to sea level and suffered repeated uplift, the early- and mid-Miocene nevertheless was a time of global sea-level rise (Haq *et al.*, 1987), thus creating accommodation space. Furthermore, certain structures such as Gebel el Zeit have been affected by major subsidence. However, this has resulted in drowning, relatively shallow-marine Aquitanian 'Nukhul' carbonates being overlain by deeper-marine Burdigalian to Serravallian (Rudeis–Kareem) marls. Clearly, carbonate production has not been sufficiently rapid to compensate for subsidence. While this inefficiency may admittedly be explained by many factors, it is possible that the latitude of the northwestern Red Sea region during the early- and mid-Miocene, although sufficiently favourable for coral growth, nevertheless was not constantly warm and thus not particularly favourable for high benthic carbonate production. A somewhat comparable situation exists today in the Gulf of Suez (27–29°N) where modern reef communities are situated at their northern limits of tolerance. Furthermore, the intracontinental setting of the rift would make its waters particularly prone to temperature changes. These short-term negative effects possibly diminish towards the Indian Ocean.

Local effects of wind on sedimentation
Because rift-type basins form topographic corridors, they favour an active wind regime which tends to be oriented parallel to the axis of the rift. This results in a certain degree of coincidence between structural (internal) and climatic (external) polarity. Today, and probably in the past, the strongly predominant north-north-

westerly wind, particularly active in the Gulf of Suez, influences shallow-water movement and thus reef growth and transport of their detritus. Had this been the case during the Miocene one would logically expect somewhat better development of reefs at the northern ends and thicker accumulations of detrital carbonate, and possibly even siliciclastic sands, at the southern end of shallow marine platforms. While this is not striking, at least on the platforms examined, there is some indication of preferential accumulation of carbonate sands at the southern end both of Abu Shaar and, especially at Zug al Bohar. Furthermore, at this latter locality, reefs are best developed towards the northern end of the block.

Where Precambrian crystalline basement has emerged above Miocene sea level, due partly to the axial plunge of certain blocks (Abu Shaar–Esh Mellaha), this local source of terrigenous material was exported by coastal currents. At Abu Shaar, cross-bedding directions within the arkosic sands and gravels clearly indicate a north-west provenance, probably from the emergent Esh Mellaha block, and thus longshore transport.

Other possible climatic effects: sporadic terrigenous input
The considerable peripheral and local relief has favoured input of coarse terrigenous materials. However, these discharges have not prevented the development of carbonate platforms due partly to structural relief (blocks) which favoured crestal carbonate production. It may also have been due to the sporadic nature of the siliciclastic input due to the absence of permanent fluviatile systems. Large-scale drainage systems, limited by the proximity of the rift shoulders, may have been further reduced by semi-arid climates favouring flash-flooding. Following each flood event, the temporary turbid waters cleared rapidly favouring renewed carbonate sedimentation.

CONCLUSIONS

Shallow marine sedimentation within the peripheral parts of the north-western Red sea has been strongly influenced by the morpho-structural setting, both with respect to the periphery and axis of the rift, and especially by the various structural properties of individual tilt-blocks. These latter include angle and direction of tilt, plunge of the main axis of the block as well as its vertical relief, these parameters influencing the relative importance of carbonate and siliciclastic facies and their geometry with respect to the footwall crest of the block. Miocene tilt-block platforms of the north-western Red Sea are also characterized by an important synsedimentary disintegration occasioned by repeated tectonic adjustment. Oblique fault-systems crossing the tilt-block and reactivation of bounding faults have caused repeated collapse of the carbonate platforms and the formation of coarse slope deposits.

Factors other than syn-rift tectonics also have influenced sedimentation. The relatively thin (less than 150 m during 10 m.y.) carbonate platforms may be due to several factors including their peripheral location; rising mid-Miocene global sea level initially flooded the axial parts of the rift and the crests of the more peripheral tilt-blocks were flooded relatively late in the mid-Miocene. However, the drowning of mid-Miocene platforms (Gebel el Zeit) situated closer to the axis suggests that carbonate production was not dynamic; it may have been limited by the cooler waters relating to relatively high palaeolatitudes.

The Middle Miocene tilt-block platforms of the north-western Red Sea and south-western Gulf of Suez have much in common with other rift settings: the Lower Cretaceous of north Spain (Rosales *et al.*, 1994), the Lower Cretaceous of both the west Atlantic and the Campos Basin of Brazil, the Jurassic of north Mexico, and others. However, they also have their own peculiarities, many of which are discussed in this volume.

ACKNOWLEDGEMENTS

The authors wish to thank the following colleagues whose constructive reviews have enabled considerable improvement of the original manuscript: Dan Bosence, Paul Enos, Wolfgang Schlager and James Lee Wilson. Their help is greatly appreciated. This research was financed by the French oil companies ELF and TOTAL and by the E.C. (Brussels) 'Science Program' in the form of the project RED SED. The financial help is gratefully acknowledged.

Chapter D6
The tectono-sedimentary evolution of a rift margin carbonate platform: Abu Shaar, Gulf of Suez, Egypt

N. E. Cross, B. H. Purser and D. W. J. Bosence

ABSTRACT

Abu Shaar is a dolomitized Miocene, mixed carbonate-siliclastic platform outcropping on the crest of a fault block in the south-western margin of the Gulf of Suez, Egypt. The Miocene platform rests unconformably on Precambrian basement and is cut to the east by a basinward-dipping and throwing extensional fault; Miocene strata are tilted to the western margin of the basin. Up-section decreases in dip, successive westward thickening of strata and onlap of beds on the hangingwall dip slope indicate that fault-block rotation occurred before and during deposition of this Langhian-aged syn-rift platform.

Miocene depositional morphology has been exhumed by Quaternary erosion to reveal platform-margin buildups and slopes in the footwall, and gently sloping open-marine to restricted-marine platform-top facies in the hangingwall dip slope. Hangingwall areas are characterized by coarse, shallow marine sandstones derived from exposed basement of the northern Esh el Mellaha fault block and the basin margin Red Sea Hills. The southern margin passes into a transfer zone. The last Miocene carbonates to be deposited on the platform-top are restricted marine carbonates and signal the onset of evaporative conditions in the Gulf.

The large, three-dimensional exposures of Abu Shaar have enabled the recognition and platform-wide correlation of nine depositional sequences which provide a spatial and temporal framework for syn-rift platform evolution. Three depositional sequences show evidence of syndepositional fault-block rotation with synchronous footwall uplift, and erosion and hangingwall deepening and flooding. These depositional sequences are characterized by thickening down the hangingwall ramp, onlap on to the hangingwall and footwall unconformities overlain by aggradational shelf-margin reefs and steep slopes off the footwall block. Where fault throw reduces into the transfer zone, footwall uplift is less, slopes are less steep, and shelf-margin reefs are thicker and more progradational. Two depositional sequences have platform-wide occurrence but show no thickness changes and are interpreted to result from regional (tectonic or eustatic) sea-level changes. Two other depositional sequences are restricted to the steep slopes of the footwall margin and are generated by submarine fault-related, scarp retreat. They are downlapping wedge-shaped sequences in which shelf-margin reefs pass down-dip into slope deposits. They do not have equivalents in the transfer zone margin or the hangingwall and therefore are not generated by fault-block rotation.

Facies models are proposed for the footwall and hangingwall regions of the platform and comparisons are made with other tilt-block carbonate platforms.

INTRODUCTION

The Red Sea rift system and its northern sub-basin, the Gulf of Suez, is a young extensional basin comprising numerous rotated fault blocks (Figure D6.1). In the south-west of the Gulf, fault blocks are overlain by early to middle Miocene (Aquitanian–Langhian) sediments which crop out on Gebel Zeit, Esh el Mellaha and Abu Shaar (Figure D6.1). These large, three-dimensional

Sedimentation and Tectonics of Rift Basins: Red Sea–Gulf of Aden. Edited by B.H. Purser and D.W.J. Bosence. Published in 1998 by Chapman & Hall, London. ISBN 0412 73490 7.

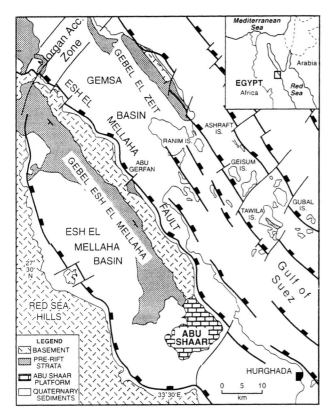

Figure D6.1 Geological setting of Abu Shaar within the south-west Gulf of Suez (after Bosworth, 1994).

outcrops provide a detailed record of the nature and evolution of syn-rift facies and depositional sequences (Purser et al., this volume).

Abu Shaar is a 10 km × 10 km mixed carbonate–siliciclastic platform developed on the south-east footwall and hangingwall dip slope of the Esh el Mellaha fault block, close to the south-west margin of the Gulf of Suez (Figure D6.1). The platform is limited to the east by the Esh el Mellaha fault and steep (40°) slopes which descend below the coastal plain, while to the south-west, the subhorizontal platform top passes into the subsurface below the western Quaternary alluvium of the Esh el Mellaha basin (Figure D6.1). This relatively simple platform morphology reflects the underlying structural framework; Abu Shaar is an exhumed fault-block carbonate platform whose steep north-eastern and south-eastern slopes, together with the subhorizontal plateau, record original Miocene depositional morphology. The stratigraphy of the Abu Shaar platform is well exposed in several deeply incised wadis which dissect the footwall of the block (Wadis 1 to 8; Figure D6.2). Of these, Wadi Bali'h cuts 4 km across the block exposing sections from the faulted eastern platform margin through to the western hangingwall dip slope.

Esh el Mellaha and the nearby fault block of Gebel el Zeit have been studied since the beginning of this century (Gregory, 1906; Hume, 1916; Madgewick et al., 1920). Subsequent oil exploration in the Gulf of Suez has stimulated numerous regional studies of its tectono-stratigraphic evolution (Garfunkel and Bartov, 1977; Sellwood and Netherwood, 1984; Webster and Ritson, 1984; Montenat et al., 1986; Richardson and Arthur, 1988; Purser et al., 1990; Purser and Philobbos, 1993; Patton et al., 1994; Schütz, 1994; Bosworth, 1995). Most studies outline the main attributes of these early rift structures, notably the presence of Precambrian basement and overlying pre-rift Mesozoic and early Tertiary sediments. North-west–south-east oriented fault blocks formed a structural template which strongly influenced the distribution of the syn-rift Miocene sediments which unconformably overlie the pre-rift strata. However, due to the lack of age-diagnostic biota and widespread dolomitization, the precise age of the Abu Shaar platform is not well constrained. Lithological correlation with dated adjacent areas suggests a Burdigalian to Langhian age (Cofer et al., 1984; see discussion in Bosworth, 1995); a Langhian to early Serravallian age is proposed by Plaziat et al. (this volume).

Various aspects of the Abu Shaar platform have been studied by previous workers, including platform stratigraphy and sedimentology (Burchette, 1988; James et al., 1988), microfacies (Cofer et al., 1984), carbonate/siliciclastic interactions (El Haddad et al., 1984), stromatolites (Rouchy et al., 1983; Monty et al., 1987) and diagenesis (Aissaoui et al., 1986; Coniglio et al., 1988; Clegg et al., this volume). In view of this number of previous studies, particularly those of Burchette (1988) and James et al. (1988), the necessity of an additional contribution could be questioned. While the previous works provide an overview of the carbonate sedimentology and stratigraphy they are nevertheless limited in both the extent and the depth of their treatment of the Abu Shaar platform. Many studies have focused on the well exposed north-eastern, footwall margin where a series of wadis expose Miocene reefs and platform slopes. Moreover, the sections sited on the footwall consist almost entirely of carbonate facies, and previous work has exaggerated the importance of carbonates over siliciclastic facies which dominate parts of the hangingwall dip-slope stratigraphy. A sequence stratigraphic treatment of these sections has not been previously attempted. Following an extensive programme of detailed fieldwork, this study provides a three-dimensional study of the sedimentary facies, depositional sequences and stratigraphic evolution of this exhumed fault-block platform. This contribution is complemented by other, more specialized studies of aspects of the Abu Shaar platform that are not considered here; reef palaeoecology by Perrin et al. (this volume), slope sedimentation by Purser and Plaziat (this volume), synsedimentary seismite deformation by Pla-

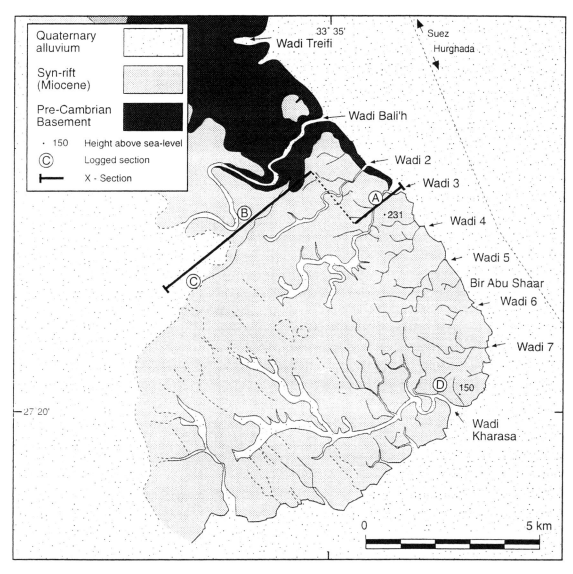

Figure D6.2 Geological map of the Abu Shaar platform illustrating wadi sections, locations of logs and cross-section (Figure D6.8).

ziat and Purser (this volume) and platform dolomitization by Clegg et al. and by Purser (this volume).

Because of the relatively complete three-dimensional exposure, and easy access, the Abu Shaar platform is unique in providing a well constrained case-study for fault-block sedimentation as modern topography is essentially inherited from the original marine morphology. The north-eastern slopes of Abu Shaar, represent original platform margin slopes where palaeodepths can be measured, the three-dimensional geometry of depositional sequences can be studied as can the three-dimensional stratigraphic and structural evolution of what is interpreted here to be a syn-rift carbonate platform.

Understanding the relationships between sedimentary facies, depositional architecture and extensional tectonism, has obvious economic implications. The Abu Shaar platform provides an important on-shore analogue for potential reservoir facies not only in the Gulf of Suez (Kulke, 1982; El Hilaly and Darwish, 1986; Darwish and Saleh, 1990) but also for other Miocene carbonates in arid climate, rift basin settings (cf. Sun and Esteban, 1994). It also serves as a predictive model which can be applied to other tilt-block carbonate platforms where tectono-sedimentary relationships are less well constrained than those developed on Abu Shaar.

STRUCTURAL SETTING

Regional

The Red Sea rift and the Gulf of Suez, were initiated in the latest Oligocene–early Miocene (Bosworth *et al.*, this volume). However, whereas the Red Sea continues to extend, the extension in the Gulf of Suez decreased through the Miocene (Courtillot *et al.*, 1987a). Following an extensive period of pre-rift erosion, the northern parts of the Red Sea were affected by a series of north–south, 040° and 120° oriented fault systems, prior to the development of the present-day (north-north-west–south-south-east) 140° rift axis (Montenat *et al.*, 1986). Partially transcurrent fault-bound basins, with little vertical movement were filled with non-marine siliciclastic sediments of late Oligocene age (Montenat *et al.*, this volume; Sellwood and Netherwood, 1984). Subsequent extension, mainly perpendicular to the N140 axes, define a series of fault blocks, including Esh el Mellaha, and half-graben during the early Miocene.

Propagation of the Red Sea rift and its associated widening during the Oligo-Miocene extended northwards to the Aqaba transcurrent fault system (Freund *et al.*, 1968; Patton *et al.*, 1994; Bosworth, 1995). Further transcurrent movements along this north-north-east–south-south-west Aqaba system since the Serravallian–Tortonian marks the boundary between the active Red Sea rift and the failed Gulf of Suez rift (Bosworth, 1995). Esh el Mellaha and adjacent fault blocks, although today situated within the south-west Gulf of Suez, exhibit early sedimentary styles comparable to those of the Miocene of north-western Red Sea (cf. Purser and Philibbos, 1993).

Local

The Esh el Mellaha and Gebel Zeit blocks are bound to the north-east by major (140°) oriented normal faults (Figure D6.1). Well, outcrop and seismic data from the neighbouring Gebel Zeit block indicate that these major bounding faults tend to be listric at depth (Bosworth, 1995; this volume). The complementary fault system which borders the south-western side of the Esh el Mellaha block probably coincides with the main escarpment marking the south-western boundary of the rift adjacent to the Red Sea Hills (Figure D6.1). On Abu Shaar, Miocene or pre-Miocene tilting of the Esh el Mellaha fault block is clearly expressed by an angular unconformity where syn-rift Miocene marine carbonate platform sediments rest on pre-rift Precambrian basement (Figure D6.2).

The amount of fault-block rotation controlled the depth and gradient of the hangingwall dip slope and as such influenced the local bathymetry of the transgressive Miocene sea that flooded the Esh el Mellaha fault block. This in turn constrained the degree to which Abu Shaar was isolated from the western shoreline of the present-day Red Sea Hills (Figure D6.1) and therefore the supply of siliciclastic sediments from these peripheral basement outcrops. Both geophysical data (Bosworth, 1995) and projection of the pre-Miocene surface at outcrop, suggest that the pre-rift/syn-rift surface descends to depths exceeding 1 km to the west within the Esh el Mellaha basin (Figure D6.1). Although this present thickness of strata results in part from post-Miocene movements, these seem to be minor based on the subhorizontal (<1°) dip of the uppermost bedding surfaces on Abu Shaar. Water circulation within this early elongate gulf would have been determined both by its depth and also by any structurally controlled communications with the more open-marine waters of the axial parts of the rift system. While structural geometries preclude any major obstruction to the south-east, closure towards the north-east may have been controlled by the Morgan Accommodation Zone at the north-east end of the Esh el Mellaha fault block (Figure D6.1). However, this structural barrier, even if it existed, implies that the inferred bathymetry to the south-west of Abu Shaar extended northwards for at least 70 km. The Esh el Mellaha basin (Bosworth, 1995; formerly Tarbul syncline: Montenat *et al.*, 1986; Burchette, 1988) therefore was of sufficient dimensions to permit free-water circulation and probably normal marine salinities during the early- to mid-Miocene.

In addition to its well-documented south-westerly tilt, the Esh el Mellaha and Abu Shaar fault blocks also plunge to the south-east. Basement attains a present-day altitude of about 300 m near the southern end of Esh el Mellaha, descending to approximately 200 m in the northern parts of Abu Shaar, from where it drops below the Quaternary desert plain in the vicinity of Bir Abu Shaar (Figures D6.2, D6.3). The decrease in footwall altitude reflects the presence of an interbasinal transfer zone at the south-eastern end of the Esh el Mellaha border fault (Hurst, 1987; Gawthorpe and Hurst, 1993). Onlap at the base of the Abu Shaar platform, from both the south-east and the south-west, indicates that this structural geometry existed during the Miocene transgression. Sections through the carbonate platform therefore thin towards the north-east of Abu Shaar where Precambrian basement of Esh el Mellaha rises higher than the adjacent platform. In this area, a Miocene footwall island, of at least 50 m in altitude, marked the north-eastern boundary of the submerged Abu Shaar platform. During the Miocene this strike-trending topographic high provided an important source area for siliciclastic input to the Abu Shaar platform (Figure D6.3).

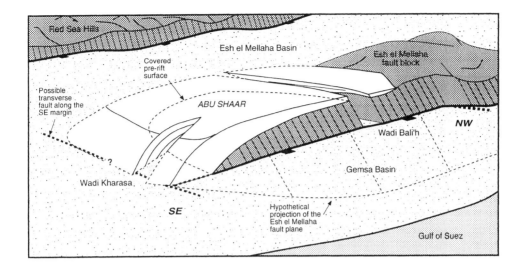

Figure D6.3 Oblique aerial photograph and geological sketch illustrating rifted margin to Gulf of Suez with Red Sea Hills (background), the Esh el Mellaha block and border faults, and Abu Shaar platform, Quaternary alluvium and coastline.

At the north-eastern end of Abu Shaar (Wadi Treifi) a right-lateral offset of approximately 6 km suggests a cross-strike fault (040°) separates the Abu Shaar platform from the adjacent Esh el Mellaha block. In Wadi Treifi (Figure D6.2), Miocene conglomerates composed of basement and Eocene chert pebbles infill a palaeo-valley suggesting syn-tectonic deposition. An early Miocene fault is also thought to have separated the Esh el Mellaha and Abu Shaar fault blocks because of differences in their structural and stratigraphic histories.

On Esh el Mellaha the Precambrian basement and overlying pre-rift Cretaceous and Eocene rocks are tilted towards the south-west at an angle approaching 8° (Bosworth, 1995) while at Abu Shaar the Cretaceous and Eocene are not exposed, the Precambrian basement surface is lower, and the Miocene is inclined to the south-west at an angle of between 1 and 4° (average 2°).

The structural setting of the Esh el Mellaha–Abu Shaar fault block has obvious implications for the sedimentary evolution of the Abu Shaar carbonate platform and its associated siliciclastic facies. Miocene bathymetry and sedimentation were influenced by this underlying tectonic framework. Specifically, open-marine facies surrounding the footwall crest of Abu Shaar, together with the inclined nature of the pre-Miocene hangingwall dip slope substratum, imply that sedimentation, notably in its initial transgressive stages, did not occur on a subhorizontal platform-top but rather on an inclined ramp (Burchette, 1988). This hangingwall dip slope, inclined towards the south-west, probably passed well below wave-base into an open-marine environment in the Esh Mellaha basin. The Miocene tilt-block platform was limited on the opposing north–east side by a sharp break in slope, coincident with the footwall scarp of the Esh Mellaha fault, and to the south-east by an unexposed cross-strike fault or transfer zone (Figure D6.3).

REGIONAL SYN-RIFT STRATIGRAPHY

The stratigraphy of the Gulf of Suez is summarized in Figure D6.4 and has been largely established from subsurface work in the axial parts of the Gulf. Plaziat *et al.* (this volume) discuss the stratigraphy in more detail together with the problems of dating rift margin exposures and integrating these with the axial areas.

The Abu Zenima Formation (Chattian–Aquitanan) consists of up to 100 m of continental to shallow marine clastics (Sellwood and Netherwood, 1984) and is thought by many to represent sedimentation at the onset of Oligo–Miocene rifting in the Gulf of Suez (Chenet and Letouzey, 1983; Montenat *et al.*, 1988; Richardson and Arthur, 1988; Montenat *et al.*, this volume). Unequivocal evidence for rift initiation is represented by the Nukhul Formation (Aquitanian–Burdigalian) which consists of a variety of fluvial to shallow-marine facies (Saoudi and Khalil, 1986) up to 700 m thick (Richardson and Arthur, 1988). The succeeding Rudeis Formation (Burdigalian–Langhian) also represents deposition during a period of rifting (Evans and Moxon, 1986) when a well-defined fault block/half-graben basin morphology strongly controlled sedimentation. During the Rudeis, basinal areas were dominated by *Globigerina* marls up to 1000 m thick (Richardson and Arthur, 1988), whereas on approximately time-equivalent footwall highs, including Abu Shaar, shallow marine platform carbonates and coarse siliciclastics were deposited. These sediments tend to exhibit a progressive change from open-marine facies near the base to more restricted facies at the top (Purser *et al.*, 1990). The latter appears to represent a widespread change preceding regional evaporite deposition (Belayim, South Gharib and Zeit formations) during the middle and late Miocene (Orszag-Sperber *et al.*, this volume).

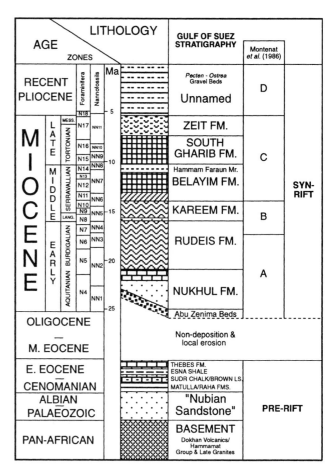

Figure D6.4 Summary stratigraphy of the Gulf of Suez (after Bosworth, 1994; Montenat *et al.*, 1986).

Local basement (pre-rift) stratigraphy

The platform carbonates and siliciclastics of Abu Shaar rest unconformably on a Precambrian basement of calc-alkaline metavolcanics and pyroclastics (Dokhan Group) and metasediments (Hammamat Group). Further north-west on Esh el Mellaha, these are intruded by granites and overlain by Mesozoic (Nubian) sandstones and Eocene (Thebes) cherty limestones. Granites comprise much of the Red Sea Hills along the rift margin, some 15 km to the south-west of Abu Shaar. These igneous, volcanic and sedimentary basement rocks had considerable footwall relief during the Miocene and therefore constituted an important source of siliciclastic material.

Local Miocene (syn-rift) stratigraphy

On Abu Shaar, platform carbonates and siliciclastics attain a maximum thickness of 120 m in Wadi Kharasa. This Miocene sedimentary cover has been assigned to the Rudeis Formation by previous workers (Cofer *et al.*, 1984; Burchette, 1988; James *et al.*, 1988). The Rudeis Formation is a lithostratigraphic term generally applied

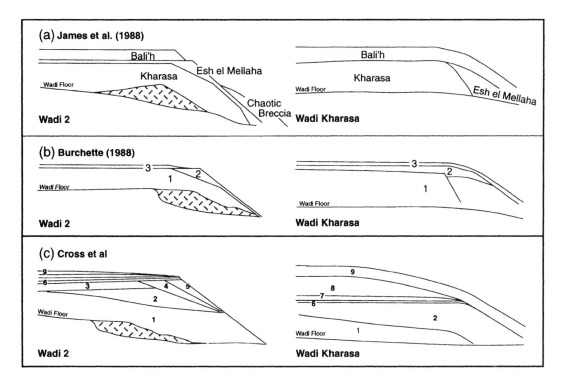

Figure D6.5 Previous stratigraphic subdivisions of the Miocene carbonates of Abu Shaar based on sections in Wadi 2 (Figure D6.2) and Wadi Kharaza. (a) Members of James et al. (1988). (b) Numbered sequences of Burchette (1988).

to the subsurface basinal facies in the Gulf of Suez. The platform sediments of Abu Shaar correlate with informal Group B of Montenat et al. (1986; this volume) (Figure D6.4).

James et al. (1988) subdivided the stratigraphy of the Abu Shaar platform into four informal members: the Kharaza, Esh el Mellaha, Bali'h and Chaotic Breccia Members (Figure D6.5). The Kharaza, Esh el Mellaha and Bali'h Members are broadly consistent with the three sequences (1–3) described by Burchette (1988) (Figure D6.5). James et al. (1988) define the four members as follows.

The **Kharaza Member** is volumetrically the most important member, consisting of various open-platform carbonate facies on the footwall and mixed carbonate/siliciclastic facies on the hangingwall dip slope. This member represents platform initiation with onlapping surfaces progressively blanketing fault-generated basement topography.

The **Esh el Mellaha Member** consists of coralgal reef and associated fore-reef facies developed along the eastern footwall following fault-induced platform margin collapse.

The **Bali'h Member** displays a distinct change from open to restricted carbonate facies in the upper platform stratigraphy as a prelude to regional evaporite deposition in the mid- to late-Miocene.

The **Chaotic Breccia Member** is a complex unit of breccia and bedded sediment which overlies the fore-reef beds of the Esh el Mellaha Member along the eastern margin. According to James et al. (1988), this member represents collapse of platform-top sediments following evaporite dissolution.

The Kharaza, Esh el Mellaha, Bali'h and Chaotic Breccia Members are bounded by well-defined, angular unconformities and subhorizontal disconformities which are generally accompanied by major facies shifts.

DEPOSITIONAL SEQUENCES, FACIES AND ENVIRONMENTS

More detailed stratigraphic analysis following extensive fieldwork by the present authors has shown that the units recognized by previous workers (Burchette, 1988; James et al., 1988) can be further subdivided into nine unconformity and disconformity-bound depositional sequences. The divisions of previous authors (above) can be recognized within these depositional sequences so that the members have been retained here, albeit with minor correlation changes (cf. Figures D6.5 with D6.6–D6.8). The recognition of more detailed platform-wide depositional sequences provides a new, detailed and genetic, three-dimensional picture of platform evolution.

The recognition of depositional sequences within the footwall platform margin is based on well-defined unconformities and disconformities (sequence boundaries). Each of the depositional sequences shows dis-

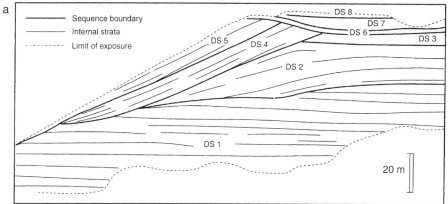

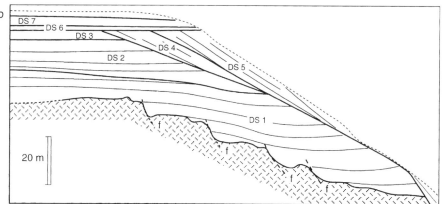

Depositional sequences, facies and environments 279

Figure D6.7 Photograph and interpreted geological sketch of Wadi Kharasa (eastern face) illustrating stratigraphic geometries, depositional sequences (DS) and bounding surfaces. (For location see Figure D6.2.)

tinctive geometries and facies which allow them to be correlated around the entire north-east and south-east platform margin (Cross, 1996). A progressive lateral change in geometry and facies occurs within each of the sequences, however, a three-dimensional correlation is facilitated by the close spacing of wadi sections around this platform margin (Figure D6.2).

The members and depositional sequences exposed along the eastern footwall platform margin can be traced for a distance of approximately 1–2 km into the platform before wadi sections rise, and beyond the outcrops towards the south-west or where basement rises to the surface. However, Wadis Bali'h and 3 expose the most complete section across the platform (Figure D6.8) and the Bali'h Member can be directly correlated into the hangingwall dip slope. South-westerly correlation of other units is hindered by the presence of the intervening basement which formed a footwall island

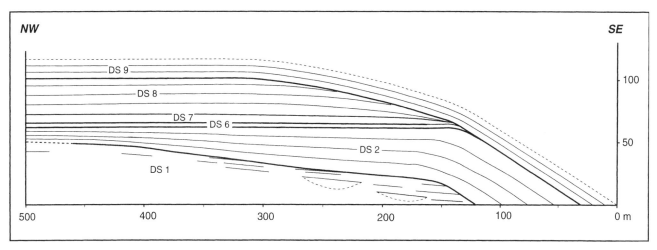

Figure D6.6 Photograph and interpreted geological sketch of (a) Wadi 3 (southern face), and (b) Wadi 2 (northern face) illustrating stratigraphic geometries, depositional sequences (DS) and bounding surfaces. Note Precambrian basement outcropping in Wadi 2. (For locations see Figure D6.2.)

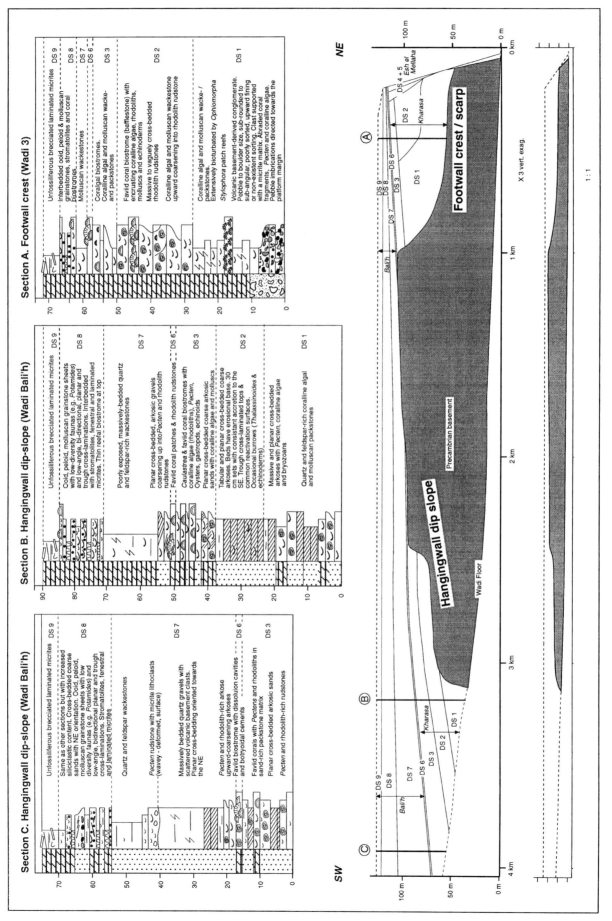

Figure D6.8 Composite cross-section through Wadi Bali'h and Wadi 3 (for locations see Figure D6.2), illustrating across platform correlation of depositional sequences (DS) described in text. Facies and lithologies of depositional sequences are described in logs A, B and C (for time equivalent log D from south of platform see Figure D6.9). True to scale section at base of figure.

during mid-Miocene times. Obtaining precise and complete correlation of depositional sequences and their boundaries between the hangingwall dip slope and the footwall in Wadi Bali'h, is achieved firstly by hanging the stratigraphy from the base of the Bali'h Member and then by the extrapolation of stratigraphic thicknesses, altitudes and dips of bounding surfaces from known points along the footwall margin (Figure D6.8). Furthermore, in spite of some lithological differences between the two sides of the fault block, there are characteristic facies and sequence geometries which facilitate the correlation of these units. The correlation of depositional sequences between the north-east and south-east platform margins and the south-west hangingwall dip slope has been achieved by walking out and correlating each side of each wadi section with the aid of large photomosaics (Cross, 1996) and is an obvious prerequisite to understanding the three-dimensional evolution of the platform.

Platform sediments exposed on Abu Shaar comprise a great variety of carbonate and siliciclastic facies many of which have been described and interpreted by Cofer *et al.* (1984), Burchette (1988), and James *et al.* (1988). Each of the James *et al.* (1988) four members are described and then for a detailed treatment are subdivided into newly defined, local, depositional sequences (DS 1 to DS 9) whose character generally varies according to their structural position.

The Kharaza Member

The Kharaza Member rests unconformably on pre-rift Precambrian basement (Figures D6.(b) and D6.8). On the footwall of the block the lower bounding surface of the Kharasa Member is a low-angle, north-easterly, basinward sloping, angular unconformity (Figure D6.6(b)). It is locally terraced and cut by small synsedimentary extensional faults (e.g. Wadi 2, Figure D6.2). Along the footwall margin, the basement is highly fractured and filled with micrite containing planktonic foraminifera and brecciated basement pebbles (cf. Burchette, 1988). Across the hangingwall dip slope the unconformity has a profile which steps down towards the south-west. A gentle slope of approximately 1° is interrupted by two steps in Wadi Bali'h each of which is approximately 20 m in height, where the unconformity slopes at up to 30° (Figure D6.8).

On the north-east footwall the Kharaza Member varies from 0 m to a maximum exposed thickness of *c.* 100 m in an eastward direction. To the south-east, the top of the Precambrian passes into the subsurface north of Wadi Kharaza (Figure D6.2) and thus the base of the member is not exposed. This member can be subdivided into three separate unconformity/disconformity-bounded depositional sequences (DS 1 to DS 3) (Figures D6.6–D6.8).

Depositional Sequence 1

In the footwall exposures in north-eastern Abu Shaar the lowest part of the Kharaza Member (DS 1) is up to 70 m thick and has an aggradational geometry with subhorizontal strata which onlap the footwall crest towards the south-west (Figure D6.6). The base of DS 1 is marked by a basal conglomerate composed of subangular basement clasts. This clastic base, up to 15 m thick, passes rapidly up into carbonate facies dominated by highly bioturbated (*Ophiomorpha*), mollusc and coralline algal wackestones and packstones with minor thickets of the coral *Stylophora* (Plate 20(a)).

On the hangingwall dip slope the lowest exposed part of DS 1 consists of strata dipping towards the south-west at up to 4° (Figure D6.8). In this position, DS 1 has a maximum exposed thickness of 20 m and onlaps the basement towards the north-east (Figure D6.10). As with the footwall side, the base of this depositional sequence is marked by a basement-derived conglomerate. This facies passes rapidly up into a mixed carbonate and siliciclastic succession of interbedded quartz and feldspar-rich, bioclastic grain-/rudstones and coarse arkoses.

In Wadi Treifi (see above) the lowest parts of the conglomerates include several spectacular palaeosols which pass up into marine siliciclastics and carbonates. The lower non-marine facies thin southwards where they are thought to correlate with 10 m of non-marine conglomerates (DS 1) on the north-eastern entrance of Wadi Bali.

Depositional Sequence 2

On the north-eastern footwall exposure the base of DS 2 is marked by a low-angle unconformity which slopes down eastwards off the platform margin (Figures D6.6, D6.8). It passes south-westward as a subhorizontal disconformity and is truncated approximately 0.6 km from the platform margin by the westerly inclined lower bounding surface of DS 3. DS 2 has a relatively consistent thickness of 25 m at the north-east platform margin with an eastward progradation which downlaps its lower bounding surface (Figure D6.6). Away from the platform margin it has an aggradational geometry and rests conformably on DS 1 (Figures D6.10, D6.11). DS 2 is dominated by coralgal facies with faviid and poritid baffle- and framestones (Figure D6.8). These occur either as massive beds with occasional large-scale, sigmoidal shaped reefal units oriented towards the platform margin, or thinner biostromal units. Coralline algal-rich facies are best displayed in thick, vaguely cross-bedded rhodolith rudstones (Plate 20(b)).

On the hangingwall dip slope (Figures D6.8, D6.10) the base of DS 2 is locally channelised with a basement pebble and intraclast conglomerate fill (Wadi Bali'h). DS 2 is approximately 15 m thick with distinctive coarse arkoses (Figure D6.8, section B). These comprise a number of tabular and planar, cross-bedded sand sheets (Plate 20(c)). Individual beds are up to 4 m thick with generally sharp, planar, locally erosional bases and are often several hundred metres in lateral extent. Cross-bedding occurs in cosets of approximately 30 cm in thickness which have a relatively consistent accretion direction to the south-east.

To the south-east, DS 2 thickens to 45 m in Wadi Kharaza, and shows a progressive increase in coralgal reef facies together with an aggradational and progradational geometry (Figure D6.7). A zonation of coral genera and morphologies reflect the transition from back-reef to lower reef-front through intermediate zones as a function of palaeobathymetry (Perrin *et al.*, 1995; Perrin *et al.*, this volume).

Depositional Sequence 3
In exposures of the north-eastern footwall the base of DS 3 is marked by a sharp, planar, low-angle unconformity surface which slopes to the south-west away from the platform margin (Figures D6.6, D6.8). DS 3 increases from 7 to 15 m in a south-westerly direction (Figure D6.8). In the platform interior, this depositional sequence consists of subhorizontal strata which onlap and downlap DS 2. Coralgal facies of DS 3 are broadly similar to those described in DS 2. The inclined sequence boundary which marks the base of DS 3 in the north-east of the platform is not present at the south-east margin in Wadi Kharasa.

On the hangingwall dip slope the lower bounding surface of DS 3 is inclined to the south-west at approximately 2° (Figures D6.8, D6.10, D6.12) and truncates the more steeply dipping upper strata of DS 2 at a low angle. This inclined surface is also onlapped by the lower strata of DS 3 in a north-easterly direction (Figure D6.10). In Wadi Bali'h, DS 3 thickens towards the south-west from 11 m to 17 m over a distance of less than 1 km and is dominated by coarse bioclastic (rhodolith and *Pecten*) rudstones with occasional faviid coral biostromes (Figures D6.8, D6.10). Coral biostromes are best developed adjacent to the basement onlap (Figure D6.12) where they are interbedded with conglomerates. They have an elongate (north-east/south-west) profile wedging out to the south-west. Coralgal facies on the hangingwall dip slope are comparable to those developed along the footwall except for the notable absence of columnar *Porites*. Some display dissolution cavities lined with early marine cements with fissures, desiccation structures and internal sediments which are interpreted as top DS 3 karstic features (for details see Purser, this volume).

On the hangingwall dip slope there is a marked lateral variation from more carbonate-rich facies to more siliciclastic-rich facies (section C, Figure D6.8) in the south-west (Wadi Bali'h) extremities of the Kharaza Member.

Interpretation of the Kharaza Member
Depositional Sequence 1
In the footwall the marked angular unconformity at the base of the DS 1 represents the syn-rift unconformity with erosion of the underlying Precambrian basement and possibly pre-rift strata. The basinward dipping unconformity surface in the footwall and its hangingwall dip-slope equivalent suggests that relative sea-level fall on the footwall was in response to extensional north-east–south-west faulting and fault-block rotation.

Following this initial tectonic phase, the generation of marine accommodation space for the deposition of DS 1 on both the footwall and the hangingwall dip slope must have occurred during a platform-wide relative sea level rise. Overall, the aggradational geometries of DS 1 indicate platform sedimentation was able to keep up with this relative rise.

Conglomeratic facies represent the initial erosion of the footwall crest prior to carbonate platform development. They indicate alluvial to beach environments adjacent to the basement fault block and are similar to present-day sediments occurring along the Esh el Mellaha block (cf. Roberts and Murray, 1988b). In Wadi Treifi, the isolated occurrence of Eocene chert cobbles indicates either transport from the west of the Esh Mellaha–Abu Shaar ridge or a previous local occurrence of pre-rift Eocene, that has since been eroded.

On the footwall, the rapid vertical transition to micrite-rich facies indicates deposition within relatively low-energy, sheltered conditions probably in a back-reef setting (cf. interpretation of the Esh el Mellaha Member, below).

Sedimentation in the hangingwall slope was in a mixed siliciclastic and carbonate setting. Facies exposed in Wadi Bali'h indicate that both spatially and temporally, carbonate–siliciclastic transitions occur over relatively short distances (metre scale). Carbonate facies represent a background of shallow-marine, *in situ* carbonate production while siliciclastics represent periodic influxes. Well-developed cross-bedding oriented towards the south-east indicates a north-west provenance and longshore transport parallel to the crest of the fault block. The deposition of carbonate facies was precluded by each new phase of siliciclastic supply which buried the carbonate interval. Following siliciclastic deposition, reduction in supply allowed stabilization of the substrate and growth of carbonate producers.

Deposition of DS 1 seems to have occurred during a phase of tectonic inactivity. The occurrence of micritic carbonates and the absence of widespread platform margin facies at the footwall extremity of the block indicates that initially the platform extended beyond the present bordering fault zone (Burchette, 1988). This may have been achieved where the preceding conglomerates bridged the fault scarp and adjacent basin (Burchette, 1988).

Depositional Sequence 2
A relative sea-level fall and platform margin erosion on the footwall is indicated by the unconformity surface at the base of DS 2 and the correlative disconformity in the hangingwall slope. Accommodation for DS 2 was generated by a relative sea-level rise which was accompanied by a marked platformward-shift in facies along the footwall margin. Facies relationships and stratal geometries indicated that the platform margin responded to sea-level rise by back-stepping followed by basinward progradation (Figure D6.6(a)). The abundance of a diverse coralgal framework along much of the footwall margin is indicative of a shallow-marine, high-energy shelf margin environment. Rhodolith rudstones represent deposition in low-relief shoals along the platform margin and across the platform top within a relatively high-energy current-swept environment.

Deposition of DS 2 on the hangingwall dip slope occurred within a mixed carbonate and siliciclastic setting similar to that inferred for DS 1. The context of planar and tabular cross-bedded arkoses indicate the migration of sand sheets in a tidally influenced off-shore setting. The unidirectional palaeocurrent pattern indicates that strong tidal currents were probably accentuated by the narrow gulf produced by the block and basin topography in the Esh el Mellaha basin (cf. Roberts and Murray, 1988a).

The low-angle platform relief (DS 1) around the south-east of Abu Shaar (Wadi Kharaza) allowed the growth of a more progradational (DS 2) platform margin (cf. Hurst, 1987; Burchette, 1988; Gawthorpe and Hurst, 1993). The platform-wide correlation of DS 2, its constancy of thickness on the hangingwall dip slope, and its lower sequence boundary suggests that relative sea-level change at this time was regional rather than in response to local fault-block rotation.

Depositional Sequence 3
Along the north-east footwall, a relative sea-level fall is inferred by the erosion surface seen at the base of DS 3. This is best displayed where DS 3 oversteps DS 1 and DS 2 (Figure D6.8). The facies comprising DS 3 are similar to those of the preceding sequence with a coralgal reef rimming the platform in the north-east and an emergent basement island shoreward of the platform margin. Mixed carbonate and siliciclastic facies present on the hangingwall dip slope are similar to those of DS 2 and DS 3, albeit with the widespread development of coral biostromes (Figure D6.9).

Following the broad expanse of shallow marine sediments which filled accommodation space in DS 2, DS 3 shows significant thickening of strata towards the south-west. This wedge-shaped accommodation space, filled with shallow-marine facies is considered to have resulted from fault-block rotation generating a rise in relative sea level prior to or during DS 3 sedimentation. This is supported by the erosion surfaces and basal onlap around the footwall island in the northern parts of the platform in Wadi Bali'h (Figures D6.8, D6.10). Laterally extensive coral biostromes on the hangingwall dip slope indicate that block rotation was slight and that sedimentation was able to keep up with relative sea-level rise.

The absence of DS 3 in Wadi Kharasa reflects the south-easterly decrease in fault displacement and therefore block rotation towards the transfer zone. This spatial relationship produces a complex stratigraphy whereby DS 3 developed above an erosion surface in the north-east of the platform while sedimentation continued with the coeval development of DS 2 in the transfer zone setting to the south.

The Esh el Mellaha Member

The Esh el Mellaha Member consists of two depositional sequences (DS 4 and DS 5) which are confined to the north-east footwall platform margin between Wadi Bali'h and Wadi 7 (Figures D6.2, D6.6, D6.8). The correlation of the Esh el Mellaha Member from the platform margin across the platform top by James *et al.* (1988) is unsupported by our fieldwork as its upper surface is always truncated so that no equivalent platform-top strata are preserved. The James *et al.* (1988, their Fig. 10b) platform top unit of the Esh el Mellaha Member correlates with our DS 6 (base of the Bali'h Member; Figure D6.8).

The Esh el Mellaha Member also lacks a correlative equivalent along the south-eastern platform margin (Figure D6.7). In Wadi Kharasa, James *et al.* (1988) correlate the Esh el Mellaha Member (their Fig. 10a) with steeply dipping fore-reef slope strata. We have traced these reefal slopes around the entire platform edge and found them to correlate with the fore-reef slopes of DS 2 (Figures D6.6, D6.7). Burchette's (1988) correlation of this Member (his 'sequence 2') to the south-east margin is also not supported. His 'sequence 2' in Wadi Kharaza correlates with our DS 7 and DS 8 in the Bali'h Member (Figures D6.5, D6.8). In this respect, the three Members of James *et al.* (1988) do not strictly correspond with the three sequences of Burchette (1988) (Figure D6.5).

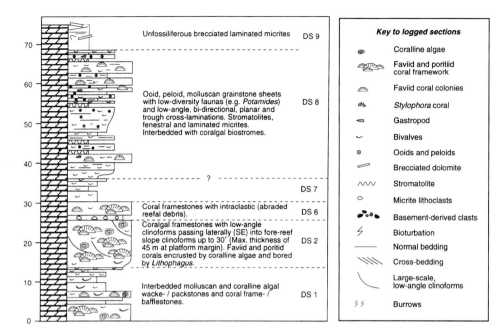

Figure D6.9 Log D (Figure D6.2) through section in eastern face of Wadi Kharasa (Figure D6.7) illustrating facies, lithologies and depositional sequences described in text and Figure D6.8 (including key to Figure D6.8).

Depositional Sequence 4
At the eastern platform margin, DS 2 and DS 3 are truncated by an angular unconformity at the base of DS 4 (Figures D6.6, D6.8). The shape of this sequence boundary is laterally variable, since in Wadi 2 it is sharp and planar while in Wadi 3 it has a listric profile (cf. Figures D6.6(a) and (b)). This sequence is wedge-shaped, thinning (from 15 to 0 m) both platformward and basinward. Where accessible, DS 4 is composed of massive coralgal framestones, although locally it displays internal bedding which is inclined and tapers towards the platform margin at up to 25°. Bedding shows a progradational geometry which downlaps the lower sequence boundary (Figure D6.6).

Depositional Sequence 5
The lower bounding surface of this sequence has a similar geometry to the base of the preceding sequence. It is sharp and planar rather than concave-up and is locally eroded into terraces (Wadi 5, Abu Shaar). The basal surface of DS 5 passes down slope and is bevelled at shallow levels at about 25° (Figure D6.6(b)). At lower levels the surface cuts through both DS 1 and the Precambrian basement and passes into the subsurface of the coastal plain at about 80°.

DS 5 is composed of thin clinoform wedges which downlap DS 4 and steepen towards the platform margin (Figure D6.6). The facies grade downslope from *in situ* coralgal reefs to a mixture of fore-reef blocks and finer bioclastic grain/rudstones. Mollusc, *Halimeda* or coralline algal clasts all dominate at different levels (Plate 21(a)). Coarse redeposited facies consist of chaotically and unsorted blocks (centimetre to metre across) of reefal sediment in a finer-grained matrix (Plate 21(b)). Bed contacts are either planar or concave-up, erosional and internal truncation surfaces are common. Finer grainstone facies occasionally display inverse grading above sharp and locally channelled surfaces. Beds often show lens-shaped geometries producing an intense interdigitation of facies. These slope deposits are dealt with in detail by Purser and Plaziat (this volume).

Interpretation of the Esh el Mellaha Member
The development of planar, steeply inclined sequence boundaries at the bases of the Esh el Mellaha Member has been previously attributed to fault-related margin collapse (Burchette, 1988; James *et al.*, 1988) and this view is supported by our work (Figures D6.11, D6.13(a)). Facies deposited on the fore-reef slope of the Esh el Mellaha Member can be broadly grouped into two categories based on depositional textures. Coarse coralgal/reef-rock rud-/floatstones represent gravity induced mass-transport debris deposited during local episodes of reef-front and fore-reef slope instability. Finer bioclastic grain-/rudstones are the result of grain-flows in which the platform top provided a constantly replenished supply of loose sediment readily available for off-platform transport (Figure D6.13(b)).

Following platform margin collapse to form the surface at the base of DS 4, several further collapse events are indicated by internal (DS 4) truncation surfaces, and the base of DS 5. Once initial collapse

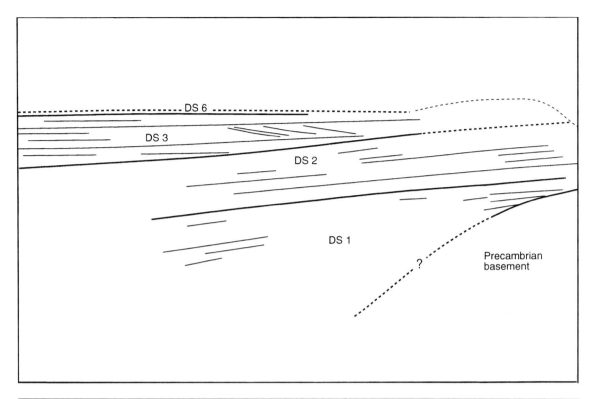

Figure D6.10 Stratigraphic geometries and depositional sequences in Wadi Bali (located adjacent section B, Figure D6.8). Note onlap on to Precambrian basement and south-westward thickening of DS 3.

had occurred, the platform margin was increasingly susceptible to further instability by the inclined sediment wedges. The accommodation space represented by these units does not necessarily result from a change in relative sea level.

286 Tectono-sedimentary evolution of rift margin carbonate platform

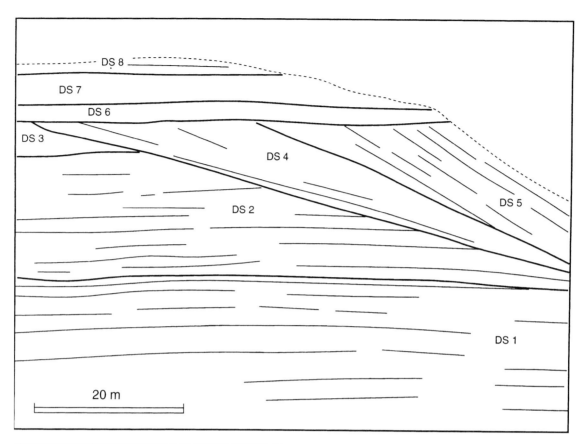

Figure D6.11 Detail of Figure D6.6(b) illustrating stratigraphic geometries and depositional sequence relationships of DS 1 to 8 at the footwall platform margin.

Whereas the sequence boundary at the base of DS 5 obviously post-dates DS 4, the exact timing of the lower, more steeply inclined surface is more difficult to constrain. The more steeply sloping basal surfaces of DS 4 and DS 5 both truncate the basement and the north-east margin of DS 1 and maybe superimposed on earlier erosion surfaces related to the tectonically active phase prior to DS 3. Truncation of the platform and basement along the margin fault zone has removed peripheral facies in DS 1 producing an anomalous situation whereby muddy, back-reef facies now constitute the seaward edge of platform (see DS 1) as in Read's (1985) 'erosional escarpment-type rimmed-shelf margin'.

The occurrence of the Esh el Mellaha Member is restricted to the steep, fault-controlled north-east platform margin (Figures D6.6–D6.8). To the south-east (Wadi Kharaza) the Esh el Mellaha Member is absent, reflecting the change towards a platform margin away from the main active north-west–south-east fault system to a margin of lower relief. It seems likely that in this south-east position the border fault was blind and produced a low amplitude flexure rather than a fault scarp at the surface (Hurst, 1987; Burchette, 1988; cf. Hurst and Surlyk, 1984).

The absence of the Esh el Mellaha Member within the hangingwall dip-slope stratigraphy (Figure D6.8) implies that footwall margin collapse was not in direct response to fault-block rotation. The section across the platform (Figure D6.8) indicates that a post DS 3 relative sea level rise would have resulted in some sedimentation in the hangingwall dip-slope area. Similarly if sequence boundaries at the base DS 4 and DS 5 developed in response to relative sea level fall then platform-wide correlative erosion surfaces would be expected in both the hangingwall dip slope (Wadi Bali'h) and south-east platform margin (Wadi Kharasa).

The localized footwall distribution of DS 4 and DS 5 precludes regional relative sea level change as a controlling factor in their development. If DS 5 developed during a relative sea level rise following platform margin collapse then platform-wide correlative equivalents would be expected. These are not conclusively identified but may be represented by the upper levels of DS 2 and DS 3 in the south-east (Wadi Kharaza) and south-west (Wadi Bali) respectively rather than the development of a new depositional sequence (DS 4).

The Bali'h Member

The Bali'h Member can be sub-divided into three platform-wide depositional sequences (DS 6 to 8) (Figures D6.6–D6.8, D6.11). On the footwall, the base of the Bali'h Member (base of DS 6) is marked by an erosion surface, which today dips gently to the south-west at approximately 1°, and subsequent beds that overstep DS 3, DS 4 and DS 5. Locally, this sequence boundary has channels up to 1 m deep and 5 m in width that are filled with intraclastic breccia-conglomerates. This surface represents a considerable break in platform evolution because some, if not all, of the allochthonous platform slope deposits of DS 4 and 5 must have been derived from a platform top, of which nothing is preserved (Figure D6.11, D6.13).

Depositional Sequence 6
On the footwall, DS 6 has a relatively constant thickness of 2.5 m and comprises a massively bedded, aggradational sequence composed of a single coral biostrome. On the south-east face of Wadi 2 (Figures D6.6(b), D6.11) it displays low-angle progradation towards the platform margin. At the margin it consists of almost entirely columnar *Porites* framestone but passes into the platform interior as faviid framestones and bafflestones rich in coralline algae and molluscs (Figures D6.8, section A, B, and C; Figure D6.10).

This sequence is particularly distinctive because in the footwall it oversteps earlier sequences and is preferentially mineralized with white calcite/dolomite and black iron oxides. The cements develop layers up to 20 cm thick on top of the biostrome and also occur as botryoidal linings to cavities following coral dissolution (Clegg *et al.*, this volume). Cements are often interlayed with geopetal micrite within such cavities.

On the hangingwall dip slope, a 2 m thick coral biostrome of similar geometry and facies forms a correlative equivalent. It is well exposed in Wadi Bali'h (Figures D6.8, D6.12) and its upper surface forms a prominent light grey plateau with faviid coral colonies and patch reefs up to 5 m wide and 1 m in height within a surrounding rhodolith rudstone.

Depositional Sequence 7
DS 7 is underlain by a planar, locally erosional disconformity surface which slopes gently towards the south-west (Figures D6.8, D6.12). DS 7 shows marked thickening from the footwall and across the hangingwall dip slope. In the north of the platform, DS 7 thickens from 4 m close to the platform margin to 30 m, 4 km to the south-west (Wadi Bali'h; Figures D6.8, D6.12).

At the platform margin, this sequence is composed of beige-coloured bioclastic wackestones with a low quartz and feldspar content (~5%). However, towards the platform interior it becomes progressively more siliciclastic-rich with sub-angular quartz and feldspar grains comprising up to 30% floating in a micrite matrix. In the hangingwall dip slope (Wadi Bali'h), the base of this sequence consists of a variable thickness of cross-bedded *Pecten*-rich gravels and rudstones (Figure D6.8, sections B, C). These are thinnest where they onlap the basement

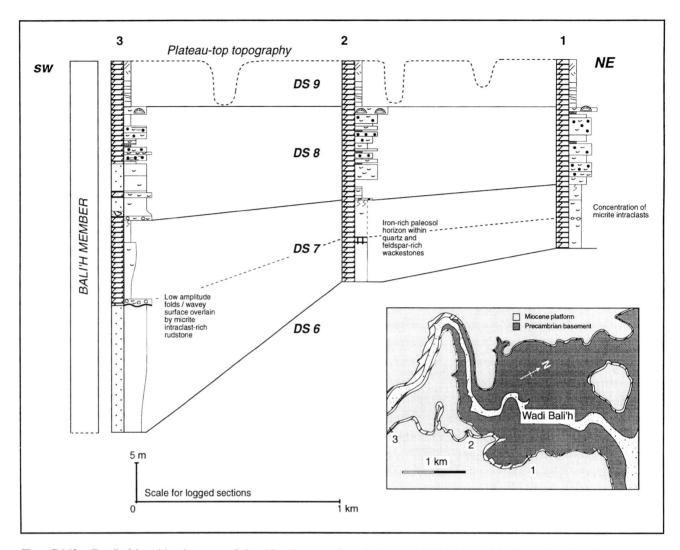

Figure D6.12 Detail of depositional sequences 7, 8 and 9 and sequence boundaries exposed in hangingwall dip-slope of Wadi Bali showing south-westward thickening and nature of hangingwall sequence boundaries. For location see map inset.

and thicken to 15 m over a distance of 1 km to the south-west. Cross-bedding indicates no consistent accretion direction. This *Pecten*-rich unit is overlain by massive to crudely cross-bedded coarse arkosic sands which indicate a north-easterly palaeocurrent direction. In the south-west part of Wadi Bali'h, DS 7 is split into two units by a deformed surface overlain by an intraclastic conglomerate (Section C at 40 m, Figure D6.8). This surface marks a change in the hangingwall dip slope from coarse siliciclastic gravels below to fossiliferous carbonates above and, as such, marks an important surface within DS 7. Across the hangingwall dip slope this surface can be traced towards the north-easterly platform margin where it passes into an iron-rich palaeosol horizon. However, a correlative equivalent of this surface does not exist to the east of the basement island within the footwall stratigraphy so in this chapter it is not taken as a sequence boundary. An alternative view is presented in Purser *et al.* (this volume) and Plaziat *et al.* (this volume) who interpret this surface as a sequence boundary and correlate it with the erosion surface at the base of our DS 6 in the footwall.

Depositional Sequence 8
Along the footwall platform margin the base of DS 8 is marked by a low-angle disconformity which slopes towards the platform interior at less than 2° (Figures D6.8, D6.12) and is represented by a hardground surface with borings of *Botula*. DS 8 thickens from 4 m close to the footwall platform margin to 25 m, 6 km to the south-west (Wadi Bali'h) (Figure D6.8). Even though the eastern edge of the platform is truncated, apart from coral biostromes there is little evidence of significant platform margin buildups or sand shoals. However, along the southern margin of Abu Shaar,

thicker coral biostromes form clearly defined platform margin buildups.

DS 8 comprises thinly bedded aggradational strata which are laterally continuous for 6 km across the platform top. The most characteristic facies include: laminated and fenestral micrites, stromatolites, various peloid, ooid, restricted mollusc, miliolid/alveolinid grainstones and coralgal biostromes. Oolitic and peloidal grainstone facies, often with *Potamides* gastropods and fenestral textures, occur as laterally continuous sheets across the platform and display small-scale, planar and trough cross-lamination (ripples) (Plate 21(c)). Stromatolites (up to 50 cm thick) form semi-continuous sheets, some comprising 10 cm diameter columns up to 20 cm high, or form secondary domes within larger domes up to 1 m high and 2 m in diameter (Plate 21(d)). This sequence also includes several coralgal biostromes which attain a maximum thickness of 4 m at the platform margin (e.g. Wadi 3) and thin towards the south-west. The siliciclastic component of these facies increases towards the south-west and the north-east. Planar, cross-laminated, fine quartz sands and marls are the predominant sediment types within the south-western part of the platform.

Several metres of talus deposits at the base of the footwall platform margin are probably the equivalent of this sequence since they exhibit similar restricted facies to those described from the platform top. Such facies include the stromatolites described by El Haddad (1984), Monty *et al.* (1987) and James *et al.* (1988), together with the distinctive fore-reef slope pisolites described by Aissaoui *et al.* (1986), James *et al.* (1988) and Purser (this volume).

Interpretation of the Bali'h Member
There is a marked change in facies from the Kharaza and Esh Mellaha Members to the Bali'h Member. Platform-rimming reefs and fully marine bioclastic sands and gravels are no longer dominant (except along the southern margin) and an upward transition through depositional sequences 6, 7 and 8 records increased restriction and evidence of peritidal conditions and elevated salinities.

Depositional Sequence 6
James *et al.* (1988) place this depositional sequence at the top of the Esh el Mellaha Member and refers to it as a 'reef veneer'. Although it shares facies similarities with the underlying Esh el Mellaha Member (DS 4 and DS 5) it is separated by a major erosion surface (Figures D6.11, D6.13), has a different geometry (Figure D6.8), and its origin is not associated with the platform margin collapse represented by the Esh el Mellaha Member.

The planar erosion surface at the base of the sequence corresponds to a platform-wide relative sea level fall in which the platform top (Kharaza and Esh el Mellaha Members) underwent peneplanation and local incision. Subsequent relative sea level rise of only a few metres was associated with aggradation and limited progradation of the reef of DS 6 in a shallow-marine setting. On the hangingwall dip slope the laterally extensive equivalent of the poritid biostrome confirms platform-wide shallow marine conditions. This unit therefore acts as a marker horizon indicating an original horizontal surface which is now rotated by about 1°.

Haematite and calcite mineralization within the footwall extremity of this sequence has been attributed to a karstic exposure surface (El Haddad *et al.*, 1984; Aissaoui *et al.*, 1986; James *et al.*, 1988; Burchette, 1988). This interpretation seems unlikely because the presence of calcite and haematite cements in dissolved corals within otherwise completely dolomitized carbonates indicates the mineralization event was post-dolomitization (Clegg *et al.*, this volume). These same cements also occur within fractures cross-cutting DS 9 in

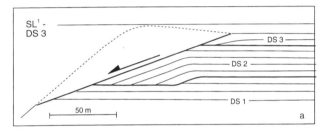

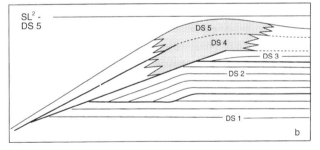

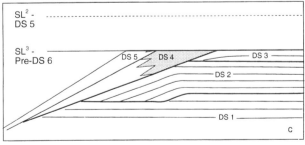

Figure D6.13 Sketches based on outcrops exposed in Wadi 3 (Figure D6.6) to illustrate evolution (a–c) of depositional sequences 3, 4, 5 and 6 on the footwall of Abu Shaar. (a) Fault-related slump scar truncating DS 1, 2, 3, note no relative sea-level change from SL^1 is required to generate this erosional surface; (b) deposition of DS 4 and 5 with platform margin reefs and fore-reef slopes, requiring a relative rise in sea level (SL^2); (c) relative sea-level fall to SL^3 to truncate platform top equivalents of DS 4 and 5 prior to deposition of DS 6 on DS 3, 4 and 5.

Wadi 3. Recent work has revealed a microlayering of dolomite and calcite whose isotopic properties suggest a hydrothermal origin (Clegg et al., 1995) and that this is probably post-Miocene. The local occurrence of mineralization along the footwall, referred to previously as evidence for possible 'tectonic tilting' by James et al. (1988), also supports a hydrothermal origin associated with the margin fault-zone.

Depositional Sequence 7
At the platform margin, the base of DS 7 coincides with a relative sea level fall which appears to be in response to fault-block rotation since DS 6 is tilted towards the south-west and DS 7 shows a marked south-westerly thickening. A subsequent rise in relative sea level was again associated with platform sedimentation as indicated by the aggradational geometry of DS 7.

Facies within this sequence reflect deposition in a shallow-marine environment. The footwall area comprises bioclastic wackestones indicative of an inner platform setting which become more siliciclastic to the west. As with some of the earlier sequences, it is likely that the platform margin facies are not preserved and have been truncated by subsequent fault movement (see above interpretation of Esh el Mellaha Member).

To the west of the basement island, sand-rich, cross-bedded, *Pecten* and rhodolith rudstones reflect deposition in a series of shoreface bars. In the south-west, cross-bedded, coarse sands indicate a provenance from the south-west rift margin at a time when the Esh el Mellaha basin must have been nearly filled. A southwest provenance for siliciclastics in this sequence not only indicates the development of a subhorizontal shelf but precludes the existence of a rimmed shelf margin in the subsurface of the Esh el Mellaha basin. The development of a shallow mixed carbonate-siliciclastic shelf attached to a continental margin is comparable to fringing platforms of the present-day Gulf of Suez (Purser et al., 1987; this volume).

Depositional Sequence 8
The base of DS 8 corresponds to a minor sea level fall prior to a relative rise in sea level. The south-westerly thickening of DS 8 also implies synsedimentary rotation of the Abu Shaar platform.

Facies comprising DS 8 reflect platform-wide deposition in peritidal environments. The grainstone sand sheets, extending across the platform top are not limited to the high-energy platform margin environment. These relatively thin, locally mounded, small-scale, cross-bedded facies indicate a platform interior sand shoal environment. However the associated low-diversity–high abundance molluscan faunas indicate that circulation was periodically restricted. The absence of structures reminiscent of emergence and desiccation also suggest that stromatolites developed within a sub-tidal environment. The progressive increase in siliciclastic facies towards the south-west reflects the proximity of the westerly basement outcrop source areas and reduced bathymetry within the Esh el Mellaha basin.

The Chaotic Breccia Member

This member comprises one depositional sequence, DS 9.

Depositional Sequence 9
The base of DS 9, along with much of the sequence, is poorly exposed on Abu Shaar. The relationship between DS 9 and the underlying sequences can only be seen in Wadi Kharasa where the platform margin of DS 2, 6, 7 and 8 are eroded (Figure D6.7). In most areas it is an aggradational sequence which varies in thickness from 5 m in the north of the platform to 15 m along the south-east margin.

DS 9 consists entirely of unfossiliferous, laminated dolomicrites which are often folded and brecciated (Figure DS 6.9). Individual folds, which may also affect underlying, marine sediments of DS 8, attain 10 m in amplitude, notably in the south-west parts of the platform. In thin-section, these unfossiliferous dolomicrites exhibit pseudomorphs of gypsum and celestite.

Interpretation of Depositional Sequence 9
Evidence of a relative sea level fall prior to the deposition of DS 9 is displayed at the margin of Wadi Kharaza where erosion occurs of DS 2, 6, 7 and 8 (Figure D6.7). Sequence relationships indicate that DS 9 developed following a subsequent relative sea level rise allowing it to prograde over the truncated margin of DS 8.

The dominance of unfossiliferous micrite indicates deposition within a low-energy, restricted environment. The widespread occurrence of this characteristic laminite facies over the entire platform and possibly some of the footwall slopes, indicates regional restriction prior to the onset of late Miocene evaporitic sedimentation (Orszag-Sperber et al., this volume). A detailed treatment of this facies and the nature of its deformation is given in Plaziat and Purser (this volume).

DISCUSSION

Tectonic controls on the development of the Abu Shaar platform

Several lines of evidence indicate that synsedimentary tectonism was an important factor for controlling the distribution and evolution of facies, and depositional sequences on the Abu Shaar platform.

Tectonic setting

The overall control on the distribution of facies and environments exerted by fault block/half-graben rift basin morphologies has been well documented, particularly for siliciclastic depositional systems (e.g. Leeder and Gawthorpe, 1987; Gawthorpe and Hurst, 1993; Leeder and Jackson, 1993). Abu Shaar is a Miocene carbonate platform which developed in the south-west Gulf of Suez during the later stage of active rifting at a time when fault block/half-graben rift morphology strongly controlled sedimentation (e.g. Burchette, 1988; Purser *et al.*, 1990). Clearly the location of the Abu Shaar platform on the Esh el Mellaha and Abu Shaar fault blocks, together with its overall geometry, implies a close tectono-sedimentary relationship.

Thickness variations

On a regional scale, the Abu Shaar platform occurs in a stratigraphic section which is known from seismic evidence to increase in thickness considerably from north-east to south-west across the Esh el Mellaha basin (Bosworth, 1995). In the north-east of Abu Shaar, footwall platform thickness is approximately 80 m, whereas to the south-west, Bosworth (1995) calculates a maximum thickness of 1600 m in the hangingwall of the Esh el Mellaha basin.

On a more local scale, north-east to south-west, thickening across the platform of depositional units occurs three times within the Abu Shaar stratigraphy (Figure D6.8). The south-westerly thickness increases in DS 3, DS 7 and DS 8 are achieved by repeated thickening of individual strata, and in the case of DS 3, north-easterly onlap. Since all of the sediments on Abu Shaar have been deposited at, or close to sea level, such thickness variations confirm repeated synsedimentary tilting of the fault block and platform. Episodes of synsedimentary tilting are minor and probably come towards the end of the main phase of rifting represented by the major Precambrian–pre-Miocene unconformity (cf. Burchette, 1988). DS 2 and DS 6 show no thickness changes in this north-east to south-west direction and the lower and upper surfaces of DS 1 and DS 9 respectively, are not preserved and thicknesses cannot be observed.

All depositional sequences that are laterally continuous around the eastern footwall margin of Abu Shaar also show thickness increases along strike towards the south-east indicating increased accommodation space, due to reduced footwall uplift, towards the transfer zone.

Tilting of beds

Platform strata on Abu Shaar have extremely gentle dips to the south-west. However, the extensive outcrops and wadi sections make it possible to physically correlate stratigraphic contacts and measure from oriented photomontages changes in dip from the base of the section to the top. Lower sequences (DS 1 and 2) of the Kharaza Member dip at up to 4° to the south-west in the hangingwall dip slope and these decrease to 3° in DS 3 (Figure D6.8). The uppermost units of the Bali'h Member (DS 7 to 9) dip at 1–2° to the south-west. These upward decreases in dips are visible over the 5–6 km outcrop on Abu Shaar. This pattern of upward-decreasing south-westerly dips confirms minor, but repeated, tilting of the Abu Shaar fault block during the Miocene.

Sequence boundaries

The recognition of nine unconformity/disconformity-bound depositional sequences indicates that the evolution of the Abu Shaar platform was interrupted by relative sea level changes. With the exception of DS 4 and DS 5, sequence boundaries identified on Abu Shaar correlate to relative falls in sea level and erosion of the platform margin.

Carbonate platforms sited on footwall crests commonly develop unconformities which relate to episodes of relative sea-level fall during footwall uplift (e.g. Leeder and Gawthorpe, 1987; Rosales *et al.*, 1994; Pickard *et al.*, 1994). However, since relative sea level fall can also be conditioned by more regional sea level changes it is unwise to correlate all of the Abu Shaar sequence boundaries with phases of footwall uplift.

The thickness variations described above indicate that accommodation for DS 3, DS 7 and DS 8 was generated by fault-block tilting. This would also imply that the lower bounding surfaces of these three sequences represent erosion associated with footwall uplift. Moreover, the correlative equivalents of these surfaces in the hangingwall dip-slope stratigraphy correspond to coeval deepening during relative sea level rise (cf. Bosence *et al.*, 1994; Gawthorpe *et al.*, 1994). This relationship is confirmed by the correlation of footwall unconformities with tilted flooding surfaces on the hangingwall dip slope (Figures D6.8, D6.10, D6.11).

Fault-related platform margin collapse

Local platform margin collapse represented by the Esh el Mellaha Member (DS 4,5) is associated with the steep fault-controlled escarpment along the north-easterly footwall. Episodes of marginal collapse at this time cannot be related to rotation because no accommodation space was generated on the hangingwall dip slope. Furthermore, there is no evidence of tilting after the collapse event since DS 6, the base of the Bali'h Member, is a shallow-marine biostromal coral unit of similar thickness and facies which has platform-wide constant thickness and bathymetry. Despite significant footwall erosion, all the evidence indicates that this was not a response to fault-block rotation. Although the slump scars at the bases of DS 4 and DS 5 are not necessarily eroded scarps of the main platform border fault, their occurrence along the north-east fault-bounded platform margin suggests that synsedimentary movements may have served as an important triggering mechanism on an already steep escarpment-type margin.

The lower scarp surface which truncates DS 1 and Precambrian basement equally is considerably steeper and is interpreted as the remnant of a synsedimentary fault scarp.

Platform stratigraphic geometries

The distinction between ramps and rimmed shelves and their progradation and aggradational characteristics can be used to identify an underlying tectonic control (Gawthorpe *et al.*, 1989). On Abu Shaar a relationship exists between the spatial occurrence of sequence geometries and their position with respect to the underlying fault block. Depositional sequences can be characterized into three groups based on their geometry: aggradational, progradational and local progradational wedges.

Aggradational sequences are the dominant group and emphasize how the topography/bathymetry of the underlying fault block controlled the lateral expansion of the platform along the north-east footwall margin. These sequences, rather than representing aggradation of a rimmed shelf along the footwall, represent the back-reef remnants of a platform margin that developed to the north-east of the present-day outcrops. Synsedimentary truncation by the block-bounding fault precluded the long-term progradation of the platform margin and resulted in the truncated escarpment and the preservation of back-reef facies adjacent to the faulted margin (cf. Burchette, 1988). Aggradation of sequences on the dip slope of the block was facilitated by the progressive increase in subsidence and therefore accommodation space towards the Esh el Mellaha basin. Unlike many published models (e.g. Leeder and Gawthorpe, 1987; Burchette and Wright, 1992), the development of a dip-slope platform margin on the Esh el Mellaha fault block was prevented by regular siliciclastic influx which infilled basin and dip-slope accommodation.

Although progradational sequences are not preserved along much of the footwall margin of Abu Shaar, a reduction in both the relief and the effects of synsedimentary truncation allowed their development along the south-east margin. Aggradational sequences on the north-east platform margin become gradually more progradational towards the south-east (e.g. DS 2). This relationship reflects the more ramp-like geometry reflecting the slope of the basement into the subsurface at a position where the fault block passes into a transfer zone in the south-east (Wadi Kharasa). The transfer zone not only promoted progradation but the reduced footwall uplift, emergence and erosion which accompanied synsedimentary fault movements in the north-east, also allowed greater aggradation of sequences.

Distribution and evolution of carbonate facies

There are major differences between the carbonate facies of the four members described above. Along the footwall (Figure D6.11(a)) the Kharaza Member is characterized by platform margin coralgal reefs which form a broken fringe and well-developed reef slopes in the south-east of the platform (Wadi Kharasa). Back-reef areas are characterized by marine molluscan and coralline algal-rich shoals with lower energy algal, molluscan and foraminiferal wackestones and packstones around the footwall crest. The hangingwall dip-slope area (Figure D6.11(b)) is characterized by biostromal reef units interbedded with shallow, open-marine bioclastic rudstones, grainstones and packstones and shallow-marine siliciclastics (see below).

The Esh el Mellaha Member is limited to the eastern footwall margin where the facies are initially similar to the coralgal reefs of the underlying Kharaza Member. Subsequently, redeposited carbonates occur on the platform slopes which, from their composition (corals, coralline algae, *Halimeda* and molluscs), must have been derived from an upslope shallow-marine platform margin. This is not preserved as the basal Bali'h Member truncates the tops of the Esh el Mellaha slope deposits. Clearly, open-marine conditions continue at this time in the footwall area.

Facies of the Bali'h Member contrast markedly with those of the Kharasa and Esh el Mellaha Members. Basal coral biostromes in DS 6 pass up into mollusc-rich shallow-marine bioclastic sand and gravels of DS 7 which grade laterally into shallow-marine siliciclastic facies in the south-west. The succeeding DS 8 shows clear evidence of restriction with the occurrence of ooid and peloidal sand sheets, restricted molluscan faunas and stromatolites interbedded with coralgal biostromes. These changes occur across the entire platform so that restricted marine conditions affect both the footwall as well as the hanging wall dip-slope area. This restriction increases into the Chaotic Breccia Member of DS 9 with platform-wide, unfossiliferous laminated micrites and scattered evaporite pseudomorphs. There is no evidence of a normal marine margin to the east or to the west at this time.

The upward restriction seen through DS 7, DS 8 and DS 9 probably corresponds with the early stages of evaporite deposition in the Belayim Formation (Orzsag-Sperber, this volume).

While the overall distribution of carbonate facies on the Abu Shaar platform is closely related to the morphology of the underlying fault block (Figure

D6.11), the nature of carbonate producers and therefore much of the platform facies were influenced by climatic factors.

Distribution and evolution of siliciclastic facies

The composition of siliciclastic facies is relatively consistent and reflects the proximity and abundance of pre-rift igneous and volcanic basement source areas. However, there are significant variations in palaeocurrent directions through the Kharasa and Bali Members in the hangingwall dip slope. The Kharaza Member (Wadi Bali) consists of a series of arkosic sands and gravels, each grading up into more carbonate-rich facies (Figure D6.14(b)). Progradation of sliliciclastics is towards the south-east, parallel to the strike of the hangingwall dip slope and indicates a supply from the north-east of Abu Shaar. In the Bali Member, coarser, *Pecten*-rich, arkosic sands of DS 7 have more variable cross-bedding directions to between the south-east and south-west. Higher in the same sequence, arkosic gravels indicate a supply from the south-westerly rift margin. The provenance of these particular sands, from the south-westerly rift margin, implies that the Esh el Mellaha basin was largely filled and was a shallow-marine area at this time.

Many of the depositional sequences show marked lateral changes in siliciclastic content (Figure D6.11(b)). Within the Kharaza Member, arkosic gravels and sands are an important component in the north-eastern part of Abu Shaar (Wadi Bali) situated relatively close to the Esh Mellaha source. Within the Bali'h Member, there is a marked increase in both conglomerates, sands and marls towards the west parts of Abu Shaar where terrigenous facies predominate. These later variations also confirm a western supply from the periphery of the rift.

Synsedimentary deformation

Synsedimentary deformation structures have been described from a number of horizons from both footwall and hangingwall locations on Abu Shaar (Plaziat *et al.*, 1990; this volume). Synsedimentary slumps, debris flows and slide scars on the footwall slopes are related to platform margin instability and may have a variety of triggering mechanisms. However their presence on the hangingwall dip-slope strata (brecciation, folding, thrusting and fissuring) are considered to be of seismic origin by Plaziat *et al.* (1990; this volume) which may relate to the fault movements associated with block rotation demonstrated above.

Comparison with other tilt-block carbonate platforms

Facies models for the development of carbonate platforms in fault block/half-graben settings were originally proposed by Leeder and Gawthorpe (1987). Examples have since been published on the syn-tectonic evolution of carbonate depositional within extensional basins including: the Lower Carboniferous of northern England (e.g. Ebdon *et al.*, 1990), and Ireland (Pickard *et al.*, 1994) (e.g. Hurst and Surlyte, 1984), the Lower Jurassic of the Tethyan region (Santantonio, 1994), the Lower Cretaceous of northern Spain (Rosales *et al.*, 1994) Oligo–Miocene of Sulawesi, Indonesia (Wilson and Bosence, 1996) and the Miocene of Malta (Dart *et al.*, 1993). These examples have largely confirmed the tectono-sedimentary model of Leeder and Gawthorpe that footwall areas are often escarpment or accretionary platform margins with shelf margin buildups and footwall-derived redeposited facies while hangingwall dip slopes may be ramp margins or ramps evolving into rimmed shelf margins. The attached platform of Abu Shaar indicates that siliciclastic sediments derived from the rift border fault may fill the hangingwall basin so that it maintains a ramp-like profile or fills completely with restricted-marine or continental facies.

Examination of the large, three-dimensional outcrop of Abu Shaar shows that the stratigraphy and depositional sequences of tilt-block platforms have an important three-dimensional aspect that has not been obvious in earlier models. As well as the footwall to hangingwall stratal thickening and decrease in dip which was well documented by Rosales *et al.* (1994) important along-strike thickening, decrease in depositional slopes, and increase in progradational geometries is shown where the Abu Shaar tilt-block passes laterally into a transfer zone. These changing stratal geometries reflect a reduction in fault throw as the fault tip is approached in the transfer zone and complements. This contrasts with the stratal geometries in clastic systems in transfer zones recently documented by Gawthorpe and Hurst (1993).

The asymmetric and diachronous nature of depositional sequences on tilt-block carbonate platforms is emphasized in this chapter, and by Bosence *et al.* (1994) and Rosales *et al.* (1994), and demonstrates the differing facies response to footwall uplift and hangingwall subsidence on rotating tilt-blocks, and the distinctive wedging of depositional sequences responding to tilt-block rotation. Tectonic quiescence and regional subsidence was shown to favour highstand progradation of shallow footwall facies into hangingwall dip slopes. This study extends these concepts to a three-dimensional view of depositional sequences where the tilt-block passes into a transfer zone but also shows that depositional sequences can be used to demonstrate that not all sea-level changes on tilt-block platforms are due to block

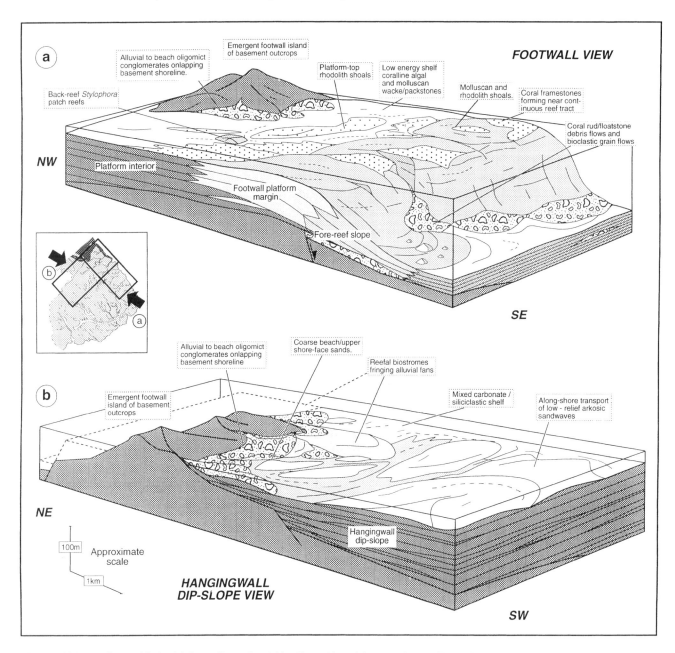

Figure D6.14 Facies models for (a) footwall margin of Abu Shaar (viewed from south-east) illustrating the relationship between the footwall platform margin and the underlying fault block, and (b) hangingwall dip-slope of Abu Shaar (viewed from south-west) where gentle slopes of hangingwall are filled with mixed carbonate-siliciclastic facies.

rotation. Because platform carbonates build up to sea level they preserve previous horizontal surfaces within their stratigraphy. Therefore thickness changes, and repeated thickness changes can be used to demonstrate rotation which results in synchronous footwall uplift and erosion (generating a lowstand systems tract), and hangingwall subsidence (generating a transgressive and highstand systems tract). Conversely, regional sea level falls and rises generate parallel-sided depositional sequences which pass over the footwall area. It is important to note that regional (tectonic or eustatic) sea-level rise is required to generate any carbonate stratigraphy over the footwall of tilt-blocks as fault movement results in footwall uplift, emergence and cessation of shallow-water carbonate production.

CONCLUSIONS

1. Abu Shaar is a mixed early-middle Miocene carbonate/siliciclastic platform that developed on the footwall and hangingwall dip slope of a basement fault-

block with a basinward dipping fault and landward dipping strata on the south-western margin of the Gulf of Suez.

2. The Miocene stratigraphy of Abu Shaar has been subdivided into nine unconformity-bound depositional sequences which subdivide the previous lithostratigraphic members of James et al. (1988) and sequences of Burchette (1988). This provides a genetic classification of stratal geometries and surfaces which extends in three-dimensions over the exposed part of the platform. The depositional sequences provide a time-frame and genetic stratigraphy for detailed discussion of the sedimentological and tectonic evolution of the platform.

3. Dips within the syn-rift strata are gentle and fan out towards the marginal Esh el Mellaha basin with lowest beds dipping at around 4° and upper units dipping at about 1° to the south-west. In the footwall, steep syndepositional slopes (40°) descend off the footwall into the Miocene gulf to the east. Miocene strata thicken westerly into the hangingwall and thin to the footwall where a basement island persists, with onlapping syn-rift through Miocene times. The thickness changes, in shallow-marine carbonates and siliciclastics, and the upward decrease in dip towards the hangingwall indicate that the fault-block was periodically rotated during the Miocene. The pivot of rotation is thought to be some distance to the east of the present-day footwall scarp.

4. The nature of the block margin fault zone exerted a major control on sequence geometries and thicknesses. Along the steep, fault-controlled footwall margin in the north-east of Abu Shaar, sequences are thin and aggradational. Towards the lower relief, south-easterly margin, correlative equivalents are thicker and more progradational, recording the presence of an intrabasinal transfer zone.

5. The carbonate facies of Abu Shaar evolve from an early stage of fully marine carbonate facies, often with fringing and barrier reefs around the entire footwall crest with gentle slopes into the hangingwall sea and steep, reef-rimmed slopes into the footwall and transfer zone to the east and south respectively. Later stages show evidence of restriction of circulation with the appearance of peritidal facies indicative of increased salinity and lower hydraulic energy (e.g. ooid sand sheets, stromatolites and peloidal and micritic facies and restricted molluscan faunas) with some marine interfingering (e.g. coralgal biostromes and bioclastic facies). This vertical change is probably related to regional restriction and evaporative conditions in the Gulf of Suez during the Serravallian.

6. The composition of siliciclastic sediments and crossbedding directions record both the provenance and nature of the evolving depositional morphology of the Abu Shaar platform. In the Kharaza Member, the predominance of arkosic gravels and sands with south-east cross-bedding directions indicates a northeast provenance, probably from the neighbouring Esh el Mellaha range. This contrasts with the coarse gravels of Precambrian basement rocks whose northeast progradation indicates transport from the rift periphery. The transport of peripheral-derived siliciclastics during sedimentation of the Bali member confirms the gradual filling of the basin and thus the evolution from an initial open-marine ramp in a subhorizontal shelf attached to the rift periphery.

7. The footwall margin has two depositional sequences interpreted to be infills of erosionally modified, submarine slump-scars driven by both fault-related movement and by associated relative sea-level changes.

ACKNOWLEDGEMENTS

Initially work for this project was undertaken by the French group GENEBASS and this has been extended by an international group joined by researchers from the University of London (Royal Holloway). This combined research has been financed by the European Community 'Science' project RED SED, British Petroleum and Elf-Aquitaine. The authors gratefully acknowledge this support and thank T. Burchette for his review and suggestions that have improved our initial manuscript.

Chapter D7
Miocene coral reefs and reef corals of the south-western Gulf of Suez and north-western Red Sea: distribution, diversity and regional environmental controls

C. Perrin, J.-C. Plaziat and B. R. Rosen

ABSTRACT

Coral reefs developed within the Gulf of Suez–northern Red Sea region during a relatively brief time interval corresponding to the maximum of the Middle Miocene worldwide marine transgression which was also associated with phase of a warm global climate. The reefs occur in mixed carbonate–siliciclastic sequences which belong to the marine Miocene syn-rift unit (Group B or Upper Rudeis–Kareem Formations) and possibly extend from Langhian to early Serravallian age. Studied reef locations extend along the western coast of the Gulf of Suez and north-western Red Sea from north of Hurghada to the Abu Ghusun–Ras Honkorab area.

Reef distribution and development within the region appear to be highly controlled by the tectono-sedimentary regional setting of the early rift system. In particular, the structural framework strongly controlled the regional palaeotopography which in turn constrained the transport of siliciclastic sediments and the circulation of regional and local open-marine waters. Coral reefs typically occur both on structural and sedimentary palaeohighs (horsts and fronts of fan deltas, respectively), and preferentially developed on the north-east side of fault blocks facing the open-marine waters of the axial rift zone. Reef growth was generally favoured at locations sheltered from the main siliciclastic supply from the south-west rift shoulder, while transport of terrigenous material was concentrated within palaeo-depressions (half-graben and graben). Coral assemblages appear to be restricted to small-sized fringing reefs, thin biostromes, or shallow marine facies with scattered coral colonies. In most areas, the steep north-east facing footwall slopes seem to have limited reef progradation.

The coral fauna shows an entirely Miocene Mediterranean affinity with no apparent faunal overlap with the Miocene reef coral fauna from the neighbouring Indo-Pacific. Material collected from the studied areas includes 16 genera and 27 species, a similar richness to Middle Miocene reef coral faunas from the western Mediterranean. Coral reefs and their coral fauna became definitively extinct within the Gulf of Suez–northern Red Sea region during the Serravallian, as a result of an increasing restriction of marine conditions which eventually led to evaporite sedimentation. This was directly related to the gradual closure of the Suez Isthmus, probably due to both eustatic and tectonic movements, which finally isolated the Gulf of Suez–Red Sea from the Mediterranean basin. Thus coral faunas became extinct in this region much earlier than in the Mediterranean generally, where reef corals occurred until the end of the Miocene.

INTRODUCTION

The Miocene coral reefs of the Gulf of Suez and north-western Red Sea developed during a relatively brief period of time coincident with the Middle Miocene worldwide maximum transgression during the late Burdigalian to Langhian, around 16 Ma. This period is known to have been climatically favourable for coral reef development not only within the Mediterranean reef province but also in higher palaeolatitudes, including Poland, Japan, southern Australia and New Zealand (Adams et al., 1990; Plaziat, 1995). Thus the Miocene coral reef belt was wider than today. Within the Gulf of Suez–Red Sea area, reef development and distribution

Sedimentation and Tectonics of Rift Basins: Red Sea–Gulf of Aden. Edited by B.H. Purser and D.W.J. Bosence. Published in 1998 by Chapman & Hall, London. ISBN 0412 73490 7.

was favoured by a variety of environmental conditions including availability of substrates, sedimentation rates and open-marine circulation which were, in turn, strongly dependent on the regional syn-rift tectono-sedimentary setting.

Miocene coral reefs have been reported by numerous authors along the Gulf of Suez and Red Sea coasts since the beginning of the century (Blanckenhorn, 1901). However, most of the more recent Miocene studies of this region have focused on the relationship between the geodynamics of the rift system and sedimentation (e.g. Sellwood and Netherwood, 1984; Scott and Govean, 1985; Montenat et al., 1986a, 1988; Purser et al., 1990; Plaziat et al., 1990; Hughes et al., 1992; Purser and Philobbos, 1993). More local works on the best exposed platforms have concentrated on geological history and platform stratigraphy (Burchette, 1988; James et al., 1988) with more detailed studies of particular facies such as stromatolites and evaporites (Monty et al., 1987; Orszag-Sperber et al., 1994). Hence, in spite of their economic potential, coral reefs themselves have remained relatively poorly known. In particular, there has been no attempt to investigate their composition and internal structure, nor the regional and local controls on reef geometry and development. Moreover, the taxonomy of the corals has been overlooked for nearly a century, the last paper on the subject being that by Gregory (1906).

This chapter discusses various sites where Miocene carbonate sequences include reefs. It focuses on regional and local environmental controls on reef distribution and development, throughout the area. The regional significance of the coral fauna and its relationship to the Mediterranean Miocene reef province is also discussed.

Following the initial work of the French GENEBASS group, a regional sedimentological study of the Egyptian syn-rift Miocene series was carried out by a research team from Université de Paris-Sud, Orsay, as part of the EC RED SED Programme. The latter project also included more specialized work on platform geometry, sequence stratigraphy and diagenesis of the Gebel Abu Shaar fault-block, undertaken by researchers from the Universities of Paris-Sud, London and East Anglia and from IGAL (Paris).

Reefs developed on the Abu Shaar and Esh Mellaha fault blocks provide the best exposed three-dimensional outcrops. Therefore detailed palaeoecological analysis using qualitative and quantitative approaches involved detailed fieldwork (by C.P.) (Perrin et al., 1995). This study is part of a NERC project (Palaeozonation and Quantitative Palaeobathymetry of Caenozoic Rifts). The detailed results of this work are beyond the scope of this chapter and will be presented elsewhere. Fieldwork on the Abu Shaar–Esh Mellaha reefs and taxonomic identifications of their coral fauna were carried out by C.P. For locations situated south of Hurghada the occurrence and geometry of reefs, and their relationship to the structural framework, were recorded by a field team here represented by J.C.P under the RED SED Programme. The coral fauna in this case was identified by B.R.R.

AGE AND REGIONAL SETTING OF REEF DEVELOPMENT

Stratigraphy

Regional stratigraphy of the Gulf of Suez and northern Red Sea has been largely based on the lithostratigraphy of offshore deposits. In the Gulf of Suez, the Neogene series is subdivided into the Gharandal Group (Abu Zenima, Nukhul and Rudeis Formations) and the Ras Malaab Group (Kareem, Belayim, South Gharib and Zeit Formations). These correspond to the GENEBASS stratigraphy defined from onshore sections by Montenat et al. (1986a, 1988) which consists of four groups (A to D) separated by regional angular unconformities which record the major stages of rifting.

Miocene coral reefs belong to the syn-rift Miocene marine Group B which includes both carbonate and siliciclastic deposits. Biostratigraphic data obtained from planktonic foraminifera and calcareous nannoplankton occurring within the lower marls indicate that Group B reefal sedimentation may range from Langhian to Lower Serravallian in the Gulf of Suez (Plaziat et al., this volume).

Tectono-sedimentary regional setting

Within the northern Red Sea and Gulf of Suez area, Group B deposits overlie the protorift sediments of Group A and the pre-rift series, or rest directly on the Precambrian basement. The base of Group B is marked by a major angular unconformity recording the second major phase of rifting at the end of the early Miocene, which is well-known as the Mid-Clysmic Event (16.5 Ma). The contrasted palaeotopography resulting from normal faulting and block tilting at that time (Sellwood and Netherwood, 1984; Scott and Govean, 1985; Montenat et al., 1986a, 1988; Evans, 1988; Richardson and Arthur, 1988) favoured local reef development during the eustatic highstand. The fault-blocks, which formed substrates for fringing reefs, also acted as morpho-structural barriers protecting reefs from excessive terrigenous supply from the rift shoulder. Sands and gravels were deflected by blocks which were generally elongated parallel to the coast (Purser et al., this volume) and detrital inputs were transported through a variety of structural depressions. This system favoured reef development along the seaward faces of emergent basement blocks whose Group A continental cover had been partly eroded. However, some reefs also developed on sedimentary palaeohighs such as Middle Miocene fan-delta fronts. Morpho-structural isolation from the Mediterranean basin is another major tectono-sedimentary event which led to a regional restriction of marine conditions throughout the Red Sea rift system, initiating deposition of the Middle to late Miocene evaporites (Group C).

Regional distribution of Miocene coral reefs and coral facies

Miocene reef corals (*sensu* Rosen, 1984, p. 204) have been known from Egypt since the latter part of the last century (Fuchs, 1883; Felix, 1884, 1903, 1904; Gregory, 1898, 1906). Miocene coral reefs and reefal limestones (Figure D7.1) have been reported from the southern Cairo area, north-west of Suez, from various localities along both coasts of the Gulf of Suez, from the Maqna area (south-east of the Gulf of Aqaba and in the north-east area of the Red Sea), and along the western coast of the Red Sea from Hurghada to Port Sudan (ibid. and also Scott and Govean, 1985; Ott d'Estevou *et al.*, 1986a,b; Prat *et al.*, 1986; Thiriet *et al.*, 1986; Jarrige *et al.*, 1986; Roussel 1986; Roussel *et al.*, 1986; Montenat *et al.*, 1986b, 1990; Burollet, 1986; Le Nindre *et al.*, 1986; Purser *et al.*, 1990, 1994). Considering the size of this literature on syn-rift Miocene deposits, there is a surprising lack of knowledge about reef structure and composition. Older works up to the early part of this century were mainly devoted to coral descriptions, while more recent studies have been concerned with rift tectonics and their relationship to sedimentation. There has also been a general tendency to apply the terms 'reef', 'reefal' or even 'reefoid' to virtually any lithified biogenic carbonate deposits, and even to terrigenous material containing only scattered coral fragments! This makes it difficult to gain a realistic picture of the overall distribution of 'true' coral reefs within the area, and in the present account, therefore, we have concentrated only on areas which have been studied by at least one of the authors.

The present study is not a complete and consistent evaluation of all Miocene coral reefs occurring between Suez and Berenice. Field observations and related sampling vary in degree of detail and tend to be discontinuous. The studied reefs extend over 500 km along the southern Gulf of Suez and north-western Red Sea coast from north of Hurghada to the Abu Ghusun–Ras Honkorab area (Figure D7.2).

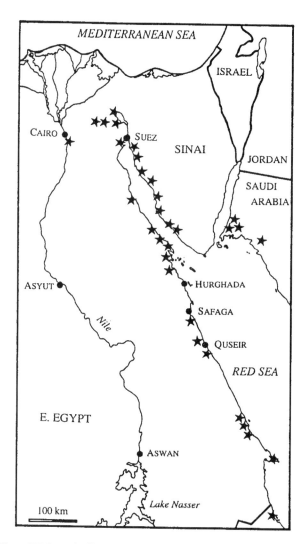

Figure D7.1 Distribution of coral reefs and reef coral facies (stars) within the Gulf of Suez–northern Red Sea region, based on compilations from literature (see text), from the Egyptian Miocene reef coral collections of the Natural History Museum (London) and from our own coral collections.

OCCURRENCE, GEOMETRY AND DEVELOPMENT OF CORAL REEFS

North of Hurghada area

Local tectono-sedimentary setting
The Gebels Esh Mellaha and Abu Shaar, although both corresponding to the same major fault-block tilted to the south-west, exhibit differences in their Middle Miocene carbonate facies. The rotation axis of the block pitches to the south-east leading to the gradual deepening of the pre-Miocene basement towards the southern end of the block. This strongly favoured the development of mixed siliciclastic–carbonate sequences and coral reefs on Abu Shaar during the Middle Miocene transgression. This is indicated by the gradu-

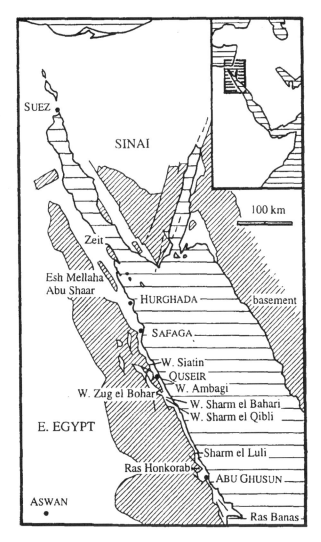

Figure D7.2 Location map of the studied reef sites.

ally increasing thickness of Miocene marine footwall sequences from Wadi Bali to Wadi Kharasa, and also by the occurrence of best-developed coral reefs on the south-east margin of the platform.

The Esh Mellaha structural block is bounded on its north-east side by a system of clysmically oriented (N140) normal faults which separate Esh Mellaha from the north-eastern neighbouring block of Gebel el Zeit, by the structural depression of the Gemsa Plain (see Figure D6.1 in Cross et al., this volume). This area was continuously subsiding during the Middle Miocene and was characterized by pelagic marls (Prat et al., 1986). The occurrence of coral reefs along the north-eastern side of these fault blocks is partly controlled by the presence of these open-marine conditions to the East.

On its western side, the Esh Mellaha–Abu Shaar block is bordered by the Mellaha Basin extending westwards to the Red Sea Hills. These hills form the main external rift scarp, which is bounded by a system of normal faults. The Mellaha Basin has been filled with marine Miocene sediments, the siliciclastic content of which gradually decreases towards the south-east, including the south-western Abu Shaar ramp. The existence of this western basin which was open to the south and connected by straits between basement islands to the open-marine waters of the axial rift zone, at least south of the Esh Mellaha–Abu Shaar barrier, is of major importance for reef occurrence and distribution. In particular, the presence of open-marine waters within the Esh Mellaha Basin, at least during the first part of the mid-Miocene transgression, enabled reefs and coral assemblages to develop on the south-west ramp of the Abu Shaar block. In addition, the depth of the Esh Mellaha Basin indirectly controlled the extent of siliciclastic supply from the western Red Sea Hills to the Abu Shaar ramp, and was therefore probably a factor limiting reef growth in that area.

The Precambrian basement, which forms the elongated crest of Gebel Esh Mellaha, and which also outcrops in the northern parts of Gebel Abu Shaar, consists of granitoids, volcanic and metasedimentary rocks. The Middle Miocene transgression never reached the highest crest of the Precambrian basement of Esh Mellaha. This formed a north-west–south-east chain of elongated islands during mid-Miocene time. These would have acted as a protective barrier blocking longshore siliciclastic inputs from the north-west, on the north-east and south-west sides of the blocks and hence favoured reef growth.

Gebel Abu Shaar
Reef geometry and distribution within depositional sequences
The general north-west–south-east polarity of Gebel Abu Shaar is modified in detail by several cross-faults with minor lateral movements. The morphology of the north-eastern footwall slope was therefore not strictly linear but broken by promontories and embayments of Precambrian basement, with a lateral topographic irregularity of about 100–200 m. This palaeomorphology has had strong influence over geometry and thickness of slope deposits (Purser and Plaziat, this volume) and coral reefs. Modern erosion has tended to adjust the platform margin according to the line of the Clysmic fault-scarp, hence favouring preservation of reefs fringing the structural embayments.

Repeated synsedimentary tilting during the mid-Miocene has had obvious effects on the overall platform geometry and development, and is expressed through unconformities forming major sequence boundaries. South-west rotation of the block tended to create accommodation space on the hangingwall ramp, maintaining relatively open-marine conditions in spite of

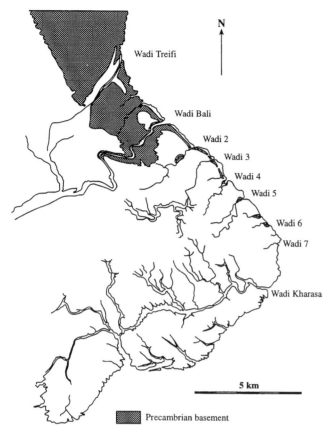

Figure D7.3 Northern Hurghada area, geological map of the Gebel Abu Shaar with location of the main wadis.

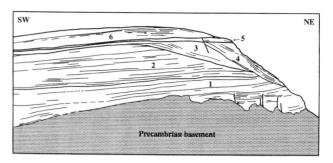

Figure D7.4 Northern Hurghada area, Gebel Abu Shaar, example of sequence geometry on the footwall margin (Wadi 2) showing the tilted erosional surfaces and the depositional sequences (1,2,3,4,5,6) traced from photomosaic: 1–2, Kharasa Member; 3–4, Esh Mellaha Member; 5, 'Reef Veneer' of James et al. (1988); 6, first depositional sequence of the Bali'h Member (compare with Cross et al., this volume).

repeated terrigenous inputs and aggradation. This resulted in mixed carbonate–siliciclastic sedimentation including coral reefs and scattered coral patches, often associated with rhodolith facies and coralline algal crusts. However, in the upper half of the series, environmental conditions within the whole Red Sea basin gradually became more restricted. The very shallow facies of the south-western ramp ended, at the top of the sequence, with subtidal microbial laminites which are known from Esh Mellaha to the Abu Ghusun area. These cover the summit of the Abu Shaar platform but, although they contain rare pseudomorphs of evaporitic minerals, there is no definite evidence that gypsum developed on the platform itself.

By contrast, synsedimentary tilting tended to reduce accommodation space on the footwall margin. While sequence boundaries on the south-west ramp correlate with tilted flooding surfaces, they also correspond to marked erosional surfaces on the footwall margin (Figure D7.4). Although coral reef growth was presumably favoured by the nearby open marine conditions of the axial zone, and by the limited siliciclastic supply in this area, the general steepness of the north-eastern footwall slopes appears to have been an important factor limiting reef progradation.

While the south-west sedimentary tilting has resulted in contrasting variations of accommodation space between the east-north-east and west-south-west parts of the platform, palaeomorphology of the basement plunging to the south-east has led to a north-west–south-east gradient in accommodation space. This led to maximal reef development to the south-east, in the area of Wadi Kharasa, and together with the lower-angle slopes of the underlying substrate here, this also favoured reef progradation.

James et al. (1988) subdivided Mid-Miocene sedimentation on Abu Shaar into four major successive units of the Rudeis Formation, respectively called the Kharasa Member, Esh Mellaha Member, Bali'h Member and Chaotic Breccia Member. The first three members are broadly equivalent to the three sequences of Burchette (1988). They also correspond to depositional sequences defined by Cross et al. (this volume): their DS1 to DS3 to the Kharasa Member, sequences DS4 and DS5 to the Esh Mellaha Member, DS6 to DS8 to the Bali'h Member, while the Chaotic Breccia Member correlates with their sequence DS9. Note, however, that there are major differences between these two sets of authors concerning their correlations and general geometry of individual sequences (Cross et al., this volume, for discussion). Moreover, some of these correlations are not supported by our own work (Figure D7.4).

Except for the final Breccia Member which consists of subtidal microbial laminites, each sequence includes, at least locally, coral reefs and/or open-marine facies with scattered coral colonies. On the platform scale, however, development of coral assemblages and, hence, reefs, may have been very local and occurrence of coral reefs on the Abu Shaar platform has been largely overestimated in

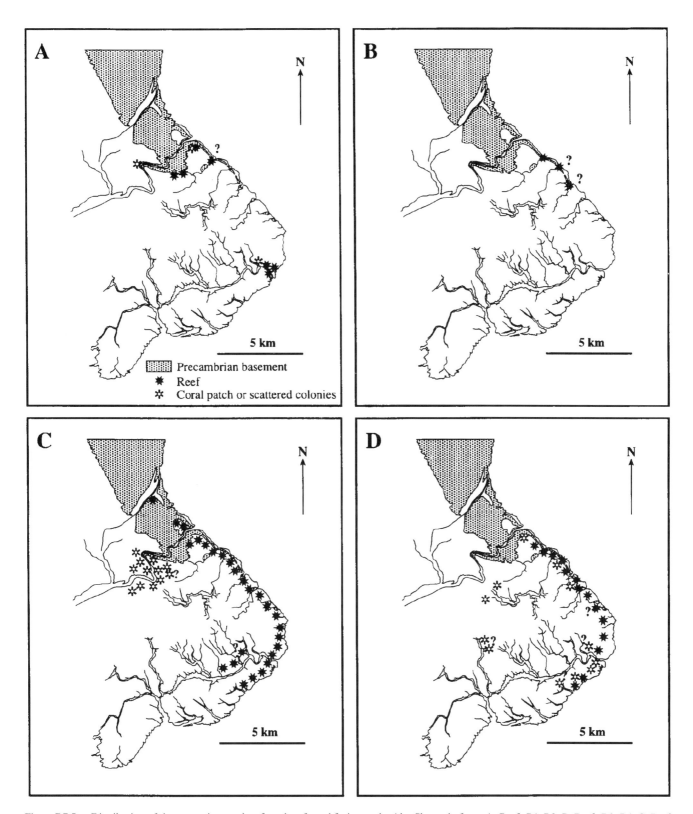

Figure D7.5 Distribution of the successive coral reefs and reef coral facies on the Abu Shaar platform. A. Reefs R1–R2. B. Reefs R3–R4. C. Reef R5. D. Reefs R6–R8.

previous stratigraphical and sedimentological studies (Figure D7.5).

Mid-Miocene reefal and associated sedimentation followed local continental filling of palaeodepressions, in particular at the northern end of Abu Shaar (palaeovalley of Wadi Treifi) and at the entrance of Wadi Bali (north face). The Kharasa Member includes two platform-wide depositional sequences (DS 1 and DS 2 of Cross *et al.*, this volume), with both carbonate and terrigenous facies. Although these sequences contain coral reefs (R1 and R2), buildup development appears to be restricted to local areas (Figure D7.5(a)). These reefs include progradational biostromes occurring on the western side of the Precambrian palaeo-islands and rare scattered *Stylophora* clusters on the south-west hangingwall. Reefs preserved on this south-west dip slopes indicate the persistence of open-marine conditions in the western parts of the platform. Although a thin coral reef outcrops at the front of Wadi Bali, equivalents of these buildups are absent along the north-eastern footwall scarp where they never developed or are not preserved – possibly due to syn-Miocene faulting.

The Esh Mellaha Member consists of both reefs (R3 and R4) and coarse bioclastic facies organized in two sequences along the north-east footwall from Wadi Bali to the north of Wadi Kharasa (Figure D7.4), while they seem to be absent in Wadi Kharasa. Their lenticular geometries and their sharp inclined basal surfaces on the platform margin were both conditioned by previous erosion (and not collapse) of the platform front (Figure D7.4). Observations concerning reef assemblages and internal structure remain incomplete due to inaccessibility of the outcrops (Figure D7.5(b)). However, reefs in these two sequences consist of thin lenticular units of coral framestones or scattered colonies on the palaeo-slopes, while on the platform top, dense coral frameworks are rare and often replaced by bioclastic facies.

The Esh Mellaha Member is topped by a major platform-wide erosional surface truncating the underlying sequences of Esh Mellaha and Kharasa Members. This subhorizontal discontinuity records a major relative sea level fall resulting from both local tilting of the fault block and eustatic sea level change. Subsequent rise of sea level favoured the widespread growth of a coral reef (R5) fringing the north-eastern footwall and extending as coral patches across the subhorizontal dip-slope surface (Figure D7.5(c)). This reef corresponds to the first sequence of the Bali'h Member and correlates with both the 'Reef veneer' of James *et al.* (1988) and with the depositional sequence DS 6 of Cross *et al.* (this volume).

Overlying sequences of the Bali'h Member (DS 7 and DS 8 of Cross *et al.*, this volume) begin with terrigenous-rich sediments in which scattered coral colonies occur.

Subsequent conditions generally appear to have been less favourable for coral reefs. Three brief episodes of reef growth, however, occur over most of the platform (Figure D7.5(d)). They consist of thin biostromal reefs (R6, R7 and R8) displaying locally dense coral framework which is associated with an open-marine molluscan fauna containing *Periglypta* and *Lithodomus*. The presence of these reefs confirms optimal marine conditions contrasting with episodes of terrigenous sedimentation. The latter include numerous stromatolitic horizons generally associated with oolites and a restricted marine biota such as small *Potamides* and ? *Cerastoderma* (*Cardium*) or even *Acetabularia* algae. Sequence geometry and facies distribution within the Bali'h Member reflect a widespread uniformity of the sedimentary system, although a subtle east–west gradient still persisted as shown by a gradual westward decrease of organisms, possibly related to increasing siliciclastic content towards the rift margin. The initial south-west dip slope or ramp has thus accreted to a very shallow, subhorizontal shelf which probably extended to the western rift margin. Moreover, evolution of the Bali'h Member records alternating normal and restricted marine conditions which can be related to regional environmental changes leading to the final restriction of the Gulf of Suez–Red Sea Basin.

Coral reefs and reef assemblages
Patterns of reef palaeozonation were assessed using methods developed by Perrin *et al.* (1995) together with mapping of reef assemblages, and careful prior analysis of internal reef structure. In particular, on reef palaeo-slopes, attention was focused on qualitative and quantitative surveys on the same reef palaeosurface. Modern surface slopes are not always concordant with the original reef palaeoslopes, and may cut across several reef units. This explains the major differences between the reef palaeozonations presented below, and those described by James *et al.* (1988). For example, for their 'Reef veneer' (R5), they recorded coral framework extending downslope from the reef crest for 75 m, and believed this could be used as a direct indication of original palaeodepths. In fact, their profile cuts through several major sequences of R5, R4 and probably R3. Our own reef surveys shows that coral framework does not occur below a palaeodepth of 20–25 m.

Coral reefs fringing the north-eastern footwall
Coral reefs fringing the north-eastern footwall are mainly represented by the fringing reefs of the Esh Mellaha Member (R3 and R4) and by the basal reef of the Bali'h Member (R5; Figure D7.5(b), (c)). These reefs are exposed on the preserved part of the footwall Miocene cover, above the fault scarp, along most of the north-eastern platform margin. In the northern part of

the platform (Wadi Bali), the two oldest reefs (R1 and R2) are represented by only one single individual reef body (Figure D7.5(a)), a few metres thick, which is dominated by faviid corals (*Montastraea mellahica*, *Montastraea* sp. and *Tarbellastraea*), and *Stylophora* patches associated with encrusting coralline algae.

The basal reef of the Bali'h Member (R5) comprises 11 to 14 coral species on the frontal part of the platform where it prograded towards the north-east or to the east depending on the local morphology of the platform margin. In the northern part of the platform, at Wadi Bali, these coral assemblages outcrop discontinuously as a result of subsequent major erosion, and consist of scattered faviid coral colonies in growth position encrusted by coralline algae. Southwards, the outcrop is much better preserved and frequently shows a downward continuity from the reef front to the reef palaeoslope. Shallowest zones of the reef are characterized by a faviid-dominated assemblage (*Favites neglecta*, *Montastraea mellahica*, *Montastraea* sp., *Tarbellastraea*, for the most common species), gradually passing downslope to a *Porites*-dominated assemblage (columnar or branching forms of *Porites collegniana*, often associated with microbial crusts) and/or to a *Caulastraea* assemblage. Further downslope, massive colonies of *Porites* are dominant and are themselves replaced downwards by a mussid-dominated assemblage (*Acanthastraea*, *Lithophyllia*, *Mussismilia*) associated with coralline algal and bryozoan crusts.

Fringing coral reefs prograding
to the south or south-east
Fringing coral reefs prograding to the south-east or to the south (R1, R2 and R5) occur in the southern parts of the platform (at Wadi Kharasa and further to the southwest), where they prograded to the south or to the south-east (Figure D7.5). In Wadi Kharasa, the two oldest reefs (R1 and R2) are about 15 and 20 m thick respectively, and show a complete profile from reef slope to back-reef facies. Back-reefs are characterized by rhodolith rudstones or floatstones and/or by dense frameworks of branching corals (*Caulastraea* sp. and *Stylophora* with some branching *Porites*). The general palaeozonation pattern shows faviid and *Stylophora* assemblages, sometimes with a *Caulastraea* sp.-dominated assemblage on the reef flat. On the reef-front, coral framework is mainly built by branching *Porites* and faviid corals (*Montastraea*, *Favites*, *Tarbellastraea*), while reef slopes are clearly dominated by biodetrital facies (coral rudstones and floatstones), except in some areas where an *Acanthastraea*-dominated assemblage occurs.

On the south-eastern margin of the platform (southwest of Wadi Kharasa), only the front and external flat of the second reef are preserved. They contain a branching *Porites* assemblage and a faviid-dominated zone (including *Montastraea mellahica*, *Tarbellastraea* sp., *Favites neglecta* and minor *Favia*), with scattered colonies of *Acanthastraea* 'seawards'. A similar pattern is found in the basal fringing reef of the Bali'h Member (R5) with occurrence of both the faviid and branching *Porites* assemblages.

Coral reefs developed west
of the basement palaeo-island
Westward-prograding reefs (R1 and R2) developed on the south-west side of the basement palaeo-island in the northern part of the platform (Figure D7.5(a)). Two successive reef units (R1 and R2) can be seen outcropping over an area about 100 m across in Wadi Bali. In addition, just above the Precambrian basement, a thin bioclastic layer, including a few massive faviid colonies in growth position, is interbedded with coarse conglomerates of volcanic pebbles and boulders.

The lower reef has a maximum thickness of 5 m and is interbedded with flat tidal-channel fills which have truncated the coral framework. Reef assemblages are largely dominated by faviid corals (*Montastraea mellahica*, *Montastraea* sp., *Favia* sp., *Tarbellastraea* cf. *ellisiana*) associated with coralline algal crusts. Towards the basement palaeo-island, the reef facies grades laterally into bioclastic packstones rich in molluscs and coralline algal fragments. The thickness of the upper reef varies from about 4 to 10 m. A facies rich in rhodoliths and coralline algal crusts commonly occurs at the reef base and passes gradually upwards into coral framework. The latter is relatively diverse and dominated by *Stylophora*, and faviid corals (*Caulastraea* and the same species as above) associated with some minor forms and coralline algal crusts. Eastwards, coral framework is replaced by bioclastic sands with fragments of *Stylophora*. The reef is overlain by packstone layers showing an upward-increase in siliciclastic content, and containing abundant coralline algal fragments and molluscs. The uppermost layers are siliciclastic sandstones with low-angle planar foreset bedding and birds-eye structures typical of beach deposits. Both reefs are rapidly replaced westwards on the hangingwall ramp by mixed siliciclastic–bioclastic sands including occasionally small-sized patches of *Stylophora*, 1 to 3 m thick.

Reef blankets
Three thin biostromal reef blankets (R6, R7 and R8; Figure D7.5(d)) are interbedded within stromatolitic and other restricted shallow facies of the Bali'h Member. The lower reef unit (R6) occurs on the north-east footwall and is very discontinuous (<2 m thick) as a result of differential erosion at its top. Reef facies are

characterized by their poorly preserved framework or by scattered coral colonies, by the dominance of *Acanthastraea* locally associated with minor massive *Porites* or *Caulastraea*, and by the local abundance of the bivalve *Lithodomus*. The second reef extends as a thin biostrome across the whole platform. Its thickness varies from one to a few metres. Thickness of the reef front increases gradually along the eastern margin, particularly on the southern margin where it locally reaches 10–12 m. Reef framework consists of massive faviid corals, massive and branching *Porites* and dominant *Acanthastraea*, associated with encrusting coralline algae and a molluscan fauna including *Periglypta* and numerous boring bivalves (*Lithophaga*, Pholadidae).

Coral patches

Small coral patches and clusters of a few massive colonies are scattered in a mixed siliciclastic–bioclastic sand which outcrops continuously over the subhorizontal Abu Shaar plateau (Figure D7.5(c)). The basal part of the sandy layer (approximately the lowest metre) locally comprises abundant coralline algal fragments with both encrusting and branching rhodoliths and coralline algal crusts, which together form a coralline algal pavement colonized by coral colonies. The size of coral patches is about 0.5–1 m thick and 1–2 m across, and there is a gradual spectrum from isolated coral colonies, through colony clusters to true coral patches. Coral framework is filled with a muddy carbonate matrix. No pattern was found either in the frequency and size of patches and clusters, or in the distribution of coral taxa. These are represented by faviid corals (mainly *Montastraea mellahica*, *Tarbellastraea* cf. *ellisiana* and *Favia* sp., associated with *Montastraea* sp., *Favites neglecta*, *Favites neglecta* var. *minor*) and large massive colonies of *Porites*.

Gebel Esh Mellaha

The north-eastern footwall margin of Gebel Esh Mellaha is characterized by a highly diverse facies. Locally, where the footwall ridge has been disrupted by cross-faults, such as in Wadi Gerfan, coarse discharges of terrigenous sands and conglomerates formed shallow-marine fan deltas, the tops of which are often colonized by small patch-reefs similar to those of Wadi Sharm el Bahari (see below). In the southern area, the steep footwall escarpment is overlain by discontinuously outcropping reef talus. These sedimentary slopes are dipping at 25°–30° to the north-east and rise up to about 50 m above the present-day surrounding alluvial plain. Only medium and deep parts of the reef palaeoslopes are preserved, suggesting the original existence of a shallow reef fringing the Precambrian basement.

Reef facies consist of scattered coral colonies within a packstone biodetrital matrix including coarse coral fragments. The coral fauna is largely dominated by Faviidae, especially *Montastraea mellahica*, *Favia* sp., *Favites neglecta* and *Tarbellastraea*, associated with less common massive *Porites* and solitary *Lithophyllia*. Minor and rare components include *Acanthastraea*, *Mussismilia*, *Caulastraea*, agariciid corals and coralline algal crusts, together with small-sized azooxanthellate-like corals. *Stylophora* was found only as fragments of branches. The deepest parts of the reef palaeoslopes contain some corals in growth position together with coral fragments and overturned colonies, indicating that coral growth occurred in deeper palaeodepths than on Abu Shaar reef slopes. This may be related partly to more efficient protection of the Esh Mellahar reefs against siliciclastic input from the rift margin as suggested by the lower siliciclastic content of reef facies and associated sediments compared to Abu Shaar. Downslope, reef facies are overlain by large domal stromatolites with numerous *Lithodomus* borings up to 2.5 cm in diameter, and reworked azooxanthellate-like solitary corals, both indicative of prevailing normal marine conditions.

Safaga area (Figure D7.6)

In the Safaga area, the pre-Miocene basement forms very high Clysmic-oriented ridges reaching a present-day elevation of more than 800 m. These are rimmed by narrow, mixed siliciclastic–carbonate platforms on their north-east margins. Carbonate sequences including reefs occur in the upper part of these platforms and outcrop at 290 m above present-day sea level. Slope geometry suggests submarine palaeotopography corresponds approximately to present-day relief, equivalent to a palaeodepth range of about 300 m. Platform sedimentation began with low-stand terrigenous slope deposits, a few tens of metres in thickness. These lower, almost pure, terrigenous sequences are locally overlain by biodetrital carbonates containing large reef olistoliths. These have been derived from a reef-rimmed platform which developed on a terraced terrigenous platform during the subsequent mid-Miocene sea-level rise. Reef sequences are 10 to 15 m thick and outcrop along the north-east margin of the fault block where they are locally interrupted by submarine detrital flows filling palaeodepressions. An erosional terrace of the basement, formed during the sea-level high-stand, clearly shows that reefs were directly fringing the steep basement slopes, without any lagoon or back-reef. This relatively simple type of reef is similar to the fringing reefs of Gebel Esh Mellaha.

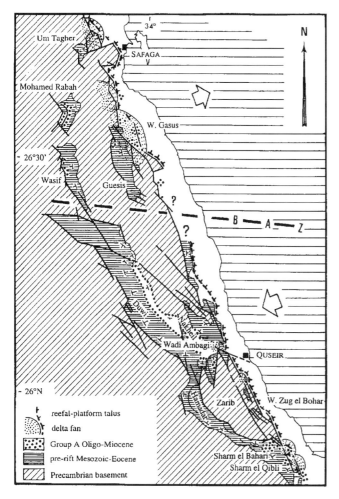

Figure D7.6 Distribution of reefs and carbonate-platform talus from north of Safaga to South of Quseir. Reef platforms developed in front of palaeohighs: basement fault-blocks, delta-fans of early mid-Miocene age (B1) or pre-rift sedimentary outcrops (Cretaceous–Eocene). B-A-Z, Brothers Accommodation Zone.

Area south of Quseir (Figure D7.6)

The studied area south of Quseir extends from north of Wadi Ambagi to the southern Wadi Sharm el Qibli. In the northern part of this region, reefs are developed on the north-east side of clysmically-oriented basement fault blocks, while to the south, the reef substrate consists of submarine sedimentary palaeohighs formed by antecedent Gilbert-type fan deltas (e.g. at Sharm el Bahari). Accumulation of these coarse conglomeratic fan deltas has been directly influenced by the tectonically controlled palaeotopography which constrained transport of sediments (for details see Purser et al., this volume). At Wadi Sharm el Qibli, reefs also fringed a palaeohigh made up of residual Oligo-Miocene continental sediments not removed by the subsequent early Miocene erosion.

Wadi Ambagi (Figure D7.7)

In this area, the tectonic framework is complicated by a major N110-oriented fault. The geometry and development of the platform has been strongly constrained by the underlying morphology of the fault scarp and also by the presence of a small hangingwall fault block of Eocene limestone (Figure D7.7; see also Plaziat et al., this volume). The spectacular relief of this horst gives the misleading impression that the block morphology alone determines the east-facing orientation of the reef front. In fact, the reef developed on mixed carbonate–siliciclastic platform sediments showing an overall progradation towards the north-north-east but with its final talus facing the east-south-east. The occurrence of coral fragments within channel fills intercalated within the lower clays points to an older coral fauna which is not preserved *in situ*.

The youngest coral reef developed on the edge of a subhorizontal massive carbonate bed within which there is an abundant molluscan fauna and debris of the crab *Daira*, associated with rare corals. The reef is about 100 m wide and mainly represented by its fore-reef talus containing downslope-transported reef blocks. At the platform edge, *in situ* coral framework consists mainly of columnar *Porites*, while a more diverse coral fauna occurs within the reef blocks of the palaeoslope. This includes *Favites profunda*, *Tarbellastraea* sp. and *Lithophyllia gigas*. A terminal bioclastic reef talus, partly replaced by gypsum (Orszag-Sperber et al., 1986), contains a diverse reefal malacofauna which is not preserved within the upper massive reef carbonate. In contrast, another talus facies is exclusively characterized by the mytilid mollusc *Brachidontes* associated with other minute bivalves. The composition and very low diversity of this fauna reflects the onset of Red Sea marine restriction. At the same time, the coral reef was overlain by sands and stromatolites prior to burial by onlapping gypsum. This exceptional situation (with overlying gypsum) shows that the whole Wadi Ambagi area was affected by renewed subsidence before deposition of the Upper Miocene evaporites. A similar situation also exists further north at Wadi Siatin, where poritid and faviid corals occur within a horizontal marine carbonate facies.

South Wadi Ambagi–Wadi El Aswad

In this area, a clysmically oriented (N140) basement fault block is rimmed on its south-eastern side by pre-rift Cretaceous–Eocene sediments. These formed complex palaeohighs bordering the mid-Miocene coast, in front of which a mixed terrigenous–carbonate platform developed. Three individual coral reefs are interbedded within the platform sequences (Purser et al., this volume), but coral skeletons are poorly preserved within the reef-core facies. However, the composition of coral

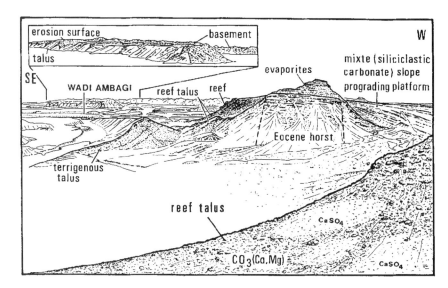

Figure D7.7 Relationships between substratum palaeotopography, detrital platforms and reef growth; west of Quseir, south and north of Wadi Ambagi. Talus along the basement blocks relates to a narrow detrital terrace fringed by coral reefs which have been subsequently eroded. The Eocene horst was completely buried by a detrital shelf prograding to the north-east when the reef was growing. The reef talus was locally covered by quartz sands and gravels but remained outcropping when the late-Miocene gypsum onlapped and covered this relief. A diagenetic replacement of the biogenic carbonate locally led to a chalky sulphatized facies.

fauna has been inferred from a collection of much better preserved coral fragments which are reworked within a coarse gravel slope deposit, exposed at the foot of the platform front. This coral fauna appears to be relatively diverse and includes *Porites collegniana*, *Favites neglecta*, *Favites neglecta* var. *minor*, *Montastraea mellahica*, *Tarbellastraea ellisiana*, *Tarbellastraea* sp., *Thegioastraea* cf. *multisepta*, and *Lithophyllia* sp.

Wadi Asal–Wadi Zug El Bohar area (Figure D7.8)

The eastern side of the major basement block of Ras Zarib is subdivided into a series of small horsts and graben which are bounded by Clysmic (N140) and oblique (N115) oriented faults (Figure D7.6). The crests of these secondary fault blocks were submerged relatively late in the mid-Miocene, and hence reef growth occurred mainly on their eastern flanks. Small inner graben were flooded when reefs were growing on the platform margin, but coarse terrigenous materials were transported through these structural depressions, preventing flourishing coral assemblages from colonizing these areas. However a faviid coral assemblage developed, probably during a brief episode of reduced siliciclastic supply (Figure D7.8(d)).

Reef sequences fringing the north-east side of the fault blocks have been subdivided into three distinct coral reefs (R1, R2 and R3). The whole reef complex rests on terrigenous slope deposits. These consist of a lower coarse lag containing boulders of Nubian sandstone and granite which is overlain by sandstones and gravels locally rich in coralline algal fragments and rhodoliths. Initial dipping of this sedimentary slope was about 10° to the north-east, but later increased to about 20° as a result of syn- and post-reefal tilting (Figure D7.8).

The oldest reef (R1), which is the best developed and reaches about 25 m in thickness, shows an overall progradation of 350 m towards the north-east. It consists of massive carbonate especially rich in encrusting coralline algae, large mytilid bivalves, echinoids, crabs and coral colonies, intensively bored by *Lithodomus* and clionid sponges. Reef builders include a rather diverse coral fauna with *Porites* cf. *collegniana*, *Stylophora regulata*, *Tarbellastraea ellisiana*, *Tarbellastraea* sp., *Montastraea mellahica*, *Thegioastraea diversiformis*, *Thegioastraea multisepta*, *Favites neglecta* var. *minor*, and *Aquitanophyllia grandistellae*. This reef, crossed by neptunian dikes, is truncated by a normal Clysmic fault (N140) showing a 10 m throw, the steep plane of which is bored by large *Lithodomus* (Figure D7.8(c)). A slight (3 m) reactivation of this fault also occurred after the development of the youngest reef (R3).

A sandy unit is deposited above and in front of the first reef (R1), and includes a second reef (R2) which shows an apparent biostromal geometry. This resulted from reef truncation by a synsedimentary fault-system before the development of R3 (Figure D7.8(a)–(c)). In detail, 5 m thick, the R2 may be subdivided into three lenticular units interbedded with gravels. Reef-building assemblages are dominated by massive faviid and mussid corals (in particular *Aquitanophyllia grandistellae*), while near the reef top, columnar *Porites* colonies are associated with large rhodoliths. Neptunian dikes,

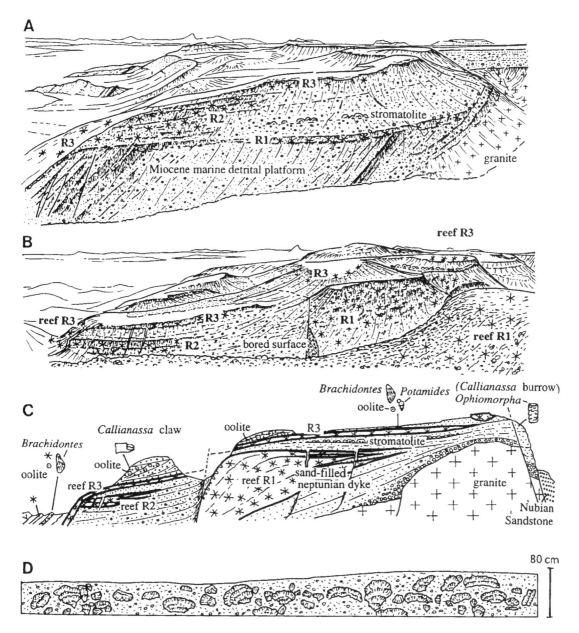

Figure D7.8 Fringing reefs in a syn-rift tectonic setting south of Wadi Zug el Bohar. The reef facies developed on the eastern side of narrow fault blocks consisting of Precambrian granite and succeeded to a detrital marine sloping platform showing synsedimentary tectonic features. A. Tilted platform edge with a granite substratum. Stromatolites and other local restricted marine facies developed in the back-reef depression. On the other hand, a final oolite veneer belongs to the same transitional strata as the *Brachidontes* and *Potamides*, illustrating the general restriction of the Red Sea basin before evaporite sedimentation. B. Maximal development of the first reef showing, in the background, the stromatolite facies developed in the back-reef depression. The faulted front was bored by bivalves prior to its burial by clastic sediments which are themselves affected by the fault reactivation. The later reefs are represented by thinner biostromes associated with faulting and final tilting towards the rift axis. C. Interpretative synthetic cross-section. D. Coral-rich bed typical of a reefal episode within the inter-horst corridors subjected to excessive detrital influx. This type of *in situ* reef coral assemblage is only preserved in the very best outcrops while poor preservation of similar facies might lead to misleading interpretation as a dense coral-framework.

0.5–1 m wide, indicate fracturing related to limited tilting. The structural relationships between R1 and R2 suggest a relative sea level fall of about 5 m probably related to minor tectonic uplift of the adjacent block rather than to eustatic sea level change. However, the overlying strata containing the youngest reef (R3) suggest a subsequent gradual sea level rise of about 15 m that may have been due to eustatic change.

R3 is a 1 m thick biostromal blanket, overlying both R1 and R2, and gradually thinning towards the highest part of the platform (Figure D7.8(a)–(c)). Reef-builders consist mainly of scattered massive coral colonies. This reef is tilted towards the north-east and envelops the truncated front of R2. It also extends to the west across lagoonal deposits which have filled a depression that was probably formed by tilting of R1. These deposits consist of quartz sands, oolites, potamid and mytilid molluscs, and dome-shaped stromatolites (but few open-marine organisms) (Figure D7.8(a)–(c)). Nearby sections, located about 2 km to the south-east, indicate that the lagoon with stromatolites extended laterally behind a local ridge formed by the nontilted reef, R2. This shows that restriction of lagoonal environments may be brought about by the presence of a local barrier of sedimentary and/or tectonic origin. *Botula* borings in the sandstone sill suggest that this represents a very shallow-marine surface.

The overlying terrigenous and oolitic sands form a prograding lenticular ridge of eastward-dipping, cross-bedded sands, locally reaching about 10 m in thickness. A particular facies characterized by reworked corals occurs within a channel cutting through the unfossiliferous, uppermost platform deposits. This may record either a fourth relative sea level rise, or a late reworking of reef facies during the initial phase of marine restriction. On the platform front, between the marine talus and the Miocene gypsum, there are inclined terrigenous layers containing oolites, mytilid bivalves and scattered coral fragments. These were derived from scouring and light erosion of the platform top, probably during the relative sea level fall associated with Miocene evaporite precipitation.

Sharm el Bahari–Sharm el Qibli area (Figure D7.9)
The hills bordered by these two wadis and the western valley enable the Miocene series to be studied in three dimensions. Palaeohighs reaching about 35 m above the present-day wadi floor were formed by subaerial erosion of the tilted, Oligo-Miocene, proto-rift series (Group A). This palaeotopography controlled the distribution of subsequent syn-rift coarse delta fans (B1). Today, both proto-rift and syn-rift formations are slightly tilted to the east. Several reef units outcrop in the wadis Sharm el Bahari and Sharm el Qibli. They developed preferentially on the palaeohighs formed by the Oligo-Miocene hills or the syn-rift delta fans (Figure D7.9).

At Sharm el Bahari, a biogenic limestone developed on top of a submarine conglomeratic fan which prograded towards the south-east (Figure D7.9, sections I and II). The lenticular coral-rich facies reaches its maximum thickness of 20 m towards the north-west and thins gradually downslope towards the south-east. A bioclastic slope dips at about 30° to the south-east, following the underlying fan slope. Patchy reef framework only occurs at the top of the hill and seems to have been built mainly by encrusting and branching coralline algae with scattered massive coral colonies, intensively bored by Lithophaginae. The particularly poor preservation of these corals may give misleading estimates of their relative abundance. Downslope, the massive carbonate facies is gradually replaced by a bioclastic packstone with scattered coral fragments, interbedded within terrigenous marine slope deposits. This relief was gradually buried by siliciclastic deposits followed by oolitic sands and then stromatolitic ridges and domes. These are overlain by horizontal microbial laminites and, further to the east, by Middle to late Miocene evaporites.

At Sharm el Qibli, three successive fringing reefs occur on the western side of a tilted palaeohigh of Group A deposits, and fringe the narrow marine platform slopes deepening towards the south-west (Figure D7.9, sections III and IV). The dips of these slopes decrease gradually upwards in the platform sequence and the surface on which the youngest reef grew is subhorizontal. The initial reef (R1) developed on a gravel layer a few metres thick which caps the palaeohigh. These gravels, which now outcrop over the platform top, indicate that, at that time, the palaeohigh was buried under the same marine fan delta. The latter is cut by the wadi and better exposed on its northern side (Figure D7.9). The R1 fringing reef is followed by two successive buildups (R2 and R3) which are developed on a later terrigenous unit and record a slight westward shift of location relative to R1. Both reefs R2 and R3 appear to occur somewhat lower, topographically, than R1, due to a late westward tilting related to a major fault bordering the Ras Zarib basement block. The coral fauna is represented both by branching and massive colonies including dominant *Stylophora regulata* associated with *Porites collegniana*, *Montastraea alloiteaui*, *Tarbellastraea ellisiana*, *Mussismilia provincialis* and *Lithophyllia michelotti*. Reef palaeoslopes locally consist of siliceous debris flows rich in molluscs and coralline algae, dipping to the south-west. This shows that relative heavy terrigenous supply is not necessarily incompatible with reef growth, provided that siliciclastic inputs are sporadic.

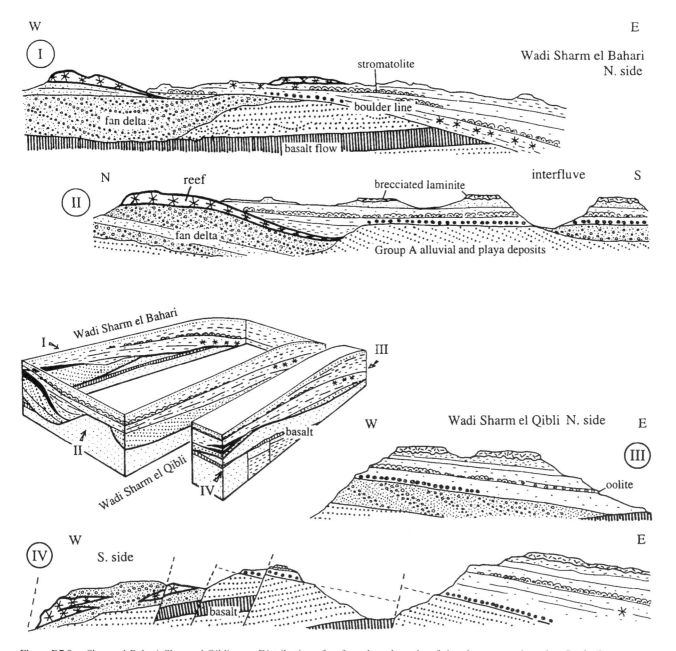

Figure D7.9 Sharm el Bahari–Sharm el Qibli area. Distribution of reef corals and coral reefs in a lower tectonic setting. Reef substrata are palaeohighs built by detrital sediments accumulated within corridors between the local early-Miocene highs. A contrasted erosional surface carved hills and valleys which then controlled transport and accumulation of the marine mid-Miocene coarse detrital Gilbert-type fan-deltas. Sharm el Bahari and Sharm el Qibli reefs developed on these sedimentary submarine palaeohighs, with slope deposits dipping southwards or eastwards according to the nearby residual depressions. Subsequent differential subsidence induced a fan-shaped geometry of strata dipping eastwards to the rift axis. These sediments, in particular the stromatolite facies record marine ecological restriction although a few open-marine episodes favoured brief reef coral development (but no coral reefs) on the initially low-angle slope, dipping of which has been later on increased by local tectonics. The uppermost laminites are folded and brecciated during a seismic-triggered sliding, which may reflect an episode of this post-depositional tilting.

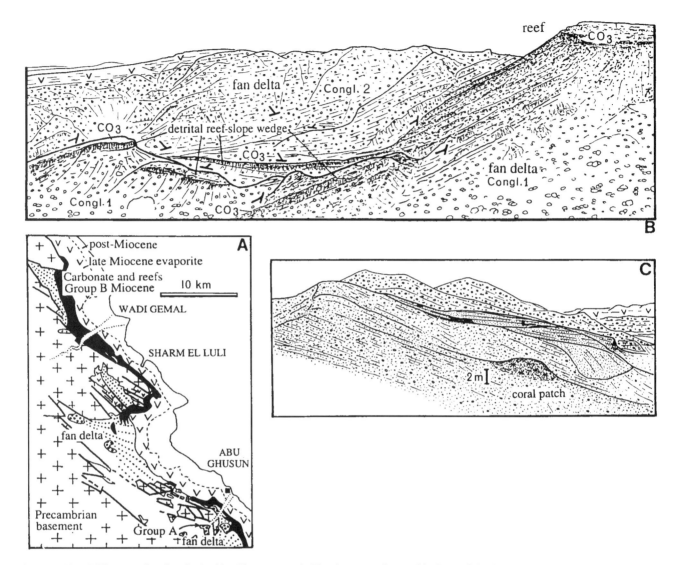

Figure D7.10 Different reef settings in the Abu Ghusun area. A. The almost continuous fringing reef developed on the Group B detrital positive reliefs filling the oblique structural corridors near Ras Honkorab or on the eastern side of the basement horsts (Sharm el Luli, north-west of Abu Ghusun) and also on Early Miocene continental Group A residual reliefs (south of Abu Ghusun). B–C. Geometrical relationships between reef development and coarse alluvial fans, south of Abu Ghusun. B. The coral-rich facies onlapping a detrital shelf (1) was locally buried by later conglomerates (2). Carbonate wedges containing abundant coral debris are the best preserved reef coral facies. C. However, a very local, small lens consisting of *in situ* corals may be interpreted as a coral patch developed on the gravel slope during a decrease of detrital influx.

In this area, a subsequent minor relative sea level fall of about 12 m is expressed by a lower horizontal surface associated with a boulder lag. This erosional discontinuity separates the lower marine sequences containing the coral reefs from an upper marine series gradually onlapping the underlying submarine relief. This upper series contains stromatolite levels which are locally well developed and reflect the onset of marine restriction. The overall geometry of the depositional sequences, which show a marked eastward thickening without any clear gradient in grain-size, suggests basinwards differential subsidence. The only clear facies change within this low-angle platform slope is recorded by the limitation of open-marine biota to the eastern parts of the platform. Here, there are rhodolith-rich facies, lenticular layers containing coral fragments and scattered colonies, and levels with a diverse marine molluscan fauna. These assemblages clearly contrast with the more restricted unfossiliferous sands and clays with large-sized stromatolites which dominate the western parts of the platform.

Abu Ghusun–Ras Honkorab area (Figure D7.10)

Wadi Abu Ghusun coincides with a major cross-fault system. South of the wadi, the Precambrian basement is downfaulted, and reef facies overlie a coarse Gilbert-delta fan which displays steep slopes with clastics and coarse interbedded wedges containing sulphatized coral blocks (Figure D7.9). North of the wadi, a clysmically oriented fault block formed a continuous barrier, 10 km long. Along its axial side, a complex of fringing reefs developed on a narrrow terrace. Similarly, some 10 km to the north, the structural high of the easternmost fault block of Ras Honkorab favoured flourishing reef growth on its north-east side (e.g. at Sharm El Luli, see below). In both localities, however, coral reefs, although occurring in zones sheltered from high siliclastic sedimentation, developed on narrow terrigenous shelves. Subsequent deformation has changed the structural context of each area, resulting in very different outcrop conditions. An important uplift affected the reef of Abu Ghusun which was almost completely broken up during the Miocene and is represented mainly by reefal olistoliths (Purser et al., 1996, and this volume). In contrast, the reefs located north-east of Ras Honkorab occur at the base of the basement slope.

Abu Ghusun
The Abu Ghusun platform displays synsedimentary tectonic features suggesting that uplift occurred relatively early. Within the uppermost part of the platform, where coral colonies can be seen in growth position, subvertical neptunian dikes, filled with sandstones, record the early stages of platform dislocation which mainly affected its frontal zone. On the eastern platform margin, talus deposits 10 m thick, including reefal olistoliths, overlie the Precambrian basement and the tilted Oligo-Miocene continental series. This geometry has possibly resulted from the truncation of the front of a narrow shelf, by a mid-Miocene fault scarp. The plurimetric olistoliths, reefal components of which vary from block to block, are inclined in all directions (Purser *et al.*, 1996). A particularly large olistolith, 22 m long, occurs near the footslope forming a prominent outcrop which could easily be confused with a patch-reef. Within this large carbonate block, inconsistent orientations of corals, together with the occurrence of the five different taxonomic assemblages, enable at least ten individual smaller-sized blocks to be distinguished. The similar growth-orientation of coral colonies within each single block suggests that these blocks are an agglomeration of olistoliths located at the base of a gully and that it was derived from a coral reef which consisted of various reef-building assemblages. The most common corals include columnar *Porites*, which is dominant at the platform top, *Thegioastraea*, *Tarbellastraea* and *Favites*. The overall coral richness, however, is similar to that observed in other mid-Miocene reefs of the region, while the molluscan fauna includes frequent large-sized cowries, *Conus* and *Lithodomus*.

Sharm el Luli
At Sharm el Luli (Figure 7.10(a)), synsedimentary tectonics, consisting of repeated tilting, have given rise to an overall downstepping geometry of the three successive reef units (R1, R2 and R3), but the tectonics are also expressed as sedimentary discontinuities, olistoliths and slumping structures (Purser *et al.*, 1996). The minimal amount of relative uplift, about 20 m, can be estimated from the difference in altitude between the first and last reef unit. The Precambrian basement is encrusted by laminated, partly brecciated microbialites, this being the only evidence here of the early Miocene, pre-evaporite, restricted marine phase (Orszag-Sperber *et al.*, this volume). Terrigenous sands with coralline algal, bryozoan and molluscan debris form two massive beds totalling 30 m in thickness and overlain by the first coral reef, R1. These basal terrigenous deposits are dipping 25° to the north-east. This material prograded to the north as shown by the 10° dip of its internal oblique stratification. This indicates that these gravelly sands were supplied by the western rift border and transported around the end of the block. They accreted northwards, building a narrow platform, and obliquely filled the depression along the basinwards side of the block, so forming the substrate of the first fringing reef. This reef prograded towards the north-east. Massive coral colonies in growth position are scattered within a muddy carbonate sand and include *Porites*, *Thegioastraea*, *Tarbellastraea* and *Favites*. The reef-slope facies consists of fine-grained, mixed siliciclastic–bioclastic sands. The present-day inclination of the top of the reef-platform has resulted from subsequent tilting of about 10° to the north-east, as confirmed by inclined geopetal cavity-infillings. The reef front is overlain by siliciclastic sands dipping at 20° towards the north-east, suggesting an original sedimentary slope of less than 10°.

The second reef (R2) developed on the north-eastern margin of R1 and the elevation of its upper surface is about 10 m lower, topographically. The geopetal infillings within reef cavities are statistically dipping at 5° to the north-east, confirming that uplift of R1 is essentially due to tilting. The slight dip of R2, together with neptunian dikes and large reefal olistoliths on the frontal slopes, indicate synsedimentary tectonic movements. A facies rich in plate-like (2 cm thick) colonies of *Tarbellastraea* is a characteristic assemblage of this particular reef. Platey coral facies may indicate reduced light intensity (Insalaco, 1996) perhaps suggesting growth in somewhat deeper water than the other coral facies in this region. The present-day elevations of the slightly tilted reef flat, and the base of the adjacent olistolith, show a difference of about 10 m. This does not seem to be related to sedimentary instability but rather suggests a dislocation due to the partial seismic

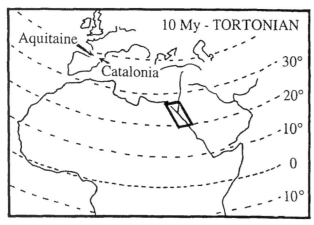

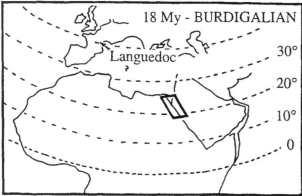

Figure D7.11 Evolution of palaeolatitude in the studied area from early Miocene to late Miocene times, after Ricou (1994).

liquefaction of the underlying sands (Plaziat and Purser, this volume).

Reef R3 is subhorizontal and its overall geometry seems to record a palaeo-sea-level a few metres lower than the post-tilting crest of R2. This slight sea-level fall could explain the *in situ* reworking of many coral colonies within R3, some of which show three different orientations of their geopetal infillings of borings (Purser *et al.*, 1996). Molluscs including *Periglypta* and *Conus* are locally abundant. An asymmetric depression about 5–10 m deep is oriented parallel to the reef front and is located in the middle of the R3 reef-flat, the eastern part of which is characterized by abundant coral colonies in growth position. The steep western slope of this depression is inclined towards the north-east and corresponds to a synsedimentary fault-plane bored by *Lithodomus*, *Aspidopholas* and *Cliona*. The opposite slope rises gently to the east up to the same level as the western reef-flat and represents a reef-flat surface which has been tilted by this clysmically oriented synsedimentary fault. The development of R3 therefore has to be regarded as syn-tectonic in spite of the overall subhorizontal geometry of the reef. In addition, the western part of this reef is affected by clastic-infilled dikes with a N130–135 orientation.

Beach boulders and cobbles of reefal carbonates fill the deeper part of the intra-reefal structural depression. Later oolitic and terrigenous sands contain a nearly monospecific molluscan fauna consisting of abundant *Brachidontes* and rare *Chlamys* associated with very small bivalves. This fauna records the onset of pre-evaporitic restriction. A similar mytilid bivalve facies also covers the eastern reef talus. The latter is dipping at 25° to the north-east where it is overlain by a blanket of microbial laminites capping the highest parts of the reef-flat and the reef-front. Marls with coral fragments, together with subhorizontally bedded, unfossiliferous sandstones, fill the local depressions of the tectonically generated palaeotopography, and are also sealed by 3 m of laminated carbonates. These first stages of marine restriction are followed by gypsum deposits.

Discussion

Comparison of the syn-reef tectonic history relating to the neighbouring Abu Ghusun and Sharm el Luli reefs demonstrates that differences in amplitude of local structural movements may completely modify reef preservation. In addition to the relatively spectacular reefs formed in front of structural blocks, other coral buildups situated both to the south-west of the Ras Honkorab block and to the north of Sharm el Luli (at Wadi Gemal) are developed on coarse alluvial fans. The geometry of these reefs often consists of subhorizontal biostromes and palaeoslopes with coarse coral fragments, or less frequent coral-patches overlying gravel and boulder slopes at a palaeodepth of about 20 m. Reef organisms include faviid corals, *Stylophora*, molluscs (*Lima*, *Cypraea*), crabs (especially *Daira*) and echinoids (*Echinometra mathaei*, cf. Soudet in Montenat *et al.*, 1986b and Negretti *et al.*, 1990). Some of the most external reef talus deposits which are directly overlain by evaporites, are partly replaced by gypsum (Orszag-Sperber *et al.*, 1986), making detailed study of peri-reefal talus (e.g. at Ras Honkorab) difficult. However, when all the reef occurrences in the region are considered together, it appears that reef development generally succeeded a relatively long phase of coarse siliciclastic sedimentation. In particular, at a locality situated 5 km north of Sharm el Luli, sediments occurring immediately below the reefs have been dated as Upper Langhian by Philobbos *et al.* (1993). The coarse clastic substratum supporting small coral patches was taken to represent the beginning of the middle Miocene, since it discordantly overlies Group A sediments dated at Ras Honkorab (Montenat *et al.*, 1986a) and at Safaga (Thiriet *et al.*, 1986) as Aquitanian. The main reefal constructions are therefore probably Langhian–Serravallian in age.

Table D7.1 Determinations of Miocene coral species collected from the various studied reef locations and previous records of coral species in other localities from western Mediterranean: L, Languedoc and C, Catalonia (data from Chevalier, 1962); A, Aquitaine basin, SW France (data from Cahuzac and Chaix, 1993)

	Egyptian reefs	Western Mediterranean localities
POCILLOPORIDAE	*Stylophora regulata*	
	Stylophora cf. raristella	C?
	Madracis cf. decaphylla	
AGARICIIDAE	*Pavona cf. banoensis*	
PORITIDAE	*Porites collegniana*	A
	Porites cf. collegniana	
	Porites sp.	
FAVIIDAE	*Caulastraea* sp.	
	Favia sp.	
	Favites neglecta	L; C
	Favites cf. neglecta	
	Favites neglecta var. *minor*	
	Favites (Dictyoastrea) profunda	
	Montastraea alloiteaui	
	Montastraea cf. alloiteaui	
	Montastraea mellahica	L; C
	Montastraea cf. mellahica	
	Montastraea sp.	
	Tarbellastraea ellisiana	A
	Tarbellastraea reussiana	L; C; A
	Tarbellastraea spp.	
	Thegioastraea diversiformis	
	Thegioastraea cf. multisepta	
MUSSIDAE	*Mussismilia provincialis*	
	Acanthastraea sp.	
	Aquitanophyllia grandistellae	
	Lithophyllia patula	
	Lithophyllia gigas	L; C?
	Lithophyllia michelotti	
	Lithophyllia sp.	
DENDROPHYLLIIDAE	*Balanophyllia* sp.	
	Dendrophyllia cf. colonjoni	

DISCUSSION

The coral fauna

Composition and diversity

A check list of coral species is given in Table D7.1. Specimens from Gebel Abu Shaar and Esh Mellaha are housed at the Laboratoire de Paléontologie, Muséum National d'Histoire Naturelle (Paris) while corals collected from the other localities, south of Hurghada are deposited in the Department of Palaeontology, the Natural History Museum (London). These two collections were identified in separate studies (see Introduction) and possible conspecifics and synonyms between them have yet to be checked. In most studied localities, corals are preserved in a dolomitized muddy carbonate matrix and skeletons are generally totally dissolved and/or replaced by dolosparite and dolomicrosparite (Purser, this volume). As already mentioned, in some places (e.g. Abu Ghusun area, see above), coral reef facies are completely or partly replaced by diagenetic gypsum, making coral determinations impossible at species or even generic level, and analysis of reef assemblages hazardous. Although coral diagenesis can vary greatly even within the same specimen and can affect estimation of species richness, the collections presented here appear to be reasonably comprehensive for a primary evaluation of coral diversity within the region.

The taxonomy of Miocene corals from the Mediterranean biogeographical province is based mainly on Chevalier's works (1954, 1962). Although several more detailed revisions of some taxa have been made since that time (Oosterbaan, 1988, 1990), a complete taxonomic revision of the Miocene Mediterranean corals is still needed, based on recent advances in systematics of modern scleractinian corals. For the Miocene coral fauna from the Gulf of Suez–Red Sea region in particular, previous taxonomic work is found only in the older literature. This consists of the short papers by Gregory (1898, 1906) on coral collections from the Suez area and from Gebels Esh Mellaha and Abu Shaar, and by Felix (1884, 1903, 1904) on corals from Sinai, the

Suez area and Middle Egypt. Moreover, most of the species described in these works are now considered invalid and/or lack adequate descriptions or illustrations. It will be difficult to revise them without making direct comparisons with these authors' original specimens. In the meantime, most of the coral species from the present study can be regarded, by default of such revision, as new records for the area, though taxonomic work is still at a preliminary stage.

The studied coral fauna comprises 16 genera and 27 species, of which only three species can be regarded as non-reef dwellers. The apparent low diversity of non-reef corals probably reflects a sampling bias towards coral reef facies. All the reefal species belong to the Miocene Mediterranean coral fauna, i.e. all corals identified here to species level are previously known only from the Miocene of the Mediterranean region. Some problematic specimens remaining unidentified at species level, however, eventually emerge as species previously unknown in the Mediterranean province, or even completely new species.

Since coral reefs are widely regarded as high diversity ecosystems, this coral fauna may appear relatively poor, especially when compared with Recent Indo-Pacific reefs. However, because at any one time the foremost factors affecting local taxonomic richness are the size of the global and the regional pools of taxa (Rosen, 1981; Perrin *et al.*, 1995), the richness of the present studied reefs should first be compared with the global mid-Miocene and Mediterranean coral faunas. For reef genera, these are 79 and 32, respectively, (Rosen, 1988, Table 2) compared with modern totals of 109 and zero (Veron, 1993). Thus the global generic figure for the mid-Miocene is considerably lower than the modern global total, and the mid-Miocene total for the Mediterranean is itself only a rather small proportion of its contemporaneous global total. The mid-Miocene Red Sea fauna amounts to half the Mediterranean total and, hence in its proper context, is not really as poor as might at first be thought.

Variations in species and generic richness within the different sites (Table D7.2) are highly dependent on several sampling factors.

1. The depth and detail of investigations were not equal at each individual site. The areas of Gebels Abu Shaar and Esh Mellaha, which both show the highest species and generic richness, were studied in much more detail than the other areas, with palaeoecological and taxonomic surveys of reefs and associated facies carried out by a coral specialist.
2. Sampling and evaluations of coral assemblages were also strongly influenced by the conditions of outcrop preservation, and by the state of diagenesis and weathering of coral skeletons. In particular, corals in some areas appear to be much better preserved as reworked fragments in reef talus and palaeoslopes than within reef core facies. In such cases, it is not possible to make proper estimates of composition of the different reef-building assemblages and relative abundances of taxa. Only the overall coral fauna can be assessed for the locality as a whole. Selective preservation of some taxa also needs to be taken into account, since fragile and branching corals in particular are more likely to be underestimated or simply not recorded. This might apply in some areas to genera like *Stylophora* and *Caulastraea* and possibly also *Acropora*.

Table D7.2 Distribution of the identified coral genera within the studied reef sites

		Esh Mellaha	Abu Shaar	Wadi Ambagi	Wadi el Aswad	Wadi Zug el Bohar	Sharm el Qibli	Abu Ghusun
POCILLOPORIDAE	*Stylophora*	+	+			+	+	
	Madracis		+					
AGARICIIDAE	*Pavona*	+						
PORITIDAE	*Porites*	+	+	+	+	+	+	+
FAVIIDAE	*Caulastrea*	+	+				?	
	Favia	+	+					+
	Favites	+	+	+	+	+		+
	Montastrea	+	+		+	+	+	
	Tarbellastraea	+	+	+	+	+	+	+
	Thegioastrea				+	+		+
MUSSIDAE	*Mussismilia*	+	+				+	
	Acanthastrea	+	+					
	Aquitanophyllia	+	+			+		
	Lithophyllia	+	+	+	+		+	+
DENDROPHYLLIIDAE	*Balanophyllia*	+	+					
	Dendrophyllia		+					

3. For several reasons including both 1. and 2. above, it was not possible to make equally good collections at each locality.

Bearing these problems in mind, the taxonomic records shown in Tables D7.1 and D7.2 provide a reasonable preliminary picture for coral distributions within the mid-Miocene reefs along the Egyptian coast of Gulf of Suez and Red Sea. At most localities, the coral fauna is largely dominated by *Porites* and faviids, or sometimes the less common, closely related mussids. This general pattern characterizes most Tertiary reefs of the Mediterranean coral province (Chevalier, 1962, 1977) but seems also to apply to at least some Miocene reef coral faunas in the Indo-Pacific realm, too (McCall *et al.*, 1994).

General palaeobiogeographical context
During the Middle Miocene, the region of the Gulf of Suez and northern Red Sea was connected to the main basin of the Mediterranean through a wide, open-marine seaway in the Suez area, while to the south, it was probably separated from the Indian Ocean by a land barrier (Adams *et al.*, 1983; Steininger and Rögl, 1984). Reef corals, as other tropical marine organisms, were distributed within three marine biogeographical provinces (the Mediterranean and the two present-day coral reef provinces of tropical Atlantic and Indo-Pacific) separated at that time by land or cross-latitude marine barriers. As far as reef corals are concerned, no faunal overlap seems to occur at the Miocene time between these three provinces, although this widely accepted pattern should be re-evaluated by an overall taxonomic revision of Miocene scleractinians. While the western Atlantic/Caribbean region seems to have been isolated from the Mediterranean at the beginning of the Neogene, open marine seaway with the Indo-Pacific through the Middle East area became gradually interrupted during the early Miocene as a result of the progressive tectonic emergence of a land barrier (see McCall et al., 1994 for discussion).

Faunal history
The exact timing of the divergence of Mediterranean open-marine faunas from those elsewhere within the Miocene differs according to particular authors but is widely thought to have happened during the Aquitanian or the Burdigalian (Chevalier, 1962, 1977; Adams *et al.*, 1983; Rögl and Steininger, 1983, 1984; McCall *et al.*, 1994). However, divergence of Indian Ocean and Mediterranean reef-coral (and echinoid) faunas seems to have begun prior to complete physical closure of marine seaways through the Middle East. It should be noted, however, that the Lower Miocene coral fauna from the Makran area in southern Iran described by McCall *et al.* (1994) is entirely of Indo-Pacific affinity and does not share any species with coral assemblages of the same age in the Mediterranean nor with the mid-Miocene coral fauna presented here.

By contrast, within the Mediterranean region, Egyptian corals show strong affinities with the Middle Miocene coral faunas of Languedoc (southern France), Catalonia (Spain) and the Aquitaine Basin (south-west France). Taxonomic richness of the mid-Miocene Egyptian fauna, both at generic and species levels, is similar to that of late Burdigalian–early Langhian reefal facies from Languedoc (20–25 species; 10–11 reef genera) and Catalonia (23–31 species including 7 nonreef-coral species; 19–21 genera) (Chevalier, 1962). The Middle Miocene fauna from south-west France seems to display a higher taxonomic richness, comprising 46–47 species and 23–24 genera, according to Chaix (in Cahuzac and Chaix, 1993). It should be noted, however, that this coral assemblage appears to occur near the northern limit of reef-coral development in this region (Figure D7.11) and actually consists of a mixture of reef- and nonreef forms, the latter representing roughly half (26 species and 14 genera) of the total fauna. The precise amount of faunal overlap between the Egyptian corals and those from south France and Spain is difficult to establish at species level on current knowledge because it is likely that some forms may prove to be synonyms when the respective faunas are eventually revised. At generic level, their overlap is 11 genera.

General biodiversity of Egyptian Middle Miocene reefs

As has already be mentioned for corals, the other groups of organisms occurring within the reefs (i.e. echinoids, molluscs, forams, etc.) show an apparently low relative diversity. This is not, however, a feature of the Egyptian Miocene alone, but may also apply to the Mediterranean faunas as a whole, as might be revealed by a critical review of taxonomic records from Mediterranean reef facies of different ages in the Miocene. Of particular importance is that these faunal lists often represent assemblages both from reefs and peri-reefal environments or are even based on facies which are interpreted as reefal just because they contain fragments of corals (including both reef and/or non-reef corals).

The Middle Miocene Egyptian molluscan fauna associated with coral reefs is characterized by a fairly low diversity and usually a small size of individuals. Strombidae seem to be absent and the Conidae and rather small Cypraeidae are nowhere frequent. Boring Lithophaginae are abundant but also of limited size. This fauna, however, is typical of open-marine waters with, in addition to the above-mentioned molluscs, large *Periglypta*, the echinoid *Echinometra*, the nautilid *Aturia aturi* and the crab *Daira*. Larger foraminifera are not known within reef-cores but were only observed within bioclastic facies where rare *Borelis* and *Amphistegina* occur. Moreover, it appears that even within the non-reef facies, which are the richest in molluscs, the fauna is often almost monospecific. This is the case for the *Chlamys (Aequipecten) submalvinae* (Blanckenhorn) levels which are abundant on the Abu Shaar platform and are associated with layers bearing the small oyster *Cubitostrea* cf. *fimbriata*.

It is clear that precise estimates of Egyptian Miocene coral reef biodiversity are limited by our present-day knowledge of these faunas. It should be stressed that comprehensive comparisons with other coral reef faunas from the Mediterranean and Indian Ocean will be impossible as along as taxonomic lists rigorously established from coral reefs *sensu stricto* are lacking, and until faunal lists of the last century are critically revised.

As the rather low biodiversity of the mid-Miocene Egyptian coral reefs does not seem to be confined locally or regionally to the Gulf of Suez–Red Sea basin, it cannot be related to unfavourably low temperatures or to unsuitable lagoon-like environments within the studied region. Although this fairly low diversity might reflect absence of direct exchange with the major ocean systems (Atlantic or Indo-Pacific), diversity is not greatly different from a hypothetical richness of contemporaneous Mediterranean faunas.

Local and regional controls on reef geometry and development

Tectonic control

Tectonics exerted both indirect and direct major controls on reef growth in three different ways.

1. The structural framework resulting from the initial stages of syn-rift tectonics provided the general antecedent palaeotopography which determined which sites on palaeohighs were favourable for reef development, and which areas had open marine waters (Figure D7.12).

2. The horst and graben pattern also constrained transport of terrigenous sediments through structural depressions, while palaeohighs protected reefs growing on their north-eastern sides from siliciclastic supply from the western rift-margin. This general palaeomorphology, together with the episodic transport of terrigenous sediments along structurally controlled pathways, is similar to that found today in the Gulf of Suez and the Gulf of Aqaba where coral reefs are fringing both the narrow shorelines and the fronts of alluvial fans (Roberts and Murray, 1984, 1988; Purser *et al.*, 1987).

3. Synsedimentary tectonics, which are mainly expressed by tilting of fault blocks, controls relative sea-level changes and hence the resulting accommodation space (Figure D7.12).

The regional rift setting has thus had a long-lasting influence on reef location. In these circumstances, it is often difficult to distinguish eustatic and tectonic roles in relative sea-level changes.

Siliciclastic sedimentation

Although terrigenous sedimentation may locally have prevented reef growth, co-occurrence of siliciclastic sedimentation and coral reefs was favoured both by palaeomorphology and regional climatic conditions. The warm and probably sub-arid climate prevailing during mid-Miocene times prevented intense chemical weathering of basement rocks and production of fine terrigenous sediment. High-water turbidites were rare. When climatic conditions induced episodes of coarse silicilastic sedimentation, their influence on coral reef development was secondarily confined by the structurally controlled palaeotopography. This sedimentation pattern does not appear to have inhibited reef development overall, as the sediments built shallow-water shelves and prevented reef growth only for brief intermittent periods. Intervening times favouring reef growth lasted much longer. Massive siliciclastic supply from the western rift border and its longshore transport on the western flank of the fault

Figure D7.12 Diverse types of mid-Miocene reef settings observed along the southern Gulf of Suez–north-western Red Sea coast. Final tilting is removed and a given stage of reef growth has been illustrated in some cases. A. On a tilted basement block, with its crest and eastern side protected from terrigenous sedimentation. A local environmental restriction on the back-reef platform seems to have favoured stromatolite development in this area but the widespread occurrence of the restricted marine facies at that time suggests a regional marine restriction phase. B. Reef development on a detrital shelf, geometry and sedimentation of which were influenced by a pre-rift strata horst. Reef location is not directly dependent on the buried horst but is related to the position of the platform edge. This low-setting and the post-reef subsidence may explain the final evaporite overlap. C. Synsedimentary tectonics induces relative sea-level changes between the successive reefs. The frontal part of the first reef is faulted. The fault-plane is then bored and buried by subsequent back-reef terrigenous deposits contemporanous to the growth of the second fringing reef. Occurrence of synsedimentary earthquakes is shown by open wedges within the later reef facies. D. Another type of syntectonic reef development with limited tilting of the reef core but with truncation of the reef front. Extensional fissure opening (syndepositional neptunian dikes) is related either to reef faulting and tilting or to seismic instability. E–H. Various expressions of reef growth on sedimentary frontal slope of a terrigenous platform. Steepness, tectonic instability (earthquakes) and rigidity (early or late lithification) are the main controlling factors. Low-angle slopes enable shallow reef-flat and deeper coral patches to develop (E). A steeper slope favours frontal dislocation (tilted reef blocks) and talus bevelling (F). Steep but stable setting induces a well-developed talus and continuous rimmed-platform affected by local open-fractures (G). Major tectonics and seismic instability of steep edges may be characterized by abundant olistoliths which may be the only trace of a former reef-front (H).

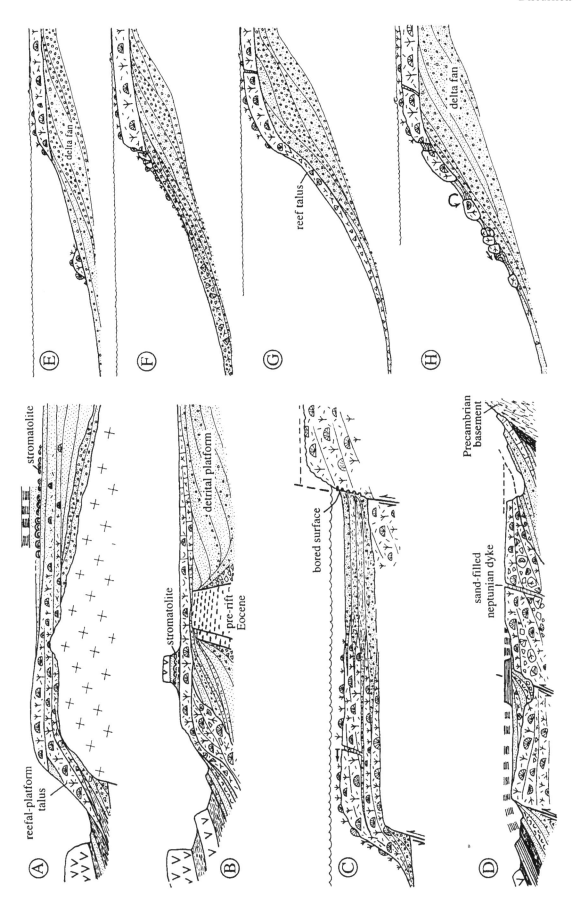

blocks nevertheless tended to limit reef development to the west (e.g. on the western ramp of the hangingwall of the Abu Shaar tilted-block).

Relative sea-level changes
On a regional scale, relative sea-level changes controlled the connection between the Gulf of Suez–Red Sea basin and the main open-marine basin of the Mediterranean, and hence the stability of open-marine waters within the axial zone of the rift (see below). On a platform scale, relative sea-level changes controlled the local extent of carbonate sedimentation and hence the location and development of coral reefs. Sea-level changes were caused both by the local tilting of fault blocks and by local and/or regional subsidence combined with eustatic changes (Figure D7.12). Synsedimentary tilting of fault blocks created accommodation space on the hangingwall slope and may have resulted in a local emergence and/or major erosion on the upper footwall deposits (e.g. Gebel Abu Shaar; see Cross *et al.*, this volume). Subsequent reef growth on the footwall margin was therefore necessarily associated with a relative sea-level rise resulting from subsidence and/or eustatic rise. Individual reef development and geometry were both directly controlled by the rate of relative sea-level rise and by the palaeomorphology of the antecedent reef substrate. In particular, the general steepness of the eastern footwall scarps prevented extensive reef progradation and favoured development of narrow fringing reefs. Regional and/or local sea level falls not only caused emergence and erosion, but also renewal of terrigenous sedimentation favouring new reef growth.

Increasingly restricted conditions
Increasing restriction of marine conditions through the mid-Miocene is indicated by the overall development of a shelf facies which is characterized by a very low abundance and taxonomic richness of marine organisms, association with stromatolites (Bali'h Member on Abu Shaar platform referred to the South Gharib Formation, cf. Plaziat *et al.*, this volume), with an upward progression to unfossiliferous microbial laminites (e.g. Chaotic Breccia Member) prior to evaporitic sedimentation. However, restriction of marine conditions took place gradually and episodically as indicated by alternation of these conditions with times when open marine sedimentation favoured periodic reef growth. This was the result of gradual closure of the Isthmus of Suez due to tectonic movements probably acting together with high frequency eustatic sea-level changes. Further implications of this gradual restriction are that it prevented flourishing reef growth and well-developed reef-building assemblages to thin biostromal blankets and eventually led to the regional extinction of coral faunas.

CONCLUSIONS

The distribution, composition and development of coral reefs within the Gulf of Suez and northern Red Sea are dependent on both global and regional factors. Global controls were related mainly to the major worldwide transgression of the Middle Miocene, whose maximum highstand is recorded at around 16 Ma, and the warm global climate prevailing at that time.

By contrast, regional and local environmental controls on coral reefs are strongly related to the tectonosedimentary setting of the rift system. In particular, the tectonically controlled palaeomorphology determined the availability of substrates suitable for coral reefs, influenced open-marine water palaeocirculation, and constrained transport of siliciclastic sediments. In addition, there was relatively little fine-grained terrigenous input because the regional climate was warm and sub-arid. These regional conditions favoured the development of small fringing reefs rather than large reef complexes, barrier-reefs or even large patch-reefs. Stratigraphical and geometrical relationships between reefs within the same platform were controlled by the interaction of eustatic and tectonic movements which also influenced the distribution of suitable substrates.

By determining the topographical relationships with neighbouring basins, rift tectonics exerted an indirect influence on the available species pool of corals and associated open-marine organisms within the area. The Miocene coral belong to the fauna of the Mediterranean coral reef province and there appears to be no faunal overlap at species level with the Miocene of the Indo-Pacific. Species richness of corals, though apparently poor, seems to be of the same order as reef coral assemblages of the same age in western Mediterranean, according to our present knowledge of the Eygptian Miocene fauna and the state of Tertiary coral taxonomy. Associated faunas also show an apparent taxonomic richness comparable to that of their Mediterranean equivalent. Therefore, the elongate shape of the Miocene Gulf of Suez–northern Red Sea basin, and the probably narrow seaway in the Suez area, do not appear to have limited faunal exchanges between this basin and the Mediterranean and open marine faunas seem to have been in biogeographical continuity. It should be noted, however, that a similar situation exists today: the elongate and relatively narrow basin of the Red Sea is connected to the Indian Ocean by the narrow and shallow seaway at Bab el Mandeb and yet it has one of the highest recorded coral diversities west of Australia and Indonesia (Sheppard, 1987).

The Mediterranean coral fauna and associated coral reefs disappeared from the Gulf of Suez–northern Red Sea region during the Serravallian. This is earlier than the final disappearance of reef-corals from the Mediterranean basin as a whole, at the end of the Miocene and reflects early regional isolation resulting from the gradual tectonic closure of the Suez area seaway.

ACKNOWLEDGEMENTS

Research by C.P. was undertaken while in receipt of a Natural Environment Research Council (UK) project on 'Palaeozonation and Quantitative Palaeobathymetry of Cenozoic Reefs' (Grant GR3/8143, to C.P., B.R.R., and Dan Bosence of Royal Holloway, University of London). Fieldwork by J.C.P. was undertaken with funds from the EC 'RED SED' Programme. These research funds are gratefully acknowledged. B.R.R. also wishes to thank Marie-Laure Caparros and Margaret Hughes for their help with taxonomical work on the corals.

Chapter D8
Miocene periplatform slope sedimentation in the north-western Red Sea rift, Egypt

B. H. Purser and J.-C. Plaziat

ABSTRACT

Syn-rift tectonics in the north-west Red Sea are expressed by multiple fault blocks whose surfaces are generally tilted. This highly irregular morphology was formed during the early Miocene and progressively buried during the early- and mid-Miocene transgression when most sediments were deposited on inclined surfaces. Five examples of sedimentation on steep footwall escarpments of tilted Precambrian basement blocks are discussed. Only the uppermost parts of slopes are exposed and the discussion relates to the proximal segments (10–250 m) of more extensive slopes (possibly exceeding 1000 m), most of which are buried below the Quaternary plains.

Sediments deposited on these slopes include both carbonates and siliciclastics, the latter derived from adjacent pre-Miocene, substratum. The siliciclastic sediments are related to basement relief and to the frequency of cross-faults which tend to concentrate terrigenous supply. The relatively steep inclinations, averaging 20° but locally attaining 40°, are maintained by the heterogeneous textures and angular nature of the components, many of which are derived from platform-edge reef complexes. The numerous sedimentary features illustrated in this contribution record several major processes. Sediment production may be local, generally in the form of corals, molluscs, red and green algae, and stromatolites. However, most sediment is derived from the shelf and top of the slope, generally situated close to Miocene sea level, and delivered to the talus via mass- or grain-flows. The sediments are not significantly modified by burrowing organisms and are frequently affected by an active submarine cementation. They are constantly remobilized during periods of tectonic instability resulting in the formation of steep, by-pass surfaces below which sediments, including numerous olistoliths, may locally onlap the slope. Miocene erosion of the slopes and platform margins ranges in scale from kilometric blocks to smaller (200 m) lenticular units whose gravitational displacement leaves lunate scars on the footwall margin. Smaller-scale erosion includes the formation of gulleys and planar erosion surfaces abraded by the continual down-slope movement of sediments. Proximal slope environments are often transit zones for sediments which are ultimately stabilized lower on the slope.

The mid-Miocene sediments of the north-western Red Sea clearly illustrate the factors initiating platform and slope sedimentation. Relief has been created not by sedimentary aggradation but mainly by tectonic processes in the form of tilted fault blocks whose crests and flanks determine sediment type and geometry. Their understanding and prediction depends essentially on the understanding of the synsedimentary structural framework. Their occurrence also closely relates to their position with respect to the structural axis or periphery of the rift.

INTRODUCTION

Rift basins are characterized by multiple phases of normal faulting and the creation of numerous structural blocks (Montenat et al., this volume). Because these blocks are formed and reactivated during sedimentation they exert an important influence on sediment composition and geometry. Since the initiation of the Red Sea during the late Oligocene, bathymetric relief has been an important factor influencing sedimentation. Although relatively modest during the initial proto-rift phase (Orszag *et al.*, this volume), it was accentuated during the subsequent early Miocene transgression which drowned a very irregular bathymetric relief dominated by the presence of high (possibly exceeding 1000 m) and relatively steep (10°–40°) slopes.

This early Miocene relief expressed the major phase of rifting in this part of the basin (Bosworth *et al.*, this volume). Deep (3000 m) structural depressions were filled with early- and mid-Miocene marls and late-Miocene evaporites. They are bordered by structural highs which persisted throughout much of the Miocene. This early relief has been frequently reactivated and influenced Pliocene and Quaternary sedimentation, and today is expressed both on land and in the marine bathymetry. Marine slopes are thus a permanent feature of the rift sedimentation.

This contribution discusses only mid-Miocene (Langhian–Serravallian), shallow-marine slope processes. The morpho-tectonic relief created by early Miocene extension served as a substratum for mid-Miocene sedimentation and, although sedimentation continued into the late Miocene, it became evaporitic during the Serravallian (Orszag-Sperber *et al.*, this volume).

The multiple early Miocene blocks which constitute many of the coastal hills of the north-west Red Sea (Figure D8.1) and south-west Gulf of Suez region generally are composed of Precambrian basement covered by a veneer of pre-rift Mesozoic and Paleogene sediments, the principal oil reservoirs within the Gulf of Suez. Most of these structural units are elongate parallel to the axis of the rift – the so-called 'Clysmic direction' (Hume, 1921), and are tilted either towards the axis or to the periphery of the rift. Furthermore, their main axes generally plunge and thus the structural culmination of most blocks is of limited dimensions (Purser *et al.*, this volume). In other words, the substratum for early- and mid-Miocene sedimentation is almost everywhere inclined and the shallow-marine highs did not possess the subhorizontal morphology frequently associated with carbonate platforms. Therefore many sediments are slope deposits with typical clinoform bedding (Figure D8.2). This implies that sediment production, in particular carbonate, was limited to a relatively small surface at the platform top; this is especially the case where the structural culmination remained above Miocene sea level. In this case, carbonate production was confined to a system of narrow fringing reefs probably comparable to those of the modern Gulf of Aqaba (Dullo and Montaggioni, this volume). This situation was prevalent during the relatively early stages of the Miocene transgression and with time, slope accretion, particularly on relatively low-angle dip-slope ramps, gradually

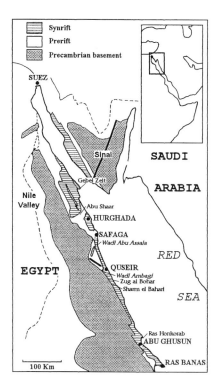

Figure D8.1 Map of the north-western Red Sea and south-western Gulf of Suez showing the distribution of areas studied.

developed a horizontality, as is described on Abu Shaar by Cross *et al.* (this volume). Steeper footwall slopes, on the contrary, tended to retain angles generally exceeding 15°. These constitute the principal subject of this contribution.

General attributes

The mid-Miocene slope deposits of the north-west Red Sea (Egypt) are often characterized by relatively steep angles (Figure D8.2); while 15° is typical, they may locally attain 40°, and the presence of numerous horizontal geopetal structures clearly indicate only slight post-Miocene tilting. The slope deposits exhibit many gravitational effects and they are characterized by coarse (sand and gravel) sedimentary textures often with a silty matrix. Reefal olistoliths, many exceeding 10 m in diameter, are frequent and their presence indicates instability of the upper parts of the slope.

Miocene slope sediments have five principal sources:

- Many slopes grade upwards into massive reefs which have grown either on the footwall crest or have fringed the steep flanks of emergent blocks. These reefs produce coarse detritus, including numerous olistoliths.
- Oolitic and bioclastic sands produced on the platform may be swept over the well-defined edge to accumulate on the slope. However, the quantity concerned is

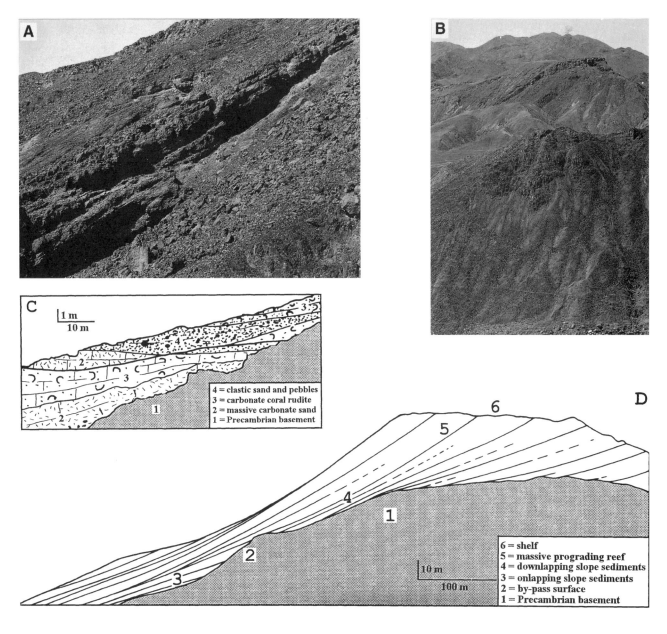

limited, presumably because of limited subhorizontal substrates favouring oolite development.
- Both calcareous and argillaceous muds are decanted on to the slopes. However, the sources of these fine suspensions may be various, including the platform culmination.
- Autochthonous marine benthos living on the slopes have contributed significant amounts of coarse carbonate. They include corals, molluscs, red and green algae, and domed stromatolites.
- Noncarbonate detritus derived from the Precambrian crystalline basement, together with cobbles of pre-rift Eocene chert, are major constituents particularly where the slope has developed against an emergent basement block (Figure D8.2).

Correlation between different slopes and with contemporaneous platform and basin deposits

In principle, sedimentary sequences deposited on slopes should record events affecting the shallower platform environments (Reijmer and Evaars, 1991). Prograding reefs may be directly correlated with their coarse proximal detritus. However, where slope relief is relatively high, locally attaining 250 m (at outcrop, near Safaga, see below), it is rarely possible to trace any given unit to its platform source due to the presence of numerous discontinuities. These unconformities may have subvertical relief and were submarine escarpments over which sediment was by-passed during movement down-slope. The nature and origins of these unconformities, a characteristic feature of these mid-Miocene slope deposits, will be discussed as their presence tends to preclude direct correlation with platform sequences. In some cases, notably at Abu Shaar, the overall sedimentary evolution from open-marine facies rich in coral and other biogenic debris, to peritidal environments with oolite and green algae, is recorded both on the platform and in the slope deposits enabling a general correlation to be made. However, slope deposits rarely record the more localized events affecting the platform top, e.g. modest fluctuations in relative sea level. Furthermore, the numerous angular discontinuities are often very local and many can not be correlated laterally along the same slope. Their development and limited extension reflect slope instability generally conditioned by tectonic movements.

Visible relief of slope deposits at outcrop is often in the order of 50 m, attaining a maximum of about 250 m at Wadi Abu Assala (Safaga, Figures D8.1, D8.2). The lower parts of the slopes and adjacent basins are masked by the Quaternary plain. However, subsurface data in the northern Red Sea (Miller and Barakat, 1988) and in the southern Gulf of Suez (Bosworth, 1994a) indicate relief locally exceeding 2000 m. While this relief may be less in the peripheral areas of the rift, it is nevertheless clear that the exposed slope deposits, discussed in this contribution, represent only the upper, proximal parts of a more extensive system. Because no wells are situated close to outcrop, the precise dimensions of any given slope have not been established.

These mid-Miocene slope deposits are a significant element of rift sedimentation and record the early stages of platform development. Because many are directly connected to reefs and beach deposits they record absolute depths of specific ecological and diagenetic phenomena. Furthermore, because processes of formation are generally dominated by mass-flows and other collapse events, they record the relative tectonic instability of the rift.

THE MID-MIOCENE SLOPES AND THEIR SEDIMENTS

The Miocene platforms have been discussed by Montenat *et al.* (1986), Purser *et al.* (1987, 1990, this volume), Burchette (1988), James *et al.* (1988), Purser and Philobbos (1993), and Cross *et al.* (this volume) and all contributions note the presence of platform slope deposits. However, information concerning the sedimentary properties of these facies and their processes of deposition are relatively limited. Although virtually all structural blocks exposed along the north-west Red Sea and south Gulf of Suez coasts possess slope deposits, the present authors have selected five (Figure D8.1) which are considered to be representative. While some lie directly on high peripheral basement relief (Safaga), others (such as Abu Shaar and Esh Mellaha) flank individual blocks separated from the rift periphery by major structural depressions. The slopes examined are generally oriented towards the axis of the rift. Those situated near the Gulf of Suez rest on steep footwall escarpments, the blocks being tilted towards the rift margin. Others in the north-west Red Sea have developed on hangingwall dip slopes, the blocks being tilted

Figure D8.2 Slope deposits at Wadi Abu Assala (Safaga). A. Lower slope deposits onlapping Precambrian basement. B. Panorama to south-west of Wadi Abu Assala showing clinoforms in front of high basement relief. C. Sketch of main facies illustrated in photo A. D. Profile showing geometry of the clinoforms which onlap and downlap the Precambrian basement. E. Panorama of clinoforms downlapping basement, north of Wadi Abu Assala.

towards the rift axis. Their opposing footwall scarps are generally buried in coarse terrigenous sediments derived from the rift periphery (discussed by Purser *et al.*, this volume). Therefore the structural context and synsedimentary tectonics differ, and these variations influence the general character of the respective slope deposits.

Wadi Abu Assala, Safaga

This area was examined initially by Thiriet *et al.* (1986) and by Bosworth *et al.* (this volume, and for location) and is characterized by high (833 m) Precambrian basement relief which is terraced locally at about 250 m. The terrace coincides with the development of narrow (200–500 m) fringing carbonate shelves composed of massive reefs (Figure D8.2(e)). Although the top of the shelf (platform) is subhorizontal, it is composed almost entirely of massive, slightly concave clinoform beds dipping towards the south-east. This slope relates to a major east–west cross-fault approximately coincident with the modern wadi. The clinoform deposits thus overly a fault scarp which was modified during the mid-Miocene transgression to form a somewhat irregular substratum inclined at about 15° (Figure D8.2).

These slope deposits locally comprise two major depositional sequences (Figure D8.2). The lower half is a 10–20 m thick series of massively bedded, mixed carbonate and siliciclastic detritus which onlaps the inclined basement surface. This onlap sequence contains abundant, coarse coral debris which must have been derived from a higher reef. The higher parts of the talus, on the contrary, downlap both the inclined basement substratum and the preceeding onlap sequence. The uppermost parts of this upper sequence are the direct extension of the massive reef which culminates the slope and, in common with the basal onlapping sequence, contains large reef olistoliths and coral debris.

The two sequences are best developed within embayments in the inclined basement substratum, and are absent on intervening promontories. These depositional sequences may be interpreted in two ways:

1. The lower, onlapping sequence, may be the product of a lower sea level, the upper (downlapping) unit recording stabilization and lateral accretion during maximum sea level.
2. Both sequences are the product of a common, high sea-level stand clearly recorded by the clinoforms constituting the narrow reef-shelf (Figure D8.3).

The eventual contemporaneity of the two sequences is caused by a local steepening of the substratum which became too steep to support the downlapping sands and gravels favouring the creation of a by-pass zone. Detritus exported over the by-pass stabilized further down the slope to onlap the basement. With the

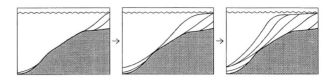

Figure D8.3 Sketch showing interpretation of contemporaneous onlapping and downlapping clinoforms.

resulting decrease in slope, the by-pass zone was gradually buried and the contemporaneous downlapping sequence prograded over the bathymetrically lower onlap sequence. Support for this second hypothesis comes from the presence of a marked steepening of the lower part of the Precambrian substratum, and by the presence of abundant reef detritus within the lower, onlapping sequence. This detritus clearly indicates the existence of *in situ* reefs above, the sole trace of which is the massive reef complex, clearly the source of the second, downlapping sequence. A similar situation is present at other localities, notably at Abu Shaar and also around the deeper flanks of the Bahama platforms (Grammer *et al.*, 1993) where coarse Quaternary detritus derived from the peripheral escarpment and the platform edge during relatively low sea-level onlap the steep by-pass escarpment.

Lateral variations in slope stratigraphy

Slope deposits, attaining a maximum thickness (measured normal to bedding) of about 80 m, vary in composition and geometry around the palaeoembayment cut by the Abu Assala cross-fault. The steeply inclined scarp, bevelled during sea-level rise, was not a simple linear surface but consisted of a series of secondary embayments and narrow promontories probably resulting from coastal erosion of the highly fractured and faulted basement. These morphological irregularities controlled to a considerable degree the nature of slope sedimentation and thus its lateral variations.

1. The major depressions (embayments), possibly coinciding with the intersection of two fault systems (Bosworth *et al.*, this volume) favour deposition of coarse conglomerates composed of rounded pebbles and cobbles of basement, often with a sandy matrix, mixed with scattered biogenic debris. The embayment probably linked to a fluviatile system comparable with the 'sharms', a characteristic element of the modern Red Sea coastline (Plaziat *et al.*, this volume).
2. Regular surfaces or weakly developed promontories, somewhat removed from the linear influx of terrigenous detritus, favoured the development of reefs and their associated clinoforms.
3. Local, sharply defined ridges, although favouring reef development near sea level, were essentially by-pass

zones at greater depth. They were populated by spectacular stromatolitic encrustations developed on steep substrates extending vertically for at least 50 m (Figure D8.15(a)).

Although these three major facies, together with other important variations within the slope deposits, are approximately contemporaneous, the stromatolitic unit developed relatively late (Thiriet *et al.*, 1986), and appears to record the initial effects of pre-evaporite restriction.

Wadis Ambagi and Aswad, Quseir

Carbonate platforms in these areas (Figure D8.1) were examined in detail by Roussel *et al.* (1986) and by Purser *et al.* (this volume, also for localities). A narrow shelf fringing relatively low (100 m) basement relief between Wadis Ambagi and Aswad, is bordered by a nearly continuous slope averaging 50 m in height. This 'Ambagi Shelf' is much simpler tectonically and morphologically than that at Safaga. Although directly anchored on Precambrian basement it is not flanked by high relief as it is situated several kilometres from the edge of the rift.

The block on which is developed the fringing reef and associated clinoform beds is tilted towards the axis of the rift. Its eastern edge formed a substratum for reefal sedimentation, its landward (western) border remaining slightly above Miocene sea level. The steep, linear fault scarp bordering the eastern edge of the shelf (Figure D8.4) was slightly bevelled by marine erosion and onlapped by open-marine sands and gravels. A major reef-terrace has constituted the principal source for adjacent slope deposits across which it has prograded some 250 m.

The slope deposits, inclined at angles averaging 15° (Figure D8.4), include many angular discontinuities discussed in detail by Purser *et al.* (this volume). In spite of these complications there appears to be a general stratigraphy comprising the following slope facies.

1. The massive, subhorizontal reef (no. 1) has furnished coarse detritus which forms well-bedded clinoforms locally attaining 150 m in thickness. These bioclastic gravels are rich in coral debris and scattered

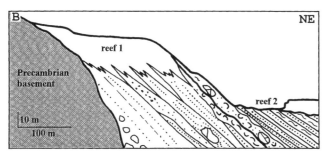

Figure D8.4 Slope deposits at Wadi Ambagi, Quseir. A. General view of shelf slopes and downstepping reefs. B. Sketch showing main sedimentary units shown in photo A; reef 1 fringing steep fault-scarp progrades across its own talus. C. Detail of slope deposits showing mainly grain-flows in the lower part and coral-rich debris-flow (mass-flow) above (man gives scale).

olistoliths, and mainly comprise well-bedded grain-flows. Locally a number of muddy mass-flow deposits occur. These carbonate slope sediments are diluted by large amounts of terrigenous sands and gravels which have been transported across the narrow shelf from the adjacent basement. Unlike Wadi Abu Assala (Safaga), carbonates and siliciclastics are intimately mixed. This mixing appears to be the result of the relatively simple morphology (at least on a local scale) of the shelf, there being only one embayment (discussed below). The absence of fault/fracture controlled drainage systems on the topographically low hinterland, has favoured a dispersed terrigenous supply to the narrow shelf and slope.

2. The upper part of this mixed carbonate–siliciclastic unit is marked by a major (10 m thick) debris-flow (Figure D8.4(c)) which can be traced along the Ambagi slope for several kilometres. This massive unit is composed of coral debris mixed with olistoliths, its irregular base eroding the unit below. This debris-flow does not contain interbedded sands or muds and would appear to result from a single event.

3. The debris-flow unit is overlain continuously by relatively fine, well-bedded bioclastic and siliciclastic sands (Figure D8.4(b)). These exhibit numerous ripples and local discontinuities with downlapping beds. These relatively fine slope deposits, typical of both hydrodynamic and tectonic stability, have prograded laterally for about 200 m prior to truncation and burial by a second, topographically lower, reef complex.

The regularity of the Ambagi Shelf is interrupted near its southern end (Wadi el Aswad) by a clearly defined palaeovalley or canyon (Figure D8.20(b)) whose depth was about 35 m. The axis of this depression plunges across the shelf and was probably eroded along a minor north-north-east–south-south-west cross-fault. It is filled firstly by a massive debris-flow containing reef olistoliths and pebbles of Precambrian basement, followed by well-bedded bioclastic sands which drape the flanks of the depression.

Wadi Abu Ghusun

Located near the south-eastern extremity of the Egyptian Red Sea coast (Figure D8.1), the area of Abu Ghusun has been mapped by Montenat *et al.* (1986) and Philobbos *et al.* (1993). The general structural and sedimentary dynamics of the Abu Ghusun platform are discussed by Purser *et al.* (this volume). The platform is traversed by a major north-east–south-west fault-line depression which is followed by the modern Wadi Abu Ghusun. Both the north-eastern slopes of the platform and the sides of the fault-line depression are flanked with Miocene slope deposits exhibiting many indications (Figure D8.5) of synsedimentary tectonic instability.

The north-eastern flanks of the platform

Spectacular slope deposits rising to an altitude of about 120 m dip at an angle of about 20°, and pass below subhorizontal late-Miocene evaporites towards the east (Figure D8.6(a)). The relatively thin (75 m) talus is heterogeneous and composed essentially of noncarbonate terrigenous materials derived from the adjacent Precambrian substratum which forms the crest of the block, notably towards its north-western, up-plunge end. The predominant slope facies (Figure D8.6) include the following.

- lensoid bodies of unsorted conglomerate composed of rounded cobbles of basement volcanics emplaced during sporadic mass-flow events;
- mixed carbonate–siliciclastic sands with scattered corals, molluscs and echinoids, exhibiting clearly defined planar bedding, emplaced by repeated grain-flow processes;
- thin (1–3 m), discontinuous beds of relatively pure, micritic carbonate. Textures and well-developed bedding indicate that this facies probably has been deposited from suspension directly onto the slope;
- numerous large olistoliths, perhaps the most characteristic elements within the slope complex (Figure

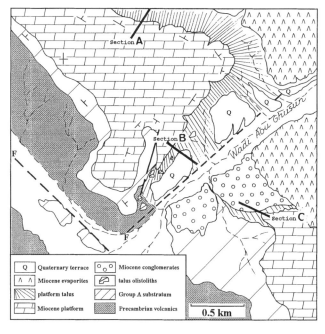

Figure D8.5 Simplified geological map of the Wadi Abu Ghusun area (after Montenat *et al.*, modified). The Miocene platform is flanked by slope deposits (note dip directions) which contain many large olistoliths recording the demise of the platform. Note location of sections illustrated in following figures (D8.7–D8.10).

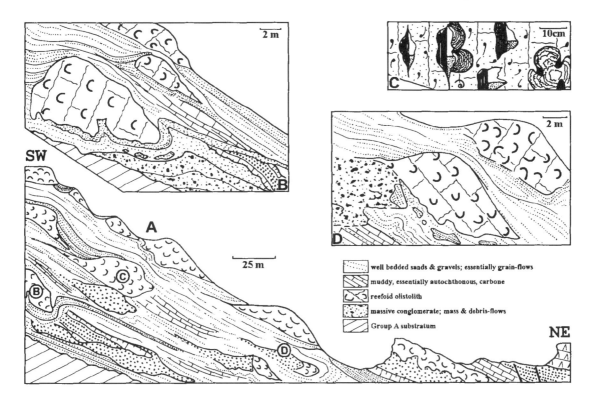

Figure D8.6 Section A across the north-eastern slope of Abu Ghusun platform. A. General panorama showing main slope units and location of detail sketches. B. Large reefal olistoliths which have slid on conglomeratic slope beds deforming them. C. Detail of olistolith showing two generations of geopetal filling (see photo, Figure D8.18(b) recording history of the olistolith. D. Reefal olistoliths, conglomeratic mass-flow and sandstone clasts. Most of these features record the disintegration of the platform.

D8.6(b)). Most are blocks of reefoid carbonate within which individual corals exhibit a common orientation.

This primary growth polarity has been rotated on the slope with repeated turning being recorded by multiple geopetals (Figure D8.6(c)). In addition to reef-derived olistoliths there also exist several large (15 m) blocks of beach-rock each composed of well-bedded basement pebbles and coral debris. These blocks not only confirm emergence of the crest but also explain the rounding of the ubiquitous basement materials. In a number of examples the sliding of large olistoliths down the palaeoslope has crumpled underlying beds on the downside of the block (Figure D8.6(b)). Olistoliths are not concentrated within any given horizon, their dispersion, together with the ever-present basement materials, clearly indicate constant destruction of the platform probably due to tectonic instability.

The flanks of the north-east–south-west fault-line valley
The Abu Ghusun platform is dissected by a 1 km wide depression. This is interpreted as a Miocene palaeovalley because of the orientation of the marine slope deposits and because of its partial filling by marine conglomerates. The steeply inclined (15°–30°) marine talus bordering both sides of the palaeovalley rest against subvertical palaeoescarpments which cut the Precambrian basement, proto-rift clastics and the earlier parts of the Miocene carbonate platform (Figure D8.7, D8.8). It is not surprising, therefore, that slope deposits are composed essentially of olistoliths, the finer sediments being an intimate mixture of silty carbonate and siliciclastics.

In common with most mid-Miocene slope deposits, those bordering the fault-line depression include the following facies.

- The predominant sediment type is massive, 2–5 m thick beds with wispy traces of highly deformed bedding, composed of unsorted pebbles and cobbles of angular carbonate and subrounded pebbles of Precambrian basement, generally within a sandy matrix (Figure D8.7(b)). These are interpreted to be of mass-flow origin.
- Parallel-bedded sands and gravels, individual beds sometimes being characterized by well-sorted, specific granulometry. These are generally dominated by siliciclastic grains and these inclined beds may downlap on to local, subhorizontal surfaces (Figure

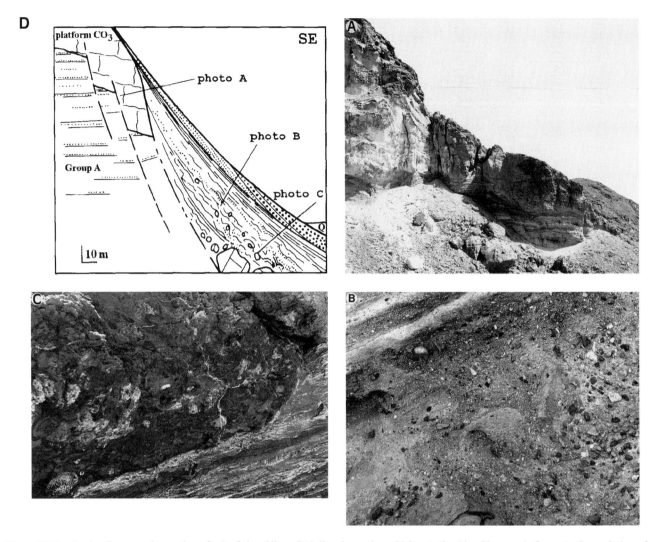

Figure D8.7 Section B across the northern flank of the oblique fault-line depression which cuts the Abu Ghusun platform. A. General view of synsedimentary faults and slopes. B. Mass-flow deposit composed of mixed detrital carbonate (from platform) and basement pebbles. C. Large reefal olistoliths with gypsified corals, embedded in grain-flow sands.

D8.16(a)). These sediments, often with marked basal surfaces, may be composed of relatively fine, well-sorted sand typical of grain-flow sedimentation.

- Piles of large (10–20 m) olistoliths (Figures D8.8(a)), each composed of cemented corals and reefoid debris, occur on both sides of the depression. Their disorganized arrangement and considerable sizes suggest that they have accumulated at the foot of a submarine escarpment, confirmed by the proximity (200 m) of the *in situ* platform. The reef-derived blocks are embedded in a fine silty matrix with vague lamination suggesting environments situated below wave-base. Adjacent sediments contain numerous massive coral colonies often concentrated within specific beds. This coarse reef detritus, which grades laterally into laminated sands (Figure D8.8(c)), records the sporadic disintegration of the non-lithified parts of the reef platform above.

The highly varied textures and bedding types visible at Wadi Abu Ghusun could be mistaken as indicating varying distance from the platform source, the olistoliths and mass-flow deposits being more proximal than the laminated sands and silts. However, all sediments have been deposited at a distance of only 200 m from the palaeoescarpment bordering the fault-line depression. Dip angles and bedding direction of all facies are very similar clearly indicating a common source. The important variations must therefore be controlled by fluctuations affecting the supply, possibly relating to repeated movements in the adjacent fault-line scarps bordering the depression.

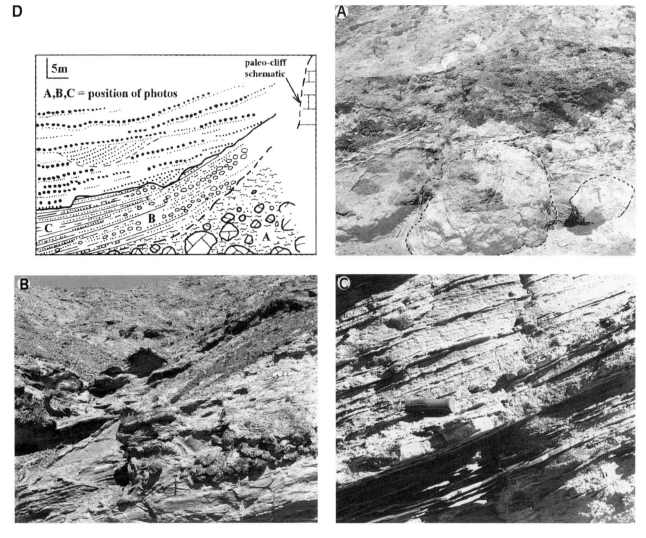

Figure D8.8 Section C across the southern flank of the fault-line depression; locations of photos shown in sketch. A. Proximal slope materials situated near base of submarine cliff; note abundant olistoliths. B. Finer slope deposits with massive coral colonies dispersed in well-bedded silts. C. Laminated grain-flow silts and sands (knife gives scale). Notice that the distance between photos A and C is only 50 m.

The Abu Shaar platform

Probably the most spectacular mid-Miocene platform of the Red Sea region, Abu Shaar has been studied by many workers and is discussed by Cross *et al.* and by Purser *et al.* (this volume). The elongate block of Precambrian basement is some 20 km from the periphery of the rift and, in common with others in the region, is tilted towards the periphery. It has been buried progressively by onlapping mid-Miocene sediments which have been tilted repeatedly, the gently dipping (3°) dip slope contrasting with the steep footwall escarpment bordering the north-eastern side of the platform (Figure D8.9). This escarpment is buried by a series of spectacular talus deposits exhibiting numerous angular discontinuities constituting a relatively complicated lithostratigraphy discussed by Prat *et al.* (1986), Burchette (1988), and James *et al.* (1988). Together with the subhorizontal platform-edge deposits, the series has now been subdivided into nine depositional sequences (Figure D8.9(a)) by Cross *et al.* (this volume). While these are clearly expressed along the higher parts of the footwall escarpment, the presence of subvertical by-pass surfaces interrupt the continuity of sediments deposited on the slope (Figure D8.9(b)) making correlation between platform and slope sequences difficult.

The footwall slope series may be examined in two parts.

The upper slope and platform edge (Figure D8.9(a))

The north-east dipping beds, discussed by Cross *et al.* (this volume), comprise sequences 2–6 which clearly record fluctuating slope conditions relating to changes in relative sea level. These sequences are all truncated by

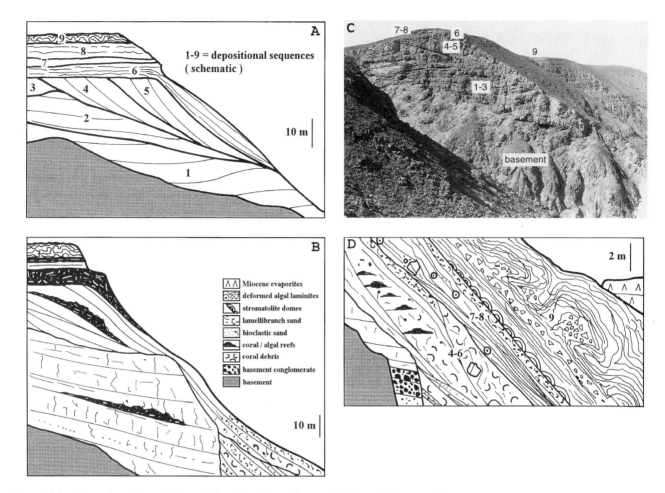

Figure D8.9 General attributes of footwall slope deposits, north-east Abu Shaar tilt-block platform. A. Simplified stratigraphy of upper slopes showing numerous discontinuities and nine depositional sequences defined by Cross *et al.* (this volume). B. Upper parts of slope deposits showing major discontinuities and bypass surfaces. C. Panorama of footwall escarpment (section III-2) showing erosion surfaces and depositional sequences. D. Details of lower parts of slope showing the possible slope equivalents of platform depositional sequences; notice numerous erosional surfaces which complicate slope stratigraphy.

the 5–6 sequence boundary and thus are not directly correlatable with sediments on the lower parts of the slope. As noted by Cross *et al.*, these platform edge, upper slope deposits, are essentially coral–algal reefal facies whose massively bedded debris tends to downlap on to the truncated preceding unit.

The lower slope (Figure D8.9(d))

These sediments have received considerably less attention from other workers. Dipping at angles averaging 30°, these well-bedded sediments exhibit numerous angular discontinuities which could be recorded as sequence boundaries. They are probably the result of repeated, relatively localized, slope collapse. However, in spite of these complications, there does exist a general slope stratigraphy which can be correlated in very general terms with that developed on the platform and along the slope.

1. A basal, subhorizontal, massively bedded carbonate with scattered coral debris is enriched in subangular boulders of basement in its lower parts (Figure D8.9(d)). This is considered to be the equivalent of sequence 1; it onlaps and progressively buries the steep footwall escarpment of Precambrian volcanic basement. This unit is truncated by a subvertical surface, probably of tectonic origin, which formed a submarine escarpment against which the succeeding unit has been stacked.

2. The main unit, averaging 25 m in thickness, is dominated by a massive carbonate whose coral, red algal and molluscan debris float in a muddy matrix. These are interpreted as mass-flow deposits and often have an erosional base (Figure D8.10). They may be separated by well-bedded grainstones composed mainly of molluscan debris lithified by an abundant fibrous marine cement. This unit, in general, is characterized by the abundance of coral debris and is probably the equivalent of sequences 5 or 6 of the platform margin and top.

3. The onlapping beds above are similar to those immediately below and also include mass-flow and

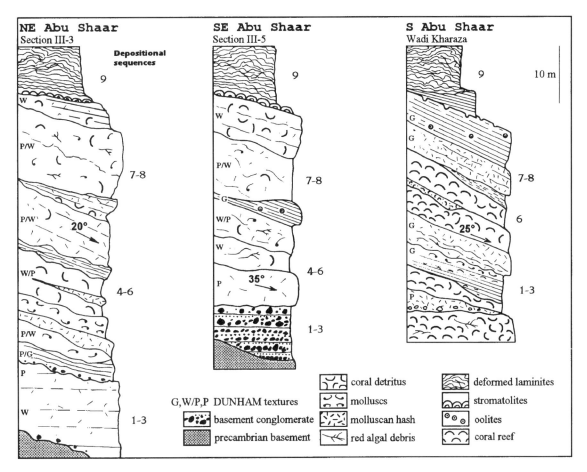

Figure D8.10 Sections around the north-eastern and southern flanks of Abu Shaar platform showing main lithological packages and lateral variations. Mass-flow units predominate.

bedded grain-flow units. The latter, however, are locally rich in oolitic grains swept from the platform and are therefore suggestive of sequences 7 and 8. The upper part of this unit is often rich in dwarfed mytilids indicating restriction, also common within sequence 8 on the platform.

4. The top of the preceding unit is marked by a planar, highly lithified, inclined surface on which is developed a series of spectacular stromatolitic domes. These stromatolites are discussed in the following section and have formed in an unusual context. They are frequently bored by marine bivalves and associated with rare solitary corals.

5. The uppermost slope unit, perhaps the most spectacular (Figures D8.9, D8.19(f)), is a 15 m thick laminated carbonate which is intensely folded and brecciated. This unfossiliferous dolomicrite contains scattered pseudomorphs of gypsum and celestite and has generally been interpreted as a collapse breccia resulting from the dissolution of Miocene evaporites (Burchette, 1988; James et al., 1988). However, the development of major folds, some of which attain 10 m

in amplitude, within which breccias are locally concentrated, is not typical of collapse. In these slope deposits these deformations could be interpreted as being the consequence of gravity sliding. Although this may indeed be the case, the correlation with sequence 9 on the platform where fold-structures and breccias are also well developed, clearly mitigates against a simple gravitational origin. This intense deformation, discussed in detail by Plaziat and Purser (this volume), is attributed to local sliding, both on the slope and on the platform, triggered by Miocene seismic (earthquake) shock.

These five lithostratigraphic units occur on many parts of the footwall slope (Figure D8.10). However, they are not continuous. In common with other Miocene slope deposits, notably at Wadi Abu Assala (Safaga), the footwall surface is not a simple escarpment determined by a single bounding fault. Although relatively linear, it nevertheless is modified by a number of embayments, the largest being that at wadi III-3 (Figure D8.11). At this point, Precambrian basement advances some 200 m to the east. This irregularity is related to a north-north-east–south-south-west cross-fault which in-

332 Miocene periplatform slope sedimentation

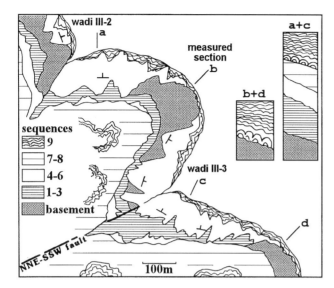

Figure D8.11 Simplified map of the footwall periphery of Abu Shaar showing local palaeoembayments and their effects on slope sedimentation. The thickest packages are localized within the embayments (sections a and c) while the thinnest are on the promontories. Embayments have been formed by local footwall collapse.

tersects the north-north-west–south-south-east footwall scarp. The embayment existed during mid-Miocene sedimentation as is confirmed by marked lateral changes in slope stratigraphy and thickness (Figure D8.11). While the embayment coincides with a 60 m thick series of multiple mass-flow deposits, it also includes scattered pebbles of pre-rift Eocene chert derived from the opposite, hangingwall slope of the block. On the adjacent promontories the lower slope units are limited to the stromatolitic unit which is developed directly on the steep basement escarpment where it is overlain by the brecciated laminite. Clearly, the cross-fault and related embayment have favoured the funnelling of both carbonate and noncarbonate detritus down the footwall slope.

The Esh Mellaha platform

The Esh Mellaha block is the longest (65 km) continuous structural unit within the north-western Red Sea (Prat *et al.*, 1986; Bosworth *et al.*, 1995; Purser *et al.*, this volume) and is situated immediately to the north of Abu Shaar from which it is separated by a north-east–south-west transfer fault which has slightly offset the two blocks. In common with Abu Shaar and other blocks in the region, it is tilted towards the periphery of the rift. Tilting of the block to the west has resulted in a marked morpho-structural asymmetry, the north-eastern (footwall) flank being marked by a steep escarpment (Figure D8.12) against which mid-Miocene marine sediments are stacked. However, this syn-rift cover is preserved only locally. Most of the Esh Mellaha range consists of Precambrian basement, notably along the footwall escarpment, the syn-rift cover having been removed by Miocene and later erosion.

The footwall escarpment, although in general linear, in detail is modified by a number of shallow embayments giving a lunate morphology to the border of the block (Figure D8.12(a)). This somewhat irregular edge is draped locally by steeply (30°) inclined talus deposits (Figure D8.12(b)) which are preserved both in the embayments and on adjacent promontories, notably on the south-east extremity of the block. The general stratigraphy of the slope deposits is similar to that at Abu Shaar.

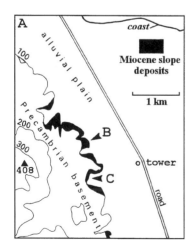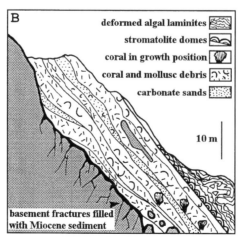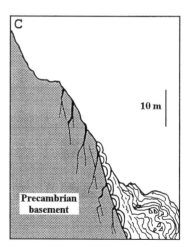

Figure D8.12 General attributes of footwall slopes, south-eastern Esh Mellaha tilt-block. The thickest slope packages are often localized on the promontories and the thinnest in the embayments where most of the sequence has been removed by Miocene erosion; stromatolites encrust Precambrian basement.

- Subhorizontal, massively bedded muddy carbonates onlap the steep footwall surface. These may be the equivalents of depositional sequence 1 at Abu Shaar.
  1. The essential part of the deposits exposed on the south-eastern corner of Esh Mellaha (Figure D8.12(b)) are massively bedded carbonates often rich in coral debris, these being correlated with units 4–6 at Abu Shaar. Although most of these sediments have been deposited *en masse*, the presence of faviid coral colonies in subvertical growth position clearly implies a less violent style of sedimentation. These open-marine slopes frequently contain cavities filled with subhorizontal geopetal muds confirming the perenity of the original sedimentary dip (Figure D8.18(a)).
  2. These massive slope deposits are limited by a planar, highly lithified surface generally encrusted with domed stromatolites identical to those at Abu Shaar (James *et al.*, 1988).
  3. The slope complex, in common with that at Abu Shaar and many other blocks on the north-west Red Sea coast, terminates with a 10–20 m thick, unfossiliferous algal laminite deformed into folds and highly brecciated.

This stratigraphic sequence, as at Abu Shaar, is characterized by marked lateral variations along the slope. However, in contrast to that at Abu Shaar, the *thinnest* sequence is developed within the lunate embayments where subvertical surfaces limiting Precambrian basement are encrusted with domed stromatolites (Figure D8.12(c)), the earlier, coral-rich slope unit being absent. On adjacent promontories, however, the sequence is complete. These very rapid variations clearly imply Miocene erosion of the pre-stromatolite units, their removal, within the embayment, leading to a thin, incomplete slope deposit. The arcuate form of the embayment together with its modest dimensions, are indicative of gravity collapse possibly during a single event. This may have been facilitated by the presence of a highly fractured basement substratum and the proximity of a major fault system whose movements may have triggered gravitational sliding.

This situation differs from that at Abu Shaar where the *thickest* slope deposits occur within the embayment. At Abu Shaar, the embayment coincides with the intersection of a cross-fault and a major bounding-fault, this intersection probably favouring local erosion and enlargement of the structurally controlled embayment. This latter was formed relatively early, prior to deposition of the first slope sequences. At Esh Mellaha the embayments do not appear to coincide with the presence of cross-faults and are most probably the result of local collapse favoured by local basement failure.

This occurred somewhat later than at Abu Shaar, subsequent to the deposition of the initial slope deposits. However, further north along the footwall escarpment (at Esh Mellaha) in the vicinity of Wadi Gerfan, a local embayment, comparable to that at Abu Shaar, coincides with the intersection of cross-faults favouring the creation of a relatively wide (5 km) embayment filled with a thick sequence of marine conglomerates and carbonates.

These five examples (Wadi Ambagi, Safaga, Abu Ghusun, Abu Shaar and Esh Mellaha) demonstrate the main structural, morphological and stratigraphic aspects of mid-Miocene slope sedimentation. There also exist other examples, many discussed by Purser *et al.* (this volume). It is recalled that the preceding descriptions concern only the relatively steep ($>10°$) slope deposits whose sedimentary clinoforms are readily visible in the field. However, most syn-rift sediments have been deposited on slopes, many less than 5°, and consequently are less readily discernible. These have not been included, although they are discussed by Cross *et al.* and by Purser *et al.* (this volume).

SPECIFIC SLOPE FEATURES AND SEDIMENTARY DYNAMICS

The preceding section gives a relatively general picture of mid-Miocene slope facies and their tectonic context. However, it is important to examine these sediments in greater detail, permitting a better understanding of the processes involved thus enabling comparison with slope sedimentation in other sedimentary basins (Figure D8.13).

Subvertical, somewhat polished escarpments flanking the numerous wadis enable the detailed study of sedimentary structures. The series of clinoform beds characterizing these slope deposits indicates that constructional processes exceed erosion. However, the slope deposit may represent only a minor part of the sedimentary history of the slope environment; the major part may have been eroded or redeposited lower down the slope as indicated by numerous traces of synsedimentary erosion. Therefore, in considering slope sedimentation it is necessary to consider the two opposing effects of creation and destruction in order to deduce a balanced picture of slope sedimentation. This is especially the case with these Miocene examples because only the uppermost parts (maximum 250 m) of a considerably more extensive (possibly 1000 m) slope, are exposed. In this context, it is logical that destructive processes may be very significant and that the proximal slope environment is little more than a transit zone for sediment being exported further down the slope.

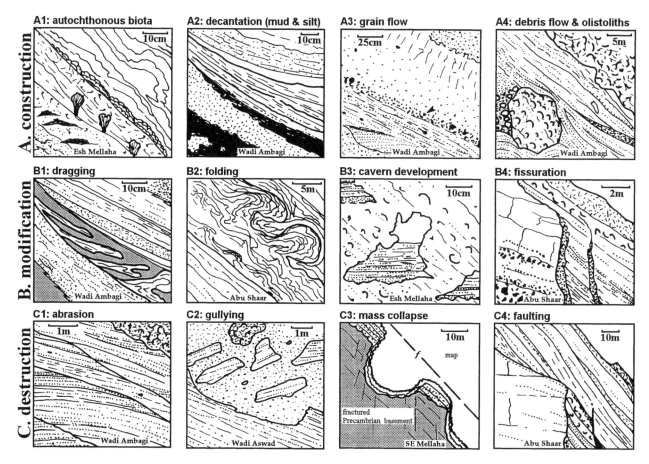

Figure D8.13 Graphic summary of the main sedimentary features observed on the mid-Miocene slopes in the north-western Red Sea.

Constructional features

The bulk of the sediment deposited on the slope is derived from higher elevations on the slope and, especially, from communities fringing the platform. Certain ecological communities living on these slopes have created sediment on the slope itself while others are delivered by oceanic waters.

Autochthonous faunas and floras (Figure D8.14)

Because many slopes may be followed up-dip to palaeosea-level (generally recorded by beach deposits or fringing reefs), a precise depth may be attributed to certain inhabitants of the slope. For example, it is shown below that stromatolites are known to have flourished at depths of 50 m. Space does not permit detailed analyses of the varied slope communities.

Corals

These are often the predominant component at the top of the slope, notably of sequences 4–6 at Abu Shaar (Perrin *et al.*, this volume). However, scattered colonies of massive faviids, in subvertical growth position, occur down the slope to palaeodepths of about 50 m, notably on the south-eastern corner of the Esh Mellaha platform. Rare solitary corals occur directly below stromatolites both at Abu Shaar and at Esh Mellaha, where they also lived at depths of about 50 m.

Molluscs
Although an important component of slope detritus, it is difficult to demonstrate an autochthonous origin. Two exceptions may be cited. The first concerns the uppermost (youngest) bed of several open-marine slopes, situated immediately below the characteristic stromatolite horizon, notably at Abu Shaar. This mollusc-rich bed(s) is composed almost entirely of mytilids (Figure D8.14(b)). These are not preserved in growth position but their extreme abundance on the slope and rareness on the platform, suggests an autochthonous origin. A second, very unusual molluscan community comprises only nautiloids (Figure D8.14(a)). Found only at one locality (Wadi Treifi, 2 km north of Wadi Bali), these small (2 cm), possibly immature nautiloids were crammed into a cavern some 50 cm in diameter where they appear to have been the sole, but numerous (several hundred) occupants.

Figure D8.14 Autochthonous biological slope communities. A. Lenticular mass of small nautiloids (coin measures 2.5 cm). B. Mytilid lamellibranchs which are very abundant near the top of the slope sequence; although statistically horizontal, the slope of the bed (indicated by line) is about 35°. C. Pisolite beds at Abu Shaar showing homogeneous composition. The pisolites are formed by microbial and diagenetic processes discussed in detail by Purser (this volume). D. Rhodophyte construction at Abu Shaar situated at palaeodepths of about 50 m on the slope (coin measures 2.5 cm).

Coralline algae

These are important contributors to slope sedimentation, generally in the form of detritus. However, red algae may also form small (1–2 m thick) *in situ* constructions notably on the steep footwall slopes of Abu Shaar (Figure D8.14(d)) where they lived at depths ranging from 45 to about 60 m. Their morphologies include both laminated encrustations and massive colonies with short (1–2 cm), generally massive branches. Within any given construction, the detrital fraction is generally dominant and this may exhibit a geopetal disposition (Figure D8.14(d)) confirming the semi-permanent nature of the constructed framework.

Green algae. *Halimeda*, a fairly rare constituent within the platform facies, is locally abundant on the upper slopes, notably at Abu Shaar where it occurs in small pockets in wadi III-3. Its limited distribution would suggest that it favoured the upper slope environment. However, the scarceness of *Halimeda* debris within the grain-flows and other slope deposits indicates that it was not a significant contributor to the sediment budget.

Pisolites/oncoids

Subovoid pisolites averaging 1–2 cm in diameter occur on the upper slopes of the Abu Shaar platform where they have been recorded by El Hadded *et al.* (1984), Aissaoui *et al.* (1986), James *et al.* (1988) and by Burchette (1988). This enigmatic facies occurs only at Abu Shaar where it is restricted to the upper parts of the slope, about 10 m below the top of the talus of sequence 6 (Cross *et al.*, this volume). The pisoids are limited to a distinct bed (Figure D8.14(c)) which extends laterally along the slope for about 1 km (in the vicinity of Bir Abu Shaar). Its rather vague bedding is inclined at about 15–20°, although locally may be horizontal. Seemingly of microbial origin, the irregularly laminated cortex of individual pisolites generally coats a molluscan fragment. The cortex is composed both of microcrystalline as well as fibrous carbonate, the latter being more typical of marine diagenetic processes (discussed by Purser, this volume). The pisoid bed attains a maximum thickness of about 1.5 m, being composed only of pisoids there being no other biogenic detritus in spite of the fact that adjacent beds are rich in bioclasts. The homogeneous nature of these sediments (Figure D8.14(c)) strongly suggests that they have formed on the slope, confirmed by their apparent absence on the platform above. Bedding attitudes and general stratigraphic occurrence (it may be traced up the slope over a distance of about 10 m) indicate that the pisoids are neither a soil nor a cave deposit; they have little in common with the pisolite beds of the Permian Capitan Complex (New Mexico). They could have formed at relatively shallow (about 10 m) depths on local ledges developed on the slope. Whatever their origin, they are limited to the slope environment.

Stromatolites

Although volumetrically unimportant, an extensive stromatolitic unit is one of the most spectacular slope facies. Developed on steep, locally vertical, substrates (Figure D8.15), this unit occurs on several slopes along the Red Sea coast: Ras Honkorab, Wadi Ambagi, Abu Shaar and Esh Mellaha, where it invariably occupies a common stratigraphic position marking the top of open marine talus and immediately preceding the unfossiliferous, brecciated algal laminites and associated evaporites.

In addition to their unusual environmental occurrence, the morphology of the stromatolites is also unusual. The basic unit is a banana-shaped protuberance averaging 15 cm in length and 5 cm in diameter, whose downwards curvature reflects gravitational influence. In cross-section (Figure D8.15(e)) each unit has a well-developed asymmetry. The preferential growth on the upper surface of the protuberance is thought to relate to an algal phototrophic response, the result being a fairly regular convex upper/outer surface of the 'banana' and an irregular lower/inner surface. Adjacent stromatolitic protuberances are closely packed laterally but generally have centimetre-sized spaces below each unit, these latter being partially filled with detrital sediment or fibrous submarine cement. The closely packed mass of downward-curved protuberances forms a continuous sheet encrusting the substratum, the latter frequently being Precambrian basement whose rocky surface possibly favoured these encrustations. The sheets are modified into broad, weakly defined domes averaging 2–3 m in diameter and about 30 cm in amplitude (Figure D8.15(b)). They are often bored by marine organisms and may be associated with rare solitary corals.

The stromatolitic sheets, exposed at the base of exposed slopes about 100 m below the equivalent platform sediments (also locally stromatolitic) may be

Figure D8.15 Stromatolites which frequently encrust steep slopes around Miocene platforms. A. General view of stromatolite encrustation at Wadi Abu Assala whose vertical extension (arrows) is at least 40 m. B. Typical morphology of stromatolites, the 10 cm wide, closely packed columns (arrow) comprising low amplitude (50 cm) domes averaging 2–5 m in diameter (Esh Mellaha). C. External morphology showing downward-curved lobes (Wadi Abu Assala). D. Another morpho-type composed of subvertical ridges (Wadi Abu Assala). E. Polished slab (approximate growth position) showing the internal structure comprising closely packed downward-sloping lobes (Abu Shaar).

Specific slope features and sedimentary dynamics 337

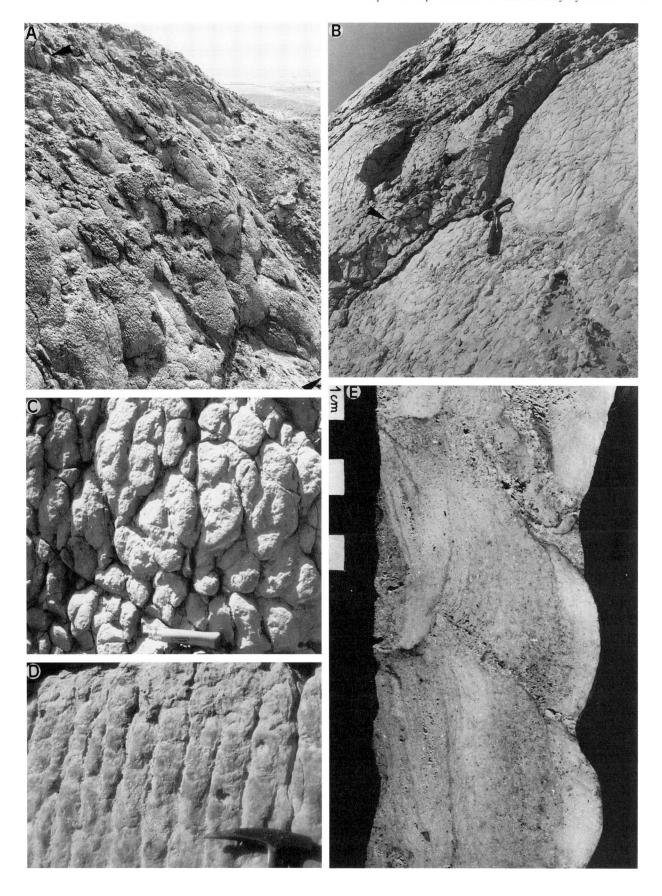

followed up the slope without interruption for distances of 20–50 m (Figure D8.15(a)), above which they are truncated by Miocene or Quaternary erosion. Their vertical extension clearly implies growth in water depths attaining (and possibly exceeding) 50 m, the absence of notches or other phenomena suggestive of sea level lowering tending to preclude peritidal conditions.

The possibility that these unusual encrustations result from diagenetic processes (cf. James et al., 1988) is not supported by our work. However, these Miocene examples do contain laminae of fibrous crystalline carbonate which may also line intra-stromatolitic cavities. The gravitational effects clearly expressed by these bodies, their internal lamination and general morphologies are all typical attributes of stromatolites. Their relatively deep (at least 40 m) environment and form are somewhat comparable to the microbial crusts described by Dullo et al. (1990) from the modern Sudanese shelf of the Red Sea.

Materials derived from the platform
Most sediment comprising the Miocene slopes must have originated on the upper parts of the slope, notably the reefs, or the periphery of the shelf. Sediment has been transported and deposited by two main mechanisms.

Mass-flows (Figure D8.16(c)). Massive beds, generally with sharp or erosional bases, composed of unsorted gravels, sands and muds, are interpreted as mass or debris-flow deposits. They are the predominant facies on many slopes, notably at Abu Shaar and Esh Mellaha. Generally fairly coarse in texture, they often contain massive coral colonies and olistoliths mixed with biogenic rudstone. They may include pebbles and cobbles of Precambrian basement and a finer terrigenous sand fraction which may predominate. Individual mass-flows averaging 1–5 m in thickness can rarely be traced laterally for more than several hundred metres. However, the presence of numerous erosional surfaces within most slope sequences (see below) suggests that individual mass-flows initially could have been more extensive, this being the case on the Ambagi shelf where an individual debris-flow can be traced along the slope for about 5 km.

Mass-flow sedimentation may result from oversteepening of the slope or may be triggered by seismic shock, both systems probably existing. However, the lateral continuity of at least one example (Ambagi shelf) would suggest seismic control.

Grain-flow. Slope deposits include numerous, relatively thin (5–10 cm) beds of sand or gravel exhibiting vague, planar stratification. They may have a fairly sharp base and a weakly expressed upwards decreasing granulometry (Figure D8.16(b)) with concentrations of pebbles composed of Precambrian rocks or pre-rift Eocene chert, marking the base. Although sometimes resembling turbidite beds, the presence of burrows within these beds (Figure D8.17(a)) and the absence of terminal lamination suggests that grain-size evolution may be the result of a diminishing supply rather than gravitational sorting.

Micro-onlap sequences
Grain-flow deposits very frequently show onlapping relationships with respect to the underlying erosion surface (Figure D8.16(b)). Beds exhibiting onlapping relationships may range in thickness from a few centimetres to about 1 m, although larger-scale onlapping is often present immediately below steep by-pass surfaces (Figure D8.2). This mode of accumulation seems to be typical of steep slopes, i.e. sediments originate from above rather than from below as is the case with the more classical 'onlap' sequences. Detritus, delivered by gravitational processes, is stabilized locally where the slope decreases in angle, the deposit gradually accreting back up the slope progressively burying the by-pass surface.

Olistoliths
These blocks may be isolated, or multiple when they occur within debris-flows. Generally composed of reef materials, often with corals in growth position, they may attain 10–15 m in diameter, notably at Abu Ghusun (Figure D8.19(d)). The delivery of certain large olistoliths to the slope has been accompanied by modifications of associated slope deposits. At several localities the sliding of the block has crumpled the sedimentary substratum (Figure D8.17(d)). Following stabilization, certain olistoliths have obstructed subsequent detritus which has been piled against the up-slope side of the block (Figure D8.17(c)).

Together with *in situ* production of biogenic carbonate, the delivery of mixed carbonate and terrigenous materials from above contributes to slope accretion. However, the presence of numerous erosional discontinuities indicates that much (probably most) detritus has accumulated lower on the slopes and that the proximal slopes, visible at outcrop, are mainly transit zones.

Modification features

Sediment stabilized on the slope may be modified *in situ* by various mechanical, biological or diagenetic processes.

Mechanical effects
The upper surfaces of certain beds may be modified by sliding of the overlying bed. Thus, the surface may exhibit small-scale (centimetre) drag effects (Figure D8.17(a),(b)). On a larger scale (metric) gravitational

Figure D8.16 Non-biological slope construction phenomena. A. Downlapping grain-flows at Wadi Abu Ghusun. B. Grain-flows (a), graded mass-flows (b), onlapping microsequences (c) and erosional surfaces (d), at Wadi Ambagi. C. Mass-flow at Wadi Abu Ghusun. D. Mass-flow with numerous carbonate clasts (a), and deformation (b) at base of grain-flow (c) at Wadi Ambagi.

movement results in slump-like folding, individual folds sometimes attaining 10 m in amplitude. This is particularly the case when the sediments are fine-grained, notably at Safaga (Figure D8.18(c)) and, especially, with the laminated dolomicrites which form the uppermost unit of many mid-Miocene slope and platform deposits (Figure D8.19(f)). In the latter case, the sharply defined planar base has acted as a sole on which the folded mass

Figure D8.17 Modifications affecting Miocene slope deposits. A. General view of slope showing burrows (a), and drag-effects (b), as well as onlap microsequences and erosion surfaces. B. Detail of drag or slump effect shown on photo A. C. Olistolith of Miocene carbonate with coarse debris piled on up-slope side. D. Large reefal olistolith which has deformed preceding slope deposits during its emplacement (hammer gives scale); all photos from Wadi Ambagi.

has been slightly displaced. The folded material appears to have been a mixture of lithified and friable beds with both plastic stretching and highly brecciated beds.

Rarer mechanical effects include the formation of neptunian dikes. These generally measure 5–20 cm in width and may be traced vertically throughout the entire slope sequence. They may be comparable to the fissures observed by Grammer *et al.* (1993) on the deep slope deposits flanking the Bahamian escarpments and are possibly the initial trace of slope failure. The formation of these fissures is not associated with a lining of submarine cement indicating the rapidity with which they have been filled with sediment.

Biological effects
Burrowing is fairly rare, probably because of the relatively unfavourable substrate which was constantly

Figure D8.18 Modifications affecting Miocene slope deposits. A. Dissolution cavities filled with subhorizontal geopetal carbonate often with marine microfaunas (Esh Mellaha). B. Constructional cavities in reefal olistolith showing two generations of geopetal filling, the first vertical, the second horizontal (arrows). They record the mobilization of the olistolith. C. Slump-folds and minor faults in argillaceous slope deposits at Wadi Abu Assala. D. Laminated grain-flow deformed by emplacement of overlying mass-flow (Wadi Ambagi).

moving, buried or eroded. At Wadi Ambagi, burrows developed within certain marine beds (Figure D8.17(a)) indicate that these beds are not a single event; others emanating from specific surfaces suggest periodic slowing of slope sedimentation.

Diagenetic effects

Many slope deposits are highly cemented with fibrous, isopachous carbonate typical of submarine lithification. This affects grainstones, particularly those situated in the youngest (outer) part of the marine slope immedi-

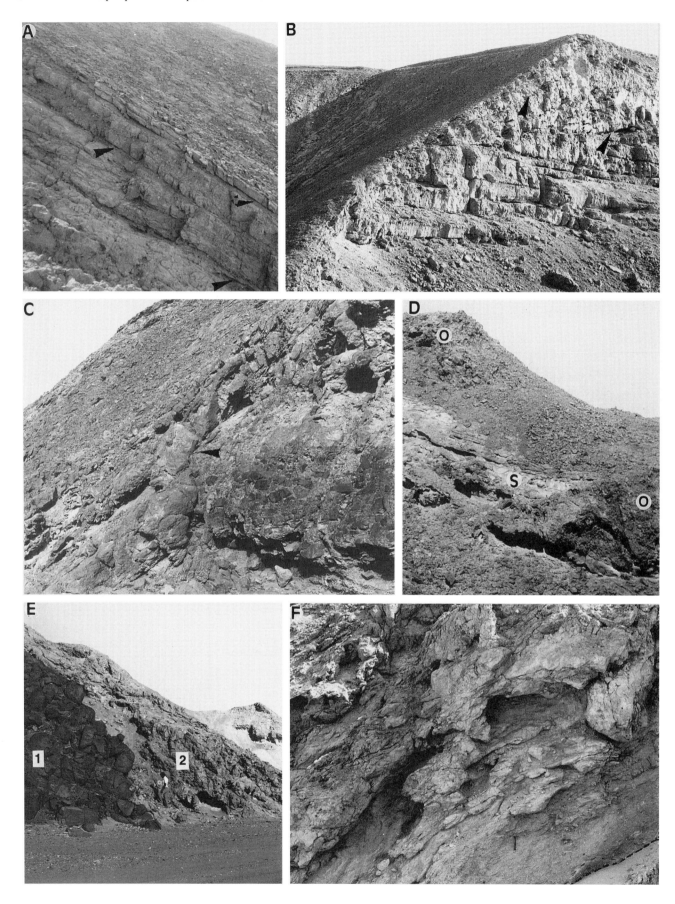

ately preceding the stromatolitic horizon (Aissaoui et al., 1986). These cements and related diagenetic features (dolomitization) are discussed by Clegg et al. and by Purser (this volume). Early lithification of slope deposits is also confirmed by the presence of molluscan and annelid borings which are especially common on by-pass surfaces. Similar diagenetic phenomena have been noted by Grammer et al. (1993) within the deep slope deposits flanking the Bahamian platforms.

The presence of small (5–30 cm) caverns, many of which appear to be the result of Miocene dissolution, probably are related to sporadic emergence of the platform. These caverns, generally filled with subhorizontal, laminated geopetal silts containing globigerinid foraminifera and other micro-organisms, indicate the original angles (up to 40°) of slope (Figure D8.18(a)). However, in some cases, notably at Sharm el Luli (Ras Honkorab) these geopetal fillings are slightly (10°) inclined, recording the tilting of the reefs and associated talus.

Destructional features

The very limited thickness (10–20 m) of many slope deposits probably results from several processes including relatively modest rates of production on the upper slope and narrow shelf above. However, as already noted, it probably results mainly from the constant erosion and destruction of these relatively unstable upper slope deposits, illustrated by the following.

Large-scale gravity collapse
On a large scale, the presence of embayments several hundreds of metres in width, probably indicates massive, localized collapse of the slope and platform margin, notably at Esh Mellaha and Abu Shaar (Figures D8.11, D8.12). As already noted, the stratigraphic gaps associated with these embayments confirm gravitational collapse. In other examples, slope and platform edge sediments have been removed by synsedimentary faulting or by more generalized collapse of the margin following movements of the adjacent bounding faults. This is the case at Abu Shaar (Cross et al., this volume), and at Zug al Bohar where the fault planes have been exposed to marine boring. Because many slope deposits are stacked against footwall escarpments, they are situated close to bounding fault systems whose rejuvenation is suggested by angular sequence boundaries and by the existence of small synsedimentary faults sealed by slope deposits. This instability, together with the basinwards slope of the bedding-planes, indicates the precarious nature of many (probably most) slope deposits.

Truncation surfaces
On a somewhat smaller (metric) scale, most slope accumulations exhibit many truncation surfaces of varying nature. Inclined grain-flow deposits may be cut by relatively planar erosion surfaces (Figures D8.17(a), D8.19(a)); these small-scale angular unconformities recording down-slope movement of sediment whose transit has abraded the friable substratum.

Escarpments
Small, subvertical escarpments, 0.5–10 m in height, are common. They may affect a given bed (Figure D8.20(a)), recording the dislocation of the bed, sometimes confirmed by the presence of numerous clasts. More important scarps, often subvertical, may be bored by marine organisms confirming the early lithification of the substratum. These escarpments, many of which are tectonic in origin, record zones of sediment by-pass. The foot of the abraded surface may be buried in onlapping detritus which progressively buries the surface.

Gulleys and canyons
Where cross-faults intersect the boundary fault, subsequent (submarine) erosion may create a relatively large-scale embayment discussed in the preceding chapter (D7). In addition, lenticular sand bodies whose main axes are more or less parallel to the dip of the slope, occur in relatively small (1–10 m) gulleys eroded on the slope (Figure D8.20(b),(c)). While the factors determining their initial occurrence are unknown, they have probably been enlarged by the funnelling of sand down the gulley. Clearly exposed examples occur at Wadi el Aswad (Figure D8.20(b)) and near the entrance to Wadi Kharasa (Abu Shaar, Figure D8.20(c)).

In conclusion, most large exposures of slope sediments exhibit a multitude of sedimentary and diagenetic features (Figure D8.21) which have recorded the many constructional and destructional processes associated with these Miocene slope environments.

DISCUSSION AND CONCLUSIONS

This contribution concerns a specific aspect of platform sedimentation, namely the peripheral slopes. Their

Figure D8.19 Destructional features affecting slope deposits. A. Lower slopes at Abu Shaar (section III-3) showing sedimentary dip and series of low-angle erosional surfaces (arrows). B. Upper slopes at Abu Shaar (section III-2) showing multiple truncation surfaces (arrows), and palaeoembayment in shadow to left. C. Truncated sequence (basal Kharaza beds) forming steep bypass surface (arrow) onlapped by slope deposits (Abu Shaar). D. North-eastern slope of Abu Ghusun platform showing large olistoliths (o) with sands piled on their upslope sides (s). E. Dark Precambrian volcanics (1) onlapped by massive debris flows resulting from the collapse of the upper slope and platform edge (Wadi Gemal, north Ras Honkorab). F. Brecciated and folded microbial laminites which have slid on basal surfaces (dotted line), footwall at Abu Shaar.

344 Miocene periplatform slope sedimentation

Figure D8.20 Destructional features affecting slope deposits. A. Minor scarp (below hammer) formed by detachment of slope beds which have broken into clasts (arrows); Wadi el Aswad, Quseir. B. Gulley cutting Precambrian basement (lower left) filled with multiple mass-flows (in axis) and grain-flows on adjacent slopes (Wadi Ambagi). C. Gulley filled with bedded carbonate sand at entrance to Wadi Kharasa, Abu Shaar. D. Large fissure (3 m, arrows showing opposing walls) cutting slope deposits, filled with coarse sands at Wadi el Aswad.

creation, accretion and destruction obviously are closely linked to the history of the entire platform, as discussed in other contributions to this volume (Bosworth *et al.*, Cross *et al.*, and Purser *et al.*).

Specific characteristics of mid-Miocene platform slopes

The platform slopes examined in this contribution represent the earliest phases of platform and slope evolution. These Miocene examples are all situated near the periphery of the rift and therefore reflect this

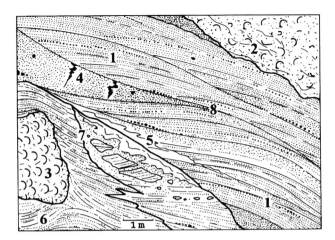

Figure D8.21 Field-sketch of slope deposits showing typical association of various construction, modification and destructional features at Wadi Ambagi: 1, bedded grain-flows; 2, unbedded mass-flows; 3, reefal olistolith; 4, burrows; 5, drag-effect; 6, sediments deformed by deposition of large olistolith; 7, escarpment formed by detachment of slope bed which has broken into clasts; 8, abrasion surface onlapped by grain-flow.

particular structural setting. Those situated closer to the axis of the rift, as in other rift systems, are generally limited to the subsurface and have not been studied in detail. A single exception crops out at Gebel el Zeit (Purser *et al.*, and Bosworth *et al.*, this volume), a structural block which had considerable bathymetric relief during the early- and mid-Miocene. However, because this block is situated in the axial zone of the rift, it subsided rapidly and shallow-marine facies are very thin (20 m). This tilted block is buried below pelagic marls and late-Miocene evaporites. Although these parallel-bedded, globigerinid-rich marls exhibit few sedimentary features typical of slopes, they nevertheless contain highly mixed and reworked microfaunas giving inverted stratigraphic sequences discussed by Plaziat *et al.* (this volume). The reworking of these relatively deep marine sediments could have been favoured by seismic instability affecting an inclined substratum.

The Miocene slopes examined in this text concern only the upper, proximal segment of a more extensive talus most of which is buried below adjacent Quaternary plains. This limitation must be kept in mind for it possibly explains many of the sedimentary properties of these deposits, notably the predominance of erosional phenomena which tend to impart a relatively heterogeneous character to these proximal deposits.

These slope deposits, although attached to structural blocks which may be situated some distance (up to 20 km) from the rift margin, are composed of both carbonate and terrigenous materials, the latter being derived locally. The crest of most blocks, especially along the up-plunge segments of the footwall escarpment, were generally situated above palaeosea-level and thus were sources for platform and slope detritus. These 'young' platforms, although at first sight unfavourable for siliciclastic sedimentation, nevertheless can be dominated by noncarbonate facies.

Slope creation

The mid-Miocene slopes of the north-western Red Sea are sedimentologically 'young' and represent the initial stages of slope development. This context demonstrates clearly the factors responsible for initial slope development, an aspect rarely considered in the literature. The relief of the Miocene slopes has a structural origin, the steepest slopes generally being related to steep footwall escarpments. In other well-known examples such as the clinoforms composing the Triassic platforms of the Dolomites (Bosellini, 1984), the Cambrian ramps of the Appalachians (Read, 1982; Barnaby and Read, 1990), the Bahamian platforms and their deep flanking deposits (Schlager *et al.*, 1976; Mullins, 1985; Grammer *et al.*, 1993) and many others, relief favouring the formation of clinoforms is generally attributed to sedimentary aggradation, typical of the steepened ramps of Read (1982) and the Type III platform margins of Wilson (1974).

The tectonic creation of Miocene slopes in the Red Sea rift system is important, for at least two reasons: firstly, their prediction in the subsurface may be aided by considering the synsedimentary tectonic framework, sometimes observed directly on seismic profiles; secondly, the structural frame and related sediment geometry may persist for relatively long periods of geological time. Although attenuated by sedimentary accretion and modified by relative sea-level fluctuation, structural geometry is frequently rejuvenated, notably in distensive tectonic regimes. Thus, in the north-western Red Sea, the early to mid Miocene framework controlling the occurrence of mid-Miocene slopes, has been rejuvenated during the Pliocene and Quaternary, and even today influences the distribution of modern reefs within the south-western parts of the Gulf of Suez (Orszag-Sperber *et al.* and Plaziat *et al.*, this volume). In other words, the structural frame has persisted over a period of about 20 m.y.

Slope accretion

Miocene slope sedimentation involves both carbonate and siliciclastic components whose relationships depend on the relief and structural architecture of the basement crest or hinterland. Where basement has been moderately elevated (50 m) above Miocene sea-level, detritus has been dispersed along the adjacent shelf and the slope is an intimate mixture of carbonate and noncarbonate materials. However, where a high basement has been affected by cross-faulting, the resulting erosional depression concentrates and funnels the terrigenous detritus across the shelf and down the slope, leading to a separation of carbonate and siliciclastic materials.

The angles of slope deposits are often high, averaging 20° but locally attaining 40°. This elevated angle is probably favoured by the very poor sorting and highly angular nature of the slope derbis (Kirkby, 1987). Bedding surfaces, especially at Abu Shaar, are generally planar suggesting that the foot of the slope is considerably lower than the present day outcrops. Others, notably at Safaga, tend to be slightly concave-up suggesting proximity of the toe of the slope. Slopes are frequently characterized by contemporaneous downlap and onlap relationships (Figure D8.2). The latter are particularly frequent immediately below steepened by-pass surfaces which have been buried by sediment delivered from above, possibly in a manner comparable to that described by Grammer et al. (1993) for sediments accumulating at the foot of the Bahamian platform escarpment where they have been emplaced during high sea-level stand.

Basal (toe) relationships are never visible within the Miocene Red Sea slopes. The tops, on the contrary, are clear. Generally speaking, there is only very modest aggradation and clinoform beds often extend up to a subhorizontal surface, the shelf/platform resulting from progradation of the clinoforms. Where limited aggradation has occurred, notably at Abu Shaar, there is generally an erosional discontinuity separating the slope and the subhorizontal platform sediments.

The north-western Red Sea deposits do not record the detailed sea-level events affecting the platform above, as been the case in the Northern Calcareous Alps (Reijmer and Evaars, 1991) and in the Bahamas (Mullins, 1985; Grammer et al., 1993). However, a very general correlation between platform top and slope stratigraphy has been established at Abu Shaar, based on oolitic and other peritidal allochems. Apart from local backstepping (Abu Shaar) or downstepping (Ras Honkorab), the highly discontinuous nature of slope sediments, in general, tends to mask the effects of sea-level movement. These discontinuities record erosion and by-passing of sediment towards the deeper parts of the slope. As a result, the proximal slope deposits of the north-western Red Sea tend to be thin (20–50 m). While these modest accumulations may be conditioned by the limited production on the very narrow platform culmination, they probably are due to the instability of sediments in these steep, tectonically unstable settings.

Slope destruction

As already noted, the mid-Miocene slopes exhibit a multitude of erosional features: steep by-pass surfaces, angular unconformities and erosional discontinuities, mass-flow sedimentation, olistoliths, breccias and larger-scale collapse of the slope and platform periphery. Erosional or by-pass margins (Read, 1982) are not uncommon in the geological record: Ladinian of the Dolomites (Bosellini, 1984), lower- and mid-Cambrian of the Appalachians (Barnaby and Read, 1990), Devonian of the Canning Basin (Playford and Lowry, 1966; Kerans et al., 1986), Bahamian Platforms (Mullins, 1985; Grammer et al., 1993). Although the origins of erosional margins are rarely apparent, certain authors attribute platform edge collapse to seismic instability (Cook, 1983; Bosellini, 1984). The creation of coarse breccias, typical of many platform slope deposits, has been ascribed to collapse either during periods of lower sea level (Read, 1982) or to seismic instability. Their widespread and repeated occurrence on mid-Miocene slopes of the north-western Red Sea probably reflects seismic instability relating to rift dynamics, collapse leading to platform edge erosion and, presumably, the accumulation of thick piles of coarse detritus at the foot of the slope (Burchette, 1988).

The evolution of mid-Miocene slopes

The platforms and slopes exposed within the coastal areas of the north-west Red Sea and south Gulf of Suez, although initially developed during the mid-Miocene, have been rejuvenated during the Pliocene, spectacular Pliocene and Quaternary slope deposits occurring on the islands of Gubal, Shadwan and Giftun (discussed by Orszag-Sperber et al., this volume). This late Neogene and recent relief is accentuated by diapirism of late Miocene evaporites whose mobilization and geometry are often related to the location of the initial, early-Miocene extensional faults. Slopes and platforms located near the periphery may continue to rise above sea level where they will be dismantled by subaerial erosion. Others situated closer to the axis subside. If regional climatic conditions favour rapid carbonate production, sedimentation may compensate or exceed subsidence resulting in the development of major platforms and slopes. This positive evolution may be more likely in the intermediate zones of moderate subsidence where shallow-water conditions and rapid organic production favour perpetuation of platforms and slopes whose location and geometry may retain an intimate relationship with the constantly rejuvenated structural frame.

Chapter D9
The tectonic significance of seismic sedimentary deformations within the syn- and post-rift deposits of the north-western (Egyptian) Red Sea coast and Gulf of Suez

J.-C. Plaziat and B. H. Purser

ABSTRACT

Seismites are a typical attribute of the north-western Red Sea rift sediments. The chronological distribution of these sedimentary expressions of tectonic instability depends on diverse factors including sedimentary texture, early lithification, and proximity of structural or sedimentary slopes. In spite of these local factors, a clear relation exists between increasing frequency of palaeoseismites and the tectonically controlled stratigraphic boundaries of the major phases of syn-rift sedimentation. Seismites in sediments having horizontal disposition include mainly *in situ* deformations while slope and near-slope deposits show diverse sliding structures related to intraformational liquefaction. These are especially abundant and spectacular even on low-angle ramps due partly to their proximity to the steep slopes characteristic of rifts.

INTRODUCTION

This contribution does not attempt to demonstrate the seismic origins of all synsedimentary deformation structures recorded in the Miocene sediments of the region; this would require detailed discussion of each structure and its particular setting. However, a previous synopsis of the Egyptian syn-rift seismites has been made by Plaziat *et al.* (1990) and this will be utilized as a base to interpret the various syndiagenetic deformations, including dololaminite breccias, which have been the subject of several divergent hypotheses. The authors agree with various suggestions that interpretation of these structures be made with caution and therefore our interpretations follow a detailed analysis of each outcrop. Unfortunately, space does not permit extended descriptions and discussion.

In addition to our own detailed studies (Plaziat and Ahmed, 1994; Purser et al., 1994), the reader should consider Ambraseys and Sarma (1969); Lowe (1975); Youd (1975); Tinsley et al. (1985); El Isa and Mustafa (1986); Owen, 1987; Vittori *et al.* (1991); Guiraud and Plaziat (1993); Ricci Luci (1995), and Plaziat and Ahmamou (1997). In addition to these studies, many workers have interpreted certain deformational structures as seismites but have failed to offer critical evidence.

Since the introduction of the notion of 'seismite' by Seilacher (1969), the definition has been enlarged considerably to include many types of deformation and resedimentation phenomena. The term now covers all sedimentary deposits exhibiting a deformational structure or remobilization fabric which can be attributed to earthquake shock whose epicentre is more or less distant from the site in question (Figure D9.1). In common with turbidites, the definition was founded initially on objective study of bedded deposits having synsedimentary deformation structures. The proposed origin of the deformation was then based on interpretation of the deformational mechanisms and chronology of the sediment reorganization. However, contrary to turbidites, it is not this mechanism that has provided the name 'seismite' but rather the cause that has triggered the mechanisms. Earthquake destabilization of subaqueous deposits may have resulted either in a slight bedding disturbance or a complete dispersion of the sediment particles; debris and grain flows, as well as turbidity currents may be triggered by earthquake destabilization. It has even been suggested that the flysch turbidites are

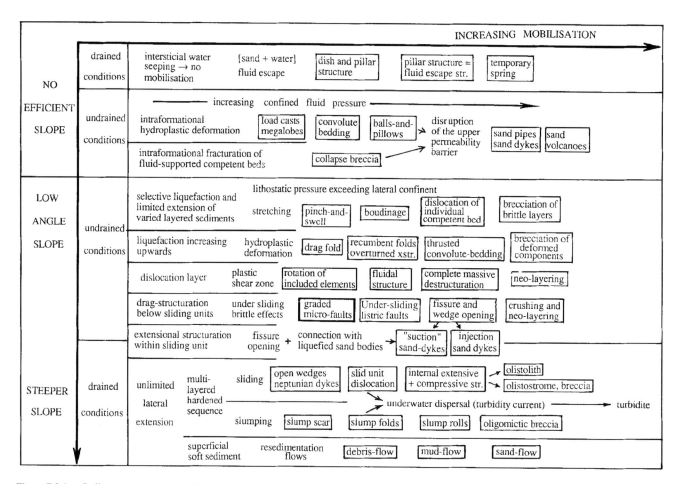

Figure D9.1 Sedimentary structures which may be observed in seismites (framed) resulting from post-depositional liquefaction of at least part of a given water-saturated deposit. These structures are arranged in two important gradients: increasing slope and increasing lateral or vertical mobilization. Influence of sedimentary and diagenetic lithology is implied although not fully expressed. Because other causes may also induce liquefaction, these deformational structures and remobilization fabrics are not all specific for palaeo-earthquakes. However, most of these features are suggestive, or are possible indications which should be integrated in any field study when deciphering palaeoseismicity.

not necessarily a primary deposit but are a form of seismite (Kastens, 1984; Plaziat, 1994).

To identify a seismite it is therefore necessary to demonstrate that the mechanical modifications have resulted from earthquake destabilization. Because there exists no direct way of identifying palaeoseismicity, interpretations essentially involve elimination of all other potential causes of destabilization. Thus, it is necessary to situate each 'potential seismite' within a detailed regional context (morphology, diagenetic state, depth of burial, etc.), before interpreting its origin. Furthermore, before interpreting a remobilized sediment as a 'seismite' the deformed beds should be described objectively: an *in situ* breccia, a sand layer with loadcasts or convolute bedding, a carbonate mud with open and subsequently closed fissures, a slightly brittle bed dislocated by sliding, an olistolith, a debris flow or even a turbidite, etc.

Within the syn-rift Upper Oligocene–Miocene and in the post-rift Pliocene and Quaternary sediments of the north-western Red Sea and Gulf of Suez coasts, numerous seismites have been identified by the authors (Plaziat *et al.*, 1990; Plaziat and Ahmed, 1994; Purser *et al.*, 1994). One would logically consider that a rift setting be exceptionally favourable for sedimentary remobilization as a result of repeated synsedimentary tectonic activity: not only repeated movements along normal and strike-slip faults but also by the associated steep topographies. The authors confirm that the temporal distribution of seismites is not haphazard and that their frequency and most extensive expressions may help characterize the periods of more intense tectonic activity.

DEFORMATION PROCESSES AND MORPHO-STRUCTURAL INFLUENCE ON SEISMITE TYPES

Deformational process will not be examined in detail, a summary of which is given on Figure D9.1. The type of destabilization (deformation, dislocation, grain remobilization, etc.) of sedimentary units affected by earthquakes, in particular, vary according to the diagenetic state of the sediment and to its position with respect to a slope (Plaziat and Ahmed, 1994; Plaziat and Ahmamou, 1997). Most seismite deformations are the result of a more or less achieved liquefaction of a particular sediment horizon situated near the sedimentary surface. The local collapse of unstable blocks, the direct fissuration of lithified beds by lateral shearing waves, are subaerial processes which play only a minor role within subaquatic or phreatic environments. On the other hand, reorganization of component grains in water-saturated soft sediments (closer packing) leads to an expulsion of interstitial water as a result of diminished porosity (Figure D9.2). The potentially most favourable sediments for this reorganization occur at a shallow depth where compaction by sedimentary loading has not expulsed interstitial waters and where lithification is absent or localized. This liquefaction results in a more or less viscous state which tends to liberate the liquefied sediment and its cover from gravity-stabilizing forces (friction) and enables the action of previously inhibited tractional forces. The increasing water/sediment ratio leads to different rheological states: a hydroplastic state which involves limited intergranular mobility permits only deformation of the bedding. It evolves into a liquid state which enables flow of a high-density fluid, even on a very weak slope. Finally, the fluidized state is characterized by fluid turbulence enabling an extreme mobility leading, for example, to sand and water injection (Lowe, 1975, and other references in Guiraud and Plaziat, 1993). Cross-cutting injection and internal 'load' deformation demonstrate that fluid escape generally developed within a multilayered series and was initially blocked by an impermeable cover. These undrained conditions favoured overpressure which, in turn, helped more sliding of the overlying strata by reducing the stabilizing effects of gravity.

In horizontal beds liquefied layers reorganize *in situ* as a function of their respective density and cohesiveness (e.g. load-casts, balls and pillows, collapse breccias) but if a local slope is present, even if very slight, lateral gravitational forces also operate. A horizon having high water content, completely liquefied, may flow on a slope less than one degree unless the overlying beds are too rigid or laterally blocked. If the slope is sufficient, or where the subhorizontal surface is situated close to a depositional slope, this type of obstacle is easily ruptured: lateral force will result in buckling, folding or breaking of more competent beds and the underlying lubricant facilitates the displacement and dislocation of

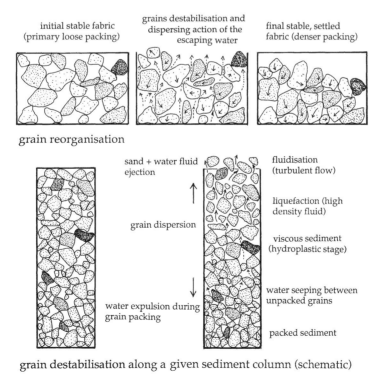

Figure D9.2 Sketch summarizing the processes involved in liquefaction induced by grain reorganization in soft, water-saturated sediments.

its cover. In this type of sliding the rheological properties of each bed will determine the style of deformation both within the displaced mass and in the underlying beds, the latter being more or less dragged by movements of the overlying mass.

Thus, it is clear that the rift fault blocks have topographies which favour the movement of superficial, even partially lithified deposits: perfectly horizontal surfaces are rare and steep scarps associated with half-graben and their bordering fault scarps are frequent (Figure D9.3). In the associated fan deltas with steep frontal slopes, the loose granular material is at its highest instability and susceptability to liquefaction, factors which also influence, even at some distance, the remobilization of subhorizontal sediments.

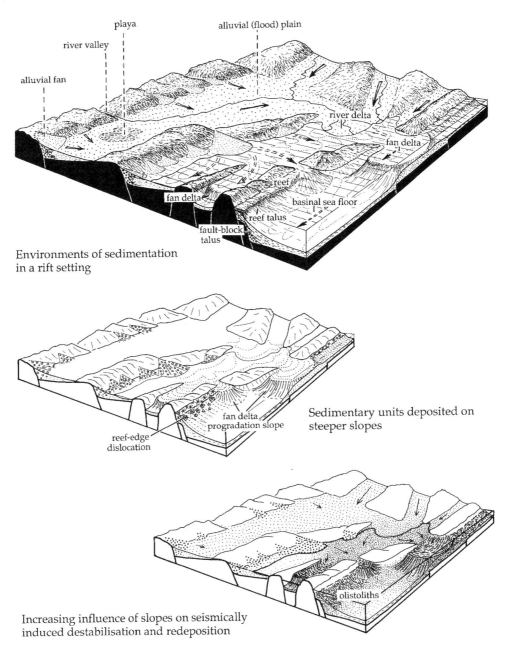

Figure D9.3 Combined influences of sedimentary environment, slope angle, and lithology on the local susceptibility to seismic destabilization in a marginal rift setting. Block-faulting is a basic cause of inclined sedimentary substratum and thus leads to sedimentary facies highly prone to deformation and reorganization during reactivation of syn-rift faults.

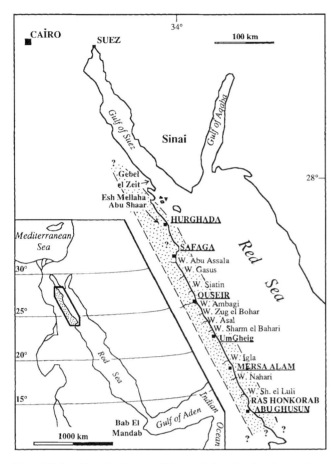

Figure D9.4 Map showing localities studied and regional extension of the Middle Miocene, C1 brecciated laminite (dotted).

CHRONOLOGICAL DISTRIBUTION OF SEISMICITY IN THE SYN-RIFT SERIES

The post-Eocene syn-rift strata of the north-western Red Sea and western Gulf of Suez (Figure D9.4) have been subdivided into four major tectono-sedimentary units (Figure D9.5) (Montenat et al., 1986; Plaziat et al., this volume). Group A (late Oligocene–early Miocene), essentially of continental origin, is associated with initial faulting and strike-slipping within a larger rift basin (Ott d'Estevou et al., 1989b). Termed 'proto-rift' (Orszag-Sperber and Plaziat, 1990), this tectonic stage is not preceded by a period of general doming but has been associated with local highs developed along the borders of relatively shallow graben and half-graben. These local reliefs were elevated progressively by increased tilting (Orszag-Sperber et al., this volume). The marine Group B (early–mid Miocene) is associated with major axial subsidence and follows a major tectonic reactivation (post–Nukhul and mid-Clysmic events) giving rise to the development of the well-known Clysmic blocks (Purser et al., this volume). The essentially evaporitic Group C (mid–late Miocene) records the filling of the deep, axial depressions resulting from the preceding subsidence. Finally, the 'Pliocene' and early Quaternary Group D deposits are contemporaneous with renewed peripheral uplift and rejuvenation of certain Miocene blocks together with important halokinesis and glacio-eustatic sea-level fluctuations.

The authors review seismites in chronological order.

The seismites of continental Group A1

Debris and mud-flow deposits are frequent on alluvial fan slopes. These massive, poorly sorted facies, although influenced by gravity and high viscosity, should not be confused with remobilized deposits of seismic origin. The bedding of materials is locally affected by post-depositional deformation indicative of seismic destabilization, the most diagnostic seismite criteria being load-casts, slumps and megalobes affecting horizontal deposits (Plaziat et al., 1990). These structures are most frequently developed within the upper parts of graben fillings suggesting increased seismic activity (Orszag-Sperber and Plaziat, 1990). The lithology of these terminal deposits tends to be more favourable for seismic liquefaction: upper channelized materials are better sorted and somewhat finer than the generally coarse conglomeratic textures typical of basal Group A. However, the most abundant load-casts are not restricted to sand layers but include mudstones and muddy sands.

The seismites of marine, restricted Group A2

The localized marine ingression which resulted in local precipitation of a gypsum unit exceptionally lies conformably on the continental A1 unit at Wadi Gasus (Safaga). A progressive tilting of the half-graben may be deduced from the formation of megalobes affecting the upper continental deposits (conglomerates and mudstones) followed by several slumps in the subaquatic, grey muds and sands intercalated between the beds of evaporite, one of which has slid. A more massive sand and gravel resedimentation unit fills the depression, including plurimetric olistoliths of consolidated marl (Figure D9.6). Thus, the usual major angular unconformity between Groups A and B is replaced locally (at Wadi Gasus) by a progressive tilting associated with increased seismic activity.

The seismites of marine Group B

The uplifted blocks tend to favour remobilization of marine sediments deposited on basement highs but also within the sedimentary slopes developed progressively in prograding platforms. Subhorizontal deposits, however, also occur, notably in the upper parts of Group B when

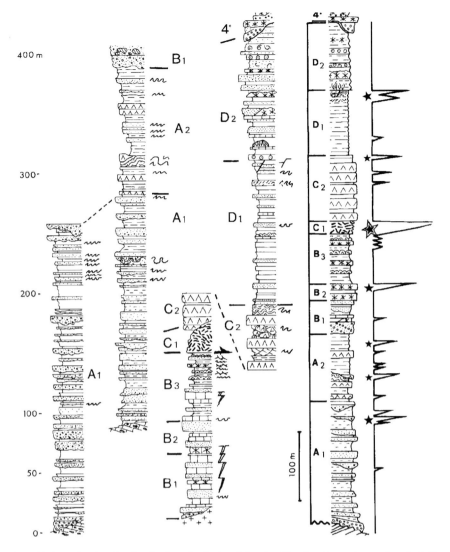

Figure D9.5 Lithostratigraphic logs showing the distribution of seismite-rich units. The first four columns (on the left) are based on sections at Wadi Hamadat, Wadi Gasus, Wadi Bali and Wadi Siatin, respectively. These four sections illustrate the varied responses to seismic events (symbols adjacent to log) according to varying lithologies and geological settings. The ubiquitous deformation of the Middle Miocene laminites (Unit C1) suggests that favourable lithology is the controlling factor; the organic-rich laminae interbedded within lithified dololaminites (stromatolites) constitutes a fragile fabric which has favoured widespread crumbling and brittle dislocation. The right-hand column summarizes the stratigraphic distribution and relative intensity of seismite deformation structures within the syn-rift sediments of the Egyptian Red Sea coastal region.

structural relief has been levelled. It is, therefore, of considerable interest that deformational structures such as load-casts are more frequent within these locally emergent horizontal beds (Figures D9.7, D9.13(c), Wadi Asal). In addition, conglomeratic fan deltas with steep sedimentary slopes are affected locally by swarms of vertical sand-dikes (Sharm el Bahari, Abu Ghusun, Figure D9.8(b)(c)).

Coarse terrigenous deposits characterize the peripheral slopes of many platforms (Wadi Ambagi, Abu Ghusun, Abu Shaar) and include numerous reef-derived olistoliths (Figures D9.9, D9.18) generally embedded in lenticular sand and gravel units. Although of multiple

Figure D9.6 Mud olistoliths in a coarse sandy, redeposited material marking the transition between tilted Group A evaporites and the horizontal B1 marine conglomerates: Burdigalian, Wadi Gasus, south Safaga.

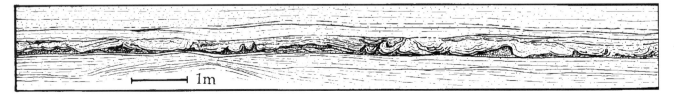

Figure D9.7 Example of *in situ* (static) seismites in the upper part of Group B. This hydroplastic load-cast deformation affecting a sandy mud is the lowermost of a series of seven deformed levels observed at this site. This structure records the first of a series of seismic shocks possibly emanating from a nearby cross-fault: south Wadi Asal.

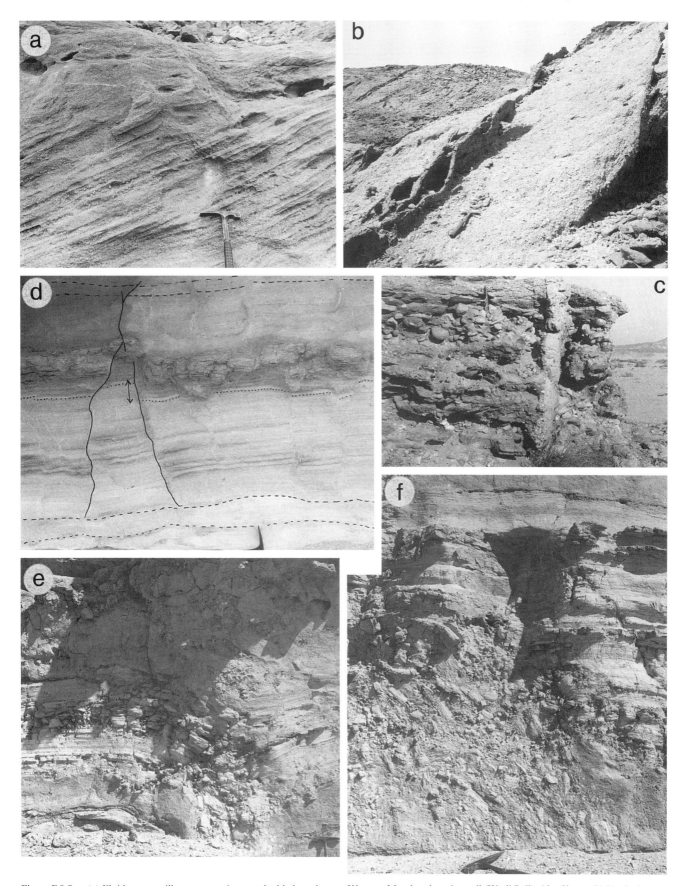

Figure D9.8 (a) Fluid-escape pillar structure in cross-bedded sandstone: Kharasa Member, hangingwall (Wadi Bali), Abu Shaar. (b) Vertical sandstone dikes cutting a conglomeratic fan-delta: upper B3 unit (Plaziat *et al.* this volume), S. Wadi Abu Ghusun. (c) Sandstone dike in marine conglomeratic fan-delta at Wadi Sharm el Bahari. (d) 'Graded faults' in a fine, laminated sand situated immediately below the slid, brecciated C1 laminite; this type of deformation, induced by the dragging effect of the oversliding laminite, may affect underlying deposits to depths of several metres. (e), (f) Basal and internal brecciation of the laminite (C1) unit indicating the mechanical nature (sliding) of this type of brecciation: Wadi Zug al Bohar.

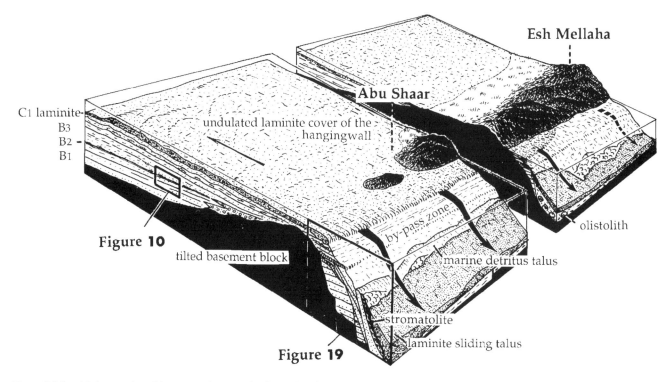

Figure D9.9 Major stratigraphic contrast between the footwall and hangingwall cover of a basement tilt-block. The weaker slope of the hangingwall suggests a lower susceptability to earthquake destabilization compared to the steeper footwall escarpment, the latter generally characterized by multiple unconformities and a well developed talus. Sliding following erosion has formed talus wedges below the by-pass escarpment.

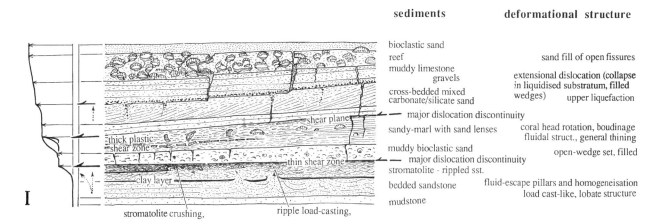

Figure D9.10 Seismites on the low-angle hangingwall slope of Abu Shaar platform. Submarine sliding and limited internal dislocations affect a 10–20 m thick lithologically varied sequence. These sedimentary and diagenetic variations strongly influenced the type of deformation.

origins, their great abundance associated with sliding structures and mass-flows in the above-mentioned localities (notably at Abu Ghusun), strongly suggest seismic origin (Purser *et al.*, this volume).

Open fissures descending more than 20 m from specific surfaces, notably in Wadi Sharm el Luli, Wadi Ambagi, and Wadi Bali at Abu Shaar, have been filled with mud or sand containing clasts of adjacent beds, and gravels derived from the upper walls and from the collapse of immediately overlying beds (and thus are intraformational). These express distensional movements affecting the series of beds situated above or below a sheared level (Figure D9.10, Plate 22). Such shear-induced dislocation may either be a sedimentary

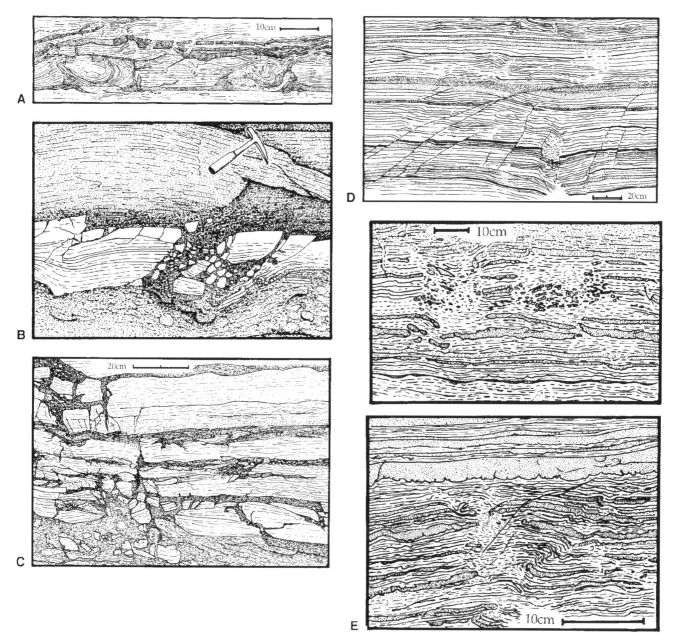

Figure D9.11 (A) Associated lobes (Macar's 'pseudonodules'), boudinage structures and intensely convoluted material illustrating the considerable influence of lithology on the type of structure produced. Sand beds may either be dislocated into boudins (brittle behaviour) or intruded into a dislocation fracture (fluid behaviour), or folded (hydroplastic behaviour). The intercalated sandy mud is generally deformed hydroplastically but cross-cutting fractures and micro-brecciation indicate a somewhat brittle behaviour; upper Group B, Sharm el Bahari. (B) Collapse brecciation due to the failure of the substratum. The intrastratal location of this deformation is indicated by the sandy nature of the breccia matrix; lateral flow is inferred from the fragments included within the sand; upper part of Group B, south Wadi Asal. (C) Complex association of liquefaction-induced intrastratal deformations. The boudinage affecting mud layers probably preceded the multilevelled complete liquefaction which, in turn, caused vertical dislocation and collapse-brecciation during failure of the substratum. This progressive disorganization is the complex scenario expected to occur during the repeated shocks of seismic waves during a major earthquake; the diverse facies comprising a formation is involved successively in the deformation according to their susceptibility to liquefaction, to their ductile or brittle behaviour, and to the nature of the adjacent materials and their rheological properties; upper part of Group B, Wadi Sharm el Bahari. (D) Graded faults and collapse breccia within a thinly bedded silt and sand deposit. The intraformational deformation results from a bed-on-bed dislocation and sliding. Overburden possibly explains the association of normal and reverse faulting and a local brecciated wedge: the final effect is a general thinning with extension of the median, deformed part. Microload casts and disturbed bedding of top layers demonstrate the post-burial origin of this typical seismite (cf. Seilacher, 1969); upper part of Group B, Wadi Siatin. (E) Pinch-and-swell, boudinage and crumbly dispersion characteristic of the differential behaviour of more competent mud and more liquidized sand within an extensional setting. The purely *in situ* (static) intrastratal deformation of (1) contrasts with the lateral asymmetry (recumbent folds and microthrusts) of a subsequent shear-deformation in (2). The discontinuties, stretching and boudinage do not result from sliding; upper part of Group B, laminated unit, Wadi Siatin.

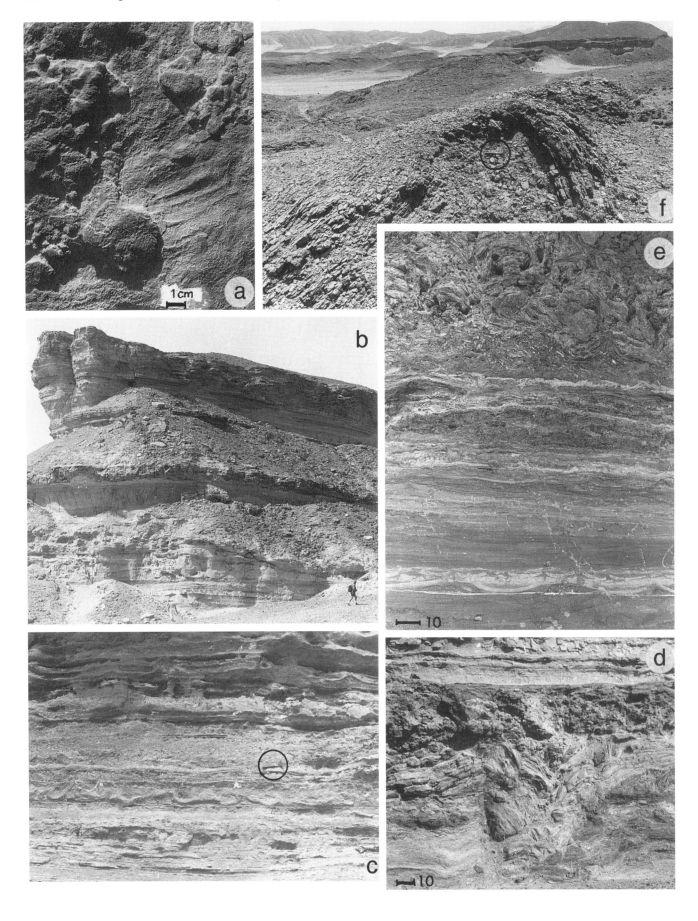

break (diastem) reactivated as a shear-plane or be expressed as a more diffuse zone developed within a thick bed which has become plastic during pounding by earthquakes. These large-scale structures are difficult to interpret without taking into account the underlying and overlying beds. They are generally confused with faults and thrusts mainly because they are rarely completely exposed in the smaller outcrops. Fortunately this type of fissure-filling is frequently associated with adjacent deformational structures exhibiting horizontal displacement including folded bedding, normal and reverse faults, deformed sheared layers, boudinage, or brecciation characteristic of seismic liquefaction (Figure D9.11). Thus, the causal relation between fissure-opening and intraformational dislocation can generally be demonstrated.

In situ deformation (load-casts, pinch and swell, etc.), without lateral displacement, are frequently associated with sliding and horizontal shear structures developed during subsequent sliding. These deformation structures generally have been acquired during a common seismic event when specific processes affected beds with different lithologies and diagenetic properties within the same sedimentary sequence (Figure D9.11(c),(e)). However, these composite structures may rather be developed progressively during increased fluidity resulting from progressive increase in pressure of fluid escape: porewater tends to migrate upwards and progressively fluidizes the upper, more susceptible facies during and following the arrival of seismic waves, as shown by sand pipes and the resulting sand volcanoes that continue to function for a relatively long time after the impact of an earthquake which triggered deep sand liquefaction.

Within a given vertical section one may observe many different types of seismite although their frequence and style vary according to local topography and facies. For example (south bank, Wadi Asal, Figure D9.12) several subhorizontal beds with basal load-casts record *in situ* deformation (Figure D9.12(c)). Somewhat higher (5 m), a 2 m thick sequence of mud and sand beds has been deformed by large load-casts and subsequent sliding resulting in structures including fissure systems, graded faults and breccias. However, there appears to be no link between the uppermost folded and brecciated carbonate laminite and the horizontal substratum from which the folds are dissociated (Figure D9.15). On the contrary, at Abu Shaar, a tilted, high-relief submarine block during Group B sedimentation (Cross *et al*, this volume), there appear to be few *in situ* deformations; instead, the platform cover is characterized by frequent dislocation structures including subvertical, sediment-filled fissures (see above), intraformational slide-breccias, subhorizontal dislocations affecting successive beds via steps or ramps or metre-sized, fluidized and sheared beds. Stromatolitic beds appear to have been particularly susceptible to fragmentation creating vuggy breccias which are usually mineralized with ferric oxides and silica (Abu Shaar, Wadi Asal). Certain horizons affected by horizontal intraformational movements are difficult to characterize; their fluidal structure, subparallel to bedding, is only locally associated with rotation of individual constituents such as coral colonies (Figure D9.10) and stomatolite domes (Figure D9.14). Finally, the upper (B3) marine strata at Abu Shaar are affected by low amplitude folds which become increasingly pronounced towards the base of the brecciated dololaminite (Group C1) which caps the platform and forms piles up to 30 m thick on the footwall slope.

This brecciated dololaminite is a sedimentary unit extending along much of the north-western Red Sea margin. It is probably of cyanobacterial origin and includes alternating laminae rich in organic matter and locally preserved moulds of evaporite minerals. This laminite, deposited in restricted subtidal environments, is the initial expression of isolation of the Red Sea before evaporite precipitation. Its large, compressional folds associated with brecciation (Figures D9.13, D9.16) have been interpreted in several ways. They were regarded initially as giant stromatolites (El Haddad *et al*., 1984; James *et al*., 1988; Burchette, 1988), while the breccias, locally attaining 20 m in thickness, have been widely interpreted as the products of collapse following evaporite dissolution (Coniglio *et al*., 1988; Burchette, 1988). The disorganized brecciated fabric has also been

Figure D9.12 (a) Horizontal groove-marks on a slightly indurated clast of the C1 breccia. The varied degrees of lithification explain the association of disharmonic folding and brecciation within the same sedimentary unit, Wadi Asal. (b) Section through the Middle Miocene (B3 to C1), south Wadi Asal. The Group B3 restricted marine deposits (few molluscs and frequent stromatolites) terminate in desiccated silts with mud cracks (e) situated below the C1 dololaminite, the latter being partly brecciated and folded (photo f). (c) Two load-cast horizons in continental mudstones situated several metres below outcrop shown in (b). The presence of a series of stratiform deformations extending about 30 m below the top of the folded laminite (photo f) may record a single or multiple earthquakes. (d) Brecciated dolomicrite with marked undulations situated between a flat, basal slip-surface and an undisturbed cover. These relationships suggest a distinct sliding event during B3 sedimentation. (e) Transition from the perfectly preserved desiccation chips and the coarse breccia. A rippled sand unit included several distinct destructured levels including shear-levels. However, the main discontinuity is situated at the top of the irregularly laminated bed (neo stratification). The basal breccia is a fragmented material filled by microbreccia. This facies grades into the coarser, upper breccia (6) via an undulating transition surface. (f) A surficial fold within the upper part of the C1 laminite. The cores of these folds are composed of a coarse, highly crushed breccia. These folds are interpreted as sliding-induced swells which have moved along the intraformational shear zone. Note that the long axis of this feature is oriented N 150°, i.e. close to the orientation of the rift axis.

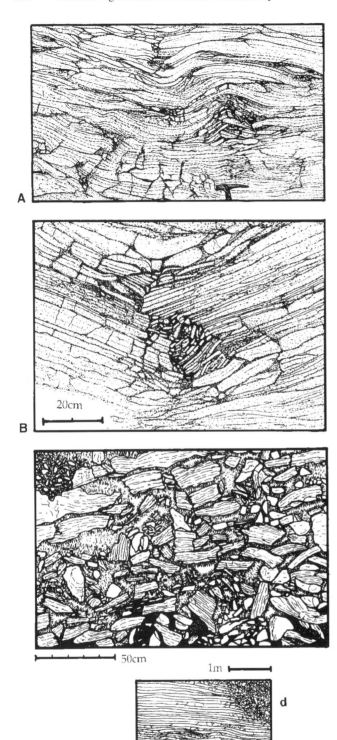

Figure D9.13 (A) and (B) Incipient brecciation in the Middle Miocene (C1) laminites. Fissures have developed during folding of the partially lithified carbonate. Internal dislocations parallel to the bedding favour differential creeping (disharmonic folds) which, in turn, causes axial (hinge) brecciation. Folding and the resulting breccia are obviously related to a general sliding as the substratum is not deformed except in rare levels; south Wadi Asal. (C) The vuggy fabric of the brecciated laminite favoured growth of celestite crystals. This mineralization (b2) is localized between the lowermost breccia with micro breccia matrix (b1), and the upper, weakly deformed laminite (c). The complex remobilization of this basal breccia is expressed by a wavy limit and upwards injection of fine-grained breccia (d) interpreted as being the result of a caterpillar-motion within the basal shear-zone limiting the laminite unit (a). During lateral creeping, the minute particles formed by crushing have filtered through the interclast pores and, in turn, have favoured basal lubrication of the sliding unit.

identified as a calcrete-type paleosol by El Aref (1993). Since 1988 the present authors have interpreted these brecciated and folded laminites as being the result of sliding (Plaziat *et al.*, 1990; Purser *et al.*, 1994). The thickness and extent of this deformation (southern Gulf of Suez to Abu Ghusun) are thought to reflect a major seismic event. During sliding, this brecciated/folded laminite has deformed the underlying open-marine beds comprising part of the Abu Shaar gently sloping ramp, the uppermost parts of the marine series (locally rich in coral) exhibit low amplitude (1–2 m) drag-fold undulations whose wavelengths range from several tens to several hundreds of metres (Figure D9.17). This is also obvious at Sharm el Bahari. On the other hand, the south- and north-east-facing escarpments of the Abu Shaar plateau (Figure D9.18) exhibit local thinning and removal of certain beds of the uppermost marine facies (B3 in Sequence 8 of Cross *et al.*, this volume). More plastic horizons are flattened and squeezed between rigid beds.

Figure D9.14 (a) Inclined and near-vertical clasts associated with narrow doming of the fine, lower breccia demonstrate the plastic rheological behaviour of the overall brecciated unit which creeps on the most finely broken basal materials. This sliding is also associated with folds whose main axes are approximately perpendicular to the sliding direction: C1 breccia, south Wadi Asal. (b) Dislocated stromatolitic unit showing incipient fracturation and clast rotation within a facies most susceptible to shear stress. Complete brecciation is often the final structure which may be mineralized (iron and silica): B3 unit south-west of Wadi Kharasa, Abu Shaar. (c) Various brittle behaviour in the C1 laminite. Brecciation is more or less obvious depending on the relative thickness of the lithified layers. Note that the biggest (thickest) clasts have been rotated along oblique shear zones. The pseudocontinuity of the thinnest layers expresses the asymmetric nature of folding and faulting which is characteristic of these internal shear-zones: south Wadi Asal. (d) Associated breccia and weakly deformed parts of the C1 laminite showing the maximum degree of heterogeneity observed within this slid unit. Note that the larger flat clasts tend to rise to the left, i.e. the southwards sliding direction: Abu Shaar, south-west of Wadi Kharasa. (e) Complex basal brecciation of the C1 laminite, south Wadi Asal. The lowermost, vaguely bedded, crushed unit is the most affected by sliding to the right. The breccia, which tends to coarsen upwards, grades into less dislocated beds at the top.

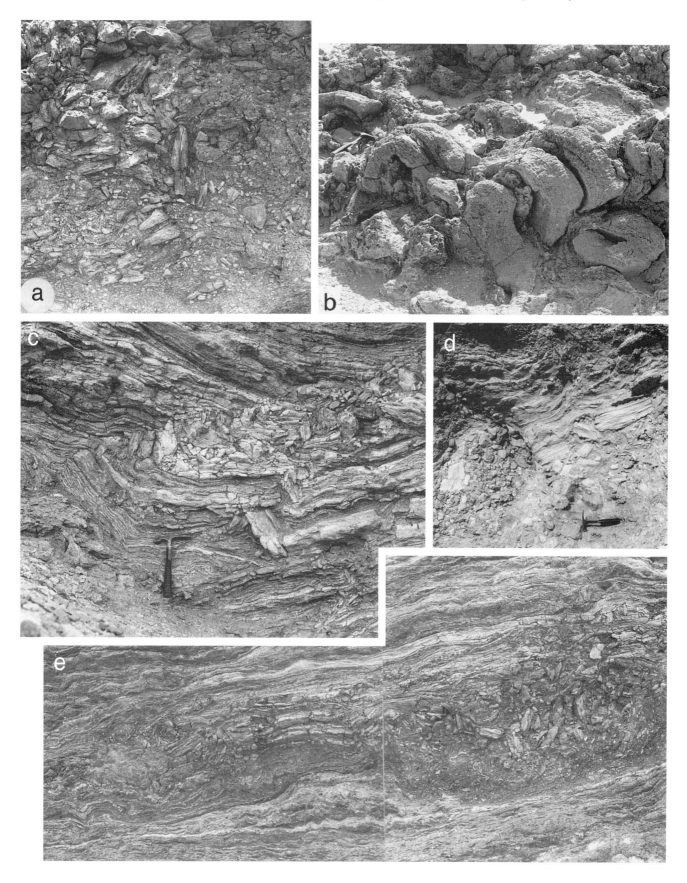

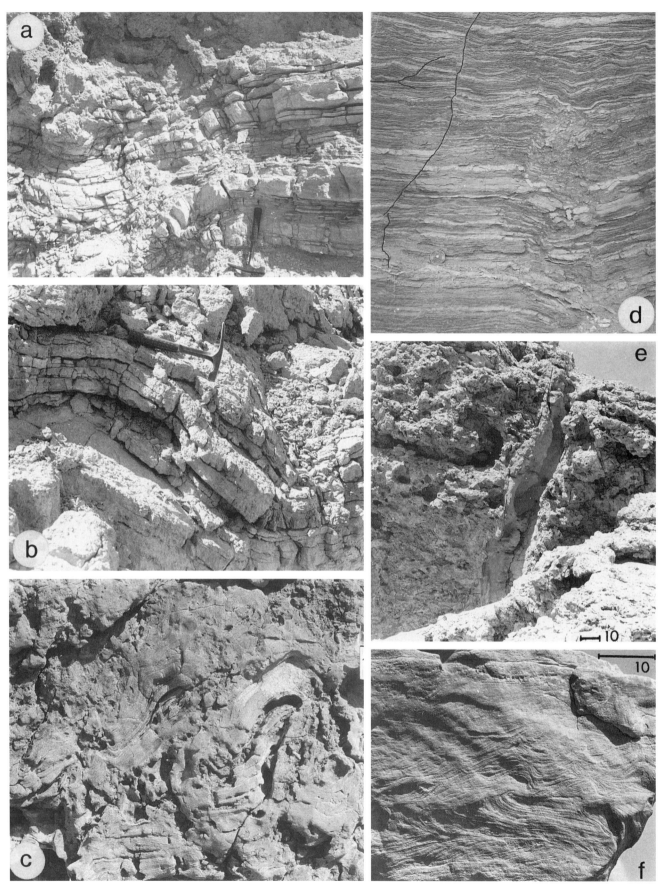

Figure D9.15 (a) Folded and brecciated laminite coating basement relief situated immediately below the lower Miocene A2 evaporite; Ras Honokorab. (b) Folded beds within the upper C1 brecciated laminite situated below the Group C, Upper Miocene evaporites; south Wadi Ambagi, Quseir. (c) Local ductile disharmonic behaviour within the lower (A2) brecciated laminite. This demonstrates that the laminite was composed both of lithified and plastic sediment; Ras Honkorab. (d) Associated microreverse fault and vertical wedge of collapse breccia (extensional dislocation) characteristic of limited sliding of the B3 unit; Wadi Siatin. (e) Quartz sand dike filling a subvertical fracture cutting a mid-Miocene reef at Sharm el Luli. (f) Hydroplastic deformation of a ripple cross-bedded oolitic sand. This convolute structure is associated with other *in situ* deformations including fluid-escape pillars: B3, Bali Member, western Abu Shaar platform.

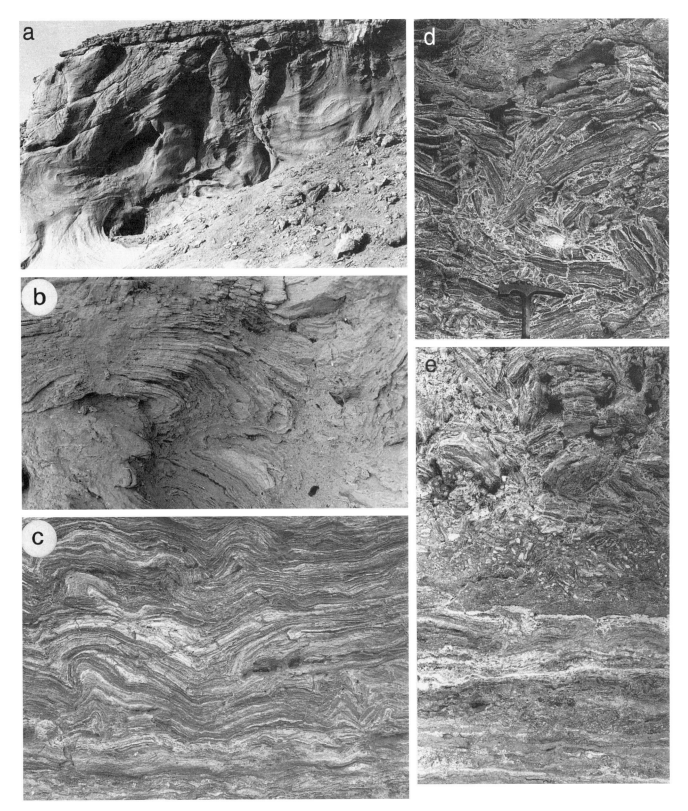

Figure D9.16 (a) Mega-convolute bedding (3 m in thickness) within Group D 'Pliocene' sands, south Wadi Igla (interpretation in Plaziat and Ahmed, 1994). This hydroplastic deformation grades upwards into a more consolidated, bedded sand cover exhibiting multiple thrusting. These relationships demonstrate intraformational liquefaction rather than slump-folding. (b) Vertically stacked folds affecting C1 laminites associated with completely homogenized muds. The stacked folds, in the absence of overburden, suggest that the structure results from a horizontal shear-zone; south-west of Wadi Kharasa, Abu Shaar. (c) Horizontal fissuration planes cross-cutting folded laminite with pseudo-stromatolitic domes and swells. The brittle behaviour is less obvious than the ductile disharmonic folding, although both brittle and ductile effects are intimately associated. Organic matter-rich laminae intercalated between the light-grey dolomicrite laminae may have acted as a lubricant which favoured interlaminae motion; C1 laminite, south Wadi Asal. (d) Vuggy laminite breccia with celestite crystals lining the interclast voids. The contorted aspect results from the internal creeping and clast rotation within a sliding shear-zone; south Wadi Asal; i.e. detail of the transition zone between the lower, autochthonous sand deposits and the upper laminite breccia (C1), south Wadi Asal. The original bedding is progressively destroyed with alternating crushed beds and lenses of dislocated beds. The apparent neolayering ends abruptly below a massive microbreccia (1 mm clasts in a mud matrix) which grades into the upper coarse breccia but also fills the interclast voids below. This fine material probably acted as a plastic (viscous) sole favouring sliding of the laminites and drag-dislocation of underlying beds; shear-stress is dispersed via several dislocation planes and plastic shear-zones within the laminite and its substratum.

Although at Abu Shaar steep-slope sliding towards the south and east, i.e. towards the axis of the rift, alone is not sufficient proof of seismic causes, the thickness (several tens of metres) and regional extent (350 km) of these deformations, generally on very weak slopes (Figure D9.19), are difficult to explain other than in terms of seismic destabilization. The remobilization could possibly be favoured by a lowering of sea level, especially in view of the fact that they immediately precede the deposition of Group C evaporites. However, it is not likely that a mechanism relating to sea-level lowering initiated these ubiquitous movements, even on low-gradient surfaces.

Because these deformed laminites probably reflect a major seismic event, one may suspect that the many and varied seismites within the underlying Group B sediments, are also the products of this same high-magnitude earthquake, their intraformational localization depending mainly on the susceptibility of particular sedimentary fabrics. At several localities there is, in fact, a clear relationship between the breccia and the immediately underlying dislocations. However, there also exists a considerable number of seismites which are separated from the deformed laminites capping Group B by several metres of comparable lithologies that are undeformed. The more localized and laterally discontinuous nature of these lower deformations suggests that they are the expression of seismic events whose magnitudes were markedly lower than the event affecting the laminites capping the sequence. An increase in the magnitude of earthquakes towards the major environmental change affecting the region (Group B–C transition) would support a tectonic cause for this basin-wide change (Orszag-Sperber *et al.*, this volume).

The seismites of Group C evaporites

Because of their weathered state most outcrops of Group C evaporites are unfavourable for detailed study. Only in exceptional sections such as those near Quseir and Gebel el Zeit can one observe details (Figure D9.20). At these localities intraformational folds occur within gypsum which are not necessarily of seismic origin. Nevertheless, folds associated with small-scale thrusting as well as bedded megabrecciated gypsum interbedded with plastic marls indicate that important

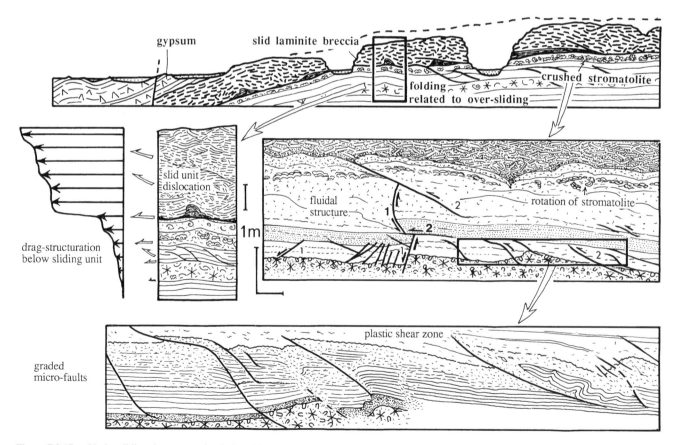

Figure D9.17 Under-sliding destructuration induced by the southwards displacement (to the left) of the C1 laminite cover. The dragging which affects the uppermost 10 m of marine (B3) strata is expressed by intraformational faulting ('graded faults'), multilevel brecciation and low-amplitude folding (middle right). Extensional faulting (1) is modified by subsequent compressive thrusting (2) during the smae sliding event. Southern edge of Abu Shaar.

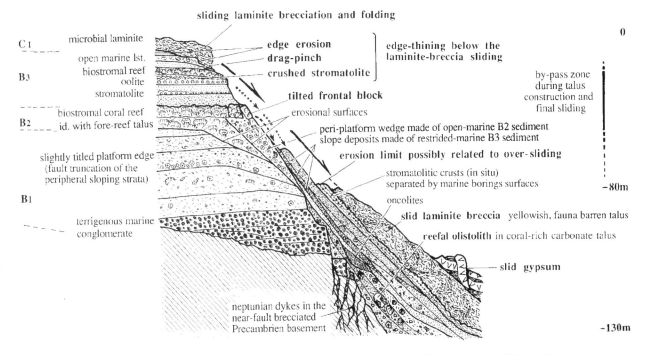

Figure D9.18 Slope deposits and associated seismites on the steep (35°) footwall of the Abu Shaar and Esh Mellaha tilt-blocks.

intraformational sliding was frequent near the top of the formation at Gebel el Zeit and in Wadi Siatin. There also exist bedded microbreccias suggesting remobilization of laminated evaporites (Wadi Ambagi).

In certain wells in the Gulf of Suez, terrigenous sand and debris flows indicate repeated displacement on slopes (Rouchy et al., 1995; Orszag-Sperber et al., this volume). These resedimented materials, including centimetre-sized evaporite debris, have been interpreted as the result of slope instability triggered by seismic shocks.

Seismites in the post-evaporite (Pliocene–Quaternary), Group D sediments

The Pliocene sediments exposed along the north-western coast of the Red Sea between Quseir and Abu Ghusun comprise two distinct formations discussed by El Haddad et al. (1984) and by Orszag-Sperber et al. (this volume). A lower unit composed of relatively fine-grained clastics with rare dwarf molluscan faunas indicative of restricted environments, is succeeded by a somewhat coarser sandy or calcareous facies locally rich in oolite, red algae, coral and oyster beds. The abrupt change from the restricted facies to shallow open-marine sediments coincides with a 5–10 m thick unit exhibiting

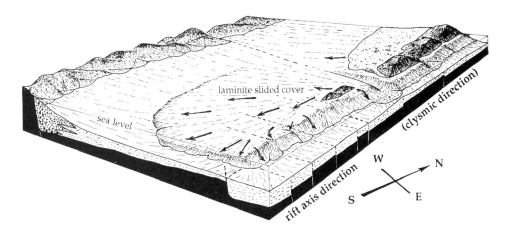

Figure D9.19 Seemingly opposed sliding of the laminite cover (C1) on the plateau of Abu Shaar tilt-block; in fact, the general southward direction of sliding results from the combined influence of the south-west dip-slope and the north-east footwall slope.

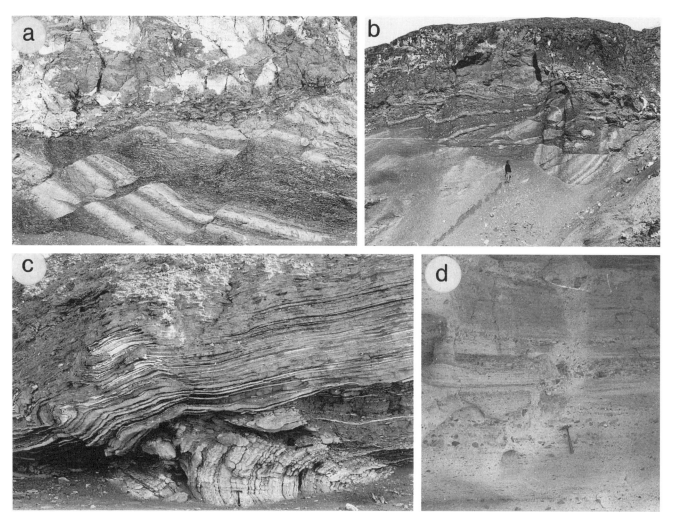

Figure D9.20 (a) and (b) Intraformational dislocation in muds situated within the C2 gypsum, north. Gebel el Zeit. Faulting and block rotation affecting the layered muds forms the substratum for the sliding of the upper gypsum unit. Displacement may have been triggered by seismic instability. (c) Folded and truncated laminated gypsum situated below a thick, coarse terrigenous unit intercalated within the Upper Miocene C2 evaporites at north Wadi Ambagi, Quseir. While slope oversteepening may have caused this sliding, seismic triggering can not be discounted. (d) Sand wedge in Pleistocene fluviatile sediments at Wadi Nahari. Limited modification and collapse of the coarse materials suggests that a fissure was immediately filled and closed, rather than resulting from fluid escape.

large-scale deformational structures related to liquefaction and sliding processes. In the region of Mersa Alam, these deformations include breccias which may be stratiform or locally associated with synsedimentary faulting (Wadi Igla), and plastic folds in a metre-thick, fine sands (Figure D9.16(a)) resulting from liquefaction (Plaziat and Ahmed, 1994). Individual outcrops of these strata having limited vertical dimension (2–10 m), it is more difficult than in the Miocene deposits to prove that faulting and brecciation are limited to intraformational levels. Therefore, in a few cases (Wadi Igla) it can not be demonstrated that they are deep tectonic faults or that they bend and terminate into a local intraformational slide. Whatever the precise geometry, the stratiform distribution of these deformational structures can be readily correlated from one valley to another over a distance of about 50 km. This would suggest that breccias and associated fluidal structures in subhorizontal deposits are probably of seismic origin.

Discrete *in situ* hydroplastic deformational structures, including load-casts, occur below the major facies change (at Um Gheig and south of Safaga, Figure D9.21(a)). Thus, there is a certain similarity between the sequence of deformations affecting the sediments of Groups A, B and C; the frequency and scale of deformational structures increase towards the top of both sedimentary sequences, each of which is followed by a major change in sedimentary environment (Purser et al., 1990).

For Quaternary strata it is somewhat more difficult to identify with certainty the exact causes of distensional, landslide deformations affecting continental and

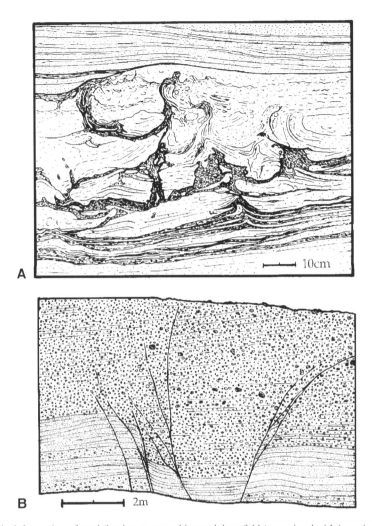

Figure D9.21 (A) Hydroplastic deformation of sand (load-casts, stretching and drag-folds) associated with intrusion of silts into dislocation cracks; 'Pliocene', south Safaga. (B) Pre-lithification faulting and tilting in upcurrent-sloping fluvial sediment within a 'Pliocene' terrace. This type of deformation suggests tilting superficial listric faulting. Because of the coarse texture of the materials and the large-scale tilting it may be a classical rift-tectonic faulting rather than a sliding-induced deformation.

marine sediments. The important fluctuations of sea level and contrasting rainy phases may have been partly responsible. However, the occasional existence of fissures filled with sand or pebbles, both within reef-terraces and in fluviatile gravels (Wadi Nahari, Figure D9.20(d)), in many respects comparable to those occurring in the Miocene Group B, is suggestive of seismic origin.

Marine and continental Quaternary deposits tend to be relatively coarse-grained (corals, boulders, gravels) which are not particularly prone to seismic deformation in spite of the fact that structural movements continued, notably during the early Quaternary. Modern earthquakes are relatively rare and of low magnitude suggesting a general, recent stabilization of the rift periphery (Plaziat *et al.*, this volume).

DISCUSSION: RELATIONSHIPS BETWEEN SEISMITES AND TECTONIC HISTORY

The creation of structural relief during the evolution of the rift system is clearly recorded by sediment geometry and facies. To this record may be added that of seismites, bearing in mind that their development also depends on the susceptibility of sediment relative to liquefaction. Thus, for both sedimentological (granulometry or early lithification) and topographic (absence of slopes) reasons, certain sedimentary units are considerably less favourable than others in registering the effects of palaeo-earthquakes. Within the rift context of the Red Sea–Gulf of Suez, there exist both favourable factors such as frequent synsedimentary slopes, and unfavourable factors such as the frequency of coarse, poorly sorted materials (Figure D9.3). However, in spite of these potentially unfavourable factors, the syn- and post-rift sediments of the north-western Red Sea and Gulf of Suez are characterized by the abundance of seismites and, locally, by their volumetric importance.

The interest of seismites lies particularly in their chronological distribution: generally speaking, the outcropping reactivated fault-planes do not provide much information concerning the absolute chronology of tectonic activity. Final movements often mask earlier displacement and faults sealed by specific sedimentary horizons are rare. Within this rift system, seismic events sufficient to deform sediments (seismites) occur throughout the history of the basin. However, a marked increase in frequency and magnitude of seismic events appears to coincide with tectono-stratigraphic boundaries and major changes in sedimentation. For example, there is a marked change in environment (restricted, marine ingression) related to block faulting and tilting at the end of Group A, and there are major changes in environment at the end of Groups B and C as well as within Group D. The most important seismically related deformational structures coincide with the onset of mid-Miocene evaporite sedimentation thus suggesting a structural control. Even though the development of local, kilometre-sized grabens is responsible for limited Group A evaporite deposits, it is the development of a structural sill north of the Gulf of Suez that initiated Belayim–South Gharib (Group C) evaporite sedimentation throughout the Gulf of Suez and much (if not all) the Red Sea (Orszag-Sperber *et al.*, this volume). Therefore, coincidence with one or several earthquakes of exceptional magnitude would not be surprising.

While the expected high frequency of seismic activity in a rift setting is recorded by the syn-rift sediments, the authors admit that stable cratonic settings are not necessarily devoid of major earthquakes. In the pre-rift Nubian Sandstones of Egypt (Plaziat and Purser, 1995), and the Tawila Sandstones of Yemen (unpublished); favourable fluviatile sands include spectacular seismites, comparable with the deformed beds deposited in a strike-slip setting of Early Cretaceous Sandstones of Nigeria (Guiraud and Plaziat, 1993). However, the authors insist on the basic difference between the respective deformational styles: sand liquefaction (load-casting, convolute folding and injection) are characteristic of subhorizontal fluviatile basins whereas shear dislocations, sliding and brecciation associated with contrasted reliefs are more typical of rift settings.

CONCLUSION

Within the rift, different types of morpho-structural domains developed simultaneously. High reliefs with steep footwall slopes certainly are the most susceptible to slumping and delivery of olistoliths triggered by seismic liquefaction of underlying sediments. Although necessarily very important in the organization of talus deposits, seismic destabilization is difficult to prove. Thus, the most convincing seismites are recorded from subhorizontal dip slopes and from structural depressions or corridors (sediment transfer zones) between structural highs. These ubiquitous records of palaeo-seismicity complement the structural history of rifting and may confirm the tectonic origin of major environmental changes within the basin.

SECTION E
SYN-RIFT CARBONATE DIAGENESIS

The two chapters in this section examine the complex diagenetic history of Miocene syn-rift carbonate sediments from the north-western Red Sea and western Gulf of Suez. This complexity is shown to be related to the varying primary mineralogy of the sediments, climate, steep depositional slopes, overlying evaporites, hydrothermal fluids related to extensional faults and high heat flow throughout this rift basin margin.

E1 Syn-rift diagenesis of middle Miocene carbonate platforms on the north-western Red Sea coast, Egypt
B.H. Purser

E2 The dolomitization and post-dolomite diagenesis of Miocene platform carbonates: Abu Shaar, Gulf of Suez, Egypt
N. Clegg, G. Harwood and A. Kendall

Chapter E1
Syn-rift diagenesis of Middle Miocene carbonate platforms on the north-western Red Sea coast, Egypt

B. H. Purser

ABSTRACT

The mid-Miocene carbonates exposed along the north-western coastal region of the Red Sea and southern Gulf of Suez, although relatively thin (less than 150 m), are highly modified by a series of Miocene diagenetic phases. Developed under near-surface conditions, the general diagenetic evolution began with marine cementation of slope deposits. This was followed by massive dolomitization and dissolution of aragonitic debris. The geometry of the dolomitized strata does not appear to be related to specific sequence boundaries or faults and its pervasive nature, negative oxygen isotopes and associated hydrocarbons all suggest dolomitization from heated basinal sea-waters. A later diagenetic phase relating to the penetration of sulphate-rich brines into the dolomite substratum is expressed by a polyphased diagenesis involving dolomite dissolution, aragonite precipitation and dolomitization, celestite cementation and dissolution. Large volumes of secondary sulphate have replaced the dolomite substratum.

The phases of diagenesis, most of which were related to the lateral and vertical movement of large volumes of marine waters, were probably stimulated by variations in density aided by a relatively high temperature gradient. The delivery of these marine waters to the interstitial pore-system was facilitated by the inclined nature of the sediment/water interfaces. These factors, and the resultant diagenesis, are a natural consequence of rift basins.

INTRODUCTION

The Red Sea–Gulf of Suez rift contains up to 5000 m of syn-rift sediments, a large part of which are early to mid-Miocene pelagic marls of the Rudeis–Kareem formations which are not exposed on the western periphery. These thick basinal sediments record subsidence during the major phase of rifting. Early to mid-Miocene, shallow-marine carbonates are thinly developed (20–30 m) on actively subsiding blocks such as Gebel el Zeit (Figure E1.1) (Prat et al., 1986; Bosworth, 1995) but attain thicknesses of up to 150 m at outcrop where they characterize the culminations of syn-rift fault blocks (Figure E1.2). Probably of Langhian–Serravallian age (Philobbos et al., 1993; Plaziat et al., this volume), these mid-Miocene carbonate platforms constitute parts of the coastal Red Sea Hills, both in the south-western parts of the Gulf of Suez and in the north-western Red Sea. Similar carbonates also occur in Sudan (Schroeder et al., this volume). Each carbonate platform, formed on a tilted, early Miocene fault block, is flanked, generally along its basinwards side, by spectacular slope deposits (Purser and Plaziat, this volume) which record Miocene submarine relief. This relief is often rejuvenated by repeated movement along adjacent border-faults and by structural tilting (Cross et al., this volume). Marked synsedimentary relief therefore is one of the major attributes of mid-Miocene rifting.

Open marine carbonate sedimentation is replaced progressively during the Serravallian by restricted peritidal facies often dominated by laminated micrites and stromatolites. These pass into massive evaporites which attain 2000 m within the deeper parts of the basin (Richardson and Arthur, 1988). However, bathymetric relief persisted during evaporite sedimentation. At outcrop, evaporites onlap the flanks of mid-Miocene platforms such as Abu Shaar (Rouchy et al., 1983) (Figure E1.2).

Sedimentation and Tectonics of Rift Basins: Red Sea–Gulf of Aden. Edited by B.H. Purser and D.W.J. Bosence. Published in 1998 by Chapman & Hall, London. ISBN 0412 73490 7.

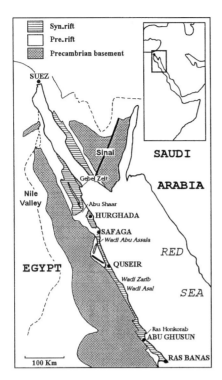

Figure E1.1 Simplified geological map of north-western Red Sea and Gulf of Suez showing localities studied.

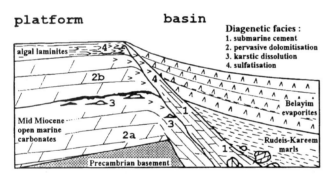

Figure E1.2 Schematic relationships between the structural, stratigraphic and four main diagenetic phases discussed in this contribution.

Virtually all Neogene carbonates in the north-western Red Sea are strongly modified by multiple diagenetic processes. Dolomites are ubiquitous. However, early marine cements, celestite, secondary calcite and silicate cements, are also frequent. Because extensive fieldwork (Montenat et al., Cross et al., Purser et al., this volume) has shown that most Miocene carbonates exposed along the periphery of the rift have never been buried more than several hundred metres (and possibly less), their diagenetic properties have been acquired under near-surface conditions.

Previous work

Diagenesis of mid-Miocene carbonates of the north-western Red Sea has received only limited attention and most studies are restricted to the Abu Shaar block. Aissaoui et al. (1985) have considered the nature and origins of aragonitic fabrics in general, with brief reference to the Miocene aragonites at Wadi Asal. Aissaoui et al. (1986) have also examined the multiple diagenetic phases affecting the Miocene carbonates at Abu Shaar, stressing the importance of both the early marine cements and, especially, the dolomitization of virtually all platform and slope carbonates. These studies were continued by Coniglio et al. (1988) demonstrating the strong predominance of light oxygen and carbon isotopes. A comprehensive summary of the diagenetic properties of a series of carbonate platforms was prepared by Purser (1987) noting the considerable variety of minerals including polyphased dolomites, secondary gypsum, celestite, aragonite and secondary silica associated with the transition from mid-Miocene carbonates to late-Miocene evaporites. The importance of secondary sulphates was first noted by Orszag-Sperber et al. (1986). Reference to specific diagenetic aspects of Miocene carbonates has been made by Rouchy et al. (1985) and by Youssef (1989a,b) concluding a bacterial origin of lenticular calcite bodies occurring within late Miocene evaporites. Sun (1992) and Sun and Esteban (1994) have shown the potential of sulphate-rich waters to dissolve aragonitic bioclasts, possibly explaining the widespread vuggy porosity typical of many Red Sea Miocene platforms. Finally, the detailed description of Miocene dolomites and their isotopic properties, at Abu Shaar, by Clegg et al. (this volume) should be considered in conjunction with the present, more general, evaluation of the diagenetic evolution of mid-Miocene carbonates.

Objectives of this contribution

Pervasive diagenetic modification of Miocene carbonates and, probably, the siliciclastic sediments (Darwish and El Araby, 1993), is one of the major attributes of these syn-rift sediments. This short contribution evaluates the possible relationships between diagenesis and rift geodynamics: are the multiple diagenetic phases the result of multiple subaerial exposure during repeated lowering of relative sea level, or do they result from deeper-seated movements of phreatic waters? These diagenetic processes influence the petrophysical character of syn-rift sediments and thus determine the reservoir potential of these rocks; the possible relationships between basin dynamics and reservoir distribution are considered.

The region studied, while including a limited number of platforms (Figure E1.1) nevertheless covers a considerable portion of the rift (400 km). More importantly, it includes platforms located near its periphery (Zug al Bohar, Ras Honkorab), others situated at mid-distance between the periphery and the axis (Gebel Abu Shaar), as well as Gebel el Zeit situated near the axis of the Gulf of Suez (Purser et al., this volume). Because each area has a somewhat different sedimentary and tectonic history, notably Gebel el Zeit, the choice enables limited comparison of the various diagenetic properties and their position within the rift.

THE PRINCIPAL DIAGENETIC ATTRIBUTES OF THE MID-MIOCENE CARBONATES

The diagenetic setting

The general nature and intensity of diagenesis depends on the chemical composition of the interstitial waters, on their movement and renewal, and on their temperature. Other secondary factors relating to the above include substrate mineralogy, permeability of the sediments affected, and presence or absence of organic materials, including bacteria. These obvious controls should be kept in mind when considering the geological framework within which the mid-Miocene carbonates have been diagenetically modified.

The morpho-structural framework
The creation of a marked bathymetric relief by extensional faulting is expressed, in this area, by numerous, generally elongate highs. This relief has strongly influenced the sedimentary pattern (Purser et al., this volume) favouring the development of carbonate platforms bordered by depressions filled with basinal Rudeis–Kareem marls, and subsequently by late-Miocene evaporites. The rift is flanked by a high peripheral escarpment much of which was acquired during the Pliocene and Quaternary.

The stratigraphic context
Mid-Miocene carbonates, rich in aragonitic components, were deposited on and around structural highs with peripheral slopes. Sediments deposited on the platforms are interrupted by numerous discontinuities enabling their subdivision into depositional sequences (Burchette, 1988; Cross et al., this volume), some of which are related to emergence during repeated tilting of the blocks. The Miocene carbonates pass upwards into mid to late Miocene evaporites reflecting a regional increase in basinal salinities.

Thermal gradient
Rifts are generally characterized by relatively high thermal gradients, that in the southern Gulf of Suez being in the order of 1.7°F per 100 ft (W. Bosworth, personal communication). This gradient could have been considerably higher during phases of maximum rifting. This relatively high vertical gradient also varies laterally being highest in the vicinity of major faults (Bosworth, personal communication). Hot springs occur on both sides of the Suez rift.

Hydrology (Figure E1.3)
All previously mentioned factors have a potential influence on the composition and movement of interstitial waters and thus on diagenesis. The submarine relief with its relatively steep slopes forms a barrier to marine currents and waves and favours the injection of sea water into the permeable substratum (Buddemeir and Oberdorfer, 1986). The entry of marine waters via an inclined surface, in principle, should be recorded by the preferential development of cements near the entry surface, these cements decreasing towards the interior as movements of interstitial waters slow. Horizontal sea floors, on the contrary, facilitate the movement of currents and the underlying interstitial waters therefore tend to stagnate.

Exposure of the platform culmination during periods of relative sea-level lowering, in principle, favour the development of shallow, fresh-water lenses. The resulting diagenesis would be recorded by the dissolution of aragonitic debris and by the precipitation of calcite cements whose geometry is closely related to the exposure surface; such diagenetic effects tend to be stratiform.

The change from normal marine to hypersaline conditions during the Miocene has important diagenetic implications. Because this change was also associated with inclined sea floors, the denser brines would have displaced the less saline interstitial fluids leading to a chemical disequilibrium, particularly with respect to the less stable aragonite (Sun, 1992). Furthermore, the formation of multiple secondary basins by rift-related block-faulting would tend to compartmentalize the Miocene waters (Figure E1.3). Thus, waters situated

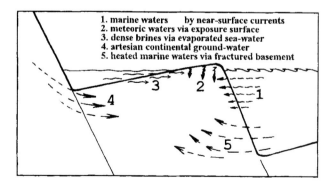

Figure E1.3 The five potential hydrological systems which may possibly influence syn-rift diagenesis of carbonate sediments deposited on carbonate platforms.

within graben, separated from the oceanic mass by intervening structural barriers, logically would evolve either towards hypersalinity or towards less saline waters, depending on climate and meteoric influx. Resulting differences in density would stimulate movement of interstitial waters through intervening barriers.

Relatively high peripheral relief, under humid climatic conditions, would favour an artesian system by which groundwaters would discharge into peripheral basins to be mixed with sea waters, possibly improving their diagenetic potential.

Finally, relatively high thermal gradients, typical of rift systems, would stimulate movement of any warm interstitial waters whose buoyancy would favour an upward and, possibly, lateral migration in a manner comparable to that envisaged by Saller (1984). The presence of steep submarine slopes would favour the input of cool, dense basinal waters into a heated substrate. These warm marine waters could subsequently rise via faults and fractures, the latter being abundant within the Precambrian crystalline basement. These waters may have penetrated the sedimentary cover, either along faults, or in a dispersed manner via the extensive fracture system.

Whatever the mechanism, there exists a series of important factors, most relating to rift geodynamics, which favour the movement of various interstitial waters and thus a high diagenetic potential. However, before describing the suite of diagenetic products, it is important to consider briefly the possible criteria for the recognition of one or other hydrodynamic model. In principle, waters entering a sedimentary sequence from a given point or surface will leave a diagenetic imprint whose lateral (or vertical) variations will be related to this surface. Thus, sea water entering an inclined substrate will favour a high degree of cementation (or dolomitization) near to this surface. Meteoric waters entering the exposed platform during periods of sea-level lowering will create a lenticular diagenetic body closely related to this surface, as noted by Sun (1992). Finally, waters penetrating via a fault-plane may form diagenetic bodies which are oblique to the sedimentary stratification.

On the other hand, a diagenetic style which can not be related to a given sedimentary or structural surface suggests that the fluids or other controlling factors, either affected the sedimentary mass as a whole (burial/pressure) or have been delivered (in the case of fluids) via multiple conduits. This could have been the case, for example, where formation waters penetrated the sedimentary cover via a complex fracture system, or where lateral artesian flow, relating to peripheral relief, affected the entire sedimentary column.

General diagenetic properties and diagenetic sequence

Within the mid-Miocene carbonates the various diagenetic phases exhibit two, basically different, geometries; some phases are related directly to specific surfaces, others, notably the regional dolomitization, affect virtually all carbonates exhibiting only minor lateral or vertical variations. The dolomite mass is never stratiform and appears to be a regional phenomenon recorded as far south as Sudan (Schroeder et al., this volume). The presence of an important evaporite formation which onlaps carbonate slopes has led to the premature conclusion (Purser, 1987) that dolomitization in the Gulf of Suez is related to late Miocene brines. However, there exists neither a geometric relationship between the dolomite and the carbonate/evaporite interface, nor an isotopic signature reflecting an evaporitic source (Aissaoui et al., 1986; Clegg et al., this volume).

The relatively complicated diagenetic sequence (Purser, 1987) involves four major phases, each of which has a number of secondary events:

1. An early fibrous cement affects virtually all grainstones, notably those deposited on the platform slopes. It is closely related to peripheral slopes of many Miocene platforms in the north-western Red Sea.
2. Dolomitization and dissolution of aragonitic skeletons, intimately associated events, affect most carbonates on the north-western Red Sea coast including the early marine cements. Both dolomitization and dissolution are pervasive on the Abu Shaar platform, there being few visible petrographic variations or systematic isotopic changes (Clegg et al., this volume).
3. Dissolution of dolomite occurs on several scales. Karst cavities, 1–5 m in diameter, at Abu Shaar and at Wadi Asal, are related to specific discontinuities, i.e. are stratiform, while sparry calcite cement, although volumetrically unimportant, is widespread.
4. A series of diagenetic minerals and fabrics relating to highly saline waters include replacement of dolomite by secondary sulphates, gypsum cements and diagenetic silicate typical of sulphate-rich waters. Frequent in the upper part of the mild-Miocene carbonates, these diagenetic effects are related stratigraphically to the carbonate–evaporite transition.

Early marine cements

Borings (generally *Lithodomus*) which affect both stratiform discontinuities and certain fault planes (Figure E1.4(a)) demonstrate the early lithification of Miocene

Principal diagenetic attributes of mid-Miocene carbonates 373

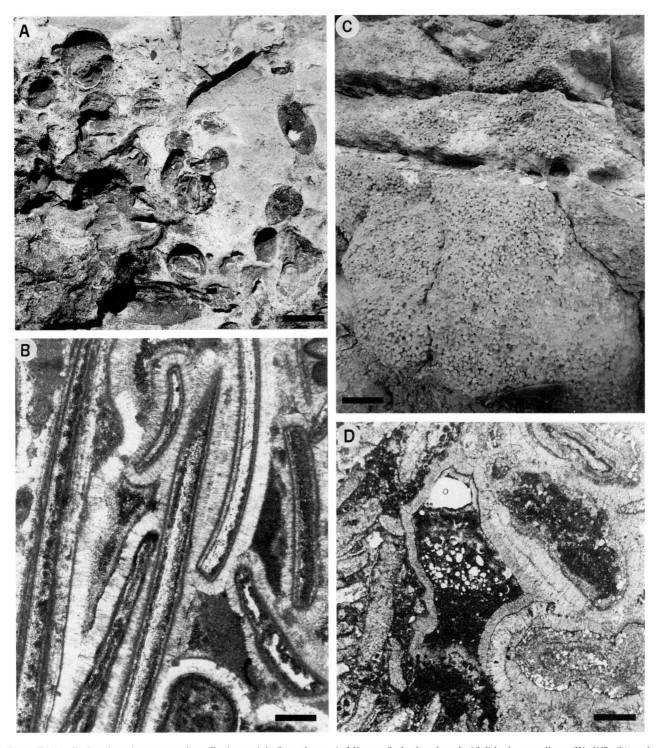

Figure E1.4 Early submarine cementation affecting periplatform slopes. A. Miocene fault-plane bored with lithodome molluscs (Wadi Zarib), scale 2 cm. B. Isopachous fibrous cement now completely dolomitized (Abu Shaar), scale 350 μm. C. Outcrop of pisolite beds; note hammer-head at bottom (Abu Shaar), scale 10 cm. D. Detail of submarine cement and internal, void-filling sediments (Abu Shaar), scale 500 μm.

sediments. Lithification is also confirmed in the field by the abundance of olistoliths (not all of which are reefal) within most slope deposits.

Fibrous cements

Borings penetrating grainstone textures may truncate fibrous cements whose fabrics are typical of many modern submarine cements (Shinn, 1969; James and Ginsburg, 1979; Sandberg, 1985). Fibrous cements within the Miocene carbonates at Abu Shaar are discussed in detail by Aissaoui et al. (1986) who has distinguished a series of morphological variations including a botryoidal morphology typical of certain modern aragonitic cements. Other, less fibrous crystal fabrics, are suggestive of a magnesian calcite precursor. However, the possibility of two primary mineralogies (aragonite and magnesian calcite) is not supported by the subsequent diagenetic behaviour of these early phases as they are both replaced by dolomite (Figure E1.4(c), (d)). All cements retain their initial fabric. On the contrary, all aragonitic debris (e.g. corals and molluscs) is strongly affected by dissolution while rhodophyte debris, originally composed of magnesian calcite, is perfectly preserved. Because all the cements are preserved (rather than dissolved) they also were probably composed of magnesian calcite. Furthermore, on a global scale, aragonite cements are rare within Tertiary limestones, deposited in essentially 'calcitic seas' (Sandberg, 1985).

Pisolites
Fibrous magnesian calcite cements are highly developed within the platform peripheries at Zug al Bohar and at Ras Honkorab (Figure E1.1) where their presence is reflected by the relatively high densities of slope deposits. They are particularly well developed within the coarse pisolitic rudstone (Figure E1.4(c)), a well-known facies within the footwall slope deposits at Abu Shaar (El Haddad et al., 1984; Aissaoui et al., 1986; Burchette, 1988; James et al., 1988; Cross et al., this volume). Slightly elongate pisoliths averaging 1 cm in diameter, are concentrated within a single bed varying in thickness from 1 to 4 m. This very distinct bed, dipping at about 25° in conformity with surrounding slope deposits, is composed almost uniquely of pisoliths. Its upper surface is locally bored (Aissaoui, 1986). The enigmatic origin of the pisoliths, already discussed by several authors (op. cit.), may be sedimentary and/or diagenetic. They have never been observed on the platform culmination and appear to be limited to the peripheral slopes which, together with their very homogeneous textures (i.e. only pisoliths), indicates an autocthonous origin. Their inclined but stratiform disposition and their isopachous, typically phreatic cementation (Figure E1.5(a)) suggest they are neither cave pearls nor vadose pisolites analagous to those of the Permian Capitan complex of New Mexico (Dunham, 1969).

Individual pisoliths at Abu Shaar have a somewhat flattened nucleus. The nuclei are generally composed of clotted micrite (Figure E1.5(b), (c)) occasionally associated with microbal structures (Aissaoui et al. 1986). (Figure E1.6(b)) and only rarely a mollusc or bryozoan fragment. The cortex, which generally attains 0.5–1 mm in thickness, is composed of multiple, isopachous layers of fibrous carbonate separated by thinner, micritic laminae (Figure E1.5(c), (d)). Contact between the two cortical fabrics is rarely flat, the micritic fabric tending to penetrate the adjacent fibrous lamina. Furthermore, both fabrics often exhibit a small-scale domal morphology. These phenomena, including the clotted micrite nuclei, all suggest microbial influence. Cortices of individual pisoliths, generally darkened by the presence of micrite laminae, are followed by a more translucent, isopachous layer of fibrous cement common to adjacent grains. This cement, generally devoid of micritic laminae, is overlain by fine, dolomitized internal sediment (Figure E1.4(d)) occasionally containing foraminifera and other marine debris.

The pisoliths at Abu Shaar therefore include two distinct components. The sedimentary nucleus and cortex appear to be essentially of microbial origin. Their development on a relatively steep submarine surface can best be explained by the presence of a microbial substrate which tended to stabilize the grains, nevertheless permitting their occasional movement facilitating cortical growth. Experimental studies by Buczynski and Chafetz (1991) utilizing organic substrates have shown that spherical grains composed of both micritic and fibrous carbonate do not involve significant movement. The second, predominantly fibrous, cement while possibly associated with microbial activity, seems to express the diminishing effects of organic substrates. This brief episode of organo-diagenetic sedimentation is preceded and followed by open-marine slope sedimentation.

Micritic cements
Microcrystalline cements affect certain grainstones deposited on the platforms. These unlaminated, isopachous cements are comparable to the magnesian calcite cements typical of many modern reefs and beach-rocks. They are relatively uncommon within the slope deposits. The selective development of fibrous and micritic cements respectively, in different parts of the platforms, may reflect differences in water temperatures and salinities, the coarser fibrous calcites, developed at depths attaining 50 m on the slopes, being conditioned by cooler waters and normal marine salinities.

Hydrodynamics
The clearly expressed field relationships between slope sedimentation and a high degree of submarine cementation suggest cause and effect. Because fibrous cementation affects only the slopes (Figure E1.2), and fails to penetrate more than 10 m into the adjacent, subhorizontal platform strata, it forms a dense external carapace similar to that observed on certain Pacific atolls (Aissaoui et al., 1986) and other oceanic platforms such as Belize (James and Ginsburg, 1979). The peripheral

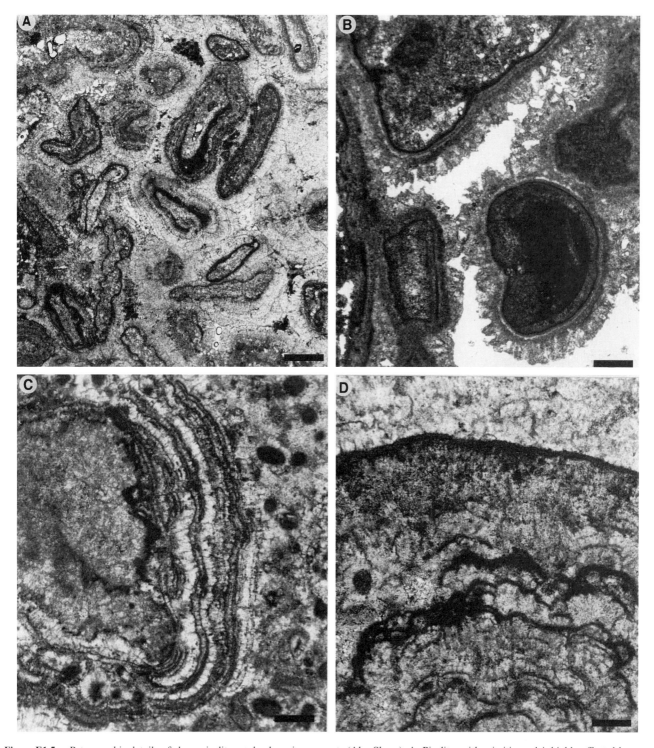

Figure E1.5 Petrographic details of slope pisolites and submarine cements (Abu Shaar). A. Pisolites with micritic nuclei, highly affected by dolomitized fibrous cement, scale 1 cm. B. Pisolites, micrite nuclei and irregularly laminated cortex possibly of microbial origin, scale, 350 μm. C. Pisolite showing clotted micrite nucleus and alternating micritic and crystalline cortical structure, scale 150 μm. D. Irregular, discontinuous nature of cortical laminae suggestive of microbial influence, scale 25 μm.

occurrence of this diagenetic facies indicates centripetal influx of marine waters favoured by the inclined nature of the substrate, i.e. a rift-related submarine morphology.

Petrophysical implications
Because these early marine cements have reduced primary porosity, the least porous carbonates are located around the peripheries of north-western Red

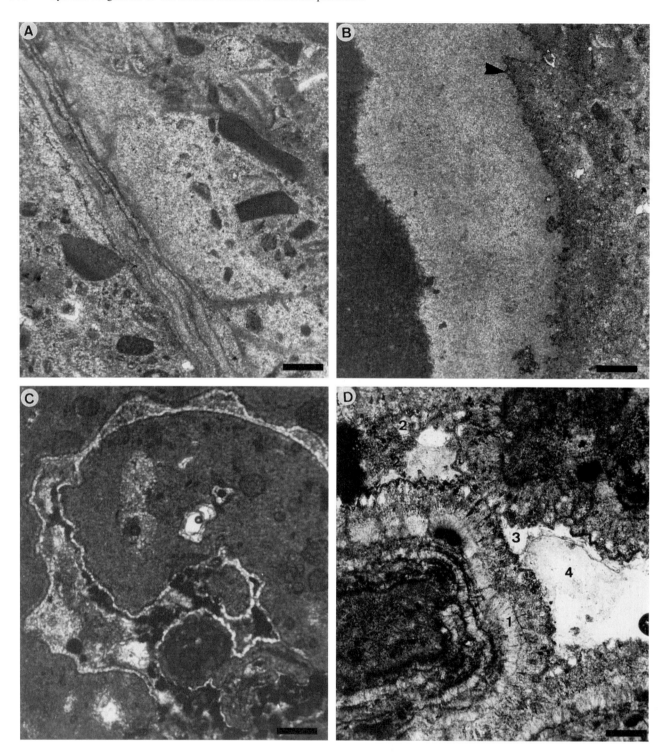

Figure E1.6 Various petrofabrics in dolomite (slope deposits, Abu Shaar). A. Branched microchannel within various microcrystalline dolomite fabrics, scale 100 µm. B. Microcrystalline fabrics; oxidation of contact (arrow) suggests multiple phases, scale 100 µm. C. Dissolved gastropod lined with dolospar overlain by dolomitized internal micrite, scale 500 µm. D. Diagenetic sequence; 1 = fibrous cement, 2 = internal sediment, 3 = dolospar, 4 = sulphate cement, scale 100 µm.

Sea platforms. Inversely, throughout most platform interiors, of which Abu Shaar perhaps is the best example, most sediments, irrespective of their mineralogies and textures, are porous and permeable. However, these final properties are also the result of several additional diagenetic processes discussed below.

Polyphased dolomitization and dissolution

On most platforms, notably Abu Shaar, virtually all carbonates have been replaced by dolomite. On others, notably Gebel el Zeit, situated closer to the subsident rift axis, dolomitization is only partial. Within any given platform, dolomite fabrics are homogeneous and there are no apparent vertical or lateral variations, the only differences visible at outcrop reflecting variations in primary sedimentary properties. Dolomitization is massive and pervasive. Most dolomites have preserved, often with remarkable detail, the primary skeletal structures and the fabric of the early fibrous cements (Figure E1.4(b), (d)). This fabric-preserving dolomite generally has replaced magnesian calcite debris, the aragonitic components having been dissolved. Common to many dolomites in general, the temporal association of dolomitization and dissolution of aragonite is a well-known phenomenon, dissolution often determining the petrophysical properties of the dolomite (Purser et al., 1994). However, it is not always clear whether dissolution preceded or was contemporaneous with dolomitization. Within the Miocene carbonates of the north-western Red Sea, microcrystalline dolomite has replaced all sedimentary micrite and magnesian calcite bioclasts as well as the peripheral envelopes of aragonitic debris, notably the molluscs and corals, the central parts having been dissolved (Figure E1.6(c)). These secondary voids are generally lined by a fringe of limpid dolospar (Figure E1.6(c)).

These petrographic relationships (dolomicrite–dissolution–dolospar) may be interpreted in two manners. Firstly, the contrasting dolomicrite–dolosparite fabrics, separated by a phase of dissolution, may suggest two phases of dolomitization. However, because the process of dolomitization involves dissolution of the precursor mineral (calcite or aragonite), often on a nanoscopic scale, and the precipitation of another (dolomite), it may result in secondary porosity, especially if dissolution exceeds dolomite precipitation. Furthermore, dolomite crystallinity depends to a considerable degree on the nature of the pre-dolomite fabric; dolomite 'replacing' micrite frequently gives a microcrystalline dolomite while dolomite growing into pore-space nearly always results in dolospar. Thus, a second hypothesis involves contemporaneous dolomitization, dissolution and dolospar growth as a more or less continuous process. This latter hypothesis, in most cases, is preferred.

In other, not infrequent examples, two or more distinct phases of dolomitization may be demonstrated, notably where a zone of pyrite or manganese oxide corrodes the initial dolomite phase (Figures E1.6(b), E1.7(d)) which is overlain by a subsequent dolospar. Finally, within dolomitized micrites there are frequent discontinuities which appear to be caused by internal erosion, and these may be accentuated by thin pyritic staining of the discontinuity (Figure E1.6(b)). The discontinuities separate slightly different dolomite fabrics suggesting multiphased dolomitization.

These petrographic considerations are not necessarily irrelevant to the interpretation of dolomitization processes based on carbon and oxygen isotope data, discussed by Aissaoui et al. (1986), James et al. (1986) and by Clegg et al. (this volume). The wide, seemingly disorderly scatter of isotopic signatures (discussed below) could reflect a number of factors, including bulk sampling involving the mixing of several dolomite phases.

Dolomite petrotypes: comparison of Abu Shaar and Gebel el Zeit

Dolomite fabrics discussed above and by Aissaoui et al. (1986) and by Clegg et al. (this volume) concern samples from the Abu Shaar platform. Material collected by Roussel (1986) between Wadi Sharm el Bahari and Quseir (Figure E1.1) also exhibit fabrics similar to those described from Abu Shaar. However, dolomites from Gebel el Zeit (Figure E1.1) differ in several respects from the more peripherally located (Abu Shaar, Quseir) platforms. While dolomites at Abu Shaar are dominated by microcrystalline, fabric-preserving textures, those at Gebel el Zeit frequently are coarsely sparitic, often fabric-destructive dolomite (Figure E1.7).

The dolospars at Gebel el Zeit may be zoned, notably where they fill secondary voids (Figure E1.7(c)). Normally characteristic of void-filling dolosparitic cements, the same zoned spar also appears to be 'replacing' molluscan debris (Figure E1.7(a), (b)). This would suggest that the 'replacement', in fact, is preceded by a dissolution front permitting free crystal growth into nanoscopic voids. The resulting dolomite crystals, which may attain 1 mm in size, may have slightly curved faces typical of late, hydrothermal cements. However, they rarely exhibit undulose extinction.

Relationship between dolomitization and late Miocene evaporites

The main phase(s) of dolomitization, both at Abu Shaar and at Gebel el Zeit, preceded evaporite sedimentation and thus are genetically unrelated to these late-Miocene brines. This is readily demonstrated by the following diagenetic sequence. Petrographically, dolomicrites are followed closely by dolospars, the latter developed in secondary voids where they may often be associated with gypsum cements (Figures E1.6(d), E1.7(d)). The latter (gypsum) are invariably the final diagenetic phase being separated from the older dolospars by a zone of framboidal pyrite or manganese oxide, reflecting reducing conditions. At Gebel el Zeit, dolomitized Nukhul Formation carbonates are overlain by about 50 m of

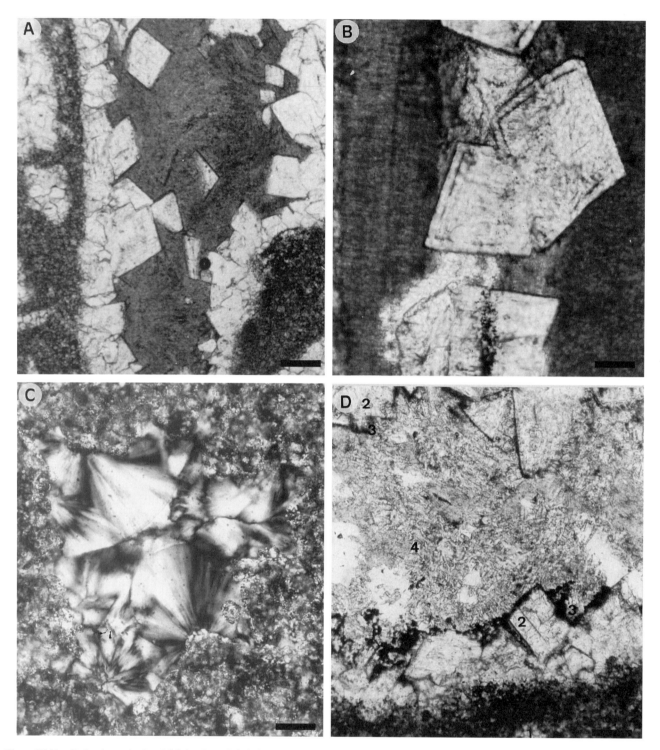

Figure E1.7 Dolomite and related fabrics from Gebel el Zeit. A. Large (1 mm) dolomite crystals replacing lamellibranch (central part of photo), scale 500 μm. B. Detail of dolomite showing zoned nature of crystals replacing mollusc, scale 200 μm. C. Fibrous quartzine filling void in dolomite matrix, scale 100 μm. D. Diagenetic sequence in void; 1 = dolomicrite; 2 = corroded dolospar; 3 = iron oxide; 4 = anhydrite; scale 200 μm.

finely bedded globigerinid marls, locally rich in organic matter, in turn overlain by the Belayim Formation evaporites (Figure El.2). The zone of framboidal pyrite and manganese oxide separating dolomite and gypsum cements probably was acquired during the subsequent burial of the dolomite within the organic-rich marls, residual porosity later being filled by gypsum cement. That the sulphate-rich water was a relatively late diagenetic event is also demonstrated by the presence of fibrous quartzine which follows dolospar development (Figure E1.7(c)). Finally, on a somewhat larger scale, large olistoliths eroded from an early Miocene fault scarp at South Zeit and deposited within the Rudeis–Kareem marls, are composed of crystalline dolomite. They are bored by marine annelids and heavily encrusted with iron and manganese oxides, the latter probably formed under reducing conditions within the marls. It is difficult to imagine the penetration of brine-related waters through the relatively impermeable marly cover and their subsequent entry into the underlying carbonate. Massive dolomitization therefore seems to have occurred prior to Miocene evaporite deposition and thus the parental water movement favouring dolomitization must have been conditioned by other factors. However, although the brines are not an important factor in mid-Miocene dolomite formation, they are nevertheless associated with limited quantities of much later dolomite formed during evaporite sedimentation.

Mineralogical and isotopic properties of the dolomite

The composition of dolomites at Abu Shaar, discussed by Aissaoui et al. (1986) and by Clegg et al. (this volume) is fairly constant, ranging from 49 to 50.5 mol % of magnesium carbonate. Oxygen and carbon isotope values, measured on bulk samples of pure dolomite, have a considerable range. The $\delta^{18}O$ and $\delta^{13}C$ are highly negative, the oxygen attaining $-11\%_0$ PDB (Clegg et al., 1995). Predominance of light oxygen isotopes tends also to preclude dolomitization from brines (see above), unless these signatures have been reset by hydrothermal fluids, as has been suggested by Coniglio et al. (1988). However, the role of meteoric or even mixed waters is also doubtful (Clegg et al., 1995; Clegg et al., this volume), in view of the excellent preservation of most (but not all) detrital feldspars. The most logical explanation of the relatively negative oxygen isotope values relates to temperature, as proposed by Clegg et al. (this volume). This interpretation is supported by comparable negative $\delta^{18}O$ values in dolomitized limestones of Valinginian age encountered at ODP site 639 on the Iberian continental margin of the eastern Atlantic (Loreau and Cross, 1988). The $\delta^{18}O$ values ranging from -8 to -10 from sparry, fracture-filling dolospars and can not be explained in terms of meteoric waters and are interpreted as being the products of hydrothermal waters.

The $\delta^{13}C$ values in dolomites from Gebels Abu Shaar and Zeit (Clegg et al., 1995) are also negative, probably due to the presence of hydrocarbons. Together with the presence of hydrocarbons in fissures and fault planes in well-cores (W. Bosworth, personal communication), these values tend to support the hypothesis of upward migration of the dolomitizing fluids.

Hydrological model

Dolomitization from hot waters implies upwards and, perhaps, lateral flow of interstitial waters. These have two potential sources: artesian flow of continental groundwaters relating to high peripheral relief, or convectional movement of warm marine waters. The presence of hydrocarbons and the large quantities of magnesium required for dolomitization strongly favour the second hypothesis. However, the composition of the dolomitizing fluids and the timing of dolomitization relative to the late Miocene evaporites, have been interpreted in several manners. Most authors (including Clegg et al., this volume) have suggested dolomitization from brines whose isotopic signatures have been reset by elevated temperatures. While this may indeed be the case, it is not supported by the diagenetic sequence discussed in the previous paragraph. Dolomitization from normal sea waters having an oxygen isotopic signature of -10 would require temperatures of about 85°C; dolomitization from brines would require somewhat higher temperatures. Marked bathymetric relief during early and mid-Miocene rifting when Precambrian basement and pre-rift sediments were probably exposed to deep marine waters along submerged footwall escarpments (Figure E1.3), probably facilitated the entry of cool sea waters into the fractured basement. Subsequent warming stimulated by a relatively high thermal gradient, resulted in their decrease in density and subsequent upwards movement and lateral migration into the Miocene, syn-rift cover. The wide scatter of negative $\delta^{18}O$ values may reflect either the mixing of several dolomite phases, or temperature fluctuation of the dolomitizing fluids. Preservation of associated detrital feldspar is due possibly to the marine waters being saturated in potassium (and other elements) during their passage through the fractured basement. Finally, many outcrops of Miocene dolomite, notably at Abu Shaar (Wadi Kharasa), are heavily stained with iron and manganese oxides whose constituents probably were derived from underlying basement rocks.

Anastamosing microchannel systems (Figure E1.6(a)) are frequent within the dolomitized Miocene carbonates, both at Abu Shaar and at Gebel el Zeit. Individual microchannels measure several hundred microns in width, their lateral walls not being clearly defined because of progressive crystallization fabrics affecting the adjacent matrix. Channels tend to be somewhat irregular and may be traced across a given sample. Their two-dimensional geometry suggests that they were initially developed as fractures affecting a weakly consolidated sediment. Subtle variations in adjacent crystal fabric, and iron staining, although not necessarily related directly to dolomitization, nevertheless indicate the passage of interstitial waters through the sedimentary sequence.

Petrophysical properties of the dolomites

Most Miocene dolomites are highly porous. However, individual voids generally correspond to primary porosity, notably in the grainstones (Figure E1.8(b)), or to secondary porosity created by dissolution of aragonitic bioclasts (Figure E1.8(a)). The well-known inter-rhombic pore systems, often associated with sucrosic dolomites (Purser *et al.*, 1994), are rare. Dense dolomite fabrics generally correspond to carbonates which have been well cemented by fibrous calcite (Figure E1.8(c)). In other words, the petrophysical properties of the dolomites are essentially inherited from the sedimentary and predolomite diagenetic fabrics; dolomitization appears not to have created porosity. However, dolomitization is associated genetically with large-scale dissolution of aragonitic bioclasts and the possibility exists that the carbonate required for large-scale do-

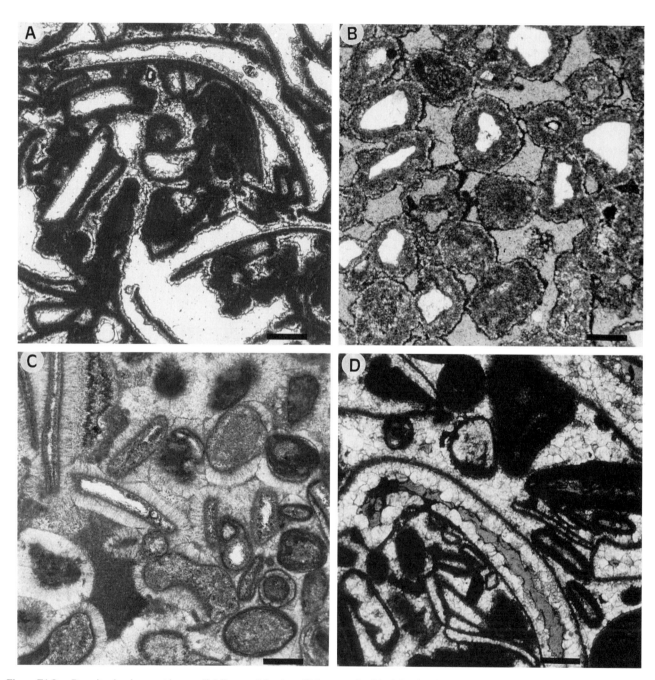

Figure E1.8 Porosity development in syn-rift Miocene dolomites. Unless stated, all is dolomite. A. Porosity (white) is primary (intergranular) and secondary (dissolution of bioclasts), scale 100 µm. B. Porosity (grey in photo) is all primary, intergranular, unmodified by dolomitisation, scale 50 µm. C. Dolomite whose low porosity reflects a high degree of submarine cementation, scale 50 µm. D. Dense dolomite whose porosity is modified by dolospar cementation, scale 100 µm.

lomitization was acquired, at least in part, by dissolution of certain sedimentary constituents.

Karstic dissolution

Large-scale dissolution with individual caverns sometimes exceeding 1 m, occur on several of the platforms studied and are generally situated directly below certain discontinuities, notably those marking the top of reef bodies. These sediment-filled cavities are interpreted to be karstic in origin. They are not abundant and there are no laterally continuous karst horizons. The most spectacular cavities are formed within a series of patch reefs situated 50–60 m above the floor of Wadi Bali, near the top of the Kharasa Member (Cross et al., this volume). At locality 'A', situated on the south-eastern side of the wadi, large faviid coral colonies have been partly dissolved, the resulting caverns being lined with a centimetre-thick layer of fibrous, botryoidal cement (Figure E1.9), probably of marine origin (cf. Aissaoui et al., 1986). The coral substrate, and the vug-lining cements are dolomitized. In addition, scattered dolomite crystals 2 mm in diameter are scattered over the smooth surface of the cement (Figure E1.9(b)) suggesting multiphased dolomitization.

At locality 'B' situated near the top of the cliff limiting the north side of Wadi Bali, the same reefal unit is cut by a series of metre-sized caverns lined with dolomitized fibrous cement. The caverns are filled with bedded, dolomitized silts cut by a second cavern system also filled with laminated internal sediment (Figure E1.10). The geometry of this multiphased cavern system is not totally clear. However, their fairly regular, subvertical walls suggest that they are enlarged fractures possibly created by Miocene seismic instability or, possibly, by cave collapse. That they are modified by dissolution is indicated not only by their width (2 m) but also by the selective removal of adjacent coral colonies, resulting voids subsequently being enlarged to form a tubular network lined with dolomitized botryoidal cement. The internal sediments which fill most caverns locally exhibit a system of polygonal desiccation cracks. The cavern system is capped by cross-bedded arkosic sands situated close to the top of depositional Sequence 3 of Cross et al. (this volume).

The angular discordance marking the top of depositional Sequence 3, clearly visible along most of the north-eastern, footwall scarp of the Abu Shaar block, is affected locally by metre-sized cavities filled with marine sediment. Their presence, together with the cavities described above, indicates lowering of relative sea level. Because similar surfaces have not been observed on other platforms, their origin at Abu Shaar probably relates to local structural movement. Other cavities developed near the top of Sequence 6 at north-east Abu Shaar, are discussed by Clegg et al. (this volume).

Pervasive and local karstic dissolution, common to many Miocene carbonate platforms in the north-western Red Sea region, poses a number of problems concerning the nature of the aggressive waters and their hydrology. The mid-Miocene climate probably was relatively dry and thus unfavourable for karst development. However, aridity may have been lower on the somewhat elevated rift periphery, as indicated by the important quantities of coarse terrigenous material delivered to the basin and the existence of extensive meteoric aquifers can not be excluded. Furthermore, in spite of widespread dissolution, sparitic calcite cementation is very moderate and most carbonates are very porous. This would suggest that the dissolved carbonate either was incorporated into the dolomite or that it was flushed out of the system.

Polyphased diagenesis relating to the carbonate–evaporite transition

Mid-Miocene open-marine carbonates, exhibiting the diagenetic features discussed in the preceding paragraphs, grade upwards into unfossiliferous dolomicrites containing scattered gypsum pseudomorphs. This progressive restriction culminates in the precipitation of a thick unit of mid–late Miocene evaporites (Figure E1.2). The evaporite formation, discussed in detail by Orszag-Sperber et al. (this volume), can be conformable on the underlying dolomites. In most cases, however, evaporites onlap inclined slope deposits. In both cases, (conformity and onlap) the transition from dolomite to evaporites is associated with a considerable variety of diagenetic features, most of which relate to the penetration of brines into the dolomitic substrate. This penetration was probably facilitated by the density of the brines and by the inclined nature of the carbonate substrate.

Cavity formation

The dolomite–evaporite transition is frequently associated with relatively large cavities which may attain 1 m in diameter (Figure E1.11 (a), (b)). At Wadi Zarib and Wadi Asal, these cavities are common and may have several possible origins. The lowermost cavities occur within a folded, brecciated dololaminite where their origin results mainly from mechanical processes relating to palaeoseismicity, discussed by Plaziat and Purser (this volume). Cavities formed higher in the sequence are the result of dissolution which has been interpreted as karstic (El Aref, 1993).

The dissolution cavities situated close to the dolomite–evaporite transition, and also within the 10 m thick Um Gheig dolomite situated near the top of the evaporites at Wadi Asal, are lined with laminated

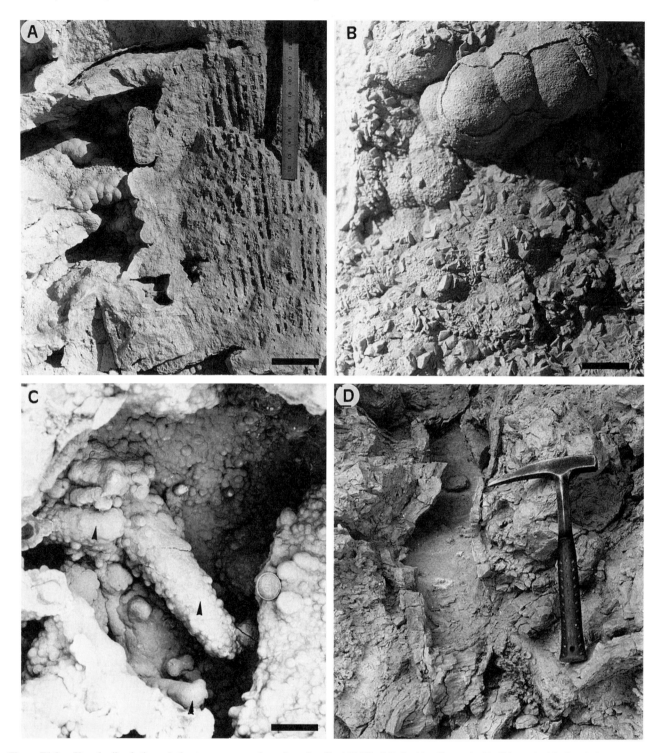

Figure E1.9 Karstic dissolution relating to a sequence boundary, locality 'A', Wadi Bali, Abu Shaar. A. Faviid coral with dissolution cavity partially filled with dolomitized botryoidal cement, scale 4 cm. B. Detail of A, dolomitized botryoidal cement and later phase of large dolomite crystals, scale 1 cm. C. Cavity in coral; mollusc boring (arrows) selectively preserved and coated with cement, scale 2 cm. D. Karst cavity, locality 'B', Wadi Bali, showing secondary cavern (see Figure E1.10).

speleothems (Figure E1.11) interpreted by El Aref (op. cit.) as calcrete. Their karstic origin, however, probably was not associated with meteoric waters; the dry climatic conditions indicated by the laminated dolomites and by the associated evaporites, tend to preclude this possibility. Dissolution most probably resulted from the penetration of sulphate-rich brines (Sun, 1992; Sun and Esteban, 1994) into a fractured dolomite substratum.

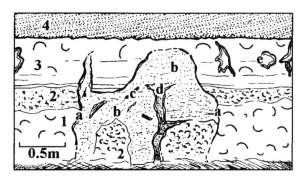

Figure E1.10 Field-sketch showing the multiple karstification at locality 'B', Wadi Bali, Abu Shaar. 1–4, sedimentary sequence; (a) wall of major cavern lined with cement; (b) cavern filling; (c) clasts of cement; (d) secondary cavern lined with botryoidal dolomite and filled with sediment; (e) isolated caverns formed by dissolution of corals.

Cavity fillings

Cavities may be lined by finely laminated cements exhibiting mamillary morphology. These speleothems are often composed of fibrous aragonite (Aissaoui, 1985) attaining 10 cm in thickness (Figure E1.11(b), Plate 23). Characterized by negative oxygen isotopes they have been interpreted by Aissaoui (op. cit.) as the products of meteoric diagenesis. The precipitation of non-marine aragonitic speleothems (rather than calcite) probably reflects the relatively high magnesium saturations of interstitial waters resulting from dissolution of the dolomite substratum. It may also have formed from relatively hot waters, as suggested by the strongly negative (−10) values of the aragonite.

The aragonite speleothems are also replaced by dolomite (Figure E1.11(c), Plate 23) and, in most cases, cavities lined with finely laminated dolomite exhibit fibrous ghost structure. The characteristic undulatory extinction of these dolomite cements, reminiscent of baroque dolomite, may be conditioned by the radiating fabric of the precursor aragonite rather than a deformed crystal lattice.

Certain cavities, notably at Wadi Asal, are lined with large (10 cm) celestite crystals (Figure E1.12(c)) whose growth probably was favoured by the presence of strontium-rich brines resulting from the dissolution or dolomitization of aragonite speleothems. The celestite has been coated locally with laminated cement (Figure E1.12(c)), following which the celestite has been dissolved (Figure E1.12(a)). The hollow pseudomorphs (looking rather like honeycomb), in turn, have either been broken and reworked into the cave-fillings, or have been filled with dolomitized silts (Figure E1.12(d)).

Secondary sulphate and pinnacle formation

At Ras Honkorab, Wadi Asal, Um Gheig and Wadi Zarib the primary evaporites grade laterally into white powdery secondary anhydrites, gypsum cements and calcite. The secondary nature of the sulphates is readily demonstrated by the presence of characteristic bedding structures which are common to both the dolomite substratum and the sulphate. It is confirmed by the sulphate composition of certain fossils including corals (Figure E1.13(c)). The contact between the precurser dolomite and the secondary sulphate is fairly sharp but irregular, being expressed in the field by a marked colour contrast between the grey dolomite and the white sulphate (Figure E1.13 (a), (b)). On a microscopic scale, the transition is considerably more complicated. The dolomite, at distances of 0.5–1 m from the white secondary sulphate, includes a variety of diagenetic modifications including calcitization (dedolomite), precipitation of fibrous quartzine and other secondary silicates (Figure E1.11(d)) typical of sulphate waters. The pervasive replacement of the dolomitic host-rock by gypsum or anhydrite is also characterized by a fine, often intense fracturing of bioclasts and matrix probably reflecting volume changes relating to dolomite–anhydrite–gypsum transformations.

On a larger scale, field studies immediately to the south of Wadi Um Gheig and in Wadi Zarib have indicated a very irregular morphology comprising a series of conical hills averaging 20–50 m in altitude. These are well developed within areas of outcropping evaporites where they are reminiscent of certain karst morphologies (El Aref, 1993). Examination of individual conical hills (Figures E1.14, E1.15 (a)) shows that a carapace of massive sulphate envelops a subvertical dolomite core. The extremely irregular limit of the dolomite has been interpreted as karst by El Aref (op. cit.). However, the contact between the gypsum carapace and the dolomite core, although clear, nevertheless is gradual; it is diagenetic. Associated with many of the minerals already discussed, it also is characterized by nodular anhydrite (Figure E1.15 (b)) formed within the adjacent dolomite, the resulting diagenetic structure being comparable to that of certain nodular sabkha anhydrites.

In certain areas, notably between Quseir and Wadi Asal, the eastern limits of the outcropping Miocene carbonates coincide with a series of anomalous outliers. Composed of unfossiliferous dolomite, they form 100 m high conical hills (Figure E1.15 (d)) which dominate the surrounding evaporites. The 'pinnacles' are generally rich in cavities lined with botryoidal aragonite (Wadi Zarib) and other speleothems or brecciated dolomicrite. These pinnacles are interpreted as residual dolomite which has been exposed following Quaternary erosion of their evaporitic carapaces.

The diagenetic phenomena associated with the carbonate–evaporite transition are numerous and their diagenetic evolution complicated. These modifications have been conditioned by the penetration of sulphate-rich brines into a brecciated dolomite host-rock.

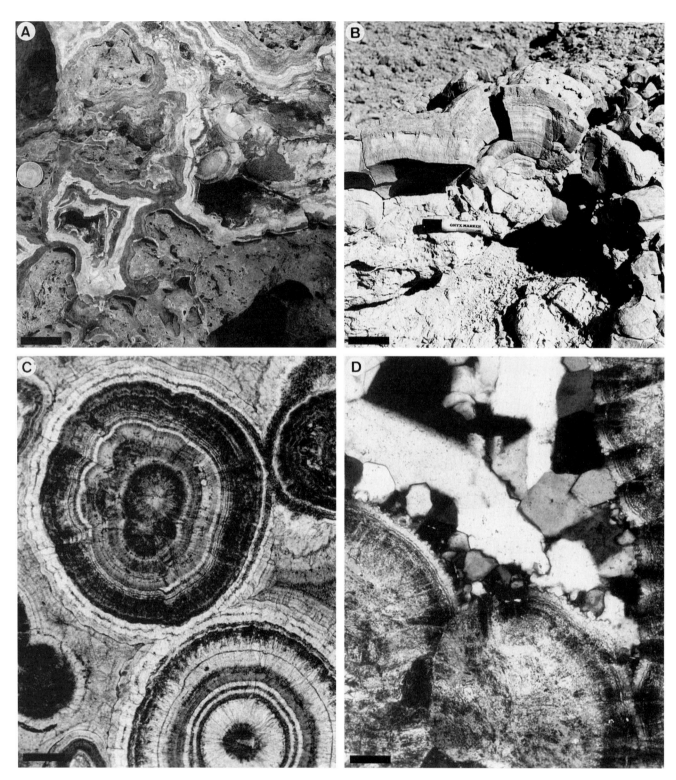

Figure E1.11 Cavern development and speleothems relating to dolomite–sulphate transitions, Wadi Zarib. A. Vugs in dolomite, lined with aragonite (dark) and dolomite (white) cement, scale 2 cm. B. Thick aragonitic cavern-filling exposed by erosion of dolomite substratum, scale 3 cm. C. Dolomite cave pearls from cavern, scale 100 μm. D. Dolomite (probably replacing aragonite ooids) overgrown by diagenetic quartz, scale 50 μm.

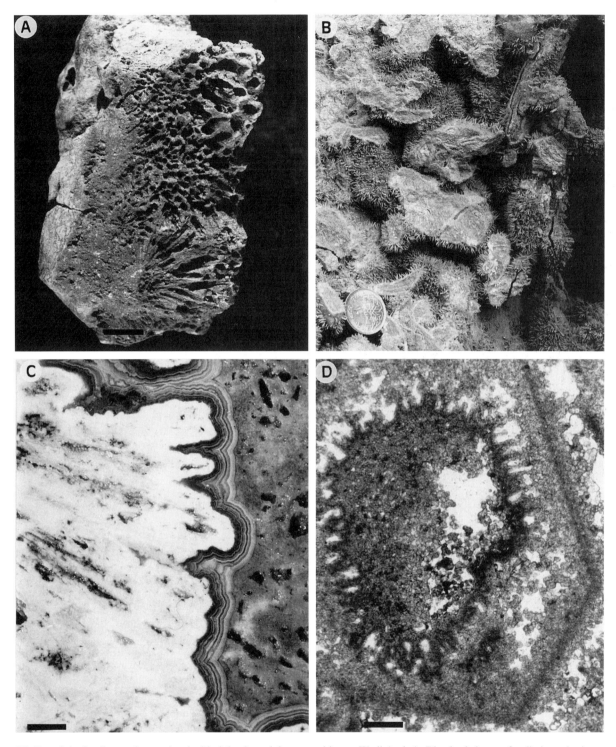

Figure E1.12 Celestite diagenesis associated with dolomite–sulphate transition at Wadi Asal. A. Dissolved cluster of radiating celestite crystals, scale 1 cm. B. Celestite crystals in vuggy dolomite breccia, coin measures 2 cm. C. White pseudomorph (probably celestite), laminated cement and internal sediment, all dolomite, scale 5 mm. D. Dissolved celestite crystal, coated with microbial micrite and filled with internal sediment, all dolomite, scale 500 µm.

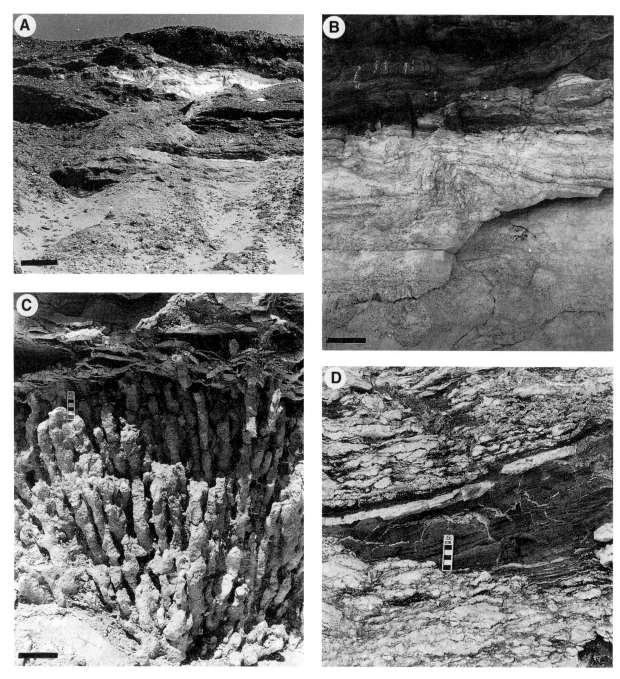

Figure E1.13 Field aspects of dolomite–sulphate transition. A. White lense of secondary sulphate and calcite in grey dolomite host-rock, Wadi Asal, scale 2 m. B. Detail of A showing sedimentary bedding and progressive contact between secondary sulphate and dolomite substratum, scale 25 cm. C. Coral colony completely replaced by sulphate (Quseir), scale 10 cm. D. Lateral transition between laminated dolomite (dark grey) and secondary sulphate (white), Quseir.

DISCUSSION

The multiple diagenetic phases affecting a relatively thin (less than 150 m) series of syn-rift carbonates has been examined briefly; geochemical studies would no doubt reveal additional diagenetic minerals. It may also be noted that these same Miocene sediments are well known for their small but frequent strata-bound sulphide ore deposits (El Aref and Amstutz, 1983). Clearly, syn- and post-sedimentary diagenetic features are abnormally frequent.

Four main diagenetic phases have been identified, although this is probably an oversimplification. Certain stratigraphic levels are marked by multiple diagenetic

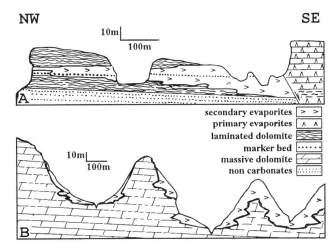

Figure E1.14 Sections showing relationships between secondary sulphates and dolomitic substratum. A. Wadi Asal showing lateral penetration of secondary sulphate (see Figure E1.13A). B. Um Gheig, showing formation of 'pinnacle' dolomite by replacement by secondary sulphate.

events; diagenesis, in one form or another, is complex, dolomites in particular being polyphased. However, it is perhaps surprising that the ubiquitous dolomites rarely exhibit recrystallization fabrics and the dolomites are generally characterized by their relatively 'immature' fabric; dolomite cements are not strongly developed and the dolomites generally have preserved a high porosity and permeability. The main exception concerns the mid-Miocene carbonates at Gebel el Zeit where Nukhul carbonates exhibit a fairly dense crystalline fabric. These open marine carbonates, although incompletely dolomitized, have well-developed dolomite cement.

The important differences in dolomite fabrics between the more peripheral platforms such as Abu Shaar and the more axial highs such as Gebel el Zeit, are not completely understood. The incomplete dolomitization of the more basinal block (Zeit) may be related to its relatively rapid burial beneath the pelagic Rudeis–Kareem marls. More peripheral platforms, which have never been buried, have been affected by dolomitizing fluids during a longer period resulting in a higher degree of diagenetic alteration.

The four main phases of diagenetic activity (submarine cementation, dolomitization, karstic dissolution and sulphatization, cf. Figure E1.2) are each related to a somewhat different platform morphology. The initial submarine cementation essentially affects the slopes where it preceeds dolomitization, the latter affecting most of the platform, including the cemented slopes. The generalized, pervasive nature of the replacement dolomite and associated dissolution of aragonitic debris, unrelated to any sequence boundary or major fault, suggests that warm dolomitizing fluids have risen through a diffuse system of fractures. The rather poorly developed karst system(s), on the contrary, is related to certain sequence boundaries. However, these karstic features can not be correlated from block to block and thus probably result from lowering of relative sea level during tilting and uplift of the block. Finally, the complex diagenetic activity relating to sulphate-rich brines formed during mid–late Miocene evaporite sedimentation, attains its maximum where evaporites onlap an inclined carbonate substratum.

CONCLUSION

The mid-Miocene carbonates deposited on and around the numerous tilt-block platforms of the north-western Red Sea and southern Gulf of Suez are characterized by an exceptionally high degree of diagenetic modification; primary mineralogies are rare. The frequent submarine cementation, massive polyphased dolomitization, dissolution and sulphatization require major changes in the composition of the interstitial waters conditioned by efficient hydrological systems.

Four major diagenetic phases (systems) affect many platforms, their distribution within any given platform being either stratiform or pervasive. Submarine cementation, the first diagenetic phase, although stratiform, mainly affects the steeply inclined peripheral slopes. Massive dolomitization, probably associated with widespread dissolution of aragonitic skeletons, on the contrary, is pervasive; it is not related to any given stratigraphic unit or boundary. Post-dolomite karstic dissolution, relatively unimportant volumetrically, is associated with specific sequence boundaries. Sulphatization and silicification fabrics are related to the transition from mid-Miocene dolomites to mid–late Miocene evaporites. Although essentially stratiform, sulphate diagenesis (like submarine cementation) affects periplatform slopes.

The mineralogical composition and geometry of the major diagenetic products constrains to a certain degree the hydrological models envisaged. Submarine cements, best developed along the basinwards slopes of many platforms, probably reflect input of sea water by oceanic currents. This delivery was probably favoured by the inclined morphology of the sea floor and, possibly, by differences in density between the cooler sea water and the slightly warmed interstitial fluids. Pervasive dolomitization and dissolution (non-karstic), on the contrary, are not geometrically related to the periplatform slopes and required a large-scale hydrodynamic drive. The two most obvious possibilities concern either regional artesian flow conditioned by high peripheral relief, or hydrothermal buoyancy of heated interstitial marine waters. Because of the great quantities of

Figure E1.15 Field aspects of 'pinnacle' formation by sulphatization. A. Dolomite chimney (arrow), exposed on top of conical hill, enveloped in sulphate (south Um Gheig). B. Detail of A, showing formation of nodular sulphate in dolomite host-rock. C. Dolomite hills enveloped in sulphate (s), (south Um Gheig). D. Conical hill ('pinnacle') resulting from Quaternary erosion of sulphate carapace (Wadi Zarib).

magnesium required, together with the exceptionally negative (−10) oxygen isotopic signatures, the latter model is preferred. Finally, the sulphate-related diagenesis affecting the outer flanks of many mid-Miocene platforms, was conditioned by a major change from normal to hypersaline waters. Their penetration into the dolomite substratum was probably favoured both by the inclined nature of the latter and by the higher density of the highly saline fluids.

The hydrological models envisaged involve submarine slopes and varying sea-water compositions due either to local restriction related to the formation of numerous half-graben basins, or to regional hypsersalinity associated with the development of mid–late Miocene evaporites. Interstitial water-flow could also have been stimulated by thermal buoyancy favoured by high thermal gradients. All these conditions are satisfied by the structural and thermal attributes of the rift.

ACKNOWLEDGEMENTS

The author wishes to thank Dan Bosence, Jean-Paul Loreau and Andy Hedburg for useful reviews and suggestions which have considerably improved the initial manuscript. Discussions with Francois Arbey notably concerning silicate diagenesis, and photos offered by Djaffar Aissaoui are much appreciated. Fieldwork was financed by the French Petroleum Institute (IFP) and by an Economic Community 'Science Program'. These grants are gratefully acknowledged.

Chapter E2
The dolomitization and post-dolomite diagenesis of Miocene platform carbonates: Abu Shaar, Gulf of Suez, Egypt

N. Clegg, G. Harwood and A. Kendall

ABSTRACT

Abu Shaar is a mixed siliciclastic/carbonate fault-bounded platform. Aragonite bioclasts have dissolved, but calcitic bioclasts and the sediment matrix are mimetically replaced by dolomite. Four types of water have been suggested to be involved in the dolomitization: meteoric, normal marine, hypersaline and hydrothermal. Well-preserved unstable calcic feldspars within dolomites suggest that waters undersaturated with respect to these feldspars, including surface meteoric waters, were never present in sufficient quantities to result in mixing-zone dolomitization, nor have such waters been responsible for precipitating pre- or syn-dolomitization meteoric cements. Stable isotopic values of the dolomites extend over a range of $\delta^{18}O$ 2 per mil to −10 per mil VPDB and $\delta^{13}C$ 3 per mil to −11 per mil. Since meteoric waters are unlikely to be involved, it is probable that the negative oxygen isotope values represent the influence of higher temperatures on isotope fractionation, probably due to heated waters exploiting basement fractures. Light carbon isotopes can be attributed to the influence of hydrocarbons. The small number of positive oxygen isotope values indicates that a surface normal marine or hypersaline sea water has also been involved. Mixing of these two waters in the platform deposits has resulted in almost complete dolomitization of the carbonates.

Post-dolomitization hydrothermal cements are present along the footwall margin. They are largely calcitic, often preceded by iron oxides, and contain thin bands of dolomite. Distribution of the cements suggests that they were precipitated from a fluid sourced from the footwall fault. Stable isotopes suggest that this fluid was heated, contained abundant organic carbon and had mixed with surface sea water. A second type of calcite cement, preceded by an evaporite residue, shows vadose textures throughout the slope. These vadose cements and cements from the centre of the platform have similar oxygen and carbon stable isotope values to the hydrothermal cements, and may therefore have a similar origin.

GEOLOGICAL SETTING

Abu Shaar is a tilted horst, situated 25 km to the north of Hurghada, on the western side of the Gulf of Suez, forming part of the north-west/south-east trending fault system that dominates the rift structure (Figure E2.1). A full description of the structure and stratigraphy of the Gulf of Suez is available in Schutz (1994) and Patton et al. (1994). Approximately 100 km² of the block is exposed, which consists of outcrops of Precambrian volcanics overlain by mid-Miocene carbonates (James et al., 1988). The highly fractured Precambrian volcanics are exposed only along the north-eastern part of the footwall margin, due to the general south-west tilt of the block. Abu Shaar is dissected by sub-parallel wadis, which cut the block perpendicular to the footwall bounding fault (Figure E2.1). The coincidence of these wadis with steps in the eastern margin suggests that they are exploiting north-east/south-west trending faults. These wadis provide excellent access to the majority of the platform sequence (Figure E2.2).

The Abu Shaar carbonates are thought by James et al. (1988) to have been deposited during the early to mid-Miocene, but age assessments are imprecise. Sedi-

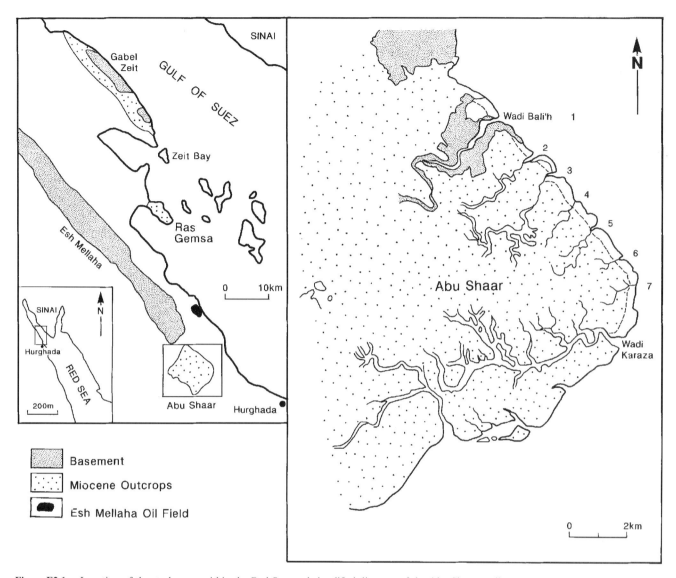

Figure E2.1 Location of the study area within the Red Sea, and simplified diagram of the Abu Shaar wadi system and geology.

ments are marine but show a gradual restriction in their upper portion, terminating with a series of dolomicrites with numerous gypsum pseudomorphs (for stratigraphic and sedimentologic details see Cross *et al.*, this volume). The carbonates form a platform sequence with multiple fringing and patch reefs towards the margin, with associated bioclastic material, including coralline algae and molluscs. Towards the hangingwall the carbonates grade into cross-bedded siliciclastics, which contain abundant quartz and well-preserved feldspar clasts. Investigation of cross-bedding directions by El Haddad *et al.* (1984) and Purser *et al.* (this volume) indicates that clastics were sourced both from the Esh Mellaha block to the north-west, and from the western periphery of the rift. In the upper portion of the platform, the carbonates grade from open marine to restricted deposits, which contain stromatolites, oolites and open marine incursions with corals and coralline algae. The sedimentology and stratigraphy of the area has been studied in detail by El Haddad *et al.* (1984); Burchette (1988); James *et al.* (1988); Assaoui *et al.* (1993); Purser *et al.* (1993); Cross *et al.* (this volume) and Purser *et al.* (this volume).

METHODS

Extensive thin section and geochemical analyses have been conducted on the Abu Shaar carbonates. A large range of samples were collected throughout the study area, approximately 300 in total from 25 locations. These included vertical sections throughout the platform, individual beds traced from the footwall margin to the hangingwall deposits, bioclasts and calcite cement samples. Approximately 200 thin sections were made,

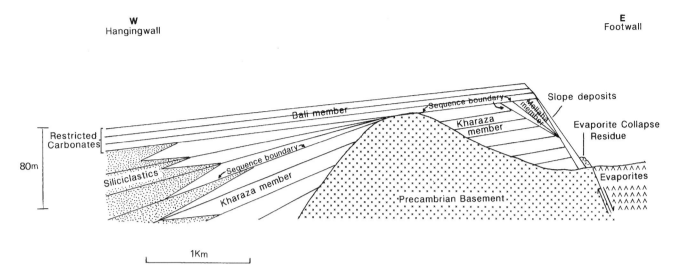

Figure E2.2 Simplified east–west (footwall to hangingwall) cross-section through Abu Shaar. Detailed sections through the platform are presented in Cross *et al.* (this volume).

which were all stained with Alizarin Red S and potassium ferricyanide, to distinguish calcite, ferroan dolomite and dolomite, using the method of Dickson (1966). Polished sections were made of 70 samples, which were analysed with cathode luminescent and fluorescent microscopes. Ten samples were investigated using Scanning Electron Microscopy, including etched, polished and fractured examples of each sample.

Geochemical analysis of both bulk rock and individual components concentrated on carbon and oxygen stable isotopes, trace element analysis and X-ray diffraction. All oxygen and carbon stable isotope values quoted are compared against the PDB standard. Approximately 136 samples in total were analysed for stable isotopes: 90 dolomites and 36 calcites (all calcites are post-dolomite cements). Total rock dolomite samples weighed approximately 6 mg, and individual dolomite components and calcite samples weighed approximately 2 to 6 mg. The sample preparation method of McCrea (1950) adapted by Dr Paul Dennis at the University of East Anglia was used for all samples. Whole rock samples were crushed in an agate pestle and mortar, and individual components were drilled out with a dentist's drill. The resultant powders were ashed in a PT 7300 RF plasma barrel etcher for four hours at 200 W forward power. The samples were then reacted in carbonate reaction vessels with 102% orthophosphoric acid (specific gravity 1.92 at room temperature) at 55°C for dolomite samples and 25°C for calcite samples. The gas collected was analysed in a VG SIRA series II stable isotope ratio mass spectrometer, with duel inlets and triple collectors, fitted with a 10-port auto sampler manifold. All samples were compared against a Carrara marble 90–200 μm size fraction standard.

Trace elements were analysed for approximately 50 dolomite and 20 calcite samples. Sample weights were 20–30 mg for dolomite bulk-rock samples, 2–10 mg for individual dolomite components, and between 5 and 25 mg for calcite samples; again all calcite samples are post-dolomite cements. Samples were weighed accurately, and dissolved in 5% Aristar acetic acid. The dissolved samples were filtered, any residue was weighed and subtracted from the original weight (this was only necessary for four siliciclastic rich samples). Blank samples were prepared to act as an internal standard, which underwent the same processes as the carbonate samples. Spectrosol standard solutions diluted with acetic acid were used to produce the calibration curves. The samples were analysed in a Thermo Jarrel Ash Polyscan 61E Inductively coupled plasma argon emission spectrometer (ICP) with autosampler.

X-ray diffraction analysis concentrated on bulk-rock samples, due to the quantities of material needed. Samples were crushed with an agate pestle and mortar. One hundred samples were analysed to investigate stoichiometry, using the method of Lumsden and Chimahusky (1980), and 10 samples, largely of the evaporite residue, were analysed over 0 to 60 degrees, to identify mineralogy. All samples were analysed in a Phillips PW 1170 X-ray diffractometer.

DIAGENESIS

Pre-dolomitization diagenesis at Abu Shaar appears to be limited to the micritic lithification of the platform as a whole and the precipitation of marine cements in the slope deposits. These marine cements have been investigated by Aissaoui *et al.* (1986). They were precipitated from marine water in intergranular voids, are often associated with internal sediment and probably represent synsedimentary diagenesis.

Dolomitization is the dominant diagenetic process that has affected the platform; the majority of carbonate has been replaced by micritic and mimetic dolomite. This dolomitization was previously studied by Coniglio *et al.* (1988), and is extended in this current research. Due to the significance of this process at Abu Shaar the majority of this chapter is devoted to the dolomitization of the platform.

Previous studies of the area offer three possible explanations for this dolomitization.

1. Mixing of meteoric and marine waters (Coniglio *et al.*, 1988).
2. Dolomitization by hypersaline brines (Sun, 1992).
3. Hydrothermal resetting of a hypersaline dolomite (Coniglio *et al.*, 1988).

Dolomitization models for Abu Shaar must take into consideration the complete micritic replacement (fabric preserving) of the carbonates, and the widespread negative carbon and oxygen stable isotopes. In the Gulf of Suez, Mediterranean, and Iraq, Miocene dolomites are largely interpreted as evaporitic in origin (Buchbinder, 1979; Oswald *et al.*, 1990; Oswald *et al.*, 1991; Sun, 1992, 1994), which together with the significant evaporite deposits in the Gulf of Suez (Schutz, 1994), suggests that an evaporitic origin for the Abu Shaar dolomites would be expected.

Post-dolomitization diagenesis is limited to the precipitation of cements, which are largely calcitic, in voids in the slope and the upper platform deposits. Voids in the central platform also contain post-dolomite cements but these are uncommon. These cements will be described and interpreted under three headings: hydrothermal cements, vadose (Wadi 3) cements and central platform cements.

Dolomitization

Petrography

Abu Shaar carbonates have undergone complete micritic dolomitization, often preserving the fine microstructure of bioclasts. Aragonitic bioclasts including corals have been completely dissolved, whereas the structure of calcitic bioclasts, particularly those initially composed of high magnesian calcite such as coralline algae, are perfectly preserved, despite complete replacement by dolomite. Preservation is commonly so perfect that the cellular structure of the coralline algae is retained (Figure E2.3) and without staining the extent of dolomitization cannot be determined in thin sections. This micritic replacement is due to abundant nucleation sites having been present in these fine-

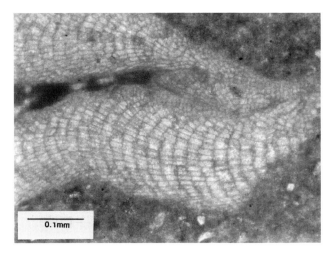

Figure E2.3 Preservation of the cellular structure of coralline algae, despite complete replacement by dolomite.

grained high magnesian calcites, and rapid recrystallization (Sibley and Gregg, 1987; Bullen and Sibley, 1984). Low magnesian calcite bioclasts, such as oysters, and fibrous marine cements in the slope, commonly retain their texture and exhibit fine preservation, comparable to that of the coralline algae (Figure E2.4). Despite the fine replacement of the majority of the bioclasts by dolomite, it is common to find a small proportion of bioclasts still partially preserved as calcite (determined by Alizarin Red S staining) (Figure E2.5). This partial calcitic preservation occurs across the whole platform.

The carbonate matrix is replaced by a fine micrite in which small rhombs commonly exhibiting cloudy cores and clear rims and approximately 10 to 40 μm in size, can be identified (Figure E2.6). Rhombs with multiple zones are rarely present. Dolomites luminesce a moderate bright orange under CL, with occasional faint zoning. This zoning corresponds to the clear rim/cloudy centre partitioning of the rhombs, with cloudy centres of the rhombs being brightest. The dolomite glows a moderate bright green under fluorescent light, which according to Dravis and Yurewicz (1985) indicates the presence of organic material. All lithologies appear to have been dolomitized similarly: the only significant geographic change in dolomite petrography observed is that several dolomite outcrops in hangingwall deposits exhibit minor (less than 10%) preservation of the original (?) calcitic matrix (Figure E2.7).

Most dolomite replaces original carbonate, but locally up to 5% occurs as a cement, forming a partial lining to voids and nucleating on replacive dolomite crystals (Figure E2.8). The dolomite cement crystals are inclusion-free, due to their direct precipitation from the dolomitizing fluids.

Stable isotope geochemistry

The carbon and oxygen stable isotopes of bulk rock samples, bioclasts and cements, all of which have been dolomitized, show a wide range of values: $\delta^{18}O$ from 0.5 to −10 per mil VPDB and $\delta^{13}C$ from 2.5 to −10.5 per mil (Figure E2.9). These are similar to the values found by Coniglio *et al.* (1988) for individual clasts, bioclasts and

Figure E2.4 Dolomitized fans of fibrous marine cements. The fibrous nature of these cements is retained as inclusions, which would be lost if these cements had suffered recrystallization.

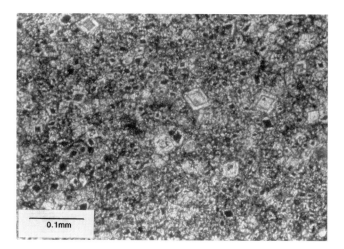

Figure E2.6 Dolomitized fine-grained sediment matrix. The size of rhombs varies between 0.01 and 0.04 mm. This photograph is representative of the sediment matrix from across the whole platform.

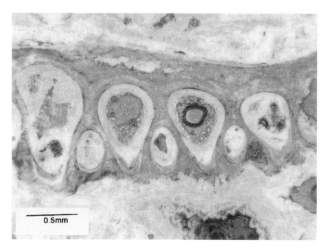

Figure E2.5 Partial preservation of the original calcite in a mollusc fragment, despite complete dolomitization of the surrounding sediment, as detected by Alizarin Red S staining (calcite is represented by the dark coloured material in the photograph). Preservation of calcite is not usually as good as this, occurring in a small proportion of the bioclasts across the whole platform.

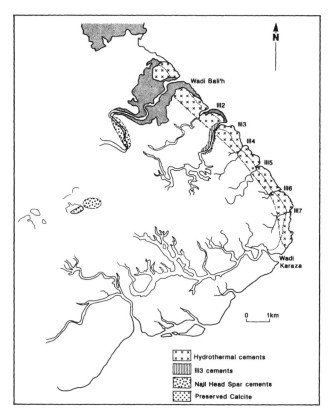

Figure E2.7 Map of the distribution of post-dolomitization cements, and pods of preserved calcite in the sediment matrix, within Abu Shaar.

bulk-rock analyses. Coniglio *et al.* (1988) found no discernible patterns or trends in the carbon and oxygen stable isotope values and other variables. They found significant isotopic variations between samples from the same hand specimens. This present study is focused on the geographic and stratigraphic variations between dolomite samples, in an attempt to trace fluid flow through the platform.

Figure E2.8 Dolomite crystals lining a void, which have nucleated on dolomitized sediment. The limpid, inclusion-free nature of these crystals, compared to the inclusion-rich dolomite, represents direct precipitation from the dolomitizing fluid rather than replacement of a precursor carbonate.

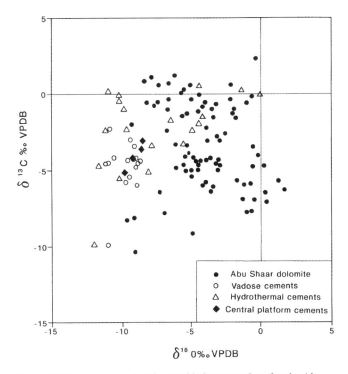

Figure E2.9 Oxygen and carbon stable isotope values for the Abu Shaar dolomites, hydrothermal cements, vadose cements and central platform cements.

Discussion of the stable isotopic values

Coniglio *et al.* (1988) suggested that the negative oxygen isotope values could be due to the influence of surface meteoric waters. Modern groundwater values for $\delta^{18}O$ of −6.5 per mil, measured from springs along the footwall of Abu Shaar in this present study, support this hypothesis, although this does not account for the very negative $\delta^{18}O$ values (as low as −10 per mil) common in the dolomites. Aissaoui *et al.* (1986) also suggested the influence of meteoric waters, as a possible explanation for the dissolution of aragonite bioclasts. They concluded that this dissolution had a karstic origin, based on the local development of Miocene soils (dolocretes) at Wadi Kharasa (Figure E2.1), and the iron oxide mineralization present along the footwall margin. This interpretation is seemingly inconsistent with the excellent preservation of feldspars throughout the platform. Feldspar grains and feldspar-rich basement clasts exhibit no corrosion (Figure E2.10) within dolomites, excluding those affected by modern weathering (exposed surfaces and immediately adjacent to open joints). Most of the feldspars are albite, but more calcic plagioclases up to anorthite are definitely present (determined by X-ray diffraction). It is suggested that the very presence of, and the perfect preservation of these highly unstable anorthite grains indicates that no significant amounts of surface meteoric water have passed through the dolomitized sediments. Otherwise they would exhibit some degree of corrosion as do those affected by modern weathering.

A significant volume of meteoric water would have been needed to dissolve aragonite (up to 20% of the original sediment) and mix with sea water to dolomitize almost the entire platform if these alterations were

Figure E2.10 An example of the preservation of feldspars seen across the platform despite complete dolomitization of the surrounding sediment. Feldspars occasionally show alteration, but are generally unaffected by the diagenetic fluids that have interacted with the platform.

products of surface meteoric diagenesis. The iron oxide mineralization seen along the footwall margin suggested by Aissaoui et al. (1986) to indicate the involvement of meteoric waters appears to be sourced from the footwall fault, and will be interpreted as of hydrothermal origin. These hydrothermal waters may indeed have had a meteoric source but would have been so modified by their interaction with the hot basement that they must now be considered as a distinct water type.

The negative oxygen isotopic values in the Abu Shaar dolomites might also be explained by their precipitation from non-meteoric waters at high temperatures. The tectonic setting of Abu Shaar within a rift makes this explanation a distinct possibility. The high geothermal gradient in rifts is commonly associated with hydrothermal circulation of sea and meteoric waters. Upwelling of these heated waters could have dolomitized the Abu Shaar platform and would have been responsible for the depleted oxygen isotopic values in the dolomites. Using the calculations of Land (1983), the most negative oxygen isotopic values seen at Abu Shaar ($\delta^{18}O$ −10 per mil VPDB), would require water with a temperature of approximately 100°C (assuming an initial water composition of $\delta^{18}O$ 0 per mil (sea water)). Most of the oxygen stable isotope values for Abu Shaar dolomites ($\delta^{18}O$ −6 per mil), however, would have been created from sea water with a temperature of 60°C. Present-day deep brines in the Red Sea (Discovery Deep), which have already mixed with colder sea water, have temperatures ranging from 44°C to 56°C (Craig, 1969). If similar brines had been involved in the dolomitization of Abu Shaar, they could account for the bulk of the oxygen stable isotope values.

The considerable spread of oxygen and carbon stable isotope values implies that dolomitization involved either more than one type of water or a water which varied considerably in temperature. Dolomite samples with positive oxygen isotope values suggest the involvement of a surficial normal marine or a hypersaline sea water. Sun (1992) suggested that many porous dolomites (including those from the Miocene of the Gulf of Suez) formed by the interaction with evaporated sea water brines that had precipitated gypsum in the absence of pre-existing carbonates. These brines are undersaturated with respect to aragonite and if placed in contact with carbonates are capable of producing moldic aragonite porosity without precipitating calcite cement, and dolomitizing the host sediment. At Abu Shaar aragonite dissolution is not associated with pre- or syn-dolomitization calcite cement and from the evidence of feldspar preservation cannot be attributed to meteoric diagenesis, leaving Sun's interpretation as the only viable mechanism. This inferred brine is thus a more probable candidate for causing the regional dolomitization of the platform. Brines would have been expected to have entered the platform during and after formation of the late Miocene evaporite sequence of the Red Sea–Gulf of Suez.

Different parts of the range of dolomite isotopic values are therefore to be explained by the involvement of different waters: sea-water brines and hydrothermal waters of unknown origin which had been drawn down into the basement, heated and driven upwards into the platform. Unlike Coniglio et al. (1988), who in one of their interpretations of Abu Shaar dolomites used a two-stage dolomitization model involving a later resetting by hydrothermal waters, we believe that the near perfect mimetic replacement of bioclasts by dolomite suggests that no dolomite recrystallization occurred. If, as we suggest, two waters were responsible for dolomitizing the Abu Shaar carbonates they must have acted simultaneously. The range of dolomite isotopic values is therefore to be attributed to water mixing, rather than to varying degrees of resetting by a second dolomitization event.

We believe the very depleted carbon isotopic values of some Abu Shaar dolomites (down to −10 $\delta^{13}C$) indicate the involvement of hydrocarbons during dolomitization. The Esh Mellaha oil field lies only 10 km to the east of the platform and was supplied with hydrocarbons that migrated westward from the Gemsa Trough (Salah and Al Sharhan, 1996). The Gemsa Trough could also have supplied hydrocarbons to the Abu Shaar platform (Figure E2.1). Alteration of evaporites in the nearby Gabel el Zeit and Ras Gemsa fault blocks, directly link hydrocarbon migration to dolomitization. Here calcium sulphates have been altered to calcium carbonates by sulphate-reducing bacteria in the presence of hydrocarbons. The resultant calcite and aragonite have subsequently undergone partial dolomitization (Youssef, 1989a,b; Pierre and Rouchy, 1988). Youssef (1989a) identified two dolomites at Ras Gemsa: one pre-dating alteration of sulphates, with isotopic values averaging $\delta^{18}O$ 1.05 and $\delta^{13}C$ −3.2 per mil and a second type post-dating sulphate alteration with isotopic values averaging $\delta^{18}O$ −6.8 and $\delta^{13}C$ −24.05 per mil. The first type involved dolomitization by a hypersaline brine; the second by a heated brine with entrained hydrocarbons. We would interpret the waters responsible for dolomitization at Ras Gemsa to be similar to those causing dolomitization at Abu Shaar. Pierre and Rouchy (1988) suggested the negative oxygen isotopes at Ras Gemsa were due to the influence of heated continental groundwaters, but which did not exceed 80°C (the upper limit for sulphate reduction). It is possible that continental groundwaters may have been involved, but the aridity of the area and the proximity of the Gulf of Suez suggest a heated normal marine water or a brine is more likely to be responsible, as was suggested by Youssef (1989a).

Diagenesis 397

There is a geographic variation in carbon isotopic values at Abu Shaar; moving from west to east (hangingwall to footwall) $\delta^{13}C$ becomes progressively heavier (Figure E2.11). Values do show considerable scatter, but when footwall, central and hangingwall zones are plotted separately, an increase in $\delta^{13}C$ values towards the footwall can be discerned (Figure E2.11). This geographic pattern is seemingly independent of lithologic and stratigraphic control. If depleted carbon isotopic values are attributed to the involvement of hydrocarbons, the more negative values of the hangingwall indicate this area was closer to the hydrocarbon source. Hydrocarbons are interpreted to have been moved with heated dolomitizing fluids exploiting basement fractures, originating in the deeper portions of the rift including the Gemsa Trough. The clastic sediments adjacent to the hangingwall fault are still highly permeable and would have formed a conduit for the heated fluids and hydrocarbons into the centre of the platform. However dolomitization of the entire platform by a single plume of heated water exploiting the hangingwall fault appears doubtful, because the intense fracturing of the basement (which can be observed at Wadi Bali'h and wadis further to the south) should have exposed the entire platform to these heated fluids. We therefore suggest the depleted oxygen isotope values obtained across the whole platform imply that heated waters entered the platform from below across its entire width (Figure E2.11).

The variations in $\delta^{13}C$ observed across the platform, could be interpreted as the result of a mixing of seawater derived brines with the heated waters rising throughout the platform. In this interpretation marginal areas of the platform were subjected to greater amounts of infiltrating brines generated in the evaporitic Red Sea (i.e. towards the footwall). This variation in the amount of brines mixing with the rising heated waters created the $\delta^{13}C$ footwall to hangingwall gradation. This model at its simplest would generate $\delta^{13}C$ values that fall along a mixing line between the two endmembers; but this does not occur (Figure E2.9). If the mixing model is valid then departures from the ideal mixing line must be attributed to variations in the composition of one or more of the waters and variations in the water/rock interaction. The sea-water derived brines could have varied in response to the degree of evaporation to which they were subjected, whereas the hydrothermal waters could have varied in response to: 1. the original isotopic composition of the parent water, 2. the degree to which it interacted with the basement, 3. its temperature, and 4. its hydrocarbon content.

The absence of any discernible variation in $\delta^{18}O$ values, however, casts doubt on the mixing model requiring a different explanation for the $\delta^{13}C$ variation. The depleted $\delta^{18}O$ values in footwall dolomites suggest heated waters also exploited faults proximal to the footwall and the eastern bounding fault of the Abu Shaar block. However, they do not exhibit the associated depleted $\delta^{13}C$ values in the hangingwall deposits. This suggests that the heated waters that exploited faults proximal to the footwall contained smaller quantities of hydrocarbons. At first sight this would seem to be improbable because the source of hydrocarbons was probably the Gemsa trough, lying to the east of Abu Shaar (Salah and Al Sharhan, 1996; Figure E2.1). It would therefore be expected that the majority of hydrocarbons would have exploited the footwall fault. At Ras Gemsa hydrocarbons have been utilized by bacteria to reduce sulphate evaporites. Hydrocarbons migrating up the Abu Shaar footwall faults will also have been in contact with Upper Miocene evaporites and thus could have been partially removed from the heated waters by similar bacterial mediation. We speculate therefore (because evaporites adjacent to Abu Shaar, replaced or not, are not exposed) that the variation in $\delta^{13}C$ in dolomites across the Abu Shaar platform is a product of a variation in the amount of hydrocarbons supplied to the platform controlled by sulphate reduction, itself controlled by proximity to Upper Miocene evaporites.

Trace element geochemistry and dolomite stoichiometry
Attempts to discern fluid movement through the platform by using variations in dolomite stoichiometry and

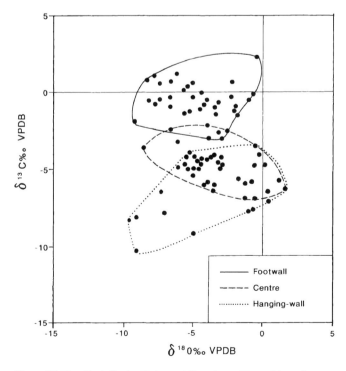

Figure E2.11 East (footwall) to west (hangingwall) partition of carbon stable isotope values for Abu Shaar dolomites.

trace element composition have been unsuccessful. Dolomite stoichiometry varies around a value of 50% $CaCO_3$; any variation being within the margin of error of the method employed. Fifty samples have been analysed for all major and minor trace elements: iron, manganese, strontium, sodium, zinc, silicon, aluminium, lead, copper, nickel, barium, vanadium, potassium and cadmium. All elements exhibit a wide variation in values (for details see Clegg, 1996), but no correlation has been found with stratigraphic or geographic location or with lithology. Aluminium positively correlates with silicon, which probably represents contamination by clastic material. Iron and manganese also positively correlate, but are of no use for identifying fluid flow through the platform.

Post-dolomitization diagenesis

Hydrothermal cements

Iron oxide and calcite cements are located along the footwall (E) margin of the Abu Shaar platform, occupying voids in the slope and upper portions of the platform deposits (Figures E2.7, E2.12). The calcitic portion of the cements extends 100 m laterally into the platform and iron oxides extend about 200 m (Figure E2.12). The distribution of the cements clearly suggests that they were sourced from the footwall fault. This represents the continued exploitation of the footwall fault, which probably supplied fluids to the platform throughout its diagenetic history. Iron oxides partially fill large cavities (up to 50 cm in diameter), are up to 3 cm thick and usually preceded the calcite. Iron oxides are a replacement of a precursor calcite rather than a direct precipitate (Figure E2.13). Iron and manganese oxyhydroxides are currently being precipitated in the Atlantis and Tethis deeps of the Red Sea (Butuzova et al., 1990) and are precipitated from springs along the western coast of Sinai (Alan Kendall and Gill Harwood, personal communications). These are formed by oxidation of Fe^{2+} and Mn^{2+} supplied by hydrothermal brines. A similar hydrothermal source is believed to be responsible for the iron oxides distributed along the footwall margin of Abu Shaar.

The calcite cements are up to 30 cm thick and are best exposed in the reef at the top of the footwall escarpment (Figure E2.14(a),(b)). Cements are less thick in adjacent slope sediments due to the smaller void size, and in platform sediments due to distance from the source. Cements are usually vein fills, but also occupy intergranular voids. Slope cements are isopachous, but equivalent cements in the platform are occasionally stalactitic and are sometimes intercalated with internal sediment (Figure E2.14(a)). Calcite cements consist largely of laminated fibrous botryoidal cements (Figure E2.14(c)), with thin layers of micritic dolomite in the oldest cements (Figure E2.14(d)). It has been impossible to definitively identify the dolomite as a cement or as a replacement of an earlier carbonate cement. However, the dolomite layers do suggest that dolomitization was sporadic and, in part, occurred almost simultaneously with precipitation of the calcite cements. Dolomite brightly fluoresces, indicating the presence of organic

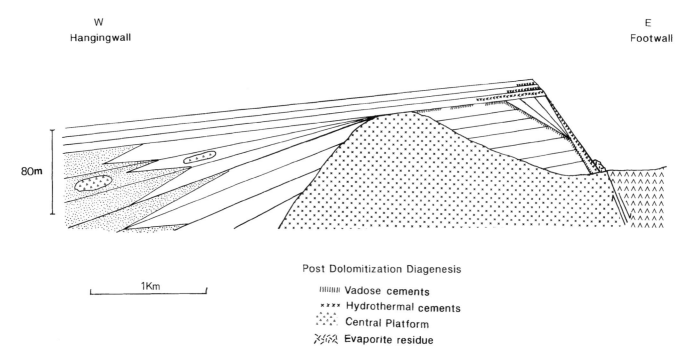

Figure E2.12 East (footwall) to west (hangingwall) cross-section through Abu Shaar, detailing the distribution of post-dolomitization cements.

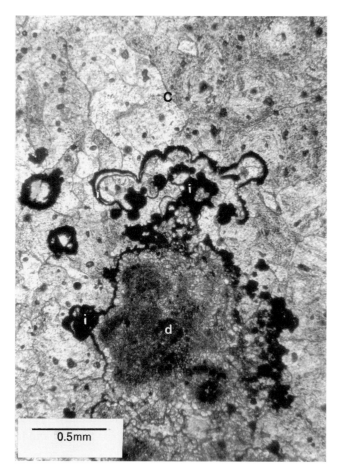

Figure E2.13 Iron oxide replacement of calcite hydrothermal cements. These iron oxides cut across calcite crystal boundaries and therefore must be a replacement rather than a direct precipitate (d = dolomitized sediment, c = calcite cements, i = iron oxides).

matter. This organic matter may also demonstrate a link between hydrothermal waters, entrained hydrocarbons and dolomitization.

Geochemistry of the hydrothermal cements
The small volume of dolomite in these cements has meant that they could not be separately analysed. Calcite samples were deliberately chosen to avoid contamination with dolomite.

Oxygen isotopic values of the calcite cements are in the range $\delta^{18}O$ 0 to −11 per mil VPDB (Figure E2.9), the more negative values suggesting that the precipitating waters were hot. Because of their geographic restriction to the east of the Abu Shaar block, these fluids were probably basinal waters drawn down into the basement, rather than continental waters supplied from the west. Negative carbon isotopes of the calcite in the range $\delta^{13}C$ 1 to −10 per mil suggest the influence of hydrocarbons (Figure E2.9). The low $\delta^{13}C$ values of these cements are not present in the surrounding dolomitized sediments. Thus at the time these cements were precipitated either sulphate reduction of evaporites was not occurring, or the light carbon mobilized by sulphate reduction migrated to produce these cements.

The heavy (0 per mil) $\delta^{18}O$ seen in the samples (Figure E2.9) may indicate precipitation from waters contaminated by marine-derived brines – a similar interpretation to that suggested for the dolomites. The presence of vadose calcite cements in the upper portions of the platform indicates that at least the upper part of the platform was locally exposed. Mixing of the surface and heated waters must have occurred at a lower level, perhaps within the slope deposits, which are both permeable and porous. However there must have been a compositional change in one of the waters involved, because dolomitization in these cements is minor.

Relatively high concentrations of trace metals such as zinc, lead, iron and manganese may be present in the calcite cements but exhibit a wide range of values: zinc 0–1500 ppm, lead 0–3000 ppm, strontium 150–4500 ppm, iron 50–200 ppm and manganese 0–100 ppm. This variation probably reflects a pulsed nature of the fluids from which these cements were precipitated. Fluctuations may also represent the mixing with marine or evaporitic waters, producing the lower trace element values as well as the more positive oxygen and carbon stable isotope values. High strontium values reflect the presence of aragonite, also seen by X-ray diffraction; although preserved aragonite is present only in small quantities, the majority has probably been converted to calcite.

Vadose (Wadi 3) cements
These calcitic cements occur only in Wadi 3 (Figure E2.1), where they are largely confined to voids in the slope deposits and along the major marginal unconformity between the Kharaza and Mellaha member of James *et al.* (1988) (Figure E2.12). Hydrothermal cements are also present in dolomites from this Wadi but are found in much higher elevations of the upper portions of the platform and slope deposits (Figure E2.12). The cements are largely vadose, with frequent stalactites and evidence of cave pools (Figure E2.15(a)). They are commonly found in association with dolomitic internal sediment, and dolomite and calcite clasts (up to 3 cm long), which have been transported from the upper portions of the platform through voids in the lithified slope deposits.

Cements are usually crystalline, unlike the hydrothermal cements which are fibrous. They are not associated with iron oxides; however, a residue rich in iron and manganese is commonly present (Figure E2.15(b)). This occurs (1) within voids in the slope deposits, (2) along the unconformity separating the Kharasa member from the Bali'h and Mellaha members, and (3) discontinuously within the platform deposits for

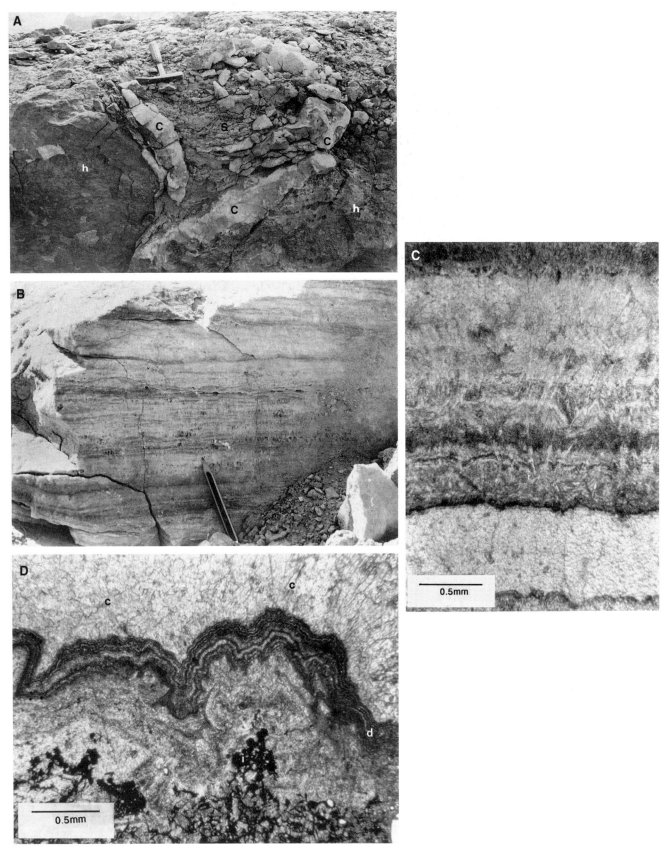

Figure E2.14 **A.** Calcite hydrothermal cements (c) lining a 50 cm diameter void, with a later infilling of remobilized dolomite sediment (s) (h = host rock). **B**. Laminated, calcite hydrothermal cements in outcrop. These cements often show this laminated texture, consisting of fibrous cements, which have been picked out by later weathering. **C**. Calcite hydrothermal cement in thin section. This fibrous texture, with occasional layers of randomly oriented crystals (r) forms the majority of these cements in the upper platform deposits. **D**. Fine bands of micritic dolomite, separated by fine calcite layers, within the calcite hydrothermal cements. This dolomite is probably a replacement of calcite rather than a direct precipitation (c = calcite, d = dolomite, i = iron oxides).

Figure E2.15 **A.** Vadose cements from the entrance to Wadi 3. Stalagmites and stalactites are common throughout the slope, and large voids such as the one in the photograph often show parallel lines, representing palaeowater levels in these voids (p). **B.** Evaporite residue overlain by laminated sediments. The residue has suffered partial erosion and resedimentation of residue clasts (c). This section forms part of the fill to the major unconformity that separates the Kharasa and Mellaha members. This photograph is representative of the residue that occurs in voids throughout the Wadi 3 slope deposits.

1 km towards the hangingwall (Figure E2.12). This is similar to the evaporite residue described by James et al. (1988), which overlies the slope deposits to the south of Wadi 5. X-ray diffraction analysis of the residue shows it to consist mainly of halite, sulphates and quartz, with no indication of the iron which provides the red/orange colouring. The residue is often commonly found on the roofs of voids, forming the core to calcitic stalactites. The geographical distribution of the residue suggests that it was transported from above and behind the platform margin, yet there are no deposits in this area which could produced this residue. The presence of halite and sulphates suggests that it may have been a solution product of evaporites, which existed above the platform but have been subsequently eroded/dissolved. The calcite cements and residue are commonly found associated with late-stage sulphate cements, again suggesting a relationship between these cements and evaporite dissolution and re-precipitation.

Geochemistry of the vadose cements
The calcitic portion of the cements has a comparable isotopic signature to the majority of the hydrothermal cements (Figure E2.9), but they do not show evidence supporting any mixing with normal marine/hypersaline sea water as seen in the hydrothermal cements. A similar hydrothermal source would therefore be expected. Trace elements are also similar to the bulk of the hydrothermal cements: zinc 20–200 ppm (one exceptionally high value has 3300 ppm), lead 0–250 ppm, strontium 0–300 ppm, iron 0–650 ppm, manganese 0–300 ppm, again suggesting a similar origin to the hydrothermal cements.

The vadose nature of these cements and their distribution suggests a meteoric source but their stable isotopes and trace element values suggest a source similar to the hydrothermal cements. The only hypothesis that accounts for both the geochemistry and their distribution/morphology is that hydrothermal fluids rose to the west of the margin, cooled or perhaps became denser by dissolution of the overlying evaporites, and moved down and to the east through the footwall deposits.

Central platform cements
These cements consist of calcite crystals, approximately 0.5 cm thick with well-developed obtuse rhombohedral terminations (nail-head spar) (Figure E2.16(a)). They are located in the centre and back portions of Wadi Bali, where they partially fill voids, within clastic and carbonate rich units (Figures E2.7, E2.12, E2.16(b)). Their distribution does not show any stratigraphic control. Stable isotope values for these cements are in the range: $\delta^{18}O$ −8.5 to −10 per mil VPDB and $\delta^{13}C$ −3 to −5 per mil. These values plot with the hydrothermal cements and suggest a similar source (Figure E2.9). It is likely that these cements are sourced from faults or fractures within the central portion of Wadi Bali'h.

CONCLUSIONS

A three-stage model for the diagenesis of the Abu Shaar platform is presented in Figure E2.17.

Dolomitization

Dolomitization of Abu Shaar carbonates involved surface sea-water-derived brines and waters that had been drawn down into the basement and driven upwards into the platform. The basement-derived waters were hot and entrained hydrocarbons. The lack of recrystallization fabrics in the dolomite suggests no resetting of the stable isotopes occurred and that both waters entered the platform simultaneously and mixed. Brines also dissolved aragonite bioclasts. This interpretation explains how aragonite could be completely dissolved while unstable feldspars are preserved. We assume that the hydrothermal fluids had come into equilibrium with feldspars in the basement. The majority of aragonite dissolution probably preceded complete dolomitization of the carbonates because dolomite cements line aragonite moulds.

The hydrothermal waters changed composition from hangingwall to footwall. The waters exploiting hangingwall faults had much more negative $\delta^{13}C$ than those exploiting faults towards the footwall. The higher footwall $\delta^{13}C$ values are attributed to sulphate reduction in evaporites adjacent to the footwall fault, oxidizing the hydrocarbons and trapping the light $\delta^{13}C$ in the carbonates that replace the sulphate. Mixing of these heated waters with the surface hypersaline sea water and differing rock/water interactions has resulted in the wide spread of $\delta^{18}O$ values and the footwall to hangingwall increase in $\delta^{13}C$.

Post-dolomitization diagenesis

Hydrothermal cements along the footwall margin record the last stages of dolomitization. These cements are sourced from the footwall bounding fault and although largely calcitic contain layers of dolomite. These bands suggest that dolomitization was pulsed and in part the result of heated waters exploiting basement faults. Stable isotopes demonstrate that the waters from which these cements precipitated also involved surface normal marine/hypersaline sea water, mixing with the heated waters exploiting the footwall fault. Calcitic cements found at Wadi 3 show similar geochemical signatures to these cements, but are vadose throughout the whole slope deposits. These cements may have been sourced

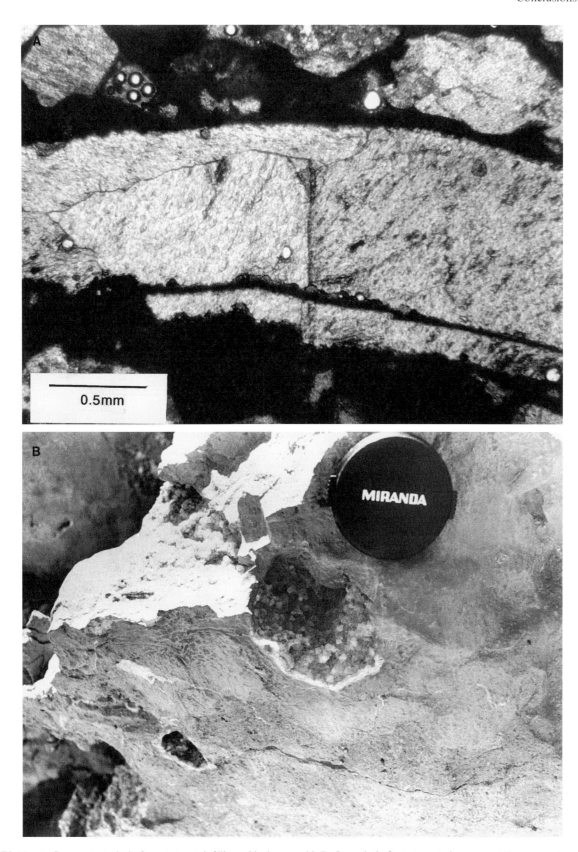

Figure E2.16 **A.** Sparry, central platform cements infilling a bioclast mould. **B.** Central platform cements in outcrop. These cements are not confined to any particular layer within these deposits and coat only a small proportion of the voids.

404 Dolomitization and post-dolomite diagenesis of Miocene platform carbonates

Dolomitization of Abu Shaar by Normal Marine/Hyposaline and Hydrothermal waters. This includes removal of large amounts of aragonite by the hypersaline brine

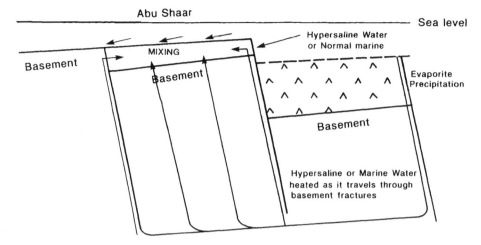

Emplacement of hydrothermal cements during last stages of dolomitization along footwall margin.

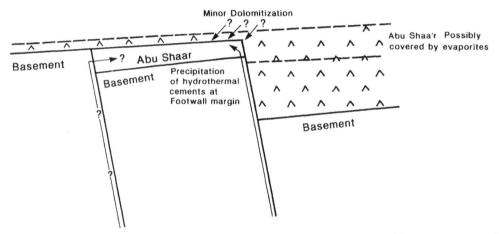

Dissolution of overlying evaporites. Deposition of solution residue on top and within the slope deposits and precipitation of the III3 and possibly Nail head spar cements

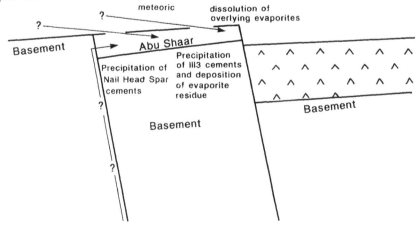

Figure E2.17   Three-stage model for dolomitization and post-dolomitization diagenesis of Abu Shaar.

from above and behind the margin, perhaps from faults within the platform. They were formed later than the hydrothermal cements as they show no evidence of dolomitization and were precipitated at a time when the platform was exposed. A third type of cement found in the central platform deposits of Wadi Bali'h is again geochemically similar to other calcite cements. Precipitation of this cement probably represents a similar origin to the others, probably due to heated waters exploiting central platform faults.

Hot water dolomitization

The involvement of temperature in dolomitization is usually limited to burial dolomitization. In the surface and near surface environment, the involvement of normal marine/hypersaline sea water and meteoric (continental groundwaters) are usually invoked to explain dolomitization (Machel and Mountjoy, 1986; Hardie 1987; Purser et al. 1994). In the case of Abu Shaar, raised geothermal gradients have resulted in convection of waters, probably drawn down into the basement from the Gulf of Suez. This has created a thermobuoyancy drive capable of moving the waters through the platform, and may have favoured dolomitization due to the increase in temperature (increased temperature favours dolomitization; Machel and Mountjoy, 1986). Although temperatures are lower, perhaps a similar situation is occurring in the Florida continental shelf; Fanning et al. (1981) report dolomitization occurring in carbonates in the Florida shelf based on a decrease in Mg^{2+} and increase in Ca^{2+} in waters exiting the shelf deposits. These waters also show elevated temperatures, raised to 40°C, associated with heating in the shelf, which results in the convection of waters entering the shelf at depth. Machel and Mountjoy (1986) state that normal marine sea water is capable of dolomitization, and that it is the hydrologic parameters which may determine if massive dolomites will form, i.e. a mechanism capable of transporting dolomitizing fluids through carbonates. The increased heating associated with raised geothermal gradients in the Gulf of Suez due to rifting is therefore extremely important in the dolomitization of Abu Shaar.

Dolomitization of Miocene deposits in the Red Sea and Mediterranean is often attributed to the action of hypersaline sea waters (Sun, 1992; Oswald et al., 1990; Oswald et al., 1991) and on a global scale also (Sun and Esteban, 1994). At Abu Shaar hypersaline waters are probably responsible for dolomitization, but in this case are driven by heating, not density. The result of the heating is to produce dolomites with negative $\delta^{18}O$ from a hypersaline water. Negative $\delta^{18}O$ is often attributed to the involvement of meteoric waters, such as the suggestion of Coniglio et al. (1988) that dolomitization of Abu Shaar was in a mixing zone between meteoric and marine water. Pierre and Rouchy (1988) suggest that continental groundwaters were involved in the sulphate reduction of Ras Gemsa and South Zeit, approximately 70 km to the north of Abu Shaar, based on negative $\delta^{18}O$ results. The results of this investigation support the idea that heating is perhaps responsible for these light isotopic values as suggested by Youssef (1989a) for Ras Gemsa evaporites, as it can also explain the movement of these fluids into these deposits and the transport of hydrocarbons which are responsible for the sulphate reduction seen at both Ras Gemsa and South Zeit.

ACKNOWLEDGEMENTS

We thank Bruce Purser, Fabien Orszag-Sperber, Dan Bosence, Claire Noble and Nigel Cross for their assistance in the field. Laboratory analyses were conducted with the aid of Stephen Bennett, Alex Etchells, Paul Dennis and Liz Rix. Paul Dennis and Julian Andrews helped with interpretation of the geochemical results.

Section F
Evaporites and salt tectonics

During the mid to late Miocene the Red Sea and Gulf of Suez became isolated from the world's ocean and up to 2 km of salt is deposited within the basin. This may reach 4 km thick in salt diapirs. The three chapters in this section firstly review the nature and origin of these evaporites and then studies from the Egyptian and Yemeni margins illustrate the important controls these evaporites have on the post-evaporite sedimentology and basin evolution.

During the production of this section on evaporites Gill Harwood from the University of East Anglia died. We dedicate this section to Gill Harwood in recognition of her contribution to carbonate and evaporite studies worldwide, and as a valued colleague and friend.

F1 A review of the evaporites of the Red Sea–Gulf of Suez rift
F. Orszag-Sperber, G. Harwood, A. Kendall and B.H. Purser

F2 Post-Miocene sedimentation and rift dynamics in the southern Gulf of Suez and northern Red Sea
F. Orszag-Sperber, B.H. Purser, M.Rioual and J-C Plaziat

F3 Salt domes and their control on basin margin sedimentation: a case study from the Tihama Plain, Yemen
D.W.J. Bosence, M.H. Al-Aawah, I. Davison, B.R. Rosen, C. Vita-Finzi and E. Whitaker

Chapter F1
A review of the evaporites of the Red Sea–Gulf of Suez rift

F. Orszag-Sperber, G. Harwood, A. Kendall and B. H. Purser

ABSTRACT

The evaporites of the Red Sea–Gulf of Suez rift system occur at many stages during the evolution of the rift. earliest Miocene (Aquitanian–Burdigalian) evaporites formed within relatively shallow proto-rift depressions. However, the main mid–late Miocene evaporite formations, locally attaining 2500 m in thickness, were deposited in structural lows formed during the major phase of rifting subsidence. These evaporites began to form during the Serravallian and thus are somewhat older than the Messinian evaporites of the Mediterranean. The two areas must have been disconnected during the late Miocene global sea-level low. Field evidence reveals that the onset of Miocene evaporite deposition within the north-western parts of the Red Sea was not correlative with global sea-level lowering. Structural movements peripheral to the rift system are thought to have generated a barrier located to the north of the Gulf of Suez.

Post-Miocene evaporites also occur both in Pliocene sediments of the Gulf of Suez and in the Quaternary, the latter in small depressions situated on the landward sides of fringing reefs. These evaporites, located near the rift periphery, occur within a post-rift setting. Finally, hot brines occurring today within the deep oceanic parts of the rift may possibly represent an evaporite model whose geodynamic context is that of rift oceanization (drift). Therefore, the various evaporite units within the Red Sea have developed during different stages of rift evolution and are not restricted to the earliest stages.

Evaporites deposited within an active rift system may be characterized by subaquatic facies because the irregular, structurally controlled coastal high relief areas preclude the development of sabkha-type evaporites. The extensional regime also favours shallow-burial, halokinetic remobilization which influences sea-floor morphology and the subsequent sedimentary history of the basin.

INTRODUCTION

Important evaporite sequences are commonly associated with rift basins, notably those situated at low palaeolatitudes (Kinsman, 1975; Rona, 1982). The inherent relationships between evaporite sedimentation and the geographically and chemically restricted environments, which exist during the initial stages of a rift basin, and ocean formation, have been demonstrated by many studies of which the Atlantic is an example (Belmonte et al., 1965).

The favourable combination of climate, drainage and morphology, necessary for massive evaporite development, is favoured by narrow, elongate depressions supplied with sea water via a vertically or laterally restricted connection with the open ocean.

When restriction is terminated by rift opening and/or deeping (Evans, 1978), evaporite sedimentation is succeeded by open marine facies and, because of the low latitude setting, these are frequently calcareous. Thus, one may expect in rift basins a sequence evolving from continental, brackish marls and carbonates with dolomite, anhydrite (or gypsum), and eventually more soluble Na or K salts, and finally, marine carbonates (Evans, 1978). This well-known sequence has been shown to exist in the lower Cretaceous of the west Africa and Brazilian margins (Reyre, 1966, 1984;

Fonseca, 1966), and in the Triassic of the North Atlantic margins, notably in Morocco (Van Houten, 1977).

The Miocene evaporites of the Red Sea and Gulf of Suez have been well documented, particularly by petroleum exploration in the basin. Heybroek (1965) was one of the first to note the relationships between these evaporites and their structural context. Academic research has been concerned with offshore seismic and DSDP drilling campaigns (Ross et al., 1970) and the R.V.J. Charcot Transmerou Cruise (1983) which proved the extension of Miocene evaporites across much of the Red Sea (Pautot et al., 1986). It has been shown that the offshore Miocene evaporites, which include both sulphates and halite salts, may attain 2500 m in thickness but that the somewhat thinner Miocene in onshore wells are dominated by sulphates.

Studies of Miocene evaporites in the Gulf of Suez have been published by Hassan and El Dashlouti (1970); Ross (1972); Ross and Schlee (1973); Khedr (1984); Kulke (1982). Other, more recent publications have noted the relationships between evaporite development and their structural setting: Mulder et al. (1975); Sellwood and Netherwood (1984); Barakat and Miller (1984); Richardson and Arthur (1988); Evans (1988); Orszag-Sperber and Plaziat (1990); Orszag-Sperber et al. (1994), and others. Detailed work, including petrology, diagenetic evolution and environmental factors relating to the formation of Miocene evaporites in the Red Sea and Gulf of Suez, are by Friedman (1972); Stoffers and Kühn (1974); Rouchy et al. (1983, 1985, 1995); Youssef (1986, 1989a,b); Orszag-Sperber et al. (1986); Purser et al. (1987), Wali et al. (1987), El Aref (1992).

The post-evaporite strata are often affected by evaporite-related tectonics. Seismic profiles of these halokinetic structures are illustrated by Miller and Barakat (1988); Rioual et al. (1995), El Anbaawy et al. (1992) and by Bosence et al. and Orszag-Sperber et al. (this volume).

This chapter summarizes the essential attributes of evaporites outcropping along the north-western Egyptian coast of the Red Sea and the Gulf of Suez, as well as specific aspects of evaporites encountered in offshore wells and imaged on seismic profiles. Their age, depositional environments and especially, their relationships to rift evolution are discussed.

This review is not limited to the middle Miocene evaporites which, although the most important, are not the only evaporitic event. Evaporites occur in the Pliocene and also in the Pleistocene and Holocene, where their structural setting differs from that of their older (Miocene) counterparts. These post-Miocene evaporites have been discussed very briefly by Friedman et al. (1973); Kushnir (1981), Purser et al. (1987a); Plaziat et al. (1995); El Haddad et al. (1983/84). Also, the thick post-Miocene evaporites of the Danakil

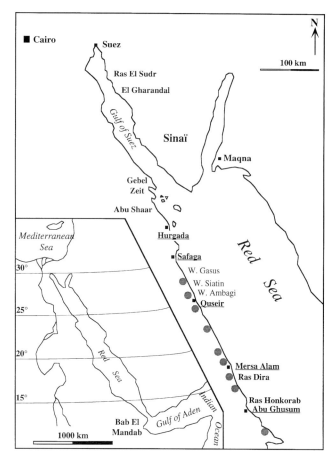

Figure F1.1 Location of the areas discussed in the text. The filled circles are Pleistocene lagoons.

depression (Eritrea) have been discussed by Holwerda and Hutchinson (1968); Hutchinson and Engels (1970); Lowell and Genik (1972).

We do not discuss the diagenesis of the evaporites, although sulphates frequently replace carbonates, including reefs, during the infiltration of brines into permeable substrata (Purser et al., 1987a) and the replacement of sulphates by carbonates related to bacterial sulphate reduction in presence of organic-rich sediments (Rouchy et al., 1985). These features are often associated with the Miocene series of the Gulf of Suez and the north-western Red Sea coast, although they are not specific to this period.

THE TEMPORAL DISTRIBUTION OF THE EVAPORITES (MIOCENE–RECENT)

The structural evolution of the Gulf of Suez as well as the north Red Sea is well documented (Bosworth, Montenat et al., Purser et al., this volume, and references therein). These authors indicate that the beginning of rifting is late Oligocene, and that proto-rift sedimentation is generally continental (group A,

Montenat et al., 1986; Plaziat et al., this volume). Subsequent to marine flooding in the Gulf of Suez and the Red Sea (Group B, Rudeis Formation, see Plaziat et al., this volume) thick evaporites were deposited and followed by 'Pliocene' marine strata. This general Miocene–Pliocene stratigraphic sequence extends from the Gulf of Suez to Ethiopia (Crossley et al., 1992).

By contrast, rifting in the Gulf of Aden began early in the Oligocene (Hughes et al., 1992) and was followed by Sheba mid-oceanic ridge formation in the late Miocene. The major evaporite unit in this region (Gulf of Aden) is of Eocene age (Taleh or Rus formation) and thus, is pre-rift: it reflects regional marine regression favouring sabkha formation (Bossellini, 1989; Watchorn et al., this volume).

Major evaporites of Miocene age therefore are confined to the Red Sea and Gulf of Suez. Because these contrast with contemporaneous open-marine conditions in the Gulf of Aden, the presence of a bathymetric barrier at the entrance to the Red Sea (Bab el Mandeb) is highly probable.

The stratigraphy of the Gulf of Suez and Red Sea areas, together with the occurrence of the major evaporitic units is given in Figure F1.2. This table also includes onshore equivalents cropping out, mainly along the north-western (Egyptian) coast of the Red Sea and Gulf of Suez (Montenat et al., 1986) and data from Crossley et al. (1992), Bossellini (1989) and Beydoun (1992) concern the others parts of the Red Sea. This table merits the following remarks.

- Older, somewhat thinner evaporitic units, deposited before the main evaporitic period, also outcrop locally along the north-west Red Sea coast. These early (proto-rift in Plaziat et al., 1990) evaporites have received less attention and are considerably thinner than the main mid-Miocene evaporites. These early evaporites overlie the continental sediments (Group A, Montenat et al., 1986, 1990; Orszag-Sperber and Plaziat, 1990; Montenat et al., this volume). This early evaporite unit does not appear clearly in wells, although Richardson and Arthur (1988) and Hughes et al. (1992) note the presence of evaporites within the Nukhul Formation (Ghara Member) in the Gulf of Suez. They considered these lower evaporites to be absent from the Red Sea.
- The principal evaporite unit has not been precisely dated and is probably middle to late Miocene. The earliest of these evaporites belong to the Kareem Formation (Markha Member).
- This evaporite unit is overlain by up to 1000 m of marine 'post-evaporitic' series. However, both outcrop and subsurface studies (by the authors) indicate the existence of evaporitic units within these Pliocene 'post-evaporite' beds. Their presence is particularly clear in the area immediately west of Gebel Zeit, and on seismic profiles in the Gulf of Suez.
- Relatively thin (2–8 m) Quaternary evaporites are widespread along the north-western Egyptian Red Sea coast where they have been deposited in shallow coastal lagoons (Plaziat et al., 1995, and this volume).
- It is clear therefore, that evaporitic sedimentation within the Red Sea rift system is not a single event; there exist at least four evaporitic bodies ranging in age from early Miocene to Quaternary, which have been deposited under variable structural settings and oceanographic conditions that have changed during the evolution of the rift. These changes will be discussed in detail.

AGE OF THE MIOCENE EVAPORITE FORMATIONS

Early Miocene evaporites

The pre-rift Cretaceous–Eocene carbonates of the Red Sea are overlain unconformably by reddish proto-rift clastics (300 m) of late Oligocene–early Miocene age (Group A, Montenat et al., 1986). These are located within structural depressions oriented obliquely (N40, N120) with respect to the Red Sea axis. Continental siliciclastic facies are overlain by thick units of well-bedded gypsum, each unit being separated by green marls, silts and diatomites (Orszag-Sperber and Plaziat, 1990). The marls contain Mediterranean foraminifera of Aquitanian age at Wadi Gasus (south Safaga, Figure F1.1). The evaporites are overlain by Burdigalian marine sediments (Montenat et al., 1986; Thiriet, 1987). Possible correlative evaporites are referred to as the Ghara Member (Nukhul Formation) by Hughes et al. (1992) corresponding to the N5 zone of Blow are found in the Gulf of Suez.

These lowest evaporites of Aquitanian–early Burdigalian age, have no known equivalents within the Mediterranean region.

The mid to late Miocene evaporites

The above-mentioned early Miocene evaporites are followed unconformably by marine clastics and carbonates probably of Langhian to Serravallian in age (Group B, Montenat et al., 1986; Plaziat et al., this volume). This open marine phase is followed by widespread evaporitic sedimentation in the mid to late Miocene.

The EGPC Stratigraphic Committee (1964) and the National Subcommittee (1974) have defined a standard Miocene lithostratigraphy that is now used extensively in the Gulf of Suez. The main evaporite unit begins with the Kareem Formation, upper part of the early Miocene

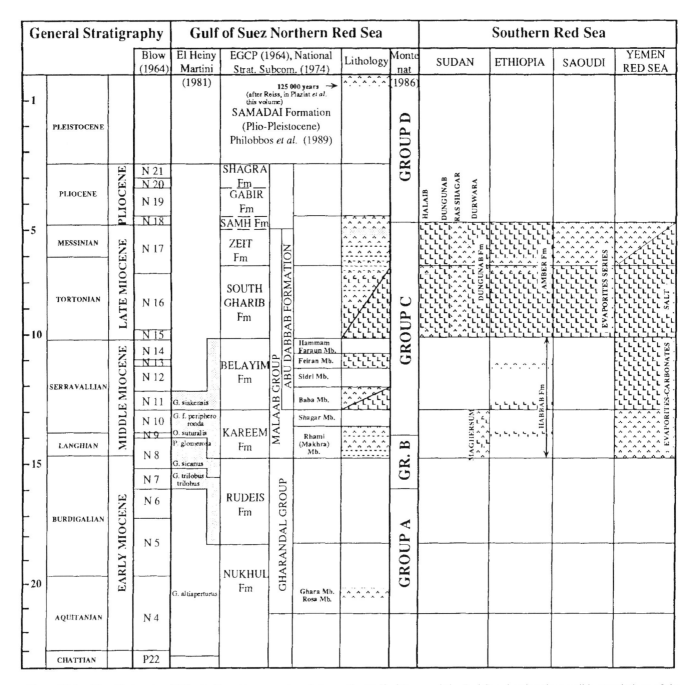

Figure F1.2 Simplified regional lithostratigraphic comparison between the Gulf of Suez and the Red Sea showing the possible correlations of the Miocene evaporite formations.

(Langhian–Serravallian in Figure F1.2). Evaporites continued through the middle–late Miocene as the Belayim, South Gharib and Zeit Formations, and are separated by widespread shaly beds. At outcrop, these evaporites were termed 'Group C' by Montenat et al. (1986).

Figure F1.2 indicates possible age correlations between the evaporite formation in the Red Sea and Gulf of Suez. The base of Markha Member of the Kareem Formation overlies the Rudeis Formation whose top has been dated as zone N9 of Blow (*Orbulina suturalis*, Langhian). In Sinai, these evaporites are overlain by the Shagar shales which contain zone N9 microfauna and nannoplankton indicative of zone NN5 (El Heini and Martini, 1981; F. Sullivan, in Scott and Govean, 1985). In the DSDP sites 225 and 227, black shales intercalated in evaporites contain *Discoaster quinqueramus*, zone Martini NN11, Serravallian (Boudreaux, 1974). All these faunas have Mediterranean affinities.

Ouda and Masoud (1993) indicate that the Kareem and Belayim Formations cover the stratigraphic interval from the datum of first appearances of *Orbulina universa*, to the datum of disappearance of *Globorotalia mayeri* and *G. siakensis*, which is equivalent to the Serravallian.

Thus, foraminifera and nanoplankton as well as the macrofauna collected both from outcrops and wells in the Gulf of Suez and the north-western Red Sea, clearly point to a Mediterranean origin for strata from the Nukhul Formation to the evaporitic Belayim Formation, indicating at least episodic connections during evaporite deposition.

The principal evaporite mass began to develop at the end of the Langhian and therefore does not correlate with the well-known Messinian 'salinity crisis' of the Mediterranean. Nevertheless, this correlation is widely admitted in the literature; for example, for Stoffers and Kühn (1974b), 'the waterfall descending from the Strait of Gibraltar must be held responsible not only for the up to two kilometer thick evaporite sequence in the Mediterranean but also for the three to four kilometer thick evaporites in the Red Sea'; Ross et al. (1973) and Coleman (1974b) have correlated their reflector 'M' with a well-known 'reflector S' in the Red Sea, both reflectors supposedly marking the Upper Miocene termination of evaporitic sedimentation and, thus, the Miocene–Pliocene boundary. The upper part of the evaporite series is poorly dated. However, it is generally accepted (Cox, 1929; Heybroek, 1965) that the marine transgression marking the end of evaporitic sedimentation occurred in the Pliocene and probably corresponded to isolation from the Mediterranean and opening to Indian Ocean waters. The change from Mediterranean to Indian Ocean appears to be based essentially on faunal affinities.

As a conclusion:

- These thick evaporites are Serravallian–Tortonian in age for their major part.
- In the northern Red Sea, evaporitic sedimentation did not cease abruptly. Onshore, at Gebel Zeit for example (Orszag-Sperber et al., this volume), as well as on certain seismic profiles in the Gulf of Suez (El Heini and Martini, 1981; Fawzy and Abdel Aab, 1984; Richardson and Arthur, 1988; Rioual et al., 1996) evaporitic horizons are interbedded within marine clastics and carbonates (Zeit Formation) of speculative late Miocene–Pliocene age.
- Therefore it was probably at the end of evaporite sedimentation that the Red Sea was cut off from the Mediterranean. Subsequently this connection was established with the Indian Ocean; according to Hughes et al. (1992) microfauna exhibit Mediterranean affinities up to (but not following) Belayim sedimentation.

- If there was a connection in the Messinian between the Gulf of Suez and the Mediterranean, this clearly creates an interesting problem because evaporitic sedimentation in the Mediterranean, at least for some authors (Hsü et al., 1973; Clauzon, 1982), was conditioned by partial desiccation following a sea-level lowering in the basin of about 1000 m. If this occurred there could not have been a Mediterranean source for evaporitic sedimentation within the Gulf of Suez and Red Sea (Figure F1.3).

However, if one retains the notion that evaporite sedimentation coincides with a lowering of sea level within the Mediterranean, and, within the constraint that delivery of important volumes of sea water are necessary for evaporite formation, there remains a major enigma, notably in the absence of Indian Ocean faunas.

Thus, it is clear that mid to late Miocene evaporites of the Red Sea and Gulf of Suez began to form somewhat earlier than those of the Mediterranean. It is beyond the

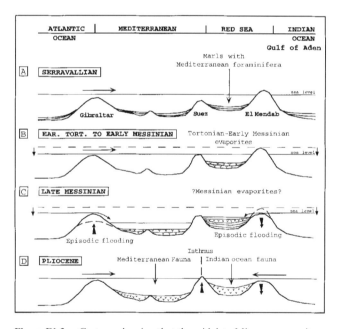

Figure F1.3 Cartoon showing that the mid–late Miocene evaporites are not Messinian in age. A. Serravallian: the fauna in the Red Sea has Mediterranean affinities. The Bab El Mandab sill is high. Marl deposition in both the Mediterranean and the Red Sea. B. Early Tortonian–early Messinian: the Mediterranean continues to receive the Atlantic waters. However, the Indian Ocean does not communicate with the Red Sea. The microfauna at the base of the evaporite formation has Mediterranean affinities and thus the supply of water necessary for the deposition of 2500 m of evaporites is derived from the Mediterranean. C. During the Messinian, the Mediterranean is isolated from the Atlantic Ocean and sea level has dropped. Thus, if the evaporites of the Red Sea are Messinian it is necessary that the Red Sea receives marine waters from the Indian Ocean. This possibility is not in agreement with the published data. D. Pliocene: the Mediterranean is disconnected from the Red Sea which is open to the Indian Ocean.

scope of this chapter to make a detailed discussion of different models for the deposition of evaporites during the Messinian in the Mediterranean, and earlier in the Gulf of Suez and the Red Sea. However, in the Gulf of Suez, Richardson and Arthur (1988), Moretti and Colletta (1988) demonstrated that the restriction favouring the deposition of evaporites appears to coincide with cessation of tectonic subsidence. If so, the initial depth of the basin must be the thickness of the evaporites. Concerning the question of palaeo-water depth (Hsü, 1972; Shmaltz, 1966), the deposits in the marginal basins differ from the more axial sediments with respect to the morphology of the Gulf of Suez at this period. The necessity of a periodic influx of sea water to precipitate this volume of evaporite, together with the marine marls below the evaporite series suggest a relatively deep depositional environment in the central basins.

Outcrop sections show that evaporites deposited on the top of the blocks and in the depression between them (Wadi Ambagi) indicate a difference in level of about 150 m, which may be the depth minima for the deposition of the marine evaporites, at least in marginal areas.

MIOCENE EVAPORITIC FACIES AND SEDIMENTATION

Early Miocene evaporites

Outcrops of this unit have been studied in Sudan, at Marsa Shinab and Gebel Sagun (Montenat *et al.*, 1990) where 30 m of coarsely crystalline and laminated gypsum exhibit frequent slump structures. At Ras Honkorab, on the Egyptian Red Sea coast (Figure F1.1), approximately 10 m of bedded gypsum and marls overlie the Precambrian basement. These evaporites occur within a well-defined graben whose adjacent horst culminated some 100 m above the gypsum (Figure F1.4(a). The gypsum passes laterally into subaquatic algal laminites which encrust the steep flanks of the graben with a depositional relief of at least 50 m. These relationships indicate that the evaporites were deposited in water depths of several tens of metres, and not in a sabkha environment.

These early evaporites are well exposed at Wadi Gasus (near Safaga), and attain 20 m in thickness (Plate 24(a)). Lithologically, these evaporites are generally highly weathered and only locally can one observe laminations and traces of subvertically elongate selenite crystals, confirming subaquatic origin. They comprise three individual beds separated by marls with marine microfaunas and diatomites. Each gypsum bed, 2–6 m in thickness, is composed locally of laminated gypsum detritus. Certain laminae show erosional basal contacts

Figure F1.4 Outcrops of various gypsum units. A. Evaporites (arrows) Group A (Aquitanian), at Ras Honkorab (Red Sea), deposited in a half-graben. Basement to right. B. Unconformable contact (a) between open-marine Group B (b) and evaporitic Group C (c) sediments at Wadi Ambagi, south-west Quseir. C. Massive selenite layers (Group C), with diagenetic spherulite (arrow). Wadi Siatin.

suggesting a detrital slope sedimentation. Marl intercalations include blocks of gypsum attaining 20 cm in diameter. These features suggest reworking of these evaporites on a slope. This may have constituted the flanks of an early-Miocene structural depression. The Nukhul evaporites are known only from certain wells located in sub-basins within the southern parts of the Gulf of Suez where they appear to pass laterally into carbonates. They occur as secondary anhydrites, and are thought to indicate shallow-marine and possibly sabkha conditions (Hughes et al., 1992).

Mid to late Miocene evaporites

As indicated on Figure F1.2, this formation is composed essentially of sulphates and halite. In the Gulf of Suez, well logs and outcrops in Sinai suggest that parts of the Belayim Formation are composed mainly of anhydrite while the upper parts (South Gharib Formation) are dominated by halite (Hassan and El Dashlouti, 1970).

However, this division does not appear to be constant; well data from Saudi Arabia suggest inverted relationships while in Sudan and Ethiopia, anhydrite and halite are interbedded through the evaporite unit (Beydoun, 1992).

It is difficult to correlate and compare the lithologies of outcrops with those encountered in the subsurface, mainly because they are not deposited in the same environment: in the basin (where mainly halite exists) and on the periphery (where they occur essentially as sulphates). Also, the basinal evaporites are often converted into anhydrite after burial.

Outcrop data

Sections (Figure F1.5) have been studied by the authors at Wadi Siatin (north Quseir) (Figure F1.4(c)), Wadi Ambagi (south Quseir), Gebel Zeit, and in Sinai (Ras Sudr). Evaporites are often discordant on Group B carbonates (Figure F.4(b)). Globally, these sections expose bedded gypsum, often transformed into ala-

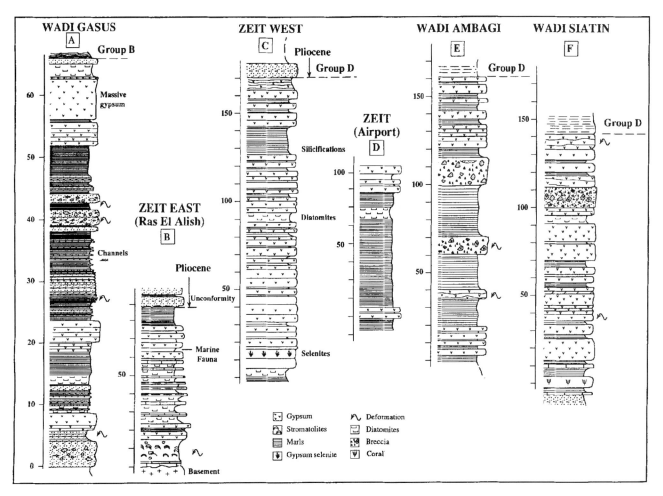

Figure F1.5 A. The earlier Aquitanian evaporites (Group A). Section at Wadi Gasus. This section shows intercalations of marine marls and diatomaceous layers in the evaporites. Slumps and olistoliths suggest the instability of the region during the deposition. B–F. The middle–late Miocene evaporites (Group C). Note deformations and breccia at Wadi Ambagi and Wadi Siatin.

bastrine, which alternate with greenish marls, and local dolomitic limestone. The lithologies are varied (Figure F1.6): selenite crystals (sometimes of metric size), gypsified algal laminites, diagenetic nodules of gypsum and various silicificated nodules are present. Intercalated greenish marls contain a poor marine fauna including diatomites (at Gebel Zeit) in which there are fishes of shallow-marine origin (Gaudant and Rouchy, 1986). No sedimentary or diagenetic criteria for sabkha environments are present although these are often mentioned in the literature (El Haddad, 1984).

In western Sinai, all the evaporites observed in the area from Ras Sudr to north of Abu Rudeis, are gypsum: halite facies were either not deposited on the margins of the Gulf of Suez, or, more likely, have been subsequently dissolved.

The basal Kareem evaporites outcrop east of Wadi Gharandal and at Wadi Gharandal where they overlie the Hawara Formation. The latter are sand-dominated, with hummocky to swaley cross-stratification and contain rare seismite structures. Evaporites are dominantly nodular in fabric, but with large (metric-scale) palmate gypsum crystals.

The Belayim evaporites at Wadi Gharandal are also characterized by palmate gypsum crystals (Figure F1.6(d)). Some possess an internal fill of poikilitic coarse secondary gypsum, perhaps after a former halite crystal core. This indicates deposition of the gypsum at brines nearing halite saturation; replacement of crystal centres by halite is common in sulphate-evaporite sequences immediately underlying halite facies (Hovorka, 1992).

Underlying these evaporites (South Wadi Nahila Quarry), are sandy, carbonate-rich marls with abundant echinoid and oysters fauna, indicating normal marine conditions prior to evaporite deposition. Although the marl-evaporite boundary is slumped, it appears that palmate, bottom-growth gypsum was the initial sulphate facies, indicating rapid isolation of the depositional basin and rapid evaporation to sulphate salinities. Brine depths in which these crystals grew were probably a few metres, and these depths must have been maintained during growth of the stacked crystals.

Thus, the evaporites examined by the authors in south-western Sinai and the western margin of the Red Sea are pure or near-pure sulphate. All identifiable depositional fabrics demonstrate growth in shallow pools of gypsum-saturated brines, with periodic increases to near-halite salinity. Even where textures are not clear, the purity of the sulphates argues for subaqueous sulphate deposition.

Stable isotopes of gypsum (Laboratory of Hydrogéologie, University Paris-Sud), show that the δ values ($11.1 < \delta^{18}O < 14.30$; $21.2 < \delta^{34}S < 23.90$) lie within the range expected for Miocene sulphates of marine origin. This is in agreement with Rouchy et al. (1995).

Attia et al. (1995) examined sulphur and oxygen isotopes from fluid inclusions and suggested that the Middle Miocene gypsum of the eastern coast of the Gulf of Suez grew from mixed sea and non-marine waters, having formed in a subsidiary basin separated from the main Gulf of Suez trough. We lack field arguments to support this interpretation.

Well data

Evaporites were found in the leg 23 drilling in the Red Sea (Stoffers and Kühn, 1974) on sites 225, 227 and 228, near the axis of the Red Sea. They have been described in the DSDP publications. The sedimentary sequence includes four units, the deeper being composed of anhydrite and halite interbedded with black shales. There exist two anhydrite units, the lower being nodular with a dolomitic matrix cut by fissures filled with organic matter, the upper consisting of finely laminated anhydrite. Stoffers and Kühn (1974) consider that undulating laminations, nodular gypsum and anhydrite are characteristic of sabkha deposits and deep-basin/shallow deposits.

The following remarks are based mainly on the studies of industrial wells and interpretations of Hassan and El Dashlouti (1970), Miller and Barakat (1988), Richardson and Arthur (1988), Mitchell et al. (1992), Hughes et al. (1992), Rouchy et al. (1995) and others.

The lithologies recorded in these sections are variable, but facies do not appear to be specific to any given formation. While in the Belayim, South Gharib and Zeit, halite is often recorded, notably within the deeper axial parts of the rift, gypsum seems to predominate towards the periphery of the basin. Ghorab et al. (1969) have recorded potassic salts in the South Gharib Formation (well Abu Shaar NE1, Gebel El Zeit 1, Ras El Bahar 3) which indicates that a supersaline phase was reached locally.

Within the **Rahmi Member** anhydrite predominates. Barakat and Miller (1984) record halite beds intercalated within anhydrite, associated foraminiferal assemblages indicating that these evaporites have been deposited in fairly deep environments (Hughes et al., 1992). In the Red Sea, correlative evaporites were composed mainly of halite, with limited anhydrite intercalations. These lateral variations between the Gulf of Suez and the Red Sea may suggest that brines delivered to the Red Sea had already precipitated their sulphates within the Gulf of Suez.

The **Belayim Formation** (30 to 1000 m) records the beginning of the main evaporite mass. This unit comprises both anhydrite and halite, with clastic sand intercalations, suggesting sub-littoral environments.

Miocene evaporitic facies and sedimentation 417

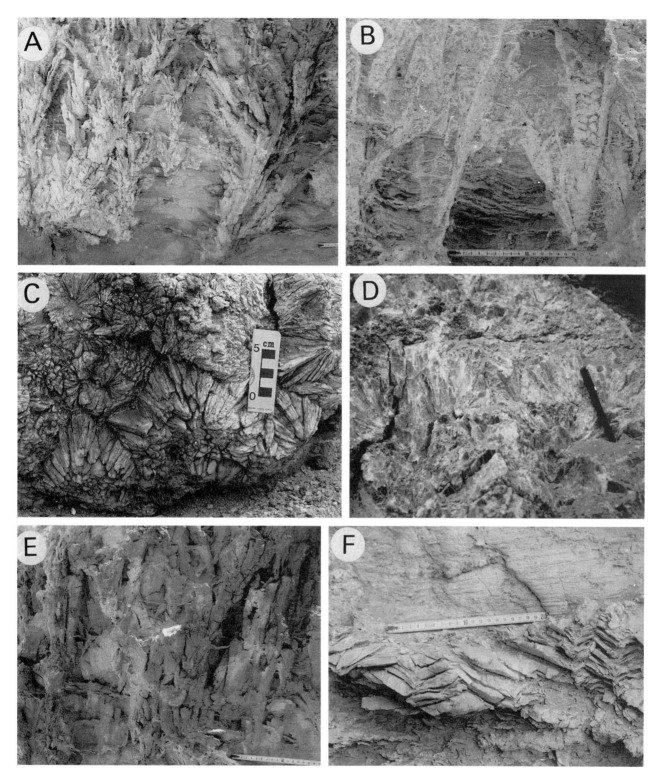

Figure F1.6 Facies of the Group C evaporites. A. Large chevron-type selenite crystals growing through stratified green marls, Wadi Siatin. B. Selenite twins in laminated carbonate-clay material, Wadi Siatin. C. Gypsum crystals growing from a nucleation cone. These diagenetic structures are often found in the vicinity of fractures, Zeit section. D. Palmate gypsum crystals, Wadi Gharandal, Sinai. E. Vertically oriented prismatic selenite crystals, Wadi Ambagi. F. Aggregate of gypsum twin-crystals in stratiform position.

We studied a well core (Tawila 2, also described in Rouchy et al., 1995) situated to the west of the island of Tawila which penetrated the Shagar Formation characterized by the zone of *Orbulina suturalis*, this mass of evaporites therefore having been deposited between the Serravallian and the Pliocene. It included the Belayim and part of South Gharib Formation composed of alternating clastics, anhydrite and halite, overlain by the upper part of the South Gharib and Zeit Formations. These latter units are dominantly halite with minor intercalations of anhydrite and siliciclastics. The Belayim, in particular, illustrates the existence of what appears to be massive anhydrite. However, on polished surfaces, this anhydrite is seen to be finely laminated, individual laminae being composed of anhydrite crystals. The anhydrite also exists in the form of breccia, the contact between *in situ* anhydrite and anhydrite breccia being locally erosive. Coarse sands associated with the breccias exhibit frequent oblique stratifications (Figure F1.7). These attributes are indicative of reworking of the evaporites (Rouchy et al., 1995). They are often associated with nodular anhydrite, this latter probably of diagenetic origin.

The **South Gharib** includes anhydrites with shale intercalations notably towards the periphery of the basin, halite being more frequent towards its centre. These evaporites have been interpreted as being shallow marine in origin.

Zeit evaporites include alternations of anhydrite, halite and clastics (shales and sandstones), possibly indicative of shallow-marine environments whose salinities fluctuated.

Thus, the sedimentary and diagenetic properties of evaporites from outcrops and wells indicate that sulphates and chlorides have been formed in subaquatic environments whose precise depths are difficult to interpret. Parts of the evaporites, notably the halite, could have been precipitated in relatively deep waters, as suggested by their structural context discussed in the following section. Although noted in the literature, we have not observed any features suggesting the existence of sabkha palaeoenvironments, although these may well have existed locally in peripheral parts of the rift.

RELATIONSHIPS BETWEEN MIOCENE EVAPORITE SEDIMENTATION AND RIFT STRUCTURE

The early Miocene evaporites

This gypsum unit is related closely to the earliest phases of rifting. These evaporites are restricted to late-Oligocene graben where they overlie continental proto-rift sediments, both at Ras Honkorab and at Wadi Gasus (Orszag-Sperber and Plaziat, 1990; Plaziat et al., 1990; Montenat et al., this volume). In the latter locality, the evaporites have been displaced and deformed by synsedimentary gravity processes. These early evaporites are thought to have been located in structural depressions isolated from open-marine environments by intervening structural highs. In other words, these early evaporites are the expression of proto-rift tectonics and early Neogene transgression. Infiltration of marine waters during the initial stages of the Burdigalian transgression led to the formation of evaporites. This structural setting may be compared with that of the Danakil depression in Eritrea (Figure F1.8). This depression (Holwerda and Hutchinson, 1968; Hutchinson and Engels, 1970; Barberi et al., 1972) is the northern prolongation of the Afar depression. It is almost completely filled with sediment ranging in age from early Miocene to Quaternary, including up to 5000 m of evaporites. At a depth of 610 m evaporites have been dated as 125 000 years and at 73 m as 76 000 years (Hutchinson and Engels, 1970).

The uppermost part of the South Gharib and Zeit Formations in the Red Sea may correlate with the deepest Danakil salt. The uppermost part of the Danakil evaporites are much younger than the top of the Red Sea salts. Today, in the region of Dallol, several centimetres of evaporites are being precipitated per year (Hutchinson and Engels, 1970).

According to Hutchinson and Engels (1970), the distribution of evaporite and siliciclastic facies within the Danakil graben exhibits an east–west salinity gradient, the bitterns being concentrated within the axis of the half-graben. Because lower salinity evaporites are located along the slope, water supply has probably been derived from the adjacent Red Sea passing over, or through, the emerging Danakil horst.

Comparison between the Gulf of Suez and the Red Sea graben and their filling by evaporites and the tectonic evolution of the Danakil depression is speculative. However, the Danakil depression could be the result of asymmetric subsidence of continental blocks marginal to newly opened rift and thus forming local restricted basins favouring the accumulation of thick evaporites (Hutchinson and Engels, 1970). In both cases, the evaporites are the sedimentary expression of the initial marine transgression.

The mid–late Miocene evaporites

This thick (up to 2500 m) evaporite complex has been formed under specific geodynamic conditions involving the north-eastern movement of the Arabian Plate (Cochran, 1983b; Le Pichon and Gaulier, 1988; Rihm et al., in this volume). Crossley et al. (1992) note that the evaporites, at least in part, have been deposited in deep depressions, isolated from the Indian Ocean (Figure

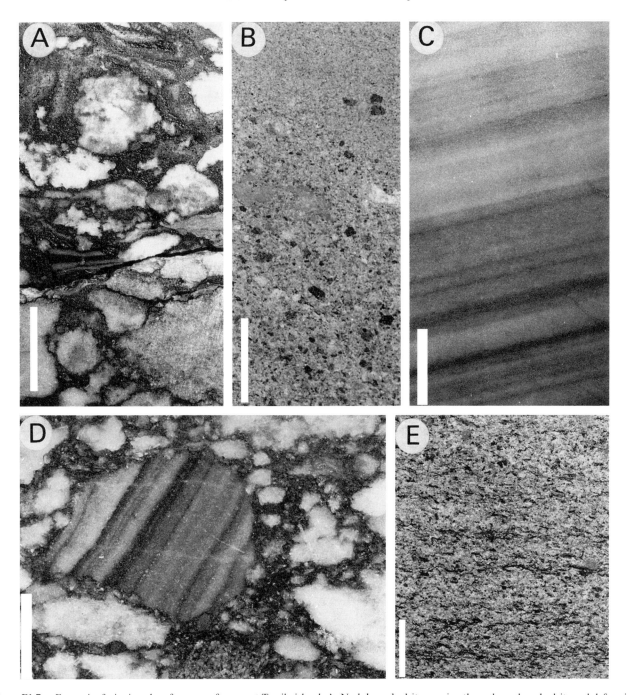

Figure F1.7 Evaporite facies in subsurface cores from west Tawila island. A. Nodular anhydrite growing through sandy anhydrite and deforming the laminations of the host sediment. Bar: 5 cm. B. Coarse and fine sands separated by an erosional structure. Sands consist of fine-grained anhydrite. Bar: 1 cm. C. Laminated fine-grained anhydrite. Bar: 2 cm. D. Breccia composed of nodular and layered anhydrite fragments. Bar: 1 cm. E. Layered sands composed of anhydrite and dolomite debris. Bar: 1 cm.

F1.9). Their precipitation has also been facilitated by global sea-level lowering at ?10 m.y. (Serravallian) according to Coletta *et al.* (1988), Evans (1988), Richardson and Arthur (1988). These latter authors note that the evaporites were deposited after the main phase of mid-Miocene extension, at least in the Gulf of Suez (see also Bosworth, this volume). They also note that evaporite facies in the Gulf of Suez (Figure F1.9) have a 'tear drop' geometry (Schmalz, 1966; Hsü, 1972) indicating that connection with the parental Mediterranean persisted at least until the end of the Serravallian (10 m.y.) at the top of the Belayim Formation, and that the rift system opened to the Indian Ocean at the beginning of the Pliocene (5.3 m.y.). Thus, the switch of supply from the Mediterranean to the Indian Ocean must have occurred during the late Miocene.

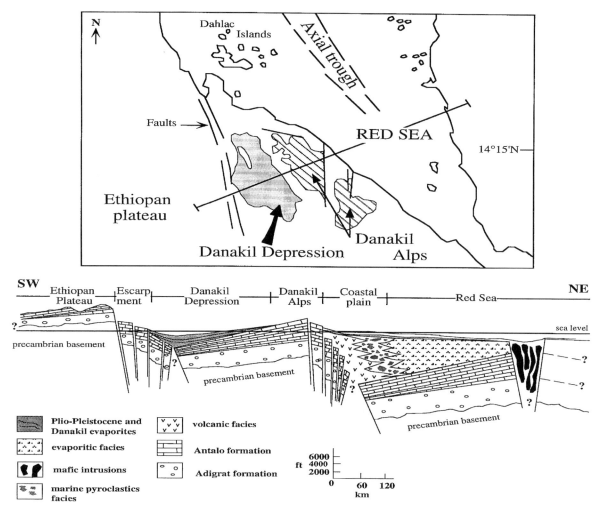

Figure F1.8 Danakil depression. A – location of the area, B – section after Hutchinson and Engels (1970).

However, as already noted, any evaporites of Messinian age deposited within the Red Sea rift must have been related to Indian Ocean waters, because the Mediterranean at that period was a subaerial topographic depression (Hsü et al., 1973; Cita, 1980). The return to normal marine conditions was the result of global sea-level rise near the end of the Messinian (Richardson and Arthur, 1988), leading to flooding from the Indian Ocean.

Orszag-Sperber et al. (1994) have indicated that tectonic control, if not the only factor, has played an important role in determining the onset of evaporitic conditions, and that global sea-level lowering was less relevant. These conclusions are based on field studies at Gebel Abu Shaar (south-west Gulf of Suez) where open-marine (Miocene) carbonates, deposited on the platform, grade upwards into oolitic and stromatolitic carbonates associated with dwarf potamids and other molluscs, indicating shallow, partially restricted environments (Figure F1.10). These peritidal facies are followed conformably by unfossiliferous dololaminites containing gypsum and selenite pseudomorphs and black silicifications. These features clearly record increasing restriction on the top and on the slopes of the platform. Total thickness of these transitional (open marine to dololaminites) is about 20 m, indicating a moderate sea-level rise. In other words, restriction ultimately leading to the precipitation of a thick, middle to late Miocene evaporite series, cannot be due to lowering of sea level, at least in its initial stages.

Evans (1988) suggested that restriction increased considerably after 11 m.y. as indicated by the thick South Gharib Formation halite. This increase was thought to be due to the proposed eustatic sea-level lowering at least at about 10.5 m.y. (Haq et al., 1987). At Gebel Abu Shaar, massive evaporites do not occur on top of the platform, although the presence of strontium-rich residues suggest that they may well have existed previously (as noted by Clegg et al., 1995, and this volume). On other platforms located on the north-

Relationships between Miocene evaporite sedimentation and rift structure 421

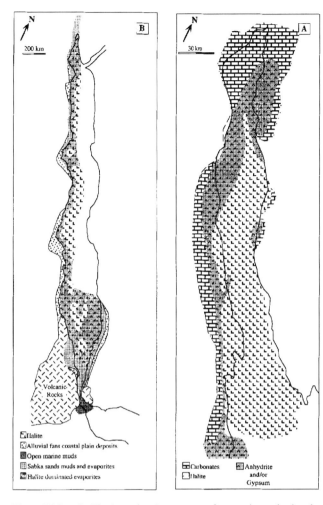

Figure F1.9 A. The 'tear drop' geometry of evaporite and related facies in the Gulf of Suez, showing the distribution of the evaporites during the middle–late Miocene (after Richardson and Arthur, 1988). B. Distribution of the middle–late Miocene evaporites in the Gulf of Suez and the Red Sea (after Crossley *et al.*, 1992).

western (Egyptian) Red Sea coast, including Wadi Ambagi (Figure F1.4), evaporites onlap periplatform slopes and also overlie its culmination. In common with Abu Shaar, the transition from open-marine to evaporite conditions includes intermediate facies, locally with mytilid bivalves indicating progressive restriction. The transition occurs at a present altitude of about 150 m.

Thus, there is no evidence that evaporite deposition was the consequence of a lowering of mid-Miocene sea level. The initial restriction must have been provoked by tectonic movements which progressively restricted the circulation of the Red Sea waters with respect to their Mediterranean source. Prior to the deposition of the evaporites, the basin was tectonically isolated. Restriction occurred during a relatively high sea level, probably Serravallian in age, before the classical Tortonian and Messinian lowering took place.

Certain workers (Plaziat *et al.*, 1990, and this volume) have indicated tectonic instability during evaporite sedimentation suggesting that at least part of the evaporite series is resedimented. Also, Rouchy *et al.* (1995) have described resedimentation features in cores from the Tawila well (Gulf of Suez). Many (perhaps most) outcrops along the north-western coast of the Red Sea exhibit spectacular slump features and large evaporite olistoliths. This instability was probably favoured by at least two factors. Firstly, many of the evaporites were deposited in relatively deep structural depressions. These would have been bordered by slopes, and this tectonic setting would have favoured slumping. Secondly, the Neogene sediments of the north-western Red Sea exhibit many deformational structures which affect virtually all types of lithologies including evaporites. They have been interpreted by Plaziat *et al.* (1990) to have triggered by paleoseismic activity.

Evaporites at passive margins

Red Sea rifting is considered by many (Rona, 1982; Evans, 1978) as a model for the formation of the South Atlantic. Relations between evaporite deposition and the hydrographic restriction associated with the initial stages of rifting have already been noted. Typically, evaporites are the initial sediments resulting from the interaction of structural setting (rifting), climate and global sea level. Thus, evaporites recording the initial stages of marine transgression can be chronological markers for the creating of new oceans (Atlantic), as has been noted by Evans (1978).

The relationships between evaporite sedimentation and Red Sea rifting differ according the period of deposition. While the older (early Miocene) evaporites indeed mark the initial phase of marine ingression, the main (middle–late Miocene) mass certainly does not. These latter evaporites follow an important period of open-marine sedimentation involving the formation of widespread early Miocene carbonate platforms (Purser *et al.*, and Cross *et al.*, this volume). These evaporites began to form some time during the end of the Langhian, the Serravallian and probably the Tortonian, i.e. almost 10 m.y. after the first marine transgression which may have occurred during the late Oligocene within certain parts of the rift (Abu Ouf and Geith, this volume).

Although the factors leading to the onset of middle–late Miocene evaporite sedimentation throughout the Red Sea may indeed be relative to rift dynamics, they can not be regarded as expressing opening of the proto-Red Sea. A possible tectonic control must be sought not in the opening, but rather in the well-developed rift basin. Because Miocene waters were initially supplied via the Mediterranean, isolation from that northern source must be due either to eustatic lowering of sea level or a disconnection caused by the formation of a

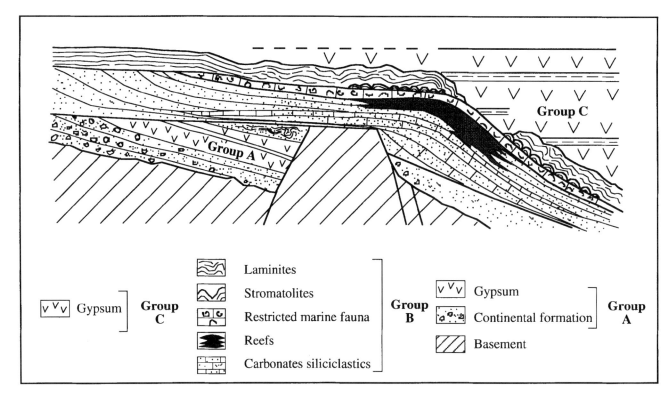

Figure F1.10 Simplified profile (without scale) showing the geological relationships between the various evaporite formations. Explanations in the text.

structural barrier to the north of the Gulf of Suez (Orszag-Sperber *et al.*, 1994).

EVAPORITE TECTONICS

Seismic surveys in the Gulf of Suez and Red Sea have demonstrated the widespread, vertically continuous, distribution of the mid to late Miocene evaporites (Mitchell *et al.*, 1992). In the Gulf of Suez, seismic profiles show considerable variations in the thickness of the evaporites which occur on two scales: a regional thickening towards the axis of the Gulf of Suez, where evaporites may attain 2500 m in structural depressions (Heybroeck, 1965); and, secondly, as a series of localized thickening and thinnings clearly the result of post-depositional movements.

These halokinetic features and their related sedimentary effects are discussed further by Bosence *et al.* and Orszag-Sperber *et al.* (this volume). These are observed mainly in the southern parts of the Gulf of Suez where spectacular north-west–south-east elongate diapirs are related to rejuvenation of underlying extensional faults (Figure F1.11, and see Figures F2.7, F2.8 in Orszag-Sperber *et al.*, this volume). These structures may be expressed as modern bathymetric ridges, many supporting living reefs. These diapiric structures also occur in the Red Sea (Orszag-Sperber *et al.*, Bosence *et al.*, Carbone *et al.*, this volume).

The association of salt diapirism and rift tectonics is well established. Those occurring in the northern part of the Red Sea (Mart and Ross, 1987) and in the southern Gulf of Suez have maximum burial depths of 1000 m, differing considerably from the great burial depths (3000–6000 m) in others basins (Yemen, in Davison *et al.*, 1994). The minimal burial depths of the diapiric features in the northern Red Sea appear to be best explained by the extensional tectonic setting, where faults underlie diapirs seen on seismic profiles in the south-eastern Gulf of Suez, affect post-Miocene (Plio-Quaternary) sediments (Rioual *et al.*, 1996).

Maps prepared by Mart and Ross (1987) and by Rioual in Orszag-Sperber *et al.* (this volume) demonstrate remobilization of evaporites along normal faults. Certain evaporite 'intrusions' constitute walls measuring 4 km in width and 30 km in length (Mart and Ross, 1987, Fig. 3).

Onshore, the diapir at Al Salif (Yemen, Red Sea) discussed by El Anbaawy *et al.* (1992), Davison *et al.* (1994) and Bosence *et al.* (this volume) is an asymmetric structure extending some 4 km above the Miocene salt layer; it displaces part of the sedimentary cover. Recent vertical displacement, based on dates of Quaternary reefs formed on the diapir, are in the order of 4.6 mm/year, implying that vertical displacement (from a depth of 4 km) has required 14 m.y., and thus began relatively soon after evaporite sedimentation.

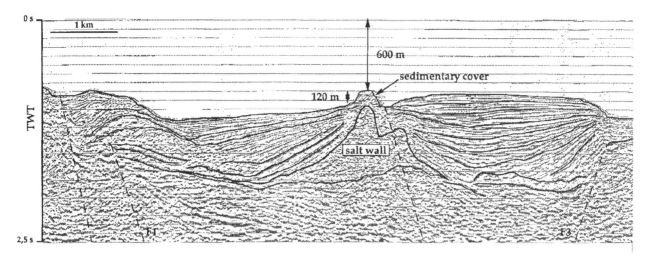

Figure F1.11 Evaporite tectonics: example of a salt wall growing through post-Miocene sedimentary cover (south Gulf of Suez).

On the north-western (Egyptian) coast of the Red Sea, a series of small, linear diapirs occur at Ras Dira (Orszag-Sperber *et al.* this volume, Plate 25) where they are composed only of gypsum at the surface, and changes in dips of carbonates and siliciclastics occurring at Ras Shagra and Ras Greifat are related to halokinetic movements (Orszag-Sperber *et al.*, this volume, Figure F2.20, Plate 25).

POST-MIOCENE EVAPORITES

Pliocene evaporites

On seismic profiles and in wells, the Zeit Formation includes evaporitic horizons intercalated within carbonates or siliciclastics whose ages have not be determined. On outcrops on the west of Gebel el Zeit several tens of metres of gypsum with local teepee structures (Plate 24(c)) associated with oolitic carbonates, sandy marls and dwarf molluscs indicate shallow-marine environments with locally emerged sabkha conditions (see Figure F2.15 in Orszag-Sperber *et al.*, this volume).

Although the age of these surface evaporites is not established, Prat *et al.* (1986) has included them within the Miocene. However, thick conglomerates are intercalated between the main (Belayim) evaporites and these younger evaporites which clearly post-date the main Miocene evaporites formation.

In Saudi Arabia, Le Nindre *et al.* (1986) mention the existence of gypsum intercalations within Pliocene marls, at the Maqna area.

Quaternary evaporites

The thick (5000 to 7000 m) series of evaporites within the Danakil depression (Eritrea) has been noted (Figure F1.8). Although the lower parts are Tertiary in age, the upper parts are dated as Pleistocene and possibly even Recent. Their parental waters are thought to have infiltrated from the Red Sea through the intervening structural block.

Plaziat *et al.* (1995 and this volume) record a series of Quaternary evaporitic lagoons on the Egyptian shoreline of the Red Sea–Gulf of Suez, where subaquatic gypsum has been deposited within small depressions (Figure F1.1).

These evaporites comprise 2–8 m thick units of well-bedded gypsum with thin dolomitic intercalations (Figure F1.12). Gypsum occurs in the form of 1–5 cm crystals of selenite which have grown on the sediment–brine interface. Locally, crystals form disorganized masses (Figure F1.12), affecting regular selenite growth and algal mats. That these evaporites are of marine origin is confirmed by their isotopic properties (δ^{18}O and δ^{34}S). These highly restricted facies are locally underlain by domed stromatolites which cap underlying open-marine sediments.

The evaporitic lagoons are barred by reefs of Eemian (5e) age, and the gypsum has been dated as 123 ± 10 k.y. by Choukri and Reys (in Plaziat *et al.*, 1996). Excavations on the landward side of the fringing reefs indicates erosion of the reef during a lower sea level followed by subsequent rise, flooding and infilling by subaquatic gypsum.

Present-day coastal sabkhas are small and relatively rare along the coast of the north-west Red Sea. However, they are somewhat better developed in Sudan and Saudi Arabia. Their absence, in spite of favourable climatic conditions, is due to an unfavourable topography; an irregular coastal relief relating in part to rift tectonics, is bordered by submarine scarps and relatively deep coastal waters. These morphologies tend not to

Figure F1.12 Pleistocene evaporites along the Egyptian Red Sea coast (see Figure F1.2 for location of the Quaternary lagoons). A. Wadi Samadai area. Lenses of gypsum alternating with very thin, parallel beds of dolomitized carbonate and local algal laminations in which small twin-crystals of gypsum have grown (hammer gives scale). B. Close-up of A (scale bar is 5 cm). C. Selenite crystals deforming algal laminations at Wadi Samadai (scale bar is 5 cm). D. Close-up of C showing interstratified carbonates and clays cut and deformed by growing gypsum crystals (scale bar is 5 cm). E. Detail of B showing lenses of gypsum growing in carbonate matrix (scale bar is 1 cm).

favour the rapid coastal accretion normally associated with sabkha plains.

The axial brines

The deep axial zone of the Red Sea, formed during post-Miocene ocean spreading, is bordered by fault scarps cutting both Miocene evaporites and their overlying post-Miocene cover (Degens and Ross, 1969). The axial zone, whose average depth is approximately 1800 m, is deepened locally to 2807 m, these local depressions being floored by basalts, thin (0.5 m) pelagic marls, and pools of brine whose saturation may attain 270% (Blanc and Anchutz, 1995). A number of hypotheses have explained the origin of these brines which are discussed by Blanc et al. (this volume).

DISCUSSION AND CONCLUSIONS
Multiplicity of evaporite phases

This review has shown that evaporites, an important facies within the Red Sea and Gulf of Suez rift, have multiple origins and structural settings.

The oldest evaporites (early Miocene, Group A) are localized within tectonic depressions generated during the early stages of rifting and thus formed before, or during, the initial phases of marine flooding corresponding to the classical peri-Atlantic situation where evaporites are the first expressions of a new ocean. However, these early evaporites are relatively thin (30 m) and localized; they do not represent the main evaporitic phase.

The better known mid–late Miocene (Belayim, South Gharib, Zeit Formations, Group C) evaporites post-date an important period of open-marine, early Miocene sedimentation. They locally attain 2500 m in thickness and fill deep, open-marine depressions created by extensional tectonic movements. Other, younger evaporites, occur within post-Miocene sediments, notably in the Gulf of Suez, where they correspond to a relatively late phase of rift evolution.

Furthermore, thin (8 m) discontinuous evaporitic units occupy small depressions landward of Pleistocene fringing reefs where they are situated within a late- or post-rift context. The presence of these evaporites shows a brief lowering of the sea during the 5e high sea level stage.

The Danakil depression is filled by 7000 m of Tertiary to Recent evaporites which have accumulated in a restricted basin during the early stages of the Red Sea development (Hutchinson and Engels, 1970).

Finally, hot brines are forming locally today within the deep axial trough. These potential basinal evaporites, closely associated with hydrothermal waters, are forming within the oceanic trough created by the opening of the Red Sea; thus, they coincide with the more advanced stage of basin evolution.

If all the evaporite units are considered, it is clear that their formation is not the result of any one geodynamic, structural or climatic factor. Although the Miocene evaporite units are closely related to their structural framework, they do not all correspond to the initial phases of rifting. On the contrary, precipitation of the main mid–late Miocene evaporite body seems to have been favoured by geographic isolation from the world ocean by structural movements situated to the north of the Gulf of Suez.

Relationships between the Red Sea and Mediterranean Miocene evaporites

It has been suggested that the Gulf of Suez–Red Sea evaporites are contemporaneous with those of the Mediterranean, having been deposited in a restricted extension of this latter basin (Hsü et al., 1978). This hypothesis appears to lack foundation because the main evaporites of the Red Sea began to form during the Serravallian, and possibly even during the Langhian, and thus are considerably older than the Messinian evaporites of the Mediterranean. Although evaporitic sedimentation in the Gulf of Suez–Red Sea probably continued into the Messinian – and even into the Pliocene – it is difficult to imagine a direct linkage with the Mediterranean at this time in view of the major drop in sea level generally accepted to have occurred during the Messinian, which would have isolated it from the Gulf of Suez. The late Miocene evaporites of the Red Sea–Gulf of Suez have faunas similar to those from the Indian Ocean and a southern link to a marine source seems likely at this time.

The factors determining mid–late Miocene evaporite formation in the Red Sea

Because the main Belayim–South Gharib–Zeit Formation (Group C) evaporites locally contain Serravallian microfaunas with Mediterranean affinities, they must have been sourced from Mediterranean waters, at least in their early stages. Restriction leading to evaporite precipitation was probably created by a structural sill situated to the north of the Gulf of Suez. That this restriction was the result of global sea-level lowering, is not supported by field evidence. The later, Messinian, phases of evaporite sedimentation are probably related to Indian Ocean sources immediately prior to complete opening, probably during the Pliocene. Thus, Miocene evaporite sedimentation within the Red Sea–Gulf of Suez rift system was controlled essentially by structural adjustments not within the rift itself but relating to the formation of structural sills at each extremity of the rift basin.

The importance of halokinetic movements

Although evaporites may be deposited in extensional rift settings, compressive foreland basins (Arabo-Persian Gulf) and in stable cratonic areas (Anglo-Paris basin), their subsequent remobilization and diapirism appears to be favoured by extensional tectonics. Thus, the Gulf of Suez–Red Sea rift system is characterized by salt tectonics which influence marine morphology and thus sedimentation. The distribution and geometry of many Quaternary and Recent reef systems within the southern Red Sea and in the Gulf of Suez appear to be controlled by underlying salt diapirs.

Possible relationships between evaporite facies and their structural context

As noted above, evaporites are known to form under various structural settings. The Miocene (and younger) evaporites of the Red Sea have been precipitated, almost exclusively, under subaquatic conditions: sabkha-type evaporites are rare. The lack of sabkhas is probably due to the morphology of coastal areas which today are generally elevated above Neogene and Quaternary sea levels and dissected by drainage systems relating to the uplifted rift margins. Thus, the flat, low relief coastal areas favouring sabkha formation and progradation are uncommon, at least along the northern parts of the Red Sea and Gulf of Suez. This situation contrasts with that of the Arabo-Persian Gulf, both during the Neogene and today, where a more stable, low relief, cratonic foreland favoured widespread sabkha development.

ACKNOWLEDGEMENTS

We would like to express gratitude and thanks to Dan Bosence who greatly improved the text, and Tadeuz Peryt for his review. This research was supported by the Centre National de la Recherche Scientifique, Elf-Aquitaine and Total-Compagnie Française des Pétroles (GENEBASS) and the EEC (SCIENCE Program).

Chapter F2
Post-Miocene sedimentation and rift dynamics in the southern Gulf of Suez and northern Red Sea

F. Orszag-Sperber, B. H. Purser, M. Rioual and J.-C. Plaziat

ABSTRACT

Field studies and seismic profiles along 400 km of the Egyptian coast cover important segments of the northern Red Sea and Gulf of Suez rift. Because the post-Miocene (Pliocene) coincides with the partial oceanization of the Red Sea as a consequence of transform movements along the Aqaba fault system, the region and stratigraphic level considered are keys to the understanding of rift evolution.

The considerable thickness (1000 m) of constant shallow-marine post-Miocene sediments indicates continued post-Miocene subsidence. Within this general context, three distinct regions are recognized, each characterized by a specific tectonic and post-Miocene sedimentary record.

1. The northern and central parts of the Gulf of Suez are characterized by relatively homogeneous terrigenous and evaporitic sedimentation undisturbed by post-Miocene tectonics.

2. The southern Gulf of Suez, between the Morgan accommodation zone and the Aqaba transform, on the contrary, exhibit marked lateral variations between siliciclastics and carbonate sedimentation. These changes are controlled by reactivation of Miocene fault blocks and related diapirism of the Middle to (?)Late Miocene evaporites. In addition, spectacular southward gravity sliding on the evaporite substratum, facilitated by the accommodation space created by the opening of the Red Sea, has led to extensional collapse and formation of secondary post-Miocene basins north of the Aqaba transform zone.

3. The northern Red Sea also includes numerous post-Miocene movements, both parallel and oblique to the axis of the rift, relating to rejuvenated Miocene faults and evaporite remobilization. These movements, in common with the Gulf of Suez, occur on a local scale. However, unlike the Gulf of Suez, the post-Miocene series is also affected by a regional dip toward the axis of the rift and by the frequent discharges of basement-derived conglomerates, these structural and sedimentary features reflecting the opening of the Red Sea and uplift of the rift periphery.

The post-Miocene sediments indicate that the northern Red Sea has attained a post-rift, passive margin stage, that the southern Gulf of Suez, affected by recent extensional movements, has not yet attained this post-rift stage, while the central and northern parts of the Gulf of Suez are still tectonically stable.

INTRODUCTION

Tectonic and stratigraphic framework

The structural evolution of the Red Sea–Gulf of Suez rift system has been described by many authors including Heybroek (1965), Garfunkel and Bartov (1977) and subsequently by Sellwood and Netherwood (1984), Montenat et al. (1986), Colletta et al. (1988), Richardson and Arthur (1988), Purser et al. (1990). These contributions are reviewed in preceding contributions of this volume, notably those of Montenat et al., Purser et al., and Plaziat et al.

The marked structural configuration, created during the phase of maximum subsidence and rifting (early Miocene), although dominated by N140 ('Clysmic') fault system, also included numerous north–south, north-east–south-west faults oriented obliquely to the

Sedimentation and Tectonics of Rift Basins: Red Sea–Gulf of Aden. Edited by B.H. Purser and D.W.J. Bosence. Published in 1998 by Chapman & Hall, London. ISBN 0412 73490 7.

axis of the rift. This highly irregular Miocene morphology was subsequently attenuated by a thick series (2000–5000 m) of middle to upper Miocene Rudeis–Kareem pelagic marls and Miocene evaporites.

Synsedimentary deformation frequently affects these evaporites and suggests structural rejuvenation and, possibly, continued subsidence. These syn-rift evaporites, locally attaining depositional thicknesses of 2000 m as shown by seismic profiles (Heybroek, 1965; Richardson and Arthur, 1988; Orszag-Sperber *et al.*, this volume), are an important element in the subsequent tectonic and sedimentary history of the basin. They are the subject of this contribution.

Field studies by Montenat *et al.* (1986), Philobbos *et al.* (1989), Purser and Philobbos (1993) and Rioual *et al.* (1996) have shown that the various structural components, including both north-west–south-east blocks, and many oblique cross-faults, have been rejuvenated since their initial, early Miocene creation, indicating continued extensional movements. In addition, the South Gharib–Zeit evaporite series (Group C) (Figure F2.2) has been remobilized, at least locally, by halokinetic movements coinciding with the rejuvenation of the underlying early Miocene fault system. The post-evaporite beds, locally exceeding 1000 m in thickness, are affected by these faults and halokinetic domes. Furthermore, along most of the north-western Red Sea coast, these deposits are affected by moderate structural dips (5° to 20°) oriented towards the axis of the rift.

This structural setting which affects the post-evaporite sediments is considered here in terms of rift

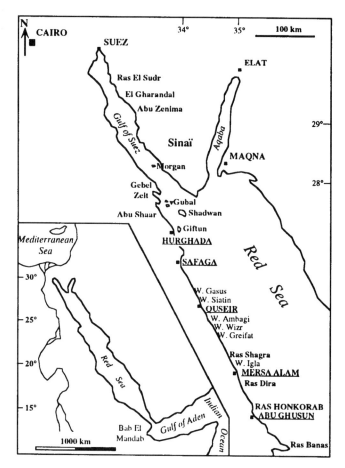

Figure F2.1 Location of the study in the Gulf of Suez and the northern Red Sea.

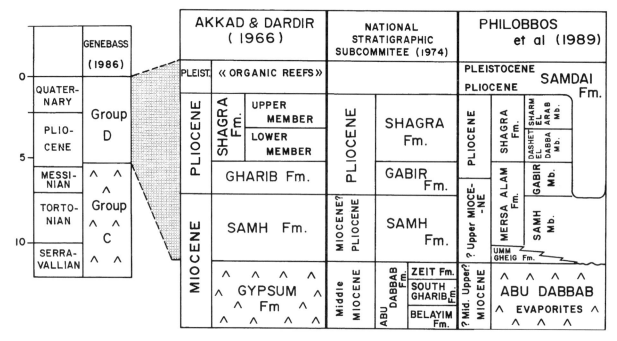

Figure F2.2 General stratigraphic correlations between outcropping and offshore post-Miocene series with a focus on the 'Pliocene' series.

evolution. More precisely, the question concerning both the temporal and spacial transition from the syn-rift to the post-rift phase, and the eventual stratigraphic position of a post-rift unconformity, are important. For many researchers (Colletta et al., 1988; Richardson and Arthur, 1988; Bosworth, 1995), the peripheral parts of the northern Red Sea region attain a post-rift passive margin stage at the end of Miocene evaporite deposition. Davison et al. (1994) also indicate this for the southern Red Sea. The post-Miocene opening of the axial zone relating to movement along the Aqaba transform, does not affect the Gulf of Suez segment of the rift. This has led some workers to consider the Gulf of Suez as an 'aborted rift', i.e. that the entire Gulf of Suez basin has been a passive margin since the deposition of the middle–late Miocene evaporites, the top of which are considered as the post-rift unconformity.

The post-evaporite, Pliocene and Quaternary sediments and their structural deformation, therefore are fundamental to the understanding of rift geodynamics, notably in the northern Red Sea, the Aqaba transform zone and the southern Gulf of Suez, which are considered as individual segments of the highly varied Red Sea–Gulf of Suez rift system.

Age of the post-evaporite deposits

The post-Miocene evaporite sediments are poorly dated. Of little economic interest to oil companies in the Gulf of Suez, post-evaporite formations are generally grouped under the term 'Plio-Quaternary'. At outcrop, the varied deposits are split in many formations and members whose precise relations are described by Akkad and Dardir (1966), Abu Khadra et al. (1984), Philobbos and El Haddad (1983), Philobbos et al. (1989). Their relations are illustrated on Figure F2.2. They have been amalgamated into a single 'Group D' by Montenat et al. (1986).

Along the Egyptian coast of the Red Sea, the uppermost Miocene evaporites include laminated muds which grade upwards into restricted marine silts, there generally being no perceptible discordance. These post-evaporite siliciclastics and the succeeding open-marine siliciclastics and carbonates, generally have the same 5°–20° dips whereas they are truncated near the coast by subhorizontal and more gentle sloping terraces whose numerous U/Th dates (see Plaziat et al., this volume) indicate Pleistocene ages. A similar situation exists on the islands in the southern Gulf of Suez (Figure F2.1). The angular discordance at the base of the Pleistocene terraces coincides with a general marked diagenetic change, the inclined, underlying series being highly cemented. This structural and diagenetic limit has been conventionally considered to mark the boundary between the Pliocene and the Quaternary. A clear-cut change in the fossils contrasts with the Miocene fauna; in particular, the molluscs and corals show a drastic taxonomic replacement between the pre-evaporite (mid-Miocene) faunas without Acropora and those of the post-evaporite ('Pliocene'), which are Acropora dominated. This change, which reflects the opening of the Red Sea to the Indian Ocean, was also described by Cox (1929) with respect to the molluscs (strombids, Pinctada, Tridacna, etc.).

Despite a dramatic lack of biostratigraphic data, these structural, sedimentary and biological arguments strongly suggest that most of the marine sediments deposited between the Miocene evaporites and the dated Pleistocene terraces belong to the Pliocene. The only puzzling question is the precise limits of this 'Pliocene'.

Thus, within an essentially lithostratigraphic context, this contribution describes the important sedimentary and structural variations which characterize the 'Pliocene' strata. The variations in time and space are interpreted in terms of rift evolution, with particular emphasis on the possible transition from syn-rift to aborted or post-rift settings.

'PLIOCENE' SEDIMENTATION AND TECTONICS

Both structural and sedimentological variations discussed in the following paragraphs confirm a well-established structural subdivision of the region. The fundamental separation between the northern Red Sea and the Gulf of Suez is further subdivided, within each of these major rift basins, into segments, each separated by accommodation zones. The Gulf of Suez comprises three major segments (Coletta et al., 1988; Steckler et al., 1988; Bosworth, 1995, and this volume) (Figure F2.3(a)). Individual structural blocks in the northern zone are tilted towards the south-west, and are not considered in this contribution. Those of the median zone are tilted toward the north-east, being separated by the Morgan accommodation zone from the third, southern region whose constituent blocks are tilted towards the south-west.

The Red Sea is also segmented into zones initially based on the location of heat-flow anomalies (Makris and Henke, 1992; Rihm and Henke, this volume), as well as the distribution of ancient lithospheric fracture zones (Makris and Henke, 1992) (Figure F2.3(b)), some of them favouring fluviatile drainage into the basin (Mitchell et al., 1992).

The central Gulf of Suez region

The northern and central regions of the Gulf of Suez are less complicated structurally than the southern region

430 Post-Miocene sedimentation and rift dynamics

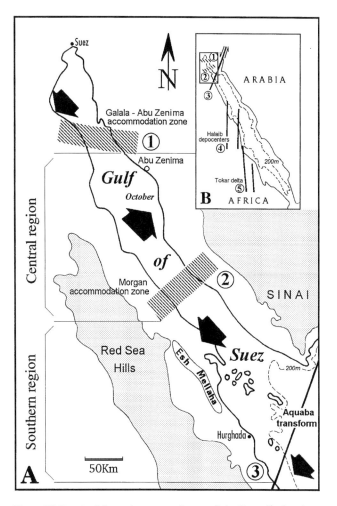

Figure F2.3 A. Schematic structural map of the Suez rift showing three asymmetrical provinces separated by two accommodation zones, after Colletta et al. (1988) and Bosworth (1994). Arrows: dips of the major tilted blocks. B. The Red Sea also is segmented, along major Precambrian shear zones which have acted as major conduits for the sediments into the basin (depocentres of Halaib and Tokar). After Mitchell et al. (1992).

(Colletta et al., 1988; Richardson and Arthur, 1988). Seismic profiles indicate few faults and virtually no mobilization of the well developed middle–(?)late Miocene evaporites. North-west–south-east oriented Miocene blocks, however, are well defined and these tend to separate the region into two major north-east and south-west sub-basins. The evaporite and post-evaporite cover, although thinning over these blocks, suggests their progressive burial and does not exhibit the repeated changes in thickness normally associated with rejuvenated fault movements.

The relatively thick (1000 m) post-Miocene fill, composed essentially of siliciclastics with local evaporitic intercalation (Fawz and Abdel Aab, 1984), is referred to the Pliocene. Seismic profiles show few or poorly defined sequences, being characterized by regular, parallel reflectors suggestive of low energy, shale and sand alternations. Well logs indicate only rare carbonate units. The central Gulf of Suez is characterized by both structural and stratigraphic simplicity since the Miocene.

The southern Gulf of Suez

Examination of outcrops along the south-west coast of the Gulf of Suez and the offshore islands of Gubal, Tawila, Shadwan and Giftun Kebir (Figure F2.1) has been complemented by the interpretation of seismic data (thanks to the Compagnie Générale de Géophysique) and a new high resolution seismic survey, between the islands of Shadwan and Giftun Kebir (Gasperini et al., 1994). Previous studies have indicated considerable post-Miocene (Zeit Formation) tectonic activity and subsidence (Moretti and Colletta, 1987; Richardson and Arthur, 1988). This instability has been confirmed by the present authors. These Pliocene and Quaternary movements are possibly related to the adjacent Aqaba transform (Garfunkel et al., 1981) (Figure F2.3(a)) and the region is characterized by the presence of numerous north-west–south-east oriented domal and anticlinal structures visible both on seismic profiles and at outcrop. These structures reflect both rejuvenation of the pre-evaporite fault blocks, and remobilization of the evaporites, the two deformational mechanisms often affecting a given block. In addition to these local structural movements which are visible on north-east–south-west sections (i.e. transverse with respect to the rift axis), an important regional tectonic change occurs as one approaches the Aqaba transform as seen on seismic lines parallel to the axis of the rift, as well as in the modern bathymetry of the southern Gulf of Suez.

Structural features within the sedimentary cover

Both industrial and high resolution seismic profiles record various degrees of halokinetic movement (Gasperini et al., 1994; Rioual et al., 1996) of South Gharib and Zeit evaporites. Although not clear on seismic profiles, these movements appear to be related genetically to the rejuvenation of pre-evaporite blocks. This relationship is visible at Gubal Island where outcropping diapiric gypsum cutting bedded dolomites, overlies a Precambrian high, demonstrated by wells (Figure F2.4(b)). The post-evaporite Gubal structure has a moderately developed north-west–south-east axis. This typical Miocene (Clysmic) trend is also confirmed by the geometry of the diapirs and evaporite ridges and walls deduced from seismic profiles (Figure F2.5). A similar north-west–south-east polarity has been demonstrated by Mart and Ross (1987, fig. 3) within the northern part

'Pliocene' sedimentation and tectonics 431

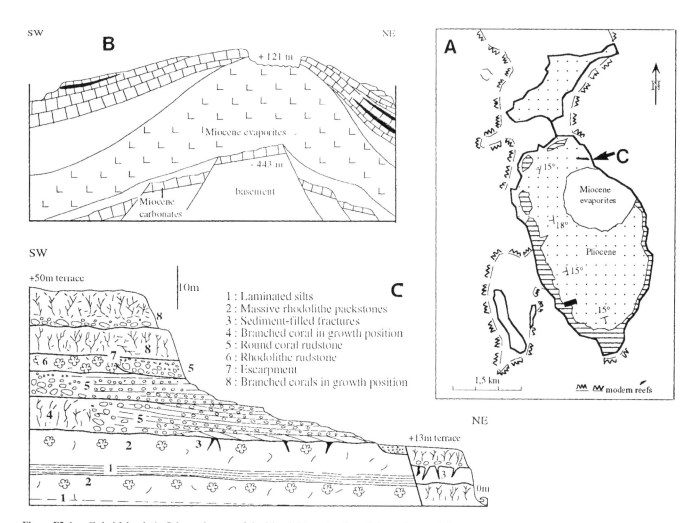

Figure F2.4 Gubal Island. A. Schematic map of the island. Note the dips of the main post-Miocene carbonate beds (dotted) showing the domal structure of the island, related to halokinetic movements. B. Schematic sketch (without scale) showing the consequence of the rejuvenation of a fault-block leading to the triggering of salt movement (gypsum at outcrop), which induce numerous discontinuities in the Pliocene series, on the flanks of the dome (after outcrop, seismic line and a well). C. Detail of B, showing the numerous synsedimentary discontinuities and reef facies. (See A for location of the section.)

of the Red Sea where certain evaporite walls, 4 km in width, have been traced laterally for 30 km.

The north-west–south-east orientation of other islands including Ashrafi, Shadwan and Giftun, as well as the onshore structure of Gebel el Zeit and Ras Gemsa, is parallel to the rift axis. These structures all include moderate dipping (5°–20°) post-Miocene sediments. Furthermore, each structure is clearly asymmetric, having a steep north-east, presumably footwall, escarpment, relating to adjacent north-west–south-east oriented faults. Hangingwall dip slopes are invariably inclined towards the south-west, in common with those of older, early–mid Miocene blocks, confirming that post-Miocene tilting, accentuated and otherwise modified by evaporite remobilization, is genetically related to fault rejuvenation (Figure F2.6).

In addition to north-west–south-east oriented post-Miocene structures, there exist a number of structures lacking clear polarity, some of which may be related to the intersection of major north-west–south-east and secondary north-east–south-west or north–south cross-faults. This would seem to be the case of the south-eastern extremity of Abu Shaar where the 'Pliocene' strata are strongly deformed. Other surface structures in the vicinity and south of Hurgada are highly faulted domes suggestive of near surface diapirism. At Shadwan island, diapiric evaporites coincide with the presence of the north-east–south-west oriented faults, this being particularly well exposed near the northern end of Shadwan (Figure F2.6).

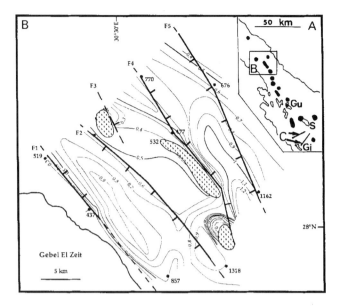
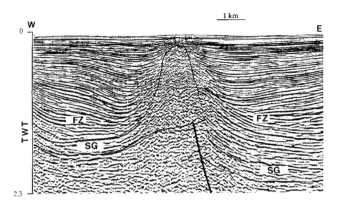

Figure F2.5 A. Location of diapiric structures in the southern Gulf of Suez (interpretation of seismic survey lines). B. Detail: isopach map of the post-Zeit formation (Pliocene and Quaternary) showing that the diapir structures are associated with the main faults. Diapiric structures, and faults (F) N120 oriented, delimit depocentres. Equidistances of curves: 0.1 s; (1318) = depth of the South Gharib Fm in the wells; Gi–Giftun Island; Gu–Gubal Island; S–Shadwan Island. C. Location of the high resolution seismic survey lines (Gasperini et al., 1993). (See Figure F2.7 and F2.9.)

Figure F2.6 Diapiric structure triggered by the rejuvenation of the block. SG: South Gharib Formation; FZ: Zeit Formation (southern Gulf of Suez).

Stratigraphic and sedimentary expression of local syn-'Pliocene' deformation

Structural movements during post-Miocene sedimentation are evident both on seismic profiles in the southern area of the Gulf of Suez and at outcrop, on the islands of Giftun, Gubal, Shadwan and on the western coast of the Gulf.

Seismic profiles, notably those traversing the various halokinetic structures, exhibit numerous stratal discontinuities and thickness variations which record the modifications of sea-floor morphology induced by the subjacent halokinesis (Figure F2.7). Local sequences with onlapping, radially dipping beds, record periods of structural and bathymetric elevation. Other sequences, sometimes better expressed in the upper half of the evaporitic cover, downlap on to centripetally inclined beds, recording periods of evaporite withdrawal or dissolution. Some diapirs are associated with a vertical change in seismic facies. The lower parts of the intruded cover are characterized by a regular velocity of flat, parallel reflectors, suggestive of quiet water, possibly argillaceous sedimentation. Near the top of the sedimentary cover, they are replaced by a massive seismo-facies with vague reflectors indicating fairly steep radial dips. These upper facies are probably massively bedded carbonates whose dips are typical of periplatform slopes. The culmination of the platform may coincide with a modern bathymetric high populated with living reefs (Figure F2.8).

High resolution seismic profiles (Figure F2.9) between the islands of Shadwan and Giftun Kebir show

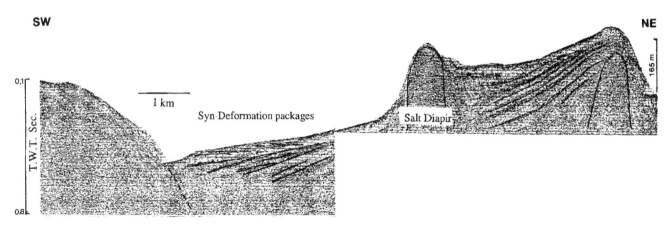

Figure F2.7 High resolution seismic line. The diapir (salt-wall) separates two minor basins, having numerous discontinuities. For location see Figure F2.5C.

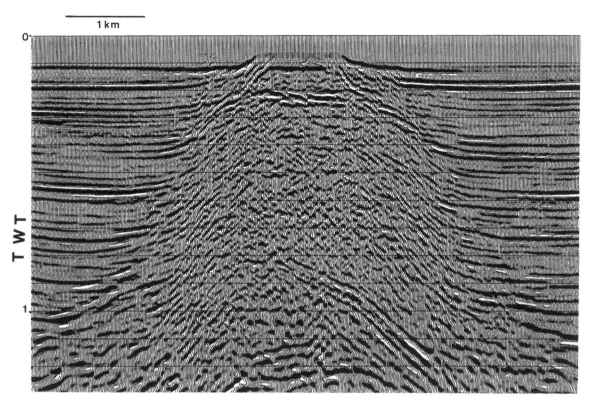

Figure F2.8 Close-up of part of a seismic section showing a diapir with the development of carbonate platform with radially oriented slopes situated near the culminations.

numerous discontinuities and other modifications within the near-surface (?Quaternary) sediment. These north-east–south-west profiles traverse major north-west–south-east oriented structural depressions, the flanks of which exhibit numerous slump structures, erosional truncations and olistoliths, all typical of a relatively recent instability. This is particularly clear along the escarpment flanking the north-eastern coast of Giftun Kebir where the sub-recent sedimentary cover is stacked, almost directly, against the subvertical submarine escarpment, there being virtually no slope deposits relating to the scarp (Figure F2.9).

Outcrops: in contrast to the lower and mid-Miocene series (Rudeis–Kareem Formations) the post-Miocene sediments are invariably of shallow-marine origin. Fluctuation of relative sea level is recorded by multiple, stacked reefs typical of the sedimentary cover situated near the culmination of most Pliocene structures, including diapirs at Gubal and Shadwan islands. Alternating with shallow subtidal sediments and other facies rich in birdseye structures, they are particularly well developed along the eastern side of Giftun Kebir where a total of 13 reefs and beaches occur within a 100 m section.

Periodic lowering of relative sea level is also recorded by spectacular cliffs which truncate reefs, notably along the north-eastern side of Gubal island (Figure F2.4(c)). At Gubal, an 8 m high escarpment is flanked by coarse reef-derived conglomerate which grades rapidly into parallel-bedded gravels and sands with gentle (3°) sedimentary dips inclined towards the periphery (north-east) of the structure. This well-sorted carbonate detritus has probably been deposited in beach and proximal slope environments.

The existence of bathymetric slopes surrounding these structures is recorded by rapid changes in thickness both of the shallow-marine carbonates and the locally developed marly horizons, clearly visible on both the north-east and south-west flanks of the Gubal high and on the northern slopes of Shadwan. It is also demonstrated by the geometry and bedding of certain shallow-marine facies, notably oolitic sands and associated reefs, both at Gubal and Giftun Kebir. On both islands, the 5–10 m thick, white oolitic limestones are concentrated around the south-eastern flanks of the high, the spectacular foreset bedding prograding mainly towards the south. This polarity may reflect the influence of wind-driven surface currents. At Giftun Kebir, these oolitic and other sands downlap on to parallel bedded carbonates suggesting shallow slope conditions (Figure F2.10(b)).

Gently inclined bathymetric slopes are confirmed by the lateral changes in these shallow platform sediments, notably along the south-western side of Giftun Kebir (Figure F2.10). A 20 m thick oolitic unit located near

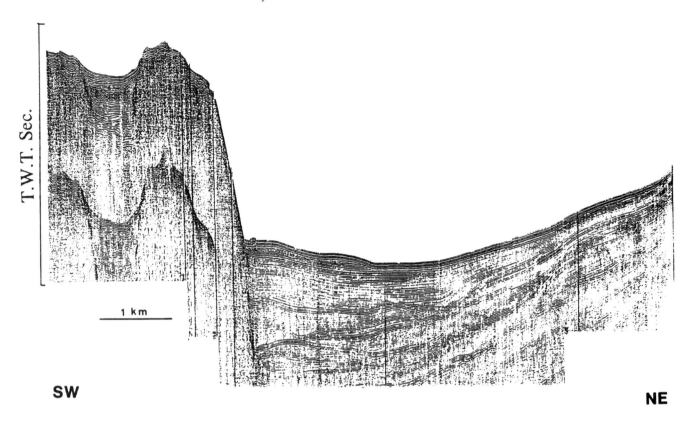

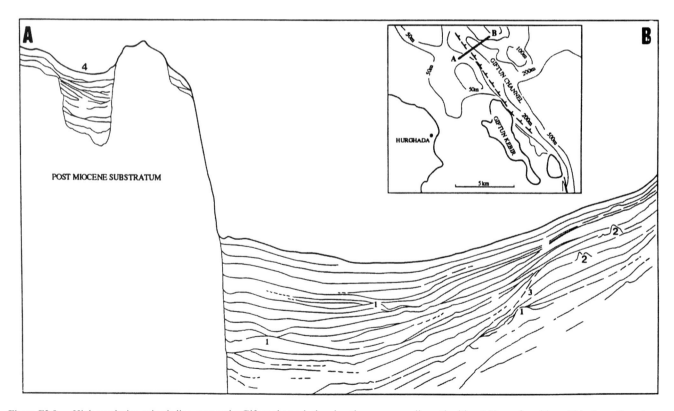

Figure F2.9 High resolution seismic line, across the Giftun channel, showing the numerous discontinuities. 1 Unconformities within the sedimentary succession; 2 – olistoliths (?); 3 – erosional truncations; 4 – channel.

'Pliocene' sedimentation and tectonics 435

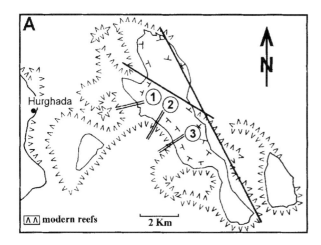

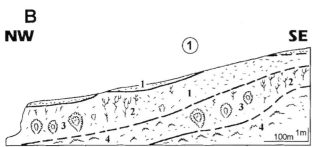

1: Coral and algal debris, muddy matrix (back-reef)
2: Branched corals in growth position (inner-reef)
3: Massive Porites in growth position (outer-reef)
4: Encrusting red algae (reef-front)

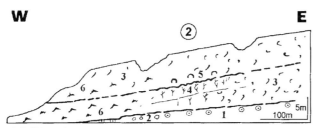

1: Oolitic grainstone
2: Conglomerate (oolitic Lst.)
3: Molluscan pack/wackestones (back-reef)
4: Branched corals (reef)
5: Massive corals (reef)
6: Red algal framestone (reef-front)

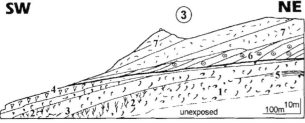

1: Molluscan pack/wackestones (back-reef)
2: Branched corals (reef)
3: Red algae
4: Pinna biostrome
5: Laminated sand with birdseyes
6: Downlapping oolitic sands
7: Massive molluscan pack/wackestone

Figure F2.10 Giftun Island. A. Map showing the divergent dips (15° to 20°) of the carbonate beds indicating the domal structure of the island. B. Sections across the western flanks of Giftun Kebir Island, showing the numerous discontinuities and lateral changes in facies, indicating repeated movements of relative sea level. (1) North-western flank (probably Quaternary), covered with multiple reefs; (2) western flank upper part of Plio-Quaternary series; (3) south-western flank middle part of series showing peri-platform facies and downlapping oolitic sands.

the culmination of the structure grades laterally into biogenic facies including small coral reefs and a spectacular lens of large *Pinna* lamellibranchs.

Instability is recorded by the development of local sedimentary sequences at Gubal where each of four 10 m thick sequences grades upwards from laminated marl and rippled sands through massively bioturbated lime muds culminating either in massively bedded carbonates rich in corals, red algae and pectinids, or in a cross-bedded oolite. The basal marly facies changes rapidly in thickness, suggesting bathymetric relief. Instability is recorded also by characteristic deformation of the sequence top; unlithified carbonate has been slightly remobilized, creating small (10 cm), asymmetric flow-structures (Figure F2.11(a),(c)). This instability is confirmed by larger-scale deformation of oolitic sands whose foreset bedding has slumped (Figure F2.11(d)). These deformations are particularly frequent within the post-evaporite sediments on Gubal Island and probably record repeated seismic shocks relating to fault movements in the basement. Synsedimentary structural movements are recorded in the post-Miocene cover both by the presence of normal faults sealed by shallow-marine sediments, and by the presence of well-defined sediment-filled fractures (Figure F2.11(e)), truncated by 'Pliocene' erosion surfaces.

Three post-Miocene structures in the vicinity of Hurghada, whose domal form and multidirectional dip pattern are suggestive of underlying diapirism (Figure F2.12), have a 'Pilocene' cover composed essentially of shallow-marine siliciclastics. In common with their calcareous equivalents on the offshore islands, these siliciclastics also record considerable instability of relative sea level. A section measured immediately to the north-west of the Sheraton Hotel comprises multiple 5–10 m thick mixed siliciclastic–carbonate sequences, each generally characterized by bioturbated or cross-bedded clastics in the lower parts and laminated sands or muds with birdseye structures at the top, the sequence being limited by a sharply defined discontinuity overlain by lithoclasts. Instability is confirmed by several structures (folds, dislocations, neptunian dikes), typical of seismic instability (Plaziat and Purser, this volume).

On a larger scale, two domal structures, Gebel Dishshit (10 km south of Safaga) and Sharm el Arab

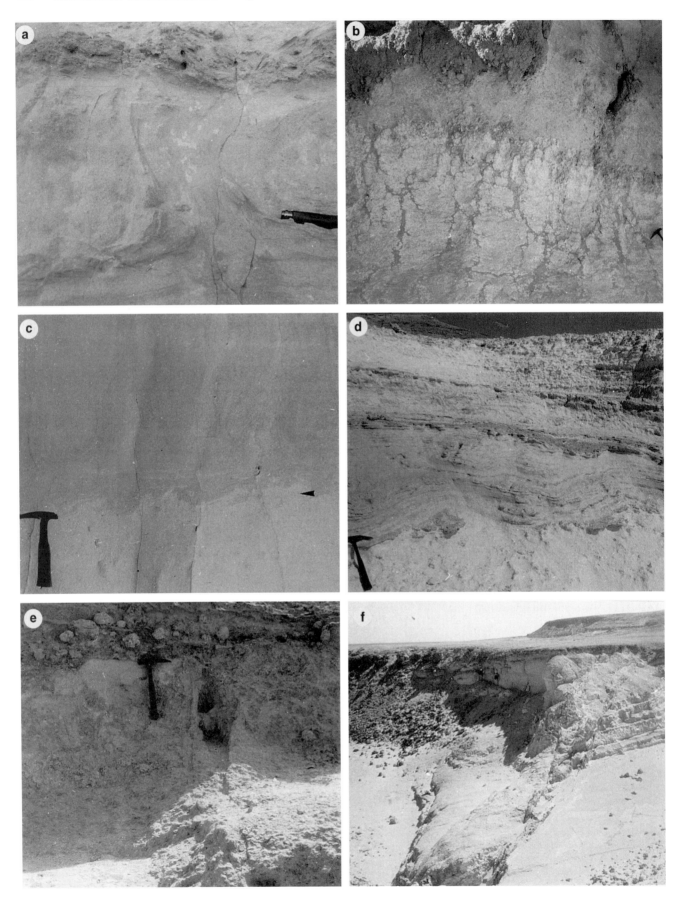

'Pliocene' sedimentation and tectonics 437

Figure F2.11 Tectonic and sedimentary deformations affecting the Pliocene carbonates on Gubal Island. (a) Fluidification structure in oolitic sands; (b) synsedimentary fissuration of rhodophyte reef directly above a diapir; (c) flow structure (arrow) at upper limit of minor carbonate sequence; (d) deformation affecting top of massive carbonates and base of overlying laminites; (e) synsedimentary fracture filled by sediments and sealed by rhodolite rudstones; (f) north-west–south-east trending fault on the north-eastern side of Gubal Island sealed by Pleistocene terrace.

(35 km south of Hurghada) expose post-Miocene sections which include large blocks (1–3 m) of 'Pliocene' sandstone and plurimetric submarine escarpments recording major gravitational collapse. At both structures the sequence also includes large-scale (20–30 m high) clinoform beds composed of siliciclastic sands and gravels associated laterally with reefal carbonate platforms, these relationships being typical of shallow-marine fans (Figure F2.13).

Regional tectonics and their sedimentary expressions
As already noted, there exists an important difference between the post-Miocene sections exposed on the islands and those examined on the nearby continental shoreline; while the former (islands) are totally carbonate, the latter are composed essentially of relatively coarse siliciclastics, including local conglomerate levels composed of rounded Precambrian basement whose source almost certainly is the nearby (20 km) western periphery of the rift (Plate 25(b)).

This major north-east–south-west variation is best explained in terms of relative proximity to these peripheral detrital sources. However, the distance between the purely calcareous series on Giftun Kebir island and the predominantly terrigenous sequence at the Sheraton (Hurghada) section is only 7 km. While these sections may not be strictly contemporaneous, both are relatively thick (150 and 200 m) and probably overlap in time. It is unlikely that this very rapid transition occurred on a shallow, horizontal sea floor and probably is the result of a 'Pliocene' depression between the continental shoreline and the Giftun Kebir high. This depression trapped continental detritus facilitating the development of carbonates on the offshore bathymetric high whose positive relief is confirmed by numerous reefs and beach deposits noted above.

Further to the north, in the vicinity of Gebel el Zeit, mapping by the authors has indicated the presence of post-Miocene evaporites (Figure F2.14). The south-west

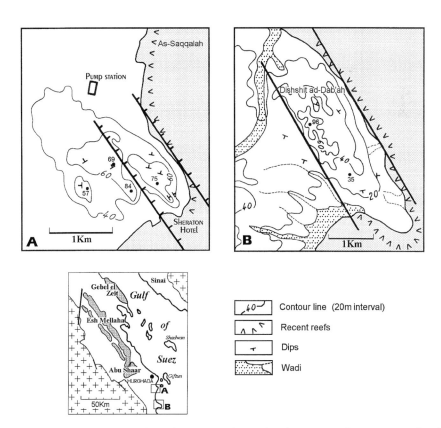

Figure F2.12 Structures south of Hurghada. Morphology of two post-Miocene domal structures (dips: 5° to 20°) of carbonate and siliciclastic beds. A. As Saqqalah area; B. Dishshit Dabah area.

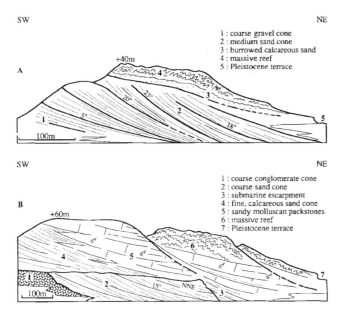

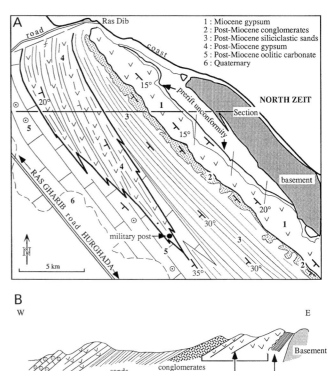

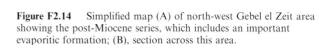

Figure F2.13 Simplified sections across the flanks of the domal structures south of Hurghada. A. Gebel Dishishit (location Figure F2.12) showing two successive cones on which developed a Pliocene or early Quaternary reef. B. Sharm el Arab (35 km south of Hurghada), showing multiple cones and related carbonates. Note that the style of sedimentation is different from that of the nearby Giftun island.

Figure F2.14 Simplified map (A) of north-west Gebel el Zeit area showing the post-Miocene series, which includes an important evaporitic formation; (B), section across this area.

dip slopes of the north Zeit structure expose a thick section of middle (?) to late Miocene evaporites composed of massive sulphate beds which dip westwards at angles of 15°–20°. This structural dip also affects the post-evaporite conglomerates as well as the succeeding terrigenous sands and gravels, these latter attaining a thickness of about 150 m. In the southern parts of the area these siliciclastics pass upwards into oolitic carbonates with scattered corals and oysters while to the north, in the vicinity of Wadi Dara, they are overlain by laminated gypsum whose bedding and teepee structures differ strongly from the most massive marine Miocene evaporites to the east. All clastics, carbonates and evaporites exhibit a constant south-west dip, similar to that of the underlying Miocene series, confirming an important post-Miocene tilting of the Zeit block. These relationships also indicate the existence of post-Miocene evaporitic conditions within a post-Miocene structural depression.

Southern Gulf of Suez seismic profiles parallel to the axis of the Gulf of Suez indicate major changes within the post-Miocene series as one approaches the northern Red Sea (Figure F2.15). In the axial parts of the Gulf, between north-eastern Zeit and Gubal highs, the post-evaporite fill, affected locally by a series of diapiric intrusions, has a relatively simple structure with moderate dips. Seismic facies typically comprise closely spaced, flat reflectors which are generally continuous for several kilometres, typical of well-bedded, calm water deposition. There are few major discontinuities, apart from those close to diapiric structures.

In the zone situated between Gubal and Shadwan islands respectively, there is a progressive change from the regular seismic facies noted above, and a more hummocky facies with irregular, discontinuous reflectors, typical of carbonates; also lenticular sand bodies and small reefs occur. This facies coincides with the presence of multiple discontinuities locally associated with large clinoforms prograding towards the south or south-east. This zone is limited to the south-east by a sudden change in both structural and sedimentary style coincident with a north–south line extending across the Gulf of Suez from the north-western extremity of Shadwan island. This line coincides with a clearly defined bathymetric scarp visible both to the north and the south of Shadwan (Figure F2.16). To the west of this scarp, modern water-depths are about 75 m. To the east, on the contrary, the bathymetry descends rapidly to about 600 m forming the large, moderately deep Shadwan Embayment situated to the north-west of the Aqaba transform, i.e. within the southernmost Gulf of Suez.

The escarpment limiting Shadwan Embayment to the north coincides with a system of north–south faults with downthrows to the south-east. These faults appear to

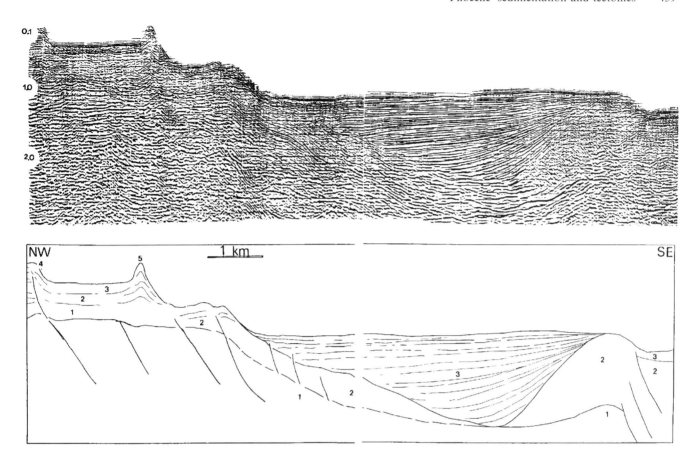

Figure F2.15 North-west–south-east seismic profile across the southern part of the Gulf of Suez showing the post-Miocene collapse of the platform margin. 1 – Top Miocene evaporites; 2 – post-evaporite slope deposits; 3 – post-collapse secondary basin filling; 4 – edge of Suez platform; 5 – reef pinnacle.

bottom-out on the top of the Miocene evaporites. Thus, above the top evaporite reflector, the post-Miocene cover is strongly deformed being cut by numerous normal faults. The present bathymetry is also irregular and comprises several, clearly defined pinnacles whose radially inclined reflectors are suggestive of reef talus. Located close to the north–south escarpment limiting the Suez Platform, these pinnacles may have developed by rapid vertical reef growth during the collapse and subsidence of the substratum. The collapse zone exhibits an irregular, discontinuous seismic facies comparable to that of the peripheral zone of the Suez Platform.

The Shadwan Embayment coincides with the presence of a post-evaporite secondary basin limited to the north-west by the structural slope leading up to the edge of the Suez Platform, and to the south-east by a well-defined block. This latter, flanked by steep escarpments, is underlain by the reflector marking the top of the evaporites which is not cut by faults limiting the block, i.e. the faults, in common with those limiting the Suez platform, bottom-out on top of the Miocene evaporites.

The structural and stratigraphic relationships show that the southern parts of the Gulf of Suez have been affected by a major gravity collapse and lateral sliding of 'Pliocene' blocks on a Miocene evaporite sole. Their displacement and rotation have created secondary basins, expressed in the modern bathymetry of the Shadwan Embayment. The secondary basin is characterized by closely spaced, flat, continuous reflectors suggestive of well-bedded sediments. A systematic decrease in dip and onlapping relationships of each reflector, indicates sustained rotation of the limiting (south-east) block. These gravity features have occurred relatively close to (35 km) but, nevertheless, north of the Aqaba transform zone; the superficial faults are not the direct expression of the plate boundary. Moreover, the fault escarpment limiting the Suez Platform, oriented close to north–south, is not parallel to the north-north-east–south-south-west Aqaba transform. This north–south trend is a frequent cross-fault direction in the Gulf of Suez and northern Red Sea (Montenat et al., this volume). These large-scale gravity displacements, lubricated by the evaporite substratum, were probably facilitated by accommodation space created by the extension of the northern Red Sea, and triggered by repeated seismic events relating to transform movements.

440   Post-Miocene sedimentation and rift dynamics

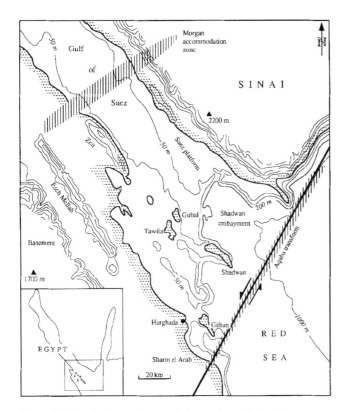

Figure F2.16 Bathymetric map of the southern Gulf of Suez showing the abrupt passage from 80 to 1000 m within the southern Gulf of Suez. Note that the boundary with the Red Sea is not the N20 Aqaba transform, but the north–south, 200 m bathymetric curve.

Although seismic coverage does not extend to the Aqaba transform, i.e. northern Red Sea, bathymetry suggests that the vertical component of this deformation is considerably less than that produced by the gravity collapse to the north. In other words, both structurally and morphologically, the Aqaba transform fault does not strictly mark the limits between the Red Sea and the Gulf of Suez, as the topographic boundary is located some 35 km to the north-west.

The north-western Red Sea

The post-Miocene series are well-exposed in most major wadis between Safaga and Mersa Alam (300 km to the south) and have been examined by many workers including Akaad and Dardir (1966), Philobbos and El Haddad (1983), Thiriet (1987), Philobbos et al. (1989), Soliman et al. (1993) and others. The lithostratigraphic schemes of some of these studies are outlined in Figure F2.2. In general, the Miocene evaporites are conformably overlain by restricted, shallow-marine siliciclastics, there being no perceptible tectonic break, at least within these peripheral parts of the rift. In some areas, such as Safaga (Thiriet, 1987) and Wadi Samadai (Philobbos et al., 1989), Miocene evaporites are overlain by continental sediments including fluviatile channels and palaeosols with roots. The overall evolution from fluviatile deposits or essentially terrigenous, restricted marine sediments (Mersa Alam Formation) through open-marine mixed carbonate–siliciclastic deposits (Shagra Formation) to coarse conglomerates (Samadai Formation), is fairly widespread. However, this transition is locally modified by structural movements relating both to east–west cross-faulting (Philobbos et al., 1993) and to movements of the underlying Miocene evaporites.

The marine 'Pliocene' deposits at all localities examined are composed of very shallow marine facies, often with birdseye structures and beach-type bedding, alternating with palaeosols, reflecting repeated emersions (Figure F2.17). These sediments also include numerous deformation structures such as fluidification features, folding, sliding and stratiform brecciation, thought to record seismic instability (Plaziat et al., 1990; Plaziat and Ahmed, 1994; Plaziat and Purser, this volume).

Local tectono-sedimentary features

Aerial photographs and detailed fieldwork reveal that 'Pilocene' strata (200–300 m in thickness) form a highly irregular coastal zone 1–2 km in width. Unlike their equivalents in the Gulf of Suez, these strata are affected by a general structural dip of 10°–20° towards the axis of the rift. However, these moderate dips are often modified locally by varied structural features detailed below. Tectonics generally occurred during, and influenced, post-Miocene sedimentation. The following features have been mapped.

a. Wadi Abu Dabbab (98 km south of Quseir)
A very moderate flexure affects the Shagra Formation and is overlain by a local buildup of reefal carbonates with some siliciclastics, the lenticular reef and peripheral carbonates measuring about 1–5 km in width and 100 m in thickness (Figure F2.18). This sedimentary buildup has locally modified the regional north-east structural dip (preserved below).

b. Ras Shagra (20 km north of Mersa Alam)
This local structural feature is intersected by the present shoreline (Figure F2.19) and appears to be a faulted dome whose landward flanks dip at 10°–30° to the west and north-west. The western flank, forming spectacular coastal cliffs, comprises three lensoid reef bodies and associated rhodolite facies. Intercalated sands with birdseye structures and laminated ilmenite concentrations indicate repeated beach deposition. These littoral carbonates and fine siliciclastics pass laterally (to the west) into coarse conglomerates with local buildups of larger oysters. Their relationships suggest local carbon-

'Pliocene' sedimentation and tectonics 441

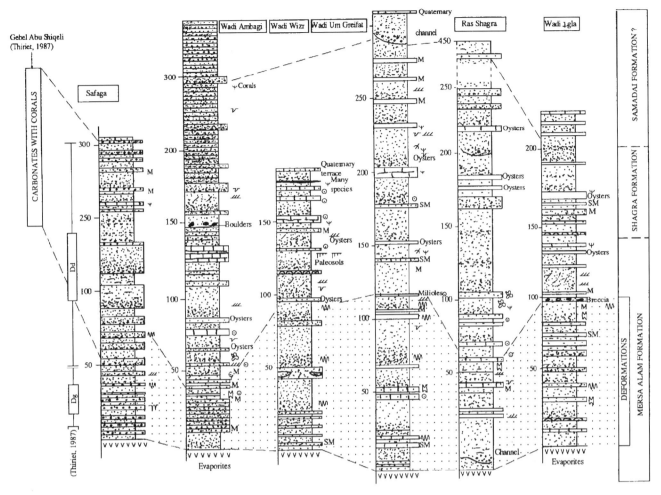

Figure F2.17 Lateral lithological variations within the Pliocene sediments of the Red Sea. There is a clear change between the bottom part, with many synsedimentary deformations, and the upper part, characterized by a more marine sedimentation. For location of the sections, see Figure F2.1. Y: corals; M: molluscs; sm: small molluscs (restricted environment)

ate buildups on the culmination of a structural dome, the coarse clastics being localized in the structural lows.

c. Wadi Igla (12 km north of Mersa Alam) (Figure F2.20)

The shoreline 1 km south of Wadi Igla is intersected by a clearly defined, east–west trending anticlinal ridge affecting the post-Miocene series. The major wadis coincide with open synclines flanking the ridge, both Wadi Igla and an unnamed wadi, 2 km to the south, terminating in deep littoral embayments or sharms. The stratigraphic sequence exposed in both wadis is composed essentially of dolomites containing rare molluscs and oolites of the basal (Samh) member of the Mersa Alam Formation and appears to be laterally constant. However, following a spectacular synsedimentary deformation (Plaziat and Ahmed, 1994) the succeeding open-marine facies exhibits marked lateral changes. At Wadi Igla, situated within the depression, sediments of the Shagra Formation consist mainly of stacked conglomerate fans separated by thin (3–5 m) reefal bodies. A similar section, composed of massively bedded conglomerates, occurs in the wadi located within the next depression some 2 km to the south. On the intervening structural high, the equivalent section is composed essentially of reefal carbonates; conglomerates are rare. In common with the previous localities, the syn-Pliocene formation of an east–west structural ridge has influenced surface drainage and shallow marine funnelling of coarse siliciclastic sediments along the oblique depressions (with respect to the rift axis), the intervening structural high favouring local carbonate buildups.

d. Wadi Um Greifat (65 km north of Mersa Alam) (Figure F2.21)

The regional (north-east) dip affecting the post-Miocene strata along most of the north-western Red

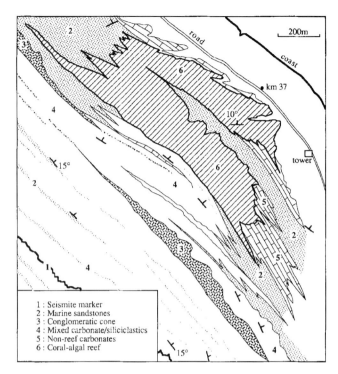

Figure F2.18 Map of Wadi Dabbab area, showing the development of a lenticular reef within an essentially siliciclastic shallow marine regime.

Sea coast, is modified by a complicated structural pattern involving domal and elongate anticlines limited by coastal cliffs presumably related to a major, clysmically oriented fault parallel to the shoreline. This structure is cut by a north-east–south-west fault separating a simpler high axis to the south-east from a more complex dome and syncline northern structure. Such structural features suggest underlying diapirism, possibly related to the intersection of major basement faults. All structures affect the Pliocene strata but not the Quaternary terraces. The sediment geometries suggest that the structures were active during Pliocene times.

A high east–west cliff cutting across the southern anticline exposes three superimposed reefs, each limited by an erosion surface below which cavities associated with dissolution of corals indicate periodic emergence. The discontinuities are offlapped by siliciclastic sands with spectacular clinoform bedding. These sand bodies, which do not completely bury the reefs, are localized around both the east and west flanks of the reefal buildups indicating terrigenous sedimentation around the shallow marine flanks of an emergent carbonate high.

e. Ras Dira (27 km south of Mersa Alam)

Quaternary terraces and alluvial cones form a narrow (5 km) coastal plain (Plate 25(a), (b)) which is affected locally by a series of small conical hills composed of deformed Miocene gypsum. Individual evaporite hills are arranged along a N120 line suggesting structural control. The highest (71 m) as well as the smaller metric reliefs pierced the 'Pliocene' strata and induced concentric deformation below Quaternary sediments, confirming the diapirism.

North-western Red Sea tectonics and its influence on sedimentation

These examples demonstrate the local influence of structural movements probably relating to both rejuvenation of underlying north-west–south-east Miocene blocks as well as to upward mobilization of the Miocene evaporites. That these post-Miocene structures are relatively common is indicated by the numerous lateral variations within these 'Pliocene' strata (Figure F2.18). The lower parts of these series (Mersa Alam Formation)

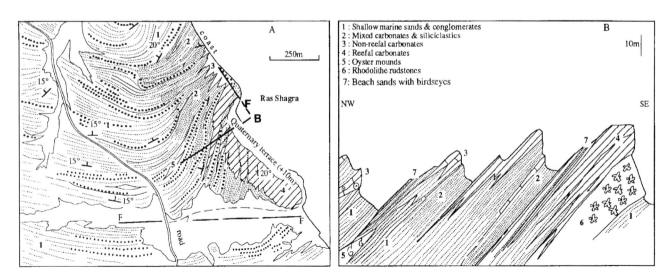

Figure F2.19 Ras Shagra area. A. Map showing the relationships between shallow marine siliciclastics (sands and conglomerates) and reefal carbonates (F = fault). B. cross-section.

'Pliocene' sedimentation and tectonics 443

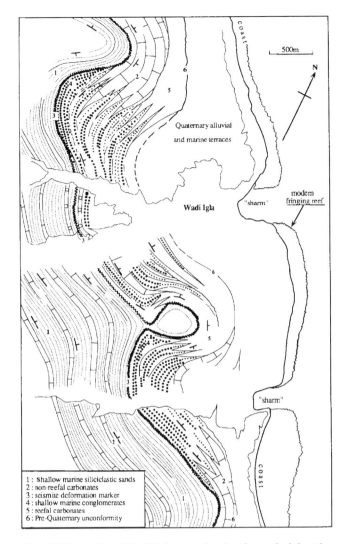

Figure F2.20 Map of Wadi Igla area, showing the marked domal structure which has influenced the facies distribution following a seismite deformation event.

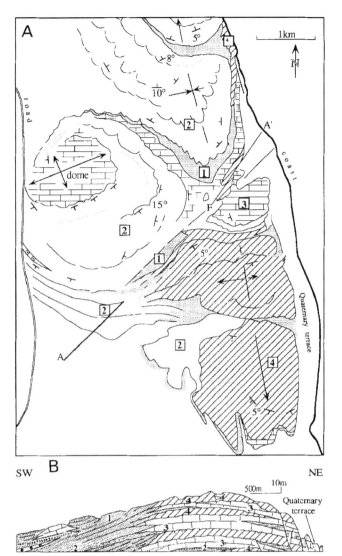

Figure F2.21 A. Map of Wadi Um Greifat area whose structural style is suggestive of evaporite doming. The lenticular shallow marine siliciclastics are stacked against the reef high. AA′ : Cross-section. B. Section across Um Greifat showing the relationships between Pliocene siliciclastic and carbonate sedimentation and the development of an anticlinal structure. 1 – marine siliciclastic sands; 2 – mixed carbonates and siliciclastics; 3 – non-reefal carbonates; 4 – reefal carbonates.

generally comprise fine sands, reddish and greenish mudstones and occasional limestones with dwarf (restricted marine) lamellibranchs. However, as one proceeds upwards, sediments become more variable; the open-marine facies of the upper Mersa Alam Formation (Gabir Member) and, especially, the Shagra Formation may be dominated by terrigenous gravels and conglomerates (as at Wadi Um Gheigh) or be composed mainly of shallow marine carbonates rich in red algae and corals (as at Wadi Wizr). Finally, the uppermost conglomeratic Samadai Formation, probably Pleistocene in age, is generally separated from the underlying Shagra Formation by an angular discordance (El Haddad et al., 1993). However, these coarse sediments are not always present and open-marine carbonates may occupy much of the 'Pliocene'. The five localities previously examined in detail suggest that the major sedimentary variations are related to cross-trending structural features which traverse the predominant north-west–south-east polarity of the rift.

The regularly bedded, fine-grained, restricted marine facies are terminated abruptly by a highly deformed stratiform unit often affecting 10 m of sediments. Interpreted as the result of a major 'Pliocene' seismic event, these deformations involve fluidification of unconsolidated sands, sliding and associated or brecciation of more hardened layers. These seismites record movements along a buried fault parallel to the rift axis, as they may be followed along the coastal zone for at least 80 km between Wadi Wizr and Wadi Igla. This episode of deformation coincides with a marked change in facies, separating restricted marine clastics below from open-marine carbonates and siliciclastics above.

In addition to tectonically induced discontinuities within the 'Pliocene' series, a major angular discordance separates the Samadai and Shagra Formations from the subhorizontal Pleistocene and younger terraces discussed by Plaziat et al. (this volume). This unconformity is especially evident where horizontal Quaternary reef terraces overlie relatively steeply inclined 'Pliocene' such as Um Greifat or Ras Shagra. However, this discordance is also present along most of the north-western Red Sea coast where the post-Miocene sediments generally dip (5°–20°) towards the axis of the rift, which differs from the complex Pliocene setting of the Gulf of Suez.

DISCUSSION AND CONCLUSIONS

This review of 'Pliocene' sedimentation concerns an extensive region (400 km, from Ras Gharib in the north to Mersa Alam in the south) including two significant segments of the rift system, the Gulf of Suez and the north-western Red Sea. It treats a stratigraphic interval – the 'Pliocene' – with considerable importance in terms of rift dynamics for it coincides with the main activity of the Aqaba transform (Bayer et al., 1988; Lyberis, 1988) and the southern oceanic separation of the African and Arabian plates. Within the Red Sea domain, the 'Pliocene' therefore encompasses the transition from rift to post-rift (drift) geodynamics. A comparison of structural and sedimentary variations along this regional north-west–south-east transect may be summarized as follow.

1. The Miocene substratum

Important syn-rift movement during the early and middle Miocene created a system of tilted blocks whose main axes were oriented either parallel or oblique to the north-west–south-east rift axis. These movements, often of considerable amplitude (>1000 m), affected both the north-western Red Sea and the Gulf of Suez. They are recorded by a relatively constant stratigraphy (Plaziat et al., this volume) which varies depending more on the position of each block with respect to the rift periphery than to its situation within the Red Sea or Suez segments (Purser et al., this volume).

2. Major differences between Miocene and post-Miocene tectonics and sedimentation

Because of the considerable early and mid-Miocene morpho-structural relief, syn-rift sediments are characterized by important changes in facies and thickness between the structural highs and adjacent lows which were partially filled with up to 2000 m of fine, often pelagic clays. Post-Miocene structural movements, although frequent, have generally concerned reactivation of Miocene faults resulting in rotation of the Miocene substratum. These movements not only influenced the location and thickening of carbonate and siliciclastic sedimentary units within the 'Pliocene' cover, but also explain their thickness between fault blocks and diapirs.

The comparison of facies and thickness has important implications in terms of subsidence rates and rift geodynamics. Admittedly difficult to generalize, the thickness of Miocene sediments is often in the order of 100–200 m on the structural culminations and, as already noted, may exceed 2000 m in adjacent lows (Richardson and Arthur, 1988; Bosworth, 1995). Post-Miocene sediments both in the Gulf of Suez and in the north-western Red Sea, at outcrops, average about 200–400 m. However, on offshore seismic profiles and wells, they attain 1500 m (Miller and Barakat, 1988). The time interval for Miocene sedimentation is about 18 Ma, that for the Pliocene, approximately 3 Ma. It would seem therefore that the rates of post-Miocene sedimentation are generally greater than those of the Miocene (Purser and Philobbos, 1993). In terms of subsidence, consideration must be given to the facies concerned. The thickest Miocene strata, developed in structural lows within the Gulf of Suez, are middle-Miocene Rudeis–Kareem Formations including marls whose pelagic character indicates relatively deep environments before and during their deposition. Thus, these facies do not necessarily reflect synsedimentary subsidence. Pliocene facies, on the contrary, are mainly shallow-marine siliciclastics and carbonates. Their considerable thickness (1000–1500 m) attained during a relatively short time interval therefore implies important subsidence. Moretti and Colletta (1987) and Richardson and Arthur (1988) demonstrated that an increased rate of tectonic subsidence characterizes a one million year episode beginning at 6.4 Ma, within the Suez rift as a result of oblique strike-slip motion relating to a change in the structure of the Aqaba transform zone. Subsequently, the tectonic subsidence is supposed to be reduced or null, suggesting that the Plio-Quaternary subsidence reflects only stratigraphic accumulation.

Miocene tectonics involved vertical movement along faults oriented both parallel and oblique to the rift axis. Pliocene tectonics, in addition to rejuvenation of the Miocene fault system, also included important halokinetic movements which generated low reliefs locally influencing Pliocene sedimentation. These movements result in domes and faulted anticlinal structures which are characteristic of the Pliocene. Buried evaporite remobilization depends on a number of factors. Shallow burial (less than 100 m), notably in the peripheral part of the rift, would suggest that burial is of less importance than extensional reactivation of the Miocene faults; these extensional movements appear to favour evaporite ascension. Diapirism and other halokinetic movements, typical of the Red Sea and southern Gulf of Suez, are rare within the central and northern Suez region, in spite of the fact that these latter areas also include Miocene blocks and fault systems. Lack of diapirism suggests that the basement faults have not been rejuvenated during the Pliocene.

3. Pliocene tectonics and sedimentation

There is a similarity between the post-Miocene sequences in the southern Gulf of Suez and northern Red Sea regions. At outcrop there tends to be an evolution from relatively fine siliciclastic sediments lying conformably on Miocene evaporites, to essentially carbonate facies towards the top. Exclusively carbonate sedimentation is restricted to the top of certain structural highs, notably in the southern Gulf of Suez. Siliciclastics, derived from the periphery of the rift, are deposited in the intervening lows. Many zones of Pliocene siliciclastic discharges coincide with zones of earlier syn-rift clastic sedimentation, thus indicating a common structural control from faults and corridors oriented obliquely to the axis of the rift.

There are also marked differences both in tectonic style and sedimentary response, which characterize each of the major provinces.

The **northern and central Gulf of Suez**, examined on seismic profiles, are characterized by predominantly fine siliciclastics and evaporitic sedimentation whose constant distribution reflects the absence of halokinetic and other structural movements.

The **southern Gulf of Suez**, on the contrary, is characterized by widespread halokinetic movement which generally exhibits a north-west–south-east orientation relating to rejuvenation of older Miocene fault systems. These produce positive bathymetry which has favoured the development of small but numerous carbonate platforms. Increasing predominance of carbonate facies towards the south of the Gulf of Suez, and especially in the vicinity of the escarpment limiting the Shadwan Embayment, reflects more favourable conditions for reef growth and related carbonate production.

The escarpment marking the edge of the Suez Platform has been created fairly recently because the bounding fault cuts about 500 m of post-evaporite strata. These important movements are related to gravitational sliding of post-Miocene sediments on the Miocene evaporites. The displacement and rotation of blocks creates secondary Pliocene–Quaternary basins which includes the 35 km wide Shadwan Embayment situated to the north-west of the Aqaba transform fault. These various structural movements – block tilting, halokinetic remobilization and gravity sliding – all confirm the instability of the southern Gulf of Suez, only part of which can be related to the direct influence of the Aqaba transform fault. These movements, together with a relatively rapid Pliocene sedimentation, indicate that the region continues to be affected by extensional tectonics related to rifting.

These post-Miocene movements, as already noted, are reflected in the mineralogical composition and geometry of the Pliocene sediments. On a regional scale, the transition between siliciclastic and carbonate facies occurs both across and parallel to the Gulf of Suez axis. On a south-west–north-east cross-profile (Figure F2.22), siliciclastic sediments characterize the peripheral zone of the rift, the axial parts being protected by multiple structural traps, escape terrigenous contamination. The predominance of shallow-marine carbonates within the axial parts of the basin is unusual but confirmed on the north-west–south-east axial profile. Clastics and evaporites occupy the northern, proximal part of the basin, and grade southwards into massive carbonates towards the edge of the Gulf of Suez. This geometry is typical of many carbonate platform settings. However, this carbonate relief, with its erosional distal margin, is located near the axis of a structural depression rather than on the flank of the basin (Figure F2.22).

The Pliocene sequences outcropping in the region of Hurghada, although dominated by siliciclastic facies, do not include thick conglomerates typical of the Samadai Formation of the north-western Red Sea coast. Furthermore, the post-Miocene formations, within the Gulf of Suez, in areas somewhat removed from local struc-

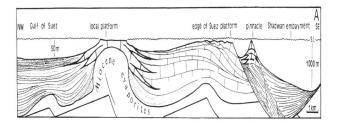

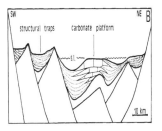

Figure F2.22 General interpretation of the structure of the Gulf of Suez, based mainly on seismic lines. A. Interpretation of a seismic line parallel to the axis. Carbonates are located on the top of diapirs and on the south-eastern edge of the Suez platform. The depressions are filled with siliciclastics. Compare this sketch with the seismic line in Figure F2.15. B. Simplified section perpendicular to the rift axis showing the peripheral shoulders bordering the rift graben. Note that the Gulf of Suez, which contains several carbonate platforms, is a depression bordered by important reliefs (2200 m in the Sinai, 1700 m for the Red Hills).

tural highs (i.e. to the west and the north-west of Gebel el Zeit) are not affected by the distinct regional dip, typical of the north-western Red Sea region.

The **north-western Red Sea region**. The Aqaba transform fault separating the Gulf of Suez and the Red Sea segments of the rift has been active throughout the Pliocene (Lyberis, 1988). However, accommodation space generated by this opening has favoured gravity sliding within the southern parts of the Gulf of Suez, creating the escarpment of the Suez basin. Bathymetric maps do not indicate any major morphological change coincident with the Aqaba transform zone situated about 40 km south-east of the edge of the Suez escarpment.

The post-Miocene strata within the north-western Red Sea, as already noted, differ in two respects from that of the southern Gulf of Suez. Firstly, its upper parts frequently include massive conglomerates (Samadai Formation) composed of Precambrian basement materials suggesting active erosion possibly related to uplift of the rift periphery. Secondly, the post-Miocene strata, including the Samadai conglomerates, are affected by a general dip towards the axis of the rift. This north-east dip must be related either to axial subsidence or to the peripheral uplift, more probably to a combination of both.

Seismic profiles confirmed by DSDP (Ross and Schlee, 1973), indicate that the escarpments created by axial spreading cut both 'reflector S' (top Miocene evaporites) and the greater part of the post-evaporite 'Pliocene' cover. This implies that Pliocene sedimentation corresponds to the final phases of rifting, immediately preceding local oceanization of the axial zone (incipient sea-floor spreading). The unconformity terminating the post-Miocene series and predating the early Quaternary terraces, created by regional axial subsidence, and/or by peripheral uplift, may therefore be regarded as the post-rift unconformity, the importance of which is discussed by Bosence *et al.* and by Plaziat *et al.* (this volume). Although structural movements continued within the peripheral parts of the north-western Red Sea (uplifted and tilted Pleistocene terraces), including uplift of the shoulders (basement mountains), these may be considered as post-rift, relating to isostatic adjustment of a passive margin.

4. Post-Miocene sedimentation and non-tectonic factors

Both Miocene and Pliocene sediments are closely linked to rift evolution and thus to tectonics. However, eustatic and climatic factors also play important roles. The eustatic component is difficult, if not impossible to evaluate not only because of lack of dates, but also because of the presence of numerous discontinuities most of which appear to express seismic instability. Furthermore, the amplitude of post-Miocene structural movements, frequently exceeding several hundred metres, far exceeds that proposed by the Pliocene sea-level curves of Haq *et al.* (1987). All measured sections indicate that post-Miocene sedimentation remained close to contemporaneous sea level; sedimentation infilled available accommodation space and continuously compensated rise in relative sea level. Thus, although local depositional sequences do occur, no regional sequence stratigraphic organization has been noted.

Climatic factors, although insufficiently studied, clearly have been important in determining mineralogy and textures of the 'Pliocene' series, the lower parts of which (Samh Member) are composed of fine siliciclastics with few feldspars (Soliman *et al.*, 1993). However, the upper parts (Gabir and Shagra) include numerous mass-flow deposits rich in feldspars and amphiboles. These appear to have been due both to relatively arid climatic conditions and, in contrast with the major terrigenous discharges of the Halaib and Tokar regions (Sudan), to the small catchment areas on the peripheral shoulder of the rift.

5. Conclusions

The 'Pliocene' sediments of the Gulf of Suez and the northern Red Sea region record considerable subsidence as well as important structural movements (faulting plus halokinesis) relating to extensional rift tectonics as already noted by Heybroek (1965). Lateral variation in style and intensity recorded by these sediments show that the region comprises three stages of rift evolution.

The central and northern Gulf of Suez, characterized by post-Miocene stability and homogeneous 'Pliocene' sedimentation, are in a less advanced stage relative to the evolution of the south Gulf of Suez. The southern Gulf of Suez, between the Aqaba transform and the Morgan accommodation zone, characterized by extensional structural movements and considerable Pliocene subsidence, but lacking generalized basinwards tilting, appears to be in a syn-rift stage. That this situation has not necessarily terminated is shown by the Quaternary movement affecting a number of its component structures. Conversely, the Red Sea has attained a passive margin situation, the regional tilting of Pliocene strata following the near-oceanization of the axial zone, followed by Quaternary erosion and subsequent deposition of horizontal Late Pleistocene terraces exhibiting very minor changes in altitude, resulting in a clearly defined post-rift unconformity.

ACKNOWLEDGEMENTS

The constructive criticism of Dan Bosence and Ian Davison improved the manuscript. We thank the Compagnie Générale de Géophysique (France) for access to seismic data and useful discussions. The authors acknowledge the efficient help of Lucas Gasperini, Bernard Gensous and Michel Tesson during the high resolution seismic campaign and also the Egyptian General Petroleum Corporation (EGPC) for technical help and for permission to publish seismic profiles. The authors express their gratitude for the financial support of the EEC (SCIENCE Program).

Chapter F3
Salt domes and their control on basin margin sedimentation: a case study from the Tihama Plain, Yemen

D. W. J. Bosence, M. H. Al-Aawah, I. Davison, B. R. Rosen, C. Vita-Finzi and E. Whitaker

ABSTRACT

This chapter examines the Pliocene and Quaternary post-rift sediments overlying, and deformed by, the intrusion of salt diapirs on the coastal plain of Yemen in the south-eastern Red Sea. The coastal plain region has been dominated by siliciclastic deposits (Al Milh Sandstone) derived from erosion of the Yemen rift shoulder escarpment and deposited in a range of fluvial, lacustrine, aeolian and shallow-marine environments. These have accumulated on and around, and have been intruded by, the Salif Evaporite which forms a 3 km high salt wall that outcrops at the Salif salt mine. The Pliocene to Holocene Kamaran Limestone is deposited on and seaward of the Salif salt wall and thickens offshore through Kamaran Island to offshore wells. The salt wall is considered to have acted as a barrier to sediment transport from the rift margin resulting in ponding of siliciclastic sediments on the coastal plain and offshore carbonates.

The upper surface of the Salif salt diapir underwent differential sea-floor dissolution as the salt uplifted into the shallow-marine environment. Less soluble rafts of gypsum and dolomite formed upstanding ridges surrounded by locally derived conglomerates. Coral patch reefs and carbonate sands and gravels accumulated within a few metres of the dissolved upper surface to the salt. Analysis of the coral fauna indicates that it comprises a number of taxa tolerant of increased levels of turbidity but only a few taxa are tolerant of increased salinity. A thin cover of sediment and strongly stratified water column is thought to explain the unusual occurrence of a fully marine fauna close to the dissolved salt.

Carbon and strontium isotopic dating of the limestones indicates older (Pliocene–Pleistocene) tilted beds on the flanks of the diapir and younger (Holocene) reef limestones on top of the diapir today occurring at an altitude of 16–20 m above sea level. These relationships reflect those seen on a seismic section through the diapir, indicating that the rise of the salt is punctuated by periods of subsidence.

INTRODUCTION

The syn-rift stratigraphy of the southern Red Sea basin is dominated by salt domes, walls and pillows which rise from Miocene salts and occasionally outcrop on the Tihama Plain of Yemen. Recent seismic (Heaton *et al.*, 1995) and structural (Davison *et al.*, 1996) studies show that salt walls and pillows intrude some 2 km of Miocene to Recent basin margin strata and divide the south-eastern basin margin of the Red Sea into a series of margin-parallel structural zones, salt walls, pillows and minibasins. The northern Tihama Plain (Figures F3.1, F3.2) provides the only outcrops in Yemen of these salt domes which comprise halite, gypsum/anhydrite, dolomite and shale of the salt diapirs, deformed post-salt siliciclastic and evaporitic sediments, and sediments of the overburden of the salt domes. These outcrops around the salt domes are also the only onshore record of Miocene to Pliocene sedimentation in the south-eastern Red Sea. They have responded in a complex fashion to the rising salt domes (Figure F3.3) and reveal the most recent history of the style and rates of salt dome movement. This contribution focuses on the effects the salt domes have had on rift margin

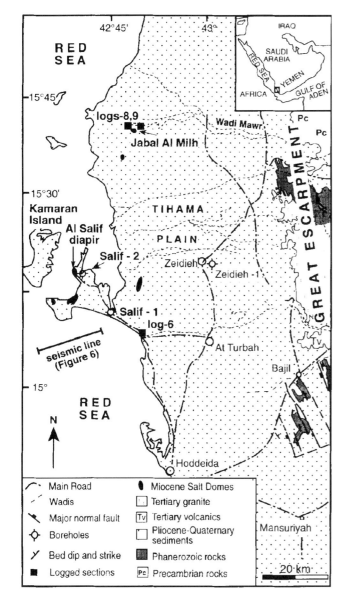

Figure F3.1 Location and geological setting of study area illustrating location of logs 6, 8 and 9 (Figure F3.5), seismic line (Figure F3.6), and location of log 6 (Figure F3.7).

sedimentation. Data for this analysis come from a new stratigraphic analysis and classification of the main basin margin lithologies together with detailed fieldwork on sedimentary rocks in the deformed diapir walls, and Pliocene and Quaternary sediments which have accumulated over the rising, and sometimes subsiding, salt domes. The diapirs studied at outcrop occur around Al Salif and Al Milh (Figure F3.1) and are both sites of commercial salt extraction.

The major results of this study on the effects of salt domes on basin margin sedimentation are that:

1. The Salif salt wall provides a barrier to the basinward transport of rift margin siliciclastics which are ponded behind the salt wall while carbonate facies accumulate on the crest and basinward of the salt wall. The shoreline has been located over the salt wall since at least the early Pliocene.
2. The upper surface of the salt dome underwent differential sea-floor and subaerial dissolution generating a complex topography of upstanding ridges of gypsum/anhydrite and dolomite with local scree deposits of these lithologies and a lower undulose halite surface overlain by gypsum conglomerates.
3. Patch reefs develop within a metre of the dissolved salt surface. These patch reefs are dominated by branching frameworks of the coral *Galaxea fascicularis* which is unexpected as today it is not known to be a salinity tolerant genus. These patch reefs are thought to form in response to intensely stratified, but turbid, marine waters.
4. Mapping and dating of sediments over the salt wall indicate an early phase of reef growth which was followed by uplift, tilting and erosion, prior to subsidence, further reef growth and final uplift through to the present day. Seismic evidence of marginal onlapping reflectors indicates that upward movement of the salt dome was also interrupted by periods of subsidence.

GEOLOGICAL SETTING OF SOUTH-EASTERN RED SEA SALT DOMES

Stratigraphic setting

The southern Red Sea rift opened to marine environments soon after the main phase of Afar plume magmatism between 26 and 32 Ma (Davison *et al.*, 1994; Menzies *et al.*, this volume). Hughes and Beydoun (1992) suggest from micropalaeontological evidence that brackish and marginal marine environments characterized the earliest rift sediments of this area during the late Oligocene. In the early Miocene deep marine conditions are evident (Hughes *et al.*, 1991). These are followed by mid to late Miocene salt deposits which were formed soon after rifting of the basin margin continental crust ceased in this area (Heaton *et al.*, 1996; Davison *et al.*, 1996). Extension was subsequently localized at the centre of the Red Sea. Deposits of this early phase of the rift basin do not outcrop onshore in western Yemen except that post-salt clastics occur in small vertical or overturned sections on, within or adjacent to salt domes (Figure F3.1; Davison *et al.*, 1996). However, because of petroleum exploration in this area, seismic sections tied to well data are available of the post-salt deposits and have been integrated with our fieldwork (Figure F3.2; Heaton *et al.*, 1995).

These Miocene strata may be classified lithostratigraphically as part of the informal Baid Formation of the Jizan Group described from southern Saudi Arabia

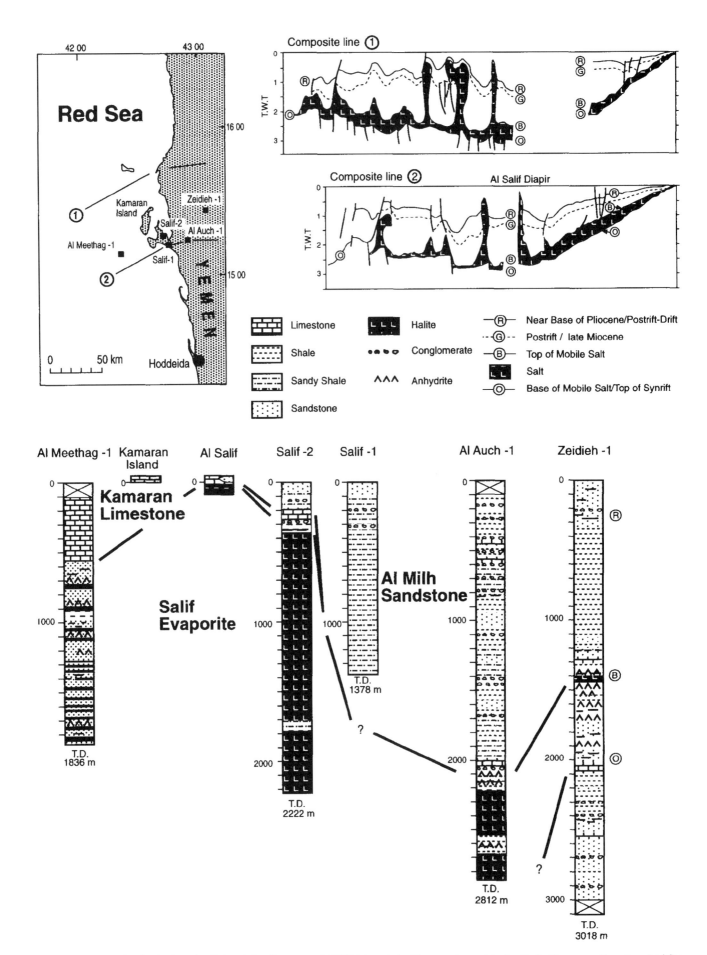

Figure F3.2 Regional seismic and well data (after Heaton *et al.* 1995) integrated with our outcrop data for the Salif area, south-eastern Red Sea.

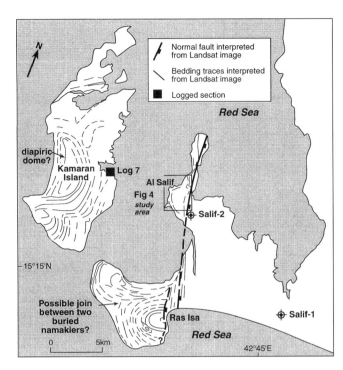

Figure F3.3 Tracing from Landsat satellite image indicating bedding of cover rocks and faults over Salif diapir together with location of log 7 (Figure F3.7) on Kamaran Island (after Davison et al., 1996).

(Blank et al., 1986). The Baid Formation is a unit of mixed lithologies within the dominantly volcaniclastic and volcanic strata on the Jizan Group. In northwestern Yemen three distinct lithological units can be recognized at the surface and in the subsurface. These units have not been assigned formal lithostratigraphic names in Yemen or in Saudi Arabia because complete stratigraphic sections are not exposed, and details of their composition in the subsurface have not been published. Despite this an informal classification of these sediments is useful as their occurrence and nature provide information on the sedimentological evolution of this basin margin. The names we use follow previous informal usage in the area, where possible, and all the names are taken from the best known outcrops of these units. The three main lithotypes in this area are here informally named the Salif Evaporite, the Al Milh Sandstone and the Kamaran Limestone.

Salif Evaporite

A lower, westward-thickening evaporitic unit is strongly deformed into a series of salt domes, salt walls and salt canopies (Heaton et al., 1995). These outcrop in the salt quarries at Al Salif (42°4′ E, 15°15′ N) and also at Jabal Al Milh (42°4′ E, 15°40′ N) (Figure F3.1) and have been penetrated by numerous wells in the southern Red Sea (Heaton et al., 1995; Hughes and Beydoun, 1992). The evaporite layer passes eastwards across the Tihama Plain from at least 600 m of halite at Al Auch-1 to about 500 m of interbedded shales, sands and anhydrite at Zeidieh-1 before thinning on to the basin margin (Figure F3.2). This evaporite unit has been described informally as the 'Salif evaporite' by Abu Khadra (1982), El Anbaawy et al. (1992) and Davison et al. (1996); the 'massive salt zone' (Mitchell et al., 1992), the 'salt' (Hughes and Beydoun, 1992), or the 'main evaporite group' and the 'mobile salt' by Heaton et al. (1995). At outcrop the Salif Evaporite is intensely deformed, with vertical bedding and steeply plunging isoclinal folds, in salt domes which are at least 1500 m wide at Salif and at least 250 m wide at Jabal Al Milh (Davison et al., 1996). The strata at Salif comprise mainly halite with thin anhydrite interbeds with one 100 m (minimum tectonic thickness) gypsum/anhydrite unit and at least one unit of dolomite, gypsum/anhydrite, shale and fine sandstone (Figures F3.4, F3.5) up to 20 m thick which is repeated by folding within the Salif diapir (Figure F3.4). These sediments have been recently described in some detail by El Anbaawy et al. (1992) and we have little to add to their facies descriptions and interpretations. Wells at Al Meethag-1, Al Auch-1, and Zeidieh-1 also penetrate beds of anhydrite (Figure F3.2). From seismic evidence the maximum undeformed stratigraphic thickness of the salt is between 1.5 and 2 km thick but it reaches up to 4 km in salt diapirs (Heaton et al., 1994). The Salif diapir has risen over 3 km with respect to the main salt layer (Figure F3.6). In the Zeidieh-1 well the Salif Evaporite is underlain by at least 1000 m of sands, sandy shales and conglomerates (Heaton et al., 1995). The upper surface of the Salif Evaporite is either an unconformity with the Al Milh Sandstones in the eastern part of the Salif Peninsula, or the Kamaran Limestone as seen in the western part of the Salif peninsula (Figure F3.4). We have not found fossils in the Salif Evaporite. Grolier and Overstreet (1978) report a middle to late Miocene microflora from the early vertically oriented sediments at Salif and a similar age is suggested by El Anbaawy et al. (1992) and in the offshore by Hughes and Beydoun (1992) from regional comparisons with Egyptian evaporites of the South Gharib and Zeit Formations. Heaton et al. (1995) indicate ages of c. 11–17 m.y. for the 'mobile salt' of this area which we presume to be based on micropalaeontology from their wells. Hughes and Filatoff (1995) in their recent review of the Tertiary rocks of the Red Sea coast of Saudi Arabia also equate the main evaporite unit (their Mansiyah and Ghawwas Formations of the Maqna Group) with the South Gharib and Zeit Formations of Egypt.

Al Milh Sandstone

Siliciclastic sediments deformed by, and therefore younger than the salt, crop out in a number of exposures

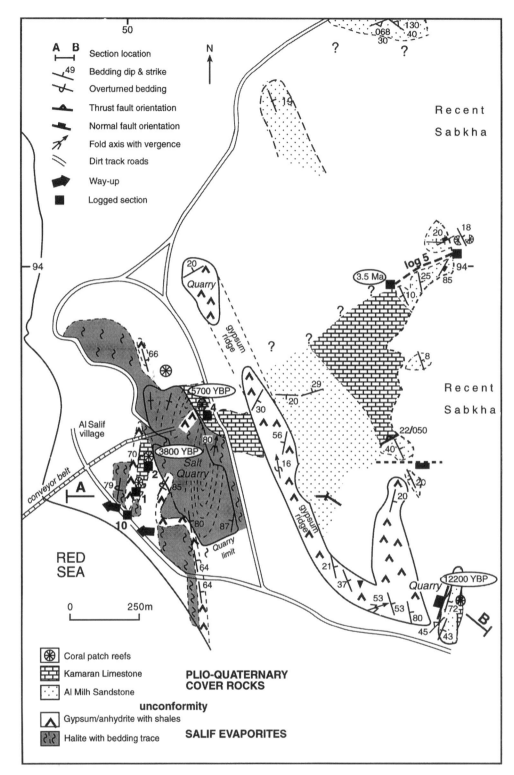

Figure F3.4 Geological map of Salif area showing location of logs, location of dated samples (Appendix) and cross-section A–B in block diagram (Figure F3.10).

at Jabal Al Milh (42°50′ E, 15°40′ N) and also unconformably overlie that salt on the eastern side of the Salif Peninsula. In the subsurface up to 2000 m of siliciclastic deposits overlie the Salif Evaporite in the Al Auch-1 well and 1300 m of siliciclastics in the Zeidieh-1 well (Figure F3.2). These are labelled 'Abbas unit' of the 'post salt clastics' by Hughes and Beydoun (1992). The largest outcropping sections of this unit are a number of

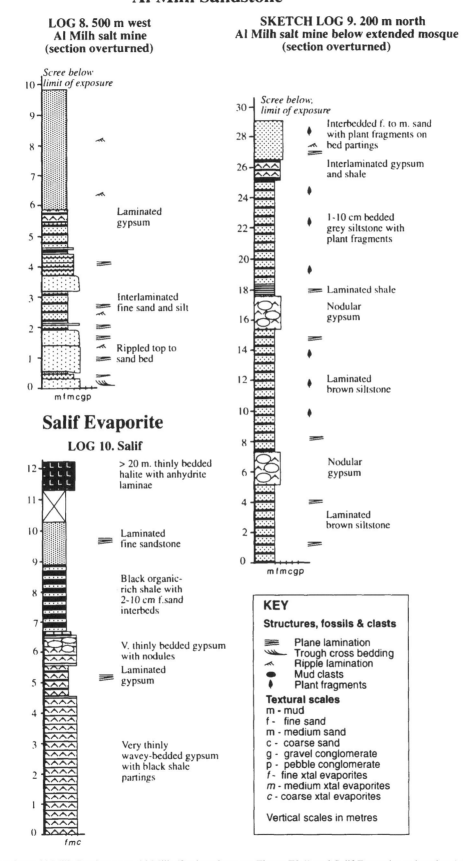

Figure F3.5 Logs from Al Milh Sandstone at Al Milh (for locations see Figure F3.1) and Salif Evaporite at location 1 (Figure F3.4).

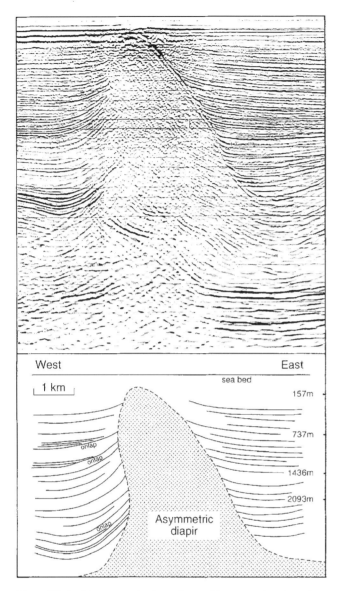

Figure F3.6 East–west seismic section and interpretation through Salif salt dome (modified after Davison et al., 1996). For location of line see Figure F3.1.

isolated hills around the salt mine at Jabal Al Milh (Davison et al., 1996) and other outcrops are also described here from an unamed hill 0.5 km due west of the salt mine and at Jabal Al Milh to the south of the salt mine (Figure F3.5). These isolated outcrops are locally overturned and deformed by the intruding salt diapir. Beneath the 'extended mosque' (Plate 27B) an overturned section exists which is presented in Figure F3.3. The section comprises 29 m of interbedded siltstones, shales and sandstones with some nodular gypsum horizons. These deposits are interpreted to have formed in continental environments, mainly alluvial flood plain and lacustrine, with some inland sabkhas represented by the nodular gypsum beds. The section to the west of the salt quarry is also overturned and tightly folded and one accessible 10 m portion of this has been logged (Figure F3.5). Here a section of coarse to fine sandstones with a mud-flake conglomerate and thin gypsum interbeds are interpreted as three fining-upward fluviatile cycles. At the foot of the hill beneath the castle at Jabal Al Milh, south of the salt mine, there is a 100 m thick section of brecciated gypsum with 5–6 m of sandstone on top. No way-up data have been found here and it is not clear whether or not this section relates to the Salif Evaporites or the younger Al Milh Sandstones.

At Salif various sections and small sand quarries in siliciclastic sands to the west of the salt mine which are correlated with the exposures at Al Milh because they unconformably overlie the salt. Thus, the exposed clastics at Salif are clearly post-salt and most are unlithified. One quarry section has been logged (log 5, Figure F3.7) and comprises 9 m of silts and fine sands interpreted to be of fluviatile origin. A similar section is exposed in the roadside quarry in the south-west of the Salif peninsula (Figure F3.4) with basal coarse sands and scattered pebbles of metamorphic basement rocks. Both these locations expose the junction with overlying Kamaran Limestone which is conformable at the former and unconformable at the latter location. The siliciclastic deposits of the Tihama Plain are the present-day expression of the Al Milh Sandstone and comprise alluvial fan sands and gravels derived from the Yemen Highlands (Figure F3.1), aeolian dunes, and coastal siliciclastic beach sands (e.g. log 6, Figure F3.7), sabkhas and muddy, mangrove shorelines.

No age-diagnostic fossils have been found in outcrops of these sandstones and conglomerates, the only fossils being badly altered ostracods and scattered plant fragments. The sedimentary facies of the sections examined represent fluviatile, lacustrine and sabkha environments where fossils of biostratigraphic value are not expected. On the Salif peninsula the sands interfinger laterally with the Kamaran Limestone dated by ^{14}C as 3700 and 3900 BP (see Appendix for details of ^{14}C dating) and in the south-east of the peninsula are overlain by 3.5 Ma ($^{87}Sr/^{86}Sr$ dating, see Appendix) and 12 200 BP Kamaran Limestone. Heaton et al. (1995) state that in the subsurface 'biostratigraphic data are absent because the depositional environments were either hypersaline or fluviatile'. However their figures 3 and 6 indicate a <11 m.y. (above marker horizon B) for this siliciclastic unit. Hughes and Beydoun (1992) indicate a Pliocene to Pleistocene age for their 'post-salt clastics' which we interpret to be equivalent to this unit.

Kamaran Limestone
This unit comprises those limestones resting unconformably on the salt domes in western Salif together

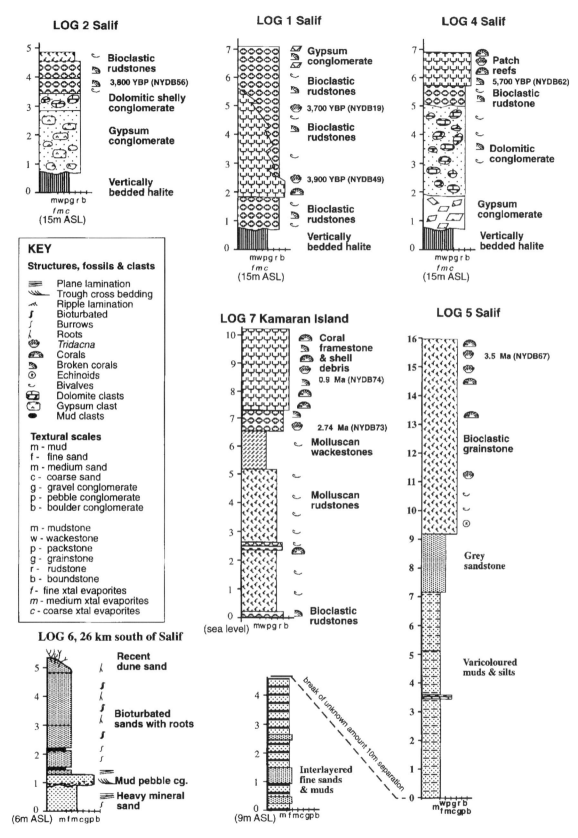

Figure F3.7 Sedimentary logs of sections overlying salt at Al Salif (logs 1, 2, 4 and 5, for locations see Figure F3.4), and Kamaran Island (log 7, Figure F3.3) and coastal section to the south of Salif salt dome (log 6, Figure F3.1).

with those exposed on Kamaran Island. In both cases the top of the limestone is the present-day land surface. Wells drilled through the extensive, deformed sections of limestone on Kamaran Island encounter saline water and a Landsat image (Figure F3.3) indicates domed strata suggesting that the island is underlain by salt. A similar situation exists further north in the Farasan Islands of southern Saudi Arabia where diapiric structures have been mapped and salt is penetrated at depths of 45–154 m (Macfadyan, 1930; Bantan, 1995). The Kamaran Limestone outcrops are overlain by Quaternary beach sands and scree so that the original thickness is unknown. Ten metres of bioclastic and reefal limestones are exposed in cliffs to the south of Kamaran harbour below the old hospital (log 7, Figure F3.7) and this may represent the most complete section exposed on the island. There are also scattered outcrops of bioclastic grainstones and rudstones suggesting up to 30 m of section in strata. El Anbaawy (1993) records a maximum of 28 m in his 'Kamaran reef limestones'.

In the subsurface, and offshore, the top 500 m of Al Meethag-1 well encounters limestones (Heaton et al., 1995) which we interpret to be lateral equivalents to the limestones of Kamaran and Salif (Figure F3.2). Hughes and Beydoun (1992) label these limestones as the 'Meethag unit' of their 'post-salt clastics'.

Cox (1931) records a large number of molluscs from the 'Reef Limestone' of Kamaran; many of these are long ranging and a number still live in the Indo-Pacific region and he suggests that these beds 'cannot be much older than Pleistocene' (Cox, 1931, p.4).

Carbon and strontium isotope dating indicates ages from 3700 BP to 3.5 Ma (Appendix). This is consistent with ages of < 5 m.y. given in Heaton et al. (1995) from the Al-Meethag well. The limestones therefore span the Pliocene and Quaternary periods. Grolier and Overstreet (1978) report late Pliocene microfauna from marine sediments overlying the evaporites at Salif.

Structural setting of Al Salif Diapir

The Al Salif diapir is an asymmetric structure which has risen up to 3 km above the main Miocene halite source layer (Figure F3.6). It lies very close to the present-day Red Sea coastline and our mapping (Figure F3.4) tied to Landsat analysis (Figure F3.3) indicates that the Al Salif peninsula is produced by a linear north–south trending salt diapir, or salt wall, bounded by a normal growth fault on the eastern margin. The diapiric wall continues offshore for several kilometres (Figures F3.2, F3.3). Stratigraphic geometries on the seismic section (Figure F3.6) indicate that diapirism began soon after deposition of the main salt interval, and growth of sedimentary strata into the synclines on either side of the diapir can be observed in the late Miocene to Quaternary strata (Figure F3.6). Up-dip truncations and onlap towards the diapir indicate upward salt movement as well as periods of subsidence (Figure F3.6). Overburden reflectors break up close to the top of the diapir (above 0.5 sec TWT) suggesting that rapid facies variations or pervasive faulting has occurred. This is consistent with our field observations that the vertical salt within the diapir is overlain unconformably by gently folded overburden of variable lithologies which are cut by numerous faults (Figure F3.4). Holocene Kamaran Limestone is now raised above sea-level indicating that salt is still available for supply into the diapir and that the diapir is still actively growing.

Two separate partial ring structures on the Ras Isa peninsula are interpreted to be namakiers (surface salt flows which have coalesced) with an approximately east–west contact (Figure F3.3). A similar deformation style is seen on Kamaran Island (Figure F3.3) where the Kamaran Limestone (Figure F3.7, log 7) dips at various orientations (our observations). Similar salt canopies have been described in the offshore area by Heaton et al. (1995).

Discussion

The spatial and temporal distribution of these three major lithologies of the Jizan Group in north-west Yemen provide information on a number of aspects of the evolution of this basin margin during the Neogene to Quaternary. The Salif Evaporite thins into the basin margin and halites are replaced by anhydrite (plus gypsum at outcrop). Above the salt there are two main lithologies which are developed on either side of the Salif salt ridge. Inshore from the salt ridge the Al Milh Sandstone dominates and this is shown in the exposures around Jabal Al Milh to have been deposited in fluviatile, lacustrine and inland sabkha environments. To the west and offshore from the salt ridge where persistent shallow-marine conditions have resulted in the reefal platform carbonates of the Kamaran Limestone. Therefore the shoreline has maintained its present-day location at least back to the Pliocene and seems likely to have been controlled by the underlying salt ridge. This control can be seen today because the Quaternary sediments on Salif peninsula are dominated by siliciclastic beaches, sabkhas and an alluvial plain on the eastern side and abutting a prominent ridge of gypsum (Figure F3.4), and by reefal limestones and raised beaches on the western side. However, away from the barrier of the salt ridge siliciclastic sediments make up the entire Quaternary section (Figure F3.7, log 6).

SEDIMENTOLOGY OF SALIF COVER ROCKS (KAMARAN LIMESTONE AND AL MILH SANDSTONE)

Upper surface of salt diapir

The subvertical Salif Evaporite within the diapir is unconformably overlain by up to 50 m of the Al Milh Sandstone and Kamaran Limestone, which are subhorizontal or tilted (Figure F3.4), and are the main subject of this chapter. The unconformity is a smooth, sharp, undulose surface of halite which we interpret to be a dissolution surface (Figure F3.7, Plate 27(a)). Surface waters have produced a spectacular karst and pot-hole system on top of the Al Salif diapir, with one crevasse proved by drilling to reach 20 m deep.

Where the unconformity surface involves boudins and rafts of sulphate, carbonate and shale lithologies within the halite these stand up as ridges (Figure F3.4, Plate 27(c),(d)). Outcrops to the east of the quarry comprise folded and sheared nodular gypsum which forms the resistant north–south ridge to the Salif peninsula ('Gypsum ridge' of Figure F3.4, Plate 27(c)). Exposures to the west of the quarry are smaller but display a spectacular unconformity surface where the boudins or rafts of sulphates, carbonates and shales stand up as two smaller ridges, one repeated by folding, with a relief of at least 6 m (Figures F3.4, F3.8). These ridges are partially exhumed by present-day weathering and are avoided by the salt miners, leaving residual ridges within the quarry (Plate 27(c)).

Because the Quaternary continental and marine strata rest on this erosion surface and onlap on to the ridges this unconformity is considered to represent the dissolution surface where the Miocene salt encountered marine and/or phreatic waters when it neared the surface. Indications of the environment within which the upper surface of the salt was dissolved are revealed in the immediately overlying deposits which are discussed in detail below.

Sedimentology of the Kamaran Limestone at Salif and Kamaran

The sediments overlying the salt have been examined and logged (Figures F3.4, F3.7) at a number of locations around the quarry at Al Salif and near the port on Kamaran Island (Figure F3.3). Sections to the west of the gypsum ridge are characterized by the Kamaran Limestone resting unconformably on the halite surface (logs 1, 2, 4; Figure F3.7) while sections to the east of the ridge, while not reaching down to the salt surface, expose the Al Milh Sandstone which is overlain by the

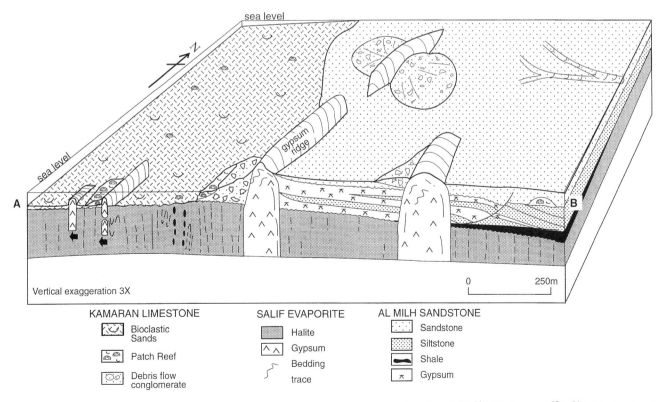

Figure F3.8 Diagram illustrating east–west cross-section of Salif salt wall with cover rock geology (with ^{14}C BP dates and $^{87}Sr/^{86}Sr$ Ma dates) and environments and sedimentary facies reconstructed for about 4000 BP.

Kamaran Limestone (log 5, Figure F3.7). A similar section, but with an unconformity below the limestone is exposed in the roadside quarry to the south-east of Salif (Figure F3.4). The following facies are recognized in the Kamaran Limestone.

Gypsum conglomerate

The first sediment to be deposited above the salt is usually a basal, matrix- or clast-supported gypsum conglomerate. This unit thickens, and clast size increases, eastwards towards the gypsum ridge (logs 2, 1 and 4, Figure F3.7) where it is particularly well developed on the eastern edge of the salt quarry (Plate 28(a)). Here angular pebbles, cobbles and boulders (up to 1.5 m across) of sheared gypsum (and some anhydrite) form a chaotic poorly sorted conglomerate with a finer gypsiferous matrix.

This facies is interpreted to be a subaerial scree deposit formed adjacent to the north–south ridges of sheared gypsum. Locally and adjacent to other ridges of sulphate and evaporite (e.g. log 2, Figure F3.7) there are other less extensive exposures of this conglomerate. From field relations the diapir surfaces must have been dissolved to produce the gypsum ridge prior to localized scree deposition.

Shelly, dolomite conglomerate

Stratigraphically above the gypsum conglomerate are thinner (20–80 cm) conglomerates with well-rounded to angular pebbles and cobbles of dolomite together with coarse molluscan debris (Figure F3.7, Plate 28(b)). The dolomite clasts are bored by bivalves and sponges and encrusted by oysters, barnacles and serpulids.

The size and local source of the dolomite clasts together with their boring and encrustation all indicate a high-energy marginal marine environment such as a beach. Similar facies are found on the present-day beaches at Salif. The occurrence of shelly, dolomite conglomerates above the gypsum conglomerates indicates a marine transgression over a shoreline of gypsum boulders. In this situation the gypsum clasts would be expected to have dissolved and evidence for this may be the rapid vertical change from gypsum clasts to carbonate clasts. However, locally, adjacent to the middle ridge of the three gypsum ridges (Figure F3.4) angular clasts of gypsum occur in a matrix of shell gravel indicating marine conditions; the pebbles are not bored. We are not aware of any previous records of this association, or of bored gypsum clasts (personal communication, R. Bromley, 1994, University of Copenhagen), and suggest that gypsum pebbles could only be preserved in marine waters if they were very rapidly buried and isolated from the waters of marine salinity indicated by the preserved faunas.

Bioclastic rudstones

These deposits occur either above the shelly, dolomitic conglomerates, or, immediately overlying eroded halite at log 1 (Figure F3.7; Plate 27(d)), or, as the first marine carbonate facies in the quarry to the south-east of Salif (Figure F3.4). The 50 to 200 cm thick rudstones contain clasts of molluscs (particularly ribbed oysters, *Tridacna* and pectinids); corals are common, as are coralline algae (including rhodoliths), barnacles, dolomite pebbles and siliciclastic grains (feldspars, quartz and ferromagnesium minerals) derived from the basement outcrops of the Great Escarpment (Figure F3.1). Corals and *Tridacna* have been dated using ^{14}C and ^{87}Sr/^{86}Sr (see Appendix) and ages range from 2.75 Ma (1.8–3.65) on Kamaran Island (log 7, Figure F3.6) to 3800 BP (± 200) at Salif (log 2, Figure F3.7).

The diverse association of marine skeletal fragments indicates normal marine conditions together with locally derived dolomite clasts and siliciclastic grains from the rift scarp some 60 km to the east. The coarse texture, association of grains and context all indicate high-energy, shallow-marine, shoreline conditions similar to those occurring on beaches and shoreface environments off the west coast of the Salif peninsula today. Palaeoecological analysis of the corals based on their present-day habitats indicates that the corals in the rudstones come from a variety of habitats but all within the 5–20 m depth range; some of these corals present are tolerant of turbid waters.

Patch reefs

Small, 1 to 2 m thick, and up to 10 m across, reefs are common on the western side of Salif (Figures F3.4, F3.7, Plate 28(c),(d)) particularly on the western edge of the gypsum ridges which would have faced the open sea (Figure F3.8). The reef frameworks are unexpectedly found to occur within 20 to 30 cm of the dissolved upper surface of the salt. *In situ* coral frameworks (Plate 28(c),(d)) are surrounded by coral debris and the bioclastic rudstone facies. The corals are dominated by *Galaxea fascicularis*, particularly in the lower sections of reefs or in small reefs and coral patches. This coral forms a closely spaced, but delicate branching, thicket rather than a rigid framestone. Corallites rise some 20 to 30 cm from the coenosteum (Plate 28(d)) in a form known as '*anthophyllites*' (Chevalier, 1971) with nestling bivalves and associated *Tubipora*. Upper sections of the reefs have a more diverse coral assemblage with *Acropora, Goniopora, Stylophora, Turbinaria, Favia, Fungia, Coscinaraea,* and *Porites* in addition to *Galaxea* and *Tubipora* together with coralline algae and bivalves (ribbed ostreids, *Tridacna*, pectinids and carditids).

Both the setting and the composition of these reefs are unusual. The close proximity of the corals to the dissolved halite surface would suggest that the ambient waters would have been of higher than normal salinity and that this might be reflected in the lower diversity of the lower *Galaxea* dominated reef-building communities. In addition to the monotypic stands of *Galaxea* the growth-form *anthophyllites* is also unusual for this coral. The normal growth form of *Galaxea*, which is common throughout the Indo-Pacific region as a reef-building coral, is a cushion shape with the corallites largely contained within a lightly calcified coenosteum (Chevalier, 1971; Sheppard *et al.*, 1986). At Salif the corallites are extended as isolated branches some 20 to 30 cm above a basal coenosteum. The only previous records of branched growth forms are by Chevalier (1971) who describes the form '*anthophyllites*' from the present-day Red Sea which has corallites extending some 6.5 cm above the level of the coenosteum. Chevalier (1971) reports this form occurring only in turbid areas and makes no comment on salinity levels. Similarly, Sheppard and Sheppard (1991) describe *G. fascicularis* as being extremely abundant in turbid waters. Compilations of salinity-related coral communities from the Indian Ocean and Arabian Gulf (Sheppard *et al.*, 1992) indicate that *Galaxea* is not a salinity tolerant genus and does not extend into the elevated salinities of the Arabian Gulf. Our conclusion is that, although the reefs are in close proximity to the upper surface of the salt, they were in some way isolated from more saline waters and that our current knowledge of the ecology of *Galaxea* suggests that the low diversity and unusual growth-form may relate to nearshore turbidity associated with siliciclastic sediment supply. However, it should be noted that the extremely long corallites of the Salif corals have not, to our knowledge, been described from modern environments. The association of corals in the patch reef indicates water depths of 5–20 m; a number are known to be tolerant of turbid waters but only one (*Porites* cf. *compressa*) possibly indicating increased salinities.

Bioclastic grainstones
Coarse bioclastic grainstones occur mainly to the east of the Salif peninsula as a crudely bedded 6–7 m thick unit (log 5, Figure F3.7). Allochems include molluscan, echinoid and coral debris while *Tridacna* occurs locally. The grainstones have 2–3 cm subvertical, walled burrows which are interpreted as thallasinoidean shrimp burrows. Up-section coral debris and large gastropod moulds are more common and the section ends with scattered, large *in situ* corals (*Cyphastrea serailia* and *Oulophyllia crispa*) and *Tridacna*. Specimens of *Tridacna* were collected for dating and gave a $^{87}Sr/^{86}Sr$ age of 3.5 Ma (2.2–4.3 Ma, early Pliocene) making this section considerably older than the late Quaternary raised beaches and reefs on the western side of Salif.

These grainstones have marine allochems and a coarse, mud-free texture indicating moderate to high hydraulic energy. They are interpreted to have formed in a shallow shoreface or shoal environment and terminate with a horizon colonized by corals and *Tridacna*. The corals are today found in calm clear waters in the 10–40 m depth range and they are also tolerant of salinities up to 46 ppt.

Sedimentology of Al Milh Sandstones at Salif

Interlayered fine sands and muds
This facies occurs at the base of the exposed siliciclastic section to the east of Salif (log 5, Figure F3.7). Here, unlithified or poorly lithified fine sands and muds are interlayered on a millimetre to centimetre scale. Within this are interbedded, 20 to 50 cm thick, fine-grained sands. These are plane-laminated and in one case the sand has deformed into ball and pillow structures. In the roadside quarry to the south-east of Salif (Figure F3.4) 75 cm of a similar facies is developed, but with thin (10–20 cm) beds of white micrite. The only biota is represented by thin sandy burrow-fills.

The fine-grain size, planar lamination, local micrite interbeds, and near-absence of biota suggest a low-energy fresh or brackish water lake or coastal lagoon.

Varicoloured muds and silts

This 7 m thick unit occurs as interlayered fine sands and muds (log 5, Figure F3.7). Silts and muds are interlayered and vary in colour through greens, greys, pinks, browns and reds which are interpreted to represent varying oxidation states of iron. One 10 cm thick bed of green-grey sparry dolomite occurs midway up the section. These structures and textures together with evidence of varying iron oxidation all suggest deposition in a quiet-water, non-marine overbank environment.

Grey sands
Two metres of grey, well-sorted fine sand occur midway up log 5 (Figure F3.7). These sands lack any structures or biota, and within the context of an overlying shallow-marine grainstones, could have formed in a number of fluvial, aeolian or coastal environments.

Coarse arkosic sand and conglomerate
Coarse feldspathic sands occur at localities 63 (1.8 m) and 98 (5 m) (Figure F3.4) and as a 30 cm thick bed at the base of log 6 (Figure F3.7) to the south of Salif. These unfossiliferous sands have decimetre-scale plane bedding and trough cross-stratification. Palaeocurrents are to the north-west at locality 98 (Figure F3.4) and in the sands at log 6. Minor erosion surfaces occur within the sands which are lined with small pebbles of Mesozoic

and Precambrian basement lithologies, together with locally derived mud pebbles (log. 6, Figure F3.6). These deposits are interpreted as distal fluviatile sands deposited from rivers discharging from the Great Escarpment some 60 km to the east (Figure F3.1) and are similar to the fluviatile sands being deposited in wadis and as sheets over the Tihama Plain today.

Heavy mineral sands
Plane-laminated, well-sorted, medium sands rich in heavy minerals occur at the base of the sea-cliff (log 6; Figure F3.7) to the south of Salif. This suite of textures and structures and the occurrence of similar sands on the present-day beaches in this area indicate moderate- to high-energy beach environment. The likely source for the heavy minerals are the Tertiary Yemen Volcanic Group which are being eroded in the escarpment mountains (Figure F3.1) today.

Bioturbated fine sand
The upper part of the sea-cliff to the south of Salif (log 6; Figure F3.7) comprises fine- to medium-grained sands with occasional plane laminations and burrows with an inter-laminated mud. These pass up-section into poorly sorted, medium sands which are well bioturbated by roots and rhizcretions and in turn are overlain by present-day dune sands stabilized by shrubs. These upper sands are seen to be of aeolian origin and a similar origin is interpreted for the lower units except where fluvial reworking is indicated by burrowed mud layers.

DISCUSSION CONCERNING THE ROLE OF SALT DIAPIRISM IN CONTROLLING BASIN MARGIN SEDIMENTARY EVOLUTION

Basin margin siliciclastic continental sediments

The cover rocks to the Salif diapiric wall indicate an eastern area dominated by continental siliciclastic sediments (Al Milh Sandstone) and a western shallow-marine carbonate province of the Kamaran Limestone. This small-scale study from Salif reflects the broad-scale lithological pattern presented earlier from regional borehole data (Figure F3.2). To the east of the salt wall there are fluvial sands and conglomerates, fluvial overbank deposits and lacustrine or brackish lagoonal sediments. The coarser fluvial sands are probably younger than the finer-grained siliciclastic units as they pass laterally or are overlain by Pleistocene carbonates (^{14}C dates of 5700 and 12 200 BP; see Appendix and Plate 27). The overbank and lacustrine/lagoonal deposits pass up into limestones dated as Pliocene. The occurrence of the clastics mainly to the east of the salt wall, their palaeocurrents (to the north-west) and their clast composition (basement clasts from the Great Escarpment) all indicate that the salt wall acted as a north–south barrier and trap for sediment transport from the east. Davison *et al.* (1996) calculate that there is enough buoyancy force in extensional tectonic regimes for salt to rise some 50–200 m above the sea floor. The ponding of shoreline clastics behind salt ridges is seen on seismic sections through Cretaceous syn-rift deposits of the Reoconcavo basin (Davison, personal observation).

Laterally equivalent strata away from the ridge (log 6; Figure F3.7) are also clastic dominated and, although undated, this section records an earlier regressive shoreline (passing from beach through fluviatile to aeolian deposits) which formed at a relatively higher sea-level than that of today. The heavy-mineral rich sand on the present-day beaches to the south of Salif derived from the Tertiary Yemen Volcanic Group contrast in their mineralogy with the quartz and feldspar rich arkosic sediments found around Salif which are probably derived from the granites and Precambrian basement further north on the Great Escarpment (Figure F3.1).

Shallow-marine carbonates isolated from siliciclastics by salt ridge

While some thin carbonate units occur to the east of the salt ridge within the thicker fluvial sediments the central and western sections are dominated by the shallow-marine Kamaran limestones. Log 5 from the east of Salif (Figure F3.7) indicates alluvial plain siliciclastics transgressed by early Pliocene seas which deposited shallow-marine grainstones and then reefal carbonates dated at 3.5 Ma. These are similar in age to the Pliocene reefal rudstones and framestones of Kamaran Island and the 500 m of limestones at the top of Al Meethag-1 and suggest a widespread area of shallow-marine carbonate shelf existed at this time in the southeastern Red Sea. All the sediments examined at outcrop were deposited in shallow-marine, often reefal environments and coral faunas suggest water depths of 5–20 m. There is no evidence of any deeper facies or rapidly shallowing-upwards sections as might be expected over a rising diapir. This is probably because the exposed sections are too small (< 10 m) to record larger-scale events.

Although the Pliocene to Pleistocene limestones on Kamaran are not seen in contact with the Salif evaporites, reports of saline groundwater on Kamaran Island and the dome-shaped overburden structures on the island (Figure F3.3) indicate that the limestones overlie salt (cf. El Anbaawy, 1993). This pattern is consistent with the data from Al Meethag well which penetrated 500 m of near-surface limestone dated as post 5 Ma by Heaton *et al.* (1995). The environment of accumulation of the underlying sands in this well is not known.

The persistence of the siliciclastic/carbonate boundary in the Salif region since the Pliocene is considered to be controlled by the location of the north–south salt ridge at Salif. The salt ridge formed a positive structure causing ponding of siliciclastic sediments, deflection of transport pathways and the rapid lateral transition from clastics to carbonates in an otherwise low-relief area. This interpretation is supported by the seismic profile (Figure F3.6) through the diapir which indicates that reflectors cannot be correlated from the west to the east of the diapir. Well-defined sub-parallel reflectors in the east change to less continuous weaker reflectors in the west with three distinct onlap surfaces. These features suggest changes in lithological character and patterns of accumulation either side of the diapir.

Patch reef colonization over salt diapir

The younger Quaternary limestones rest directly on the Salif evaporites, but adjacent to the resistant gypsum/evaporite ridges and are found over older subaerial gypsum conglomerates (logs 2, 5; Figure F3.7; Plate 27). This suggests initial Quaternary emergence and dissolution of the crest of the salt ridge followed by a relative sea-level rise during which the shallow-marine dolomite conglomerates with bored pebbles, bioclastic grainstones, rudstones and patch reefs accumulated. This sequence of facies (e.g. logs 1, 2, and 5; Figure F3.7) indicates a transgressive event followed by post 3700 BP (late Holocene) uplift.

The establishment of the reefs immediately over the dissolved surface of the salt is unusual. The well-preserved reef frameworks, although initially of low diversity, develop a fully marine aspect within a metre of the top of the salt. At this time the salt was probably being dissolved, with solutions passing through the earlier deposited sediments raising sea-floor salinity. The sediments deposited over the salt are coarse and clean and could not have formed a permeability barrier isolating the salt from the colonizing marine communities. An alternative explanation for the juxtaposition of the presumed normal marine reefs and the dissolving salt is that the reefs formed later in time and were lowered on to a dissolving surface of the salt, possibly in the subsurface. However, the near-perfect preservation of delicate branching coral frameworks above the salt (Plate 28(c), (d)) indicates no disruption or breakage of the coral framework that would have occurred if the reefs had been disturbed after deposition.

We are left with the conclusion that the reefs did indeed grow closely above the dissolving salt, however only 5 of the total 21 coral taxa occurring at Salif are known to be tolerant of (i.e. not restricted to) increased salinities. The current information on the ecology of *Galaxea* suggests that the monotypic stands and low diversity of the reefs may relate to shallow turbid coastal waters rather than increased salinity levels. However, it should be noted that the extreme branch elongation seen in the Salif *Galaxea* has not been reported from other modern environments even though *Galaxea* is common in the Red Sea (Sheppard and Sheppard, 1991). Presumably present-day analogues may be occurring in the Red Sea today but we have no knowledge of any reports of exposed sea-floor salt seeps in this area. We conclude that the Salif coral fauna is not a salinity tolerant fauna but that it shows a number of taxa that are tolerant of turbid waters and that there is also an indication of tolerance to lowered water temperatures.

Present-day analogues of coral reefs growing over salt domes comes from the Gulf of Mexico where Jurassic salt penetrates the sea floor to form saline brines and seeps (Rezak *et al.*, 1985). On one of these domes (West Flower Garden Bank) the sea floor is colonized by corals and coralline algae at water depths of 50–100 m and coral thickets are concentrated along fault scarps originating from deformation within the salt cover and the sea floor. Coral thickets also surround the brine pools and this appears to be facilitated by an intense stratification of the sea water. Rezak *et al.* (1985) record a salinity change from 200‰ to 36.7‰ over just 2 cm in the water column on the East Flower Garden Bank. While stratification of water would be less likely in the shallow sites of the Salif reefs these remarkable data from the Gulf of Mexico hint at a possible explanation of coral patch reefs growing immediately over a dissolved salt surface.

Tectono-sedimentary evolution of salt dome cover rocks

The limestones have been dated using ^{14}C and $^{87}Sr/^{86}Sr$ isotopes from *in situ* corals and *Tridacna* shells (see Appendix for details and ranges). A pattern of ages is apparent, with the youngest dates (3700–5700 BP) coming from limestones on the crest of the diapir and older dates (12 200 BP to 3.5 Ma) from the limestones on the eastern and western margins of the diapir. The oldest reefal limestones are considered to have formed soon after the salt diapir impinged on the sea floor during the Pliocene and Pleistocene or to have been flooded after an earlier period of emergence and coastal plain sedimentation as occurred to the east of Salif (log 5, Figure F3.7). The oldest date in the west on Kamaran is 2.75 Ma and the oldest date to the east of Salif is 3.5 Ma. Both these earlier limestones have been subsequently deformed by tilting and faulting and are covered by later coastal plain siliciclastics (Figure F3.8). Because the 12 200 BP reefal limestones are topographically lower (7–8 m) than the youngest limestones on the crest of the diapir there must have been a phase of at least 13–

14 m of subsidence between 12 200 and 5700 BP (for the difference in dates and heights of samples 25 and 62, see Table F3.2, Appendix).

During the late Holocene (5700 to 3700 BP) shallow seas covered the crest of the Salif diapir and shallow-water carbonate facies accumulated immediately over the dissolved upper surface of the salt. Since 3700–3900 BP the diapir has moved upwards by at least 17 to 18 m at rates of between 3.6 and 5.4 mm/yr (ages and altitudes from samples 19 and 62, Table F3.2, Appendix) if it assumed that the reefs formed at or just below sea level and discounting effects of compaction and eustatic sea-level changes. Interestingly, this date coincides with the first records of human settlement in this area, as shell middens from Ras Isa (Figure F3.3) have similar ^{14}C ages (personal communication, C. Phillips, 1994, University College London) indicating that fishing communities used the emergent islands or shoreline over the rising salt dome.

In summary, the mapping and dating of the shallow-water reef limestones indicate two phases of reef growth and uplift, interrupted by a phase of subsidence. We are unaware of other records of salt dome subsidence and suggest that it may also be a larger-scale process as onlap surfaces from the seismic profile of the Salif diapir (Figure F3.6) suggest a period of uplift and deformation followed by subsidence, base level rise, sediment accumulation and onlap.

CONCLUSIONS AND IMPLICATIONS

1. This chapter and our related contribution concerning the structural evolution of the Salif salt dome (Davison et al., 1996) represent one of the first detailed studies of the sedimentary cover over the Miocene salt domes in the Red Sea (also Orszag-Sperber et al., this volume) and indicate that Miocene evaporites are overlain by Pliocene and Quaternary carbonate and siliciclastic sediments.

2. A review of the lithostratigraphy of this basin margin from new outcrop data integrated with previously published well and seismic information indicates that three major lithotypes are present which are informally named the Salif Evaporite, the Al Milh Sandstone and the Kamaran Limestone. The middle to late Miocene Salif Evaporite thins on to the basin margin and is deformed into a range of diapiric structures, some of which reach the surface as in the Salif salt wall. The Al Milh Sandstone is mainly continental and initially thickens basinward but is locally ponded behind the Salif salt wall and is thinly developed offshore. The Kamaran Limestone occurs over the Salif salt wall and then thickens offshore through Kamaran Island and to the Al Meethag-1 well.

3. The Salif salt dome is the exposed section of a north–south salt wall that has risen some 3000 m from the regional top salt horizon. It is intensely deformed and folds are picked out by deformed and boudinaged rafts of sulphate evaporites and dolomites. The cover rocks are deformed by extensional growth faults.

4. The Al Milh Sandstone around Salif comprises sands, muds and conglomerates produced by erosion and transport of clasts from the Precambrian, Mesozoic and earlier Tertiary volcanics of the Great Escarpment rift shoulder. The facies represent deposition in fluvial wadis and overbank environments, beaches, aeolian dunes and lakes or lagoons of the precursor Tihama Plain. The shoreline has been located around the Salif salt wall since at least the early Pliocene.

5. A range of shallow-marine carbonate and evaporite lithologies make up the Kamaran Limestone which was deposited on top of the salt and adjacent to ridges of carbonate and sulphate lithologies left by sea-floor dissolution. Immediately overlying the salt are subaerial scree deposits of locally derived gypsum which are overlain by transgressive shelly dolomite conglomerates, bioclastic grainstones, rudstones, and patch reefs.

6. Patch reefs are largely constructed by an unusual form of the coral *Galaxea* with exceptionally long (20–30 cm) corallites. Although these unusual *Galaxea* frameworks together with a fauna of 24 other coral taxa are preserved within a metre of the dissolved surface of the salt it is unlikely that the ambient marine waters were especially saline as the majority of the corals present are not known to be tolerant of increased salinities. It is probable that the bottom waters at Salif were salinity stratified. However, a number of the corals (including the elongate form of *Galaxea*) are known today from turbid shoreline waters. Independent evidence for increased turbidity at Salif comes from the siliciclastic-rich sediment surrounding the corals.

7. Dating of the limestones by ^{14}C and ^{87}Sr/^{86}Sr, combined with their stratigraphic position, orientation and altitude indicate that earlier tilted, Pliocene limestones occur on the eastern and western flanks of the salt wall and that the youngest Quaternary limestones rest uncomformably and horizontally on top of the salt. This suggests an early phase of uplift and erosion followed by subsidence, Quaternary limestone formation, and the final uplift of the salt dome which is continuing today.

8. The wider implications of this work are that salt walls may form barriers to separate clastic from carbonate facies in rift basin–margin environments; that reefs may immediately colonize salt domes penetrating the sea floor; and that salt domes subside as well as move upwards.

ACKNOWLEDGEMENTS

This work was carried out with funding from a Royal Society Expeditions grant, British Petroleum and EC Science grant 'RED SED' which are gratefully acknowledged. We also thank the Governor of Hurghada and the managers of the Salif Salt Company who enabled us to carry out fieldwork in Salif and Kamaran during 1991 and 1993.

APPENDIX

Dating the Salif cover rocks

Diagenetically unaltered mollusc shells and corals were collected from logged sections in the field for isotope dating. All samples were checked petrographically in the laboratory and all heavily bored, recrystallized, or neomorphosed shells were discarded. The samples were cleaned of any matrix or cement mechanically by scraping, drilling and finally in an ultrasonic bath using distilled water. Corals were fragmented to remove debris and cements within the skeletal chambers and were checked for aragonitic composition using X-ray diffraction (Vita-Finzi, 1991). *Tridacna* shells were sectioned and acetate peels were made to check for pristine, unaltered shell material. Initially all samples were dated using ^{14}C but a number were found to be too old for this technique and were subsequently dated using ^{87}Sr/^{86}Sr isotope dating. The methods and results for these two techniques are given below. Care should be exercised in comparing the results from these two techniques because they are effective for different time ranges and have very different values of precision. Our ^{14}C technique has a maximum range back to 15 000 BP beyond which counts are similar to background values. ^{87}Sr/^{86}Sr isotope dating has far less precision and accuracy but is a useful technique throughout the late Cenozoic and is based on the evolving strontium ratio with time (Figure F3.9). It has a precision varying from 0.1 to 1 m.y. depending on the age of the sample and its position on the isotope ratio curve (Figure F3.9).

Strontium isotope dating – methods and results

Original calcitic *Tridacna* shells were selected for ^{87}Sr/^{86}Sr isotope dating (Faure, 1977; De Paolo, 1986)

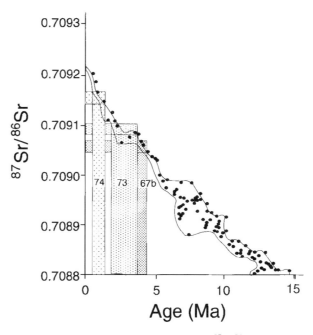

Figure F3.9 Graph illustrating evolution of ^{87}Sr/^{86}Sr in marine waters from the Miocene to the present day (after Beets, 1992; Hodell *et al.*, 1991) and the graphical technique used to establish geological age from the ^{87}Sr/^{86}Sr ratio of marine fossils (*Tridacna*) from Salif and Kamaran.

on the 5 collector thermal ionization mass spectrometer (VG 354) in the Royal Holloway Radiogenic Isotope Laboratory. Table F3.1 lists the samples, their altitudes above sea level, and ^{87}Sr/^{86}Sr ratios for material collected from the Salif and Kamaran areas. No suitable material was found in the diapiric sediments or from the Al Milh area.

To obtain ages from strontium isotope ratios it is necessary to compare analysed ratios from unaltered marine fossils with ratios obtained from chronostratigraphically constrained samples. For this comparison we use data from Beets (1992) and Hodell *et al.* (1991) from the Gulf of Aden at DSDP sites 588, 588A and 722 (Figure F3.9). The spread of published ratios and dates have been enveloped to show minimum and maximum strontium ratios and dates. Hodell's results have been standardized to those of Beets (1992) to correct for their SRM standard value of 0.710235. Our analysis of the standard SRM 987 gave a mean value of 0.710250 ± 23 (N = 36). No standardization was required for our data

Table F3.1 Samples and ^{87}Sr/^{86}Sr ratios of fossils from Salif area

Sample (height ASL)	Material (facies)	^{87}Sr/^{86}Sr (± 0.000011)	Age (Ma) (Min-Max)	Epoch (stage)
Salif sample 67b (log 5, Fig. F3.6) (29 m)	*Tridacna* (Bioclastic grainstone)	0.709057 (−0.709046 +0.709068)	3.5 (2.25–4.7)	Pliocene (Zanclean)
Kamaran sample 73 (log 7, Fig. F3.6) (7 m)	*Tridacna* (Bioclastic rudstone)	0.709092 (−0.709081 +0.709103)	2.74 (1.8–3.65)	Pliocene (Piacenzian)
Kamaran sample 74 (log 7, Fig. F3.6) (8.5 m)	*Tridacna* (Patch reef)	0.709160 (−0.709149 +0.709171)	0.9 (0.5–1.3)	Pleistocene

Table F3.2 Details of samples and ^{14}C dates from fossils from Salif area

Sample	Height ASL (metres)	Material (facies)	Age BP	Age range
NY DB 19 (UCL-300) (Log 1, Fig. F3.7)	20	*Tridacna* (patch reef)	3 700	±250
NY DB 25 (UCL-303) (Log. 63, Fig. F3.4)	6–8	*Tridacna* (bioclastic rudstone)	12 200	±400
NY DB 49 (UCL-380) (Log. 1, Fig. F3.7)	16.5	misc. bivalves (bioclastic rudstone)	3 900	±400
NY DB 56i (UCL-379) (Log. 2, Fig. F3.7)	19	coral (bioclastic rudstone)	3 800	±200
NY DB 62 a(UCL-381) (Log. 4, Fig. F3.7)	21	coral (patch reef)	5 700	±250

because our results are within six decimal places of Beets' standard value. Examples of our ratios are also given (Table F3.1) together with their analytical errors to show how age ranges are obtained graphically. Ages are subsequently quoted as the midway figure in Ma.

The strontium ratios (Table F3.1) indicate a range in ages from Pliocene to Pleistocene for samples from Salif and Kamaran. The Salif sample 67b with an age of 3.5 Ma (29 m height ASL) is from the outcrop of Kamaran Limestones to the east of Salif (Figures F3.5, F3.6). These sections dip at 10° and 25° and are considered to have a faulted contact with the adjacent sands and limestones with much younger ages (5700, 12 200 BP see below). The latter sites are topographically lower at 21 m and 6–8 m respectively indicating an inverse or offlapping stratigraphy for the limestones at Salif. However, dates from within the one section at Kamaran indicate upward younging as would be expected although the vertical accretion rates are low for reefal limestones at 0.8 m/m.y.

^{14}C dating

Samples (25 g minimum) were ^{14}C dated at the carbon 14 dating laboratory at University College London (Vita-Finzi, 1991). Ages are given in conventional years BP together with an age range derived from the standard error of counts. Sample details and results are given in Table F3.2.

Samples 19, 49 and 56 all give statistically identical dates and all come from outcrops of horizontally bedded bioclastic rudstones or patch reefs just above the dissolved salt surface from the western side of the salt quarry (Figures F3.4, F3.7). They represent the youngest strata dated and the most recent uplift event of the salt dome. Sample 62 comes from a coral patch reef in horizontally bedded sediments just to the east of the salt quarry (Figure F3.4,) and indicates a slightly older (5700 BP) colonization event on the salt dome as it impinged on the sea floor. The slightly higher altitude of this sample compared with those to the west of the quarry is considered to represent an original topographic high which would have resulted in earlier sea-floor colonization of corals followed by later colonization of lower marginal areas at 3700–3900 BP. The oldest sample dated by ^{14}C dating (sample 25) gives 12 200 BP and this comes from a similar marginal position off the crest of the dome similar to the dipping Pliocene reefal carbonates from log 5 (Figure F3.4). Sample 25 is from subhorizontal bioclastic rudstones from the small quarry (loc. 63, Figure F3.4) where underlying siliciclastic strata dip at up to 75° to the west-south-west. These topographically lower, but older rudstones, must have formed at only a few metres of water depth and therefore represent an earlier submergent phase of the salt dome compared with the younger 5700–3700 BP reefal phase discussed above. Because it is topographically lower than the younger reefs this 12 200 BP event is thought to be an older shallow-marine phase followed by possible uplift and then submergence before the most recent uplift phase which raised the youngest reefs to their present altitude.

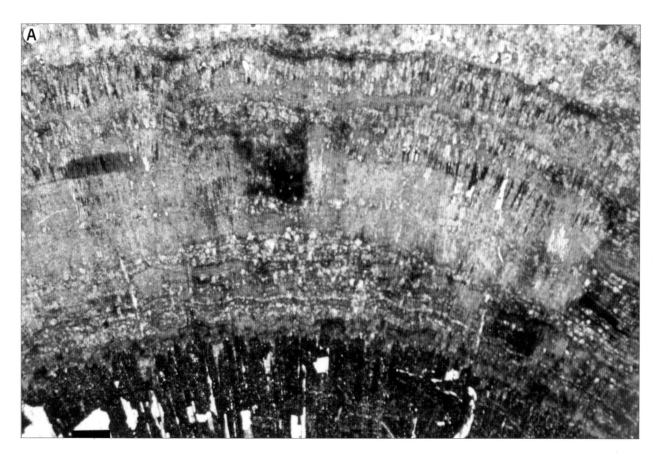

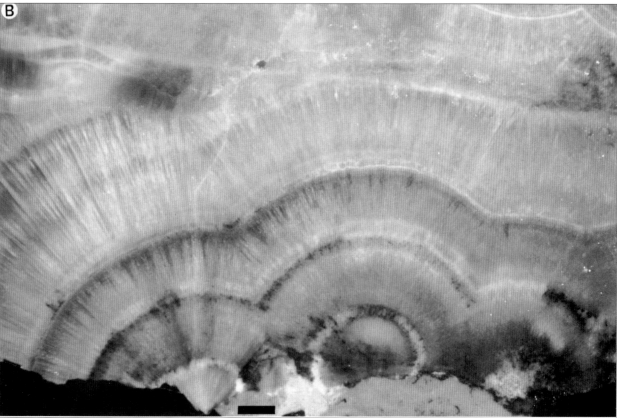

Plate 23 Diagenesis of aragonitic speleothems, Um Gheig (after Aissaoui, 1984).
A. Fibrous aragonite cement replaced by dolomite; aragonite = black; calcite = pink; dolomite = beige. Dolomite appears to be replacing the calcite, its selective replacement of certain laminae possible reflecting minor variations in the composition of the calcite; scale 50 m. B. Polished surface of aragonite speleothem; scale 5 mm.

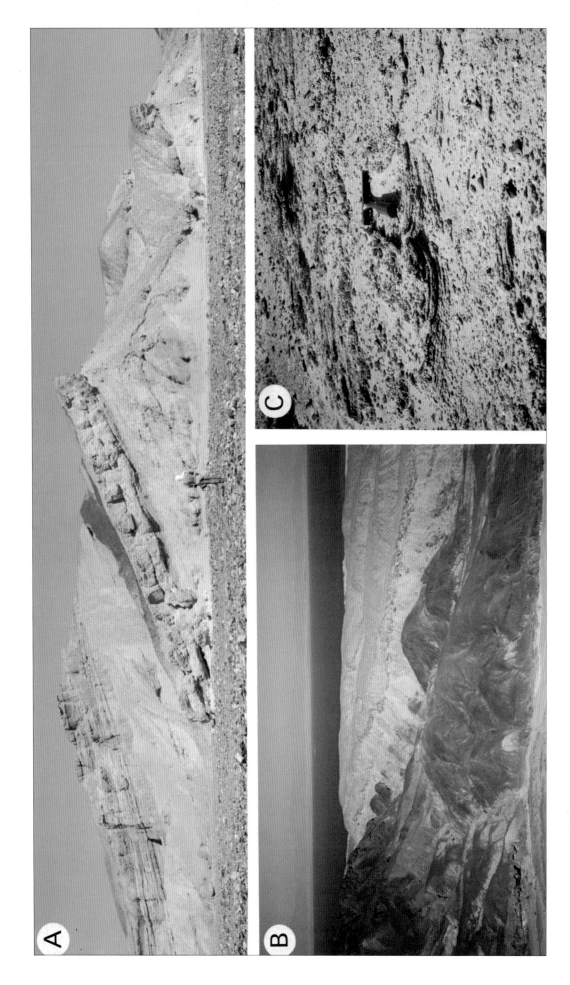

Plate 24 Evaporites, NW Red Sea. A. Outcrop at Wadi Gasus, showing the massive beds of Group A gypsum, intercalated in green marine marls. B. Gebel el Zeit (eastern part) showing the contact between evaporites (white) (group C-gypsum) with an eastward dip, and the basement (black). C. Pliocene evaporites: teepee structure western part of Gebel el Zeit. (see Figure F1.1 for location).

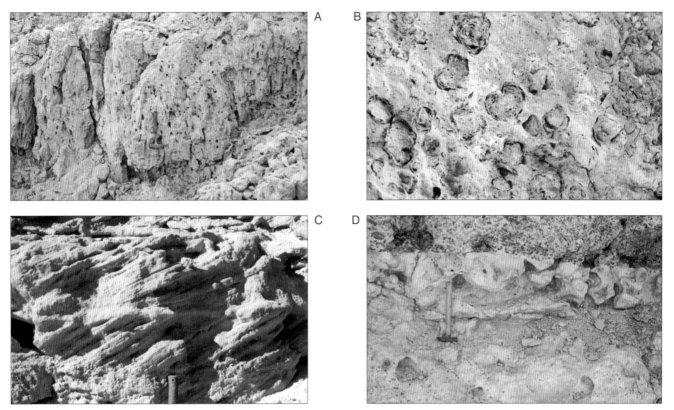

Plate 20 A. Reef framestone of branching coral *Stylophora* exposed in Wadi 3 in depositional sequence 1 (hammer for scale). B. Rhodolith rudstone with concentric crustose rhodoliths set in coarse bioclastic packstone matrix, Wadi Kharasa (50 cm field of view). C. Tabular cross-stratified, coarse-grained arkose in depositional sequence 2, Wadi Bali'h, section B (4 cm diameter hammer handle). D. Truncated reef at top of DS 4 (Wadi 3).

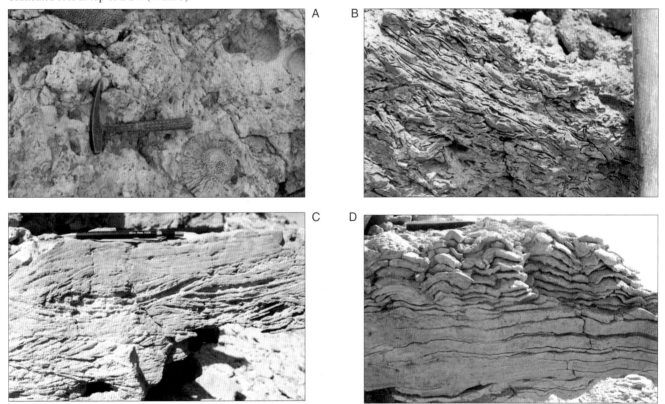

Plate 21 A. Faviid coral floatstone with chaotically arranged corals from slope facies in depositional sequence 5 (small wadi south of Wadi 2). B. Molluscan grainstone from slope facies in depositional sequence 5 (small wadi south of Wadi 2). Original depositional slope to right (4 cm diameter hammer handle). C. Bidirectional trough cross-laminated ooid grainstone from platform margin in depositional sequence 8 (Wadi 5). (12 cm pen). D.Cryptalgal laminites passing up to stromatolite domes (laterally linked) in dolomicrite. Depositional sequence 8, Wadi Kharaza (20 cm hammer head). Abu Shaar platform, SW Gulf of Suez.

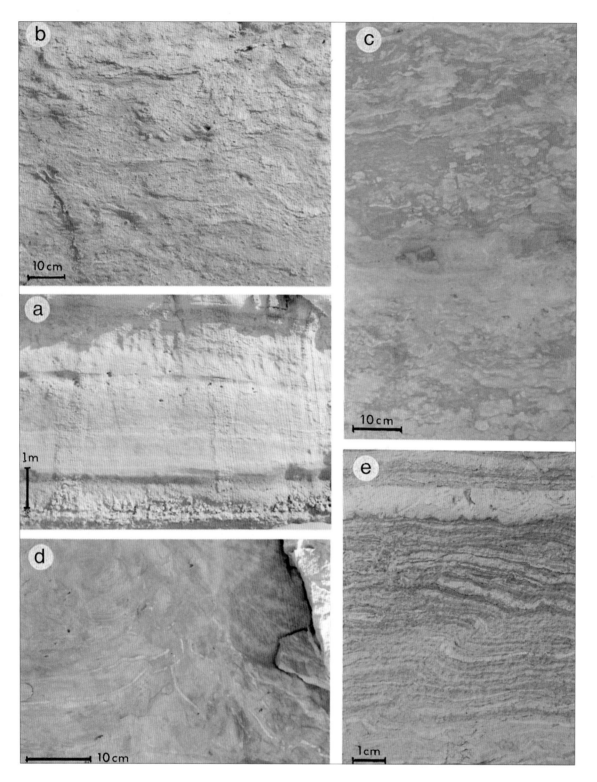

Plate 22 Seismite structures. (a) Massive beds in the Middle Miocene Kharasa Member. The apparent absence of layering in these pink coloured detrital and bioclastic beds is interpreted as resulting from deformation of interbedded muddy sands and muds within a 5 m thick shear-zone; Wadi Bali, Abu Shaar hangingwall. (b) Lensoid fluidal structure with microfolds characteristic of the pink coloured lower part of Kharasa Member. This shallow marine facies has been affected by the lateral displacement; Wadi Bali, Abu Shaar. (c) Detail of photo a showing the lumpy structure in the upper muddy unit. Proximity to the stretched level (see photo b) suggests that this unusual structure results from the shearing and deformation of a layered mud. (d) Fluid-escape pillar structure in bedded oolite. The bent and dislocated muddy layers indicate the transitional (semiplastic) state of the liquefied sand: hydroplastic folds grade into more fluidized materials with mud clasts, located within the axial part of the structure: SW Wadi Bali, Abu Shaar. (e) Laminated mud and silts exhibiting two deformational styles. The lobate loadcast-like and boudinage structures represent *in situ* (static) deformation interpreted as being the consequence of vertical compression exceeding lateral confinement pressure. The mud layers behaved as plastic, more cohesive beds relative to the yellow silts. An asymmetric folding and microfaulting shows that this deposit suffered a subsequent intraformational shear also responsible for stretching. Both deformations result from the shearing due to lateral sliding of the overlying C1 laminite; Wadi Siatin, Egyptian Red Sea Coast.

Plate 25 Post-Miocene sediments, NW Red Sea. A. Gubal Island showing typical depositional sequence ranging from laminated carbonate silts at the base (1), grading upwards into bioturbated carbonate muds (2) and terminating in cross-bedded oolitic sands (3). The sequence is unconformably overlain by a Quaternary reef-terrace (4) at an altitude of 40 m. B. View westward from Giftun Kebir Island showing shoreline immediately south of Hurghada and elevated rift periphery. The post-Miocene carbonates in foreground, only 7 km from the periphery, have virtually no siliciclastics suggesting that a structural depression separates the Giftun structure from the nearby peripheral sources.

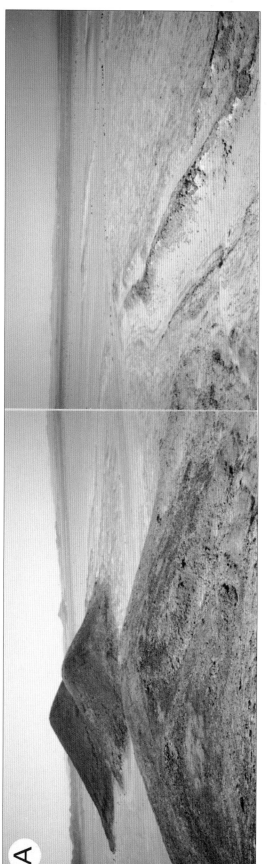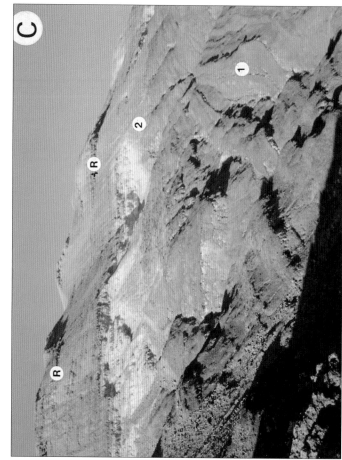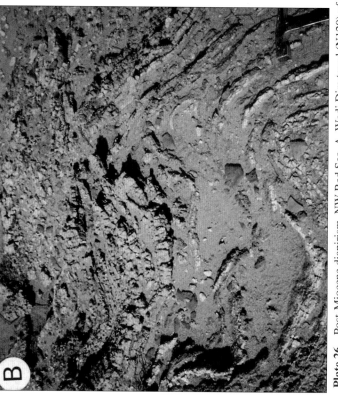

Plate 26 Post-Miocene diapirism, NW Red Sea. A. Wadi Dira trend (N120) of diapirs. S. of Marsa Alam, NW Red Sea Coast. Note the deformation of the gypsum beds around the hills. B. Detail of the deformation of gypsum. C. South-west coast of Shadwan Island showing diapir of Miocene gypsum. 1 – Miocene gypsum; 2 – top of the diapir; R – Pliocene reefs.

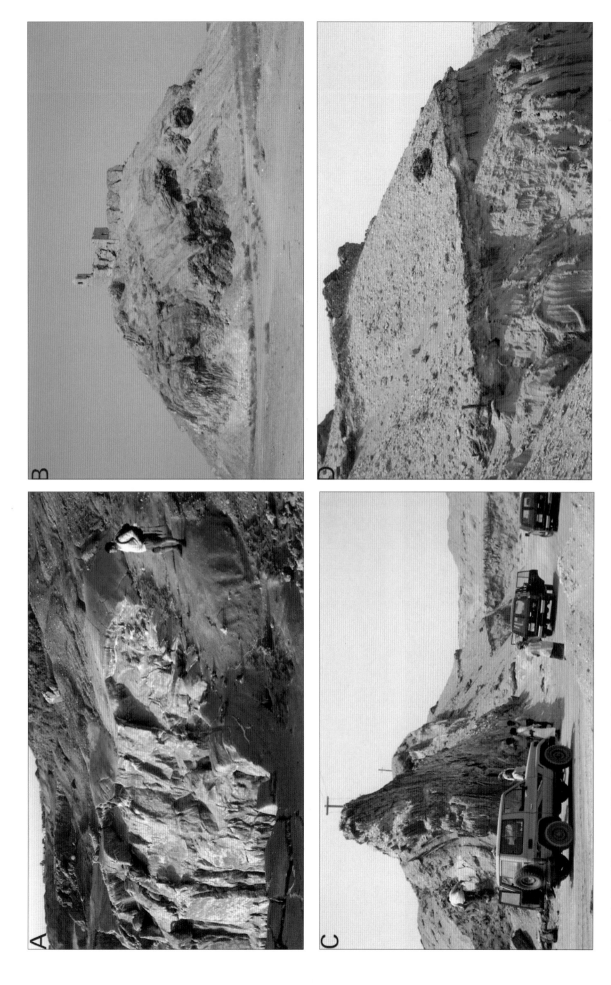

Plate 27 Post-Miocene diapirism, W. Yemen. A. Upper dissolved surface of salt and overlying Quaternary deposits on eastern edge of Salif quarry. B. Jabal Al Milh with extended mosque and overturned and folded section of overburden (log 9, Figure F3.5). C. Vertical bedded sulphates, carbonates, siliciclastics forming ridge (log. 10, Figure F3.5) through halite dissolved to a lower level. The pre-Quaternary unconformity above the salt is shown as a gently sloping resistant level above the two black vehicles to the right. D. Pre-Quaternary unconformity over vertically bedded halite (detail of Plate 27C). Hammer head rests on halite erosion surface. Lowest Quaternary bed has abundant gypsum and dolomite pebbles, followed by 6 m of bioclastic rudstones dated at 3700 to 3900 BP at 16 to 22 m above sea level.

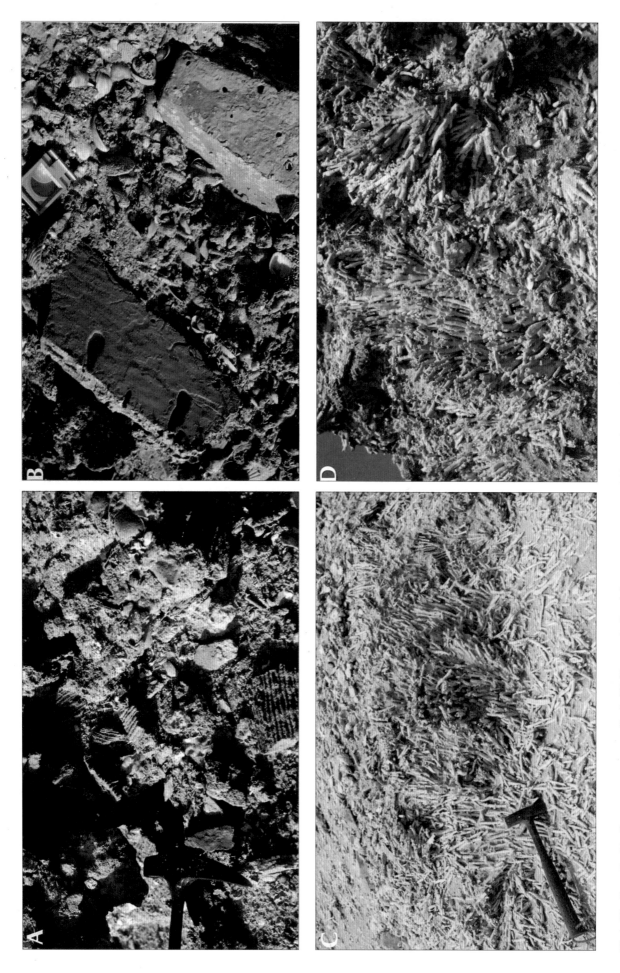

Plate 28 Post-diapiric Quaternary cover, on diapir, W. Yemen. A. Basal Quaternary gypsum pebble conglomerate with matrix containing marine molluscs, (detail of lowest Quaternary bed in Plate 27D). Note differential dissolution ridges on surfaces of pebbles. B. Bored and encrusted pebbles in dolomite conglomerate with shelly matrix, (detail of lowest Quaternary bed in Plate 27D). C. Patch-reef with *in situ* thicket of coral *Galaxea fascicularis* overlain by bioclastic rudstone (log. 1 Salif, Figure F3.7). D. Detail of *Galaxea fascicularis* with associated molluscan fauna (log 4, Figure F3.7) illustrating the lower coenosteum and elongate branches of these colonies (gastropod lower right 4 cm across.)

Section G
Post-rift axial sediments and geochemistry

Incipient oceanic troughs, over 2 km deep, became established in the late Miocene in the central and southern Red Sea. Cored sediments from these axial troughs are examined in this section in three related contributions arising from recent cruises by Italian and French research vessels. These axial sediments are shown to be diverse in their nature and origin and are strongly affected by early marine diagenesis.

G1 Axial sedimentation of the Red Sea Transitional Region (22°–25°N): pelagic, gravity flow and sapropel deposition during the late Quaternary
 M. Taviani

G2 Sedimentation, organic geochemistry and diagenesis of cores from the axial zone of the southern Red Sea: relationships to rift dynamics and climate
 P. Hofmann, L. Schwark, T. Brachert, D. Badaut, M. Rivière and B.H. Purser

G3 Metalliferous sedimentation in the Atlantis II deep, Red Sea: a geochemical insight
 G. Blanc, P. Anschutz and M.-C. Pierret

Chapter G1
Axial sedimentation of the Red Sea Transitional Region (22°–25° N): pelagic, gravity flow and sapropel deposition during the late Quaternary

M. Taviani

ABSTRACT

The Red Sea is an under-supplied marine rift basin. Study of sediments based on core and dredge samples from the Transitional Region (22°–25° N) show that offshore Quaternary sedimentation is predominantly pelagic-carbonate oozes with $CaCO_3$ contents up to 70 to 80%. Interruptions of this standard sedimentation style are due to climatically driven hydrological anomalies connected to sea-level lowstands (hard layers) and pluvial phases (sapropels). By using the last glacial/interglacial cycle as a reference, it is apparent that the entire basin changes from organic to inorganic carbonate factories in tune with sea-level fluctuations, and that these events affect sedimentary processes and CO_2 sinks differently. While climate seems to be the leading factor in forcing sedimentation patterns in the offshore Red Sea, structure primarily controls the infilling of embryonic oceanic troughs (e.g. Nereus Deep) which, besides hydrothermal metal-enriched sediments, receives a significant input through mass gravity transport, mainly pelagic-calciturbidites. Various kinds of pelagic carbonates (ranging from hardgrounds to friable pteropod limestones) and mixed MORB volcanoclastic–carbonate rocks formed on seamounts and slopes of axial trough segments.

INTRODUCTION

The study of the Red Sea rift significantly contributes to the understanding of the sediment-infill history of recently formed segments of oceanic troughs. Thus, an appreciation of processes active within such an embryonic ocean basin will ultimately provide data for modelling early oceanic sedimentation patterns. The

Deep Sea Drilling Project Leg 23 drilled six sites in the Red Sea and this provides most of the documentation at present available on the nature of the offshore syn-rift sediments resting above the Miocene evaporites (Ross et al., 1973). A summary of the main sedimentary patterns of the axial zone of the Red Sea rift combining DSDP and well evidence is given by Stoffers and Ross (1977) and Coleman (1993). Information on the uppermost (Quaternary) part of the basin's deep-sea infill are found in a few specifically sediment-oriented contributions (e.g. Milliman et al., 1969; Stoffers and Ross, 1977) while most deal with palaeoceanographic aspects (e.g. Degens and Ross, 1969; Locke and Thunell, 1988). For more than two decades much attention has been devoted to the metal-enriched sedimentation which imprints so much of the deposits accumulating within the deepest parts of the Red Sea (e.g. Degens and Ross, 1969; Bignell et al., 1976; Bignell, 1978; Guennoc and Thisse, 1982; Bäcker, 1982; Thisse and Guennoc, 1983; Karbe, 1987; Blanc et al., this volume). Thus, in spite of its obvious geodynamic importance, knowledge of the sedimentary evolution of the axial zone of the Red Sea is not as advanced as one might expect for such an important rift basin.

The aim of this chapter is to provide information on the main characteristics of axial sedimentation in the north-central part of the Red Sea basin. I do not discuss here the metalliferous sediments. Data presented here were obtained from coring and dredging of the Red Sea between 22°–25° N and originate from the MR79 and MR83 oceanographic cruises carried out aboard the Italian cableship *Salernum* and RV *Bannock* respectively (Figure G1.1(a)). Location and attributes of samples discussed in this chapter are given in Table G1.1.

Sedimentation and Tectonics of Rift Basins: Red Sea–Gulf of Aden. Edited by B.H. Purser and D.W.J. Bosence. Published in 1998 by Chapman & Hall, London. ISBN 0412 73490 7.

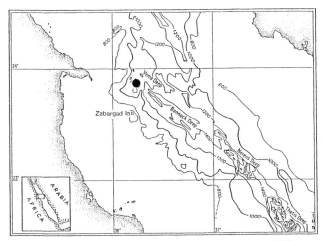

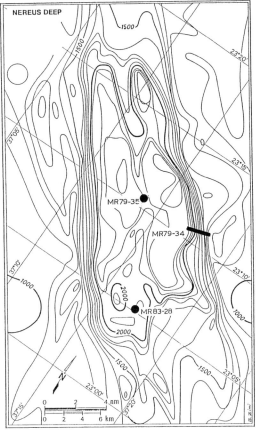

Figure G1.1 Transitional Region of the Red Sea. (a) General bathymetric map of the study area. Circle is the location of core MR79-46, adopted as the standard reference for non-metalliferous deep-sea sedimentation for the Transitional Region. (b) Detailed bathymetry of Nereus Deep, the northernmost segment of truly oceanic trough in the Red Sea; note complex topography, steepness of walls and location of cores MR79-35 and MR83-28 with turbidites discussed in the text and dredge station MR79-34 providing various types of hardgrounds and pelagic limestones.

GEOLOGICAL AND HYDROLOGICAL SETTING OF THE STUDY AREA

The marine portion of the Red Sea rift is a branch of the extensional system comprising the continental East Africa Rift and the Gulf of Aden Rift (Freund, 1970; Coleman, 1993). The Red Sea basin is an elongated depression 2000 km long, 250–450 km wide, bounded by the uplifted shoulders of the Arabian and Nubian plates. Morphologically, we can distinguish a shallow and wide main trough and a deep (up to 2500 m) and narrow segmented axial trough (Coleman, 1974). The hydrological budget of the present day Red Sea is characterized by evaporation exceeding run-off, resulting in a net loss of 2 m/year (Siedler, 1969). Balance is kept by water exchanges with the Indian Ocean via the shallow (∼137 m: Werner and Lange, 1975) Bab-el-Mandab sill. Bottom temperature (21°C) and salinity (42‰) are considerably higher than in oceans at comparable depths (Morcos, 1970; Edwards, 1987). Active hydrothermalism occurs in the deepest depressions and is accompanied by formation of brines and precipitation of metals (Degens and Ross, 1969; Bignell et al., 1976a; Bäcker, 1982). Temperature and salinity of such depressions may reach values considerably higher than those of brine-free environments (Degens and Ross, 1969; Karbe, 1987).

The Red Sea today shows different stages of oceanization with rifting propagating northwards (Cochran, 1983b). True sea-floor spreading began in the southern part of the basin c. 4 m.y. ago (Cochran, 1983b) causing the formation of a wide (5–30 km) and deep (up to 2500 m) axial trough (Braithwaite, 1987; Coleman, 1993). North of about 21° N oceanization has been suggested as limited sea-floor spreading is by discrete cells (Bonatti et al., 1984; Bonatti, 1985). The most recent, and most noteworthy truly oceanic segment, accompanied by emplacement of MORB basalt, is the Nereus Deep (or Trough) at 23°–23°30′ N. North of 25°N, the axial trough is more subdued since the basin is still under diffuse extension (Cochran and Martinez, 1988) with only some incipient, ultra-slow sea-floor spreading volcanic zones (Guennoc et al., 1990).

The Transitional Region (or Zone) is a key sector of the Red Sea basin between 20°–25° N, at the intersection of an oceanic ridge and a continental rift (Cochran, 1983b; Bonatti et al., 1984; Marshak et al., 1992). This region is characterized by a complex morphology where segments of axial troughs at different evolutionary stages intercalate with seamounts and unoceanized areas of the main trough (Pautot, 1983; Bonatti et al., 1984). Nereus Deep (Figure G1.1(b)) is a 40 km long, 12 km wide, U-shaped trough where emplacement of oceanic crust through sea-floor spreading is presently taking place (Bonatti et al., 1984). The Nereus Deep is a small-propagating oceanic rift which began opening at its southern tip about 2–3 m.y. BP but only later, about 1 m.y. BP, at its northern end (Bonatti et al., 1984). Dredging from its axis recovered fresh basalt with a MORB affinity (Bonatti et al., 1984). Today the trough

Table G1.1 Location and main attributes of core and dredge samples from the Red Sea Transitional Region: NRSS = normal Red Sea sediments, HL = hard layer, S = sapropel, MS = metal-enriched sediment, T = turbidite, HG = hardground, PL = pteropod limestones, VB = volcanoclastic-carbonate breccia

Core #	Lat. N	Long. E	Water depth (m)	Recovery (cm)	Notes
MR 79 3	23°43'50	36°13'00	925	240	NRSS
MR 79 16	23°04'95	37°55'97	983	107	NRSS, HL
MR 79 17	23°10'92	37°05'72	1128	253	NRSS, HL
MR 79 20	23°21'23	37°01'07	1379	456	NRSS, S, HL
MR 79 22	23°14'24	37°28'30	937	326	NRSS, HL
MR 79 29	23°10'96	37°13'19	2446	480	MS, T
MR 79 32	23°07'75	37°18'57	2216	463	MS
MR 79 35	23°11'46	37°14'70	2452	246	MS, T
MR 79 36	23°19'67	36°27'27	723	303	NRSS, S, HL
MR 79 37	23°29'65	36°40'13	1782	326	MS, NRSS, S
MR 79 39	23°37'19	36°55'48	1330	475	NRSS, HL
MR 79 42	23°46'89	37°09'84	860	377	NRSS, HL
MR 79 46	23°53'75	36°26'80	1224	526	NRSS, S, HL
MR 83 28	23°05'00	37°17'00	2211	678	MS, T
Dredge #	Lat. N	Long. E	Water depth (m)		Notes
MR 79 34	23°10'89	37°16'25	1749		HG, VB, PL
	23°11'89	37°17'84	1172		
MR 79 40	23°50'18	36°50'93	710		HG
	23°49'96	36°51'79	934		
MR 79 41	23°49'32	36°49'58	722		HG
	23°49'70	36°49'69	578		

is the site of intense metal-enriched sedimentation and brine discharge (Bignell and Ali, 1976; Bignell et al., 1976a; Bonatti et al., 1986). The accumulation rate of metal-enriched sediment in core MR83-28 is about 50 cm/10^3 years, i.e., comparable to other oceanized cells to the south (Taviani, 1984).

AXIAL ZONE SEDIMENTS

Definition and composition of normal Red Sea sediments: Quaternary biogenic calcareous oozes

The Red Sea basin is a sediment-starved rift. The Quaternary Red Sea is largely an arid land-locked basin. Terrigenous input to basinal sedimentation is subordinate and reflects the region's dominantly arid conditions which prevent significant fluvial discharges, at least for most of the Quaternary. Detrital mineral phases include quartz, feldspars, dolomite, clay minerals and micas (e.g. Besse and Taviani, 1982; Hofmann et al., this volume). Aeolian transport may be an important source of particles but their impact on deep-sea sedimentation has never been correctly evaluated. Atmospheric dust mesh-samples collected over the offshore north-central Red Sea reveal a modest load of fine particles of dominantly clay minerals (kaolinite, smectites), carbonate, quartz and palygorskite (Tomadin et al., 1989). Under-supply by terrigenous particles is partly balanced by pelagic sedimentation due to accumulation of tests of planktic organisms. Normal Red Sea Sediment (NRSS, thereafter) typically contains up to 70–80% biogenic $CaCO_3$ and is a structureless pteropod–globigerina–nanno-ooze (e.g. Degens and Ross, 1969). NRSS is dominantly a mixture of both low-Mg calcite (coccoliths and planktic forams) and aragonite (thecosomatous pteropods) mineral phases. Due to Red Sea hydrological peculiarities (high S and T, and deficiency in nutrients), and its geographic (desert-enclosed basin) and geometric (shallow-silled) configuration, calcareous plankton have a relatively low diversity by comparison to other tropical–subtropical oceans. Coccolithophores show an uneven distribution in the modern Red Sea and a few species (*Gephyrocapsa oceanica, Emiliana huxleyi, Umbellosphaera irregularis, G. ericsonii*) dominate the floral assemblages (e.g. McIntyre, 1969; Okada and Honjo; 1975, Winter et al., 1979). This skewness is reflected by the dominance of a small number of coccolithophorid taxa in the skeletal populations of the NRSS clay fraction (McIntyre, 1969; Winter et al., 1979). Similar considerations apply to planktic forams as well. Rather typically, late Quaternary living and skeletal planktonic foram assemblages are conspicuously dominated by few taxa (e.g. Berggren and Boersma, 1969; Behairy and Yusuf, 1984; Ivanova, 1985), among which *Globigerinoides sacculifer, G. ruber, Globigerinella siphonifera* and *Orbulina universa* are prominent in forming the sandy fraction of the NRSS. Notably, *G. sacculifer* and *O. universa* attain an abnormally large size in the Red Sea and this fact has sedimentological repercussions; in fact, winnowing of NRSS by bottom currents produces a coarse sand with a high percentage of *G. sacculifer* and *O. universa* tests, often associated with pteropod shells because of a preferential removal of smaller forams (like *G. ruber*, globigerinids and juveniles). Similar sediments occur on the margins of the Nereus Deep and may eventually be lithified as pteropod–globigerina limestones (see discussion below). Thecosomatous pteropods are also responsible for the coarser fraction of the NRSS and are relatively diverse (e.g. Herman, 1971; Almogi-Labin, 1982; Almogi-Labin *et al.*, 1986; Ivanova, 1985). Subordinate to the pelagic holoplanktic component, NRSS includes other biogenic skeletal particles, i.e. meroplanktic mollusc larvae, benthic forams, ostracods, gastropods, bivalves and decapods. Based on preservation of aragonitic pteropod shells, Almogi-Labin *et al.* (1986) pointed out that aragonite preservation is best during glacial stages, at times of relative low sea levels, while interglacial conditions led to more active dissolution and precipitation of high-Mg calcite cement. Accumulation rates of NRSS are generally 3–10 cm/10^3 years (Degens and Ross, 1969; Ivanova, 1985).

This NRSS composition is valid for the late Quaternary under standard (i.e. temperature and salinity similar to the present day) Red Sea hydrological conditions. During periods of climatically driven, highly disturbed hydrological conditions NRSS, formation and deposition stopped and the basin turned to completely different sedimentation style, as is discussed below.

Although the Red Sea is a semi-enclosed sea whose deeper parts are only tens of kilometres away from the shore, its deep-sea pelagic sedimentation shares all the fundamental attributes found in open ocean-based biogenic oozes (Pickering *et al.*, 1989). Accumulation of typical NRSS is taking place from at least isotopic stage 5 or 6 (last 150 000 years: Schoell and Risch, 1976; Reiss *et al.*, 1980; Ivanova, 1985) but probably since the Pliocene judging from DSDP results (Stoffers and Ross, 1977).

Deep-sea sedimentation in the Transitional Region of the Red Sea

Sediment distribution from cores and grabs indicates what in the Transitional Region NRSS, is typical Quaternary sedimentation over the entire main and axial troughs at depths below 600 m. The rain of pelagic tests obviously also affects the deeper oceanic troughs but this input is largely obscured by the overwhelming precipitation of metal-enriched sediment and gravity transport processes (see below). I have selected core MR79-46 (see Figure G1.1(a)) for location and Table G1.1 for attributes) as my standard reference to illustrate the main sediment types and their palaeoceanographic significance. Core MR79-46 has a continuous record back to ~70 000 years (isotope stage 4) as shown by its correlation with other Red Sea cores based on stable isotope stratigraphy (Taviani, 1983).

Most of the core is NRSS. Detrital mineralogical phases, detected by X-ray diffractometry (Philips Cu Ka), include quartz (average 15%), feldspars (up to 15%), dolomite (5%), clay minerals and micas (D. Besse and M. Taviani, unpublished). NRSS sediment supply approximates rate of accumulation (~5 cm/10^3 years), although this is not necessarily true at times of strong dissolution/lithification. The NRSS accumulation rates are comparably high but still within the world's ocean values for similar sediments (1 to 6 cm/10^3 years: Garrison, 1993; Ross 1995).

This general trend of calcareous pelagic supply is interrupted by two significant palaeoceanographic basinal events. These are sapropels and hard layers which are related to different palaeoceanographic conditions.

Sapropel deposition

Black shales and sapropels have long been known to occur in the Plio-Pleistocene sedimentary record of the Red Sea basin, marking times of bottom-water stagnation (Herman, 1965a; Stoffers and Ross, 1977). The latest of such events took place ~10–11 000 years BP) and is marked by a 2–3 cm thick dark-greenish layer which can be seen in the offshore sedimentary record at depths greater than 800 m (Herman, 1971; Besse and Taviani, 1982; Locke and Thunell, 1988; Taviani, 1995a; Hofmann *et al.*, this volume). Compositional characteristics of the sapropel in core MR79-46 (Figure G1.2) have been described by Besse and Taviani (1982) and Taviani (1995a). Biogenic carbonates are significantly less than in NRSS while terrigenous particles (especially clay minerals) are higher. Total organic carbon is >2%, which is one order of magnitude higher than NRSS (Figure G1.2; Taviani, 1995a).

The formation of this sapropel is intriguing as its deposition followed a period of generalized hypersaline conditions which affected the Red Sea basin during the last glacial. At 18–21 000 years BP (last glacial maximum) a 120 m global sea-level fall must have seriously affected the water exchanges of the shallow-silled Red Sea with the Indian Ocean. This severe reduction led to a rise in the Red Sea salinity to more than 50‰ (e.g. Reiss *et al.*, 1980; Almogi-Labin, 1982; Thunell *et al.*, 1988). The consequences of these high salinity conditions were dramatic and normal marine life (including coral reefs: Taviani, this volume) was probably exterminated and only a few euryhaline species survived (Taviani *et al.*, 1992). Among the latter were the thecosomatous pteropod *Creseis acicula* and the foram *Globigerinoides ruber* which, in fact, are often associated with the lithified layers marking the time of hypersaline conditions (Degens and Ross, 1969). Such conditions

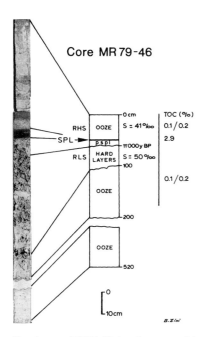

Figure G1.2 Gravity core MR79-46 showing a complete stratigraphy including NRSS (pteropod–globigerina–nanno-ooze), hard layers and sapropels: TOC = total organic carbon; pspl = protosapropel; chronology (y BP) and palaeosalinities (s) are discussed in the text.

prevailed until ~11 000 years BP (e.g. Degens and Ross, 1969; Locke and Thunell, 1988; Brachert and Dullo, 1990). With the advancing deglaciation and related sea-level rise, the hydrological regime of the Red Sea changed again. The end of the high salinity period is marked by deposition of the sapropel whose origin is the result of widespread stagnation in the basin (Figure G1.3). It has been proposed that such stagnation was caused by a wet phase (rainy period) in contrast to the previously dry period (Rossignol-Strick, 1987; Thunell et al., 1988; Taviani, 1995a). According to Rossignol-Strick (1987) the stagnation was density driven and was probably triggered by an input of fresh water into the basin. Enhanced run-off into the Red Sea basin was via entry points such as the now inactive Khor Baraka (the Sudanese delta, DSDP Site 228: Ross et al., 1973).

It is, however, quite clear that the concentration of sea water during the glacial period acted as a preparatory phase leaving the Red Sea basin filled with dense and saline water (Taviani, 1995a). It has also been suggested that the penetration itself of normal saline 'oceanic' water from Bab-el-Mandab during the deglaciation, besides marking the end of the saline domain, would have been capable of bringing about basinal stagnation (Taviani, 1990). Flooding of the basin by water of slightly lower salinity than the dense glacial Red Sea waters would have caused the required contrast of density for the formation of sapropels (Taviani, 1990). In such a case, organic matter contained in the sapropel should be mostly of marine origin. However,

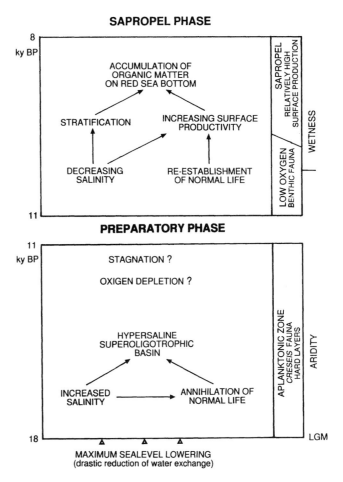

Figure G1.3 Cartoon showing main palaeoceanographic basinal events and their links to general climate. Preparatory phase: time span 18–11 ka BP (LGM = Last Glacial Maximum) as deduced by the Red Sea deep-sea core record. Sapropel phase: time span 11–8 ka BP.

preliminary data on the type of organic matter found in sapropels from the southern Red Sea mainly reflect terrestrial sources (Hofmann et al., this volume) suggesting transport from a strengthened run-off from the mainland.

Hard layers and genesis of pelagic limestones

The steep walls of the axial troughs and seamounts of the Red Sea are sites of extensive submarine cementation leading to the formation of different types of pelagic limestones which record one or more lithification/dissolution events. Deep-sea lithification produces various kinds of indurated carbonates, mostly at the expense of NRSS soft sediment. These have long been known from the Red Sea and various mechanisms of formation have been suggested (Natterer, 1898; Gevirtz and Friedman, 1966; Milliman et al., 1969; Friedman et al., 1971; Friedman, 1972; Milliman, 1977; Brachert and Dullo, 1990; Brachert, 1994). Interglacials, characterized by relative high sea-level conditions, are times of inorganic precipitation of high-Mg calcite (10–

14 mol % of magnesium), while aragonite cements preferentially precipitate during glacial periods, under conditions of more restricted circulation and increased salinity (Milliman, 1977; Almogi-Labin et al., 1986). Formation of such carbonates is not unique to the deep Red Sea and similar carbonates occur in the Mediterranean (Groupe Escarmed, 1986; Allouc, 1990) and Bahamas (Mullins, 1985; Wilber and Neumann, 1993). The most widespread type of lithification is represented by pelagic ooze cemented by aragonitic and high-Mg calcite cements (e.g. Friedman et al., 1971). On the sea floor they represent more or less regular and continuous ledges, not dissimilar from those described by Brachert and Dullo (1991) from Red Sea fore-reef slopes off Sudan. They easily break up during sampling and this is likely to be responsible for the 'hard layers' found in sediment cores at specific stratigraphic intervals of the late Quaternary. Their recurrence within cores proves that, during the Quaternary, the Red Sea witnessed many periods of $CaCO_3$-inorganic precipitation leading to extensive submarine lithification (Herman, 1965b; Degens and Ross, 1969; Olausson, 1971; Milliman, 1977; Hofmann et al., this volume). A critical revision of the genetic problems connected to the formation of the lithified layers is beyond the scope of the present article and only a few points are underlined here. Slabs of such more or less friable limestone have been previously reported as occurring in the Transitional Region (Bonatti et al., 1984). Their heavy isotopic composition for both stable oxygen and carbon ($+6.6 \delta^{18}O$ and $+3.7 \delta^{13}C$: Bonatti et al., 1984) is generally interpreted as a signal of glacial high-salinity conditions (e.g., Deuser and Degens, 1969; Milliman, 1977).

Limestones recovered from the steep flanks of the Nereus Deep and slopes of seamounts comprise: 1. micritic hardgrounds (Figure G1.4(a)), 2. pteropod–globigerina limestones (Figure G1.4(b)) and 3. volcanoclastic–pelagic carbonate (Figures G1.5(a),(b), G1.6(a),(b)).

1. Micritic hardgrounds occur on seamounts and slopes. They are occasionally encrusted by subfossil deep-sea epifauna, mostly serpulid polychaetes and corals (Figure G1.4(a)). Multiple dissolution/lithification events as well as staining by Mn-Fe oxides are normally recognizable and probably indicate relative antiquity of such hardgrounds; encrusting epifauna also indicates a prolonged history of exposure on the sea bottom since the preservation state of calcareous organisms is very variable and ranges from glossy, well-preserved skeletons to highly corroded, recrystallized ones on the same surface (Figure G1.4(a)).

2. Pteropod–globigerina limestones are by far the commonest product of deep-sea submarine lithification (hardgrounds) in the Red Sea. Some clearly form through lithification of NRSS accumulated on the slopes (Figures G1.4(b), G1.5(c), G1.6(c),(d)). There is evidence that such deposits, while still unlithified, may be winnowed by weak bottom currents and that this fabric may be inherited by the limestone. In fact, some of the slabs dredged from the Nereus Deep show a fabric where the fine fraction is lacking and the elongated tests of the thecosomatous pteropod Cresis acicula show a distinct orientation in response to unidirectional currents (Figures G1.5(c), G1.6(c)). Polarity of pteropods has been observed also in other sectors of the Red Sea axial zone (e.g. Chen, 1969). Relatively strong bottom currents capable of forming ripple marks and sand dunes have been documented in the axial zone of the southern Red Sea and related to internal waves developed at brine–sea water interfaces (Young and Ross, 1977). 'Winnowed' pteropod–globigerina limestones indicate that bottom currents are (at least occasionally) active along the edges of troughs and seamounts, as also testified by the hardgrounds colonized by serpulids and corals discussed above. These limestones are often very friable and are weakly cemented by aragonite. Friability, lack of very intense lithification and recrystallization, compositional and petrographic similarities with last glacial hard layers recovered in sedimentary cores, are good indications that such limestones are mainly late Pleistocene in age.

3. Volcanoclastic–pelagic carbonate breccias (Figures G1.5(a),(b), G1.6(a),(b)); these unusual sedimentary rocks have never been reported from the Red Sea. They occur at active spreading centres, such as the Nereus Deep. Fragmental volcanic material is MORB-basalt in composition. Underwater photos and dredging from the Red Sea axial trough have documented the common occurrence of volcanic rock fragments (Young and Ross, 1977; Eissen et al., 1989). Axial volcanism in the Red Sea oceanized troughs is by extrusion of pillow lavas (Young and Ross, 1977; Bonatti et al., 1984; Eissen et al., 1989) which is normally regarded as a quiet process (Bonatti, 1967). Although hydrovolcanic explosions are theoretically possible, since maximum hydrostatic pressure in the Nereus Deep (maximum water depth 2500 m) is c. 250 bars (Cas and Wright, 1988), basalt fragmentation probably is due to brecciation along fault scarps, as suggested by Eissen et al. (1989). Although obviously Quaternary, the exact age of these mixed carbonate breccias is not easily to assess. However, preservation of pteropods is only by moulds and the presence of high-Mg cement points to an interglacial time of formation (Almogi-Labin et al., 1986). Furthermore, basalt must have been extruded some time age since its clasts now occur at shallower depths compared to present-day extrusions.

Figure G1.4 Examples of hardgrounds. (a) Micritic hardground dredged between 700 and 900 m from station MR79-40; observe that fresh detachment surfaces are clear while surfaces witnessing prolonged exposure on the sea bottom are blackened by Fe-Mn oxides; multiple events of $CaCO_3$ precipitation/dissolution are represented in this hand sample; various generations of encrusting benthic organisms form an almost continuous coverage over the exposed surface of the hardground: subfossil serpulid polychaete (*Metavermilia* and *Protula* among others) tubes (S), with various degrees of preservation, while highly worn and recrystallized corals (C) are probably juveniles attributable to *Guynia* sp. (H. Zibrowius, personal communication, 3 May 1996). (b) Pteropod-limestone, more friable than the sample of Figure G1.5(a), was dredged deeper on the steep slope of the Nereus Deep (MR79-34); note monospecific aggregates of the thecosomatous pteropod *Creseis acicula* (P) which is known to be tolerant of very high (50‰) salinities.

Gravity flow deposits

Gravity and piston cores collected from the Nereus Deep indicate the importance of gravitational processes, both in the volume and the recurrence of resedimented beds (Taviani and Trincardi, 1992). These sediments are represented by normally graded pelagic carbonate tur-

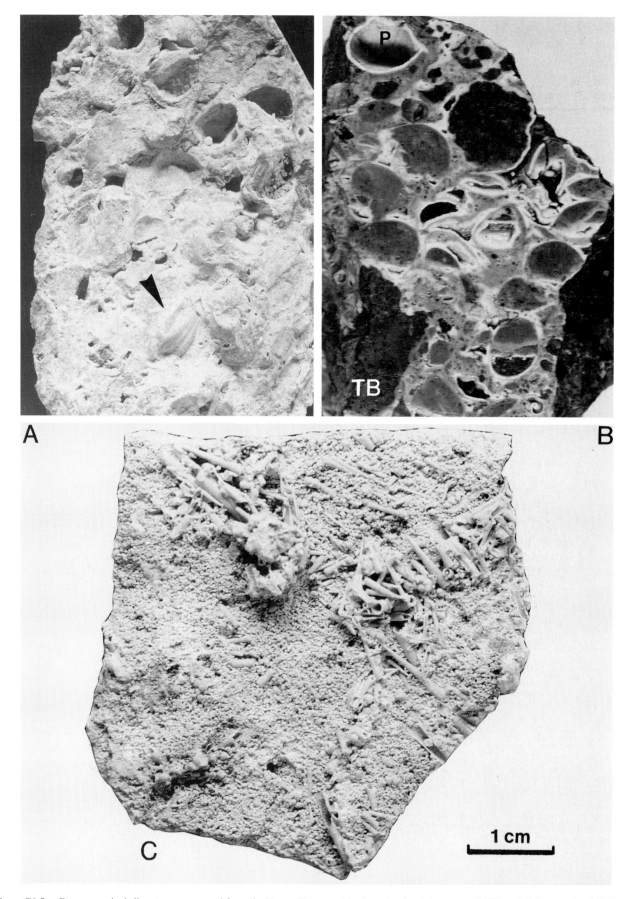

Figure G1.5 Deep-sea pelagic limestones recovered from the Nereus Deep, east flank, water depth in excess of 1100 m (all from station MR79-34). (a) Pteropod-calcareous breccia; note (arrow) richness in large thecosomatous pteropods (*Diacria quadridcutata*) which are preserved only as moulds in a dominantly high-Mg-cemented limestone. (b) Section of previous sample showing that the limestone is actually a volcanic-(MORB-basalt)–pteropod–calcareous breccia; note that both biogenic and volcanoclastic components are of similar (comparably large) size suggesting a sorting process by gravity before a later stage of lithification; observe also dissolution of pteropods (P = *Diacria*) leaving pseudomorphic vugs. (c) Slab of globigerina pteropod-limestone with evidence of winnowing (MR79-34); pteropods are *Creseis acicula* (note their polarity probably reflects bottom currents before lithification); forams are mostly *Globigerinoides* spp. and *Orbulina*; cement is aragonite.

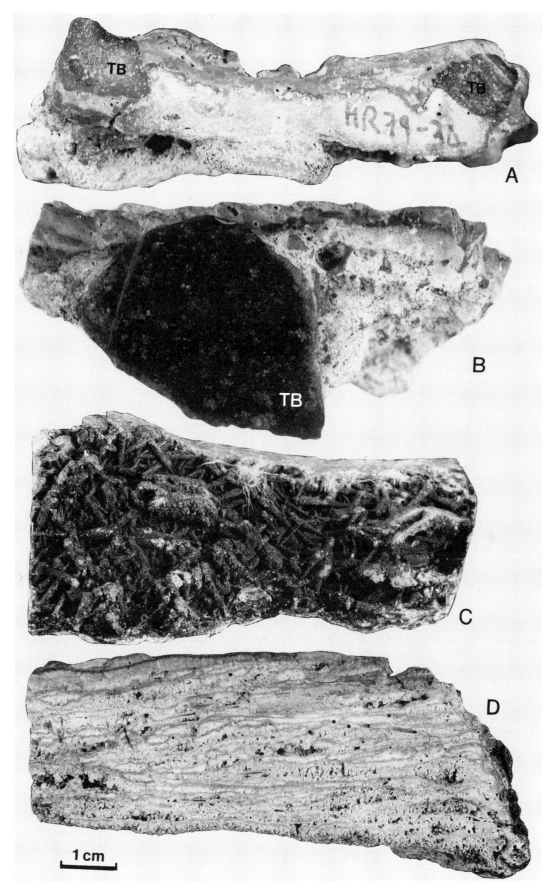

Figure G1.6 Carbonate rocks dredged from the steep walls of the Nereus Deep at depths in excess of 1100 m (all from station MR79-34). (a) Volcanic–calcareous breccia: subangular pebbles of MORB-tholeiitic basalt (TB) are encased within a high-Mg calcite cemented hardground. (b) As before, note the large basaltic pebble. (c) Blackened and highly corroded pteropods (*Creseis acicula*) cemented over the hardground exposed surface: observe imbricated fabric of the cone-shaped shells and lack of matrix (winnowing of other pelagic components before lithification; compare Figure G1.5(c). (d) Laminar calcareous crust (mostly high-Mg calcite), vugs represent solution of former pelagic tests (forams and pteropods).

bidites and, in places, by MORB-volcanoclastic/carbonate turbidites which form discrete centimetre to decimetre thick layers. Their visibility within the sedimentary sequence is strongly enhanced by their being emplaced within bright varicoloured metalliferous sediments (Plates 29 and 30). Red Sea axial trough calciturbidites distinctly remobilize only pelagic material, which is generally very well sorted. I have no evidence of shallow-shelf particles within axial calciturbidites. Although sorting may be the result of gravitational particle setting, it may well represent remobilization of winnowed NRSS, such as described above. The volcaniclastic component may result from more than one process. Fragmentation of MORB-basalt has been previously discussed while dealing with volcaniclastic–pelagic carbonate breccias and is genetically related to fault brecciation. I consider this also to be the simplest explanation for the production of basalt clasts of core MR79-35 that have then been moved downslope by gravitational processes (Figure G1.7). The alternative possibility of primary debris flows of hyaloclastic lavas, similar to those from young seamounts of the East Pacific Rise discussed by Londsdale and Batiza (1980), is not supported by texture (lack of hyaloclastic rims on basaltic pebbles) and composition (mixture with pelagic carbonates) of our Red Sea deposits. These gravity flow deposits are likely to have been triggered by the elevated seismicity within the trough (Fairhead and Girdler, 1970; Daggett et al., 1986). Earthquakes with magnitude > 5 are not infrequent in the Red Sea region and the high seismicity of the region is outlined also by the frequency of microearthquakes (e.g. Boulos et al., 1987). The presence of thin, repetitive turbiditic events is not limited to the Nereus Deep. In fact, downslope remobilization of unstable sediment from the shoulders of rift cells is probably common. Hofmann et al. (this volume) provide examples of similar deposits from the southern part of the basin. The 14 m long piston core 338 collected by RV *Marion Dufresne* within Commission I basin (latitude 19°17.3 N – longitude 38°56.4 E, 2108 m water depth) contains many thin (millimetre to a few centimetres) terrigenous and calciturbidites. A conservative estimate of turbidite accumulation within Commission I Basin, based on silty and sandy fractions alone, shows that coarse to relatively fine-grained turbidites account for as much as one fifth of the entire cored sequence.

DISCUSSION

The glacial to interglacial changeover and organic vs. inorganic switching

The orbitally controlled cyclical nature of sea-level changes during the Quaternary acts as a major control on the sedimentary style of the Quaternary Red Sea. It is important to emphasize that the Red Sea changed its sedimentary and oceanographic behaviour at the same time as sea-level change. During the last 25 000 years, NRSS are quantitatively and temporally the standard style before the last glacial maximum and since 11 000 years BP. NRSS is basically the result of prolific biogenic calcification by pelagic organisms. Conversely, hard layers are the signature of a totally different basinal behaviour where biological carbonate factories are almost completely closed down and inorganic carbonate precipitation replaces the biogenic one but at a lesser scale. The meaning of such on/off switching can be better appreciated when we consider that not only the pelagic carbonate factories fit this scheme but also much of the benthic system as well (Taviani, this volume). The cyclicity of alternate factories of completely different carbonate efficiency is repetitive for a major part of the Quaternary history of the Red Sea. The Red Sea is a sizable basin whose CO_2 sink properties (large-scale calcification vs. small-scale lithification) fluctuate rapidly being in phase with sea-level changes.

Structural and climatic control on Red Sea offshore sedimentation

Climate and structure both exert controls on rift sedimentation but an important problem concerns their relative importance in controlling sedimentary facies distribution. Claims on the importance of climate in offshore Red Sea sedimentation are numerous (e.g. Herman, 1965b; Degens and Ross, 1969; Bignell, 1976; Deuser et al., 1976; Rossignol-Strick, 1987; Thunell et al., 1988; Taviani, 1995a; Hofmann et al., this volume) and the relation between deep-sea sedimentation and tectonics is discussed by Hofmann et al. (this volume).

For modelling purposes, it is evident that one fundamental step forward would be to identify consistent and predictable attributes which are unequivocally linked to either structure or a change in climate (Lambiase and Bosworth, 1995). The marine Red Sea rift is of critical importance in such an analysis because of its intermediate stage of rifting, ideally located between the continental East Africa Rift and the Gulf of Aden, whose oceanization is more advanced.

Evaluation of the most important factors controlling sedimentation in continental rifts has been pursued by Lambiase and Bosworth (1995). These authors consider that structure is the primary factor in controlling sedimentary patterns while climate, although relevant on a local scale, is superimposed. A similar analysis can be attempted to check the relative importance of structure and climate in controlling the late Quaternary offshore sedimentation of the Transitional Region; possibly the best area in the Red Sea marine rift to model sedimentation. Many different structural situations coexist there, ranging from an incipient axial trough with all its major attributes (young and rugged topography, seismicity, deepness, volcanism) to accommodation zones (more subdued topography, relative aseismicity, lack of volcanism), seamounts and so on. It must be emphasized that the varied nature of the Pleistocene climate, with its rapid waxing and waning of ice caps (and concomitant sea-level fluctuations), makes the Quaternary shallow-silled Red Sea somewhat unique for some of the sedimentary patterns observed.

The present study consists in evaluating the inferences of structure and climate on the deep-sea sediments described earlier for the axial zone. Climate seems to be the leading factor in defining the type and quantity of materials making up the late Quaternary succession, and also in controlling the timing of switching styles through the glacio-eustatic sea-level changes. Structure is obviously important in creating the topography of the basin (Figure G1.7), but similar products, basically oozes with variable quantities of siliciclastics, sapropels and pelagic limestones can also be produced under tectonic scenarios other than rifts. The importance of climate in shaping the sedimentary succession is better exemplified by the rate of terrigenous particles exported to the deep sea. Enormous volumes of siliciclastics are trapped on the alluvial plains of both sides of the rift but there are no mechanisms to convey a significant amount of siliciclastics to the offshore axis. The regional aridity prevents such mechanisms as more vigorous streams and rivers that might otherwise be efficient at supplying detritus to the shelf. Furthermore, coral reefs, themselves strongly controlled by climatic conditions (temperature), may act as further barriers for siliciclastics. Sediment starvation in the Red Sea rift decreases at times of climatically driven changes when run-off is more active. On the other hand, the analysis of gravitational sedimentation offers an opposite view since structure is dictates the style of sedimentary-infilling of young oceanic troughs by means of topography (extreme steepness of the trough), trigger mechanism (axial quakes) and volcanic processes (Figure G1.7). In respect of gravity flow processes, climate seems to play a minor role, mainly by modulating the quantity of terrigenous siliclastics available for the turbidites which is likely to be greater at times of lower sea levels.

ACKNOWLEDGEMENTS

I am indebted to: Fabio Trincardi, John Reijmer, Andre Droxler, Thomas Brachert and Bruce Purser for useful discussions on gravity flow and carbonate sedimenta-

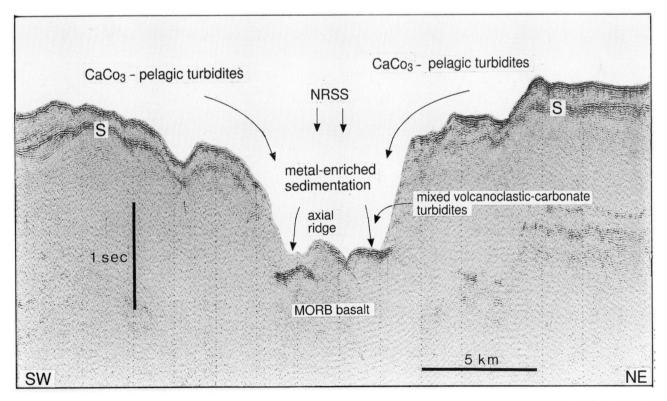

Figure G1.7 Schematic representation of main processes contributing sediment to a Red Sea incipient oceanic trough based on data from the Nereus Deep. 30 kJ seismic profile is across the northern part of Nereus Deep; S = reflector marking top of the Miocene evaporites.

tion, Helmut Zibrowius for determination of deep-sea corals and serpulid polychaetes, the late Lucien Leclaire, Pierre Clement and Jean-Pierre Giannesini for their friendly cooperation in studying *Marion Dufresne* core 388; Narriman Taviani, Ylenia Foschi and Alessandro Remia, for technical work, Luciano Casoni for art work, Vladimiro Landuzzi for X-ray diffractometry, Gabriele Marozzi for photographic work. Thanks are due to Captain, officers, crew and colleagues aboard cableship *Salernum* (cruise MR79) and RV *Bannock* (cruise MR83). I am grateful to Dan Bosence and Bruce Purser for editorial help and critical comments. Funding by CNR and from EC programme RED SED, through grants SC1*CT91-0719 and ERB SC1*CT92-0814 are gratefully acknowledged. This is IGM scientific contribution No. 1021.

Chapter G2
Sedimentation, organic geochemistry and diagenesis of cores from the axial zone of the southern Red Sea: relationships to rift dynamics and climate

P. Hofmann, L. Schwark, T. Brachert, D. Badaut, M. Rivière and B. H. Purser

ABSTRACT

Systematic study of sedimentary structures, textures, clay mineralogy, sedimentary dynamics, organic matter and lithified carbonate crusts within a 16 m core from the Suakin Deep (central Red Sea, depth 1969 m), is interpreted in terms of rift dynamics and global climatic change. The presence of 12 turbiditic sand bodies composed of planktonic skeletons mixed with material eroded from deep submarine scarps bordering the axial zone may relate to instability associated with the opening of the rift. More subtle expressions of rift dynamics include Fe and Mn oxides and certain smectitic clays precipitated from hydrothermal fluids arising within the axial zone.

Clay minerals and other silicates (quartz, feldspars, amphiboles and mica) are related to terrigenous sources during periods of delta activity on the Sudanese and adjacent shorelines. The geochemical properties of four organic-rich 'sapropels' also indicate terrigenous sources, the autochthonous marine contribution being minor. These geochemical studies also indicate that organic components, although supplied via oxic waters, have been deposited under highly saline, reducing conditions, suggesting a stratified water column. Associated aragonite crusts have positive $\delta^{13}C$ ($+3‰$) and $\delta^{18}O$ ($+6‰$) isotopic properties confirming high salinities which have also favoured the precipitation of aragonitic spherulites and other diagenetic fabrics.

Comparison of this core from the Suakin Deep with published data from the axial zone confirms the widespread distribution of sapropels and lithified crusts whose age is in the order of 15 000 to 20 000 y BP. Furthermore, the presence of similar crusts overlain by sediment enriched in organic matter, also dated as 15 000 y BP, from the Sudanese Shelf (600 m) indicates that these phenomena are not confined to the axial zone. Cores from the Gulf of Aden, on the contrary, do not contain sapropels or carbonate crusts.

Conditions favouring the development of sapropels and carbonate crusts are related to elevated bottom-water salinities. Because the sapropels and crusts are dated as 15 000–20 000 y BP, they correspond to the end of the last glacial period. In agreement with published data, the authors conclude that the lowering of global sea level has resulted in the partial separation of the Red Sea and Indian Ocean by the Bab-el-Mandab sill, thus stimulating overall restriction in the Red Sea. Subsequent sea-level rise, possibly aided by continental run-off, has resulted in a stratified water column with reducing sea-floor conditions favouring preservation of sapropels. Thus, the sediments of the axial zone of the Red Sea record both rift dynamics and global climatic change.

INTRODUCTION

The 2000 km long Red Sea rift, limited along most of its length by peripheral continental scarps, is bordered by coastal plains and extensive but relatively deep (500 m) submarine shelves. These latter are terminated by a major marine escarpment which borders the axial trough whose morphology relates to the relatively recent opening of the Red Sea. The trough contains sediments whose properties record both rift dynamics and the precipitation of minerals from warm hydrothermal brines discussed by Blanc et al. (this volume).

Sedimentation and Tectonics of Rift Basins: Red Sea–Gulf of Aden. Edited by B.H. Purser and D.W.J. Bosence. Published in 1998 by Chapman & Hall, London. ISBN 0412 73490 7.

The post-Miocene sediments deposited within the axial trough are generally laminated, with relatively bright colours and have varied mineralogies. These attributes are further enhanced by the presence of heavy metals and other potentially economic minerals whose existence has stimulated research. The list of publications and research reports concerning the axial zone, its geophysical, igneous, hydrothermal and sedimentological properties, is considerable and has been reviewed by Degens and Ross (1969), Blanc *et al.*, and by Taviani (this volume).

In view of the very numerous publications concerning the sediments of the axial zone, the need for a further publication could be questioned. The present contribution is considered necessary for two reasons. Firstly, this volume, the first treating the entire sedimentary record of the Red Sea–Gulf of Aden rifting, would be incomplete were it not to include the sediments of the axial zone. More importantly, this contribution discusses several specific facies typical of these axial environments, namely the organic-rich ('sapropel') horizons, the very characteristic turbiditic sand bodies, and the better-known carbonate crusts. While relating certain aspects of these three facies to their structural context, factors other than rift dynamics, notably climatic fluctuations, are stressed. The relative importance of both tectonic and climatic controls on deep-marine sedimentation in the Red Sea are confirmed by comparison of these axial sediments with others situated on the shallower (500–1000 m) Sudanese shelf and in the oceanic environments of the Gulf of Aden.

The geological setting of the axial zone

Seismic profiles indicate that the bathymetry of the Red Sea is closely related to its structural framework. According to Ross and Schlee (1973), typical profiles across the basin include three major segments (Figure G2.1).

1. Peripheral shelves extend seawards from the Sudanese and Saudi Arabian shorelines for distances of 30–120 km. These have a fairly regular relief modified locally by bathymetric highs relating to evaporite diapirism. These shelves generally slope gently towards the axial zone. Average depths are in the order of several hundred metres, their sediments being predominantly calcareous. Seismic profiles show a gently undulating sedimentary series overlying a clearly defined 'reflector S' marking the top of the Miocene evaporites.

2. The shelves are bordered towards the axis of the Red Sea by a 'marginal zone' of irregular relief relating to a closely spaced system of faults. This zone is limited by a marked break in slope generally situated at depths of 500–1000 m, which descends towards the axis.

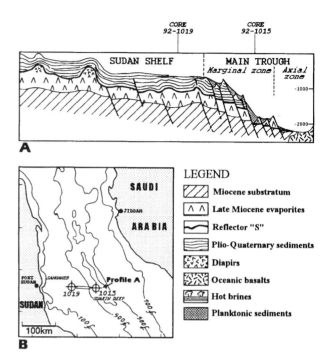

Figure G2.1 Location of material studied: A, schematic profile, after Ross and Schlee, 1977 (modified) across the Sudanese Shelf showing the major stratigraphic and morphostructural components; B, map of the Central Red Sea showing the location of cores studied.

3. The 'main trough' has average depths of about 1800 m. It includes many local highs and depressions. It is important to note that 'reflector S', confirmed by DSDP drilling to mark the top of thick, presumed Miocene evaporites, is truncated by the recent escarpment bordering the 'main trough' (Figure G2.1). This reflector is not present within the axial zone where thin Quaternary sediments overlie basalts. Thus, within local depressions such as 'Atlantis II' sediments dated as 23 000 BP lie on basalt. The lateral continuity of 'reflector S', marking the upper surface of Miocene evaporites, indicates that they probably extended across the entire Red Sea basin. The absence of evaporites only within the 10–30 km wide axial zone (Ross and Schlee, 1977) indicates that lateral separation of the two margins is a post-evaporite event and thus probably of Pliocene and Quaternary age.

Distribution of material studied

The essential part of this research is based on core 92-1015 (Figure G2.1, 19° 37.55 N; 38° 37.71 E) collected by the French research vessel *Marion Dufresne* during the 1992 'RED SED' cruise. The core site is located near the foot of the main axial escarpment. This core has been chosen for study because of its 16 m length and, especially, for its spectacular lithological variations. The core was taken in a water depth of 1969 m within the Suakin Deep, some 150 km off the Sudanese coast (Figure G2.1(b)), and is compared with others from the adjacent Sudanese shelf (core 92-1019) and with cores collected in the Gulf of Aden.

THE MINERALOGICAL AND TEXTURAL PROPERTIES OF THE AXIAL ZONE OF THE RED SEA

General properties of sediments in the axial zone

These are fairly well known thanks to numerous oceanographic cruises generally relating to the study of the metalliferous deposits and associated hot brines (Degens and Ross, 1969; Stoffers and Ross, 1977). Although sedimentary facies vary from core to core, nevertheless four major facies exist.

1. The mineralogy involves varying admixtures of carbonates (generally magnesian calcite) and silicates, the latter tending to be more important within the lower parts of the Quaternary. Silicate sands are rich in feldspars (32% of bulk mineralogy; Stoffers and Ross, 1977) and quartz. Subsidiary heavy minerals and micas confirm a detrital origin, probably in part from metamorphic sources. Clay minerals include illite, kaolinite and chlorite, mainly of detrital origin. Palygorskite and montmorillonite are very common and also appear to be detrital. Pink or dark grey horizons generally are enriched in organic matter, or in Fe and Mn oxides, the latter relating to hydrothermal activity, discussed by Blanc et al. (this volume).

2. Multiple hard layers (crusts) are widespread as very thin (several millimetres) horizons interbedded within soft clays. These crusts, composed mainly of aragonite, have been studied in considerable detail by Herman (1965a), Gevirtz and Friedman (1966) and by Milliman et al. (1969). They are also discussed in detail in this contribution.

3. Dark layers of organic-rich sediment ('sapropels') are present and their origins and preservation have been the subject of considerable debate (Herman, 1965b; Besse and Taviani, 1982; Rossingnol-Strick, 1987; Locke and Thunnell, 1988; and Taviani, this volume). Their geochemical properties are presented in this contribution.

4. Sedimentary structures, notably bedding, are often well developed. Laminations are expressed by contrasts in colour and by the presence of silt and fine sand composed mainly of planktonic skeletons and quartz. Graded bedding is a common feature (Taviani, this volume) and these generally involve a transition from relatively coarse pteropod sands to finer globigerinid silts. Erosion surfaces and other discontinuities are frequent and, together with the presence of clasts and graded bedding, indicate resedimentation, notably along the steep flanks of the axial zone.

Clearly, the Quaternary sediments of the axial zone have various mineralogical compositions and origins which may be summarized as follows:

- detrital silicates and dolomite, probably derived from peripheral deltas or via aeolian means;
- autochthonous carbonates produced by planktonic microfaunas, siliceous skeletons being relatively rare;
- physico-chemical precipitations of magnesian calcite and aragonite, from marine waters;
- precipitations from hydrothermal waters, notably the various metal oxides;
- diagenetic precipitates (carbonate cements, pyrite, etc.) from modified sea waters;
- mechanical reworking of underlying pre-Quaternary substrates, notably along the steep, fault-bound axial scarps, giving rise to lithic fragments.

Some of these products are related directly to rift geodynamics, including the suite of minerals precipitated from axial hydrothermal waters and the numerous resedimentation phenomena. Others, such as the axial muds, are related to bathymetry and thus to the rift framework. Still others are related indirectly to rift morphology and the interaction of this morphology with certain hydrographic movements; the Bab el Mandab sill is probably a structural element which influences the influx of oceanic waters into the geographically confined Red Sea basin, notably during periods of lower sea level. Finally, other properties of the axial sediments are totally independent of rift tectonics. These include fluvio-detrital or aeolian supply of silicates and organic material from coastal sources, notably during periods of elevated humidity.

Several specific aspects of axial geochemistry and diagenesis are considered in detail by Blanc (this volume) while turbiditic facies and associated oceanic muds are discussed by Taviani (this volume). The present authors, while illustrating the more typical sedimentary aspects of core 92-1015, present detailed analyses and interpretations of the organic fraction as well as the nanofabrics associated with the characteristic aragonitic crusts. These contributions supplement the numerous published, often more generalized studies of axial zone sediments.

Specific properties of axial sediments in core 92-1015

General aspects of the core
Situated near the foot of the axial scarp at water depths of 1969 m (Figure G2.1(a)), this 16 m core exhibits a subtle vertical gradation in colour from dark-beige between 16 and 5 m (below the sediment surface) followed by greenish-beige between 5 m and the sediment surface. There also exist several pinkish-brown horizons, notably near the base. These changes in colour

have not been correlated with any significant change in mineralogy but probably indicate minor variations in the concentration of the oxidized iron or manganese.

Core 92-1015, although generally fine-grained (muds and silts) includes about 12 distinct sand horizons plus many silty laminations, the sand–silt layers totalling about 25% of the core sequence (Figures G2.2, G2.3). The sand fraction is composed mainly of globigerinid foraminifera, pteropods and micromolluscs together with quartz, feldspar and mica grains. The muddy intervals also contain dispersed ('floating') skeletal and detrital grains, and have a massive aspect probably due to bioturbation. The core includes a number of dark-grey horizons enriched in organic matter (see following section), those situated between 5.5 and 6.5 m lying almost directly on limestone crusts, the remainder of the core being unlithified.

Mineralogy
The core is an intimate mixture of carbonate and silicate minerals, the former tending to increase upwards (Table G2.3). This same upwards increase in carbonate has been noted in other parts of the Red Sea (Ross and Schlee, 1973). It may reflect climatic changes on the adjacent continents, notably the transition from the more humid glacial to present desert climates with reduced input of detrital clays and silts.

The carbonate fraction
The carbonate content of the analysed samples ranges from 18 to 73%. The lowest carbonate values generally occur in the darker, brown to black, muds. High carbonate fractions are associated with lithified crusts (samples MD 921015-4-113-18) and pteropod-rich basal parts of turbiditic units (samples MD 921015-4-31-32), both of which are located in the upper part of the core.

Skeletal carbonate grains include calcareous nanoplankton (Figure G2.4), calcareous dinoflagellate cysts and pelagic gastropods as well as numerous globigerinid forams and pteropods notably within the sandy-textured turbidite beds. Those associated with the calcareous crusts range in age from 13 830 to 20 640 years (Table G2.1), and probably represent the flora and fauna of the glacial Red Sea. This granular fraction is accompanied by an important mud fraction whose biogenic nanoconstituents (Figure G2.5) are impossible to identify; they could, at least in part, be of continental origin. Certain nanoparticles are composed of aggregates of acicular aragonite crystals (see following section), these probably being marine precipitates. A minor dolomite fraction is probably of detrital origin, most Tertiary sediments outcropping on the continental shore being composed of dolomite.

The non-carbonate fraction
The locally predominant coarse silicate fraction includes sand to silt-grade quartz, feldspars and micas as well as various unidentified coloured minerals. The subangular form of the quartz and feldspar grains and their considerable size (up to 1 mm) confirm their detrital origin.

A limited number (11) of samples (Figure G2.2) have been analysed by X-ray diffraction and studied by Transmission Electron Microscope. The quantitive estimates for major elements have been made by SEM-EDS on bulk samples, in order to detect possible anomalies in the vertical distribution of minerals, notably in the vicinity of the carbonate crusts and sapropels.

Clay mineralogy
All samples analysed by X-ray diffraction (see Appendix), both in the axial core (92-1015) and from the

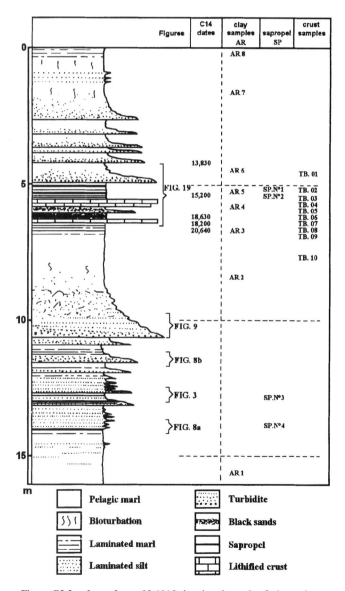

Figure G2.2 Log of core 92-1015 showing the major facies and distribution of samples.

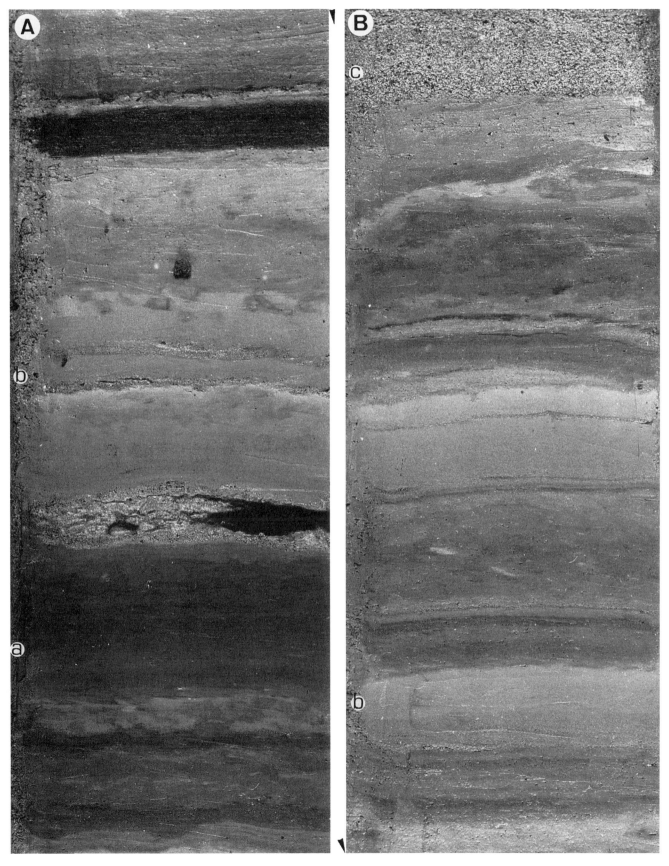

Figure G2.3 Photos of lower part of core 92-1015 (depth 13.0–12.6 m) showing (a) sapropel N° 3, (b) laminated marls and silts, and (c) thin turbidite sand. Photos A and B are continuous, scale 1:1.

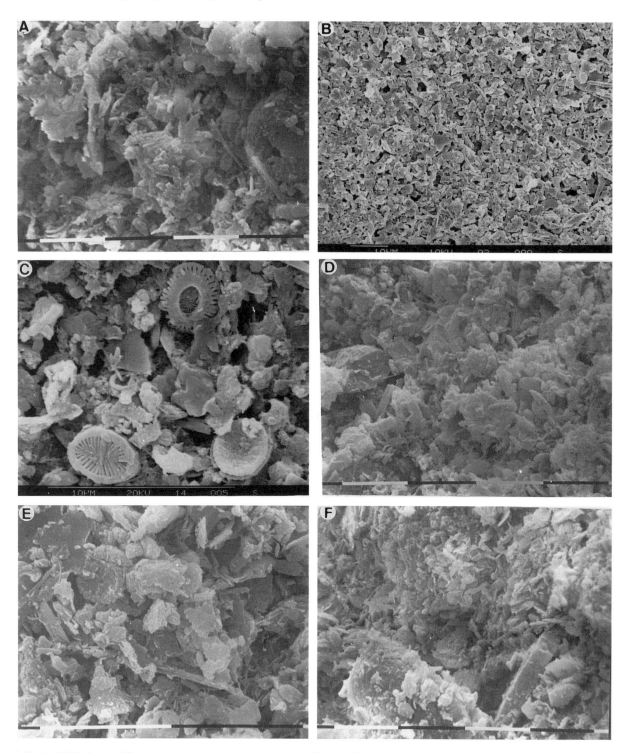

Figure G2.4 SEM photos of fine, essentially carbonate, fraction in cores 92-1015 (Suakin Deep) and 92-1019 (Sudanese Shelf). Notice that virtually all components are of biogenic origin. A. Core 92-1015, 1.2 m, mud and silt-sized bioclastic debris of unidentified origins; scale = 10 μm. B. Core 92-1015, 4.7 m, well-sorted bioclasic silt, top part of turbidite unit (sample TB 01), scale 10 μm. C. Core 92-1015, 5.7 m, slightly lithified carbonate mud composed essentially of nanoplankton and fine, bioclastic debris; (sample TB 05), scale 10 μm. D. Core 92-1015, 7.5 m, fibrous crystals possibly of chemical origin, and biogenic debris; scale 10 μm; E. Core 92-1015, 13.4 m, microbioclastic silt and scattered coccoliths; scale 10 μm. F. Core 92-1019, 1 m, microbioclastic silt, scattered coccoliths and coarser debris; scale 10 μm.

Sudanese Shelf (core 92-1019) contain smectite (25–50%), chlorite (10–20%), attapulgite (10–15%), illite (10–30%) and kaolinite (10–30%). The Hofmann-Klemen test showed the simultaneous presence of montmorillonite and beidellite. In addition, there also exist traces of an irregularly interstratified clay mineral.

Table G2.1 ^{14}C datations and C and O isotopic properties of cores 92-1015 and 92-1019, central Red Sea. (Analyses by Laboratoire, d'Hydrologie & Géochimie Isotopique, Université de Paris Sud, Orsay)

Core depths (m)	Material	^{14}C age (y BP)	^{13}C(‰PDB)	^{18}O(‰PDB)
Core 92-1015, Suakin Deep				
−4.2	turbiditic sand	13 830 ± 690	+3.82	+5.23
−5.3	lithified crust	15 200 ± 280	+4.47	+6.99
−6.2	lithified crust	18 630 ± 890	+4.17	+6.58
−6.3	lithified crust	18 200 ± 565	+4.17	+6.18
−6.5	lithified crust	20 640 ± 1010	+3.52	+5.46
−10.5	turbiditic sand	–	+2.49	+4.58
Core 92-1019, Sudanese Shelf				
−0.5	lithified crust	14 970 ± 380	+4.22	+6.99

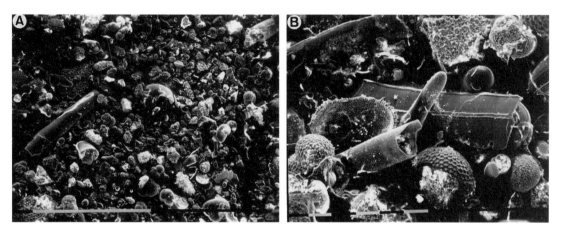

Figure G2.5 SEM photos of coarse fraction (mainly turbiditic) in core 92-1015. A. Fine planktonic and benthic biogenic detritus and terrigenous siliciclastic grains; depth 0.5 m, scale 1 mm. B. Pteropod, globigerinid foraminifera and siliciclastic grains, base turbidite; depth 4 m, scale 100 μm.

Scattered quartz and amphiboles are also present, and TEM study reveals that the finest fraction is dominated by smectite and the coarser by chlorite. The morphology of the particles confirms the presence of two distinct types of smectite, flakey particles (Figure G2.6(a)) and particles having rolled peripheries. A third type of smectite, composed of minute sheets set on an amorphous substrate (Figure G2.6(c)), has been noted only in samples from the axial zone. In addition, palygorskite was noted in all samples, occurring either as 1–2 micron crystallites, or as masses of fibres (Figure G2.6(b)), the latter suggestive of aeolian transport. Manganese and iron oxides (goethite, Figure G2.6(d)), or ferrihydrite are relatively abundant in the axial core (92-1015).

The homogeneous vertical distribution of clay mineral types together with the nanoscopic properties indicated by TEM suggest that they are essentially of detrital origin. An exception concerns a specific (third) smectite fraction, limited to the axial core, and thus probably of diagenetic origin. The iron and manganese oxides, most abundant within the axial core, are hydrothermal products, discussed in detail by Blanc *et al.* (this volume).

Vertical changes in mineralogy and chemistry

Although the number of samples analysed is limited, there appears to be a fluctuation in the relative abundance of non-carbonate minerals which coincides with core depths of 5.5–6.5 m, being near sample AR4 (Table G2.2). This anomaly is best expressed in terms of Al-Fe and Al-Mn ratios (Figure G2.7(a),(b)), the Al being located mainly in the detrital clay minerals, the Fe and Mn in oxides formed from axial hydrothermal solutions. These ratios reflect the relative importance of detrital and authigenic supplies. The marked decrease both in Fe and Mn (relative to Al) at core depths of about 6 m is coincident with the development of carbonate crusts and local enrichments in organic carbon. These mineralogical anomalies, reflecting a relative increase in hydrothermal Fe and Mn, can be explained in two ways:

- either they may record an increase in hydrothermal activity within the axial zone;
- or they may reflect a decrease in the supply of detrital clays during a constant hydrothermalism.

At the beginning of the last glaciation the ratios Si-Al and Al-K (Figure G2.7(c)), following a minimum,

486 Sedimentation, organic geochemistry and diagenesis of cores

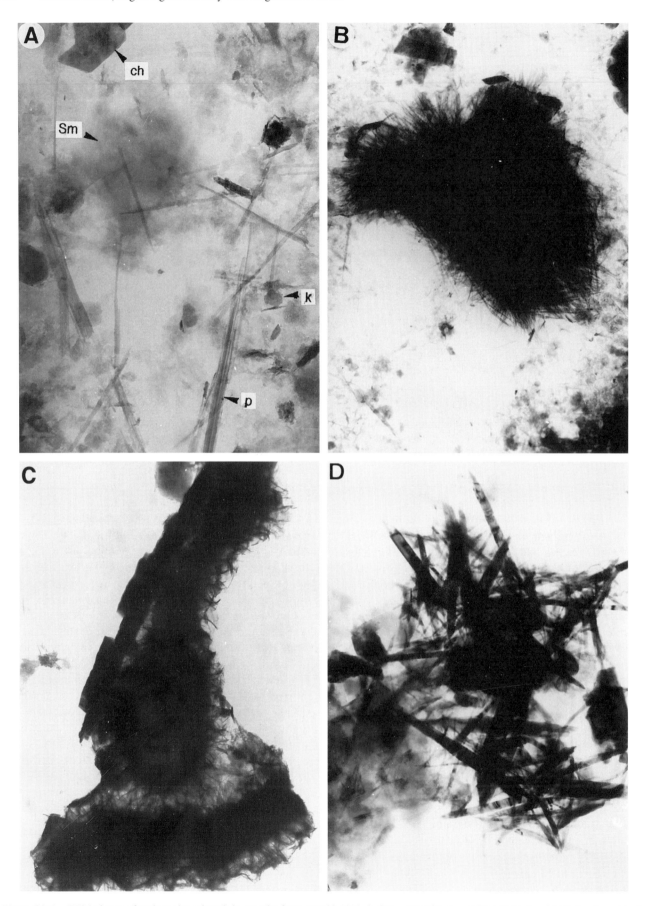

Figure G2.6 TEM photos of various clay minerals in samples from core 92-1015. A. Sm = smectite, p = palygorskite, k = kaolinite, ch = chlorite (scale: 1 μm : 5.4 cm). B. Aggregate of playgorskite crystals (scale: 1 μm : 1.08 cm). C. Flakes of smectitic clay mineral on amorphous substrate (scale: 1 μm : 1.25 cm). D. Geothite crystals (scale: 1 μm : 4.35 cm).

Table G2.2 Major-element compositions of bulk sample were determined by scanning electron microprobe (SEM) fitted with an energy dispersive spectrometer (SEM PHILIPS 505 with LINCK System).

Core	Core 1015								Core 1019		
Depth (mbsf)	15.67	8.65	6.92	6.22	5.67	4.39	1.70	0.23	6.42	3.65	0.13
Sample	AR1	AR2	AR3	AR4	AR5*	AR6	AR7	AR8	AR9	AR10	AR11
SiO_2	32.22	33.70	23.57	21.09	35.41	25.37	25.98	30.97	27.29	17.67	25.91
TiO_2	0.78	0.98	0.64	0.65	0.83	0.47	0.61	0.81	0.55	0.39	0.37
Al_2O_3	11.65	10.95	9.00	7.51	11.93	9.69	10.40	12.62	9.57	6.91	10.93
FeO	7.32	6.93	7.28	8.20	9.90	8.81	7.32	8.45	4.01	2.87	4.55
MnO	0.65	0.91	0.98	2.06	1.27	1.28	0.55	1.35	Tr	Tr	Tr
MgO	5.19	5.15	5.14	4.28	5.34	5.15	5.38	4.93	5.56	5.57	4.95
CaO	38.05	36.64	49.45	51.38	32.59	43.67	46.21	33.73	48.74	63.13	48.23
Na_2O	1.21	1.83	0.98	1.57	0.78	1.93	1.05	2.64	1.73	1.22	1.76
K_2O	1.99	1.81	1.70	1.33	1.95	1.57	1.74	1.95	1.45	1.27	1.63
S	0.29	0.26	0.29	0.29	Tr	0.31	0.24	0.25	0.23	0.24	0.20
Cl	0.66	0.84	0.95	1.65	Tr	1.37	0.53	2.30	0.87	0.73	1.47
P_2O_5	Tr	Tr	Tr	Tr	Tr	0.36	Tr	Tr	Tr	Tr	Tr
TOTAL	100.00	100.00	100.00	100.00	100.00	100.00	100.00	100.00	100.00	100.00	100.00
Si/Al	2.35	2.61	2.22	2.38	2.52	2.22	2.12	2.08	2.42	2.17	2.01
Al/K	5.42	5.58	4.89	5.23	5.65	5.69	5.51	5.99	6.10	5.05	6.21
Al/Fe	2.24	2.23	1.74	1.29	1.70	1.55	2.00	2.10	3.37	3.39	3.38
Al/Mn	25.08	16.80	12.76	5.07	13.09	10.51	26.30	12.96			

* ≤ 0.63 mm fraction.

increase progressively between 20 000 and 15 000 y BP, Al decreasing with respect to Si and K decreasing with respect to Al. Thus, detrital sedimentation becomes slightly enriched in Si and less rich in K, possibly reflecting a more humid, hydrolysing climate during which the alteration of micas is expressed by the loss of Al and K and the proportional development of higher Si ratios. These products of a relatively humid, early glacial phase, are eroded and delivered to the basin during the dryer glacial maximum.

Carbon dating of the upper parts of core 92-1015 (Table G2.1) permits a general estimation of sedimentation rates. These appear to be markedly slower (18 cm/1000 years) between 5.3 and 6.5 m (coincident with both the carbonate crusts and the non-carbonate mineral anomalies) relative to sedimentation rates (93 cm/1000 years) between 5.3 and 4.2 m, immediately above. These estimations, involving somewhat lower rates of sedimentation within the zone of mineral anomalies, would imply slower input of detrital clays

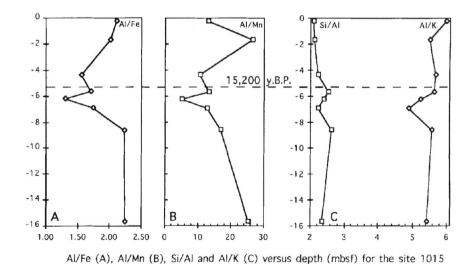

Al/Fe (A), Al/Mn (B), Si/Al and Al/K (C) versus depth (mbsf) for the site 1015 (Suakin deep sea plain).

Figure G2.7 Bulk analyses of main elements in core 92-1015. In general, Fe and Mn occur in oxides precipitated from axial hydrothermal waters while Al, Si, and K are mainly from detrital clays and silicate sands. Note the anomaly reflecting greater continental supply in the vicinity of 15 000 y BP.

and thus tend to confirm the second hypothesis, i.e. that the variations observed are the result of external, possibly climatic, influences, rather than fluctuations in axial hydrothermal activity. It is important to note that these mineralogical anomalies in core 92-1015 are dated, ranging in age from 20 640 to 15 200 BP (Table G2.1) and thus are coincident with the last glacial period.

Sedimentary structures

Throughout much of the core, bedding is fairly well developed and relates both to marked changes in colour (Figures G2.8 and G2.10) and to granulometry (Figure G2.8). Bedding surfaces may be flat or irregular and, together with sharp granulometric changes, reflect temporary breaks in sedimentation. Virtually all sand and silt beds are characterized by sharp basal surfaces (Figure G2.8(b)); fine massive or laminated muds are sharply overlain by relatively coarse sands. These latter may contain mud clasts, lithic fragments and other heterogeneous siliciclastic and biogenic detritus. Sands nearly always fine upwards into laminated silts (Figures G2.8, G2.9) before passing into massive muds. These sedimentary phenomena are typical expressions of resedimentation. The coarse sand textures are generally massive in aspect and may include a minor mud fraction while finer sands and silts are laminated and muddy. These features are all indicative of turbidite sedimentation rather than grain flow, and represent redeposition of pelagic material at the foot of a marked escarpment. Similar resedimentation phenomena have been recorded by Taviani and Trincardi (1992) and Taviani (this volume).

The turbidite horizons, of which there are at least 12 in this 16 m long core (Figure G2.2), vary in thickness, granulometry and composition. Those situated in the lower parts of the core (15 to 13 m) are relatively thin (2–10 cm) and fine-grained (Figures G2.4, G2.8). The lowermost turbidite (Figure G2.8(a)) sharply overlies a 2 cm thick dark layer enriched in organic matter. The sand comprising the turbidite is dominated by *Globigerina* with subsidiary pteropods, micromolluscs and unidentified microbioclastic debris. Silicate grains include silt-sized, subangular quartz, feldspar grains and mica of detrital origin. As one progresses up the core, turbidite horizons tend to thicken and particles are coarser. This tendency attains a maximum at 12 m where beige muds are sharply overlain by coarse sands and gravels comprising numerous lithic fragments and relatively large pteropods (Figure G2.9). The lithic debris includes fresh, angular fragments of hard limestone and subrounded pieces of slightly lithified, dark-grey siliceous siltstone. Blackened grains of unknown origin are also common. The bulk of this detritus appears to have been derived from a partially lithified substratum. Because of its relatively coarse grain-size (some clasts attaining 2 cm) and remoteness from shorelines (150 km), it is highly probable that this material has been derived locally from the adjacent slope (fault scarp) bordering the axial zone. The median parts of the 1.2 m thick turbidite body are medium-grained sands dominated by globigerinid foraminifera and by highly worn pteropods (Figure G2.9(b)). Most skeletal debris is opaque and internal chambers are filled with carbonate cement. The uppermost, laminated part of the turbidite bed is composed of silt-sized quartz and feldspar grains with frequent biotite and muscovite flakes. Biogenic constituents (mainly globigerinids) are a secondary constituent. In sum, this turbidite exhibits a vertical gradation both in grain-size and in composition (Figure G2.9) whose variations reflect a decreasing transport potential typical of turbulent flow.

A second series of closely spaced turbidite sands occurs within the top 6 m of core 92-1015. These sands differ from the lower units both in terms of texture and composition. They tend to be relatively fine, being dominated by globigerinid formaminifera and microgastropods (Figure G2.5) with subsidiary pteropods and benthic foraminifera. Very fine quartz silt, possibly of aeolian origin, is a minor fraction (less than 5%) and mica is very rare. These sands therefore are composed of relatively pure biogenic carbonate.

Sedimentary dynamics

The turbidite sands as a whole are composed mainly of planktonic skeletons, and these same organisms also occur within associated muds (Figure G2.5). The sand bodies, locally enriched in elements derived from the underlying substratum, have been redeposited at depths approaching 2000 m. Because adjacent slopes culminate at depths of about 1000 m, it is clear that the turbidites are generated well below wave-base. The component skeletal debris has been deposited both on the peripheral shelf and on the adjacent slopes. Feldspar, mica and other minerals including an important clay fraction, have probably been delivered by deltas on the Sudanese coast. Aeolian transport has probably supplied both the fine detrital silicates and dolomite. These dispersed elements, combined with a more abundant planktonic supply of carbonate skeletons, are subsequently remobilized down the axial slopes to accumulate within the deeper axial zone of the Red Sea.

The mechanism triggering resedimentation on the steep axial slopes could be gravity collapse. However, given the dense system of faults visible on seismic profiles crossing the periphery of the shelf (Ross and Schlee, 1973), and seismic instability near the axial part of the rift, it is highly probable that turbidite sedimentation is a direct expression of Red Sea opening.

Figure G2.8 Turbidite sands in core 92-1015. A. Sapropel N° 4 (a) sharply overlain by sand (b) which grades into laminated silt (c), core depth −14 m. B. Light-grey, massive muds (a) sharply overlain by sands (b) grading upwards into laminated silts (c); core depth −11.5 m. Scales for both photos are 1:1.

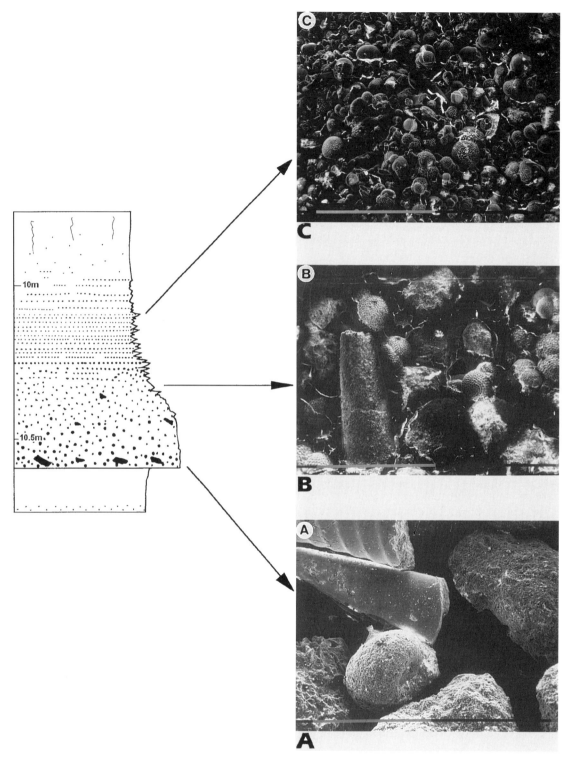

Figure G2.9 Turbidite body situated between 10 and 11 m in core 92-1015 showing typical vertical gradation from massive gravels to laminated sands and silts. A. Base of turbidite with clasts eroded from underlying substratum, coarse siliciclastic (quartz, feldspar) sands and worn pteropods. B. Coarse sands composed of worn pteropods and large globigerinid forams. C. Fine sand composed essentially of globigerinid forams. (Scale on all photos is 1 mm.)

In addition to the turbidites, fine siliciclastic sands, almost devoid of biogenic detritus, are also an important fraction of certain organic-rich horizons. This is particularly true for the dark-grey to black horizon situated at −5.2 m (Plate 31(a)). This dark, laminated unit is composed mainly of angular, silt-sized grains of quartz, feldspar and mica and many other coloured grains forming a grain-supported fabric with a muddy matrix. The latter includes particles of organic material. The fine amorphous pyritic fraction is significant (about 10%), its presence probably being responsible for the dark colour and somewhat crystalline aspect of the sediment. The microfaunal content of this horizon is low compared to that of the turbidite horizons and these fine muddy sands do not have the attributes of typical turbidites. Their siliciclastic constituents, obviously of continental origin, must have been supplied in relatively high amounts during periods of delta activity. Their intimate association with organic matter would also imply an essentially continental origin for this latter, confirmed by geochemical analyses discussed in the following section.

ORGANIC GEOCHEMICAL PROPERTIES AND ORIGINS OF SAPROPELS

A total of 43 samples were analysed for organic carbon, carbonate and total sulphur contents (Table G2.3). In addition, Rock-Eval analysis was performed on 11 samples while the characterization of aliphatic hydrocarbon fractions of the soluble organic matter was conducted on 25 samples.

Organic geochemistry of core 92-1015

Sulphur content
The total sulphur content of most samples is low (less than 0.6%, Figure G2.10(b)). An exception is four black clay samples that contain between 1.40 and 3.60% total sulphur. These samples are situated at two distinct stratigraphic levels at approximately 13–14 m and 5–6 m core depths respectively.

Organic matter content
The organic content of most sediment samples varies between 0.1 to 0.5% total organic carbon (TOC; Table G2.3 and (Figure G2.10(c)) and, hence, lies in a range typical for marine sediments deposited under oxic conditions. Exceptions are four sapropelic black clay layers, which contain between 1.66 and 2.14% organic carbon. These are also characterized by elevated sulphur contents (Figure G2.11: SP 1 to SP 4), indicative of deposition under anoxic bottom-water conditions (Berner, 1984, 1989; Stein, 1991). Two additional black clay layers in the lower part of the core (between 12.5 and 12.6 m) and a brown clay layer in the upper part (4.30 m) also contain relatively high amounts of organic carbon (approximately 0.9% TOC), but do not contain elevated sulphur values. They fall in the range of fine-grained, normal marine sediments (Figure G2.11). The organic carbon and carbonate contents of the sediments with low amounts of TOC show no correlation (Figure G2.12). However, the four organic-rich sapropelic layers contain relatively low amounts of carbonate (< 35%).

Type of organic matter
The organic matter of 11 samples (TOC between 0.29 and 2.14%) from the upper (0.32 m and 4.30 to 5.46 m) and the lower part (12.53 to 14.06 m) of the core is dominantly of terrigenous origin. The kerogens of the sediments are characterized by hydrogen indices (HI) between 14 and 150 mg HC/g TOC, oxygen indices (OI) between 98 and 828 mg CO_2/g TOC, and Tmax values of on average 419 °C (Table G2.3) these values being typical for degraded organic matter and for immature type III kerogens. The best kerogen quality (HI from 92 to 150) is associated with the four organic-rich sapropelic layers (Figure G2.13, SP 1 to SP 4). It appears that preservation conditions for organic matter were favourable during deposition of the sapropels, as indicated by the high carbon and elevated sulphur contents of these sediments (Figure G2.13).

Saturated hydrocarbon distribution patterns
The saturated hydrocarbon fractions of the analysed extracts consist mainly of *n*-alkanes (Figure G2.14). Typical is a bimodal *n*-alkane distribution with maxima at *n*-C_{18} and *n*-C_{31} (Figure G2.14(a)). The *n*-alkane distribution patterns are dominated by the long-chain alkanes with an odd over even predominance. Carbon preference indices vary between 1.4 and 3.8 for the *n*-C_{25} to *n*-C_{32} range. The *n*-C_{27}, *n*-C_{29} and *n*-C_{31} alkanes are the major compounds. Several samples are also characterized by a relatively high abundance of the *n*-C_{22} alkane when compared to the *n*-C_{21} and *n*-C_{23} alkanes (Figure G2.14(b)) while isoprenoid hydrocarbons are sparse with pristane and phytane representing the major compounds of this group. Traces of steroid hydrocarbons occur in most samples (*note*: the solvent was checked carefully to exclude contamination). Triterpenoids are also present in low concentrations and are dominated by $C_{31\ \beta\beta}$ homohopane. Samples containing relatively high amounts of $C_{31\ \beta\beta}$ homohopane are also always characterized by a high relative abundance of the C_{35} *n*-alkane (Figure G2.14(b)), which may indicate a common source for both compounds. The sapropelic organic-rich layers generally contain relatively highest amounts of C_{35} *n*-alkane, and of the $C_{31\ \beta\beta}$ homohopane when compared to the *n*-C_{34} alkane (Figure G2.15(b), (c)). The *n*-alkane distribution patterns of these layers are also dominated by the *n*-C_{27}, *n*-C_{29} and *n*-C_{31} alkanes and contain only small amounts of short-chain alkanes in the range of *n*-C_{15} to *n*-C_{20} (Figure G2.14(b)). Two samples (921015-4-81-84 and 921015-8-45-46. Figure G21.5(a)) display *n*-alkane patterns indicative of mature oils with an envelope extending from *n*-C_{15} to

492 Sedimentation, organic geochemistry and diagenesis of cores

Table G2.3 Carbonate, organic carbon and total sulphur content of the studied samples and results from Rock-Eval pyrolysis of selected organic-rich samples

Sample	mbsf	TOC (%)	$CaCO_3$ (%)	S_{tot} (%)	HI (mg HC/g TOC)	OI (mg CO_2/g TOC)
MD 921015-1-32-33[1,2]	0.32	0.37	42	0.19	14	828
MD 921015-2-36-33	1.86	0.29	47	0.19		
MD 921015-2-95-98	2.46	0.21	44	0.24		
MD 921015-3-47-48	3.47	0.29	38	0.20		
MD 921015-3-68-70[1]	3.67	0.23	34	0.25		
MD 921015-3-130-131[1,2]	4.30	0.88	52	0.17	37	382
MD 921015-4-31-32[1,2]	4.81	0.21	73	0.14		
MD 921015-4-62-65[1,2]	5.14	0.66	35	0.54	48	373
MD 921015-4-65-70[1,2]	5.17	1.74	28	2.39	130	209
MD 921015-4-81-84[1,2]	5.32	0.45	23	0.23	24	264
MD 921015-4-87-90[1,2]	5.38	0.52	32	0.43	43	340
MD 921015-4-94-98[1,2]	5.46	2.10	18	3.60	134	98
MD 921015-4-113-118[1,2]	5.65	0.16	70	0.14		
MD 921015-4-143-144	5.93	0.13	57	0.13		
MD 921015-5-22-23	6.22	0.13	69	0.15		
MD 921015-5-52-54	6.53	0.13	55	0.18		
MD 921015-5-54-60	6.57	0.15	50	0.22		
MD 921015-5-60-65	6.62	0.11	56	0.12		
MD 921015-5-70-71	6.70	0.12	53	0.15		
MD 921015-5-130-131	7.30	0.21	50	0.16		
MD 921015-6-25-26[1,2]	7.75	0.21	60	0.18		
MD 921015-6-86-87	8.35	0.18	56	0.17		
MD 921015-6-136-137	8.86	0.15	41	0.13		
MD 921015-7-35-36[1,2]	9.35	0.13	37	0.12		
MD 921015-7-95-96	10.35	0.14	44	0.24		
MD 921015-8-45-46[1,2]	10.95	0.20	61	0.25		
MD 921015-8-104-105	11.55	0.16	32	0.11		
MD 921015-9-25-26[1,2]	12.25	0.29	31	0.14		
MD 921015-9-53-54[1,2]	12.53	0.94	41	0.31	70	465
MD 921015-9-54-56[1,2]	12.55	0.34	39	0.16		
MD 921015-9-56-58[1]	12.57	0.94	30	0.14	76	321
MD 921015-9-58-62[1]	12.60	0.29	40	0.13		
MD 921015-9-101-104[1,2]	13.02	0.24	64	0.28		
MD 921015-9-104-105[1,2]	13.04	2.14	26	1.40	150	171
MD 921015-9-105-107[1]	13.06	0.23	52	0.15		
MD 921015-9-127-128	13.27	0.16	53	0.15		
MD 921015-10-75-76[1,2]	14.02	0.24	41	0.16		
MD 921015-10-50-55[1,2]	14.06	1.66	31	1.83	92	231
MD 921015-10-55-57[1]	14.09	0.28	49	0.14		
MD 921015-10-57-61	14.25	0.20	40	0.12		
MD 921015-10-135-136	14.85	0.27	57	0.15		
MD 921015-11-15-16[1,2]	15.15	0.25	44	0.16		
MD 921015-11-70-71	15.70	0.16	42	0.13		

[1] Samples analysed by gas chromatography. [2] Coupled gas chromatograpy/mass spectrometry.

n-C_{35} (Figure G2.14(c)). It appears that these samples were impregnated by a mature oil. (Contamination during drilling is possible but improbable.) However, the triterpenoid and steroid distribution patterns of these samples do not display a mature signature, indicating that the impregnating oil consisted mainly of n-alkanes.

Sources and preservation of organic matter

Marine vs. terrigenous sources of organic matter
Fluctuation in the relative amount of marine versus terrigenous organic matter is indicated by the variation of the n-$C_{(15+17)}$/n-$C_{(27+39+31)}$ ratio (Figure G2.15 (a)). The n-C_{15} and n-C_{17} alkanes are derived mainly from algal and cyanobacterial sources (Volkman and Maxwell, 1986 and references therein), while n-C_{27}, n-C_{29} and n-C_{31} alkanes originate mainly from waxes of higher plants from terrigenous sources (Volkman and Maxwell, 1986 and references therein). All samples are

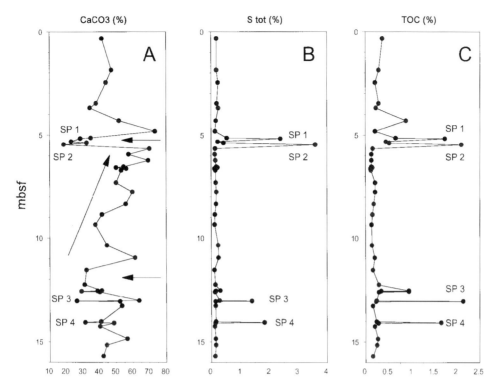

Figure G2.10 Vertical distribution of carbonate (A), total sulphur (B) and organic carbon content (C). Depths are indicated in metres below sediment surface and sapropels (SP) 1 to 4 are indicated.

dominated by n-alkanes derived from terrigenous sources. The highest input of terrigenous-derived organic matter generally occurs in the sapropel layers, which is also confirmed by the low hydrogen indices from the Rock-Eval pyrolysis for the kerogens of these samples. An interval with increased marine input is present in the lower part of the core between -11 m and -15 m. This interval is interrupted by the deposition of organic-rich layers that are characterized by high terrigenous input.

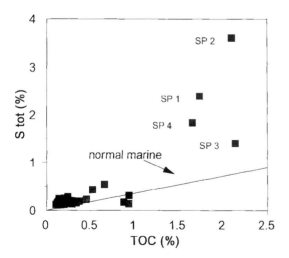

Figure G2.11 Relationship between total sulphur and organic carbon content. The trend indicated for normal marine sediments is from Berner (1984).

Bacterial contributions

The contribution of bacterial biomass is confirmed by the relative abundance of $C_{31\ \beta\beta}$ homohopane compared to the C_{34} n-alkane (Figure G2.15(b)). This triterpenoid hydrocarbon is thought to be derived mainly from bacteria living under oxic conditions (Simoneit, 1986 and references therein). As noted above, the striking covariance of the $C_{31\ \beta\beta}$ homohopane/n-C_{34} and n-C_{35}/n-C_{34} ratios (Figure G2.15(b), (c)) suggests that both compounds are derived from the same biological precursor. Generally, the highest contributions of bacterial biomass are restricted to organic matter-rich layers located in two intervals in the upper and lower part of the core (Figure G2.15(b), (c)).

Preservation conditions

The pristane/phytane ratio has been used extensively as an indication for redox conditions in sedimentary environments. Both compounds can be derived from the phytol side-chain of the chlorophyll (Didyk et al., 1978), where reducing conditions favour the formation of phytane, and oxidizing conditions lead to the preferential formation of pristane. A rigorous application of the pristane/phytane ratio requires that both products be derived from the same precursor, which is difficult to demonstrate (ten Haven et al., 1987) especially without complementary compound-specific isotope determinations. The pristane/phytane ratio is

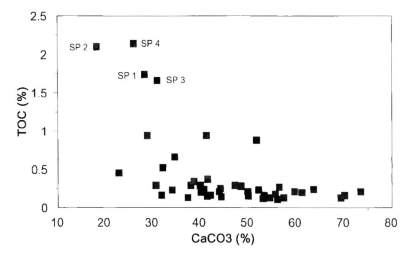

Figure G2.12 Diagram showing the relationships between organic carbon and carbonate content.

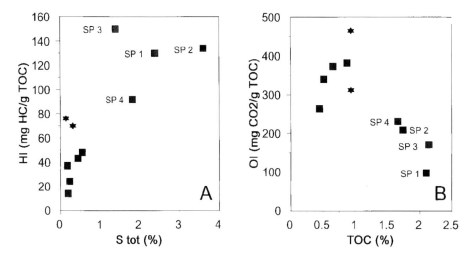

Figure G2.13 A. Relationships between hydrogen index and sulphur content. B. Relationship between organic carbon content and oxygen index. Organic matter-rich samples MD 9210159-53-54 and MD 92101559-56-58 are indicated with star symbols.

also affected by mixing of organic matter from different sources. Terrigenous organic matter is usually oxidized during transport prior to deposition, thus carrying an oxic signature, even when deposition takes place under anoxic conditions. Salinity also has an effect on the pristane/phytane ratio (ten Haven et al., 1987). Sediments deposited under strictly hypersaline conditions display extremely low (<0.2) pristane/phytane ratios. All these factors complicate the rigorous interpretation of the observed pristane/phytane ratio of the analysed extracts. It is worth noting, however, that most of the organic matter is characterized by pristane/phytane ratios below unity (Figure G2.16(a)), even though a large percentage of the organic matter appears to be derived from terrigenous sources. This implies that the marine components of the organic matter probably were deposited or formed either in a hypersaline environment and/or under reducing conditions.

Salinity
An indication for hypersalinity is also given by the high abundance of the n-C_{22} alkane in many samples. Ten Haven et al. (1988) reported a similar dominance of the n-C_{22} alkane over the C_{21} and C_{23} alkanes from several extracts and oils derived from organic matter of sediments deposited under hypersaline conditions. They defined on an empirical basis the R22 index ($2 \times n\text{-}C_{22}/(n\text{-}C_{21} + n\text{-}C_{23})$), where ratios above 1.5 are thought to be indicative of organic matter deposited in hypersaline environments. Recent investigations by Barbe et al. (1990) of modern evaporite environments in Bonmati Salina (south-east Spain) showed a predominance of the n-C_{22} alkane in the lipid composition of samples obtained from the lower salinity carbonate domain in the salina where branched hydrocarbons predominated. Extremely high concentrations, with n-C_{22} being the dominant lipid of the hydrocarbon

Organic geochemical properties and origins of sapropels   495

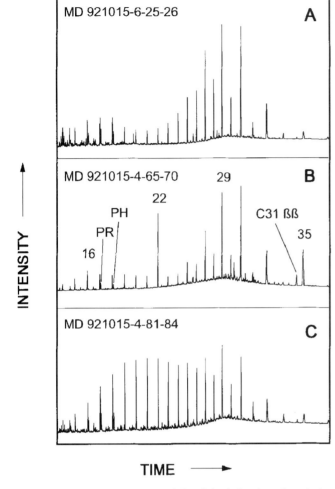

Figure G2.14 Chromatograms of the aliphatic fraction of a pelagic mud (A), a sapropel layer (B) and a sample impregnated by a mature oil of unknown origin (C).

fraction, were detected in samples obtained from higher-salinity gypsum/carbonate facies. Samples from this environment of the salina also showed high abundances of odd-numbered n-alkanes in the range of n-C_{27} to n-C_{33} demonstrating that hypersaline depositional environments are often likely to contain a significant fraction of allochthonous organic matter, as noted in the Red Sea samples (Figure G2.14(b)). If the empirical parameter is applied to core 92-1015, two periods with elevated salinity can be distinguished (Figure G2.16(b)). The lower interval with high R22 ratios reaches from approximately 12.6 to 15.5 m and an upper interval defined by only one sample at 5.17 m (MD 921015-4-65-70). However, it is questionable whether this salinity parameter can be applied successfully to the sediments studied since many other indicators for hypersaline conditions reported by ten Haven et al. (1988) are not present.

Conditions for sapropel formation in core 92-1015

Sapropels (TOC > 1%) were deposited as four distinct events (SP 1–SP 4, Figure G2.10). The TOC and total sulphur contents of the sapropel layers are similar to those recorded for Quaternary organic-rich sediments of the Black Sea deposited under anoxic conditions (Leventhal, 1983). However, anoxia did not extend through the entire water column since all sapropel layers are characterized by relatively high amounts of the $C_{31\ \beta\beta}$ homohopane, a compound which is derived from bacteria or cyanobacteria living under oxic conditions (Ourisson et al., 1979). Therefore, a ventilated upper water column probably persisted continuously during deposition. The development of anoxia probably was not caused by a high bioproductivity in the upper water column since the organic matter that accumulated in the sapropel layers is derived mainly from terrigenous sources with only a minor contribution from autochthonous marine organic matter. Furthermore, reduced dilution by autochthonous carbonate does not appear to be the cause of the deposition of carbonate-poor and organic matter-rich sapropelic layers; TOC values calculated on a carbonate-free basis (Figure G2.17) retain elevated amounts of organic matter in two intervals in the lower and upper part of the core.

Data from the upper part of core 92-1015 appear to record conditions similar to those described in the central and northern Red Sea by Locke and Thunnell (1988). Elevated salinities are indicated for the SP 1 horizons by R22 values greater than 1.5 and positive oxygen isotope values (Table G2.1) This interval may correspond to the glacial maximum (around 18 000 y BP) when elevated salinity developed during a drop in sea level in the order of 130 m (Bard et al., 1990) which, in turn, led to partial isolation of the Red Sea basin. This rise in salinity coincides with the deposition of organic-rich sediments. The organic matter is highly degraded and derived mainly from terrigenous sources which, in turn, may indicate higher rates of erosion along the basin edges caused by sea-level drop. Elevated bottom-water salinities extended to approximately 11 000 y BP (Locke and Thunnell, 1988) and are also indicated by the isotopic composition of the carbonate crusts formed during this period. Elevated salinities favoured the accumulation of organic matter in a salinity-stratified water body. The return to lower surface-water salinities during this period, as recorded by the return of less salinity-tolerant foraminifera (Locke and Thunnell, op.cit.; Rossignol-Strick, 1987), is probably related to more humid climatic conditions between 8000 and 12 000 y BP, as recorded in lacustrine sediments of northern Africa (Ritchie et al., 1985). Anoxic bottom-water conditions, in conjunction with elevated salinities, are also recorded for the lower part of

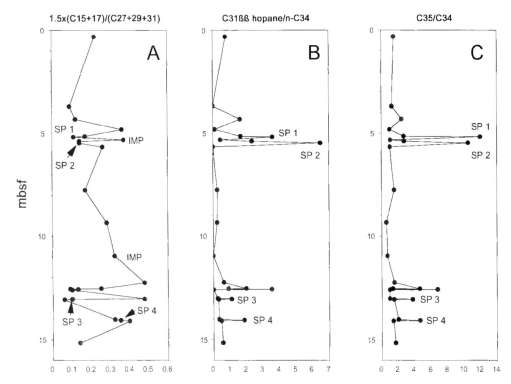

Figure G2.15 Variation of selected molecular parameters for discrimination of the sources of organic matter, details of which are discussed in the text. (IMP = impregnated sample.)

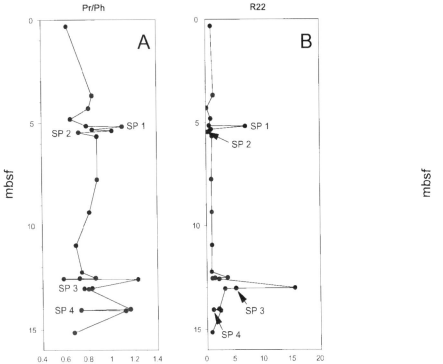

Figure G2.16 Variation of selected molecular palaeoenvironmental indicators, discussed in the text.

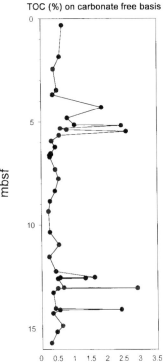

Figure G2.17 Variation of organic carbon content calculated on a carbonate-free basis.

the core. However, since dates for the lower sapropels (SP 3 and SP 4) are not available, possible correlation with a sea-level low-stand, perhaps in conjunction with a glacial maximum, must remain speculative.

DIAGENETIC PROPERTIES AND ORIGIN OF LITHIFIED CRUSTS

Lithified crusts, situated between 5.5 and 6.3 m in core 92-1015, comprise a series of very thin (1–3 mm) laminae of relatively hard limestone separated by soft carbonate mud. The crusts are part of a very characteristic 2 m thick unit comprising laminated brown or dark-grey muds relatively rich in organic matter and, locally, in silt-sized siliciclastic detritus (Figure G2.18) with very rare planktonic skeletons. Crusts range in age from 20 640 y BP at the base of 15 200 y BP at the top, i.e. over an interval of 80 cm (Table G2.1) and thus are associated with relatively low sedimentation rates averaging 6.8 cm/1000 years.

These fragile crusts are generally broken into coarse clasts during coring. However, in spite of their disruption it is possible to reorient the fragments which reveals that the lithified laminae have smooth upper and irregular lower surfaces, the latter due to numerous protruding pteropods. The morphological contrast between upper and lower surfaces reflects a diagenetic fabric which is denser towards the top of the crust, in this respect resembling the shallow-water submarine hardgrounds of the Persian Gulf described by Shinn (1969). While certain crusts are massive, others exhibit a clear lamination related to porous horizons alternating with denser laminae. Conspicuous skeletons include the conical pteropod *Creseis acicula* which are generally oriented parallel to the internal lamination. Crusts may also be associated with 2–10 mm sized, aragonitic aggregates and discontinuous platy chips, these localized diagenetic features having much in common with the larger-scale 'lumps' often associated with shallow-marine hardgrounds.

Diagenetic fabrics and mineralogical composition of the crusts

The basic fabric consists of blunt triplets of aragonite fibres (maximum length 50 μm, length/width ratio being 7:1) arranged in a highly porous network of interlocking spherulites (Figure G2.19(c)). The spherulites have an outer diameter of about 40 μm. Within the core of each rosette the radial pattern is more irregular while the outer parts consist of clearly defined crystals (Figure G2.19(f)). At the periphery, crystals have a width of 4–5 μm. Crystals of adjacent rosettes are closely intergrown but growth is blocked by biogenic or other obstacles around which crystals tend to adapt to residual pore-space (Figure G2.20(b)).

Peloidal textures, 20–30 μm in diameter, also occur. They exhibit rims of dentate magnesian calcite typical of peloidal cements in reefs (MacIntyre, 1985), or epitaxial fibrous aragonite. Others lack overgrowth. Individual peloids consist of loosely packed fragments of fibrous aragonite or cog-shaped aggregates of dentate magnesian calcite. The latter tend to encrust the fibrous elements or exhibit imprints of aragonite crystals which they clearly post-date.

A xenotopic, cryptocrystalline matrix composed of submicron to μm-sized nanograins may either be finely dispersed between aragonite fibres or may fill porosity between larger idiomorphic crystals (Figures G2.19(e), G2.20(a)). Aragonite prisms (1–2 μm) which float within this matrix are interpreted as syntaxial overgrowths on aragonite nuclei; they tend to overgrow and envelope cryptocrystalline nanograins. Although on a larger scale, similar diagenetic overgrowths of echinoderm rim-cements occur in shallow-marine environments (James and Bone, 1989). This cryptocrystalline matrix may be the dominant constituent (Figure G2.19(d), (e)) within spherulites or other allochems.

Volumetrically, skeletal grains and non-carbonate particles generally constitute less than 20% of the rock volume which is dominated by carbonate precipitates. Angular terrigenous particles lacking cement overgrowths possibly represent wind-blown material (Figure G2.20(c)). Skeletal grains include pelagic gastropods (Figure G2.21), calcareous spheres interpreted as algal

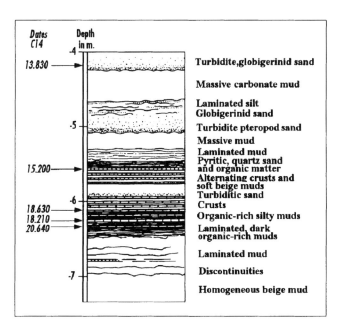

Figure G2.18 Log showing ages (^{14}C) and relationships between organic-rich muds and lithified crusts in core 92-1015, Suakin Deep.

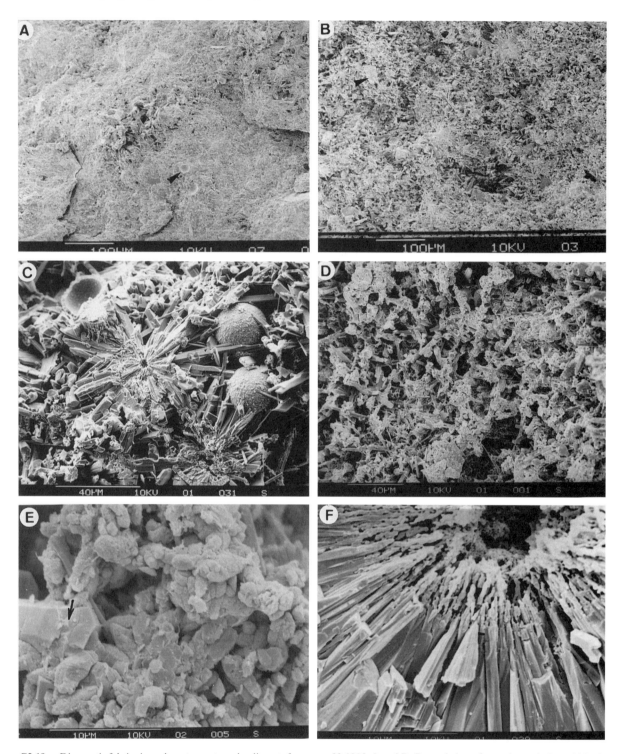

Figure G2.19 Diagenetic fabrics in carbonate crusts and sediments from core 92-1015. A and B. General view of porosity variation within the crust; recognizable sedimentary particles (arrows) float in a crystalline matrix (sample TB 04). C. Diagenetic spherulite composed of radiating aragonite fibres whose growth is obstructed locally by sedimentary particles; note hollow centre of spherulite, discussed in text (sample TB 05). D. Aggregates of cryptocrystalline carbonate dispersed in fibrous, diagenetic aragonite (sample TB). E. Detail of cryptocrystalline fabric composed of submicron-sized carbonate nanograins. These are locally incorporated into fibrous aragonite crystals (arrow), (sample TB 08). F. Detail of centre of aragonitic spherulite showing hollow centre and possible traces of dissolution (TB 05).

cysts (Figure G2.19(c)), debris of calcareous nanoplankton (Figure G2.20(e)), rare planktonic foraminifera and pteropods. Microgastropods and pteropods generally have rims of fibrous aragonite (Figure G2.21(a)) which has an irregular thickness (50–100 μm) and exhibits lobate contacts with the adjacent carbonate matrix.

Diagenetic properties and origin of lithified crusts 499

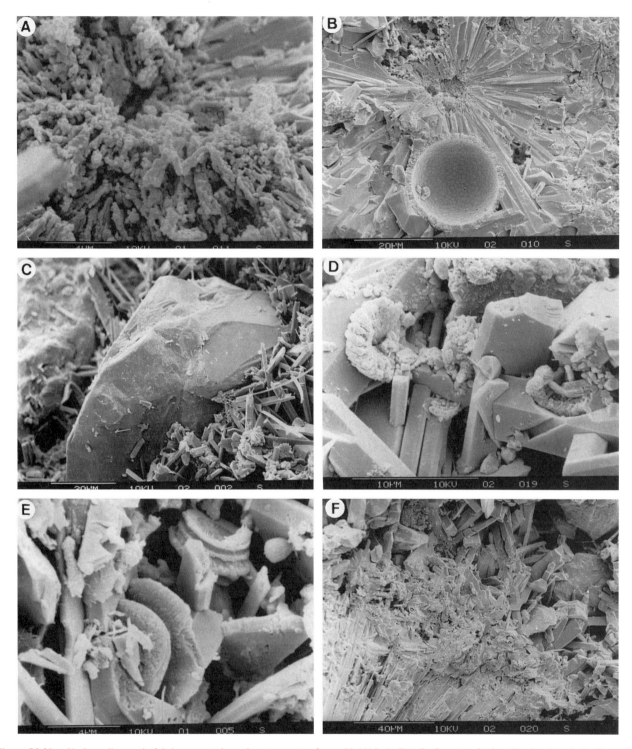

Figure G2.20 Various diagenetic fabrics present in carbonate crusts of core 92-1015. A. Detail of centre of spherulite showing chain-like arrangement of nanograins (sample TB 07). B. Spherulite whose radial growth is obstructed locally by calcisphere, the latter with thin isopachous fringe of cement (sample TB 03). C. Angular sediment particle (possibly quartz) embedded in diagenetic matrix (sample TB 08). D. Coccolith with significant diagenetic overgrowths which partly englobe aragonite fibres (sample TB 03). E. Details of coccolith preserved within an essentially diagenetic, fibrous fabric (sample TB 07). F. Fibrous aragonite with concentration of the fine debris around edge of spherulite (sample TB 07).

Cement on calcareous algal spheres is minor and is composed of μm-sized crystals disposed obliquely with respect to the surface of the cyst. There are marked variations in the style of preservation of coccoliths which either have clean surfaces or show significant diagenetic overgrowths (Figure G2.20(d), (e)).

Discussion: origin of the crusts

As noted by Gevirtz and Friedman (1966) and by Milliman *et al.* (1969), pteropods typically exhibit linings of fibrous aragonite which may completely fill intraparticle voids. Although the sediment is not grain-supported in the classical sense (Dunham, 1962), fringes of aragonite also occur on the outer surfaces of pteropods where they interlock with the cryptocrystalline matrix or with fibrous spherulites (Figure G2.19(c)). These relationships raise the question concerning the origin and timing of these fabrics.

Basically, there are two possible explanations concerning the growth of aragonite cements on the external surfaces of pteropods. Firstly, the shells may have been embedded within and supported by a fine carbonate matrix. During crystal overgrowth the matrix was displaced. Interpenetration and overpacking of small allochems near the outer margins of cements tend to support this interpretation. Based on stable isotope data from the Shaban Deep, Stoffers *et al.* (1990) have suggested that cementation is related to microbial activity within the sediment column. However, a significant overpacking of particles was not observed to occur at the compromise boundaries of adjacent cemented pteropods or spherulites. Furthermore, compromise boundaries tend to be linear and thus typical of cavity fillings. The second possibility is that crystal cements were formed at the sediment–water interface, i.e. before burial.

Aragonite cements, characteristic of pteropods and microgastropods, are absent on other biogenic particles. Furthermore, epitaxial aragonite does not occur in calcareous algal spheres but these may be embedded within cement fringes formed on adjacent pteropods. These selective relationships suggest that aragonite cement growth is controlled by the substrate. However, spectacular stellate crystal growths also represent additional *in situ* aragonite growth. Neev and Emery (1967) and Druckmann (1981) have described stellate aragonite precipitates ('whitings') from the Dead Sea which seem comparable in fabric to the spherulites of the Red Sea. Thus, one can not discount the possibility that the spherulites have formed during a two-stage process involving nucleation (as 'whitings') within the water column and subsequent diagenetic growth on the sea floor. This may explain the intergrowths between adjacent spherulites or matrix nanoparticles.

Laminae encrusted by tubular microfossils have been described by Brachert (1994). This growth results in a distinct cryptic microbial lamination of certain crusts suggesting that the millimetric lamination reflects sedimentary rhythmicity punctuated by breaks in deposition. Sedimentary breaks are also indicated by the upwards decrease in porosity in certain laminae, this phenomenon being comparable to hardgrounds in Quaternary sediments of the Mediterranean (Aghib *et al.*, 1991). This vertical change in fabric within each lamina is considered to involve several processes. It begins with the accumulation of particles, including sperulites, their subsequent lithification by aragonitic overgrowth, terminating by the infiltration of fine detrital carbonate subsequently lithified by cryptocrystalline aragonite. Thus, the two principal fabrics (porous, spherulitic and dense cryptocrystalline) reflect the relative importance of one or other of these parameters during sea-floor accretion of the crust.

The nucleus of many spherulites commonly is a microvoid several microns in diameter (Figure G2.19(f)). It may also be made of aggregates of chains of submicron-sized, elongate nanoparticles (Figure G2.20(a)). Although the significance of these nuclei is not fully understood, they may be due to original

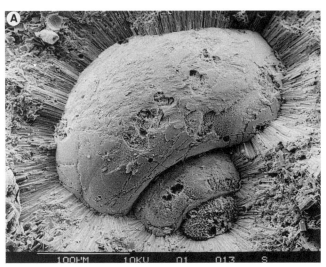

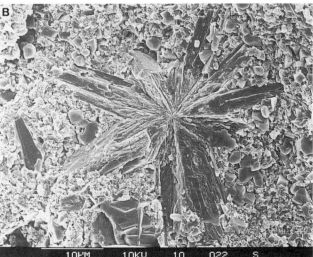

Figure G2.21 A. Pelagic microgasteropod on which has developed a thick layer of fibrous, epitaxial aragonite cement (sample TB 05). B. Microspherulite in soft sediment.

microbial aggregates. Calcareous microbial precipitates generally are composed of magnesian calcite. However, depending on the culture conditions, aragonite spherulites can develop (Buczynski and Chafetz, 1991) and these are similar in form and size to the material from the Red Sea.

As already noted, there is a strong similarity between the magnesian calcite peloids of the Red Sea and reef peloids, the latter precipitating from normal sea water (Friedman et al., 1974; Macintyre, 1985; Montaggioni and Camoin, 1993). According to laboratory studies, similar peloids result from bacterial activity (Chafetz, 1986; Chafetz and Buczynski, 1992). Thus, irrespective of the site of nucleation (sea-water column or sediment surface), it is possible that much of the carbonate comprising the crusts, including both sperulites and peloids, has microbial origins.

Palaeooceanographic control of crust formation

The oceanographic record of the late Pleistocene–Holocene of the Red Sea (Locke and Thunnell, 1988; Almoghi-Labin et al., 1991; Taviani, this volume) may explain the formation of these deep marine crusts. Principal parameters involved are glacio-eustatic fluctuations of sea level and climate-related precipitation. According to the model, global sea-level lowering of at least 130 m during the last glacial maximum resulted in the partial isolation of the Red Sea from the Indian Ocean, present depths at the Bab el Mandeb sill being 137 m (Taviani this volume, a). This restriction resulted in high salinities recorded both by the low content of calcareous nanoplankton and by the very positive stable oxygen isotopes of the plankton (Thunnell et al., 1988) and the crusts (Table G2.1). These positive values ($\delta^{18}O + 6\text{\textperthousand}$), reflecting high salinities, together with the ^{14}C ages of 15 200–20 640 y BP indicate that the crusts formed very close to the last glacial maximum.

COMPARISONS WITH OTHER AREAS OF THE RED SEA AND GULF OF ADEN

Axial zone of the Red Sea

Cores from other areas of the axial zone of the central and northern Red Sea (Taviani this volume, a) appear to have much in common with core 92-1015 from the Suakin Deep. Those situated near the foot of peripheral scarps or in adjacent depressions, contain numerous turbidite sands (Taviani and Trincardi, 1992) composed mainly of planktonic skeletons. Their presence confirms the regional importance of resedimentation processes within these deep environments which appear to be the major cause of relatively high rates of sedimentation within the rift axis.

The presence of lithified crusts is also widely reported (Herman, 1965a,b; Gevirtz and Friedman, 1966; Milliman et al., 1969; Brachert and Dullo, 1991) from the central and northern Red Sea. Their ages, averaging 15 000 years, are comparable to those obtained from crusts in core 92-1015 demonstrating the regional character of these hardgrounds. This age has led a number of workers, including Deuser and Degens (1969) and Milliman (1977), to relate this phase of marine diagenesis to overall restriction of the Red Sea during the lowering of glacial sea level. The present authors have given further evidence supporting this conclusion.

Similar organic-rich sediments have also been recorded in other parts of the Red Sea. Locke and Thunnell (1988) have identified two sapropels containing 1–1.8% TOC at water depths of 1275 m in the central Red Sea (19°34′ N–38°59′ E). These intervals contain similar amounts of carbonate as the upper sapropels (SP 1 and SP 2) from core 92-1015 and also have a relatively high quartz content probably indicating that the organic matter was derived mainly from continental sources. These sapropels have also been dated as 8000–11 000 y BP for an upper interval and about 18 000 y BP for a lower. Locke and Thunnell (op. cit.) reconstructed the development of the Red Sea for the last 18 000 years based on stable isotopes and palaeontological evidence, concluding that restriction occurred during the last glacial maximum during which low sea-level stand probably led to the isolation of the Red Sea from the Indian Ocean. As a result, elevated salinities affected the entire water column (Kolla and Biscaye, 1977). When the connection between the Red Sea and Indian Ocean was re-established during deglaciation, surface-water salinity of the Red Sea began to decrease, as evidenced by the return of *G. sacculifer* and *G. aequilateralius*. However, deep-water salinities remained elevated until at least 11 000 y BP. The decrease of salinity in the surface waters may have caused a density stratification with a lower water-mass characterized by higher salinity. Furthermore, humid conditions possibly relating to intensification of the southwestern monsoon over Africa and Arabia (Street and Grove, 1979; Ritchie et al., 1985; Cullen, 1981; Prell and Van Campo, 1986; Rossignol-Strick, 1987) between 8000 and 12 000 y BP, may have increased continental run-off further amplifying salinity stratification. This prevented the formation of new bottom-waters, leading to oxygen-deficiency and thus ideal conditions for sapropel preservation.

Sudanese Shelf

Cores taken on the Sudanese Shelf during the 1992 RED SED cruise, while having certain properties in common with sediments examined in core 92-1015 (axial zone), also exhibit marked differences. Core 92-1019, taken at a depth of about 600 m, some 100 km east-south-east of Port Sudan, also has well developed aragonitic crusts at core depths of 0.5 m whose petrographic fabrics are similar to those seen in crusts from the axial zone (Brachert, 1996). Carbon-dating indicates that the crusts in the core from the Sudanese Shelf are similar in age (about 15 000 years) to those dated from the axis. Furthermore, both the axial and the shelf cores have similar O and C isotopes, the positive O ($+6.99‰$) and C ($+4.22‰$) values (Table G2.1) indicating precipitation of diagenetic carbonate from relatively saline (restricted) marine waters. These saline waters, common both to the axial and shelf environments, confirm the overall restriction of the Red Sea at about 20 000–15 000 y BP. The presence of crusts on the shelf, situated at some considerable distance (50 km) from the axial zone, clearly reduces the possibility of hydrothermal influence on their formation. Situated at much shallower depths than the underlying Miocene evaporites, they also preclude the possibility that the elevated salinities were the result of evaporite dissolution.

The lithified crusts (hardgrounds) encountered on the shelf (core 92-1019) coincide with a change in sedimentation (Plate 32) separating massive, light-grey, bioturbated lime muds below, from vaguely bedded, dark-grey, organic-rich muds above. These darker facies overlying the hardground pass gradually upwards into light-grey carbonate muds. The presence of organic matter on the shelf, although in lower concentrations relative to the 'sapropel' horizons in the axial core, nevertheless demonstrate that its distribution within the rift is not confined to the rift axis. The stratigraphic association between the organic-rich horizons and the diagenetic crusts, both in the axis and on the shelf, would imply a common, basin-wide, climatically controlled cause.

Sediments within cores from the Sudanese Shelf also differ in many respects from those of the axial zone. The lithified crusts in cores 92-1019 (shelf) and 92-1015 (axis) have comparable ages (about 15 000 years). However, they do not occur at similar depths; situated at about 6 m within the axial core, they occur at 0.5 m within the core taken on the shelf. These differences in core-depth of what appears to be a common horizon could suggest considerably slower rates of sedimentation on this part of the shelf. This is not altogether surprising. In these relatively shallow (300–600 m) shelf areas, there are few turbidites (Brachert, 1996) which appear to be characteristic of near-axial environments.

In general, cores from the shelf are massive, medium-grey, bioturbated lime muds and bedding is rare. SEM studies indicate that the carbonate mud fraction is composed of unidentifiable, silt-sized skeletal debris (Figure G2.4(f)) plus scattered globigerinids and accessory benthonic foraminifera. There is also a significant amount (5–10%) of fine, thin-shelled molluscan debris, in this respect these hemipelagic facies differing from their axial equivalents.

Mineralogical studies and chemical analyses indicate only traces of Fe and Mn oxides (Table G2.2) these latter being typical of the axial zone where their precipitation is related to hydrothermalism. Clay minerals, on the contrary, are very similar indicating that they are largely independent of hydrothermal waters, probably being essentially of detrital origin.

Gulf of Aden

Cores taken offshore Djibouti at water depths of 1500–2000 m during the 1992 RED SED cruise differ in many respects from those of the Red Sea. Sediments are characterized by a medium- to dark-grey colour, being somewhat darker than those of the axial and shelf areas of the Red Sea. The Gulf of Aden cores consist of a relatively monotonous sequence of massive, silty muds exhibiting bioturbation; bedding is rare. Furthermore, no lithified crusts nor organically enriched ('sapropel') horizons were encountered suggesting that these latter phenomena are specific to Red Sea environments. This observation has obvious implications concerning the origins of the crusts and associated organic matter, discussed in the following section. Finally, cores from the Gulf of Aden did not include turbiditic horizons.

DISCUSSION AND CONCLUSIONS

This multidisciplinary study has shown that the 16 m of sediment in core 92-1015 recovered from the axial part of the central Red Sea, records a series of tectonic and climatic events. In common with other parts of the Red Sea, this core is characterized by numerous (12) turbidite sands composed essentially of planktonic skeletons mixed with material eroded from the structural scarps bordering the axial trough. These resedimented materials appear to be confined to the axial parts of the rift where they may record seismic instability related to rifting. Additional expressions of rift geodynamics are relatively subtle. Axial, hydrothermal waters have contributed Fe and Mn oxides and possibly modest amounts of neoformed smectite, but these chemical products appear to be confined to the axial zone.

In core 92-1015 the content of organic matter is generally low. Exceptions are two intervals situated in the upper and lower parts which contain between 0.45% and 2.14% organic carbon. Four organic-rich layers also contain elevated amounts of sulphur which was probably deposited under anoxic bottom-water conditions. The distribution of organic matter in the upper sapropels is similar to that described by Locke and Thunnell (1988), suggesting that widespread sapropel formation probably occurred between 8000–12 000 y BP and 18 000 y BP. The age of the lower sapropels (SP 3 and SP 4) is still unknown.

Organic matter in core 92-1015 was derived mainly from terrigenous sources there being relatively little autochthonous marine organic matter. Finally, two periods of elevated surface-water salinity are recorded by high R22 values. One of these periods corresponds to the last glacial maximum at around 18 000 y BP when the connection between the Red Sea and Indian Ocean was reduced to depths of less than 10 m (present depth is approximately 137 m) and thus severely restricted.

The formation of sapropels and carbonate crusts are related to a common factor, namely glacial sea-level minimum. The lithological break expressed by the aragonite crusts and upper sapropels has much in common with certain ancient hardgrounds and related sequence boundaries. They have been correlated and dated (15 000–20 000 years) throughout much of the Red Sea where their widespread distribution appears to have been determined by elevated salinities. These hypersaline conditions, clearly recorded both in the geochemical properties of the sapropels and the strongly positive C and O isotopes of the aragonitic crusts, although related indirectly to glacial sea-level lowering, are caused by the presence of a structural barrier – the Bab-el-Mandab sill; sediments at comparable depths within the Gulf of Aden do not appear to record these sea-level fluctuations. In other words, the hardgrounds within the Red Sea, typical expressions of sequence boundaries, are peculiar to that basin in spite of the fact that they are conditioned by global sea-level oscillation.

ACKNOWLEDGEMENTS

The authors express their gratitude to Monsieur J. Balut of TAAF (Paris) whose collaboration facilitated the organization of the 1992 *Marion Dufresne* cruise 'RED SED' which provided the material for this research. Thanks are also extended to the Laboratoire d'Hydrologie et Géochimie Isotopique, Université de Paris Sud (Orsay) for ^{14}C dates and isotope data, to Y. Raguideau, M. Helmer, and J. Didelot of Orsay, J. Clément and J-P Gianessini (Museum National d'Histoire Naturelle, Paris), A. Matari and B. Spitthoff (Cologne), U. Krautworst, G. Ritschel and K. Schuchmann (Mainz). The authors also thank Alan Huc of the Institut Français du Pétrole for reviewing and improving the manuscript. This research was financed by EC project 'Science Program RED SED'.

APPENDIX: ANALYTIC METHODS

Clay mineralogy

X-ray diffraction was carried out on the fraction < 2 μm using a SEIMANS K4 generator with Co anticathode, quartz monochromator in front position and CPS 120 INEL. Diffractionel software was used for XRD analysis. Deflocculation of clays was done by successive washing with distilled water following removal of carbonates with 0.2 N HCl. The clay fraction (< 2 μm) was separated by sedimentation and deposited on a glass slide from which an oriented aggregate of clay particles was made.

Three types of oriented preparations were made: air-dried, heated to 500°C for 2 h; and ethylene glycol saturation. X-ray identification was made according to the position of the 001 series of basal reflections. Samples were Ca-saturated with an excess of 1 N CaCl$_2$. Following multiple treatments the clays were washed with distilled water to remove excess CaCl$_2$ and glycolated in order to compare their expandability with the natural state. The smectites were identified using the Hofmann-Klemen test. The Li-saturated samples were heated to 250°C for 24 h and saturated with ethylene glycol. Following this treatment, only beidellite gave the expanded spacing at 17.1 Å while montmorillonite is not expanded.

The nature and quantity of major elements, based on bulk analyses, were made by SEM microprobe fitted with an energy dispersive spectrometer (EDS). The relative errors are less than 3% for elements having concentrations between 5% and 10%, and 20% for elemental concentrations of less than 5%. The detection limit is 0.1%. The weights were recalculated to 100% (oxide values in % by weight).

Transmission electron microscopy (TEM), although providing information concerning sizes and shapes of clay minerals, does not completely characterize a mineral type (Beutelspacher and van der Marel, 1968; Sudo *et al.*, 1981). In diffraction mode, TEM studies of individual particles give additional information useful in determining the crystal structure (Gard, 1971; Wilson, 1987). Single crystal spot patterns have pseudohexagonal symmetry (indexed as hkO) compatible with three-dimensional lattice (chlorite, illite, kaolinite). Ring patterns obtained from thin clay particles are indicative of a turbostratic disorder (smectite). The distinction of dioctahedral and trioctahedral networks may be made by measuring the 060 (or 06.33) reflection.

Organic geochemistry

The sampling strategy, firstly, was to collect evenly spaced samples at intervals of approximately 50 cm for routine analysis (TOC, carbonate and sulphur contents) in order to document the distribution of these basic fractions over the entire length of the core; and secondly, to sample sediment layers that appeared to be sapropelic on visual inspection. The sampling of each sapropelic layer was complemented by two additional samples, above and below the sapropel respectively. The samples ranged from green and brown marls to black calcareous clays. In addition, one lithified carbonate crust (921015-4-113-118) was studied in detail. A total of 43 samples were analysed for organic carbon, carbonate and total sulphur contents (Table G2.3). Rock-Eval analysis was performed on 11 organic-rich samples. Characterization of the aliphatic hydrocarbon fractions of the soluble organic matter was made on 25 samples by gas chromatography and on 20 samples by gas chromatography/mass spectrometry (Table G2.3).

Total carbon and sulphur determinations were made with a LECO CS-225 instrument which was also used to determine total organic carbon (TOC) following removal of carbonate with HCl. Carbonate contents were calculated by weight difference and expressed as percent $CaCO_3$. Rock-Eval pyrolysis was performed according to the method described by Espitalié et al. (1977) using a Rock-Eval plus instrument by Vinci Technologies. A SOXTEX apparatus was used to extract the ground sediment samples with a mixture of dichloromethane and methanol (99/1 v/v). Total extracts were separated into saturated hydrocarbons, aromatic hydrocarbons and heterocompounds by medium pressure liquid chromatography (Radke et al., 1980). The aliphatic hydrocarbons were then analysed using a Hewlett-Packard 5890 II gas chromatograph equipped with a HP5 methylphenyl silicon capillary column of 50 m length, 0.2 mm internal diameter and a 0.33 μm film thickness. Following cool on-column injection, the oven temperature was programmed to rise from 70 to 140°C at 10°C/min and to 300°C at 5°C/min. Helium was used as a carrier gas. The gas chromatography of selected aliphatic hydrocarbon fractions was complemented by gas chromatography/mass spectrometry on an HP 5890 II gas chromatograph coupled with an HP 5989A mass spectrometer. The oven was programmed to rise from 70 to 140°C at 10°C/min and to 300°C at 3°C/min, with experimental conditions identical to those outlined above.

Chapter G3
Metalliferous sedimentation in the Atlantis II Deep: a geochemical insight

G. Blanc, P. Anschutz and M.-C. Pierret

ABSTRACT

The 23 000-year-old sediments of the Atlantis II Deep lie on a basaltic substratum and are overlain by a hot NaCl–rich brine pool. In order to assess a geochemical model based on the sedimentological and chemical processes leading to the formation of the Atlantis II hydrothermal deposit, we have examined the chemical composition and the XRD data from 120 samples from two cores that sampled the entire sediment sequences of the Atlantis II Deep. A normative quantification of the temporal and spatial distribution of major mineral species (sulphides, sulphates, carbonates and oxides) enabled us to propose new lithological sequences that differ from those proposed by Bäcker and Richter (1973). These sequences are defined by genetic units labelled 1, 2, 3a–f, 4 and U and L for the western and south-western basins, respectively. The units are characterized by specific mineral associations deriving from hydrothermal inputs, biogenic and detrital sedimentation and from intensive secondary transformations. The supplied biogenic and hydrothermal particles acted as precursors for stable phases that accumulated within the sediment and participated with several chemical processes involving acid–base and redox reactions. Thus, we defined a Fe-Mn redox cycle relating to the removal of Mn^{++} and the formation of ferrihydrite that incorporated silica in its structure. Ferrihydrite has been partly transformed into goethite or haematite as a function of the temperature and the metal content of the solution in equilibrium. Lepidocrosite and akaganeite acted eventually as precursors for these oxides. The reactions of carbonate dissolution and oxidation of dissolved H_2S by Fe-oxides acted as the main pH buffer of the brine during the discharge of acid hydrothermal fluid, and permitted the precipitation at the equilibrium of carbonate solid solutions and occasionally anhydrite, respectively. The geochemical mechanisms leading to the formation of the Atlantis II ore deposits proceeded in different environments during the history of the deep. The precipitation of sulphides and sulphates by quenching effect are located in the vicinity of discharge sites of the hydrothermal fluid. If Zn, Fe and H_2S are in excess in reduced solution, ZnS and FeS_2 could, however, precipitate at the equilibrium far from the hydrothermal fluid springs. The precipitation of Fe and Mn-oxides operated at the sea water–brine interface. The authigenic carbonates were formed in the upper part of the reduced lower brine where the concentrations of dissolved Fe and Mn were high and where CO_3^- was realized by the dissolution of biogenic carbonates. The Fe-Mn redox cycle worked in the lower brine. The Fe and Mn-oxides were transformed into more stable phases after their settling into the sediment. The formation of carbonate solid solutions and pyrite of unit 1 resulted from reactions of early diagenesis relative to the bacterial activity. Faunal associations showed that unit 4 is contemporaneous with unit U. Mineral associations indicate that unit 3f (Mn-oxides) is contemporaneous to unit L (anhydrite and Fe-oxides). Lateral bathymetric variations and changes of the hydrodynamic regime within the brine pool explained the differences in mineralogy and thickness between the correlated units. The proposed geochemical model includes chemical processes that acted at temperatures lower than 250°C and pH range from 4 to alkaline pH of sea water during the entire hydrothermal history of the Atlantis II Deep.

INTRODUCTION

In the submarine environment, the spreading centres are favourable locations for the venting of hydrothermal solutions and the formation of mineral deposits (Von Damm, 1990; Rona and Scott, 1993). Hydrothermal research at sea-floor spreading centres began in the mid-1960s with the discovery of hot metalliferous brines and sediments pounded in deeps along the axis of the Red Sea, which is a divergent plate boundary between the African and Arabian peninsulas separating at a slow rate of about 2 cm/year (Charnock, 1964; Miller, 1964; Swallow and Crease, 1965; Miller et al., 1966; Bischoff, 1969; Degens and Ross, 1969).

Among the 18 deeps within the median valley of the Red Sea (Bignell et al., 1976a,b; Pautot et al., 1984), the Atlantis II Deep with 60 km^2 surface is the largest brine-filled closed basin located near 21°25′ N and 38°05′ E in the axial rift of the Red Sea (Figure G3.1). The rift zone is covered or bordered by a thick Miocene evaporite sequence which contributes to the formation of the dense NaCl-brines that accumulate in the depression (Shanks and Bischoff, 1977; Zierenberg and Shanks, 1986; Dupré et al., 1988). The Atlantis II Deep is divided into four sub-basins called North (N), East (E), South-West (SW) and West (W) (Figure G3.1). They are separated by bathymetric highs that do not extend above 2000 m depth contour limiting the top of the brine pool. Numerous studies have shown that the ≈5 km^3 pool of brine is stratified into two distinct, dense, and convective layers of differing temperature and chlorinity (Turner, 1969; Bäcker and Schoell, 1972; Voorhis and Dorson, 1975; Schoell and Hartmann, 1973; 1978; Hartmann, 1980). Temperature and salinity records from the RED SED cruise (September 1992) reveal a new stratification in the hydrothermal brine system of the deep. This stratification consists of the pre-existing lower and upper convective layers (e.g. LCL and UCL1), with enhanced temperatures up to 66°C and 55°C, respectively, and two additional upper convective layers named UCL2 and UCL3, respectively, which formed more recently and have distinct temperature, salinity and thickness (Blanc and Anschutz, 1995).

Cooling of the high temperature inflowing brine by mixing into the pre-existing brine pool results in precipitation of metalliferous sediment. The depositional environment of the Atlantis II Deep contrasts with those of high temperature vents on spreading centres where metal sulphide is precipitated due to thermal quenching near the vent, but the majority of metal transported to the sea floor is dispersed by the buoyant plumes where it is oxidized and eventually deposited as the highly diluted basal metalliferous sediment that commonly overlies oceanic basalt. In contrast, the brine pool confines the major part of the metalliferous sediment to deposition within the Atlantis II Deep and has resulted in the formation of a very large sedimentary ore deposit. More than 600 sediment cores have been collected within the Atlantis II Deep in order to evaluate the economic potential of the deposits. The total estimated reserves of dry, salt-free metalliferous sediment are approximately 94×10^6 metric tons (Mt), including 1.9 Mt Zn (average grade 2.1%), 0.4 Mt Cu (0.46%), 5400 t Co (59 g/t), 3750 t Ag (41 g/t), and 47 t Au (0.51 g/t) (Guney et al., 1988). This hydrothermal deposit remains the largest sulphide accumulation discovered on the sea floor and has an economic potential whose tonnage rivals the largest volcanic sulphide deposit in the world (Boldy, 1977). A 10–30 m thick layer of metalliferous sediment occurs at the bottom of the brine pool, overlying tholeiitic basalts (Bäcker and Richter, 1973). The metalliferous sediments are fine-grained (≤2 μm), delicately banded silicates, sulphides, sulphates, oxides and carbonates. A general lithostratigraphy of the metalliferous sediments has been established by Bäcker and Richter (1973). The bottom to the top of sedimentary series from the N, E and W basins are as follows. 1. The basal sediment of the detrital-oxide-pyrite (DOP) zone consists predominantly of biogenic-detrital marl that directly overlies basalt. Toward the top of this zone, there are pyrite and oxide layers. 2. The lower sulphide zone (SU1) is composed predominantly of dark red-brown, Fe-rich clays that are interlayered with black to violet, very fine-grained,

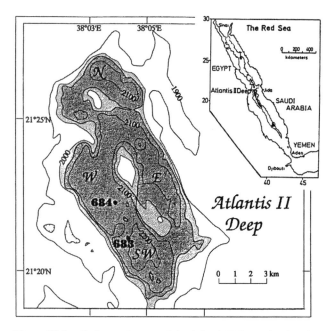

Figure G3.1 Bathymetric map of the Atlantis II Deep showing the location of cores 683 and 684. The 2000 m contour line corresponds approximately to the top of the brine-filled basins (grey area). The dark area, bounded by the 2050 m contour, shows the extent of the lower brine.

sulphide-rich sediments. 3. The central oxide zone (CO) consists of very fine-grained to amorphous Fe oxyhydroxide, and low amounts of Fe-rich phyllosilicate. 4. The upper sulphide zone (SU2) is similar in most characteristics to SU1 zone. It consists of interlayered sulphides and Fe-rich phyllosilicates. The silicates of the SU2 zone, however, are typically green in contrast to the red to brown colours observed in the SU1 zone. 5. The amorphous silicate zone (AM) is the youngest zone and is being formed today. These very young sediments average ≈95 wt% of interstitial brine and contain abundant X-ray amorphous material, including Fe oxides and silica.

The lithostratigraphy in the SW basin is different from other basins. The SW basin CO zone contains a greater proportion of haematite. The sulphide–oxide–anhydrite zone (SOAN) is characterized by disrupted sequences of haematite, Fe-rich phyllosilicates, sulphides, and anhydrite. Slumping and resedimentation are common features. Above the SOAM zone is the oxide–anhydrite zone (OAM), which consists of haematite and anhydrite with admixed Fe-rich clays and sulphides. This is characterized by breccias and intrabasinal turbidites. The upper sulphide–amorphous silicate zone (SAM) is still being deposited and it is distinguished from the AM zone by higher base-metal sulphide content.

Although these lithostratigraphic sequences have been generally referred to in many studies, they remain schematic, and cannot be used as a pattern of the metalliferous sedimentation of the Atlantis II Deep. In this chapter, we present an overview of mineralogical and chemical results and attempt to give the general pattern in space and time of the sedimentological and chemical processes leading to the formation of the Atlantis II ore deposits. For this purpose, a high depth resolution of the subsampling and the normative procedure for quantifying the distribution of major minerals have been made on two cores (683 and 684) that were collected during the 'HYDROTHERM' cruise (May 1985) (Blanc et al., 1986; Blanc, 1987). Both cores reached the basaltic substratum and recovered the almost entire undisturbed sedimentary sequence present in the SW and W basins, respectively. Intensive chemical and mineralogical studies of these cores enabled us to redefine precisely the lithological units. Considering the mineral associations of the sedimentary sequence and by comparison with available experimental data, we deduced the chemical processes which prevailed during the formation and transformation of the Atlantis II sediments. The temporal and spatial variation of these chemical processes enables us to propose a geochemical model of metalliferous sedimentation during the hydrothermal history of the Atlantis II submarine ore deposit.

MINERALOGY OF THE LITHOLOGICAL UNITS

The new lithologic description presented in Figure G3.2 can nevertheless be compared readily to that of Bäcker and Richter (1973). The thicknesses of units 1, 2, 3 and 4 are comparable; the average thicknesses of the facies DOP, SU1, CO, SU2 were calculated from several hundred cores collected in the W, E, and N basins (Urvois, 1988). Core 684, where allochthonous material does not occur, is therefore representative of the entire sedimentary sequence of the Atlantis II Deep. Because the complete sedimentary series has rarely been recovered in the SW basin, average thicknesses of lithologic facies could not be calculated in this basin (Urvois, 1988). Disturbances only occur within the two lower metres of the bottom part of core 683. Thus, core 683 with two units (lower and upper units), is representative of the entire sedimentary series from the SW basin, and

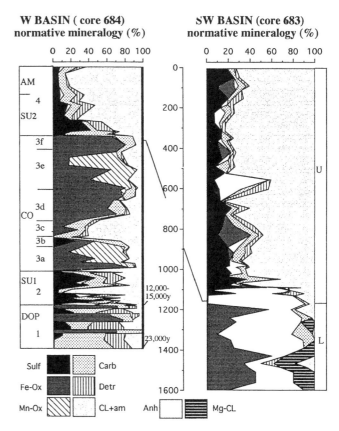

Figure G3.2 Vertical distribution of minerals in cores 683 and 684, expressed as weight percent of dry and salt-free sediments. The ages (core 684) were deduced from intercore correlation within the Red Sea using absolute abundance patterns of planktonic fauna. Lithological units defined in this study are named 1, 2, 3a–f, and 4 (core 684) and U and L (core 683). Corresponding facies proposed by Bäcker and Richter (1973) are shown for core 684: detrital-oxide-pyrite (DOP), lower sulphide (SU1), central oxide (CO), and amorphous silicate (AM). Sulf: Sulphides; Fe-Ox: iron oxides and oxihydroxides; Mn-Ox: manganese oxides and oxihydroxides; Anh: anhydrite; Carb: carbonates; Detr: detrital silicates; CL+am: clay minerals and amorphous Fe and Si compounds; Mg-CL: magnesian phyllosilicates.

is useful to give further insight into the past hydrothermal activity of the Atlantis II Deep.

Core 684

Core 684 from the W basin is divided into four main units (Figure G3.2).

1. Unit 1 (bottom to 1155 cm) consists of 2.3 m of multicoloured sediments that directly overlie the basaltic substratum (Blanc, 1987; Anschutz, 1993). This unit is composed of mixed carbonate and silicate layers separated by three iron oxyhydroxide (goethite or lepidocrocite mixed with siderite) interlayers. The primary calcite and aragonite, which normally compose the observed foraminifera and pteropod tests, are rare or absent (Anschutz and Blanc, 1995a). The recognizable major carbonate species consist of various solid solutions that will be defined subsequently. Biogenic amorphous silica originating from radiolaria and diatoms is associated with calcedony and quartz. Clay minerals (kaolinite, illite and chlorite) are abundant and can account for more than 25 wt% of the sediment. These clay minerals are considered generally as detrital clays inherited from Tertiary sedimentary series (Badaut, 1988). From the bottom to 1305 m, pyrite represents less than 2 wt% of the solid fraction, whereas its amount reaches 20 wt% near the top, where a minor amount of sphalerite occurs (<2 wt%).

2. Unit 2 is dominantly characterized by sulphide minerals (Figure G3.2). The S, Zn, Cu and Fe concentrations indicate that the cumulated content of sphalerite, pyrite, and chalcopyrite is up to 50 wt%. The distribution of the three mineral species is not uniform. Two sulphidic layers are recognized, from 1150 to 1100 cm and from 1066 to 1000 cm, which are separated by a thin layer of clay-rich sediment. Anhydrite is common at the bottom of unit 2. Carbonates in this unit are siderites which contain $MnCO_3$ in solid solution. Detrital silicates, dominantly quartz and feldspar, and biogenic amorphous silica (diatoms and radiolaria), may comprise more than 20 wt% of the sediment, as estimated by visual observations, but very few calcareous tests were found (Anschutz and Blanc, 1993a). The top of unit 2 is marked by a manganosiderite and calcite-rich layer.

3. Unit 3 is characterized by the dominance of Fe and Mn oxides and oxyhydroxides (Figure G3.2). Goethite, haematite, manganite, groutite and todorokite are the main mineral species, with variable proportions through the unit. Subunit 3a (1000–875 cm) contains dominantly goethite and manganite, which account for 80 wt% of the total sample. Mn concentration shows that the amount of manganite increases from 14 wt% at the bottom to 60 wt% at the top of this subunit. Unit 3b (875–835 cm) is a thin sulphide subunit, without Mn^{3+} or Mn^{4+} minerals, but which contains manganosiderite.

Haematite also occurs in this unit. Unit 3b is overlain by carbonate and iron silicate-rich layers (unit 3c). The sediment from 750 to 600 cm (unit 3d) consists essentially of goethite (60–80 wt%) and Ca, Fe and Mn carbonates (from 6 to 20 wt%). From 600 cm to 415 cm (unit 3e), the goethite becomes enriched in Mn. At shallower depths, the separate phases groutite and goethite occur, followed upward by manganite and todorokite. The core section from 415 to 335 cm (unit 3f) contains mostly goethite (50–70 wt%) with minor amounts of haematite (10–20 wt%), manganosiderite and no calcite. The uppermost samples of this unit contain 3 wt% anhydrite.

4. Unit 4 (335 cm to top of the core) is a facies characterized by sulphides, clays, and poorly crystallized material which is considered as Fe- and Si-bearing products (Figure G3.2). At the base of unit 4, the collected samples are composed of manganeous siderite (50 wt%), sulphide (16 wt%), and Ca-sulphate (3 wt%), followed upwards by a ZnS-rich interval. The amount of ZnS regularly decreases upwards from 20 wt% to 4 wt%. Quantities of FeS_2 and CuFeS decrease similarly to ZnS, but with lower amounts. At the top of the core (140–0 cm) the sulphides are X-ray amorphous and therefore, are most likely iron monosulphides (Brockamp et al., 1978; Zhabina and Sokolov, 1982; Pottorf and Barnes, 1983). Samples of unit 4 contain up to 5 wt% anhydrite. In this unit, Fe–Mn and Ca carbonates range from 3 to 15 wt%. The Fe- and Si-bearing poorly crystallized portion comprises up to 85 wt% of the solid fraction.

Core 683

Two major units were defined in the SW basin (Figure G3.2).

1. The lower unit (unit L) consists of anhydrite (12 to 70 wt%), talc and serpentine (up to 28 wt%) and Fe-oxides (20 to 60 wt%). It can be divided into two subzones: a magnetite-rich facies (bottom to 1365 cm) in which talc is a major component and which contains coarse basaltic fragments at 1465 cm, and a haematite-rich facies (1365 to 1180 cm) in which the anhydrite content has the maximal values. The haematite-rich facies contains carbonates (<11 wt%).

2. The upper unit (unit U) is characterized by the presence of sulphides and abundant clay and Fe- and Si-bearing compounds (Figure G3.2). The lower part from 1180 to 1100 cm has a low sulphide content (<2 wt%), mostly ZnS, and is essentially composed of anhydrite (78 to 90 wt%). At 1090 cm, sulphides become an important component and reach the highest value of both cores (55 wt%). The anhydrite content decreases significantly. The upper 10 m of the core has a relatively constant sulphide amount (7 to 20 wt%) except at 700 cm, 155 cm, and in the upper 50 cm, where the ZnS

content is higher (from 19 to 25 wt% of total sulphide minerals). Fe-oxides and anhydrite appear sporadically in the upper part of unit U.

Temperature and mineral precipitation

A large range of temperatures (i.e. from 250°C to 500°C) have been deduced from rare paragenesis with copper sulphides (Pottorf and Barnes, 1983) and fluid inclusions in sulphate minerals (Oudin et al., 1984; Ramboz et al., 1988) in the metalliferous sediments of the Atlantis II Deep. On the other hand, clay and carbonate minerals, which are major minerals, indicate crystallization temperatures in the range between 50°C and 130°C (Badaut et al., 1985, 1990; Badaut, 1988; Decarreau et al., 1990), and between 35°C and 80°C (Blanc, 1987; Zierenberg and Shanks, 1988), respectively. Investigations on the maturation of the organic matter in the Atlantis II Deep also indicated a low temperature history for the Atlantis II Deep system (Simoneit et al., 1987). Calculation of the solubility of NaCl in the interstitial brines shows that the temperature never exceeded 200°C in the Atlantis II system (Anschutz and Blanc, 1993b). These results indicate that the temperature in the sedimentary system of the Atlantis II Deep never exceeded 250°C during the hydrothermal history.

Metalliferous sedimentation and diagenetic processes

Considering the quantitative distributions of minerals (sulphides, sulphates, carbonates and oxides) occurring in each lithological unit, we attempt to define the physical and chemical parameters which prevailed during the formation of the Atlantis II ore deposits.

Sulphides

1. In the lower part of the unit 1, the pyrite amount is lower than 2% (Figure G3.3). Above 1305 cm, the increase in pyrite up to 20 wt% suggests an enhanced biological production of H_2S associated with a high input of reactive Fe in the system. The low $\delta^{34}S$ of pyrite within unit 1 ($-35‰$ to $-20‰$) is consistent with a bacterially mediated sulphate reduction (Kaplan et al., 1969; Shanks and Bischoff, 1980). However, 7 to 20 wt% of pyrite is extremely high for marine sediments. Because the abundance of microbiological mediated pyrite is limited mostly by the amount of iron oxides available, it never exceeds a few per cent, even in the most anoxic and organic matter-rich marine sediments (Jorgensen, 1977; Berner, 1981; Elsgaard and Jorgensen, 1992). Thus, pyrite contents up to 20 wt% must be related to an additional supply of reactive iron oxide, over the detrital input. This additional supply can originate only from hydrothermal inputs. Sphalerite observed at the top of the unit 1 is probably derived

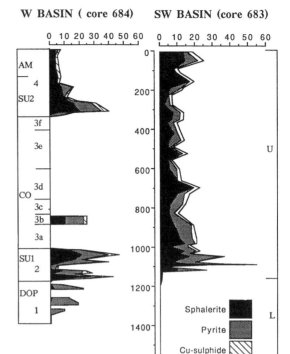

Figure G3.3 Sulphide distribution in cores 683 and 684.

from the reaction of dissolved Zn with bacteriological reduced sulphur. Therefore, the Atlantis II Deep was under the influence of hydrothermal activity during deposition of the upper part of the unit 1.

2. The Fe, Zn, and Cu sulphides constitute up to 55 wt% within the units 2, 4, and U (Figure G3.3) and show uniform $\delta^{34}S$ values of about $+5.4‰$ (Shanks and Bischoff, 1980; Zierenberg and Shanks, 1988). Such isotopically heavy values are comparable to those ($+0.9‰$ to $+6.35‰$) determined at the East Pacific Rise hydrothermal sites (Von Damm, 1990) and indicates that the sulphur does not result from the biological reduction of dissolved sulphate. Thus, high inputs of metals and H_2S-rich saline hydrothermal fluids are responsible for the accumulation of sulphides that spread over the entire area of the Atlantis II Deep. The coupled transport of metals and H_2S in saline solutions occurs only at high temperature, when the sulphide solubility and the chloride complex stability are reached (Barnes, 1979). Thus, the sulphide minerals precipitated out of equilibrium through quenching effect.

Generally, the wurtzite precipitates when $(Zn+Fe)/H_2S < 1$ (Scott and Barnes, 1972; Honnorez et al., 1985). Wurtzite has never been detected in the Altantis II sediment, whereas sphalerite is abundant. This observation suggests that the metals were always in excess in the hydrothermal fluids. However, high contents of sphalerite and pyrite occur in the lower parts of the units 2, 4 and U (Figure G3.3). This association suggests that similar physical and chemical conditions occurred for the formation and conservation of these sulphides. At temperatures lower than 300°C, and even in the largely over-saturated solutions, the nucleation kinetic of the pyrite is extremely low (Berner, 1970; Luther, 1991; Schoonen and Barnes, 1991). The formation of the pyrite (FeS_2) is derived from the transformation of the iron monosulphide (FeS). FeS is more soluble than FeS_2 and a very reducing solution enriched in Fe and H_2S is required for its precipitation. In such a solution, also enriched in Zn, sphalerite can be precipitated. Temperatures from 60°C to 150°C represent a broad interval of the estimated temperatures of the brine during the settling of the sulphide units (Anschutz and Blanc, 1993b). Considering these temperatures and activities of suphur and zinc in the range between 10^{-1} and 10^{-9}, and between 10^{-1} and 10^{-7}, respectively; thermodynamic computation shows that the oxido-reduction conditions of the brine correspond to the stability field of the pyrite (Anschutz, 1993). In these conditions, the sphalerite precipitated through quenching effect, but also at equilibrium if the amount of dissolved sulphur is in excess. FeS precipitated initially through a quenching effect and then was transformed into pyrite whereas the Fe in excess accumulated in solution. The transformation of FeS into pyrite is favoured by the occurrence of intermediate sulphur species, such as polysulphites, polythionates or thiosulphates (Schoonen and Barnes, 1991). These species result from the oxidation of H_2S in contact with iron-oxides that are simultaneously formed. Because FeS is predominantly transformed into pyrite, and not into marcasite, the pH of the system was probably higher than 4 (Murowchick and Barnes, 1986). At the top of the units 4 and U, amorphous iron monosulphides are not associated with pyrite. This suggests extremely reduced conditions in the recent brines where the intermediate species cannot be formed.

The copper sulphides yield a low amount of high temperature minerals such as chalcopyrite and cubic cubanite. It is generally lower than 3 wt%, rarely up to 8 wt%. Chalcopyrite is the main Cu-bearing mineral and has been described as rounded and fissured particles up to 35 μm in size (Pottorf and Barnes, 1983), whereas most of the sulphides occur as less than 2 μm grains in the Atlantis II sediment. This suggests mechanical transport of previously precipitated particles within the internal plumbing of the Atlantis II Deep, before the mineralizing fluid reached the point of discharge. Thus, the discrepancy in the temperatures noted previously, may be explained.

Sulphates
Distribution of sulphate species is presented in Figure G3.4. Anhydrite ($CaSO_4$), gypsum ($CaSO_4$, $2H_2O$) and bassanite ($CaSO_4$, $0.5H_2O$) were determined by XRD.

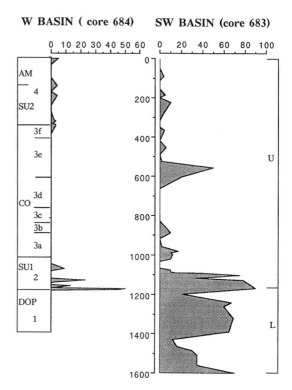

Figure G3.4 Calcium sulphate distribution in cores 683 and 684.

However, the hydration of calcium sulphates is relative to the temperature and salinity of the solution. Thermodynamic considerations on the water activity of the lower and interstitial brines suggest that gypsum and bassanite result from precipitation during the desaltation of sediment needed for analysis (Blanc, 1987; Monnin, 1989; Anschutz, 1993). Hence, anhydrite is the Ca-sulphate stable with the *in situ* conditions of the Atlantis II brines.

A hot hydrothermal fluid in equilibrium with the anhydrite has a very low concentration of dissolved sulphate and becomes undersaturated by cooling. Thus, the accumulation of anhydrite within the Atlantis II sediment implies an additional supply of dissolved sulphates in the medium. This additional supply can only be derived from either sulphide oxidation or sea water by mixing or molecular diffusion. The oxidation of the sulphur minerals or dissolved H_2S leads to the formation of intermediate sulphur species, which are metastable in reducing environments (Boulègue, 1978). The thiosulphate ($S_2O_3^=$) is the most stable species for closed-neutrality pH (Goldhaber, 1983; Moses *et al.*, 1987). The oxidizing agent can be either oxygen or oxidized dissolved and crystallized species. On the other hand, anhydrite is rapidly formed when Ca-rich, hot hydrothermal fluids are mixed with sulphate-rich sea water (Janecky and Seyfried, 1984). In the unit L of the SW basin, anhydrite is associated with iron oxides. Considering the contributions of the hydrothermal brine and sea water, a reactional model including both sulphate supplies can be proposed with two processes, as follows.

Process 1 – anhydrite precipitation using sulphate from sea water

$$SO_4^= + Ca^{2+} \longrightarrow CaSO_4$$

$$14Fe^{2+} + 7/2O_2 + 14H^+ \longrightarrow 14Fe^{3+} + 7H_2O$$

$$Fe^{3+} + 3H_2O \longrightarrow Fe(OH)_3 + 3H^+$$

Process 2 – anhydrite precipitation after sulphide oxidation

$$FeS_2 + 14Fe^{3+} + 7H_2O \longrightarrow 15Fe^{2+} + S_2O_3^=$$
$$+ 14H^+ + 2O_2$$

$$S_2O_3^= + O_2 + H_2O + Ca^{2+} \longrightarrow CaSO_4 + 2H^+$$

Anhydrite occurs at the base of each sulphide unit, corresponding to transition layers between oxide and sulphide deposits. Its content ranges from 34 to 90 wt% at the base of unit U, and reaches values up to 50 wt% at the base of unit 2, whereas only 3 wt% of anhydrite occurs at the base of unit 4 (Figure G3.4). The increase in anhydrite could be explained by process 1 when sediments were in contact with oxidizing sea water. However, process 2 started when reducing conditions were established in the brine that covered the deposit and allowed the conservation of sulphate and sulphite minerals.

Anhydrite occurs sporadically within the sulphide units with various proportions between 2 and 50 wt%. Because the $SO_4^=$ and H_2S stability fields yield distinct oxido-reduction states, the coprecipitation of sulphates and sulphides cannot exist at equilibrium. However, $SO_4^=$ is not transformed into H_2S in a reducing environment because of the very low kinetic rate of the reaction at temperatures lower than 250°C (Ohmoto and Lasaga, 1982; Sato, 1992). This explains the occurrence of $SO_4^=$ in the present-day lower brine and of anhydrite within the sulphite units. The solubility product of the anhydrite could be reached if the brine temperature and the $SO_4^=$ and Ca supplies increased.

Carbonates

In unit 1, the layers enriched in Ca-carbonates consist of secondary products such as Mn-calcite, Ca-siderite and Mg-calcite (Figure G3.5). Between 1187 and 1173 cm, kutnahorite is the predominant carbonate. The detrital silicate-rich layers also includes ankerite. Ca-siderite occurs in the goethite and lepidocrosite layers of unit 1. Rhodocrosite is recognized in the goethite-rich sediments of units 3c, d, and f, and also at the top of the unit 1. In unit 3d, ankerite and Mn-calcite occur sporadically. Carbonates are absent within the Mn-oxide layers.

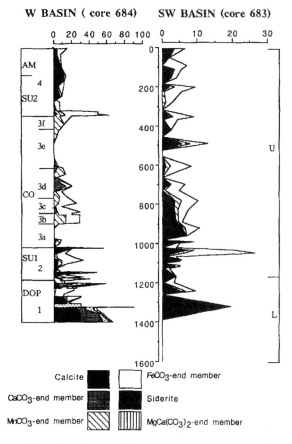

Figure G3.5 Distribution of the carbonate species in cores 683 and 684. The solid solutions are represented by their end-members.

However, the carbonate content of the metalliferous sediments is characterized by solid solutions of Fe and Mn carbonates with various composition. Nevertheless, the Fe endmember is predominant. These solid solutions, which are secondary products, imply the occurrence of dissolved Fe and Mn during carbonate formation. The chemical processes leading to carbonate formation will be summarized in the following paragraphs.

Unit 1

The processes proposed for carbonate formation of unit 1 required early diagenetic reactions relative to the bacterial activity. Reduced Fe and Mn can be released in solution through two independent pathways: (a) the bacterial reduction of Fe and Mn oxides occurring in the sediments; (b) the autocatalytic reduction of the Fe and Mn oxides by H_2S that is released in the interstitial solution during the bacterially mediated sulphate reduction.

Through the pathway (a), Fe and Mn oxides are reduced because they are electron acceptors during the organic matter oxidation (Berner, 1981; Pedersen and Price, 1982; Balzer, 1982; Aller and Rude, 1988). The oxidant consumption via the bacterial activity follows a reaction sequence relative to the available oxidants such as oxygen, nitrates, manganese oxides, iron oxides or oxyhydroxides, and sulphates (Stumm and Morgan, 1981; DeLange, 1986). The bacterial reduction of oxygen, sulphates and nitrates releases CO_2 and HS^- in the interstitial medium, whereas that of Fe and Mn oxides gives HCO_3^-, Mn^{++}, and Fe^{++} ions. The CO_2 production induces the destabilization of the biogenic calcite and aragonite (Emerson and Archer, 1990). Thus, high supplies of organic matter associated with an enhanced biological activity lead to the production of cations which can then be precipitated as secondary carbonates. In these conditions, the nature of the formed carbonate is related to the pre-existing amount of Fe and Mn oxides in the sediment.

Through pathway (b), the oxides can be reduced by autocatalytic reaction involving free H_2S (Aller, 1980; Aller and Rude, 1988; Goldhaber and Kaplan, 1974; Canfield, 1989). Therefore, enhanced bacterial activity increases the rate of Fe^{++} and Mn^{++} release.

The chemical composition of the carbonates from unit 1 can be explained by these two early diagenetic pathways acting simultaneously in an environment supplied with biogenic carbonates, organic matter and oxides. However, the high contents of Fe and Mn that average 7.48% and 1.38%, respectively within the sediment of unit 1 suggest that the hydrothermal activity worked long before the deposit of metalliferous sediment *sensu stricto*. Because Fe is more rapidly oxidized than Mn (Stumm and Morgan, 1981), Fe content is higher than that of Mn and the increase of Mn content at the top of unit 1 indicates the installation of a reducing condition induced by the formation of a permanent brine pool system.

Some mineral parageneses are surprising and require explanation.

(i) **Siderite is associated with pyrite**. Pyrite is an insoluble mineral that controls the dissolved concentration in Fe by rapid reaction with low concentration in H_2S. Siderite generally occurs within H_2S and pyrite-free reduced environments where the Ca/Fe ratio is low (Berner, 1981). However, the siderite and pyrite association in unit 1 could be explained by the presence at one place of two microenvironments having distinct characteristics. Within the first, at the proximity of a sulphate-reducing bacteria, iron is reduced by H_2S and immediately reacts with another sulphide to give pyrite. Within the second, the iron oxides are reduced by bacteria, Fe^{++} is removed in solution and then precipitated as siderite. These coupled processes were already described to explain the formation of intertidal concretions while the Fe oxide reduction exceeds that of sulphate.

(ii) **Siderite is associated with goethite and lepidocrosite from unit 1**. The occurrence of lepidocrosite indicates an oxidizing environment where dissolved Fe^{++} tends to be oxidized. However, the association with the Fe^{++}-bearing siderite suggests bacterial reduction of the Fe oxides followed by the formation of Casiderite using biogenic carbonates present within the sediment. Thus, although carbonates and oxyhydroxides are not precipitated within the same environment (carbonates in the reduced sediment and oxyhydroxides in the overlying oxidized brine), they nevertheless accumulate in the same area.

Units 2, 3, 4, L and U

The high sedimentation rate of the metalliferous sediments and the occurrence of a hot metal-rich brine during these deposition periods have considerably diluted the organic matter supplies (Simoneit *et al.*, 1987; Blanc *et al.*, 1990) and induced an abiotical media (Ryan *et al.*, 1969). Thus, early diagenetic reactions involving bacteria cannot explain the formation of carbonates within units 2, 3, 4, L, and U. Alternatively, the autocatalytic formation of Fe and Mn carbonates involved either (a) the direct precipitation from the brine or (b) the oxidation of sulphides.

(a) Balance calculation has shown that a large part of the biogenic calcite was dissolved during its deposition in the Atlantis II Deep (Anschutz and Blanc, 1995b). The dissolution of the calcite could result from a decrease of pH due either to the oxidation of Fe^{++} and Mn^{++} at the brine–sea water boundary, or the precipitation of sulphides in the brine. The rhodocrosite, the siderite and solid solutions between these two endmembers, are less soluble than calcite (Lippmann, 1980). Thus, these secondary phases probably precipitated at the place of calcite dissolution. Furthermore, the contribution of the biogenic carbonate to the composition of secondary carbonates is supported by their carbon isotopic composition (Zierenberg and Shanks, 1988). Consequently, calcite acted as a pH buffer through a consumption of protons of the system. These considerations lead us to propose a pattern of chemical reactions as follows.

$$\left.\begin{array}{l} O_2 + Fe^{++} + Mn^{++} + 2H_2O \longrightarrow FeOOH \\ \quad + MnOOH + 4H^+ \\ Me^{++} + H_2S \longrightarrow MeS + 2H^+ \end{array}\right\} \text{acidification}$$

$$CaCO_3 + 2H^+ \longrightarrow Ca^{++} + H_2CO_3 \quad \}\text{neutralization}$$
$$xFe^{++} + (1-x)Mn^{++} + H_2CO_3 \longrightarrow$$
$$(Fe_x Mn_{1-x})CO_3 + 2H^+$$

(b) During the oxidation of iron sulphide into oxide, siderite is the most stable phase, thermodynamically (Stumm and Morgan, 1981, p.433). The most enriched carbonate-bearing layers are found in the sulphide units. The predominance of the Fe^{++} endmember suggests that the environment was enriched in Fe^{++} with respect to Mn^{++} and deficient in reduced sulphur with respect to the content in dissolved Fe^{++} supplied by the hydrothermal fluid. The carbonate layer located at the top of unit 2 (e.g. 1000 cm) could be interpreted as the boundary between the sulphide and carbonate stability fields. In the goethite and haematite-bearing units (3d and 3f), the predominant carbonate is the rhodocrosite (Figure G3.5). This observation indicates that the dissolved Fe had a low concentration because it was controlled by oxide phases having a very low solubility. Mn^{++} could accumulate in solution and then be precipitated as rhodocrosite. In spite of the high content in iron oxides, the occurrence of Mn-carbonate suggests, however, a reduced environment. Conversely, the carbonates are absent within the Mn-oxides facies. Thus, the medium should be more oxidizing without dissolved iron.

Iron and manganese oxides and oxyhydroxides
These mineralogical species occur in each unit of the Atlantis II Deep. The ferric species are ferrihydrite, goethite, lepidocrosite, haematite, and magnetite. The manganiferrous species are manganite, groutite, and todorokite. Their distribution is shown in Figure G3.6. In this chapter, three points will be discussed: (i) the formation and transformation of the ferrihydrite; (ii) the formation of iron oxides; and (iii) the formation of manganese oxides.

Formation and transformation of the ferrihydrite
The ferrihydrite is found predominantly in units U and 4. We will attempt to answer the following questions: How can its formation be explained? Which are the factors allowing its conservation? Which are the parameters involved for its diagenetic transformation to geothite and haematite?

The formation of the ferrihydrite ($5Fe_2O_3$, $2H_2O$) is well documented in the literature (Towe and Bradley, 1967; Chuckrov et al., 1973; Schwertmann and Fischer, 1973; Karim, 1984; Taylor, 1984). Assuming the Fe^{++} oxidation, $Fe(OH)_2$ is formed first, then $Fe(OH)_3$. The monomers of ferric hydroxides ($Fe(OH)_3$) aggregate ($nFe(OH)_3$) and then give ferrihydrite (Hazemann, 1991). In the Atlantis II Deep, the ferrihydrite formation could result from coupled chemical reactions illustrated in Figure G3.7. Thus, the $Fe(OH)_3$ precipitation is

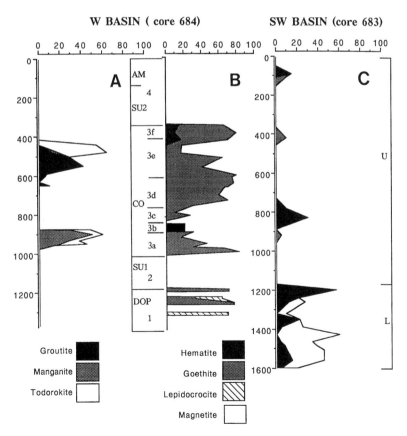

Figure G3.6 Distribution of iron and manganese, oxides and oxyhydroxides in cores 683 and 684. A: Mn-oxides in core 684; B: Fe-oxides in core 684; C: Fe-oxides in core 683.

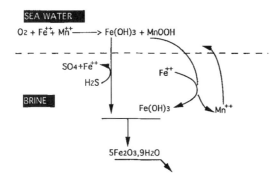

Figure G3.7 Iron oxidation–reduction cycle in the Atlantis II brine and in deep Red Sea water.

possible using either the dissolved O_2 at the brine–sea water boundary:

$$2Fe^{++} + 1/2O_2 + 5H_2O \longrightarrow 2Fe(OH)_3 + 4H^+ \quad (1)$$

or the Mn-oxides into the anoxic brine pool

$$Fe^{++} + MnOOH + H_2O \longrightarrow Fe(OH)_3 + Mn^{++} \quad (2)$$

Reaction (1) has been described experimentally (Chuckrov et al., 1973; Schwertmann and Thalmann, 1976; Karim, 1984). Considering the dissolved Fe and Mn profile in the present-day brine pool, Danielsson et al. (1980) and Hartmann (1985) proposed reaction (2). Mn-oxides only occur in sub-units 3a and 3e (Figure G3.6(a)), whereas Mn-carbonates occur in the other oxidizing units (Figure G3.5). Therefore, reaction (2) could mainly explain the absence of Mn-oxides and the formation of $Fe(OH)_3$. At the brine–sea water boundary, several phases are formed, such as Fe and Mn-oxides. The model proposed by Danielsson et al. (1980) for the present-day brine (reaction 2) could also be used to explain the absence of Mn-oxides in the major part of the oxidizing unit. Mn^{++} is removed by reaction (2) and can be oxidized again, as shown in Figure G3.7.

The ferrihydrite is less stable than goethite and haematite for a large range of temperature and pressure. Thus, its conservation in the Atlantis II sediment implies specific chemical environments.

Preservation of the ferrihydrite before its deposition. In the presence of a ferric phase, such as ferrihydrite or iron hydroxide, H_2S is immediately oxidized through the following reactions (Canfield, 1989; Stumm and Sulzberger, 1992).

$$4(5Fe_2O_3, 2H_2O) + 5H_2S + 70H^+ \longrightarrow 40Fe^{++} + 5SO_4^= + 76H_2O \quad (3)$$

$$8Fe(OH)_3 + H_2S + 14H^+ \longrightarrow 8Fe^{++} + SO_4^= + 20H_2O \quad (4)$$

Reactions (3) and (4) are irreversible for temperatures lower than 250°C (Ohmoto and Lasaga, 1982). Thus, the presence of ferrihydrite within the sediment suggests a deficit of H_2S with respect to iron oxides. These reactions act as pH buffer of the brine during the discharge of acid hydrothermal fluids. The sporadic occurrence of sulphate in the sulphide units appears to be explained by these reactions that produced $SO_4^=$ in solution. Therefore, iron participated to an oxidation–reduction cycle in which oxygen from sea water and several chemical components from the hydrothermal fluid are involved. The deposition of the ferrihydrite became effective when the entire H_2S content was consumed (Figure G3.7).

Preservation of the ferrihydrite after its deposition. In a natural environment, the ferrihydrite tends to be transformed into well-crystallized products, such as goethite and haematite. In units U, 4 and 3c, Si contents are higher than 8 wt% and the ferrihydrite bears a part of this silica (Badaut, 1988; Anschutz and Blanc, 1995c). Experiments have shown that Si increases the stability of the hydrated oxides because of the creation of Si-O-Fe bonds (Carlson and Schwertmann, 1981; Cornell et al., 1987). Thus, the presence of Si in the ferrihydrite could explain its preservation within the Atlantis II sediment.

Goethite and haematite occur in units 1, 3, U and L (Figure G3.6(b), (c)). Bischoff (1969) proposed that these two minerals are genetically linked ($Fe_2O_3 + H_2O \longrightarrow 2FeOOH$) in the Atlantis II Deep. However several studies in natural surface environments (Feitknecht and Michaelis, 1962; Langmuir, 1971; Fischer and Schwertmann, 1975; Schwertmann and Murad, 1983; Cornell and Giovanoli, 1987; Cornell et al., 1987; Cornell, 1988) have shown that they can be formed from a common original phase, the ferrihydrite. Alternatively, the lepidocrosite (γ-FeOOH) and akaganeite (β-FeOOH) could also act as original phases. The lepidocrosite is found within the sediment of unit 1 (Figure G3.6(b)) and the akaganeite was recognized only in the pounded particles of the present brine (Holm et al., 1982; 1983). These two minerals spontaneously precipitated by oxidation of Fe^{++} using dissolved O_2 (lepidocrosite) or by hydrolysis in Cl–rich solution (akaganeite) (Figure G3.8). They are unstable and tend to be transformed into goethite (Murray, 1979). On the other hand, the ferrihydrite and the amorphous iron oxides, previously named 'limonite' (Bäcker and Richter, 1973), are ubiquitous in the young sediments (Badaut, 1988). Thus, a large part of iron oxides occurring in the Atlantis II sediment originated from the transformation of the ferrihydrite (Figure G3.8). However, numerous experiments have shown that these transformations depend on temperature, pH and chemical components of the system (Lewis and Schwertmann, 1979; Karim, 1984; Cornell et al., 1987; Cornell, 1988; Fischer and Schwertmann, 1975; Stiers and Schwertmann, 1985; Cornell and Giovanoli, 1987; Schwertmann

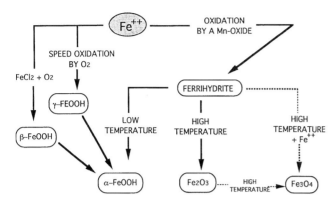

Figure G3.8 Scheme showing various chemical pathways for the formation of iron oxides in the Atlantis II Deep. α-FeOOH: geothite; β-FeOOH: akaganeite; γ-FeOOH: lepidocrosite; Fe_2O_3: haematite; Fe_3O_4: magnetite.

and Murad, 1983). Considering these experimental results, a chemical pattern for the precipitation of iron oxides can be proposed (Figure G3.8) and some conditions of the environment where they formed, can be defined as follows.

The haematite occurs in unit L (Figure G3.6(c)). If it derived from ferrihydrite, its occurrence indicates high temperatures in the SW basin. However, the occurrence of associated anhydrite suggests that the temperature did not exceed 250°C. Nevertheless, the haematite is an important end-product of the reaction at a temperature from 90°C (Lewis and Schwertmann, 1979). Conversely, temperatures lower than 50°C preferentially induced the appearance of goethite. Thus, the association of haematite and goethite within unit 3f (Figure G3.6(b)), where Si, Zn, and Cu have low contents, could be explained by a slight increase of the environment temperature (over 50°C) inducing the ferrihydrite transformation.

The haematite from unit 3b and from 750–900 cm in unit U is associated with Zn and Cu sulphides (Figure and G3.6(b),(c) and G3.3). The high content in Zn and Cu in the brine could increase the stability of the ferrihydrite, and the nucleation of the haematite.

Unit 3 consists predominantly of goethite and Mn-oxides (Figure G3.6(a), (b)). The incorporation of Mn into the ferrihydrite certainly favoured the formation of goethite at a temperature lower than 50°C. The oxidation of Fe^{++} by Mn-oxides was probably the main process for the ferrihydrite precipitation (Figure G3.8). Unit 3d is enriched in Mn-goethite, and the Mn content increases in the goethite until the formation of two distinct phases, goethite and groutite in unit 3e (Figure G3.6(a), (b)).

Formation of iron oxides
Here the following questions are considered. What information can be deduced from the occurrence of the goethite and lepidocrosite layers of unit 1? What information can be deduced from the distribution of ferriferous phases of unit 3? Which are the processes responsible for the formation of iron oxides from the unit L?

Oxyhydroxides occur in three layers (1295–1278 cm; 1240–1200 cm; 1173–1155 cm) interbedded into the biodetrital sediments of unit 1 (Figure G3.6(b)). Because these layers extend over the entire deep (Bäcker and Richter, 1973; Clément and Giannésini, 1992), they accumulated under a large brine pool. The lepidocrosite (γ-FeOOH) is the main mineral in the older layers. Experiments showed that the precipitation of lepidocrosite involving the oxidation of a $FeCl_2$ solution at 25°C, preferentially replaces those of the goethite and ferrihydrite in several conditions. These conditions are: low sulphate and carbonate contents (Carlson and Schwertmann, 1990), high Cl/Fe ratio (Taylor, 1984), low Si concentration (Schwertmann and Thalmann, 1976; Karim, 1984), rapid oxidation of Fe^{++} (Murray, 1979). By comparison, the formation of the lepidocrosite in the Atlantis II Deep could result from a rapid mixing of a Fe-rich and Si-poor brine with sea water. This implies that during the first hydrothermal discharge, the Atlantis II sediment of unit 1, could have been overlain by an oxidizing brine, largely mixed with sea water. At the alkaline pH of the sea water, a dissolution/precipitation reaction generally changes the lepidocrosite into goethite (Schwertmann and Taylor, 1972):

$$\gamma - FeOOH + OH^- + H_2O \longrightarrow Fe(OH)^-_{4(aq.)} \longrightarrow \text{goethite} \quad (5)$$

Thus, the conservation of the lepidocrosite in unit 1 suggests that pH of the interstitial brines remained acid from this period of deposition.

Iron oxides and oxyhydroxides are predominant in units 3c, 3d, 3f and between 1000 and 970 cm in unit 3a (Figure G3.6(b)). The Mn-oxides are absent in these layers (Figure G3.6(a)). The goethite layers of unit 3 are relatively enriched in Cd (10 to 350 ppm), Pb (85 to 750 ppm), Zn (0.16 to 0.73%) and Cu (0.05 to 0.5%). These elements are readily adsorbed at the surface of oxides and oxyhydroxides. Thus, the goethite of unit 3 was probably in contact with a metal-rich environment. This agrees with the transformation of ferrihydrite to oxides after its deposition into an metal-rich brine.

At the base of core 683 (SW basin), the magnetite is associated with anhydrite and talc. This facies occurs only in the deepest area of the SW basin whereas at shallower depths, the basaltic basement is covered by haematite (Hackett and Bischoff, 1973; Zierenberg and Shanks, 1983). Thus, it appears that the conditions needed for the precipitation of magnetite have existed only in the central part of the SW basin. The magnetite is generally formed in a reducing environment, whereas anhydrite is stable in an oxidizing environment. However, anhydrite and magnetite can precipitate from the same solution at temperatures lower than 250°C (Ohmoto and Lasaga, 1982). At temperatures higher than 250°C, the anhydrite and dissolved sulphate react either with dissolved Fe^{++}, or with magnetite directly, to form sulphides (Ohmoto and Rye, 1979). Zierenberg and Shanks (1983) found few occurrences of chalcopyrite associated with magnetite and anhydrite and proposed a minimum temperature of 300°C for the formation of magnetite. However, no sulphide has been found in unit L, and we saw that chalcopyrite probably precipitated in the internal plumbing before the discharge of hydrothermal fluid into the brine. Thus, we propose that the reaction of the transformation of haematite into magnetite: $Fe^{++} + Fe_2O_3 + H_2O \longrightarrow Fe_3O_4 + 2H^+$ given by Hackett and Bischoff (1973) and Zierenberg and Shanks (1983), probably acted at temperatures lower than 250°C (Fig. G3.8).

Formation of the Mn-oxides

This section will focus on the following questions: which are the processes leading to the nature, and the distribution, of the Mn-oxides and oxyhydroxides?

Manganite (γ-MnOOH), groutite (α-MnOOH) and todorokite (MnO_2) are predominant in units 3a and 3e (Figure G3.6(a)). In an oxidizing medium, the Mn(III) species (i.e. manganite and groutite) are thermodynamically less stable than the Mn(IV) species (todorokite, for instance) (Bricker, 1965). However, manganite can exist at the metastable state long after its precipitation (Hem and Lind, 1983; Diem and Stumm, 1984). In a reducing Fe^{++}-rich medium, these species are unstable (cf. reaction 2). Thus, the occurrence of these two Mn-oxide-rich layers can be related to periods of Fe^{++} disappearance in the medium. The direct precipitation of these minerals using oxygen and Mn^{++} in solution is a slow reactive process, rarely observed in natural environments. Experiments in oxidizing environments show that only the hausmatite (Mn_3O_4) and the feitknechtite (β-MnOOH) precipitate spontaneously at temperatures lower and higher than 25°C, respectively; and then changed irreversibly into manganite and todorokite, respectively (Giovanoli et al., 1976; Hem and Lind, 1983). Thus, slight changes of temperature could lead to the formation of manganite or todorokite in the Atlantis II sediments. The formation of the groutite is still unknown, however, a solid solution of goethite–groutite has been recognized in marine sediments (Varentsov et al., 1989) and experimentally synthesized in an oxidizing Fe^{+++} and Mn^{++} solution (Stiers and Schwertmann, 1985). The contents in Ba (1500 to 3700 ppm) and in Sr (300 to 800 ppm) are comparable to those determined in the manganese nodules (Nicholson, 1992), and thus do not indicate an anomalously enriched environment with respect to normal oceanic water. From these above considerations, one can conclude that an oxidizing environment prevailed during the formation of the Mn-oxides from the Atlantis II Deep.

Consequently, the spatial distribution of recent Mn-oxide deposits gives information on the geometry of the brine pool which covered the sediment during the deposition of units 3a and 3e. Using our lithological data and those given by Bischoff (1969), Bäcker and Richter, (1973); and Clément and Giannésini, (1992) a map of the Mn-oxide distribution has been prepared (Figure G3.9). The recent Mn-oxide deposits extend

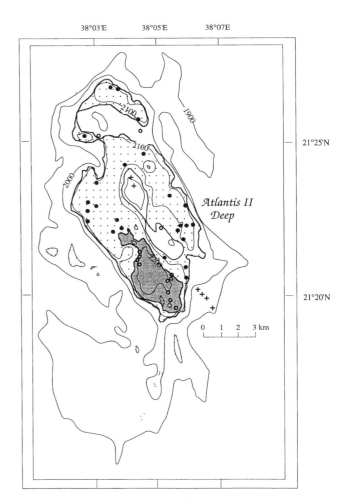

Figure G3.9 Bathymetric map showing the lateral distribution of the Mn-oxide facies (dotted area) within the Atlantis II Deep. This map was compiled using data from Bischoff (1969), Bäcker and Richter (1973), Clément and Giannesini (1992) and this study. Dark dots: location of cores within Mn-oxides have been recognized; open dots: location of cores without Mn-oxides; crosses: core locations where present-day accumulation of Mn-oxides occurs. Dark dots are all in depths between 2180 and 2070 m (dotted area). Below 2180 m, Fe-oxides occur only in the south-western basin (grey area). This lateral varation of oxide facies results mostly from a bathymetric effect inducing oxidation changes of the brine. The grey area was covered by an Fe^{++}-rich brine; the dotted area was covered by a brine depleted in Fe^{++}, but enriched in Mn^{++}.

over the rim of the deep, the central sill and the Chain Deeps. These areas are presently covered either by the upper brines or by the solutions of the overlying transition zone. These solutions yield concentrations in Mn considerably higher than those of Fe (Danielsson et al., 1980; Hartmann, 1985).

The Mn-oxide units of the CO facies are found in cores from the W, N, and E basins located between 2050 and 2100 m, approximately, and in cores collected from the surrounding deeps, at greater depths (N basin and west rim of the W basin) (Figure G3.9). The cores collected at depths exceeding 2050 m show thin CO facies and consist of goethite and detrital minerals. A relatively thick Mn-oxide layer occurs in the N basin. This indicates that the N basin could act as a marginal basin, comparable to the present chain deeps. Thus, the distribution of the present and older Mn-oxide deposits suggests that the Atlantis II sediments were covered by a thin brine during the deposition of units 3a and 3e. Because inflow of the hydrothermal fluid was probably low, this brine was mainly oxidized, only the deepest area of the SW basin being covered by a reduced brine. A transition zone occurs between the goethite and Mn-oxide facies. This transition zone consists predominantly of both Fe and Mn-oxides and is occasionally associated with Mn-goethite, when the Mn-oxides become predominant. Thus, the redox state of the brine and its stratification changed progressively during the history of the Atlantis II Deep.

GEOCHEMICAL HISTORY OF THE ATLANTIS II DEEP

Considering the chemical processes discussed, a geochemical model can be proposed to explain the mineral distribution in space and time of the Atlantis II Deep.

Sequential evolution of the W basin deposits

Unit 1
Deposition of unit 1 is divided in two temporal steps (Figure G3.10). The oldest biodetrital sediments from the W basin (23 000 yr BP) (Anschutz and Blanc, 1995a) were deposited directly on the basalt, in contact with the normal deep sea water of the Red Sea. Biogenic and detrital sedimentation continued to predominate, but an additional input of metals (Mn and Fe) modified the nature of unit 1 sediments. The oxides have been changed into secondary products by biogeochemical reactions involving bacterially mediated sulphate reduction and oxidation of organic matter. The high amount of authigenic pyrite and carbonates indicates a large input of metals, and enhanced bacterial activity. The iron oxyhydroxide layers interbedded in unit 1 cover the entire deep. These layers indicate periods during which the hydrothermal input was higher than that of biogenic and detrital particles. The formation of lepidocrosite, which is enriched in the oldest layer, indicates an oxidizing environment. Because this mineral rapidly reacts with H_2S resulting from the bacteria activity, it probably acts as an Fe source for the formation of diagenetic pyrite, before and between the Fe oxyhydroxide layers. Part of goethite and lepidocrosite was changed into siderite by reduction. The Mn and Fe contents increase from the bottom to the top of unit 1. Considering the calculated accumulation rates, we can conclude that the W basin has been filled by a brine pool since 19 000 BP, and that the Atlantis II Deep was therefore influenced by hydrothermal activity long before 11 000 BP (Anschutz and Blanc, 1995a, b). This dates the deposition of the metalliferous sediment (unit 2) that succeeded unit 1.

Unit 2
A general framework for the formation of unit 2 is presented in Figure G3.11. During this period of deposition, the background accumulation of biogenic and detrital particles was minor, the medium became sterile, and the hot reduced hydrothermal inflows increased drastically. This inflow of hydrothermal brine was first discharged into an oxidizing environment which progressively became more reducing. This hypothesis is sustained by the mineralogical observation showing the following temporal sequence: Fe-oxides and

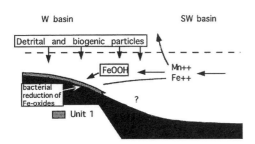

Figure G3.10 Schematic representation of deposition of unit 1.

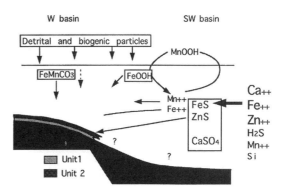

Figure G3.11 Schematic representation of deposition of unit 2.

anhydrite; Fe and Mn-carbonates, and anhydrite; sulphides. This mineralogical sequence occurring within each sulphide unit and can be explained as follows:

- Occurrence of Fe-oxides, only. The dissolved metals are supplied into the brine by the hydrothermal inflow. Because of the reducing condition of the brine, the metals, Fe and Mn in particular, spread over the entire deep. Oxides precipitate at the brine–sea water interface, in contact with dissolved oxygen, and in the brine pool. However, Mn-oxides are finally reduced and therefore are absent within the sediment.
- Occurrence of anhydrite. The hot and Ca-rich hydrothermal fluid is mixed with the brine in which dissolved sulphate occurs. The anhydrite precipitated by quenching effect. Part of dissolved sulphate resulted from the oxidation of H_2S dissolved in the inflowing hydrothermal fluid. This oxidation is favoured by the presence of Fe and Mn-oxides. The anhydrite content is high near the spring, whereas it decreases with distance.
- Occurrence of carbonates. The increase in carbonate content at the base of the sulphide units indicates an enhanced supply of metals into a reducing environment.
- Occurrence of sulphides. Sulphides precipitate by quenching effect near the spring and are brought subsequently into the surrounding basins as particles. The coupled transport of dissolved metals and H_2S implies high temperature of the fluid at the discharge point.

The high content of anhydrite and sulphides within unit 2 of the core 684 suggests that the discharge point of the hydrothermal fluid was probably located near or in the W basin during this deposition.

Unit 3

Figure G3.12 illustrates the general framework for deposition of unit 3. The drastic change of the mineralogy between units 2 and 3 indicates that the media was deficient in dissolved H_2S, whereas it was still supplied by dissolved metals. The transition zone between a reduced metal-rich brine and oxygenated sea water is the main site of precipitation. Precipitation was controlled mostly by the establishment of a redox cycle involving Fe and Mn (Figure G3.7). Fe precipitated as oxides, and Mn as carbonates.

The occurrence of Mn-oxides in unit 3a indicates that the brine was enriched in dissolved Mn with respect to Fe during this period of deposition. The sediment of the W basin was covered by a solution having characteristics comparable to those of the present upper brines or of the overlying transition zone. To explain this hydrological situation, it seems that the hydrothermal discharge into the brine pool was low during this period. In this case, the temperature of the brine probably decreased and convection ceased. Consequently, the molecular diffusion favoured the development of a thick transition zone between the brine and sea water.

During deposition of unit 3b, the brine was supplied by a fluid enriched in dissolved H_2S. Because no oxidation of sulphides into anhydrite occurs, the hydrothermal spring must have been located in a reducing environment.

Unit 3c is characterized by the incorporation of silica into the ferrihydrite and in the amorphous ferric products. During deposition of unit 3d, the silica input was low and the Fe-Mn redox cycle dominated. The brine of the W basin was therefore reduced but deficient in dissolved H_2S, and the amount of dissolved Fe remained in excess with respect to the Mn-oxides formed at the brine–seawater interface. The hydrogeochemistry of the brine during the deposition of unit 3e is comparable to that of unit 3a. The occurrence of haematite in unit 3f implies that the temperature of the brine increased.

Unit 4

The general framework for the formation of unit 4 is shown in Figure G3.13. The transition zone between units 3 and 4 is mineralogically comparable to that of units 1 and 2. However, the sulphates and sulphides are present but in less quantity. After the total consumption of dissolved H_2S, most metals precipitated as oxides and carbonates. The high silica content favoured the formation of ferriferous clays (Anschutz and Blanc, 1995c). Thus, the discharge point of the hydrothermal fluid was probably distal from the W basin.

Sequential evolution of the SW basin deposits

Unit L

The metalliferous sediments of unit L directly overlie the basalt (Figure G3.12). The unit L consists predominantly of Fe-oxides and anhydrite. These minerals precipitated from a fluid enriched in metals and calcium,

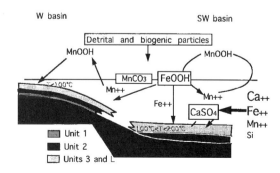

Figure G3.12 Schematic representation of deposition of units 3 and L.

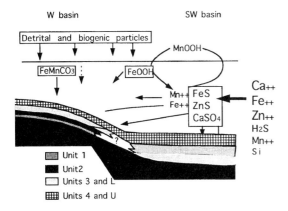

Figure G3.13 Schematic representation of deposition of units 4 and U.

but deficient in H₂S. The anhydrite is formed mainly by a quenching effect near the discharge point of the hydrothermal fluid, and the redox cycle involving Mn-oxides and Fe^{++} permitted the formation of ferrihydrite. The deeper part of the SW basin was probably filled by a hot reduced brine, deficient in H₂S. This environment favoured the change from ferrihydrite to haematite. The bottom of core 683 is characterized by the occurrence of magnetite that must be related to a local thermal event. However, the absence of sulphides suggests that the temperature should not exceed 250°C.

Unit U

The increased amount of anhydrite within the lower part of unit U suggests that the discharge point of the hydrothermal fluid was close to site 683 (Figure G3.13). The rapid development of the sulphide unit succeeding the anhydrite facies, was probably due to an increase of the H₂S concentration in the inflowing brine. Thus, the change of the sulphide content in unit U can be related to the temperature of the solution. Indeed, the inflowing and pre-existing brines should have been occasionally in thermal equilibrium near the spring. In this case, the precipitation of sulphides was possible by equilibrium reactions in the brine that cooled gradually. The metastable sulphur species, such as thiosulphates, enabled the monosulphides of iron to be transformed into pyrite. These species derived from the oxidation of H₂S in contact with Fe-oxides which are predominant in unit U. The redox cycle between Mn-oxides and dissolved iron should be effective because the deficit of H₂S with respect to metals dissolved in a reduced brine. This mechanism of reactions probably worked during deposition of the entire unit U.

Spatial evolution of the Atlantis II deposits

The faunal associations recognized in cores 683 and 684 suggest that unit 4 of the W basin is contemporaneous with unit U of the SW basin (Anschutz and Blanc, 1993a). This correlation is also supported by their mineralogy. Thus, unit L probably corresponds to the upper part of unit 3. However, differences in (a) mineralogy, and (b) thickness, of the sedimentary sequences occur between the correlated units. These differences could be explained by lateral bathymetric variations and by changes of the hydrodynamic regime within the brine pool.

(a) Unit L contains Fe-oxides and anhydrite, whereas unit 3 consists of oxides and oxyhydroxides of iron and manganese. The occurrence of anhydrite in the SW basin is related to the proximity of the inflowing thermal fluids. The brine of the W basin, cooler and distal from this source, was undersaturated with respect to the calcium sulphates, and was not supplied with anhydrite precipitated in the SW basin. The difference in nature of the Fe-oxides between the two basins can be explained in terms of lateral thermal gradient, the SW basin being hotter than the W basin. The absence of Mn-oxides in core 683 shows that the brine was thick and reduced in the SW basin whereas the W basin was covered only by the upper part of the brine. The mechanism of reactions working simultaneously in both environments is presented in Figure G3.14. We propose that the redox cycle involving Fe and Mn acted in these two places.

The formation of the magnetite facies at the base of core 683 implies a local thermal event. However, the formation of haematite facies within unit 3f of the W basin could be considered as a distal consequence of this thermal pulse. In this case, the base of the core 683 should be correlated with the unit 3f.

Units 1, 2, 3a–e have been recognized in the W, E and N basins, and on the rim of the SW basin. Considering the previous lithological data (Bäcker and Richter,

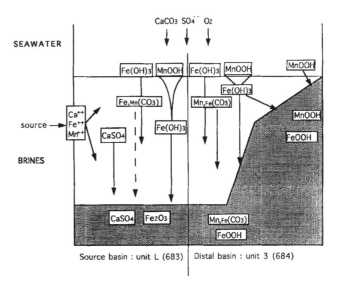

Figure G3.14 'Schematic representation of the alternative evolutions of chemical components at the brine–sea water interface.

1973), we propose that the extension of the characteristic lithology of the SW basin corresponds to a basaltic flow which buried and destroyed the pre-existing units 1, 2 and 3 a–f. This basaltic flow corresponds to the basement of the central part of the SW basin.

The occurrence of haematite at the base of unit U (SW basin) suggests that the brine temperature of the SW basin might have been higher than that of the W basin during deposition of unit 4. The temperatures deduced from the oxygen isotopic compositions of clays support this hypothesis (Badaut et al., 1990; Decarreau et al., 1990). The ferrous smectite (100–130°C) is recognized only in the SW basin whereas the nontronite (about 70°C) is predominant in the W basin (Badaut et al., 1988). This difference in clay minerology indicates also that basin SW sustained more reducing conditions than the others. Because of greater depth of the SW basin, the thickness of the brine pool of the SW basin was always higher than that of the W basin. Thus, the variation of the brine thickness could induce changes in chemical processes in the W basin, whereas those of the SW basin were not affected.

Section H
Post-rift coastal sediments, reefs and geomorphology

The five chapters in this section focus on the Quaternary post-rift sediments and geomorphology of the basin margin. As such the section complements Section G on the axial post-rift sediments.

The contributions cover a wide range of topics from the sedimentology of the modern carbonate sediments – particularly the reefs – to the uplifted Quaternary reef-terraces that fringe the basin, and an analysis of the relationship between geomorphology and structural processes on the uplifted south-western Red Sea margin in Yemen.

H1 Present-day sedimentation on the carbonate platforms of the Dahlak Islands, Eritrea
 F. Carbone, R. Matteucci and A. Angelucci
H2 Quaternary marine and continental sedimentation in the northern Red Sea and Gulf of Suez (Egyptian coast): influences of rift tectonics, climatic changes and sea-level fluctuations
 J.-C. Plaziat, F. Baltzer, A. Choukri, O. Conchon, P. Freytet, F. Orszag-Sperber, A. Raguideau and J.-L. Reyss
H3 Post-Miocene reef faunas of the Red Sea: glacio-eustatic controls
 M. Taviani
H4 Modern Red Sea coral reefs: a review of their morphologies and zonation
 W.-C. Dullo and L. Montaggioni
H5 Tectonic geomorphology and rates of crustal processes along the Red Sea margin, north-west Yemen
 I. Davison, M. R. Tatnell, L. A. Owen, G. Jenkins and J. Baker

Chapter H1
Present-day sedimentation on the carbonate platform of the Dahlak Islands, Eritrea

F. Carbone, R. Matteucci and A. Angelucci

ABSTRACT

The Dahlak archipelago is part of a carbonate shelf, separated from the Eritrean coast by the Massawa Channel, where present-day carbonate sediments accumulate on and around uplifted blocks of Pleistocene 'Dahlak Reef Limestone'. They are the uppermost part (about 100 m thick) of a shallow-water sedimentary complex, that overlies an evaporite sequence of Miocene age with a measured maximum thickness of over 3000 m. The physiography of the Dahlak Reef Limestone outcropping on the islands and also that of present-day depositional environments are related to recent, still active, tectonic movements and recent sea-level changes. The complex sea-floor topography prevents the development of any organized facies belts from the inner shelf to the outer margin. Widespread records of a late Holocene higher sea-level are found all around the archipelago: fossil wave-cut notches 1–3 m above present sea level; dried-up intra-island lagoonal areas; upward shoaling subsurface sequences in sabkha and lagoonal areas; moribund or dead fringing reefs; palimpsest character of many bioclastic sediments.

Structural features and sedimentary facies distribution divide the Dahlak shelf into two main units: a north-west unit, including islands and islets east of the North Massawa Channel, the largest of which is Harat; and a south-east unit, including the large island of Dahlak Kebir and, to the east, the Dahlak Bank.

In the north-western unit, the distribution of depositional facies is closely related to a tectonic pattern which originated from the intersection of fault systems, including the Danakil Rift and ring-like fault systems at the edges of bathymetric depressions resulting from the collapse of the cover rocks due to mobilization or near-surface dissolution of salt domes.

Present-day sedimentation is controlled by three major morpho-structural features:

1. Emerging structural highs, forming a number of islands, islets and rocks, sometimes with intra-island lagoons, mangrove swamps and wide supratidal flats which, in places, pass into sabkhas.
2. Shallow submerged structural highs, with flat tops and separated from surrounding deeper floors by step faults.
3. Deep structural lows, arranged in troughs and ovoid depressions characterized by muddy sedimentation, with abundant faecal pellets and assemblages of pteropods, planktonic foraminifers and calcareous nanoplankton.

The south-eastern unit, informally named Dahlak Kebir shelf, consists of a large uplifted block of Dahlak Reef Limestone, bordered westward by the deep narrow tectonic trough of the South Massawa Channel and eastward by the steep edge of the axial trough of the Red Sea. Along the south-western edge of this shelf sedimentation is influenced by the largest island of the archipelago and also by a system of step faults, oriented north-west–south-east parallel to the axis of the Massawa Channel and by a series of deep troughs with predominantly muddy sediments. The eastern sector of the shelf, particularly the Dahlak Bank area, is a wide, shallow-water plateau consisting of emergent flattened Pleistocene coral reefs and several shoals, composed of storm-thrown, coarse, carbonate sand and coral debris.

INTRODUCTION

The Dahlak archipelago is located in the southern Red Sea, off the Eritrean coast (15°40′N, 40°10′E). It consists of 130 low flat islands, islets and rocks, covering about 3000 km^2. The largest island, Dahlak Kebir, has a surface area of over 758 km^2 and a maximum elevation of 48 m.

This archipelago lies on a wide carbonate shelf, separated to the west from the Eritrean coast by the deep tectonic trough of the Massawa Channel. Eastward, it is bounded by a steep continental slope which runs parallel to the axial trough of the Red Sea (Figure H1.1). The islands and the present-day depositional substratum of the shelf consist of a sequence of Plio-Pleistocene shallow-water limestones, the Dahlak Reef Limestone (Nir, 1971).

The morpho-structural setting of the shelf depends on several factors: (i) the present sea-floor spreading of the Red Sea; (ii) salt tectonics, involving the underlying (more than 3000 m thick) evaporite series of Upper Miocene age; and (iii) recent eustatic sea level changes. However, the extremely diversified physiography of the shelf is mostly a reflection of the combination of different tectonic trends.

The archipelago has been the focus of a number of investigations, by the authors and others from the Dipartimento di Scienze della Terra, Università 'La Sapienza', Roma, and the Italian Consiglio Nazionale delle Ricerche. Fieldwork took place in the 1970s and in the early 1980s (Matteucci, 1974; Angelucci et al., 1975, 1978, 1981, 1982, 1985; Bono et al., 1976, 1983; Conforto et al., 1976; Belluomini et al., 1980; Civitelli and Matteucci, 1981; Fumanti, 1983a, b; Carboni et al., 1993). After a long interval, the research, now including extensive echo-sounding profiling and bottom sampling, was resumed in 1994 and 1995, in conjunction with the Eritrean Ministry of Marine Resources.

ENVIRONMENTAL PARAMETERS

Data on the climatic regime of the archipelago were obtained from: (i) the general account of the Red Sea of Edwards (1987) and Sheppard et al. (1992); (ii) data collected between 1901 and 1942 at the weather stations of Nakhra and Massawa (Fantoli, 1966), and (iii) direct observations made during the field surveys by Angelucci et al. (1978, 1985).

The archipelago has a megathermal climate, with mean annual temperatures above 25°C and mean, monthly and seasonal minimum values above 15°C. The thermoregulating effect of the sea is evident at Massawa and on the islands. Here, while annual temperatures are on average high, thermal excursions are moderate.

Mean annual rainfall is below 250 mm, with peaks in December and February, the low total indicating an arid climate. Comparisons of temperatures and rainfalls yield very low, monthly and annual, aridity indexes (De Martonne, 1926; Thornthwaite, 1948). Evaporation is consequently very high and run-off scarce or absent (Angelucci et al., 1978).

Temperature, salinity and pH were measured in the open sea, nearshore, in shallow waters and relatively restricted lagoonal environments within the archipelago over a long period. Water temperature generally reflects seasonal changes. The mean surface temperature varies from 26°C in February to 31°C in August. In April, salinity is 40.2‰ in open-marine environments and 42.7‰ in the innermost part of the eastern lagoon of Isratu island (Bono et al., 1983), as opposed to a mean annual value for Red Sea of 38‰ (Edwards, 1987). Water pH, measured in the open sea, is 7.8 just below the surface and 7.7 at a depth of about 50 m.

Currents are generally weak and less intense to the south, near the strait of Bab El Mandeb. In all seasons, the average water flow trends north-west–south-east, parallel to the Red Sea axis. During summer, the southwestern monsoon causes surface water to flow northward, whereas, from October to May, the north-eastern monsoon produces a reverse flow. This flow pattern is locally very variable, as a result of eddies which, in many places, are superimposed on the general pattern. The eddies vary both in intensity and location moving water from the central Red Sea towards the coast. Although this flow is generally weak, it may increase in shallow waters near reefs and shoals.

Wave motion is usually moderate (waves of up to 1.5 m high), but it may become strong (waves of up to 4 m) in some periods of the year and particularly in August.

Tidal excursions, measured at different locations in the archipelago, range from 0.50 to 1.20 m, compared with 0.90 m recorded between the archipelago and Kamaran island (Edwards, 1987).

GENERAL GEOLOGICAL SETTING

The separation of the African and Arabian plates (giving rise to the Red Sea axial trough by strike-slip) took place in successive stages. It began in an area of thinned continental crust formed during the Pan-African orogeny (about 600 m.y. BP), and continued by wrench-faulting tectonics until the commencement of the present stage, characterized by progressive sea-floor spreading from south to north (Makris and Rihm, 1991).

On the evidence of axial linear magnetic anomalies, sea-floor spreading began around 5 m.y. ago (Roeser, 1975; Izzeldin, 1987; La Brecque and Zitellini, 1985). The SONNE 53 geophysical survey (Egloff et al., 1991; Makris and Rihm, 1991) in the southern Red Sea

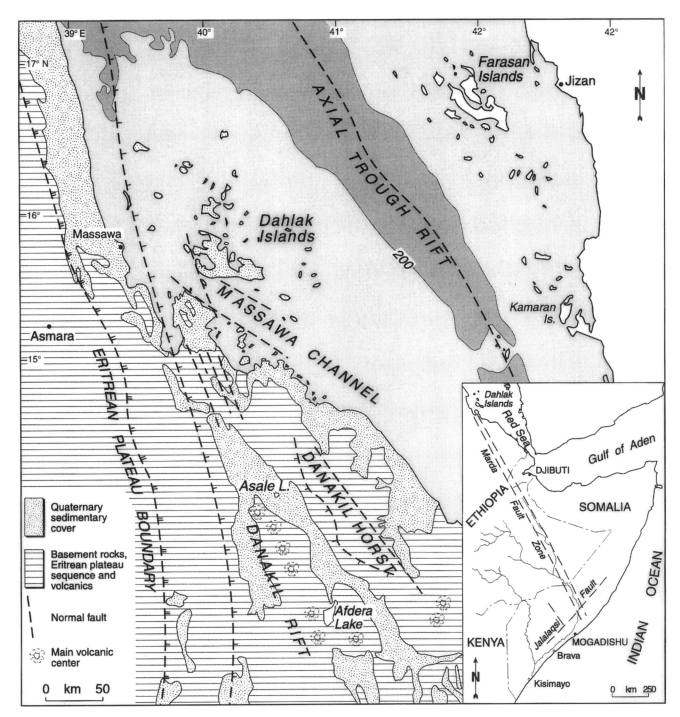

Figure H1.1 Tectonic sketch of the Eritrean coast, southern Red Sea.

identified oceanic crust outcropping in the Sudanese axial trough (Suakin Deep). In most of the Sudanese offshore areas, the same survey reported salt domes rising directly from the sea floor. The continent–ocean transition is clearly identified along the western flank of the central Red Sea. Here, a considerable portion of old oceanic crust, far from the Red Sea axis, underlies pre-evaporite sediments (Egloff et al., 1991). The sedimentary sequence below the salt sequence is extremely thick (over 2–3 km in some places), and can be assigned to the early Miocene, according to the stratigraphic data of Bunter and Abdel Magid (1989) and Beydoun (1989). This suggests a fairly recent age for the underlying oceanic crust.

During the Plio-Pleistocene, sea-floor spreading created the axial trough. This is separated from shelf areas by large, down-faulted blocks, some of which cover many kilometres (Izzeldin, 1987). Faulting of the shelf is more widespread in the southern Red Sea, where the sequence is often folded and injected with salt diapirs.

Many petroleum prospecting wells have been drilled on the African coast of the Red Sea. These have reached, but not entirely passed through, more than 3000 m of evaporites of likely Miocene age (Ross and Schlee, 1973; Angelucci et al., 1981), which are overlain by shallow-water skeletal limestones.

Numerous wells drilled by AGIP in the Dahlak Islands, during the late 1930s, provided a major contribution to the understanding of the subsurface stratigraphy. Thanks to these wells, AGIP's geologists defined three superimposed lithological units on Dahlak Kebir Island. The uppermost unit (outcropping) consists of the Dahlak Reef Limestone (Figure H1.2), whose thickness does not exceed 100 m. An unconformity separates this unit from the middle, of presumed upper Miocene–Pliocene age, made up of alternations of laminated clays, marls, anhydrites and sands. A 10–20 m thick sequence of crab-bearing clays occurs about 60–70 m above the bottom of this unit. The total thickness varies from 80 to 300 m over short distances. This implies a period of subaerial erosion prior to deposition of the overlying reef limestones. The crab-bearing beds are recorded in all wells, and are used as a correlation marker. The lowest unit is a thick salt sequence of Miocene age (penetrated by the Suri 7 well to 2250 m). Here, predominantly massive salt is associated with laminated clays, gypsum and anhydrite.

The mobilization of the evaporite sequence and uplift of salt diapirs occur in many places along the southern Red Sea coast; at Farasan Islands domal structures are connected with the occurrence of salt rocks below a thin cover of reef limestones (Bantan, 1995), while along the north coast of Yemen salt diapirs crop out, actively deforming Quaternary marine sedimentary cover (Davison et al., 1996; Bosence et al., 1996).

Deposition of the carbonate sequence outcropping on the Dahlak shelf and part of the Eritrean coast is presumed to have been initiated during a period of high sea-level stand, favouring the growth of luxuriant reefs and the deposition of related facies.

Studies of the coral faunas of the Dahlak Reef Limestone, exposed on islands and along the Eritrean coast (Montanaro Gallitelli, 1939, 1943, 1972) have assigned a Pleistocene age to them.

Radiometric dating of a small number of samples has given a variety of ages. Nir (1971), using radiocarbon dating, found values of $16\,400 \pm 400$ and $28\,600 \pm 700$ BP for limestones outcropping on Entedebir Island (6–7 and 19 m above msl). By contrast, from samples of the same island and Dahlak Kebir (Ghubbet Mus Nefit), at 1 to 5 m above msl, Conforto et al. (1976) obtained two groups of ^{230}Th dates: one of 120 ka BP and the other ranging from 160 ka to 170 ka BP. The 120 ka group corresponds to a period of high sea level, during the last interglacial (isotope substage 5e) and is consistent with regional data on raised reefs along the west coasts of the Red Sea and the Gulf of Aden (Faure et al., 1973, 1980) and on Zabargad Island, off the Sudanese coast (Hoang and Taviani, 1989). The older group of dates coincides with a lowstand of sea level during a glacial period (isotope stage 6), according to conventional curves of sea-level change, such as that for the Huon Peninsula, New Guinea (Bloom et al., 1974; Aharon and Chappell, 1986). The latter finding might indicate vertical tectonic movements, thereby validating the field observations of Belluomini et al. (1980) of a salt rock outcrop underlying the Dahlak Reef Limestone, outcropping at Dahlak Kebir near Gembeli village.

The Dahlak Reef Limestone consists of a sequence of shallow-water limestones of predominantly coralgal facies (Figure H1.2), outcropping on many islands. At Dahlak Kebir, a sequence approximately 30 m thick comprises reefal facies, intercalated with coquinas very rich in bivalve and gastropod shells, typical of back-reef environments. Similar sequences are observed in many places in the archipelago, suggesting less local variability of depositional environments in the Pleistocene, and thus, probably, a greater lateral structural continuity of the Dahlak shelf.

LOCAL STRUCTURAL PATTERN

Data on the regional structural setting of the archipelago are available in Facca (1965), Frazier (1970), Lowell and Genik (1972), Cochran (1983b), and Bunter and

Figure H1.2 Pleistocene Dahlak Reef Limestone along the northeastern coast of Ghubbet Mus Nefit. Whole massive coral colonies scattered in coarse skeletal debris.

Abdel Magid (1989). Frazier (1970) placed the horst and graben pattern of the Dahlak shelf in the more general framework of the tectonic history of the Eritrean coast.

As reported by Facca (1965), the first structural interpretation of the archipelago was by Migliorini (unpublished reports of AGIP Mineraria), who described a range of dome-shaped structures on Dahlak Kebir Island that he attributed to salt diapirs.

The Dahlak shelf is a wide shallow-water area broken by horsts and graben, and separated from the Eritrean coast by the Massawa Channel. The distribution of sedimentary facies on the shelf is closely tied to its tectonic pattern. The occurrence of ancient carbonates on the islands, as well as the physiography of present-day depositional environments are attributable to recent and still active tectonic movements (Figure H1.3). These movements gradually segmented the major horst, which extended from the edge of the Red Sea axial trough of the Eritrean coast and has a counterpart in the Farasan archipelago, off the Saudi coast (Sheppard, 1986; Bantan, 1995).

The present tectonic pattern of the area is dominated by different fracture systems, the most prominent of which is the Danakil Rift, cutting the northern portion of the archipelago from the Eritrean coast. The Danakil Rift is assumed to represent the northern extension of the Marda Fault zone (Boccaletti *et al.*, 1991), an aborted rift system extending to the Somali coast of the Indian Ocean (Figure H1.1). A north-west–south-east trending fracture system, associated with more recent spreading of the Red Sea and widely distributed in the islands, must have been the main determinant for the development of the trough in the South Massawa Channel and also a series of aligned depressions, the most obvious of which is Ghubbet Mus Nefit (Plate 33(a)). Other variably oriented fracture systems locally associated with the main trends, include a south-west–north-east trending system, particularly evident in the northern sector of the archipelago. On many islands, surface morphology is governed by a closely spaced net of small-displacement step-faults. The occurrence of fissures and active faults indicates continuing tectonic movements (Figure H1.3), reflected in the numerous earthquakes in the area (Al-Amri, 1994). Locally, annular fracture systems, collectively aligned with the north-west–south-east system and formed by the dilation and collapse following dissolution of the salt substratum, are very significant (Figure H1.4(a)). Frazier (1970) suggested that deep elliptical depressions (e.g. Ghubbet Mus Nefit) formed when the overburden was pierced by salt diapirs, whereas gently anticline settings (e.g. Schumma island) are salt domes which still have their sedimentary cover. These elliptical depressions show a typical step-fault pattern along their edges

Figure H1.3 Evidence of the dense net of normal faults characterizing the northern sector of the Ghubbet Mus Nefit coast. Exposed fault-planes are well preserved, confirming active faulting.

(Plate 33(c)), often with a gentle centrifugal dip of the strata. Where the dip gave rise to only small differences in elevation (some tens of metres at the most), it was capable of diversifying depositional environments during the most recent sea-level changes: some depressed areas are now occupied by very shallow lagoons, sabkhas and deflation plains.

Sea-level fall during the last glacial period caused the exposition of the sea-floor and subsequent weathering and erosion of Pleistocene coral limestones. These features are particularly clear in areas poorly affected by tectonics, like the eastern Dahlak Kebir shelf (Dahlak Bank), where a gently sloping sea floor 0 to 30 m deep is punctuated by closed depressions (maximum 20 m deep) and by emergent relics of fossil coral buildups, sculpted during subaerial exposure. This pattern is less evident in northern and western portions of the archipelago, where recent tectonics partially hide the effects of negative changes of sea level. Therefore the origin of terrace-like morphologies, shown between 0 and 40 m depth by echo-sounding profiles, is difficult to ascribe.

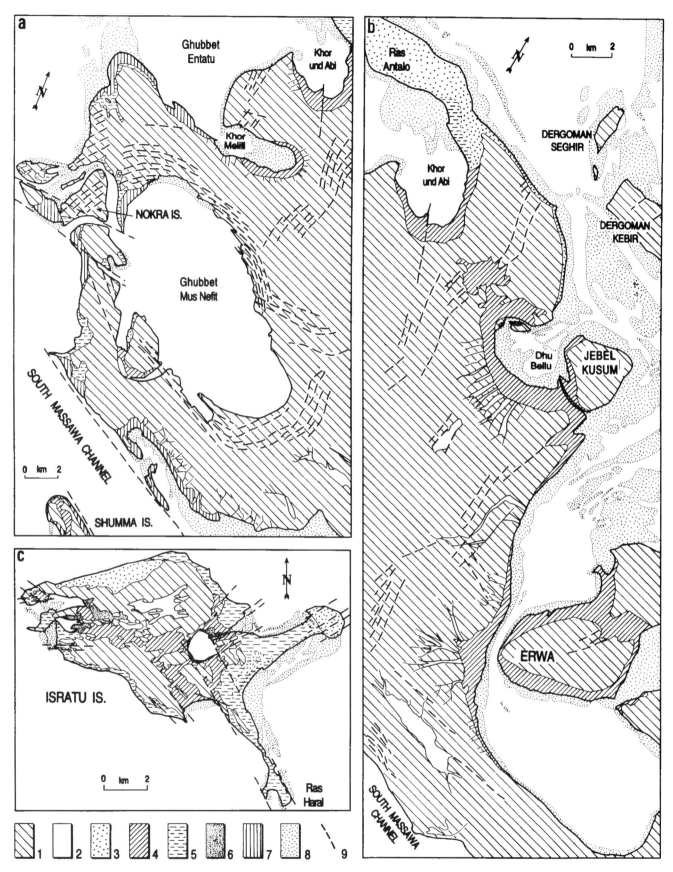

Figure H1.4 Facies patterns, related to morphostructural setting in some Dahlak Islands: (a) Eastern Dahlak Kebir Island; (b) East coast of Dahlak Kebir Island; (c) Emergent structural high of Isratu Island. (1) Dahlak Reef Limestone. (2) Eluvium covering wide hollows subject to deflation. (3) Emergent marine terraces covered by thin veneers of aeolian carbonate sand. (4) Coastal and inner sabkha areas. (5) Accretionary beach ridges and coastal dune fields. (6) Mangrove swamps. (7) Flat hard bottom covered by coral carpet. (8) Sandy bodies forming beach ridges, coastal spits, bars and tidal deltas. (9) Main faults.

SEDIMENTARY PATTERNS

The Dahlak shelf includes a number of depositional units with contrasting facies patterns (Figure H1.5). Two main sectors may be identified: a southern one, with the extensive shallow-water area of the Dahlak Kebir shelf; and a northern one, where sea-floor morphology is more diversified, with numerous bathymetric highs of variable size, separated from adjacent basins by step-faults. The overall sedimentary pattern is very diverse, with varying carbonate sediment production and storage rates. These variations depend not only on the morphologies and bathymetric locations of the depositional areas but, more importantly, on recent changes in relative sea level and therefore on increase or decrease of accommodation rate.

Dahlak Kebir shelf

This sector of the archipelago consists of a large uplifted block of Dahlak Reef Limestone. Westwards, it is bounded by the deep tectonic trough of the South Massawa Channel and eastwards by the steep rectilinear edge of the Red Sea axial trough, these two margins showing different tectonic configurations and depositional features (Figure H1.5, cross-section B).

South-western shelf edge

On this margin, sedimentation is influenced by the presence of the largest island of the archipelago, Dahlak Kebir, and by the north-west–south-east fracture system which opened the Massawa Channel (Figure H1.6, profile 1). This fracture system is also associated with deep bathymetric lows, including Ghubbet Mus Nefit (Figure H1.4(a), Plate 33(a)), and with the narrow, elongate, tectonic pillar of the Dahlak Reefs (Figure H1.6, profile 2). As reported by Frazier (1970), this structural situation is closely related to rift tectonics and local collapse of limestones resting on top of the salt diapirs. Therefore sedimentation is very variable along this edge, depending on local physiography and wave energy. Skeletal sands dominate in both intertidal zones and deeper waters (Figure H1.8 (c)). However, increasing depth leads to a higher muddy fraction, which contains an abundant epipelagic faunal assemblage (Figure H1.8(h)). Over the abraded flats along the coast or around islets and emerging rocks, coral communities are scattered and only occasionally form a coral carpet (*sensu* Geister, 1983). In most cases, the substratum is hard and abraded, with rare and isolated settlements of massive corals, often encrusted with corallinaceans or covered by *Sargassum*. The outer margin of the Dahlak Reefs, facing the Massawa Channel, is a poorly developed fringing reef (Plate 34(b)). At a depth of 15–20 m, the reef terminates against bioclastic sediments. In this area, the reef crest displays the richest and most flourishing coral assemblage in the area, dominated by massive corals (poritids and faviids). In the shallow waters of the back reef, the coral cover decreases rapidly, to be replaced by seaweed meadows.

Inner shelf

The shelf east of Dahlak Kebir Island (Figure H1.4 (b)) is a broad shallow area of quiet waters, with irregular isopic facies belts unrelated to any fault pattern. The innermost area, bordering the coast, has a series of embayments, fringed by wide intertidal and supratidal belts (Plate 33(b)). Beach facies with very diverse morphology and development are mostly associated with aeolian deposits of skeletal sands or coquinas (mainly strombids and arcids). The coquinas result from intense deflation which removes the fine fraction. The beaches protect extensive flats, sporadically flooded by the sea during exceptional tides and storms. These areas commonly resemble coastal sabkha (Plate 34(a)), with evaporation and desiccation producing polygonal cracks, teepee structures and curled mud chips (Davies, 1970; Purser, 1973; Hardie, 1977). Mangrove thickets are found in limited areas characterized by a hard and fissured substratum veneered by mud. The coastal embayments of Dahlak Kebir show the muddy carbonate sediments typical of sheltered environments with poorly oxygenated waters. Sediments are often covered by *Thalassia*, *Syringodium* and *Sargassum* with dense populations of epiphitic soritid forams, and green algae (*Caulerpa* and *Microdictyon*) are also abundant in places. The belt from the low-intertidal to the high-subtidal, bordering these lagoons to the east, consists of highly bioturbated muddy and skeletal sands. Wide portions of the sea floor show burrows and mounds of varying size originated from callianassids, constantly modelled and flattened by tides.

Sediment distribution by tidal currents produces an extremely diverse bottom morphology, forming sandy bars and shoals (Plate 33(d)) which sometimes pass into tidal deltas. An intricate net of channels ensures sea-water exchange, even in the innermost areas.

Outer shelf

This sector, enclosing part of the Dahlak Bank, is a wide shallow-water area (Figure H1.6, profile 2), whose seaward side is marked by the 20 m isobath. The sea-floor morphology is irregular, partly because of numerous small islands and sandy shoals (Plate 34(c)). The present facies distribution appears to be controlled by a palaeomorphology formed during the last relative low-stand of sea level. The emerging portions of the islands and islets are raised reefs. Large amounts of skeletal sand are stored on extensive abraded flats around these islands, occasionally forming emerging sandy cays 2–3 m high.

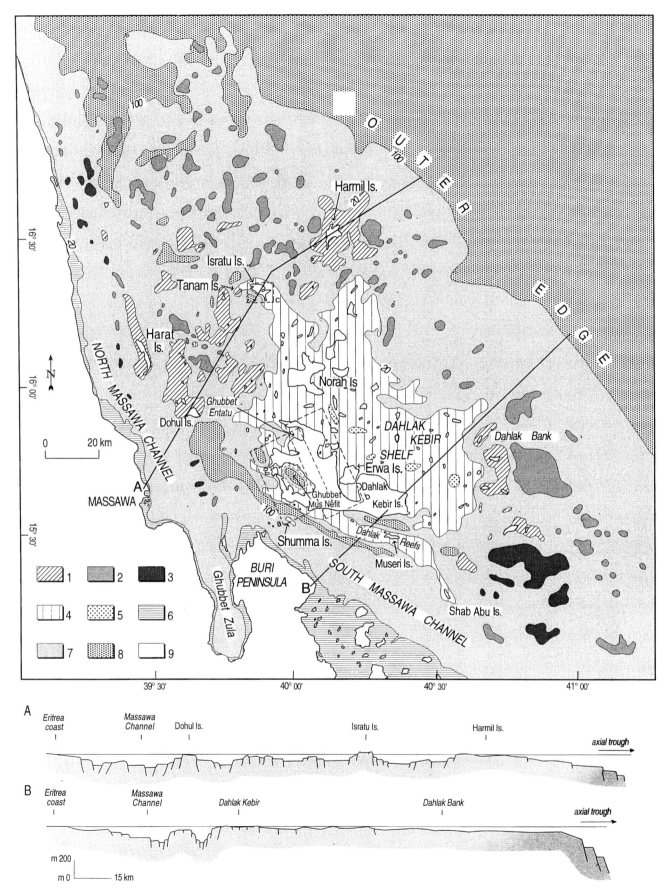

Figure H1.5 Regional distribution of the main sedimentary units: (1) Emergent structural highs. (2) Submerged shallow structural highs. (3) Structural highs submerged below wave base. (4) Shallow Dahlak Kebir shelf. (5) Sea floor depression inside the Dahlak Kebir shelf. (6) Shallow coastal shelf. (7) Undifferentiated deeper sea floor of the Dahlak Islands shelf. (8) Basin areas deeper than 100 m. (9) Supratidal areas. A. Highly schematic cross-section of the northern Dahlak Islands shelf. B. Highly schematic cross-section of the southern Dahlak Islands shelf.

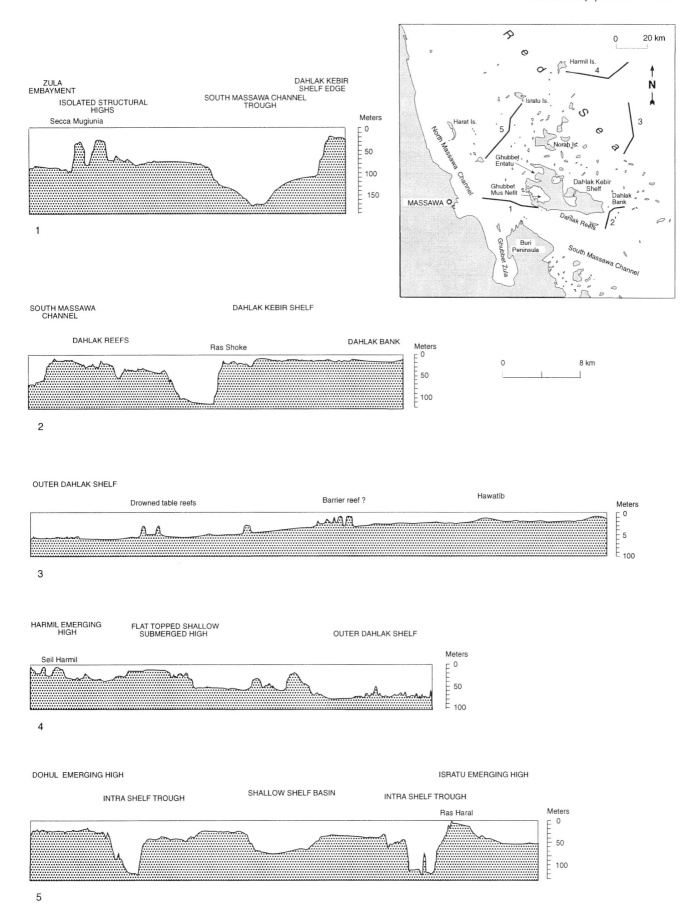

Figure H1.6 Selected echo-sounding profiles summarizing the main topographic features of the Dahlak Islands shelf.

Throughout the area, extensive bioclastic sediments originate not only from erosion of the emerged fossil reefs but also from intense bioerosion of skeletal material of indigenous communities (mainly bivalves, corals and echinoids). Transport of this material by waves and currents creates a cover of sand, with grain-size differing from place to place and where bivalve communities dominate (Figure H1.8 (a), (d)).

The living corals predominantly form small fringing reefs, generally along the outer edges of abraded flats. Small patch reefs, massive coral heads and knolls are also present on the inner part of the abraded flats, generally occupied by phanerogam meadows.

North-eastern shelf edge

The outermost facies belt of the Dahlak Kebir shelf (below the 20 m isobath) is generally a planar surface sloping gently to the 100 m isobath, where a sudden increase in slope marks the edge of the deep axial trough of the Red Sea (Figure H1.6, profile 3). A distinctive feature of this zone is the occurrence of isolated, circular pinnacles, 200–500 m wide, rising abruptly from the submarine plain to within 5–10 m of the water surface. These may be karst relics of one sector of the Dahlak Reef Limestone, weathered during periods of lowstand of sea level in the last glacial stage, or bioconstructions. In the latter case, they may include both fossil and living coral build-ups with flat tops, like the table reefs of Ginsburg and Schroeder (1969), quoted in James (1983). The absence of living corals on the tops of some of these build-ups and the presence of encrustations of red algae and *Sargassum* on dead coral, may indicate drowning.

The sediments on the outer portion of the shelf are usually skeletal sands, with a gradual increase in muddy fraction and pelagic assemblages (pteropods, planktonic foraminifers and nanoplankton) to seaward (Figure H1.8 (f)). The benthos consists mainly of echinoids and rotaliform macroforaminifers, including *Operculina* and *Heterocyclina* (Accordi *et al.*, 1994).

Emergent structural highs

North of Dahlak Kebir, almost all islands and islets of the archipelago are related to isolated, uplifted blocks (Figure H.5, cross-section A; Figure H1.6, profiles 4, 5). On major islands, the physiography of the depositional environments is controlled by tectonics: wide uplifted areas, with exposures of Dahlak Reef Limestone, are separated by depressed areas, with arid plains subject to deflation, sabkhas and vast tidal flats. On islands such as Isratu (Figure H1.4(c)) and Harmil, intra-island lagoons and pools are relics of wide arms of the sea, which extended into the islands until recently. This assumption is corroborated by the occurrence of fossil sea-level notches, and ancient beach deposits in the innermost parts of the islands. The sediments in these intertidal and supratidal environments are variable, depending on physico-chemical parameters and on hydrodynamic energy. In the lagoons, the type of sediment is closely connected with water flow, which often changes the water completely during tides, removing the fine sediment fraction and oxygenating the environment. Oligotypic facies with small turriculate gastropods and algal mats characterize the innermost tidal flats and mangrove swamps. Minimum water movement is recorded at these points of transition between lagoon bottoms and sabkha areas where dark and fetid lime muds, typical of reducing environments, occur.

In this area of the archipelago, present coral colonization is mainly at the edges of the abraded shelves bordering the islands, forming small but well-developed fringing reefs. On the shelves small patch reefs compete with *Thalassia* or *Sargassum* meadows in shallow-water environments. The distribution and zonation of corals is mainly controlled by sea-floor morphology, tidal streams, and even less by the effects of prevailing winds. In the fringing reef lying along the northern coast of Tanam Island Civitelli and Matteucci (1981) recognized, moving seawards from the inshore up to 7 m depth, four coral assemblages: (i) *Stylopora pistillata* and calcareous algae, (ii) *Acropora pharaonis* and *Lobophyllia corymbosa*, (iii) *Porites*, and (iv) sparse corals. This zonation is partially comparable to that suggested by Rosen (1971, 1975) for the shallow-water coral communities of the Indo-Pacific area, hydrodynamism being the most important zoning factor, even if most of the fringing reefs of the archipelago lack the highest energy zones.

Along the shorelines, sedimentation is characterized by bioclastic sands (Figure H1.8 (b)), which generally build beaches, shoals and coastal spits. The generalized progradation of the shoreline occasionally creates sandy beach ridges, which develop around emerging relics of Pleistocene reefs. The composition and grain-size of these sands is very variable, reflecting changes in the benthic communities and sea-water energy.

Shallow submerged structural highs

These tectonically delineated bathymetric highs, 5–20 m deep, often have a flat submerged top (Figure H1.6, profile 5). The facies pattern varies, depending on different parameters, the most important of which is bathymetry. On these wide wave-swept areas, *Thalassia* and *Sargassum* blankets usually consolidate a sandy substratum, where corals are represented only by scattered heads and knobs of poritids and faviids, and patch reefs and coral thickets are infrequent. In some instances, the tops of the highs are sediment-starved, and dead corals, largely encrusted with corallinaceans, often almost cover the substratum. This may be related to rapid drowning, possibly after a stand in intertidal or supratidal environments caused the death of the reef community. The sediment on the tops of these structures has very variable composition and grain-size (Figure H1.8 (e)), depending on local carbonate-producing organisms *in situ* and on the hydrodynamic energy at the water–sediment interface. Notable proportions of carbonate mud appear only in areas where deposition is below the current and wave action.

Deep structural lows

A particular feature of deposition on the northern and western Dahlak shelf is the occurrence of numerous, deep bathymetric depressions (Figure H1.5). These are generally ellipsoidal or form narrow elongate troughs, surrounded by, but separated from, shallow waters by step-faults (Figure H1.6, profile 5). These depressions, which may exceed 150 m depth, are attributable to tectonics, particularly to interaction between the regional fracture system (mainly north-west–south-east trending) and local salt tectonics, and have a dramatic impact on the depositional pattern of the shelf. Even if they have only limited connection with the open sea, given their physiography (e.g. the inner Ghubbet Mus Nefit) or location inside the shelf (the Ghubbet Entantu, or the trough of Isratu), they host sediments with a significant and diversified epipelagic assemblage (Figure H1.8 (g)) of planktonic foraminifers (*Gallitellia vivans, Globigerina bulloides, Globigerinita glutinata, Globigerinoides ruber, Gl. bilobus, Hastigerina aequilateralis, H. siphonifera, Orbulina universa, O.* sp. 'small'), pteropods (*Creseis acicula, C. virgula, Hyalocyclis striata, Diacria quadridentata, Cavolinia longirostris, C. uncinata, C. inflexa, Atlanta plana*) and calcareous nanoplankton (*Gephirocapsa oceanica, Umbilicosphaera sibogae, Helicosphaera kamptneri, Siracosphaera* sp., *Calcidiscus cfr. leptoporus*), which normally cannot be preserved on winnowed skeletal sandy bottoms (Carboni *et al.*, 1993).

The micritic and very fine bioclastic material deposited in these depressions generally results from bioerosion processes, characteristic of most of the shallow-water environments. The fine sediment produced is transported in suspension to the depressions by wave and tidal currents whose intensity creates a variable depth limit, below which particles are deposited (Figure H1.7). High production of biodetrital micrite from numerous carbonate organisms is present in Ghubbet Entatu (Angelucci *et al.* 1982). The central depression of the Ghubbet Entatu, with depth ranging from 30 to 40 m, receives fine sediment from the inner lagoon of Kor Melill (Figure H1.4 (a)), where tidal flow ensures complete daily recharge of water. The high tide supplies clear water to the inner part of the lagoon, whereas the low tide removes cloudy water, rich in suspended sediment.

DISCUSSION AND CONCLUSIONS

The depositional pattern of the Dahlak shelf is strongly influenced by tectonics. The combination of this and other geological and physico-chemical parameters regulates the distribution of depositional facies over the shelf. Among these parameters, the most crucial are: (i) the physiography of the substratum, reflecting subaerial modelling during the last glacial sea-level lowstand; (ii) the hydrodynamic regime, and (iii) organogenic sediment production and sedimentation rates. The main geological features of the Dahlak platform are as follows.

1. **Structural setting**. This reflects the segmentation of a wide, shallow-water carbonate unit of Pleistocene age. The tectonic pattern results from the combination of fracture systems, which are in part inherited, including ancient wrench-faulting tectonics of the crust (Makris and Rhim, 1991) and in part linked to recent spreading of the Red Sea. These processes are coupled with local tectonics, related to salt diapirism (Frazier, 1970). The result is a series of bathymetric highs and deep depressions connecting the Massawa Channel with the edge of the axial trough of the Red Sea.

2. **Facies distribution**. The sedimentary facies pattern is quite different from that of a rimmed carbonate shelf (Ginsburg and James, 1974), typically bordered by continuous to semi-continuous rim or barrier along the shelf margin, which restricts water circulation to form an inner low-energy lagoon. The morpho-structural setting of Dahlak islands inhibits the development of wide low-energy shallow-water environments which might have facilitated the deposition of carbonate muds. Pure lime-mud sediments occur only in the innermost sheltered shallow-water areas (Figure H1.7, profile 3) and in tectonic depressions where suspended sediment flows beneath the limits of wave and current action (Figure H1.7, profiles 1, 2). In these intrashelf depressions, the micritic and fine bioclastic sediment is rich in planktonic skeletons (pteropods, foraminifers and calcareous nanoplankton), associated with variable percentages of bioclastic material. The depressions act as sedimentary traps for fine particles generated by bioerosion and held in suspension by the hydrodynamic regime. They also provide 'sanctuaries' for the preservation of a relatively diversified epipelagic assemblage, present in all the waters of the archipelago, but not preserved in environments with a stronger hydrodynamic regime.

Skeletal sediments are the most widespread. Bioclasts generally originate from molluscs (bivalves are the dominant constituent), corals, bryozoans (also lunuli-

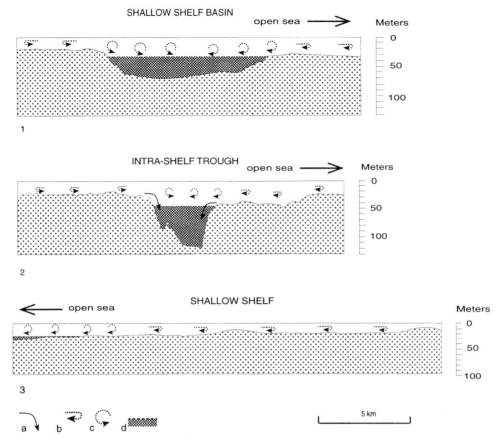

Figure H1.7 Schematic representation of relationships between bathymetric profiles and movement of lime mud particles caused by waves and tidal currents: (1) muddy fraction transport and sedimentation in a shallow shelf basin, northern Dahlak shelf area. (2) Muddy fraction transport and sedimentation in a deep structural low, northern Dahlak shelf area. (3) Muddy fraction transport and sedimentation in the Dahlak Kebir shallow shelf. The arrows indicate the main transport mechanisms of sediment: (a) skeletal material; (b), (c) lime mud particles; (d) area where fine sediment fraction is deposited.

forms), echinoderms, balanids, foraminifers, crustaceans, polychaetes and red algae. Frequently there is a sharp distinction between the components of the living communities (identified from shells and fragments) and those making up the skeletal sediment (which is highly bioeroded, encrusted and sometimes dark stained). Some of the bioclastic sediments covering the Dahlak shelf do not reflect present-day depositional environments and are regarded as palimpsest or relic sediments (Swift et al., 1971).

Figure H1.8 Main sediment types in the Dahlak Islands archipelago. Scale bar = 1 mm in all figures. (a) Bivalve sand – Unsorted and sub-rounded lithoclasts and bioclasts mainly deriving from erosion of emergent Dahlak Reef Limestone, Intertidal zone of Tor Island, Dahlak Bank. (b) Bivalve sand – Abraded fragments of bivalve and gastropod tests mixed with scarce benthic foraminifers. Surf zone of the southern bay of Harmil Island. (c) Balanid/bivalve sand – Coarse and unrounded balanid, bivalve and gastropod remains locally form the substratum for bryozoan and benthic foraminifer communities. Depth 24 m, Seil Amber, South Massawa Channel. (d) Bivalve muddy sand (granular fraction > 62 μ) – Bivalves filled with skeletal sand, showing early cementation at the sediment–water interface. Depth 18 m, Dhu-I-kuff Island, outer Dahlak Kebir shelf. (e) Mollusc/foraminiferal sand – Unsorted bivalve and gastropod remains are associated with well-developed benthic foraminifers, of which *Rotaliina* and *Miliolina* are abundant, and *Textulariina* are scarce. Depth 36 m, Dohul-Isratu islands profile (Figure H1.6,5), northern Dahlak Islands shelf. (f) Pteropod/foraminiferal muddy sand (granular fraction > 62 μ) – Unsorted skeletal remains and faecal pellets from the coarse-grained fraction, in addition to abundant pteropods and planktonic foraminifers. Benthic foraminifers (*Rotaliina* and *Miliolina*) are also present. Depth 68 m, Harmil Island-outer edge profile (Figure H1.6,4), northern Dahlak Islands shelf. (g) Pteropod muddy sand (granular fraction > 62 μ) – The coarse fraction mainly consists of hardened ovoid faecal pellets and peloids, loose or cemented together to form 'grapestones'. Pteropod shells and very thin echinoid spines are the most common bioclasts. Depth 50 m, Isratu-Harmil islands profile, northern Dahlak Island shelf. (h) Pteropod mud (granular fraction > 62 μ) – The sediment contains abundant hardened ovoid faecal pellets, derived from mud-feeding animals. Among pteropods and heteropods the genera *Creseis*, *Cavolinia* and *Atlanta* are very common. Depth 70 m, Secca Mugiunia–Erc Abdulla Abu Madda bar profile (Figure H1.6,1), South Massawa Channel.

Shallow-water skeletal sands are distributed over and around emerging structural highs and wave-swept areas, where they form beach ridges, sandy shoals and sandy cays. In supratidal environments, deflation and aeolian accumulation of skeletal sands are the main control of sedimentation. Evaporation locally creates sabkha conditions in depressed areas inside the islands and along the coasts.

As a result, the general sedimentary framework of the archipelago is characterized by a series of facies belts which do not simply trend parallel to the shelf margins but, depending on depth and morphology of the bottom, show complex, bathymetrically related patterns. A sequence of clean skeletal sands, muddy sands and muds rich in epipelagic remains appears both as a graded sequence from the shelf areas to the outer edge and as concentric belts in the intrashelf depressions, although the latter show a less rich and diversified pelagic component.

3. **Coral colonization**. The distribution and zonation of living corals show the general characters pointed out in various places along the western coast of Red Sea (Braithwaite, 1982, 1987; Head, 1987). The coral cover of the shallow sea floors is irregularly distributed, conditional upon morphology, tectonic stability of the substratum and location with respect to the dominant water flow. Fringing reefs, patch reefs and coral carpets are encountered everywhere around the islands and on the tops of shallow-submerged highs. However, these build-ups consist chiefly of a thin, superficial coral envelope, covering an eroded substratum. Coral community diversification and distribution are closely connected to local water energy and morphology (Civitelli and Matteucci, 1981). No large reefal complexes have been discovered, even in places with favourable structural setting and physiography (Dahlak Reefs).

4. **Present-day evolutionary trend**. The effects of recent tectonic activity and the last eustatic sea-level change are demonstrated in the archipelago by morphological and sedimentological features which define a stadial evolutionary trend. The fall of sea level below the shelf edge, during the Wisconsin interstadial, is reflected by sea-floor morphology particularly in areas poorly affected by tectonics. A recent relative stand of sea level just above its present position (1–3 m) is demonstrated throughout the archipelago by: (i) fossil wave-cut notches at variable distances from the present shoreline and on the borders of inner depressions on major islands. They lie at the same elevation on most of the islands visited; (ii) present intra-island lagoons, which are interpreted as relics of more extensive sea arms; (iii) ancient beachrocks and lagoonal deposits within present sabkha areas, forming the substratum, and (iv) present sabkha deposits overlying swamp facies rich in gastropod (cerithiid) and mangrove remains. A general sedimentary progradation is responsible for the development of beach ridges, shoals and sandy cays along the coasts. This stand of sea level just above the present one can be correlated with eustatic sea-level change. Holocene marine deposits, 1–3 m above present sea level, are reported on many coasts of East Africa and the Red Sea.

ACKNOWLEDGEMENTS

We acknowledge the co-operation of the researchers from Consiglio Nazionale delle Ricerche and from Dipartimento di Scienze della Terra, Università 'La Sapienza', Rome, who have investigated the archipelago of the Dahlak Islands for more than a decade and who processed the multiple data required for this study. We are particularly indebted to our colleagues G. Accordi, F. Chiocci, J. Pignatti and P. Tortora, with whom we initiated a new cycle of studies in 1994. We are grateful to the Eritrean Ministry of Marine Resources for authorizing us to start new investigations. Finally, we would like to thank Dr J.C. Hillmann, Assistant Head of Research, Research and Environment Division of the Eritrean Ministry, for helping us to resolve many of the problems connected with our recent marine survey. Also we thank C.J.R. Braithwaite and the volume editors B. Purser and D. Bosence for critically reviewing the original manuscript.

Chapter H2
Quaternary marine and continental sedimentation in the northern Red Sea and Gulf of Suez (Egyptian coast): influences of rift tectonics, climatic changes and sea-level fluctuations

J.-C. Plaziat, F. Baltzer, A. Choukri, O. Conchon, P. Freytet, F. Orszag-Sperber, A. Raguideau and J.-L. Reyss

ABSTRACT

This contribution is based on new observations of the relations between continental and marine deposits along the 800 km of northernmost coast of the Red Sea. One hundred U/Th dates of the marine units (coral reefs, molluscs, echinoids, gypsum) together with 20 site descriptions and age discussions enable an interpretation of sediment relations with palaeoclimatic and glacio-eustatic sea-level changes. Isotope stage 5e is studied in detail and shows high frequency changes, but the study also concerns the overall Early Pleistocene to Holocene evolution. Stability is a general attribute of this part of the eastern African coast with the exception of the southern Gulf of Suez. North of the Aqaba transform fault the Gulf appears to be still in the rifting stage.

INTRODUCTION

The Egyptian coasts of the north-western Red Sea and Gulf of Suez are characterized by a series of spectacular Quaternary reef-terraces which, until recently, have been little studied. The oldest works (Newton, 1899, 1900, Sandford and Arkell, 1939, Butzer and Hansen, 1968, Veeh and Giegengack, 1970; Gvirtzman and Buchbinder, 1978) have shown that sediments and fossil communities are typical of the Indo-Pacific (Figures H2.10, H2.11), often occurring as reefs ranging in altitude from +1 to +42 m.

These terraces are examined here for possible effects of Quaternary rifting despite the fact that a number of authors regard this area as entering or within the post-rift phase. In addition, this contribution examines the relationships between alluvial discharges and sea-level fluctuations as these are both influenced by glacio-eustacy and the associated changes in average rainfall.

Research by the present authors since 1986, is supported by more than 110 U/Th and ^{14}C dates (Reyss et al., 1993; Plaziat et al., 1996). Our first general results on continental and marine deposits were published separately by Freytet et al. (1993) and Plaziat et al. (1995). The present contribution is the first attempt to synthesize these studies.

This study endeavours to discriminate the respective influences of tectonics, climate and eustatic sea-level changes on various coastal deposits (reefs, alluvial fans; lake and river deposits) which are related to Quaternary chronology by means of dated reefs and salina gypsum. Thus, after a general discussion of the present-day sedimentary conditions, the effects of glacio-eustatic cycles at these tropical latitudes and innovations in dating methods we present a synthesis of our research in two parts: firstly, short accounts of local field records have been selected in order to illustrate the prevailing tectonic, alluvial or reef influences on the different marine and continental Quaternary settings. Fifteen littoral and wadi sections with maps illustrate these various sedimentary settings and demonstrate the overall influence of glacio-eustatic changes, especially during the middle and late Pleistocene cycles. A minor sea-level fall during the well-documented 5e high-stand is indicated. Late Pleistocene tectonics is revealed to be important only in the southern Gulf of Suez, which points to a rift-type activity of the Gulf while the Red Sea coast is almost stable since middle Pleistocene times. Secondly, sediment accumulation in the Suez isthmus acted as a northern barrier during the entire Quaternary

Sedimentation and Tectonics of Rift Basins: Red Sea–Gulf of Aden. Edited by B.H. Purser and D.W.J. Bosence. Published in 1998 by Chapman & Hall, London. ISBN 0412 73490 7.

538 Quaternary marine and continental sedimentation

but did not prevent Mediterranean lagoonal molluscs from settling in Gulf lagoons and bays where they were associated with Indo-West Pacific mangrove and reef faunas.

GEOGRAPHIC AND CLIMATIC CONTEXT

Within the Red Sea rift system which comprises 3800 km of African coastline between 13° and 30° N, a series of local studies of Egyptian outcrops have been made over 850 km, between Wadi Lahami (north of Ras Banas) and Ismailia (Figure H2.1). This coastal margin of the Eastern Desert is characterized by a hyper-arid climate with an alleged 5 mm average annual rainfall. This does not necessarily fall every year at a given place and therefore does not reflect the real influence of running water within a wadi basin. The relatively narrow coastal plain (generally less than 1 km to 10 km, and 20–30 km wide south of the Gulf of Suez) is limited by a continuous basement relief, the so-called Red Sea Hills, a littoral escarpment which culminates at between 1000 and 2200 m at less than 40 km from the shore (Figure H2.2). The individual drainage basins thus have a very small extent: a few square kilometres to a maximum of 1800 km² (Wadi Gemal). Even during periods of greater humidity, the discharges via numerous, but relatively short wadi systems have been limited. These have resulted mainly in alluvial cones and terraces, the coarseness to which is a function of the proximity to the peripheral relief. Today, wadis appear to be carrying little coarse detritus, the occasional flash-floods having mainly erosional effects. The wadi transport potential is limited essentially to sands and silts that are also derived from the reworking of aeolian material infiltrated into the older coarser materials bordering the wadis. This fine material, which tends to make the desert soils impermeable, encourages surface water-flow in the form of sheet-floods which extend to the shoreline during brief periods of catastrophic flooding which may result in human casualties.

Wind action is seasonally important but the volume of displaced sand is modest, mainly because of the limited input of fine-grained sediment in the alluvial fans

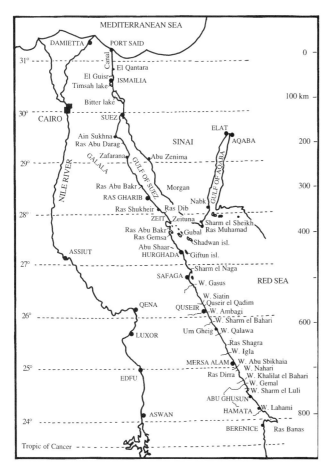

Figure H2.1 Location map of localities.

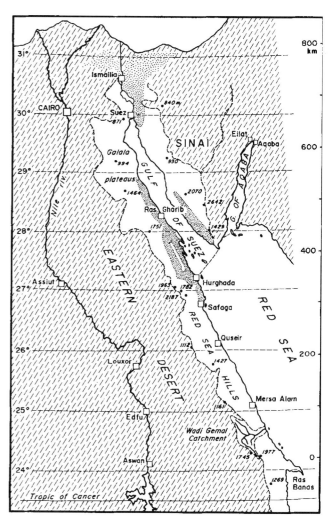

Figure H2.2 The catchment areas (in white) of Red Sea–Gulf of Suez basin. Altitudes in metres. Dotted: sand and gravel plains.

and because of the predominant north to south (Shamal) wind, parallel to the shoreline and hinterland relief. Thus, winds have no access to the sands of the Eastern Desert. Aeolian processes are characterized by a high degree of deflation and black pebble lags tend to exaggerate the visual importance of even the thinnest gravel beds. It has also probably contributed to the erosion of depressions which may extend below sea level such as north of Ras Shukheir. Sand transport is sufficient to form small shadow dunes on the lee-sides of rocks which may extend up to 50 m above the coastal plain. In flat alluvial areas, sands are trapped by vegetation in form of nebkhas.

Vegetation is very scarce but tends to increase towards the south, notably in the form of grassy shrubs and scattered acacias. Subsurface flow in the wadis and vegetation density are related to heavier dews and more frequent rains south of Quseir, which thus tend to modify locally the overall hyper-arid aspect of the rocky desert. The Suez Isthmus also benefits from a slightly more humid climate and modern irrigation. In spite of these damping effects, dune fields are more extensive in the north, notably east of the Canal.

Coral reefs form an almost continuous fringing belt between the Gulf of Suez and Ras Banas. However, these are generally narrow, with frequent breaks or 'sharms' situated at the mouths of second-order wadis whereas major wadis have formed fan deltas which have displaced coral growth some distance from the shoreline. The formation of sharms has been attributed to fluvial erosion during the last glacial sea-level lowstand (Gvirtzman et al., 1977). These deep embayments have not yet been filled by later terrigenous sediments due to very low supply rates. This is just one example that suggests that Quaternary climatic changes have strongly influenced littoral sedimentation. Therefore, it is logical to compare the respective roles of tectonic relief and climate when considering the Quaternary history of the rift.

The coastal region is only locally fringed by cliffs, notably between Ain Sukhna and Ras Abu Darag and in the vicinity of Safaga. Elsewhere, low alluvial plains alternate with low cliffs (5–30 m high) cut into Quaternary reefal carbonates; the latter often developed on Pleistocene alluvial cones or Pliocene carbonate and terrigenous halokinetic domes (Um Gheig, South Hurghada, Ras Shukheir, Ras Gharib) and, much more rarely on Miocene dolomites (Ras Dib, Ras Gemsa). Therefore, the coastline appears not to be a simple faultline morphology, in spite of the linear nature of the shoreline. This morphological diversity of settings underlines an important problem concerning tectonic activity during the Plio-Quaternary, and thus, the syn- or post-rift setting of the present rift margin.

CHRONOLOGY OF QUATERNARY DEPOSITS AND PROBLEMS OF DATING

The oldest Quaternary deposits comprise continental alluvial fans (Butzer and Hanzen 1968; Freytet et al., 1993) principally made of debris flows, which may be related to superimposed alluvial systems which are unconformably stacked. This coarser piedmont material constitutes alluvial cones with radii ranging from 1 to 15 km, as well as the fluvial (river and wadi) material entrenched into the Neogene substratum and preceding Quaternary deposits. A relative chronology of the interdigitated continental and marine deposits has been established (Freytet et al., 1993) which demonstrates the repeated succession of planar erosion surfaces covered with subhorizontal fanglomerate beds, and phases of linear incision. These narrow valleys are filled with basal debris-flow sediments and subsequent fluvial deposits locally associated in their upper part with travertines (Figure H2.3).

Four fanglomerate formations have been defined in the region south of Quseir (Figure H2.4). The latest three are associated with reef-terraces, suggesting the possible correlation of glacio-eustatic sea-level drop with climatic variations, notably humidity increase, which was especially effective at latitudes around 25° N.

Each arid, interglacial reef-terrace was followed, during a period of anaglacial cooling, by planar erosional reworking of part of the upper alluvial fan resulting in deposition of a new fanglomerate (Figure H2.3). This suggests a subarid climate with contrasting seasons favouring debris-flows and sheet-floods. These deposits are subsequently cut by valleys 2–15 m deep, reflecting a greater erosional potential. The glacial maximum is another period of aridity while the following cataglacial warming and renewed humidity progressively supplies the fluviatile drainage system. The most humid episode also favoured development of travertines, mostly within abandoned channels and ponds of the braided network. The rise of base-level probably also influenced alluvial deposition due to reduction of stream velocities. The new sea-level highstand favoured marine erosion of the prominent parts of the coastal alluvial cones and the construction of fringing reefs on a narrow terrace during a new phase of aridity.

Repeated phases of erosion and alluvial deposition clearly have destroyed or masked some of the smaller deposits, favouring selective preservation of the more widespread sub-arid cones. However, one can occasionally observe the entrenched traces of the most humid phases (fluvial deposits) previously underestimated or disregarded (Freytet et al., 1993).

Within the marine deposits, several dating methods have been used. Biostratigraphy has not furnished

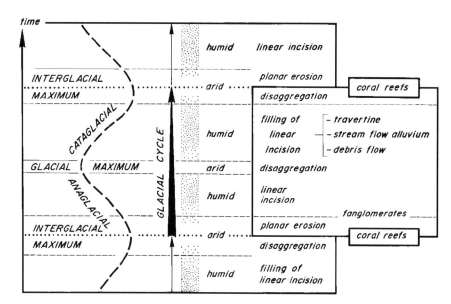

Figure H2.3 Climatic and sedimentary evolution during a glacial cycle between two high-stand reefs represented by a schematic sea-level curve (after Freytet et al., 1993). Disaggregation is a permanent process but it is especially common when erosion is limited by aridity.

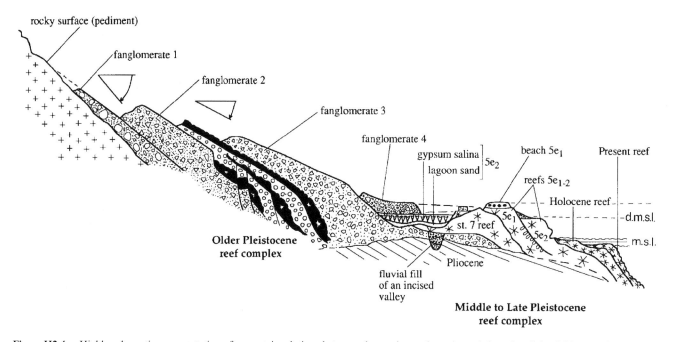

Figure H2.4 Highly schematic representation of geometric relations between the marine and continental deposits of the African Red Sea coast. Tectonic tilting and faulting occur through the Pleistocene. The Holocene high stand notch, and division of the stage 5e reef are only locally observed. (m.s.l. = mean sea level; d.m.s.l. = derived mean sea level).

precise ages because there have been few changes within the documented molluscan faunas since the last glaciation (Taviani, 1982; Taviani, this volume b). A sandy facies containing minute *Clypeaster*, *Laganum* (Figure H2.11(j), (k)) and *Chlamys vasseli* (Shagra Fm) is generally referred to as a 'Late Pliocene or early Quaternary' age and may possibly be assigned to the Lower Pleistocene according to a Sr isotope date of 1 million years at Gubal island (Taviani and Dullo, unpublished).

Radiometric dates have been the principal base for chronology (Figure H2.5). Several dates are relatively old and difficult to localize in the sedimentary systems (Butzer and Hansen, 1968; Veeh and Giegengack, 1970), while the more recent dates (Andres and Radtke, 1988; Hoang and Taviani, 1991; El Moursi, 1992) give precise

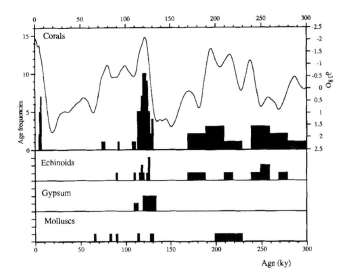

Figure H2.5 Summary diagram of the U/Th dates produced during the research programme sponsored by the PNRCO, by Choukri and Reyss (Centre des Faibles Radioactiviltiés, Gif-sur-Yvette). Age frequencies are number of samples compared with the $\delta^{18}O$ SPECMAP stack plotted against age, using the time-scale developed by Imbrie et al. (1984).

but sometimes geologically questionable information. Dates presented here have been made systematically through several sections and cross-checked (Reyss et al., 1993; Choukri, 1994; Choukri et al., 1995; Plaziat et al., 1995). For the lower Holocene levels, both ^{14}C (shells) and U/Th (coral, echinoid) dates are found to be in complete agreement. For the older, Late Pleistocene reef-terraces, only U/Th dates exist. These have been derived mainly from aragonitic or slightly calcitised corals (<5% calcite), aragonitic gastropods (Terebralia, Melanoides) and several bivalves (Tridacna) (Figure H2.5). In addition, a new approach using echinoid spines has given encouraging results. Assuming that the diagenetic change of the calcitic echinoid skeleton from porous stereome to massive monocrystalline calcite occurs very early in diagenesis, it was proposed that the resulting radiometric age should be very close to the time of deposition. This has been confirmed by comparing coral and echinoid spine U/Th ages from the same strata (Choukri, 1994; Choukri et al., 1995).

Another new method has been the U/Th dating of subaquatically precipitated gypsum, formed in sea-water salinas closed by a permeable sediment barrier. $^{234}U/^{238}U$ ratios close to 1.15 (which is the usual coral value), gives a convincing argument for the selection of sea-water derived gypsum. Concordance of ages derived from associated corals and gastropods confirms the reliability of this method.

Thus we base our chronology on a large suite of radiometric dates (Figure H2.5) compared with previous studies. However, we lack dates older than 300 000 years BP (no Sr isotope dates). Most of the dates refer to the Late Pleistocene isotope stage 5e highstand, around 123 ka BP, known in northern Europe as the Eemian and in the Mediterranean as the Eutyrrhenian stage. We had to reject a few aberrant measurements because the numerous dates obtained from the same coral-terrace have provided puzzling results. Most of these misleading dates are related to an artificial decrease in age caused by late precipitation of diagenetic aragonite cement affecting the aragonitic skeletons of Pleistocene corals situated within the Holocene and present-day splash zone (Figure H2.6). This is demonstrated by the striking correlation between present altitudes and age anomalies (inconsistency), within a given reef-terrace. In the vicinity of exposed promontories such as Ras Shagra and Sharm el Naga (Figure H2.6), the lowest levels may be diagenetically rejuvenated by more than 50%.

Ages considered to be too old have also been noted, especially within gastropods. On the contrary, Tridacna gives ages which are generally too young. We nevertheless suggest that, because of the limited discrepancy between dates based on Tridacna shells and corals, the results are meaningful providing that their use is limited to the distinction of one or other of the last three interglacial maxima.

Prehistoric artefacts, notably flint tools made from Eocene cherts, have been found at a number of localities and, although not furnishing precise dates, nevertheless are useful for assessing the minimum ages of certain terraces (Montenat, 1986). In addition a typical Acheulean tool has been observed on the last interglacial reef-terrace at Ras Barra (Figure H2.7(a), (b)). Abundant flint detritus spread at several localities on the Miocene Abu Shaar plateau may record Upper Paleolithic or Neolithic occupation but marine shell clusters and a midden with flint tools detected south of Bir Abu Shaar (Figure H2.7(d), (l)) give ^{14}C ages of 5345-6073 years BP. The associated Terebralia, a typical mangal gastropod, gives useful information concerning the Neolithic climatic optimum (see below and Plaziat, 1995).

Data from selected local studies

The sites, discussed below in some detail, have been selected from a larger set, as the best representatives of the various local tectonic, climatic and sea-level controls on sedimentation, from which certain conclusions will be discussed.

A. Southern Gulf of Suez: Gebel el Zeit region from Ras Gemsa (S) to Ras Gharib (N) (Figure H2.8)

This is one of the areas which clearly express the effects of neotectonic uplift by up to 12 m of the reef-terrace formed during the last interglacial period.

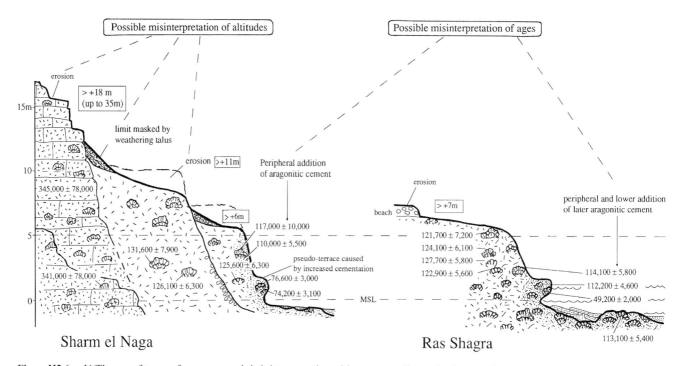

Figure H2.6 U/Th ages of two reef sequences and their interpretation with respect to diagenetic changes of coral skeletons on cliff shores exposed to sea water splash, spray and impregnation. The respective derived Mean Sea Levels are deduced from altitudes of characteristic sedimentary surfaces modified by erosion and levelling below slope deposits. Note the frequent age inversions with respect to sample altitudes and the vertical extension of every reef construction. Where a normal, about 8 m altitude of stage 5e derived MSL is established (Ras Shagra), the abnormally low-age values (114–49 ka) cannot be referred to the younger 5c, 5a because these relative high-stands were only several tens of metres below present sea level.

Ras Dib (Figures H2.9, H2.10 (a), (c))
Immediately to the south of Ras Dib a narrow reef-terrace is developed against Neogene sediments. The reef-front has been totally eroded seawards of a clysmically-oriented (N140°) fault which has lowered the eastern block by several metres. The reef-core (Figure H2.10) composed of perfectly preserved corals and molluscs and dated as isotope stage 5e, lies on cemented marine gravels and sands including blocks of recrystallized coral whose present altitude is 11 m above mean sea level; although rising progressively westward towards the land. This seaward slope also affects a discontinuity, relating to a pause in reef-growth situated some 2 m above its base. This surface coincides with a local external deposit of terrigenous gravels. Renewed growth by 1.5–3 m high colonies of *Porites* which attain altitudes of +17/18 m along the seaward margin, but rise towards the interior to about +20.5 m.

Excellent outcrop conditions here favoured reliable interpretation of the altitudes measured and the method we use to identify palaeo-sea level. The recommended reference-level necessary to reconstruct the palaeo-sea level (and thus, to measure tectonic effects) is mean sea level (MSL). In reefal environments (cf. Van de Plaasche, 1986), this level coincides with the culmination of the subhorizontal epi-reefal terrace (Figure H2.12) which approximately constitutes the top of the reef-flat. Today, along the northern Red Sea coast, this surface lies some 0.8 to 1.5 m below the sand slope (berm) and the sand ridge piled by storm waves, on the upper beach but depends on the local exposure to wave energy.

Thus, in the best outcrops of the last interglacial reefs such as at Ras Dib, a clear-cut exhumed relief of the coral growth zone was not found to be representative of MSL because the reef-flat was not constituted by fused coral heads but has been levelled by detrital sediments burying the uneven coral colonies in growth position (Figure H2.9). This coral surface may be 0.5–2 m lower than the palaeo-MSL. Here, the difference between +17 m (coral heads) and +20.5 m (bioclastic/beach accumulation) illustrates that the potential error is not negligible. An estimation of +19 m for the derived (= interpreted) MSL. appears to be more likely, especially if one considers the exposed nature of the promontory. Accordingly, the term 'mean sea level' (MSL) will be used in this contribution in the restrictive sense of 'derived mean sea level' without discussion of the altitudinal significance of the different facies observed and measured (coral heads, beach-rock, uppermost beach deposits) as well as the conditions of the present and fossil wave exposure.

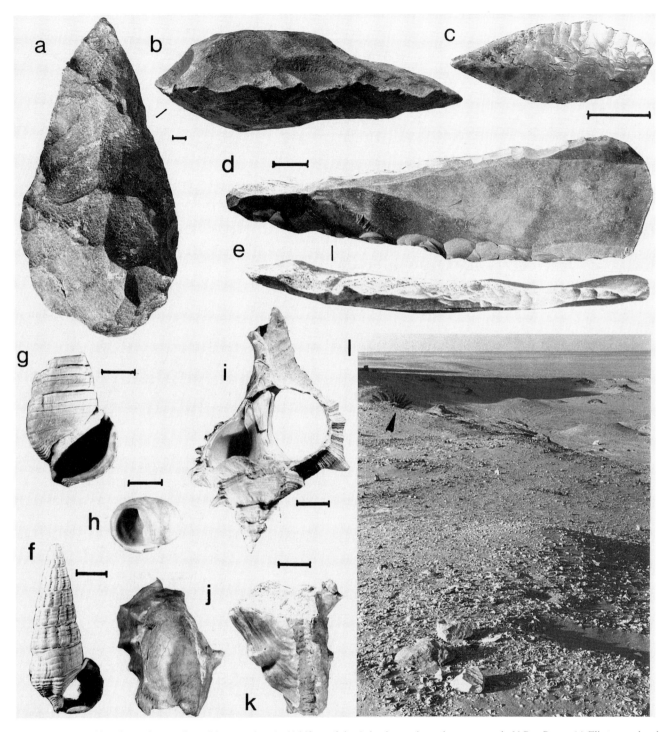

Figure H2.7 Prehistoric artefacts and nutrition remains: (a, b) biface of the Acheulean culture, basement rock, N Ras Barra. (c) Flint arrowhead isolated on an alluvial terrace of Wadi Kharasa, N Hurghada. (d)–(l) Midden on the south-eastern foot of Abu Shaar plateau, N Hurghada. (d), (e) flint knife. (f), (g) *Terebralia palustris*, pierced and broken. (h) *Nerita*. (i) *Murex*. (j), (k) *Saccostrea cuccullata*, the mangrove oyster with *Avicennia* roots prints. (l) the midden site, with a head of palm-tree (arrow) within the local depression of an ancient spring of sulphate water (gypsum crust). Scale bar: 1 cm.

Andres and Radtke (1988) obtained a figure of +15 m for the height of this reef based on the barometric values for this outcrop or an adjacent site. Although not in contradiction with our levelling results, this altitude should not be used to characterize either the reef flat altitude (their '+15 m niveau über Meer Riff') or the derived MSL because it corresponds with the basal surface of coral growth.

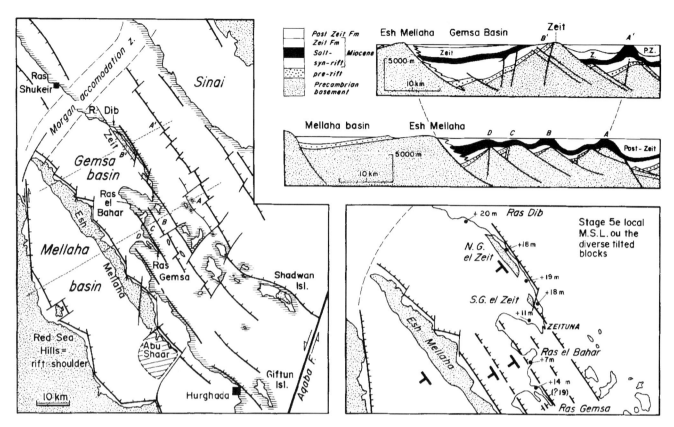

Figure H2.8 The southern Gulf of Suez. Tectonic map and sections after Bosworth (1995) slightly modified. The tilted blocks have been secondarily faulted and the evaporitic Miocene has laterally migrated and accumulated in halokinetic structures above the crest of tilted blocks. Thus a rejuvenated block-tilting may be easily confused with a recently reactivated halokinetic doming, except for the Gebel el Zeit basement shoreline.

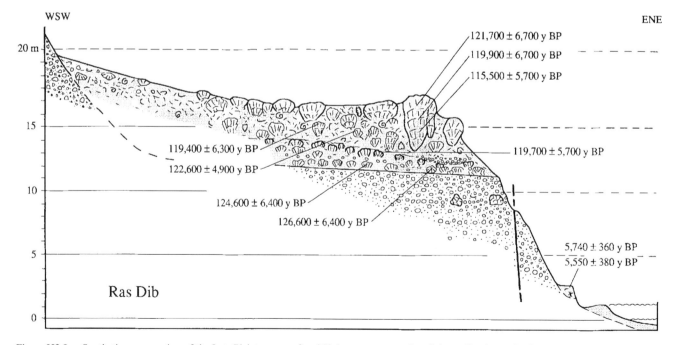

Figure H2.9 Synthetic cross-section of the Late Pleistocene reef and Holocene emergent beach immediately south of Ras Dib lighthouse. Altitude levelling and U/Th dates of aragonitic corals are related to Present MSL.

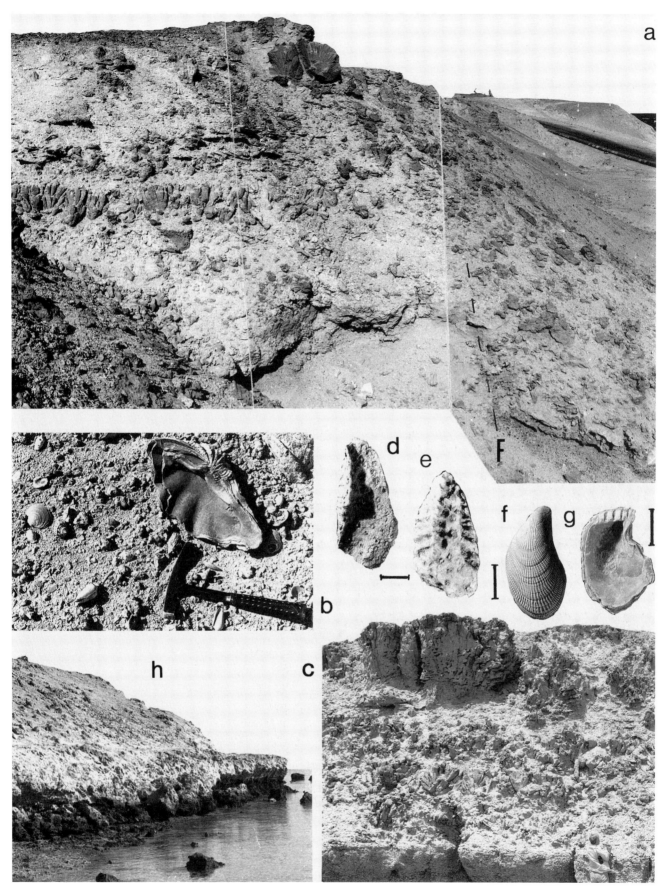

Figure H2.10 Last Interglacial reefs. (**a**)–(**c**) Ras Dib site. The diverse superimposed reefal facies belong to the same stage 5e reef growth. The giant *Porites* of the upper external part of the outcrop suggest that erosion removed the frontal part of the reef. The conglomeratic substratum is locally lowered by a normal fault (F). (**b**) Mollusc shells on the detrital slope, with *Tridacna, Codakia, Strombus, Turbo, Dentalium* in a loose matrix. (**d**), (**e**) *Dendostrea foliacea* with their substrate mould of *Acropora* branches. (**f**) *Septifer excisus*. (**g**) *Isognomon nucleus*. Scale bars: 1 cm (**h**) N Mersa Alam. A littoral profile of the Late Pleistocene reef. The whole outcrop belongs to only one reef but cementation of the lower 2 m produces a bench or a pseudo-terrace by differential erosion.

Ten samples of corals from Ras Dib have been dated (U/Th). The average age of 121.16 ± 6.5 ka (115–126.6 ka) agrees closely with the age obtained from an echinoid spine from the same site (119.6 + 9.3). Another methodological contribution of the sample-clustering method is produced when one relates the dated coral samples to their respective altitudes (Figure H2.9). This allows an appreciation of the significance of ranges of error of the radiometric method.

+17 m: 123.6 (+6.5 − 6.2); 121.7 (+6.9 − 6.5) ka BP
+16.5 m: 119.9 (+6.8 − 6.5) ka
+15 m: 117.7 (+6.8 − 6.4); 115.5 (+5.9 − 5.6) ka
+14 m: 119.7 (+5.9 − 5.6) ka
+12.5 m: 124.6 (+6.6 − 6.2) ka
+12 m: 126.6 (+6.6 − 6.3) ka

Although all values are within the range of the possible measurement error of the median age (127.6 − 114.6 ka), it is significant that the oldest dates were obtained from both the highest and the lowest samples. Within the 3 m vertical section of a single coral head there is also a progressive inversion of radiometric dates from base to top: 115.5, 119.9 and 121.7 ka. These dates are essentially the same and within the margin of error. The large number of dates obtained at each site, while appearing wasteful, is important to the interpretation of stratigraphically complicated sites. Concerning the history of the reef-terrace at Ras Dib, the interrupted reef growth during the 5e isotopic stage is revealed to be of limited importance; samples located on either side of the sedimentary discontinuity have statistically identical ages, close to average. The calcitised coral boulders preserved in the underlying marine deposit possibly belong to isotopic stage 7 or an earlier highstand.

Towards the south (Figure H2.8)
The fault scarp affecting Precambrian basement forms part of the Quaternary coast and scattered remnants of the same 5e reef-terrace (127.6 ka + 4.7 − 4.5) are at the same altitudes as at Ras Dib (+12/13 m at the base and +17.5/18 m at the top of the coral terrace). Further south, within the structurally (and topographically) low area separating the two Zeit basement mountains, where Miocene and Pliocene rocks form the substratum, a Quaternary section occurs which is similar to that published by Andres and Radtke (1988). The reef-flat has been re-measured at +19 m and a horizon with coral blocks forming a terrace overlying coarse, possibly continental gravels, has been estimated at +36 m (compared with the +50 m of the previous authors and their +65 m terrace which has not been observed by us). This high level, undatable by radiometric methods, has been considered as Early Pleistocene or even Pliocene. The ages of the Andres and Radtke '+15 m terrace' as 94.0 − 140.0 ± 20 ka are not in contradiction with our data.

In front of the south (small) Gebel el Zeit (Figure H2.8) South of the extensive oil installations, a gently sloping plateau is developed on a larger reef-flat cut by a long spectacular ravine (Figure H2.13). This reef overlies marine gravels (+6.3 − 7.5 m) situated on the outer slope of a tilted Pliocene block which rises to +15 m and is overlain by a first reefal formation of branched *Acropora* and *Stylophora* dated at 185.2 (+12.8 − 11.5) and 186.0 (+10.6 − 9.7) ka BP. This lower coral facies is buried below a reef-core facies which is reinforced by colonies of top-shaped colonies of *Porites*. Within the seaward parts of the reef-core, between +8 and +12 m, the coral ages range from 121.5 to 208 ka: 121.5 (+6.1 − 5.8), 124.7 (+4.2 − 4) and 131.1 (+4.3 − 4.2) ka associated with other massive corals giving ages of 197 (+10 − 9) and 208 (+12 − 11) ka. These dates suggest a stage 7 reef, here represented by the underlying branched corals. In the landwards part of the reef, *Porites* colonies between +15.4 and +17 m, all belong to the stage 5e highstand. Dates range from 122.7 (+5.8 − 5.5), 129.1 (+4.1 − 4), 131.4 (+2.6 − 2.5), 120.3 (+6.4 − 6.1), 122.5 (+5.7 − 5.4). In addition, an echinoid spine gave a similar age of 117.6 (+6.8 − 6.4) ka BP.

It is impossible to separate the two successive coral communities on diagenetic properties, although their respective ages indicate two distinct episodes separated by the glacial isotopic stage 6. Both units occur at the same altitude and the only explanation of the perfect preservation of corals belonging to the isotopic stage 7 would be from the protection offered by the overlying stage 5e reef during the last glacial lowstand.

Further south, adjacent to abandoned buildings, a trench reveals a sand containing perfectly preserved molluscs situated below reef 5e (corals dated at 125.6 (+4.5 − 4.4), 119.3 (+6.5 − 6.2), and an echinoid as 127.2 (+8 − 6) (Figure H2.11(a)). The reef does not appear to be higher than +10 m, but it is covered with slope rubble.

The repeated tilting of the Zeit blocks during Miocene and Pliocene times thus was continued at least to the Quaternary 5e stage. However, the similar altitudes of reef corals of 5e and 7 stages indicate that tilting was not a continuous process but appears to have paused at least between the stage 7 reef growth and uplift of the 5e reef.

The western hangingwall side of Gebel el Zeit (Figure H2.8)
Field study is limited to the southern coastal outcrops plus a sample of sand rich in molluscs collected along

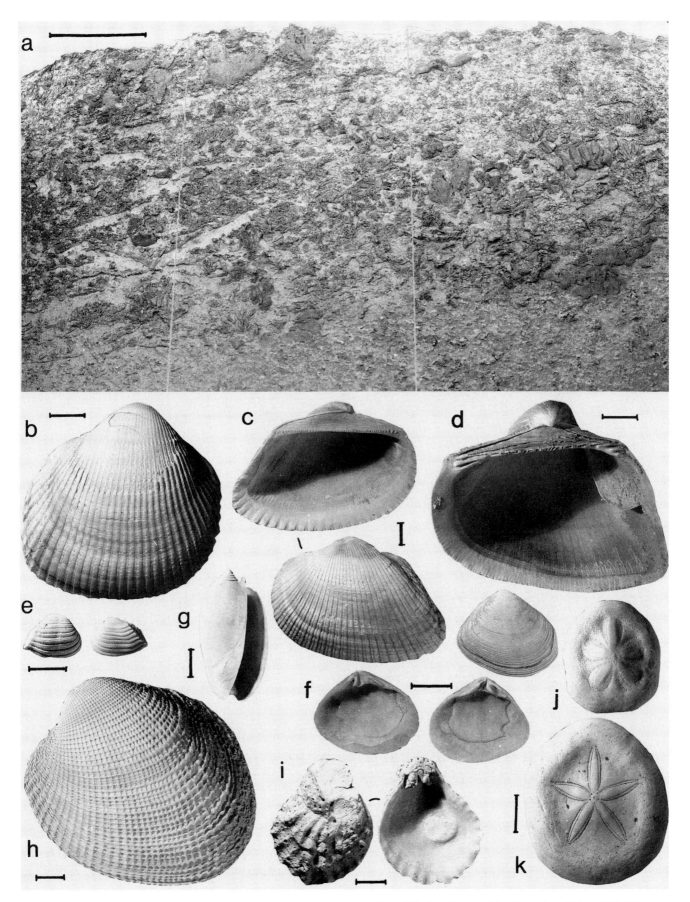

Figure H2.11 Pleistocene reef and open-marine sand fauna. (**a**) Excavated section of the Last Interglacial reef, southern Gebel el Zeit. The bioconstruction is rich in corals developed on a coquina sand deposited in an infralittoral, protected environment characterized by *Fulvia*. (**b**) *Fulvia papyracea*. (**c**) *Anadara erythraeonensis*. (**d**) *Cucullaea cucullata*. (**e**) *Corbula acutangula*. (**f**) *Atactadea glabrata*. (**g**) *Oliva bulbosa*. (**h**) *Periglypta reticulata*. (**i**) *Plicatula plicata*. (**j**) *Clypeaster reticulatus*. (**k**) *Laganum depressum*. Scale bars: 1 cm.

548 Quaternary marine and continental sedimentation

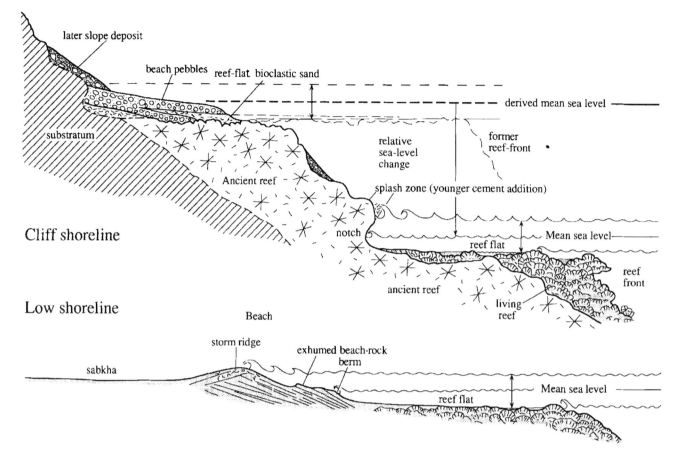

Figure H2.12 Schematic representation of the reference levels used for the measurement of relative sea-level change: location of the present mean sea level, estimation of the most likely location of the palaeo-(derived) MSL and measurement of the vertical distance between these inferred levels.

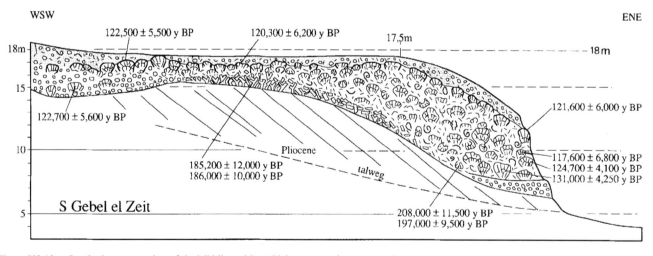

Figure H2.13 Synthetic cross-section of the Middle and Late Pleistocene reef-terrace on the eastern flank of the Southern Gebel el Zeit. *In situ* and reworked corals from a late Middle Pleistocene (isotopic stage 7) terrace have been separated by dating.

the track north-east of the airfield, at an altitude of approximately +15 m.

West of the Zeituna oil-company base
On the shore of Zeit Bay immediately to the west of the police post, a gravel terrace is situated at +10–12 m (Figure. H2.8, H2.14), containing *Terebralia, Cerasto-*

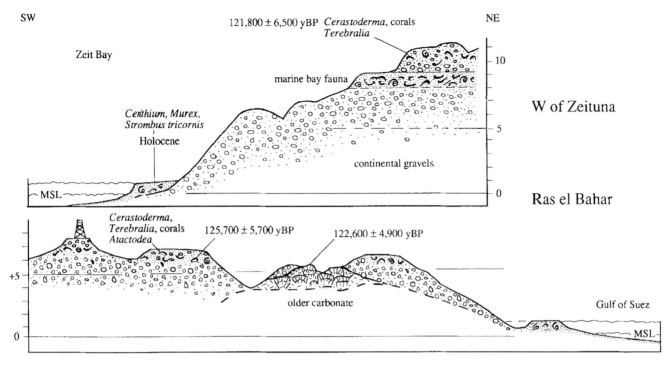

Figure H2.14 Late Pleistocene beach clastics with a mixed malacofauna bringing together components from exposed beach (*Atactodea*, corals), mangrove substrate (= mangal: *Terebralia, Nerita*) and protected lower intertidal (*Cerastoderma*) environments. Altitudes at Ras el Bahar seems to be unchanged since stage 5e times but a 3 m uplift is likely west of Zeituna.

derma, Nerita and more open-marine molluscs and coral debris dated at 121.8 (+ 8.6 − 6.4) ka. These gravels cap a 1 m thick marine sand (+ 9 m) rich in molluscs overlying thick continental gravels. In addition, a probable Holocene + 1 m terrace occurs, which is rich in molluscs.

Further south in Zeit Bay, near Ras el Bahar and Ras Gemsa, the more internal environments belong to variable settings. At the landward end of Ras el Bahar, a terrace with pebbles situated between + 5 and + 8 m (Figure H2.14) includes fauna derived from an open, high-energy shoreline: *Atactodea* (Figure H2.11(f)), *Nerita* (Figures H2.7(h)) and coral debris. These are mixed with molluscs typical of more sheltered environments (*Potamides conicus, Cerastoderma glaucum*, and *Terebralia palustris* (Figures H2.20, H2.25) and overlie a bay sand with marine molluscs. This coarse-grained coastal complex is interpreted as a spit deposit supplied by the southward longshore wave transport and stabilized within the lee-side of the promontory. A sample of *Stylophora* gave an age of 125.7 (+ 5 − 4.8) ka indicating that the area has not been affected by uplift since the 5e stage. This is also confirmed by an adjacent reef facies located at + 5 to + 6 m dated as 122.6 (+ 5 − 4.8) ka. The cliffs of Miocene gypsum and overlying post-evaporite (Pliocene ?) series forming the Ras Gemsa Peninsula are cut by local marine terraces with corals along their eastern side (Figure H2.15). A sandy facies rich in *Anadara* associated with other open-marine shells including scattered *Cucullaea* (Figure H2.11(d)) and corals is situated between + 10.3 and + 13.5 m. These deposits are dated as 124.7 (+ 6.7 − 6.4) ka and 125.5 (+ 5.8 − 5.5) ka and are adjacent to an undated upper (19.6 m) gravel deposit with rare, worn, molluscan debris. Nevertheless, it is clear that the Gemsa Peninsula has been uplifted (tilted ?) since the 5e highstand by at least 5 m, and possibly up to 12 m if we assume that the gravels represent a Last Interglacial beach deposit.

Esh Mellaha

Further south, towards Esh Mellaha, no reefal facies have been noted. Their absence may reflect a decrease in altitude of 5e terraces, from north-east to south-west, in relation to western tilting of Neogene blocks (Figures H2.8, H2.32). These tilted blocks include Gebel el Zeit which has been uplifted by 10 to 13 m (difference between the usually accepted eustatic reference level of + 6 to + 8 m for 5e MSL and the present altitudes at Ras Dib and south Zeit), while Ras Gemsa was uplifted in the order of + 5 to + 8 m (and, possibly, + 11 m). On each of these blocks, the subsiding areas would have resulted in the disappearance of coastal deposits below Holocene continental sediments. The axis of rotation theoretically coincides with the localities where 5e terrace materials are situated at the normal eustatic (+ 6/8 m) levels – this

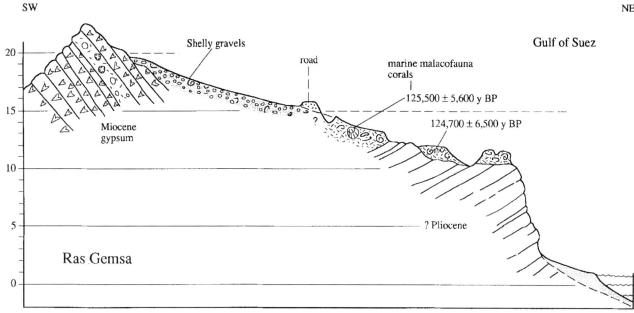

Figure H2.15 Synthetic cross-section at the north-eastern base of Ras Gemsa peninsula. The upper open-marine terrace with dated corals and molluscs, suggests a 6 m, post-5e uplift which may be extended up to 11 m if the gravels with shell clasts are considered as beach deposits of the same high sea level.

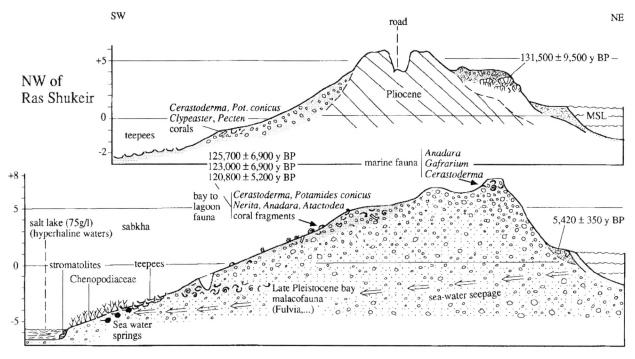

Figure H2.16 Synthetic cross-sections of the Pleistocene and Holocene marine deposits associated with a Pliocene crest and its southwards coarse detrital prolongation north-west of Ras Shukeir. The depression below sea level more likely results from deflation rather than from modern subsidence, because the Late Pleistocene reef and beaches are still close to their normal altitude.

being the case for Ras el Bahar – or close to points that are slightly uplifted (+3/5 m) like the end of Zeit Bay (for the South Zeit block).

Between Ras Shukheir and Ras Gharib (Figure H2.1) North of the Zeit region this depression possibly exhibits a similar structural behaviour (Figure H2.16). A narrow, dipping Pliocene section (outcropping along the roadside) forms a N140 structural crest which supports reefal facies plastered along the seawards surface, at an altitude of +4 m.

A coral which gives an age of 131.5 (+9.9 − 9.2) ka is overlain by 1 to 2 m of beach cobbles. Towards the south, along the crest, a terrace with cobbles and marine molluscs (*Anadara* and rare *Cerastoderma*) are probable equivalents of this 5e reef. These beds occur at +7.5 m. To the west of the structural ridge, the depression is occupied by a sabkha surrounding modern saline lakes whose surface is situated at 5.9 m below present MSL.

Between the structural crest and the sabkha, a gravel terrace with *Stylophora* debris and molluscs including *Anadara*, *Nerita*, *Potamides* and *Cerastoderma* (Figure H2.20) covers the gentle slope. At an altitude of +5.7 and −2.7 m two coral samples gave ages of 120.3 (+5.3 − 5.1), 125.7 (+7.1 − 6.7) and 123 (+7.1 − 6.7) ka indicating that these beds are also 5e in age, although situated topographically lower than the marine terrace with *Anadara*. Below these *Cerastoderma* gravel beds, which are comparable to those at Zeit Bay and Ras el Bahar, are subtidal sands with marine shells. It would seem, therefore, that the complete 5e sequence is tectonically lowered to the west with respect to the coastal terrace which may be stable or slightly uplifted.

The linear disposition of the uplifted Holocene beds parallel to pre-existing Clysmic rotational faults, suggest a renewed structural tilting of this area rather than a doming. Rejuvenation of the Miocene faults predates the Holocene upper beach (5.4 ka BP coral by U/Th) situated about 25 m landwards and about 1 m above the present beach berm.

At Gemsa Peninsula (Figure H2.8) a similar structural tilting has been interpreted by Ibrahim *et al.* (1986) to be the result of Holocene, halokinetic doming. The age of this deformation was based on a supposed 'Flandrian' age of the coral facies which the present authors have shown to be markedly older (Pleistocene 5e) and our data do not support the proposed age and tectonic process for this uplifted reefal facies (Plaziat *et al.*, 1995).

Deposits older than 5e have not been examined in detail, either because they are essentially continental alluvial fans (south of Zeit, north of Ras Shukheir), or because they consist of a thin reefal veneer underlying 5e reefs (south Zeit, stage 7). There are virtually no older reef-terraces preserved along the coast of Gebel el Zeit, suggesting that multiple rejuvenation of faults occurred repeatedly during the Quaternary. This contrasts with a relative stability since the beginning of the Holocene.

B. The area of Quseir el Qadim (Figure H2.1, H2.17)

This site is the most extensive known exposure of Quaternary evaporites of the Egyptian coast. It clearly demonstrates the relationships between Quaternary salina evaporites and reef-terraces. This 1200 m long depression is elongated parallel to the coast and measures 400–700 m in width. It has been connected to the sea via a 150 m wide, 500 m long channel. The depression has developed at the end of a modern talweg cut into Miocene evaporites and Plio-Quaternary conglomerates. It is partially lined with a thin cover of modern sandy alluvium which forms part of a more extensive sabkha (+1 m) which is flooded during rains.

The coastline is rocky, being formed by Pleistocene reefal limestone terrace situated between +7 and +8 m. It is covered by a gravel beach-deposit rising to +9.5 m. The reef, dated as 5e (131 + 6.6 − 6.3 ka) with rare corals dated as 121.1 (+6.2 − 5.9) and 131 (+6.6 − 6.3) ka overlies older Pleistocene calcitized coral limestones notably including echinoid spines that gave doubtful ages greater than 300 ka. Within the depression, the fossiliferous cap of outliers made of older reefal limestone, together with scattered remains below the gypsum terrace, clearly show that the older reef and conglomerate substratum is overlain by sands and gravels containing open-marine molluscs such as *Cucullaea* (Figure H2.11(d)), *Glycymeris*, *Lambis*, *Conus*, *Fasciolaria*, etc. (Figure H2.25(f)). This marine bed is locally overlain by sands containing large *Anadara* (Figure H2.11(c)) and *Dosinia* (Figure H2.20(f)). These are covered all over the basin, just below the salina gypsum, by a horizon with *Potamides conicus* and a dwarf mytilid (*Brachidontes variabilis*) plus rare *Melanoides tuberculata* (Figure H2.25(g)(h)). The faunal succession thus reflects a progressive restriction which has culminated with the precipitation of laminated, subaquatic gypsum (Figures H2.17, H2. 25(b)). Near the northern and southern extremities of the depression, as well as along the landward edge, the lower marine gravelly sands, deposited around the periphery of the bay, contain *Nerita* and *Terebralia* indicating that a mangrove fringed the bay during the period of maximum flooding. These open-marine facies fill irregularities on the basal erosion surface, being situated between +1 and +6 m; their upper limits attain +6.5 and +6.6 m of positive features and only +1.5 m within adjacent depressions. Furthermore, the gypsum also drapes the palaeotopography, thinning over highs and thickening in lows. Its top height varies from +2 to +6.8 m. The bedded gypsum is overlain locally by cracked unbedded gypsum considered to be of sabkha origin, culminating at +7.6 m. Gypsum crusts of pedological origin, associated with alluvial sediments, occur up to heights of +8.8 m. The subaquatic bedded gypsum is clearly of marine origin as indicated by isotopic properties (δO^{18} and δS^{34}) (Orszag-Sperber *et al.*, this volume) and by $^{234}U/^{238}U$ ratios that are close to sea water and to Quaternary aragonitic corals. However the pedogenic crusts suggest a continental contribution of dissolved Miocene sulphate.

552 Quaternary marine and continental sedimentation

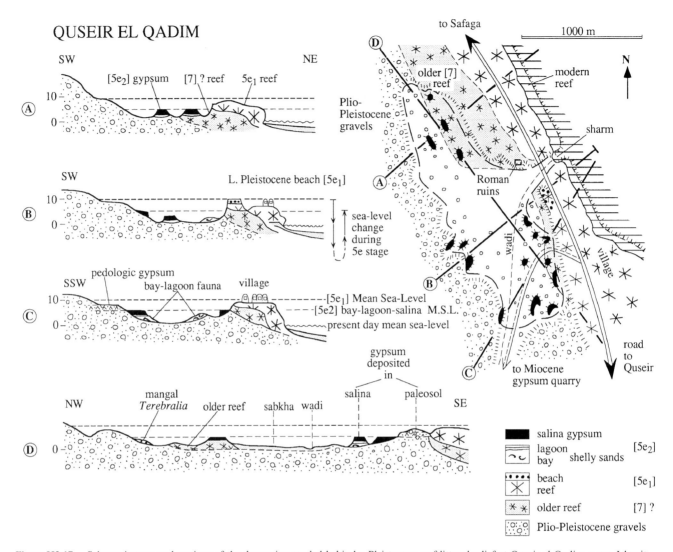

Figure H2.17 Schematic map and sections of the depression eroded behind a Pleistocene reef littoral relief at Quseir el Qadim, a pre-Islamic settlement. A village-hotel today covers the last interglacial reef terrace. The post-reef depression is lined with marine sands (with open-marine fossils including corals and mangal molluscs) and subsequently filled with subaquatic gypsum up to about 3 m below the previous reef MSL. The local, within 5e erosion of the Plio-Pleistocene substratum reached the same depth to that of the Last Glaciation lowest-stand.

In sum, landwards of the +9/10 m reef terrace of 5e age the MSL is estimated to have been slightly uplifted by 1 to 4 m. A series of younger marine beds have been deposited within a depression eroded in the softer, landwards parts of the reefal substratum (composed of older weathered reefs and loose gravels) during a brief drop (at least by 8–10 m) of the stage 5e highstand (Figure H2.18). The erosion is interpreted as resulting from a eustatic lowering because this same sedimentary sequence is observed all along the 300 km of the north-west Red Sea coast. The enclosed marine and subsequent lagoonal facies ($5e_2$) therefore have been deposited during a renewed sea-level rise of about +7 m which flooded the depression. This 'khor', i.e. a desert estuary or lagoon, had a depth of about 10 m and was flooded by the infragypsum transgression whose euhaline waters were systematically drained and refilled by tides. However, the entrance to the lagoon has subsequently been narrowed and closed probably by the deposition of sands and gravels supplied by longshore transport, thus progressively restricting the lagoon and finally, isolating a gypsum salina. The ages of the infragypsum marine molluscs and corals (see above) as well as that of the gypsum (125.9 + 8.8 − 8.2 ka and 128.5 + 6.4 − 6.1 ka) are very close to those of the pre-erosional outer reef. Thus, it is clear that the entire sequence, including the gypsum, is part of 5e isotopic substage (= Eemian) rather than one of the 5c or 5a, 100 and 80 ka BP substages. The sea-level fluctuations recorded by the Quaternary deposits at Quseir el Qadim are thus interpreted as the result of a very high frequency oscillation whose precise age is impossible to determine, being within the margin of error of the dating method. Other gypsum salinas which occur along the

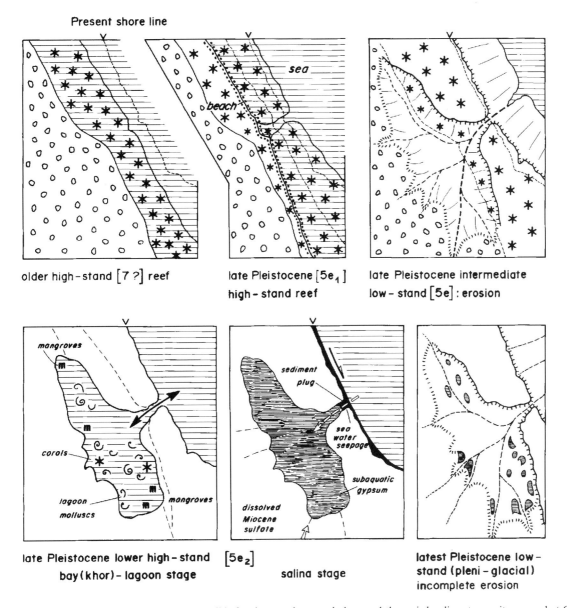

Figure H2.18 Cartoon of the successive stages responsible for the complex morphology and the varied sedimentary units exposed at Quseir el Qadim. Reconstruction based on dates and levelled transects in Figure H2.17.

southern coast corroborate this glacio-eustatic interpretation.

C. The Sharm el Bahari area (Figure H2.19)

Thirty five kilometres south of Quseir, this site exhibits a sequence slightly different from that discussed above. Typical of a region little affected by structural uplift, this area is associated with the development of a Pleistocene 'sharm' which preceded the modern Sharm el Bahari. A $5e_1$ reef dated as 130.8 (+4.9 − 4.7) and 121.1 (+7.2 − 6.8) ka culminates at +7 m, but is covered by beach gravels and shell beds to a height of +8.9 m. Rare *Terebralia* and oysters exhibiting moulds of mangrove roots indicate that a mangal probably benefited from the input of fresh water as indicated by the presence of the melanid (*Thiara voumanica* Bgt) (Figure H2.20(b)). This gastropod is known only from non-marine Pleistocene deposits of Somalia (Nardini, 1933) and lives today in various localities between Syria and Zanzibar. The present valley does not include Quaternary gypsum beds but, along its southern flank, an older reef culminating at +8.75 m is cut by a fluviatile erosional trench partially filled by a lower reef-terrace composed of corals 126.2 (+7.2 −6.8 ka), with *Tridacna* and echinoids (128.1 + 6.5 −6.3 ka). These are overlain by sands rich in open-marine molluscs, grading upwards to gravels with *Terebralia* (Figure H2.20(d)) and mangrove oysters plus *Isognomon ephippium* (Figure H2.20(c))

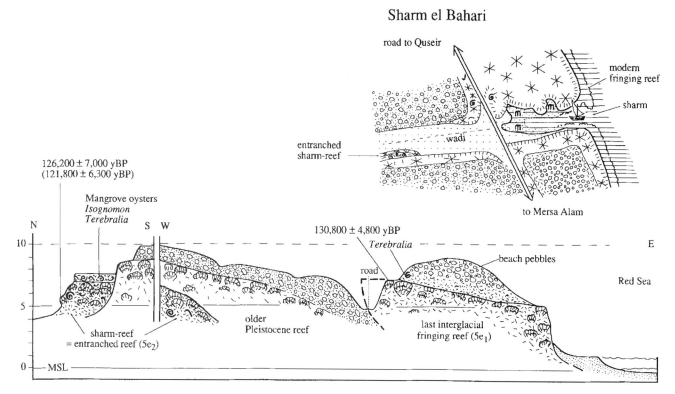

Figure H2.19 Schematic map and sections of the Pleistocene deposits outcropping in the wadi valley of Sharm el Bahari. m: Present-day mangal composed of small *Avicennia* mangrove trees associated with *Potamides conicus*. The east–west section shows relations between the three reefal units and terrigenous terraces. The entrenched location of the lower last-interglacial reef and beach ($5e_2$) better seen in north–south section is interpreted as a paleo-sharm.

which characterize an entrenched mangal community situated about 1 km landwards of the modern mangroves. These reefal and mangal deposits define a palaeosharm similar to that existing today. The modern sharm is one of the longest examined on the Egyptian Red Sea coast. It appears to coincide with an ephemeral wadi which was observed following one of the rare rains on the distant mountains (26 November 1989) to erode and transport blocks of asphalt from the coastal highway. A comparison between the molluscan faunas of the Pleistocene and the present-day mangals indicates that the Egyptian mangal community was more plentiful and diverse during the last interglacial period, while being fringed by a coral reef similar to that of today. This is the only Pleistocene sharm observed by the authors.

D. The Wadi Qalawa area (Figures H2.20(a), H2.21, H2. 25(e))

The peculiarity of this site concerns the presence of a lacustrine limestone situated between the second ($5e_2$) marine flooding and gypsum beds. The ($5e_1$) reef and its beach cobbles terrace attain $+7$ m while the adjacent subaquatic gypsum does not exceed $+5.6$ m. The latter extended several hundred metres behind the reef terrace and rises progressively to about $+10$ m. This slope may be the result of structural tilting. Along the south side of the valley the bedded evaporites attain $+8$ m. Their sandy basal facies with *Potamides conicus* and *Melanoides* overlies an intermediate horizon locally rich in large smooth-shelled oysters and *Anadara* which caps an older, pre-5e reef culminating at $+8.5$ m. However, within the axis of the palaeovalley, a muddy sand with small and large *Dosinia* represents the second $5e_2$ marine (lagoonal) transgression at $+5$ m (Figure H2. 21). A probably brief emergence is reflected by ferruginized root traces and by secondary, twinned gypsum crystals. The overlying carbonate mud includes a basal, thin (7 cm) discontinuous layer with *Potamides*, *Terebralia* and *Melanoides* followed by a bedded white carbonate mud with *Lymnaea* (Figure H2.25(e)), *Hydrobia*, and charophytes, associated with scattered *Potamides*, *Melanoides* (Figure H2.20(a)) and stromatolitic crusts with an isolated crab claw and a *Pupa* shell. This is a typical lacustrine deposit indicating variable salinity ranging from mesosaline to nearly fresh-water. The overlying gypsum unit rises to $+10$ m. It is relatively pure but poorly stratified, showing numerous root-traces. It is probably not the equivalent to the salina gypsum facies situated lower and closer to the sea, but rather is a

Figure H2.20. (a) Changing environment in two superimposed carbonate layers: a lower saline lake episode with Potamids (p) is succeeded by a freshwater (to oligohaline) layer, with *Melanoides tuberculata* (m) and a *Lymnaea* (l). Wadi Qalawa. (b)–(e) Mangal molluscs. Wadi Sharm el Bahari. (b) *Thiara voumanica*. (c) *Isognomon ephippium*. (d) *Terebralia palustris*. (e) *Gafrarium pectinata*. (f) *Dosinia erythracea*. (g) *Cerastoderma glaucum*, S Zeit, W Zeituna. (h), (i) Holocene reef terrace, completely emergent at low tide. The coral heads are abraded by several tens of centimetres and covered with a gravelly beach-rock (r). N Ras Shagra.

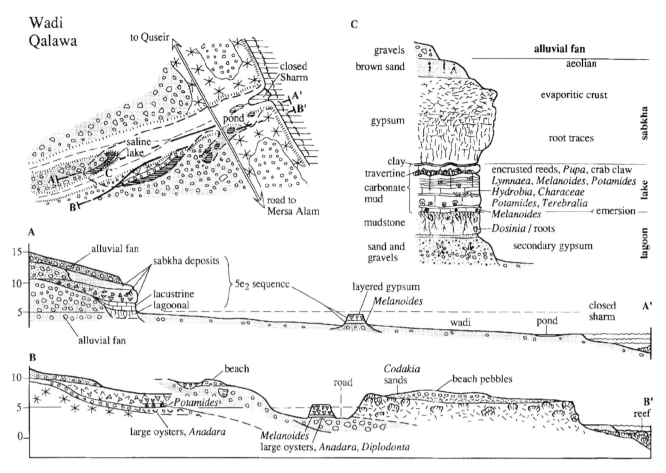

Figure H2.21 Schematic map and sections of the lower Qalawa valley. The present reef shows an intermediate type between a continuous fringing reef and a sharm: the deep axial valley groove is not filled by coral construction but its access to boats is blocked by a shallow reef barrier. The two Pleistocene reefs are associated with an intercalated alluvial fan, excavated by the wadi prior to the Last Interglacial marine-to-salina (5e$_2$) ingression. Gypsum occurs over the lacustrine unit in the internal sabkha depression (see sections A and B) at a higher altitude than the salina gypsum visible in more seaward sequences. The axial lacustrine carbonates underlying this internal gypsum (A and C) therefore probably correlate with salina isolation in this unique setting.

continental product situated 2 m higher than the salina water-table, where, together with the associated lacustrine facies, it has resulted from dissolution and reprecipitation of Miocene carbonate and gypsum outcropping inland, during a more humid period which followed the 5e$_2$ transgression. These lacustrine and pedogenic gypsum deposits may record the beginning of the stage 5e to stage 4 regression.

However, it is also conceivable that the underlying lagoonal (*Terebralia* bed) and lacustrine (*Lymnaea*, *Melanoides*, *Potamides* beds) deposits record a major fresh-water supply into isolated basin synchronous with the salinas.

Our interpretation for the latter origin of this sulphate is supported by the fact that the gypsum crust developed in an adjacent continental (alluvial fan) gravel extending landwards to altitudes of +10.75 m. This type of soil-crust has also been observed at a number of sites, e.g. 2 m higher than the bedded gypsum at Quseir el Qadim, while similar massive gypsum with root-traces has been noted north of Ras Shagra (Figure H2.22) at altitudes varying from one depression to another over distances of a few hundred metres. Therefore, it is logical to suspect the continental origin of these perched sulphate facies, especially in view of the extensive Miocene gypsum deposits outcropping at short distances (< 1 km) inland.

E. The Wadi Nahari area (Figure H2.23)

This locality has been described as a compound fan system (Freytet *et al.*, 1993; Baltzer *et al.*, 1994). The earliest boulder conglomerates have a dip which markedly exceeds that of the alluvial terraces, strongly suggesting tectonic tilting. Alluvial-fan gravels are deposited on an erosion surface which cuts Pliocene beds deformed by halokinetic movement (Figure H2.22). A limited rejuvenation (several metres doming) affects the basal discordance. This doming has resulted in the removal of the apex of the alluvial cone. Below the alluvial fan, perched about 10 m above the modern

Chronology of Quaternary deposits and problems of dating 557

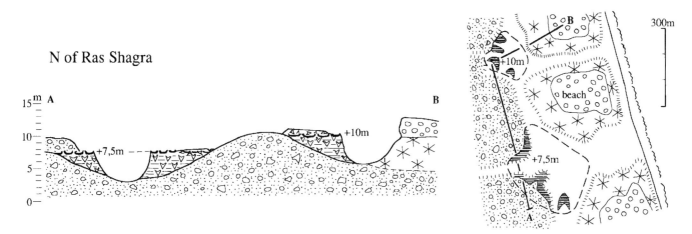

Figure H2.22 Schematic map and section of gypsum salinas N Ras Shagra. In spite of a back reef relief similar to the khor–lagoon–salina basins, the two subaquatic gypsum lenses are not interpreted as being marine evaporites but as sabkha deposits filling independent continental depressions fed by freshwater enriched in dissolved Miocene sulphate. They have different altitudes and no fossiliferous marine/lagoonal precursory bottom-layers.

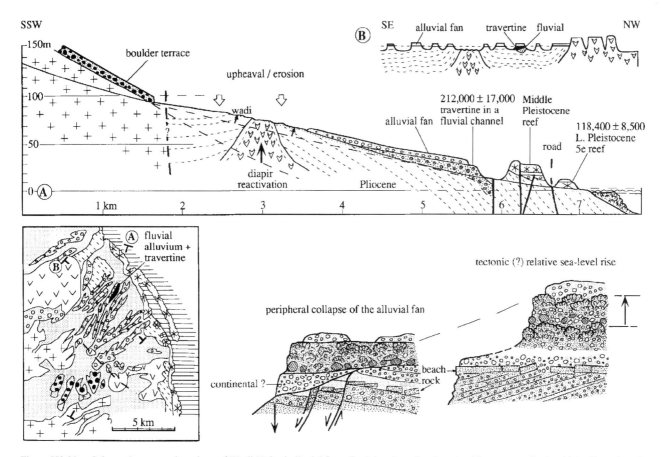

Figure H2.23 Schematic map and sections of Wadi Nahari alluvial fans, fluvial and marine deposits. The gypsum diapir which affects the Pliocene tilted deposits has been recently reactivated as the wadi longitudinal profile is not regularly concave. The local fluvial channel with associated travertine fill referred to stage 7, is a very small sediment body (map after Freytet *et al.*, 1993). It is covered by an extensive, fan-shaped alluvial sheet probably more recent than the perched, tectonized Middle Pleistocene reef. The last interglacial reef, between road and sea, however, is not uplifted. This locality expresses fully the complexity of time-correlation between contrasted sedimentary units, in a long-living tectonic setting.

wadi, a local lithified alluvium is cut by sharp cliffs. These gravels and conglomerates have been transported by a river within a narrow valley that deposited relatively well-sorted material with frequent oblique bedding. They are laterally associated with a mud and carbonate unit containing travertine-encrusted reeds. The calcareous facies thickens laterally (5 m) and contains *Melanoides tuberculata* at the base. These shells have given a U/Th age of 212 (+20 −15) ka. It is therefore possible that the linear incision and subsequent fluvial infilling, corresponding to humid episodes, represent the anaglacial and the cataglacial phases respectively of the isotope stage 8, or possibly even the preceding stage 10 lowstand. The fluvial formation plunges and is cut out by a fault. A reef-terrace, uplifted at +26 m, replaces it seaward. It could not be dated but can not be younger than stage 7. A sandy gravel with beachrock formation, below a reef unit, may represent the prolongation of the fluviatile facies but, in spite of a comparable lithification, and its modification by faults, there is no definitive proof of lateral continuity. The faults are sealed by a double coral terrace overlying beach deposits with imbricated clasts and heavy mineral concentration. This reef superposition is the only record of a relative sea-level rise by about 2 m and may result from tectonics or gravity collapse. Between the coastal road and the shore, the usual Late Pleistocene littoral sequence outcrops: below the beach horizon at +7.5 m, the reef (+6 m) has been dated at 118.4 (+9 − 8.3) ka BP based on a sample of echinoid spines. It thus corresponds to stage 5e. A gypsum deposit is localized within the depression situated behind the reef has been dated at 131.9 (+21.9 −18.6) ka.

A thin modern alluvial fan covers the Pliocene eroded plain near the margin of the Precambrian basement.

Other neighbouring sites (Wadi Um Gheig, Wadi Greifat) also exhibit an alternation of fluviatile beds deposited in linear depressions, and coarse fan deposits overlying planar erosion surfaces. However these have not been dated.

F. The Wadi Igla area (Figures H2.1, H2.24)

This medium-sized wadi terminates on the coast as a sharm. Continental and marine deposits cut by this wadi, have been studied intensively (Freytet *et al.*, 1993 Ahmed *et al.*, 1994; Baltzer *et al.*, 1994; Plaziat *et al.*, 1995). Between the basement relief and the coast, the 6 km long valley exposes four alluvial terraces. They have no relation with the overhanging unit composed of metre-sized boulders and located in the prolongation of a rocky sediment surface in the upstream area (piedmont). Both the sediment, a mass-flow with large blocks, and its substratum slope (4°) suggest structural tilting towards the east. The age of this first fan deposit is uncertain, it may be Pliocene or early Pleistocene. The other alluvial terraces have slopes of 1.4 −1.7°.

Four kilometres seaward, the highest +42 m Pleistocene reef R4 (Figure H2.24) caps a mixed marine unit lying on the Pliocene with a marked angular unconformity. This series is composed of carbonate slope deposits (R1–R3) interbedded in coarse unfossiliferous sands and gravels of fluviatile origin. Closer to the sea but west of the road, the flanks of the incised valley are also composed of prograding units of peri-reefal carbonates and coarse terrigenous sediment with oblique bedding. Changes in sea level are not easy to decipher but were probably limited whereas the overall lowering towards the east of both the reefs and clastics suggests a subsequent tilting in the order of several degrees including the capping R4 reef. Between the older marine series (R1–R3) and the uppermost reef (R4) a muddy palaeosol occurs with root-traces and nodules. At this level, a synsedimentary tectonic event is recorded by a fault sealed by the marine gravel beds. A seaward downthrow of 1.5 m is compensated by the thickened continental unit.

The first deposits of R4 unit are beach gravels and scattered boulders (50 cm) covered by an extensive reef-terrace whose fresh-looking corals are in fact highly recrystallized. This terrace is covered by alluvial fan gravels. Towards the sea, a series of later faults has lowered this R4 reef below wadi level. A coarse terrigenous unit buries this Early to Middle Pleistocene reef complex and underlies the younger littoral reef and beach outcrops levelled at around +10 m. A series of dates indicates that this lower terrace is in fact composite. The seaward parts of the terrace give ages of 117.6 (+4.4 − 4.2), 115.6 (+5.8 − 5.5) and 118 (+5.6 − 5.3) ka BP but other dates suggest that these external parts are piled against diagenetically altered reef carbonates whose aragonitic corals and echinoids range from 191 to 358 ka BP. Within a 10 m long depression, preserved on this old coastal terrace that is 1.5 m lower than the +9.5 m beach associated with the 5e reef, a gypsum-cemented sand lens occurs which is rich in dwarf *Terebralia* with associated *Nerita*, *Clithon*, *Brachidontes* and *Melanoides*. This brackish malacofauna (Figure H2.25 (g),(h)) indicates that a typical mangal developed in a progressively restricted lagoonal environmental at the end of the episode $5e_2$ as the *Terebralia* give an age of 114.5 (+7.9−6.8) and 125.5 (+12.1−11) ka BP.

The slopes of the three lowest fluvial terraces within the valley are not different from that of the modern wadi. Their altitude in the east link them to a palaeo-sea level corresponding to the lower (younger) reefs (+1/+10 m). This suggests structural stability since the beginning of the upper Pleistocene.

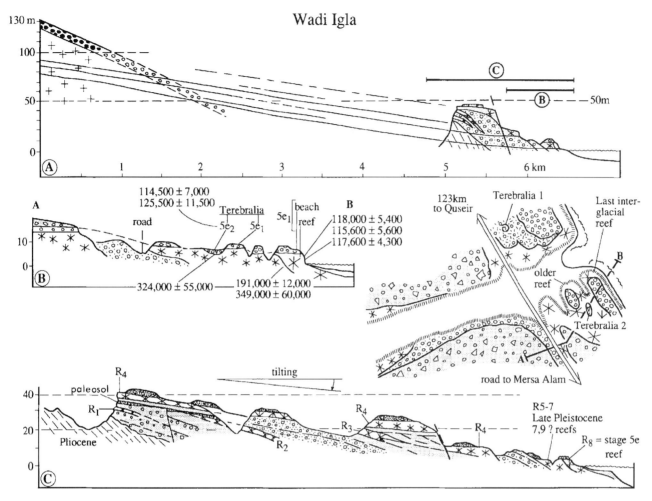

Figure H2.24 Longtitudinal section of the Pleistocene deposists of Wadi Igla and schematic map of the near-mouth area. This is one of the most complex sections, rich in uplifted carbonate reef-units (undated) associated with coarse terrigenous gravels that are repeatedly faulted (c). The lower, littoral reef complex (B) gives unusually old dates at the same +10 m altitude $5e_1$ reefs fringe stages 7 and 9 reefal facies. Together with the inland lagoon and mangrove (*Terebralia*) patch ($5e_2$), this suggests that tectonic uplift ceased before the end of Middle Pleistocene.

G. The Wadi Khalilat el Bahari area (Figures H2.1, H2.26, H2.27)

This location has much in common with that of Wadi Igla; both areas show the relationships between raised reef-terraces and terrigenous continental materials. However, a more simple fan-architecture suggests a different geomorphologic setting. Initial studies include those of Purser *et al.* (1987), Plaziat *et al.* (1989), and M'Rabet *et al.* (1991). The exceptional outcrop conditions, notably those of the older Pleistocene reefs, occur along a shallow wadi which cuts a 4 km long, 2–4 m high, continuous wall (Figure H2.26a). The two reef complexes are separated by a thick, pink alluvial-fan. The older reef complex, at least mid-Pleistocene in age, is situated between +30 and +42 m of altitude. It overlies an ancient alluvial fan composed of the same pink granitic materials. The sea has transgressed over this poorly sorted deposit in two distinct steps. An initial sandy deposit in the form of a pre-littoral ridge has a relief of about 4 m. It developed on a shallow marine terrace cut in the preceding cone. Strong wave action is reflected in well-sorted sands with concentrations of heavy minerals localized on either side of a coastal spit (Figure H2.26(b),(c)). An initial reef (R1), formed on the seawards flank of this sand ridge. On the opposite side, relatively fragile, branching and cup-shaped corals occupied a protected embayment. The reef and ridge have been truncated and lowered (by several metres) by a series of small faults prior to being covered by a gravel unit. This structural lowering of the peripheral parts of the cone may have facilitated the marked marine transgression, covering the ridge with discontinuous corals and coralline algae crusts (R2) which also developed behind the flooded ridge. During a period of subsequent sea-level stability, beach pebble deposits (now uplifted between +42 and +32 m) gradually filled the lagoonal depression, extending beyond the reef front.

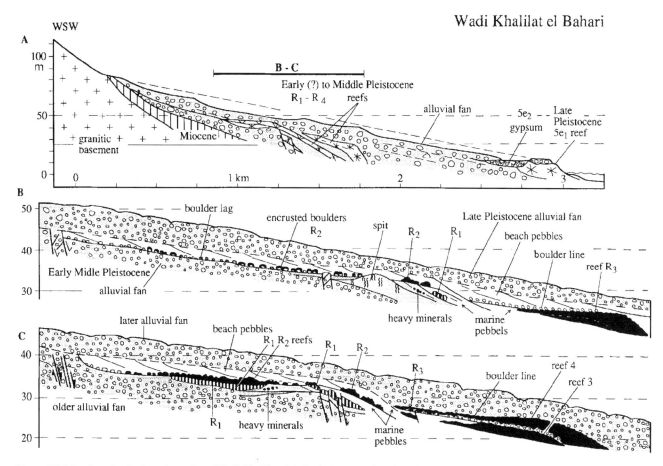

Figure H2.26 Longitudinal sections along Wadi Khalilat el Bahari. A: general section showing the separated reef distribution and alluvial-fans. B,C: parallel sections of the upper (Early to Middle Pleistocene) reef complex and alluvia, on both sides of the wadi trench. Left side (B) shows a more simple sequence, with a clear sand-spit relief and the only R3 reef being covered by a thick unit of beach pebbles (in place of R4). Right side (C) reveals a more complete reef sequence, outcropping 150 m to the south. Faulting is clear and the R4 reef develops above a boulder marker-line. Beach pebbles (in white) have been laterally replaced by R4 directed buried below the Middle Pleistocene upper alluvial fan. The rapid changes, southwards, suggest a transitional setting between an alluvial fan prominent coastline and an inter-alluvial fan embayment (Figure H2.27).

The new reef R3 which is situated seawards of R2, and about 5 m lower, corresponds to a minor lowering of relative sea level interpreted here as a minor eustatic drop. Finally, a rise of about 3 m resulted in a capping of R3 by an additional reel-flat (R4), whose levelled coral-head morphologies and coralline encrustation indicate that it developed up to sea level. In this respect it differs from the preceding R1–R2 reefs in which irregular, subtidal growth-morphologies are filled with gravels. These (R1–R2) reefs have been termed 'immature reefs' by Plaziat et al. (1989) and may illustrate either the instability of relative sea level or the sudden beach growth. This instability would reflect, in part, tectonic adjustments. However high frequency eustatic sea-level changes can not be excluded, especially between R3 and R4 constructions. Influence of the fan morphology appears to be indirect as the reef-terraces formed within the relative depression between two alluvial cones (Figure H2.27). Tectonics and eustatic events therefore had to compromise with the sedimentary evolution (spit development, bay filling) in reef-terrace construction.

Figure H2.25 (a) Late Pleistocene hills isolated by erosion on a marine-to-lagoon sand coating the Pleistocene gravel. These lagoon to salina deposits ($5e_2$) are excavated behind the higher Late Pleistocene ($5e_1$) reef and pebble beach, Wadi Abu Sbikhaia. (b) Marginal gypsum thinning on to the Early Pleistocene gravel substrate with an intermediate marine shell fauna. Quseir el Qadim. (c) Laminated gypsum, with spaced terrigenous mud intercalations, deposited in salina water and (d) massive gypsum with tubular root-traces, characteristic of a sabkha environment, Wadi Abu Sbikhaia. (e) Lacustrine limestone with a displaced *Potamides*, associated with *Lymnaea*. Wadi Qalawa. (f) Marine fauna of the bay-stage preceding the sulphate salina. *Glycymeris pectunculus, Anadara erythraeonensis, Gafrarium pectinatum, Plicatula plicata, Morula*, Quseir el Qadim. (g) Small *Terebralia palustris, Potamides conicus, Melanoides tuberculata, Brachiodontes variabilis* and a *Callianassa* claw, indicative of lagoon mangal, Wadi Igla. (h) The mytilid *Brachiodontes variabilis* associated with *Potamides conicus* and *Cerithium scabridum*. More restricted lagoon environment, Quseir el Qadim.

562   Quaternary marine and continental sedimentation

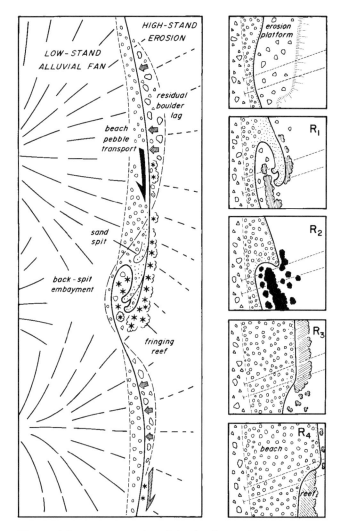

Figure H2.27 Reconstruction of the shoreline evolution at Wadi Khalilat el Bahari. Right: cartoon of the facies changes resulting from erosion, spit growth, tectonics and sea-level changes. The dashed lines show the respective location of right and left banks of the wadi. Left: a tentative generalization during R1 stage.

An important coarse alluvial fan deposit separates these undated older (R1–R4) reefs from the younger reef-terrace bordering the present shoreline. Vague bedding within this fan (Figure H2.26(a)) indicates a rather steep progradation slope, probably influenced by about 100 m glacio-eustatic lowering of sea level during isotopic stage 6. Within this coarse detrital fan, with 5 m blocks of Precambrian basement, certain finer beds are cemented with carbonate preserving root-traces. This indicates a somewhat humid climate (Freytet *et al.*, 1994) and is attributed to the anaglacial phase prior to the penultimate glaciation. Within a notch cut into this cone, a 5e reef, dated as 119.9 (+5.8 − 5.5) and 124.2 (+4.9 − 4.7) ka BP exhibits a reef-flat which, in common with (R1–R4) older reefs, is gently inclined (+6 to +4 m) towards the sea. It is overlain by a relatively thick gravel beach deposit culminating at +8.75 m,

dated by scattered coral heads at 115.7 (+6.4 − 6.1) ka BP.

Behind the Pleistocene beach, a shallow depression with gypsiferous sands indicates the existence of a sabkha perched at +7 m. These sands are possibly an equivalent of the gypsiferous sand overlying a bed with *Potamides conicus* and *Nerita* noted at +4.5 m some 2 km further north, the latter being typical of the lagoon-salina infill of $5e_2$ age.

In the splash range of waves from the present beach a narrow terrace is commonly situated at +2.1–1.3 m which cuts the 5e reef. This terrace is attributed to a differential erosion of the 5e reef material infilled by aragonite cements in the Holocene and present-day splash-zone. The age of the cemented corals reflects this subsequent lithification, with a stage 5e age lowered to 93.9 (+4.5 − 4.2) ka, in common with other 5e localities situated on exposed promontories (Sharm el Naga, Ras Shagra, Figure H2.6).

The gypsum deposits are generally overlain by coarse alluvial fan gravels as in other palaeosalina localities. This renewed cone sedimentation has been attributed to the more humid anaglacial period of the last glaciation. The volume of this discharge appears to be relatively moderate as it only locally buries the 5e terrace and does not reach the present shoreline. However, this assumes little subsequent erosion of these deposits situated about 130 m above during the pre-Holocene lowstand.

As at Wadi Igla, Wadi Khalilat el Bahari shows two principal reef complexes recording Pleistocene sea-level highstands. The older is uplifted, faulted and inclined towards the sea and probably corresponds to the mid- and possibly early-Pleistocene and may represent several interglacial periods (stages 9, 11, 13, etc.) during which respective highstands have been considered as lower than the present-day sea level. The second complex is documented only by the 5e reef, but it is suspected that the stage 7 highstand reef has not locally developed when compared with Wadi Igla and other localities where the 7 and 5e reefs occur at the same altitude.

The stability of the region after the mid-Pleistocene is confirmed by the Holocene reef altitude at Wadi Gemal: here, an abrasional surface cuts the coral platform dated at 5.8 ± 0.25 ka BP (cf. Figure H2.20(h),(i)). This high, derived MSL (+1 m) has been dated from six localities between Safaga and Ras Banas, giving ages ranging from 6410 to 5390 years BP. The notches visible on many rocky coasts do not reflect different Holocene relative sea levels.

H. The Wadi Abu Shikhaia area
(north of Wadi Samadai) (Figure H2.28)

This site is unusual, not only because it has no dated 5e reef-terrace, but also because of a series of much older low-lying reefs, overlain by the subaquatic gypsum precisely dated from the last interglacial (5e) stage.

The highest reef-terrace (+8.5 m) is covered by a beach gravel deposit up to +11.5 m. This complex is interpreted as representing the 5e stage, mainly on the basis of identical morphology and lithology with the nearby dated 5e sequences. However, lower samples from this terrace, including coral, *Tridacna* and echinoid spines, have given ages varying between 259 and 435 ka. As there are no topographically higher reefs at this site, it is therefore probable that early- and mid-Pleistocene reefs have not been uplifted in this area and remain in a position similar to that of the stage 7 reef at Wadi Igla. Within the depression, which is elongated parallel to the shoreline, situated immediately landwards of the 5e reef-terrace, the layered gypsum formation formed under salina conditions attains a thickness of 4 m. It culminates with teepee structures at a height of +7.8 m. Thus, there appears to be a difference in level of about 3.7 m between the respective MSL of $5e_1$ beach terrace and $5e_2$ salina. This is comparable to the 2.7 m difference at Quseir el Qadim, 150 km to the north, and in other intermediate outcrops. The recent collapse of part of the gypsum cliff exposed an exceptional unweathered section (Figure H2.28) from which five dates have been established: 113.4 (+6.4 − 6.1), 135.3 (+15 − 13), 130.5 (+10 − 9), 131 (+9 − 8.5) and 120.5 (+16 − 14) ka BP. Isotope ratios $^{234}U/^{238}U$ indicate a marine source for the uranium of these sulphates. At another southern locality, Sharm el Luli, this gypsum formation has also been dated as 126.1 + 7.6 − 7.1 ka BP and the adjacent +9 m peri-reefal coquina as 131.3 + 10.1 − 9.3 ka BP.

The Wadi Abu Sbikhaia locality indicates that certain parts of the coastline have contrasting early Quaternary histories but nevertheless have a similar Late Pleistocene–Holocene evolution. This reinforces the general impression of relative structural stability since the mid-Pleistocene. The slightly elevated (< +1 m MSL) Holocene reef, at this locality, is dated as 6100 + 270 years BP.

REGIONAL SYNTHESIS AND DISCUSSION

A. Climatic evolution reflected in the sedimentary sequences

It is possible to associate several sedimentary units with different climatic and sea-level conditions.

1. The Plio-Quaternary coarse fanglomerate beds

At a number of localities (north and south of Hurghada, Wadi Nahari for instance) continental boulder beds overline marine Pliocene formations affected by haloki-

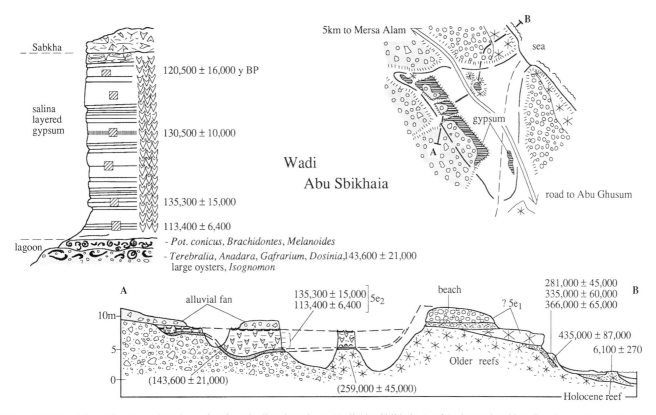

Figure H2.28 Schematic map and sections of reefs and salina deposits at Wadi Abu Sbikhaia. Reef $5e_1$ is not dated but directly overlies much older reef carbonate deposits (undated stage 7). A 3 m thick salina gypsum sequence has been precisely dated. The $5e_2$ lagoonal ingression is expressed by a basal mangal malacofauna, replaced upwards by the restricted lagoon molluscs.

netic and other rift tectonics. The age of these boulder facies is not yet known. They occur near the foot of the basement escarpment, notably at the opening of oblique (east–west) fault-related valleys. These fanglomerates, with a muddy matrix containing metre-sized blocks, indicate a climate having marked seasonal contrasts with periods of violent flooding. While it is possible that these coarse discharges are of early Quaternary age they could also be of late Pliocene age (cf. the Villafranchian conglomerates of the margins of the Mediterranean), reflecting the more humid start to the anaglacial phase of one of the first glaciations.

2. Old alluvial cone deposits

These form important local accumulations markedly less coarse than the preceding facies and lack visible relationships with marine deposits. By analogy with better constrained data from recent alluvial fans they are interpreted as the products of early Pleistocene, glacio-eustatic sea-level falls. For example, within the upper parts of Wadi Khalilat el Bahari, the superposed cones. (separated seawards by the older Pleistocene reefs) have been distinguished only landwards by a discontinuity within the progressive upward decrease in the grain-size of the coarser constituents. No other alluvial unit documents the glacio-eustatic cycles of the Early Pleistocene revealed by marine isotopic analyses (Figure H2.29). No reef units represent the numerous highstands supposed to be within 20 m below present MSL belt that now outcrops in most of the uplifted sites. This is a general problem to be discussed later.

3. The mid-Pleistocene alluvial cones and reef terraces

Coral beds give ages of 180–300 ka, suggesting that they represent highstand stages 7 and 9 although one should be careful interpreting these older dates. The age of the terrigenous sediments has not been established precisely.

The isotopic stage 7-reef often constitutes the eroded substratum for the 5e reef, their similar altitude complicating the separation of these two constructed units. It is therefore logical to refer the older uplifted (+30/45 m) reefs south of Quseir to some of the five (isotopically proven) highest stands of the Middle Pleistocene. We would suggest that the biological diversity of these older reefs appears to be lower than that of 5e (and subsequent Holocene reefs) although specific differences have not been established. Within this elevated reef complex, reef-terraces separated by 2–5 m change of relative sea level have been noted at several localities; these seem to correspond to very high-frequency sea-level fluctuations, rather than express discontinued highstands separated by lower, glacial phases.

Fluvial and associated travertine deposits indicate periods of permanent water-flow. These are chronologically comparable to the mid-Pleistocene Saharan lakes (Gasse *et al.*, 1987; Petit-Maire *et al.*, 1989). It is not

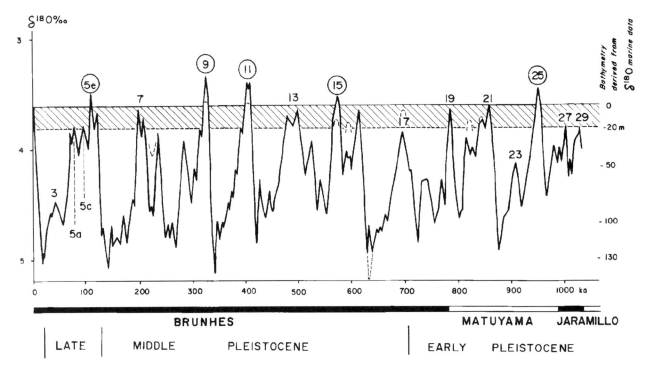

Figure H2.29 Eustatic sea-level fluctuations deduced by Soepri Hantoro (1992) from oxygen isotope data of Shackleton *et al.* (1990). Above 0 m line the isotope stages are supposed to have been originally above the present MSL (encircled) and others, between 0 and −20 m, are tectonically uplifted in the Pleistocene of Egypt). This diagram shows that at least 9 high-stands are liable to generate emergent reefs during the last million years (more than 15 for the whole Quaternary).

certain that all river deposits belong to the stages 8 to 7 cataglacial phase, but the 212 000 BP date of the *Melanoides* layer at Wadi Nahari is in full agreement with this interpretation. A pause in alluvial dynamics during the highstand is related to a phase of marked aridity. The fanglomerate cone deposition rather seems to have been formed during the subsequent sea-level fall under seasonally contrasted conditions intermediate between sub-arid and humid. This climatic improvement in the desert marks the beginning of an anaglacial (cooling) phase prior to the more local linear incisions which are logically related to the more rainy period preceding a glacial maximum (Freytet et al., 1993).

4. The marine deposits of the Last Interglacial period

Our current research has been concentrated on these sediments partly because of their considerable complexity and also because these marine facies are susceptible to reliable U/Th dating.

The very minor uplift which appears to have affected the eastern Egyptian coast over the last 200 000 years enables precise study of the marine deposits of the highest (5e) level, but not of the subsequent (5c, 5a, 3), less elevated highstands (Figure H2.29), even in the uplifted sites of Gebel el Zeit.

Establishing the amplitude of the possible tectonic uplift is a complex problem.

1. The consistent definition of a derived mean sea level is already difficult (Figure H2.6). It requires excellent outcrop conditions enabling not only identification of the top of a reef construction, but also the upper beach deposits or at least the reef-flat as exposed at low tide. Erosion of unlithified beach and reef-flat sediment, and subsequent sedimentary cover by slope screes may modify these surfaces by 1–3 m.
2. It is uncertain whether the usual definition of MSL of the highest stand of stage 5e at +6 m is justified. Our data suggest that, in the Red Sea, it may rather have been situated close to +8 m. Whatever the definition, this emergent shoreline has remained at a relatively constant altitude along more than 600 km of the north-western Red Sea coast. Although this sea level has been determined from deposits associated with fringing reefs, the authors stress that bioconstructed carbonates are not necessarily precise sea-level indicators. Reef framework top may descend to depths of several metres in incipient constructions (immature reefs). Only the encrusted and sediment-levelled subhorizontal reef-flat can be used to characterize mean sea level in this region.

Stage 5e deposits can be subdivided into two episodes whose ages are so close that they can not be separated effectively by radiometric methods (Figures H2.30, H2.31).

Initially, sea level possibly reached an altitude of +8 m and fringing reefs developed along most of the coastline, as is the situation today. The associated pebble beach was initially supplied by erosion of older alluvial fan deposits and by a limited input from the fluvial system. This is notably obvious south of the mouths of major wadis because oblique wave transport towards the south spread these pebbles along a few kilometres of the coast into beach terraces attaining +10 m in altitude. It is the lower part of the thicker terraces that records the ancient mean sea level, their top being heaped up during storm events.

Subsequently, a brief lowering of sea level is expressed by the erosion of depressions at least 10 m deep, situated immediately landwards of the cemented reef-terrace (Figure H2.31). The linear eastern limit of these depressions, observed at numerous localities, can not be explained in terms of Clysmic faulting. The depressions are related to a system of subparallel valleys oriented normal to the shoreline which have cut across the reef-terrace. The second-order branches, developed parallel to the coastline, have eroded the softer and weakly indurated materials located immediately behind the rectilinear reef-terrace (e.g. Quseir el Qadim). It is not possible to establish the amplitude of this sea-level drop, but it is noted that the last glacial lowstand which exceeded −120 m, did not deepen these valleys significantly. There are no other known sedimentary expressions of this erosion, the resulting alluvial fans being necessarily situated below modern sea level. These major coastal depressions are filled with basal marine facies such as gravels with molluscs typical of protected embayed environments, including *Anadara*, *Dosinia*, *Gafrarium*, large smooth oysters, associated locally with mangal molluscs such as *Terebralia*, *Nerita*, *Isognomon*, and mangrove oysters. At other localities, characterized by a more active water circulation, an open-marine molluscan fauna is associated with scattered corals (Quseir el Qadim) or even reefal trottoirs (Sharm el Bahari). These more or less enclosed embayments (sharms or khors) communicated with the open sea via a narrow channel, constricted by the 5e reef-terrace outcropping at that time about 3 m above sea level. Without any apparent change in sea level, faunas indicate a progressive, important restriction affecting all the bays, limiting faunas to *Potamides conicus*, *Cerithium*, a small *Dosinia*, a *Hydrobia* and the dwarf *Brachidontes variabilis*. A further increase in salinity led to the precipitation of laminated gypsum. This salinity, obviously exceeding 115 g/litre, nevertheless fluctuated, permitting the occasional presence of *Potamides* (which can not withstand salinities in excess of 100 g/litre (cf. Plaziat, 1993). The morphological factors determining a marine source and high salinities are debatable. The lagoonal basins progressively closed by a permeable barrier (Plaziat et al., 1995) may

566 Quaternary marine and continental sedimentation

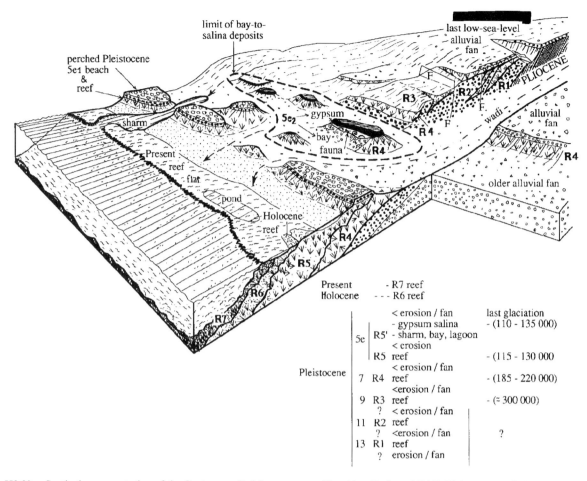

Figure H2.30 Synthetic representation of the Quaternary Red Sea sequence. The older, Early to Middle Pleistocene reefs (R1–R4) are highly schematic (see Figure H2.24). The present geometric relationships are the result of both tectonics and eustatic sea-level changes. The Late Pleistocene 5e deposits must be split in two different stages ($5e_1$ reef and beach; $5e_2$ bay to gypsum–salina ingression). The emergent Holocene reef (R6) is referred to an Holocene optimum. R7 is the modern living reef.

result from an increase in longshore detrital supply as a greater humidity would have increased seasonal flooding and pebble delivery (monsoon enhancement). It is likely that the beginning of anaglacial cooling would result in increased southerly atmospheric circulation and rainfall, notably at latitudes close to the tropics. Heavier rains would have increased dissolution of Miocene gypsum which commonly forms the nearby hinterland areas. Thus, it is possible that the salinas have been supplied both by marine and by saline continental waters, the former seeping through a permeable coastal barrier ridge. Unpublished geochemical data do not contradict this interpretation and the presence of gypsum soil crusts situated somewhat higher than the salina (subaqueous) evaporites, suggest underflow of continental waters enriched in dissolved sulphate.

5. Emergence during the Last Glaciation and the preceding cooling episodes

No traces have been found of the 5a, 5c highstands because of insufficient uplift. One is thus obliged to group the entire post 5e Pleistocene events into a single episode, admitting that important climatic fluctuations during the last 100 000 years must have occurred.

Renewed sheet-transport and reactivation of alluvial fans as well as erosion of valleys bordered by alluvial terraces located at altitudes lower than those of pre-5e, record a more recent, relatively elevated rainfall. Nevertheless, one should not consider only the obvious effects of erosion and transport; arid or hyper-arid brief episodes probably existed, but are only rarely expressed. The carbonate crusts (calcretes) which developed locally on the 5e reefs (Freytet *et al.*, 1994), indicate a relative and sporadic aridity.

The volume of post-5e detritus deposited within the region studied is very moderate and valley erosion did not exceed several metres. Therefore, it is clear that the modern 'sharms' reflect permanent conditions unfavourable to coral growth, both during periods of highstand and lowstand, rather than being the result of major erosion during a single (final) glacial lowstand. The permanent nature of the exits of many medium and

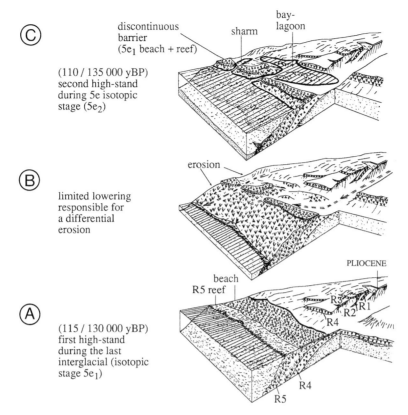

Figure H2.31 Three-stage reconstruction of the Late Pleistocene littoral evolution relating to high frequency limited sea-level changes. Based on Figure H2.30.

large wadi systems is a corollary; the deepening of valleys implies export of detritus beyond the shoreline, forming alluvial or marine fans which today are situated below sea level.

6. The Holocene optimum

There are two types of evidence for a climatic optimum during a higher Holocene highstand. Firstly, reef facies are found everywhere above present sea level. Secondly, traces of human occupation reflect in part the presence of a flourishing mangal extending as far north as the southern Gulf of Suez. The common factors shared by these optima (reefs, beaches, middens) is their age, which ranges from 6400 to 5400 years BP (13 out of 17 ^{14}C dates).

Slightly emergent reefs are cut by an abrasion surface which suggests a derived MSL situated about 0.5–1 m above the present reef-flat (Figures H2.20 (h),(i)). The only exceptions are at Gebel el Zeit, where a beach situated at +2 m suggests recent uplift in agreement with the major post 5e movements, and at the very exposed site north of Ras Abu Bakr, where the sandy beach deposit with few coral fragments situated at +4 m, has an age of 5800 BP.

At the southernmost point of the studied area (north of Wadi Lahami) an epireefal sand yielded an exceptionally rich molluscan fauna with few corals dated at 4820 + 220 BP. The favourable climate suggested by this biodiversity is also reflected in the former northwards extension of mangals within the Gulf of Suez. Their malacofauna suggests conditions more favourable than those of mangrove assemblages living in the modern Red Sea but similar to those of the Indian Ocean (Gulf of Aden and Gulf of Oman, cf. Plaziat, 1995). Oysters fixed to *Avicennia* roots, and *Terebralia* have been consumed in quantity by Neolithic populations who abandoned clusters of broken shells at the south-eastern periphery of Gebel Abu Shaar, dated as 5365, 5645, 5675 and 6073 years BP. A 20 m long midden is located close to a vanished water source (Figure H2. 7 (d)–(l)). The proximity to Bir (well) Abu Shaar, which has been reactivated today, indicates a permanent water supply from the adjacent fractured basement block, and thus does not imply a markedly greater humidity than that prevailing today. It is nevertheless probable that the abundance of flint artefact detritus on the Abu Shaar plateau indicates the existence of limited pastures that have since been eliminated by hyper-aridity.

B. Local and regional variations

Having considered some of the general conclusions we will now evaluate the more local characteristics of Quaternary sedimentation, which generally relate to tectonics.

1. Varying degrees of tectonic deformation

a. Northern Gulf of Suez (Figure H2.1)

Some 6 km north of Suez city the stage 7 marine sands outcropping at an altitude of about +6 m indicate that the isthmus has not been significantly uplifted or lowered since the upper Pleistocene. The Holocene brackish deposits at the periphery of the Bitter Lake (with *Cerastoderma*, *Potamides*, *Corbicula* and *Melanoides*) confirm that there has been a very limited recent uplift. South of Suez the 5e marine terrace indicates a derived MSL situated between +6 and +7 m (north Ras Abu Darag; 15 km north of Zafarana) indicating that the coast between the Galala–Abu Zenima and the Morgan (Miocene) accommodation zone is also relatively stable.

b. Southern Gulf of Suez between Ras Gharib and Ras Gemsa (Figures H2.8, H2.32)

This area contrasts with the adjacent sites as it is characterized by a considerable degree of Quaternary instability. Block faulting of the coastal zone north of Ras Shukheir has been recorded by the creation of narrow post-Pliocene horst and graben. On the other hand the 5e beach-deposits culminating at +8 and +5 m are not abnormal.

The most important structural effects occur along the flank of Gebel el Zeit. Uplift in the order of 11–13 m (with respect to a reference level of +6 or +8 m) of the 5e terrace indicates that a post-Miocene fault bordering the footwall of the block has been rejuvenated since 120 000 BP. The Holocene beach is also uplifted by 1–1.5 m, its initial height being about +1 m. Around Zeit Bay, into the hangingwall, uplift of the 5e terrace decreases westwards and finally disappears, suggesting that the Zeit block has been tilted during the Quaternary (Figure H2.32). At Gemsa Peninsula, which represents a Miocene gypsum anticline located on the crest of another block, the 5e terrace has also been uplifted by at least 5 m (and possibly 11–13 m), again suggesting repeated tilting rather than late Pleistocene halokinetic movement (Figure H2.15).

The eastern border of the Esh Mellaha and Abu Shaar blocks appears to be devoid of 5e marine sediment, suggesting that in this more peripheral setting (with respect to the rift axis) the high marine terrace is either replaced by continental sediments or, more likely, is buried below the recent alluvial fan deposits during a recent phase of tilting.

c. Red Sea coast from Hurghada southwards

In the vicinity of Hurghada, 5e reef and beach deposits are situated at about +7 m and do not reflect any movement since 120 000 years BP. Late Pleistocene stability is also typical of much of the Red Sea coast; derived MSL varies between +5 and +9 m, with few exceptions (Sharm el Naga: +11 m, north Ras Shagra: +12.5 m, and Wadi Abu Sbikhaia: +11.5 m). These exceptions indicate limited uplift in the order of 3/5 m or 5/6.5 m depending on the reference level envisaged.

If one considers present altitudes of the 5e subaquatic gypsum (+7/9 m), stability of the Red Sea coast appears even clearer, suggesting that uplift at the rift periphery ceased south-east from the Aqaba transform zone during the last 120 000 years. However, this recent regional stability follows a mid-Pleistocene history of severe uplift in the order of 25–40 m notably in the vicinity of Safaga, Ras Shagra, Wadi Igla, Wadi Nahari and Wadi Khalilat el Bahari.

In sum, the structural behaviour of the southern Gulf of Suez differs markedly from the general evolution of the northern Gulf at Suez as well as along the coast of the north-western Red Sea. This would suggest that the recent structural movements in the southern Gulf of Suez are related to the strike-slip activity of the Aqaba transform fault. However, a precise location of modern earthquakes (Figure H2.33) shows that the more active faults are located along the axis of the Gulf of Suez; not

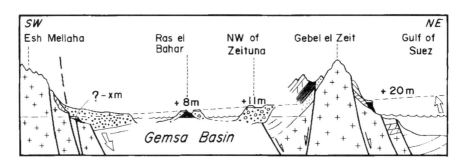

Figure H2.32 Tectonic interpretation of the east–west change in altitude of the Late Pleistocene 5e southern Gulf of Suez shore deposits (reef-beach). A general tilting accounts for the eastward increasing uplift.

only in the southernmost half-graben but also between the Morgan and the Galala–Abu Zenima accommodation zones. This suggests a modern reactivation of rifting north-west of the Aqaba–Levant fault.

Continued tilting of blocks within the Gulf of Suez contrasts with the more stable situation in the north Red Sea, and suggests that the northward propagation of rifting currently affects the south of Suez region which has not yet attained the post-rift phase.

Along most of the north-western Red Sea coast, the regional dip of 5–20° towards the basin axis, affecting Pliocene formations as well as the parallel but more gentle slope of Pleistocene alluvial deposits and early reef-terraces, seems to record a post-rift differential axial subsidence of the Red Sea.

Relationships between Quaternary continental and marine sedimentation and the hinterland morphology

Compared with the Sudan and Eritrean coastal regions (Mitchell *et al.*, 1992) the limited dimensions of the Egyptian Red Sea river catchment areas explain the relatively small contribution of deltas and alluvial fans to Quaternary sedimentation. Modern aridity also accentuates this already limited detrital supply. However, during Quaternary times increased rainfall favoured the development of alluvial fan hundreds of metres of kilometres in radius and attaining thicknesses of at least 10 m.

Marine sediments deposited during periods of high-sea level benefited from the relief created by the alluvial cones, and small reefs developed on narrow shelves cut into these gravel. Coastal erosion contributed gravels and especially boulders to adjacent infralittoral areas where they were colonized by corals and red algae. In this setting sea-level rise is expressed by initially increasing grain-size (alluvial fan gravels, beach pebbles, infralittoral boulders), followed by reef construction. Regression during sea-level stability is expressed by progradation of beach pebbles or gravels, rather than by reactivation of alluvial fans. This confirms that the reefs formed during the highest sea level and coincided with periods of aridity. During this phase, wadis were a

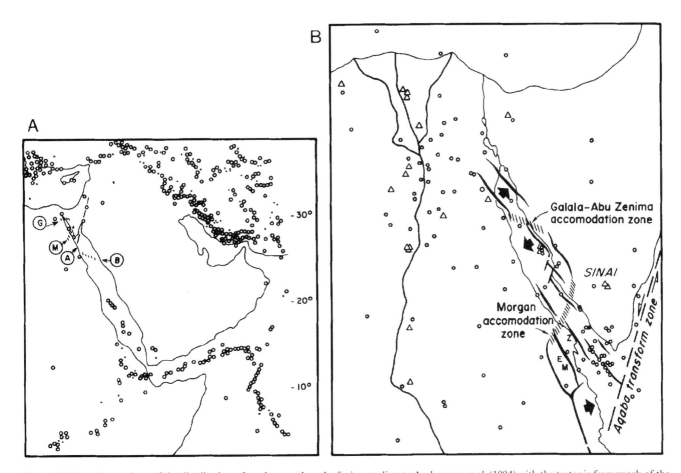

Figure H2.33 Comparison of the distribution of modern earthquake foci according to Ambraseys *et al.* (1994) with the tectonic framework of the northern Red Sea and Gulf of Suez after Bosworth (1995). A: Aqaba transform fault; B: Brothers Island accommodation zone; EM: Esh Mellaha; G: Galala-Abu Zenima a.z.; M: Morgan a.z.; Z: Gebel el Zeit. Notice the weak seismicity of the northern Red Sea and the location of epicentres along the axial faults of the segmented Gulf of Suez. (A) Instrumental period (1964–1992), o: magnitude equal to or exceeding 5: lower magnitude. (B): △: pre-instrumental period earthquakes; o: 1989–1992.

limited source of new detrital material as is demonstrated by the lateral, longshore decrease in thickness and grain-size away from major wadi mouths.

Eustatic lowering of sea level is expressed by a down-stepping of 5e terraces, but this is not always clear and generally only of 1–3 m amplitude. It suggests that the lowering rate was not constant and with high-frequency variations reflecting fluctuations during a climatic 'degradation', i.e. progressively more humid conditions.

The subsequent incision of valleys by increased fluviatile discharge has been interpreted as being related to the end of anaglacial (cooling) periods (Freytet *et al.*, 1993). If one accepts that sea level, at that moment, was close to minimum, it coincides with the maximum slope gradient and thus with increased erosive potential resulting in valley incision of the oldest cones, and reefs. Thus, it seems inevitable that the formation of sharms has involved several glacio-eustatic oscillations rather than a single Würm lowering (cf. Gvirtzman *et al.* 1977). Our observations at Wadi Sharm el Bahari confirm the existence of a stage 5e palaeosharm. However, small sea-level falls have also resulted in minor erosion, as documented during the brief lowstand which subdivides the 5e interglacial maximum. The resulting shallow waters of the sea transgressed during the subsequent sea-level rise and was a condition favouring its closure and fill by evaporitic precipitation.

Conversely the larger wadis are not associated with 5e gypsum deposits. This may reflect the excessive width of the pre-existing valley which prevented closure by coastal sand bars, or because alluvial supply exceeded incision. In the case of high alluvial supply, a fan-delta could develop. The arcuate form of certain reefs suggests that they have grown on such a drowned alluvial cone or on a detrital spit attached to the bulge of an ancient alluvial fan, from which it tends to extend in a seaward south direction. This has been demonstrated at Wadi Nakari for the modern reef and coastal lagoon (M'Rabet *et al.*, 1991) and also in the uplifted, Pleistocene reef complex at Wadi Khalilat el Bahari (Figure H2.27; Plaziat *et al.*, 1989).

It is clear, therefore, that the importance and geometry of terrigenous discharges is strongly influenced by the size of the catchment area which varies considerably from one wadi to another. The regional distribution of 5e evaporite salinas (from Ras Banas to Safaga) also reflects this wadi diversity (Figure H2.34). However, they did not develop further north possibly because large gravel discharge systems dominate much of this region. In our opinion their limitation to the southern Egyptian coast does not reflect specific structural controls. Indeed, it should be stressed that temporary 5e fluctuations of sea level are clearly expressed north of Safaga; at Ras el Bahar where a poorly defined detrital terrace occurs with *Terebralia* (5e2) situated 1.5 m below the highest 5e1 beach. Furthermore, within the carbonate terrace at Sharm el Naga (north Safaga, Figure H2.6) a subvertical surface with large subtidal *Lithodomus* borings separates a more lithified reefal facies (5e1, dated as 126 and 131.6 ka) culminating at +11 m, from another (5e2) reef which culminates at +6.5 m (126.6 ka). In a very different setting, on the flank of the modern, deep depression north of Ras Shukheir (Figure H2.16), two gravel terraces situated at +7.5 m and +5.5 m have different molluscan faunas respectively dominated by *Anadara* and by *Cerastoderma* and may represent the successive 5e highstands.

Different sedimentary discontinuities were noted between superimposed Pleistocene reef growth stages. At Ras Dib, our dates indicate that the interruption of coral growth occupied a relatively short, unmeasurable time interval and was probably a local event (Figure H2.9). At south Zeit (Figure H2.13), a minor discontinuity corresponds to a simple change between a branched *Stylophora* and *Acropora* association and the usual overlying reef-terrace with massive *Porites* and faviids but without any apparent diagenetic difference. These are nevertheless interpreted as expressing two different highstands (7 and 5e) separated by a 100 m sea fall but this can only be resolved by accurate radiometric dates.

Certain gravel horizons have also been regarded as discontinuities of glacio-eustatic origin, although their true significance is highly variable. At Wadi Abu Sbikhaia, a reef dated at >300 ka is separated from the supposed 5e reef by a small gravel bed whose characteristics are not different from a comparable layer located between reefs 5e and 7 at Wadi Igla.

3. The age of the Suez Isthmus and the Gulf isolation from the Mediterranean domain (Figure H2.34)

The presence of the Mediterranean cardid (cockle) *Cerastoderma glaucum* (Figure H2.20(g)) associated with *Potamides conicus* within the Late Pleistocene sediments between north Ras Shukheir and Hurghada has stimulated re-examination of the problem concerning the presence or absence during Quaternary times of a biogeographical barrier between the Mediterranean and Indian Ocean provinces (Plaziat *et al.*, 1985). Now confined to the canal and the Port of Suez, *C. glaucum* had never been recorded from older Indo-Pacific sediments and might be interpreted as having been introduced at the time of the 'Lessepsian migration' (Por, 1978) but in the opposite direction to this (not towards the Mediterranean).

Between south Zeit and Hurghada the 5e beach gravels with *Cerastoderma glaucum* deposited in sheltered environments (bays, lee coasts) also contain Indo-Pacific molluscs, in particular *Gafrarium* and *Terebralia palustris* (Figure H2.20(d)(e)). This suggests that the

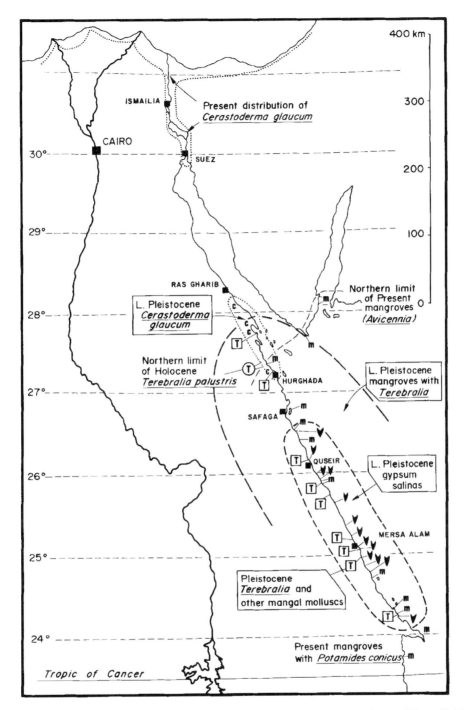

Figure H2.34 Palaeogeographic distribution of characteristic molluscs along the African Quaternary shores of Egypt. T: *Terebralia palustris* (see Figure H2.20), this mangal Potamid reached the southern Gulf of Suez during the Late Pleistocene 5e stage and the Holocene optimum. It is not restricted to the lagoon semi-enclosed depressions of the Red Sea coast. m: The present-day *Avicennia* mangal with *Potamides conicus* (Figure H2.24). c: *Cerastoderma glaucum* (Figure H2.24) in the Late Pleistocene of southern Gulf of Suez.

southern Gulf of Suez was a meeting (mixing) zone for the different biogeographic populations, possibly reflecting episodic free-water circulation between the two seas. However, the near-total absence of other typical Mediterranean species (except *Potamides conicus*) implies an efficient barrier separating these biogeographic provinces.

The authors now consider that the absence of complete faunal mixing is due to the presence of the intervening sedimentary relief created by the Nile, which attains +15 m at El Guisr, north of Ismailia. Thus, the isolated Mediterranean species reflect discontinuous colonization comparable to that between certain Mediterranean lagoons and the Saharan lakes (Gasse *et al.*,

1987; Plaziat *et al.*, 1991; Plaziat, 1993). Within the Gulf of Suez, this had not occurred before the upper Pleistocene (stage 5e) while in the Sahara it happened repeatedly earlier in the Quaternary. However, in both cases transportation must have been effected by subaerial means (probably migratory birds).

Thus, according to the available data, the Suez Isthmus has acted as a barrier at least since the Pliocene, preventing the mixing of faunas in spite of episodes of climatic compatibilities. The present altitude of stage 7 shell-beds situated at +6 m north of Suez town, and the fluviatile molluscs (*Unio, Etheria*) of the El Guisr sill (33 000 ± 735 years BP) also suggests a relative stability.

CONCLUSION

Integrated studies of the Quaternary deposits along the Egyptian African coast have led to a number of surprising conclusions including relative tectonic stability over the last 200 000 years (Figure H2.35). This stability (with the exception of the southern Gulf of Suez) favoured a detailed study of glacio-eustatic sea-level variations.

During the Late Quaternary, the north-western coast of the Red Sea appears to be an area of vertical stability while the adjacent shoulders of the rift are reputed to be actively rising and its axis to continue to deepen and widen during an incipient phase of ocean-floor spreading. Thus, it is suggested that the relatively narrow coastal plains may approximately coincide with a transition belt between uplift and subsidence. However, it is clear that within this overall framework, some local blocks have continued to undergo vertical movements throughout Quaternary times; Early to Middle Pleistocene reefs have been uplifted by up to 30–40 m. Furthermore, in the southern Gulf of Suez, basement blocks have continued to rotate until the late Pleistocene and even during the Holocene. Rifting has not, therefore, totally ceased. This is especially true if one extrapolates the effects of a relatively short time interval (Late Quaternary, less than 0.2 my) to a scale of several million years which would result in important geomorphological and bathymetric changes, especially because continental relief is only moderately affected by erosion in arid or hyper-arid climates. However, one should not interpret all Quaternary alluvial sedimentation as a function of this modern analogue; one can not limit the

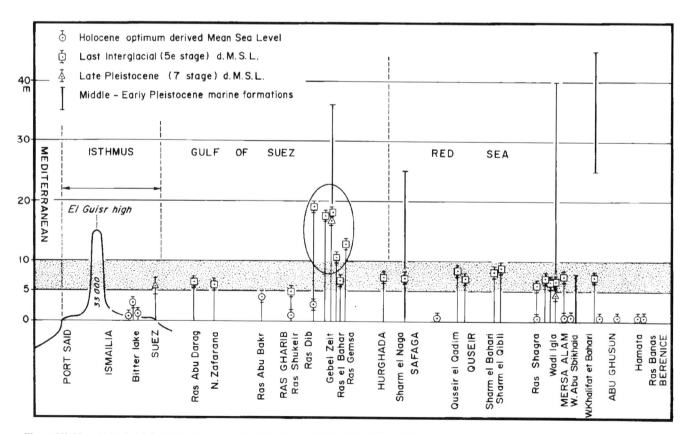

Figure H2.35 Altitude of the derived mean sea level in the studied localities. The relative stability of the rift margin during Late Pleistocene and Holocene times is documented from most of the dated sites. The contrasting, uplifted 5e reefs and beaches of the southern Gulf of Suez are encircled.

importance of the anaglacial and cataglacial humid phases to brief flood events. The presence of coarse alluvial fan deposits, in spite of limited catchment areas, suggests relatively long periods of relatively high seasonal rainfall which occurred between the phases of highest and lowest sea-level stands.

The influence of the rift tectonics inherited from the Miocene, has almost certainly diminished. Nevertheless, it is of some importance to note that the geometry and distribution of many sedimentary units (reefs, fluvial cones) are related today, to the relief of these Neogene blocks. The recent deposits tend to mask the topography created by Miocene rifting, these older structural units also being modified by post-Miocene halokinetic movements which affect both Pliocene and early Quaternary cover.

The present north-western Red Sea is often associated with tectonically controlled morphologies and flourishing reef systems. This is true, but the reefs in general are not developed on fault blocks (with the exception of Gebel el Zeit region). Egyptian Quaternary reefs rather took advantage of detrital accumulations and illustrate in a spectacular manner the potential of reef communities to colonize repeatedly the changing coastal morphologies resulting from coarse siliciclastic progradation. This region probably is one of the most favourable for the study of late Pleistocene reefs, partly due to its mummifying arid conditions which have favoured preservation of primary mineralogy and reef structure and because it has numerous spectacular outcrops freshly cut by littoral and wadi cliffs.

ACKNOWLEDGEMENTS

This research benefited from EEC founding (program SCIENCE) and from the French Programme d'Etude des récifs coralliens (PNRCO). In the field, help from our colleagues from Assiut University has been essential. We also thank sincerely the editors, and especially E. Davaud and D. Bosence, for their valuable help in improving the manuscript.

Chapter H3
Post-Miocene reef faunas of the Red Sea: glacio-eustatic controls

M. Taviani

ABSTRACT

The Red Sea maritime rift has been discontinuously colonized by coral reefs since the Miocene. The Tethyan Miocene period of coral growth was interrupted when hypersaline conditions became established in the basin. The second period of reefal recolonization, started from the Indian Ocean, is poorly known in its Pliocene part but Pleistocene reefs are abundant throughout the entire basin. The Quaternary reefs appear strictly controlled by the Pleistocene cyclicity of ice ages which in turn modulate with the basin's hydrological conditions. It is proposed that the shallow-silled Red Sea basin was cyclically affected by more or less severe biotic turnovers as a consequence of periods of high environmental stress (temperature, salinity) during glaciations.

The fact that the last interglacial (isotope substage 5e) coral reef fauna differs from the modern biota inhabiting the Red Sea is interpreted as a consequence of a basinal extinction event at the peak of the last glaciation due to a rise in salinity. Periods of increased humidity (rainy phases) are represented by marine benthic biota as well as proven by the occurrence in the Red Sea of early Holocene mangals, at present absent or strongly reduced in size and complexity.

INTRODUCTION

The Red Sea region has been a site of prolific, albeit discontinuous, coral reef growth since the Miocene. This is to say that coral reefs accompanied the evolution of the Red Sea rift since its incipient stages as a fully maritime system beginning during the Middle Miocene (Gregory, 1906; Buchbinder, 1979; James et al., 1988; Perrin et al., this volume). However, the history of coral reef colonization has been marked by numerous and often dramatic setbacks. In fact, although latitudinal, geographic and overall climatic conditions have been largely favourable for coral reef growth in the Red Sea during the last 16 m.y., the environment within the basin has at times been unsuitable for such a complex tropical marine ecosystem.

It is convenient to subdivide the history of coral reef growth in the Red Sea into two periods. In Miocene times, the Red Sea embayment was connected to the Tethyan domain via a Mediterranean seaway (e.g. Montanaro Gallitelli, 1939, 1941; Coleman, 1974 a, b, 1993; Por, 1989). Therefore, the Miocene coral populations were biogeographically related to Mediterranean faunas. These reefs appear only seldom to be of significant size and their biodiversity is comparably low (e.g. Purser et al., 1990, Sun and Esteban, 1994). They represent a typical 'arid land-locked subtropical–temperate setting' where full reef growth is limited by a combination of periods of increasing salinity and, perhaps, low temperatures allowing only the formation of shortlived fringing reefs of modest size (Sun and Esteban, 1994). Examples of such coral build-ups are included in the Burdigalian to Serravallian carbonates outcropping along the Gulf of Suez and Red Sea (e.g. Carella and Scarpa, 1962; El Haddad et al., 1984; Monty et al., 1987; Burchette, 1988; Purser and Hötzl, 1988; Purser et al., 1990; Tewfik et al., 1992; Darwish and El Azabi, 1993).

In this period, the maritime environments of the basin were dominantly controlled by its tectonic setting which played a decisive role in dictating its hydrological conditions and, ultimately, the kind of ecosystems inhabiting the Red Sea. At times, periods of variable restricted circulation brought about increases in salinity. The marginal marine biota were exterminated when conditions of very high salinity became established at the end of the Middle Miocene c. 12 m.y. ago leading to extensive deposition of salt (Stoffers and Kühn, 1974; Stoffers and Ross, 1977; Braithwaite, 1987).

The second phase is the post-Miocene recolonization of the basin after this major saline period, when the Red Sea opened to the Indian Ocean, sometime in the Pliocene when marine biota recolonized the basin from the south (Coleman, 1974a,b; Braithwaite, 1987; Por, 1989).

The aim of the present chapter is twofold. Firstly, to focus on the second phase of reefal colonizations (post-Miocene), with special regard to the late Pleistocene. The re-examination of available evidence shows that most post-Miocene reefs are likely to be Pleistocene in age. Secondly, to comment upon the Quaternary faunal evolution documented in the Red Sea reefal and associated ecosystems and their climatic implications. It is suggested that glacial periods negatively affect coral-reef biota as has been the case during the last glaciation. It is suggested that wet and arid climatic phases are also accompanied by faunal rearrangements of specialized shallow-marine biota.

The Plio-Pleistocene problem

Many authors dealing with the post-Miocene evaporite coastal stratigraphy of the Red Sea margin, refer to the general term 'Plio-Pleistocene' for their chronology (e.g. Fuchs, 1878; Cox, 1931).

Precise biostratigraphic dating of the age of post-Miocene sequences is in many cases inadequate and a serious source of biostratigraphic error might be derived from pre-World War II papers. These early geological reconnaissance surveys of the Red Sea–Gulf of Aden coastal areas predated the resolution of modern biostratigraphy and the introduction of accurate isotopic dating. A further complication stems from the intrinsic nature of tropical carbonate marginal (shallow) marine facies which typically lack fossils of biostratigraphic value.

Pliocene reefs

References to 'Pliocene' reefs in the Red Sea region are scattered in the literature (e.g. Montanaro Gallitelli, 1939, 1973; El Shazly et al., 1974; Purser and Hötzl, 1988; Purser et al., 1993; Rioual et al., 1996). Limiting our analysis to some modern examples, it appears that most of these Pliocene age assessments are not substantiated by reliable biostratigraphy. Thus, El Shazly et al. (1974) refer the 'old coral reefs' outcropping on the island of Zabargad, offshore Egypt at latitude 23°N, to the Pliocene. These coral limestones, with no biostratigraphically useful fossils but containing extant species of molluscs, were originally attributed to the Plio-Pleistocene by Moon (1923), and to the Quaternary by El Shazly et al. (1967), Bonatti et al. (1983), and Hoang and Taviani (1991). Montenat et al. (1988) indicate that reefal carbonates contribute to their 'Group D sedimentary sequence' of the northern Red Sea and Gulf of Suez. In their scheme, a Pliocene age is proposed for these rocks although in the text (p. 170) these authors, while discussing the faunal content, refer these carbonates to the Plio-Pleistocene. Some support for a Pliocene age of the Ifal Formation (Midyan region of Saudi Arabia), which includes some reef facies, derives from foraminiferal data obtained from underlying echinoid sands but even this chronology is somewhat uncertain (Purser and Hötzl, 1988). Purser et al. (1993) attribute a Pliocene age to coral reef carbonates exposed at Umm el Gerifat (south of Quseir, Egypt) unconformably overlain by Pleistocene deposits. These carbonates have an Indo-Pacific fauna and their age assignment is based only on their comparably high diagenetic modification and tectonic disturbance (Purser et al., this volume). Based on Sr-dating, Bosence et al. (this volume) produce the only ascertained evidence of Pliocene reefs where coral reefs are seated on top of a salt dome in Yemen.

It appears that common criteria used for Pliocene age assignment of marine rocks of the Red Sea margins are: 1. their interposition between the top of the Miocene and terraced coral reefs for which a Quaternary age is widely accepted; 2. the degree of tectonic deformation (mostly folding and faulting), and 3. the severity of lithification and recrystallization. None of these criteria is of strict time-stratigraphic merit as they reflect the local geologic history of a given site. In fact, 'old-looking' Pleistocene reefal-carbonates occur in tectonically active areas as, for example, Zabargad (Bonatti et al., 1983), Sinai (Gvirtzman and Buchbinder, 1978; Strasser et al., 1992), Giftun, Shaker and Gubal islands (personal observations).

In summary, with a few exceptions, most of the post-Miocene reefal limestones observable along the entire Red Sea basin may very likely be Quaternary in age.

Quaternary coral reefs

As discussed above, the most appropriate interpretation at present is to attribute to the Pleistocene period the vast majority of post-Miocene coral reef (and reef-associated) facies outcropping in the Red Sea. The Pleistocene lasted about 1.8 m.y. (Van Couvering, 1995) and has been punctuated by about 20 sea-level fluctuations which can be correlated with variations in ice volumes. The Plio-Pleistocene oxygen isotopic record (Tiedemann *et al.*, 1994) shows the number of Pleistocene relative highstands far exceeding the five or so uplifted coral reefs directly dated from the Red Sea. These occur in less than 0.5 m.y. and only cover part of the Brunhes epoch. The coral reef legacy of the largest part of the Pleistocene (c. 1.2 m.y.) is therefore unaccounted for. It seems logical to conclude that many older occurrences of reef carbonates may belong to this 'missing' part of the Pleistocene. This may explain the advanced diagenesis of certain coral reefs (compare Al Sayari *et al.*, 1984; Dullo, 1986). Repetitive subaerial exposures of the carbonates will trigger many diagenetic events which will have taken place under climatic conditions punctuated alternatively by arid and more humid phases (e.g. Deuser *et al.*, 1976; Conchon *et al.*, 1994; Freytet *et al.*, 1994; Szabo *et al.*, 1995). Furthermore, considering that the Red Sea is an active rift basin, the duration of the Quaternary is sufficient for reefs to be affected by tectonic processes which have been used as an additional argument for their antiquity.

Clearly, the comparably 'old' Pleistocene coral reefs and associated facies, are in strong need of detailed work to establish their ages and evolution through time. Hopefully, a larger application of Sr-dating to such carbonates will significantly clarify their chronology as is the case for reefs developed on salt domes in Yemen (Bosence *et al.*, this volume).

Modern Red Sea coral reefs are largely of the fringing type, although small barrier reefs and atolls are also known to occur in this basin (e.g. Loya and Slobodkin, 1971; Mergner and Schumacher, 1974; Angelucci *et al.*, 1975; Gvirtzman and Buchbinder, 1978; Roberts and Murray, 1988a, Montaggioni *et al.*, 1986; Head, 1987; Guilcher, 1988; Dullo and Montaggioni, this volume). They grow almost continuously along the rift margin, bordering suitable areas of mainland with little supply of terrigenous silt and offshore islands from the extreme south and north to the Gulfs of Suez and Aqaba. Growth of true coral reefs stops at about 29°N. Palaeontological data from the western side of Sinai (Hammam Faraun) reveals that this was the northernmost limit of reefs during the last interglacial as well (Taviani *et al.*, 1995). The location, growth and morphology of the reefs are the result of an interplay between topographic features, terrigenous input and glacio-eustatic erosion (Gvirtzman *et al.*, 1977; Plaziat *et al.*, 1989; Dullo and Montaggioni, this volume). It is worth mentioning that modern coral reefs are known to have been rapidly uplifted by a few metres above present sea level as a response to seismic shocks as recorded for the Shadwan 1969 earthquake (Kebeasy, 1990).

The distribution of Pleistocene coral reefs largely follows the morphology and zonation patterns of their modern counterparts and, thus, are normally represented by narrow fringing reefs. Their preservation varies from site to site (Figures H3.1, H3.2). The best preserved Pleistocene reefs belong to the last interglaciation (see below) and excellent exposures are found along the Egyptian coast, especially on the tilted block of Gebel Zeit (Figures H3.1(c)(d), H3.2(b)), Wadi Khalilat el Bahari (Plaziat *et al.*, 1989), Zabargad island (Figures H3.1(b), H3.2(a)) (Taviani, unpublished) and Saudi Arabia (Dullo, 1987, 1990).

Well-preserved coral reefs from the last interglacial expose a complete sequence of reef facies landwards from the sea as:

1. reef slope
2. reef crest
3. reef flat
4. back-reef lagoon (where many subfacies are distinguishable)
5. shore (sandy-gravelly, beach-rock, rocky)
6. backshore (sabkha)

Facies 2 to 4 are those commonly represented in the Pleistocene reef systems of the Red Sea (Figures H3.2). Well-preserved upper reef slope facies are uncommon, as is the case of true shoreline deposits. To the author's knowledge the best examples are to be found at Gebel Zeit (where facies 6 is also documented) and Zabargad island. Eurihaline to mesohaline lagoonal (with *Potamides* (= *Pirenella*) *conicus* (Blainville, 1826), *Cerastoderma glaucum* (Poiret, 1789), *Anadara* spp. among others) and mangal (mangrove environments) facies are known from the Egyptian coastal section (Plaziat *et al.*, 1995; Plaziat *et al.*, this volume).

Terraced Quaternary coral reefs

Raised 'coral beaches' have long been known to occur along the entire Red Sea–Gulf of Aden region (Issel, 1869, Walther, 1888; Bullen-Newton, 1900; Hall and Standen, 1907; Nardini, 1934, 1937; Abrard, 1942; Selli, 1944, 1973; Said, 1962, 1990a; Sestini, 1965; Montanaro Gallitelli, 1973; and many others). The application, since the late 1960s, of dating techniques, especially uranium-series dating of coral aragonite (e.g. Veeh and Giegengack, 1970) has dramatically enhanced the stratigraphic resolution allowing discrimination among the various mid-late Pleistocene units. Previous attempts to establish radiometric ages of raised coral reefs through ^{14}C dating proved of limited value because of the age of the reefs being infinite and geologically inconsistent (Nesteroff, 1960; Berry *et al.*, 1966; Hötzl *et al.*, 1984; also Behairy, 1983).

In certain sectors of the Red Sea region, characterized by comparably high uplift rates, up to five major raised coral reef units, often arranged in terraces, occur. This is the case of the Afar region, Sinai, offshore tectonic

Figure H3.1 Early (?) Pleistocene back-reef limestone (Old Coral Reef Fm) from the Evaporite Valley on Zabargad Island (Egypt, Red Sea). The coral rock is very recrystallized and most of its palaeontological content dissolved; the arrow indicates a mould of a *Strombus* gastropod (S). B. Example of a late Pleistocene fringing reef. A distinct coral terrace (arrow) developed on the peridotitic bedrock at Zabargad Island (southern side, Turtle Beach, see Hoang and Taviani, 1991). Its altitude is about +15 m above msl. C. Panoramic view of raised terraces at Gebel Zeit (Egypt); their age is Eemian (last interglacial) and often represent complete reef sequences (from shore to upper reef slope). D. Detail of the Eemian coral reefs of Gebel Zeit (Egypt). The reefal build up (ER) representing upper fore-reef facies transgressed over continental gravels and sands.

islands (e.g. Tiran, Zabargad, The Brothers), the Egyptian coast at Râs Gemsah and Gebel Zeit. These coral units correspond to high-stands correlatable with sea-level fluctuations triggered by the accumulation and disruption of Pleistocene ice caps. From the stratigraphically oldest to the youngest these are as follows.

1. ***c*. 400 ka coral reefs.** The oldest dated coral terrace is a veneer with scattered corals, probably representing a former shoreline as shown by its mollusc content (Bosworth and Taviani, unpublished observations) is found at Gebel Zeit (north-western Egypt); a single TIMS U/Th age (coral sample) of 426 ka BP has been obtained (Bosworth and Taviani, 1996). This terrace is about 42 m above present msl and represents a high-stand correlatable with isotope stage 11 of the oceanic oxygen isotope curve.

2. **330–290 ka coral reefs.** Uranium-series dated units of such an age have been detected in Sinai (Kronfeld *et al*., 1982; Gvirtzman *et al*., 1992; Gvirtzman, 1994), Egypt (El Moursi, 1993) and possibly, although undated, Saudi Arabia (Dullo, 1990). These coral reefs correlate with isotope stage 9.

3. **170–250 ka coral reefs.** Raised coral reefs of such an age are widespread in the Red Sea and correlate with isotope stage 7. They occur in Sinai (Gvirtzman, 1994), Egypt (El Moursi, 1993; Choukri *et al*., 1995), The Brothers and Zabargad (Hoang and Taviani, 1991), Saudi Arabia (Dullo, 1990), Dahlak (Conforto *et al*., 1976), and Afar (Hoang *et al*., 1974).

4. **115–135 ka coral reefs.** Uplifted beaches and coral reefs of this age are among the commonest landforms of the Red Sea coastline. They correlate with the last interglacial (Eemian or Sangamonian, isotope substage 5e). A large quantity of U-series dates of aragonite scleractinian coral and *Tridacna* and sea-urchin calcite makes the Red Sea–Gulf of Aden the best dated basin in the world for this time-span. These last interglacial reefs have been positively identified in Sinai

Figure H3.2 A. Shelly gravel of a former Eemian shoreline at Zabargad Island (Egypt). Fossils (arrows) are intertidal gastropods, namely *Planaxis sulcatus* (Born, 1778) (P) and *Nerita sanguinolenta* Menke, 1829 (N). B. Detail of an upper fore-reef facies with large *in situ* stony corals (e.g. *Favia, Platygyra, Porites, Goniastrea* and *Favites*); intracoral sediment is unlithified and contains a very diverse fossil fauna including shells of typical fore-reef gastropods (e.g. *Neritopsis radula* (L. 1758); Eemian terrace, Gebel Zeit (Egypt). C. Paired giant clam (*Tridacna maxima* Roeding, 1798) preserved *in situ* from an Eemian terrace at Wadi Gemal island (southern Egypt); facies is the transition between reef flat and back-reef lagoon. D. *In situ Dendropoma maximus* (Sowerby, 1825) vermetid gastropods from an Eemian terrace at Wadi Gemal island (southern Egypt); these attached shells are excellent palaeoenvironmental indicators since they normally inhabit the seaward face of reef flat, reef edge and uppermost reef flat between 0–2 m.

(Kronfeld *et al.*, 1982; Gvirtzman *et al.*, 1992; Strasser *et al.*, 1992; Gvirtzman, 1994), Tiran island (Goldberg and Beith, 1991), Egypt (Veeh and Giegengack, 1970; Hoang and Taviani, 1991; Andres and Radtke, 1988; Reyss *et al.*, 1993; El Moursi, 1993; Choukri *et al.*, 1995; Plaziat *et al.*, 1995, 1996), Sudan (Hoang *et al.*, 1995), Saudi Arabia (Dullo, 1990), Dahlak (Conforto *et al.*, 1976), Afar and Gulf of Aden (Labou *et al.*, 1970; Faure *et al.*, 1973, 1980; Hoang *et al.*, 1974, 1980). Based on field and faunal evidence, these last interglacial deposits outcrop in Yemen and offshore islands (El-Anbaawy, 1993), Yemen (Bosence *et al.*, this volume), Eritrea (Selli, 1973, Montanaro Gallitelli, 1973) and Djibouti (Abrard, 1942). These raised reefs and shelly terraces generally occur at altitudes of between 5–10 m above present MSL. Because it is generally accepted that during the last interglacial sea level peaked *c.* 6–7 m higher than today, 5e interglacial reefs have been widely used as reliable tracers of vertical movements. From the screening of the available data, it seems that most of the Red Sea region has been reasonably stable during the last 100 000 years, including 'tectonic' islands such as The Brothers and Zabargad, offshore Egypt, (Hoang and Taviani, 1991). Notable exceptions are the Afar triangle (Faure, 1975; Faure *et al.*, 1980), Tiran island (terrace at +40 m, 146 ± 16 ka: Goldberg and Beith, 1991), Gebel Zeit and Râs Gemsah blocks in Egypt (Andres and Radtke, 1988; Reyss *et al.*, 1993; Plaziat *et al.*, 1995; Bosworth and Taviani, 1996), and reefs above salt domes from Yemen (Bosence *et al.*, this volume).

Besides these uplifted terraces, there is ample documentation of Pleistocene coral terraces still submerged, representing relative sea-level lowstands (e.g. Brachert and Dullo, 1990; Gvirtzman, 1994) although on subsiding coastlines highstand terraces are drowned as well, as is the case in Tawila and Shaker islands in the southern Gulf of Suez (personal observation).

5. **Holocene coral reefs.** Emergent Holocene corals (younger than 6000–7000 years BP, isotope stage 1) have been documented from Sinai (Gvirtzman et al., 1992; Gvirtzman, 1994), the Jordanian side of the Gulf of Aqaba (Bouchon et al., 1981), Egypt (Plaziat et al., 1995, 1996; Montaggioni, personal communication, 1992), Saudi Arabia (Behairy, 1983) and Yemen (Bosence et al., this volume). Such corals are often associated with an erosional surface and typically occur at an altitude of +0.5 m and may indeed represent a short-lived higher-than-present Holocene seastand dated at c. 5.5–6.5 ka (Gvirtzman, 1994; Plaziat et al., 1995; Bosence et al., this volume).

LATE QUATERNARY FAUNAL TURNOVERS

The Last Interglacial: differences between Eemian and Recent faunas

Given the faunal richness of last interglacial (isotope substage 5e) terraces, an excellent opportunity exists to assess the effect of the last glaciation on coral-reef growth in the Red Sea region. To date, a significant palaeontological and zoological database is available on molluscs while information on other groups, including scleractinian corals, is more scanty. Taviani (1982) discussed the subtle but highly significant differences that exist between reef-associated mollusc faunas from the Last Interglacial and the Recent in the Red Sea–Gulf of Aden rift basin.

Isotope substage 5e reefal complexes differ from their modern counterparts because they contain taxa which are now: (i) completely extinct, (ii) locally extinct, (iii) rare within the Red Sea basin or have shifted more to the south. Limiting our discussion to some of the commonest, larger and easily classifiable taxa (Figure H3.3), this argument can be supported by the following observations.

Case (i) is best represented by the keyhole limpet *Diodora impedimentum* Cooke, 1885 (synonym: *Capulina ruppelli* var. *barroni* Bullen-Newton, 1900: see Mienis, 1981). This species occurs in 5e coral reefs throughout the entire Red Sea basin, including the Djibouti part of the Gulf of Aden and is often abundant (Bullen-Newton, 1900; Abrard, 1942; personal observation). In spite of extensive research, no specimens of this characteristic gastropod have been found in modern reefs and we conclude that the taxon did not survive the last glaciation.

Case (ii) is illustrated by other conspicuous components of the Red Sea Eemian fauna, i.e. *Columbella turturina* Lamarck, 1822, and *Rhinoclavis vertagus* Linné 1767. Their fossil distribution parallels that of *D. impedimentum* and are equally common in the last interglacial reefs. These taxa thrive today in the Indo-West Pacific region (Houbrick, 1978; Cernohorsky, 1972). However, both species have been completely eradicated from the Red Sea region. Fossils belonging to *C. turturina* are consistently found in Eemian deposits and were first noticed by Issel (1869). *R. vertagus* is less abundant than *C. turturina* but is widespread in the Eemian of the Red Sea–Gulf of Aden region (Hall and Standen, 1907; Jousseaume, 1888, 1930; Nardini, 1934; Barash et al., 1983; and personal observations). Within this same category are other taxa, among which are *Cerithium madreporicolum* Jousseaume, 1930, *Conus litteratus* Linné, 1758 and *Cucullaea cucullata* (Roeding, 1798). The first species is occasionally found in fossil localities throughout the entire Red Sea (Egypt, Zabargad Island, where it is common; Saudi Arabia at Sharm Obhor, personal observation) and Gulf of Aden (Jousseaume, 1930, Abrard, 1942); the second species is known in the Eemian of Zabargad Island (personal observation). Both species are unknown from the modern Red Sea (Houbrick, 1992; Wils, 1986); the latter occurs in 5e back-reef sandy deposits of the Egyptian coast at Quseir, Râs Gemsah, and Hammam Faraun (Plaziat and Taviani, unpublished data). *Cucullaea cf. cucullata* also occurs in the pre-Eemian Pleistocene of Giftun Kebir, Egypt (within the *Pinna*-unit: Taviani, 1995c) but is at present absent from the Red Sea, while it is known from Aden (Oliver, 1992).

Case (iii) is based on the observation that some species were exceedingly abundant in Eemian times but are rare today in Red Sea reefs. The common cowrie *Cypraea moneta* Linné, 1758, inhabits shallow back-reef environments, and can be collected, often by the hundreds, on many 5e reefs, representing one of the commonest fossils in the Eemian of Aqaba, Gulf of Suez and northern Red Sea, Zabargad Island, Jeddah area etc. (literature and unpublished data) but is rare in the Red Sea basin at present (Foin and Ruebush, 1969; Mienis, 1971; O'Malley, 1971; Barletta, 1972). The same applies to *Oliva bulbosa* Roeding, 1798, whose shells are very common in 5e deposits as far north as Aqaba, while today this species is scarce in the Red Sea and often found much more to the south. A similar case is thought to occur with the bivalve *Corbula taitensis* Lamarck, 1818 (= *C. acutangula*, Issel, 1869) which appears commonly associated with Eemian back-reef deposits. However, its occurrence within older Pleistocene deposits (Farasan Islands: Cox, 1931, and Egypt: personal observations) as well as in the modern Red Sea (Oliver, 1992), makes it of lesser palaeobiogeographic value.

Similar extinction trends appear among corals as well. Three species of scleractinians, i.e., *Turbinaria peltata* (Esper, 1794), *Cycloseris vaughani* (Boschma,

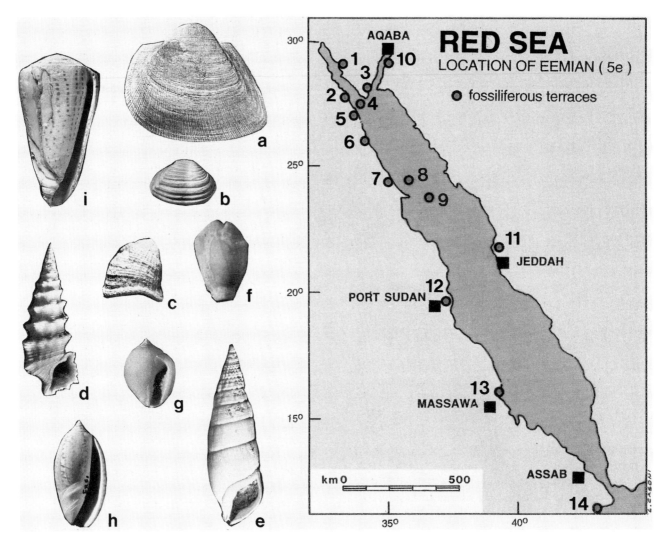

Figure H3.3 Some representative Eemian molluscs discussed in the text: a = *Cucullaea cucullata* (Roeding, 1798), ×0.8; b = *Corbula taitensis* Lamarck, 1818, ×2; c = *Diodora impedimentum* Cooke, 1885, ×1.5; d = *Cerithium madreporicolum* Jousseaume, 1930, ×1.5; e = *Rhinoclavis vertagus* (Linné, 1767), ×0.8; f = *Cypraea moneta* Linné, 1758, ×0.8; g = *Columbella turturina* Lamarck, 1822, ×1.5; h = *Oliva bulbosa* Roeding, 1798, ×0.8; i = *Conus litteratus* Linné, 1758, ×0.4. Map shows the location of most relevant Eemian outcrops whose mollusc fauna have been used in the present study. *Egypt:* 1 = Hamman Faraun; 2 = set of terraces of Râs Gharib, Râs Gemsah and Gebel Zeit; 3 = set of terraces nearby Râs Muhammad and Sharm el Sheikh; 4 = terraces of Gubal (Jubal) and Tawila islands; 5 = terraces around Hurghada; 6 = terraces around Quseir; 7 = terraces nearby Berenice; 8 = Wadi Gemal island; 9 = Zabargad island. Jordan: 10 = Yamanie; Aqaba. Saudi Arabia; 11 = Sharm Obhor, Jeddah. Sudan: 12 = nearby Port Sudan and Marsa Fijia. Eritrea: 13 = Massawa (from Selli, 1973). Djibouti: 14 = terraces nearby Djibouti (from Abrard, 1942, and A. Bonfitto, unpublished, 1993).

1923) and *Pavona minuta* Wells, 1954, have been recorded from an Eemian reef terrace at Sharm al Harr (Saudi Arabia) but do not seem to occur in the modern Red Sea coral fauna (Dullo, 1987, and C. Dullo, *in litt.*, 8 March 1996).

Impact of the last glacial age on coral reef growth

The idea that Pleistocene coral reefs of the Red Sea have a fauna somewhat different from the present one is not new. Many new taxa belonging to molluscs, scleractinian corals and echinoderms, apparently exclusive to the Pleistocene 'raised coral reefs' have been described by authors (e.g. Issel, 1869; Jousseaume, 1888, 1930; Bullen-Newton 1900; Fourtau, 1914; Cox, 1931; Nardini, 1934, 1937; Selli, 1973; Montanaro Gallitelli, 1973; Dollfus and Roman, 1981). The occurrence of extant Indo-Pacific species no longer living in the Red Sea–Gulf of Aden has also been underlined in the past and their likely palaeobiogeographic meaning is discussed by Nardini (1934), Abrard (1942), and Selli (1944, 1973). Admittedly, many of these taxa are misinterpretations and their correct taxonomic position is still in need of systematic revision. However, the basic observation of important faunal differences between Pleistocene and modern Red Sea reefs was correctly laid down in the past.

The faunal differences existing between the last (isotope substage 5e) and the present interglaciation (isotope stage 1) coral reefs can be best interpreted as a basin-wide extinction event(s) which took place during the last glaciation. As documented by faunal and oxygen stable isotope data, periods of high salinity were established in the Red Sea during the last glaciation. Such 'saline' phases were a consequence of strongly diminished water exchanges with the Indian Ocean at the Bab-el-Mandab shallow sill (c. 137 m present-day depth) during Pleistocene periods of relative sea-level lowering (e.g. Degens and Ross, 1969; Ivanova, 1985; Locke and Thunell, 1988). The latest hypersaline event took place during the last glacial maximum and ended about 11 ka BP (e.g. Ivanova, 1985; Locke and Thunell, 1988; Brachert and Dullo, 1990). Salinity rose to 50‰ (e.g. Reiss et al., 1980; Almogi-Labin, 1982; Locke and Thunell, 1988; Thunell et al., 1988). Prolonged periods of high salinity affecting the whole of the basin are lethal to stenoecious organisms and the overall Red Sea fauna was at that time very likely annihilated (Por, 1971, 1978; Gvirtzman et al., 1977; Taviani et al., 1992). However, excess salinity was not the only factor causing the proposed extinction. In fact, extinctions and geographic displacements also affected taxa living outside the Red Sea (Taviani, 1995b). It is likely that such biotic perturbations were a response to a combination of factors which included salinity, temperature drops (Ivanova, 1985) and disruption of the internal organization of reefs due to sea-level changes. By approaching quasi-evaporitic conditions, the semi-enclosed Red Sea basin (Thunnell et al., 1988) greatly amplified a global phenomenon of stress on reef ecosystems of the entire Indo-West Pacific tropical belt during the last glaciation. This caused significant faunal turnovers and local extinctions (Taviani, 1995b).

This agrees with the view of Gvirtzman et al. (1977) that the present Red Sea fauna is the result of a Holocene recolonization of the basin. Similar extinction events may also have taken place during previous glaciations but with the lack of proper faunal data, we cannot confirm such a hypothesis. We only note that the Middle Pleistocene reefs (undated) of Khalilat el Bahari, Egypt, show a strange low-diversity mollusc and coral assemblage (Plaziat et al., 1989).

It is worth mentioning that faunal disturbances at times of basinal isolation are linked to sea-level variations during the Quaternary and have been invoked to explain the biogeographic evolution of the Red Sea fauna (Foin, 1972; Por, 1978; Dollfus and Roman, 1981; Barash et al., 1983).

The Holocene and late Pleistocene wet phases and their faunal signature

In a recent paper Plaziat et al. (1995) call attention to the climatic significance of late Pleistocene and Holocene optima which led to an invasion of Indian Ocean mangals within the Red Sea. Such an expansion is well documented by mangal associated malacofaunas, including the gastropod *Terebralia palustris* (Linné, 1767) found along the Egyptian coastline up to the southern Gulf of Suez (south Gebel Zeit: Plaziat, 1995). Radiometric dating of such fossil faunas indicates two periods of mangrove growth. The oldest occurrence is related to the last interglaciation (isotope substage 5e) while the younger refers to the Holocene, c. 5 ka BP (Plaziat, 1995). Today, mangroves of the Red Sea region are deprived of such accompanying and easily preservable faunas (Plaziat, 1995; Taviani, 1995a). This absence is a consequence of the extreme aridity of the whole region which began less than 5000 years ago (Taviani, 1995a, with references therein). However, independent onshore and offshore geological data as well as archaeological evidence suggests that there were periods considerably wetter than today with many rainy phases punctuating the late Quaternary climate of the region (e.g. Anton, 1984; Rossignol-Strick, 1987; Stanley and Warne, 1993; Gasperini et al., 1994; Freytet et al., 1994; Szabo et al., 1995; Taviani, 1995a). The latest rainy period initiated some 11 ka BP and lasted until 5 ka BP (e.g. Anton, 1984; Szabo et al., 1995), correlates with expansion of brackish mangals in the Red Sea region (Plaziat, 1995). These mangals then shrunk with the onset of the present phase of aridity.

The myth of Quaternary seaways between the Red Sea and the Mediterranean

Since the last century, claims have been made of possible interconnections via Suez between the Red Sea and the Mediterranean during the Quaternary (Por, 1978). The recent record of marine faunas containing the Mediterranean cockle *Cerastoderma glaucum* in sediments from the last interglacial (isotope substage 5e) bays of the Egyptian coastline between Râs Gharib and Hurghada, has been taken as an argument for such a connection (Plaziat et al., 1995). However, the assemblage containing this 'Mediterranean' element is otherwise exclusively Indo-West Pacific in affinity, e.g. *Anadara ehrenbergi* (Dunker, 1868), *Cardites rufa* (Laborde and Deshayes, 1834), *Chama brassica* Reeve, 1847, *Gafrarium pectinatum* (Linné, 1758), *Dosinia erythraea* Roemer, 1860, *Monilea obscura* (Wood, 1828), *Nerita polita* Linné, 1758. The presence of *Cerastoderma* cannot be taken as evidence of a direct connection with the Mediterranean since its occurrence may be explained by passive dispersal mechanisms, perhaps by migratory birds, as is the case in Saharan saline lakes (Plaziat, 1991, 1993). Recent fieldwork north of Ismailia shows that a fluviatile barrier existed between the two basins and that the previous assumption of a late Pleistocene connection with the Mediterranean is unsubstantiated (Plaziat et al., this volume).

CONCLUSIONS

1. The overall history of coral reefs in the Red Sea maritime rift is one of discontinuous growth controlled by tectonic and climatic changes. The first Miocene phase of coral growth was largely controlled by the rift's tectonic evolution and ended when hypersaline conditions were established at the end of this period. The second phase (post-Miocene) is poorly known in the Pliocene and these older reefs are better attributed to the Quaternary.

2. The evolution of Quaternary reefs is largely controlled by climate, specifically by Pleistocene glacio-eustatic sea-level fluctuations which in turn modulate with the basin's hydrology. It is postulated that the shallow-silled Red Sea basin experienced a series of more or less completed biotic recolonization events from the Indian Ocean. These faunal incursions follow periods of high environmental stress (temperature, salinity) which could bring about the destruction of stenoecious biota as was the case during the last glacial period. Sea-level highstand conditions (equivalent to interglacial and warm-interstadial periods) are times of healthy coral growth which in a time-span of a few thousand years (<10 ka) may recover a high level of internal organization and diversity as proven by the modern Red Sea coral reefs. In contrast, sea-level lowstands affect coral reef growth negatively depending upon the amount of sea-level drop, its duration, and climatic conditions. Rates of sea-level fall and rise should be taken into account with respect to their control on coral reefs. Rapid vertical sea-level displacement intuitively limits a full physical and biological coral-reef development by imposing continuous environmental disturbance. A slow rate of sea-level change or, better, prolonged sea-level stillstands promote full space occupancy by coral reefs and the establishment of complex biotic relationships among associated fauna.

3. Significant faunal turnovers are also observed in the latest Quaternary of the Red Sea region in response to periods of increased humidity (rainy phases). This is testified by the occurrence in the Red Sea of early Holocene mangals, which at present have disappeared or are strongly reduced in size and complexity. The data presented in this chapter supports the view that even very 'stable' ecosystems like coral reefs or rain forests may react quickly to environmental changes by reorganizing themselves (Stanley, 1995, for an overview).

ACKNOWLEDGEMENTS

This chapter benefits from field evidence gathered during many years of research in the Red Sea basin. The author acknowledges financial support from EC grants SC1*CT91-0719 and ERB SCI*CT92-0814 (RED SED Programme). I am also indebted to: GRSTS of Florence for organizing expeditions in Jordan (1974–1978) and Saudi Arabia (1977–1978) also providing logistic support in Zabargad in 1979 and 1980; CNR (Italian National Research Council) for funding research in Egypt and offshore islands in 1979, 1980 and 1983; William Bosworth (Marathon Petroleum Egypt) for the organization of fieldwork in Zabargad, Sinai and Gebel Zeit (1993–1994); Skipper Renato Marchesan and the crew of motor boat *Ernesto Leoni* for their cooperation in surveying Gubal, Tawila, Shaker and Zabargad Islands in 1991, 1992 and 1993; Michele Miele, Italian Embassy Scientific Attaché in Cairo for helping in logistics. I also thank J.-C. Plaziat and W.-C. Dullo for sharing information on Red Sea Pleistocene coral reefs. Thanks are also due to William Bosworth, Georg Heiss, Lucien Montaggioni, Bruce Purser, Paolo Colantoni, Mara Marchesan, Paul Crevello, Antonio Bonfitto, Rita Impiccini, Marco Oliverio, Marcello Quarantini, W.-C. Dullo, F. Baltzer, Massimo Salmi, Alessandro Olshki, Paolo Notarbartolo, Renato Marchesan, Ilaria Marchesan for their help in the field. I am indebted to Gino Zini (drawing), Paolo Ferrieri (photos), Ylenia Foschi and Alessandro Remia (text and figure preparation). Thanks are due to Marco Roveri for comments on an earlier draft of the chapter and to Dan Bosence whose many suggestions helped to improve the text. This is IGM scientific contribution n. 1022.

Chapter H4
Modern Red Sea coral reefs: a review of their morphologies and zonation

W.-C. Dullo and L. Montaggioni

ABSTRACT

Modern reefs in the Red Sea exhibit different controls on their formation. The majority of the reefs belong to the fringing type. These reefs grow close to the mainland and are absent in wadi mouths. Their arrangement and orientation are strongly controlled by the siliciclastic input from the hinterland. In addition, discharge events due to ephemeral precipitation may cause local but serious damage of the reef-building assemblages. However, the general pattern of reef arrangement follows the morphology of the coastline, which itself portrays the tectonic framework of the Red Sea. Barrier reefs and even atolls also occur, predominantly in the central and southern part of this young ocean. Both reef types show a strong control by the tectonic framework of rift-related origin. The outlines of the reef crest and the orientation of the foreslopes follow the tectonic pattern. Spectacular drop-offs are widespread and represent predominantly fault planes of horsts and graben structures parallel to the rift. Furthermore rift-related salt diapirism occurs, which influences reef shapes, indicated by circular and semicircular outlines. Besides these controls, sea-level changes are a trigger as well. They are documented in onshore and offshore terraces and their present-day position can be attributed to the last glacial–interglacial cycle. These different controls along with the two major reef types and their typical zonation patterns are discussed briefly.

INTRODUCTION

Since its opening in Tertiary times, the Red Sea has been fringed by coral reefs of different proliferation (Cochran 1983b; Purser and Hötzl, 1988). Due to the rift-related origin of the Red Sea, the gross reef physiography seems to have been primarily controlled by plate tectonics rather than by eustasy. However, in relation to this major control, salt diapirism and siliciclastic input are additional factors which may influence present-day morphology and setting.

The first reports concerning Red Sea coral reefs date back to the *post mortem* publication of Forskal (1775), who visited the eastern coast, while the work of Klunzinger (1879) focused mainly on the classification of scleractinian corals from the western coast. The first descriptions of living reefs from a geological point of view were published in the classical work of Johannes Walther (1888) and almost twenty years later, Hume (1906) described continuous fringing reefs along the south-western shore of the Gulf of Aqaba. Modern sedimentological studies were initiated by the work of Nesteroff (1955) around the Farasan Islands. A decade later Friedman (1968) provided a comprehensive inventory on the geology and geochemistry of reefs, carbonate sediments and waters in the Gulf of Aqaba, which is valid for many reefs fringing the Red Sea. Since this classical study a variety of papers has been published dealing with different aspects of the living reefs among which only the most recent ones are cited including references (Montaggioni et al., 1986; Piller and Pervesler, 1989) and reviews by Head (1987) and Sheppard and Sheppard (1991).

The invention of small submersibles has opened a new era of reef research. Fricke and Schuhmacher (1983) and Fricke and Landmann (1983) were the first to study reefs in the Gulf of Aqaba and the Sinai. Subsequently, these investigations were extended into

Sedimentation and Tectonics of Rift Basins: Red Sea–Gulf of Aden. Edited by B.H. Purser and D.W.J. Bosence. Published in 1998 by Chapman & Hall, London. ISBN 0412 73490 7.

the central Red Sea by Dullo *et al.* (1990) and Brachert and Dullo (1990, 1991). Beside the living reefs, geologists have been attracted by the raised Quaternary reefs, recorded at various topographic levels (Dullo, 1990; Gvirtzman, 1994; Strasser *et al.*, 1992).

The margins of the Red Sea and its two northern gulfs have had different tectonic evolutions (Purser and Hötzl, 1988; Roberts and Murray, 1983) and therefore different controls on reef formation since the early rifting stages to the present day. In connection with this tectonic control, siliciclastic input and salt diapirism are other important mechanisms influencing the setting of modern reefs in this desert-enclosed young ocean.

The present chapter discusses briefly the different controls on reef growth in the Red Sea and reef morphology, as well as important aspects of zonation patterns from selected sites (Figure H4.1).

SILICICLASTIC CONTROL

In the northern end of the Red Sea, especially in the Gulf of Aqaba but also in the Gulf of Suez, the marine shores are flanked by rugged mountains which attain 2000 m altitude at a distance less than 20 km from the present coast. While this steep morphology continues below sea level in the Gulf of Aqaba down to 1850 m (Mergner and Schuhmacher, 1974), the Gulf of Suez is shallow with an average bathymetry of 70 m, being almost filled with sediments. The contrasting morphol-

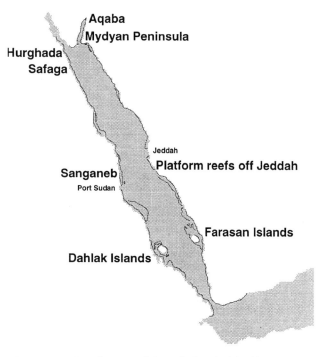

Figure H4.1 Locality map of the reefs sites cited in this paper.

ogies are caused by the different origins of the two gulfs. The Gulf of Suez cuts through the stable Arabo-Nubian shield and represents an aulacogen, while the Gulf of Aqaba is the active continuation of the Red Sea rift, formed as a consequence of transform movements along the Aqaba–Levante structure (Garfunkel *et al.*, 1981).

Coral reefs in the northern Red Sea and the two gulfs are mainly of the fringing type. In the Gulf of Aqaba reefs are attached almost directly to the coast (Mergner and Schuhmacher, 1974) while the development of even small lagoons is an exception. This steeply inclined morphology with small reefs growing directly on the Precambrian basement is also a typical feature of the deeper foreslope environment (Fricke and Landmann, 1983; Fricke and Schuhmacher, 1983). Due to a much shallower environment in the Gulf of Suez, there are reefs transitional from fringing reefs towards barrier reefs, especially along the south-eastern coast near the entrance to the Red Sea.

In general the fringing reefs form a narrow band along the shore (Plate 35(a)), which is only interrupted in areas of wadi mouths (= *sharm* in Arabic). These locations of ephemeral discharge follow the erosional patterns developed during the last glacial maximum, when sea level was more than 120 m lower than today (Gvirtzman *et al.*, 1977). They are pronounced features along the Gulf of Aqaba and most parts of the Red Sea, while they are only subordinate around the Gulf of Suez. When sea level dropped below 70 m during most of the glacial time, nearly the entire gulf floor was emergent.

There is, however, a notable equilibrium between carbonate and siliciclastic sedimentation in front of large, relatively old alluvial fans (Plate 33(b)), where reefs are developed along the seaward margins of fans (Roberts and Murray, 1983). The reef framework protects the siliciclastic sediments from being eroded and redistributed by marine processes. Most of these fans originated during the Pleistocene (Gvirtzman *et al.*, 1977; Hayward, 1982) and are capped with Holocene sediments (Al Sayari *et al.*, 1984).

Locally, the belt of modern reefs situated in front of alluvial fans on which they developed, exhibit the shape of barrier reefs due to the diverse topography of the substratum. Present-day drainage patterns on the fans even control the small-scale morphology of these reefs, indicated by small channels and furrows which display the same pattern and orientation as the drainage system on land (Gvirtzman *et al.*, 1977; Dullo, 1990). In addition, the pools on broader reef-flats are related to channels which extend from wadis into the reef.

The fact that reefs do not flourish or even colonize large coastal re-entrants and small cuts or channels through the fringes and barriers, suggests that land-derived siliciclastics are still transported through these pathways. Simple sediment traps in the fringing reefs near Aqaba have confirmed this fact, having up to 20% siliciclastics (Schuhmacher et al., 1995). These channels are also the prominent pathways to export skeletal grains provided by the reef carbonate 'factory'. Submersible observations (Fricke and Landmann, 1983; Brachert and Dullo, 1991) have revealed that large volumes of sediment are actively moving down these pathways, sometimes causing submarine erosion.

Modern sediment flux to the coast is governed by the arid climate and is therefore ephemeral. Sudden sediment transport can cause major damage to the living reef environment, although much of the sediment is transported via the existing pathways. Severe damage is rare and occurs in the order of hundreds of years (Hayward, 1982). However, some floods may be so violent that larger areas of reefs become inundated and covered by siliciclastics up to cobble size (Dullo, 1990). Because of the localized sediment delivery to reefs through the adjacent terrestrial drainage patterns and varying flood intensities, sediment inundation is generally incomplete and local. Therefore, recolonization of damaged reef areas may occur subsequently to the flooding event. Although cobbles seem to be difficult to recolonize, coralline algae and scleractinians have high potential to settle on these unstable substrates and in doing so, they even stabilize this specific sediment (Dullo and Hecht, 1990). Numerous examples can be seen in the uplifted terraces, where reefs grow directly on cobble-sized flood sediments (Dullo, 1990). Therefore, siliciclastic sedimentary processes are essential to form the substratum for subsequent reef growth (Hayward, 1982).

In combination with these ephemerally transported coarse siliciclastics, fine-grained material is spread around the reef. However, in general the sediment load is too small and the periods of turbid sea water too brief, to influence reef growth except around the distributional channels.

MODERN REEFS OF THE AQABA–EILAT GULF AREAS

Biozonation of fringing reefs

The following natural environments can be distinguished on the well-developed reef areas, from the shore seawards (Gabrié and Montaggioni, 1982) (Figure H4.2).

1. The Beach Zone. Quartzofeldspathic material collected by nearby wadis from surrounding mountains are partly trapped onshore, forming poorly developed beaches, about 20 m in width. Fauna includes molluscs (*Littorinidae*, *Ostreidae*) and crustaceans (crabs, cirripeds).

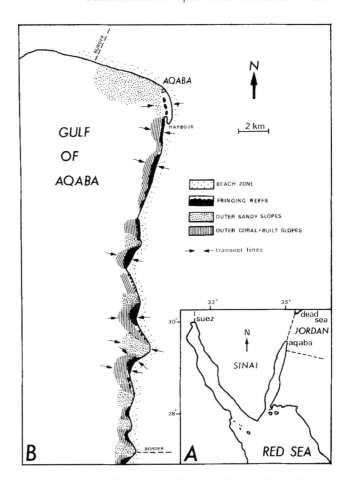

Figure H4.2 Distribution of the fringing reefs and adjacent depositional zones in the Gulf of Aqaba, Jordan coast.

2. The Back-reef Zone (Plate 35(c)). Present only in the widest, central, parts of reefs, the back-reef zone is a sandy depression between the beach and the reef-flat, having a maximum width of 40 m and a depth of 1.5 to 2 m below mean sea level. Bottom sediment is colonized by scattered coral heads (*Stylophora, Seriatopora, Platygyra, Millepora*), alcyonarian colonies (*Lithophyton, Cladiella, Sinularia*) and seagrass beds (*Halophila, Halodule*).

3. The Reef-flat Zone (Plate 35(d)). The zone, about 20 m wide, is a dead coral pavement, with a 0.5–1 m rear-reef step. Coral communities are composed of small-sized colonies (*Stylophora, Seriatopora, Acropora*, mainly) associated with hydrocorals (*Millepora*). The reef front forms a nearly vertical drop-off, 2–4 m high, characterized by the occurrence of *Millepora* and branching red algae.

4. The Outer Slope Zone. Two main types of outer slopes (Figure H4.3) can be defined according to the nature of substrates: loose sedimentary slopes, coral-built slopes.

The first type of slope is a sandy talus found in the vicinity of embayments. Dense mats of scattered patches

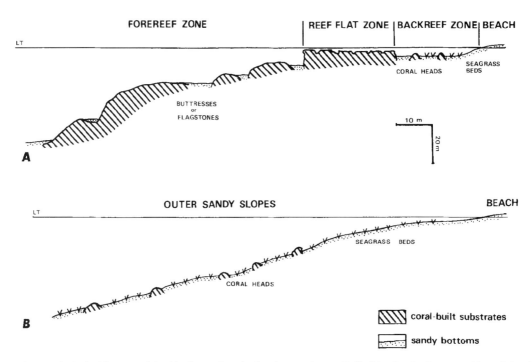

Figure H4.3 Main morphological features of the fringing reefs and related outer slopes. Gulf of Aqaba, Jordan coast (A. well-developed reefs; B. embayment areas).

of *Halophila* plants cover the bottom down to 50 m. No coral framework has developed at depths of 1–50 m. The second type of slope is best developed close to headlands (Figure H4.4), where it occurs in the form of a coral-built fore-reef. The upper part of the slope (0–20 m) is dominated by branching growth forms (*Stylophora, Seriatopora, Acropora, Echinopora*). The lower fore-reef zone (20–50 m) exhibits prolific massive growth-forms (*Montipora, Astreopora, Porites*, and various *Faviidae*). A series of sandy pools generally remains conspicuous between 5 to 10 m; at depths of 15 to 50 m, only small pockets of sediment accumulate on the coral-built surfaces.

Classification of reefal sediments from the Gulf of Aqaba resulted in the definition of several facies (Gabrié and Montaggioni, 1982). The beach comprises a terrigenous, well-sorted, medium-sand facies. The backreef zone is typified by a mixed terrigenous-coral, fine-sand facies and, on the basis of foraminiferal composition, a milliolid-peneroplid subfacies. The reef-flat zone exhibits a coralgal, poorly sorted coarse-sand facies, with a homotrematid subfacies. Along the outer slope, various facies are related to the nature of substrate and water depth: (i) terrigenous well-sorted fine-sand facies, close to wadi mouths; (ii) coralgal, poorly sorted coarse-sand facies and acervulinid-homotrematid subfacies, on the upper fore-reef areas (0–20 m); (iii) coral–molluscan–foraminiferal, poorly sorted medium-sand facies, enriched in amphisteginids, along the lower fore-reef areas.

Figure H4.4 Fore-reef buttress zone dominated by *Acropora* colonies. Fringing reef near Sharm Mujjawan, Gulf of Aqaba, Saudi Arabia.

Origin of reef morphologies

As emphasized by Gvirtzman *et al.* (1977) and Gvirtzman (1994), since the middle Pleistocene, reef-building organisms have repeatedly recolonized the same localities during each successive sea-level rise along the Aqaba–Eilat Gulf coasts. Each recolonization stage occurs in the form of a thin (less than 8 m), narrow biogenic pavement, formed on the reef-flat or, more frequently, on the outer slope of an older reef, uplifted during episodic tectonic activity. Therefore, the development of reef systems in the northern Red Sea cannot be explained by their relative youth (from 10 000 to 13 000 years BP according to Gvirtzman et al., 1977). Gvirtzman et al. ascribed what they called limited development of coral reefs in the Gulf of Aqaba–Eilat to the lack of vitality of hermatypic corals in these relatively high latitudes. This hypothesis cannot be confirmed by our own observations although on the reef-flat itself, oligospecificity and reduced growth of coral assemblages occur. In contrast, the coral assemblages of the outer slopes are highly diversified and very dense and linear growth rates of corals are comparable to those of the Indian Ocean (Heiss, 1994). Furthermore, evaluations of the carbonate budget also indicate prolific reef growth (Heiss, 1996). Thus, the stability of the morphological characteristics of these reefs suggests that their evolution is controlled both by geological and ecological factors. Pre-existing topography has played a major role: the absence of continental shelf and the very steep inclination of submarine slopes has always limited the seaward net accretion. Further more, the past climate of the Red Sea during the entire Holocene was generally similar to that of today, i.e. very arid (annual rainfall at Aqaba 25 mm; average yearly evaporation 2000 mm), which determines a negative hydrologic balance, and consequently hypersaline waters (41.5 to 43‰) and periods of exceptional emersions, detrimental to coral life. During the glacial periods, the climate was cooler and more humid (Gvirtzmann et al., 1977) and intense erosion took place, creating deep channels and spreading alluvial sediments into the centre of the bays. The tectonic activity which characterizes this region is also responsible for mass mortality of coral assemblages following sudden and irreversible uplifts, the latest dating about 2200 years BP (Montaggioni, unpublished).

The most remarkable characteristic of outer reef-flats along the Aqaba coasts is the vertical drop-off and the absence of spur-and-groove structures which are typical of most reef formations. Spur and grooves are features which form in response to hydrodynamic actions (Roberts et al., 1975) and which generally develop on reef fronts in exposed or very exposed conditions. The calm conditions in the Gulf of Aqaba would therefore be the likely cause of the absence of spur-and-groove structures. The existence of spurs and grooves on the reefs of the west coast of the gulf (Friedman, 1968; Sneh and Friedman, 1980) probably results from the differential effect of the dominant winds, which blow predominantly from the north and north-east, generating marine currents which are relatively stronger along the western shores.

The two types of outer slope represent different stages with a common origin. The stage reached by an outer slope is in part determined by the proximity of terrestrial water run-off. The local absence of any buildup in close proximity to the wadis may be explained by the occasional fresh-water run-off and input of terrigenous sediments, detrimental to coral growth. On the contrary, rocky headlands are probably more favourable to a lush coral growth.

TECTONIC CONTROL

Since the late Oligocene, the sedimentary evolution of the Red Sea basin and its margins has been controlled tectonically (Purser and Hötzl, 1988; Bosworth 1994a). The orientation of fringing, barrier reefs and even the scattered atoll-like structures parallel to the coastline and to major fault structures demonstrates this very clearly. All reefs in the Gulf of Aqaba and most reefs in the northern and central Red Sea are oriented parallel to the main axis of the rift.

In addition to the orientation the almost vertical foreslopes in the shallow water of up to 50 m relief, and its exposed cliffs, demonstrate a strong tectonic control of the reef morphology. The deeper foreslopes, 90 m below sea level, exhibit even higher tectonic cliffs (Brachert and Dullo, 1990).

Furthermore, the numerous faults affecting certain onshore terraces of both siliciclastic sediments and carbonates of marine origin also indicate this tectonic control. Spectacular examples crop out especially in the southern part of the Gulf of Aqaba, where Eemian reefs are uplifted 20 m above present sea level and older Pleistocene reefs occur at levels of about 98 m (Dullo, 1990). This tectonic displacement is still active as it caused the most recent earthquake in this region in November 1995. Examples of fault displacement of young sediments are seen in numerous coastal localities (Briem and Blümel, 1984).

Uplifted Holocene reefs

In the Gulf of Suez and along the Egyptian Red Sea coast, exposed Holocene reef-flats occur near Hurghada and Safaga and Giftun islands. They vary in altitude between 0.2 and 0.5 m above the present-day reef-flats. North of Hurghada, however, and on the adjacent islands of Tawila and Gubal, no elevated Holocene reef-flats are present (personal observation). ^{14}C-dates (Table H4.1) indicate that the emergent reefs are between 3900 and 5500-years-old. These reefs are of two types: (i) stepped, horizontal terraces fringing the continental coastline, and (ii) gently seaward-sloping surfaces around islands. This strongly suggests sharp, rapid uplift, probably related to rifting, and progressive tilting, presumably controlled by evaporite tectonics or block rotation (Bosworth, 1994a).

Table H4.1 Radiocarbon dates of scleractinian corals, exposed reef flats, Gulf of Suez (Egypt)

Laboratory code	Altitude above present mean low tide level (m)	Material dated	Mineralogy (A = aragonite; Hmc = high magnesian calcite)	Conventional age (years BP)
LGQ 648	0.40	*Acropora*	A	5580 ± 140
LGQ 649	0.40	*Acropora*	A	5320 ± 130
LGQ 650	0.40	*Acropora*	A	3950 ± 130
LGQ 651	0.45	*Leptastrea*	A/Hmc	5500 ± 130
LGQ 652	0.45	*Faviid*	A/Hmc	3930 ± 130
LGQ 653	0.45	*Faviid*	A/Hmc	4650 ± 150
LGQ 654	0.20	*Acropora*	A	4920 ± 140
LGQ 659	0.20	*Favites*	A/Hmc	4960 ± 140

Similarly, exposed remains of former reef-flats, coral heads and beach-rocks are found along the eastern coast of the Red Sea, from Aqaba (Jordan) to Shu'ayaba (Saudi Arabia), at altitudes ranging from 0.50 to 1.80 m above mean low tide level (Montaggioni, personal observation). Their ages vary between 2000 and 6200 years BP (Table H4.2). These terraces are thought to have resulted from differential uplift along coast-parallel faults and subsequently suffered wave erosion. This is in accordance with previous works from the Jordan and Saudi Arabian coasts in the Red Sea (Hötzl et al., 1984).

Given the distribution pattern of the Holocene reef exposures, it may be assumed that the latitude of Hurghada (i.e. entrance to the Gulf of Suez) coincides with two tectonic provinces, namely a southern region which is relatively rising and a northern region which, at least today, is relatively stable or subsiding (Bosworth, 1994a).

Recent demise of reef assemblages

In addition to exposed reefal deposits, undisturbed *in situ* dead coral heads, pavements and molluscan assemblages were found on modern reefs, throughout the eastern Red Sea shorelines. These features are located at depths ranging between 0.20 and 1.20 m below present-day mean low tide level, which is above the expected depth range of these organisms. ^{14}C-dates indicate they were flourishing reefs, between 4600 and approximately 2000 years BP (Table H4.3). At present, these deposits occur in various reef zones, mainly in back-reef zones. The coral build ups are commonly covered by mangrove-bearing muddy sands. The molluscan assemblages are typical of back-reef endofauna (Lucinids such as *Codakia divergens*, Cerithids); which are fossilized in the form of shelly floatstone crusts, 2–7 cm thick, exposed at low spring tide. The demise of the coral and molluscan communities is believed to have resulted from recent, low-amplitude, vertical movements, i.e. uplifting or tilting according to the sites considered, thereby modifying the original depth range of the relevant biological assemblages.

Table H4.2 Radiocarbon dates of scleractinian corals, exposed reef flats and associated deposits, eastern coast of the Red Sea (Jordan and Saudi Arabia)

Laboratory code	Location	Altitude above mean low tide level (m)	Material dated	Mineralogy (A = aragonite; Hmc = high magnesian calcite)	Conventional age (years BP)
Gak 11405	South of Jeddah (reef islet)	0.60	*Porites*	A	2140 ± 100
Gak 11410	Shu'ayaba (Saudi Arabia) 20°50'N; 39°20'S)	1.20	Beach-rock	A/Hmc	3160 ± 110
Gak 11411	- idem -	1.80	Beach-rock	A/Hmc	2950 ± 130
Gak 9434	Gulf of Aqaba (29°26'42"N; 34°58'02"S)	0.50	*Faviid*	A	5600 ± 150
Gak 9436	- idem -	0.50	*Faviid*	A	5820 ± 120
Gak 9437	- idem -	0.55	*Faviid*	A/Hmc	6210 ± 130
Gak 9438	- idem -	0.45	*Faviid*	A/Hmc	5570 ± 280

Table H4.3 Radiocarbon dates of coral and molluscan samples from dead reefal pavements, Jordan and Saudi Arabia

Laboratory code	Location	Depth below mean low tide level (m)	Material dated	Mineralogy (A = aragonite; Hmc = high magnesian calcite; Lmc = low magnesian calcite)	Conventional age (years BP)
Gak 11401	Aqaba	1.00	coral pavement	A/Hmc	1860 ± 100
Gak 11403	Aqaba	1.20	coral pavement	A/Hmc	1500 ± 130
Gak 11404	Saudi Arabia Shu'ayaba swamps	0.40	coral pavement	A/Hmc	3400 ± 110
Gak 11406	Eliza shoals, Jeddah, Saudi Arabia	0.10	Molluscan floatstone	A/Hmc/Lmc	2720 ± 100
Gak 11409	Shu'ayaba reef, Saudi Arabia	0.70	coral pavement	A/Hmc/Lmc	2780 ± 110
Gak 11413	Old King's Palace, North of Jeddah, Saudi Arabia	0.10	coral pavement (*Goniastrea*)	A/Hmc	4620 ± 110
Gak 11414	- idem -	0.10	skeletal floatstone	A/Hmc	2280 ± 80
Gak 11415	- idem -	0.10	coral pavement (*Porites*)	A/Hmc	3300 ± 120

Tectonic control of the shape of Sanganeb Atoll

Sanganeb Atoll is located in the central Red Sea, offshore Sudan. It is an almost closed structure, open only to the leeward western-side. The elongate shape has a length of 5.5 km and a width of almost 2 km. Maximum depth of the lagoon is a little less than 60 m (Figure H4.5). The trend of the two margins in the east and the west run parallel to the major structural patterns in the central Red Sea at 010°.

This atoll displays steeply inclined or even vertical slopes (Figure H4.6) which differ significantly on their leeward and windward margins. Windward margins have a prominent terrace at about 90 m and below 95 m the foreslope is almost vertical (Figure H4.7). The leeward margin, in contrast, is subdivided by a number of small terraces each covered with sand. The deeper wall again is almost vertical.

Below 120 m on leeward and windward margins, the steep foreslopes are characterized either by flat surfaces or vertical spurs and grooves having a spacing which may range between 2 and 3 m. The walls display a rugged surface (Figure H4.7) due to numerous ledges (Brachert and Dullo, 1990). These ledges are only a few centimetres thick and protrude about 25 cm horizontally from the cliffs. They are covered with loose sediment derived from the living reef in the shallow water. The internal structure of these ledges exhibits a fabric, characterized by frequent borings with subsequent infilling, lithification and renewed boring. This leads to an almost obliteration of the primary skeletal fabric of deep-water benthic organisms such as azooxanthellate corals, bryozoans and serpulids. A detailed description of the facies patterns and the mode of ledge rock formation is given by Brachert and Dullo (1990, 1991).

These submersible investigations were limited to 200 m water depth due to the diving limit of the submersible. Therefore, continuation of the steep slope down into the basin was checked by means of boomer seismics (500–2000 Hz). A more detailed description of the slope morphology including the seismic pattern in the lagoons is given by Reijmer *et al.* (submitted) (Figure H4.8). Boomer profiles oriented perpendicular to the margins show a continuation of the steep slope down to about 750 m. However, the windward margin displays several internal steps below 250 m, which are interpreted as fault-related displacements because the steep wall apparently is a fault plane. The leeward

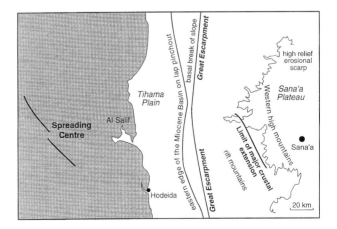

Figure H4.5 Sketch of the Sanganeb Atoll, showing the deep slopes investigated by submersible. After Brachert and Dullo, 1990.

margin shows a similar steep slope down to about 750 m (Figure H4.8), but has no internal steps. Displacements, however, are seen within the lagoon sediments where they form a series of westward-dipping steps, which are obviously caused by normal faults and microhorst structures.

The tectonic pattern of combined horst and graben structures, which show a general trend of 010°, also applies to the overall framework for the atoll. It represents a tectonic horst, which is subdivided into minor steps by normal faulting (Figure H4.9). The continuous faulting and very gentle tilting towards the west is indicated by minor reef growth on the leeward (west) margin. This tectonic control is also seen in the barrier-reef system running parallel to the coastline off Sudan, where similar steep slope morphologies occur (Brachert and Dullo, 1990).

The modern offshore knoll reef system of Jeddah area

In the vicinity of Jeddah and along the major part of the western coast of Saudi Arabia, coral reefs occur as two morphologically distinct systems: the fringing reef and the offshore reef platform. The former constitutes an almost continuous belt along the shore and often has wide back-reef zones. About 3 km offshore, the marginal shelf area consists of a shallow (average depth 20 m) regular platform, covering a surface area of 800 km². It is bound both seawards and shorewards by near-vertical escarpments marking the edge of the Red Sea trough. Attention has been drawn to the steep, ocean-facing topography with rapid drops to depths of 400–800 m (Behairy and El Sayed, 1983). These observations suggest that the offshore platform, together with the main coastal alignments and the coast-paralleling trench deeper than 350 m, have been defined by large-scale faults (Guilcher, 1982; Coleman, 1977). The platform is occupied by a system of scattered reefs, rising from inter-reefal sandy bottoms. Seismic and drilling data reported by Berry *et al.* (1966) from the nearby Sudanese reef-platforms indicate that the modern central Red Sea reefs are plastered on to the surfaces of Pleistocene reef limestones that are about 200 m thick which, in turn, lie on the eroded surface of a Tertiary marine series. The total thickness of Holocene reef growth is uncertain (Braithwaite, 1982). Based on physiographical and biological criteria, the knoll reef system can be divided into the following zones (Figure H4.10).

1. The offshore reef bodies. The tops of the flattened tops of the reef bodies consist of poorly zoned reef-flats mainly occurring as a pavement with scattered coral colonies (Plate 36(a)). The coral community of the innermost part is dominated by *Stylophora pistillata*. Coralline algae are represented by massive branching and crustose forms (*Lithothamnium, Spongites*). The soft sediments deposited on top of reef frameworks form either scattered and ephemeral sandy pockets, 2–10 cm thick or larger bodies, up to 1 m thick, which are likely to be retained permanently within wide ponds.

Figure H4.6 Steep slope at −40 m on the windward margin of Sanganeb atoll. Note divers for scale.

Figure H4.7 Vertical wall at −195 m on the windward margin of Sanganeb atoll, spacing of the channels is 2 m.

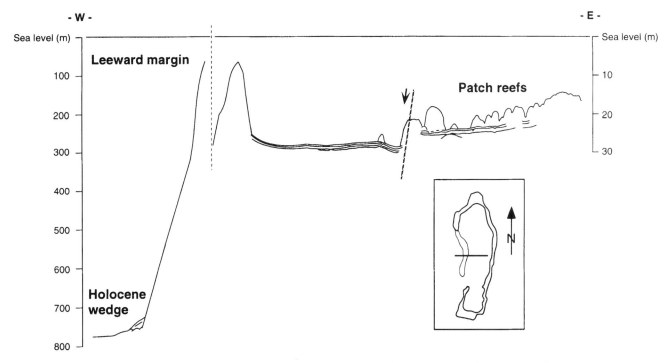

Figure H4.8 Interpretation of a boomer line through the leeward margin of Sanganeb atoll. Note the different scaling in bathymetry. Overall length of the section is 0.9 nautical miles.

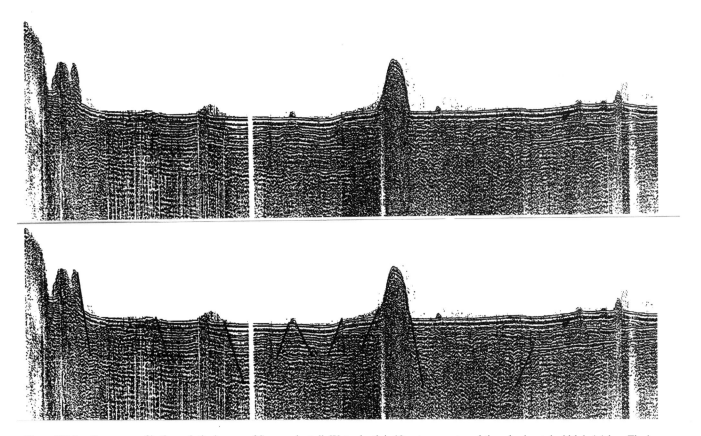

Figure H4.9 Boomer profile through the lagoon of Sanganeb atoll. Waterdepth is 10 m two way travel time, horizontal width is 1.1 km. The lower interpreted profile exhibits several faults.

The upper parts of the steeply sloping reef flanks are typically settled by the hydrocoral *Millepora*. Some scleractinians are also characteristic of this zone (*Pavona, Goniastrea, Leptoria, Hydnophora, Oulophyllia, Symphyllia*). Skeletal material occurs as sediment pockets infilling cavities in the coral-built framework.

The base of the reef flanks consists of a gentle sloping sandy talus (15–38 m) with scattered protruding framework having up to 1 m relief. This zone is characterized by a high species diversity. Rocky outcrops have a patchy cover of coral species belonging to the genera *Stylocoeniella, Acropora, Astreopora, Alveopora, Pachyseris, Podabacia*, associated with alcyonarians.

2. The offshore inter-reef areas. They form sandy bottoms occupied by scattered to densely packed coral knobs and patches. Some coral forms are linked closely to this type of sandy biota (*Psammocora, Acropora scandens, Cycloseris, Fungia, Siderastrea*).

3. The fringing fore-reef zone. This consists of a vertical, 20–25 m drop-off, changing seawards into a gentle sandy slope followed by a sandy plain. The top of the fore-reef exhibits biological communities which are similar to those mentioned from the offshore upper reef flanks. The corals which dominate this zone are *Millepora, Pavona, Goniastrea, Leptoria, Acropora* which colonize the margins of grooves. In deeper waters, large heads of *Porites*, associated with *Montipora, Favia, Favites, Platygyra* and alcyonarians are encountered. At the foot of the walls, under overhangs, the dominating corals are *Acropora, Astreopora, Alveopora, Echinophyllia* and *Pachyseris*. The great homogeneity in the community composition and distribution of both the fringing reef and the offshore reef platform reflects the lack of lateral changes in ecological conditions between the two systems. Only the foraminiferal assemblages, dominated by miliolids and peneroplids, appear to be markedly different in composition.

A general facies model was proposed based on the major representative sedimentary types (Montaggioni et al., 1986). The offshore knoll platform system has no significant lateral facies zonation. This is consistent, to a great extent, with the physiographical and ecological attributes of the offshore platform. Knolls and table reefs with poorly zoned surfaces are symmetrically distributed and occur along the seaward edge and the landward edge of the platform. The absence of lateral zonation may be ascribed both to the relatively low diversity of the reef-associated biota which lack the required ecological spectrum, and to the apparently homogeneous water-energy conditions which prevail across the platform. However, just as the offshore buildups are ecologically zoned with respect to depth, distinct sedimentary facies are vertically discernable. These comprise a combined reef-top and upper reef-flank facies (i.e. *Tubipora*-encrusting foraminiferal-bryozoan) and a combined lower reef-flank and inter-reef facies (i.e. molluscan/free-living foraminiferal) respectively. This general zonation pattern is also valid for the Sanganeb atoll.

SALT DIAPIR CONTROL ON REEFS

While extensional tectonics control most of the overall present-day arrangement and orientation of the reefs in the Red Sea, there is also a control due to salt diapirism, which may follow the major tectonic lineaments (north-north-west–south-south-east). Salt diapirism as a control on reef formation is most obvious in high quality maps or aerial photographs. The reefs show always

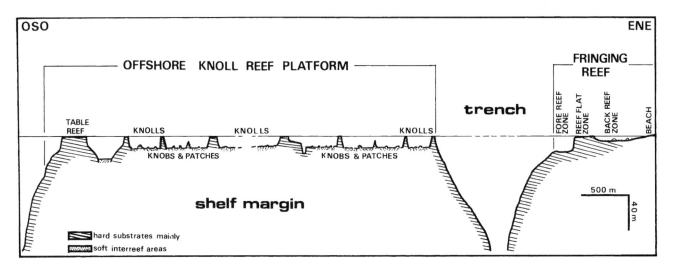

Figure H4.10 Physiography and morphology of the modern reefs from the Saudi Arabian coast, north of Jeddah.

circular or semicircular outlines of their reef crest. Good examples are the reefs south of the Midyan peninsula, where the Gulf of Aqaba enters the Red Sea and along the Egyptian coast. Other beautiful examples are the two islands groups in the southern Red Sea, the Farasan Islands (Figure H4.11) and the Dahlak archipelago (see Carbone *et al.*, this volume).

These two archipelagos are located in the southern part of the basin, where the desert-enclosed 'ocean' reaches its maximum width of 360 km. The water depth around the islands does not exceed 100 m. The semicircular geometries of the reefs, which mainly follow semicircles affecting nearby coast-lines, measure several tens of kilometres. Their shape suggests ascending salt domes partly buried by sediment.

The Farasan Islands comprise a more or less uniform reef-flat with faulting and uplift, strongly influenced by salt diapirism. The exposed reef limestones onshore are mainly of Eemian age (Dabbagh *et al.*, 1984) and are elevated less than 15 m above present sea level. These reef limestones, underlain by marly limestones, pass down into yellow and green marly limestones, which show intercalations of clay rich in fish remains and diatoms (Dabbagh *et al.*, 1984). They 'onlap' dome-shaped gypsum and anhydrite, which are uppermost Miocene in age (Skipwith, 1973). The islands themselves exhibit a tectonic pattern similar to that of the Sanganeb atoll. Graben and horst structures run parallel to the main Red Sea graben, although regional faulting of the Farasan Bank was responsible for the delimitation and arrangement of the islands. The present-day fracturing of the islands is due to diapirism and thus, partly causes gravitational sliding (Dabbagh *et al.*, 1984).

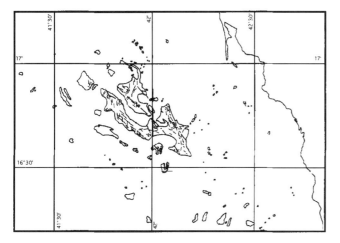

Figure H4.11 Sketch map of the Farasan Islands. Note the frequent semicircular shapes of the coastline on the small reef islands, which have been caused by diapirism. Major tectonic lineaments are indicated. After Dabbagh *et al.*, 1984.

SEA-LEVEL CONTROL AND REEF TERRACES

Besides tectonic displacements and uplift, there are several records of reef formations due to sea-level changes. Most obvious are those occurring on land as terraces or fossil wave notches. Although there is still ongoing tectonic activity which caused the recent demise of coral assemblages, there is a prominent highstand of sea level during early Holocene (Friedman, 1968; Gvirtzman, 1994). This sea level occurs 1 m above the present sea level and has been dated to 5.2 ka BP (Gvirtzman, 1994) which is isotope stage 1.1 (Pisias *et al.*, 1984).

Older highstands occurred during the period of the last interglacial. In general, during isotope stage 5 (80–125 ka BP) several reef-terraces were formed, and those of Eemian age (isotope stage 5.5: 120–125 ka BP) being the most widespread along the Red Sea coast (e.g. Dullo, 1990). They occur up to 20 m above present sea level. However, younger onshore terraces also occur at about 7 m above present sea level, which correspond to isotope stage 5.1 and 5.3 (Gvirtzman, 1994). This elevation is normally ascribed to reefs of the substage 5.5 (Bloom *et al.*, 1974). Thus, this abnormally high elevation reflects the tectonic overprint affecting ancient sea-level markers. In addition there are also terraces present along the whole Red Sea coast, which are attributed to the isotope stages 7.1 (206 ka BP) and 9.3 (310 ka BP mostly referred to in literature as older than 250 ka). These terraces are elevated to around 15 m and 30 m above present sea level; however, tectonic displacement may result in different altitudes (Dullo, 1990).

Submarine morphologies of the fringing and barrier reefs exhibit several prominent steps among which those at about −20 m, −60 m and −90 m are the most prominent. According to Gvirtzman (1994) the submarine terrace at −20 m was formed during isotope stage 5.2 (91.0 ± 6.8 ka BP).

The deeper terrace situated at about −60 m was formed during isotope stage 4.2 (64.0 ± 6.4 ka BP), while for the next deeper terrace no dates are currently available (Gvirtzman, 1994). However, one may speculate in accordance with other observations in the Indian Ocean (Dullo *et al*, 1997) that the prominent terrace level at about 90 m was formed during the late isotope stage 3 (37–28 ka BP). On top of the two reef-terraces located at about −90 m and −60 m small framework buildups began to grow during conditions of rising sea level after the last glacial maximum, and they now exhibit clear morphologies of drowned reefs (Plate 36 (b)) presumably due to meltwater pulses.

CONCLUSIONS

The arrangement and morphological shape of the different types of modern coral reefs in the Red Sea depend strongly on the tectonic framework of rift related origin. The general orientation of the reefs follows the overall pattern of rift-related faults. This is frequently documented by extremely steep and almost vertical foreslopes. Furthermore, many atoll-like structures result from graben and horst tectonics. Siliciclastic input controls mainly the fringing reefs, which are interrupted near wadi mouths due to the distribution pattern of run-off waters. In combination with rift related tectonics there is a pronounced control by salt diapirism, especially in the southern part of the Red Sea around the two major island groups and in the north the entrance of the two gulfs into the Red Sea. Eustatic sea level changes have influenced the present-day bathymetrical morphology of the reefs, especially the submarine terraces as well as those onshore. Furthermore, the last sea-level lowstand during maximum glaciation has caused erosion and karstification, still seen in present-day reef geometries. However, in conclusion the major control appears to be the rift-related geodynamics.

ACKNOWLEDGEMENTS

Fieldwork was financially supported by the Deutsche Forschungsgemeinschaft 'Evolution of Reefs', coordinated by Professor E. Flügel (Erlangen), the Mission Océanographique Française au Moyen-Orient, led by Professor J. Jaubert (Nice), and by the RED SED programme of the European Community, led by Professor B. Purser, which is gratefully acknowledged. Critical reading and reviewing of the first draft were done by the editors as well as by T. Scoffin which is gratefully acknowledged.

Chapter H5
Tectonic geomorphology and rates of crustal processes along the Red Sea margin, north-west Yemen

I. Davison, M. R. Tatnell, L. A. Owen, G. Jenkins and J. Baker

ABSTRACT

The geomorphology of an extended continental margin of the Red Sea is examined in north-west Yemen where there are tight constraints on the temporal evolution of the margin from the start of plume volcanism at 30.5 Ma through to the present day. Red Sea rifting was preceded by impingement of the Afar plume which produced a large range of mountains reaching up to 3.7 km above MSL. A new drainage system has evolved from an elevated flat plateau since plume-related volcanism and uplift began. The main wadis have deep incised valleys which run almost perpendicular to the Red Sea coastline and major extensional faults, indicating that gravity is the main control on their orientation. The rate of surface uplift in the region since 30.5 Ma has averaged at least 0.08 mm/year as pre-volcanic Late Cretaceous/Tertiary marine sandstones are now situated at an elevation of up to 2.65 km above MSL. The resultant denudation was sufficient to produce enough sedimentation to continually fill the accommodation space created by subsidence and extension along the margins of the Red Sea. Up to 8 km of clastic sediments and evaporites have been deposited since early Miocene times (approximately 24 Ma) in the Red Sea. This material indicates that an average of 3 km of denudation has taken place across the Yemen Highlands, implying that the original volcanic rock volume may have been up to three times larger than that preserved at the present day.

INTRODUCTION

Active extensional tectonics produces new topography rapidly, controlling erosional and depositional processes, and strongly affecting the nature of drainage and slope development. There is very little published work on large-scale geomorphological analysis of extensional continental margins (Ollier, 1985; Gilchrist and Summerfield, 1990; Gilchrist et al., 1994; Kooi and Beaumont, 1994; Steckler and Omar, 1994) and this is the first study of its kind which concentrates on the landscape evolution of the eastern Red Sea margin. This study examines a 100 km long (north–south) segment of the active eastern margin of the Red Sea in north-west Yemen, eastwards from the Red Sea spreading centre to the drainage divide with the great desert of Arabia, which lies approximately 300 km east of the Red Sea coastline (Figures H5.1, H5.2). The aim of the chapter is to quantify geomorphological, tectonic and magmatic processes at this spreading margin, and to describe the tectonic geomorphology using detailed 1:50 000 topographic maps and observations of the surface geology. Our study is integrated with accurate dating of the flood volcanic province (Baker et al., 1996a) and associated plutons (Zumbo et al., 1995; S. Blakey, unpublished), which provide important controls on the rates of crustal processes associated with Red Sea opening. Constraints on the timing and magnitude of extension and denudation will be discussed, showing the influence of tectonics and rock type on landscape evolution from 30.5 Ma through to the present. This quantitative approach aims to improve the understanding of the landscape evolution at an actively extending margin situated above a decaying mantle plume, that has produced the youngest major continental flood volcanic province in the world which is associated with an oceanic spreading centre.

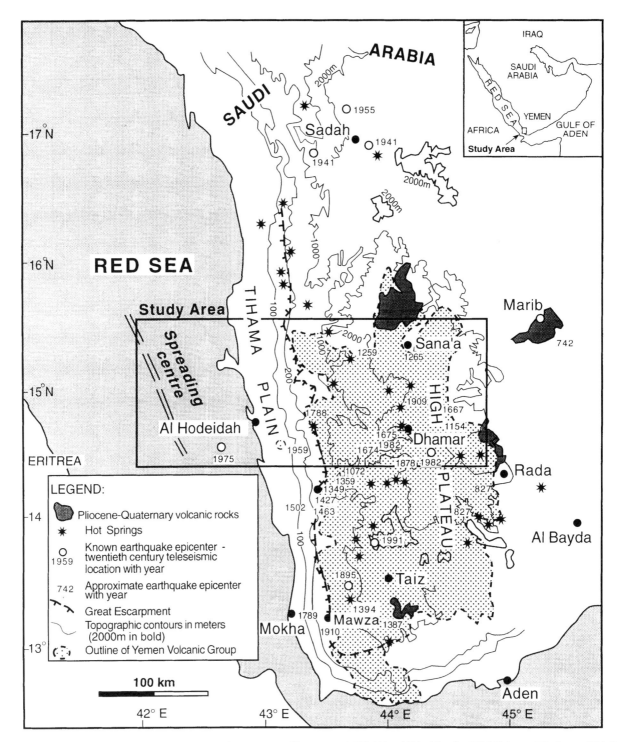

Figure H5.1 Topography of western Yemen. Data from topographic maps at 1:250 000 published by the UK Ministry of Defence, 1986. Hot spring locations from El Shatoury et al. (1979) and earthquake epicentres from Ambraseys and Melville (1983).

GEOLOGIC HISTORY

The initial rifting of the Red Sea was related to the emplacement of the Afar plume, which caused extensive flood volcanism between 30.5 and 26.5 Ma in the study area followed by synchronous crustal extension, intrusive magmatism and uplift (Menzies et al., 1992;

Davison et al., 1994; Baker et al., 1996a). Restricted volcanism has continued to the present day with the most active area on the Sana'a Plateau and in the Dhamar area (Manetti et al., 1991; Figure H5.1), where the latest eruption occurred in 1937 (Macdonald, 1972; Ambraseys et al., 1994). Many earthquakes with un-

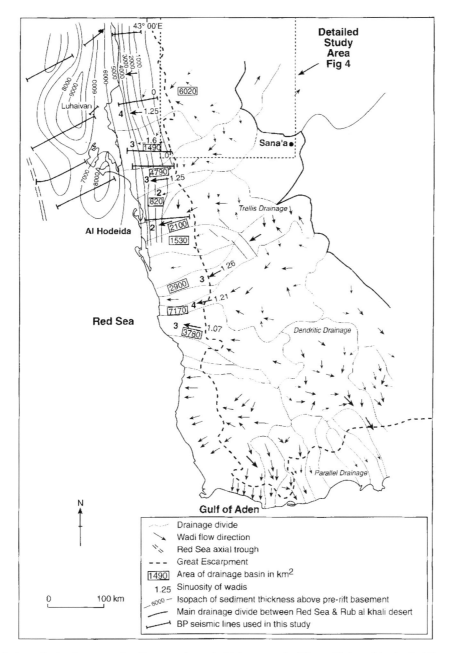

Figure H5.2 Major drainage basin shapes, total sedimentary isopach thickness on the Tihama Plain and in the Red Sea and river sinuosity in western Yemen.

known focal mechanisms have been recorded in Yemen in historical times and several thousand people were killed in Dhamar (Ms = 5.7) in 1982 (Plafker *et al.*, 1978; Ambraseys and Melville, 1983; Ambraseys *et al.*, 1994). The present topographic expression of the rift-flanks (maximum elevation *c.* 3.7 km) is the result of the interaction of magmatic and structural processes which produced rock and surface uplift (*sensu* England and Molnar, 1990), and surface processes which modify the topography and denude the upper crust. Major extension only began after the main flood volcanic phase, and accompanied the intrusion of the granite plutons from 27.5 to 20.9 Ma, but probably continued at least until 16.1 Ma, the age of the youngest intrusive dikes.

Prior to Red Sea rifting palaeocurrents in the fluvial sandstones of the Tawilah Group of the Cretaceous to early Tertiary age are all directed towards the east and north east in north-west Yemen (Al Subbary, 1990, Al Subbary and Nichols, 1991, Al Subbary *et al.*, 1993, and this volume). All major wadis near the Red Sea now flow approximately westwards into the Red Sea. Hence, the drainage observed at the present day has evolved in response to the creation of the Red Sea depression produced by crustal extension between approximately

27.5 and 10 Ma (Davison *et al.*, 1994). There is probably no influence of any pre-existing drainage on the present-day pattern as the area was covered with a > 2 km thick carapace of volcanic rocks erupted in a short period of time between 30.5 Ma and 26.5 Ma (Baker *et al.* 1996a). Rifting ceased in this area around late Miocene times as flat-lying sedimentary and volcanic rocks of this age lie unconformably over the rotated fault blocks (Davison *et al.*, 1994). Present-day extension of the onshore region is now thought to be minimal, as lavas with K/Ar dates of less than 10 Ma are subhorizontal and unfaulted (Huchon *et al.* 1992). Extension is now presumed to be located nearer the propagating spreading centre in the Red Sea which is situated approximately 60 km offshore. Extension was dominated by domino-style normal faults with a regular spacing of approximately 2–8 km and an extension factor (β) of at least 1.6–1.8 calculated from bedding cut-off angles (Davison *et al.*, 1994). This is regarded as a minimum value as a large, but unaccountable, component of extension is taken up in the granite plutons which may attain 25 km width in a south-west–north-east direction.

Contemporary denudation is strongly controlled by precipitation, which is a function of altitude, with < 80 mm/year on the Tihama Plain and to the far east bordering the Arabian desert, to > 800 mm/year over the mountains of the high plateau (Remmele, 1989). Rainfall occurs as torrential cloudbursts, which produce highly erosive floods with rapidly deposited wadi sediments. A bank section in Wadi Mawr attests to this, where a post-1968 plastic oil container was found buried below a 3 m thickness of coarse-grained wadi conglomerate in a location approximately 3 km west of the Great Escarpment on the Tihama Plain. Vegetation comprises scrubs and grasses, with irrigated areas providing lush orchards and grain fields. Agricultural terracing and irrigation channelling has been developed over the last two thousand years up to the peaks of the highest mountains, so that present-day erosion rates are probably less than the effective rates prior to human habitation.

STRATIGRAPHY

The Precambrian basement comprises gneisses and schists intruded by granites of Pan-African age (500–800 Ma, McCombe *et al.*, 1994). This is unconformably overlain by a sedimentary sequence which is approximately 1 km thick, dominated by Jurassic limestones and shales (Amran Group) in the lower part, and Cretaceous to early Tertiary sandstones (Tawilah Group) in the upper part (Figure H5.3). These are overlain by a thick sequence of acid and basic lavas of the Yemen Volcanic Group which reach up to at least

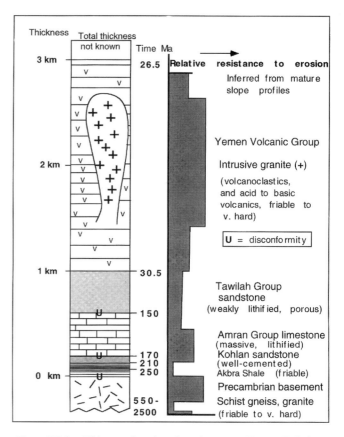

Figure H5.3 Litho-stratigraphy of north-western Yemen. Relative strengths of rocks are estimated from average rock slopes on mature valley profiles. The Precambrian basement, lower Amran Formation, and Tertiary granite lithologies often have subvertical profiles and indicate they are the most resistant to erosion.

2 km in thickness at the present day (Baker 1996), and probably attained 4–5 km in thickness before any erosion took place (see later). The presence of marine Tawilah sediments just below the base of the lavas (Al Subbary *et al.*, this volume), as well as no marked erosional periods within the sequence, and little evidence for major eustatic sea-level changes (> 100 m), suggest that uplift during this time (Eocene–Oligocene?) was minimal. The relative resistance of the individual stratigraphic units can be estimated from the average mature slope angles for each lithology in the largest valleys (Figure H5.3). These indicated the basement, Tertiary granites and the lower part of the Amran Group to be most resistant to weathering with slope dips of greater than 60°, the Tawilah Group, and the Volcanic Group have similar slope profiles (30°) and the upper part (200 m in thickness) of the Amran Group is the weakest section of the stratigraphy (10° slopes) (Figure H5.3).

No syn-rift sedimentary rocks have been found within the mountains of northern Yemen, but post-rift deposits of Miocene to Recent age reach thicknesses of greater than 3414 m on the Tihama Plain (proved by the Abbas-1 well, Heaton *et al.*, 1996, Bosence *et al.*, this volume). These sedimentary rocks provide information on the rates of deposition, and hence rates of transfer of sediment from the eroding mountains during the latter part of their history. The thick section of sedimentary strata below the Tihama Plain indicate that denudation was dominantly produced by fluvial erosion with an efficient and rapid transport of sediment from the mountains.

GEOMORPHOLOGY OF NORTH-WEST YEMEN

Methods of study

A section of north-west Yemen was studied by field mapping aided by the interpretation of LANDSAT TM imagery, aerial photographs and with reference to previously existing geological maps (Ministry of Oil and Mineral Resources, 1992, Kruck, 1983). Topographic maps (Series Y.A.R. 59, Overseas Survey Directorate, Ordnance Survey, 1985 at a scale of 1:50 000 with a 40 m contour interval) were used to examine drainage patterns and densities and to construct base-level (streamline) surface maps using the method of Filosofov (1960, 1970) with the Horton-Strahler stream ordering system (Strahler, 1964). Geomorphological indices were calculated using river sinuosity (Schumm, 1963), drainage density (stream length/area in m/m^2) and valley depth: width ratios (Bull and McFadden 1977). Seismic profiles provided by BP and the Yemen Ministry of Oil and Mineral Resources and borehole data (Hughes and Beydoun, 1992; Mitchell *et al.* 1992; Heaton *et al.* 1996) were used to estimate sediment ages and volumes deposited in the Red Sea and on the Tihama Plain.

Geomorphologic characteristics of north-west Yemen

General description
In north-west Yemen vegetation is sparse and therefore the geomorphology is essentially a function of tectonics, contrasting rock types, and geomorphologic processes coupled with climatic change throughout late Cenozoic times. Differential tectonic activity and the duration of land-form development can be assessed using base-level surfaces, morphologies of fault scarps and mountain fronts, the geomorphology of alluvial fans, and the characteristics of drainage systems.

The geomorphology of drainage systems is strongly related to tectonics and geological structures and lithology. Drainage basin shapes, which are strongly influenced by uplift and structural controls, allow a determination of, and comparison between, different tectonic settings. The degree of stream incision is a function of rock type, stream power and the rate of tectonic uplift. Bull and McFadden (1977) showed that the ratio of valley width:valley depth is a good indicator of tectonic uplift. Stream characteristics can be compared with the experimental work of Ouchi (1985) and the observations of Adams (1980) to identify zones of differential vertical crustal displacements.

A transect extending between 15°00′ to 15°45′ N and 43°10′ to 44°00′ E covering an area of approximately 100 km (north–south) × 100 km (east–west) was studied in detail to examine the relationship between tectonics and geomorphology (Plate 37). This region was chosen because the geology of the area is reasonably well known (Kruck, 1983; Kruck *et al.*, 1984; Davison *et al.*, 1994), although much of it has been mapped using remote sensing. The region contains a complete stratigraphic section through the Phanerozoic sedimentary rocks and Cenozoic volcanic rocks in Yemen. Four distinct topographic zones are present (Plate 37). These include the following.

1. A highly-eroded **Precambrian basement** to the north of 15°30′ N has a major east–west trending contact with Mesozoic sedimentary rocks to the south. This contact forms an impressive scarp and represents an erosion surface where the Phanerozoic cover has been flexed upwards, and preferentially stripped from the basement to the north. There is no clear surface expression of any structural discontinuity along this boundary. The area to the north has a high drainage density on a relatively flat-lying less-elevated (*c.* 1000 m above MSL) landscape, in comparison to the Mesozoic succession farther south which has an average elevation greater than 2000 m ASL.

2. The **Tihama Plain** is a 40–75 km wide north–south trending plain which rises gently from the Red Sea to an altitude of approximately 200 m at the foot of the Great Escarpment. Only the major wadis flow across the Tihama Plain, and some of these have been canalized for irrigation purposes in their upper sections.

3. The **Great Escarpment** trends north-north-west–south-south-east and can be traced from southern Yemen northwards (Figure H5.1) for over 1000 km to the east of Makkah in Saudi Arabia and periodically farther northwards for approximately 1000 km to the Gulf of Aqaba. The escarpment rises from an altitude of approximately 200 m on the Tihama Plain to more than 1000 m above sea level. Its origin is complex, sometimes formed by a number of extensional fault terraces (e.g. Jabal Al Dhamir, Plate 39) which gradually step down towards the Tihama Plain, while elsewhere (e.g. Jabal Bura) it is produced by resistant granite intrusives and porphyritic acid lavas which are less denuded than the surrounding rocks. Individual fault scarps bordering the Tihama Plain appear to have undergone little erosion (1–2 km) from their original positions (e.g. Bajil area, Davison *et al.*, 1994). Well and seismic data indicate that marine transgressions and regressions, dominated by

clastic sediments with evaporites, oscillated between the centre of the Red Sea and the Great Escarpment since late Miocene times (Heaton *et al.*, 1996), which suggests little retreat of the escarpment. This is in contrast to fault scarps along some parts of the Gulf of Suez and northern Red Sea where the scarp is believed to have retreated by up to 100 km since the Miocene (Steckler and Omar, 1994). The resistant granites and coarse-grained acid volcanic rocks are probably responsible for the lack of retreat. Six major wadis trending east–west and north-west–south-east dissect this escarpment within the study area. These have a spacing of between 10 and 40 km.

4. The **Highlands** of north-western Yemen form an extensive elevated plateau averaging approximately 2200 m above sea level, with a maximum altitude of 3660 m Jabal an Nabi Shuyab. This area of high relief corresponds to the present outcrop of Tertiary and Quaternary volcanic rocks. Several distinct areas can be defined based on drainage patterns and topography (Figures H5.4 and H5.5, Plate 37(b)).

(i) The **Sana'a Plateau** is approximately 2500 m in altitude to the south and west, with a shallow basin centred near Sana'a at an altitude of 2200 m, with centripetal drainage. This basin has up to 80 m of Quaternary clastic sediments and tuffs (J. Mather, personal communication). Strombolian cones and associated lava flows of Quaternary age are common throughout the region.

(ii) The **western Sana'a plateau mountains** are steeply dissected and characterized by trap-type topography produced by resistant lava flows, with a superimposed dendritic drainage. Ancient valley-fill deposits are present within this region and deeply eroded stratified scree slopes are common along many of the valleys. Fourth- and fifth-order streams flow westwards and west-north-westwards to the Tihama Plain. These mountains are separated from the lower rift mountains by a large erosional scarp which reaches up to approximately 1 km of relief (Plate 37(b)). This scarp is highly irregular and appears to be a simple erosional feature which is cutting back into the Sana'a Plateau. The scarp is being actively eroded and large landslides are preserved along most of its length. However, the scarp has only retreated some 20 km inland from the zone of major crustal extension defined by the eastern limit of significantly rotated pre-rift strata (Figure H5.4).

(iii) The **rift mountains** are also deeply incised and have an average elevation of approximately 1500 m. They are dominated by trellis-pattern drainage comprising west-south-westward draining ephemeral streams with north-north-west and south-south-east flowing consequent streams, whose orientation is strongly controlled by domino-style rotational fault blocks with a north-north-west–south-south-east strike (Plate 37(b), Plate 38). Little superficial sediment is present on

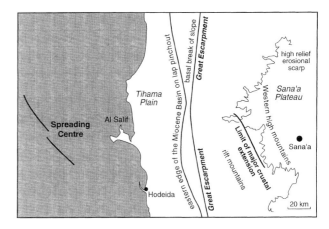

Figure H5.4 Summary of major tectonic and topographic features along the Red Sea rift margin. Note the escarpment along the western border of the Sana'a Plateau is only 20 km from the extended zone, which indicates the amount of retreat of the drainage basins since extension ceased.

the valley slopes, but wadi deposits up to 40 m thick are present, and are currently being actively eroded. Some drainage domains are isolated, with the deepest parts of these basins corresponding to the greatest down throw on the major normal faults.

(iv) The **Granite Mountains** on the western border of the highlands, with an average elevation of 2000 m, are drained with a dendritic pattern on the western margin. This region has the steepest river gradient and slopes in the study area.

Drainage basin characteristics
Drainage basin size in the study area varies from 820 km^2 to 7190 km^2 with the larger basins drained by sixth-order streams and the smaller basins drained by third-order streams (Horton-Strahler scheme, Figure H5.2). A first-order river is a river with no tributaries, a second-order river is one which has tributaries of first-order streams, and a third-order stream is one which has tributaries of second-order streams (at the scale of observation of 1:50 000 topographic maps). There is an even spacing of drainage outlets along the Red Sea margin which has produced a broadly even thickness of sediments deposited on the Tihama Plain reaching approximately 4 km at the Red Sea coastline (Figure H5.2). Length to width ratios (L:W) of the drainage basins vary from 8:1 to 1.5:1 (Figure H5.2). The largest drainage basins are bottle-shaped, with the neck at their lower western reaches where drainage constriction is controlled by the highly resistant granite mountains and also possible differential uplift.

Drainage patterns
Rifted basins generally produce a trellis-type drainage with high sinuosity rivers at the large scale of observation, but lower sinuosity at a smaller scale. This is

because major normal faults develop perpendicular to the regional dip of the landscape. Hence, small rivers follow fault plane directions for long distances until either they manage to break through the normal fault footwall into the adjacent hangingwall drainage basin, or pass through transfer zones. Trellis-pattern stream paths will develop if the rate of footwall uplift is greater than the stream incision rate. The drainage basins in Yemen exhibit some areas of high extension where trellis-stream paths are developed due to fault barriers (Figures H5.5, H5.6). First- and second-order wadis follow the fault trends (Figures H5.5, H5.6), whereas third- and sixth-order wadis are only occasionally deflected from a course which is controlled by gravity and the regional topographic slope down to the Red Sea coastline (Figure H5.5). The minimum vertical displacement rate of the major normal faults in the study area is equivalent to 0.11 mm/year (2 km throw in 17.5 m/year maximum period of extension from 27.5 Ma to 10 Ma, Davison *et al.*, 1994) and provides an important constraint on the minimum rate of incision of the major third- to sixth-order wadis, which must be greater than the vertical displacement in the fault footwall. Footwall uplift may have been as great as 25% of the total fault displacement in a thin elastic crust with no sediment fill in the newly created half-graben (cf. Kusznir *et al.*, 1991), which is equivalent to 0.03 mm/year of vertical footwall uplift. The first- and second-order wadis must have incision rates lower than this. There are no discernible topographic offsets across the major fault planes at the present day, but wadi deposits situated 10 km west of Sana'a exhibit normal faults with up to 5 m of displacement attesting to continued local tectonic activity.

The eastern part of the area is characterized by dendritic drainage (centripetal on the Sana'a Plateau) superimposed on the Tertiary and Quaternary volcanic rocks. In the rift mountains a trellis drainage is controlled by the domino fault blocks (Figure H5.5). A dendritic drainage dominates from the Great Escarpment on to the Tihama Plain (Figure H5.4, H5.5). In the northern area a dendritic drainage is superimposed on the exposed Precambrian basement.

All the larger wadis are of low sinuosity with values ranging between 1.07 to 1.6 (Figure H5.2). The larger

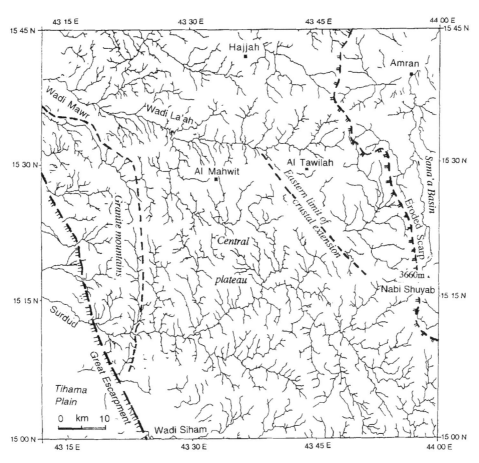

Figure H5.5 Drainage pattern for north-western Yemen with first-order streams removed. Note the change from east to west, from dendritic drainage (volcanic plateau) to trellis (central zone of rotated domino fault blocks) to a dendritic style of drainage (Tertiary granites of the Great Escarpment). Wadis traced from 1:50 000 topographic maps.

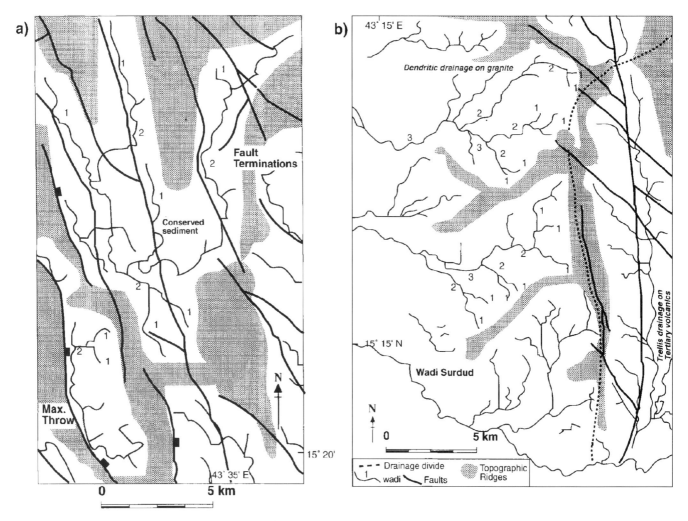

Figure H5.6 (a) Drainage patterns and extensional domino faulting in the central area. Note some first-order streams manage to cut through small (< 1 km throw) normal faults suggesting they were incised more rapidly than smaller faults which grew at slower extension rates. Numbers along wadis refer to stream order. (b) Drainage patterns along the granitic batholith in the south-west of the study area. Note the distinct change from north-south trending trellis to east-west trending dendritic drainage across the lithological boundary. The location of these areas are shown on Plate 37.

wadis have average stream gradients of 1:70 in the study area, with the gradient increasing to 1:30 in their upper reaches (last 30 km).

Streamline surfaces
Streamline maps are constructed by joining points of equal elevation on adjacent streams of the same order, based on the Horton-Strahler scheme.

Streams of different orders will effectively lower topography to a level which is reflected by the tectonic setting of a region. By constructing base-level surface maps for streams of different orders, it is possible to identify current and past regions of tectonism in geologically active areas. Zuchiewicz and Oaks (1993) have shown this to be a particularly successful technique in areas such as the Basin and Range Province, USA. The method has helped identify zones of active faulting, regional warping and migrating zones of differential uplift.

Not surprisingly, the first-order streamline surface closely resembles the corresponding topographic contours rather than the geological map boundaries (compare Plate 37(a) and Figure H5.7). The first-order stream pattern and topography is controlled by the domino faulting in the centrally extended zone (Figure H5.7). This suggests that first-order stream incision rate was smaller than the vertical slip rate of the major normal faults. Unfortunately, only the minimum slip rate on the faults can be determined at 0.11 mm/year (average of 2 km of slip in maximum time span of approximately 17.5 m/year). The streamline contours are aligned parallel to prominent topographic features such as the Great Escarpment and the Basement/Mesozoic boundary.

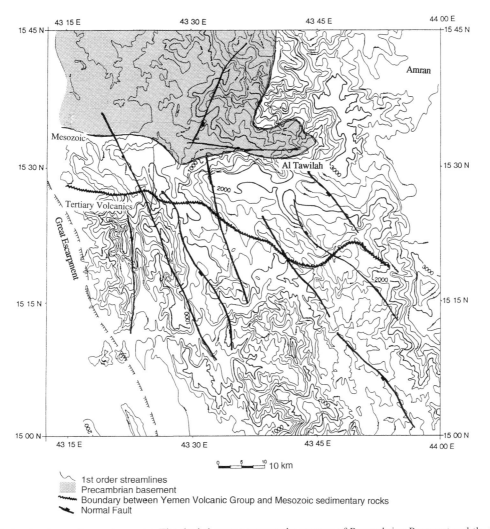

Figure H5.7 First-order stream-line contour map. The shaded areas represent the outcrop of Precambrian Basement and the thick sinuous east–west trending line represents the approximate boundary between the outcrops of Mesozoic and Tertiary volcanics.

The second-order streamline surface has some resemblance to the present-day topographic contours, but has a clear correlation to the bedrock geology of north-west Yemen (compare Plate 37 and Figure H5.8). The Tertiary granite intrusive within the Great Escarpment and the faults that cross-cut this body are clearly reflected in the streamline contours (Plate 37, Figure H5.8). In the centrally extended zone, the major drainage divide between the north and south drainage domains is picked out. The extensional domino faults nearest the Tertiary granite are reflected in the streamline surface maps, but farther east, within the central zone of extension, the influence of tectonics on second-order drainage pattern is not observed.

The third- and fourth-order streamlines are generally oriented north–south, but this pattern is disrupted by two transverse depressions that relate to the courses of the present Wadi La'ah and Wadi Surdud (Plate 37, Figure H5.9). The Wadi Surdud depression trends north-east–south-west, whereas the Wadi La'ah depression trends west-north-west–east-south-east, with the Wadi Siham depression in the far south sector of the study area dissecting the Great Escarpment trending north-west–south-east. In the remainder of the centrally extended zone the streamline contours strike very approximately east–west with no internal features, which suggests that the domino faults played little or no part in the growth of large-scale geomorphological features within this zone, and that gravity forces were dominant. The most closely spaced isobases delimit the lithological boundary between the Precambrian basement and the preserved Mesozoic succession, and also between the volcanic plateau in the east and the Mesozoic cover.

Valley depth : width ratios

Valley depth to width (D:W ratio) ratios should indicate the degree of maturity of the landscape with lower ratios

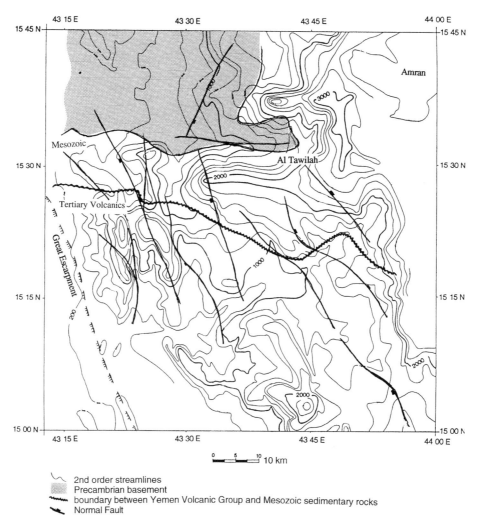

Figure H5.8 Second-order stream-line contour maps. The shaded areas represent the outcrop of Precambrian Basement and the thick sinuous east–west trending line represents the approximate boundary between the outcrops of Mesozoic and Tertiary volcanics.

indicative of higher maturity; or the hardness of the rocks, with harder rocks having higher D:W ratios. The D:W ratios have been calculated along the two major Wadis La'ah and Surdud every 5 km along the Wadi courses and within their major catchments (Figure H5.10). The D:W ratio decreases steadily downstream along Wadi La'ah and Wadi Mawr, but the ratio is highly variable along Wadi Surdud, with the highest ratios at the granitic escarpment reaching 0.5. The main influence on this parameter is probably lithology. A comparison of Plate 37 and Figure H5.10 shows that the more resistant lithologies have higher ratios. The very high D:W ratios across the Granite Mountains, which reach up to a maximum of 50×10^{-2} in Wadi Surdud is probably indicative of the resistant lithologies. The major valley networks within the study area and their associated streams have average D:W ratios which vary between 4.5×10^{-2} and 28×10^{-2}.

Drainage densities

Drainage density reflects the amount of surface water run-off and the ease of initiation of new stream channels and relief. High densities should be produced in weak, impermeable rocks in semi-arid environments where flash flooding is common. Drainage density (total stream length/unit area) has been measured throughout the study area using 1:50 000 topographic maps (Figure H5.11) with averages taken in 4×4 km square grids. Stream density is highest (>9 km/km^2) in the following areas.

(a) along the boundary between the Precambrian basement and the preserved Mesozoic succession to the north;

(b) at the contact between the Yemen Volcanic Group and the underlying Tawilah sandstone in the centrally extended zone;

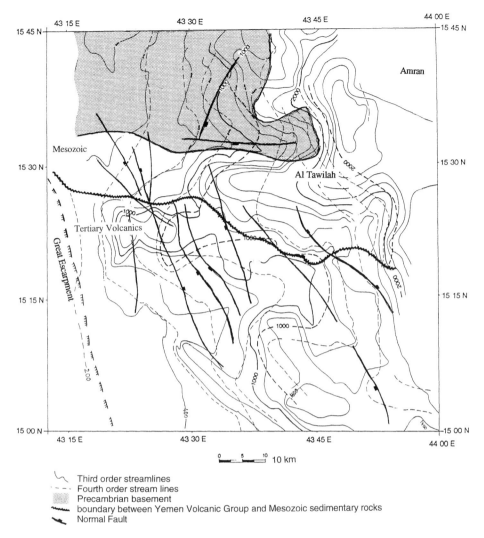

Figure H5.9 Third- and fourth-(broken lines) order stream-line contour maps. The shaded areas represent the outcrop of Precambrian Basement and the thick sinuous east–west trending line represents the approximate boundary between the outcrops of Mesozoic and Tertiary volcanics.

(c) at the acid volcanic centres and Tertiary basalts and ignimbrites towards the south of the area;

(d) along Tertiary granite intrusives of the Great Escarpment and the Tertiary basalts at the entrance of Wadi Surdud to the Tihama Plain.

The lowest drainage densities (1–2 km/km^2) are on the flat plateaux of Amran limestone, Tawilah sandstone and Tertiary volcanics in the north-east of the study area (Plate 37, Figure H5.11), and near to the Tihama Plain where seepage is probably great.

In summary, variation in drainage density is dependent mainly on the bedrock lithology. However, generally higher densities are developed in the extended fault block terrain (rift mountains) due to the increased relief produced by faulting.

RATES OF CRUSTAL PROCESSES

Magmatic eruption and intrusion rates

^{40}Ar/^{39}Ar dating of lowermost flood basalts has established that the onset of flood volcanism in the study transect occurred at 30.5–29.2 Ma (Baker et al. 1996a). A switch from exclusively basaltic volcanism to bimodal basalt–rhyolite volcanism took place at 29 Ma, and uppermost rhyolitic ignimbrites yield plateau ages on feldspar mineral separates of 26.7–27.0 Ma (Baker et al. 1996a; Chiesa et al., 1989). Elsewhere in western Yemen the flood volcanic pile was erupted over a marginally longer period from 30.9–29.5 Ma (Baker et al. 1996a). The vertical accumulation rate of magma at the surface averages 0.5 mm/year, if it is assumed that 2 km of lava was erupted in 4 m/year.

The precise ^{40}Ar/^{39}Ar ages enable some crude estimate of eruption recurrence rates to be made. Lower basalt flows were erupted every 10–100 k.yr., whereas the bimodal upper flows were perhaps only produced every 100–500 k.yr (Baker *et al.* 1996a). However, the decline in eruption rate is somewhat misleading. The voluminous rhyolitic volcanic rocks and intrusive granites, are largely the product of extensive fractional crystallization from basaltic parents (Baker, 1996), and large amounts of intrusive and/or cumulate rocks, at least equivalent to 3–10 times the volume of erupted rhyolitic volcanic rocks and intruded granites, must have been added to the lithosphere at this stage.

While the preserved flood volcanic pile was produced in a period of <4–5 m.y., $^{40}Ar/^{39}Ar$ plateau and Rb-Sr isochron ages for intrusive rocks from the study transect and elsewhere in western Yemen range from 27.5–16.1 Ma. It is unclear exactly what these intrusive rocks represent; the feeder systems of now eroded flood volcanic flows or perhaps merely the feeder systems of less voluminous intraplate volcanism triggered by the onset of crustal extension.

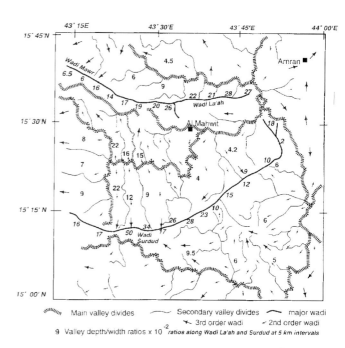

Figure H5.10 Map of valley depth : width ratios ($\times 10^{-2}$) for the Wadis Surdud and La'ah and surrounding tributaries.

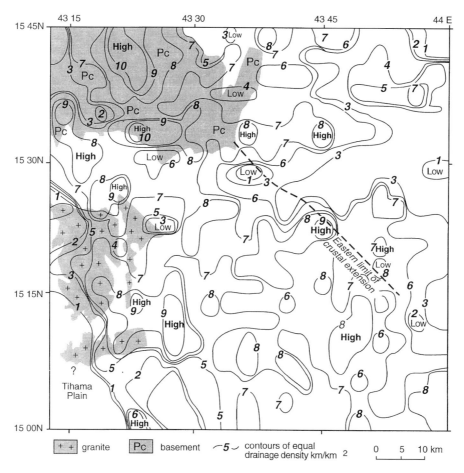

Figure H5.11 Drainage density map for the study area. Contours were constructed using averages over a 4×4 km square grid. Note the high densities at the contact between Phanerozoic sedimentary rocks and Precambrian basement, and the low density over the north-western plateau area composed of flat-lying permeable Amran limestone and Tawila sandstone.

Granite intrusion rates are not known, but the span of granite ages (27.5–20.9 Ma, S. Blakey, unpublished) gives a probable maximum time span 7.6 m.y. for individual intrusions. The granites have vertical walls in the upper 3 km of the crust indicating large-scale extension must have been taking place during granite emplacement, with emplacement accommodated by an equivalent amount of crustal extension. Considering the width of the granites reaches 25 km measured in the direction of crustal stretching (north-west–south-east), and that the width of the extended exposed onshore zone in approximately 75 km, the granite intrusions could represent a maximum of 50% of the total crustal extension. However, it is unlikely that the granites are continuous at depth throughout the crust and the volume proportion of granite is probably far less than their relative proportion indicated by surface outcrop.

Extension rate

It is still difficult to bracket the exact duration of extension. Probably the best indicators of the age of maximum extension are the intrusive ages of the granites. The granites were erupted over an approximate time span of 7.6 m.y. from Jabal Dubas (the oldest at 27.52 ± 0.27 Ma to Jabal As Sharqi 20.88 ± 0.18 Ma (whole rock-biotite mineral pairs, 2 sigma error, S. Blakey unpublished). Most of the crustal extension probably occurred during this time, so the minimum extension period is 7.6 Ma (minimum range of granite intrusion ages) and the maximum extension period could be as great as 17.5 Ma. (from 27.5 Ma, oldest granite extrusions and youngest pre-extension basalts to 10 Ma date of basalts unconformably overlying fault blocks). This gives extension rates of 6.5 mm/year to 2.7 mm/year, respectively, over the 110 km wide extended onshore zone, which has been stretched by β factor averaging 1.7 (strain rate of 1.3×10^{-15}) calculated from the bedding cut-off angles of the domino faults (Wernicke and Burchfiel, 1982). This is similar in magnitude to the average half-spreading velocity of 55–60 mm/year estimated for the southern Red Sea and Gulf of Aden spreading rates since 20 Ma (Sahota, 1990).

Surface uplift rate

Impact of the Afar plume at approximately 31 Ma (Menzies et al., this volume) is postulated to be the predominant influence over uplift and magmatism in the region. However, major extension only began once the Afar plume magmatism had ceased, and is related to regional stresses that produced extension in a north-east–south-west direction during massive granite emplacement and domino faulting (Davison et al., 1994).

Shallow-marine sedimentary rocks of the Tawilah Group situated 100 m below the base of the Yemen Volcanic Group are now at elevations of up to 2650 m ASL on the Sana'a Plateau indicating maximum surface uplift of 2420 m since impingement of the plume at 30.5 Ma or shortly before, as sea level was approximately 130 m higher than the present day at 30 Ma (Haq et al., 1987). This implies an average surface uplift of 0.08 mm/year since this time if the land surface was around sea level at 30.5 Ma. The surface elevation was partly produced by Oligocene basalt–rhyolite magmatism. Volcanism continued through to the Quaternary with a pile of conformable lavas and volcaniclastics reaching an exposed maximum thickness of c. 2–3 km (Davison et al., 1994). This should have increased the surface elevation by 600 m if local isostatic subsidence developed due to loading (assuming Airy isostasy and a crustal density of 2750 kg/m^3). Thinning of the lithosphere, buoyancy above the Afar plume, magmatic underplating and flexural upwarp probably account for the additional elevation.

Melt modelling and lithospheric thicknesses

Oligocene flood and intraplate basalts have an isotopic signature that requires them to have been derived from the upwelling of the Afar plume and the asthenosphere (Baker et al., 1996b). Modelling of the rare earth element data for these rocks requires them to be the product of mixing of melts from garnet- and spinel-facies mantle, with some 10–60% of the melt derived from spinel-facies mantle (Baker, 1996). As the transition from spinel- to garnet-facies mantle occurs at 60–80 km (McKenzie and O'Nions, 1991), and Pan-African lithosphere is typically 100–120 km thick (Mooney et al., 1985), substantial thinning of the lower part of the lithosphere must have taken place in order to permit ascending plume and asthenospheric mantle to ascend into the spinel-stability field and undergo decompression melting. Replacement of cold lithospheric mantle with plume or asthenospheric material would have been a major factor in driving surface uplift (White and McKenzie, 1989). As upper crustal extension post-dated flood volcanism in Yemen then the mantle plume must have played an important role in driving lithospheric thinning.

As a guide to the amount of uplift produced by lithospheric thinning we can predict what the surface uplift would be for a thinning factor $\beta = 1.5$. In the absence of upper crustal extension this would produce a surface uplift of approximately 1000 m in an air-loaded case (assuming standard lithospheric thermal properties, using the equation in Summerfield, 1991, p. 99). A further component of local peak uplift may have been driven by the isostatic response of differential erosion on rift flanks, as the region uplifted (England and Molnar

1990). The highly dissected topography of the rift highlands has valleys with a local relief up to 1800 m and valley height–width ratios of $10–50 \times 10^2$. Using the method described by Gilchrist et al. (1994), and knowing the geometry and volume of denuded rock, and assuming Airy isostasy and maximum uplift of the peaks produced by differential erosion can be calculated to be approximately 1 km. This will only occur where the relief is largest in the granite area, and along Wadi La'ah. As there is little topographic relief on the Sana'a Plateau (400 m) remaining uplift on the plateau must have occurred either by crustal flexure or possibly by magmatic underplating, although the crustal thickness is thought to be normal (35 km) below the Sana'a Plateau (Egloff et al., 1991, Makris et al., 1991).

Studies by Menzies et al. (1992) and Gallagher et al. (1996) have shown that the apatites in the Precambrian basement rocks, 10 km north of Al Mahwit were totally annealed at 17 Ma with sharp 16 μm track lengths indicating cooling was rapid and that the rocks cooled through a blocking temperature of approximately 110°C at this time. This indicates the absolute minimum age of the onset of denudation (either tectonic or erosional). Unfortunately, there are still no constraints on the timing of the maximum age of the onset of denudation, other than it is probably older than the fission track closure ages at 17 Ma. If we assume a 55°C/km rifting geotherm with 110°C closure temperature for apatite, this gives approximately 2 km of denudational cooling in the last 17 Ma, which implies a denudation rate of 0.11 mm/year over this period.

Limited uplift probably started around the same time as Afar plume magmatism which began at 30.5 Ma (Baker et al., 1996a), as there are extensive laterite deposits developed at the top of the Tawilah Group (Davison et al., 1994; Al-Kadasi, 1994). These deposits are not present east of the Balhaf Graben (F. Watchorn, personal communication 1994), and there is no evidence of erosion of these sediments. This indicates that subaerial uplift was restricted to an area which extends 400 km eastward from the spreading centre of the Red Sea, and that the amount of vertical uplift was restricted to probably something in the region of tens to a hundred metres so that there was no visible erosion during this time.

Sedimentation rate

The sedimentation rates of drainage basins may be determined by calculating the amount of sediment deposited in the basins over a known time interval. Seismic reflection data have been used to constrain the volume of sediment which was deposited into the studied portion of the Red Sea Basin and on to the Tihama Plain since rifting began (Figure H5.2, H5.12).

There are some large potential errors in this calculations, but such estimates are important in quantification of basic surface processes operating in sedimentary basins, i.e., denudation and deposition. Possible errors are summarized as follows.

1. The amount of evaporite within the post-rift section is estimated from well and seismic information. Ten wells are currently available, the Al-Meethag-2 well contains up to 63% evaporite and 37% clastic sediment, compared to the Abbas-1 well onshore which contains only 12% evaporite and 88% clastic sediment (documented in lithological columns constructed from drill-cuttings and well logs in Heaton et al., 1996). An estimate of 12% of evaporites in the post-rift section was assumed for the Tihama Plain and a 60% volume of evaporites was assumed for the offshore area. These evaporites will have been mainly derived externally from the World ocean system (as the eroded material from the Yemen Highlands only contains on average 2 wt% Na_2O (Baker, 1996) and cannot be included in the estimates of sediment budget. It is estimated that the calculated volume of evaporites could lead to an error of $\pm 20\%$ in the calculation of the detrital sediment volume in the Red Sea.

2. The accurate porosity/depth relationships of the sediments in the Red Sea are not available due to the poor well control (10 wells in the offshore and Tihama Plain study area), and it is estimated that this could lead to an error of $\pm 10\%$, which is the expected maximum deviation from standard porosity depth relationships below 0.5 km depth.

3. The depth to the top of the basement on seismic reflection profiles is not easy to pick as the reflector is weak and discontinuous (Figure H5.12). Furthermore, the seismic velocity of the section is difficult to estimate due to the rapid variation in total evaporite thickness. It is estimated that this velocity error may be up to $\pm 10\%$ judging from the variation in calculated interval velocities on the seismic sections along the margin.

4. The proportion of eroded clastic and intruded or extruded volcanic rocks in the syn-rift sequence identified on the seismic sections is not known. An estimate of 50% clastic and 50% volcanic rocks has been assumed.

5. A small amount of carbonate sediments is also present in the section, and have been included with the clastic sediments in the sediment budget calculation.

This gives a possible error of approximately 30% (using the square root of the sum of the errors squared) in the calculated volume of sediments in this part of the Red Sea. Despite the difficulties in calculating sediment volume, a sediment budget has been calculated as this provides important first-order constraints on the volume of eroded material derived from the rift shoulder. A transect was defined where detailed topographic maps were available at 1:50 000 scale with contour intervals of

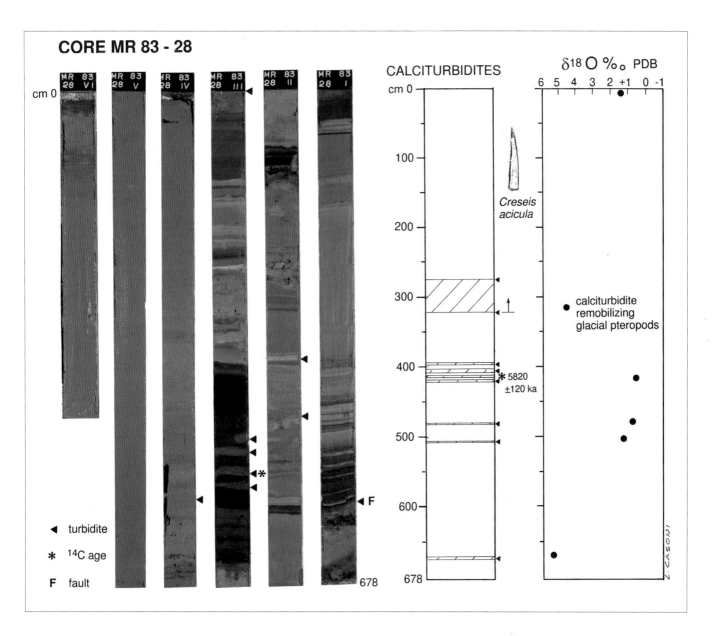

Plate 29 Example of calci-turbidites in piston core MR83-28, Nereus Deep, water depth 2211 m. Individual turbidites range from 3 to 40 cm in thickness. Thin turbidites, remobilizing only pelagic carbonates, are well-sorted carbonates containing large forams (mostly *G. sacculifer*, *O. universa*) and pteropods. The turbidites show normal grading. Remobilization of glacial carbonates, mostly overgrown thecosomatous pteropods (*Creseis acicula*), is evidenced by the inverse stratigraphy of oxygen stable isotope composition. In fact, the thickest calciturbidite (c. 300 cm core depth) clearly contains carbonates older than 11 000–12 000 years BP as indicated by its heavy $\delta^{18}O$ composition (compare Deuser and Degens, 1969) which overlies a sequence of thin pelagic calciturbidites ^{14}C-dated at 5820 + 120 ka (total carbonate).

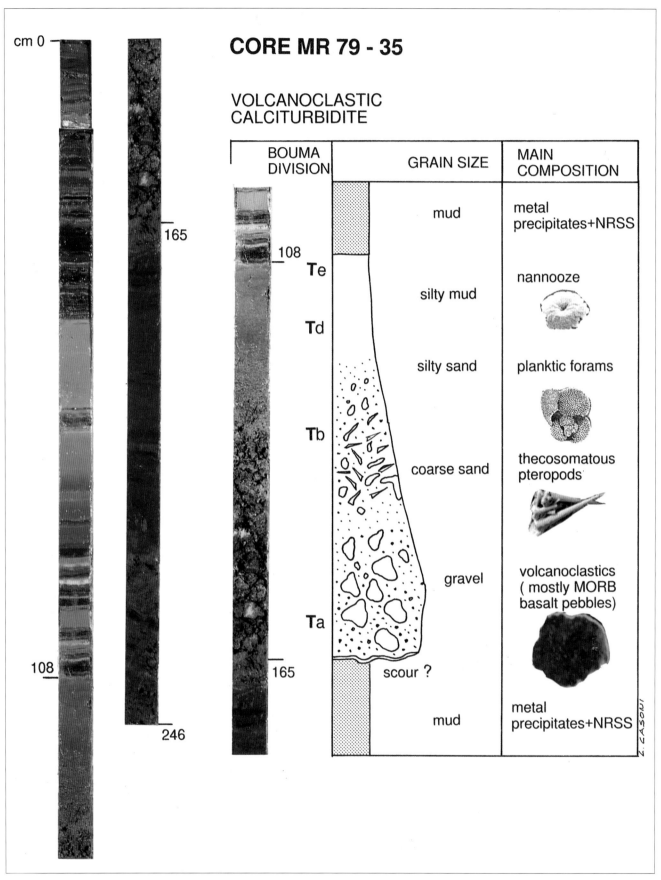

Plate 30 Mixed volcanoclastic-calcareous turbidite in gravity core MR79-35 from the Nereus Deep, water depth 2452 m. Pebbles from the basal unit (Ta) are angular to subangular pebbles of MORB-tholeiitic basalt (Rossi and Rossi, 1982); source material is within Nereus Deep, where emplacement of MORB-basalt is documented at shallower depths (compare Figure G1.6b, Plate 29a, b).

Plate 31 Core 92-1015 from the Suakin Deep (-1969 m); A, core-depth -3.6 m; B, core-depth -5.3 m. The aragonitic crust (a) is overlain by sapropel N°2 (b) which grades upwards into silty marls (c) which are interrupted by thin turbiditic sand layers (d). (scale approx. 1:1).

Plate 32 Core 92-1019 from the Sudanese Shelf (depth 600 m); A, core-depth -6 m; B, core-depth -0.5 m. Bioturbated carbonate muds (a) are locally lithified (b), especially in B (0.5 m) where the crust is dated as 15 000 BP. This crust (in B) is overlain by dark, organically enriched muds (c) which grade up into massive carbonate. Note the similarity with certain ancient hardgrounds. (scale 1:0.75).

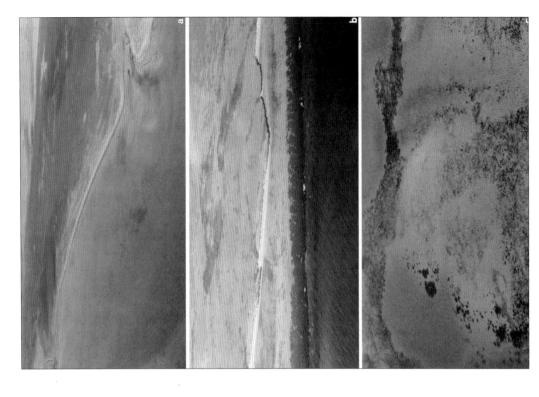

Plate 33 Oblique aerial views of some sedimentary belts of the Dahlak Kebir shelf, Eritrea. (a) Ghubbet Mus Nefit basin, produced by collapse of Dahlak Reef Limestone resting on top of a Miocene salt dome on the south-western side of the Dahlak Kebir shelf edge. Note, on the left side, the ring-like, normal fault systems and the deep channels linking the basin with the open sea. (b) Dhu Bellu lagoon, north-east Dahlak Kebir coast. Observe the widespread, shallow, subtidal sand accumulations (light blue) and the lagoonal channels (dark blue). (c) Typical tectonic pattern of Dahlak Kebir Island. Interaction of normal fault systems generating step-like morphology along the edge of Mus Nefit depression. (d) Detail of a tidal channel showing the morphology of a tidal bar along the eastern coastal belt of Dahlak Kebir, near Erwa Island. Colour changes reflect the water depth and the distribution of sand bodies.

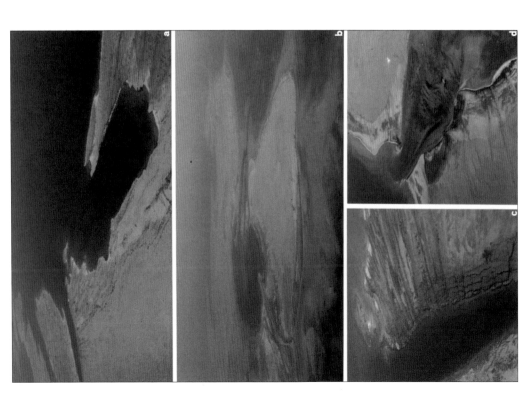

Plate 34 Oblique aerial views of some sedimentary belts of the Dahlak Kebir shelf. (a) Typical view of the wide supratidal flat of the north-east Dahlak Kebir coastal belt. The brown colour marks a coastal sabkha area where ground-water table fluctuation and evaporation is intense. (b) Elongate fringing reef along the south-western Dahlak Kebir shelf edge. The dark belt, behind the wave breaker zone, is mainly made-up of living corals; the inner, light-green area consists of skeletal sand and rubble with scattered sea grass and coral patches (dark blots). (c) Coral-capped hard bottom rising from a depth of about 5 m in the inner shelf belt of Dahlak Kebir Island. Dark areas correspond to living corals, buff-coloured ones to skeletal sand bottoms.

Plate 35 A. View over the fringing reef in the southern Gulf of Aqaba near Haql, Saudi Arabia. The erosional features of the last glacial low stand of sea level are visible. B. Equilibrium coexistence between siliciclastic input and reef growth. Note the undisturbed fringing reef in front of the wadi fan. South of Haql, Gulf of Aqaba, Saudi Arabia. C. Back reef zone showing living *Stylophora pistillata* and larger areas covered with loose carbonate sand. Fringing reef near Sharm Mujjawan, Gulf of Aqaba, Saudi Arabia. D. Reef flat formed by several microatolls composed of *Porites*. Numerous noncalcareous algae grow on the barren flat. Fringing reef, Sahrm al Harr, Northern Red Sea, Saudi Arabia.

Plate 38 Highly dissected rotated fault block terrain in Wadi Lahima, W. Yemen. Location shown on Plate 37.

Plate 39 The Great Escarpment, Jabal Ad Dhamir, W. Yemen, looking southward with the Tihama Plain in the lower right.

Plate 35 A. View over the fringing reef in the southern Gulf of Aqaba near Haql, Saudi Arabia. The erosional features of the last glacial low stand of sea level are visible. B. Equilibrium coexistence between siliciclastic input and reef growth. Note the undisturbed fringing reef in front of the wadi fan. South of Haql, Gulf of Aqaba, Saudi Arabia. C. Back reef zone showing living *Stylophora pistillata* and larger areas covered with loose carbonate sand. Fringing reef near Sharm Mujjawan, Gulf of Aqaba, Saudi Arabia. D. Reef flat formed by several microatolls composed of *Porites*. Numerous noncalcareous algae grow on the barren flat. Fringing reef, Sahrm al Harr, Northern Red Sea, Saudi Arabia.

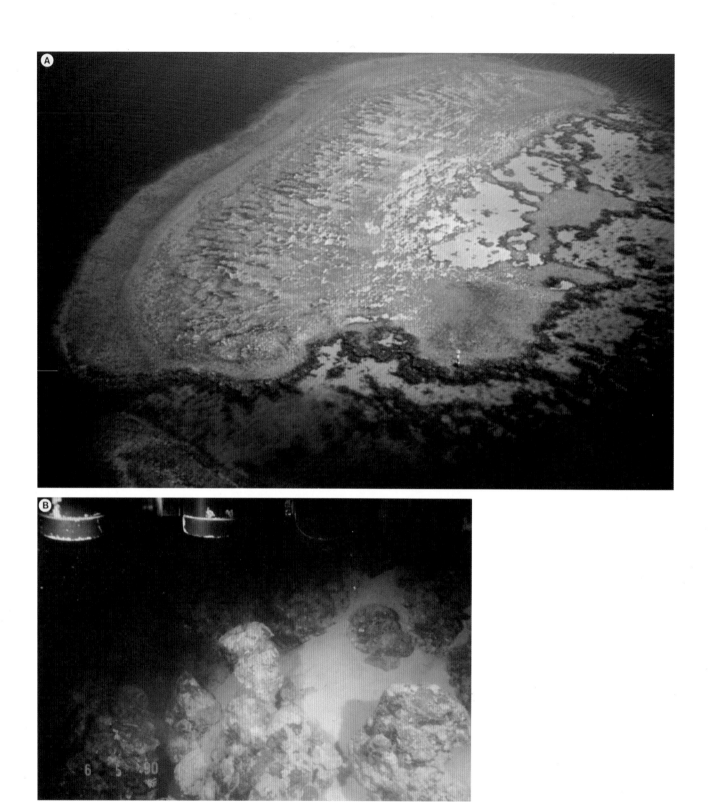

Plate 36 A. Aerial view of a table reef characterized by a poorly zoned reef flat. (North of Jeddah, offshore platform). Maximum width of the reef flat: 200 m. B. Drowned reef morphology on top of the -60 m terrace at Khor Shinab, Sudan. The shape of the ancient shallow-water corals are perfectly displayed although presently overgrown by crusts of coralline algae and sessile foraminifera.

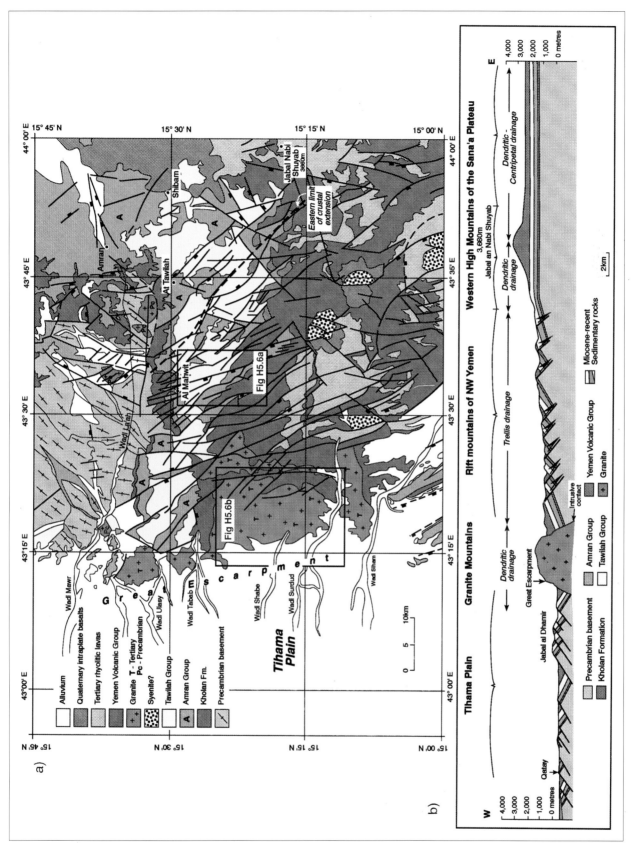

Plate 37 (a) Geological map across north-west Yemen based on field mapping (Davison *et al.*, 1994) and remote sensing on Landsat colour-enhanced images. (b) A composite geological cross-section across the study area, showing the main geomorphological regions. Location shown on Figure H5.1.

Plate 38 Highly dissected rotated fault block terrain in Wadi Lahima, W. Yemen. Location shown on Plate 37.

Plate 39 The Great Escarpment, Jabal Ad Dhamir, W. Yemen, looking southward with the Tihama Plain in the lower right.

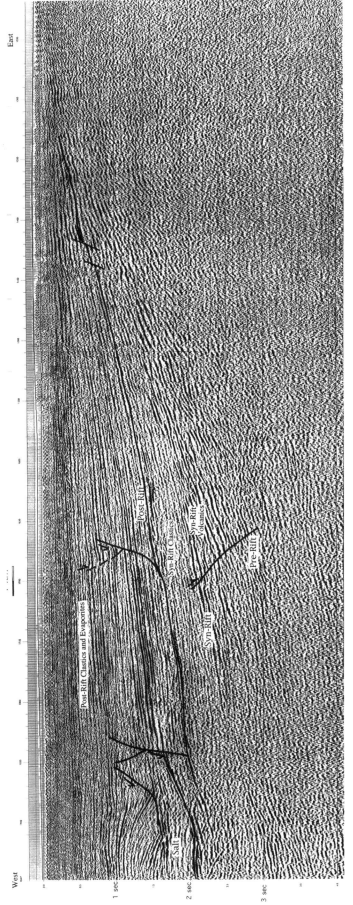

40 m (Figure H5.1, UK Ministry of Defence, 1985). On a typical two-dimensional cross-section in the transect area at the latitude of Dhamar it was estimated from the well data and the seismic section that the total area of approximately 300 km^2 of clastic sediment was deposited out to the Mid-Ocean axial trough in the Red Sea (Figure H5.13). When the calculated evaporite volume is removed from the total sediment volume it appears that there is a greater amount of clastic sediment deposited on the Tihama Plain than in the Red Sea (Figure H5.13). Average vertical clastic sedimentation rates in the Red Sea are estimated at 0.08 mm/year since the early Miocene (2 km average sediment thickness/25 m.y.) with a maximum sedimentation rate of 0.32 mm/year (8 km maximum thickness of sediment/25 m.y.). In a two-dimensional section, the rate of transport from the Highlands into the Red Sea and Tihama area is equal to 1.2×10^5 km^2/year (area of the Red Sea Basin filled with clastic sediments 300 km^2/period of deposition 25 m.y.).

Denudation rate

If the deposited clastic sediment area in the section (300 km^2) is divided by the length of the drainage catchment from the Great Escarpment to the watershed with the Arabian Desert (100 km distance) then the average denudation over the drainage area can be calculated as 3000 m since rifting began at approximately 27.5 Ma (denudation rate = 0.17 mm/year, Figure H5.13). This volume of eroded sediment is more than twice the calculated rock volume along the transect in the Yemen Highlands that lies above sea level (not including the sediment-filled Tihama Plain). This implies that the amount of volcanic rock denuded from above the present topography is approximately 3 ± 1 km in thickness, and this is almost twice the average thickness of the present preserved pile of the Yemen Volcanic Group. This amount of rapid denudation is also supported by reset fission track ages, which have been obtained from basement rocks 10 km north of Al Mahwit (Menzies et al., 1992). In this area all of the Amran, Tawilah and Volcanic Group rocks have been eroded which reach a minimum total thickness of 3 km again suggesting this amount of denudation.

Figure H5.12 Seismic section across the Tihama Plain. Large rotations of 15° can be observed at deep syn-rift levels, overlain by shallow-dipping post-salt reflectors. Most post-salt reflectors converge towards a common pinch-out near to the eastern end of the seismic line. The location of the seismic section is shown on Figure H5.2.

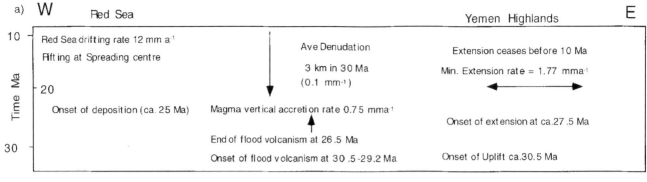

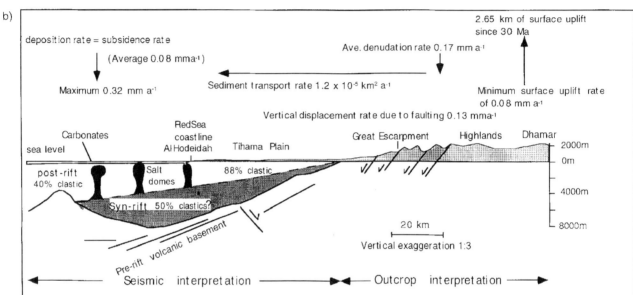

Figure H5.13 Summary diagram of vertical tectonic strain and rates, erosion and deposition rates in north-western Yemen. (a) Timing and rates of tectonic, denudation and depositional activity in north-western Yemen. (b) Cross-section of western Yemen outlining the topography and tectonic structures, showing rates of processes. Partly based on seismic line shown in Figure H5.12.

DISCUSSION

Great Escarpment

The small amount of erosion of the Great Escarpment and its relative linear form would suggest that it is a young feature produced by rapid differential uplift. Steep slopes and high river gradients across the granitic mountains support this assertion. The Escarpment does not appear to have undergone a large retreat since rifting began. This is probably due to the resistant nature of the capping volcanic rocks. It does not exhibit any recent flat irons (e.g. Plate 38) and is highly dissected due to extensive erosion which suggests that it is a mature feature. The seismic reflection profile (Figure H5.12) does not show any major onlapping sequence of Miocene to Recent sediments progressively across the Tihama Plain. The Miocene pinch out mapped on the Tihama Plain (interpreted by R. Heaton, BP exploration, Figure H5.4), indicates that the maximum possible retreat of the escarpment since Pliocene times is 10–20 km (Figure H5.4).

Drainage patterns and sediment transfer

Present-day drainage patterns are controlled by the dominant lithological boundaries and structure. From east to west, there is a progressive change from centripetal/dendritic (Sana'a Plateau) to superimposed dendritic (Western Sana'a Plateau Mountains) to trellis (centrally extended zone, Yemen Rift Mountains) and dendritic drainage (towards the Tihama Plain on the Great Escarpment, Plate 38). The central zone is characterized by slower uplift on the footwalls of the fault blocks than the Great Escarpment, and is dominated by north-west–south-east trending faults which control the recent trellis drainage. At the Great Escarpment faults have maximum throws of 4 km which is calculated to be the equivalent to a vertical slip rate of 0.22 mm/year assuming a maximum 17.5 m.y. period of extension. Vertical displacement rates in the Central Highlands are not constrained, but faults in this area have a maximum throw of 2 km suggesting that rates may be lower (0.11 mm/year). Major wadi incision rates were greater than footwall uplift rates. However, minor wadis follow the extensional fault trends indicating that their incision rates were lower than footwall uplift rates.

The internal drainage basins are bounded by topographic highs related to faults. Sediment is at present being conserved within small valleys, but these storages are short-lived features of the landscape. Active headward erosion of streams in adjacent basins is cutting back and will eventually break through the drainage divides. There is significant Recent denudation with the sediment being rapidly transported to the Tihama Plain, and active headward erosion of the valleys producing 1000 m high scarps along the western margin of the Sana'a Plateau (Figure H5.4). The present-day scarp is being eroded and major landslides occur along this topographic break.

It is interesting to speculate what the Yemen margin will look like in 100 m.y. time. At current denudational rates the margin would be reduced to sea level within the next 30 m.y. in the absence of any further uplift. However, denudation rates will diminish as topography reduces, and it will be considerably longer before the margin is eroded to sea level. Once the pre-rift sedimentary rocks of the Kholan, Amran and Tawilah Group are removed it will be difficult to observe evidence for extension in the Precambrian basement, and it will probably be assumed that the crustal extension ended at the Great Escarpment, whereas crustal extension extends some 75 km east of this.

CONCLUSIONS

It is estimated that up to 6 km, with an average of about 2 km, of siliciclastic sediments have been deposited in the south-eastern Red Sea since erosion and deposition began following plume volcanism.

If this material was directly derived from the adjacent Red Sea mountains, which appears very likely, it implies that an average thickness of 3 ± 1 km of material has been eroded from the Yemen Highlands in the transect area. Hence, the original volume of volcanic rocks in the Afar plume province may have been at least twice as large as the present-day preserved volume, as only 1–1.5 km thickness remains at the present day. The large amounts of material apparently eroded from western Yemen suggests that a large part of the flood volcanic record has been removed. Magma vertical addition rates at the surface may have been as great as 0.66 mm/year if a total volcanic thickness of 3 km was erupted in approximately 4–5 m.y.

The Red Sea Highlands have been uplifted to 3660 m maximum elevation (Jabal Nabi Shuyab) from an elevation close to sea level in the last 30 m.y. due to the effects of the Afar plume. This maximum uplift occurs eastward of the zone of upper crustal extension. Calculations suggest that this uplift is due to lithospheric heating, magmatic addition to the crust, the surface extrusion of lavas and crustal flexure. In addition, differential erosion of the area is calculated to have caused up to a maximum of 1 km of local peak uplift. The present-day elevation of pre-volcanic marine sediments at 2650 m above sea level indicates a minimum surface uplift rate of 0.08 mm/year over this time span, if most of the uplift occurred since the first lava eruption. The drainage pattern which has evolved from a flat-lying elevated volcanic plateau is dominated by fourth- to sixth-order wadis which dissect the Yemen Highlands almost perpendicularly to the main tectonic grain imposed by large-scale extensional domino faulting. Hence, denudation rates in large wadis outstripped vertical strain rates imposed by normal faulting and regional uplift.

Wadi traces in the highly extended areas are trellis type because footwall uplift rates on normal faults were more rapid than incision rates of first- and second-order wadis. Third- to sixth-order wadi incision rates were greater than footwall uplift rates (minimum rate = 0.04 mm/year), so that straighter wadi courses developed. Extension increased drainage density by increasing the tortuosity of the stream system, and increasing topographic gradients.

All the vertical rates of crustal processes (surface uplift, denudation, magmatic accretion) that have been measured are significantly less than 0.5 mm/year, whereas horizontal extension rates are measured at several millimetres per annum. These are thought to be representative long-term rates, whereas the short-term rates may be much more rapid.

ACKNOWLEDGEMENTS

I. Davison and L.A. Owen would like to thank the Royal Society and the RED SED grant from the European Union to B. Purser with D. Bosence as UK co-ordinator for financial help. Matthew Tatnell held a student grant at Royal Holloway on the Basin Evolution and Dynamics MSc course at Royal Holloway when this work was undertaken. Gareth Jenkins completed some of this research as an BSc undergraduate project sponsored by the Department of Geology at Royal Holloway. Joel Baker held a Commonwealth Scholarship during his study for a PhD at Royal Holloway. Mike Ellis, Richard Collier and Dan Bosence are thanked for their thoughtful reviews.

REFERENCES

Abbate, E., Bigazzi, G., Norelli, P. and Quercioli, C. (1993a) Fission-track ages and the uplift of the northern Somali plateau, in *Geology and Mineral Resources of Somalia and Surrounding Regions* (eds E. Abbate, M. Sagri and F.P. Sassi), Istituto Agronomo Oltremare, Firenze, Relazione e Monografia, **113**, pp. 369–378.

Abbate, E., Bruni, P., Fazzuoli, M. and Sagri, M. (1988) The Gulf of Aden continental margin of Northern Somalia: Tertiary sedimentation, rifting and drifting. *Memorie Società Geologica Italiana*, **31** (1986), 427–445.

Abbate, E., Bruni, P. and Sagri, M. (1993b) The northern Somalia Daban basin in the Gulf of Aden rift system: basin fill, architecture and structural evolution, in *Geodynamics and Sedimentation of the Red Sea–Gulf of Aden Rift System*, (eds E.R. Philobbos and B.H. Purser), Geological Society of Egypt, Special Publication, **1**, pp. 363–382.

Abbate, E., Ficarelli, G., Pirini Raddrizzani, A., Salvietti, Torre, D. and Turi, A. (1974) Jurassic sequences from the Somali coast of the Gulf of Aden. *Rivista Italiana Paleontologia e Stratigrafia*, **80**, 409–478.

Abdallah, A.M., Adindani, A. and Fahmy, N. (1963) Stratigraphy of the Lower Mesozoic rocks, western side of the Gulf of Suez, Egypt. *Geological Survey and Mineral Resources Department, Egypt*, Paper. No. **27**, 23 pp.

Abdel Khalek, M.L., Abdel Wahed, M. and Sehim, A. (1993) Wrenching deformation and tectonic setting of the Northwestern part of the Gulf of Aqaba, in *Geodynamics and sedimentation of the Red Sea–Gulf of Aden Rift system*, (eds E.R. Philobbos and B.H. Purser), Geological Survey of Egypt. Special Publication, **1**, pp. 409–445.

Abdelghany, O., Piller, W.E. and Toleikis, R. (1996) Nummulitide Foraminiferen (Gattung Planostegina) im Unter- und Mittelmiozän der Parathethys und des mediterranen Raumes und ihre stratigraphische Bedeutung. *Sediment*, **96**, Wien, Abstr. Vol. p. 1.

Abou Khadrah, A. and Wahab, S.A. (1984) Petrography and diagenesis of the Samh Formation and younger sediments, North Mersa Alam, Red Sea, Egypt. *Journal African Earth Sciences*, **2**, 277–286.

Abou Ouf, M.A. and Geith, A.M. (this volume), Sedimentary evolution of early rift troughs of the central Red Sea margin, Jeddah, Saudi Arabia, in *Sedimentation and Tectonics of Rift Basins: Red Sea–Gulf of Aden*, (eds B.H. Purser and D.W.J. Bosence), Chapman & Hall, London, pp. 135–145.

Aboul-Basher, H.M. (1979) Control of manganese mineralization in relation to the tectonic history of the Red Sea coastal deposits. MSc thesis, Faculty Science, University Khartoum, 165 pp.

Abrard, R. (1942) Mollusques Pléistocenes de la Côte Francaise des Somalis. *Archives du Museum National d'Histoire Naturelle*, Paris, **18** (Ser. 6), 5–105.

Abu Khadra, A.M. (1982) Sedimentological evolution of the Yemen Arab Republic. *Bulletin Faculty of Science, Sana'a University*, Sana'a, **2**, 39–55.

Abu Khadrah, A.M. and Darwish, M. (1986) On the occurrence of raised beach sediments in the Hammam Faraun area, Sinai, Egypt. *Arabian Gulf Journal of Scientific Research*, **4**, 1159–1175.

Abu Khadrah, A. and Wahab, S.A. (1984) Petrography and diagenesis of the Samh Formation and younger sediments, North Mersa Alam, Red Sea, Egypt. *Journal of African Earth Sciences*, 277–286.

Abul Nasr, R.A. (1990) Re-evaluation of the Upper Eocene rock units in west central Sinai, Egypt. *Ain Shams University, Earth Science Series*, **4**, 234–247.

Accordi, G., Angelucci, A., Carbone, F. et al. (1994) *Present-day marine sedimentary environments of Dahlak Islands archipelago (Eritrea, Red Sea), Marine survey of February 1–12, 1994, Final Report*, Centro di Studio per il Quaternario e l'Evoluzione Ambientale del CNR, Rome.

Adams, C.G., Gentry, A.W. and Whybrow, P.J. (1983) Dating the terminal Tethyan event. *Utrecht Micropaleontological Bulletin*, **30**, 273–298.

Adams, C.G., Lee, D.E. and Rosen, B.R. (1990) Conflicting isotopic and biotic evidence for tropical sea-surface temperatures during the Tertiary. *Palaeogeography, Palaeoclimatology, Palaeoecology*, **77**, 289–313.

Adams, J. (1980) Active tilting of the United States midcontinent: geodetic and geomorphic evidence. *Geology*, **8**, 442–446.

Aghib, F.S., Bernoulli, D. and Weissert, H. (1991) Hardground formation in the Bannock Basin, Eastern Mediterranean. *Marine Geology*, **100**, 103–113.

AGIP (1977–1984) Exploration data (Surface geology, geochemistry, geophysics and subsurface well results) for the AGIP permit areas in the Gulf of Aden region, PDR Yemen.

Aharon, P. and Chappell, J. (1986) Oxygen isotopes, sea level changes and the temperature history of a coral reef environment in New Guinea over the last 10^5 years. *Palaeogegraphy, Palaeoclimatology, Palaeoecology*, **56**, 337–379.

Ahmed, E.A., Soliman, M.A. and Essa, M.A. (1994) Sedimentology and evolution of the Quaternary sediments, NW Red Sea, Egypt, *Geological Society Egypt*, Special Publications, **1**, 295–320.

Aissaoui D.M. (1985) Botryoidal aragonite and its diagenesis. *Sedimentology*, **32**, 345–361.

Aissaoui, D.M., Coniglio, M., James, N.P. and B.H. Purser. (1986) Diagenesis of a Miocene reef-platform: Jebel Abu Shaar, Gulf of Suez, Egypt, in *Reef Diagenesis*, (eds J.H. Schroeder and B.H. Purser), Springer-Verlag, 112–131.

Akaad, M.K. and Dardir, A. (1966) Geology and phosphate deposits of Wassif-Safaga area, *Geological Society of Egypt*, Paper 36, 35 pp.

Akamaluk, T. (1989) Gravity field and the crustal structure of East Africa. *Hamburg Geophysical Einzelschrift*, **A92**, University, Hamburg.

Akkad, S.E. and Dardir, A. (1966) Geology of the Red Sea coast between Ras Shagra and Mersa Alam, with short notes on results of exploratory work at Gabal el Rusas lead and zinc deposits. *Geological Survey of Egypt*, Memoir 3, 1–67.

Al Amri, A.M.S. (1994) Seismicity of the south-western region of the Arabian Shield and southern Red Sea. *Journal of African Earth Sciences*, **19** (1/2), 17–25.

Al Kadasi, M. (1994) Temporal and spatial evolution of the basal flows of the Yemen Volcanic Group. Unpublished PhD thesis, Royal Holloway University of London, 301 pp.

Al Sayari, S.S., Dullo, C., Hötzl, H., Jado, A.R. and Zöttl, J.G. (1984) The Quaternary along the coast of the Gulf of Aqaba, in *Quaternary Period in Saudi Arabia* (eds A.R. Jado and J.G. Zöttl) Springer, Wien, pp. 32–47.

Al Shanti, A.M.S. (1966) Oolitic iron ore deposits in Wadi Fatima between Jeddah and Mecca, Saudi Arabia, *Saudi Arabian Directorate General Mineral Resources Bulletin*, **2**, 51 pp.

Al Sharhan, A.S. and Kendall, C.G.S. (1944) Depositional settings of the upper Jurassic Hith anhydrite of the Arabian Gulf – an analog to Holocene evaporites of the United Arab Emirates and Lake MacLeod of Western Australia. *American Association of Petroleum Geologists Bulletin*, **78**, 1075–1096.

Al Subbary A.K. (1990) Stratigraphic and sedimentological studies of the Tawilah Formation, Al-Ghiras area, northeast Sana'a, Yemen Arab Republic. Unpublished MSc thesis, University of Sana'a.

Al Subbary, A. (1996) The sedimentology and stratigraphy of the Cretaceous to early Tertiary Tawilah Group, western Yemen. Unpublished PhD thesis, Royal Holloway University of London, 316 pp.

Al Subbary, A.K. and Nichols, G.J. (1991) Cretaceous–Early Tertiary pre-rift sediments, Yemen. *British Sedimentological Research Group Meeting, Edinburgh*, Abstract Vol., 65.

Al Subbary, A.K., Nichols, G.J. and Bosence, D.W.J. (1993) Cretaceous to Tertiary pre-rift fluvial to shallow marine sediments in Yemen, in *Tectonics and Sedimentation of Red Sea–Gulf of Aden Region* (eds E. Phillobos and B.H. Purser) Geological Society of Egypt Cairo, Special Publication No. 1, pp. 383–407.

Al Subbary, A.K., Nichols, G.J., Bosence, D.W.J. and Al Kadasi, M. (this volume) Pre-rift doming, peneplanation or subsidence in the southern Red Sea? Evidence from the Medj-zir Formation (Tawilah Group) of western Yemen, in *Sedimentation and tectonics of rift basins: Red Sea–Gulf of Aden* (eds B.H. Purser and D.W.J. Bosence), Chapman & Hall, London, pp. 119–133.

Ali Kassim Mohamed (1991) Oligo-Miocene sedimentation in the Boosaaso and Qandala Basin, Gulf of Aden, NE Somalia, in *Geologia del Basamento Italiano – Convegno in onore di Tommaso Cocozza*, Siena, Abst. Vol., 87.

Ali Kassim Mohamed (1993) I bacini Oligo-Miocenici della Somalia nordorientale; evoluzione sedimentaria e strutturale. PhD thesis, IV Ciclo, Dip. Sc. Terra Università di Siena; Biblioteca Nazionale, Firenze, Roma.

Allam, A. (1988) A lithostratigraphical and structural study on Gebel el Zeit area, Gulf of Suez, Egypt. *Journal of African Earth Sciences*, **7**, 933–44.

Allan, T.D. (1970) Magnetic and gravity fields over the Red Sea. *Philosophical Transactions Royal Society London*, **A 267**, 153–180.

Allen, J.R.L. (1965) A review of the origin and characteristics of recent alluvial sediments. *Sedimentology*, **5**, 89–191.

Allen, J.R.L. (1983) Studies in fluviatile sedimentation: Bars, bar complexes and sandstone sheets (low sinuosity braided streams) in the Brownstones (L. Devonian, Welsh Borders). *Sedimentary Geology*, **33**, 237–293.

Aller, R.C. (1980) Diagenetic processes near the sediment–water interface of Long Island Sound. II – Fe and Mn. *Advances in Geophysics*, **22**, 351–415.

Aller, R.C., and Rude, P.D. (1988) Complete oxidation of solid phase sulfides by manganese and bacteria in anoxic marine sediments. *Geochimica Cosmochimica Acta*, **52**, 751–765.

Allouc, J. (1990) Quaternary crusts on slopes of the Mediterranean Sea: A tentative explanation for their genesis. *Marine Geology*, **94**, 205–238.

Almogi-Labin, A. (1982) Stratigraphic and paleoceanographic significance of late Quaternary pteropods from deep-sea cores in the Gulf of Aqaba (Eilat) and northernmost Red Sea. *Marine Micropaleontology*, **7**, 53–72.

Almogi-Labin, A., Hemleben, C., Meischner, D. and Erlenkeuser, H. (1991) Paleoenvironmental events during the last 13,000 years in the Central Red Sea as recorded by Pteropoda. *Paleoceanography*, **6**, 83–98.

Almogi-Labin A., Luz B. and Duplessy J.-C. (1986) Quaternary paleo-oceanography, pteropod preservation and stable-isotope record of the Red Sea. *Palaeogeography, Palaeoclimatology, Palaeoecology*, **57**, 195–211.

Altherr, R., Henjes-Kunst, F. and Baumann, A. (1990) Asthenosphere versus lithosphere as possible sources for basaltic magmas erupted during formation of the Red Sea: constraints from Sr, Pb and Nd isotopes. *Earth and Planetary Science Letters*, **96**, 269–286.

Altherr, R., Henjes-Kunst, F., Puchelt, H. and Baumann, A. (1988) Volcanic activity in the Red Sea axial trough – evidence for a large mantle diapir? *Tectonophysics*, **150**, 121–133.

Altichieri, L., Angelucci, A., Boccaletti, M., Cabdulaqaadir, M.M., Arush, M., Piccoli, G. and Robba, E. (1982) Preliminary study on the Paleogene Formation of Central Somalia (Hiraan, Galgadund, Mudug and Nugaal Regions). *Quaderni di Geologia della Somalia*, **6**, 183–214.

Ambraseys, N. and Melville. E.C.P. (1983) Seismicity of Yemen. *Nature*, **303**, 321–323.

Ambraseys, N.N., Melville, C.P. and Adams, R.D. (1994) *The Seismicity of Egypt Arabia and the Red Sea*, Cambridge University Press.

Ambraseys, N.N. and Sarma, S. (1969) Liquefaction of soils induced by earthquakes. *Seismology Society of America, Bulletin*, **59**, 651–664.

Anderton, R. (1976) Tidal shelf sedimentation: an example from the Scottish Dalradian. *Sedimentology*, **23**, 429–458.

Andres, W. and Radtke, U. (1988) Quartäre Strandterrassen an der Küste des Gebel Zeit (Golf von Suez/Ägypten). *Erdkunde*, **42**, 7–16.

Angelier, J. (1979) Néotectonique de l'arc égéen. *Mémoires Société Géologique Nord*, **13**, 1–418.

Angelier, J. (185) Extension and rifting: the Zeit region, Gulf of Suez. *Journal of Structural Geology*, **7**, 605–612.

Angelier, J. and Bergerat, F. (1983) Systèmes de contraintes et extension continentale. *Bulletin Centre Recherche Exploration Production Elf Aquitaine*, **7**, 137–147.

Angelucci, A., Befani, G., Biagi, P.F. et al. (1978) Geological framework of Tanam, Wusta and Isratu in the Dahlak Islands (southern Red Sea). *Geologica Romana*, **17**, 345–388.

Angelucci, A., Boni, C.F., Bono, P. et al. (1982) Il Ghubbet Entatu nell'arcipelago delle Isole Dahlak (Mar Rosso): un esempio di sedimentazione carbonatica. *Bollettino della Società Paleontologica Italiana*, **21** (2/3), 189–200.

Angelucci, A., Boni, C.F., Bono, P. et al. (1985) L'arcipelago delle Isole Dahlak nel Mar Rosso meridionale: alcune caratteristiche geologiche. *Bollettino della Società Geografica Italiana*, **Ser. XI**, Vol. II, 233–262.

Angelucci, A., Civitelli, G., Funiciello, R., Mariotti, G., Matteucci, R., Passeri, L., Pialli, G., Praturlon, A. and Sirna, G. (1975) Preliminary report on the carbonate sedimentation at the Dahlak islands (Red Sea, Ethiopia). *Geologica Romana*, **14**, 41–61.

Angelucci, A., Matteucci, R. and Praturlon, A. (1981) Outline of geology and sedimentary environments of the Dahlak Islands (southern Red Sea). *Bollettino della Società Geologica Italiana*, **99**, 405–419.

Anschutz, P. (1993) Genèse et évolution géochimique des sédiments métallifères de la fosse Atlantis II (Mer Rouge). Thèse Université Louis Pasteur, Strasbourg.

Anschutz, P. and Blanc, G. (1993a) L'histoire sédimentologique de la fosse Atlantis II (mer Rouge). Les apports de la micropaléontologie. *Compte Rendue Academie Science*, Paris, Série II, **317**, 1303–1308.

Anschutz, P. and Blanc, G. (1993b) Le rapport NaCl/eau des boues minéralisées de la fosse Atlantis II (Mer Rouge). Calcul de la teneur en halite des sédiments et implication sur la paléotempérature du milieu. *Compte Rendue Academie Science*, Paris, Série II, **317**, 1595–1600.

Anschutz, P. and Blanc, G. (1995a) Diagenetic evolution of the DOP facies from the Atlantis II Deep (Red Sea): evidence of early hydrothermal activity. *Oceanologia Acta*, **18**, 105–112.

Anschutz, P. and Blanc, G. (1995b) Chemical mass balances in metalliferous deposits from the Atlantis II Deep, Red Sea. *Geochimica et Cosmochimica Acta*, **59**, 4205–4218.

Anschutz, P. and Blanc, G. (1995c) Geochemical dynamics of the Atlantis II Deep (Red Sea): Silica behavior. *Marine Geology*, **128**, 25–36.

Antkinson, R.J., Posner, A.M. and Quirck, J.P. (1968) Crystal nucleation in Fe (III) solutions and hydroxide gels. *Journal Inorganic Chemistry*, **30**, 2371–2381.

Anton, D., 1984. Aspects of geomorphological evolution; paleosols and dunes in Saudi Arabia, in *Quaternary Period in Saudi Arabia*, (eds A.R. Jado and J.G. Hotzl), pp. 275–296.

Arafa, A.A. (1982) Calcareous nannofossils from the Kareem Formation (Middle Miocene) Gulf of Suez area, Egypt. *Neues Jahrbuchs für Geologie und Paläontologie. Monathefte*, 449–455.

As Saruri, A.L. (1995) *Lateral lithologic change of the Habshiyah formation (middle Eocene) of the southern Arabian peninsula (Yemen)*. Proceedings of Rift Sedimentation and Tectonics in the Red Sea–Gulf of Aden region. Sana'a, Yemen, 23–31 October p 58.

Attia, O.E., Lowenstein, T.K. and Wali, A.M.A. (1995) Middle Miocene gypsum, Gulf of Suez: marine or non marine? *Journal of Sedimentary Research*, **A65**, 4, 614–626.

Azzaroli, A. (1958) L'Oligocene e il Miocene della Somalia, Stratigrafia, Tettonica, Paleontologia (Macroforaminiferi, Coralli, Molluschi). *Palaeontologica Italica*, **52**, 1–142.

Azzaroli, A. and Fois, V. (1964) Geological outlines of the Northern end of the Horn of Africa, in: *Proceedings XXII. International Geological Congress*, New Delhi, Sect. **4**, 293–314

Bäcker, H. (1982) Metalliferous sediments of hydrothermal origin from the Red Sea, in *Marine Mineral Deposits* (eds P. Halbach and P. Winter), Verlag Gluckauf Gmbh, Essen, 102–136.

Bäcker, H., Lange, K. and Richter, H. (1975) Morphology of the Red Sea Central Graben between Subair Islands and Abul Kizaan. *Geologie Jahrbuch*, **D13**, 79–123.

Bäcker, H. and Richter, H. (1973) Die rezente hydrothermal-sedimentäre Lagestätte Atlantis II Tief im Roten Meer. *Geologie Rundschau*, **62**, 697–741.

Bäcker, H. and Schoell, M. (1972) New deeps with brines and metalliferous sediments in the Red Sea. *Nature Physical Science*, **240**, 153–158.

Badaut, D. (1988) Les argiles et les composés silico-ferriques des sédiments métallifères de la fosse Atlantis II (Mer Rouge). Formation et diagenèse des dépôts. Thèse Docteur des Science, Université Paris Sud, Orsay.

Badaut, D., Besson, G., Decarreau, A. and Rautureau, R. (1988) Occurrence of a ferrous trioctahedral smectite in recent sediments of Atlantis II Deep, Red Sea. *Clay Mineralogy*, **20**, 1–16.

Badaut, D., Blanc, G. and Decarreau, A. (1990) Variation des minéraux argileux ferrifères, en fonction du temps et de l'espace, dans les dépôts mètallifères de la fosse Atlantis II en Mer Rouge. *Comptes Rendue Academie Science* Paris, **310**, 1069–1075.

Badley, M.E., Egeberg, T. and Nipen, O. (1984) Development of rift basins illustrated by the evolution of the Oseberg feature, Block 30/6, offshore Norway. *Journal Geological Society of London*, **141**, 639–649.

Badley, M.E., Price, J.D., Rambech Dahl, C. and Agdestein, T. (1988) The structural evolution of the northern Viking Graben and its bearing upon extensional modes of basin formation. *Journal Geological Society of London*, **145**, 455–472.

Baker, J., Menzies, M. and Snee, L. (1994) Stratigraphy, $^{40}Ar/^{39}Ar$ geochronology and geochemistry of flood volcanism in Yemen. *Mineralogical Magazine*, **58A**, 42–43.

Baker, J., Snee, L. and Menzies, M. (1996) A brief Oligocene period of flood volcanism in Yemen; implications for the duration and rate of continental flood volcanism at the Afro-Arabian triple junction. *Earth and Planetary Science Letters*, **138**, 39–55.

Baker, J.A. (1996) Stratigraphy, geochronology and geochemistry of Cenozoic volcanism in western Yemen. Unpublished PhD thesis, Royal Holloway University of London, 386 pp.

Baker, J.A., Thirlwall, M.F. and Menzies, M.A. (1996b) Sr-Nd-Pb isotopic and trace element evidence for crustal contamination of a mantle plume: Oligocene flood volcanism in western Yemen. *Geochimica Cosmochimica Acta*, in press.

Baldridge, S., Eyal, Y., Bartov, Y., Steinitz, G. and Eyal, M. (1991) Miocene magmatism of Sinai related to the opening of the Red Sea. *Tectonophysics*, **197**, 181–201.

Ball, J. (1939) Contributions to the geography of Egypt. Bulletin of Ministry of Finance, Survey and Mines Department, Cairo, Egypt. 308 pp.

Bally, A.W. (1981) Atlantic type margins, in *Geology of Passive Continental margins*. (eds A.W. Bally, A.B. Watts, J.A. Grow, W. Manspeizer, D. Bernoulli, C. Schreiber and J.M. Hunt), American Association Petroleum Geologists Bulletin Course Note Series 19.

Bally, A.W. (1982) Musings over sedimentary basin evolution. *Philosophical Transactions of the Royal Society London*, **A305**, 325–328.

Baltzer, F., Conchon, O., Freytet, P. and Purser, B.H. (1993) Climatic and tectonic evolution recorded by Plio-Quaternary sedimentary terraces and fans along the Egyptian coast of the Red Sea, *Geological Society Egypt*, Special publication, **1**, 321–342.

Balzer, W. (1982) On the distribution of iron and manganese at the sediment/water interface: thermodynamic versus kinetic control. *Geochimica Cosmochimica Acta*, **46**, 1153–1161.

Bantan, R. (1995) Geology of Farasan Islands, Red Sea, Southwest Saudi Arabia, in *Rift Sedimentation and Tectonics in the Red Sea–Gulf of Aden Region*, (ed. D. Bosence), San'a, Yemen, 23–31 October 1995, 8–9.

Barakat, M.G., Darwish M. and El Barkooky, A.N. (1986) *Lithostratigraphy of the Post Carboniferous–Pre-Cenomanian clastics in west central Sinai and Gulf of Suez, Egypt.* Proceedings of the Eighth. Exploration Seminar. E.G.P.C., 380–405.

Barakat, M.G., Darwish, M. and El Outefi, N.S. (1988) *Eocene tectono-stratigraphy and basin evaluation in the Gulf of Suez Petroliferous Province.* Proceedings of the Nineth. E.G.P.C. Exploration Seminar, Cairo, 22 pp.

Barakat, J. and Miller, P.M. (1984) *Geology and petroleum exploration, Safaga concession, Northern Red Sea, Egypt.* Proceedings of the Seventh E.G.P.C., Exploration Seminar, Cairo.

Barash, A., Danin, Z. and Yaron, L. (1983) The genus *Rhinoclavis* (Gastropoda: Cerithiidae) in the Red Sea. *Annali del Museo Civico G. Doria*, **85**, 95–117.

Barbe, A., Grimalt, J.O., Pueyo J.J. and Albaiges, J. (1990) Characterisation of model evaporitic environments through the study of lipid components. *Organic Geochemistry*. **16**, 815–828.

Barberi, F., Giglia, G., Tazieff, H. and Varet, J. (1972) Tectonic significance of the Afar (or Danakil) depression. *Nature*, **235**, 144–147.

Bard, E., Hamelin, B., Fairbanks, R.G. and Zindler, A. (1990) Calibration of the ^{14}C time scale over the pase 30,000 years using mass spectrometric U-Th ages from Barbados corals. *Nature*, **345**, 405–410.

Barletta, G. (1972) Malacofauna del Mar Rosso: V. I Cipreidi delle Dahlak meridionali. *Conchiglie*, **8**, (1–2), 8–14.

Barnaby, R.J. and Read, J.F. (1990) Carbonate ramp to rimmed shelf evolution: Lower to Middle Cambrian continental margin, Virginian Appalachians. *Geological Society of America Bulletin*, **102**, 391–404.

Barnes, H.L. (1979) Solubilities of ore minerals. in *Geochemistry of Hydrothermal Ore Deposits* (ed. Barnes H.L.), Wiley-Interscience. New York, pp. 404–460.

Barnes, S.V. (1976) Geology and oil prospects of Somalia, East Africa. *American Association Petroleum Geologists Bulletin*, **60**, 389–413.

Bartov, Y., Lewy, Z. and Steinitz, G. (1980) Mesozoic and Tertiary stratigraphy, paleogeography and structural history of the Gebel Areif in Naga Area (Eastern Sinai). *Israel Journal of Earth Science*, **29**, 114–139.

Bartov, Y., Steinitz, G., Eyal, M. and Eyal, U. (1980) Sinistral movement along the Gulf of Aqaba – its age and relation to the opening of the Red Sea. *Nature*. **285**, 220–221.

Basahel, A.N., Bahafzalla, A., Jux, U. and Omara, S. (1982) Age and structural setting of a Proto-Red Sea embayment. *Nues Jahrbuch für Geologie und Palaeontologie*, **8**, 456–468.

Baumgartner, T.R. (1974) Tectonic evolution of the Walvis Ridge and west African Margin, South Atlantic Ocean. Thesis Oregon State University.

Bayer, H.J. Hötzl, H., Jado, A.R., Ruscher, B. and Voggenreitter, W. (1988) Sedimentary and structural evolution of the northwest Arabian Red Sea margin. *Tectonophysics*, **153**, 137–151.

Beach, A. (1986) A deep seismic reflection profile across the northern North Sea. *Nature*, **323**, 53–55.

Beadnell, H.J.L. (1905) The relations of the Eocene and Cretaceous systems in the Esna–Aswan reach of the Nile

Valley. *Quarterly Journal of the Geological Society of London*, **61**, 667–678

Beets, C.J. (1992) Calibration of late Cainozoic marine strontium isotope variations and its chronostratigraphic and geochemical applications. Thesis, Vrije Universiteit, Amsterdam, 133 pp.

Behairy A.K. and El Sayed M.K. (1983) Bathymetry and bottom relief beyond the reef complex off Jeddah, Red Sea. *Journal Faculty Marine Science Jeddah*, **3** (1404 H), 73–80.

Behairy A.K and Yusuf, N. (1984) Distribution of some planktonic foraminifera species in deep-sea cores from the Red Sea and their relation to eustatic sea-level changes. *Palaeogeography, Palaeoclimatology, Palaeoecology*, **46**, 291–301.

Behairy, A.K.A., (1983) Marine transgressions in the west coast of Saudi Arabia (Red Sea) between mid-Pleistocene and present. *Marine Geology*, **52**, M25–M31.

Belluomini, G., Esu, D., Manfra, L. et al. (1980) Gasteropodi dulcicoli e terrestri nell'isola di Dahlak Kebir – testimonianze di una fase umida olocenica nelle Isole Dahlak, Mar Rosso. *Bollettino Malacologico*, **16**, 369–390.

Belmonte, Y., Hitz. P and Wenger, R. (1965) The salt basins of the Gabon and he Congo (Brazzaville): a tentative paleogeographic interpretation. in *Salt Basin around Africa*. The Institute of Petroleum, London, Elsevier Publishing Company, pp. 55–74.

Ben Menahem, A., Nur, A. and Vered, M. (1976) Tectonics, seismicity and structure of the Afro-Eurasian junction – the breaking of an incoherent plate. *Physics and Earth Planetary International*, **12**, 1–50.

Berckhemer, H., Baier, P., Bartelsen, H. et al. (1975) Deep seismic soundings in the Afar region and on the Highland of Ethiopia, in *Afar Depression of Ethiopia*, (eds A. Pilger and A. Roesler), Interunion Commission on Geodynamics. Scientific Report, **14**, Schweizerbart, Stuttgart, I, pp. 66–79.

Berggren, W.A. and Boersma, A. (1969) Late Pleistocene and Holocene planktonic foraminifera from the Red Sea, in *Hot Brines and Heavy Metal Deposits in the Red Sea*, (eds E.T. Degens and Ross, D.A.), Springer-Verlag, New York, pp. 282–298.

Berhe, S.M. (1986) Geologic and geochronologic constraints on the evolution of the Red Sea–Gulf of Aden and Afar depression. *Journal African Earth Science*, **5**, (2), 101–117.

Bernard, P.C., Thompson, S., Bastow, M.A., Ducreuz, C. and Mathurin, G. (1992) Thermal maturity development and source rock occurrence in the Red Sea and Gulf of Aden. *Journal of Petroleum Geology*, **15**, 173–186.

Berner R.A. (1970) Sedimentary pyrite formation. *American Journal Science*, **268**, 1–23.

Berner R.A. (1981) Kinetics of weathering and diagenesis, in 'Kinetics of Geochemical Processes', (eds A. Lasaga and R. Kirkpatrick), *Reviews in Mineralogy*, **8**, 111–134.

Berner, R.A. (1984) Sedimentary pyrite formation: an update. *Geochimica Cosmochimica Acta*, **48**, 606–615.

Berner, R.A. (1989) Biochemical cycles of carbon and sulfur and their effect on atmospheric oxygen over Phanerozoic time. *Palaeogeography, Palaeoclimatology, Palaeoecology*, **75**, 7–122.

Berry L., Whiteman A.J. and Bell S.V. (1966) Some radiocarbon dates and their geomorphological significance, emerged reef complex of the Sudan. *Zeitschrift für Geomorphologie*, **10**, 119–143.

Berthelot, F. (1986) Etude thermique du Golfe de Suez dans son contexte geodynamique. Thesis, Université Pierrie Marie Curie, Paris VI, 190 pp.

Besse, D. and Taviani, M. (1982) The last Quaternary sapropelitic level in the Red Sea: its micropaleontological–mineralogical characteristics and paleoceanographic significance. Proceedings INQUA XI Congress, Moscow 1982, Abstracts, **1**, 36.

Betton, P.J. and Civetta, L. (1984) Strontium and neodymium isotopic evidence for the heterogeneous nature and development of the mantle beneath Afar (Ethiopia). *Earth Planet Science Letters*, **71**, 59–70.

Beutelspacher, H. and van der Marel, H.W. (1968) *Atlas of electron microscopy of clay minerals and their admixtures*. Elsevier, Amsterdam.

Beydoun, Z.R. (1964) *The stratigraphy and structure of the Eastern Aden Protectorate*. Overseas Geology and Mineral Resources, Supplement Series, 5, Her Majesty's Stationary Office, London, 107 pp.

Beydoun, Z.R. (1966) Geology of Arabian peninsula. Eastern Aden Protectorate and part of Dhufar. *United States Geological Survey Professional Paper*, **560-H**, 1–48.

Beydoun, Z.R. (1969) Note on the age of the Hadhramaut Arch, southern Arabia. *Overseas Geological Mineral Resources*, **10**, 236–240.

Beydoun, Z.R. (1978) Southern Arabia and northern Somalia: comparative geology. *Philosophical Transactions of the Royal Society, London*, Series A, 267–292.

Beydoun, Z.R. (1988) *The Middle East: Regional Geology and Petroleum Resources*. Scientific Press, Beaconsfield, UK.

Beydoun, Z.R. (1989) The hydrocarbon prospects of the Red Sea–Gulf of Aden: a review. *Journal of Petroleum Geology*, **12**(2), 125–144.

Beydoun, Z.R. (1991) Middle East hydrocarbon reserves enhancement, 1975–1990; significant developments and implications for potential future large increases. *Journal of Petroleum Geology*, **14** (1), Supplement 1, ii–iv.

Beydoun, Z.R. (ed.) (1992) The Red Sea–Gulf of Aden, II. *Journal of Petroleum Geology*, **15**, 121–246.

Beydoun, Z.R. and Sikander, A.H. (1992) The Red Sea–Gulf of Aden: re-assessment of hydrocarbon potential. *Marine and Petroleum Geology*, **9**, 474–485.

Bignell, R.D. (1976) Red Sea: Climate and rifting related to metallogenesis of the Red Sea. *Geological Association of Canada, Special Publication*, **14**, 181–184.

Bignell, R.D. (1978) Genesis of the Red Sea: Metalliferous Sediments. *Marine Mining*, **1** (3), 209–235.

Bignell, R.D. and Ali, S.S. (1976) Geochemistry and stratigraphy of Nereus Deep, Red Sea. *Geologisches Jahrbuch*, **D17**, 173–186.

Bignell, R.D., Cronan D.S. and Tooms, J.S (1976a) Metalliferous brine precipitates. *Geological Association of Canada*, Special Publication, **14**, 147–180.

Bignell, R., Cronan D. and Tooms J. (1976b) Metal dispersion in the Red Sea as an aid to marine geochemical exploration. *Institute Mining Metallurgy Transactions*, **84**, B274–B278

Bischoff J.L. (1969) Red Sea geothermal brine deposits: their mineralogy, chemistry, and genesis, in *Hot Brines and Recent Heavy Metal Deposits in the Red Sea* (eds E.T. Degens and A.D. Ross), Springer Verlag, Berlin, pp. 368–401.

Bishay, Y. (1966) Studies on the larger foraminifera of the Eocene (the Nile Valley between Assiut and Cairo and SW Sinai). PhD. thesis, Alexandria University.

Blakey, S., Menzies, M.A. and Thirlwall, M.F. (1994) Geochemistry of Tertiary within-plate (A-type) granitoids at a passive margin, southern Red Sea. International Association of Volcanology and Chemistry of the Earth's Interior, Ankara, Turkey (abs).

Blanc, G (1987) Géochimie de la fosse Atlantis II (Mer Rouge). Evolution spatiotemporelle et rôle de l'hydrothermalisme. Thèse de l'Université Pierre et Marie Curie, Paris VI.

Blanc, G. and Anschutz, P. (1995) New stratification in the hydrothermal brine system of the Atlantis II Deep, Red Sea. *Geology*, **23** (6), 543–546.

Blanc, G., Anschutz, P. and Pierret, M-C. (this volume) Metalliferous sedimentation in the Atlantis II deep, Red Sea: A geochemical insight, in *Sedimentation and Tectonics of Rift Basins: Red Sea–Gulf of Aden*, (eds B.H. Purser and D.W.J. Bosence), Chapman & Hall, London.

Blanc, G., Boulègue J., Baudaut D. and Stouff, P. (1986) Premiers résultats de la campagne océanographique 'Hydrotherm' (mai 1985) du Marion Dufresne sur la fosse Atlantis II (Mer Rouge). *Compte Rendue Academie Science Paris*, **302**, 175–180.

Blanc G., Boulègue, J. and Charlou, J. L. (1990) Profils d'hydrocarbures légers dans l'eau de mer, les saumures et les eaux interstitielles de la fosse Atlantis II (mer Rouge). *Oceanologia Acta*, **13**, 187–197.

Blanckenhorn, M. (1901) Neues zur Geologie und Paläontologie Aegyptens. III Das Miocän *Zeitschrift der deutsche geologische Gesellschaft*, **53**, 52–132.

Blank, H.R., Johnson, P.R., Gettings, M.E. and Simmons, G.C. (1986) Explanatory notes to the geological map of the Jizan Quadrangle, sheet 16F, Kingdom of Saudi Arabia. Ministry of Petroleum and Mineral Resources, Jeddah.

Bloom, A.L., Broecker, W.S., Chappell, J.M.A. *et al.* (1974) Quaternary sea level fluctuations on a tectonic coast: new ^{230}Th/^{234}U dates from the Huon Peninsula New Guinea. *Quaternary Research*, **4**, 185–205.

Blow, W.H. (1969) Late Middle Eocene to Recent planktonic foraminiferal biostratigraphy. Proceedings of the First International Conference on Planktonic Microfossils (eds R. Brönnimann and H.H. Renz) Brill, Leiden, pp. 199–421.

Blundell, D.J.B., Hobbs, R.W., Klemeperer, S.L., Scott-Robinson, R., Long, R.E., West, T.E. and Duin, E. (1991) Crustal structure of the central and southern North Sea from BIRPS deep seismic reflection profiling. *Journal Geological Society of London*, **148**, 445–457.

Boccaletti, M., Getaneh, A. and Bonavia, F. (1991) The Marda Fault: a remnant of incipient aborted rift in paleo African-Arabian plate. *Journal of Petroleum Geology*, **14**, 79–92.

Boeckelman, K. and W. Schreiber (1990) Sedimentology and stratigraphy of Taleh and Karkar Formations (Eocene) in Northern Somalia. *Berliner Geowissenschaftliche Abhandlungen*, **102.2**, 595–620.

Boersma, A. (1984) *Handbook of Common Tertiary Uvigerina*, Microclimates Press, New York.

Bohannon, R.G. and Eittreim, S.L. (1991) Tectonic development of passive continental margins of the southern and central Red Sea with a comparison to Wilkesland, Antarctica. *Tectonophysics*, **198**, 129–154.

Bohannon, R.G., Naeser, C.W., Schmidt, D.L. and Zimmermann, R.A. (1989) The timing of uplift, volcanism and rifting peripheral to the Red Sea: a case for passive rifting. *Journal of Geophysical Research*, **94**, 1683–1701.

Boillot, G., Mougenot, D., Girardeau, J. and Winterer, E.L. (1989) Rifting processes on the West Galicia margin, Spain, in *Extensional Tectonics and Stratigraphy of the North Atlantic Margins*, (eds A.J. Tankard and H.R. Balkwill), American Association Petroleum Geologists, Tulsa, Memoir 46, pp. 363–77.

Boldy, J. (1977) Uncertain exploration facts from figures. *Canadian Institute Mining Metallurgy Bulletin*, **70**, 86–95.

Bolli, H.M., Saunders, J.B. and Perch-Nielsen, K. (1985) *Plankton stratigraphy*, Cambridge University Press, Cambridge.

Bonatti, E. (1967) Mechanisms of deep-sea volcanism in the south Pacific, in *Researches in Geochemistry* (ed. P.H. Abelson), **2**, Wiley, New York, pp. 453–491.

Bonatti, E. (1985) Punctiform initiation of seafloor spreading in the Red Sea during transition from a continental to an oceanic rift. *Nature*, **316**, 33–37.

Bonatti, E. (1987) Rifting or drifting in the Red Sea? *Nature*, **330**, 692–693.

Bonatti, E., Clocchiatti, R., Colantoni, P., Gelmini, R., Marinelli, G. (1983) Zabargad (St. John's) Island: an uplifted fragment of sub-Red Sea lithosphere. *Journal of the Geological Society of London*, **140**, 677–690.

Bonatti, E., Colantoni P., Della Vedova, B. and Taviani M. (1984) Geology of the Red Sea transitional region (22°–25°N). *Oceanologica Acta*, **7** (4), 385–398.

Bonatti, E., Colantoni, P., Lucchini, F., Rossi, P.L., Taviani, M. and White, J. (1986) Chemical and stable isotope aspects of the Nereus Deep (Red Sea) metal-enriched sedimentation, *Memorie della Società Geologica Italiana*, **27**, (1984), 59–72.

Bono, P., Carbone, F., Ciancetti, G. *et al.* (1983) Caratteristiche idrologiche e sedimentologiche delle lagune dell' Isola di Isratu (Isole Dahlak, Mar Rosso). *Geologica Romana*, **22**, 249–270.

Bono, P., Civitelli, G. and Pialli, G. (1976) Caratteristiche sedimentologiche e idrologiche della laguna di Khor Melill (Isole Dahlak, Mar Rosso). *Geologica Romana*, **15**, 69–82.

Bosellini, A. (1984) Progradation geometries of carbonate platforms: examples from the Triassic of the Dolomites, northern Italy. *Sedimentology*, **31**, 1–24.

Bosellini, A. (1989) The continental margins of Somalia (structural evolution and sequence stratigraphy). *Memorie di Scienze Geologiche gia Memorie degli Instituti di Geologia and Mineralogia dell'Universita di Padova*, **41**, 373–458.

Bosellini, A. (1992) The continental margins of Somalia: structural evolution and sequence stratigraphy, in 'Geology and Geophysics of Continental Margins' (eds J.S. Watkins, F. Ziqiang and K.J. McMillen), American Association Petroleum Geologists, Memoir **53**, 185–205.

Bosence, D.W.J. (this volume) Stratigraphic and sedimentological models of rift basins, in *Sedimentation and tectonics of rift basins: Red Sea–Gulf of Aden* (eds B.H. Purser and D.W.J. Bosence), Chapman & Hall, London, pp. 9–26.

Bosence, D.W.J., Al-Aawah, M.H., Davidson, I., Rosen, B., Vita-Finzi, C. and Whittaker, L. (this volume) Salt domes and their control on basin margin sedimentation: a case study from the Tihama Plain, Yemen, in *Sedimentation and tectonics of rift basins: Red Sea–Gulf of Aden* (eds B.H. Purser and D.W.J. Bosence), Chapman & Hall, London, pp. 450–467.

Bosence, D.W.J., Cross, N., Hardy, S. and Purser, B.H. (1994) Depositional sequences and sequence boundaries from a Miocene syn-rift carbonate platform. International Sedimentological Congress, Recife, Brazil, Abstract volume, p. B2.

Bosence, D.W.J., Nichols, G., Al-Subbary, A-K., al-Thour, K. and Reeder, M. (1996) Syn-rift continental-marine depositional sequences. Tertiary, Gulf of Aden, Yemen. *Journal of Sedimentary Research*, **66**, 766–777.

Bosence, D.W.J., Pomar, L., Waltham, D. and Lankester, T.G. (1994) Computer modelling a Miocene carbonate platform. Spain. *American Association Petroleum Geologists Bulletin*, **78**, 247–266.

Bosence, D.W.J. and Waltham, D. (1990) Computer modelling the internal architecture of carbonate platforms. *Geology*, **18**, 26–30.

Bosworth, W. (1985) Geometry of propagating continental rifts. *Nature*, **316**, 625–627.

Bosworth, W. (1992) Mesozoic and early Tertiary rift tectonics in East Africa. *Tectonophysics*, **209**, 115–137.

Bosworth, W. (1993) Structural style and tectonic evolution of the rift basins of northeast Africa, *Geoscientific Research in Northeast Africa*, Rotterdam, Abst. Vol., pp. 213–217.

Bosworth, W. (1994a) A high-strain rift model for the southern Gulf of Suez (Egypt). in Hydrocarbon Habitat in Rift Basins. (ed. J.J. Lambiase), *Geological Society of London*, Special Publication, **80**, 75–102.

Bosworth, W. (1994b) A model for the three-dimensional evolution of continental rift basins, north-east Africa. *Geologsiche Rundschau*, **83**, 671–688.

Bosworth, W. (1995) *Structure, sedimentation and basin dynamics during early Gulf of Suez rifting*. Proceedings of Rift Sedimentation and Tectonics in the Red Sea-Gulf of Aden Region. University of Sana'a Yemen, October, p. 13.

Bosworth, W., Crevello, P., Winn, R.D. and Steinmetz, J. (this volume) Structure, sedimentation and basin dynamics during rifting of the Gulf of Suez, and northwestern Red Sea, in *Sedimentation and Tectonics of Rift Basins: Red Sea – Gulf of Aden* (eds B.H. Purser and D.W.J. Bosence), Chapman & Hall, London, pp. 77–96.

Bosworth, W., Lambiase, J. and Keisler, R. (1986) A new look at Gregory's Rift: the structural style of continental rifting. *EOS*, **67**, 577–583.

Bosworth, W. and Taviani, M. (1996) Late Quaternary reorientation of stress field and extension direction in the southern Gulf of Suez, Egypt: Evidence from uplifted coral terraces, mesoscopic fault arrays, and borehole breakouts. *Tectonics*, **16**, in press.

Bott, M.H.P. (1992) Passive margins and their subsidence. *Journal Geological Soceity of London*, **149**, 805–812.

Bott, W.F., Smith, B.A., Oakes, G. Sikander, A.H. and Ibrahim, A.I. (1992) The tectonic framework and regional hydrocarbon prospectivity of the Gulf of Aden. *Journal of Petroleum Geology*, **15**, 211–243.

Bouchon, C., Jaubert, J., Montaggioni, L. and Pichon, M. (1981) *Morphology and evolution of the coral reefs of the Jordanian coast of the Gulf of Aqaba (Red Sea)*. Proceedings Fourth International Coral Reef Symposium, Manila **1**, 559–565.

Boudreaux, J.E. (1974) Calcareous nannoplankton ranges. *Deep Sea Drilling Project, Initial Report*, **23**, pp. 1073–1099.

Boulègue, J. (1978) Géochimie du soufre dans les milieux réducteurs. Thèse Doctoral et Science, Université Paris IV.

Boulos, F.K., Morgan, P. and Toppozada, T.R. (1987) Microearthquake studies in Egypt carried out by the Geological Survey of Egypt. *Journal of Geodynamics*, **7**, 227–249.

Bown, T.M. and Kraus, M.J. (1981) Lower Eocene alluvial paleosols (Willwood Formation, Northwest Wyoming, USA) and their significance for paleoecology, paleoclimatology and basin analysis. *Palaeogeography, Palaeoclimatology and Palaeoecology*, **34**, 1–30.

Brachert, T.C. (1994) Palaeoecology of enigmatic tube microfossils forming 'cryptalgal' fabrics (late Quaternary, Red Sea). *Palaeontologische Zeitschrift*, **68**, 299–312.

Brachert, T.C. (1996) Klimastenerung von karbonatsystemem: fallbeispiele zur biofazies ozeanisches und flachmariner öcosysteme im Känozoikum. Unpublished thesis, Univerität Mainz, 238 pp.

Brachert, T.C. (submitted) *Non-skeletal carbonate production within a deep ocean basin. The 'hard layer' of the glacial Red Sea*. Proceedings Death Valley International Stromatolite Symposium, Laughlin, USA.

Brachert, T.C. and Dullo, W.-C. (1990) Correlation of deep sea sediments and forereef carbonates in the Red Sea: an important clue for basin analysis. *Marine Geology*, **92**, 255–267.

Brachert, T.C. and Dullo, W.-C. (1991) Laminar micrite crusts and associated foreslope processes, Red Sea, *Journal of Sedimentary Petrology*, **61**, 354–363.

Brahim, A., Rouchy, J.M., Maurin, A., Guelorget, O. and Perthuisot, J.-P. (1986) Mouvements halocinétiques récents dans le golfe de Suez: l'exemple de la péninsule de Guemsah. *Bulletin Societé Géologie*, France, **8** (2), 177–183.

Braithwaite, C.J.R. (1982) Patterns of accretion of reefs in the Sudanese Red Sea. *Marine Geology*, **46**, 297–325.

Braithwaite, C.J.R. (1987) Geology and paleontology of the Red Sea region, in *Red Sea* (eds A.J. Edwards and S.M. Head), Pergamon Press, Oxford, pp. 22–44.

Brannan, J., Gerdes, K.D. and Newth, I.R. (in press) Tectonostratigraphic development of the Qamar basin, eastern Yemen. *Marine and Petroleum Geology*.

Braun, J. and Beaumont, C. (1989a) Styles of continental rifting: results from dynamic models of lithospheric extension, in *Sedimentary Basins and Basin-Forming Mechanisms*, (eds C. Beaumont and A.J. Tankard) Canadian Society of Petroleum Geologists, Memoir 12, 241–258.

Braun, J. and Beaumont, C. (1989b) A physical explanation of the relation between flank uplifts and the breakup unconformity at rifted continental margins. *Geology*, **17**, 760–764.

Bray, E.A. du, Stoeser, D.B. and McKee, E.H. (1991) Age and petrology of the Tertiary As Sarat volcanic field, southwestern Saudi Arabia. *Tectonophysics*, **198**, 155–180.

Brice, S., Cochran, M.D., Pardo, G. and Edwards, A.D. (1982) Tectonics and sedimentation of the South Atlantic rift sequence: Cabinda, Angola. *American Association Petroleum Geologists Bulletin*, **66**, 5–18.

Bricker, O.P. (1965) Some stability relations in the system $Mn-O_2-H_2O$ at 25°C at one atmosphere total pressure. *American Mineralogist*, **50**, 1296–1354.

Briem, E. and Blümel, W.D. (1984) Contributions to the Quaternary geomorphology of the Ifal Depression, in *Quaternary Period in Saudi Arabia* (eds A.R. Jado and J.G. Zöttl). Springer, Wien, pp. 47–55.

Brockamp, O., Goulard, E., Harder, H. and Heydemann, A. (1978) Amorphous copper and zinc sulfides in the metalliferous sediments of the Red Sea. *Contributions Mineralogy and Petrology*, **68**, 85–88.

Brognon, G.F. and Verrier, G.R. (1966) Oil and geology in Cuannza basin of Angola. *American Association Petroleum Geologists Bulletin*, **50**, 108–158

Brown, G.F. (1970) Eastern margin of the Red Sea and coastal structures in Saudi Arabia. *Philosophical Transactions of the Royal Society London*, **A267**, 75–78.

Brown, G.F., Jackson, R.O., Bogue, R.G. and Maclean, W.H. (1962) Geologic map of the southern Hijaz quadrangle Kingdom of Saudi Arabia. United States Geological Survey Miscellaneous Geological Investigations Map I-210A.

Brown, L.F. and Fisher, W.F. (1977) Seismic stratigraphic interpretation of depositional systems: examples from Brazil rift and pull-apart basins, in *Seismic Stratigraphy–Applications to Hydrocarbon Exploration*, (ed. C.E. Payto), American Association Petroleum Geologists Memoirs 26, pp. 213–2480

Brown, R.N. (1980) History of exploration and discovery of Morgan, Ramadan and July oilfields, Gulf of Suez, Egypt, in *Facts and Principles of World Petroleum occurrence* (ed. A.D. Miall). Canadian Society Petrololeum Geologists, pp. 733–64.

Bruni, P. and Fazzuoli, M. (1977) Sedimentological observations on Jurassic and Cretaceous sequences from Northern Somalia. Preliminary report. *Bollettino Società Geologica Italiana*, **95** (1976), 1571–1588.

Bruni, P. and Fazzuoli, M. (1980) Mesozoic structural evolution of the Somali coast of the Gulf of Aden. *Accademia Nazionale dei Lincei, Atti Convegno*, **47**, (1979), 193–207.

Buchbinder, B. (1979) Facies and environments of Miocene reef limestones in Israel. *Journal of Sedimentary Petrology*, **49** (4), 1323–1344.

Buczynski, C. and Chafetz, H.S. (1991) Habit of bacterially induced precipitates of calcium carbonates and the influence of medium viscosity on mineralogy. *Journal of Sedimentary Petrology*, **61**, 226–233.

Buddemeier, R.W. and Oberdorfer, J.A. (1986) Internal hydrology and geochemistry of coral reefs and atoll slands: key to diagenetic variations, in *Reef Diagenesis*, (eds, J.H. Schroeder and B.H. Purser) Springer-Verlag, NY, pp. 91–111.

Bull, W.B. and McFadden, L.D. (1977) Tectonic geomorphology north and south of the Garlock Fault, Carolina, in *Geomorphology in Arid Regions*, (ed. D.O. Doehring) State University, New York, Binghampton, pp. 115–138.

Bullen-Newton, R. (1900) Pleistocene shells from the raised beach deposits of the Red Sea. *Geological Magazine*, **7**, 500–514, 544–569.

Bullen, S.B. and Sibley, D.F. (1984) Dolomite selectivity and mimic replacement. *Geology*, **12**, 655–658.

Bunter, M.A.G. and Abdel Magid, A.E.M. (1989) The Sudanese Red Sea, 1. New developments in stratigraphy and petroleum geological evolution. *Journal of Petroleum Geology*, **12**, 145–166.

Burchette, T.P. (1988) Tectonic control on carbonate platform facies distribution and sequence development: Miocene, Gulf of Suez. *Sedimentary Geology*, **59**, 179–204.

Burchette, T.P. and Wright, V.P. (1992) Carbonate ramp depositional systems. *Sedimentary Geology*, **79**, 3–57.

Burollet, P.F. (1986) Reconnaissance géologique de la région de Ras Benas. *Documents et Travaux Documents et Travaux, Institut Géologique Albert de Lapparent*, Paris, **10**, 171–175.

Burollet, P.F., Bolze, J. and Ott d'Estevou, P. (1982) *Sedimentology and tectonics of the Gharamul area, West of Suez gulf*. Proceedings Sixth EGPC Exploration Seminar, Cairo, Egypt, Abstract, 1–17.

Butuzova, G. Yu, Drits, V.A., Morozov, A.A. and Gorschkov, A.I. (1990) Process of formation of iron-manganese oxyhydroxides in the Atlantis II and Tethys deeps of the Red sea, in *Sediment-hosted Mineral Deposits*, (eds J. Parnell, Y. Lianjun and C. Changming). Special Publication No. 11 of the International Association of Sedimentologists, Blackwell Scientific Publications, Oxford, pp. 57–73.

Butzer, K.W. (1980) Pleistocene history of the Nile valley in Egypt and Lower Nubia, in *The Sahara and the Nile*, (eds M.A.J. Williams and H. Faure), Balkema, Rotterdam, pp. 253–278.

Butzer, K.W. and Hansen, C.L. (eds) (1968) The coastal plain of Mersa Alam, in *Desert and River in Nubia*, University of Wisconsin Press, Madison, USA, pp. 395–432.

Cahuzac, B. and Chaix, C. (1993) Les faunes de coraux (anthozoaires, scléractiniaires) de la façade atlantique française au Chattien et au Miocène. Ciências da Terra (UNL), **12**, 57–59.

Camp, V.E., Hooper, P.R., Robool, M.J. and White, D.L. (1987) The Medina eruption. Saudi Arabia: magma mixing and simultaneous extrusion of three basaltic chemical types. *Bulletin Volcanology*, **49**, 489.

Camp, V.E. and Roobol, M.J. (1989) The Arabian continental alkali basalt province: Part 1. Evolution of Harrat Rahat, Kingdom of Saudi Arabia. *Geological Society America Bulletin*, **101**, 71–95.

Camp, V.E. and Robool, M.J. (1991) Comment on 'Topographic and volcanic asymmetry around the Red Sea: constraints on rift models, by T.H. Dixon, E.R. Ivins and J.F. Brenda. *Tectonics*, **10**, 649–652.

Camp, V.E. and Robool, M.J. (1992) Upwelling asthenosphere beneath western Arabia and its regional implications. *Journal of Geophysical Research*, **97**, 15255–15271.

Camp, V.E. and Robool, M.J. and Hooper, P.R. (1991) The Arabian continental alkali basalt province: Part II. Evolution of Harrats Khaybar, Ithnayn, and Kura, Kingdom of Saudi Arabia. *Geological Society America Bulletin*, **103**, 363.

Camp, V.E., Robool, M.J. and Hooper, P.R. (1992) The Arabian continental alkali basalt province: Part III. Evolution of Harrat Kishb, Kingdom of Saudi Arabia. *Geological Society America Bulletin*, **104**, 379.

Canfield, D.E. (1989) Sulfate reduction and oxic respiration in marine sediments: implication for organic carbon preservation in euxinic environments. *Deep-Sea Research*, **36**, 121–138.

Cant, D.J. and Walker, R.G. (1978) Fluvial processes and facies sequences in the sandy braided south Saskatchewan River, Canada. *Sedimentology*, **25**, 625–648.

Canuti, P. and Marcucci, M. (1968) Microfacies Aptiens-Albiens del l'Al Medo et de Candal (Somalia). In: Proceedings 3rd African Micropalaeontological Congress.

Capaldi, G., Chiesa, S., Conticelli, S. et al. (1987a) Jabal An Nar: an upper Miocene volcanic centre near Al Mukha (Yemen Arab Republic). *Journal of Volcanology and Geothermal Research*, **31**, 345–351.

Capaldi, G., Chiesa, S., Manetti, P. et al. (1987b) Tertiary anorogenic granites of the western border of the Yemen Plateau. *Lithos*, **20**, 433–444.

Capaldi, G., Manetti, P. and Piccardo, G.B. (1983) Preliminary investigations on volcanism of the Sadah Region (Yemen Arabic Republic). *Bulletin of Volcanology*, **46**, 413–427.

Capaldi, G., Manetti, P., Piccardo, G.B. and Poli, G. (1987e) Nature and geodynamic significance of the Miocene dyke swarm in the North Yemen (YAR). *Neues Jahrbuch Mineral Abh*, **156**, 207–229.

Carboni, M.G., Casalino, R., Esu, D. et al. (1993) La componente epipelagica del Ghubbet Mus Nefit (Isole Dahlak, Mar Rosso). *Geologica Romana*, **29**, 197–211.

Carbonne, F., Matteucci, R. and Angelucci, A. (this volume) Present-day sedimentation of the carbonate platforms of the Dahlak Islands, Eritrea, in *Sedimentation and tectonics of rift basins: Red Sea–Gulf of Aden*, (eds B.H. Purser and D.W.J. Bosence), Chapman & Hall, London, pp. 527–542.

Carella, R. and Scarpa, N. (1962) *Geological results of exploration in Sudan by Agip Mineraria*. Proceedings of the Fourth Arab Petroleum Congress, Beirut, pp. 1–23.

Carlson, L. and Schwertmann, U. (1981) Natural ferrihydrites in surface deposits from Finland and their association with silica. *Geochimica et Cosmochimica Acta*, **45**, 421–429.

Carlson, L. and Schwertmann, U. (1990) The effect of CO_2 and oxidation rate on the formation of goethite versus lepidocrocite from an Fe(II) system at pH 6 and 7. *Clay Mineralogy*, **25**, 65–71.

Carver, R.E. (1971) *Procedures in Sedimentary Petrology*. John Wiley, New York. 653 pp.

Cas, R.A.F. and Wright, J.V. (1988) *Volcanic successions. Modern and Ancient*. Chapman & Hall, i-xvi + 528 pp.

Cernohorsky, W.O. (1972) *Marine Shells of the Pacific*. Volume II. Pacific Publications, Sydney, 411 pp.

Cerveny, V. and Psencik, J. (1981) *2-D Seismic Ray Package*. Charles University Prague.

Chafetz, H.S. (1986) Marine peloids: a product of bacterially induced precipitation of calcite. *Journal of Sedimentary Petrology*, **56**, 812–817.

Chafetz, H.S. and Buczinsky, C. (1992) Bacterial lithification of microbial mats. *Palaios*, **7**, 277–293.

Charnock, G. (1964) Anomalous bottom water in the Red Sea. *Nature*, **203**, 591.

Chazot, G. (1993) Evolution geochimique du magmatisme Cenozoique au Yemen: interactions entre le rift Mer Rouge–Aden et le point chaud Afar. Unpublished PhD thesis, University Lyon.

Chazot, G. and Bertrand, H. (1993) Genesis of silicic magmas during Tertiary continental rifting in Yemen. *Lithos*, **36**, 69–84.

Chazot, G., Menzies, M.A. and Baker, J. (this volume) Pre-, syn- and post-rift volcanism on the southwestern margin of the Arabian Plate, in *Sedimentation and tectonics of rift basins: Red Sea–Gulf of Aden* (eds B.H. Purser and D.W.J. Bosence), Chapman & Hall, London, pp. 50–55.

Chazot, G., Menzies, M.A. and Harte, B. (1996a) Determination of partition coefficients between apatite, clinopyroxene, amphibole, and melt in natural spinel lherzolites from Yemen: implications for wet melting of the lithospheric mantle. *Geochimica et Cosmochimica Acta*, **60**, 423–437.

Chazot, G., Menzies, M.A. and Harte, B. (1996b) Silicate glasses in spinel lherzolites from Yemen: origin and chemical composition. *Chemical Geology*. (in press).

Chen, C. (1969) Pteropods in the hot brine sediment of the Red Sea, in *Hot Brines and Heavy Metal Deposits in the Red Sea* (eds E.T. Degens and D.A. Ross), Springer-Verlag, New York, 313–317.

Chenet, P.Y. and J. Letuzey (1983) Tectonique de la zone comprise entre Abu Durba et Gebel Mezzazat (Sinai Egypte) dans le contexte de evolution du rift de Suez. *Bulletin Centres Recherche Exploration-Production, Elf Aquitaine*, **7**, 201–215.

Chenet, P.Y., Letouzey, J. and Zaghloul, E.S. (1986) *Some observations in the rift tectonics in the eastern part of the Suez rift*. Proceedings of the Seventh EGPC Exploration Conference, Cairo, pp. 18–36.

Cherchi, A., Fantozzi, P.L. and Abdirahman, H. (1993) Micropaleontological data on the Jurassic–Cretaceous sequences and the Cretaceous–Paleocene boundary in Northern Somalia (Bosaso region). *Comptes Rendues Academie des Sciences*, **316**, Serie II, 1179–1185.

Chevalier, J.P. (1954) Contribution à la révision des polypiers du genre Heliastraea. *Annales Hébert et Haug*, **8**, 105–190.

Chevalier, J.P. (1962) Recherches sur les madréporaires et les formations récifales miocènes de la Méditerranée occidentale. *Mémoires de la Société géologique de France*, N.S., XL, **93**, 1–562.

Chevalier, J.P. (1971) *Les Scléractiniaires de la Mélanésie Francais. (Nouvelle Calédonie, Iles Chesterfield, Iles Loyauté, Nouvelles-Hébrides)*. 1re Partie. Éditions de la Fondation Singer-Polignac, Paris.

Chevalier, J.P. (1977) Aperçu sur la faune corallienne récifale du Néogène, *Mémoires Bureau Recherche Geologique Minieres*, **89**, 359–366.

Chiari, R., Ferrari, M.C. and Sgavetti, M. (1994) Spectral classification of the rocks as a preliminary procedure for the lithologic interpretation of remote sensing data. An example relative to mainly carbonate and evaporitic rocks. Geology from space. *Proceeding SPIE 2320*, 24–35.

Chiesa, S., Civetta, L., De Fino, M. *et al.* (1989) The Yemen Trap Series: genesis and evolution of a continental flood basalt province. *Journal of Volcanology and Geothermal Research*, **36**, 337–350.

Chorowicz, J. and Lyberis, N. (1987) Evolution tectonique de la péninsule du Sinai: la formation des plis de l'Arc Syrien et des golfes de Suez et d'Aqaba. *Notes et Mémoires, Total-CFP*, **21**, 199–209.

Chorowicz, J., Le Fournier, J. and Vidal, G. (1987) A model for rift development in Eastern Africa. *Geological Journal*, **22**, 495–513.

Chorowicz, J., Mukonki, N.B. and Pottier, Y. (1979) Mise en évidence d'une compression horizontale liée à l'ouverture des fossés est-africains (branche occidentale), dans le seuil entre les lacs Kivu et Tanganyika. *Comptes Rendus Sommaire Société Géologie*, France, **5/6**, 231–234.

Choukri, A. (1994) Application des méthodes de datation par les séries de l'Uranium à l'identification des hauts niveaux marins de la côte égyptienne de la Mer Rouge au moyen de coraux, radioles d'oursins et coquilles. Thesis Science, University Rabat.

Choukri, A., Reyss, J.-L. and Plaziat, J.-C. (1995) Datations radiochimiques des hauts niveaux marins de la rive occidentale du Nord de la Mer Rouge au moyen de radioles d'oursin. *Compte Rendue Academie des Sciences de Paris*, ser.2, **321**, 25–30.

Chronis, G., Piper, D.J.W. and Anagnostou, C. (1991) Late Quaternary evolution of the Gulf of Patras, Greece: tectonism, deltaic sedimentation and sea-level change. *Marine Geology*, **97**, 191–209.

Chuckrov, F.V., Zvyagin, B.B., Gorskhov, A.I., Yermilova, L.P. and Balaskova, V.V. (1973) Ferrihydrite. *Izvestiiya International Geological Review*, **16**, 1131–1143.

Cita, M.B. (1980) Quand la Méditerranée était asséchée. *Recherche*, **107** (11), 26–36, Paris.

Civetta, L., La Volpe, L. and Lirer, L. (1978) K-Ar ages of the Yemen plateau. *Journal of Volcanology and Geothermal Research*, **4**, 307–314.

Civitelli, G. and Matteucci, R. (1981) La scogliera a frangia di Tanam (Isole Dahlak, Mar Rosso). *Bollettino della Società Geologica Italiana*, **99**, 517–530.

Clark, D.M. (1985), Geology of the Al-Bad quadrangle sheet 28A. Kingdom of Saudi Arabia. Directorate general of Mineral Resources Jeddah, *Open File Report, DGHR-OF-03-20*, 76 pp.

Clauzon, G. (1982) Le canyon messinien du Rhone: une preuve décisive du 'dessicated deep-basin model' (Hsü, Cita et Ryan, 1973). *Bulletin de la Société Géologique de France*, sér. 7, **24** (3), 597–610.

Clegg, N.M. (1996) The diagenesis of Miocene carbonates: Abu Shaar, Gabel Zeit and Ras Gemsa; Gulf of Suez, Egypt. PhD thesis, University of East Anglia, England.

Clegg, N. Harwood, G. and Kendall, A. (1995) Hydrothermal resetting of carbonates dolomitised by evaporitic brines: Abu Shaar and Gabal el Zeit, Gulf of Suez, Egypt. IAS Regional Meeting of Sedimentology (Aix-les-Bains), Abstract volume, p. 39.

Clegg, N.M., Harwood, G. and Kendall, A. (this volume) Dolomitization and post-dolomite diagenesis of Miocene platform carbonates: Abu Shaar, Gulf of Suez, Egypt, in *Sedimentation and tectonics of rift basins: Red Sea–Gulf of Aden*, (ed B.H. Purser and D.W.J. Bosence), Chapman & Hall, London, pp. 391–406.

Clément, P. and Giannésini, P.J. (1992) MD 29 à bord du 'Marion Dufresne', 16 octobre–28 oct. 1981. Rapport des TAAF, Museum National d'Histoire Naturelle, Paris.

Cochran, J.R. (1981) The Gulf of Aden: structure and evolution of a young ocean basin and continental margin. *Journal of Geophysical Research*, **86**, 263–287.

Cochran, J.R. (1982) The magnetic quiet zone in the eastern Gulf of Aden: implications for the early development of the continental margin. *Geophysical Journal of the Royal Astronomical Society*, **68**, 171–201.

Cochran, J.R. (1983a) Effects of finite rifting times on the development of sedimentary basins. *Earth and Planetary Science Letters*, **66**, 289–302.

Cochran, J.R. (1983b) A model for the development of the Red Sea. *American Association Petroleum Geologists Bulletin*, **67**, 41–69.

Cochran, J.R. and Martinez, F. (1988) Evidence from the northern Red Sea on the transition from continental to oceanic rifting. *Tectonophysics*, **153**, 25–53.

Cofer, C., Lee, K. and Wray, J. (1984) *Miocene carbonate microfacies, Esh Mellaha range, Gulf of Suez*. Proceedings of the Seventh EGPC Exploration Seminar, Cairo, pp. 97–115.

Coffield, D.Q. and Schamel, S. (1989) Surface expression of an accomodation zone within the Gulf of Suez rift, Egypt, *Geology*, **17**, 76–79.

Coleman, J.M. (1969) Brahmaputra River: Channel processes and sedimentation. *Sedimentary Geology*, **3**, 129–239.

Coleman, R.G. (1974a) Geologic background of the Red Sea, in *The Geology of Continental Margins* (eds C.A. Burke and C.L. Drake), Springer, Berlin, pp. 743–751.

Coleman, R.G. (1974b) Geologic background of the Red Sea, in *Initial Reports of the Deep Sea Drilling Project, 23* (eds R.B. Whitmarsh, O.E. Weser, D.A. Ross *et al.*), Washington (U.S. Government Printing), pp. 813–819.

Coleman, R.G. (1977) Geologic background of the Red Sea. Red Sea Research 1970–1975. *Mineral Resources Bulletin Jeddah 22* (C), C1–C9.

Coleman, R.G. (1984) *The Tihamat Asir Igneous Complex, a Passive Margin Ophiolite*. Proceedings of the 27th International Geological Congress, Moscow, 9, pp. 221–239.

Coleman, R.G. (1993) *Geologic Evolution of the Red Sea*, Oxford University Press, Oxford, 186 pp.

Coleman, R.G., Fleck, R.J., Hedge, C.E. and Ghent, E.D. (1977) The volcanic rocks of southwest Saudi Arabia and the opening of the Red Sea. *Mineral Resources Bulletin, Jeddah*, **22**, D1–D30.

Coleman, R.G., Gregory, R.T. and Brown, G.F. (1983) Cenozoic volcanic rocks of Saudi Arabia. *United States Geological Survey o-f rep, OF-03-93*.

Coleman, R.G., Hadley, D.G., Fleck, R.J. *et al.* (1979) The Miocene Tihama Asir ophiolite and its bearing on the opening of the Red Sea, in A.M.S. Al-Shanti (ed.) *Evolution and Mineralisation of the Arabian-Nubian Shield*, King Abulaziz University, Institute Applied Geology, Bulletin, **3**, pp. 173–187.

Coleman, R.G. and McGuire, A.V. (1988) Magma systems related to the Red Sea opening. *Tectonophysics*, **150**, 77–100.

Collier, R.E. and Gawthorpe, R.L. (1995) Neotectonics, drainage and sedimentation in Central Greece; insights into coastal reservoir geometries in syn-rift sequences, in *Hydrocarbon Habitat in Rift Basins*, (ed J. Lambiase), Geological Society, London, 80, 165–181.

Colletta, B., Le Quellec P., Letouzy, J. and Moretti, I. (1988) Longitudinal evolution of the Suez rift structure (Egypt). *Tectonophysics*, **153**, 221–233.

Conchon, O., Baltzer, F. and Purser, B.H. (1994) Enregistrement sedimentaire des variations climatiques quaternaires sur la bordure NW du rift de la Mer Rouge. *Quaternaire*, **5** (3,4), 181–188.

Conforto, L., Delitala M.C. and Taddeucci, A. (1976) Datazioni col ^{230}Th di alcune formazioni coralligene delle Isole Dahlak (Mar Rosso). *Società Italiana di Mineralogia e Petrologia*, **32** (1), 153–158.

Coniglio, M., James, N.P. and Aissaoui, D.M. (1988) Dolomitisation of Miocene carbonates, Gulf of Suez, Egypt. *Journal of Sedimentary Petrology*, **58**, 100–119.

Coniglio, M., James, N.P. and Aissaoui, D.M. (1996) Abu Shaar complex (Miocene) Gulf of Suez Egypt; deposition and diagenesis in an active rift setting, in 'Models for carbonate stratigraphy from Miocene reef complexes of Mediterranean regions' (eds E. Franseen, M. Esteban, W. Ward and J.M. Rouchy), Society Economic Paleontologists and Mineralogists, *Concepts in Sedimentology and Paleontology*, **5**, 367–384.

Cook, H.E. (1983) Ancient carbonate platform margins, slopes, and basins: platform margin and deep water carbonates. S.E.P.M. Short Course 12, 5.1–189.

Cornell, R.M. (1988) The influence of some divalent cations on the transformation of ferrihydrite to more cristalline products. *Clay Mineralogy*, **23**, 329–332.

Cornell, R.M. and Giovanoli, R. (1987) Effect of manganese on the transformation of ferrihydrite into goethite and jacobsite in alkaline media. *Clays and Clay Minerals*, **35**, 11–20.

Cornell, R.M., Giovanoli, R. and Schindler, P.W. (1987) Effect of silicate species on the transformation of ferrihydrite into goethite and hematite in alcaline media. *Clays and Clay Minerals*, **35**, 21–28.

Courtillot, V. (1980) Opening of the Gulf of Aden and Afar by progressive tearing. *Physics of the Earth and Planetary Interiors*, **21**, 343–350.

Courtillot, V. (1982) Propagating rifts and continental break up. *Tectonics*, **1**, 239–250.

Courtillot, V., Armijo, R. and Tapponnier, P. (1987a) Kinematics of the Sinai triple junction and a two-phase model of Arabia–Africa rifting, in *Continental Extensional Tectonics*, (ed. M.P. Coward), Geological Society of London Special Publication, pp. 559–573.

Courtillot, V., Armijo, R. and Tapponnier, P. (1987b) The Sinai triple junction revisited. *Tectonophysics*, **141**, 181–190.

Coutelle, A., Pautot, G. and Guennoc, P. (1991) The structural setting of the Red Sea axial valley and deeps: implications for crustal thinning processes. *Tectonophysics*, **198**, 395–409.

Coward, M.P. (1986) Heterogeneous stretching, simple shear and basin development. *Earth and Planetary Science Letters*, **80**, 325–336.

Cox, K.G. (1989) The role of mantle plumes in the development of continental drainage patterns. *Nature*, **342**, 873–877.

Cox, K.G., Gass, I.G. and Mallick, D.I.J. (1969) The evolution of the volcanoes of Aden and Little Aden, South Arabia. *Quarterly Journal of the Geological Society of London*, **124**, 283–308.

Cox, K.G., Gass, I.G. and Mallick, D.I.J. (1970) The peralkaline volcanic suite of Aden and Little Aden, South Arabia. *Journal of Petrology*, **11**, 433–461.

Cox, K.G., Gass, I.G. and Mallick, D.I.J. (1977) The Western part of the Shuqra volcanic field, South Yemen. *Lithos*, **10**, 185–191.

Cox, L.R. (1929) Notes on the post-Miocene Ostreidae and Pectinidae of the Red Sea region, with remarks on the geological significance of their distribution. *Proceedings of Malacologist Society of London*, **18**, 165–209.

Cox, L.R. (1931) The geology of the Farsan Islands, Gizan and Kamaran Islands, Red Sea. Part 2 Molluscan Palaeontology. *Geological Magazine*, **68**, 1–13.

Craig, H. (1969) Geochemistry and origin of the Red Sea brines, in *Hot Brines and Recent Heavy Metal Deposits in the Red Sea*, (eds E.T. Degens and D.A. Ross) Springer-Verlag, Berlin, pp. 208–243.

Cross, N.E. (1996) Sedimentary facies and sequence stratigraphy of a Miocene tilt-block carbonate platform, Gulf of Suez, Egypt, Unpublished PhD thesis, University of London, 268 pp.

Cross, N.E., Purser, B.H. and Bosence, D.W.J. (this volume) The tectono-sedimentary evolution of a rift margin fault-block carbonate platform: Abu Shaar, Gulf of Suez, Egypt, in *Sedimentation and tectonics of rift basins: Red Sea–Gulf of Aden*, (eds B.H. Purser and D.W.J. Bosence), Chapman & Hall, London, pp. 271–295.

Crossley, R., Watkins, C., Raven, M., Cripps, D., Carnell, A. and Williams, D. (1992) The sedimentary evolution of the Red Sea and the Gulf of Aden. *Journal of Petroleum Geology*, **15**, 157–172.

Cullen, J.L. (1981) Late Quaternary deep-ocean circulation. *Geological Society of America Bulletin*, **97**, 1106–1121.

Dabbagh, A., Hötzl, H. and Schnier, H. (1984) Farasan Islands, in *Quaternary Period in Saudi Arabia*, (eds A.R. Jado and J.G. Zöttl), Springer, Wien, pp. 212–220.

Daggett, P.H., Morgan, P., Boulos, F.K., Hennin, S.F., El-Sherif, A.A. *et al.* (1986) Seismicity and active tectonics on the Egyptian Red Sea margin and northern Red Sea. *Tectonophysics*, **125**, 313–324.

Danielsson, L.G., Dyrssen, D. and Graneli, A. (1980) Chemical investigations of Atlantis II and Discovery brines in the Red Sea. *Geochimica et Cosmochimica Acta*, **44**, 2051–2065.

Dart, C.J., Bosence, D.W.J. and McClay, K.R. (1993) Stratigraphy and structure of the Maltese graben system. *Journal of the Geological Society, London*, **150**, 1153–1166.

Dart, C., Cohen, H.A., Akyuz, H.S. and Barka, A. (1995) Basinward migration of rift-border faults: implications for facies distribution and preservation potential. *Geology*, **23**, 69–72.

Dart, C., Collier, R.E.L1., Gawthorpe, R.L., Keller, J.V.A. and Nichols, G. (1994) Sequence stratigraphy of (?) Pliocene-Quaternary syn-rift Gilbert-type fan deltas, northern Peleponnesos, Greece. *Marine and Petroleum Geology*, **11**, 545–560.

Darwish, M. (1992) *Facies developments of the Upper Paleozoic–Lower Cretaceous sequences in the Northern Galala Plateau and evidence for their hydrocarbon reservoir potentiality, Northern Gulf of Suez, Egypt*. Proceedings of the First International Conference on Geology of the Arab World, Cairo University, Cairo, 1, pp. 75–214.

Darwish, M. (1994) *Cenomanian–Turonian sequence stratigraphy, basin evolution and hydrocarbon potentialities of Northern Egypt*. Proceedings of the Second International Conference on Geology of the Arab World, Cairo University, Cairo, Egypt, 3, pp. 315–362.

Darwish, M. and El Araby, A. (1993) Petrography and diagenetic aspects of some siliciclastic hydrocarbon reservoirs in relation to rifting of the Gulf of Suez, Egypt, in *Geodynamics and Sedimentation of the Red Sea–Gulf of Aden Rift System*, (eds E. Philobbos and B.H. Purser), Geological Survey of Egypt Special Publication No. 1, pp. 155–187.

Darwish, M. and El Azabi, M. (1993) Contributions to the Miocene sequences along the western coast of the Gulf of Suez, Egypt. *Egyptian Journal of Geology*, **37**, 21–47.

Darwish, M. and Saleh, W. (1990) *Sedimentary facies development and diagenetic consideration of the hydrocarbon bearing carbonates in the Ras Fanar Field*. Proceedings of the Tenth EGPC Exploration and Production Conference, Cairo.

Davies, G.R. (1970) Carbonate bank sedimentation, eastern Shark Bay, western Australia. *American Association of Petroleum Geologists*, Memoir 13, pp. 85–168.

Davis, J.B. and Kirkland, D.W. (1979) Bioepigenic sulfur deposits. *Economic Geology*, **74**, 462–468.

Davison, I., Al-Kadasi, M., Al-Khirbash, S., Al-Subbary, A., Baker, J. *et al.* (1994) Geological evolution of the southern Red Sea rift margin: Republic of Yemen. *Geological Society of America Bulletin*, **106**, 1474–1493.

Davison, I., Bosence, D., Alsop, I. and Al-Aawah, M.H. (1996) Deformation and sedimentation around active Miocene salt diapirs on the Tihama Plain, northwest Yemen, in *Salt Tectonics* (eds I. Alsop, D.J. Blundell and I. Davison), Geological Society of London, Special Publication 100, pp. 23–39.

Davison, I., Tatnell, M.R., Owen, L.A., Jenkins, G. and Baker, J. (this volume) Tectonic geomorphology and rates of crustal processes along the Red Sea margin, NW Yemen, in *Sedimentation and Tectonics of Rift Basins: Red Sea–Gulf of Aden* (eds B.H. Purser and D.W.J. Bosence), Chapman & Hall, London, pp. 601–617.

Decarreau, A., Badaut, D., and Blanc, G. (1990) Origin and temperature formation of Fe-rich clays from Atlantis II deep deposits (Red Sea). An oxygen isotopic geochemistry approach. *Chemical Geology*, **84**, 363–364.

De Charpal, O., Guennoc, P., Montadert, L. and Roberts, D.G. (1978) Rifting, crustal attenuation and subsidence in the Bay of Biscay. *Nature*, **275**, 706–711.

De Martonne, E.M. (1926) Une nouvelle fonction climatologique: l'indice d'aridité. *La Meterologie*, 1926, 449–458.

De Paolo, D.J. (1986) Detailed record of the Neogene Sr isotopic evolution of seawater from DSDP site 590B. *Geology*, **14**, 103–106.

Degens, E.T. and Ross, D.A. (eds) (1969) *Hot Brines and Heavy Metal Deposits in the Red Sea*, Springer-Verlag, New York.

DeLange, G.J. (1986) Early diagenetic reaction in interbeded pelagic and turbiditic sediments in the Nares Abyssal Plain (Western North Atlantic): consequences for the composition of sediment and interstitial water. *Geochimica et Cosmochimica Acta*, **50**, 2543–2561.

Delfour, J. (1979) L'orogenèse panafricaine dans la partie nord du bouclier arabe, Bulletin Société Géologiqe, France, **21**, 449–56.

Deniel, C., Vidal, P., Coulon, C. *et al.* (1994) Temporal evolution of mantle sources during continental rifting: the volcanism of Djibouti (Afar). *Journal of Geophysical Research*, **99**, 2853–2869.

Deuser, W.G. and Degens, E.T. (1969) O^{18}/O^{16} and C^{13}/C^{12} ratios of fossils from the hot-brine deep area of the central Red Sea, in *Hot Brines and Heavy Metal Deposits in the Red Sea*, (eds E.T. Degens and D.A. Ross), Springer-Verlag, New York, pp. 336–347.

Deuser, W.G., Ross, D.A. and Waterman, L.S. (1976) Glacial and pluvial periods: their relationship revealed by Pleistocene sediments of the Red Sea and Gulf of Aden. *Science*, **191**, 1168–1170.

Dewey, J.F. (1982) Plate tectonics and the evolution of the British Isles. *Journal Geological Society of London*, **139**, 317–412.

Dewey, J.F. and Sengor, A.M.C. (1979) Aegean and surrounding regions: complex multi-plate and continuous tectonics in a convergent zone. *Bulletin of the Geological Society of America*, **90**, 82–91.

DGME (Department of Geology and Mineral Exploration) (1986) Geological Survey and Prospecting in the Habban-Mukalla Area PDRY 1982–1986, Peoples Democratic Republic of Yemen, Ministry of Energy and Minerals, Regional Geology, Volume 1.

Dickson, J.A.D. (1966) Carbonate identification and genesis as revealed by staining. *Journal of Sedimentary Petrology*, **36**, 491–505.

Didyk, B.M., Simoneit, B.R.T., Brassell, S.C. and Eglington, G. (1978) Organic geochemical indicators of palaeoenvironmental conditions for sedimentation. *Nature*, **272**, 216–222.

Diem, D. and Stumm, W. (1984) Is dissolved Mn^{2+} being oxidized by O_2 in absence of Mn-bacteria of surface catalysts? *Geochimica et Cosmochimica Acta*, **48**, 1571–1573.

Dixon, T.H., Ivins, E.R. and Franklin, B.J. (1989) Topographic and volcanic asymmetry around the Red Sea: constraints on rift models. *Tectonics*, **8**, 1193–1216.

Dixon, T.H., Stern, R.J. and Hussein, I.M. (1987) Control of Red Sea rift geometry by Precambrian structures. *Tectonics*, **6**, 551–571.

Dollfus, R.P. and Roman, J. (1981) Les Echinides de La Mer Rouge. Monographie zoologique et paléontologique. *Mémoires du Comité des Travaux Historiques et Scientifiques*, Section des Sciences, Paris, Bibliothèque Nationale, **9**.

Dolson, J., El-Gendi, O., Charmy, H., Fathalla, M. and Gaafar, I. (1996) Gulf of Suez rift basin sequence models. Part A. Miocene sequence stratigraphy and exploration significance in the Greater October field area, northern Gulf of Suez. *Proceedings 16th E.G.P.C Conference*, Cairo, pp. 1–8.

Dravis, J.J. and Yurewicz, D.A. (1985) Enhanced petrography using fluorescence microscopy. *Journal of Sedimentary Petrology*. **55**, 491–505.

Drooger, C.W. (1966) Miogypsinidae of Europe and North Africa, in *Proceedings of their session in Berne* (eds C.W. Drooger, Z. Reiss, R.F. Rutsch, P. Marks), Brill Leiden, pp. 51–54.

Drooger, C.W. (1993) *Radial foraminifera; Morphometrics and Evolution*. North-Holland Publ., Amsterdam.

Druckman, Y. (1981) Sub-recent manganese bearing stromatolites along the shorelines of the Dead Sea, in *Phanerozoic Stromatolites*, (ed. C.L.V. Monty), Springer-Verlag, Berlin, pp. 197–208.

Du Bray, E.A., Stoeser, D.B. and McKee E.H. (1991) Age and petrology of the Tertiary As Sarat volcanic field, southwestern Saudi Arabia. *Tectonophysics*, **198**, 155–180.

Ducci, E. and Pirini, C. (1968) Stratigraphy and micropaleontology of some Cretaceous and Lower Eocene formations from Midjiurtinia Region (Somalia). *Proceedings 3rd African Micropalaeontology Conference, Cairo*, pp. 549–559.

Dullo, W.-C. (1986) Variation in diagenetic sequences: an example from Pleistocene coral reefs, Red Sea, Saudi Arabia, in *Reef Diagenesis* (eds J.H. Schroeder and B.H. Purser), Springer-Verlag, Berlin, Heidelberg, pp. 77–90.

Dullo, W.-C. (1987) Fazies und Fossile Ueberlieferung der pleistozaenen Riffterrassen and der Ostkueste des Roten Meeres. Habilitation thesis, University of Erlangen, 247 pp.

Dullo, W.-C. (1990) Facies, fossil record, and age of Pleisotocene reefs from the Red Sea (Saudi Arabia). *Facies*, **22**, 1–46.

Dullo, W.-C., Blomeier, D., Camoin, G.F., Casanova, J., Colonna, M. *et al.* (1997) Morphological evolution and sedimentary facies on the foreslopes of Mayotte, Comoro Islands: Direct observations from submersible, in *Carbonate Platforms of the Indian Ocean and the Pacific*. (eds G Camsin and D. Bergerson), IAS Special Publication.

Dullo, W.-C., Brachert, T.C. and Moussavian, E. (1990) The foralgal crust facies of the deeper fore reef in the Red Sea: a deep diving survey by submersible. *Geobios*, **23**, 261–281.

Dullo, W.-C. and Hecht, C. (1990) Corallith growth on submarine alluvial fans. *Senckenbergiana maritima*, **21**, 77–86.

Dullo, W.-C., Hötzl, H. and Jado, A.R. (1983) New stratigraphical results from the Tertiary sequence of the Midyan area, NW Saudi Arabia, *Newsletter Stratigraphy*, **12**, 75–83.

Dullo, W.-C. and Montaggioni, L. (this volume) Modern Red Sea coral reefs, in *Sedimentation and tectonics of rift basins: Red Sea–Gulf of Aden* (eds B.H. Purser and D.W.J. Bosence), Chapman & Hall, London, pp. 589–600.

Dumont, T. and Grand, T. (1987) Caractères communs entre l'évolution précoce d'une portion de marge passive fossile (marge européenne de la Tethys ligure, Alpes occidentales) et celle du rift de Suez. *Comptes Rendus Sommaire Société Géologie*, France, Paris, **305** (II), 1369–373.

Dunham, R.J. (1962) Classification of carbonate rocks according to depositional texture. *American Association Petroleum Geologists*, Memoir 1, 108–121.

Dunham, R.J. (1969) Vadose pisolite in the Capitan Reef (Permian), New Mexico and Texas, in *Depositional Environments in Carbonate Rocks* (ed. G.M. Friedman), Society of Economic Paleontologists and Mineralogists, Special Publication 14, pp. 182–191.

Dupré, B., Blanc, G., Boulègue, J. and Allègre, C.J. (1988) Metal remobilization at a spreading centre studied using lead isotopes. *Nature*, **333**, 165–167.

Ebdon, C.C., Fraser, A.J., Higgins, A.C., Mitchener, B.C. and Strank, A.R.E. (1990) The Dinantian stratigraphy of the East Midlands: a seismostratigraphic approach. *Journal of the Geological Society, London*, **147**, 519–536.

Ebinger, C.J., Karner, G.D. and Weissel, J.K. (1991) Mechanical strength of extended continental lithosphere: constraints from the western rift system, East Africa. *Tectonophysics*, **10**, 1239–1256.

Edwards, F.J. (1987) Climate and oceanography, in *Red Sea*, (eds A.J. Edwards and S.M. Head), Pergamon Press, Oxford, pp. 45–69.

Edwards, R.A., Whitmarsh, R.B. and Scrutton, R.A. (1996) Geophysical features and geological development of the Ghana transform continental margin. *Geomarine Letters*, in press.

Egloff, F., Rihm, R., Makris, J. *et al.* (1991) Contrasting structural styles of the eastern and western margins of the southern Red Sea: the 1988 SONNE experiment. *Tectonophysics*, **198**, 329–353.

EGPC (1964) *Oligocene and Miocene rock stratigraphy of the Gulf of Suez region*. Egyptian General Petroleum Corporation, Cairo.

EGPC (1996) *Activity of oil exploration in Egypt:* 1886–1980. Eighth EGPC Exploration Seminar, Cairo.

Einsele, G. (1992) *Sedimentary Basins; Evolution, Facies and Sediment Budget*, Springer-Verlag, Berlin.

Eissen, J.P., Juteau, T., Joron, J.L. *et al.* (1989) Petrology and geochemistry of basalts from Red Sea axial rift at 18° North. *Journal of Petrology*, **30**, 791–839.

El Anbaawy, M.I.H. (1993) Reefal facies and diagenesis of Pleistocene sediments in Kamaran Island, southern Red Sea, Republic of Yemen. *Egyptian Journal of Geology*, **37** (2), 137–151.

El Anbaawy, M.I.H., Al-Aawah, M.A.H., Al-Thour, K.A., and Tucker, M.E. (1992) Miocene evaporites of the Red Sea Rift, Yemen Republic: sedimentology of the Salif halite. *Sedimentary Geology*, **81**, 61–71.

El Aref, M.A.M (1992) Petrographical and sedimentological studies on some sulfur-bearing evaporite deposits on the western side of the Gulf of Suez, Egypt. PhD thesis, University of Cairo, Egypt.

El Aref, M.A.M. (1993) Paleokarst surfaces in the Neogene succession of Wadi Essel-Sharm el Bahari, Egyptian Red Sea coast, as indication of uplifting and exposure, in *Geodynamics and Sedimentation of the Red Sea–Gulf of Aden Rift System*. (eds E.R. Philobbos and B.H. Purser), Geological Society of Egypt, Special Publication, 1, pp. 205–232.

El Aref, M.A.M. and Amstutz, G.C. (1983) Lead-zinc deposits along the Red Sea coast of Egypt; new observations and genetic models on the occurrences of Um Gheig, Wizr, Essel and Zug el Bohar. *Monograph on Mineral Deposits*, Borntraeger, Stuttgart, **21**.

El Barkooky, A.N. (1992) *Stratigraphic framework of the Paleozoic in the Gulf of Suez Region, Egypt*. Proceedings of the First International Conference on the Geology of the Arab World, Cairo University, Cairo, Egypt. (Abstract).

El Haddad, A.A. (1984) Sedimentological and geological studies on the Neogene sediments of the Egyptian part of the Red Sea. PhD thesis, Assiut University.

El Haddad, A., Aissaoui, D.M. and Soliman, M.A. (1984) Mixed carbonate-siliciclastic sedimentation on a Miocene fault block, Gulf of Suez. *Sedimentary Geology*, **37**, 185–202.

El Haddad, A., Philobbos, E. and Mahran, T. (1993) Facies and sedimentary development of dominantly siliciclastic late Neogene sediments, Hamata area, Red Sea coast, Egypt, in *Geodynamics and Sedimentation of the Red Sea–Gulf of Aden Rift System*, (eds E. Philobbos and B.H. Purser), Geological Society of Egypt, special publication 1, pp. 253–276.

El Heini, I. and Martini, E. (1981) Miocene foraminiferal and calcareous nannoplankton assemblages from the Gulf of Suez region and correlations. *Géologie Méditerranéenne*, **VIII** (2), 101–108.

El Hilaly, A. and Darwish, M. (1986) Petrographic, diagenetic and sedimentological history of the Pre-Balayim sedimentary sequences, with their reservoir characteristics, Zeit Bay Oil Field, Gulf of Suez, Egypt. *Proceedings 8th EGPC Exploration Conference, Cairo*.

El Isa, Z., Mechie, J., Prodehl, C. *et al.* (1987) A crustal structure study of Jordan derived from seismic refraction data. *Tectonophysics*, **138**, 235–253.

El Isa, Z.H. and Mustafa, H. (1986) Earthquake deformations in the Lisan deposits and seismotectonic implications. *Geophysical Journal of the Royal Astronomy Society, London*, **86**, 413–424.

Ellis, A.C., Kerr, H.M., Cornwell, C.P. and Williams, D.O. (1996) A tectono-stratigraphic framework for Yemen and its implications for hydrocarbon potential. *Petroleum Geoscience*, **2**, 29–42.

Ellis, J.P. and Milliman, J.D. (1985) Calcium carbonate suspended in Arabian Gulf and Red Sea waters: biogenic and detrital, not 'chemogenic'. *Journal of Sedimentary Petrology*, **55**, 805–808.

El Moursi, M.E.E. (El Sahyed El Sherbii) (1992) Evolution quaternaire de la plaine côtière de la Mer Rouge entre Hurghada et Marsa Alam, Egypte. Thesis Science, Marseille, France.

El Moursi, M.E.E. (1993) Pleistocene evolution of the reef terraces of the Red Sea coastal plain between Hurghada and Marsa Alam, Egypt. *Journal of African Earth Sciences*, **17**, 125–127.

El Nakhal, H.A. (1988) Stratigraphy of the Tawilah Formation (Cretaceous–Paleocene) in the Yemen Arab Republic. *M.E.R.C. Ain Shams Univ., Earth Sc. Ser.*, **2**, 161–171.

Elsgaard, L. and Jorgensen, B.B. (1992) Anoxic transformations of radiolabeled hydrogen sulfide in marine and freshwater sediments. *Geochimica et Cosmochimica Acta*, **56**, 2425–2435.

El Shatoury, M.E., Al Khirbash, S. and Othman, S.A. (1979) Analysis of lineaments in Landsat-1 photographs of Yemen and their geologic significance. *Dirsat Yamaniyyah, Journal Center Studies Research*, **3**, 3–14.

El Shazly, E.M. (1977) The geology of the Egyptian region, in *The Ocean Basins and Margins, 4A, the Eastern Mediterranean*. (eds A.E.M. Nairn *et al.*), Plenum, New York, pp. 379–444.

El Shazly, E.M., Saleeb Roufaiel, G.S. and Zaki, N. (1974) Quaternary basalt in Saint John's Island. *Egyptian Journal of Science*, **18**, 137–148.

El Shazly, E.M., Saleeb, W.S., Saleeb, G.S. and Zaki, N. (1967) A new approach to the geology and mineralisation of St. John's Island, Red Sea. *Proceedings Congress of the Geological Society of Egypt*, Session 2, Abstracts, pp. 5–8.

El Shinnawi, M.A. (1975) Planktonic foraminifera from the Miocene Globigerina Marl of Hurgada–well-134, Eastern Desert, Egypt, *Proceedings 4th African Colloquium Micropalaeontology*, Addis Ababa, 1972, pp. 199–224.

El Tarabili, E. and Adawy, N. (1972) Geologic history of the Nukhul-Baba area, Gulf of Suez, Egypt. *American Association of Petroleum Geologists Bulletin*, **56**, 882–902.

Emerson, S.R. and Archer, D. (1990) Calcium carbonate preservation in the ocean. *Philosophical Transactions Royal Society London*, **A331**, 29–40.

England, P.C. and Molnar, P. (1990) Surface uplift, uplift of rocks, and exhumation of rocks. *Geology*, **18**, 1173–1177.

Espitalié, J., Lapote, J.L., Madec, M., Leplat, P. and Bouttefeu, A. (1977) Methode rapid de characterisation des roches mère, de leur potentiel pétrolier et de leur degré d'évolution. *Revieu Instit. Francais du Pétrole*, **32**, 23–42.

Evans, A.L. (1988) Neogene tectonic and stratigraphic events in the Gulf of Suez rift area, Egypt. *Tectonophysics*, **153**, 235–247.

Evans, A.L. (1990) Miocene sandstone provenance relations in the Gulf of Suez: insights into syn-rift unroofing and uplift history. *American Association of Petroleum Geologists Bulletin*, **74**, 1386–1400.

Evans, A.L. and Moxon, I.W. (1986) Gebel Zeit chronostratigraphy: Neogene syn-rift sedimentation atop a long-lived paleohigh. *Proceedings of 8th EGPC Exploration Conference*, 1, Cairo, November 1986, pp. 251–65.

Evans, R. (1978) Origin and significance of evaporites in basins around Atlantic margin. *American Association of Petroleum Geologists Bulletin*, **62**, 223–234.

Facca, G. (1965) Etiopia, in *Enciclopedia del Petrolio*, Colombo, Roma, 4, pp. 339–359.

Fairhead, J.D. and Girdler, R.W. (1970) The seismicity of the Red Sea, Gulf of Aden and Afar triangle. *Philosophical Transactions of the Royal Society of London*, **A267**, 49–74.

Falvey, D.A. (1974) The development of continental margins in plate tectonic theory. *Australian Petroleum Exploration Journal*, **14**, 95–106.

Fanning, K.A., Byrne, R.H., Breland, J.A., Betzer, P.R., Moore W.S., Elsinger, R.J. and Pyle, T.E. (1981) Geothermal springs of the west Florida continental shelf: evidence for dolomitization and radionuclide enrichment. *Earth and Planetary Science Letters*, **52**, 345–354.

Fantoli, A. (1966) Contributo alla climatologia dell'altipiano etiopico. Regione eritrea. *Ministero Affari Esteri, Cooperzione Scientifica e Tecnica*, **LV**.

Fantozzi, P.L. (1992) Da rifting continentale a rifting oceanico: studio dell'evoluzione strutturale dei margini passivi del Golfo di Aden. PhD thesis, IV Ciclo, Dip. Sc. Terra Università di Cagliari, Torino, Siena, Biblioteca Nazionale, Firenze, Roma.

Fantozzi, P.L. (1993) From continental to ocean rifting: Evidence from northeastern Somalia (Gulf of Aden). *Proceedings European Union Geoscientists VII*, Abstract Volume 193.

Fantozzi, P.L. (1996) Transition from continental to oceanic rifting in the Gulf of Aden: structural evidence from field mapping, in Somalia and Yemen, *Tectonophysics*, **259** (in press).

Fantozzi, P.L., Abdirhaman, H.M. and Ali Kassim, M. (1993) Geological map of Northeastern Somalia. CNR-Universita di Siena.

Fantozzi, P.L. and Sgavetti, M. (this volume) Tectonic and sedimentary evolution of the Gulf of Aden continental margins; new structural and stratigraphic data from Somalia and Yemen, in *Sedimentation and Tectonics of Rift Basins: Red Sea–Gulf of Aden* (eds B.H. Purser and D.W.J. Bosence), Chapman & Hall, London, pp. 56–76.

Faulds, J.E., Geissman, J.W. and Mawer, C.K. (1990) Structural development of a major extensional accomodation zone in the Basin and Range Province, northwestern Arizona and southern Nevada; implication for kinematics models of continental extension. *Geological Society America Memoirs*, **176**, 37–76.

Faure, G. (1977) *Principles of Isotope Geology*, Wiley.

Faure, H. (1975) Recent crustal movements along the Red Sea and Gulf of Aden coasts in Afar (Ethiopia and T.F.A.I.). *Tectonophysics*, **29**, 479–486.

Faure, H., Hoang, C.T. and Lalou, C. (1973) Structure et géochronologie (^{230}Th/^{234}U) des récifes coralliens soulevés a l'ouest du Golfe d'Aden (T.F.A.I.). *Revue de Géographie Physique et Géologie Dynamique*, **XV** (4), 393–403.

Faure, H., Hoang, C.T. and Lalou, C. (1980) Datations ^{230}Th/^{234}U des calcaires coralliens et mouvements verticaux à Djibouti. *Bulletin de la Société Géologique de France*, **XXII** (6), 959–962.

Faurot, M.L. (1888) Sur les sédiments quaternaires de l'Ile de Kamaran (Mer Rouge) et du Golfe de Tadjourah. *Bulletin de la Société Géologique de France*, **16** (sér.3), 528–546.

Favre, P. and Stampeli, G.M. (1992) From rifting to passive margin: the example of the Red Sea, Central Atlantic and Alpine tethys. *Tectonophysics*, **215**, 69–97.

Fawzy, H. and Abdel Aab, A. (1984) Regional study of Miocene evaporites and Pliocene to Recent sediments in the Gulf of Suez, *Proceedings 7th Egyptian General Petroleum Corporation, Exploration Seminar*, Cairo.

Feitknecht, W. and Michaelis, W. (1962) Uber die Hidrolyse von Eisen(III)–Perchlorat-Lösungen. *Helv. Chim. Acta*, **26**, 212–224.

Felix, J. (1884) Korallen aus ägyptischen Tertiärbildungen. *Zeitschrift der deutsche geologische Gesellschaft*, **36**, 415–453.

Felix, J. (1903) Korallen aus ägyptischen Miocänbildungen. *Zeitschrift der deutsche geologische Gesellschaft*, **55**, 1–22.

Felix, J. (1904) Studien über tertiäre und quartäre Korallen und Riffkalke aus Ägypten und der Sinaihalbinsel, *Zeitschrift der deutsche geologische Gesellschaft*, **56**, 168–208.

Féraud, G., Zumbo, V., Sebai, A. et al. (1991) ^{40}Ar/^{39}Ar age and duration of tholeiitic magmatism related to the early opening of the Red Sea rift. *Geophysical Research Letters*, **18**, 195–198.

Ferertiuos, C., Papotheodorov, G. and Collins, M.B. (1988) Sediment transport processes on an active submarine fault escarpment: Gulf of Corinth, Greece, *Marine Geology*, **83**, 43–61.

Ferrari, M.C. (1993) La riflettanza spettrale VNIR come dato telerilevato e come tecnica di laboratorio per lo studio di successioni carbonatiche: applicazione ai depositi Terziari pre-rift della Migiurtinia (Somalia Settentrionale) PhD thesis, IV Ciclo, Dip. Sc. Terra Università di Cagliari, Torino, Siena, Biblioteca Nazionale, Firenze, Roma.

Ferrari, M.C., Sgavetti, M. and Chiari, R. (1996) Thematic Mapper multispectral facies in prevalent carbonate strata of an area of Migiurtinia (Northern Somalia): analysis and interpretation. *International Journal of Remote Sensing* (in press).

Filosofov, V.R. (1960) *A Short Manual of Morphometric Methods used in Search for Tectonic Structures*, Saratov University Press (in Russian).

Filosofov, V.R. (1970) Maps of isobases and residual relief (in Russian), in *Application of Geomorphic Methods in Structural Investigations*, Publishing House NEDRA, pp. 48–52.

Fischer, W.R. and Schwertmann, U. (1975) The formation of hematite from amorphous iron(III) hydroxide. *Clays and Clay Minerals*, **23**, 33–37.

Foin, T.C. (1972) The zoogeography of the Cypraeidae in the Red Sea basin. *Argamon*, **3**, 5–16.

Foin, T.C. and Ruebush, L.P. (1969) Cypraeidae of the Red Sea at Massawa, Ethiopia, with a zoogeographical analysis based on the Schilder's Regional List. *The Veliger*, **12**, 201–206.

Folk, R.L. (1968) *Petrology of Sedimentary Rocks*, Hemphill, Austin, Texas.

Folk, R.L. and Ward, W.C. (1957) Brazos River bar, a study in the significance of grain size parameters. *Journal of Sedimentary Petrology*, **27**, 3–26.

Fonseca, J.I. (1966) Geological outline of the Lower Cretaceous Bahia supergroup, in *Proceedings 2nd West African Micropaleontological Colloquium*, (ed. J.E. Van Hinte), Ibadan, pp. 49–71.

Forskal, P. (1775) *Descritiones naimalum, avium, amphibiorum, piscium, insectorum, vermium, quae in itinere orientali observit Petrus Forskal* (Post mortem autoris edit Carsten Niebuhr), Hauniae.

Fourtau, R. (1914) Notes sur les échinides fossiles de l'Egypte. *Bulletin Institute Egypt*, **7** (1913), 86–94.

Frazier, S.B. (1970) Adjacent structures of Ethiopia, the portion of Red Sea coast including Dahlak Kebir Island and the Gulf of Zula. *Philosophical Transaction of the Royal Society of London*, **A267**, 131–141.

Fredet, J.M. (1987) Tectonique et sédimentation en domaine continental: évolution du bassin paléogène d'Alès (Gard). Thesis Université Lyon I.

Freund, R. (1970) Plate tectonics of the Red Sea and East Africa. *Nature*, **228**, 453.

Freund, R., Garfunkel, Z., Zak, I., Goldberg, M., Weissbrod, T. and Derin, B. (1970) The shear along the Dead Sea rift. *Philosophical Transactions of the Royal Society of London*, **A267**, 107–130.

Freund, R., Zak, I. and Garfunkel, Y. (1968) Age and rate of the sinistral movement along the Dead Sea rift. *Nature*, **220**, 253–255.

Freytet, P., Baltzer, F. and Conchon, O. (1993) A Quaternary piemont on an active margin: the Egyptian coast of the NW Red Sea. *Zeitschrift Geomorphologisches N.F.*, **37** (2), 215–236.

Freytet, P., Baltzer, F., Conchon, O., Plaziat, J-C. and Purser, B.H. (1994) Signification hydrologique et climatique des carbonates continentaux quaternaires de la bordure du désert oriental égyptien (côte de la Mer Rouge). *Bulletin de la Société Géologique de France*, **165** (5), 593–601.

Fricke, H.W. and Landmannn, G. (1983) On the origin of Red Sea submarine canyons. *Naturwissenschaften*, **70**, 195.

Fricke, H.W. and Schuhmacher, H. (1983) The depth limits of Red Sea stony corals: an ecophysiological problem. *Marine Ecology*, **4**, 163–194.

Friedman, G.M. (1968) Geology and geochemistry of reefs, carbonate sediments and waters, Gulf of Aqaba (Elat, Red Sea). *Journal of Sedimentary Petrology*, **38**, 896–919.

Friedman, G.M. (1972) Significance of Red Sea in problem of evaporites and basinal limestones. *American Association of Petroleum Geologists Bulletin*, **56**, 1072–1086.

Friedman, G.M., Abraham, J.A., Braun, M. and Miller, D.S. (1973) Generations of carbonate particles and laminites in algal mats example from sea marginal hypersaline pool, Gulf of Aqaba, Red Sea. *American Association of Petroleum Geologists Bulletin*, **57**, 218–251.

Friedman, G.M., Amiel, A.J. and Schneiderman, N. (1974) Submarine cementation in reefs: example from the Red Sea. *Journal of Sedimentary Petrology*, **44**, 816–825.

Friedman, G.M., Schneiderman, N. and Gevirtz, J.L. (1971) Indurated hard calcium carbonate layers at the bottom of the Red Sea, in *Carbonate Cements* (ed. O.P. Bricker), Johns Hopkins University Studies in Geology, 19, pp. 116–118.

Fritsch, J., Hinz, K., von Rad, U. *et al.* (1978) Ergebnisse geophysikalischer Untersuchungen der VALDIVIA Westafrika Fahrt VA-10/1975. *Forschungsber. M78-03*, BMFT, Bonn.

Frost, S.H. (1977) Oligocene reef coral biogeography Caribbean and western Tethys. *Documents du Bureau Recherche Geologique Minieres*, **89**, 342–352.

Frostick, L.E. and Read, I. (1989) Is structure the main control of river drainage and sedimentation in rifts? *Journal of African Earth Science*, **8**, 165–182.

Frostick, L.E. and Read, I. (1990) Structural controls of sedimentation patterns and implications for the economic potential of the East African rift basins. *Journal of African Earth Science*, **10**, 307–318.

Frostick, L.E. and Steel, R.J. (1993) Sedimentation in divergent plate-margin basins, in *Tectonic Controls and Signatures in Sedimentary Successions* (eds L.E. Frostick and R.J. Steel), Special Publication International Association of Sedimentologists, 20, pp. 111–128.

Fuchs, T. (1878) Die Geologische Beschaffenheit der Landenge von Suez. *Denekschrifte Akademie Wissenschaft (Math. Nat. Kl.)*, Wien, **38**, 25–42.

Fuchs, T. (1883) Beiträge zur Kenntniss der Miocaenfauna, Aegyptens und der libyschen Wüste. *Palaeontographica*, **XXX**(2), 18–66.

Fumanti, B. (1983a) Contribution to the knowledge of the marine algae of Dahlak Islands (Red Sea, Ethiopia). *Annali di Botanica*, **41**, 95–101.

Fumanti, B. (1983b) Recent stromatolites from Dahlak Islands (Red Sea, Ethiopia). *Annali di Botanica*, **41**, 87–94.

Gabrié, C. and Montaggioni, L. (1982) Sedimentary facies from the modern coral reefs, Jordan Gulf of Aqaba, Red Sea. *Coral Reefs*, **1**, 115–124.

Gallagher, K., Menzies, M., Hurford, A. and Yelland, A. (1996) Comparative fission track data for the Red Sea volcanic and the Gulf of Aden nonvolcanic margins (in preparation).

Gard, J.A. (1971) *The Electron-optical Investigation of Clays*, Mineralogical Society, London.

Garfunkel, Z. (1988) Relation between continental rifting and uplifting: evidence from the Suez rift and northern Red Sea. *Tectonophysics*, **150**, 33–49.

Garfunkel, Z and Bartov, Y. (1977) The tectonics of the Suez rift. *Geological Survey of Israel Bulletin*, **71**, 1–44.

Garfunkel, Z., Ginzburg, A. and Searle, R.C. (1987) Fault pattern and mechanism of crustal spreading along the axis of the Red Sea from Side Scan Sonar (Gloria) data. *Annals Geophysics*, **5B**, 187–200.

Garfunkel, Z., Zak, I. and Freund, R. (1981) Active faulting in the Dead Sea rift. *Tectonophysics*, **80**, 1–26.

Garrison, T. (1993) *Oceanography*. Wadsworth Publishing Company, Belmont, California.

Garson, M.S. and Krs, M. (1976) Geophysical and geological evidence of the relationship of Red Sea transverse tectonics to ancient fractures. *Geological Society America Bulletin*, **87**, 169–181.

Gasperini, L., Gensou, E., Philobbos E., Purser, B., Reijmer, J. et al. (1994) Gulf of Suez seismic survey (Sept./Oct., 1992): scientific report. *Giornale di Geologia*, (s.3), **56** (2), 3–12.

Gass, I.G. (1977) The evolution of the Pan-African crystalline basement in NE Africa and south Arabia. *Journal of the Geological Society of London*, **134**, 129–138.

Gass, I.G. and Mallick, D.I.J. (1968) Jebel Khariz: an Upper Miocene strato-volcano of comenditic affinity on the South Arabian Coast. *Bulletin Volcanologique*, **32**, 33–88.

Gasse, F., Fontes, J.C., Plaziat, J.C., Carbonel, P., Kaczmarska, I. et al. (1987) Biological remains, geochemistry and stable isotopes for the reconstruction of environmental and hydrological changes in the Holocene lakes from North Sahara, *Palaeogeography, Palaeoclimatology, Palaeoecology*, **60**, 1–46.

Gaudant, J. and Rouchy, J.M. (1986) Ras Dib: un nouveau gisement de Poissons fossiles du Miocène moyen du Gebel Zeit (Golfe de Suez, Egypte). *Bulletin du Muséum National d'Histoire Naturelle, Paris*, 4°série, 8, section C; 4, 463–481.

Gaulier, J.M., LePichon, X., Lyberis, N. et al. (1988) Seismic study of the crust of the northern Red Sea and Gulf of Suez. *Tectonophysics*, **153**, 55–88.

Gautier, B. and Angelier, J. (1986) Distribution et signification géodynamique des systèmes de joints en contexte distensif: un exemple dans le rift de Suez. *Comptes Rendues Acadamie de Science* Paris, **303** (II), 1147–1152.

Gawthorpe, R.L., Fraser, A.J. and Collier, R.E.Ll. (1994) Sequence stratigraphy in active extensional basins: implications for the interpretation of ancient basin-fills. *Marine and Petroleum Geology*, **11**, 642–658.

Gawthorpe, R.L. and Hurst, J.M. (1993) Transfer zones in extensional basins: their structural style and influence on drainage development and stratigraphy. *Journal of the Geological Society, London*, **150**, 1137–1152.

Gawthorpe, R.L., Gutteridge, P. and Leeder, M.R. (1989) Late Devonian and Dinantian basin evolution in northern England and North Wales, in *The role of tectonics in Devonian and Carboniferous sedimentation in the British Isles*, (eds R.S. Arthurton, R.S. Gutteridge and S.C. Nolan), Yorkshire Geological Society Occasional Publication, 6, pp. 1–23.

Gawthorpe, R.L., Hurst, J.M. and Sladen, C.P. (1990) Evolution of Miocene footwall-derived coarse-grained deltas, Gulf of Suez, Egypt: implications for exploration. *American Association of Petroleum Geologists Bulletin*, **74**, 1077–1086.

Geister, J. (1983) Holozäne westindische korallenriffe: geomorphologie, önkologie und fazies. *Facies*, **9**, 173–284.

Gettings, M.E. and Stoeser, D.B. (1981) A tabulation of radiometric age determination for the Kingdom of Saudi Arabia. *United States Geological Survey Miscellaneous Document 20, Interagency Report* 353.

Gettings, E.M. (1977) Delineation of the continental margin in the southern Red Sea region from new gravity evidence. *Saudi Arabian Directorate General Mineral Resources Bulletin*, **22**, K1–K11.

Geukens, F.P. (1960) Contribution a la géologie du Yemen. *Institute Géologie de France, de L' Université Louvin*, Mémoire 21, pp. 116–180.

Geukens, F.P. (1966) Geology of the Arabian Peninsula, (Yemen). *United States Geological Survey Professional Paper*, 560-B, 23 pp.

Gevirtz, J.L. and Friedman, G.M. (1966) Deep-sea carbonate sediments of the Red Sea and their implications on marine lithification. *Journal of Sedimentary Petrology*, **36**, 143–152.

Ghorab, M.A. (1961) Abnormal stratigraphic features in Ras Gharib oil field. *Proceedings 3rd. Arabian Petroleum Congress*, Alexandria, pp. 1–10.

Ghorab, A., El Shazly, E.M., Abdel Gawad, A., Morshed, T., Ammar, A.A. and Ibrahim, A.M. (1969) Discovery of potassium salts in the evaporites of some oil wells in the Gulf of Suez region. *Geological Society U.A.R., 7th Annual Meeting.* Abstracts, Cairo.

Ghorab, M.A. et al. (1964) *Oligocene and Miocene Rock Stratigraphy of the Gulf of Suez Region*. Egyptian General Petroleum Corporation, Cairo.

Ghorab, M.A. and Marzouk, I.M. (1967) A summary report on the rock-stratigraphic classification of the Miocene non-marine and coastal facies in the Gulf of Suez and Red Sea coast. Unpublished Report, E.R. 601.

Giannerini, G., Campredon, R. and Feraud, G. (1988) Genèse des rifts intracontinentaux par propagation polygonale de transformantes continentales: exemple du triplet golfe d'Aden, mer Rouge, Afrique de l'Est. *Comptes Rendues Acadamie de Science* Paris, **306** (II), 1507–514.

Gibbs, A.D. (1984) Structural evolution of extensional basin margins. *Journal Geological Society of London*, **139**, 317–412.

Gilchrist, R., Kooi, H. and Beaumont, C. (1994) Post-Gondwana geomorphic evolution of southwestern Africa: implications for the controls of landscape development from observations and numerical experiment. *Journal of Geophysical Research*, **99**, 12,211–12,228.

Gilchrist, R. and Summerfield, M.A. (1990) Differential denudation and flexural isostasy in formation of rifted-margin upwarps. *Nature*, **346**, 739–742.

Gilchrist, R., Summerfield, M.A. and Cockburn, H.A.P. (1994) Landscape dissection, isostatic uplift and the morphologic development of orogens. *Geology*, **22**, 963–967.

Gindy, A.R. (1963) Stratigraphic sequence and structures of a post-lower Eocene section in Wadi Gasus el Bahari, Safaga district, Eastern Desert. United Arab Republic. *Bulletin of the Faculty of Sciences, Alexandria University*, **5**, 71–87.

Ginsburg, R.N. and James, N.P. (1974) Holocene carbonate sediments of continental shelves, in *The Geology of Continental Margins* (eds C.A. Burk and C.L. Drake), Springer-Verlag, Berlin, pp. 137–155.

Ginsburgh, R.N. and Schroeder, J.H. (1969) Notes for NSF Seminar on carbonate cements. Bermuda Biological Station, unpublished.

Ginzburg, A., Makris, J., Fuchs, K. and Prodehl, C. (1981) The structure of the crust and upper mantle in the Dead Sea Rift. *Tectonophysics*, **80**, 109–119.

Giovanoli, R., Feitknecht, W., Maurer, R. and Häne, H. (1976) Über die Reaktion von Mn_3O_4 mit Säuren. *Chimica*, **30**, 307–309.

Girdler, R.W. (1991) The Afro-Arabian rift system – an overview. *Tectonophysics*, **197**, 139–153.

Girdler, R.W. and Darracot, B.W. (1972) African poles of rotation. *Comments Earth Science*, **2**, 131–138.

Girdler, R.W. and Styles, P. (1974) Two stage seafloor-spreading. *Nature*, **247**, 7–11.

Girdler, R.W. and Styles, P. (1978) Seafloor spreading in the western Gulf of Aden. *Nature*, **271**, 615–617.

Godin, P.D. (1991) Fining up-ward cycles in the sandy braided-river deposits of westwater Canyon Member (upper Jurassic), Morrison Formation, New Mexico. *Sedimentary Geology*, **70**, 61–82.

Goldberg, M. and Beith, M. (1991) Jizan island: an internal block at the junction of the Red Sea rift and the Dead Sea transform. *Tectonophysics*, **198**, 261–273.

Goldhaber, M.B. (1983) Experimental study of metastable sulfur oxyanion formation during pyrite oxidation at pH 6–9 and 30°C. *American Journal Science*, **283**, 193–217.

Goldhaber, M.B. and Kaplan, I.R. (1974) The sulfur cycle, in *The Sea: Marine Chemistry*, (ed. E.D. Goldberg), J. Wiley and Sons, New York, pp. 569–655.

Goldich, S.S. (1938) A study in rock weathering. *Journal of Geology*, **46**, 17–58.

Goldschmidt-Rokita, A., Hansch, K.J.F., Hirschleber, H.B. et al. (1994) The ocean/continent transition along a profile through the Lofoten Basin, northern Norway. *Marine Geophysical Research*, **16**, 201–224.

Grammer, G.M. and Ginsburg, R.N. (1992) Highstand vs. lowstand deposition on carbonate platform margins: insight from Quaternary foreslopes in the Bahamas. *Marine Geology*, **103**, 125–136.

Grammer, G.M., Ginsburg, R.N. and Harris, P.M. (1993) Timing of deposition, diagenesis and failure of steep carbonate slopes in response to high-amplitude/high-frequency fluctuations in sea level, Tongue of the Ocean, Bahamas, in *Carbonate Sequence Stratigraphy; Recent Developments and Applications* (eds R.G. Loucks and J.F. Sarg), American Association of Petroleum Geologists, Memoir 57, pp. 107–131.

Greene, D.C. (1984) Structural geology of the Quseir Area, Red Sea Coast, Egypt. *Department of Geology and Geography Contribution*, **52**, University Massachusetts, Amherst.

Greenwood, J.E.G.W. and Bleackley, D. (1967) *Geology of the Arabian Peninsula: Aden Protectorate*. United States Geological Survey Professional Paper, 560-C, pp. 1–96.

Gregory, J.W. (1898) A collection of Egyptian fossil Madreporia. *Geological Magazine*, IV, 241–251.

Gregory, J.W. (1906) On a collection of fossil corals from eastern Egypt, Abu Roash and Sinai. *Geological Magazine*, **3**, 50–58.

Grolier, M.J. and Overstreet, W.C. (1978) Geological map of the Yemen Arab Republic. United States Geological Survey Miscellaneous Investigations Series. Map 1-1143-B.

Groupe Escarmed, 1983. Examples de sédimentation condensée sur les escarpements de la mer Ionienne (Méditerranée orientale). Observations à partir du submersible Cyana. *Revue de l'Institut Francais du Pétrole*, **38**, 427–438.

Grow, J.A., Hutchinson, D.R., Klitgord, K.D., Dillon, W.P. and Schlee, J.S. (1983) Representative multichannel seismic profiles over the U.S. Atlantic margin, in *Seismic Expression of Structural Styles* (ed. A.W. Bally), American Association of Petroleum Geologists, Studies in Geology, 15, 2.2.3.1–19.

Guennoc, P., Pautot, G., Le Qentrec, M-F. and Coutelle, A. (1990) Structure of an early oceanic rift in the northern Red Sea. *Oceanologica Acta*, **13**, 145–157.

Guennoc, P. and Thisse, Y. (1982) Genese de l'ouverture de la Mer Rouge et des mineralisations des fosses axiales. Synthése bibliographique. *Documents du Bureau Recherche Geologique Minieres*, **51**.

Guilcher, A. (1982) Physiography of the Red Sea coast and coastal waters: a summary of the present state of knowledge, with suggestions for further research. *Journal Faculty of Marine Science, Jeddah* 2 (1402 H), 27–36.

Guilcher, A. (1988) *Coral Reef Geomorphology*. Wiley, Chichester.

Guiraud, R., Issawi, B. and Bellion, Y. (1985) Les linéaments Guineo-Nubiens: un trait structural majeur à l'échelle de la Plaque Africaine. *Comptes Rendues Acadamie Sceances Paris*, **300** (II, 1), 17–20.

Guiraud, M. and Plaziat, J.C. (1993) Seismites in the fluviatile Bima Sandstones: identification of paleoseisms and discussion of their magnitude in a Cretaceous synsedimentary strike-slip basin (Upper Benue, Nigeria). *Tectonophysics*, **225**, 493–522.

Guney, M., Al-Marhoun, M.A. and Nawab, Z.A. (1988) Metalliferous submarine sediments of the Atlantis II Deep, Red Sea, *Canadian Institute Mining Bulletin*, **81**, 33–39.

Gvirtzman, G. (1994) Fluctuations of sea level during the past 400,000 years: the record of Sinai, Egypt (Northern Red Sea). *Coral Reefs*, **13**, 203–214.

Gvirtzman, G. and Buchbinder, B. (1978) Recent and Pleistocene Coral Reefs and Coastal Sediments of the Gulf of Eilat. Tenth Internaitonal Congress of Sedimentology, Jerusalem, Post-Congress Excursion Y4, pp. 163–191.

Gvirtzman, G., Buchbinder, B., Sneh, A., Nir, Y. and Friedman, G.M. (1977) Morphology of the Red Sea fringing reefs: a result of the erosional pattern of the last-glacial-low-stand sea level and the following Holocene recolonization. 2e Symposium international sur les coraux et récifs coralliens fossiles. *Mémoires Bureau Recherche Geologique Minieres*, **89**, pp. 480–491.

Gvirtzman, G., Kronfeld, J. and Buchbinder, B. (1992) Dated coral reefs of southern Sinai (Red Sea) and their implications to the Late Quaternary sea-levels. *Marine Geology*, **108**, 29–37.

Hack, J.T. (1973) Stream profile analysis and stream gradient index. *Journal Research United States Geological Survey*, **1**, 421–429.

Hackett, J. and Bischoff, J.L. (1973) New data on the stratigraphy, extent, and geologic history of the Red Sea geothermal deposits. *Economic Geology*, **68**, 553–564.

Haitham, F.M.S. and Nani, A.S.O. (1990) The Gulf of Aden rift: Hydrocarbon potential of the Arabian sector. *Journal of Petroleum Geology*, **13**, 211–220.

Hall, S.A. (1970) Aeromagnetic map of Ethiopia. University of Newcastle.

Hall, S.A. (1979) A total intensity magnetic anomaly map of the Red Sea and its interpretation. *United States Geological Survey, Saudi Arabia Mission, Project Report, SA 275*.

Hall, S.A., Andreasen, G.E. and Girdler, R.W. (1977) Total-intensity magnetic anomaly map of the Red Sea adjacent coastal areas, a description and preliminary interpretation, in 'Red Sea Research 1970–1975.' *Saudi Arabian Directorate General Mineral Resources Bulletin*, **22**, F1–F15.

Hall, W.J. and Standen, R. (1907) On the Mollusca of a raised coral reef on the Red Sea coast. *Journal of Conchology*, **12**, 65–68.

Handford, C.R. and Loucks, R.G. (1993) Carbonate depositional sequences and systems-tracts responses of carbonate platforms to relative sea-level changes, in *Carbonate Sequence Stratigraphy; Recent Developments and Applications* (eds R.G. Loucks and J.F. Sarg), American Association Petroleum Geologists, Tulsa, Memoir 57, pp. 3–41.

Haq, B.U., Hardenbol, J. and Vail, P.R. (1987) Chronology of fluctuating sea levels since the Triassic. *Science*, **235**, 1156–1167.

Haq, B.U., Hardenbol, J. and Vail, P.R. (1988) Mesozoic and Cenozoic chronostratigraphy and cycles of sea-level change, in *Sea-Level Changes: an integrated Approach* (eds C.K. Wilgus et al.), Society of Economic Paleontologists and Mineralogists, Special Publication No. 42, pp. 71–108.

Hardie, A.L. (ed) (1977) *Sedimentation on the Modern Carbonate Tidal Flats of Northwest Andros Island, Bahamas*, The Johns Hopkins University Studies in Geology, 22, Baltimore and London.

Hardie, L.A. (1987) Dolomitization: a critical view of some current views. *Journal of Sedimentary Petrology*, **57**, 166–183.

Harding, T.P. (1984) Graben hydrocarbon occurrences and structural style. *American Association Petroleum Geologists Bulletin*, **68**, 333–362.

Hardy, S. and Waltham, D.A. (1992) Computer modelling of tectonics, eustacy and sedimentation using the Macintosh. *Geobyte*, **7**, 42–52.

Hart, M.B. (1987) *Micropaleontology of carbonate environments*, Ellis Harwood Ltd., Chichester.

Hart, W.K., Woldegabriel, G., Walter, R.C. and Mertzman, S.A. (1989) Basaltic volcanism in Ethiopia: constraints on continental rifting and mantle interactions. *Journal Geophysical Research*, **94**, 7731–7748.

Hartmann, M. (1980) Atlantis II Deep geothermal brine system. Hydrographic situation in 1977 and changes since 1965. *Deep-Sea Research*, **27A**, 161–171.

Hartmann, M. (1985) Atlantis II Deep geothermal brine system. Chemical processes between hydrothermal brines and Red Sea deep water. *Marine Geology*, **64**, 157–177.

Hassan, A.A. (1967) A new Carboniferous occurrence in Abu Durba-Sinai, Egypt. *Proceedings 6th Arabian Petroleum Congress*, Baghdad, **II**, 39 (B-3), 8 pp.

Hassan, F. and El Dashlouti, S. (1970) Miocene evaporites of Gulf of Suez region and their significance. *American Association of Petroleum Geologists Bulletin*, **54**, 1686–1696.

Haunold, T.G. (1990) The new Neogene genus *Pappina* in the new family Pappinidae: polymorphine mode of chamber addition in the Bulininacea. *Journal Foraminiferal Research*, **20**, 56–64.

Haven ten, H.L., de Leeuw, J.W., Rullkötter, J.W. and Sinninghe Damsté, S.J. (1987) Restricted utility of pristane/phytane ratio as a palaeoenvironmental indicator. *Nature*, **330**, 641–643.

Haven ten, H.L., de Leeuw, J.W., Sinninghe Damsté, S.J., Schenk, P.A., Palmer, S.E. and Zumberge, J. (1988) Application of biological markers in the recognition of palhypersaline environments, in *Lacustrine Petroleum Source Rocks* (eds A.J. Fleet, K. Kelts, and M.R. Talbot), Geological Society London Special Publication, 40, Blackwell Scientific Publications, Oxford, pp. 123–130.

Hayward, A.B. (1982) Coral reefs in a clastic sedimentary environment: fossil (Miocene, SW Turkey) and modern (recent, Red Sea) analogues *Coral Reefs*, **1**, 109–114.

Hazemann, J.L. (1991) Etude cristallochimique par diffraction X et par spectroscopie EXAFS de la solution solide aFEOOH-aAlOOH. Thèse de l'Université Louis Pasteur, Strasbourg.

Head, S.M. (1987) Corals and coral reefs of the Red Sea, in *Red Sea, Key Environments* (eds A.J. Edwards and S.M. Head), Pergamon Press, Oxford, pp. 128–151.

Healy, J.H., Mooney, W.D., Blank, H.R. et al. (1982) Saudi Arabia seismic deep-refraction profile. *United States Geological Survey, Saudi Arabian Mission Rep OF-02-37*.

Heaton, R.C., Jackson, M.P.A., Bamahmoud, M. and Nani, A.S.O. (1996) Superposed Neogene extension, contraction, and salt canopy emplacement in the Yemeni Red Sea, in *Salt Tectonics; A Global Perspective for Exploration* (eds M.P.A. Jackson, D.G. Roberts and S. Snelson). American Association of Petroleum Geologists Memoir 65, pp. 333–351.

Heiss, G.A. (1994) Coral reefs in the Red Sea: Growth, production and stable isotopes. *Geomar Report*, **32**, 1–141.

Heiss, G.A. (1996) Carbonate production by scleractinian corals at Aqaba, Gulf of Aqaba, Red Sea. *Facies*, **33** (in press).

Hem, J.D. and Lind, C.J. (1983) Nonequilibrium models for predicting forms of precipitated manganese oxides. *Geochimica et Cosmochimica Acta*, **47**, 2037–2046.

Hempton, M. (1987) Constraints on Arabian plate motion and extensional history of the Red Sea. *Tectonics*, **6**, 687.

Hendrie, D.B., Kusznir, N.J., Morley, C.K. and Ebinger, C.J. (1994) Cenozoic extension in northern Kenya: a quantitative model of rift basin development in the Turkana region. *Tectonophysics*, **236**, 409–438.

Henke, C.H. (1989) Schwerefeld und Krustenaufbau der Arabischen Republik Yemen. MSc thesis, Institut Geophysics, Universität Hamburg.

Henke, C.H. (1995) Zur dreidimensionalen Modellierung und Visualisierung von Schweredaten angewendet auf das südliche Rote Meer. *Berichte aus dem Zentrum für Meeres-und Klimaforschung, Reihe C*, Nr. 7, Universität Hamburg.

Herman, Y. (1965a) Etudes des sédiments Quaternaires de la Mer Rouge. *Institut Océanographique, Annales*, **42**, 339–430.

Herman, Y. (1965b) Evidence of climatic changes in Red Sea cores, in *Means of Correlation of Quaternary Successions* (eds R.B. Morrison and H.E. Luright), 8. University of Utah Press, Salt Lake City.

Herman, Y. (1971) Vertical and horizontal distribution of pteropods in Quaternary sequences, in *The Micropaleontology of Oceans*, (ed. B.M. Funnell), Cambridge University Press, pp. 463–486.

Hermina, M., Klitzsch, E. and List, F.K. (1989) Stratigraphic lexicon and explanatory notes to the geological map of Egypt, 1:500,000. Conoco Inc. Cairo.

Heybroek, F. (1965) The Red Sea Miocene evaporite basin, in *Salt Basins around Africa*, Institute of Petroleum and the Geological Society London, Elsevier Publishing Co., Amsterdam, pp. 17–40.

Hine, A.C., Wilber, R.J., Bane, J.M., Neumann, A.C. and Lorenson, K.R. (1981) Offbank transport of carbonate sands along open, leeward bank margins, Northern Bahamas. *Marine Geology*, **42**, 327–348.

Hoang, C.T., Dalongeville, R. and Sanlaville, P. (1995) Stratigraphy, tectonics and paleoclimatic implications of uranium-series-dated coral reefs from the Sudanese coast of the Red Sea. *Quaternary International*, **31**, 47–51.

Hoang, C.T., Lalou, C. and Faure, H. (1974) Les récifs soulevés a l'ouest du golfe d'Aden (T.F.A.I.) et les hauts niveaux de coraux de la dépression de l'Afar (Ethiopie), géochronologie et paléoclimats interglaciaires, in *Les méthodes quantitatives d'étude des variations du climat au cours du Pléistocène*. Colloques Internationaux du C.N.R.S. No. 219, pp. 103–114.

Hoang, C.T., Lalou, C. and Faure, H. (1980) Ages $^{230}Th/^{234}U$ du 'Tyrrhénien' de Mer Rouge et du Golfe d'Aden. INQUA, Comptes-rendus excursion table ronde sur le Tyrrhenien de Sardaigne, Cagliari, 21–28 April, pp. 109–112.

Hoang, C.T. and Taviani, M. (1991) Stratigraphic and tectonic implications of uranium-series-dated coral reefs from uplifted Red Sea islands. *Quaternary Research*, **35**, 264–273.

Hodell, D.A., Mueller, P.A. and Garrido, J.R. (1991) Variations in the strontium isotopic composition of seawater during the Neogene. *Geology*, **19**, 24–27.

Hofmann, C., Feraud, G., Pik, *et al.* (1995) $^{40}Ar/^{39}Ar$ dating of Ethiopian Traps. EUG 8, Strasbourg, Terra Abstracts, **7**, 159.

Hofmann, P., Schwark, L., Brachert, T., Badaut, D., Rivier, M. and Purser, B.H. (this volume). Sedimentation, organic geochemistry, and diagenesis of cores from the axial zone of the southern Red Sea: Relationships to rift dynamics and climate, in *Sedimentation and Tectonics of Rift Basins: Red Sea–Gulf of Aden* (eds. B.H. Purser and D.W.J. Bosence), Chapman & Hall, London, pp. 483–507.

Holm, N.G., Dowler, M.J., Wadsten, T. and Arrhenius, G. (1983) β-FeOOH.Cl$_n$ (Akagénéite) and Fe$_{1-x}$O (Wüstite) in hot brine from the Atlantis II Deep (Red Sea) and the uptake of amino acids by synthetic β-FeOOH.Cl$_n$. *Geochimica et Cosmochimica Acta*, **47**, 1465–1470.

Holm, N.G., Wadsten, T. and Dowler, M.J. (1982) β-FeOOH (Akaganéite) in Red Sea brine. *Estudios Geol.*, **38**, 367–371.

Holwerda, J.G. and Hutchison, R.W. (1968) Potash bearing evaporites of the Danakil area, Ethiopia. *Economic Geology*, **63**, 124–150.

Honnorez, J., Alt, J.C., Honnorez-Guerstein, B.M., Lavenne, C., Muehlenbach K. *et al.* (1985) Stockwork-like sulfide mineralization in young oceanic crust: Deep Sea Drilling Project Hole 504B. *Initial Reports DSDP*, **83**, Washington DC, U.S. Gov. Printing Office, pp. 263–282.

Hötzl, H., Jado, A.R., Moser, H., Rauert, W. and Zötl, J.G. (1984) The youngest Pleistocene, in *Quaternary Period in Saudi Arabia* (eds A.R. Jado and J.G. Zöttl), Springer, Wien, pp. 314–324.

Hötzl, H., Moser, H., Rauert, W., Wolf, M. and Zötl, J.G. (1984) Problems involved in ^{14}C age determinations in carbonates, in *Quaternary Period in Saudi Arabia* (eds A.R. Jado and J.G. Zöttl), Springer, Wien, pp. 325–331.

Houbrick, R.S. (1978) The family Cerithiidae in the Indo-Pacific. Part 1: the genera *Rhinoclavis, Pseudovertagus* and *Clavocerithium*. *Monographs of Marine Mollusca*, **1**, American Malacologists, Greenville, USA.

Houbrick, R.S. (1992) Monograph of the Genus *Cerithium* Bruguiere in the Indo-Pacific (Cerithiidae: Prosobranchia). *American Malacologists, Smithsonian Contributions to Zoology*, **510**, 211 pp.

Hovorka, S.D. (1992) Halite pseudomorphs after gypsum in bedded anhydrite-clue to gypsum-anhydrite relationships. *Journal of Sedimentary Petrology*, **62**, 1098–1111.

Howell, J.A. and Flint, S.S. (1996) A model for high resolution sequence stratigraphy within extensional basins, in *High Resolution Sequence Stratigraphy: Innovations and Applications*, (eds J.A. Howell and J.F. Aitkin), Geological Society of London Special Publication, **104**, pp. 129–137.

Hsü, K.J. (1972) Origin of saline giants: a critical review after the discovery of the Mediterranean evaporites. *Earth Sciences Review*, Amsterdam, **8**, 371–396.

Hsü, K.J., Cita, M.B. and Ryan, W.B.F. (1973) The origin of the Mediterranean evaporites. *Deep Sea Drilling Project*. Initial report, Washington, **13**, Part 2, pp. 1203–1231.

Hsü, K.J., Stoffers, P. and Ross, D.A. (1978) Messinian evaporites from the Mediterranean and Red Seas. *Marine Geology*, **26**, 71–72.

Hubbard, R.J. (1988) Age and significance of sequence boundaries on Jurassic and early Cretaceous rifted continental margins. *Bulletin American Association Petroleum Geologists*, **72**, 49–72.

Huchon, P., Jestin, F., Cantagrel, J. M. *et al.* (1992) Extensional deformations in Yemen since Oligocene and the Africa-Arabia-Somalia triple junction. *Annales Tectonicae*, **5**, 141–163.

Hughes, G.W., Abdine, S. and Girgis, M.H. (1992) Miocene biofacies development and geological history of the Gulf of Suez, Egypt. *Marine and Petroleum Geology*, **9**, 2–28.

Hughes, G.W. and Beydoun, Z.R. (1992) The Red Sea–Gulf of Aden: biostratigraphy, lithostratigraphy and palaeoenvironments. *Journal of Petroleum Geology*, **15**, 135–156.

Hughes, G.W. and Filatoff, J. (1995) New biostratigraphic constraints on Saudi Arabian Red Sea pre- and syn-rift sequences. *Proceedings of the Middle East Petroleum Geosciences Conference (Geo 94)*, 517–528.

Hughes, G.W., Varol, O. and Beydoun, Z.R. (1991) Evidence for Middle Oligocene rifting of the Gulf of Aden and for Late Oligocene rifting of the Red Sea. *Marine and Petroleum Geology*, **8**, 354–358.

Hume, W.F. (1906) The topography and geology of the Peninsula of Sinai (southeastern portion). Geological Survey Egypt, Cairo.

Hume, W.F. (1916) *Report on the oil fields region of Egypt*, Egypt Survey Department, Cairo.

Hume, W.R. (1921) Relations of the northern Red Sea and associated Gulf areas to the 'rift' theory. *Proceedings of the Geological Society of London*, **77**, 96–101.

Hume, W.F., Madgwick, T.G., Moon, F.W. and Sadek, H. (1920) Preliminary geological report on the Gebel Tanka area. *Petroleum Research Bulletin Egypt*, **4**.

Hurst, J.M. and Surlyk, F (1984) Tectonic controls of Silurian carbonate. Shelf margin morphology and facies, North Greenland. *American Association Petroleum Geologists Bulletin*, **68**, 1–17.

Hurst, J.M. (1987) Syn-rift carbonate depositional patterns: Miocene, Gulf of Suez, Egypt. *American Association of Petroleum Geologists Bulletin* (Abs), **71**, 569.

Hutchinson, R.W. and Engels, G.C. (1970) Tectonic significance of regional geology evaporite lithofacies in Northeastern Ethiopia. *Philosophical Transactions Royal Society of London*, **A267**, 1181.

Ibrahim, A., Rouchy, J.-M., Maurin, A., Guelorget, O. and Perthuisot, J.-P. (1986) Mouvements halocinétoques récents dans de Golfe de Suez: l'exemple de la péninsule de Gemsah. *Bulletin de la Société géologique de France*, **2**, 177–183.

Imbrie, J., Hayes, J.D., Martinson, D.G., Macintyre, A., Mix, A. *et al.* (1984) The orbital theory of Pleistocene climate: support from a revised chronology of the marine delta peninsula 18 0 record, in *Milankovitch and Climate* (eds A. Berger J. Imbrie, J.D. Hays, G. Kukla and B. Saltzman), D. Reidel, Hingham, Massachusetts, pp. 269–305.

Inden, R.F. and Moore, C.H. (1983) Beach environments, in *Carbonate Depositional Environments*, (ed. P. Scholle), American Association of Petroleum Geologists, Memoir **33**, pp. 212–265.

Insalaco, E. (1996) Upper Jurassic microsolenid biostromes of northern and central Europe: facies and depositional environment. *Palaeogeography, Palaeoclimatology, Palaeoecology*, **121**, 169–194.

Issawi, B., Francis, M., El-Hinnawi, M. and El-Deftar, T. (1971) Geology of Safaga-Quseir coastal plain and of Mohamed Rabah area. *Annals of the Geological Survey of Egypt*, **1**, 1–19.

Issel, A. (1869) *Malacologia del Mar Rosso*, Bibliotheca Malacologia, Pisa.

Ivanova, E.V. (1985) Late Quaternary biostratigraphy and paleotemperatures of the Red Sea and the Gulf of Aden based on planktonic foraminifera and pteropods. *Marine Micropaleontology*, **9**, 335–364.

Izzeldin A.Y. (1987) Seismic, gravity and magnetic surveys in the central part of the Red Sea: their interpretation and implications for the structure and evolution of the Red Sea. *Tectonophysics*, **143**, 269–306.

Izzeldin, Y.A. (1982) On the structure and evolution of the Red Sea. PhD thesis, University Louis Pasteur, Strasbourg.

Jackson, J.A. (1987) Active normal faulting and crustal extension, in *Crustal Extensional Tectonics* (eds M.P. Coward, J.F. Dewey and P.L. Hancock), Special Publication Geological Society of London, **28**, pp. 3–17.

Jackson, J.A. and McKenzie, D.P. (1983) Rates of active deformation in the Aegian Sea and surrounding areas. *Basin Research*, **1**, 121–128.

Jackson, J.A., White, N.J., Garfunkel, Z. and Anderson, H. (1988) Relations between normal-fault geometry, tilting and vertical motions in extensional terrains: an example from the southern Gulf of Suez. *Journal of Structural Geology*, **10**, 155–70.

Jado, A.R., Hötzl, H. and Boscher, B. (1989) Development of sedimentation along the Saudi Arabian Red Sea coast. *Journal of King Abdulaziz University, Earth Sciences*, **3**, 863–888.

James, N.P. (1983) Reef environment, in *Carbonate Depositional Environments* (eds P.A. Scholle, D.G. Bebout and C.H. Moore), American Association of Petroleum Geologists Memoir, **33**, Tulsa, Oklahoma, pp. 345–462.

James, N.P. and Bone, Y. (1989) Petrogenesis of Cenozoic temperate water calcarenites, South Australia: a model for meteoric-shallow burial diagenesis of Shallow water calcite sediments. *Journal of Sedimentary Petrology*, **59**, 191–204.

James, N.P., Coniglio, M., Aissaoui, D.M. and Purser., B.H. (1988) Facies and geological history of an exposed Miocene rift-margin carbonate platform: Gulf of Suez, Egypt. *American Association of Petroleum Geologists Bulletin*. **72**, 55–572.

James, N.P. and Ginsburg, R.N. (1979) The seaward margin of Belize barrier and atoll reefs. *International Association Sedimentologists Special Publication*, **3**, 1–191.

James, N.P., Rosen, B. and M. Coniglio. (1988) Miocene platform margin reefs, Gulf of Suez, Egypt. *American Association of Petroleum Geologists Bulletin*, (Abs)., **72**, 200–201.

Janecky, D.R. and Seyfried, W.E. Jr (1984) Formation of massive sulfide deposits on oceanic ridge crests: incremental reaction models for mixing between hydrothermal solutions and seawater. *Geochimica et Cosmochimica. Acta*, **48**, 2723–2738.

Jarrige, J.J., Ott d'Estevou, P, Burollet, P.F., Icart, J.C., Montenat, C. et al. (1986) Inherited discontinuities and Neogene structure: the Gulf of Suez and NW edge of the Red Sea. *Philosophical Transactions Royal Society, London*, **317**, 129–139.

Jarrige, J.J., Ott d'Estevou, P., Burollet P.F., Montenat, C., Richert, J.P. and Thiriet, J.P. (1990) The multistage tectonic evolution of the Gulf of Suez and northern Red Sea continental rift from field observations. *Tectonics*, **9**, 441–465.

Jarrige, J.J., Ott d'Estevou, P. and Sehans, P. (1986) Etude structurale sur la marge occidentale de la Mer Rouge: le secteur du Gebel Duwi près de Quseir (Egypte). *Documents et Travaux, Institut Géologique Albert de Lapparent*, **10**, 17–127.

Jestin, F., Huchon, P. and Gaulier, J.M. (1994) The Somalia plate and the East African rift system. *Geophysical Journal International*, **116**, 637–654.

Joffe, S. and Garfunkel, Z. (1987) Plate kinematics of the Circum Red Sea – a reevaluation. *Tectonics*, **141**, 5–22.

Jones, R.W. and Racey, A. (1994) Cenozoic stratigraphy of the Arabian Peninsula and Gulf, in *Micropalaeontology and Hydrocarbon Exploration in the Middle East* (ed. M.D. Simmons), Chapman & Hall, pp. 274–307.

Jorgensen, B.B. (1977) The sulfur cycle of a coastal marine sediment (Limfjorden, Denmark). *Limnology Oceanography*, **22**, 814–832.

Jousseaume, F. (1888) Description des Mollusques recueillis par M. le Dr Faurot dans la Mer Rouge et le Golfe d'Aden. *Mémoires de la Société Zoologique de France*, **1**, 165–223.

Jousseaume, F. (1930) Cerithiidae de la Mer Rouge. *Journal de Conchyliologie*, **74**, 270–296.

Kaplan, I.R., Sweeney, R.E. and Nissenbaum, A. (1969) Sulfur isotope studies on Red Sea geothermal brines and sediments, in *Hot Brines and Recent Heavy Metal Deposits in the Red Sea*, (eds E.T. Degens and A.D. Ross), Springer-Verlag, Berlin, pp. 474–498.

Karbe, L. (1987) Hot brines and the deep sea environment, in *Red Sea, Key Environments*, (eds A.J. Edwards and S.M. Head), Pergamon Press, Oxford, pp. 70–89.

Karim, Z. (1984) Characteristics of ferrihydrites formed by oxidation of $FeCl_2$ solutions containing different amounts of silica. *Clays and Clay Minerals*, **32**, 131–184.

Karpoff, R. (1957a) Sur l'existence du Maestrichtien en Nord de Djeddah (Arabie Seoudite). *Compte Rendue Seances*, **245** (15), 1322–1324.

Karpoff, R. (1957b). Esquisse géologique de l'Arabie Seoudite. *Bulletin Société Geologique de France*, Ser 6, **7**, 653–697.

Kebeasy, R.M. (1990) Seismicity, in *The geology of Egypt*, (ed. R. Said), Balkema, Rotterdam, Brookfield, pp. 51–59.

Kennett, J.P. and Srinivasan, M.S. (1983) *Neogene Planktonic Foraminifera*, Hutchinson Ross, Stroudsburg.

Kerans, C., Hurley, N.F. and Playford, P.E. (1986) Marine diagenesis in Devonian reef complexes of the Canning Basin, Western Australia, in *Reef Diagenesis* (eds J.H. Schroeder and B.H. Purser), Springer-Verlag, pp. 357–380.

Kerdany, M.T. and Cherif, O.H. (1990) Mesozoic, in *The Geology of Egypt* (ed. R. Said), Balkema, Rotterdam, pp. 407–438.

Khalil, B. and Meshref, W. (1988) Hydrocarbon occurrences and structural style of the southern Suez rift basin. *Proceedings 9th. EGPC Exploration and Production Conference*, Cairo, Egypt, 13 pp.

Khedr, E. (1984) Sedimentological evolution of the Red Sea continental margin of Egypt and its relationship to sea-level changes. *Sedimentary Geology*, **39**, 71–86.

Kinsman, D.J.J. (1975) Rift valleys basins and sedimentary history of trailing continental margins, in *Petroleum and global tectonics*, (eds A.G. Fischer, and S. Judson), Princeton University Press, pp. 83–126.

Kirkby, M.J. (1987) General models of long-term slope evolution through mass movement, in *Slope stability* (eds M.G. Anderson and K.S. Richards), John Wiley & Sons, New York pp. 359–379.

Klitzsch, E. (1990) Paleozoic, in *The Geology of Egypt*, (ed. R. Said), Balkema, Rotterdam, pp. 393–406.

Klunzinger, C.B. (1879) *Die Korallenthiere des Roten Meeres. 2. Die Steinkorallen*, Gutmann, Berlin.

Kohn, B.P. and Eyal, M. (1981) History of uplift of the crystalline basement of Sinai and its relation to opening of the Red Sea as revealed by fission tract dating of Apatites. *Earth and Planetary Science Letters*, **52**, 129–141.

Kolla, V. and Biscaye, P.E. (1977) Distribution of quartz in the sediments of the Indian Ocean. *Journal of Sedimentary Petrology*, **47**, 642–649.

Kooi, H. and Beaumont, C. (1994) Escarpment evolution on high-elevation rifted margins; insights derived from a surface processes model that combines diffusion, advection and reaction. *Journal of Geophysical Research*, **99**, 12191–12210.

Kora, M. (1984) The Paleozoic outcrops of Um Bogma area, Sinai. PhD thesis, Mansoura University, Mansoura.

Kostandi, A.B. (1959) Facies maps of the study of the Paleozoic and Mesozoic sedimentary basins of the Egyptian region. *Proceedings UAR 1, Arabian Petroleum Congress Cairo, Congress 2*, Cairo, pp. 54–62.

Kraus, M.J. and Bown, T.M. (1986) Paleosols and time resolution in fluvial stratigraphy, in *Paleosols, their Recognition and Interpretation* (ed. V.P. Weight). Blackwells, Oxford, pp. 180–207.

Kröner, A., Greiling, R., Reischmann, T. et al. (1987) Pan-African crustal evolution in the Nubian segment of Northwest Africa, in *Proterozoic Lithosphere Evolution* (ed. A. Kröner), American Geophysical Union, Washington, Geodynamics Series, 17, pp. 235–257.

Kronfeld, J., Gvirtzman, G. and Buchbinder, B. (1982) *Geological evolution and thorium–uranium ages of Quaternary coral reefs in southern Sinai*. Proceedings Israel Geological Society, Annual Meeting 1982, Elat and Eastern Sinai, Elat, 24–27 January 1982, pp. 45–46.

Kruck, W. (1983) *Geological map of the Yemen Arab Republic: Sheet Sana'a* 1: 250,000. Federal Institute Geosciences and Natural Resources, Hannover, Germany

Kruck W., Al-Anissi, A. and Saif, M. (1984) Geological map of the Yemen Arab Republic, sheet Al-Hudaydah, 1:250,000. Federal Institute for Geosciences and Natural Resources, Hannover, Germany

Kuhnert, H. (1993) Diagenesis of corals in Miocene clastics of the Sudanese Red Sea coast, in *Geoscientific research in*

Northeast Africa, (eds U. Thorweihe and H. Schandelmeier), Balkema, Rotterdam, pp. 459–464.

Kulke, H. (1982) A Miocene carbonate and anhydride sequence in the Gulf of Suez as a complex oil reservoir. *Proceedings 5th EGPC Exploration Seminar, Cairo*, pp. 269–275.

Kushnir, (1981) Formation and early diagenesis of varved evaporite sediment in a coastal hypersaline pool. *Journal of Sedimentary Petrology*, **51**, 193–1203.

Kuss, J. (1992) The Aptian–Paleocene shelf carbonates of northeast Egypt and southern Jordon: establishment and break-up of carbonate platforms along the southern Tethyan shores. *Zeitschrift dt. geologie Ges.*, **143**, 107–132.

Kusznir, N.J. and Egan, S.S. (1990) Simple-shear and pure-shear models of extensional sedimentary basin formation: application to the Jeanne d'Arc Basin, Grand Blanks of Newfoundland, in *Extensional Tectonics of the North Atlantic Margins*, (eds A.J. Tankard and H.R. Balkwill), American Association of Petroleum Geologists Memoir **46**, pp. 305–322.

Kusznir, N.J., Marsden, G. and Egan, S.S. (1991) A flexural cantilever simple-shear/pure-shear model of continental extension, in *The Geometry of Normal Faults* (eds A.M. Roberts, G. Yielding and B. Freeman), Geological Society, London, Special Publications **56**, pp. 41–61

La Brecque, J.L. and Zitellini, N. (1985) Continuous sea floor spreading in Red Sea: an alternative interpretation of magnetic anomaly pattern. *American Association of Petroleum Geologists Bulletin*, **69**, 513–524.

Laffitte, R., Harland, W.B., Erben, H.K., Blow, W.H., Haas, W. *et al.* (1972) Essai d'accord international sur les problèmes essentiels de la stratigraphie (some international agreement on essentials of statigraphy. Internationale Ubereinkunft über die grundlagen der Startigraphie), *Comptes Rendues Société Géologique France*, 36–45.

Lalou, C., Nguyen, H.V., Faure, H. and Moreira, L. (1970) Datation par la méthode Uranium-Thorium des hauts niveaux de coraux de l'Afar (Ethiopie). *Revue de Géographie physique et de Géologie dynamique*, **12**, 3–8.

Lambiase, J.J. and Bosworth, W. (1995) Structural controls on sedimentation in continental rifts, in *Hydrocarbon Habitat in Rift Basins*, (ed. J.J. Lambiase), Geological Society Special Publication 80, pp. 117–144.

Land, L.S. (1975) Palaeohydrology of ancient dolomites: geochemical evidence. *American Association of Petroleum Geologists Bulletin*, **59**, 1602–1625.

Land, L.S. (1980) The isotopic and trace element geochemistry of dolomite: the state of the art, in *Concepts and Models of Dolomitization* (eds D.H. Zenger, J.B. Dunham and R.L. Ethington) Society of Economic Palaeontologists and Mineralogists, Special Publication No. 28, Tusla, Oklahoma, USA, pp. 87–110.

Land, L.S, (1983) The application of stable isotopes to studies of the origin of dolomite and to problems of diagenesis of clastic sediments, in *Stable Isotopes in Sedimentary Geology* (eds M.A. Arthur, T.F. Anderson, I.R. Kaplan, J. Veizer and L.S. Land), Society of Economic Palaeontologists and Mineralogists, short course No. 10, Tusla, Oklahoma, USA, 4-1-8.

Langmuir, D. (1971) Particle size effect on the reaction goethite = hematite + water. *American Journal Science*, **271**, 147–156.

Laughton, A.S. (1966a) The birth of an ocean. *New Scientist*, **27**, 218–220.

Laughton, A.S. (1966b) The Gulf of Aden. *Philosophical Transactions Royal Society, London*, **259-A**, 150–171.

Laughton, A.S. and Tramontini, C. (1968) Recent studies of the continental crustal structure of the Gulf of Aden. *Tectonophysics*, **8**, 359–375.

Laughton, A.S., Whitmarsh, R.B. and Jones, M.T. (1970) The evolution of the Gulf of Aden. *Philosophical Transactions Royal Society, London*, **267-A**, 227–266.

Le Nindre, Y.M., Garcin, M., Motti, E. and Vazquez-Lopez, R. (1986) Le Miocène du Massif de Maqna (Mer Rouge, Arabie saoudite), stratigraphie – paléogéographie. *Documents et Travaux Institut Géologique Albert de Lapparent, Paris*, **10**, 177–185.

Le Pichon, X. and Francheteau, J. (1978) A plate-tectonic analysis of the Red Sea–Gulf of Aden area. *Tectonophysics*, **46**, 369–406.

Le Pichon, X. and Gaulier, J.M. (1988) The rotation of Arabia and the Levant fault system. *Tectonophysics*, **153**, 271–294.

Leeder, M.R. (1995) Continental rifts and proto-oceanic troughs, in *Tectonics of Sedimentary Basins*, (eds C.J. Busby and R.V. Ingersoll), Blackwell Science, USA, pp. 119–148.

Leeder, M.R. and Gawthorpe, R.L. (1987) Sedimentary models for extensional tilt-block/half-graben basins, in *Continental Extensional Tectonics* (eds M.P. Coward, J.F. Dewey and P.L. Hancock), *Geological Society of London, Special Publication*, **28**, pp. 139–152.

Leeder, M.R. and Jackson, J.A. (1993) The interaction between normal faulting and drainage in active extensional basins, with examples from the western United States and central Greece. *Basin Research*, **5**, 79–102.

Leventhal, J.S. (1983) An interpretation of carbon and sulfur relationships in Black Sea sediments as indicators of environments of deposition. *Geochimica et Cosmochimica Acta*, **47**, 133–137.

Lewis, D.G., and Schwertmann, U. (1979) The influence of aluminium on the formation of iron oxides IV. The influence of [Al], [OH], and temperature *Clays and Clay Minerals*, **27**, 195–200.

Lippmann F. (1980) Phase diagrams depicting aqueous solubility of binary mineral systems. *Nues Jahrbuch Abh.*, **139**.

Lister, G.S., Etheridge, M.A. and Sismond, P.A. (1986) Detachment faulting and evolution of passive continental margins. *Geology*, **14**, 246–250.

Locke, S.M. and Thunnell, R.C. (1988) Paleoceanographic records of the last glacial/interglacial cycle in the Red Sea and Gulf of Aden. *Palaeogeography, Palaeoclimatology, Palaeoecology*, **64**, 163–187.

Lohmann, H.H., Hoffmann-Rothe, J. and Hinz, K. (1995) Argentina, in *Regional petroleum geology of the world. Part II: Africa, America, Australia and Antarctica*, (ed. H. Kulke), Borntraeger, Berlin, pp. 549–575.

Lonsdale, P. and Batiza, R. (1980) Hyaloclastite and lava flows on young seamounts examined with a submersible. *Geological Society of America Bulletin*, **96**, 545–554.

Loreau, J-P. and Cross P. (1988) Limestone diagenesis and dolomitization of Tithonian carbonates at ODP site 639 (Atlantic Ocean, West of Spain). *Proceedings of the Ocean Drilling Program, Scientific Results*, **103**, 105–143.

Lowe, D.R. (1975) Water escape structure in coarse-grained sediments. *Sedimentology*, **22**, 157–204.

Lowell, J.D. and Genik, G.J. (1972) Sea floor spreading and structural evolution of the southern Red Sea. *Bulletin American Association Petroleum Geologists*, **56**, 247–259.

Loya, Y. and Slobodkin, L.B. (1971) The coral reefs of Eilat (Gulf of Eilat, Red Sea), in *Regional Variation in Indian Coral Reefs*, (eds D.R. Stoddart and M. Yonge), Symposium of the Zoological Society of London, 28, pp. 117–139.

Luger, P., Hendriks, F., Arush, M., Bussman, M., Kallenbach, H. et al. (1990) The Jurassic and the Cretaceous of Northern Somalia: preliminary results of the sedimentologic and stratigraphic investigations. *Berliner Geowissenschaftliche Abhandlungen*, **1202**, 571–594

Lumsden, D.N. and Chimahusky, J.S. (1980) Relationship between dolomite non-stoichiometry and carbonate facies parameters. in *Concepts and Models of Dolomitization*, (eds D.H. Zenger and J.B. Dunham), Society of Economic Palaeontologists and Mineralogists, Special Publication Number **28**, Tulsa, Oklahoma, USA, pp. 123–139.

Luther III, G.W. (1991) Pyrite synthesis via polysulfide compounds. *Geochimica et Cosmochimica Acta*, **55**, 2839–2849.

Lyberis, N. (1988) Tectonic evolution of the Gulf of Suez and the Gulf of Aqaba. *Tectonophysics*, **153**, 209–221.

Macdonald, D.P. (1972) *Volcanoes*, Prentice Hall Inc, New Jersey.

Macfadyan, W.A. (1930) The geology of the Farsan Islands, Gizan and Kamaran Islands, Red Sea. Part 1 General Geology. *Geological Magazine*, **67**, 310–315.

Macfadyan, W.A. (1933) *Geology of British Somaliland*, UK Crown Agents, London.

Machel, H.G. and Mountjoy, E.W. (1986) Chemistry and environments of dolomitization – a reappraisal. *Earth Science Reviews*, **23**, 175–222.

Machette, M.N., Persounis, S.F. and Nelson, A.R. (1991) The Wasatch fault zone Utah, segmentation and history of Holocene earthquakes. *Journal Structural Geology*, **13**, 137–149.

Macintyre, I.D. (1985) *Submarine cements: the peloidal question*. Society of Economic Mineralogists and Paleontologists, Special Publication **36**, pp. 109–116.

Madgewick, T.G., Moon, F.W. and Saded, H. (1920) Preliminary report on the Abu Shaar El Quibli (Black Hill) district. *Petroleum Research Bulletin*, **6**, Government Press, Cairo, 11 pp.

Makris, J., Allam, A., Mokhtar, T. et al. (1983) Crustal structure in the northwestern region of the Arabian Shield and its transition to the Red Sea. *Bulletin Faculty Earth Science King Abdulaziz University*, **6**, 435–447.

Makris, J., Egloff, R., Jacob, A.W.B. et al. (1988b) Continental crust under the southern Porcupine Seabight west of Ireland. *Earth Planetary Science Letters*, **89**, 387–397.

Makris, J. and Ginzburg, A. (1987) The Afar Depression – transition between continental rifting and seafloor spreading. *Tectonophysics*, **141**, 199–214.

Makris, J. and Henke, C.H. (1992) Pull-apart evolution of the Red Sea. *Journal Petroleum Geology*, **15**, 127–134.

Makris, J. Henke, C.H., Egloff, F. and Akamaluk. T. (1991a) The gravity field of the Red Sea and East Africa. *Tectonophysics*, **198**, 369–382.

Makris, J., Menzel, H., Zimmermann, J. and Gouin, P. (1975) Gravity field and crustal structure of north Ethiopia, in *Afar between Continental and Oceanic Rifting*, (eds A. Pilger and A. Roesler), Schweitzerbart, Stuttgart (I), pp. 120–134.

Makris, J., Mohr, P. and Rihm, R. (eds) (1991b) Red Sea: birth and early history of a new oceanic basin. *Tectonophysics*, special issue, **198**, Elsevier, Amsterdam.

Makris, J. and Rihm, R. (1991) Shear controlled evolution of the Red Sea: pull apart model. *Tectonophysics*, **198**, 441–466.

Makris, J., Rihm, R. and Allam, A. (1988a) Some geophysical aspects of the evolution and the structure of the crust in Egypt, in *The Pan-African Belt of Northeast Africa and the Adjacent Areas*, (eds S. El Gaby and R.O. Greiling), Vieweg, Braunschweig, pp. 345–369.

Makris, J., Tsironidis, J. and Richter, H. (1991c) Heat flow density distribution in the Red Sea. *Tectonophysics*, **198**, 383–393.

Mallick, D.I.J., Gass, I.G., Cox, K.G. et al. (1990) Perim Island, a volcanic remnant in the southern entrance to the Red Sea. *Geological Magazine*, **127**, 309–318.

Manetti, P., Capaldi, G., Chiesa, S., Civetta, L., Conticelli, S. et al. (1991) Magmatism of the eastern Red Sea margin in the northern part of Yemen from Oligocene to present. *Tectonophysics*, **198**, 181–202.

Manighetti, I., Tapponier, P. and Courtillot, V. (1995) Propagation of the Arabia–Somalia plate boundary through the Gulfs of Aden and Tadjoura, into Afar. *Proceedings of 'Rift sedimentation and tectonics in the Red Sea – Gulf of Aden region'*, University of Sana'a, pp. 36–37.

Marshak, S., Bonatti, E., Brueckner, H. and Paulsen, T. (1992) Fracture-zone tectonics at Zabargad island, Red Sea (Egypt). *Tectonophysics*, **216**, 379–385.

Mart, Y. and Hall, J. (1984) Structural trends in the northern Red Sea. *Journal of Geophysical Research*, **89**, 352–364.

Mart, Y. and Ross, D.A. (1987) Post-Miocene rifting and diapirism in the Northern Red Sea. *Marine Geology*, **74**, 173–190.

Marzouk, I.M. (1988) Study of crustal structure of Egypt, deduced from deep seismic and gravity data. PhD thesis, Institut Geophysics University, Hamburg.

Matteucci, R. (1974) *Cymbaloporella tabellaeformis* (Brady) foraminifero endolitico del Mar Rosso. *Geologica Romana*, **13**, 29–43.

McCall, J., Rosen, B.R. and Darrell, J. (1994) Carbonate deposition in accretionary prism settings: early Miocene coral limestones and corals of the Makran Mountain Range in southern Iran. *Facies*, **31**, 141–178.

McClay, K.R., Nichols, G.J., Khalil, S., Darwish, M. and Bosworth, W. (this volume) Extensional tectonics and sedimentation, eastern Gulf of Suez, Egypt, in *Sedimentation and Tectonics of Rift Basins: Red Sea–Gulf of Aden* (eds

B.H. Purser and D.W.J. Bosence), Chapman & Hall, London, pp. 223–238.

McCombe, D.A., Fernete, G.L. and Alawi, A.J. (Compilers) (1994) The geological and mineral resources of Yemen. Ministry of Oil and Mineral Resources, Geological Survey and Minerals Board, Yemen Mineral Sector Project, Technical Report.

McCrea, J.M. (1950) On the isotope chemistry of carbonates and a palaeotemperature scale. *Journal of Chemistry and Physics*, **18**, 849–857.

McGuire, A.V. and Bohannon, R.G. (1989) Timing of mantle upwelling: evidence for a passive origin for the Red Sea rift. *Journal of Geophysical Research*, **94**, 1677–1682.

McIntyre, A. (1969) The Coccolithophorida in Red Sea sediments, in *Hot brines and Heavy Metal Deposits in the Red Sea*, (eds E.T. Degens and D.A. Ross), Springer-Verlag, New York, pp. 299–305.

McKenzie, D.P. (1978) Some remarks on the development of sedimentary basins. *Earth and Planetary Science Letters*, **40**, 25–32.

McKenzie, D.P., Davies, D. and Molnar, P. (1970) Plate tectonics of the Red Sea. *Nature*, **226**, 243–249.

McKenzie, D.P. and O'Nions, R.K. (1991) Partial melt distributions from inversion of rare earth element concentrations. *Journal of Petrology*, **32**, 1021–1091.

Meneisy, M.Y. (1986) Mesozoic igneous activity in Egypt. *Qatar University Science Bulletin*, **6**.

Meneisy, M.Y. (1990) Vulcanicity, in *The Geology of Egypt*, (ed. R. Said), Balkema, Rotterdam, pp. 157–172.

Menzies, M., Baker, J., Al Kadasi, M., Snee, L., Hurford, and Yelland, A. (1995) *Evolution of a Volcanic Margin: Relative and Absolute Timing of Surface Uplift, Exhumation Magmatism and Extension in Western Yemen*. Conference in Rift Sedimentation and Tectonics in the Red Sea–Gulf of Aden Region, 23–30 October 1995, Sana'a, Yemen Abstract volume.

Menzies, M., Baker, J., Bosence, D., Dart, C., Davison, I. *et al.* (1992) The timing of magmatism, uplift and crustal extension – preliminary observations from Yemen, in *Magmatism and continental breakup*, (ed. B. Storey), Geological Society of London, Special publication 68, pp. 293–304.

Menzies, M.A., Baker, J., Chazot, G. *et al.* (1966) Evolution of the Red Sea Volcanic Margin, Western Yemen. In *Large Igneous Provinces* (eds J. Mahoney and M. Coffin) American Geophysical Union Monograph.

Menzies, M., Bosence, D., El-Nakhal, H., Al-Khirbash, S., Al Kadasi, M. and Al Subbary, A. (1991) Lithospheric extension and the opening of the Red Sea: sediment–basalt relationship in Yemen. *Terra Nova*, **2**, 340–350.

Menzies, M.A., Gallagher, K., Yelland, A., *et al.* (1997) Red Sea and Gulf of Aden rifted margins, Yemen: denudational histories and margin evolution. *Geochimica et Cosmochimica Acta* (submitted).

Menzies, M., Gallagher, K., Yelland, A. and Hurford, A. (1996) Exhumation and crustal cooling history of the Yemen rift-flanks: regional fission track data from the southern Red Sea and Gulf of Aden margins. Manuscript submitted.

Menzies, M.A., Al Kadasi, M., Al Khirbash, S. *et al.* (1994a) Geology of the Republic of Yemen, In *The Geology and Mineral Resources of Yemen*, (eds D. McCombe, G.L. Fernette and A.J. Alawi), Ministry of Oil and Mineral Resources. Geological Survey and Minerals Exploration Board, Yemen Mineral Sector Project, *Technical report*, pp. 21–48.

Menzies, M.A., Yelland, A., Baker, J. *et al.* (1994b) Evolution of the Red Sea volcanic margin – a multi-isotopic approach. *United States Geological Survey Circular*, **1107**, p. 216.

Mergner, H. and Schuhmacher, H. (1974) Morphologie, okologie und zonierung von korallaentiffen bei Aqaba (Golf von Aqaba, Rotes Meer). *Helgoländer wissenschaftliche Meeresuntersuchungen*, **26**, 238–358.

Merla, G., Abbate, E., Azzaroli, A., Bruni, P., Canuti, P., et al. (1979) A geological map of Ethiopia and Somalia (1973) 1:200,000 and comment with a map of major landforms. Consiglio Nazionale delle Ricerche, Roma.

Meshref, W.M. (1990) Tectonic framework in *The Geology of Egypt*, (ed. R. Said), Balkema, Rotterdam, pp. 113–155.

Metwalli, M.H., Philip, G. and El Sayed, A.A.Y. (1978) El Morgan oil field as a major fault blocks reservoir masked by the thick Miocene salt, a clue for deeper reserves of hydrocarbons in Gulf of Suez petroleum province. *Acta Geologica Polonica*, **28**, 389–413.

Miall, A.D. (1977) A review of the braided-river depositional environment. *Earth Science Reviews*, **13**, 1–62.

Miall, A.D. (1978) Fluvial sedimentology. *Canadian Society Petroleum Geologists Memoir*, Calgary, pp. 597–604.

Miall, A.D. (1988) Architectural elements and bounding surfaces in fluvial deposits: anatomy of the Kagenta Formation (Lower Jurassic), Southwest Colorado. *Sedimentary Geology*, **55**, 233–262.

Mienis, H.K. (1971) Cypraeidae from the Sinai area of the Red Sea. *Argamon*, **2**, 13–43.

Mienis, H.K. (1981) Note on the identity of *Fissurella impedimentum* Cooke, 1885 (Prosobranchia: Fissurellidae). *Journal of Conchology*, **30**, 303–304.

Migliorini, C. (1936) *Documenti relativi alle ricerche nelle Isole Dahlac, nel bassopiano eritreo, nell'Ogadèn ed in Somalia.* A.G.I.P. mineraria, (unpublished).

Milani, E.J. and Davidson, I. (1988) Basement control and transfer tectonics in the Reconcavo-Tucano-Jatob rift, northeast Brazil. *Tectonophysics*, **154**.

Miller, A.R. (1964) High salinity in sea water. *Nature* **203**, 590.

Miller, A.R., Densmore, C.D., Degens, E.T., Hathaway J.C., Manheim F.T. *et al.* (1966) Hot brines and recent iron deposits in deeps of the Red Sea. *Geochimica et Cosmochimica Acta*, **30**, 341–359.

Miller, P.M. and Barakat, H. (1988) Geology of the Safaga Concession, northern Red Sea, Egypt. *Tectonophysics*, **153**, 123–36.

Milliman, J.D. (1977) Interstitial waters of late Quaternary Red Sea sediments and their bearing on submarine lithification. *Kingdom of Saudi Arabia, Red Sea Research 1970–75, Mineral Resources Bulletin*, 22-M, i–iii + M1–M6.

Milliman, J.D., Ross, D.A. and Ku, T.-L. (1969) Precipitation and lithification of deep-sea carbonates in the Red Sea. *Journal of Sedimentary Petrology*, **39**, 724–736.

Ministry of Oil and Mineral Resources (1992) Geological map of Republic of Yemen, scale 1:1,000,000, Mineral Exploration Board, Sana'a, Yemen.

Mitchell, D.W.J., Allen, R.B., Salama, W and Abouzakm, A. (1992) Tectonostratigraphic framework and hydrocarbon potential of the Red Sea. *Journal of Petroleum Geology*, **15**, 187–210.

Mohr, P.A. (1975) Structural setting and evolution of Afar, in *Afar between Continental and Oceanic Rifting*, (eds A. Pilger and A. Roesler), Schweitzerbart, Stuttgart (I), pp. 27–37.

Mohr, P.A. (1991) Structure of Yemeni dike swarms, and emplacement of coeval granite plutons. *Tectonophysics*, **198**, 203–221.

Mohriak, W.U., Mellow, M.R., Karner, G.D., Dewey, J.F. and Maxwell, J.R. (1989) Structural and stratigraphic evolution of the Campos Basin, offshore Brazil, in *Extensional Tectonics and Stratigraphy of the North Atlantic Margins*, (eds A.J. Tankard and H.R. Balkwill), American Association Petroleum Geologists, Tulsa, Memoir 46, pp. 577–598.

Moltzer, J.G. and Binda, P.L. (1981) Micropaleontology and palynology of the middle and upper parts of the Shumaysi Formation, Saudi Arabia. *Bulletin of the Faculty of Science, King AbdulAziz University*, **4**, 57–76.

MOMR – Ministry of Oil and Mineral Resources (1990) Magnetic anomaly map of Yemen, contour map of residual total magnetic intensity, scale 1:1.000.000, The Natural Resources Project, Sana'a.

Monnin, C. (1989) Thermodynamique des eaux naturelles concentrées: solubilité des minéraux à haute température et haute pression et coefficients d'activité des éléments en trace. Thèse Doctoral Etat. UPS, Toulose.

Montadert, L., De Charpal, O., Roberts, D.G., Geunnoc, P. and Sibuet, J. (1979) Northeast Atlantic passive continental margins: rifting and subsidence processes, in: *Deep Drilling Results in the Atlantic Ocean: Continental Margins and Paleoenvironment* (eds M. Talwani, W. Hey and W.B.F.) American Geophysical Union, pp. 154–186.

Montaggioni, L., Behairy, A.K.A., El-Sayed, M.K. and Yusuf, N. (1986) The modern reef complex, Jeddah area, Red Sea: a facies model for carbonate sedimentation on embryonic passive margins. *Coral Reefs*, **5**, 127–350.

Montaggioni, L. and Bouchon, C. (1984) Morphogenesis of a modern knoll reef platform, Saudi Arabian Shelf Margin, Jeddah area, Red Sea. *5th European Regional Meeting International Association of Sedimentologists Marseilles*, Abstracts 292.

Montaggioni, L. and Camoin, G.F. (1993) Stromatolites associated with coralgal communities in Holocene high-energy reefs. *Geology*, **21**, 724–736.

Montanaro Gallitelli, E. (1939) Ricerche sull' età di certi depositi insulari e marginali del Mar Rosso (Isole Dahlak, Penisola di Buri, dintorni di Massaua fino al Sahel). *Palaeontographia Italica*, **39**, 215–258.

Montanaro Gallitelli, E. (1941) Foraminiferi, posizione stratigrafica e facies di un calcare a 'operculina' dei Colli di Ebud (Sahel Eritreo). *Palaeontographia Italica*, **40** (n.s. 10), 67–75.

Montanaro Gallitelli, E. (1943) Coralli costruttori delle scogliere emerse di Massaua e Gibuti. Missione geologica nella Dancalia meridionale e nell' Hararino promossa dall'A.G.I.P. *Reale Accademia Italiana*, **4**, 1–76.

Montanaro Gallitelli, E. (1972) Ricerche eseguite negli anni 1941–1942 sopra i coralli delle scogliere emerse di Massaua e Gibuti. *Documentazioni Paleontologiche dell'Istituto di Paleontologia dell'Università di Modena*, **IV**, 447–519.

Montanaro Gallitelli, E. (1973) Ricerche eseguite ngli anni 1941–1942 sopra i Coralli delle scogliere emerse di Massaua e Gibuti. Accademia Nazionale dei Lincei. Missione geologica dell'AGIP nella Dancalia meridionale e sugli Altipiani Ararini 1936–1938. *Documentazione Paleontologica, Roma*, **4**, 447–519.

Montenat, C. (1986a) Un aperçu des industries préhistoriques du Golfe de Suez et du littoral égyptien de la Mer Rouge. *Bulletin Institut Français d'Archéologie orientale*, Cairo, **86**, 239–255.

Montenat, C. (ed) (1986b) Etudes géologiques des rives du Golfe de Suez et de la Mer Rouge nord-occidentale. Evolution tectonique et sédimentaire d'un rift néogène. *Documents et Travaux, Institut Géologique Albert de Lapparent*, Paris, **10**, p. 192, 12 plates.

Montenat, C., Angelier, J., Beaudouin, B., Bolze, J., Burollet, P.-F., et al. (1990) La marge occidentale de la Mer Rouge au nord de Port Soudan. *Bulletin Societe géologique France*, (8) **VI**, 435–446.

Montenat, C., Burollet, P., Jarrige, J.J., Ott d'Estevou, P. and Purser, B. (1986c) La succession des phénomènes tectoniques et sédimentaires néogènes sur les marges du rift de Suez et de la Mer Rouge nord occidentale. *Comptes Rendues Acadamie de Science Paris*, **303**, II, 213–218.

Montenat, C., Orszag-Sperber, F., Ott d'Estevou, P., Purser, B.H. and Richert, J.P. (1986d) Etude d'une transversale de la marge occidentale de la Mer Rouge: le secteur de Ras Honkorab-Abu Ghusun, in *Geological Studies of the Gulf of Suez, the Northwestern Red Sea Coasts, Tectonic and Sedimentary Evolution of a Neogene Rift*, (ed. C.P. Montenat), Documents et Travaux Institut Geologique Albert de Lapparent, **10**, pp. 145–170.

Montenat, C., Orszag-Sperber, Plaziat, J-C. and Purser, B.H (this volume) The sedimentary record of the initial stage of Oligo-Miocene rifting in the Gulf of Suez and northern Red Sea, in *Sedimentation and tectonics of rift basins: Red Sea–Gulf of Aden* (eds B.H. Purser and D.W.J. Bosence), Chapman & Hall, London, pp. 146–161.

Montenat, C., Ott d'Estevou, P., Jarrige, J.J. and Richert, J.P. (this volume) Rift development in the Gulf of Suez and NW Red Sea: Structural aspects and related sedimentary processes, in *Sedimentation and tectonics of rift basins: Red Sea–Gulf of Aden* (eds B.H. Purser and D.W.J. Bosence), Chapman & Hall, London, pp. 97–116.

Montenat, C.P., Ott d'Estevou, P. and Purser, B.H. (1986e) Tectonic and sedimentary evolution of the Gulf of Suez and the north-western Red Sea: a review, in *Geological Studies of the Gulf of Suez, the Northwestern Red Sea Coasts, Tectonic and Sedimentary Evolution of a Neogene Rift* (ed.

C.P. Montenat). Documents et Travaux Institut Geologique Albert de Lapparent, 10, pp. 7–18.

Montenat, C., Ott d'Estevou, P., Purser, B.H., Burollet, P.F., Jarrige, J.J. et al. (1988) Tectonic and sedimentary evolution of the Gulf of Suez and the northwestern Red Sea. Tectonophysics, 153, 161–177.

Monty, C.L.V., Rouchy, J-M., Maurin, A., Bernet-Rollande, M.C. and Perthuisot, J.P. (1987) Reef–stromatolites–evaporites facies relationships from the Middle Miocene: examples of the Gulf of Suez and the Red Sea, in Evaporite Basins (ed. T.M. Peryt), Springer-Verlag, Heidelberg, pp. 75–86.

Moon, F.W. (1923) Preliminary Geological Report on Saint John's Island (Red Sea), Geological Survey of Egypt, Government Press, Cairo.

Moon, F.W. and Sadek, H. (1923) Preliminary geological report on Wadi Gharandal area. Petroleum Resources Bulletin, 9, 40pp.

Mooney, W., Gettings, M.E., Blank, H.R. and Healy, J.H. (1985) Saudi Arabian seismic deep refraction profile: a traveltime interpretation of crustal and upper mantle structure. Tectonophysics, 111, 173–246.

Moore, T.A. and Al-Rehaili, M.H.A (1989) Geologic map of the Makkah quadrangle sheet 21D, Kindom of Saudi Arabia, Ministry of Petroleum and Mineral Resources, Jeddah.

Morcos, S.A. (1970) Physical and chemical oceanography of the Red Sea. Marine Biology Annual Review, 18, 73–202.

Moretti, I. and Chénet, P.Y. (1987) The evolution of the Suez rift: a combination of stretching and secondary convection. Tectonophysics, 133, 229–234.

Moretti, I. and Colletta, B. (1987) How does a block tilt: the Gebel Zeit example, Gulf of Suez. Journal of Structural Geology, 100, 9–19.

Morgan, P. (1990) Egypt in the tectonic framework of global tectonics, in The Geology of Egypt (ed. R. Said), Balkema, Rotterdam, pp. 91–111.

Morgan, P., Boulos, F.K., Hennin, S.F., El Sherif, A.A., El Sayed, A.A. et al. (1985) Heat flow in eastern Egypt, the thermal signature of a continental break up. Journal of Geodynamics, 4, 107–131.

Morley, C.K. (1995) Developments in the structural geology of rifts over the last decade and their impact on hydrocarbon exploration, in Hydrocarbon Habitat in Rift Basins (ed. J.J. Lambiase), Special Publication Geological Society of London, 80, pp. 1–32.

Morely, C.K., Nelson, R.A., Patton, T.L. and Munn, S.G. (1990) Transfer zone in the East African rift system and their relevance in the hydrocarbon exploration in rifts. American Association Petroleum Geologists Bulletin, 74, 1234–1253.

Morely, C.K., Wescott, W.A., Stone, D.M., Harper, R.M., Wigger, S.T. and Karanja, F.M. (1992) Tectonic evolution of the northern Kenyan Rift. Journal of the Geological Society of London, 149, 333–48.

Moses, C.O., Nordstrom, D.K., Herman, J.S. and Mills, S.L. (1987) Aqueous pyrite oxydation by dissolved oxygen and by ferric iron. Geochimica et Cosmochimica Acta, 51, 1561–1571.

Moustafa, A.M. (1976) Block faulting of the Gulf of Suez. Proceedings of 5th EGPC Exploration Seminar, Cairo, 19 pp.

Moustafa, A.R. and Fouda, H.G. (1988) Gebel Sufr el Dara accommodation zone, southwestern part of the Suez rift. Ain Shams University Earth Science Series, 2, 227–39.

M'Rabet, A., Purser, B.H. and Soliman, M. (1991) Diagenèsse comparée de récifs coralliens actuels et quaternaires de la côte égyptienne de la Mer Rouge, Géologie Méditerranéenne, 16 (2-3), 1989, 5–39.

Mulder, C.J., Lehner, P. and Allen, D.C.K. (1975) Structural evolution of the Neogene salt basins in the eastern Mediterranean and the Red Sea. Geologie en Mijnbouw, 54, 208–221.

Mullins, H.T. (1985) Modern deep-water carbonates along the Blake–Bahama boundary, in Deep-water Carbonates: Buildups, Turbidites, Debris Flows and Chalks, (eds P.D. Crevello and P.M. Harris), SEPM Core Workshop No. 6, New Orleans, pp. 461–490.

Murowchick, J.B. and Barnes, H.L. (1986) Marcasite precipitation from hydrothermal solutions. Geochimica et Cosmochimica Acta, 50, 2615–2630.

Murray, J.W. (1979) Iron oxides, in Marine Materials, (ed. P.J. Ribbe), American Mineralogical Society, Washington, USA, pp. 47–98.

Murray, J.W. (1991) Ecology and Paleoecology of Benthic Foraminifera, Longman, Harlow.

Nardini, S. (1933) Paleontologia della Somalia. VI. Fossili del Pliocene e del Pleistocene, 5 Molluschi marini e continentali del Pleistocene della Somalia. Palaeontographia italica, 32, Supplement 1, 169–192.

Nardini, S. (1934) Molluschi delle spiagge emerse del Mar Rosso e dell'Oceano Indiano. Introduzione e Parte I (Gasteropodi). Palaeontographia Italica, 34 (n.ser.), 171–267.

Nardini, S. (1937) Molluschi delle spiagge emerse del Mar Rosso e dell'Oceano Indiano. Parte II (Lamellibranchi). Palaeontographia Italica, 37 (n.ser), 99–152.

National Stratigraphic Sub-Committee of The Geological Sciences of Egypt (NSSC), (1974) Miocene rock stratigraphy of Egypt. Egyptian Journal of Geology, 18, 1–69.

Natterer, K. (1898) Expedition S.M. Schiff 'Pola' in das Rote Meer Noerdliche Halfte (Oktober 1895–May 1896). Denkschriften Akademie der Wissenschaften in Wien, Mathematisch-Naturwissenschaftliche Klasse, Wien, 65, 445–572.

Neev, D. and Emery, K.O. (1967) The Dead Sea: depositional processes and environment of evaporites. Bulletin Geological Survey of Israel, 41, 1–147.

Negretti, B., Philippe, M., Soudet, H.J., Thomassin, B.A. and Oggiano, G. (1990) Echinometra micocenica Loriol, échinide miocène synonyme d'Echinometra mathaei (Blainville) actuel: biogéographie et paléoécologie. Géobios, 23, 445–454.

Nelson, R.A., Patton, T.L. and Morley, C.K. (1992) Rift-segment interaction and its relation to hydrocarbon exploration in continental rift systems. American Association Petroleum Geologists Bulletin, 76, 1153–1169.

Nesteroff, W.D. (1955) Les récifs coralliens du Banc Farasan Nord (Mer Rouge): Résultats Scientifiques des Campagnes de la 'Calypso', I. Campagne en Mer Rouge. Masson et Cie, Paris.

Nesteroff, W.D. (1960) Age des derniers mouvements du graben de la mer Rouge déterminé par la méthode du C14 appliquée aux récifs fossiles. *Bulletin de la Société Géologique de France*, **20**, 415–418.

Newton, R.B. (1899) On some Pliocene and Pleistocene shells from Egypt. *Geological Magazine*, **6**, 402–407.

Newton, R.B. (1900) Pleistocene shells from the raised deposits of the Red Sea. *Geological Magazine*, **7**, 500–519, 544–560.

Nichols, G.J. and Daly, M.C. (1989) Sedimentation in an intracratonic extensional basin: the Karoo of the central Morondoava Basin, Madagascar. *Geological Magazine*, **126**, 339–354.

Nicholson, K. (1992) Contrasting minerological-geochemical signatures of manganese oxides: guides to metallogenesis. *Economic Geology*, **87**, 1253–1264.

Nio, S-D. (1976) Marine transgressions as a factor in the formation of sandwave complexes. *Geologie en Mijnbouw*, **55**, 18–40.

Nir, Y. (1971) Geology of Entedebir Island and its recent sediments, Dahlak Archipelago – southern Red Sea. *Israel Journal of Earth-Sciences*, **20**, 13–40.

Nottvedt, A., Gabrielsen, R.H. and Steel, R.J. (1995) Tectonostratigraphy and sedimentary architecture of rift basins, with reference to the northern North Sea. *Marine and Petroleum Geology*, **12**, 881–901.

Ohmoto, H. and Lasaga, A.C. (1982) Kinetics of reactions between aqueous sulfates and sulfides in hydrothermal systems. *Geochimica et Cosmochimica Acta*, **46**, 1237–1247.

Ohmoto, H. and Rye, R.O. (1979) Isotopes of sulfur and carbon, in *Geochemistry of Hydrothermal Ore Deposits*, (ed. H.L. Barnes), 2nd edn, J. Wiley & Sons, New York, pp. 509–567.

Okada, H. and Honjo, S. (1975) Distribution of Coccolithophores in marginal seas along the Western Pacific Ocean and in the Red Sea. *Marine Biology*, **31**, 271–285.

Olafsson, I., Sundvor, E., Eldholm, O. and Grue, K. (1991) Møre Margin: crustal structure from analysis of expanded spread profiles. *Marine Geophysical Research*, **14**, 137–162.

Olausson, E. (1971) Quaternary correlations and the geochemistry of oozes, in *The Micropaleontology of Oceans* (ed. B.M. Funnell), Cambridge University Press, pp. 375–398.

Oliver, P.G. (1992) *Bivalved seashells of the Red Sea*. Hemmen, Wiesbaden.

Ollier, C.D (1985) Morphotectonics of continental margins with great escarpments, in *Tectonic Geomorphology*, (eds M. Morisawa and J.T. Hack), Allen and Unwin, Boston, Massachusetts, pp. 3–25.

O'Malley, J. (1971) *Cowries of the Jeddah-Red Sea Area*, (Supplement). Of Sea and Shore Publishers, Port Gamble, Washington, USA.

Omar, G.I and Steckler, M.S. (1995) Fission track evidence on the initial rifting of the Red Sea: two pulses, no propagation. *Science*, **270**, 1341–1344.

Omar, G.I., Steckler, M.S., Buck, R. and Kohn, B.P. (1989) Fission-track analysis of basement apatites at the western margin of the Gulf of Suez rift, Egypt: evidence for synchroneity of uplift and subsidence. *Earth and Planetary Science Letters*, **94**, 316–328.

Oosterbaan, A.F.F. (1988) Early Miocene corals from the Aquitaine Basin (SW France). *Mededelingen van de Werkgroep voor Tertiaire en Kwartaire Geologie*, **25**, 247–284.

Oosterbaan, A.F.F. (1990) Notes on a collection of Badenian (Middle Miocene) corals from Hungary in the National Museum of Natural History at Leiden (The Netherlands). *Contributions Tertiary Quaternary Geology*, **27**, 3–15.

Orszag-Sperber, F., Freytet, P., Montenat, C. and Ott d'Estevou, P. (1986) Métasomatose sulfatée de dépôts carbonatés marins miocènes sur la marge occidentale de la Mer Rouge. *Compte Rendus de l'Académie des Sciences*, Paris, **302**, 1079–1084.

Orszag-Sperber, F., Harwood, G., Kendall, A. and Purser, B.H. (this volume a) A review of the evaporites of the Red Sea–Gulf of Suez rift, in *Sedimentation and Tectonics of Rift Basins: Red Sea–Gulf of Aden* (eds B.H. Purser and D.W.J. Bosence), Chapman & Hall, London, pp. 409–428.

Orszag-Sperber, F. and Plaziat, J.C. (1990) La sédimentation continentale (Oligo-Miocène) des fossés du proto-rift du NW de la Mer Rouge. *Bulletin de la Société Géologique de France* (**8**), VI, 3, 385–396.

Orszag-Sperber, F., Plaziat, J.C. and Purser, B.H. (1994) Tectonique et confinement: les dépôts évaporitiques miocènes associés au rifting de la Mer Rouge et du Golfe de Suez (Eqypte). *Comptes Rendus de l'Académie des Sciences*, Paris, **318**, II, 123–129.

Orszag-Sperber, F., Purser, B.H., Rioul, M. and J. -C. Plaziat (this volume b) Post Miocene sedimentation and rift dynamics in the southern Gulf of Suez, in *Sedimentation and Tectonics of Rift Basins: Red Sea–Gulf of Aden* (eds B.H. Purser and D.W.J. Bosence), Chapman & Hall, London, pp. 429–448.

Osman, S.H. (1992) Miocene alluvial to marine facies transition of the Mait group in the Bosaso basin (Migiurtania, Northern Somalia). *Giornale di Geologia*, **54**, 67–76.

Oswald, E.J., Meyers, W.J. and Pomar, L. (1990) Dolomitization of an upper Miocene reef complex, Mallorca, Spain: evidence for a Messinian dolomitizing Mediterranean sea. *American Association of Petroleum Geologists Bulletin*, **74**, 735 (abstract).

Oswald, E.J., Schoonen, M.A.A. and Meyers, W.S. (1991) Dolomitising seas in evaporitic basins: A model for pervasive dolomitization of Upper Miocene reefal carbonates in the western Mediterranean. *American Association of Petroleum Geologists Bulletin*, **75**, 649 (abstract).

Ott d'Estevou, P., Bolze, J. and Montenat, C. (1986a) Etude géologique de la marge occidentale du Golfe de Suez: le Massif des Gharamul et le Gebel Dara. *Documents et Travaux, Institut Géologique Albert de Lapparent*, **10**, 19–44.

Ott d'Estevou, P., Jarrige, J.J., Montenat, C., Prat, P., Richert, J.P. and Thiriet, J.P. (1987) Principaux aspects de l'évolution structurale du rift de Suez et de la Mer Rouge Nord-Occidentale (Egypte). *Notes et Mémoires, Total-CFP*, **21**, 167–197.

Ott d'Estevou, P., Jarrige, J.J., Icart, JC., Montenat, C., Henry, C. and Cravatte, J. (1986b) Observations structurales dans le secteur d'Abu Rudeis, marge occidentale du Golfe de Suez. *Documents et Travaux, Institut Géologique Albert de Lapparent*, **10**, 75–92.

Ott d'Estevou, P., Jarrige, J-J., Montenat, C. and Richert, J-P. (1989a) Succession des palaeochamps de contrainte et évolution du rift du Golfe de Suez et de la Mer Rouge nord-occidentale. *Bulletin des Centres de Recherche Exploration-Production d'Elf Aquitaine*, **13**, 297–318.

Ott d'Estevou, P., Jarrige, J.J., Montenat, C. and Richert, J.P. (1989b) Tectonic evolution and successive stages of palaeostress of the Gulf of Suez. Northern Red Sea rift. *Proceedings Symposium on the Afro-Arabian Rift System*, Karlsruhe (FRG), March 1989, Abstract.

Ouchi, S. (1985) Response of alluvial rivers to slow active tectonic movement. *Bulletin of the Geological Society of America*, **96**, 504–515.

Ouda, K.H. and Massoud, M. (1993) *Sedimentation history and geological evolution of the Gulf of Suez during the Late Oligocene–Miocene*, Geological Society of Egypt, Special Publication 1, pp. 48–89.

Oudin, E., Thisse, Y. and Ramboz, C. (1984) Fluid inclusion and mineralogical evidence for high temperature saline hydrothermal circulation in the Red Sea metalliferous sediments: preliminary results. *Marine Mining*, **5**, 3–31.

Ourisson, G., Albrecht, P. and Rohmer, M. (1979) The hopanoids: palaeochemistry and biochemistry of a group of natural products. *Pure and Applied Chemistry*, **51**, 709–729.

Owen, G. (1987) Deformation processes in unconsolidated sands, in *Deformation of Sediments and Sedimentary Rocks*, (eds M.E. Jones and R.M.L. Preston), Geological Society London Special Publication 29, pp. 11–24.

Pallister, J.S. (1987) Magmatic history of Red Sea rifting: perspective from the central Saudi Arabian coastal plain. *Geological Society of America Bulletin*, **98**, 400–417.

Pallister, J.S., Stacey, J.S., Fischer, L.B. and Premo, P.R. (1988) Precambrian ophiolites of Arabia: geologic settings, U-Pb geochronology, Pb-isotope characteristics, and implications for continental accretion. *Precambrian Research*, **38**, 1–54.

Papatheodorou, G. and Ferentinois, G. (1993) Sedimentation processes and basin-filling depositional architecture in an active asymmetric graben: Strava graben, Gulf of Corinth, Greece. *Basin Research*, **5**, 235–253.

Patton, T.L., Moustafa, A.R., Nelson, R.A. and Abdine, S.A. (1994) Tectonic evolution and stratigraphic setting of the Suez rift, in *Interior Rift Basins* (ed S.A. Landon), American Association of Petroleum Geologists Special Publication **59**, pp. 7–55.

Pautot, G. (1983) Les fosses de la Mer Rouge: approche géomorphologique d'un stade initial d'ouverture océanique réalisée à l'aide du Seabeam. *Oceanologica Acta*, **6**, 235–244.

Pautot, G., Guennoc, P., Coutelle, A. and Lyberis, N. (1984) Discovery of a large brine deep in the northern Red Sea. *Nature*, **310**, 5973, 133–136.

Pautot, G., Guennoc, P., Coutelle, A. and Lyberis, N. (1986) La dépression axiale du segment N Mer Rouge (de 25°N à 28°N): nouvelles données géologiques et géophysiques obtenues au cours de la campagne Transmerou 83. *Bulletin de la Societé Géologique de France* (**8**), II, 3, 381–399.

Pedersen, T.F. and Price, N.B. (1982) The geochemistry of manganese carbonate in Panama Basin sediments. *Geochimica et Cosmochimica Acta*, **46**, 59–68.

Perrin, C., Bosence, D.W.J. and Rosen, B. (1995) Quantitative approaches to palaeozonation and palaeobathymetry of corals and coralline algae in Cenozoic reefs, in *Marine Palaeoenvironment Analysis from Fossils*, (eds D.W.J. Bosence and P.A. Allison), Geological Society of London, Special Publication **83**, pp. 181–230.

Perrin, C., Plaziat, J.C. and Rosen, B.R. (this volume) Miocene coral reefs and corals of the SW Gulf of Suez and NW Red Sea: distribution, diversity and regional environmental controls, in *Sedimentation and Tectonics of Rift Basins: Red Sea–Gulf of Aden* (eds B.H. Purser and D.W.J. Bosence), Chapman & Hall, London, pp. 296–319.

Perry, S.K. (1986) Structural geometry and tectonic evolution of the southwestern Gulf of Suez. PhD thesis, University of South Carolina, Columbia.

Perry, S.K. and Schamel, S. (1990) The role of low-angle normal faulting and isostatic response in the evolution of the Suez rift, Egypt. *Tectonophysics*, **174**, 159–173.

Petit-Maire, N. (1989) Interglacial environments in presently hyperarid Sahara: palaeoclimatic implications, in *Paleoclimatology and Paleometerology: modern and past atmospheric transport*, (eds M. Leiney and M. Sarathein), Kluwer, pp. 637–661.

Petit-Maire, N., Reyss, J.L. and Fabre, J. (1994) Un paléolac du dernier interglaciaire dans une zone hyperaride du Sahara Malien (23°N). *Comptes Rendus Academie Science Paris*, **319**, 805–809.

Philobbos, E.R. and El Haddad, A.A. (1983) Tectonic control on Neogene sedimentation along the Egyptian part of the Red Sea coastal area, *5th International Conference on Basement tectonics*, Cairo, (Abstract).

Philobbos, E.R., El Haddad, A.A., Luger, P., Bekir, R. and Mahran, T. (1993) Syn-rift Miocene sedimentation around fault-blocks in the Abu Ghusun-Wadi el Gemal area, Red Sea, Egypt, in *Geodynamics and Sedimentation of the Red Sea–Gulf of Aden Rift System*, (eds E.R. Philobbos and B.H. Purser), Geological Society of Egypt, Special Publication **1**, pp. 115–142.

Philobbos, E.R., El Haddad, A.A. and Maharan, T.M. (1989) Sedimentology of syn-rift upper Miocene(?)–Pliocene sediments of the Red Sea area: a model from the environs of Mersa Alam, Egypt. *Egyptian Journal of Geology*, **33**, 1–2, 201–227.

Pickard, N.A.H., Rees, J.G., Strogen, P., Somerville, I.D. and Jones, G. Ll. (1994) Controls on the evolution and demise of Lower Carboniferous carbonate platforms, northern margin of the Dublin Basin, Ireland. *Geological Journal*, **29**, 93–117.

Pickering, K.T., Hiscott, R.N. and Hein, F.J. (1989) *Deep marine environments*, Unwin Hyman, London, 416 pp.

Pierre, C. and Rouchy, J.M. (1988) Carbonate replacements after sulphate evaporites in the Middle Miocene of Egypt. *Journal of Sedimentary Petrology*, **58**, 446–456.

Piller, W.E. and Pervesler, P. (1989) The northern bay of Safaga (Red Sea, Egypt): an actuopaleontological approach

I. Topography and bottom facies. *Beiträge zur Paläontologie von Österreich*, **15**, 103–147.

Pisias, N.G., Martinsen, D.G., Moore T.G. *et al.* (1984) High resolution stratigraphic correlation of benthic oxygen isotopic records spanning the last 300,000 years. *Marine Geology*, **56**, 119–136.

Plafker, G., Agar, R., Asker, A.H. and Hanif, M. (1987) Surface effects and tectonics of the 13th December 1982 Yemen earthquake. *Bulletin of the Seismological Society of America*, **77**, 2018–2037.

Plassche, O. van de (ed.) (1989) *Sea-level Research: a Manual for the Collection and Evaluation of Data*, Geo Books, Norwich.

Platel, J-P. and Roger, J. (1989) Evolution geodynamique du Dhofar (Sultanat d'Oman) pendant le Cretace et le Tertiare en relation avec l'ouverture du golfe d'Aden. *Bulletin de Societe geologique du France*, **8**, 253–263.

Playford, P.E. and Lowry, D.C. (1966) *Devonian reef complexes of the Canning Basin, Western Australia*, Western Australia Geological Survey, Bulletin **118**.

Plaziat, J.-C. (1989) Signification écologique et paléogéographique des peuplements oligotypiques de *Potamides* (Gastéropodes thalassiques). *Atti. 3e simpos. ecol. e paleoecol. della Commita bentoniche, Taormina 1985*, pp. 25–52.

Plaziat, J.-C. (1991) Paleogeographic significance of the Cardium, Potamids and Foraminifera living in intra-continental salt lakes of North Africa (Sahara Quaternary, Egypt Present lakes). *Journal of African Earth Sciences*, **12**, 383–389.

Plaziat, J.-C. (1993) Modern and fossil Potamids (Gastropoda) in saline lakes. *Journal of Paleolimnology*, **8**, 163–169.

Plaziat, J.-C. (1994) From aeolian sand deposits to deep-sea fan gravity flows, dish and pillar structure characterise seismites. *Proceedings 14th International Sedimentology Congress, Recife, F*, Abstracts, pp. 13–14.

Plaziat, J.-C. (1995) Modern and fossil mangroves and mangals: their climatic and biogeographic variability, in *Marine Palaeonvironmental Analysis from Fossils*, (eds D.W.J. Bosence and P.A. Allison), Geological Society Special Publication **83**, pp. 73–96.

Plaziat, J.-C. and Ahmamou, M. (1997) Les mécanismes à l'origine des séismites et leur signification tectonique dans le Pliocène du Saïss de Fès et de Meknès (Maroc). *Geodynamica Acta*, in press.

Plaziat, J.-C. and Ahmed, E.A. (1993) *Diversity of sedimentary expressions of major earthquakes: an example from the 'Pliocene' sandy seismites of the Egyptian Red Sea Coast*. Geological Society of Egypt, Special Publication **1**, pp. 277–294.

Plaziat, J.-C. Baltzer, F., Choukri, A., Conchon, O., Freytet, P. *et al.* (1995) Quaternary changes in the Egyptian shoreline of the NW Red Sea and Gulf of Suez. *Quaternary International, PIGC 274*, **29–30**, 11–22.

Plaziat, J.-C., Baltzer, F., Choukri, A., Conchon, O., Freytet, P. *et al.* (this volume) Quaternary marine and continental sedimentation of the N Red Sea and Gulf of Suez coast of Egypt. Influences of rift tectonics, climatic changes and sea-level fluctuations, in *Sedimentation and Tectonics of Rift Basins: Red Sea–Gulf of Aden*, (eds B.H. Purser and D.W.J. Bosence), Chapman & Hall, London, pp. 543–579.

Plaziat, J.-C., Guiraud, M., and Poisson, A. (1994) Les 'overturned cross-stratifications' correspondant à une déformation sismique: rôle de la pente sur la dissymétrie des structures. *Proceedings 15e Réunion-des Sciences de la Terre*, Nancy, p. 36.

Plaziat, J.-C., Montenat, C., Barrier, P., Janin, M.C., Orszag-Sperber, F. and Philobbos, E. (this volume) Stratigraphy of the Egyptian syn-rift; correlations between axial and peripheral sequences of the NW Red Sea and Gulf of Suez and their tectonic and eustatic controls, in *Sedimentation and tectonics of rift basins: Red Sea–Gulf of Aden* (eds B.H. Purser and D.W.J. Bosence), Chapman & Hall, London, pp. 211–222.

Plaziat, J.-C., Montenat, C., Orszag-Sperber, F., Philobbos, E.R. and Purser, B.H. (1990) Geodynamic significance of continental sedimentation during initiation of the NW Red Sea (Egypt), in 'Continental sediments of Africa', (eds Kogbe and J. Lang). *Journal of African Earth Sciences*, **10**, 355–360.

Plaziat, J.-C. and Purser, B.H. (1995) Prelithification recumbent-folding in the Nubian Sandstone of eastern Egypt: seismic versus current-drag deformation. *Proceedings of Rift sedimentation and tectonics in the Red Sea–Gulf of Aden region*, Sana'a, Abstracts, pp. 50–51.

Plaziat, J.-C. and Purser, B.H. (this volume) Tectonic significance of seismic sedimentary deformation structures within the syn- and post-rift deposits of the NW (Egyptian) Red Sea coast and Gulf of Suez, in *Sedimentation and Tectonics of Rift Basins: Red Sea–Gulf of Aden* (eds B.H. Purser and D.W.J. Bosence), Chapman & Hall, London, pp. 347–366.

Plaziat, J.-C, Purser, B.H. and Philobbos, E. (1990) Seismic deformation structures (seismites) in the syn-rift sediments of the NW Red Sea (Egypt). *Bulletin de la Société Géologique de France*, **8**(6), 419–434.

Plaziat, J.-C., Purser, B.H. and Soliman, M. (1990) Les rapports entre l'organisation des dépôts marins (Miocène inférieur et moyen) et la tectonique précoce sur la bordure NW du rift de mer Rouge. *Bulletin Société Géologique France*, **8**, VI, 397–418.

Plaziat, J.-C., Purser, B.H. and Soliman, M. (1991) Localisation et organisation interne de récifs coralliens immatures sur un cône alluvial du Quaternaire ancien de la Mer Rouge (Sud de l'Egypte). *Géologie Méditerranéene*, **16**, 41–59.

Plaziat, J.-C., Reys, J.-L., Choukri, A., Orszag-Sperber, F., Purser, B.H. and Baltzer, F (1996) U/Th chronology of the reef deposits of the Egyptian Red Sea and Gulf of Suez. *Proceedings 17th regional International Association of Sedimentologists meetings*, Sfax, Abstract, pp. 213–214.

Pomar, L. and Ward, W.C. (1994) Response of a late Miocene Mediterranean reef platform to high-frequency eustasy. *Geology*, **22**, 131–134.

Por, F.D. (1971) One hundred years of Suez Canal. A century of Lessepsian migration: retrospect and viewpoints. *Systematic Zoology*, **20**, 138–159.

Por, F.D. (1978) Lessepsian migration. The influx of the Red Sea Biota into the Mediterranean by way of the Suez canal. *Ecological Studies*, **23**, 1–228.

Por, F.D. (1989) *The Legacy of Tethys*, Monographiae Biologicae 63, Kluwer Academic Publishers, Dordrecht, Boston, London.

Pottorf, R.J. and Barnes, H.L. (1983) Mineralogy, geochemistry, and ore genesis of hydrothermal sediments from Atlantis II Deep, Red Sea. *Economic Geology Monograph*, **5**, 198–223.

Prat, P., Montenat, C., Ott d'Estevou, P. and Bolze J. (1986) La marge occidentale du Golfe de Suez d'après l'étude des Gebels Zeit et Mellaha. *Documents et Travaux Institut Géologique Albert de Lapparent*, Paris, **10**, 45–74.

Prell, W.L. and Van Campo, E. (1986) Coherent response of the Arabian sea upwelling and pollen transport to late Quaternary monsoonal winds. *Nature*, **323**, 526–528.

Prodehl, C. and Mechie, J. (1991) Crustal thinning in relationship to the evolution of the Afro-Arabian rift system: a review of seismic-refraction data. *Tectonophysics*, **198**, 311–327.

Prosser, S. (1991) Syn-rift sequences: their recognition and significance in basin analysis. PhD thesis, University of Keele, UK.

Prosser, S. (1993) Rift-related linked depositional systems and their seismic expression, in *Tectonics and Seismic Sequence Stratigraphy* (eds G.D. Williams and A. Dobb), Geological Society of London, Special Publication 71, pp. 35–66.

Purser, B.H. (ed.) (1973) *The Persian Gulf: Holocene Carbonate Sedimentation and Diagenesis in a Shallow Epicontinental Sea*, Springer-Verlag, Berlin.

Purser, B.H. (1987) Diagenése précoce des sédiments carbonatés Miocénes, NW mer Rouge. *Notes et Mémoires TOTAL, Compagnie Française des Pétroles*, **21**, 149–166.

Purser, B.H. (this volume) Syn-rift diagenesis of middle Miocene carbonate platforms in the northwestern Red Sea, Egypt, in *Sedimentation and Tectonics of Rift Basins: Red Sea–Gulf of Aden*, (eds B.H. Purser and D.W.J. Bosence), Chapman & Hall, London, pp. 369–390.

Purser, B.H., Aissaoui, D.M. and Orszag-Sperber, F. (1987a) Diagenesis and rifting: post-sedimentary evolution of Miocene carbonate sediments on the NW margin of the Red Sea, in *Genèse et évolution des bassins sédimentaires*, (eds J.L. Berthon, P.F. Burollet and P. Legrand), Notes et Mémoires TOTAL, Compagnie Française des Pétroles, Paris, pp. 145–166.

Purser, B.H., Bosence, D., Schroeder, H. and Taviani, M. (eds) (1993) *Comparative Sedimentology of the Red Sea–Gulf of Aden Rift System*. European Community Scientific report. Department of Geologie Universite de Paris-sud (Orsay).

Purser, B.H., Brown, A. and M'Rabet, A. (1994) Nature, origins and evolution of porosity in dolomites, in *Dolomites: a volume in honour of Dolomieu*, (eds B.H. Purser, M. Tucker and D. Zenger) International Association Sedimentologists, Special Publication **21**, pp. 283–308.

Purser, B.H. and Hötzl, H. (1988) The sedimentary evolution of the Red Sea rift: a comparison of the northwest (Egyptian) and northeast (Saudi Arabian) margins. *Tectonophysics*, **153**, 193–208.

Purser, B.H., Montenat, C., Orszag-Sperber, F., Ott d'Estevou, P., Plaziat, J.-C. and Philobbos, E. (this volume) Carbonate and siliciclastic sedimentation in an active tectonic setting, Miocene of the northwestern Red Sea rift, Egypt, in *Sedimentation and tectonics of Rift Basins: Red Sea–Gulf of Aden*, (eds B.H. Purser and D.W.J. Bosence), Chapman & Hall, London, pp. 239–270.

Purser, B.H., Orszag-Sperber, F. and Plaziat, J.-C. (1987) Sédimentation et rifting: Les séries néogènes de la marge nord-occidentale de la mer rouge. *TOTAL Notes et Mémoires*, **21**, 111–144.

Purser, B.H., Orszag-Sperber, F., Plaziat, J.-C. and Rioual, M. (1993) Plio-Quaternary tectonics and sedimentation in the SW Gulf of Suez and N. Red Sea, in *Geoscientific Research in Northeast Africa*, (eds U. Thorweihe and H. Schandelmeier), Balkema, pp. 259–262.

Purser, B.H. and Philobbos, E.R. (1993) The sedimentary expressions of rifting in the NW Red Sea, Egypt, in *Geodynamics and Sedimentation of the Red Sea–Gulf of Aden Rift System*, (eds E.R. Philobbos and B.H. Purser), Geological Society of Egypt Special Publication, **1**, pp. 1–45.

Purser, B.H., Philobbos, E.R. and Soliman, M. (1990) Sedimentation and rifting in the NW parts of the Red Sea: a review. *Geological Society of France Bulletin*, **8**, 371–384.

Purser, B.H. and Plaziat, J.-C. (this volume) Miocene peri-platform slope sedimentation in the northwestern Red Sea rift, in *Sedimentation and Tectonics of Rift Basins: Red Sea–Gulf of Aden* (eds B.H. Purser and D.W.J. Bosence), Chapman & Hall, London, pp. 320–343.

Purser, B.H., Plaziat, J.-C. and Philobbos, E.R. (1993) Stratiform breccias and associated deformation structures recording Neogene earthquakes in the syn-rift sediments of the Egyptian red Sea coast, in *Geodynamics and Sedimentation of the Red Sea–Gulf of Aden Rift System*, (eds E.R. Philobbos and B.H. Purser), Geological Society of Egypt Special Publication **1**, pp. 189–204.

Purser, B.H., Plaziat, J.-C. and Rosen, B.R. (1993) Miocene reefs of the NW Red Sea: relationships between reef geometry and rift evolution, in *Geodynamics and Sedimentation of the Red Sea–Gulf of Aden Rift System*, (eds E.R. Philobbos, and B.H. Purser), Geological Society of Egypt Special Publication, **1**, Cairo, pp. 89–113.

Purser, B.H., Plaziat, J.-C. and Rosen, B.R. (1996) Miocene reefs of the northwest Red Sea, in models for carbonate stratigraphy from Miocene reef complexes of Mediterranean regions. *Concepts in Sedimentology and Paleontology*, **5**, 347–365.

Purser, B.H., Soliman, M, and M'Rabet, A. (1987b) Carbonate, evaporite, siliciclastic transitions in Quaternary rift sediments of the northwestern Red Sea. *Sedimentary Geology*, **53**, 247–267.

Purser, B., Tucker, M.E. and Zenger, D.H. (1994) Problems, progress and future research concerning dolomites and dolomitization, in *Dolomites: A volume in honour of dolomites*, (eds B. Purser, M. Tucker and D. Zenger), International Association of Sedimentologists, Special Publication **21**, Blackwell Scientific Publications, Oxford, pp. 3–20.

Quennell, A.M. (1958) The structure and evolution of the Dead Sea rift. *Quarterly Journal of the Geological Society of London*, **64**, 1–24.

Quennell, A.M. (1984) The western Arabian rift system, in *The Geological Evolution of the Eastern Mediterranean*, (eds J.E.

Dixon and A.H.F. Robertson), Blackwell Scientific Publishers, Oxford, pp. 775–788.

Radke, M., Willsch, H. and Welte, D.H. (1980) Preparative hydrocarbon group type by automated medium pressure liquid chromatography. *Annals Chemistry*, **52**, 406–411.

Ramboz, C., Oudin, E. and Thisse, Y. (1988) Geyser-type discharge in Atlantis II Deep, Red Sea: evidence of boiling from fluid inclusions in epigenetic anhydrite. *Canadian Mineralogist*, **26**, 765–786.

Ramzy, M., Steer, B., Abu-Shadi, F., Schlorholtz, M., Milka, J. *et al*. (1996) Gulf of Suez rift basin sequence models. Part B. Miocene sequence stratigraphy and exploration significance in the central and southern Gulf of Suez. *Proceedings 16th EGPC conference*, Cairo, pp. 1–7.

Read, J.F. (1982) Carbonate platforms of passive (extensional) continental margins: types, characteristics and evolution. *Tectonophysics*, **81**, 195–212.

Read, J.F. (1985) Carbonate platform facies models. *American Association of Petroleum Geologists Bulletin*, **69**, 1–21.

Reading, H.G. (1986) *Sedimentary Environments and Facies*, 2nd edn, Blackwell Scientific Publications, Oxford.

Redfern, P. and Jones, J.A. (1995) The interior rifts of Yemen–analysis of basin structure and stratigraphy in a regional plate tectonic context. *Basin Research*, **7**, 337–356.

Reeder, M. (1994) Dating, subsidence and uplift studies of marine carbonates, Gulf of Aden, South Yemen. Unpublished BSc thesis, Royal Holloway, University of London.

Reijmer, J.G. and Evaars, S.L. (1991) Carbonate platforms reflected in carbonate basin facies (Triassic, Northern Calcareous Alps, Austria). *Facies*, **25**, 253–278.

Reijmer, J.G., Posewang, J. and Dullo, W.C. (1990) The Holocene morphology and sediment accumulation of Sanganeb atoll. *Atoll Research Bulletin* (submitted).

Reineck, H.E. and Singh, I.B. (1975) *Depositional Sedimentary Environments with Reference to Terrigenous Clastics*, Springer-Verlag, New York.

Reiss, Z. and Hottinger, L. (1984) The Gulf of Aqaba. *Ecological Studies*, **50**, 1–354.

Reiss, Z., Luz, B., Almogi-Labin, A., Halicz, E., Winter, A. and Wolf, M. (1980) Late Quaternary paleoceanography of the Gulf of Aqaba (Elat), Red Sea. *Quaternary Research*, **14**, 294–308.

Remmele, G. (1989) Hydroclimatic conditions in the southwest of the Arabian Peninsula (Yemen Arab Republic). *Erdkunde*, **43**, 27–36.

Ressetar, R. and Nairn, A.E.M. (1980) Two phases of Cretaceous–Tertiary magmatism in the Egyptian Eastern Desert: Paleomagnetic and K-Ar evidence. *Annals of the Geological Survey of Egypt*, **10**, 997–1011.

Reyre, D. (1966) Evolution géologique du bassin gabonais, in *Sedimentary basins of the Africa Coasts*, Association of African Geological Surveys, Paris, pp. 171–191.

Reyre, D. (1984) Caractères pétroliers et évolution géologique d'une marge passive. Le cas du bassin Bas Congo-Gabon. *Bulletin Centres Recherches Exploration et Production Elf Aquitaine*, **8**, 303–332.

Reyss, J.-L., Choukri, A., Plaziat, J.-C. and Purser, B.H. (1993) Datations radiochimiques des récifs coralliens de la rive occidentale du Nord de la Mer Rouge, premières implications stratigraphiques et tectoniques. *Comptes Rendues Académie des Sciences de Paris* (ser.2), **317**, 487–492.

Rezak, R., Bright, T.J. and McGrail, D.W. (1985) *Reefs and Banks of the Northwestern Gulf of Mexico, their Geological, Biological and Physical Dynamics*, Wiley, New York.

Ricci Lucchi, F. (1995) Sedimentological indicators of paleosismicity, in *Perspectives in Paleosismology*, (eds L. Serua and D.B. Slemmons), Association of Engineering Geologists, Special Publication 6, pp. 7–17.

Richards, M.A., Duncan, R.A. and Courtillot, V.E. (1989) Flood basalts and hotspot tracks – plume heads and tails. *Science*, **246**, 105–107.

Richardson, M.A. and Arthur, M.A. (1988) The Gulf of Suez – northern Red Sea Neogene rift: a quantitative basin analysis. *Marine and Petrology Geology*, **5**, 247–270.

Richter, H., Makris, J. and Rihm, R. (1991) Geophysical observations offshore Saudi Arabia: seismic and magnetic measurements. *Tectonophysics*, **198**, 279–310.

Rigby, J.K. and Hamblin, W.K. (1972) *Recognition of Ancient Sedimentary Environments*, Society of Economic Paleontologists and Mineralogists, Special Publication 16, 34 pp.

Rihm, R. (1984) Seismiche messungen im Roten Meer und ihse interpretation. MSc thesis, Universität Hamburg, 144 pp.

Rihm, R. (1989) Die Entwicklung des Roten Meeres abgeleitet aus geophysikalischen Messungen. *Hamburg Geophysical Einzelschrift*, **A94**, University Hamburg.

Rihm, R. (199?) Mid-ocean ridge segmentation is determined by initial rift geometry – evidence from the Red Sea. Submitted to *Terra Nova*.

Rihm, R. (1996) Die mittelozeanischen Rüeken: das grösste Vulkansysteme der Erde. *Meer und Museum*, **12**, Deutches Museum für Meereskunde und Fischesei, Stralsund, 61–68.

Rihm, R. and Henke, C.H. (this volume) Geophysical studies on early tectonic controls on Red Sea rifting, opening and segmentation, in *Sedimentation and tectonics of rift basins: Red Sea–Gulf of Aden*, (eds B.H. Purser and D.W.J. Bosence), Chapman & Hall, London, pp. 27–49.

Rihm, R., Makris, J. and Möller, L. (1991) Seismic surveys in the northern Red Sea: asymmetric crustal structure. *Tectonophysics*, **198**, 279–295.

Rioual, M., Orszag-Sperber, F. and Purser, B.H. (1995) Plateformes carbonatées plio-quaternaires dans le Golfe de Suez: rôle de la tectonique et de l'eustatisme. *Bulletin de la Société Géologique de France*, **167**, 509–516.

Rioual, M. (1996) Sédimentation et tectonique post-Miocene dans le rift du Golfe de Suez et le NW de la Mer Rouge (Egypte). Doctoral Thesis, Université de Paris Sud, 240 pp.

Ritchie, J.C., Eyles, C.H. and Haynes, C.V. (1985) Sediment and early to mid-Holocene humid period in the eastern Sahara. *Nature*, **314**, 352–355.

Roberts, G.P., Gawthorpe, R.L. and Stewart, I. (1993) Surface faulting in active normal fault zones: examples from the Gulf of Corinth fault system, *Zeitschift für Geomorphologie*, *N.F. Supplement*, **94**, 303–328.

Roberts, H.H. and Murray, S.P. (1983) Controls on reef development and the terrigenous-carbonate interface on a shallow shelf, Nicaragua. *Coral Reefs*, **2**, 71–80.

Roberts, H.H. and Murray, S.P. (1988a) Developing carbonate platforms: Southern Gulf of Suez, Northern Red Sea. *Marine Geology*, **59**, 165–185.

Roberts, H.H. and Murray, S.P. (1988b) Gulfs of the northern Red Sea: depositional setting of abrupt siliciclastic-carbonate transitions, in *Carbonate–Clastic Transitions* Doyle, (ed L.J. and H.H. Roberts), *Developments in Sedimentology*, **42**, Elsevier, 99–142.

Roberts, H.H., Murray, S.P. and Suhayda, J.N. (1975) Physical processes on a fringing reef system. *Journal of Marine Research*, **33**, 223–260.

Robertson Research International (1987) The geology and petroleum potential of the Sudanese Red Sea (Unpublished report for the Geological Research Administration of the Sudan).

Robson, D.A. (1971) The structure of the Gulf of Suez (Clysmic) rift, with special reference to the eastern side. *Journal of the Geological Society of London*, **127**, 247–276.

Roeser, H.A. (1975) A detailed magnetic survey of the southern Red Sea. *Geologisches Jahrbuch*, **13**, 131–153.

Roger, J., Platel, J-P., Cavelier, C, and Bourdillon-de-Grissac, C. (1989) Donnees nouvelles sur la stratigraphie et l'histoire geologique du Dhofar (Sultanat d'Oman). *Bulletin de Societe geologique du France*, **8**, 265–277.

Rögl, F. and Steininger, F.F. (1983) Vom Zerfall der Tethys zu Mediterran und Paratethys. Die neogene Paläogeographie und Palinspastik des zirkummediterranen Raumes. *Annalen des Naturhistorisches Museums in Wien*, (A)**85**, 135–163.

Rögl, F. and Steininger, F.F. (1984) Neogene Paratethys, Mediterranean and Indopacific seaways. Implications for the palaeobiogeography of marine and terrestrial biotas, in *Fossils and Climate*, (ed. P. Brenchley), John Wiley, Chichester, Geological Journal Special Issues 11, pp. 171–200.

Roman, J. (1979) *Une monographie des Echinides de la Mer Rouge. Principaux résultats*. Actes du Colloque Européen sur les Echinodermes, Bruxelles, 3–8 Septembre 1979, pp. 133–136.

Rona, P.A. (1982) Evaporites at passive margin, in *Geodynamics of Passive Continental Margins*, (ed. Scutton), American Geophysical Union and Geological Society of America, Geodynamics series, 6, pp. 116–132.

Rona P.A. and Scott, S.D. (1993) A special issue on sea-floor hydrothermal mineralization: new perspectives. Preface. *Economic Geology*, **88**, 1935–1976.

Rosales, I., Fenandez-Mendiola, P.A. and Garcia-Mondejar, J. (1994) Carbonate depositional sequence development on active fault-blocks: the Albian in the Castro Urdiles area, northern Spain. *Sedimentology*, **41**, 861–882.

Rosen, B.R. (1971) The distribution of reef coral genera in the Indian Ocean, in *Regional Variation in Indian Ocean Coral Reefs*, (eds D.R. Stoddard and C.M. Yonge), Symposium of the Zoological Society London, **28**, pp. 263–299.

Rosen, B.R. (1975) The distribution of reef corals. *Reports of Underwater Association* (New Series), **1**, 1–16.

Rosen, B.R. (1981) The tropical high diversity enigma – the corals' eye view, in *Chance, Change and Challenge. The Evolving Biosphere*, (eds P.H. Greenwood and P.L. Forey) British Museum (Natural History) London, Cambridge University Press, Cambridge, pp. 103–129.

Rosen, B.R. (1984) Reef Coral biogeography and climate through the late Cenozoic: just islands in the sun or a critical pattern of islands, in *Fossils and climate*, (ed. P. Brenchley), John Wiley, Chichester, Geological Journal Special Issues, 11, pp. 201–262.

Rosen, B.R. (1988) Progress, problems and patterns in the biogeography of reef corals and other tropical marine organisms. *Helgoländer Meeresuntersuchungen*, **42**, 269–301.

Rosen, B.R. and Smith, A.B. (1988) Tectonics from fossils? Analysis of reef-coral and sea-urchin distributions from late Cretaceous to Recent, using a new method, in *Gondwana and Tethys* (ed M.G. Audley-Charles, M.G. and A. Hallam), Geological Society of London Special Publications, **37**, pp. 275–306.

Rosendahl, B.R. (1987) Architecture of continental rifts with special reference to East Africa. *Annual Review of Earth and Planetary Sciences*, **15**, 445–503.

Rosendahl, B.R., Reynolds, D., Lorber, P., Burgess, C., McGill, J. et al. (1986) Structural expression of rifting: lessons from Lake Tanganyca, Africa, in *Sedimentation in the East African Rifts*, (ed. L. Frostik), Geological Society London, Special Publication, 25, pp. 29–43.

Ross, D.A. (1972) Red Sea hot brine area-Revisited. *Science*, **175**, 1455–1457.

Ross, D.A. (1995) *Introduction to Oceanography*, Harper Collins College Publishers, New York.

Ross, D.A. and Schlee, J. (1973) Shallow structure and geologic development of the southern Red Sea. *Geological Society of American Bulletin*, **84**, 3827–3848.

Ross, D.A., Whitmarsh, R.B., Ali, S.A., Boudreaux, J.E., Fleischer, R.W. et al. (1973) Red Sea drillings. *Science*, **179**, 377–380.

Rossi, P.L. and Rossi, S. (1983) Fanghi rossi del Mar Rosso. *Atti del 40 Congresso della Associazione Italiana di Oceanologia e Limnologia (IV AIOL)*, Chiavari, 1–3 Dicembre 1980 (eds R. Frache and F. De Strobel), Genova, 39-1–39-16.

Rossignol-Strick, M. (1987) Rainy periods and bottom water stagnation initiating brine accumulation and metal concentrations. 1. the Late Quaternary. *Paleoceanography*, **2**, 333–360.

Rouchy, J.M., Bernet-Rollande, M.C., Maurin, A.F. and Monty, C. (1983) Sedimentological and palaeogeographic significance of various bioconstructed carbonates associated with Middle Miocene evaporites near Gebel Esh Mellaha, Egypt. *Comptes Rendues de l'Académie des Sciences (Series 2)*, **6**, 457–462.

Rouchy, J.M., Monty, C., Pierre, C., Bernet-Rollande, M.C., Maurin, A.F. and Perthuisot, J.P. (1985) Genèse des corps carbonatés diagénétiques par réduction de sulfates dans le Miocène évaporitique du Golfe de Suez et de la Mer Rouge. *Comptes Rendus de l'Académie des Sciences, Paris*, **301**, 1193–1198.

Rouchy, J.M., Pierre, C. and Sommer, F. (1995) Deep-water resedimentation of anhydrite and gypsum deposits in the Middle Miocene (Belayim formation) of the Red sea, Egypt. *Sedimentology*, **42**, 267–282.

Roussel, N. (1986) Dynamique sédimentaire des series miocènes de la région de Quseir (Egypte), bordure NW de la Mer Rouge, Thèse 3ème cycle Université de Paris-Sud.

Roussel, N., Purser, B.H., Orszag-Sperber, F., Plaziat, J-C., Soliman, M. and El Haddad, A.A. (1986) Géologie de la région de Quseir. *Documents et Travaux, Institut Géologique Albert de Lapparent*, Paris, **10**, 129–144.

Rust, B.R. (1978) Depositional models for braided alluvium, in *Fluvial Sedimentology*, (ed. A.D. Miall), Canadian Society Petroleum Geologists Memoir **5**, pp 605–625.

Rust, B.R. and Gibling, M.R. (1990) Braidplain evolution in the Pennsylvanian South Bar Formation, Sydney Basin, Nova Scotia, Canada. *Journal Sedimentary Petrology*, **60**, 59–72.

Ryan, W.B.F., Thorndike, E.M., Ewing, K. and Roos, D.A. (1969) Suspended matter in the Red Sea brines and its detection by light scattering, in *Hot Brines and Recent Heavy Metal Deposits in the Red Sea* (eds E.T. Degens and A.D. Ross), Springer-Verlag, Berlin, pp. 153–157.

Sadek, H. (1959) *The Miocene of the Gulf of Suez (Egypt)*. Geological Survey of Egypt.

Sahota, G. (1990) Geophysical investigation of the Gulf of Aden continental margins: geodynamic implications for the development of the Afro-Arabian rift systems. Unpublished PhD thesis, University of Wales, Swansea.

Sahota, G., Styles, P. and Gerdes, K. (1995) Evolution of the Gulf of Aden and implications for the development of the Red Sea. Proceedings 'Rift Sedimentation and Tectonics in the Red Sea-Gulf of Aden Region', University of Sana'a, Yemen, October, p. 56.

Said, R. (1960) Planktonic foraminifera from the Thebes Formation, Luxor. *Micropaleontology*, **6**, 277–286.

Said, R. (1962) *The Geology of Egypt*, 1st edn, Elsevier, Amsterdam.

Said, R. (1971) The explanatory notes to accompany the Geological Map of Egypt. Geological Survey of Egypt. Paper 56.

Said, R. (1990a) Cenozoic, in *The Geology of Egypt*, 2nd edn, (ed. R. Said). A.A. Balkema, Rotterdam, pp. 451–86.

Said, R. (1990b) *The Geology of Egypt*, 2nd edn, A. Balkema, Rotterdam.

Salah, M.G. and Al Sharhan, A.S. (1996) Structural influence on hydrocarbon entrapment in the northwestern Red Sea, Egypt. *American Association of Petroleum Geologists Bulletin*, **80**, 101–118.

Saller, A.H. (1984) Petrologic and geochemical constraints on the origin of subsurface dolomite, Enewetak Atoll: an example of dolomitization by normal seawater. *Geology*, **12**, 217–220.

Sandberg, P. (1985) Aragonite cements and their occurrence in ancient limestones, in *Carbonate Cements*, (eds N. Schneidermann and P. Harris) Society of Economic Petrologists and Mineralogists, Special Publication, **36**, pp. 33–57.

Samuel, A., Harbury, N., Bott, R. and Thabet, A.M. Field observations from the Socotran platform: their interpretation and correlation to southern Oman. *Marine and Petroleum Geology*.

Sandford, K.S. and Arkell, W.J. (1939) *Palaeolithic Man and the Nile Valley in Lower Egypt*, Publication University of Chicago Oriental Institute, **46**, 105 pp.

Santantonio, M. (1994) Pelagic carbonate platforms in the geological record: their classification, and sedimentary and paleotectonic evolution. *American Association of Petroleum Geologists*, **78**, 122–141.

Saoudi, A. and Khalil, B. (1986) Distribution and hydrocarbon potential of Nukhul sediments in the Gulf of Suez. *Proceedings 7th EGPC Exploration Seminar*, Cairo, pp. 75–96.

Sato, M. (1992) Persistancy-field Eh-pH diagrams for sulfides and their application to supergene oxidation and enrichment of sulfide ore bodies. *Geochimica et Cosmochimica Acta*, **56**, 3133–3156.

Schandelmeier, H., Klitzsch, E., Hendriks, F. and Wycisk, P. (1987) Structural development of north-east Africa since Precambrian times. *Berliner Geowiss. Abh.*, **75**, 4–24.

Schandelmeier, H. and Pudlo, D. (1990) The Central African Fault Zone (CAFZ) in Sudan: a possible continental transform fault. *Berliner Geowiss. Abh.*, **A 120**, 31–44.

Schilling, J.G., Kingsley, R.H., Hanan, B.B. *et al.* (1992) Nd-Sr-Pb isotopic variations along the Gulf of Aden: evidence for Afar mantle plume-continental lithosphere interaction. *Journal of Geophysical Research*, **97**, 10927–10966.

Schlager, W. (1992) *Sedimentology and Sequence Stratigraphy of Reefs and Carbonate Platforms*. American Association Petroleum Geologists, Tusla, Continuing education course notes, Series 34.

Schlager, W., Hooke, R.L. and James N.P. (1976) Episodic erosion and deposition in the Tongue of the Ocean, Bahamas. *Geological Society of America Bulletin*, **87**, 1115–1118.

Schlische, R.W. and Olsen, P.E. (1990) Quantitative filling model for extensional basins with application to the early Mesozoic rift of eastern North America, *Journal of Geology*, **98**, 135–155.

Schlumberger (ed) (1983) *WEC Afrique de l'Quest. Part 1. Geology*. Schlumberger.

Schmaltz, R.F. (1966) Environments of marine evaporite deposition. *Minerals Industry International*, Washington, **35**, 1–7.

Schmidt, D.L., Hadley, D.G. and Brown, G.F. (1982) Middle Tertiary continental rift and evolution of the Red Sea in southwestern Saudi Arabia. *Saudi Arabian Deputy Ministry of Mineral Resources Open File Report USGS-OF-03-6*.

Schmidt, D.L. and Hadley, D.D. (1984) Stratigraphy of the Miocene Baid Formation, southern Red Sea coastal plain Kingdom of Saudi Arabia. *Deputy Ministry for Mineral Resources Technical record, USGS-TR-04-23*.

Schoell, M. and Hartmann, M. (1973) Detailed temperature structure of the hot brines in the Atlantis II Deep area (Red Sea). *Marine Geology*, **14**, 1–14.

Schoell, M. and Hartmann, M. (1978) Changing hydrothermal activity in the Atlantis II Deep geothermal system. *Nature*, **274**, 784–785.

Schoell, M. and Risch, H. (1976) Oxygen- and carbon analyses on planktonic foraminifera of core VA01-188P (Southern Red Sea). *Geologisches Jahrbuch*, **D17**, 15–32.

Schoonen, M.A.A. and Barnes, H.L. (1991) Mechanisms of pyrite and marcasite formation from solution. III. Hydrothermal processes. *Geochimica et Cosmochimica Acta*, **55**, 3491–3504.

Schroeder, J.H. (1982) Aspects of coastal zone management at the Sudanese Red Sea: Characteristics and resources, pollution, conservation and research. *Environmental Research Report, Institute Environmental Studies*, University Khartoum, 3, pp. 1–53.

Schroeder, J.H. (1985) Sparry calcite cement in Miocene coral from Khor Eit, NE Sudan. *Proceedings 5th Coral Reef Congress*, **3**, pp. 283–288.

Schroeder, J.H., and Mansour, N. (1994) Sedimentary environments in the coastal plain of the Red Sea/NE Sudan – Thematic map 1:50000. Sheet Mersa Arakiyai. Technische Fachhochschule, Berlin.

Schroeder, J.H. and Nasr, D.H. (1981) The fringing reefs of Port Sudan, Sudan: I. Morphology – Sedimentology – Zonation. *Essener Geographischearbeiten*, **6**, 29–44.

Schroeder, J., Toleikis, R., Wunderlich, H. and Kuhnert, H. (this volume) Miocene isolated carbonate platform and shallow-shelf deposits in the Red Sea coastal plain, NE Sudan, in *Sedimentation and tectonics of rift basins: Red Sea–Gulf of Aden* (eds B.H. Purser and D.W.J. Bosence), Chapman & Hall, London, pp. 190–210.

Schuhmacher, H., Kiene, W.E., Dullo, W.-Chr., Gektidis, M., Golubic, S., et al. (1995) Factors controlling Holocene reef growth: An interdisciplinary approach. *Facies*, **32**, 145–188.

Schumm, S.A. (1963) Sinuosity of alluvial rivers on the Great Plains. *Bulletin Geological Society of America*, **74**, 1089–1100.

Schüpel, D. and Weinholz, R. (1990) The development of the Tertiary in the Habban-Al Mukalla area, P.D.R. Yemen. *Zeitschrift für geologistche Wissenschaften (Berlin)*, **18**, 523–528.

Schutz, K.I. (1994) Structure and stratigraphy of the Gulf of Suez, Egypt, in *Interior Rift Basins*, (ed. S.M. Landon) American Association of Petroleum Geologists, Memoir **59**, Tusla, Oklahoma USA pp. 57–199.

Schwertmann, U. and Fischer, W.R. (1973) Natural amorphous ferric hydroxide. *Geoderma*, **10**, 237–247.

Schwertmann, U. and Murad, E. (1983) Effect of pH on the formation of goethite and hematite from ferrihydrite. *Clays and Clay Mineralogy*, **31**, 277–284.

Schwertmann, U., and Taylor, R.M. (1972) The transformation of lepidocrocite of goethite. *Clays and Clay Mineralogy*, **20**, 151–158.

Schwertmann, U. and Thalmann, H. (1976) The influence of Fe(II), Si and pH on the formation of lepidocrocite and ferrihydrite during oxidation of aqueous $FeCl_2$ solutions. *Clay Mineralogy*, **11**, 189–200.

Scott, S.D. and Barnes, H.L. (1972) Sphalerite-Wurtzite equilibria and stoichiometry. *Geochimica et Cosmochimica Acta*, **36**, 1275.

Scott, R.W, and Govean, F.M. (1985) Early depositional history of a rift basin: Miocene in western Sinai. *Palaeogeography, Palaeoclimatology, Palaeoecology*, **52**, 143–158.

Sebai, A. (1989) Datation $^{40}Ar/^{39}Ar$ du magmatisme lie aux stades precoces de l'ouverture des rifts continentaux: examples de l'Atlantique Central et de la Mer Rouge. Unpublished PhD thesis, University Nice-Sophia Antipolis.

Sebai, A., Zumbo, V., Feraud, G., et al. (1991) $^{40}Ar/^{39}Ar$ dating of alkaline and tholeiitic magmatism of Saudi Arabia related to the early Red Sea rifting. *Earth and Planetary Science Letters*, **104**, 473–487.

Seilacher, A. (1969) Fault-graded beds interpreted as seismites. *Sedimentology*, **13**, 155–159.

Selley, R.C. (1982) *An introduction to Sedimentology*, Academic Press, London, New York.

Selli, R. (1944) I caratteri e le affinità delle malacofaune quaternarie del Mar Rosso. *Giornale di Geologia*, **18**, 5–22.

Selli, R. (1973) Molluschi quaternari di Massaua e di Gibuti. Accademia Nazionale dei Lincei, Missione geologica dell'AGIP nella Dancalia meridionale e sugli Altipiani Ararini 1936–1938. *Documentazione Paleontologica*, **4**, Roma, 153–444.

Sellwood, B.W. and R.E. Netherwood (1984) Facies evolution in the Gulf of Suez area: sedimentation history as an indicator of rift initiation and development. *Modern Geology*, **9**, 43–69.

Sengor, A.M.C. and Burke, K. (1978) Relative timing of rifting and volcanism on Earth and its tectonic implications. *Geophysical Research Letters*, **5**, 419–421.

Sestini, J. (1965) Cenozoic stratigraphy and depositional history, Red Sea Coast, Sudan. *American Association of Petroleum Geologists Bulletin*, **49**, 1452–1472.

Sgavetti, M. (1992) Criteria for stratigraphic interpretation using aerial photographs: examples from the south-central Pyrenees. *American Association of Petroleum Geologists Bulletin*, **76**, 708–730.

Sgavetti, M., Ferrari, M.C., Chiari, R., Fantozzi, P.L., and Longhi, I. (1995) Stratigraphic correlation by integrating photostratigraphy and remote sensing multispectral data: an example from Jurassic-Eocene strata, Northern Somalia. *American Association of Petroleum Geologists Bulletin*, **78**, 1571–1589.

Shackleton, N.J., Berger, A. and Peltier, W.R. (1990) An alternative astronomical calibration of the Lower Pleistocene time scale based on ODP site 677, *Transaction of the Royal Society of Edinburgh, Earth Science*, **81**, 251–261.

Shanks III, W.C. and Bischoff, J.L. (1977) Ore transport and deposition in the Red Sea geothermal system: a geochemical model. *Geochimica et Cosmochimica Acta*, **41**, 1507–1519.

Shanks III, W.L. and Bischoff, J.L. (1980) Geochemistry, sulfur isotope composition and accumulation rates of the Red Sea geothermal deposits. *Economic Geology*, **75**, 445–459.

Sheppard, C.R.C. (1986) Reefs and coral assemblages of Saudi Arabia. 2. Fringing reefs in the southern region, Jeddah to Jizan. *Fauna of Saudi Arabia*, **7**, 37–58.

Sheppard, C.R.C. (1987) Coral species of the Indian Ocean and adjacent seas: a synonymized compilation and some regional distributional patterns. *Atroll Research Bulletin*, **307**, 1–32.

Sheppard, C.R.L., Price, A. and Roberts, C. (1992) *Marine Ecology of the Arabian Region. Patterns and Processes in Extreme Tropical Climates*, Academic Press, London.

Sheppard, C.R.L. and Sheppard, A.L.F. (1991) Corals and coral communities of Arabia. *Fauna of Saudi Arabia*, **12**, 3–170.

Sheridan, R.E. (1981) Recent research on passive continental margins, in *The Deep Sea Drilling Project: A Decade of Progress*, (eds J.E. Warme, R.G. Douglas and E.L. Winterer), Society of Economic Paleontologists and Mineralogists, Special Publication **32**, pp. 39–55.

Shimron, A.E. (1990) The Red Sea line – a Late Proterozoic transcurrent fault. *Journal African Earth Science*, **11**, 95–112.

Shinn, E.A. (1969) Submarine lithification of Holocene carbonate sediments of the Persian Gulf. *Sedimentology*, **12**, 109–144.

Sibley, D.F. and Gregg, J.M. (1987) Classification of dolomite rock textures. *Journal of Sedimentary Petrology*, **57**, 967–975.

Siedler, G. (1969) General circulation of water masses in the Red Sea, in *Hot Brines and Heavy Metal Deposits in the Red Sea*, (eds E.T. Degens and D.A. Ross), Springer-Verlag, New York, pp. 131–137.

Simoneit, B.R. (1986) Cyclic triterpenoids of the geosphere, in *Biological Markers in the Sedimentary Record*, (ed. R.B. Johns), Methods in Geochemistry and Geophysics, **24**, Elsevier, Amsterdam, pp. 43–100.

Simoneit, B.R., Grimalt, J.O., Hayes, J.M. and Hartmann, H. (1987) Low temperature hydrothermal maturation of organic matter in sediments from the Atlantis II deep, Red Sea. *Geochimica et Cosmochimica Acta*, **51**, 879–894.

Singer, A. (1984) The paleoclimatic interpretation of clay minerals in sediments, a review. *Earth Science Reviews*, **21**, 251–293.

Skipwith, P. (1973) The Red Sea and coastal plain of the Kingdom of Saudi Arabia. *Directorate General Mineral Resources, Technical Report TR-1973-1*.

Sneh, A. and Friedman, G.M., (1980) Spurs and groove patterns on the reefs on the northern Gulf of the Red Sea. *Journal of Sedimentary Petrology*, **50**, 981–986.

Soepri Hantoro, W. (1992) Etude des terrasses récifales quaternaires soulevées entre le détroit de la Sonde et l'île de Timor, Indonésie. Mouvements verticaux de la croûte terresre et variations du niveau de la mer, Thesis Science, Marseille-Luminy.

Soliman, M., Ahmed, E. and Purser, B.H. (1993) *Evolution of the Pliocene sediments in the NW part of the Red Sea rift, Egypt*. Geological Society of Egypt, Special Publication **1**, pp. 233–253.

Somaliland, Oil Exploration Co. (SOEC) (1954) A geological reconnaissance of the sedimentary deposits of the Protectorate of British Somaliland. Crown Agents (U.K.) 42.

Spencer, C.H. (1987) *Provisional stratigraphy and correlation of the Tertiary rocks in the Jeddah region*. Ministry of Petroleum and Mineral Resources, Deputy Ministry for Mineral Resources Jeddah.

Stanley, D.J. and Warne, A.G. (1993) Sea level and initiation of predynastic culture in the Nile delta. *Nature*, **363**, 435–438.

Stanley, S.M. (1995) New horizons for paleontology, with two examples: the rise and fall of the Cretaceous Supertethys and the cause of the Modern Ice Age. *Journal of Paleontology*, **69**, 999–1007.

Steckler, M.S. (1985) Uplift and extension at the Gulf of Suez: indications of induced mantle convection. *Nature*, **317**, 135–139.

Steckler, M.S., Berthelot, F., Lyberis, N. and Le Pichon, X. (1988) Subsidence in the Gulf of Suez: implications for rifting and plate kinematics. *Tectonophysics*, **153**, 249–270.

Steckler, M.S. and Omar, G.I. (1994) Controls on erosional retreat of the uplifted rift flanks at the Gulf of Suez and northern Red Sea. *Journal of Geophysical Research*, **99**, 12,159–12,173.

Steckler, M.S. and ten Brink, U.S. (1986) Lithospheric strength variations as a control on new plate boundaries: examples from the northern Red Sea region. *Earth and Planetary Science Letters*, **79**, 120–132.

Steel, R.J. (1993) Triassic to Jurassic megasequence stratigraphy in the northern North Sea rift to post-rift evolution, in *Proceedings of the 4th Conference on Petroleum Geology of North-West Europe* (ed. J.R. Parker), Geological Society of London, Special Publication, 299–315.

Steen, G. (1984) Radiometric age dating of some Gulf of Suez igneous rocks. Proceedings of the 6th Exploration Seminar, Cairo, March 1982, EGPC/EPEX, **1**, pp. 199–211.

Steen, G. and Helmy, H. (1982) Pre-Miocene evolution in the Gulf of Suez. GUPCO Internal Report.

Stein, R. (1991) *Accumulation of Organic Carbon in Marine Sediments*. Lecture Notes in Earth Sciences, 34, pp. 1–217.

Steininger, F.F., Bernor, R.L. and Fahlbusch, V. (1990) European Neogene Marine/continental chronologic correlations, in *European Neogene marine-continental chronology*, (eds Lindsay, E.H., Fahlbusch, V. and Mein, P.), NATO ASI Sec., A, 180, Plenum Press, New York, pp. 15–23.

Steininger, F.F. and Rögl, F. (1984) Paleogeography and palinspastic reconstruction of the Neogene of the Mediterranean and Paratethys, in *The Geological Evolution of the Eastern Mediterranean*, (eds J.E. Dixon, and A.H.F. Robertson), Blackwell Science Publishers, pp. 659–667.

Steinitz, G., Bartov, Y. and Eyal, Y. (1981) *K/Ar age determinations of Tertiary magmatism along the western margin of the Gulf of Elat*. Geology Survey Israel Current Research, pp. 27–29.

Stern, R.J. (1985) The Najd Fault System, Saudi Arabia and Egypt: a late Precambrian rift related transform system? *Tectonophysics*, **4**, 497–511.

Stern, R.J., Dixon, T.H., Golombek, M.P. *et al.* (1986) The Hamisana Shear Zone: a major Precambrian shear zone in the Red Sea hills of Sudan. *Geological Society America Abstracts*, **18**, 763.

Stiers, W., and Schwertmann, U. (1985) Evidence for manganese substitution in synthetic goethite. *Geochimica et Cosmochimica Acta*, **49**, 1909–1911.

Stoffers, P., Botz, R. and Scholten, J. (1990) Isotope geochemistry of primary and secondary carbonate minerals in the Shaban Deep (Red Sea), in *Sediments and Environmental Geochemistry, Selected Aspects and Case Histories*, (eds D. Heling, P. Rothe, U. Förstner and P. Stoffers), Springer-Verlag, Berlin, pp. 83–94.

Stoffers, P. and Kuhn, R. (1974) Red Sea evaporites: a petrographic and geochemical study, in *Initial Reports of* the *Deep Sea Drilling Project, 23* (eds R.B. Whitmarsh, O.E. Weser and D.A. Ross) Washington, U.S. Government Printing Office, pp. 821–847.

Stoffers, P. and Ross, D.A., (1977) Sedimentary history of the Red Sea, in *Kingdom of Saudi Arabia, Red Sea Research 1970–1975*, Mineral Resources Bulletin, **22-H**, i-iv + H1–H19.

Storey, B.C., Alabaster, T. and Pankhurst, R.J. (1992) *Magmatism and the Cause of Continental Break-up*, Geological Society of London Special Publication **68**.

Strahler, A.N. (1964) Quantitative geomorphology of drainage basins and channel networks, in *Handbook of Applied Hydrology*, (ed V.T. Chow) McGraw-Hill, New York.

Strasser, A., Strohmenger, C., Davaud, E. and Bach, A. (1992) Sequential evolution and diagenesis of Pleistocene coral reefs (South Sinai, Egypt). *Sedimentary Geology*, **78**, 59–79.

Street, F.A. and Grove, A.T. (1979) Global maps of lake level fluctuations since 30,000 yr. B.P. *Quaternary Research*, **12**, 83–118.

Stride, A.H. (1970) Shape and size trends for sand waves in a depositional zone of the North Sea. *Geological Magazine*, **107**, 469–177.

Stumm, W. and Morgan, J.J. (1981) *Aquatic Chemistry*, 2nd edn, J. Wiley and Sons, New York.

Stumm, W. and Sulzberger, B. (1992) The cycling of iron in natural environments: considerations based on laboratory studies of heterogeneous redox processes. *Geochimica et Cosmochimica Acta*, **56**, 3233–3257.

Sudo, T., Shimoda, S., Yotsumito, H. and Acta, S. (1981) *Electron Micrographs of Clay Minerals*, Elsevier, Amsterdam.

Sultan, M., Arvidson, R.E. and Duncan, I.J. (1988) Extension of the Najd Shear System from Saudi Arabia to the Central Eastern Desert of Egypt based on integrated field and Landsat observations. *Tectonics*, **7**, 1291–1306.

Sultan, M., Becker, R., Arvidson, R.E., Shore, P., Stern, R.J., et al. (1992) Nature of the Red Sea crust: A controversy revisited. *Geology*, **20**, 593–596.

Summerfield, M. (1991) *Global Geomorphology*, Wiley Interscience.

Sun, Q.S. (1992) Skeletal aragonite dissolution from hypersaline sea water: a hypothesis. *Sedimentary Geology*, **77**, 249–257.

Sun, Q.S. (1994) Perspective. A reappraisal of dolomite abundance and occurrence in a Phanerozoic. *Journal of Sedimentary Research*, **A64**, 396–404.

Sun, Q. and Esteban, M. (1994) Paleoclimatic controls on sedimentation and diagenesis and reservoir quality: lessons from Miocene carbonates. *Bulletin American Association Petroleum Geologists*, **78**, 519–543.

Surlyk, F. (1990) Mid-Mesozoic syn-rift turbidite systems: controls and predictions, in *Correlation in Hydrocarbon Exploration*, (ed. J.D. Collinson), Norwegian Petroleum Society, Graham Trottman, London, pp. 231–241.

Swallow, J.C. and Crease, J. (1965) Hot salty water at the bottom of the Red Sea. *Nature*, **205**, 165–166.

Swift, D.J.P., Stanley, D.J. and Curray, J.R. (1971) Relict sediments on continental shelves: a reconsideration. *Journal of Geology*, **79**, 322–346.

Szabo, B.J., Haynes Jr., C.V. and Maxwell, T.A. (1995) Ages of Quaternary pluvial episodes determined by uranium-series and radiocarbon dating of lacustrine deposits of Eastern Sahara. *Palaeogeography, Palaeoclimatology, Palaeoecology*, **113**, 227–242.

Tankard, A.J. and Balkwill, H.R. (1989) *Extensional Tectonics and Stratigraphy of the North Atlantic Margins: Introduction*, American Association Petroleum Geologists Memoir, **46**.

Tankard, A.J., Welsink, H.J. and Jenkins, Q.A.M. (1989) Structural styles and stratigraphy of the Jeanne D'Arc basin, Grand Banks of Newfoundland, in *Extensional Tectonics and Stratigraphy of the North Atlantic Margins*, (eds A.J. Tankard and H.R. Balkwill), American Association petroleum Geologists Memoir, **46**, pp. 265–282.

Tard, F., Masse, P., Walgenwitz, F. and Gruneisen, P. (1991) The volcanic passive margin in the vicinity of Aden, Yemen. *Bulletin Centres Recherché Exploration–Production Elf-Aquitaine*, **15**, 1–9.

Taviani, M. (1982) Paleontological markers in late Pleistocene raised coral reefs in the Red Sea. *Proceedings XI INQUA Congress*, Moscow, Abstract Volume 1, p. 307.

Taviani, M. (1983) Basin-wide $\delta^{18}O$ correlation of Red Sea cores (Late Pleistocene to Holocene). *Proceedings First International Conference on Paleoceanography*, Zurich 1982, Abstracts, p. 57.

Taviani, M. (1984) *Sedimentation in a Mini-propagating Rift (Nereus Trough; Central Red Sea): an Oxygen Stable Isotope Approach*. IAS, Proceedings of the Fifth European Regional Meeting of Sedimentology, Marseilles, 9–11 Avril 1984, Abstracts, 427–428.

Taviani, M. (1990) *The Sapropel at the Glacial-postglacial Transition in the Red Sea Basin: a fully Marine Event?* IAS, Proceedings of the Thirteenth International Sedimentological Congress, 26–31 August 1990, Nottingham, England, Abstracts, p. 545.

Taviani, M. (1995a) The ever changing climate: Late Quaternary palaeoclimatic modifications of the Red Sea region as deduced from coastal and deep-sea geological data. *Proceedings of the Egyptian–Italian Seminar 'Geosciences and Archeology in the Mediterranean countries'*, Cairo, 28–30 November 1993, The Geological Survey of Egypt, Special Publication 70, pp. 193–200.

Taviani, M. (1995b) Stable tropics not so stable: climatically-driven extinctions of reef-associated molluscan assemblages (western Indian Ocean, last Interglacial to Present). *IAS, International Workshop Reefs and Carbonate Platforms in the Pacific and Indian Oceans*, Sydney, Australia, 10–14 July 1995, Abstracts.

Taviani, M., (1995c) The Pinna-unit of Giftun island (Egypt, Red Sea): *Signature of a Major Catastrophic Storm Event during the Pleistocene?* Proceedings of the 'Rift Sedimentation and Tectonics in the Red Sea-Gulf of Aden Region', Sana'a, Yemen, 23–31 October 1995, Abstract Volume, p. 67.

Taviani, M. (this volume a) Axial sedimentation of the Red Sea Transitional Region, in *Sedimentation and Tectonics of Rift Basins: Red Sea – Gulf of Aden*, (eds B.H. Purser and D.W.J. Bosence), Chapman & Hall, London, pp. 469–482.

Taviani, M. (this volume b). Post-Miocene reef faunas of the Red Sea: glacio-eustatic controls, in *Sedimentation and*

Tectonics of Rift Basins: Red Sea – Gulf of Aden. (eds B.H. Purser and D.W.J. Bosence), Chapman & Hall, London pp. 580–588.

Taviani, M. and Dullo, W.C. (1993) The problem of time-distant carbonate facies overprints in tropical regions: a case from Gubal island (Egypt, Red Sea), *International Association Sedimentologists 14th regional meeting, Marrakesh*, Abstracts p. 400.

Taviani, M., Janssen, R. and Sengupta, B. (1992) *The Red Sea rift unique deep-sea biota and their significance for the basin biogenic sedimentary budget*. International Symposium on Sedimentation and Rifting in the Red Sea and Gulf of Aden, Cairo, 10–14 January 1992, Abstracts.

Taviani, M., Marchesan, M., Quarantini, M. and Salmi, M. (1995) *Reef-associated mollusk faunas of the Gulf of Suez (Egypt): last interglacial to present. Twelfth International Malacological Congress, Vigo*, Spain, 3–8 September 1995, Abstracts, pp. 333–334.

Taviani, M. and Trincardi, F. (1992) *Evidence of gravity flow deposits in a tectonically active segment of the Red Sea rift (Nereus Deep)*. Proceedings of the International Symposium on Sedimentation and Rifting in the Red Sea and Gulf of Aden, Cairo, 10–14 January 1992, Abstracts.

Taylor, R.M. (1984) Influence of chloride on the formation of iron oxides from Fe(II) chloride. II – Effect of [Cl] on the formation of lepidocrocite and its cristallinity. *Clays and Clay Mineralogy*, **32**, 175–180.

Teisserenc, P. and Villemin, T. (1990) Sedimentary basin of Gabon – geology and oil systems. *American Association Petroleum Geologists Bulletin*, **74**, 117–199.

Tewfik, N, and Ayyad, M. (1984) *Petroleum exploration in the Red Sea shelf of Egypt*. Proceedings of the sixth Exploration Seminar, Cairo, March 1982, EGPC/EPEX, 1, pp. 159–180.

Tewfik, N., Harwood, C. and Deighton, I. (1992) The Miocene, Rudeis and Kareem formations of the Gulf of Suez. Aspects of sedimentology and geohistory. *Proceedings of the Eleventh E.G.P.C. Petroleum Exploration and Production Conference*, Cairo, 1, pp. 84–112.

Thiriet, J.P. (1987) Evolution tectonique et sédimentaire de la marge occidentale de la Mer Rouge au Néogène, région de Port Safaga (Egypte). Thèse, Université Claude Bernard, Lyon.

Thiriet, J.P., Burollet, P.F., Guiraud, R., Icart, J.C., Jarrige, J.J. et al. (1985) Sur l'existence de jeux décrochants transpressifs dans la structuration précoce du Golfe de Suez et de la Mer Rouge. L'example de la région de Port Safaga (Egypte). *Comptes rendus de L'Académie des Sciences de Paris*, **301**, 207–212.

Thiriet, J.P., Burollet, P.F., Montenat, C. and Ott d'Estevou, P. (1986) Evolution tectonique et sédimentaire de la marge occidentale de la Mer Rouge au Néogène: région de Port Safaga. *Documents et Travaux, Institut Geologique Albert de Lapparent*, **10**, 93–116.

Thisse, Y. and Guennoc, P. (1983) The Red Sea: a natural geodynamic and metallogenic laboratory. *Episodes*, **3**, 3–9.

Thornthwaite, C.W. (1948) An approach toward rational classification of climates. *Geographical Review*, **38**, 55–94.

Thunell, R.C., Locke, S.M. and Williams, D.F. (1988) Glacio-eustatic sea-level control on Red Sea salinity. *Nature*, **334**, 601–604.

Tiedemann, R., Sarnthein, M. and Shackleton, N.J. (1994) Astronomic timescale for the Pliocene Atlantic $\delta^{18}O$ and dust flux records of Ocean Drilling Program site 659. *Paleoceanography*, **9**, 619–638.

Tinsley, J.C., Youd, T.L., Perkins, D.M. and Chen, A.T.F. (1985) Evaluating liquefaction potential. in *Evaluating earthquake hazards in the Los Angeles region: an earthquake-science perspective*, (ed. J.I. Zioni), United States Geological Survey professional paper, **1360**, 263–315.

Toleikis, R. (in prep.) Biostratigraphie und Paläoökologie in Miozänen Sedimenten der sudanesischen Küstenebene. PhD thesis, Technische Universität Berlin.

Toleikis, R. and Schroeder, J. (1995) Biostratigraphy and palaeoecology of Miocene sediments in the Sudanese coastal plain. *Conference in Rift Sedimentation and Tectonics in the Red Sea–Gulf of Aden Region*, 23–30 October 1995, Sana'a, Yemen, Abstract volume pp. 74–76.

Tomadin, L., Cesari, G., Fuzzi, S., Landuzzi, V., Lenaz, R., et al. (1989) Eolian dust collected in springtime (1979 and 1984 years) at the seawater-air interface of the northern Red Sea, in *Paleoclimatology and Paleometeorology: Modern and Past Patterns of Global Atmospheric Transport* (eds M. Leinen and M. Sarntheim), Kluwer Academic Publishers, pp. 283–310.

Towe, K.M. and Bradley W.F. (1967) Mineralogical constitution of colloidal 'hydrous ferric oxides'. *Journal Colloidal Interface Science*, **24**, 299–310.

Tucker, M.E. and Wright, V.P. (1990) *Carbonate Sedimentology*, Blackwell Scientific Publications, Oxford, UK

Turner, J.S. (1969) A physical interpretation of the observations of the hot brine layers in the Red Sea, in *Hot Brines and Recent Heavy Metal Deposits in the Red Sea*, (eds E.T. Degens and A.D. Ross), Springer-Verlag, Berlin, pp. 164–173.

Urvois, M. (1988) Apports de l'estimation géostatistique de l'épaisseur des unités métallifères dans la compréhension des mécanismes de mise en place des sédiments de la fosse Atlantis II (Mer Rouge). *Documents Mémoires Bureau Recherche Geologique Minieres*, **154**.

Vail, J.R. (1983) Pan-African crustal accretion in northeast Africa. *Journal African Earth Science*, **1**, 285–294.

Vail, J.R. (1985) Pan-African (Late Precambrian) tectonic terrains and the reconstruction of the Arabian-Nubian Shield. *Geology*, **13**, 839–842.

Vail, J.R. (1988) Tectonics and evolution of the Proterozoic basement of the Northeastern Africa, in *The Pan-African Belt of Northeast Africa and the Adjacent Areas*, (eds S. El Gaby and R.O. Greiling), Vieweg, Braunschweig, pp. 195–226.

Vail, P.R., Mitchum, J.R. and Thompson, S. (1977) Seismic stratigraphy and global changes of sea-level. Part 3: Relative changes of sea-level from coastal onlap, in *Seismic Stratigraphy – Applications to Hydrocarbon Exploration* (ed. C.E. Payton), American Association Petroleum Geologists, Tulsa, Memoir **26**, pp. 63–81.

Van Couvering, J.A. (1995) Setting Pleistocene marine stages. *Geotimes*, **40**, 10–11.

Van de Plassche, O. (ed.) (1986) *Sea-level Research*: a manual for the collection and evaluation of data, Geobooks, Norwich, 618 pp.

Van Houten, F.B. (1977) Triassic–Liassic deposits of Morocco and Eastern North America. *American Association of Petroleum Geologists Bulletin*, **61**, 79–99.

Van Houten, F.B., Bhattacharyya, D.P. and Mansour, S.E.I. (1984) Cretaceous Nubia Formation and correlative deposits, eastern Egypt: Major regressive-transgressive complex. *Geological Society of America Bulletin*, **95**, 397–405.

Van Wagoner, J.C., Posamentier, H.W., Mitchum, R.M., Vail, P.R., Sarg, J.F., *et al.* (1988) An overview of the fundamentals of sequence stratigraphy and key definitions, in *Sea-Level Changes: An Integrated Approach*, (eds C.K. Wilgus *et al.*), Society of Economic Paleontologists and Mineralogists, Special Publication **42**, pp. 39–45.

Varentsov, I.M., Drits, I.M., Gorhkov, V.A., Sivtsov, A.I., and Sakharov, B.A. (1989) Processes of formation on (Mn, Fe)-crusts in the Atlantic: mineralogy, geochemistry of basic and trace elements, Krylov Underwater Mountain, in *Sediment Genesis and Fundamental Problems in Lithology* (ed. V.N. Kholodov), Nauka, Moscow, pp. 58–78.

Vazquez-Lopez, R. and Motti, E. (1981) Prospecting in sedimentary formations of the Red Sea Coast between Yanbu al Bahr and Magna 1979–1986. *Saudi Arabia Deputy Ministry Mineral Resources, Jeddah, Technical Report, BRGM-TR-01-1*.

Veeh, H.H. and Giegengack, R. (1970) Uranium-series ages of corals from the Red Sea. *Nature*, **226**, 155–156.

Veron, J.E.N. (1993) *A Biogeographic Database of Hermatypic Corals; species of the Central Indo-Pacific, genera of the world*, Australian Institute of Marine Science Monograph Series, **10**, pp. 1–433.

Vidal P., Deniel C., Vellutini, P.J., *et al.* (1991) Changes of mantle sources in the course of a rift evolution: the Afar case. *Geophysical Research Letters*, **18**, 1913–1916.

Villemin, T., Angelier, J. and Bergerat, F. (1984) Tectoniques en extension et subsidence dans le Nord-Est de la France. *Annals Société Géologique Nord*, 221–229.

Viotti, C. and El Demerdash, G. (1968) Studies on Eocene sediments of Wadi Nukhul area, east coast of Gulf of Suez. *Proceedings of the 3rd African Micropaleontological Colloquium*, Cairo, pp. 403–423.

Vita-Finzi, C. (1991) First-order ^{14}C dating Mark II. *Quaternary Proceedings*, **1**, 11–18.

Vittori, E., Labrini, S.S. and Serva, L. (1991) Palaeoseismology: a review of the state-of-the-art. *Tectonophysics*, **193**, 9–32.

Voggenreiter, W. and Hötzl, H. (1989) Kinematic evolution of the southwestern Arabian continental margin: implications for the origin of the Red Sea. *Journal of African Earth Sciences*, **8**, 541–564.

Voggenreiter, W., Hötzl, H. and Mechie, J. (1988) Low-angle detachment origin for the Red Sea Rift System? *Tectonophysics*, **150**, 51–76.

Vogt, P.R., Kovacs, L.C., Bernero, C. and Srivastava, S.P. (1982) Asymmetrical geophysical signatures in the Greenland–Norwegian and Southern Labrador Seas and the Eurasia Basin. *Tectonophysics*, **89**, 95–160.

Volker, F., McGullock, M.T. and Altherr, R. (1993) Submarine basalts from the Red Sea: new Pb, Sr and Nd isotopic data. *Geophysical Research Letters*, **20**, 927–930.

Volkman, J. and Maxwell, J.R. (1986) Acyclic isoprenoids as biological markers, in *Biological Markers in the Sedimentary Record* (ed. R.B. Johns), Methods in Geochemistry and Geophysics 24, Elsevier, Amsterdam pp. 1–42.

Von Damm, K.L. (1990) Seafloor hydrothermal activity: Black smoker chemistry and chimneys. *Annual Reviews Earth Planetary Science*, **18**, 173–204.

Voorhis, A.D. and Dorson, D.L. (1975) Thermal convection in the Atlantis II hot brine pool. *Deep-Sea Research*, **22**, 167–175.

Wal, A.M.A., Eldougdoug, A.A. and Aref, M.A.M. (1987) Petrological characteristics and environmental conditions of the Miocene sulfur-bearing evaporites, Gemsa area, Gulf of Suez, Egypt. *Annals Geological Survey of Egypt*, **XVII**, 77–99.

Walther, J. (1888) Die Korallenriffe der Sinai Halbinsel. Geologische und biologische Betrachtungen. *Abhandlungen der Mathamatisch Physicalischen Classe der Königlichen Sächsischen Gesellschaft der Wissenschaften*, **14**, 439–506.

Ward, W.C. and McDonald, K.C. (1979) Nubia Formation of central Eastern Desert, Egypt – major subdivisions and depositional setting. *Bulletin of the American Association of Petroleum Geologists*, **63**, 975–983.

Warren, J.K. and Kendall, G.St.C. (1985) Comparison of sequences formed in marine sabkha (subaerial) and salina (subaqueous) settings – modern and ancient. *American Association of Petroleum Geologists Bulletin*, **68**, 1013–1023.

Watchorn, F. (1995) The role of pre-rifting structures in the tectono-stratigraphic evolution of the Gulf of Aden Margin: Hadramaut Province, Yemen, *Rift Sedimentation and Tectonics in the Red Sea–Gulf of Aden Region*, University of Sana'a Yemen, October, Abstract Volume, p. 79.

Watchorn, F., Nichols, G.J. and Bosence, D.W.J. (this volume) Rift-related sedimentation and stratigraphy in southern Yemen, in *Sedimentation and tectonics of rift basins: Red Sea–Gulf of Aden*, (eds B.H. Purser and D.W.J. Bosence), Chapman & Hall, London, pp. 163–189.

Webster, D.J. and Ritson, N. (1984) Post-Eocene stratigraphy of the Suez rift in the southwest Sinai, Egypt, *Proceedings of the Sixth EGPC Exploration Seminar*, Cairo, pp. 276–288.

Wegener, A. (1915) *Die Entstehung der Kontinente und Ozeane*, Vieweg, Braunschweig, 1. Aufl.

Weigel, W., Flüh, E.F., Miller, H., *et al.* (1995) Investigations of the East Greenland continental margin beteeen 70° and 72°N by deep seismic sounding and gravity studies. *Marine Geophysical Research*, **17**, 167–199.

Weissbrod, T. (1969) The Paleozoic of Israel and adjacent countries. Part 2, the Paleozoic outcrops in southwestern Sinai and their relation with those of southern Israel. *Geological Survey of Israel*, **48**, 1–32.

Weissel, J.K., Hayes, D.E. and Herron, E.M. (1977) Plate tectonic synthesis: the displacements between Australia,

New Zealand and Antarctica since the Late Cretaceous. *Marine Geology*, **25**, 231–277.

Weissel, J.K. and Karner, G.D. (1989) Flexural uplift of rift flanks due to mechanical unloading of the lithosphere during extension. *Journal of Geophysical Research*, **94**, 13,919–13,950.

Werner, F. and Lange, K. (1975) A bathymetric survey of the sill area between the Red Sea and the Gulf of Aden. *Geologisches Jahrbuch*, **D 13**, 125–130.

Wernicke, B. (1985) Uniform sense, normal simple shear of the continental lithosphere. *Canadian Journal of Earth Sciences*, **22**, 108–125.

Wernicke, B. and Burchfiel, B.C. (1982) Modes of extensional tectonics. *Journal of Structural Geology*, **4**, 105–115.

Wernicke, B. and Tilke, P.G. (1989) Extensional tectonic framework of the U.S. central Atlantic passive margin, in *Extensional Tectonics and Stratigraphy of the North Atlantic Margins*, (eds A.J. Tankard and H.R. Balkwill), American Association Petroleum Geologists, Tulsa, Memoir **46**, pp. 7–21.

Wessel, P. and Smith, W.H.F. (1991) Free software helps map and display data. *EOS*, **72**, (441), 445–446.

White, R. and McKenzie, D. (1989) Magmatism at rift zones: the generation of volcanic continental margins and flood basalts. *Journal of Geophysical Research*, **94**, 7685–7729.

White, R.S. (1987) When continents rift. *Nature*, **327**, 191.

Wilber, R.J. and Neumann, C.A. (1993) Effects of submarine cementation on microfabrics and physical properties of carbonate slope deposits, Northern Bahamas, in *Carbonate Microfabrics* (eds R. Rezak and D.L. Lavoie), Springer-Verlag, New York, pp. 79–94.

Wils, E. (1986) Red Sea Malacology III. Revisie: de Conidae van de Rode Zee. *Gloria Maris*, **25**, 161–206.

Wilson, J.L. (1974) Characteristics of carbonate-platform margins. *American Association of Petroleum Geologists Bulletin*, **58**, 810–824.

Wilson, J.L. (1975) *Carbonate Facies in Geologic History*, Springer-Verlag, New York.

Wilson, M.J. (1987) *A Handbook of Determinative Methods in Clay Mineralogy*, Blackie, Glasgow and London.

Wilson, M.E.J. and Bosence, D.W.J. (1996) The Tertiary evolution of south Sulawesi: a record in redeposited carbonates of the Tousa Limestone Formation, in *Tectonic Evolution of Southeast Asia*, (eds A. Hall and D.J. Blundell), Geological Society, London, Special Publication No. **109**, pp. 365–390.

Winter, A., Reiss, Z. and Luz, B. (1979) Distribution of living Coccolithosphore assemblages in the Gulf of Elat ('Aqaba). *Marine Micropaleontology*, **4**, 197–223.

Withjack, M.O. and Jamison, W.R. (1986) Deformation produced by oblique rifting. *Tectonophysics*, **126**, 99–124.

Wright, V.P. (1984) *Paleosols their Recognition and Interpretation*, Blackwells Scientific Publications, Oxford.

Wright, V.P. (1994) Losses and gains in weathering profiles and duripans, in *Quantitative Diagenesis: Recent Developments and Applications to Reservoir Geology*, (eds A. Parker and B.W. Sellwood), Kluwer Academic Publishers, Netherlands, pp. 95–123.

Wunderlich, H. (in prep.) Lithologie und Faziesanalyse an miozänen Riftsedimenten der sudanesischen Küstenebene. PhD thesis, Technische Universität, Berlin.

Yamani, M.A. (1968) Geology of the oolitic hematite of Wadi Fatima, Saudi Arabia and the economics of its exploration: PhD thesis, Cornell University Itheca, New York.

Youd, T.L. (1975) Liquefaction, flow and associated ground failure. *Proceedings of the U.S. National Conference on Earthquake Engineering*, pp. 146–155.

Young, R.A. and Ross, D.A. (1977) Volcanic and sedimentary processes in the Red Sea axial trough. Kingdom of Saudi Arabia, Red Sea Research 1970–75, *Mineral Resources Bulletin*, **22-I**, i–iii + I1–I13.

Youssef, E.A. (1986) Depositional and diagenetic models of some Miocene evaporites on the Red Sea coast, Egypt. *Sedimentary Geology*, **48**, 17–36.

Youssef, E.A. (1989a) Geology and genesis of sulfur deposits at Ras Gemsa area, Red Sea coast, Egypt. *Geology*, **17**, 797–801.

Youssef, E.A. (1989b) Some diagenetic fabrics of aragonite, calcite and native sulfur in the Miocene evaporite cycle on the Red Sea, Egypt. *Journal Mineralogy Petrology Economic Geology*, **84**, 5, 168–176.

Youssef, M.I. (1957) Upper Cretaceous rocks in Kosseir area. *Bulletin Institute Desert Egypte*, **7**, 35–54.

Zhabina, N.N. and Sokolov, V.S. (1982) Sulphur compounds in the sediments of the Atlantis II deep, Red Sea. *Marine Geology*, **50**, 129–142.

Zico, A. (1993) Late Cretaceous – early Tertiary stratigraphy of the Thermal area, East Central Egypt. *Neues Jahrbuch Geologie Palasntologie*, **3**, 135–149.

Zierenberg, R.A. and Shanks III, W.C. (1983) Mineralogy and geochemistry of epigenetic features in metalliferous sediments, Atlantis II Deep, Red Sea. *Economic Geology*, **78**, 57–72.

Zierenberg, R.A., and Shanks III, W.C. (1986) Isotopic variations on the origin of the Atlantis II, Suakin and Valdivia brines, Red Sea. *Geochimica et Cosmochimica Acta*, **50**, 2205–2214.

Zierenberg, R.A. and Shanks III, W.C. (1988) Isotopic studies of epigenetic features in metalliferous sediments, Atlantis II deep, Red Sea. *Canadian Mineralogist*, **26**, 737–753.

Zuchiewicz, W. and Oaks, R.Q. (1993) Geomorphology and structure of the Bear River Range, NE Utah; a morphometric approach. *Zeischrift für Geomorphologie, Supplement Band*, **94**, 41–56.

Zumbo, V.G., Feraud, H., Bertand, H. and Chazot, G. (1995) $^{40}Ar/^{39}Ar$ chronology of the Tertiary magmatic activity in Southern Yemen during early Red Sea rifting. *Journal Volcanology and Geothermal Research*, **65**, 265–279.

Index

Note: **Bold** page numbers refer to illustrations and *italic* page numbers to tables

Abu Durba 103, 231
Abu Ghusun area 103, **108**, **310**
Abu Ghusun Formation 150, 212, 214
Abu Ghusun platform 246
   cartoon **249**
Abu Ghusun-Ras Honkorab, reefs (Miocene) 310
Abu Imama Formation 192
Abu Rudeis 103
Abu Shaar
   facies models **294**
   Kharaza Member 281
   structure 274
   syn-rift stratigraphy 276
Abu Shaar platform 91, **217**, 242, 271
   aerial photograph **275**
   Bali'h Member 287
   Chaotic Breccia Member 290
   coral reefs **301**
   depositional sequences 277
   Esh el Mellaha Member 283
   geological map **273**
   onlap 274
   pisolites 374
   slope deposits 329
   tectonic controls 290
Abu Shaar-Esh Mellaha platforms 247
Abu Shaar tilt-block platform **250**
Abu Zenima Formation 15, 81, 146, 148, 212
Acanthastraea 303, 304
Accommodation space 12, 13, 19–23, 187, 266
   on plunging axis 254
Accommodation zones 18, 19, 23, 70, 80, 108, 171, 429
   Gulf of Suez 224, 231
Accretion of oceanic crust 48
Accretion of reefs and sands 244
Accretionary platform margins 293
Accumulation rates
   axial sediments 470
   metal-enriched sediment 469
Acetabularia 302
Acheulean tool 541
Acropora 429
Active rift models 15
Active rifting 10, 16
Aden Volcanic Line 53
Aegean Gulf 20
Aeolian dolomite 488

Aeolian quartz in axial muds 488
Afar Depression 37
Afar mantle plume 54, 55, 165, 616
African plate **224**
African Red Sea margin 48
Afro-Arabian continent 74
   plate 50
Age of the Suez Isthmus 570
Algal mats 109, 112
Algally laminated dolomicrites 257
Alkaline volcanics 53
Alluvial cones 539
Alluvial deposits 234
Alluvial discharges and sea-level fluctuations 537
Alluvial fans 20, 146, 154, 176, 561
Alluvial plain environment 107
Alluvial reefs 561
Alluvial terraces 185, 558
Al Mukalla 58
Along-strike thickening 293
Al Salif diapir, (Yemen) **449**
Amal-Zeit province 224
Amalgamated platform 245
Ambagi Shelf 243, **244**
Amphiboles, climatic implications 446
Amran Group 60, 171
Analyses of organic carbon and total sulphur axial sediments *492*
Anhydrite 81
   axial cores 510
   axial sediments 508
   nodular 383, **419**
   precipitation from hot axial waters 511
Ankerite 511
Annelid borings on by-pass surfaces 343
Annular fracture systems 527
Anoxic bottom-water conditions 491
Anoxic lacustrine carbonate 108
Antecedent drainage 19
Antithetic fault 156
Antithetic tilt-blocks 107, 116
Apatite fission track 50, 79
   analysis 53, 77
Aqaba 585
Aqaba-Dead Sea fault system 15
Aqaba transform 7, 95, 430, 444
Aqaba trend 79, 102, 214, 230
Aquitanian planktonic microfauna 214

Arabian plate 50, **224**
   magmatism 53
Arabo-African platform 147
Aragonite cements
   in axial crusts 500
   precipitation during glacial periods 472
Aragonitic crusts 497
Aragonitic speleothems 383
Arcuate embayments 333
Arid climate 7, 20, 107, 154, 176
Arkosic gravels 247
Asthenosphere 55
Asymmetric depositional sequences 293
Asymmetric graben 10
Asymmetric sedimentation 241
Asymmetry
   rift 222
   rift initiation 157
   rift sedimentation 218
   sediment composition 241
Atlantic Ocean margins 10
Atlantis II Deep 16, 505
   geochemical history 517
Atlantis II deposits 509
   spatial evolution 519
Atolls 589
   fault-controlled morphology 590
Austrotrillina asmariensis 61
Axial and peripheral sequences, correlations between 211
Axial brines 425, 506
Axial carbonates 482
Axial cores 92–101, **482**
Axial cores, distribution of minerals in **507**
Axial crusts
   diagenetic fabrics (SEM photos) **499**
   origin 500
Axial deeps 506
Axial escarpment 480
Axial plunge of a block 241
Axial sands by winnowing 470
Axial sedimentation, Red Sea 467
Axial sediments 465
   accumulation rates 470
   importance of climate 476
Axial trough 7, 37, 48, 465, 480
Axial turbidites, components (SEM photos) **490**
Axial zone 7

geodynamics 468
geological setting 480
hydrology 468
mineralogical and textural properties 481
organic geochemistry of cores 479
sediments 469

Bab el Mandab sill 3, 7, 318, 503, 583
Baba Plain 236
Bad Formation 154
Bali'h Member 277
Bandar Harshau Formation 168
Baraka suture zone 45
Barrier effect of upstream block 269
Barrier-reef system, tectonic control 590
Barrier reefs 584
Basaltic dikes 102, 237
Basaltic extrusion 7
Basaltic flows 153, 185
Basaltic lava 141
Basaltic magmatism 52, 160
Basaltic volcanism 53, 133
Basalts 144
Basement highs, Yemen 188
Basement lineaments 50
Basement structures 77, 102
Basin centre 12
Basin dynamics 77, 90
Basin margin 12
Bathymetric scarp, Shadwan 438
Beach deposits 184, 303
Belayim Formation 81, 236
  evaporites 416
Belayim province 224
Benthic foraminifera, Miocene, Sudan 197
Bimodal magmatism 52
Biostratigraphy
  Gebel el Zeit **219**
  north-western Red Sea 217
  Sudan 201, 204
Biostromal coral 291
Biostromal reefs 302
Bir Abu Shaar 274
Block rotation 247, 293
Block tilting and sedimentation 266
Blocks independent of rift periphery 257
Boosaaso Basin 63, 68, 168
Bored gypsum clasts 458
Botryoidal cement 381
Botula 288
Brachidontes 305
Braid deltas 20
Braided fluvial system 140, 178
Break-up unconformity 12, 168, 188
Breccias and intrabasinal turbidites 507
Brecciated carbonate 331
Brecciated laminite 332
Brecciation of algal laminites 358
Bregmaceros 218
Brines
  convection 506
  Red Sea axis 506
Burdigalian planktonic foraminifera 214
By-pass margins 263
By-pass surfaces 329, 338
By-pass zone 324

Calcareous nannoplankton 218, 500
Calciturbidites 476
Calcium sulphate distribution in axial cores **510**
Calcretes 172, 179
Canyons and gulleys 343

Carbon isotopes
  axial sediments 485
  geographic variation 397
Carbon-14 dates
  axial sediments 485
  from fossils from Salif area *464*
Carbonate diagenesis 367
  axial zone 479
Carbonate facies, evolution 292
Carbonate mounds 195
  Sudan **198**
Carbonate muds, Dahlak Shelf, Eritrea **535**
Carbonate platforms 6, 15, 20, 24, 95, 179
  Dahlak Islands, Eritrea 523
  demolished 254
  growth 25
  inclined 321
  initiation 6
  on diapirs 445
  syn-rift block faulting 239
  synsedimentary faulting 247
Carbonate shelf 254
Carbonate-siliciclastic platform 240
Carbonate-siliciclastic sedimentation in active tectonic setting 239
Carbonate-siliciclastic shelf 290
Carbonate-siliciclastic transitions, structural controls 263
Carbonate slope 197
Carbonate silts, axial sediments **484**
Carbonates
  precipitated from hot axial brines 511
  protected from terrigenous contamination 257
  Red Sea axial trough 482
Carlsberg Ridge 41, 48
Caulastraea 303, 304
Cavern development and speleothems **384**
Cavity fillings 383
Celestite crystals 383
Celestite diagenesis **385**
Central African Fault Zone 45, 48
Central Gulf of Suez, structural and stratigraphic simplicity 430
Central Red Sea margin 135
Centripetal evolution 112
Centripetal migration 112
Cerastoderma 302
Chalcopyrite 510
  in axial cores 508
Chalcopyrite axis, Red Sea 508
Channel-fill facies 126
Channel fills 140
Channel sedimentation 150
Channels 146
Chaotic Breccia Member 277
  Abu Shaar platform 290
Charophyte remains 215
Chattian faunas 6
Chronology of Quaternary coastal deposits, north-western Red Sea 539
Classification of continental margins 45
Clast rotation **361**
Clastic discharges 112
Clastic sedimentation rates, Red Sea 609
Clastic transit 115
Clay mineralogy axial sediments 482
Clay minerals
  analyses *139*
  axial core **486**
Cleavage **112**
Climate-controlled submarine diagenesis 502
Climates 3

Climatic and sedimentary evolution during glacial cycle **540**
Climatic control on Red Sea offshore sedimentation 476
Climatic effects, sporadic terrigenous input 270
Climatic events **471**
Climatic evolution reflected in sedimentary sequences 563
Climatic factors 4, **471**
  and sedimentation 241
Climatic influence on deep axial sedimentation 476
Clinoforms 249
Clysmic faults 105
Clysmic trend 79, 102, 214, 230
Coarse clastic discharge 253
Coastal playa 182
Coastal sabkha, Dahlak Islands 529
Coastal salina 183
Coccolithophores 469
Coccoliths in axial sediments **484**
Collapse
  breccia **360**
  peripheral reef 250
  salt diapirs 529
Compressional folds and brecciation 357
Compressional strike-slip 106, 107
Condensed sections 22
Conglomeratic valley-fill 254
Conidae 315
Conjugate continental margins 37
Constructional cavities **341**
Contemporaneity of onlap and downlap 324
Continent–ocean transitions 32
Continental and marine deposits (Quaternary) north-western Red Sea coast 537
Continental drifting 56
Continental lineaments 30
Continental rifting 73
Continental sediments 17
Coral-algal patch-reef on the culmination of cone 263
Coral colonization, Dahlak shelf 536
Coral extinction (Holocene) 579
Coral genera, distribution, north-western Red Sea 314
Coral palaeozonation 303
Coral patches, Abur Shaar 304
Coral reefs 219
  Abu Shaar platform **301**
  distribution **301**
  distribution, Gulf of Suez-northern Red Sea region **298**
  geometry and development 298
  morphologies and zonation 583
  northern Red Sea 584
  on hanging wall slope 303
Coral zonation, Dahlak Shelf, south-western Red Sea 532
Coralgal framestones 284
Coralgal framework along footwall margin 283
Coralgal mounds **194**, 195
Coralline algal crusts 303
Coralline limestone 179
Corals, palaeobiogeographical context 315
Corals (Miocene)
  composition and diversity 313
  faunal list **313**
Corridors, and shallow-marine cones 266
Crab *Daira* 315

Cross-bedded sandstone 127
Cross-fault corridor 247
Cross-faults 18, **83**, 85, **91**, 231
  as cause of disintegration of platform 263
  favouring lateral transit of siliciclastics 267
Cross-structures 80, 102
  Pliocene North-western Red Sea **443**
Cross-trending graben **105**
Crust formation in axial zone 501
Crustal domains 30
Crustal doming 14
Crustal extension 74
Crustal margin models **43**
Crustal types **39**
Cryptocrystalline matrix in crusts 497
Cypraeidae 315

Daban basin 58, 168
Dahlak Archipelago (Eritrea)
  environments 524
  sediment patterns (map) **530**
Dahlak Islands
  Eritrea 523
  Quaternary reefs 526
Dahlak Kebir 524
Dahlak Reef Limestone 524
Dahlak Reefs 529
Dahlak shelf
  bathymetric depressions 533
  horsts and graben 527
Damaged reefs 585
Danakil depression 418
Danakil Rift **525**, 527
Dead Sea transtensional systems 229
Debris-flow 20, 150, **151**, 326
  deposits 154, 338
Deccan Plateau of India 10
Deep axial sedimentation 7
Deep marine sands 482
Deep Sea Drilling Project Leg 23, 467
Deep-sea pelagic limestones **474**
Deep-sea sedimentation in Red Sea axis 470
Deformation and sedimentation 184
Deformational structures, increase towards
  top of sedimentary sequences 364
Deformed laminites 362
Density, movement of interstitial waters 372
Depocenters 93
Depositional model 90
Depositional sequences 6, 22, 24
  Abu Shaar platform 277
  and tilt-block rotation 293
Desert rose gypsum 179
Desiccation 154, 176, 182, 183, 282
Detrital dolomite 482
Detrital fans 112
Detrital overflow 261
Detrital quartz, axial zone 482
Dewatering 179, 182
Dhofar 17, 168
Diachronism 6
  north-south 157
  of rift initiation 160
Diagenesis
  and thermal gradient 371
  of Abu Shaar, three-stage model **404**
  relating to carbonate-evaporite
    transition 381
Diagenetic attributes, Miocene
  carbonates 371
Diagenetic evolution, Miocene
  carbonates 370
Diagenetic fabrics

  in carbonate crusts (SEM photos) **498**
  of axial crusts 497
Diagenetic modifications 6
Diagenetic pinnacles 383
Diagenetic properties, sequence 372
Diapiric structures, southern Gulf of
  Suez **432**
Diapirism 6
  controlling sedimentary evolution 460
  south-eastern Red Sea (Yemen) **450**
Diapirs 422
Diatomites 108, 150, 153, 214
  and fish 416
Differences between Eemian and recent reef
  faunas 579
Differences between Miocene and post-
  Miocene tectonics and
  sedimentation 444
Dike swarms 52, 54, 55
Dikes 48, 107
Dip-slope depressions, sediment traps 243
Dip-slope platform margin 292
Dip-slope ramp **19**
Discontinuous reef growth, tectonic and
  climatic changes 582
Dislocated stromatolitic unit **358**
Dispersed terrigenous supply 326
Dissolution cavities **341**
Distribution of minerals in axial cores **507**
Djibouti 54
Dolomite
  and organic matter 393, 399
  in axial sediments 482
  isotope geochemistry 394
  luminesce 393
  petrofabrics **376**
  petrography 393
  petrophysical properties 380
  petrotypes 377
  stoichiometry 397
Dolomitization 6
  and late Miocene evaporites 377
  by convection of warm sea waters 369, 405
  by hot sea water 402
  from hot waters 379
  Miocene platform carbonates 390
  polyphased 377
  post evaporite 396
  prior to Miocene evaporite deposition 379
  Sudan 201
Dolomitizing fluids, upward migration 379
Domal stromatolites 304
Dome-shaped structures, Dahlak Kebir
  island 527
Doming 115, 147
Domino-style faults 18, 230
Drag dislocation **361**
Drag effects 338
Drainage and slope development 595
Drainage basin shapes 599
Drainage densities, western Yemen 604
Drainage patterns
  and sediment transfer (Yemen) 610
  north-western Yemen 600, **601**
Drifting 7
Drowned carbonate platform 259
Drowned reefs 593
Duwi fault trend 102, 156
Duwi orientation 231

Early marine cements 370, 372
Early Miocene evaporites 411
Early rift deposits

  Gulf of Suez 148
  Sinai margin **149**
Early rift environments 5
Early rift sediments 117, 147
Early syn-rift deposits 144
  north-western Red Sea 150
Early syn-rift stage **18**
Earthquakes 568
  in Yemen 597
East African rift 106, 115
Eastern Gulf of Suez 223
Eastern margin of Red Sea, stratigraphic
  correlation **143**
Eastern Red Sea margin 48, 142
Eastern Yemen 132
East-west asymmetry, magmatic 160
East-west offsets 45
Echo-sounding profiles, Dahlak Islands
  shelf **531**
Edge of Suez platform **439**
Eemian molluscs (photos) **580**
Egyptian National Stratigraphic Sub-
  committee 212
Egyptian Red Sea coast 78
Egyptian syn-rift deposits, stratigraphy 211
El Gal Basin 69
Embryonic ocean basin 467
En-echelon structures 37
Eritrea 523
Erosional escarpment rimmed-shelf
  margin 287
Escarpment-type margin 292
Esh el Mellaha-Abu Shaar barrier 299
Esh el Mellaha basin **85**, 90, 247, 272, 274,
  291
Esh el Mellaha fault 25, 272
Esh el Mellaha-Gemsa basin **84**
Esh el Mellaha Member 277
  Abu Shaar platform 283
Esh el Mellaha platform, slope deposits 332
Esh el Mellaha tilt-block 250
Esna Shales 233
Eurasian plate **224**
Eustacy 211
Eustatic sea-level fluctuation 6
Evaporite deposition 79
  Sudan 201
Evaporite formation 25
  structural sills 426
Evaporite (Miocene), facies and
  sedimentation 414
Evaporite olistoliths 421
Evaporite phases, multiplicity of 425
Evaporite remobilization 427
  rejuvenation of fault blocks 430
Evaporite sedimentation 6
  and rift structure 418
Evaporite sequences associated with rift
  basins 409
Evaporite tectonics 422
Evaporite walls 431
Evaporites 15, 108, **108**, **113**, 146, 153, 176,
  197, 204
  along normal faults 422
  and salt tectonics 407
  at passive margins 421
  Bab el Mandeb barrier 411
  Group C 112
  in structural depressions 418
  (Miocene) Dahlak Islands, Eritrea 526
  of Red Sea-Gulf of Suez rift 409
  relations Mediterranian/Indian Ocean 413
  temporal distribution 410

Evaporitic event 102
Evaporitic facies 187
Evaporitic fills 20
Evaporitic lagoons 423
Evaporitic stage 21
Evaporitic unit, early Miocene 214
Evolution of Red Sea 45
Evolution of relief 156
Evolutionary stages 3
Evolving platform morphologies 249
Exhumed fault-block 272
Exhumed unconformity 253
Extension rate, western Yemen 607
Extensional faulting 136
Extensional regime 106
Extensional stress field 45
Extensional tectonics 105, 156
   and sedimentation 223

Fabric-destructive dolomite 377
Fabric-preserving dolomite 377
Facies model, Miocene, Sudan 205
Facies patterns and morphostructural setting, Dahlak Islands 528
Facies variability, mid-Miocene sedimentation, north-western Red Sea 263
Failed rift 12
Fan deltas 20, 24, 234, 249, 539
Farasan archipelago, Saudi Arabia 527, 593
Fartaq high 168
Fault-blocks 10, 24, 25
   and sediments 113
   evolution 82, 108
   orientation 241
   pattern 104
   rotation 9, 13, 22, **23**, 274
   tilting 291
Fault-controlled basins 61
Fault-controlled sub-basins 237
Fault-controlled thickness 188
Fault-line depressions funnel clastics 268
Fault-line erosion 253
Fault-line valley deposits 327
Fault-related uplift 24
Fault-slip 24
Fault zones 171
Faults and breccias 357
Favid corals 245
Faviid corals 302
Favites 311
Feldspars
   climatic implications 444
   preservation **395**
Fenestral micrites 289
Ferricrete 125, 127, 131
Ferricrete paleosols **128**
Fibrous quartzine 378
Fine slope deposits 326
Fission-track 7
   data 94, 229
   dating 230
   shoulder uplift 15
Flash-flood deposits 182
Flash-flood transportation 155
Flash-flooding 7
Flexural isostacy 13, 16, 20, 230
Flexure 106, 112
Flood-plain mudstones 150
Fluid-escape structure **353**
Fluidification structure **437**
Fluvial channel 131
Fluvial systems 20

Fluvio-lacustrine sediments 132
Folded and brecciated laminite **360**
Foliated metamorphic basement 104
Footwall collapse 253, **332**
Footwall-derived fans 11, 18
Footwall embayments 331
Footwall escarpment 253
Footwall highs 13
Footwall islands 274
   local source of siliciclastic detritus 267
Footwall karst 23
Footwall margin 10
Footwall slope **19**
   deposits 329
   general attributes **330**
Footwall sourced fans **19**
Footwall uplift 13, 20, 22, 23, **23**, 25, 291
   and hanging wall subsidence 293
Footwall uplift rates 611
Foreset bedding, wind-driven surface currents 433
Fracturing, relating to dolomite-anhydrite-gypsum transformations 383
Fringing carbonate shelves 324
Fringing reefs 303, 584
   biozonation 585
   Gulf of Aqaba, Jordan coast **585**
   morphological features **586**
Gabbro-granite-syenite plutons 54
Galaxea patch reefs (Yemen) 458
Gebel Abu Shaar 25
Gebel Araba 231
Gebel Dishshit 435
Gebel Duwi 81
Gebel el Zeit 80, 82, 91, 93, 103, **105**, **106**, 150, 218, 257
   block rotation 259
   deep marine facies 257
   field-views **260**
   Quaternary fault rejuvenation 551
   reef terraces 546
   structures and stratigraphic discordances **259**
Gebel Esh Mellaha, reefs (Miocene) 304
Gebel Esh Mellaha-Abu Shaar 218
Gebel Hamadat **159**
Gebel Naqara 85
Gebel Tarbul 107, **110**
Gebel Zarib 243, 261
Gemsa basin 82, 257
Gemsa Peninsula 549
GENEBASS 7, 85, 97, 240
Geochemical history of Atlantis II Deep 517
Geochemistry of hydrothermal cements 399
Geological framework, Sudan 192
Geological setting, south-eastern Red Sea salt domes 449
Geometry of syn-rift sediments 240
Geomorphological studies 7
Geomorphology 521
   methods of study 599
   Red Sea margin, Yemen 595
Geopetal muds 333
Geopetally infilled borings 312
Geotectonic settings 4
Geothermal gradient 396
German Research Foundation 210
Gharamul area 102, 104
Gharandal Group 234
Ghaydah Formation 168
Giftun Island, Plio-Quaternary cover **435**
Gilbert-type fan deltas 236
Glacial eustacy 7

Glacial periods
   negatively affecting coral-reef biota 575
   Red Sea basin filled with dense saline water 471
Glacio-eustatic changes 537
Glacio-eustatic controls on reef faunas 574
Global climatic change, recorded by deep axial sediments 479
Global sea-level
   lowering 420
   rising 241, 269
Globigerina ciperoensis 82
Globigerina marls 81, 85, **108**, 112, 142, 218
Globigerinapsis semiinvoluta 148
Globigerinid foraminifera 488
Globigerinoides 141, 197
Goethite 512
Graded faults **353**
Grain-flow 242, 326
Grain-flow deposits **244**, 328, 338
Grain-flow units 331
Grain reorganization **349**
Grain-size distribution curves 153
Granite plutons 104
   Yemen 598
Gravitary deformations **113**
Gravitational processes 111
Gravitational sliding 445
Gravity **40**
Gravity collapse 333
   and lateral sliding 439
Gravity features 148
Gravity field 37
Gravity flow deposits in axial trough 473
Gravity flows 234
Gravity models 37
Gravity sliding 331
   southern Gulf of Suez 438
Great Escarpment 121
   western Yemen 599, 610
Group A 81, 97, 102, 148
   clastics 146
   deposits **151**, **152**
   north-western Red Sea 212
Group B 97, 102, 148
   north-western Red Sea 215
   sedimentation 111
Group C 97, 102, 148
   evaporites 112, **219**
Group D 97, 102, 148
Gubal Island **431**, **437**
Gulf of Aden 9, 14, 17, 48, 54, 56, **62**, **120**, 165
   evolution **186**
   offshore 185
   sea-floor spreading 188
   stratigraphy **170**
   syn-rift **72**
Gulf of Aqaba
   transform 79
   transtensional systems 229
Gulf of Aqaba-Dead Sea 224
Gulf of Suez 9, 10, 14, 22, 25, 48, 77, 78, **84**, 97, 102, 146
   aborted rift 429
   depocentres 224
   early rift deposits 148
   geological map **225**
   isolation from Mediterranean domain 570
   Pliocene 236
   pre-rift stratigraphy 231
   radiocarbon dates, reef flats *588*

raised beaches 236
rift margin development **235**
stratigraphy **227**, 231
structural evolution 230
subsidence rates 444
syn-rift sequences **229**
syn-rift stratigraphy 233
tectonic setting **224**
timing of rifting 226
Gulleys and canyons 343
Gypsum 79, 150, 153, 176
   and celestite pseudomorphs 290
   and diatomites 414
   (Miocene) isotopes 416
   outcrops **414**
   ridge (Yemen) **452**
   salinas N Ras Shagra **557**
Gypsum/selenite crystals **417**

Habshiya Formation 172
Hadhramaut 165
Hadhramaut Group 172
Hadhramaut region **169**
Hadhramaut-Masila drainage system 187
Half-graben 9, 21, 23–5, 79
   asymmetry 231
   basins 136
   fills 15
   model 18
   structures 17
   trap or funnel clastics 260
Halimeda 197, **203**, 284, 336
Halite 79, 102
Halokinesis 112, 116
Halokinetic features 422
Halokinetic movements 112, 430
   southern Gulf of Suez 445
Hamadat corridor 147, 156
Hamadat half-graben 261
Hami Formation 168
Hammah Formation 144
Hangingwall
   basins 11, 15
   depocentres 231
   dip-slopes 19, 25, 187
   sub-basins 20, 24
   subsidence 22, 23, **23**
Haq eustatic curves 216
Hard grounds 502
Hard layers of pelagic limestones 471
Harrat Hadan 52
Heat flow data 230
Hemipelagic facies 502
Hemipelagic muds 21
Heterocyclina 532
Heterogeneous stretching **10**, 13
High resolution seismic line 434, **434**
Highstand deposits 23
Highstand systems tracts 22–5
Hinge-zone 170
Hollow pseudomorphs 383
Holocene and late Pleistocene wet phases and their faunal signature 581
Holocene coral reef terraces 579
Holocene reef terraces 587
Horizontal extension 79
Horst and graben system 112
Hot metalliferous brines 506
Hummum-Sharmah 182
Hurghada 435
Hydrocarbons 85
   and dolomitization 396
   entrained by brine 396

exploration 171, 223
   in fissures 379
Hydrodynamics, marine diagenesis 374
Hydrological budget 468
Hydrology, axial zone 468
Hydroplastic deformation **151**, **360**, **365**
Hydrothermal brine 506, 517
Hydrothermal buoyancy of heated interstitial marine waters 387
Hydrothermal calcite 396
Hydrothermal cements 398, 399
Hydrothermal history, axial brines 509
Hydrothermal ore deposit 506
Hydrothermal solutions 506
Hydrothermalism in axial depressions 468
Hypersalinity
   and sapropels 494
   of Red Sea during glaciation 502
Hyrothermal calcite 290

Ignimbrites 186
Incipient oceanic troughs 465
Inclined ramp 276
Indian Ocean 17, 58
   influx 7
Influx of clastics 114
Inheritance of Precambrian structural trends 5
Inherited relief 204
Inherited structural pattern 102, 103, 104, 116
Initial doming 114
Initial rifting 107, 215
   Oligo-Miocene 146
Inner shelf planctonics 533
In situ deformation 357
Interior lagoon 193
Internal creeping **361**
Intracontinental setting of rift 269
Intracrustal detachment surface 10
Intraformational slide-breccias 357
Intrastratal shearing 7
Iron and manganese oxides in axial zone 513
Iron-rich sandstone 127
Isolated deeps 48
Isolated platform, Sudan 197
Isostatic adjustment of passive margin 446
Isostatic compensation 7
Isotope dates, Quaternary terraces 537
Isotopes, axial sediments 485
Isotopic properties of Miocene dolomites 379

Jebel Abu Imama **194**, 204
Jebel Dhabba 179, 184
Jebel Dhabdab 182
Jebel Dyiba 195
Jebel Marun 178
Jebel Mukalla **183**, 185
Jebel Tobanam 195
Jeddah 135
Jeza Formation 172
Jizan group 143
July field **112**

Kamaran Limestone 448
Kaolinite 127
Kareem Formation 81, 236
Karstic dissolution (Miocene) 381
Karstic surfaces 23
Kerogens of axial sediments 491
Kharaza Member 216, 277
   Abu Shaar 281
Kharga Arch 15

Khor Eit 193, 195
Kinematics 116

Lacustrine facies 116, 130, 179
Lacustrine gastropods 129
Lacustrine limestones 168, 215, 554
Lacustrine mudstones 125, **126**
Lacustrine sandstones **131**
Lacustrine shales 115
Laminated dolomicrite 290
Laminated sandstones 129
Laminitic microbial unit 220
Landscape evolution 595
Large clinoforms 438
Large-scale gravity collapse 343
Last glacial impact on coral reef growth 580
Late Quaternary faunal turnovers 579
Late syn-rift model **19**
Latitudinal effects on Miocene sedimentation 269
Lava flows 144
Lepidocrocite 511
Lessepsian migration 570
Levant Shear **78**
LIP 51
   volcanism 52
Listric faults 63, **113**
Lithified crusts 7
   in axial sediments 497
   on Sudaneses shelf 502
Lithified limestone crusts 471
Lithified surface encrusted with domed stromatolites 333
Lithodomus 302, 304
Lithology, axial **483**
Lithosphere 54
Lithospheric detachment 13
Lithospheric discontinuities 45
Lithospheric thinning 10
Load-casts 351
   deformation **352**
Local source noncarbonate fraction 249
Local tectonic movements and sedimentary discontinuities 265
Longshore current 249
Longshore transport 270, 282
Low-relief setting 187
Low sea-level stands 24
Lower crust 10
Lower Rudeis Formation 148
Lowstand
   deposits 23
   periods 24
   systems tract 23
   wedges 22
Lymnaea 554

Maghersum Formation 192
Maghersum Island 193
Magmatic activity 5
Magmatic east-west asymmetry 160
Magmatic events 50
Magmatic phases 51
Magnesium calcite cements 471
Magnetic trends 37
Main phase rifting 85
Main rift events 79
Major pre-rift doming 230
Major transgressions 212
Malaab Group 234
Mangal molluscs 564
Mantle plumes 10, 53
Mantle processes 51

Marginal zone of axial trough 480
Marib Graben 41
Marine cements 282
Marine diagenesis, hydrodynamics 374
Marine fan overflows 245
Marine flooding 19, 24
Marine transgression 6
Marion Dufresne, 1992 RED SED cruise 480
Mass-flow 327, 338
Mass-flow deposits 330
Mass-flow events 326
Mass movement 20
Massawa Channel 524, 527
Maximum flooding surfaces 22
Maximum subsidence 111
Measurement of relative sea-level change **548**
Mediterranean and Red Sea evaporites 425
Mediterranean coral assemblage 6, 314
Medj-zir formation 119, 121
Mega-convolute bedding **361**
Megasequences 21
Melanoides 554, 558
Mellaha block 113
Mersa Alam Formation 440, 444
Metal-enriched sediment, accumulation rates 469
Metalliferous sedimentation 476
  dispersion 506
  in Atlantis II Deep 505
Micritic cements 374
Micritic hardgrounds 472
Microbial aggregates 501
Microbial crusts 197, 501
Microbial laminite 247
Microbial pisolites 336
Microbial structures 374
Microbial substrate stabilizes grains 374
Microchannel systems 379
Microcodium **126**, 130
Micro-onlap sequences 338
Microstructural analysis **106**
Mid syn-rift model **19**
Mid-Clysmic and Post-Nukhul merging 218
Mid-Clysmic event 95, **99**, 111, 212, 222, 236
Mid-Rudeis event 234
Mid-to-late Miocene evaporites 6, 411
Middle Miocene worldwide maximum transgression 296
Midyan area 15, 79, 142, 154
Midyan Peninsula 114
Migration of subsidence 112
Mimetic dolomite 393
Mineral deposits, Red Sea axial zone 506
Mineral paragenesis, axial brines 512
Mineral precipitation (T) in axis 509
Mineral sequence, axial cores 508
Mineralogical anomalies coincident with last glacial period 488
Mineralogy axial sediments 482
Minerals, distribution in axial cores **507**
MINOS cruise 32
Miocene carbonate platform 291
Miocene carbonate/siliciclastic platform 294
Miocene carbonates 190
  Sudanese coastal plain **200**
Miocene coral reefs 296
  regional distribution 298
Miocene depositional morphology 272
Miocene evaporites 16
  correlations of 412
Miocene fault blocks, reactivation 427
Miocene foraminifera **208**

Miocene platforms, north-western Red Sea 239
Miocene reef assemblages 302
Miocene reef settings **316**
Miocene rocks in the Sudan **192**
Miocene sedimentation, attributes influencing **241**
Miocene slope deposits, north-western Red Sea 321, 323
Miocene syn-rift platforms, north-western Red Sea 241
Miogypsinids 201
Miogypsinoides 82
Mio-Pliocene boundary 220
Mixed carbonate-siliciclastic platform 272
Models 9
Modern earthquakes 568
Modern-reef-coral fauna, entirely Indo-Pacific 315
Modern reefs, Saudi Arabian coast, physiography and morphology **592**
Moho 37
Montastraea 303
Montmorillonite 485
MORB basalt 468
Morgan Accommodation Zone **78**, 80, **84**, 231, 274
Morpho-structural asymmetry 332
Morpho-structural framework 5
Mud-flows 150
Mudstone 129
Mukalla Formation 17, 60, 179
Mukalla high 168
Mukalla-Sayhut region **167**
Multicoloured sediments, overlying basalt substratum 508
Multiple diagenetic processes 370
Multiple dissolution/lithification 472
Multiple reef bodies 246
Multiple tectonically controlled sequence boundaries 259
Multispectral satellite images 60

N2 planktonic zone 79
N4 foraminifera zone 81, 82, 204
Nail-head spar 402
Nakheil Formation 81, 147, 212
Namakiers 456
Naqara reef complex **91**
Narrowing of rift 231
Nautiloids 334
Nebkhas 539
Negative oxygen isotope values relating to temperature 379
Neogene rifting 15
Neotectonic deformation, Gulf of Suez 568
Neotectonic stability, north-western Red Sea coast 565, 568
Neotectonics
  Dahlak Islands, Eritrea 527
  Gulf of Aqaba 587
  south-western Gulf of Suez 568
Neptunian dikes 306, 340, 435
Nereus Deep 468
Nodular anhydrite 383, **419**
Nondivergent stratal geometries 187
Nonstructural controls of syn-rift sedimentation 269
Nontectonic factors 7
Normal fault systems 85
Normal Red Sea Sediment (NRSS) 469
North European rift 106
North Sea 10, 21

and Atlantic margins 4
central graben 13
North-south diachronism 157
Northern Red Sea 14, 146
  Quaternary sedimentation 537
  stratigraphic nomenclature **148**
Northern Saudi Arabia 142
Northwards rifting propagating 468
North-western Red Sea 97, 211
  early syn-rift deposits 150
  Miocene syn-rift platforms 241
  passive margin 446
  peripheral 212
  post-Miocene 441
  local tectono-sedimentary features 440
  structures 442
  Quaternary coastal deposits 539
  rates of sedimentation 444
  tectonics and its influence on sedimentation 442
Nubian sandstones **80**, 100, 130, 139, 144, 147, 243
Nukhul dolomite **260**
Nukhul Formation 15, 78, 81, **83**, 148, 234
Nullipore rocks 236

Oblique motion 41
Oblique rift 17
Oblique rift basin 17
Ocean-floor
  diagenesis 7
  spreading 5, 7, 18
Oceanic basin 12, 74
Oceanic crust 4, 32, 33, 37, **38**, 54, 56
  accretion of 48
Oceanic spreading 74
Oceanic structures **64**
Oceanization 29, **47**, 165
Oil
  in axial sediments 492
  reservoirs 261
Old structural lineaments 48
Oligocene lavas 233
Oligo-Miocene rifting, initial stages of 146
Olistoliths 153, **260**, 326, 338
Oncolites 176
Onib Hamisana suture zone 45
Onlapping detritus 343
Ooid and peloidal sand sheets 292
Oolites 197, 289, **331**
Oolitic ironstones 141
Open fissures 354
Operculina 532
Ophiomorpha 281
Orbulina 197
Organic geochemistry and origins of sapropels 491
Organic geochemistry of cores from axial zone 479
Organic matter
  content of axial sediments 491
  sources and preservation 490
Organic-rich sapropels 479
Owen fracture zone **57**
Oxygen isotopes, axial sediments 485

Palaeocliff 247
Palaeocurrents 127, 131, 132, 178, 182, 293
Palaeoembayments **332**
Palaeoflow 131
Palaeogeographic distribution of characteristic Quaternary molluscs along African shores **571**

Palaeogeographic maps **132**
Palaeogeography **133**
Palaeo-sea level identification 542
Palaeostress 105
Palaeo-valley 253
Paleosols 16, 127, **131**, 155, 232, 281, 288
Palustrine environments 150
Palygorskite 485
Pan-African basement 77
Pan-African granitic intrusions 59
Pan-African shear zones 79
Parana of Brazil 10
Parasequence boundaries 22
Partial melting 11
Passive infill of accommodation space 172
Passive margins 12
  basins 23
  north-western Red Sea 446
  sedimentation 189
  stages 21
  stratigraphy 21
Passive rifting 4, 10
Patch reefs
  colonization over salt diapir 461
  near salt surface 449
Pedestals for platforms 195
Pelagic carbonate turbidites 473
Pelagic carbonates 7
Pelagic gastropods 500
Pelagic muds 21
Pelagic sedimentation 469
Peloidal grainstone 289
Peloids, axial crusts 497
Periglypta 302
Peripheral and axial sequences, correlations between 211
Peripheral clinoforms 221
Peripheral north-western Red Sea 212
Peripheral uplift 4, 7
  western Yemen 611
  Yemen 608
Periplatform slope sedimentation 320
Peritidal conditions 289
Petrophysics, early cement 375
Pillow lavas 472
Pinctada 429
Pinna 435
Pinnacles
  formation by sulphatization **388**
  submarine, Dahlak Shelf, Eritrea 532
Pisolites, Abu Shaar platform 374
Pisolites/oncoids on slopes 336
Planktonic biozones 201
Planktonic zone N2 79
Plant fragments 129
Plate tectonic reconstruction 45
Platform amalgamation 263
Platform carbonates 247
Platform development 95
Platform margin
  Abu Shaar **286**
  collapse 291
Platform rim 195, 204
  Sudan 197
Platform siliciclastics 249
Platform slope
  characteristics 344
  Sudan **198**, 201
Platform talus 326
Platforms 182
  repeated uplift 254
  stratigraphic geometries 292
  Sudan 193

Playa deposits 150, 176, 184
Pleistocene back-reef limestone, Zabargad Island 577
Pleistocene evaporite facies **424**
Pleistocene reef and open-marine sand fauna last interglacial reef, southern Gebel el Zeit **547**
Pleistocene to Holocene evolution, north-western Red Sea coast **537**
Plio-Pleistocene Group D 112
Pliocene east-west structural ridge 440
Pliocene evaporites
  Gulf of Suez 423
  west Gebel Zeit 438
Pliocene reefs 575
Pliocene rifting, southern Gulf of Suez **445**
Pliocene sedimentation 429, 445
Pliocene subsidence 444
Pliocene tectonics 429, 445
Plume-related magmatism 11, 16
Plume-related rift basins **10**, 14
Plume-related rift models 11
Plume-related rifting 4, 17
Plume-related volcanism 9, 17, 18, 25
Pole of rotation of rift system 224
Polyphase rifting 104, 116, 156
Polyphase tectonic evolution **109**
Ponding shoreline clastics behind salt ridges 460
Porites 176, 179, 282, 303, 304, 311
Porites framestone 287
Porosity developments in syn-rift dolomites **380**
Post-depositional seismic destabilization 351
Post-dolomitization diagenesis 398
Post-dolomitization hydrothermal cements 390
Post-erosional volcanism 51
Post-evaporitic deposits 220
Post-evaporitic tectonics, southern Gulf of Suez 438
Post-evaporitic unit (Group D) 220
Post-Kareem event 212
Post-Miocene evaporites 437
Post-Miocene reef faunas 576
Post-Miocene sedimentation, Gulf of Suez/northern Red Sea 427
Post-Miocene sedimentation and subsidence 444
Post-Miocene series 446
Post-Miocene stratigraphy, north-western Red Sea **428**
Post-Miocene subsidence 427
Post-Nukhul and Mid-Clysmic merging 218
Post-Nukhul event 148, 154, 212
Post Pleistocene tectonic stability, north-western Red Sea coast 563
Post-rift
  axial subsidence 569
  coastal sediments and geomorphology 521
  environments 7
  fill 95
  magmatism 53
  north-western Red Sea 446
  passive margin stage 427
  phase 6, 184
  sedimentology 21
  stage 19
  strata 9, 12, 13
  subsidence 18
  unconformity 6, 9, 11–17, 25, 188, 429
  north-western Red Sea 446
  volcanism 18, 54

Post-salt sediment, south-western Yemen 457
Post-tectonic uplift 185
Post-volcanic subsidence 17
Post-volcanic uplift 17
Potamides 289, 302, 554
Potential hydrological systems, syn-rift diagenesis 371
Praeorbulina 197
Precambrian structural trends, inheritance of 5
Precipitation of metalliferous sediment 506
Precipitation site, axial zone metals 518
Predictive model 273
Pre-drifting restoration 71
Pre-existing continental lineaments 41
Pre-existing lithospheric structures 45
Prehistoric artefacts (photos) **543**
Pre-Nukhul event 215
Pre-rift
  doming 5, 14, 15, 119
  environments 130
  history 146
  magmatism 52
  sedimentary cover 100
  strata 9, 11, 14, 16, 17, 25, **62**
  stratigraphy, Yemen 171
  structural trends 170
  succession 58
  unconformity **245**, **260**
  unroofing 156
  updoming 16, 18
  uplift 114
  Yemen 187
  volcanics 16
  volcanism 51, 54
Pre-rift/syn-rift sediments, contact 147
Preservation of detrital feldspars 379
Prograding submarine fans 24
Propagating spreading centre 598
Propagation of syn-rift system 6
Proto-Gulf of Aden 168, 188
Proto-Gulf of Suez, rift basin 15
Proto-rift 100, 147
Proto-rift evaporites 411
Proximal fans 107
Proximal margin 10
Pteropod-globigerina limestones 472
Pteropod-globigerina sands 472
Pteropod-globigerina-nanno-ooze 469
Pteropods 488
Pull-apart basins 17, 45, **46**, 48, 77
Pure shear **10**, 13, 16, 18, 25
Pyrite in axial cores 508, 509

Qandala Basin 68
Quantification of geomorphological and tectonic processes, Yemen 595
Quaternary coastal deposits, north-western Red Sea 539, **540**
Quaternary coral reefs 575
Quaternary diapirism (Ras Dira) 442
Quaternary evaporites (NW Red Sea) 411, 423
Quaternary evolution (NW Red Sea) 564
Quaternary fault rejuvination, Gebel el Zeit 551
Quaternary Gulf of Suez 539
Quaternary Red Sea sequence, synthetic representation **566**
Quaternary reef-terraces 537, 576
  southern Gulf of Suez 541
Quaternary reefs, Dahlak Islands 526
Quaternary salina, Quseir 552

Quaternary sedimentation, northern Red Sea 537
Quaternary stage 5e, constant altitude, north-western Red Sea 564
Quseir 100
   Quaternary salina 552
Quseir el Qadim 80, 551

Radiocarbon dates
   reef flats
Gulf of Suez *588*
Jordan and Saudi Arabia *588*
Radiometric dates
   Quaternary, north-western Red Sea 540
   reef terraces, Gebel el Zeit 546
Raised beaches, Gulf of Suez 236
Ramp 20
Ramp-like geometry 292
Ramp margins 293
Ranga Formation 81, 148, 150
Ras Dib terraces 542
Ras Gemsa 549
Ras Gharib 444
Ras Honkorab 103, 146, 153, **154**, 157
   field-views 257
   geological map **255**
Ras Honkorab platforms 254
Ras Shagra 440
Ras Shukheir depression **550**
Ras Shukheir salina 551
Rates of crustal processes 595
Rates of erosion, western Yemen 609
Rates of salt diapirism 462
Reactivation 100, 103
   of Miocene fault blocks 427
Red beds 81, 140, 232, 237
Red deposits 108
Red Sea
   and Mediterranean evaporites 425
   evolution of 45
   margins 41
   stratigraphy 212
Red Sea Deeps 48
Red Sea-Gulf of Aden rift system 3
Red Sea-Gulf of Suez rift, transverse asymmetry 146
Red Sea Hills 85, 192, 240, 274, 538
Red Sea/Indian Ocean 419
RED SED Programme 190, 240, 295, 297, 319
Red terrigenous sediments 102, 115
Reddened paleosols 176
Reddish alluvial mudstones 212
Reddish continental clastics 148
Redeposited carbonates 292
Redeposited facies 20
Reef assemblages, recent demise 588
Reef biostromes 303
Reef corals, north-western Red Sea 296
Reef development 316
   age and regional setting 297
   on down-plunge end of platform 267
Reef geometry 316
Reef knolls, Saudi Arabia 590
Reef morphologies origin 586
Reef palaeoslopes 302, 304
Reef palaeozonations 302
Reef pinnacles (Post-Miocene) 439
Reef platform 590
Reef progradation 306
   footwall scarps 318
   Wadi Kharasa 300
Reef shelf **244**

Reef terraces 7, 242
   (Quaternary) and faunas (photos) **545**
   Ras Dib **542**
   sea-level control 593
Reef zonation, Saudi Arabia 592
Reefal buildup 182
Reefal carbonates 185, 242
Reefal olistoliths 321, **327**
Reefal sequences 243
Reefs **108**, 112
   and footwall crest 266
   and sea-level changes 318
   Aqaba-Eilat areas 585
   Gebel Abu Shaar 299
   olistoliths 311
   on alluvial fans 584
   on Gilbert-type fan deltas 305
   on syn-rift delta fans 308
   salt diapir control 592
   sediment inundation 585
   siliciclastic control 584
   syntectonic 312
   tectonic control 316, 587
   Wadi Ambagi 305
Reefs (Miocene)
   Abu Ghusun-Ras Honkorab 310
   biodiversity 315
   Gebel Esh Mellaha 304
   restricted conditions 318
   Safaga area 304
   Sharm el Bahari 308
   Sharm el Luli 311
   siliciclastic sedimentation 316
   south of Quseir 304
   Wadi Zug El Bohar 306
Reefs (Quaternary), Dahlak Islands 526
Reflector S 7, 446, 480
Regional unconformities 100, 104
Rejuvenation
   of fault blocks 111
   of pre-evaporite fault blocks 430
   of Precambrian fracture systems 242
Relationships between Red Sea and Mediterranean evaporites 425
Relative sea level and sedimentation 241
Relict basins 85
Relief (amplitude) of a block 241
Remobilization
   breccia **358**
   of laminated evaporites 363
Reoconcavo, Brazil 12
Resedimentation
   axial core 488
   axial zone 481
Resedimented beds 473
Resedimented evaporite 421
Reservoir facies 273
Reservoir potential of syn-rift sediments 370
Reworking of microfauna 218
Rhine graben 10
Rhizolith 176
Rhodocrosite 511
Rhodolith floatstone **199**
Rhodophyte construction, palaeodepths of 50 m **335**
Rhombic basin 77
Rhombic blocks **103**, 107, 108
Rhomboidal fault blocks 231
Rhomboidal fault pattern 231
Rift axial sands 482
Rift border faults 10
Rift development 97, 106
   Gulf of Suez 237

Rift escarpment 80
Rift evolution, three stages 444
Rift flank uplift 13, 230
Rift geodynamics, high diagenetic potential 372
Rift geometry 79
Rift histories 91
Rift initiation 78, 82, **93**
   asymmetry 157
   transverse asymmetry 159
Rift margin 112, 187
   development (NW Suez) **235**
   escarpments 94
Rift narrowing 231
Rift periphery, uplift has ceased SE of Aqaba transform 568
Rift shoulders 7, 13, 16–18, 114, 165, **169**
   uplift 15, 17, 20, 77, 80, **93**, 94
   Yemen 188
Rift stress field 95
Rift subsidence 85
Rift to drift 3, 18
Rifting
   end of, Yemen 598
   major phase 6
Rim synclines, south-western Yemen 456
Rimah Formations 172
Rimmed platform 20
Rimmed shelf margins 293
Rock-Eval analysis 491
Rock-types in outcrop, Sudan **199**
Rollover folds 63
Rollover structures 112, **113**
Rosa Member 150, 214
Rotation
   of fault-blocks 9, 13, 22, 23, **23**, 274
   of tilt-blocks 109
Rotation rates 93, **94**
Rudeis Formation 81, 234
Rus Formation 172

Sabkha 179, 182
Sabkha facies 177
Sabkhas, absence of 416, 418
Safaga area 85, **99**, 103, **104**
   reefs (Miocene) 304
Safaga region **86**
Salif diapir
   south-western Yemen 451
   structural setting 456
Salif evaporite 448
Salif salt dome (south-western Yemen), seismic section through **454**
Salif salt wall **457**, 462
Salinity
   and sapropels 471
   stratification 501
Salt, south-western Yemen 451
Salt deposits 112
Salt diapirism 20, 21, 25
   and rift tectonics 422
Salt dome cover 461
Salt domes
   and basin margin sedimentation 449
   Tihama Plain, Yemen 448
   walls and pillows (Yemen) 448
Salt ridge barrier 456
Salt subsidence (Yemen) 462
Salt tectonics 6, 116
   offshore Eritrea 526
Salt wall barrier 449
   and trap for sediment 460
Salt walls 79

Samadai Formation 440
Samalut Formation 233
Sand body in fault-line depression, North Zeit 259
Sand-dikes 352
Sanganeb atoll **589**
   seismic profiles **591**
   tectonic control 589
Sapropels **483**
   and salinity 471
   axial core 491
   deposition axial trough 470
   formation 471, 495
   from terrigenous sources 495
   in axial core **489**
   organic geochemistry and origins of 489
   organic-rich 479
Sarar Formation 168, 186
Saturated hydrocarbon in axial sediments 489
Saudi Arabia 135
   reef zonation 592
Saudi Arabian margin 154
Sea-floor spreading 16, 29, 37, 48, 54
Sea-level changes and sequence boundaries 241
Sea-level variations 185
Seamounts 472
Secondary anhydrites 383
Secondary sulphate and pinnacle formation 383
Sediment budget, western Yemen 608
Sediment by-pass 343
Sediment geometry, north-western Red Sea 216
Sediment liquefaction 349
Sediment loading 12
Sediment types, Dahlak Islands (photos) **535**
Sediment wedges 285
Sedimentary asymmetry on any given block 266
Sedimentary dikes **113**
Sedimentary dynamics, in axial zone 488
Sedimentary events **101**
Sedimentary ore deposit 506
Sedimentary patterns on Dahlak Shelf, Eritrea 529
Sedimentary slopes 204
Sedimentary structures
   in axial sediments 488
   in seismites **348**
Sedimentary thickness, western Yemen **597**
Sedimentary traps 533
Sedimentation
   and structural stability 268
   and tectonics, Yemen 187
   effects of wind 269
   on Miocene fault blocks 241
   on peripheral (western) and axial (eastern) sides of tilt-blocks **266**
   on positive and negative parts of tilt-blocks **266**
   southern Yemen 165
Sedimentation rate
   axial core 488
   of metalliferous sediments 512
   western Yemen 608
Sedimentological evolution 9
Sediments
   basinward transit 250
   exported 263
Segmented pattern 30
Seismic destabilization 362

Seismic events, deforming sediments (seismites) 365, 366
Seismic facies 438
Seismic instability 250
Seismic modelling **33**
Seismic profiles
   Sanganeb atoll **591**
   southern Gulf of Suez 432
Seismic-rich units, distribution **352**
Seismic sedimentary deformations 347
Seismicity in syn-rift series 351
Seismites 7, 107, 347
   and tectonic history 365
   deformation intensity **352**
   in Quaternary sediments 363
   on low-angle hangingwall slope **354**
Semi-arid climates 270
Semicircular geometries of reefs 593
Sequence boundaries 23, 24, 291
   can not be correlated 268
Sequence stratigraphy 9
   concepts 21, 22, 25
   half-graben 23
   methods 22
Sequential evolution, metalliferous sediments 518
Shadwan Embayment 438, 441
Shadwan Island 431
Shagra Formation 440
Shahmal wind 539
Shallow marine cones
   at exits of cross-faults 261
   Wadi Sharm el Bahari **262**
Shallow marine conglomerate cone 261
Shallow marine fans 6
Shallow marine sequences, record synsedimentary tilting 268
Shallow shelves, Sudan 195
Sharm el Arab 435
Sharm el Bahari 157, 214, **264**
   map **262, 309**
   Pleistocene deposits 554
   Quaternary 553
   reefs (Miocene) 308
Sharm el Luli, reefs (Miocene) 311
Sharm el Qibli 105, 214
Sharms 441, 539, 554, 567, 570, 584
Sharp continent-ocean transitions **34**
Shear control **44**
Shear dislocations, sliding and brecciation typical of rift settings 366
Sheba Ridge 41, 48, 54, 56
Sheet-floods 176, 538
Shelf deposits 197
Shelf-margin wedge 216
Shelves, Sudan 193
Shihr Group 165, 168, 172
Shimis sandstone 168
Shoab Ali basin 90
Shoulder uplift 58
Shumaysi Formation 135, **138**, 140
Shumaysi trough 136
Siderite 508
Siesmic section, Tihama Plain, western Yemen **609**
Siliciclastic and carbonate, transition 445
Siliciclastic control of modern reef development 584
Siliciclastic discharges, Gebel Honkorab 255
Siliciclastic facies, evolution 293
Siliciclastic sands on lee-slope 245
Siliciclastic sedimentation 6
   in oblique structural depressions 260

Siliciclastics 7, 249
   staked against the reef **443**
Sills, Sudan 195
Simple shear **10**, 16
Sinai 85, **103**, 114
Sinai margin, early-rifting deposits **149**
Sinai triple junction 79
Skeletal shelf (Dahlak) carbonates 533
Skolithos burrows 176
Slip rates 24, 25
Slope accretion 345
Slope communities (Miocene) **335**
Slope creation 345
Slope deposits 249
   Abu Ghusun 326
   Abu Shaar platform 329
   and associated seismites **363**
   Esh Mellaha platform 332
   significant of rift sedimentation 323
   visible relief 323
   Wadis Ambagi and Aswad, Quseir 325
Slope destruction 346
Slope downlap 324
Slope escarpments 343
Slope evolution 346
Slope facies stratigraphy 325
Slope fauna and flora (Miocene) 334
Slope features and sedimentary dynamics 333
Slope modification 338
Slope onlap sequence 324
Slope processes 321
Slope stratigraphy, lateral variations 324
Slope transit zone 333
Slump-like folding 339
Slump scars 20, 292
Smectite 485
Socotra 17
Solitary corals 334
Somalia 56, 168, 186
South Atlantic Creataceous rift **115**
South Atlantic rift 115, 116
South-eastern Red Sea 130
   salt domes, geological setting 449
South Gharib Formation 236
   evaporites 418
South Gharib salt 81
South Hadhramaut Arch 168, 170
Southern Arabia **133**
Southern Atlantic Cretaceous rift 20
Southern Gulf of Suez 78
   stratigraphy **80**
Southern Red Sea 16, 121
Southern Saudi Arabia 143
Southern Yemen, sedimentation 165
Spatial evolution of Atlantis II deposits 519
Speleothems 382
Sphalerite in axial cores 508
Spherulites (diagenetic) in axial crusts 497
Spreading rates 48
Spur-and-groove structures, absence of 587
Stability of the tilt-block 241
Stable isotopic values 395
Stacking patterns 22
Stage 5e deposits, north-western Red Sea coast 564
Stagnant bottom waters 7
Storm deposits 179, 183
Stratal discontinuities 432
Stratal geometries 293
Stratal surfaces 60
Stratified water column 479
Stratiform distribution of deformational structures 364

Stratigraphic correlation, eastern margin of Red Sea **143**
Stratigraphic nomenclature, northern Red Sea **148**
Stratigraphy, Southern Gulf of Suez **80**
Streamline maps, Yemen 602
Stretched continental crust 33, 37, **42**
Stretching 10, 41
Strike-slip faults 5, 45, 79, 156
   Sudan 207
Strike-slip folds 156
Strike-slip movements 41, 45, **47**, 48, 106, 156, 214
Strike-slip regime 95
Strike-slip tectonics 157
Stromatolites **91**, **108**, 216, 289, 308, **331**
   at depths of 50 m 334
   'banana-type' 336
   encrusting steep slopes **336**
   on platform slope 336
Stromatolitic domes 109
Stromatolitic encrustations 325
Stromatolitic ridges 263, **264**, 308
Stromatolitic sheets 336
Strombidae 315
Strontium isotope curve 463
Strontium isotope dating 189, 459, 463
Structural and sedimentary processes 107
Structural asymmetry 159
Structural barrier 267
Structural considerations, Sudan 207
Structural controls
   Miocene sedimentation 265
   Red Sea offshore sedimentation 476
   syn-rift sedimentation 241
Structural corridors 111, 261
Structural evolution 5
Structural framework 102
Structural highs, Dahlak Islands 532
Structural instability 259
Structural ramp 267
Structural template 272
Structures south of Hurghada **437**
Stylophora 254, 281, 303
Suakin Deep 479
Subaqueous delta system 182
Submarine cementation 471
Submarine erosion 585
Submarine escarpments 328, 437
Submarine fans 20
Submarine terraces 593
Submarine transit **105**
Subsidence 10, 12, 91
   history, Gulf of Suez 230
   migration 112
   modelling 94
Subsidence rates 77, 95
Sudan
   Miocene carbonates 190
   outcrop morphology 193
   platform slope **198**, 201
   shallow shelves 195
   shelves 193
   sills 195
   strike-slip faults 207
Sudan shelf 480
Sudanese coast, stratigraphic correlation of Miocene carbonate deposits **208**
Sudanese coastal plain 191
   Miocene carbonates **200**
Sudanese margin 153, **155**
Sudanese Miocene **203**
Sudanese shelf, diagenesis 501

Suez half-graben 91
Suez Isthmus as barrier since Pliocene 572
Suez Platform edge 445
Suez rift dynamics 94
Sulphates 102
Sulphide ore deposits 386
Sulphur isotopes, axial sediments 509
Superimposed deformations 105
Suqah trough 135
Surface uplift rate, western Yemen 607
Suture zones 48
Symmetric rift basin 10
Syndepositional deformation 215
   Plio-Quaternary 435
Syndepositional fault movement 11
Syndiagenetic deformations 347
Syn-rift
   basins 63
   carbonate platform 273
   carbonates 7
   clastics 17
   deposits **99**, 102
   diagenesis 369
   potential hydrological systems **371**
   environments 6
   magmatism 52
   megasequence 11
   Miocene coral reefs 297
   phase 3
   propagation 6
   sedimentation and tectonics 163, 269
   sedimentology 19
   sediments
     geometry of 240
     north-western Red Sea 215
     Yemen 172
   sequences, Gulf of Suez **229**
   South Gulf of Suez 446
   stage 6
   strata 9, 15, 17
   stratigraphy 6, 22
   structures 71
   talus (Miocene), NW Red Sea 326
   to post-rift, transition 431
   unconformity 5, 9, 11, 13, 14, 16, 17, 25, 117, 142, 187, 282
   volcanics 4, 14
   volcanism 51, 54
Syn-rift/pre-rift sediments, contact 147
Synsedimentary deformation 7, **110**, 155, 214, 293
Synsedimentary destruction platform 247
Synsedimentary erosion platform margin 249
Synsedimentary faulting **113**
Synsedimentary sliding 112, 247
Synsedimentary tectonic instability 326
Synsedimentary tilting 6, 249, 291
Synsedimentary truncation 292
Synthetic faults 112
Synthetic tilt-blocks 116
Syria 16
Syrian Arc 14, **78**, **98**, 230
Syrian foldbelt 159
Systems tracts 23

Talus 193
Talus-slope angles 321
Talus slopes 195
Tarbellastraea 303, 311
Tarbul syncline 274
Taurus and Zagros orogenic belts 97
Tawilah group 119, 121, 144, 172
Taxonomy, Miocene corals 313

Tayiba Formation 233
Tectonic events **101**
Tectonic evolution 97
Tectonic framework and sedimentation 276
Tectonic geomorphology 595
Tectonic instability 7
Tectonic modification and destruction of carbonate platforms 263
Tectonic setting, Gulf of Suez **224**
Tectonic stability, NW Red Sea coast 563
Tectonic stages 107
Tectonic subsidence **93**
Tectonics 211
   and geomorphology 599
   Eritrean coast, southern Red Sea **525**
   sedimentation and stratigraphy 236
Tecto-sedimentary evolution 130
   western Yemen 133
Tectono-sedimentary evolution in western Saudi Arabia 144
Tectono-sedimentary sequences 107, 115, 116
Tectono-sedimentary units 97, 100, 148, 212
Tectono-stratigraphic evolution, Eastern margin, Gulf of Suez 236
Tectono-stratigraphic sequence 237
Temperature, axial brines 509
Tension gashes 107
Terebralia 448, 553
Terrigenous clastics blocked by footwall escarpment 267
Tethyan rift 116
Thalassinoides 127, 176
Thebes Formation 81, 147, 233
Thermal subsidence 10, 13, 21, 24
Thermal uplift 12
Thinning 10
Tholeiitic basalt 107
Thorium dating of two reef sequences, interpretation with respect to diagenetic changes **542**
Thorium isotopes 7
Tihama Plain, Yemen 450, 598
Tilt-block/half-graben model 6
Tilt-blocks 7, 107, **111**, **112**, 115, 257
   carbonate platforms 20, 21, 293
   pattern 156
   platforms 242, 266
   attributes 265
   plunge 247
   rotation 109
   structures **110**
   Sudan 195
Tilted fault blocks 240
Tilting, first phase 153
Timing of rifting, Gulf of Suez 226
Topographic escarpments 34
Topography, western Yemen **596**
Trace element geochemistry 397
Transcurrent faults 74
Transfer faults 18, 108, **111**
Transfer or accommodation zones 9
Transfer zones 5, 9, 19, 24, 25, 63, 67, 69, 73
Transform faults 56, 57
Transform margin 4
Transgressive systems tracts 22, 23
Transit of terrigenous detritus 242
Transitional Region 467
   Red Sea, axial zone **468**
Transverse asymmetry
   Red Sea-Gulf of Suez rift 146
   rift initiation 159
Transverse faults 115
Trap door fault systems 231

Travertines 539
Tridacna 220, 429, 459
Tropical climates 19, 24
Truncation surfaces 343
Tufas 176
Turbid sea water too brief to influence reef
 growth 585
Turbidite sands **483**
 in axial core **489**
Turbidite sandstones 81
Turbidite sedimentation as direct expression
 of Red Sea opening 488
Turbidites 7, 168, 186, 234
 in axial sediments 488

Umm Er Radhuma Formation 172
Unconformities in slope deposits 323
Underplating 11
Undersliding destructuration **362**
University of Assiut 240
Unroofing of rift flank 238
Uplift 10, 13, 133
 and erosion 156
 at rift periphery ceased south-east from
  Aqaba transform 568
 limited accommodation space 263
Uplifted terraces 95
Upper crust 10
Upper Paleolithic or Neolithic
 occupation 541
Upward-decreasing dips 291
Upwelling of heated waters 396
Uranium dating of two reef sequences,
 interpretation with respect to
 diagenetic changes **542**
Uranium isotopes 7
Usfan Formation 135, **138**

Vadose cements 399
Valley depth:width ratios (western
 Yemen) 603
Vents on spreading centres 506
Vertical crustal displacements (Yemen) 599

Vertical relief of tilt-block 267
Volcanic flows 55
Volcanic plumes 10
Volcanic stratigraphy 14, 55
Volcaniclastic sediments 165
Volcanics 107, **120**
Volcanics of Yemen, Ethiopia and Eritrea 16
Volcanism 11, 50
Volcanoclastic-pelagic carbonate
 breccias 472
Volume of eroded sediment, Yemen 608

Wadi Abu Assala (Safaga) 324
Wadi Abu Dabbab 440
Wadi Abu Ghusun 247, **248**
Wadi Abu Sbikhaia, Quaternary 562
Wadi Ambagi Quseir 221
 reefs 305
 tilted shelf 242
Wadi Ash Sham 136
Wadi Bali 249
 depositional sequences **288**
 section **285**
Wadi Bali'h 272
Wadi el Aswad 242, **264**
Wadi Feiran 234
Wadi Gasus 150, 153, **159**, 214
Wadi Gemal 215, 562
Wadi Gharandal 236
Wadi Himmum 182
Wadi Igla 441
 Quaternary 558
 seismite deformation event **443**
Wadi Kabrit **258**
Wadi Khalilat el Bahari, Quaternary
 reefs 559
Wadi Kharasa **284**
 reef progradation 300
 reefs 303
 stratigraphic geometries **279**
Wadi Nahiri, Quaternary 556
Wadi Nukhul 214, 234
Wadi Qalawa, Quaternary 554

Wadi Sharm el Bahari, shallow-marine
 cone 261
Wadi Shumaysi 135
Wadi Treifi **253**
Wadi Um Dirra 254
Wadi Um Gheig 383
Wadi Um Greifat 441
 discontinuities offlapped by siliciclastic
  sands 442
Wadi Zug El Bohar, reefs (Miocene) 306
Wadis Gasus 217
Warm dolomitizing fluids 387
Western margin 45
Western Red Sea margin **113**
Western Yemen **132**
 tectono-sedimentary evolution 133
 topography **596**
Wind, effects on sedimentation 269
Wind-driven currents, reef growth 245
Winds 7
World Bank 4, 191, 240
Wrench faulting **106**

Yemen 14, 16, 119
 Jurassic rift 170
 margins 11
 stratigraphy 168
 streamline maps 602
 tectonic architecture 168
 vertical crustal displacements 599
 volcanic group 119, 121, 125

Zabargad Island 578
Zafarana Formation 236
Zeit Formation 81, 423
 evaporites 418
Zeit-Mellaha area **109**
Zigzag fault pattern 102, 104, 108, **109**, 115,
 116, 155
Zug al Bohar
 blocks **246**
 platform 243
 tilt-block platform **245**